Moses Fayngold and Vadim Fayngold

Quantum Mechanics and Quantum Information

Related Titles

Furusawa, A., van Loock, P.

Quantum Teleportation and Entanglement

A Hybrid Approach to Optical Quantum Information Processing

2011
Hardcover
ISBN: 978-3-527-40930-3

Gazeau, J.-P.

Coherent States in Quantum Physics

2009
Hardcover
ISBN: 978-3-527-40709-5

Bruß, D., Leuchs, G. (eds.)

Lectures on Quantum Information

2007
Softcover
ISBN: 978-3-527-40527-5

Audretsch, J. (ed.)

Entangled World

The Fascination of Quantum Information and Computation

2006
Hardcover
ISBN: 978-3-527-40470-4

Stolze, J., Suter, D.

Quantum Computing

A Short Course from Theory to Experiment

2004
Softcover
ISBN: 978-3-527-40438-4

Hameka, H. F.

Quantum Mechanics

A Conceptual Approach

2004
Softcover
ISBN: 978-0-471-64965-6

Phillips, A. C.

Introduction to Quantum Mechanics

2003
Softcover
ISBN: 978-0-470-85324-5

Zettili, N.

Quantum Mechanics

Concepts and Applications

2001
Softcover
ISBN: 978-0-471-48944-3

Merzbacher, E.

Quantum Mechanics

1998
Softcover
ISBN: 978-0-471-88702-7

Cohen-Tannoudji, C., Diu, B., Laloe, F.

Quantum Mechanics

2 Volume Set

1977
Softcover
ISBN: 978-0-471-56952-7

Moses Fayngold and Vadim Fayngold

Quantum Mechanics and Quantum Information

A Guide through the Quantum World

WILEY-VCH

The Authors

Moses Fayngold
NJIT
Dept. of Physics
Newark, NJ 07102-1982
fayngold@mailaps.org

Vadim Fayngold
vadim.research@gmail.com

All books published by **Wiley-VCH** are carefully produced. Nevertheless, authors, editors, and publisher do not warrant the information contained in these books, including this book, to be free of errors. Readers are advised to keep in mind that statements, data, illustrations, procedural details or other items may inadvertently be inaccurate.

Library of Congress Card No.: applied for

British Library Cataloguing-in-Publication Data
A catalogue record for this book is available from the British Library.

Bibliographic information published by the Deutsche Nationalbibliothek
The Deutsche Nationalbibliothek lists this publication in the Deutsche Nationalbibliografie; detailed bibliographic data are available on the Internet at <http://dnb.d-nb.de>.

© 2013 Wiley-VCH Verlag GmbH & Co. KGaA, Boschstr. 12, 69469 Weinheim, Germany

All rights reserved (including those of translation into other languages). No part of this book may be reproduced in any form – by photoprinting, microfilm, or any other means – nor transmitted or translated into a machine language without written permission from the publishers. Registered names, trademarks, etc. used in this book, even when not specifically marked as such, are not to be considered unprotected by law.

Print ISBN: 978-3-527-40647-0

Cover Design   Adam Design, Weinheim, Germany

Typesetting   Thomson Digital, Noida, India

Printing and Binding   Markono Print Media Pte Ltd, Singapore

Contents

Preface *XIII*
Abbreviations and Notations *XIX*

1 The Failure of Classical Physics *1*
1.1 Blackbody Radiation *1*
1.2 Heat Capacity *4*
1.3 The Photoelectric Effect *9*
1.4 Atoms and Their Spectra *12*
1.5 The Double-Slit Experiment *14*
Problem *19*
References *19*

2 The First Steps into the Unknown *21*
2.1 The BBR and Planck's Formula *21*
2.2 Einstein's Light Quanta and BBR *24*
2.2.1 Discussion *27*
2.3 PEE Revisited *30*
2.4 The Third Breakthrough: de Broglie Waves *31*
2.4.1 Exercise *33*
Problems *35*
References *35*

3 Embryonic Quantum Mechanics: Basic Features *37*
3.1 A Glimpse of the New Realm *37*
3.2 Quantum-Mechanical Superposition of States *39*
3.3 What Is Waving There (the Meaning of the Ψ-Function)? *42*
3.4 Observables and Their Operators *47*
3.5 Quantum-Mechanical Indeterminacy *49*
3.6 Indeterminacy and the World *53*
3.7 Quantum Entanglement and Nonlocality *58*
3.8 Quantum-Mechanical Phase Space *62*
3.9 Determinism and Causality in Quantum World *63*

3.9.1	Discussion 63
	Problems 66
	References 66

4	**Playing with the Amplitudes** 69
4.1	Composition of Amplitudes 69
4.2	Double Slit Revised I 74
4.3	Double Slit Revised II 77
4.4	Neutron Scattering in Crystals 78
4.5	Bosonic and Fermionic States 81
4.6	Path Integrals 89
	Problems 93
	References 93

5	**Basic Features and Mathematical Structure of QM** 95
5.1	Observables: the Domain of Classical and Quantum Mechanics 95
5.2	Quantum-Mechanical Operators 97
5.3	Algebra of Operators 100
5.4	Eigenvalues and Eigenstates 102
5.5	Orthogonality of Eigenstates 107
5.6	The Robertson–Schrödinger Relation 110
5.7	The Wave Function and Measurements (Discussion) 112
	Problems 116
	References 117

6	**Representations and the Hilbert Space** 119
6.1	Various Faces of a State Function 119
6.2	Unitary Transformations 121
6.3	Operators in the Matrix Form 125
6.4	The Hilbert Space 129
6.5	Operations in the Hilbert Space 135
6.6	Nonorthogonal States 142
	Problems 147
	References 148

7	**Angular Momentum** 149
7.1	Orbital and Spin Angular Momenta 149
7.2	The Eigenstates and Eigenvalues of $\hat{L}$ 151
7.3	Operator $\hat{L}$ and Its Commutation Properties 154
7.4	Spin as an Intrinsic Angular Momentum 164
7.5	Angular Momentum of a Compound System 183
7.6	Spherical Harmonics 188
	Problems 196
	References 197

8	**The Schrödinger Equation** *199*
8.1	The Schrödinger Equation *199*
8.2	State Function and the Continuity Equation *200*
8.3	Separation of Temporal and Spatial Variables: Stationary States *203*
8.4	The Helmholtz Equation and Dispersion Equation for a Free Particle *205*
8.5	Separation of Spatial Variables and the Radial Schrödinger Equation *207*
8.6	Superposition of Degenerate States *209*
8.7	Phase Velocity and Group Velocity *212*
8.8	de Broglie's Waves Revised *218*
8.9	The Schrödinger Equation in an Arbitrary Basis *222*
	Problems *226*
	References *226*
9	**Applications to Simple Systems: One Dimension** *227*
9.1	A Quasi-Free Particle *227*
9.2	Potential Threshold *232*
9.3	Tunneling through a Potential Barrier *236*
9.4	Cold Emission *241*
9.5	Potential Well *244*
9.6	Quantum Oscillator *249*
9.7	Oscillator in the E-Representation *254*
9.8	The Origin of Energy Bands *257*
9.9	Periodic Structures *260*
	Problems *269*
	References *271*
10	**Three-Dimensional Systems** *273*
10.1	A Particle in a 3D Box *273*
10.2	A Free Particle in 3D (Spherical Coordinates) *274*
10.2.1	Discussion *277*
10.3	Some Properties of Solutions in Spherically Symmetric Potential *277*
10.4	Spherical Potential Well *278*
10.5	States in the Coulomb Field and a Hydrogen Atom *281*
10.6	Atomic Currents *287*
10.7	Periodic Table *290*
	Problems *293*
	References *294*
11	**Evolution of Quantum States** *295*
11.1	The Time Evolution Operator *295*
11.2	Evolution of Operators *299*

11.3	Spreading of a Gaussian Packet	*301*
11.4	The *B*-Factor and Evolution of an Arbitrary State	*303*
11.5	The Fraudulent Life of an "Illegal" Spike	*306*
11.6	Jinnee Out of the Box	*311*
11.7	Inadequacy of Nonrelativistic Approximation in Description of Evolving Discontinuous States	*315*
11.7.1	Discussion	*316*
11.8	Quasi-Stationary States	*317*
11.8.1	Discussion	*323*
11.9	3D Barrier and Quasi-Stationary States	*324*
11.10	The Theory of Particle Decay	*327*
11.11	Particle–Antiparticle Oscillations	*331*
11.11.1	Discussion	*337*
11.12	A Watched Pot Never Boils (Quantum Zeno Effect)	*339*
11.13	A Watched Pot Boils Faster	*344*
	Problems	*350*
	References	*352*
12	**Quantum Ensembles**	*355*
12.1	Pure Ensembles	*355*
12.2	Mixtures	*356*
12.3	The Density Operator	*358*
12.4	Time Evolution of the Density Operator	*366*
12.5	Composite Systems	*368*
	Problems	*376*
	References	*376*
13	**Indeterminacy Revisited**	*377*
13.1	Indeterminacy Under Scrutiny	*377*
13.2	The Heisenberg Inequality Revised	*380*
13.3	The Indeterminacy of Angular Momentum	*382*
13.4	The Robertson–Schrödinger Relation Revised	*384*
13.5	The N–ϕ Indeterminacy	*388*
13.6	Dispersed Indeterminacy	*390*
	Problems	*394*
	References	*395*
14	**Quantum Mechanics and Classical Mechanics**	*397*
14.1	Relationship between Quantum and Classical Mechanics	*397*
14.2	QM and Optics	*400*
14.3	The Quasi-Classical State Function	*401*
14.4	The WKB Approximation	*404*
14.5	The Bohr–Sommerfeld Quantization Rules	*406*
	Problems	*409*
	References	*410*

15	**Two-State Systems** *411*
15.1	Double Potential Well *411*
15.2	The Ammonium Molecule *415*
15.3	Qubits Introduced *419*
	Problem *422*
	References *422*

16	**Charge in Magnetic Field** *423*
16.1	A Charged Particle in EM Field *423*
16.2	The Continuity Equation in EM Field *425*
16.3	Origin of the A-Momentum *427*
16.4	Charge in Magnetic Field *429*
16.5	Spin Precession *432*
16.6	The Aharonov–Bohm Effect *437*
16.6.1	Discussion *441*
16.7	The Zeeman Effect *442*
	Problems *444*
	References *445*

17	**Perturbations** *447*
17.1	Stationary Perturbation Theory *447*
17.1.1	Discussion *450*
17.2	Asymptotic Perturbations *455*
17.3	Perturbations and Degeneracy *457*
17.4	Symmetry, Degeneracy, and Perturbations *460*
17.5	The Stark Effect *462*
17.6	Time-Dependent Perturbations *465*
	Problems *471*
	References *471*

18	**Light–Matter Interactions** *473*
18.1	Optical Transitions *473*
18.2	Dipole Radiation *474*
18.3	Selection Rules *477*
18.3.1	Oscillator *478*
18.3.2	Hydrogen-Like Atom *478*
	Problems *480*
	Reference *480*

19	**Scattering** *481*
19.1	QM Description of Scattering *481*
19.2	Stationary Scattering *487*
19.3	Scattering Matrix and the Optical Theorem *490*
19.4	Diffraction Scattering *494*
19.5	Resonant Scattering *498*

19.6	The Born Approximation *501*	
	Problems *504*	
	References *505*	
20	**Submissive Quantum Mechanics** *507*	
20.1	The Inverse Problem *507*	
20.2	Playing with Quantum States *509*	
20.3	Playing with Evolution: Discussion *514*	
	Problems *522*	
	References *522*	
21	**Quantum Statistics** *525*	
21.1	Bosons and Fermions: The Exclusion Principle *525*	
21.1.1	Discussion *531*	
21.2	Planck and Einstein Again *540*	
21.3	BBR Again *542*	
21.4	Lasers and Masers *543*	
	Problems *545*	
	References *546*	
22	**Second Quantization** *547*	
22.1	Quantum Oscillator Revisited *547*	
22.2	Creation and Annihilation Operators: Bosons *548*	
22.3	Creation and Annihilation Operators: Fermions *552*	
	Problems *555*	
	References *555*	
23	**Quantum Mechanics and Measurements** *557*	
23.1	Collapse or Explosion? *557*	
23.2	"Schrödinger's Cat" and Classical Limits of QM *563*	
23.3	Von Neumann's Measurement Scheme *571*	
23.3.1	Discussion *575*	
23.4	Quantum Information and Measurements *578*	
23.5	Interaction-Free Measurements: Quantum Seeing in the Dark *586*	
23.6	QM and the Time Arrow *593*	
	Problems *595*	
	References *596*	
24	**Quantum Nonlocality** *599*	
24.1	Entangled Superpositions I *599*	
24.2	Entangled Superpositions II *601*	
24.2.1	Discussion *604*	
24.3	Quantum Teleportation *604*	
24.4	The "No-Cloning" Theorem *607*	

24.5	Hidden Variables and Bell's Theorem	*613*
24.6	Bell-State Measurements	*619*
24.7	QM and the Failure of FTL Proposals	*627*
24.8	Do Lasers Violate the No-Cloning Theorem?	*628*
24.9	Imperfect Cloning	*636*
24.10	The FLASH Proposal and Quantum Compounds	*643*
	Problems	*649*
	References	*649*

25	**Quantum Measurements and POVMs**	*651*
25.1	Projection Operator and Its Properties	*651*
25.2	Projective Measurements	*655*
25.3	POVMs	*658*
25.4	POVM as a Generalized Measurement	*664*
25.5	POVM Examples	*666*
25.6	Discrimination of Two Pure States	*670*
25.7	Neumark's Theorem	*681*
25.8	How to Implement a Given POVM	*686*
25.9	Comparison of States and Mixtures	*695*
25.10	Generalized Measurements	*697*
	Problems	*700*
	References	*701*

26	**Quantum Information**	*703*
26.1	Deterministic Information and Shannon Entropy	*703*
26.2	von Neumann Entropy	*709*
26.3	Conditional Probability and Bayes's Theorem	*711*
26.4	KL Divergence	*716*
26.5	Mutual Information	*717*
26.6	Rényi Entropy	*719*
26.7	Joint and Conditional Renyi Entropy	*721*
26.8	Universal Hashing	*726*
26.9	The Holevo Bound	*731*
26.10	Entropy of Entanglement	*733*
	Problems	*734*
	References	*735*

27	**Quantum Gates**	*737*
27.1	Truth Tables	*737*
27.2	Quantum Logic Gates	*741*
27.3	Shor's Algorithm	*746*
	Problems	*751*
	References	*752*

28	**Quantum Key Distribution** *753*	
28.1	Quantum Key Distribution (QKD) with EPR *753*	
28.2	BB84 Protocol *758*	
28.3	QKD as Communication Over a Channel *766*	
28.4	Postprocessing of the Key *769*	
28.5	B92 Protocol *776*	
28.6	Experimental Implementation of QKD Schemes *779*	
28.7	Advanced Eavesdropping Strategies *788*	
	Problems *793*	
	References *793*	

Appendix A: Classical Oscillator *795*
Reference *799*

Appendix B: Delta Function *801*
Reference *807*

Appendix C: Representation of Observables by Operators *809*
Appendix D: Elements of Matrix Algebra *813*
Appendix E: Eigenfunctions and Eigenvalues of the Orbital Angular Momentum Operator *817*
Appendix F: Hermite Polynomials *821*
Appendix G: Solutions of the Radial Schrödinger Equation in Free Space *825*
Appendix H: Bound States in the Coulomb Field *827*
Reference *829*

Index *831*

Preface

Quantum mechanics (QM) is one of the cornerstones of today's physics. Having emerged more than a century ago, it now forms the basis of all modern technology. And yet, while being long established as a science with astounding explanatory and predictive power, it still remains the arena of lively debates about some of its basic concepts.

This book on quantum mechanics and quantum information (QI) differs from standard texts in three respects. First, it combines QM and some elements of QI in one volume. Second, it does not contain those important applications of QM (helium atom, hydrogen molecule, condensed matter) that can be found almost in any other textbook. Third, it contains important topics (Chapters 13 and 20, Sections 6.6, 11.4–11.7, 24.1–24.2, 24.6, among others) that are not covered in most current textbooks. We believe that these topics are essential for a better understanding of such fundamental concepts as quantum evolution, quantum-mechanical indeterminacy, superposition of states, quantum entanglement, and the Hilbert space and operations therein.

The book began as an attempt by one of the authors to account to himself for some aspects of QM that continued to intrigue him since his graduate school days. As his research notes grew he came up with the idea of developing them further into a textbook or monograph. The other author, who, by his own admission, used to think of himself as something of an expert in QM, was not initially impressed by the idea, citing a huge number of excellent contemporary presentations of the subject. Gradually, however, as he grew involved in discussing the issues brought up by his younger colleague, he found it hard to explain some of them even to himself. Moreover, to his surprise, in many instances he could not find satisfactory explanations even in those texts he had previously considered to contain authoritative accounts on the subject.

Our original plan was to produce a short axiomatic formulation of the logical structure of quantum theory in the spirit of von Neumann's famous book [1]. It turned out, however, that the key ideas of QM are not so easy to compartmentalize. There were, on the one hand, too many interrelations between apparently different topics, and conversely, many apparently similar topics branched out into two or more different ones. As a result, we ended up adding lots of new examples and eventually reshaping our text into an entirely different framework. In hindsight, this evolution appears to resemble the way Ziman wrote his remarkable book on QM as a means of self-education [2].

The resulting book attempts to address the needs of students struggling to understand the weird world of quantum physics. We tried to find the optimal balance between mathematical rigor and clarity of presentation. There is always a temptation to "go easy on math", to "get to the subject matter quicker." We, however, feel that mathematical formalism must be treated with proper respect. While the high number of equations may understandably intimidate some readers, there is still no realistic way around it. On the other hand, replacing explanations of concepts entirely by formulas, expecting them to speak for themselves, may produce only a dangerous illusion of understanding. Hence, the way we see this book: a textbook with a human face, a guidebook by necessity abundant with equations – still not an easy reading! – but also with a thorough explanation of underlying physics. This is precisely what we feel is necessary for its intended audience – mostly college students.

There is an ongoing discussion of what would be the best way to introduce QM: doing it gradually, starting from some familiar classical concepts and showing their failure in quantum domain, or using the "shock therapy" – immediately placing the reader face to face with the most paradoxical quantum phenomena and force him/her to accept from the outset the entirely new concepts and mathematical tools necessary for their description.

We do not think that this question has a unique answer: the choice of the most suitable approach depends on many factors. The "shock therapy" method is efficient, straightforward, and saves lots of time (not to mention decreasing the size of the textbook!) when aimed at the curious and bright student – but only because such a student, precisely due to his/her curiosity, has already read some material on the topic and is thus prepared to absorb new information on a higher level. We, however, aimed at a maximally broad student audience, without assuming any preliminary acquaintance with quantum ideas. The only (but necessary!) prerequisites are the knowledge of standard classical topics – mechanics, electromagnetism (EM), and thermodynamics in physics; and linear algebra with matrices and operations on them, calculus, Fourier transforms, and complex variables in mathematics. Accordingly, our book uses gradual introduction of new ideas. To make the transition from classical to quantum-mechanical concepts smoother for the first-time reader, it starts with some familiar effects whose features demonstrate the inadequacy of classical concepts for their description. The first chapter shows the failure of classical physics in explaining such phenomena as the blackbody radiation (BBR), heat capacity, and photoeffect. We tried to do it by way of pointing to QM as a necessary new level in the description of the world. In this respect, our book falls out of step with many modern presentations that introduce the Schrödinger equation or the Hilbert space from the very beginning. We believe that gradual approach offers a good start in the journey through the quantum world.

QM is very different from other branches of physics: it is not just formal or counterintuitive – it is, in addition, not easy to visualize. Therefore, the task of presenting QM to students poses a challenge even for the most talented pedagogues.

The obstacles are very serious: apart from rather abstract mathematics involved, there is still no consensus among physicists about the exact meaning of some basic concepts such as wave function or the process of measurement [3–5]. The lack of self-explanatory models appealing to our "classical" intuition provides fertile soil for potential confusions and misunderstandings. A slightly casual use of technical language might be permissible in other areas of physics where the visual model itself provides guidance to the reader; but it often proves catastrophic for the beginning student of QM. For instance (getting back to our own experience) one of the authors, while still a student, found the representation theory almost incomprehensible. Later on, he realized the cause of his troubles: none of the many textbooks he used ever explicitly mentioned one seemingly humble detail, namely, that the eigenstates used as the basis may themselves be given in this or that representation. The other author, on the contrary, had initially found the representation theory to be fairly easy, only to stumble upon the same block later when trying to explain the subject to his partner.

When carefully considering the logical structure of QM, one can identify a fair number of such points where it is easy to get misled or to inadvertently mislead the reader. Examples of potential snags include even such apparently well-understood concepts as the uncertainty (actually, indeterminacy) principle or the definition of quantum ensembles. In this book, we deliberately zero in on those potential pitfalls that we were able to notice by looking at them carefully and sometimes by falling into them ourselves. Sometimes we devote whole sections to the corresponding discussions, as, for instance, when probing the connection between the azimuthal angle and angular momentum (Section 13.4).

There has been a noticeable trend to present QM as an already fully accomplished construction that provides unambiguous answers to all pertinent questions. In this book, we have tried to show QM as a vibrant, still-developing science potentially capable of radically new insights and important reconstructions. For this reason, we did not shy away from the issues that up to now remain in the center of lively debates among physicists: the meaning of quantum reality, the role of the observer, the search for hidden parameters, and the increasingly important topic of quantum nonlocality [3,6–8]. We also discuss some important developments such as quantum information and its processing [9–12], interaction-free measurements [13,14], and no-cloning theorem and its implications [15–18]. We devote a special chapter to a new aspect of QM, practically unknown to the broad audience: the so-called "inverse problem" and manipulation of quantum states by local changes of interaction potential or of the environment, which is especially relevant to the physics of nanostructures [19,20]. This is so important for both practical applications and developing a new dimension in our "quantum intuition" that it should become a part of all standard quantum textbooks.

Unfortunately, everything comes at a price. As mentioned in the beginning, focusing on some conceptual issues has left little room for some applications of quantum theory. In such cases, the reader can find the necessary information in the abundant standard texts [21,22], and the basic hope is that he/she will be already sufficiently equipped for reading them.

This book describes mostly nonrelativistic QM and has a clear structure. After showing the limitations of classical physics, we discuss the new concepts necessary to explain the observed phenomena. Then we formulate de Broglie's hypothesis, followed by two postulates: the probabilistic nature of the wave function and the superposition principle. It is shown then that the whole mathematical framework of the theory follows from these two postulates.

We introduce the idea of observables and their operators, eigenstates and eigenvalues, and quantum-mechanical indeterminacy with some simple illustrating examples in Chapter 3. We do not restrict to the initial concept of the wave function as a function of r and t, and show as early as possible that it could be a function of other observables. The ideas of representation theory are introduced at the first opportunity, and it is emphasized from the beginning that the wave function "has many faces."

We made every effort to bridge the gulf between the differential and matrix forms of an operator by making it clear that, for instance, Hermitian operators are those that have a Hermitian matrix. We emphasize that going from one form to another usually means changing the representation – another potential source of confusion.

We discuss the superposition principle in more detail than other textbooks. In particular, we emphasize that the sum of squares of probability amplitudes gives 1 only if the corresponding eigenfunctions are square integrable *and* the basis was orthogonalized. We introduce nonorthogonal bases as early as in Chapter 6 and derive the normalization condition for this more general case.

We did not rush to present the Schrödinger equation as soon as possible, all the more so that it is only one nonrelativistic limit of more general relativistic wave equations, which are different for particles with different spins. It expresses, among other things, the law of conservation of energy, and it follows naturally (Chapter 8) once the energy, momentum, and the angular momentum operators are introduced.

We feel that we've paid proper attention to quantum ensembles (Chapter 12). The idea of the pure and mixed ensembles is essential for the theory of measurements. On the other hand, we make it explicitly clear that not every collection of particles "in the same physical situation" forms a statistical mixture of pure states as traditionally defined, due to the possibilities of local interaction and entanglement.

In many cases, the sequence of topics reminds a helix: we return to some of them in later chapters. But each such new discussion of the same topic reveals some new aspects, which helps to achieve a better understanding and gain a deeper insight into the problem.

We are not aware of other textbooks that use both the "old" and Dirac's notation. We cover both notations and use them interchangeably. There is a potential benefit to this approach, as the students who come to class after having been exposed to a different notation in an introductory course are often put at a relative disadvantage.

We want to thank Nick Herbert and Vladimir Tsifrinovich for valuable discussions, Boris Zakhariev for acquainting us with the new aspects of the inverse problem, and Francesco De Martini for sharing with us his latest ideas on the problem of signal exchange between entangled systems. We enjoyed working with the Project Editor Nina Stadthaus during preparation of the manuscript and are grateful to her for her patience in dealing with numerous delays at the later stages of the work.

References

1 von Neumann, J. (1955) *Mathematical Foundations of Quantum Mechanics*, Princeton University Press, Princeton, NJ.
2 Ziman, J.M. (1969) *Elements of Advanced Quantum Theory*, Cambridge University Press, Cambridge.
3 Herbert, N. (1987) *Quantum Reality: Beyond New Physics*, Anchor Books, New York.
4 Bell, J.S. (2001) *The Foundations of Quantum Mechanics*, World Scientific, Singapore.
5 Bell, J.S. (1990) Against measurement, in *Sixty Two Years of Uncertainty*, (ed A. Miller), Plenum Press, New York.
6 Gribbin, J. (1995) *Schrodinger's Kittens and the Search for Reality*, Little, Brown and Company, Boston, MA.
7 Greenstein, G. and Zajonc, A.G. (1997) *The Quantum Challenge: Modern Research on the Foundations of Quantum Mechanics*, Jones & Bartlett Publishers, Boston, MA.
8 Kafatos, M. and Nadeau, R. (1990) *The Conscious Universe: Part and Whole in Modern Physical Theory*, Springer, New York.
9 Berman, G., Doolen, G., Mainieri, R., and Tsifrinovich, V. (1998) *Introduction to Quantum Computers*, World Scientific, Singapore.
10 Benenti, G., Casati, G., and Strini, G. (2004) *Principles of Quantum Computation and Information*, World Scientific, Singapore.
11 Bouwmeester, D., Ekert, A., and Zeilinger, A. (2000) *The Physics of Quantum Information: Quantum Cryptography, Quantum Teleportation, Quantum Computation*, Springer, Berlin.
12 Nielsen, M.A. and Chuang, I.L. (2000) *Quantum Computation and Quantum Information*, Cambridge University Press, Cambridge.
13 Elitzur, A.C. and Vaidman, L. (1993) Quantum-mechanical interaction-free measurements. *Found. Phys.*, **23** (7), 987–997.
14 Kwiat, P., Weinfurter, H., and Zeilinger, A. (1996) Quantum seeing in the dark. *Sci. Am.*, **275**, 72–78.
15 Herbert, N. (1982) FLASH – a superluminal communicator based upon a new kind of quantum measurement. *Found. Phys.*, **12**, 1171.
16 Wooters, W.K. and Zurek, W.H. (1982) A single quantum cannot be cloned. *Nature*, **299**, 802.
17 Dieks, D. (1982) Communication by EPR devices. *Phys. Lett. A*, **92** (6), 271.
18 van Enk, S.J. (1998) *No-cloning and superluminal signaling*. arXiv: quant-ph/9803030v1.
19 Agranovich, V.M. and Marchenko, V.A. (1960) *Inverse Scattering Problem*, Kharkov University, Kharkov (English edition: Gordon and Breach, New York, 1963).
20 Zakhariev, B.N. and Chabanov, V.M. (2007) *Submissive Quantum Mechanics: New Status of the Theory in Inverse Problem Approach*, Nova Science Publishers, Inc., New York.
21 Blokhintsev, D.I. (1964) *Principles of Quantum Mechanics*, Allyn & Bacon, Boston, MA.
22 Landau, L. and Lifshits, E. (1965) *Quantum Mechanics: Non-Relativistic Theory*, 2nd edn, Pergamon Press, Oxford.

Abbreviations and Notations

Abbreviations

BBR	blackbody radiation
BS	beam splitter
CM	classical mechanics
C_M	center of mass
C_M RF	reference frame of center of mass
e	electron
EM	electromagnetism
EPR	Einstein, Podolsky, and Rosen
F	fermi (a unit of length in micro-world, $1\,F = 10^{-15}$ m)
FTL	faster than light
IP	inverse problem
n	neutron
p	a proton or positron
PEE	photoelectric effect
QM	quantum mechanics
RF	reference frame
SR	special relativity
UV	ultraviolet (range of spectrum)

Notations

$\bar{F}$ or $\langle F \rangle$	average of F
1D, 2D, ..., n D or ND	one-dimensional, two-dimensional, ..., n-dimensional or N-dimensional

$$\mathrm{Det}(M) = \mathcal{D}(M) = \begin{vmatrix} a_{11} & a_{12} & \cdots & a_{1n} \\ a_{21} & a_{22} & \cdots & a_{2n} \\ \vdots & \vdots & \ddots & \vdots \\ a_{n1} & a_{n2} & \cdots & a_{nn} \end{vmatrix}$$

$$\text{determinant of matrix } \begin{pmatrix} a_{11} & a_{12} & \cdots & a_{1n} \\ a_{21} & a_{22} & \cdots & a_{2n} \\ \vdots & \vdots & \ddots & \vdots \\ a_{n1} & a_{n2} & \cdots & a_{nn} \end{pmatrix}$$

1
The Failure of Classical Physics

Quantum mechanics (QM) emerged in the early twentieth century from attempts to explain some properties of blackbody radiation (BBR) and heat capacity of gases, as well as atomic spectra, light–matter interactions, and behavior of matter on the microscopic level. It soon became clear that classical physics was unable to account for these phenomena. Not only did classical predictions disagree with experiments, but even the mere existence of atoms seemed to be a miracle in the framework of classical physics. In this chapter, we briefly discuss some of the contradictions between classical concepts and observations.

1.1
Blackbody Radiation

First, we outline the failure of classical physics to describe some properties of radiation.

A macroscopic body with absolute temperature $T > 0$ emits radiation, which generally has a continuous spectrum. In the case of thermal equilibrium, in any frequency range the body absorbs as much radiation as it emits. We can envision such a body as the interior of an empty container whose walls are kept at a constant temperature [1,2]. Its volume is permeated with electromagnetic (EM) waves of all frequencies and directions, so there is no overall energy transfer and no change in energy density (random fluctuations neglected). Its spectrum is independent of the material of container's walls – be it mirrors or absorbing black soot. Hence, its name – the *blackbody radiation*. In an experiment, we can make a small hole in the container and record the radiation leaking out.

There is an alternative way [3] to think of BBR. Consider an atom in a medium. According to classical physics, its electrons orbit the atomic nucleus. Each orbital motion can be represented as a combination of two mutually perpendicular oscillations in the orbital plane. An oscillating electron radiates light. Through collisions with other atoms and radiation exchange, a thermal equilibrium can be established. In equilibrium, the average kinetic energy per each degree of freedom is [1]

$$E_i = \frac{1}{2}kT, \tag{1.1}$$

where k is the *Boltzmann constant*. This is known as the *equipartition theorem*. The same formula holds for the average potential energy of the oscillator, so the total energy (average kinetic + average potential) per degree of freedom is kT. Thus, we end up with the average total energy kT per oscillation. In an open system, such equilibrium cannot be reached because the outgoing radiation is not balanced and the energy leaks out. This is why any heated body cools down when disconnected from the source of heat.

But if the medium is sufficiently extended or contained within a cavity whose walls emit radiation toward its *interior*, then essentially all radiation remains confined, and thermal equilibrium can be attained. Each oscillator radiates as before, but also absorbs radiation coming from other atoms. In equilibrium, both processes balance each other. In such a case, for each temperature T and each frequency ω there exists a certain characteristic energy density $\rho(\omega, T)$ of radiation such that the rate of energy loss by atoms through emission is exactly balanced by the rate of energy gain through absorption. The quantity $\rho(\omega, T)$ is called *spectral energy density* (the energy density per one unit of frequency range). In classical EM theory, it is determined by the corresponding field amplitudes $\mathcal{E}(\omega)$ and $\mathbf{B}(\omega)$ of monochromatic waves with frequency ω:

$$\rho(\omega, T) = \frac{1}{4}\left(\varepsilon_0|\mathcal{E}(\omega)|^2 + \frac{|\mathbf{B}(\omega)|^2}{\mu_0}\right) = \frac{1}{2}\varepsilon_0|\mathcal{E}(\omega)|^2 = \frac{|\mathbf{B}(\omega)|^2}{2\mu_0}. \tag{1.2}$$

The last two expressions in (1.2) are obtained in view of the relation $\mathbf{B} = \sqrt{\mu_0\varepsilon_0}(\hat{\mathbf{n}} \times \mathcal{E})$, where $\hat{\mathbf{n}}$ is the unit vector along the wave propagation. Note that ω is the angular frequency,[1] and all quantities involved are measured in the rest frame of the given medium.[2]

Under the described conditions, $\rho(\omega, T)$ is a *universal* function of ω and T. According to thermodynamics (Kirchhoff's law of thermal radiation), it must be the product of ω^3 and another universal function of ω/T [2,3]:

$$\rho(\omega, T) = \alpha\omega^3 f\left(\frac{\omega}{T}\right). \tag{1.3}$$

Using (1.3), one can show that the *total energy density* $\eta(T)$ of BBR is

$$\eta(T) = \int_0^\infty \rho(\omega, T)d\omega = \sigma T^4, \tag{1.4}$$

---

1) We will use throughout the book the angular frequency ω, which is ordinary frequency f (number of cycles/s) multiplied by 2π. In the physicists' jargon, the word "angular" is usually dropped.
2) Rest frame of an object is the frame of reference where the object's center of mass is at rest.

where

$$\sigma \equiv \alpha \int_0^\infty \xi^3 f(\xi) \mathrm{d}\xi, \quad \xi \equiv \frac{\omega}{T}. \tag{1.5}$$

The relation (1.4), known as the *Stefan–Boltzmann law*, is exact and has been experimentally confirmed. Figure 1.1 shows a few graphs of $\rho(\omega, T)$ obtained from experiments. But all attempts to derive the pivotal function $f(\omega/T)$ determining $\rho(\omega, T)$ and σ have failed.

By 1900 there were two half-successful attempts to derive $\rho(\omega, T)$. Their results were different due to the different models chosen to represent radiation.

The first model considered radiation as EM waves. In this model, the molecules interacting with radiation were represented as harmonic oscillators; similarly, each monochromatic component of radiation can also be considered as an oscillator with the corresponding frequency. Then, the total energy density could be evaluated as a product of the average energy $\langle E \rangle = kT$ per one EM oscillator and the number N of oscillators occupying all states with frequency ω [4]. Such an approach results in

$$\rho(\omega, T) = N(\omega)\langle E \rangle = \frac{\omega^2}{\pi^2 c^3} kT. \tag{1.6}$$

This expression is known as the *Rayleigh–Jeans formula*. Note that it does have the form (1.3). But, while matching the data at low frequencies, it diverges at high frequencies (Figure 1.2), predicting the infinite spectral density $\rho(\omega, T)$ and infinite total energy density $\eta(T)$ at $\omega \to \infty$, even at low temperatures! This conclusion of classical theory was dubbed "the UV catastrophe." Something was wrong with the classical notion of energy exchange between matter and radiation as a continuous process, especially when applied to the high-frequency range!

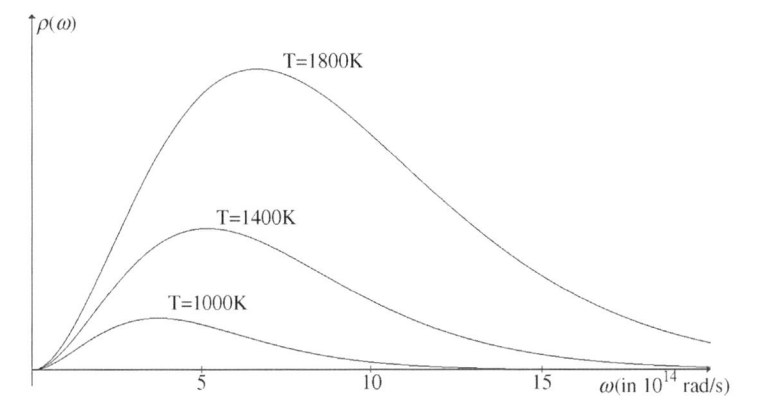

Figure 1.1 The BBR spectrum at various temperatures.

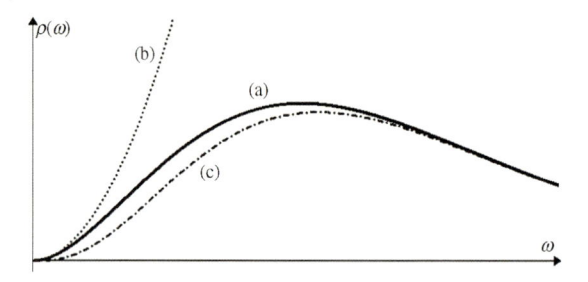

Figure 1.2 The BBR spectrum according to different approaches treating radiation as classical waves or particles, respectively. (a) BBR spectrum; (b) the Rayleigh–Jeans curve; (c) the Wien curve.

The second model suggested by W. Wien used the Newtonian view of radiation as a flux of particles. Applying to such particles Boltzmann's statistical treatment, he obtained the expression

$$\rho(\omega, T) = \text{const}\, \omega^3\, e^{-\gamma(\omega/T)}, \tag{1.7}$$

where γ is another constant. This expression also satisfies the requirement (1.3), and, in contrast with (1.6), it describes accurately the experimental data for high frequencies. However, it does not match the data at low frequencies (Figure 1.2). Something was wrong with the notion of radiation as classical particles, especially in the low-frequency range!

Thus, regardless of whether we view radiation as purely classical waves (Maxwell) or purely classical particles (Newton), either view only partially succeeds. The wave picture works well in describing low frequencies, and the particle picture works for high frequencies, but both fail to describe *all* available data. That was the first indication that the EM radiation is neither exactly waves nor exactly particles.

1.2
Heat Capacity

Heat capacity is the amount of heat dQ required to change a body's temperature T by 1 K: $C = dQ/dT$. We model the body as an *ideal gas* whose molecules do not interact with each other. The analysis for an ideal gas hinges on the number of degrees of freedom. For an atom considered as a point-like object, three mutually independent directions of its motion (or three components of its position vector) form three degrees of freedom. A diatomic molecule presents a more complex case. If it is a rigid pair of two point masses, then it has five degrees of freedom – three coordinates of its center of mass and two angular coordinates specifying the orientation of its axis. If the separation s between the two masses can change (e.g., two masses connected by a spring), then it becomes a variable s, and the total number j of degrees of freedom jumps from 5 to 6. This is the maximal number for a diatomic molecule formed from two point-like atoms. The number j here can also be determined as 3×2 (three degrees of freedom per particle times the number of particles).

But as stressed in comment to Eq. (1.1), the vibrational degree of freedom "absorbs" the energy kT, where the additional amount $(1/2)kT$ is due to the average potential energy of vibration. This can be formally described by adding and extra degree of freedom for each vibration, so that for a classical diatomic molecule we can write $j' = j + 1 = 7$.

Suppose we have a system of particles (e.g., a container with gas) in a state of thermodynamic equilibrium. The gas in this case is described by the *ideal gas equation* $PV = NkT$ [3,5,6], where P and V are the gas pressure and volume, respectively, and N is the number of molecules. For one mole of gas, that is, $N = N_A$, where N_A is the Avogadro number, we have

$$PV = RT, \tag{1.8}$$

where $R = N_A k$ is the universal gas constant.

Let us now recall the relationship between pressure P and the internal energy U of the gas, $P = (2/j')(U/V)$. Combining this with (1.8) gives

$$U = \frac{j'}{2} RT. \tag{1.9}$$

There are two different types of heat capacity depending on two possible ways of transferring heat to a system. We can heat a gas keeping it either at fixed volume or at fixed pressure. The corresponding *molar heat capacities* will be denoted as c_P and c_V, respectively. To find them, recall the first law of thermodynamics [2,3], $dQ = dU + dW = dU + PdV$, where dW is an incremental work done by the system against external forces while changing its volume by an incremental amount dV. Applying the basic definition $C = dQ/dT$, we have for the case of fixed volume $dV = 0$:

$$C \to c_V = \frac{\partial U}{\partial T} = \frac{j'}{2} R. \tag{1.10}$$

When $P =$ const, we obtain

$$C \to c_P = \frac{\partial U}{\partial T} + P \frac{\partial V}{\partial T} = c_V + P \frac{\partial V}{\partial T}. \tag{1.11}$$

By virtue of (1.8) taken at $P =$ const, this gives

$$c_P = c_V + R = \left(\frac{j'}{2} + 1\right) R. \tag{1.12}$$

The ratio

$$\gamma \equiv \frac{c_P}{c_V} = 1 + \frac{2}{j'} \tag{1.13}$$

gives us direct information about the number j'.

In the outlined classical picture, the number j' and thereby c_P, c_V, and γ are all independent of T. But this contradicts the experiments. Heat capacities of all substances at low temperatures turn out to be noticeably less than predicted and go to zero in the limit $T \to 0$. Shown in Table 1.1 are the classically predicted values of c_V, c_P, and γ for a few different substances and their experimental values at room temperature [6].

1 The Failure of Classical Physics

Table 1.1 Predicted versus observed heat capacities at $T = 293$ K (in J/(mol K)).

Gas	j	c_V		c_P		γ	
		Theory	Exp	Theory	Exp	Theory	Exp
Helium (He)	3	12.47	12.46	20.79	20.90	1.67	1.67
Hydrogen (H$_2$)	6	24.42	20.36	32.73	28.72	1.33	1.41
Water vapor (H$_2$O)	9	37.41	27.80	45.73	36.16	1.22	1.31
Methane (CH$_4$)	15	62.36	27.21	70.67	35.57	1.13	1.30

The table shows a very interesting (and mixed) picture. The measurement results almost exactly confirm theoretical predictions for monatomic gases such as helium. The calculated and measured values of c and γ are in excellent agreement for $j = 3$.

For diatomic gases ($j = 6$, $j' = 7$), however, the results are more complicated. Consider, for instance, hydrogen. Its measured value of $c_V = 20.36$ is significantly lower than the expected value 24.42. A similar discrepancy is observed for γ: the measured value is 1.41 instead of predicted 1.33. A closer look at these numbers reveals something very strange: they are still in excellent agreement with Equations 1.10–1.13, but for $j' = 5$ instead of 7. Thus, the agreement can be recovered, but only at the cost of decreasing the number j' ascribed to a diatomic molecule. It looks like two "effective" degrees of freedom "freezes" when the particles get bound to one another. Which kind of motion could possibly undergo "freezing"? Certainly not 2 out of 3 translational motions: there is nothing in the isotropic space that could single out one remaining motion. It could be either 2 rotational motions, or 1 vibrational motion. Running ahead of ourselves, we will say here that it is vibrational motion that freezes first as the gas is cooled down. Already at room temperatures, we cannot pump energy into molecular vibrations. At these temperatures, the connection between the two atoms in an H$_2$ molecule is effectively absolutely rigid.

Thus, already the experimental results for different gases at room temperature show that something is wrong with the classical picture. But the situation becomes even worse if we carry out experiments for the same gas at widely different temperatures. As an example, consider the data for molecular hydrogen (Table 1.2).

At sufficiently low temperatures, experimental values become lower than the classical prediction even after we ascribe to the diatomic molecules only five degrees of freedom instead of effective seven. For instance, the experimental

Table 1.2 Constant-volume heat capacity of hydrogen as a function of temperature (in degrees Kelvin)

T	c_V
197	18.32
90	13.60
40	12.46

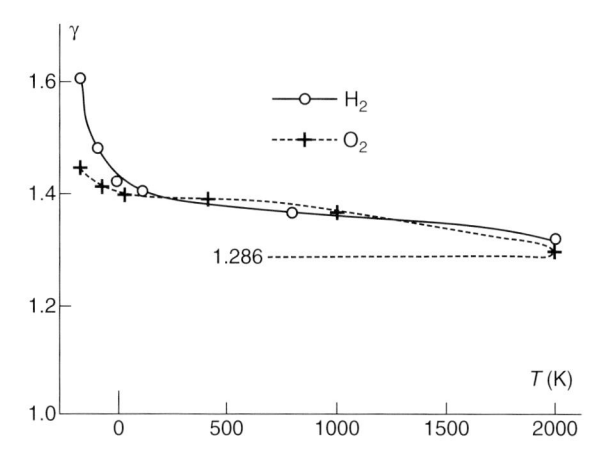

Figure 1.3 Experimental values of γ as functions of T for H_2 and O_2. The dashed horizontal line is the classical prediction for a diatomic molecule with vibrational degree of freedom ($j' = 7$). (Reproduced from Refs [3,6].)

value $c_V = 12.46$ at 40 K is significantly less than $c_V = 20.36$ measured at room temperature. It could still fall within the classical prediction, but only at the cost of reducing the number of degrees of freedom from 5 to 3. It looks as if more and more degrees of freedom become frozen as the substance is cooled down. This time we can, by the same argument as before, assume the freezing of two rotational degrees of freedom associated with the spatial orientation of the molecule.

The same tendency is observed in the measurements of γ (Figure 1.3). Contrary to the classical predictions, experiments show that γ increases with T, and this behavior, in view of Equation 1.13, can be attributed to the same mysterious mechanism of "freezing." We can thus say that decrease of c_V and increase of γ at low temperatures represent two sides of the same coin.

The general feature can be illustrated by Figure 1.4. It shows that heat capacities fall off with temperature in a step-like fashion. Each step is associated with the freezing of one or two degrees of freedom. At the end of this road, all of the initial degrees of freedom are frozen and accordingly heat capacities approach zero.

If we now increase the temperature, starting from the absolute zero, we observe the same phenomenon in reverse. As we heat a body, it effectively regains its degrees of freedom all the way back to their normal number at sufficiently high temperatures.

What causes these strange effects? And what is the "normal" number for γ to begin with? Classical physics cannot answer these questions even in the simplest case of a monatomic gas. Let us, for instance, get back to helium. We started with the apparently innocuous statement that a helium atom at room temperature has $j = 3$. But after a second thought we can ask: "Why is j equal to 3, in the first place?" After all, the helium atom consists of the nucleus and two electrons, so there are three particles in it, and therefore there must be $j = 3 \times 3 = 9$. Further, the He nucleus consists of two protons and two neutrons, so altogether we have six particles in a

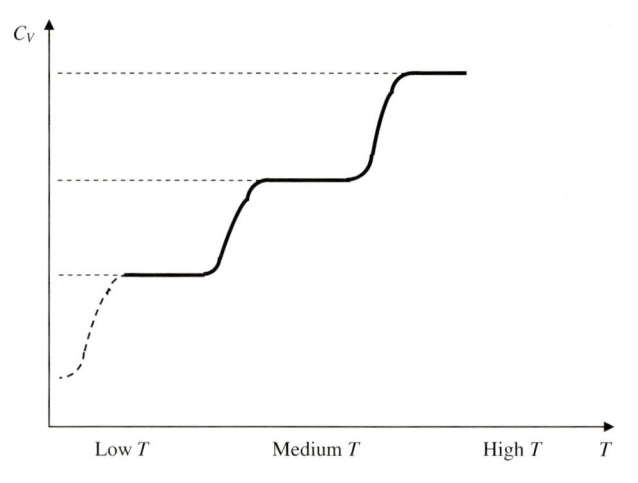

Figure 1.4 Graph of the heat capacity versus temperature for a diatomic ideal gas.

helium atom and accordingly the number j must be $j = 6 \times 3 = 18$ instead of 3. And if we also realize that each nucleon, in turn, consists of three quarks, which makes the total number of particles in the helium atom equal to 14, then the number j must be $j = 14 \times 3 = 42$. Accordingly, the theoretical prediction for the molar heat capacities for He must be $c_V = 174.6$ and $c_P = 182.9$ instead of 12.46 and 20.8, respectively. In other words, already at room temperature and in the simplest case of a monatomic gas, there is a wide (by about one order of magnitude!) discrepancy between theory and experiment. Experiment shows that *nearly all* degrees of freedom of the subatomic particles are frozen so fundamentally that they are as good as nonexisting, at least at room temperatures, and only the remaining three degrees determining the motion of the atom as a whole survive. Why is this so?

One could try to explain this by the fact that the binding forces between the electrons and the atomic nucleus are so strong that they practically stop any relative motion within an atom; this is true even more so for the protons and neutrons within a nucleus, and so on. As we go farther down the subatomic scale, the interaction forces increase enormously, thus "turning off" the corresponding degrees of freedom. But this argument does not hold. The notion about the forces is true, but the conclusion that it must "turn off" the corresponding motions is wrong.

The equipartition theorem is a very general statement that applies to *any* conservative (i.e., described by potential) forces, regardless of their physical nature or magnitude. As a simple example, consider two different types of diatomic molecules in thermodynamic equilibrium. Let each molecule be represented by a system of two masses connected by a spring, but the spring constant is much higher for one molecule than for the other. Equilibrium is established in the process of collisions between the molecules, in which they can exchange their energy. Eventually, molecules of both types will have, on average, an equal amount of vibration energy. The total mechanical energy of vibration for a spring described by

Hooke's law is $E = kA^2/2$, where k is the spring constant (not to be confused with the Boltzmann constant!) and A is the amplitude of vibration. In thermal equilibrium, we will have $E_1 = E_2$, that is, $k_1 A_1^2 = k_2 A_2^2$.

The fact that one spring is much stronger than the other will only result in the smaller amplitude of vibration for this type of molecules, that is, $A_2 = \sqrt{k_1/k_2} A_1$, and $A_2 \ll A_1$ if $k_2 \gg k_1$, but it *will not affect the energy* of vibration. Applying this to our system, we can say that the constituents of an atomic nucleus, according to the classical theory, must jitter with a very small amplitude, but with the same energy as twice the average energy of the atom's translational motion in a state of thermal equilibrium. And this energy must contribute to the observed heat capacity on the same footing as the energy of translational motion, even in the limit of arbitrarily high k. But no such contributions are evident in the observed capacities. It turns out that the word "freezing" is not strong enough and must be understood as *total elimination* of any contributions from the corresponding degrees of freedom.

It is only because the physicists did not know much and accordingly did not think much about the subatomic structure of matter 100 years ago that they could mislead themselves into believing that the existing theory at least partially accounted for the experimental observations. Strictly speaking, there was not even a partial match between the two.

What could have caused such a miserable failure of the classical picture? As we try to figure it out and go to the origins of the equipartition theorem, we realize that it was the assumption that energy exchange between the systems is a continuous process and the energy of a bound system of particles is a continuous variable. The resulting discrepancy with observations shows that there was something wrong with these classical notions of energy.

1.3
The Photoelectric Effect

The *photoelectric effect* (PEE) is the ejection of electrons from the surface of an illuminated conducting material (mostly metals). Such an effect is by itself easy to understand in the framework of classical physics. The conducting electrons in a metallic plate are bound to it by the electric forces in such a way that, while being free to move within the plate, they are not free to leave it. This can be modeled by a simple picture of an electron trapped within a potential well of macroscopic dimensions (Figure 1.5). The electron within such a well has a negative potential energy, and if its kinetic energy is not sufficiently high, it can only move within the well by bouncing off its walls, but it cannot go beyond the wall, and thus finds itself trapped.

However, when the plate is illuminated, those electrons that are sufficiently close to the surface get exposed to the EM field, which starts "shaking" them with an oscillating force. The resulting motion of the electron under such a force is well known [7,8]. For instance, if the light is monochromatic and linearly polarized, it will cause the electron to oscillate along the direction of the electric field with the frequency ω of the incident wave. In addition, the electron will start drifting along

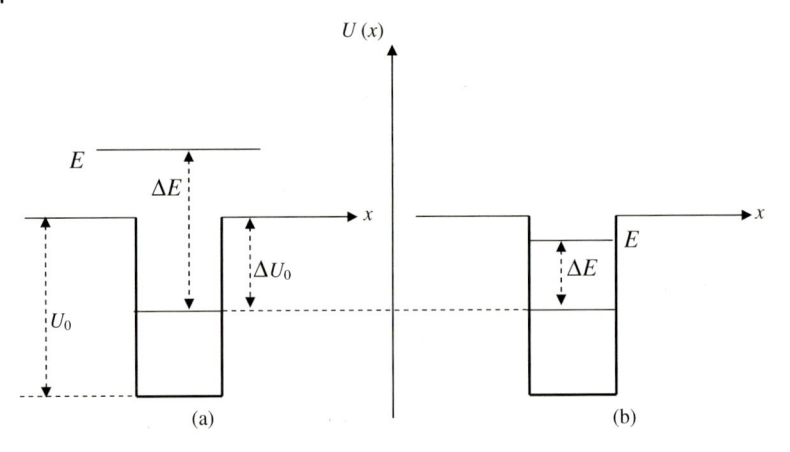

Figure 1.5 An electron in a valence band in a conducting plate can be represented as trapped within a potential well U_0. The electron can break loose of the plate only if it obtains from the environment the minimal energy equal to ΔU_0 (the work function). (a) The electron obtains the energy $K > \Delta U_0$ and becomes free. (b) The electron obtains the energy $K < \Delta U_0$. In this case, it remains trapped within the plate.

the direction of wave propagation. The kinetic energy associated with both kinds of motion stands in proportion to the light intensity I. Therefore, one can expect that at a sufficiently bright illumination the electron eventually will accumulate enough energy to quit the plate.[3] Thus, the effect itself could be easily explained. However, its specific details were in flat contradiction with theoretical predictions.

First, according to the simple picture outlined above, a sufficiently intense light beam, regardless of its frequency, must cause electron emission from the illuminated surface and should produce free electrons with accordingly high kinetic energy. On the other hand, if I is less than a certain critical value depending on the kind of material involved, no electrons will be emitted since they cannot collect enough energy to break loose from the trap (Figure 1.5b).

In fact, however, it was found that for a sufficiently high ω, no matter how dim the incident light, at least one emitted electron can be observed, and the number of emissions increases with I, but no matter how intense the light, the maximum kinetic energy of ejected electrons is the same. So instead of kinetic energy, it is the *number* of the ejected electrons that increases in proportion to I. This statement is known as the *first law of PEE* (or Stoletow's law).

---

3) The explanation seems straightforward to us, as we are looking at it retrospectively, but in fact the discovery of the PEE by Hertz, in 1887, preceded the discovery of the electron (the latter would only be discovered in 1897 by J.J. Thomson). For this reason, the nature of the phenomenon was not quite as obvious to nineteenth-century physicists as it is to us. Actually, Thomson *used* the photoeffect in his cathode ray tube experiments, which led him to identify the electron as a charged subatomic particle.

Second, the classical picture predicts that electron emission will occur regardless of the value of ω. At any ω, an electron could eventually accumulate enough energy to overcome the potential barrier if I is sufficiently high and the exposure is long enough. Moreover, according to the classical picture, low-frequency waves pump energy into electrons *more* efficiently than high-frequency ones since in each cycle the former provide more time for an electron to accelerate in one direction and thereby to attain higher speed and kinetic energy [8]. But experiments performed by Lenard in 1902 demonstrated that no matter how intense the incident light, there were no electrons emitted when ω was below a certain critical value (the *threshold frequency*), depending on the kind of material, and above the threshold frequency there was an emission, with the maximum kinetic energy of the emitted electrons increasing linearly with ω (Figure 1.6). This is the *second law of photoeffect*.

Third, classical physics predicts the existence of a certain time interval (the transition period) between the beginning of exposure and the beginning of the resulting emission. This seems natural, since according to the classical view any energy exchange is a continuous process, and it always takes a certain time for a system's energy to change. However, in experiments, electron emission started practically instantly (within less than 10^{-9} s) after the illumination. There was no

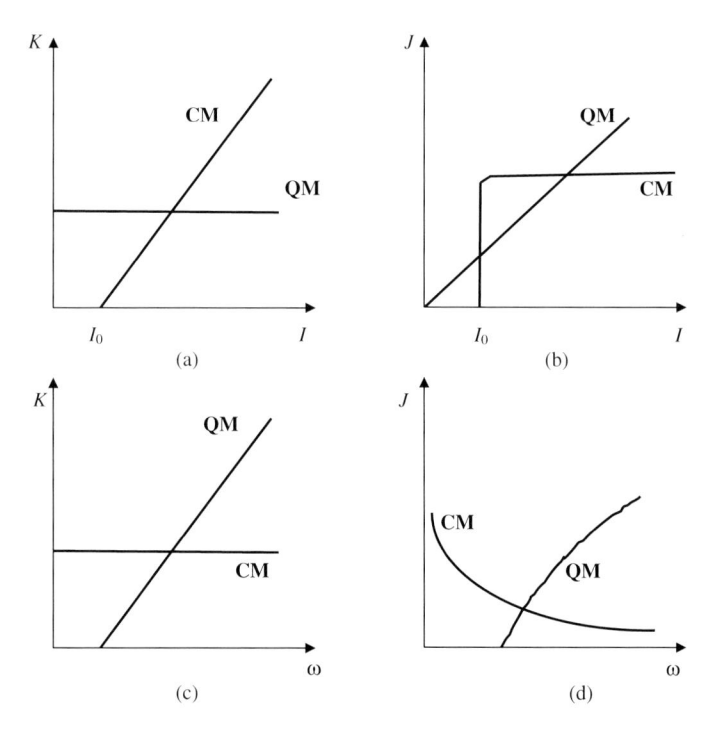

Figure 1.6 Comparison of CM and QM predictions for the basic characteristics of the photoeffect. (a) Dependence of the electrons' kinetic energy $K = K(I)$ on the intensity I of the incident light; (b) dependence of photocurrent $J(I)$ on I; (c) dependence of K on the light frequency ω; (d) dependence of J on ω.

way to reconcile this observation with the notion of interaction as a continuous process.

Summary. In the example considered here (as well as examples in the previous two sections), the notion of continuity of certain physical characteristics such as energy, intensity of monochromatic light, and so on leads to the wrong description of a real process.

1.4
Atoms and Their Spectra

According to the classical picture (the "planetary" model) based on Rutherford's experiments, an atom is a system of electrons orbiting around the nucleus like planets around the sun (hence the name of the model). This model looked simple and very compelling, all the more so that the interaction law (attraction $F \sim r^{-2}$) is mathematically the same in both cases. And some could find it philosophically attractive as well: the big is just an upscale of the small, and vice versa!

However, according to the same classical physics, such an atom cannot exist. The reason is very simple. The electrons in a Rutherford's atom seem to be in far more favorable conditions than, say, Earth's artificial satellites, which eventually spiral downward due to a small drag force in the upper atmosphere. In contrast, the electrons appear totally free of any dissipative forces: there is nothing else in the space around the nucleus. They should be ideal "planets" – no energy losses on the way!

This argument overlooks the fundamental fact that each electron carries an electric charge, and accordingly, its own electromagnetic field. Due to this field, electrons become down-spiraling artificial satellites rather than ideal planets. According to the EM theory, if an electron is moving with constant velocity, then its field just follows this motion, remaining "rigidly" attached to its "master" [7,8]. If, however, the electron accelerates, its field is getting partially detached. This is precisely what happens with an orbiting electron – orbiting involves acceleration! Accordingly, the electron must be losing its energy, which is radiated away together with the "detached" part of its field. The classical atomic electron must, in a way, move in an "atmosphere" of its own radiation field, and it must lose energy due to the "radiative friction" in this atmosphere [3,7]. As a result, very soon (in about 10^{-8} s) all electrons, having emitted a blend of electromagnetic waves of different frequencies, must fall onto the nucleus, and Rutherford's atom will cease to exist.

It seems that classical physics has come to a dead end. On the one hand, the Rutherford's experiments have shown that his planetary model is the only one possible. On the other hand, according to the ongoing experiment carried out by *Nature*, atoms are stable. And in cases when they do radiate (say, in collisions or after an optical excitation), the corresponding spectrum is discrete: one sees on a dark screen or a photographic film a set of distinct spectral lines (Figure 1.7). An atom of each chemical element has its own unique discrete spectrum.

Moreover, the features of this spectrum defied all classical notions about its possible origin. In principle, one could try to explain the discrete spectrum

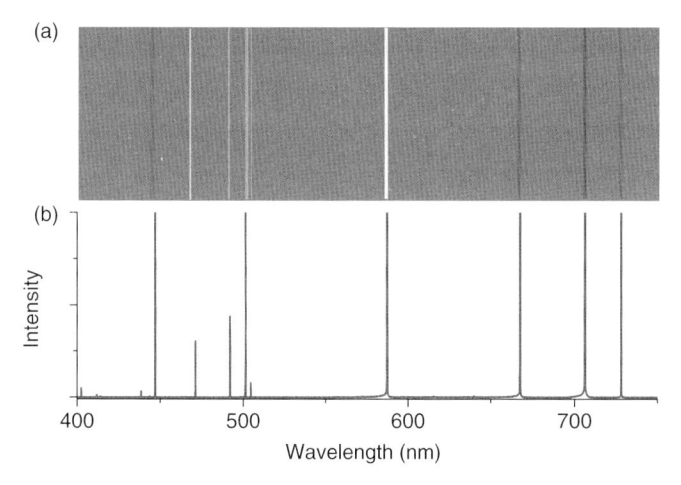

Figure 1.7 (a) Visible spectrum of a He lamp. (b) Schematic of the corresponding optical transitions of a He atom. (Courtesy Andrei Sirenko, Department of Physics, NJIT.)

classically by neglecting the continuous spiraling of the electron toward its nucleus. Suppose that spiraling is very slow as compared with the orbital frequency (it takes much longer than one complete cycle). Then the electron's motion can be approximated as periodic. In this case, its coordinates in the orbital plane can be expanded into a Fourier series [9]

$$x(t) = \sum_{n=-\infty}^{\infty} x_n e^{in\omega_1 t}, \quad y(t) = \sum_{n=-\infty}^{\infty} y_n e^{in\omega_1 t}, \tag{1.14}$$

where ω_1 is the fundamental frequency and $|n| > 1$. According to EM theory, the amplitudes x_n and y_n determine the intensity I_n of radiation with frequency $\omega_n = n\omega_1$. The possible frequencies thus form a discrete linear set $\omega = \omega_1, \omega_2, \ldots, \omega_n, \ldots$. In the same way, one can arrange the corresponding intensities I_n. However, this classical picture was also in contradiction with experiment, and even in two different ways. First, the *observed* frequencies are determined by *two* numbers m and n from the linear array of *auxiliary* frequencies $\Omega = \Omega_1, \Omega_2, \ldots, \Omega_n, \ldots$ in such a way that each ω comes as the difference of two Ω_i, that is,

$$\omega_n \rightarrow \omega_{mn} \equiv \Omega_m - \Omega_n \tag{1.15}$$

(*Ritz's combination principle*) [9]. In other words, instead of forming a linear array, the observed frequencies form a matrix

$$\omega = \begin{pmatrix} 0 & \omega_{12} & \omega_{13} & \cdots & \omega_{1n} & \cdots \\ \omega_{21} & 0 & \omega_{23} & \cdots & \omega_{2n} & \cdots \\ \vdots & \vdots & \vdots & \ddots & \vdots & \vdots \\ \omega_{m1} & \omega_{m2} & \omega_{m3} & 0 & \omega_{mn} & \cdots \\ \vdots & \vdots & \vdots & \vdots & \vdots & \ddots \end{pmatrix}. \tag{1.16}$$

A similar matrix is formed by the corresponding intensities or the oscillation amplitudes: $I_n \to I_{mn}$; $x_n \to x_{mn}$. Second, the auxiliary frequencies Ω_n do not form multiple harmonics of a certain fundamental frequency, that is, $\Omega_n \neq n\Omega_1$.

One could try to resolve these contradictions by assuming that each of the frequencies (1.16) is an independent *fundamental* frequency of periodic motion corresponding to its own degree of freedom. But then there follows an implication that even the simplest atom is a system with a huge, in principle, infinite, number of degrees of freedom, and accordingly, any substance must have an infinite heat capacity, which flatly contradicts reality. In addition, each of the fundamental frequencies must be accompanied by a set of its respective harmonics, which are absent in the observed spectra. This clash between the classical predictions and the experiment shows that something is wrong in the classical picture of electrons moving along their paths within an atom.

1.5
The Double-Slit Experiment

When Newton published his "Opticks" in 1704, he argued that light is made up of small particles – corpuscles. Although Newton's theory had some initial success, it was relatively short-lived. In 1801, Thomas Young conducted his famous double-slit experiment, which caused the whole scientific community to completely abandon Newton's corpuscular model. The experimental results bore an unambiguous signature of wave diffraction. Accordingly, we will first focus on the predictions of the classical *wave theory*.

In the simplest case, a monochromatic wave of amplitude $\mathcal{E}_0$ and frequency ω is incident at the right angle on an opaque screen with two narrow slits separated by distance d (Figure 1.8). Due to diffraction (or, if you wish, according to the *Huygens principle* [10,11]), each slit acts as an effective source of light, radiating uniformly in all directions on the other side of the slit, so we have two overlapping diverging waves from two sources. The sources are coherent since they act as "transformers" of *one* incident plane wave into two diverging ones (the so-called *wavefront-splitting* interference [10]). Denoting the wavelength as λ, we can determine the directions θ along which the waves from the two sources interfere constructively:

$$d\sin\theta = m\lambda, \quad |m| = 0, 1, 2, \ldots, \quad \lambda = 2\pi\frac{c}{\omega}. \tag{1.17}$$

More generally, we can write the expression for the sum of two diverging waves as

$$\mathcal{E}(\mathbf{r}, t) = \alpha \mathcal{E}_0 \left(\frac{e^{ik \cdot r_1}}{r_1} + \frac{e^{ik \cdot r_2}}{r_2}\right) e^{-i\omega t} \approx \alpha \frac{\mathcal{E}_0}{r} \left(e^{i\mathbf{k}\cdot(\mathbf{r}_1 - \mathbf{r})} + e^{i\mathbf{k}\cdot(\mathbf{r}_2 - \mathbf{r})}\right) e^{i(\mathbf{k}\cdot\mathbf{r} - \omega t)}$$

$$= 2\alpha \frac{\mathcal{E}_0}{r} \cos\left(\frac{kd \sin\theta}{2}\right) e^{i(\mathbf{k}\cdot\mathbf{r} - \omega t)}. \tag{1.18}$$

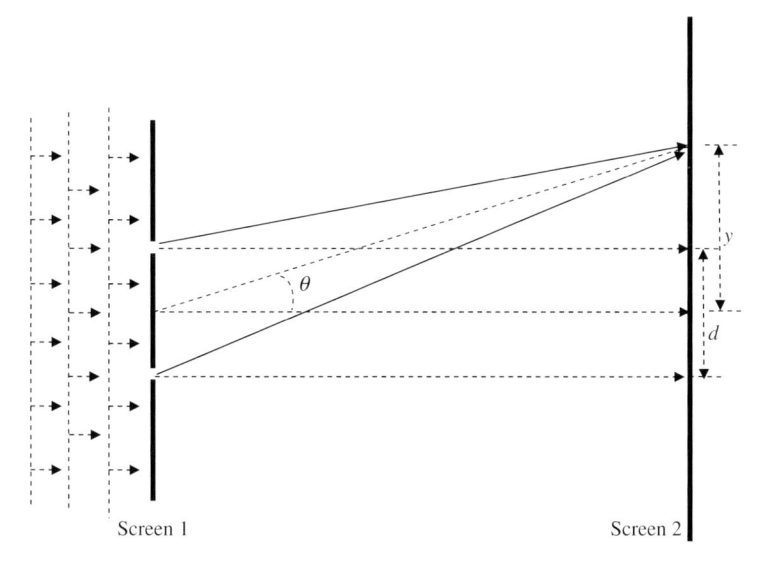

Figure 1.8 The double-slit experiment. The distance d between the slits is exaggerated for better clarity. In the actual experiment, the separation between the slits is so small that directions from them to a landing point y make practically the same angle θ with the symmetry axis.

This expression is a mathematical formulation of the Huygens principle for the case of two narrow slits. Here $\alpha \equiv 2\pi i/k$ [7], $\mathcal{E}_0$ is the amplitude of incident light (polarization ignored), r_1 and r_2 are the distances between the observation point **r** and slits 1 and 2, respectively, and the factors $\exp(ikr_j)/r_j$ describe secondary spherical wavelets coming to this point from the elements of the wavefront passing through the slits.[4] The maximal intensity will be observed along the directions θ satisfying the condition $kd \sin\theta = 2m\pi$. Since $k = 2\pi/\lambda$, this condition is identical to (1.17).

Introducing the second (observational) screen a distance $L \gg d$ away from the first one (Figure 1.8), we can write $\sin\theta \cong y/L$, where y is the distance between the observation point and the center of the screen. The resulting intensity distribution on the secondary screen is

$$I(y) \sim \mathcal{E}^2(y) \cong 4\alpha^2 \frac{I_0}{r^2}\cos^2\left(\pi\frac{d}{L\lambda}y\right), \quad r^2 = L^2 + y^2, \quad I_0 \equiv \mathcal{E}_0^2. \tag{1.19}$$

It produces alternating bright and dark fringes with the spatial period

$$\Delta y = \frac{L}{d}\lambda. \tag{1.20}$$

---

4) Strictly speaking, two narrow slits would produce two *cylindrical* waves, and we would accordingly have $\sqrt{r}$ rather than just r in the denominators of the above expressions. But this will not, for the given conditions, change the main results.

Figure 1.9 The diffraction pattern in the double-slit experiment.

This expression explains why at all we can observe spatial undulations on the screen due to the periodic undulations in the light wave, even though a typical wavelength λ in the visible region of spectrum is $\sim 0.6\,\mu k$. This is far below the minimal size we can see with the naked eye. And yet, we can see the diffraction pattern in Young's experiment! This is because of the coefficient $(L/d) \gg 1$ in Equation 1.20. In this respect, the setup with two close narrow slits and a large distance between the two screens acts as an effective magnifying glass for the wavelength, magnifying its "image" (undulation period) on the screen by the factor L/d.

The intensity distribution (1.19) for $y \ll L$ is shown in Figure 1.9. This is a typical interference picture, *regardless of the physical nature of interfering waves*. The same intensity distribution would be obtained with an acoustic wave in air or a surface wave in a pond if such a wave were incident on a large plate with a pair of narrow slits separated by a distance $d \geq \lambda$.

Now, what happens if we gradually dim the incident light? According to Equation 1.19, this will only decrease the brightness of the interference pattern. If light is truly a wave, then – no matter how faint it is – the observation screen will remain continuously, albeit less and less brightly, illuminated, and exhibit the same pattern. In other words, decrease in intensity will only result in decrease of the coefficient I_0 in (1.19). This was indeed the conclusion of Young and it was well supported by his observations. Thus, the double-slit experiment caused physicists to accept the view of light as waves in a hypothetical medium (ether).

Ironically, under a different scenario Young's experiment could have done exactly the opposite. If Young had taken high-frequency, low-intensity light sources and used short exposures and sufficiently sensitive light detectors, his work would have provided compelling evidence of the *corpuscular* nature of light. Indeed, the diffraction experiments under these conditions show gradual disintegration of the continuous pattern into granular spots. When the brightness falls beyond the visibility threshold, one can use special detectors or a scintillating screen. In the limit of nearly zero intensity, nothing will remain of the continuous undulation (1.19). Instead, one will only observe discrete and apparently random flashes on the otherwise dark screen (Figure 1.10). Remarkably, all flashes have the same brightness, or, if one uses photodetectors, their "clicks" sound equally loud (an actual

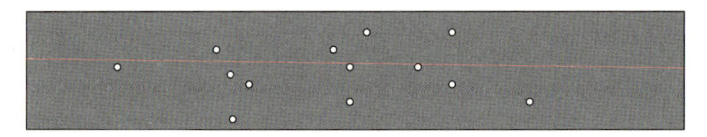

Figure 1.10 The granular pattern in the double-slit experiment at low intensities.

photodetector converts light into electric pulses, so in a real experiment one will see identical peaks on the oscilloscope screen). One must conclude then that light comes in identical portions: either one portion or nothing. This behavior has nothing to do with periodic undulations of a monochromatic wave. It looks exactly as if the screen were bombarded by identical *particles* that passed through the slits. If such measurement techniques were available to Young, then all proponents of the wave nature of light might immediately (at least, until new data were collected from experiments with high intensities or long exposures) flock toward Newton's corpuscular camp. Accordingly, the word "photon" could have become a part of scientific vocabulary about two centuries ago!

Suppose for a moment that history indeed took such a turn, forcing scientists to accept the *corpuscular* view based on the low-intensity version of Young's experiment. And, if the particle picture were accepted as the final truth for light, it would be considered equally true for electrons. So, assume that electrons are *pure particles* and consider the same double-slit experiment, but with electrons instead of light [3]. We will replace the incident wave in Figure 1.8 with a flux of electrons from a source (the "electron gun"), all having the same kinetic energy. Distance d between the slits must be much smaller in the case of electrons, so the described setup is merely a thought experiment. Real experiments with electrons have been carried out with a specially carved single crystal and used reflection from the crystal lattice rather than transmission [9,12,13], but the basic features are still the same. So we can try to visualize the electrons from the electron gun as bullets from a machine gun directed toward two slits in a steel plate. Each bullet has a chance to pass through one of the slits and hit the second screen, leaving a mark on it. In the end, we can record the resulting distribution of bullet marks on the screen. The result will be the sum of two independent contributions: one from the bullets that passed through slit 1, and the other from the bullets that chose slit 2. These predictions can be confirmed by two additional trials – one with only slit 1 open and the other with only slit 2 open (Figure 1.11). In either case, the outcome can be represented by a curve describing

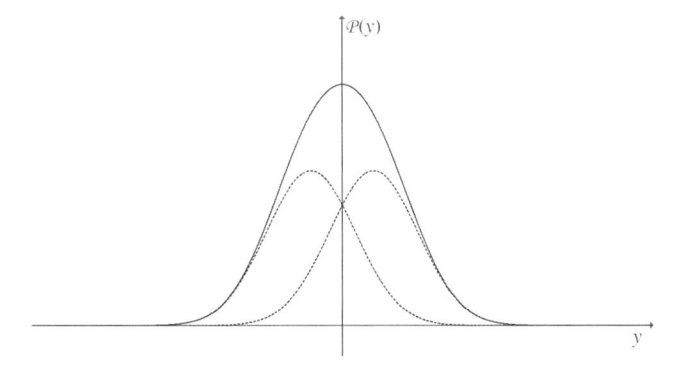

Figure 1.11 The double-slit experiment with classical particles. Each dashed curve represents statistical distribution of particle landings through the corresponding slit. The maximum of each curve is directly opposite the respective slit. The solid curve shows the net distribution with both slits open.

the density of marks on the screen. In the first case, the curve peaks right opposite the slit 1 since this is the place most likely to be hit by a bullet passing through this slit. In the second case, the corresponding curve will have its peak at the point opposite the slit 2. Then, in the case with both slits open the outcome will be just the sum of both curves: since the presence of one slit has no bearing on the probability for a bullet to pass through the other slit, the net probability (more accurately, probability density $\rho(y)$) for a bullet to hit a point y will be

$$\rho(y) = \rho_1(y) + \rho_2(y). \tag{1.21}$$

Let us now turn to the Supreme Judge – the Experiment. When carried out within the full range of parameters, it shows a very strange thing. If the flux intensity is so low that we have only one particle at a time passing through the slits, then each arrival is recorded as a point hit on the second screen (Figure 1.10), so that nothing reminds us of the diluted but still continuous pattern in Figure 1.9, characteristic of a wave. This part of the Experiment suggests that electrons are pure particles.

But there is another part that contradicts this conclusion: The *distribution* of the apparently random arrivals emerging after a long exposure is illustrated by Figure 1.12, not Figure 1.11. It describes the diffraction pattern in Figure 1.9 and is an unambiguous and indisputable signature of waves, not particles! It can only be explained as the result of *interference of waves* passing through two separate slits. In particular, it shows that there are points on the screen where particles never land, even though each particle has high probability of landing there with either slit acting separately. With both slits open, the net probability of landing at such points is *less* than that with only one slit open! The actual probability distribution is *not* the sum of the individual probabilities.

An attempt to combine both pictures on a higher level that includes all the described results leaves us with a puzzle on that level as well. We said the results obtained for low-intensity beams evidence the corpuscular aspect of matter and the ones for high intensity show its wave nature. This could produce an impression that the wave features are necessarily associated with a bunch of particles. But now we see that the

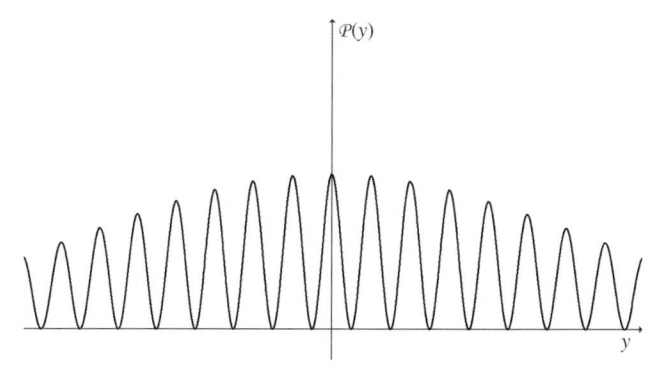

Figure 1.12 The double-slit experiment with QM particles. The diffraction pattern is the same as shown in Figure 1.9 for waves.

results are much more subtle. The *statistics* learned at long exposures shows both corpuscular and the wave aspects already at low intensities. Even if we were tempted to think that a diffraction pattern results from particles interacting with one another, this notion is immediately discarded when we turn to experiments with only one particle at a time passing through the device. It seems that under such conditions there is no room left for wave behavior! And yet, in no trial do we find a particle landing in the middle between the bright fringes. We cannot avoid the conclusion that even a single particle passing through the device has complete information telling it where it cannot land and where it can – even being encouraged to land there. Such information can exist only in a wave, in the form of instructions about directions of constructive and destructive interference impinged on it *by the geometry of both slits*. Following these instructions constitutes a wave behavior. *Each particle in these experiments interferes with itself, not with other particles.* So already a single particle shows some features of the wave behavior. And yet it crashes into the screen as a discrete unit, having nothing to do with a continuous wave. Our classical intuition is insufficient to handle the whole phenomenon.

Summary. We see a complete failure of classical physics when we attempt to apply it to the phenomena described in this chapter. In particular, we see inadequacy of the notion that an entity can be either a pure wave or a pure particle. Real objects of Nature turned out to be neither exactly waves nor exactly particles. Rather, they are something much more subtle, exhibiting sometimes one aspect, sometimes the other, depending on experimental conditions.

What is then their true nature?

The rest of the book is an attempt to describe the answer to these questions as we know it today.

Problem

1.1 Derive the Stefan–Boltzmann law using universal form (1.3) of the radiation energy density in thermal equilibrium.

References

1 Atkins, P.W. (1984) *The Second Law*, Scientific American Books, New York.
2 Massoud, M. (2005) *Engineering Thermofluids: Thermodynamics, Fluid Mechanics, and Heat Transfer*, Springer, Berlin.
3 Feynman, R., Leighton, R., and Sands, M. (1963–1964) *The Feynman Lectures on Physics*, vols. 1–3, Addison-Wesley, Reading, MA.
4 Loudon, R. (2003) *The Quantum Theory of Light*, 3rd edn, Oxford University Press, Oxford.
5 Phillies, G.D.J. (2000) *Elementary Lectures in Statistical Mechanics*, Springer, New York.
6 Frish, S.É. and Timoreva, A.V. (2007) *Course of General Physics* (in Russian), vols. 1–3, Lan Publishing House, Saint Petersburg.

7 Landau, L. and Lifshits, E. (1981) *The Field Theory*, Butterworth-Heinemann, Oxford.
8 Fayngold, M. (2008) *Special Relativity and How It Works*, Wiley-VCH Verlag GmbH, Weinheim, pp. 485–487.
9 Blokhintsev, D.I. (1964) *Principles of Quantum Mechanics*, Allyn & Bacon, Boston, MA.
10 Hecht, E. (2002) *Optics*, 4th edn, Addison-Wesley, Reading, MA, p. 385.
11 Born, M. and Wolf, E. (1999) *Principles of Optics*, 7th edn, Cambridge University Press, Cambridge.
12 Davisson, C. and Germer, L.H. (1927) Diffraction of electrons by a crystal of nickel. *Phys. Rev.*, **30** (6), 705–740.
13 Davisson, C.J. (1928) Are electrons waves? *Franklin Inst. J.*, **205**, 597.

2
The First Steps into the Unknown

As we saw in the first chapter, by the early twentieth century there were at least four different indicators that classical physics was not working at the atomic and subatomic level. First, both classical models of light – wave or particle – failed to describe BBR. Second, the classical kinetic theory failed to describe correctly the behavior of heat capacity. Third, the observed properties of PEE defied classical predictions. Finally, there was an obvious trouble with the planetary model of an atom, which involved radiation by an oscillating charge. All these examples point to the same problematic area: classical physics begins to fail whenever there are oscillation modes trying to emit or absorb energy. And, it doesn't matter whether these are EM oscillations as in the BBR and PEE, or mechanical vibrations of molecules in diatomic gases, or the electron oscillations associated with their orbital motion in an atom. Nature seems to impose some nonclassical restriction on the energies of *all* types of oscillations.

This chapter presents the venue wherein a resolution of the outlined problems may be found.

2.1
The BBR and Planck's Formula

We start with obtaining the correct formula for BBR, which was the first breakthrough leading to QM. It happened when Max Planck abandoned the classical view of energy exchange between systems as a continuous process. Planck was able to derive the correct expression for BBR under a very strange (at that time) assumption that a molecule vibrating with frequency ω can absorb (or release) energy only in separate lumps:

$$E = \hbar\omega, \quad \text{so that} \quad E_n = E_0 + nE = E_0 + n\hbar\omega, \quad n = 0, 1, 2, \ldots . \quad (2.1)$$

In other words, the allowed energy values for the oscillator form a *discrete set* of equidistant "energy levels" above the lowest possible energy E_0.

2 The First Steps into the Unknown

The crucial new characteristic here is the coefficient $\hbar$ – the *Planck constant*. Just as SR introduced the invariant speed c as a universal constant of nature, QM introduced another universal constant, h, known as *quantum of action*:

$$h = 2\pi\hbar = 6.62606876 \times 10^{-34} \text{ J s}. \tag{2.2}$$

In relativity, c plays the role of an invariant scaling parameter in the velocity space. If the velocity **v** of a particle satisfies $v \ll c$, then its motion can be described accurately by CM; otherwise, the relativistic physics takes over. Similarly, the Planck constant is another scaling parameter, representing *the least possible action in one cycle of periodic motion*. If the action S of a one-particle system, evaluated classically, satisfies $S \gg h$, then the classical approximation may be accurate; otherwise, QM takes over. Thus, the constant h determines the domain of applicability of the classical description.

Planck's next assumption was that statistical distribution of identical molecular oscillators over quantized energies (2.1) remains classical.[1] According to the classical law, the number of oscillators occupying the energy level E_n is given by the Boltzmann distribution [2], whose simplest form is

$$N_n = N_0 \, e^{-n\hbar\omega/kT}, \quad n = 0, 1, 2, \ldots. \tag{2.3}$$

Here N_0 is the number of oscillators on the lowest level (we will frequently refer to it as the "ground state"). In thermal equilibrium, most of the oscillators tend to occupy the lowest level, and the higher the level, the less its population. We want to find the *average* energy per oscillator. To this end, we wrote the above equation, taking the ground level as the reference point, that is, we set $E_0 = 0$. Denote $e^{-\hbar\omega/kT} \equiv \xi$. Then the number of oscillators in the nth state is $N_n = N_0(\xi)\xi^n$, and their total number N is

$$N = \sum_n N_n = N_0(\xi) \sum_n \xi^n = \frac{N_0(\xi)}{1-\xi}, \quad \text{so that} \quad N_0(\xi) = N(1-\xi). \tag{2.4}$$

The total energy of the system is $E_{\text{Tot}} = \sum_n N_n E_N = N_0 \hbar\omega \sum_n n\xi^n$. This sum can be easily found (Problem 2.1):

$$E_{\text{Tot}} = N_0 \hbar\omega \frac{\xi}{(1-\xi)^2} = N\hbar\omega \frac{\xi}{1-\xi}. \tag{2.5}$$

Finally, we get the sought-after expression for the average energy:

$$\langle E \rangle = \frac{E_{\text{Tot}}}{N} = \hbar\omega \frac{\xi}{1-\xi} = \frac{\hbar\omega}{e^{\hbar\omega/kT} - 1}. \tag{2.6}$$

The quantum expression (2.6) must replace the classical $\langle E \rangle = kT$, so we may avoid the UV catastrophe and get the correct formula for the BBR:

$$\rho(\omega, T) = N(\omega)\langle E \rangle = \frac{\omega^2}{\pi^2 c^3} \frac{\hbar\omega}{e^{\hbar\omega/kT} - 1} \tag{2.7}$$

This expression is in an excellent agreement with observations in all range of ω and T.

---

[1] This is the first example we look to when we need a system with some classical characteristics (i.e., a special case of classical limit of QM [1]) for deriving a QM picture of a phenomenon.

The result (2.6) applies to both the average energy of an oscillator representing an EM field of frequency ω and the average energy of an oscillator representing the corresponding molecule; both energies must be equal in thermal equilibrium. Therefore, we can also apply (2.6) to an ensemble of molecules when we want to describe the behavior of heat capacity of a substance at low T. We can then write for the fixed-volume heat capacity,

$$C_V \sim \frac{d\langle E \rangle}{dT} = T^{-2}\left(e^{\hbar\omega/kT} - 1\right)^{-2} \xrightarrow[T \to 0]{} T^{-2} e^{-2\hbar\omega/kT}. \tag{2.8}$$

This expression approaches zero at $T \to 0$, also in agreement with observations.

Note that the Boltzmann distribution (2.3) is not affected by the energy quantization – it is *only the set of possible values of energy* within *the same* distribution (2.3) that has been changed from continuous to discrete, that is, $E \to E_n$. And this alone turned out to be sufficient to change both wrong results – (1.6) and (1.7) for BBR and (1.10) for heat capacity – to correct ones, given by (2.7) and (2.8), respectively. In other words, the same Boltzmann distribution that led to the wrong results when energy was assumed continuous gives correct results if energy is discrete (summation (2.4) and (2.5) instead of integration).

How specifically does the change $E \to E_n$ do the trick? Look at Figure 2.1. Consider each energy level as a "shelf" to be filled by the corresponding number

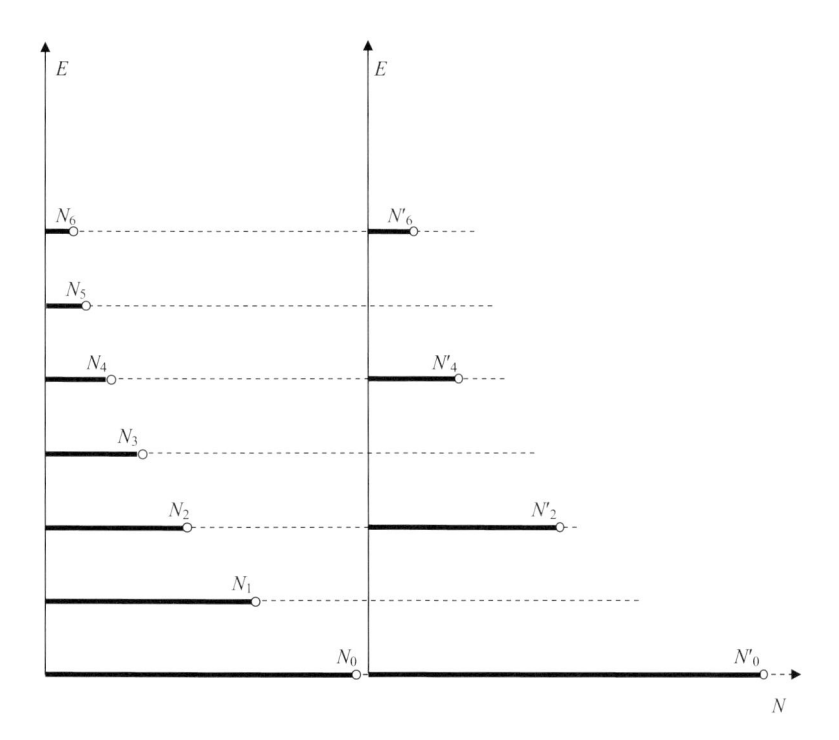

Figure 2.1 The energy levels of a harmonic oscillator and the respective populations at different quantization constants.

of particles. For a continuous energy range, we would need to distribute our N molecules over this range at the same T. A certain fraction of molecules from level E_j would have to climb up the energy scale to fill in the gap between E_j and E_{j+1}. A certain fraction of molecules from level E_{j+1} would have to go down with the same mission, but the corresponding number would be less, since N_{j+1} is less than N_j. Therefore, overall such a process would require that more energy be pumped into the system and accordingly would give both high energy density (UV catastrophe) and high ratio E/T. The lower ratio E/T and lower heat capacity are achieved if the system's energy levels are discrete (energy quantization). Thus, contradictions between classical predictions and observation both for BBR and for heat capacity are resolved in QM by the energy quantization.

What causes quantization itself? This question will be discussed in the following sections.

2.2
Einstein's Light Quanta and BBR

> The fact that beer is sold in bottled portions falls short of proving that beer actually consists of the bottled portions.
>
> *Einstein*

Here we consider a totally different approach to the BBR problem, one based on the concept of the light quanta introduced by Einstein and Bohr's rule for energy exchange between matter and quantized radiation.

After Planck had postulated the rule (2.1) for energy exchanges, Einstein took the next step. He noted that apart from BBR, PEE could also be explained correctly if electromagnetic radiation itself *actually consisted of indivisible portions* (quanta of radiation) [3]. For radiation with frequency ω, the energy and momentum of such a portion are given by

$$E = \hbar\omega, \qquad \mathbf{p} = \hbar\mathbf{k}. \tag{2.9}$$

Here $\mathbf{k} = k\mathbf{n} = (2\pi/\lambda)\mathbf{n}$ is the wave vector for the wavelength λ, pointing in the propagation direction $\mathbf{n}$.

Later, Bohr developed these ideas in his theory of an atom. When it became clear that Rutherford's model needed radical changes, Bohr proposed a new hydrogen atom model [4]. To account for the stability of atoms, he *postulated* the existence of certain selected orbits in the atomic space, on which the electron does not radiate, and accordingly, the atom remains in a *stationary* state. In the spirit of Planck's original idea, the only opportunity for the atom to radiate (or absorb) energy is to do it in a quantum jump from one stationary state to another. At these jumps, the atom changes its energy discontinuously, by one portion. Accordingly, it radiates (or absorbs) one quantum of electromagnetic energy. These are nothing else but the light quanta suggested by Einstein (and later named "photons" by Lewis); the energy of the radiated photon is equal to the energy difference between the initial and final atomic states. The radiation frequency is related to the energy difference by

$$\omega = \omega_{21} \equiv \frac{E_2 - E_1}{\hbar} = \frac{E}{\hbar}. \tag{2.10}$$

This relation, proposed by Bohr, is just the combination of Equation 2.9 and Planck's initial suggestion (2.1) about discrete energies of oscillators. If passed through a prism, each photon refracts by one fixed angle determined by frequency ω. Accordingly, a narrow line rather than a fuzzy band will appear on the screen. This is precisely what is observed in the atomic emission spectra shown in Figure 1.7. A reverse process is also possible: an atom can absorb a photon with a frequency tuned to an allowed optical transition and jump to a state with higher energy. This is an *optical excitation*. The atom does not usually stay long in an excited state; in about a few trillionths of a second it returns to its initial state, having radiated a photon of the same frequency. This gives a simple explanation of *Kirchhoff's law* of thermal radiation.

If we now combine Einstein's concept of light quanta and Bohr's rule (2.10) with Boltzmann's distribution, there follows a correct expression for the universal function $\rho(\omega, T)$ [5]. In other words, Planck's semi-empirical expression (2.7) can now be derived from basic principles of QM. Consider energy exchange between the atoms and radiation in thermal equilibrium. Each act of such exchange is an absorption or emission of a photon by an atom, accompanied by a quantum jump between atomic states m and n. We apply Bohr's rule (2.10) to this process assuming that $E_n > E_m$. In equilibrium, the population of level n is described by Equation 2.3. The absorption rate (i.e., probability w_a per atom of light absorption with transition $E_m \to E_n$ per unit time) is proportional to number of photons with suitable frequency, or equivalently, to the corresponding spectral energy density:

$$w_a = B_m^n \rho(\omega, T). \tag{2.11}$$

Here B_m^n is the proportionality constant for the given atom and given transition. Total probability of absorption (per unit volume) is the product of w_a and the number of atoms (2.3) in state m:

$$W_a = w_a N_m = B_m^n \rho(\omega, T) N_0 \, e^{-E_m/kT}. \tag{2.12}$$

Now we turn to emission. On the intuitive level, we can expect two possible mechanisms of emission. First, an excited atom can emit a photon just because it is excited and can release the corresponding energy supply under condition (2.10). This process happens of its own accord and is called the *spontaneous emission*. Its rate w'_e per atom for transition $n \to m$ is determined by the properties of the atom in question and can be denoted as

$$w'_e = A_m^n. \tag{2.13}$$

The second mechanism is the influence of external photons around the atom. Just as the photons of suitable frequency ω are readily swallowed up by an atom in a state with lower energy E_m, they also are nagging the excited atom with higher energy E_n

and *forcing* it to release its excessive energy by emitting the new photon satisfying (2.10). This kind of process is called *stimulated emission*. A photon released in such emission is in the same state as that of the photons that caused the release, and the emission rate is proportional to concentration of such photons, that is, to the radiation energy density:

$$w_e'' = B_n^m \rho(\omega, T). \tag{2.14}$$

The net emission rate per atom is

$$w_e = w_e' + w_e'' = A_n^m + B_n^m \rho(\omega, T). \tag{2.15}$$

Multiplying this by the concentration of atoms in state n given by (2.3) yields the total emission rate:

$$W_e = N_n w_e = \left(A_n^m + B_n^m \rho(\omega, T)\right) N e^{-E_n/kT}. \tag{2.16}$$

The constants A_n^m, B_n^m, and B_m^n are the *Einstein coefficients* for spontaneous and stimulated emission, respectively, and for absorption. The concept of stimulated emission was introduced by Einstein in order to satisfy the condition of equilibrium. According to this condition, the total rates of emission and absorption must be equal, $W_a = W_e$. In view of (2.12) and (2.16), this gives

$$B_m^n \rho(\omega, T) e^{-E_m/kT} = \left(A_n^m + B_n^m \rho(\omega, T)\right) e^{-E_n/kT}. \tag{2.17}$$

It is easy to see that this condition cannot be satisfied with either one of the two terms on the right when taken separately. Indeed, without spontaneous emission Equation 2.17 would reduce to $B_m^n e^{-E_m/kT} = B_n^m e^{-E_n/kT}$, which cannot hold at finite temperatures, and leaves the radiation function $\rho(\omega, T)$ arbitrary. Without stimulated emission we would have $B_m^n \rho(\omega, T) e^{-E_m/kT} = A_n^m e^{-E_n/kT}$, which cannot hold either, since at high T the absorption rate on the left would definitely overwhelm the spontaneous emission on the right. In addition, the radiation function determined from this equation would not satisfy the universal requirement (1.3). Thus, both terms are necessary for equilibrium. The physical reason is very simple: at any finite temperature the number of atoms in states with higher energy E_n is less than that with lower energy E_m. Accordingly, the rate of stimulated emission is less than that of absorption. Spontaneous emission compensates for this lower rate.

We can now turn to the *exact* equation (2.17) and consider its limit at high temperatures. *In this limit*, the radiation density increases unboundedly, so that the spontaneous emission term in (2.18) becomes negligible compared to the stimulated emission; also, both exponents approach 1. The equation in this limit can hold only if

$$B_m^n = B_n^m \equiv B, \tag{2.18}$$

that is, Einstein's coefficients for absorption and stimulated emission must be equal to each other. Then, combining the terms with unknown function $\rho(\omega, T)$ and using Bohr's condition $E_n - E_m = \hbar\omega$ gives

$$\rho(\omega, T) = \frac{A_n^m}{B_n^m} \left(e^{\hbar\omega/kT} - 1\right)^{-1}. \tag{2.19}$$

It remains to find the ratio A_n^m/B_n^m. To this end, we take the classical limit of (2.19) and require it to give the Rayleigh–Jeans formula (1.6). In this limit, observable E_n becomes continuous, meaning the discrete steps of energy change must be very small; that is, $\Delta E_{mn} = E_n - E_m = \hbar\omega \ll E_m$, $E_n \approx kT$. In other words, the classical limit is equivalent to the low-frequency limit, which corresponds to $\xi \equiv \hbar\omega/kT \ll 1$ in Equation 2.19. In this approximation, expanding the exponent into the Taylor series and retaining the first-power term, we obtain

$$\rho(\omega, T) = \frac{A_n^m}{B_n^m} \frac{kT}{\hbar\omega}. \tag{2.20}$$

Comparing this limit with the known result (1.6) for the classical domain leads to

$$\frac{A_n^m}{B_n^m} = \frac{\hbar\omega^3}{\pi^2 c^3}. \tag{2.21}$$

Putting this into (2.19) yields

$$\rho(\omega, T) = N(\omega)\langle E \rangle = \frac{\hbar\omega^3}{\pi^2 c^3} \frac{1}{e^{\hbar\omega/kT} - 1}. \tag{2.22}$$

This is the sought-after general expression for BBR. It is identical to the result (2.7,) first obtained by Planck. As another special case, it contains also the Wien formula (1.7). Indeed, in the limit opposite to $\xi \ll 1$, that is, at $\xi \equiv \hbar\omega/kT \gg 1$, Equation 2.22 reduces to

$$\rho(\omega, T) = \frac{\hbar\omega^3}{\pi^2 c^3} e^{-\hbar\omega/kT}. \tag{2.23}$$

This is identical to (1.7), with const $= \hbar/\pi^2 c^3$ and $\gamma = \hbar/k$.

Already, at this stage we can infer an interesting implication from formula (2.22). Namely, switching from the language of radiation density to the language of photons, we can evaluate the concentration of photons: dividing (2.22) by the photon's energy $\hbar\omega$ will yield the number $\nu(\omega, T)$ of photons of frequency ω per unit volume per unit frequency at temperature T:

$$\nu(\omega, T) = \frac{1}{\pi^2 c^3} \frac{\omega^2}{e^{\hbar\omega/kT} - 1}. \tag{2.24}$$

Thus, whereas the concept of light as a *classical* EM field fails to account for all properties of BBR, the concept of photons (*quantized* EM field) gives their complete explanation.

2.2.1
Discussion

It is interesting to note that all three characteristics, w_a, w_e', and w_e'' (rates of absorption, spontaneous, and stimulated emission), have classical analogues. Classically, an atom in a radiation field can be represented by an oscillator. A classical oscillator (e.g., mass μ on a spring) is described by equation

$$\mu\ddot{x} + \beta\dot{x} + kx = f_0 e^{-i\omega t}. \tag{2.25}$$

Here k is the spring constant, f_0 and ω are, respectively, the amplitude and frequency of the driving force, and β determines the dissipative force, which is proportional to the instantaneous velocity $v \equiv \dot{x}$ of the mass μ. The solution to this equation describes oscillation with the same frequency and the amplitude

$$x_0 = \frac{f_0/\mu}{\omega_0^2 - \omega^2 + i\gamma\omega}, \quad \omega_0 \equiv \sqrt{\frac{k}{\mu}}, \quad \gamma \equiv \frac{\beta}{\mu}. \tag{2.26}$$

Now, what is an absorption of a photon is, classically, an incremental increase per unit time of the oscillator's amplitude (and thereby energy) under the incident EM wave tuned to the oscillator's frequency (driven oscillator at resonance). Similarly, stimulated emission of a single photon is, in the classical limit, an incremental decrease of the oscillator's amplitude under the same force. For instance, in resonance without dissipation ($\gamma = 0$, $\omega = \omega_0$), the same force can either increase or decrease the oscillator's amplitude, depending on the phase shift between f and $\dot{x}$. If f and $\dot{x}$ oscillate in phase (e.g., you push the swing in sync with swing's velocity), then the amplitude builds up – the energy is pumped up into the system (absorption). If f and $\dot{x}$ are out of phase (you push the swing so that the force is at each moment opposite to velocity), then the swing's amplitude decreases – the swing loses energy under the action of the external force (stimulated emission). The solution of the equation for a classical oscillator driven by the *resonant force in the absence of friction* is (Appendix A)

$$x(t) = \frac{f_0}{2\mu\omega} t \sin \omega t. \tag{2.27}$$

The graph of $x(t)$ is shown in Figure 2.2. The displacement is phase shifted with respect to the force by $\pm \pi/2$ (this corresponds to the above-mentioned phase shift 0

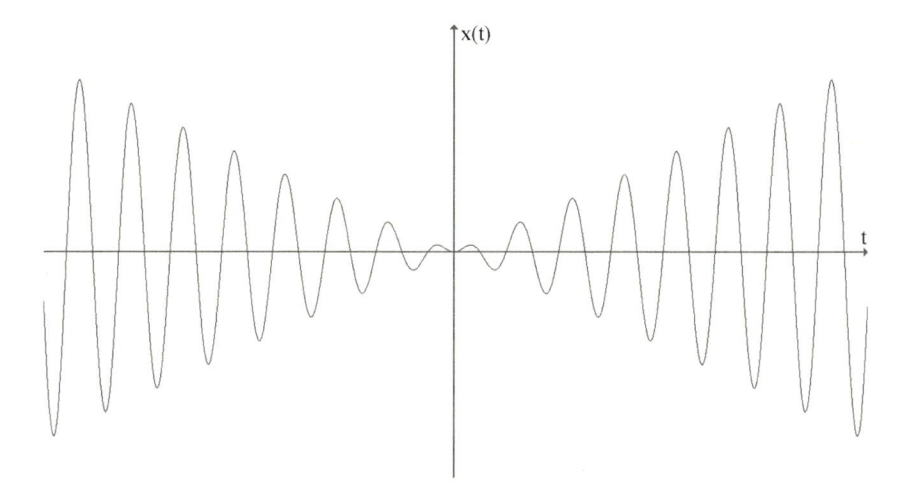

Figure 2.2 Driven oscillations at resonance without dissipation. The part of the graph showing the decrease of oscillator's amplitude and thereby its energy under the driving force corresponds to stimulated emission. The part showing amplitude (and energy) increase under the same force corresponds to resonant absorption.

or π between f and $\dot{x}$). At $t > 0$, the force leads the displacement by $\pi/2$ (is in phase with $\dot{x}$), pumping the energy into the oscillator (absorption!). At $t < 0$, the force trails the displacement by $\pi/2$ (is out of phase with $\dot{x}$), and the oscillator is losing its energy (stimulated emission!). The phase jump by π at $t = 0$ causes no discontinuity since the oscillation amplitude is zero at this moment.

In a similar fashion, we can find the classical analogy with spontaneous emission. The word "spontaneous" implies the absence of the driving force. Thus, we must set $f_0 = 0$ and, instead, excite the oscillator by a sudden hit. Alternatively, we can excite it by stretching or compressing the spring and then releasing it. The vibrations ensue, but their damping and corresponding loss of energy (emission) implies the existence of a dissipative factor. This factor is represented by a friction force $\beta \dot{x}$ in Equation 2.25. The solution of this equation with the initial condition $x = 0$ at $t < 0$ (Appendix A) is

$$x(t) = \begin{cases} x_0 e^{-(1/2)\gamma t} \sin \omega_\gamma t, & \gamma < 2\omega_0, \\ x_0 e^{-(1/2)\gamma t} \sinh |\omega_\gamma| t, & \gamma > 2\omega_0. \end{cases} \quad (2.28)$$

Here $\omega_\gamma \equiv \sqrt{\omega_0^2 - (1/4)\gamma^2}$ is the frequency of damped oscillations, which is less than the proper frequency ω_0. The amplitude x_0 is determined by the strength of the hit. The graph of this function is shown in Figure 2.3. The initial oscillations are gradually damping (spontaneous emission). In the classical scenario, they are damped due to friction; what causes this "damping" (jumping to lower energies and accordingly smaller "orbits") of an atomic electron? Here we return to the notion of "radiative friction" mentioned in Section 1.4, but now at a higher level. A deeper scrutiny shows that in QM, the role of the dissipative factor is played by *vacuum fluctuations* – an effect studied in quantum field theory. It is the birth and

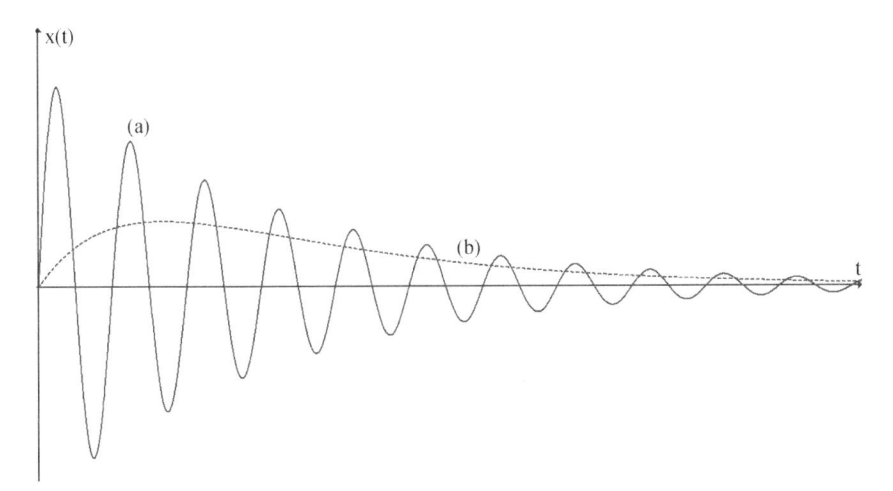

Figure 2.3 Dumped oscillations. At $t > 0$, the initially excited oscillator in the absence of driving force gradually damps out. This corresponds to spontaneous emission of initially excited atom in the absence of incident radiation.

disappearance of the so-called virtual photons, which are nagging the excited electron, forcing it to radiate. Thus, on this deeper level, spontaneous emission turns out to be not so spontaneous after all! At this level, when describing the radiation we do not distinguish between the mechanisms of spontaneous and stimulated emission [6].

And the final question: What keeps the ground level stable with no radiation from it, despite the virtual photons? In classical terms: What holds an electron in an isolated atom forever on the lowest orbit, stopping it from spiraling farther toward the nucleus, while all higher states have only finite lifetime? This question will be discussed in Section 3.5.

2.3
PEE Revisited

We now turn to the question of how the concept of photons explains all observed features of PEE described in Section 1.3 (for details of the corresponding experiments see, for example, Ref. [7]). According to the concept of quantized radiation, the incident light in this effect can be described as a beam of photons. This corpuscular aspect of light in a given situation is more relevant than the wave aspect since an electron bound to a metal plate can only be freed by absorbing a finite amount of energy exceeding a certain critical value U_0 – the *work function* characteristic of given material. It is overwhelmingly more likely for such an electron to absorb the necessary energy at once from one photon with suitable frequency than to do so in billions of steps from billions of photons with very low frequency. The latter possibility could be described classically as a continuous process, but the probability of such multistaged process is so small that it cannot be observed in known practical situations. The corpuscular aspect dominates, and an electron is freed when it absorbs a single photon with a suitable energy. All this energy is used to overcome the threshold U_0, and if there is an excess, it goes to the kinetic energy K of a released electron. Thus, we can write

$$\hbar\omega = U_0 + K. \tag{2.29}$$

This is Einstein's equation for PEE. Actually, it expresses conservation of energy by showing where the photon's energy goes. This provides a simple basis for explaining all known features of PEE.

First, we note that if ω is so low that $\hbar\omega \leq U_0$, then absorption of such a photon does not lead to an electron's release. The condition for an electron to be ejected from the surface is

$$\hbar\omega \geq U_0 \quad \text{or} \quad \omega \geq \omega_0 \equiv \frac{U_0}{\hbar}. \tag{2.30}$$

Thus, there naturally emerges the existence of the threshold frequency ω_0 corresponding to the potential threshold U_0. Even an extremely bright light will not produce any noticeable electron current if its frequency is less than ω_0.

For frequencies above the threshold, the kinetic energy of an ejected electron, according to (2.29), is $K = \hbar\omega - U_0$. In total agreement with experiments, it is a linear function of ω independent of light's intensity I_0 – just as shown in Figure 1.6c. The value of I_0 determines the number of photons in the incident beam. The more intense light will eject more electrons. Maximal kinetic energy of such electrons is determined by (2.29), but their *number* and thereby the observed electron *current* will be a linear function of I_0. This is in agreement with experimental data in Figure 1.6b.

The immediate response of the photoelectric current to the exposure also finds its explanation. Each ejection results from absorption of a single photon. This is an instantaneous quantum event rather than a continuous process. Therefore, if light in the discussed experiment reveals its particle-like aspect, then there must be practically no time delay between the action and the response, and we should expect the emergence of photoelectrons right at the start of illumination. This is precisely what is observed and what could not be explained in the classical wave framework.

Summary of Sections 2.1–2.3: The original idea of quantized energy exchange envisioned by Planck and complemented by Einstein's photon hypothesis made possible an explanation of all features of BBR and behavior of heat capacity. This was a first breakthrough toward the nonclassical physics. Then the Bohr postulates allowed researchers to explain the basic features of atomic spectra, which were totally unexplainable from the classical viewpoint. The same quantum concepts explain all features of PEE listed in Section 1.3. This was a second breakthrough.

2.4
The Third Breakthrough: de Broglie Waves

Despite its success in explaining some atomic spectra, the Bohr model left open the basic question: What is it that selects certain privileged orbits for an atomic electron? In other words, what determines the atomic stationary states?

The answer to this question comes from de Broglie's revolutionary hypothesis, one of the most important breakthroughs on the way to quantum physics. De Broglie suggested that if light waves can, under certain conditions, exhibit corpuscular characteristics, then symmetry requires that the opposite should also be true – all particles must possess wave-like characteristics, and the relationship between the wave and particle aspects must also be described by Equation 2.9. In other words, these equations must represent the universal law of nature *for all objects*. This is of such paramount importance that we rewrite them here in a slightly different form:

$$\omega = \frac{E}{\hbar}, \quad \mathbf{k} \equiv \frac{2\pi}{\lambda}\hat{\mathbf{n}} = \frac{\mathbf{p}}{\hbar}. \quad (2.31)$$

Here $\hat{\mathbf{n}}$ is the unit vector in the direction of the object's momentum. The difference from (2.9) is that it now applies to *any* physical object rather than to EM radiation alone. Therefore, the motion of a free particle with energy E and momentum $\mathbf{p}$ is, according to de Broglie, represented by a wave [8]

$$\Psi(\mathbf{r}, t) = \Psi_0 \, e^{i(\mathbf{k}\mathbf{r} - \omega t)} = \Psi_0 \, e^{i(\mathbf{p}\mathbf{r}/\hbar - Et/\hbar)}. \tag{2.32}$$

One of the motivations for de Broglie's work was the similarity in behavior of the two so-called 4-vectors: $(E/c, \mathbf{p})$, on the one hand, and $(\omega/c, \mathbf{k})$, on the other, revealed by the theory of relativity [9]. Namely, according to relativity, the phase $\mathbf{k} \cdot \mathbf{r} - \omega t$ figuring in expression (2.32) is the dot product of two 4-vectors: the "4-displacement" $s = (ct, \mathbf{r})$ and the 4-vector $k = (\omega/c, \mathbf{k})$. The first vector is the relativistic generalization of a 3D spatial displacement $\mathbf{r}$ onto the 4D space including the temporal dimension (the *space–time*, or *Minkowski space*). The second vector is a similar generalization for the frequency ω and propagation vector $\mathbf{k}$ describing a running wave. It turned out that in many respects, frequency ω behaves like time t, and $\mathbf{k}$ like displacement $\mathbf{r}$. For instance, if we switch from one reference frame to another (perform a *Lorentz transformation* [9]), then both sets of components, namely, $s = (ct, \mathbf{r}) = (ct, x, y, z)$ and $k = (\omega/c, \mathbf{k}) = (\omega/c, k_x, k_y, k_z)$, transform in the same way. They both behave as vectors in the above-mentioned 4D vector space, and their dot product is a scalar – as it should be, since it forms the total phase of the wave, and total phase is an invariant (i.e., the same for all observers) characteristic of a wave.

According to relativity, the energy and momentum of a system also form a 4-vector $p = (E/c, \mathbf{p})$ [9], which behaves in exactly the same way as the 4-vector k under Lorentz transformations. It seemed reasonable therefore to suggest that both 4-vectors may be intimately related to one another by Equation 2.31, which form the essence of de Broglie's postulate. With respect to this postulate, based on the identical relativistic properties of k and p, we can say that all QM is intrinsically relativistic.

De Broglie's second motivation was the notion that the wave aspect of particles would *naturally* admit the selection of states for a bound particle. Indeed, as we know from mechanics, the waves in a bounded volume are subject to "natural selection": only the fittest can survive. The surviving waves are those whose half-wavelength (or its integer multiple) exactly fits into the available volume along the direction of propagation of the wave (this condition weeds out inappropriate directions as well!). The simplest example is a *standing wave* on a vibrating string. The word "standing" means the absence of energy flow: we have two waves with equal amplitudes running in the opposite directions along the string, so there is no net energy transfer. Both waves are such that only an integer number of half-wavelengths fit into the string. For a string of length s, this gives

$$\lambda \to \lambda_n = \frac{2s}{n}, \quad n = 1, 2, 3, \ldots. \tag{2.33}$$

This also selects the corresponding frequencies, since the frequency f of a wave is related to its length by $\lambda f = u$, so that $\omega \to \omega_n = 2\pi(u/\lambda_n)$, where u is the wave velocity in the given medium. Thus, any selection rule for the wavelengths (λ-quantization) must automatically quantize frequencies (ω-quantization), so that only a specific discrete set of them (fundamental frequency and its harmonics) is allowed.

The ω-quantization is fundamentally important: in view of the relationship $\omega \leftrightarrow E$ expressed in Equation 2.31, it also implies energy quantization. Thus, discrete energy spectrum for a system bound in a finite region of space follows automatically from de Broglie's relations (2.31) and (2.32).

In the case of the path of a wave closing onto itself, forming a loop, only an integer number of waves can fit into the path. Imagine a snake biting its own tail and wiggling. If all adjacent wiggles are equal, then obviously only an integer number of them will fit into the length of the snake. Similarly, if you hit a ring, it will start vibrating, with only an integer number of waves running around it. For a ring of radius a, all possible wavelengths are given by

$$\lambda_m = \frac{2\pi a}{m}, \quad m = 1, 2, 3, \ldots . \qquad (2.34)$$

This condition is similar but not identical to (2.33)! The ring accommodates, apart from standing waves, the running waves of length λ_m as well, circling around and producing a circular energy flow and thereby a nonzero (but quantized!) *angular momentum*. In view of (2.31), condition (2.34) can be written as

$$a p_m = m\hbar, \qquad (2.35)$$

which can be considered as a selection rule for the azimuthal component of momentum $p_{\varphi,m} = m(\hbar/a)$. But, more importantly, this imposes quantization of the corresponding *angular* momentum **L**. The product on the left of (2.35) is the component of **L** onto the axis perpendicular to the classical orbital plane. Denoting this axis as z, we have

$$L_z = m\hbar, \qquad (2.36)$$

with integer m. As we will see later, rule (2.36) is more general than the special case (2.35) from which it was derived.

2.4.1
Exercise

As an example, let us apply the obtained quantization rules to a hydrogen-like atom. Start with the classical expression for electron's energy in the Coulomb field of the nucleus:

$$E = \frac{p^2}{2\mu_e} - \frac{Ze^2}{a}. \qquad (2.37)$$

Here Z is the number of protons in the nucleus, μ_e is the electron's mass, a is radius of its orbit, and e^2 is the shorthand for $q^2/4\pi\varepsilon_0$. From (2.37) and the expression for the centripetal force

$$\frac{\mu_e v^2}{a} = \frac{Ze^2}{a^2}, \qquad (2.38)$$

we find[2]

$$\frac{Ze^2}{a} = \frac{p^2}{\mu_e} = -2E. \qquad (2.39)$$

2) This equation is a special case of the so-called virial theorem [1], relating the average kinetic or potential energy of a system to its total energy.

Now we say that there is a de Broglie wave associated with the electron, and apply to it condition (2.35) relating p to a. Together with the first equation (2.39), this gives $p_n = n\hbar/a$ and $a = -Ze^2/2E$, which implies $p_n = -2n\hbar E/Ze^2$ (we switched back from m to n here because now n will label E rather than L). Then Equation 2.39 gives $E \to E_n = -2E_n^2 n^2 \hbar^2 / \mu_e Z^2 e^4$. Solving for E_n, we find

$$E \to E_n = -\frac{\mu_e Z^2 e^4}{2n^2 \hbar^2}. \tag{2.40}$$

This is a known expression for the energy levels of a hydrogen-like atom. Together with Bohr's rules (2.31) connecting energy levels with the radiation frequency of optical transitions, it explains quantitatively the observed spectrum of atomic hydrogen.

The energies given by (2.40) are negative since they characterize a bound electron. The lowest possible electron energy for hydrogen follows from (2.40) at $Z = 1$, $n = 1$ and equals -13.6 eV. This is the so-called ionization energy needed for the electron to break loose from its ground state, and accordingly, for the atom to get ionized. The lowest energy ($E = 0$) of the *ionized* (freed) electron is the beginning of its continuous spectrum ($E \geq 0$); it is approached "from below" by the discrete spectrum (2.40) at $n \to \infty$.

If we now put (2.40) back into (2.39), we get the expression for the radius of the *classical* orbit (the Bohr orbit) in a state characterized by quantum number n:

$$a \to a_n = \frac{\hbar^2}{\mu_e Z^2 e^2} n^2. \tag{2.41}$$

As we will see later, there is actually no such thing as a classical orbit for an atomic electron, and a_n just represents the *effective size* of a hydrogen-like atom in a state with energy E_n. For hydrogen, the smallest possible size corresponding to the ground state follows from (2.41) at $Z = 1$, $n = 1$, and is called the *Bohr radius*, but again, it should be viewed only as the effective size of an atom in this state.

Thus, the wave aspect of matter, expressed in relations (2.31) and (2.32), imposes severe restrictions on the spectra of some observables. These restrictions (quantization rules) convert some regions of continuous spectrum into discrete ones. Specifically, we have learned such rules for energy of an atomic electron and L_z-component of L.

The energy quantization works for *all bound systems*. We will learn later that for unbound but periodic systems the results of quantization are more subtle, leading to alternating regions of continuous "energy bands" and forbidden "energy gaps." As to the angular momentum, its quantization turns out to be a fundamental law, according to which any measurements of L for any object always show only discrete sets of values. All these results are the crucial points where QM comes in.

There are, however, other features of QM that are even farther away from our classical intuition. They concern the physical meaning of the wave (2.32) (what specifically is waving there?), probabilistic nature of the world, and the problem of determinism. These problems will be addressed in the next chapter and later in the book.

Problems

2.1 Derive expression (2.5) for the total energy of the system of quantized oscillators in thermal equilibrium.

2.2 For an oscillator with proper frequency ω_0, find the frequency of its damped oscillations in the presence of a dissipative force.

2.3 What would happen to a classical oscillator excited by a sudden force if the oscillator were in a viscous medium such that $\gamma > 2\omega_0$?

2.4 Imagine a fictitious universe without vacuum fluctuations but with all other laws of physics unaffected (this may be a logically inconsistent assumption!).

a) How would an excited atom behave in such a universe? (Recall that the effect of vacuum fluctuations can be formally described by the damping parameter γ in the classical model of an atom.)

b) Would thermal equilibrium be possible in such a universe?

2.5 a) Find the allowed values of de Broglie wavelength λ for a particle locked in a 1D region of size s.

b) Imagine now that s is closed into a circle of radius a. Find the quantization rule for λ in this case.

c) Compare your results for (a) and (b) and explain the difference.

(*Hint*: Compare the boundary conditions for both cases.)

References

1. Landau, L. and Lifshits, E. (1965) *Quantum Mechanics: Non-Relativistic Theory*, 2nd edn, Pergamon Press, Oxford.
2. Pathria, R.K. (2000) *Statistical Mechanics*, 2nd edn, Butterworth-Heinemann, Oxford.
3. Einstein, A. (1905) On a heuristic viewpoint concerning the production and transformation of light. *Ann. Phys.*, **17**, 132–148.
4. Bohr, N. (1913) On the constitution of atoms and molecules. Part 1. *Philos. Mag.*, **26**, 1–25.
5. Blokhintsev, D.I. (1964) *Principles of Quantum Mechanics*, Allyn & Bacon, Boston, MA.
6. Sakurai, J.J. (1984) *Advanced Quantum Mechanics*, 10th printing, Benjamin Cummings Publishing Company, San Francisco, CA.
7. McKelvey, J. (1966) *Solid State and Superconductor Physics*, Harper & Row, New York.
8. de Broglie, L. (1926) *Waves and Motions*, Gauthier-Villars, Paris.
9. Fayngold, M. (2008) *Special Relativity and How It Works*, Wiley-VCH Verlag GmbH, Weinheim, pp. 485–487.

3
Embryonic Quantum Mechanics: Basic Features

3.1
A Glimpse of the New Realm

From examples considered in the previous chapter, the first outlines of QM can be discerned. One of the basic features is discreteness of some observables. The measured values of angular momentum are always quantized. According to Equation 2.38, the component of angular momentum of any object onto a chosen axis can only have values separated by $\hbar$. We will say more about angular momentum in Chapter 7, but here, as another example, mention electron spin (*intrinsic* angular momentum unrelated to any orbital motion). Electron spin has only two possible projections onto an arbitrarily chosen direction (say, the z-direction): $S_z = \pm(1/2)\hbar$. The "+" state is usually referred to as "spin up" and the "−" state as "spin down."

The measured energy of any bound system can only have certain discrete values determined by the binding forces. Equation 2.40 for a *bound* electron in the Coulomb field is an example. However, its energy is a continuous variable at $E > 0$. The set of possible energies of a system is referred to as its *spectrum*. In QM, the spectrum can be discrete or continuous or a combination of both.

The wave aspect of matter explains also why an atomic electron in a stationary state does not radiate as prescribed by classical physics. In contrast to a tiny ball rolling along its orbit, the electron is "smeared out" over the whole orbit, like a wave on an elastic ring, which is vibrating with *all its parts at the same time*. In a way, the electron is spread out in the atomic space as music in a concert hall. Therefore, physicists nowadays rarely say "electron orbit." Rather, they say "electron configuration," "electron shell," or "electron cloud," even when referring to *only one* electron. In a state with definite energy, this cloud is axially symmetric and therefore its rotation around the nucleus produces only a steady current loop, which does not radiate.

Depicted in Figure 3.1 are a few different states of an electron in an H atom. In every such state, the electron indeed forms a sort of "cloud" around the nucleus. And, just as music in a hall, reflecting from the walls and chairs, can sound louder at some locations than at others, the electronic "cloud" in an atom can be denser at some points than at others. We will learn more about such states in Chapter 10; now we focus on other implications of de Broglie's idea.

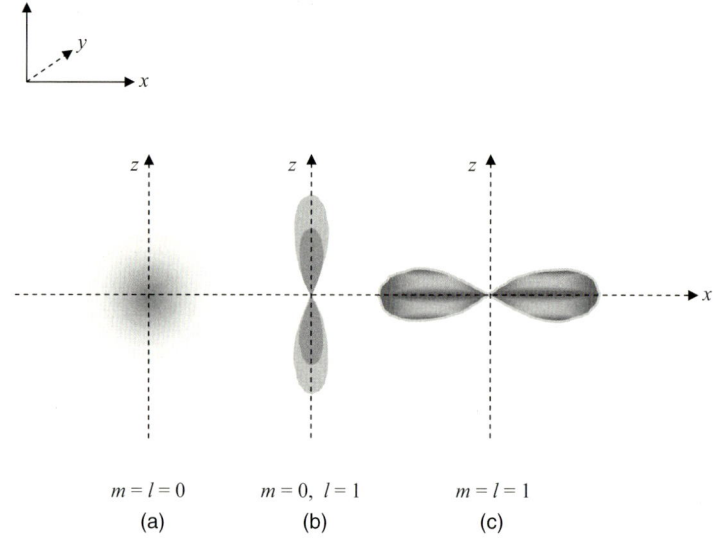

Figure 3.1 A few different configurations of a "cloud" formed by an electron in a hydrogen atom. The integers *l* and *m* are the quantum numbers determining the magnitude and the z-projection of electron's orbital angular momentum.

Equation 2.31 describes a very strange (from the classical viewpoint) symbiosis of concepts considered to be mutually exclusive: on the right we see the energy and momentum of a corpuscular object; on the left we see the characteristics of a periodic process – the frequency and wave vector of a running monochromatic wave. The equations tell us that one type of characteristic determines the other, which means that both aspects – one of *discrete things* and the other of a *process undulating in space and time* – are relevant characteristics of the same entity. So these equations express the "wave–particle duality" of the world. According to it, the wave-like or particle-like behavior as such would give us only an incomplete image of an object, working fine in some situations, but letting its counterpart take over in others. For instance, if we do find an electron at some location in an experiment designed for a position measurement, we always find the whole electron, not a part of the electron. This is a corpuscular behavior. On the other hand, there is no way of indicating the electron position *between* such measurements without actually performing a new one. All evidence indicates that the electron between position measurements exists in all locations of an extended region at once. This is a wave behavior. We express this by saying, after R. Bayerlein [1], that a microscopic object is recorded as a particle but propagates as a wave. These two properties are mutually complementary characteristics of reality (Bohr's *complementarity principle*).

From this duality there emerges, like an exotic plant from a seed, the set of properties that are alien to our "classical" experience, and all together they form the basis of QM. The wave aspect manifests itself in superposition, for instance, a quantum system existing simultaneously in two or more different states. The already mentioned "omnipresence" of a stationary electron existing almost everywhere around the nucleus is a case in point. No less puzzling is the system's sudden jump from one state to another, for instance, when we find the same electron at a definite point in a position measurement. Such a jump from many potentialities to a definite value is known as the *collapse of the wave function*. It is a discontinuous process, and in this respect it cannot even be called a "process." *It is not described by the Schrödinger (or any other) equation.* Its timing is unpredictable, and so is the outcome – a specific new state the system happened to collapse into. QM can only predict *probability* of this or that outcome. This is a radical departure from classical determinism. In the following sections, we will outline the basic quantum-mechanical concepts that superseded the classical picture of the world.

3.2
Quantum-Mechanical Superposition of States

The wave aspect of a particle enables it to exhibit the wave behavior. A wave can split into parts, for instance, by passing through a screen with two or more slits (grating), as discussed in Section 1.5 (*wavefront splitting*). Or, it can do so when passing through a beam splitter (*amplitude splitting*, Figure 3.2a). The split parts propagate independently of each other, so the wave is in a *superposition* of distinct states with different **k**. Then the split waves can be recombined using a system of mirrors (Figure 3.2b), which would produce the interference pattern on a screen.

What happens if there is only one photon passing through the device? According to QM, *each single photon* possesses wave-like properties. It can take both paths in an interferometer and then interfere *with itself* on the screen. It still lands on the screen as a particle, producing a single spot, but not anywhere. If the interference is totally destructive (two equal parts of the wave recombining at a given location are out of phase), the resulting local amplitude is zero, and the particle cannot land there. It hits primarily the spots where the parts of the wave interfere constructively

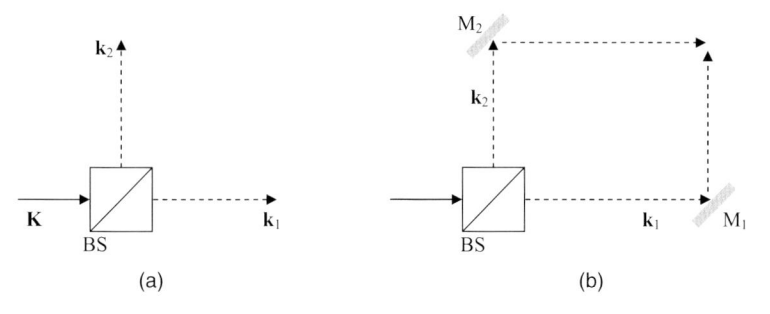

Figure 3.2 (a) Schematic of an experiment transforming a photon state with momentum **K** into superposition of states with momenta k_1 and k_2. BS – beam splitter. (b) Recombining the separated states using mirrors M_1 and M_2. The region where the recombined waves overlap will display an interference pattern. The schematic shown here is a part of the *Mach–Zehnder interferometer*, discussed in Section 23.5.

(the amplitude of the resulting wave is maximal). If all particles are also waves, they should exhibit similar behavior in similar situations.

This prediction was experimentally confirmed. Its first empirical demonstration was the electron diffraction observed by Davisson and Germer [2,3]. This experiment cannot be carried out with the traditional optical equipment (e.g., diffraction grating [4]). The reason is that according to relation $\mathbf{k} = \mathbf{p}/\hbar$, the wavelength of an incident electron with a typical kinetic energy 30–400 eV is comparable to the interatomic distance in a crystal – much smaller than the wavelength of light in the visible region of the spectrum (Problem 3.1). Therefore, Davisson and Germer used a nickel crystal instead of an optical grating. The experimental setup is shown in Figure 3.3. An electron beam is incident perpendicularly onto the face of the crystal. Most of the electrons get scattered by the surface layer of the crystal lattice acting as an optical grating. The number of electrons scattered at different angles was measured using an electron detector.

The observed distribution of scattered electrons is a typical interference pattern characteristic of waves. The positions of the diffraction maxima are found from the same relation, $m\lambda = d \sin \theta$, as in the double-slit diffraction, but now d is the lattice constant. For each energy of the beam, the value of λ determined from the measured angles $\theta = \theta_m$ is equal to the wavelength found from de Broglie's equations (2.31). The diffraction of electrons was observed also in many other experiments (see, for example, Ref. [5]), and all of them confirmed the de Broglie relations.

The validity of Equations 2.31 and 2.32 was also confirmed for other particles. Most known are the experiments with slow neutrons [6,7] (Figure 3.4), with crystals acting again as a diffraction grating. The neutron velocity must be sufficiently small (hence the term "slow neutrons") in order for the neutron wavelength to match the

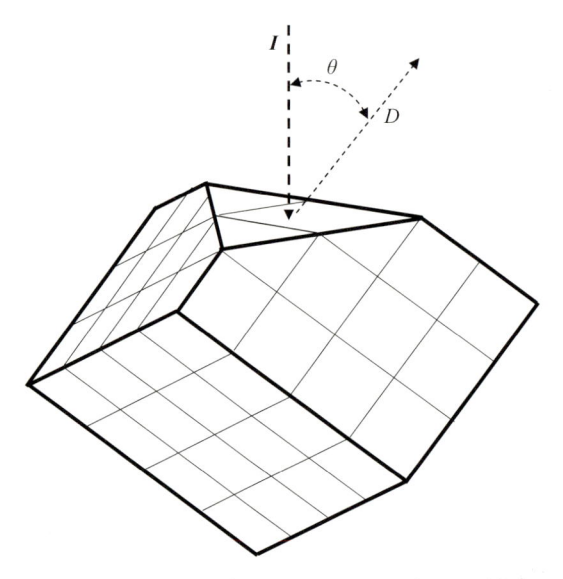

Figure 3.3 Schematic of an experiment with electron diffraction on a crystal. I – incident electron beam; D – one of the diffracted beams.

Figure 3.4 Diffraction of neutrons by a sodium chloride crystal. The diffraction pattern is similar to the one produced by X-rays with equal wavelength on the same target. Reproduced with permission from Ref. [9].

lattice constant d of the crystal. Today, the neutron diffraction is used as routinely as X-ray diffraction for characterization of crystals. Experiments exploiting the wave aspect of matter are also carried out with atoms and molecules [8]. They all demonstrate the universal nature of de Broglie waves.

All these results originate from the superposition in which a wave (particle) with definite frequency (energy) interferes with itself after taking *all* available paths at once. Generally, even for a fixed path, superposed waves may have different frequencies, for instance,

$$\Psi(x,t) = \sum_j C_j \, e^{i(k_j x - \omega(k_j))t}. \tag{3.1}$$

Equation 3.1 is a special case of Fourier series expansion of $\Psi(x,t)$. We have such a case when $\Psi(x,t)$ satisfies, for example, the *D'Alembert equation* [10,11] for a violin string, in which case $\omega(k) = uk$ with u being the wave speed along the string. A string may vibrate with only one out of the set of allowed frequencies, producing a pure tone. Its instant shape is then a sinusoid. But it also can vibrate with different allowed frequencies at once. This is a superposition described by (3.1).

If a particle can exhibit wave properties, it can also be in a superposition similar to (3.1). We will see later (Chapters 8 and 9) that this is indeed the case. For instance, a particle trapped within a region of length a behaves like a string of the same length with fixed edges and is accordingly described by an expression similar to (3.1) with one or many different frequencies. And in view of the frequency–energy connection (2.1), it follows that a single particle can be in a superposition of states with different energies.

3.3
What Is Waving There (the Meaning of the Ψ-Function)?

> Then came the Born interpretation. The wave function gives not the density of stuff, but gives rather (on squaring its modulus) the density of probability. Probability of what exactly? Not of an electron *being* there, but of the electron being *found* there, if its position is "measured".
>
> J.S. Bell, Against Measurement

Accepting that a particle can be represented by a wave and be treated as such raises a natural question: OK, but *what is it* that is waving? The answer starts with Einstein, de Broglie, and Born. According to (1.19), light intensity is proportional to the square modulus $|\mathcal{E}|^2$ of the wave amplitude. On the other hand, if light is considered as a bunch of particles, then its intensity must stand in proportion to the number of particles per unit volume (i.e., concentration) at a given locality. It follows then that $|\mathcal{E}|^2$ must be a measure of concentration of light particles. And according to de Broglie, this must be generalized to particles of *any* kind.

Well, what if we have only one particle? We cannot speak of concentration (in statistical terms) when there is only one particle. The answer to this fundamental question was suggested by Max Born [12]. According to Born, the value $|\mathcal{E}(\mathbf{r})|^2$ and, generally, $|\Psi(\mathbf{r})|^2$ is the *probability* per unit volume (*probability density*) $\rho(\mathbf{r})$ of finding the particle in the vicinity of $\mathbf{r}$. This is the famous probabilistic interpretation of the wave function. If a quantum-mechanical state is described by a function of position $\Psi(\mathbf{r}, t)$ (called a *wave function in* $\mathbf{r}$-*representation*), then

$$d\mathcal{P}(\mathbf{r}) = |\Psi(\mathbf{r})|^2 \, d\mathbf{r} \quad \text{and} \quad |\Psi(\mathbf{r})|^2 = \frac{d\mathcal{P}(\mathbf{r})}{d\mathbf{r}} \equiv \rho(\mathbf{r}). \tag{3.2}$$

Here $d\mathcal{P}(\mathbf{r})$ is the probability of finding the particle in a small volume $d\mathbf{r}$ containing the point $\mathbf{r}$. Thus, $\Psi(\mathbf{r}, t)$ describes the probability distribution associated with state Ψ. It is precisely this distribution that is shown as the "probability cloud" in Figure 3.1 for an electron in an H atom.

The probability of finding the particle in a finite region of volume V will be the integral of $|\Psi(\mathbf{r})|^2$ over this volume: $\mathcal{P}(V) = \int_V d\mathcal{P}(\mathbf{r}) = \int_V |\Psi(\mathbf{r})|^2 \, d\mathbf{r}$. If V embraces all locations allowed under given conditions (e.g., for the particle in a box, V should be the volume of this box; for a free particle, $V = \infty$), then $\mathcal{P}(V) = 1$: the particle

3.3 What Is Waving There (the Meaning of the Ψ-Function)?

must certainly be found *somewhere* within V. In such cases, we do not indicate the integration region, and, assuming that $\Psi(\mathbf{r})$ is *square integrable*, we have

$$\int |\Psi(\mathbf{r})|^2 \, d\mathbf{r} = \int \Psi^*(\mathbf{r}) \Psi(\mathbf{r}) d\mathbf{r} = 1. \tag{3.3}$$

This condition is called *normalization* and a function satisfying it is said to be *normalized*. We can represent the state as a function of some other characteristic, say, energy. Denote this function as $C = C(E)$ (the wave function in E-representation). Then its square modulus

$$|C(E)|^2 = C^*(E_j)C(E_j) = \mathcal{P}(E_j) \tag{3.4}$$

will give the probability of finding the particle in a state with energy E_j (assuming the spectrum is discrete). The net probability of finding the particle in any one of all possible energy states is

$$\sum_j \mathcal{P}(E_j) = \sum_j C^*(E_j)C(E_j) = 1. \tag{3.5}$$

For a continuous spectrum, (3.4) will take the form $|C(E)|^2 = d\mathcal{P}(E)/dE$ (probability density), and we will instead of (3.5) have the integral over the spectrum, similar to (3.3).

In some situations, the assumption of square integrability used in (3.3) is not satisfied. For example, the de Broglie wave (2.32) describing a state with definite $\mathbf{p} = \mathbf{p}'$ is not square integrable. The integral of $|\Psi|^2$ with Ψ from (2.32) diverges. This by itself does not constitute any problem, since the state (2.32) – a monochromatic wave with one fixed frequency - is only an idealization never realized in an experimental situation. Any real situation involves a group of waves (*wave packet*) within a region $\Delta \mathbf{p}$ such that all components of the group reinforce each other (interfere constructively) within a finite region $\Delta \mathbf{r}$, and oppose each other (interfere destructively) outside this region, so the resulting wave function is square integrable. But even in the idealized case (2.32), the normalization can be generalized due to the fact that the space integral of the product $\psi^*_{\mathbf{p}'}(\mathbf{r})\psi_{\mathbf{p}''}(\mathbf{r})$, while being infinite if $\mathbf{p}'' = \mathbf{p}'$, is zero if $\mathbf{p}'' \neq \mathbf{p}'$. In fact, such integral has the properties of the delta function $\delta(\mathbf{q} - \mathbf{q}')$ introduced by Dirac [13–15]. In the 1D case, it is a function of $(q - q')$ such that

$$\delta(q - q') = \begin{cases} 0, & q \neq q', \\ \infty, & q = q'. \end{cases} \tag{3.6}$$

(Since the δ-function is a routine and sometimes indispensable tool in QM, its more detailed description is given in Appendix B.) Therefore, the monochromatic function (2.32) can be normalized so that

$$\int \Psi^*_{\mathbf{p}'}(\mathbf{r}, t) \Psi_{\mathbf{p}''}(\mathbf{r}, t) d\mathbf{r} = \delta(\mathbf{p}' - \mathbf{p}''). \tag{3.7}$$

This type of normalization holds for any observable with continuous spectrum. If the spectrum is discrete (as in case (3.5)), then the normalization condition for the corresponding bound states $\Psi_j(\mathbf{r}) = \Psi_{E_j}(\mathbf{r})$ takes the form

$$\int \Psi^*_{E_j}(\mathbf{r}) \Psi_{E_k}(\mathbf{r}) d\mathbf{r} = \delta_{jk}. \tag{3.8}$$

Here δ_{jk} (Kronecker's delta) is a function of two integers j and k, which is zero when j and k are different and 1 when they are equal:

$$\delta_{jk} = \begin{cases} 0, & j \neq k, \\ 1, & j = k. \end{cases} \qquad (3.9)$$

Thus, Equation 3.8 reduces to Equation 3.3 when $j = k$. The conditions (3.7) and (3.8) tell us that the integral of the product of two wave functions corresponding to different values of the spectrum is always equal to zero. The set of functions satisfying condition (3.7) for continuous spectrum or (3.8) for discrete spectrum is called the *orthonormal basis*, by analogy with geometry, where we frequently use the sets of mutually orthogonal unit vectors $\hat{\mathbf{n}}_j$ (for more details, see Chapter 5).

These properties of Ψ, important as they are, do not answer the question: The waving of what is represented by the wave function? But they help us see its rather subtle nature. The probability amplitude Ψ_0 of the monochromatic wave (2.32) definitely does not wave. The exponential function in (2.32) is merely a complex function of $\mathbf{r}$ and t; its real and imaginary parts do wave with frequency ω, but neither of them can be directly observed. The function (2.32) by itself is not any measurable undulating entity like a water surface of a lake or an electric field in an EM wave. The latter may be described by the *real part* of an expression similar to (2.32), but only for a huge *ensemble* of photons all in one state corresponding to classical EM. The correct quantum-mechanical description of a single free particle needs the whole complex function (2.32) in its entirety.

It is easy to show that the wave function of a particle in nonrelativistic QM *must* be complex for a fundamental reason: it not only stores *all* information about the particle's state at a given moment, but also determines its evolution for all future moments. In other words, knowing $\Psi(\mathbf{r}, 0)$ determines $\Psi(\mathbf{r}, t)$. This can only be the case if $\Psi(\mathbf{r}, t)$ is a solution of differential equation of the *first order* in time:

$$i\hbar \frac{\partial \Psi}{\partial t} = \hat{H}\Psi. \qquad (3.10)$$

(An equation of a higher order would have more than one constant of integration in its general solution, so we would need extra initial conditions and knowing only the initial state would not be enough!) As we will see in the next section, the operator $i\hbar(\partial/\partial t)$ represents the particle energy, and

$$\hat{H} = -\frac{\hbar^2}{2\mu}\nabla^2 + U(\mathbf{r}) \qquad (3.11)$$

is a differential operator called the *Hamiltonian*, which for a free particle (the potential energy $U(\mathbf{r}) = 0$) reduces (up to a constant) to the Laplacian ∇^2. Equation 3.10 is the *Schrödinger equation*. Without an imaginary unit on the left, its solution for a free particle would exponentially increase in space instead of describing a wave with constant amplitude. In contrast to the D'Alembert equation (this equation is of the *second order* in time), which, for a monochromatic wave can be equally well satisfied with $\sin(\mathbf{kr} \pm \omega t)$, $\cos(\mathbf{kr} \pm \omega t)$, or $e^{i(\mathbf{k\,r} \pm \omega t)}$, Equation 3.10 with $U = 0$ can *only* be satisfied by $e^{i(\mathbf{kr} \pm \omega t)}$, which is an explicitly complex function. This result combined with (3.2) leads to another conclusion

3.3 What Is Waving There (the Meaning of the Ψ-Function)?

showing the difference between a classical and a quantum wave. Write $\Psi(\mathbf{r}, t)$ in the form

$$\Psi(\mathbf{r}, t) = \Phi(\mathbf{r}, t) e^{i\varphi(\mathbf{r},t)}, \tag{3.12}$$

with $\Phi(\mathbf{r}, t)$ and $\varphi(\mathbf{r}, t)$ being real functions of $\mathbf{r}$ and t. The de Broglie wave (2.32) is a special case of (3.12). However, whereas in a classical wave the phase may be the measurable characteristic, nothing we can measure defines the phase of a quantum wave for a single particle. For instance, in the probability density (3.2), which is a measurable characteristic, nothing is left of the phase – it drops from the square modulus (3.2). We can measure only the *phase difference* between two waves. If we have the sum of two waves (3.12)

$$\Psi(\mathbf{r}, t) = \Psi_1(\mathbf{r}, t) + \Psi_2(\mathbf{r}, t) = \Phi_1(\mathbf{r}, t) e^{i\varphi_1(\mathbf{r},t)} + \Phi_2(\mathbf{r}, t) e^{i\varphi_2(\mathbf{r},t)} \tag{3.13}$$

(as in case of the double-slit interference (1.18)), the probability density is

$$|\Psi(\mathbf{r}, t)|^2 = \left|\Phi_1(\mathbf{r}, t) e^{i\varphi_1(\mathbf{r},t)} + \Phi_2(\mathbf{r}, t) e^{i\varphi_2(\mathbf{r},t)}\right|^2$$
$$= \Phi_1^2(\mathbf{r}, t) + 2\Phi_1(\mathbf{r}, t)\Phi_2(\mathbf{r}, t)\cos(\varphi_1(\mathbf{r}, t) - \varphi_2(\mathbf{r}, t)) + \Phi_2^2(\mathbf{r}, t). \tag{3.14}$$

We see that the phase difference $\Delta\varphi(\mathbf{r}, t) = \varphi_1(\mathbf{r}, t) - \varphi_2(\mathbf{r}, t)$ explicitly enters the corresponding interference pattern, which is a measurable characteristic. But as we know from Chapter 2, even this characteristic can be measured *only for an ensemble of identical particles all in one state*, rather than for a single particle. And it is such an ensemble that forms (when sufficiently huge!) a classical wave. Thus, only the phase difference $\Delta\varphi(\mathbf{r}, t)$ between two partial waves has physical meaning. If we add an additional *overall* phase $\Lambda(\mathbf{r}, t)$ to the respective phases $\varphi_1(\mathbf{r}, t)$ and $\varphi_2(\mathbf{r}, t)$, it will drop from their difference and thus will not affect distribution (3.14). This would be equivalent to multiplying $\Psi(\mathbf{r}, t)$ by $e^{i\Lambda(\mathbf{r},t)}$:

$$\Psi(\mathbf{r}, t) \rightarrow \Psi'(\mathbf{r}, t) = e^{i\Lambda(\mathbf{r},t)} \Psi(\mathbf{r}, t). \tag{3.15}$$

It does not change the probability distribution $\rho(\mathbf{r}, t)$, including the interference pattern when $\Psi(\mathbf{r}, t)$ is a superposition (3.13). We say that $\rho(\mathbf{r}, t)$ is invariant under transformation (3.15). The transformation itself is called the *local phase transformation*. In the special case when $\Lambda(\mathbf{r}, t) = \text{const} = \Lambda$, we call it the *global* phase transformation. *All measurable characteristics of a quantum state are invariant with respect to the global phase transformation.* Some of them, like $\rho(\mathbf{r}, t)$, are also invariant with respect to the local phase transformation.

We can come to the same conclusion from another perspective. As we will see in Section 8.3, the wave function satisfies the *continuity equation*:

$$\frac{\partial \rho}{\partial t} + \vec{\nabla} \cdot \mathbf{j} = 0. \tag{3.16}$$

Here $\rho \equiv |\Psi|^2$, and

$$\mathbf{j} = \frac{i\hbar}{2\mu}\left(\Psi \vec{\nabla} \Psi^* - \Psi^* \vec{\nabla} \Psi\right) \equiv -\frac{\hbar}{\mu} \text{Im}\left(\Psi \vec{\nabla} \Psi^*\right) \tag{3.17}$$

is the *probability flux* density. Expression (3.17) illustrates the role of the quantum phase. Indeed, putting here (3.12) yields

$$\mathbf{j} = \rho \cdot \left(\frac{\hbar}{\mu}\vec{\nabla}\varphi\right). \tag{3.18}$$

This expression allows one to establish a certain analogy with fluid dynamics, where the fluid flux density is a product of mass density and velocity of the fluid element at a given location: $\mathbf{j}_\mu = \rho_\mu \mathbf{v}$. If we multiply (3.18) through by the particle's mass μ and denote the products as $\rho_\mu \equiv \mu\rho$, $\mathbf{j}_\mu \equiv \mu\mathbf{j}$, then the resulting equation will be mathematically identical to $\mathbf{j}_\mu = \rho_\mu \mathbf{v}$. The obtained new quantities ρ_μ and $\mathbf{j}_\mu$ represent the mass and flux densities of matter in a given state. Similarly, multiplying ρ and $\mathbf{j}$ by the particle's charge q, we obtain the charge and current densities in a given state. In either case, the corresponding densities satisfy the continuity relation identical with (3.16).

Expressions (3.17) and (3.18) clearly show that the flux is nonzero only when the corresponding state is described by a *complex* wave function. And it is the *gradient* of phase that is physically observable (and only indirectly, as the *average* momentum) but not the phase itself.

In CM, Equation 3.16 expresses the conservation laws for a distributed mass and electric charge, respectively. In QM, it describes the *probability conservation* (any change of $\mathcal{P}(V)$ within a volume V is associated with the nonzero probability flux *in* or *out* through the boundary of this volume). We see again that $\Psi(\mathbf{r}, t)$ generally contains only indirect information about measurable characteristics and gives direct information only about their probabilities.

The abstract nature of the wave function becomes even more evident if we consider a *system* of particles. If we have N particles, then the wave function describing the whole system is, in the case of **r**-representation, a function of at least $3N$ variables (one position vector, that is, three coordinates, for each particle):

$$\Psi(\tilde{\mathbf{r}}) = \Psi(\mathbf{r}_1, \mathbf{r}_2, \ldots, \mathbf{r}_N, t). \tag{3.19}$$

According to Born, its square modulus gives the probability density of finding one particle at $\mathbf{r}_1$, another particle at $\mathbf{r}_2$, ..., and the Nth particle at $\mathbf{r}_N$ at the moment t. The probability itself will be given by

$$d\mathcal{P}(\tilde{\mathbf{r}}, t) = |\Psi(\tilde{\mathbf{r}}, t)|^2 \, d\tilde{\mathbf{r}}. \tag{3.20}$$

Here $\tilde{\mathbf{r}}$ can be considered as a $3N$-dimensional vector with $3N$ components (x_j, y_j, z_j), $j = 1, 2, \ldots, N$, and $d\tilde{\mathbf{r}}$ is the corresponding volume element: $d\tilde{\mathbf{r}} = d\mathbf{r}_1 d\mathbf{r}_2 \cdots d\mathbf{r}_N$, with each subelement determined (in Cartesian coordinates) as $d\mathbf{r}_j = dx_j dy_j dz_j$. If the particles do not interact with each other, the above probability – as the probability of independent events – will be the product of the individual probabilities of finding each particle at a respective location, $d\mathcal{P}(\tilde{\mathbf{r}}) = d\mathcal{P}_1(\mathbf{r}_1) d\mathcal{P}_2(\mathbf{r}_2) \cdots d\mathcal{P}_N(\mathbf{r}_N)$ (we drop the time dependence here). Accordingly, the probability amplitude will be the product of the individual amplitudes:

$$\Psi(\tilde{\mathbf{r}}) = \Psi(\mathbf{r}_1, \mathbf{r}_2, \ldots, \mathbf{r}_N) = \Psi_1(\mathbf{r}_1)\Psi_2(\mathbf{r}_2) \cdots \Psi_N(\mathbf{r}_N). \tag{3.21}$$

If we want to consider $\Psi(\vec{r})$ in all these examples as a wave, it will be a wave in a 3N-dimensional space. Thus, the simple view of the amplitude Ψ as something whose square modulus represents a certain density in space becomes far less straightforward for a *system* of particles. In this case, both expressions (3.2) and (3.17) are functions of many variables satisfying the continuity equation in an abstract multidimensional space rather than in the real space. So the wave function should be considered just as a mathematical expression storing all information about the state of a system, and accordingly, we can (and frequently will) call it the *state function* as well.

3.4
Observables and Their Operators

It follows from de Broglie's postulates that observable characteristics of a system can be represented each by its specific operator. For instance, consider the de Broglie wave for a particle moving along the x-direction. Differentiate it with respect to x, then multiply through by $-i\hbar$ and use $\hbar k = p = p_x$:

$$-i\hbar \frac{\partial}{\partial x}\left(\Psi_0 e^{i(kx-\omega t)}\right) = \hbar k \left(\Psi_0 e^{i(kx-\omega t)}\right) = p\left(\Psi_0 e^{i(kx-\omega t)}\right). \tag{3.22}$$

Thus, performing the indicated operation on the de Broglie wave function returns this function multiplied by the particle's momentum. Any action performed on a function is an *operation* and is represented by the corresponding symbol named *operator*. Taking the derivative $\partial f(x)/\partial x$ can be considered as multiplying $f(x)$ by the *differential operator* $\hat{Q}_x \equiv \partial/\partial x$, so that $\hat{Q}_x f(x) \equiv \partial f(x)/\partial x$.

We write the operator *before* the function it acts upon, and $\hat{Q}f(q)$ means "the result of applying $\hat{Q}$ to function $f(q)$."

Acting on the function in (3.22) by operator $\hat{p}_x \equiv -i\hbar(\partial/\partial x)$ is equivalent to just multiplying this function by the momentum of the particle it represents. A function Ψ_q such that when acted upon by a certain operator $\hat{Q}$ it is returned unchanged, only multiplied by a numerical factor q, is called an *eigenfunction* of $\hat{Q}$, and the multiplying factor q is called an *eigenvalue* of $\hat{Q}$ (the terms come from the German *eigen*, meaning "proper"). Conversely, we say in such cases that $\hat{Q}$ *represents* observable q. The function Ψ_{q_α} is then one of the eigenfunctions, and q_α is one of the eigenvalues of this operator. The subscript α identifies specific values of q, which can be discrete or continuous within a given range. In the considered case of 1D motion, the eigenvalue of operator $\hat{p}_x$ is the observable component p_x of the particle's momentum. Therefore, this operator is called the operator of linear momentum (more accurately, of its x-component). For motion in an arbitrary direction, it is easy to show (Problem 3.5) that acting on the de Broglie wave (2.32) by operator $-i\hbar \vec{\nabla}$, where $\vec{\nabla} \equiv (\partial/\partial x)\hat{x} + (\partial/\partial x)\hat{y} + (\partial/\partial x)\hat{z}$ (with $\hat{x}, \hat{y}, \hat{z}$ being the corresponding unit vectors) is the *del* operator known in vector calculus, returns this function multiplied by the particle's momentum **p**:

$$-i\hbar \vec{\nabla} \Psi_p = p\Psi_p, \quad \Psi_p \equiv \Psi_0 e^{i(\mathbf{k}\cdot\mathbf{r}-\omega t)}, \quad \mathbf{k} = \frac{\mathbf{p}}{\hbar}. \tag{3.23}$$

In order to distinguish an operator from the observable it represents, we usually denote the operator by the same symbol capped. Thus, we write the momentum operator as

$$\hat{\mathbf{p}} \equiv -i\hbar \vec{\nabla}, \quad \text{so that} \quad \hat{\mathbf{p}}\Psi_p = \mathbf{p}\Psi_p. \tag{3.24}$$

Now take the *time* derivative of (3.23), multiply through by $i\hbar$, and use $\hbar\omega = E$. We will get

$$i\hbar \frac{\partial}{\partial t}\Psi_E(\mathbf{r},t) = E\Psi_E(\mathbf{r},t), \quad \Psi_E(\mathbf{r},t) \equiv \Psi_0\, e^{i[(\mathbf{p}\cdot\mathbf{r})/\hbar - (E/\hbar)t]}. \tag{3.25}$$

In this case, operator $i\hbar(\partial/\partial t)$ acting on a function Ψ_E describing a particle with definite energy E returns Ψ_E multiplied by E. Therefore, it can be considered as "representative" of energy and called the energy operator:

$$\hat{E} \equiv i\hbar \frac{\partial}{\partial t}. \tag{3.26}$$

Then (3.25) takes the form

$$\hat{E}\Psi_E(\mathbf{r},t) = E\Psi_E(\mathbf{r},t). \tag{3.27}$$

Using terminology introduced in the example with momentum, we say that E is one of the possible eigenvalues of $\hat{E}$, and $\Psi_E(\mathbf{r},t)$ is the corresponding eigenfunction.

Looking back at the observables $\mathbf{p}$ and E, we note that they depend on each other. For a nonrelativistic free particle of mass μ, we have $E(\mathbf{p}) = \mathbf{p}^2/2\mu$. If each observable is represented by its operator, then any relationship between them must be mirrored by the corresponding relationship between their operators. This leads to a suggestion that the energy and momentum operators, in the case of a free particle, must be related by

$$\hat{E}(\mathbf{p}) = \frac{\hat{\mathbf{p}}^2}{2\mu},$$

or, in view of (3.24) and (3.26),

$$i\hbar \frac{\partial}{\partial t} = -\frac{\hbar^2}{2\mu}\nabla^2. \tag{3.28}$$

Physically, this means that the state function describing a free particle must satisfy the equation

$$i\hbar \frac{\partial}{\partial t}\Psi(\mathbf{r},t) = -\frac{\hbar^2}{2\mu}\nabla^2 \Psi(\mathbf{r},t), \tag{3.29}$$

which is a special case of (3.10) when $U = 0$. The de Broglie wave does satisfy this equation and is thus its solution. In other words, it is a common eigenfunction of two different operators.

In the presence of forces described as the gradient of potential energy, $\mathbf{f} = -\vec{\nabla}U(\mathbf{r})$, the total energy $E = H(\mathbf{p},\mathbf{r})$ is

$$H(\mathbf{p},\mathbf{r}) = \frac{\mathbf{p}^2}{2\mu} + U(\mathbf{r}). \tag{3.30}$$

Expression (3.30) is known in CM as the *Hamilton function*. It is natural to assume that the operator representing a particle's mechanical energy in this more general case is given by (3.11) and call it the *Hamiltonian operator*. With this operator, we can generalize (3.29) onto the case when the particle is moving in a given potential field and thus arrive again, but in a different way, at the Schrödinger equation (3.10).

Mathematically, an eigenfunction $\Psi_a(\mathbf{r})$ of a differential operator $\hat{Q}(\mathbf{r})$ is the particular solution of the linear differential equation

$$\hat{Q}(\mathbf{r})\Psi_a(\mathbf{r}) = q_a \Psi_a(\mathbf{r}), \tag{3.31}$$

with an eigenvalue q_a. The set of all possible eigenvalues can be obtained by solving this equation with the corresponding boundary conditions that determine the nature of the set. For instance, a condition requiring that $\Psi(\mathbf{r}) \to 0$ at $r \to \infty$ can only be satisfied by a *discrete* set of functions $\Psi(\mathbf{r}) = \Psi_n(\mathbf{r})$ and the corresponding values q_n with integer $n = 1, 2, \ldots$. Such a condition means that the particle in the given states cannot be found far away from the origin, that is, its motion is bound within a finite region of space. An atomic electron, with its discrete spectrum, is the known example. And *in all cases* the specific eigenvalues (be they discrete or continuous) and the corresponding eigenfunctions obtained as solutions of Equation 3.31 *correctly describe the system in question*. This is another illustration of how QM gives an adequate description of reality.

The considered operators, apart from representing their respective observables, are also associated with spatial and temporal displacements. As an example, consider a small displacement $\Delta \mathbf{r}$ that takes $\Psi(\mathbf{r})$ to $\Psi(\mathbf{r} + \Delta \mathbf{r})$. For $\Delta \mathbf{r} \to 0$, this operation can be written as $\Psi(\mathbf{r} + \Delta \mathbf{r}) = \left(1 + \Delta \mathbf{r} \vec{\nabla}\right)\Psi(\mathbf{r})$, or, in view of (3.24):

$$\Psi(\mathbf{r} + \Delta \mathbf{r}) = \left(1 + \frac{i}{\hbar}\Delta \mathbf{r} \cdot \hat{\mathbf{p}}\right)\Psi(\mathbf{r}), \tag{3.32}$$

We say that operator $\hat{P}(\hat{\mathbf{p}}, \Delta \mathbf{r}) \equiv 1 + (i/\hbar)\Delta \mathbf{r} \cdot \hat{\mathbf{p}}$ performs a displacement taking $\Psi(\mathbf{r})$ to $\Psi(\mathbf{r} + \Delta \mathbf{r})$ and accordingly call it the *spatial displacement operator*. Similarly, one can show that the $\tilde{E}$ operator is associated with time displacement

$$\Psi(t + \Delta t) = \left(1 - \frac{i}{\hbar}\Delta t \cdot \hat{E}\right)\Psi(t). \tag{3.33}$$

In view of (3.26) and (3.10), the same is true for the Hamiltonian operator in the set of solutions of the Schrödinger equation:

$$\Psi(t + \Delta t) = \left(1 - \frac{i}{\hbar}\Delta t \hat{H}\right)\Psi(t). \tag{3.34}$$

The expression within parenthesis is the *time displacement operator*.

3.5
Quantum-Mechanical Indeterminacy

Another manifestation of superposition of states is indeterminacy of some physical characteristics of a system. In CM, if we fail to measure an observable to a high precision, it is due to our incompetence or to imperfection of our instruments. In QM, certain pairs of observables, such as position and momentum of a particle, cannot *in principle* be determined simultaneously. This conclusion follows immediately from de Broglie's postulate. A free particle described by de Broglie wave $\Psi_{\mathbf{p}}(\mathbf{r},t) = \Psi_0 \, e^{i(\mathbf{kr}-\omega t)}$ has a definite momentum $\mathbf{p}$. The corresponding probability distribution $|\Psi(\mathbf{r},t)|^2 = \Psi_0^2$ is uniform throughout the whole space. The particle cannot be ascribed a distinct position – it is totally *indeterminate*. On the other hand, a particle with a definite position $\mathbf{r} = \mathbf{r}'$ is described by the delta function $\Psi_{\mathbf{r}'}(\mathbf{r},t) = \delta(\mathbf{r}-\mathbf{r}')$. But $\delta(\mathbf{r}-\mathbf{r}')$ is a superposition of de Broglie waves with *all possible* $\mathbf{p}$, which interfere constructively, producing a huge splash at $\mathbf{r}'$, and interfere destructively, canceling each other everywhere else (Appendix B). Now it is the *momentum* that is totally indeterminate. Thus, a state with definite $\mathbf{p}$ has totally indeterminate $\mathbf{r}$, and a state with definite $\mathbf{r}$ has totally undefined $\mathbf{p}$. These two extreme cases suggest that, in general, an accurate measurement of one variable of the pair $(\mathbf{r}, \mathbf{p})$ comes at the cost of the other one. Position and momentum are *incompatible* (or mutually *complementary*) characteristics of an object.

Indeed, consider an intermediate case between the above two extremes, for example, a particle bound within a finite region of length a. A mechanical analogy taking account of the wave aspect of matter would be a wave on a violin string of length a. An optical analogy is a light wave between two parallel mirrors distance a apart (the *Fabry–Pérot* resonator [4,16]). Such a truncated wave can be represented by a superposition (3.1) of unbounded reiterating waves with different k such that they totally cancel each other outside region a and produce the actual shape of the wave within it. The smaller the a, the greater the needed spread Δk of the corresponding propagation numbers in such superposition. Taking a as the measure of spatial indeterminacy, $a = \Delta x$, we can evaluate the relationship between Δx and Δk using the theory of Fourier transforms [5]. As known from this theory, a function $\Psi(x)$ that is nonzero in the region of width Δx can be reproduced by a group of monochromatic waves with wave numbers in the range of width $\Delta k \sim (\Delta x)^{-1}$. Thus, Δk and Δx are mutually reciprocal, being related by $\Delta k \Delta x \sim 1$. A more rigorous treatment in Chapter 5 shows that $\Delta x \Delta k_x \geq 1/2$. In view of $\mathbf{p} = \hbar \mathbf{k}$, this translates into

$$\Delta x \Delta p_x \geq \frac{1}{2}\hbar, \tag{3.35}$$

and the same for the y and z components, respectively. This is the Heisenberg *uncertainty relationship* – probably the second (after $E = \mu c^2$) most famous relation in physics. It tells us that the product of Δp_x and Δx cannot be less than $\hbar/2$; so, the more accurately determined is x, the less accurate (better to say, more indeterminate) is p_x, and vice versa. The two opposite extremes considered in the beginning of this section are contained in (3.35) as special cases.

Similar relationships hold for some other pairs of variables. For instance, instead of considering group of waves with different k at a fixed moment of time, we can consider undulations in time for the same group at a fixed point in space. Then the same argument will show that the time interval Δt of considerable perturbation is related to the frequency spread $\Delta\omega$ by $\Delta\omega\Delta t \geq 1/2$, or, in view of $E = \hbar\omega$,

$$\Delta E \Delta t \geq \frac{1}{2}\hbar. \qquad (3.36)$$

This relationship has its own subtle points, which will be discussed in Chapter 13.

Note that indeterminacy generally does not keep us from an exact measurement of any observable, as is evident from above-mentioned examples of states with definite $\mathbf{p} = \mathbf{p}'$ or definite $\mathbf{r} = \mathbf{r}'$; it only forbids this for the whole *pair*, such as $(\mathbf{p}, \mathbf{r})$ in *the same* state. For instance, we can perform a position measurement and obtain (ideally) the *exact* outcome – a particle at $x = x_1$. In this new state, the particle will have a definite coordinate with $\Delta x = 0$. But according to (3.35), this will be a state with totally indefinite momentum, $\Delta p_x = \infty$. We cannot complement this procedure by simultaneous measurement of the particle's momentum, since the corresponding experimental setup is incompatible with the setup necessary for position measurement. The instant we discovered the wave-like aspect of particles, in which the particle's momentum is linked to the corresponding wavelength, we had to say goodbye to simultaneous coexistence of exact location *and* exact momentum of the particle. As we found more than once, in a state with definite wavelength the wave fills the whole space uniformly, and it is meaningless to ask *where* the particle associated with this wave is. In a state with definite position (e.g., an ultrashort laser pulse described by superposition (3.1)), it is meaningless to ask about the exact wavelength of the pulse.

Now we turn to the more rigorous definition of indeterminacy. For a variable in a certain state, it involves an ensemble of identical particles in this state. Imagine that the region where the state function $\Psi(x)$ is defined is chopped into very small segments all of the same length δx and centered each at $x_1, x_2, \ldots, x_j, \ldots, x_J$, respectively. Suppose we perform the position measurement in an ensemble of N particles *all in the same state* $\Psi(x)$ and find N_1 particles in the segment at $x = x_1$, N_2 of them in segment at $x = x_2$, and so on. Then we define the *average* position $\langle x \rangle$ in this state as

$$\langle x \rangle = \sum_j x_j \, \delta P(x_j) = \sum_j x_j |\Psi(x_j)|^2 \, \delta x, \qquad (3.37)$$

where $\delta P(x_j) = |\Psi(x_j)|^2 \delta x = N_j/N$ is the probability of finding a particle within a segment at $x = x_j$. Once we know the state $\Psi(x)$, we can find the average of x or of any function of x in this state, and from these find the indeterminacy. Let us define an *individual deviation* of x as

$$\Delta x_j \equiv x_j - \langle x \rangle. \qquad (3.38)$$

It is just the x-coordinate obtained in an individual trial, but measured from the average $\langle x \rangle$ rather than from the origin. Clearly, if we take the average $\langle \Delta x \rangle$ of all such deviations, we will get zero: as is evident from definition (3.38), $\langle \Delta x \rangle = \langle x \rangle - \langle x \rangle = 0$, no matter how far from $\langle x \rangle$ individual deviations may be. So $\langle \Delta x \rangle$ does not give any information about the actual spread of measured values around the average. The needed information can be obtained if we take the average of the *square* deviation (the *variance*), and then take square root of this average:

$$\langle \Delta x^2 \rangle \equiv \sum_j \Delta x_j^2 \delta \mathcal{P}(x_j), \qquad \widehat{\Delta x} \equiv \langle \Delta x^2 \rangle^{1/2}. \tag{3.39}$$

The measure of the spread so defined is called the *standard deviation*. We put (temporarily) the arc on it to avoid a possible confusion with Δx in the other part of the same equation. Using definition (3.39), the reader can easily prove the following property:

$$\langle \Delta x^2 \rangle = \langle x^2 \rangle - \langle x \rangle^2. \tag{3.40}$$

Now we can define indeterminacy of an observable in a state Ψ as *the standard deviation of this observable in a series of its measurements in this state*.

Observables a and b that cannot be measured simultaneously are *incompatible*. The corresponding operators $\hat{A}$ and $\hat{B}$ representing them do not commute; their successive application to a function gives a different result, depending on ordering, that is,

$$\hat{A}\hat{B}\Psi \neq \hat{B}\hat{A}\Psi \quad \text{or} \quad (\hat{A}\hat{B} - \hat{B}\hat{A})\Psi \neq 0. \tag{3.41}$$

The opposite is also true: If two operators representing two different observables do not commute, the corresponding observables are incompatible. We can check this rule for the already known pair x, p_x. Assuming that $\mathbf{r}$ is its own operator (this will be shown in Section 5.4) and using (3.29), it is easy to see that

$$x\hat{p}_x - \hat{p}_x x = y\hat{p}_y - \hat{p}_y y = z\hat{p}_z - \hat{p}_z z = i\hbar. \tag{3.42}$$

We want to caution the reader here about the word "uncertainty," commonly used for such characteristics as Δx or Δp_x, or generally, for standard deviations Δa and Δb of any two incompatible observables. It may produce an impression that the actual values do exist but are only *unknown* to us, probably, due to some subtle flaw in the communication between the real world and the observer; and that maybe this uncertainty can be eliminated someday if only we find out the nature of this flaw. Actually, as far as we know today from an overwhelming amount of scientific evidence, Δp_x and Δx are *real indeterminacies* in position and momentum of an object rather than the observer's uncertainty about them. These indeterminacies are the *objective characteristics* of the system, and therefore it will be more appropriate to call the uncertainty principle the "indeterminacy principle."[1]

---

1) Historically, the German *unbestimmheit* in the first formulation of the principle means "indeterminacy." The term "uncertainty" in the English formulation of this principle is the result of a sloppy translation.

3.6
Indeterminacy and the World

> If we believe that an atom consists of a nucleus orbited by electrons, the rigidity of matter is incomprehensible.
>
> David Park

The indeterminacy principle is one of paramount importance. Even the known macroscopic, so-called classical, world would not exist without it. It is *quantum-mechanical indeterminacy* that makes atoms stable. Already at this stage we can use it to answer the question raised in Section 1.4: Why can an electron live forever in the ground state, at a finite distance a from the nucleus, when classically, the truly stable state with the lowest energy would correspond to the electron merged with the nucleus, that is, to $a = 0$?

The answer is that what is true in CM may be not so in QM, and the example discussed here is just the case. The electron ground state represented by a spherical "probability cloud" in Figure 3.1a is characterized by zero angular momentum, that is, no orbiting around the nucleus. There remains only radial motion. Classically, it would correspond to the "fall–rise" cycles – falling onto and rising from the nucleus with a certain amplitude (determined by the mechanical energy E). Quantum mechanically, we have the electron wave passing by the nucleus to the opposite side as the sea wave passes around a small rock. The intermediate result is a *radial standing wave* – another example of equally weighted superposition of inward and outward motion along a radial line. And above this, we have an equally weighted superposition of such standing waves along *all radial directions at once*. The result is a *spherical standing wave*. It produces a *stationary* spherically symmetric probability distribution with effective radius a. We want to find a and E for the resulting state.

In view of the spherical symmetry of the system, we can take any radial line as a reference direction. The correct result will be obtained if we take $0.5a$ as the measure of indeterminacy of the electron position along this direction, that is, $2\Delta r = a$. If the atom collapses, $U(a)$ and Δr decrease with a. But according to (3.35), this will increase indeterminacy Δp_r of the electron's radial momentum, and consequently, the indeterminacy in its kinetic energy K. Eventually the gain in K will match the loss in U, and the "radial fall" will freeze at this stage. Thus, and ironically, it is the QM *indeterminacy* that in the final run *determines* the least possible size of an atom. This size is the trade-off between two indeterminacies – one of the radial position and the other of the radial momentum of the electron.[2]

Let us find this size. For the electron in an H atom, the total mechanical energy is

$$E = K + U = \frac{p_r^2}{2\mu_e} - \frac{e^2}{a}. \tag{3.43}$$

---

[2] We will see later that there is another factor determining the size of molecules and atoms heavier than an H atom. It comes from the *Pauli exclusion principle* (Chapter 21).

We retain only the radial component of momentum because there is no orbital motion in the considered state. From (3.35) we have $\Delta p_r \geq \hbar/2\Delta r = \hbar/a$. Since the average radial momentum $\langle p_r \rangle = 0$ (the electron cloud does not move relative to the nucleus), there follows, in view of (3.40), $(\Delta p_r)^2 = \langle p_r^2 \rangle \geq \hbar^2/a^2$. Combining this with (3.43), we find that the least possible energy in the ground state:

$$E_{min} = K_{min} + U = \frac{\hbar^2}{2\mu_e a^2} - \frac{e^2}{a}. \tag{3.44}$$

This least value can still vary as a function of a (Figure 3.5). We see immediately that a cannot approach zero since this would lead to unbounded increase of energy. The function $E_{min}(a)$ reaches its minimum at certain $a = a_0$ that can be easily found from (3.44):

$$a_0 = \frac{\hbar^2}{\mu_e e^2}. \tag{3.45}$$

This is an already familiar expression 2.41 at $n = 1$, $Z = 1$. Putting it back into (3.44) recovers Equation 2.40 at $n = 1$. Thus, we have derived here both Equations 2.40 and 2.41 for the special case $n = 1$ from the totally different perspective. The existence and stability of the minimal possible size of the atom does not even require any *circular* motion of the electron around the center! The value (3.45) for this size (the Bohr radius) determines the properties of the observed world around us. The reason our bodies do not collapse and we are still present – and the whole universe operates the way it does with its atoms protected from collapsing – is the indeterminacy principle. As mentioned above, *it is QM indeterminacy that determines the exact size of physical objects.*

Figure 3.5 The derivation of the minimal possible energy and the effective size of a hydrogen atom. Energy versus effective size of the bound electron's probability cloud.

Exercise 3.1

The derivation of minimal size and energy of H atom in Equation 3.45, albeit quantitative, is still not 100% rigorous. There was one weak point (an unspoken assumption) in this derivation. Find this assumption.

Solution:
Initially we consider a as a variable without specifying its meaning. It might be, for instance, an average radius $a = \langle r \rangle$, or it might be the distance $a = r_M$ at which the probability $\mathcal{P}(r)$ reaches its maximum. In any case, we cannot say that $U(r) = U(a) = e^2/a$. Strictly speaking, we must find the average $\langle U(r) \rangle = \langle e^2/r \rangle$ using the probability distribution yet unknown to us at this stage, and then express the result in terms of a. Using the expression e^2/a was just making an assumption without giving any justification. But the rigorous solution of the Schrödinger equation in Chapter 10 justifies this assumption.

Box 3.1 The arbitrary potential and indeterminacy

Note that the existence of a stable atomic state due to indeterminacy was shown for Coulomb potential, which is only a special case of various interaction potentials $U(r)$. This brings an interesting question: Would indeterminacy act with the same efficiency for an attractive force other than the Coulomb force? Let us narrow down this quest to a function of the form

$$U(r) = -\frac{\eta}{r^\nu}, \quad \eta, \nu > 0. \tag{3.46}$$

It has two features common with the Coulomb potential: negative singularity at $r = 0$ and decrease of the absolute value at $r \to \infty$. The same argument as before, assuming the existence of a stable state with an effective radius a, leads to

$$E_{min} = K_{min} + U = \frac{\hbar^2}{2\mu_e a^2} - \frac{\eta}{a^\nu}. \tag{3.47}$$

The problem is again about finding the minimal possible a. It is immediately seen that at $\nu \geq 2$, stable states with $E < 0$ cannot exist because in this case the barrier described by the first term of (3.47) cannot prevent the electron from merging with the nucleus [17]. So the stable bound states may exist only for $0 < \nu < 2$. For integer ν, the physical case $\nu = 1$ (Coulomb potential) is the solution of the Poisson equation in 3D space. The case $\nu = 2$ is the solution of a similar equation in 4D space (not to be confused with space–time in SR!), and so on. But according to (3.47), at $\nu \geq 2$ not even QM could make the stable ground state. We can speculate therefore that this may be one of the reasons why the space in our observable universe has only three dimensions![3]

---

3) In the *string theory*, the world is assumed to operate in 10 spatial dimensions, but only 3 of them can accommodate observable physical objects [18,19].

The atomic states can collapse even in 3D space under the overwhelming gravity in neutron stars or black holes [20–22]. In such cases, the energy needed for an increase in K due to decreasing Δr comes from the *gravitational* energy of the collapsed object. An atom well within such an object cannot be considered as isolated, and gravity is a dominating factor in this case.

Another area where QM indeterminacy plays a decisive role is particle physics. In this domain, we face the problem: How to distinguish a truly elementary particle from a compound one? Once upon a time atoms were considered elementary, as is encoded in their name ("atom" is the ancient Greek for "indivisible"). There was a brief time when the atomic nuclei were regarded as elementary. Now we know that they consist of nucleons (protons and neutrons), which can be regarded as elementary particles in many processes. This view persisted even after the discovery of neutron decay into a proton, electron, and antineutrino (*β-decay*):

$$n \to p + e + \tilde{v} \tag{3.48}$$

In spite of this process, it would be wrong to think that n is made of p, e, and $\tilde{v}$. It turns out that the relationship between "elementary" and "compound" in QM is much more subtle than in CM. One subtle point emerges when we consider an *inverse β-decay*. In this process, a proton in some heavy nuclei decays into neutron, positron, and neutrino:

$$p \to n + e^+ + v. \tag{3.49}$$

Now, the same logic would tell us that p is a compound particle consisting of n, e^+, and v. The doubly weird conclusion would be this: first, it contradicts (3.48); and second, the neutron is more massive than the proton! The latter weirdness (known as *mass defect*) can be explained by SR [23]: in the case of decay (3.49), the missing mass $\Delta \mu$ together with the corresponding energy $\Delta E = \Delta \mu c^2$ comes from the surrounding nucleons. But even so, the mere possibility of both processes (3.48) and (3.49) makes the concept of a compound particle, at best, ambiguous. The situation becomes even more embarrassing if we learn about the processes that are "halfway" between (3.48) and (3.49):

$$p + e \to n + v \quad \text{or} \quad p + \tilde{v} \to n + e^+. \tag{3.50}$$

Mathematically, all four equations are essentially the same equation, say, (3.48) with one or two terms taken to another side with the "change of sign" – converting the particle into its antiparticle. Accordingly, they describe different varieties of one process. And looking at them, we cannot say which particle is a candidate for being compound and which is elementary.

Is there a criterion for determining a truly elementary particle?

Nothing (except, perhaps, this statement) is an ultimate truth in a true science, and the fragile status of an "elementary" particle became even more evident after M. Gell-Mann and G. Zweig proposed their *quark model* [24–28]. According to it, nucleons consist of more fundamental particles – quarks. As far as we know today, quarks cannot be observed in isolation, and yet their existence is confirmed by the abundant scientific evidence.

3.6 Indeterminacy and the World

In view of the subtle nature of the concept of an elementary particle, we will restrict our quest to a more specific question: For a reaction such as (3.48), is there a way to tell whether an electron actually exists within the neutron before its decay? In other words, is there a criterion distinguishing "elementary" from "compound" for a specified type of processes? The answer is yes. And it follows from the indeterminacy principle! Let the "host" particle (e.g., n) allegedly harboring the "guest" particle (e.g., e) have the effective size r_0. If the energy $\overline{p^2} = \overline{\Delta p^2} \sim r_0^{-2}$ arising from squeezing the guest of mass μ_0 into a region with small r_0 is less than $\mu_0 c^2$ (the guest's rest energy [23]), then the guest can retain its individuality within this region and thus exist in it. Otherwise, it cannot exist there as a distinct entity.

Let us apply this to our case. We want to find out if n can be considered as a bound system $p + e + \tilde{v}$. That would be analogous to an H atom squeezed into the volume of a neutron. The effective size of a neutron is close to one Fermi (1 F $= 10^{-15}$ m) – about five orders of magnitude smaller than the Bohr radius. Even if apart from the Coulomb attraction there existed some additional stabilizing force at $r \sim 1$ F, it would not help. To see why, consider again the term $(\Delta p_r)^2 = \langle p_r^2 \rangle \geq \hbar^2/r_0^2$, but now, in view of high energy associated with this term, we have to use the relativistic $E \leftrightarrow p$ relationship

$$E^2 = \mu_e^2 c^4 + p^2 c^2. \tag{3.51}$$

Taking the average, we evaluate the second term in (3.51) as $\langle p_r^2 \rangle c^2 \geq (\hbar c)^2/r_0^2$. For $r_0 = 1$ F, this gives $\sqrt{\langle p_r^2 \rangle} c \geq 1.2 \times 10^3$ MeV. This exceeds the electron's rest energy by more than three orders of magnitude! Such an electron cannot be considered as "existing" within the neutron – it would immediately produce a firework of e, e^+ pairs, so that long before the n decays in the known way, it would "bang" in a multiparticle explosion. And it would be absolutely impossible to distinguish the initial e from its numerous clones in the e, e^+ pairs. So any notion about a single e being a part of the n before the latter decays has no physical meaning. The same is true for $\tilde{v}$. As for the proton, its rest energy is large enough, so it *could* (and actually does) fit within the given size, but it cannot make a neutron without the electron and antineutrino. All three particles emerging on the right of (3.48) in the process of β-decay *do not exist* before the decay. They are *created* in the process just as the n *disappears* in it. We face a totally new phenomenon specific of QM – the creation and annihilation of particles.

We are already familiar with one example – emission and absorption of light by atoms. A photon does not "live" within an atom before emission; it is created in the process. The atom does not "harbor" a photon after absorbing it; the photon just disappears, and the only evidence of its past existence is the increased energy (and changed angular and linear momentum) of the atom. With respect to photons, the atom can be considered as an elementary particle in the narrow sense, meaning that it does not include photons as its constituents. Similarly, n can be considered as an elementary particle in *weak interactions* – it is not a compound system of p, e, and $\tilde{v}$. And this new view of the physical nature, status, and possible transformations of elementary particles is intimately connected with the indeterminacy principle.

3.7
Quantum Entanglement and Nonlocality

Now we consider a fascinating result of superposition of states, which is found in a *system* of particles. We restrict here to the case of two particles. Suppose we have a pair of photons. Apart from its energy $E = \hbar\omega$ and momentum $\mathbf{p} = \hbar\mathbf{k}$, a photon is characterized by its *polarization* – the direction of the oscillating electric field (linear polarization) or sense of its rotation in the plane perpendicular to $\mathbf{k}$ (circular polarization) [4,16]. A circularly polarized EM wave can change the angular momentum of an atom placed in its way. This shows that polarization is the signature of the photon's intrinsic angular momentum (spin $\mathbf{s}$), which can be transferred to a photon-absorbing particle. For a circularly polarized photon, its spin projection onto $\mathbf{k}$ can be either $\hbar$ (right polarization) or $-\hbar$ (left polarization). Consider these as the basis states and represent them by symbols $|+\rangle$ and $|-\rangle$, respectively (the first example of the *Dirac notation* to be considered in Section 4.1). Denote the corresponding state functions of photons 1 and 2 as $|\Psi\rangle_1$ and $|\Psi\rangle_2$, respectively (here we only retain the spin variable). Then the possible basis states of photon 1 can be written as

$$|\Psi\rangle_1 = \begin{cases} |+\rangle_1 & \text{(right-polarized)}, \\ |-\rangle_1 & \text{(left-polarized)}, \end{cases} \qquad (3.52)$$

and similarly for the second photon, with subscript 1 replaced by 2.

Let us now turn to a *combined* state of two photons. There can be many such states. Consider one of them, specified by the condition that the net momentum and net angular momentum (and thereby its projection on any axis) are zero. The only possibility for the *net* projection onto *any* axis to be zero is that the photons have opposite spins. How can we describe such a state? By rule (3.21), it should be the product of the individual states, $|\Psi\rangle_1|\Psi\rangle_2$. But this is obtained for the case when the photons are independent. Our two photons are not independent – they are *correlated* by the condition of having opposite spins. Since their $\mathbf{k}$-vectors are also opposite due to the zero net momentum $\mathbf{k}_1 + \mathbf{k}_2 = 0$, the requirement of the zero net spin is automatically satisfied if they have the same polarization (in physical jargon, the same *helicity*): both are in state $|+\rangle$ or both in state $|-\rangle$.[4] Therefore, if we specify $|\Psi\rangle_1$ as $|+\rangle_1$ then there must be $|\Psi\rangle_2 = |+\rangle_2$, and the same for the $|-\rangle$ state. The corresponding spin function of the whole system will be either $|\Psi_{12}\rangle = |+\rangle_1|+\rangle_2$, or $|\Psi_{21}\rangle = |-\rangle_1|-\rangle_2$. But our conditions are not as detailed as to specify the spin direction of either photon. They only specify the *correlation* between these directions. Each photon knows that its polarization state must be the same as that of its partner, but neither of them knows its actual state. If a system can be in either of these two states, then it can also be in their

---

4) This should not cause any confusion! If photon 1 is right polarized with respect to $\mathbf{k}_1$ and photon 2 is right polarized with respect to $\mathbf{k}_2$, then their spins are opposite to one another, because $\mathbf{k}_2 = -\mathbf{k}_1$.

superposition. So the most general two-photon state with the zero net spin is

$$\Psi = a|\Psi_{12}\rangle + b|\Psi_{21}\rangle = a|+\rangle_1|+\rangle_2 + b|-\rangle_1|-\rangle_2, \quad (3.53)$$

where the amplitudes a and b describe the weight of the corresponding state in superposition.

Neither photon in state (3.53) can be characterized independently of its partner, and in this respect, neither of them can be singled out as a separate entity even if they are separated by voids of space. This property of state (3.53) is named *quantum entanglement*. Mathematicians prefer the term *inseparability*: expression (3.53) cannot be represented as the product of two individual states – neither of them can be factored out and thus separated from the other.

We can obtain a similar result for particles other than photons. Consider, for instance, a combined state of two particles with spin $s = 1/2$ each. In this case, the spin direction of each particle is not rigidly connected to direction of its momentum. If the *net* spin is zero, then *any* its component is also zero, which implies the opposite individual spin components along any common direction (you will find more about spin in Chapter 7). Retaining as before only characteristics of the spin state, and taking the chosen common direction as the z-axis, we can write the possible spin states of particle 1 as

$$|\Psi\rangle_1 = \begin{cases} |\uparrow\rangle_1 & \text{(spin up)}, \\ |\downarrow\rangle_1 & \text{(spin down)}, \end{cases} \quad (3.54)$$

and similarly for particle 2. The difference from Equation 3.52 is that now we write $|\uparrow\rangle$ and $|\downarrow\rangle$ instead of $|+\rangle$ and $|-\rangle$, respectively. The other difference is that now we have *one common reference direction* (say, the z-axis) for both particles, whereas the helicity of each photon was defined relative to its respective momentum. For a *combined* state with zero net spin, the same argument as before shows that such state is an *entangled superposition* similar to (3.53):

$$|\Psi\rangle = a|\uparrow\rangle_1|\downarrow\rangle_2 + b|\downarrow\rangle_1|\uparrow\rangle_2. \quad (3.55)$$

Like in case (3.53), the probability of finding the system, upon measurement, in state $|\uparrow\rangle_1|\downarrow\rangle_2$ or $|\downarrow\rangle_1|\uparrow\rangle_2$ is given by $|a|^2$ or $|b|^2$, respectively. In state (3.55) neither particle can be characterized independently of its partner, and accordingly, neither of them has its own distinct state function. Only the whole system has. We cannot represent (3.55) as a product of two separate factors describing each only one particle. Expression (3.55), like (3.53), is inseparable.

The relative weight of the two possibilities in states (3.53) and (3.55) is given by the ratio $\xi \equiv |a|^2/|b|^2$. The corresponding probabilities $\mathcal{P}_a$ and $\mathcal{P}_b$ can be expressed in terms of ξ as

$$\mathcal{P}_a = \frac{\xi}{1+\xi}, \quad \mathcal{P}_b = \frac{1}{1+\xi}. \quad (3.56)$$

Their sum is equal to 1, as it should be. When the ratio ξ is very small ($|a| \ll 1$) or very large ($|b| \ll 1$), we have a weak entanglement. This becomes obvious in the limit $\xi \to 0$ (or $\xi \to \infty$) when one of the amplitudes vanishes, and, for example, superposition (3.55) reduces to only one out of two possibilities, $|\uparrow\rangle_1|\downarrow\rangle_2$ or $|\downarrow\rangle_1|\uparrow\rangle_2$. In either of these two limits, the particles' spins remain opposite, but they are not entangled, because each particle now has its spin defined and is thereby independent of the other. The state of the system in such a limit becomes the product of individual states; that is, each of them is now represented by a separate factor. The whole state is *separable*.

Thus, the mere fact that the two spins are opposite (or same, as in case of polarized photons) does not necessarily cause the system to be entangled. It is the *absence of an individual state* of each particle in the pair (inseparability) that makes it so. The degree of inseparability (amount of entanglement) for the system of two particles reaches its maximum at $\xi = 1(|a| = |b|)$. Neglecting, for simplicity, the phase factors in the probability amplitudes, we have in this case

$$|\Psi\rangle = \frac{1}{\sqrt{2}}\left(|\uparrow\rangle_1|\downarrow\rangle_2 \pm |\downarrow\rangle_1|\uparrow\rangle_2\right). \tag{3.57}$$

Here it is equally probable to find the spin of each particle pointing up or pointing down, but in each trial each particle's spin is opposite to that of its partner. At $\xi \gg 1$, the first particle will be found with its spin up in an overwhelming amount of all trials, and each such outcome will be accompanied with the spin of the second particle pointing down. In rare cases, the first particle will be found with spin down, but again, it will be accompanied with the opposite spin of its partner. So as ξ shifts from $\xi = 1$, the correlation remains rigid but the individual state of each particle becomes increasingly defined, and therefore the entanglement gets weaker.

An important feature of entanglement is that the (anti)correlation expressed by (3.55) *does not depend on the chosen axis*. If we switch from (3.55) to another basis, say, along the *x*-direction and measure s_x instead of s_z, we will again find the individual outcomes opposite to one another.

We might be tempted to explain the entanglement by assuming that each particle in a pair emerging from the source is independent of the other, except for having its spin oriented opposite to that of its partner. In other words, each particle *knows its exact spin state*, apart from the fact that their spins are anticorrelated. This corresponds to either *a* or *b* equal to zero in the general expression (3.55). These features reiterate with each new act of pair production, except that the basis may change with each new act. For instance, the particles emerging in the next production may have their spin projections defined with respect to the *x*-axis instead of the *z*-axis; that is, one particle has its spin pointing to the right, and its partner has its spin pointing to the left. The next pair emerges with opposite spins defined with respect to some other direction and so on, so that the requirement of the zero net spin appears always to be satisfied.

This process, however, falls short of producing entangled pairs. It does satisfy the condition of the zero net spin of the pair, but only at the production stage. The condition is generally broken at the measurement stage. It only holds if we measure the spins in the same basis in which they were originally determined. If they emerged with their definite spin projections onto the *x*-axis, and we later measure the spins along the same axis, we find them opposite to one another. But if we now decide to measure spins, say, along the *z*-axis, there will be a 50% chance to find

3.7 Quantum Entanglement and Nonlocality

them both up or both down – in this case there are no correlations in the measurement results, due to independence of the individual states. This is in sharp contrast with an entangled pair, for which one gets correlated measurement outcomes *regardless* of change of basis. A state of a system of entangled particles is fundamentally different from the state of the same but independent particles that just happen to have their characteristics correlated in a certain way.

Things can be understood better through comparison. Let us compare the state (3.55) (or (3.53) with another possible state of the pair. Suppose as before that an individual electron spin component is not specified by the initial conditions, but now it does not depend on the spin of another electron. Then it has a state of its own – generally a superposition of the two basis states $|\uparrow\rangle$ and $|\downarrow\rangle$. Denoting this superposition as $|\Phi\rangle_1$ for the first particle and $|\Phi\rangle_2$ for the second, we have

$$|\Phi\rangle_1 = \alpha_1|\uparrow\rangle_1 + \beta_1|\downarrow\rangle_1, \qquad |\Phi\rangle_2 = \alpha_2|\uparrow\rangle_2 + \beta_2|\downarrow\rangle_2. \tag{3.58}$$

The two states here are totally independent of each other. Now consider the combined state

$$|\tilde{\Phi}\rangle = |\Phi_1\rangle|\Phi_2\rangle = (\alpha_1|\uparrow\rangle_1 + \beta_1|\downarrow\rangle_1)(\alpha_2|\uparrow\rangle_2 + \beta_2|\downarrow\rangle_2). \tag{3.59}$$

Multiplication gives

$$|\tilde{\Phi}\rangle = c_1|\uparrow\rangle_1|\uparrow\rangle_2 + c_2|\uparrow\rangle_1|\downarrow\rangle_2 + c_3|\downarrow\rangle_1|\uparrow\rangle_2 + c_4|\downarrow\rangle_1|\downarrow\rangle_2, \tag{3.60}$$

where

$$c_1 = \alpha_1\alpha_2, \qquad c_2 = \alpha_1\beta_2, \qquad c_3 = \beta_1\alpha_2, \qquad c_4 = \beta_1\beta_2. \tag{3.61}$$

Apart from familiar products $|\uparrow\rangle_1|\downarrow\rangle_2$ and $|\downarrow\rangle_1|\uparrow\rangle_2$, describing the states with opposite (antiparallel) individual spins, we see here the products $|\uparrow\rangle_1|\uparrow\rangle_2$ and $|\downarrow\rangle_1|\downarrow\rangle_2$, describing the states with parallel spins – both up or both down. The presence of the additional terms means that *there is no correlation between the particles* – they are generally independent of each other. Measuring the spin of one particle bears no effect on the state of the other one, and vice versa. In the case when the superpositions (3.58) are equally weighted, there is an equal (25%) chance for any one out of the four possible outcomes.

Mathematically, the absence of entanglement in state (3.60) is manifest in its separability. Expressing the coefficients c_j, $j = 1, 2, 3, 4$, as the products (3.61) and working backward from (3.60), one can represent it as the product (3.59) of two separate factors $|\Phi\rangle_1$ and $|\Phi\rangle_2$. The state (3.60) is *separable*, and therefore not entangled. In contrast, the state (3.55) describes rigid correlation – the spin measurements can only produce the antiparallel doubles, so that *measurement of only one spin automatically gives the result for another one*.

An entangled pair may also have a *nonzero* net spin. That is fine as long as the spin value obtained for one particle determines exactly the value for the other regardless of the choice of basis in which spin is to be measured. The same idea also holds for observables other than spin.

We will discuss some implications of entanglement in Section 4.5, and some other aspects of this phenomenon in Chapters 23–26.

3.8
Quantum-Mechanical Phase Space

Here we consider QM indeterminacy from the viewpoint of the phase space. In classical statistical physics, we use the phase space to describe an ensemble of particles. For a particle's motion in one dimension, the corresponding phase space is two-dimensional: one dimension representing coordinate q and the other the corresponding momentum p. At each moment, the state of the particle is uniquely defined by its position (q, p) on the *phase diagram*. As time evolves, both q and p change, and the point on the diagram traces out a trajectory. For a harmonic oscillator, the trajectory is an ellipse (Figure 3.6a). In this case, the classical action for one period can be written as $S = \oint p(q) dq$ [29]. Here S gives the area enclosed by the loop, and its value is uniquely determined by the oscillator's energy. Since in CM the latter is a continuous parameter, so is S. Quantum mechanically, however, this area, as we will see in Section 9.6, can only change by units h (Figure 3.6b). This results in the energy quantization for any system performing periodic motion (the Bohr–Sommerfeld quantization condition [5,12,13,17])

$$S(E) = \oint p(q) dq = \left(n + \frac{1}{2}\right) h, \quad n = 0, 1, 2, \ldots. \tag{3.62}$$

We will have some specific relevant examples later in the text (Sections 9.6 and 14.5).

Since in QM a particle cannot have simultaneously definite p and q, its position on the diagram is fuzzy, filling out a patch whose area, according to (3.35), cannot be less than $(1/2)\hbar$. The shape of the patch depends on the state (Figure 3.7). For instance, it can be stretched into an infinitely narrow (and accordingly, infinitely long) stripe, say, parallel to the q-axis at $p = p_0$; this would correspond to a state with definite momentum p_0 and totally indefinite position $-\infty < q < \infty$. Or it could

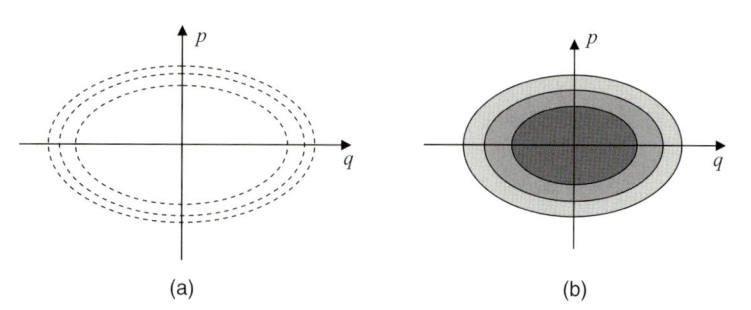

Figure 3.6 (a) Trajectories of a classical harmonic oscillator in the phase space. The size of each ellipse is a continuous function of oscillator's energy $E(p, q)$, where p and q are coordinate and corresponding momentum, respectively. (b) Trajectories of a quantum oscillator in the phase space. They look like those in (a) but have quite different meaning. A quantum oscillator with a fixed energy is described by the *whole area* within the corresponding ellipse, rather than by the ellipse itself. Also, in contrast with CM, the area of an ellipse can only change by a fixed quantity $\Delta A = 2\pi\hbar$ (is quantized). Since there is one-to-one correspondence area – energy – the energy is also quantized together with the area.

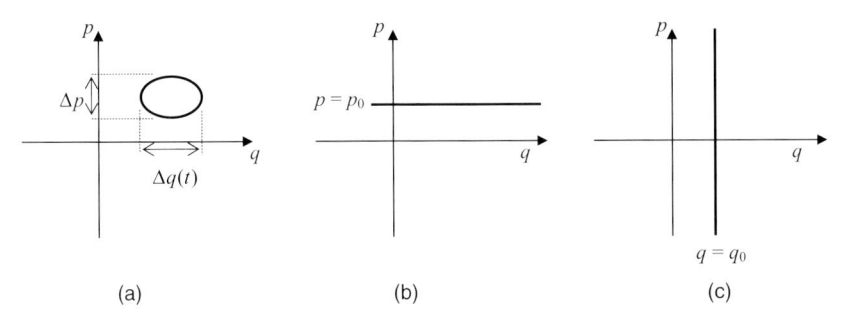

Figure 3.7 Granular phase space. Incompatibility of observables p and q is reflected in the granular structure of the corresponding phase space: in contrast with CM, in QM no state can be represented by a single point on the phase diagram. Instead, the phase space is quantized into cells or granules with the least possible area $\Delta A_{min} = (1/2)\hbar$. The shape of the granules is not specified and depends on a state of the system. (a) A possible state of a free particle described by a wave packet; whereas Δp is a fixed characteristic of the given packet, $\Delta q(t)$ is generally a function of time (the packet's shape evolves); (b) a state with sharply defined $p = p_0$; (c) a state with sharply defined $q = q_0$.

form a similar vertical stripe at $q = q_0$; that would represent a state with definite position and totally undefined momentum $-\infty < p < \infty$. These are two extremes of a shape corresponding to a wave packet with finite Δp (Chapter 11). An interesting thing about quantum phase space is that while its area representing a system cannot be less than $(1/2)\hbar$, it can change continuously with time during the evolution of a wave packet, or it can change by discrete units $h = 2\pi\hbar$, as in (3.25) for a bound system. Some states represented by an ellipse with the minimum possible area $(1/2)\hbar$ in the phase diagram are called *squeezed states* [30–32].

3.9
Determinism and Causality in Quantum World

> God does not play dice.
>
> A. Einstein

> Einstein, stop telling God what to do.
>
> N. Bohr

3.9.1
Discussion

In CM, all characteristics of a system, for example, the exact position and momentum of a point particle at some moment of time, as well as the forces acting on it, completely determine its future behavior. The sequence of all future states of a system – its evolution – is predetermined by its current state and environment. This statement is known as *classical determinism*.

Inherent in classical determinism is predictability of the future behavior of a system. A soccer player can estimate beforehand the trajectory of a soccer ball from its instant position and velocity and act accordingly. Astronomers can predict the future positions of celestial bodies (in particular, solar and lunar eclipses) with astounding accuracy. This predictability can also be extended back in time, to recover the past history of an object from its current state and known environment. For instance, we can trace out the planetary motions in our solar system back to a distant past, an ability that enabled historians to pinpoint the dates of many events by comparing concurrent eclipses described in some chronicles with those calculated by the astronomers.

In QM, the deterministic (predictable) description of the world is restricted. The *state function* of a system changes *continuously* between measurements, and this change is described by the Schrödinger equation. This evolution can be exactly predicted – it is fully deterministic. But the wave function as such is *not a dynamic variable*; it only carries information about *probabilities* of measurement results. Therefore, generally only *probability* of a certain event can be predicted, not the event itself. Due to QM indeterminacy, a system may even not possess a definite value of a certain characteristic, the exact knowledge of which, from a classical viewpoint, may be absolutely essential to describe its evolution. This shows that unpredictability is linked with the indeterminacy principle.

The probabilistic nature of the state function leads to another profound implication. The function (and the state of a system it describes), while changing continuously between measurements, can change abruptly and unpredictably (collapse to another, quite different, state) in the process of measurement. This follows from the Born interpretation combined with de Broglie's postulates. Consider the already mentioned thought experiment – a position measurement for a single free particle with definite momentum. In such a state, it is equally likely (in practical terms, equally unlikely, in view of the vastness of space) that the particle be found within a cubic meter at the North Pole, precisely, as it would within a cubic meter at the center of Andromeda's Nebula, about 2.5×10^6 light years from us. The particle, roughly speaking, is "smeared out" uniformly over the whole universe. Suppose now that we performed the measurement and found the particle (by an extraordinary whim of chance) in a detector D at a definite location in Brookhaven National Laboratory. Once the particle is in this detector, it is not anywhere else. This means a dramatic change of the initial state: the probability cloud spread over the whole universe instantaneously collapses into a state all within D (see more about this in Chapter 23). Such an event challenges our "classical" intuition. Nothing *material* can shrink down from huge to small in no time, yet the wave function does. This is another illustration that the wave function is by itself not an observable property, nor does it give *direct* information about any such properties. The de Broglie wave did *not* predict that the particle was going to wind up in D. It only gave us the *probability* (and a vanishingly small one at that) of finding the particle there. Accordingly, the corresponding measurement, as mentioned before, is *not* described by the Schrödinger equation.

Many people, including Einstein, Schrödinger, and Bohm, considered this as evidence of the incompleteness of QM. However, all attempts to find a deeper reality – *hidden parameters* of a sort that eventually determine the measurement results – have failed. Moreover, as has been shown by John Bell [33,34], the predictions of standard QM differ from those of all its known modifications incorporating hidden parameters, and the difference can be checked experimentally. And, all the experiments have unequivocally confirmed conventional QM, according to which the world is intrinsically probabilistic. Chance turned out to be an inherent element of reality. But, on the other hand, its role and its rule are restricted: in any state the *probability* of any possible outcome (if not the outcome itself) of a measurement is predictable. Without such predictability, the world would be an indescribable mess; there would be no laws and no life, let alone an intelligent life and science.

One of the most striking features of QM is the role of measurement. In classical physics, the "observer" (or the measuring device) is largely redundant. Nature doesn't notice (at least, in principle) if the device is there or not. An ideal classical measurement can reveal the pre-existing characteristics of the system or verify their expected values without perturbing the system. In QM, the device is just as important as the object under observation. The quantum measurement can "create" a new reality by dramatically changing the state of a measured system. This property has much to do with the existence of the minimal quantum of action (2.3). Indeed, a measurement involves interaction between the studied system and measuring device. Classically, this interaction can be made arbitrarily small, thus preserving the measured characteristic from undesired perturbations. According to QM, however, the action associated with the interaction energy cannot be less than h; therefore, if the system itself is sufficiently small, the significant perturbations cannot be avoided, and accordingly, the exact outcome of measurement generally cannot be predicted.

For instance, an electron in an atom has an indeterminacy in its position $\Delta q \cong a$, where a is the atomic size. One could ask *where exactly* the atomic electron is at a given moment of time, but this would be a meaningless question unless and until one performs a position measurement, by, say, illuminating the atom with gamma photons and recording the characteristics of a scattered photon. The recording could, in principle, narrow down the indeterminacy from a to a far smaller value. However, in the process of such scattering, the electron would undergo uncontrollable exchange of momentum with the photon, which would radically change the initial state. The electron could get a "kick" sufficient for breaking loose of the atom (ionization). And the more accurate its position measurement, the harder the kick, for a very simple reason: a photon can pinpoint the electron's location only to the accuracy comparable with the photon's wavelength. A more accurate position measurement requires more energetic photons with shorter wavelength. In determining the electron's position to accuracy higher than the atomic size a, the energy gain by the electron by far exceeds the ionization energy. Thus, we will get a far more "finely" defined *instant* position of the electron at the moment of measurement, but this happens at the cost of the very same state we wanted to measure. The atom will be destroyed (more politely, ionized), with sharply increased indeterminacy in the electron's momentum. This is in stark contradiction with the classical picture ("the planetary model") at the end of Chapter 2, according to which the electrons orbit the atomic nucleus along definite trajectories, with the values of both position and momentum being the exact functions of time.

Problems

3.1 Find the wavelength of an electron moving with kinetic energy of 100 eV.

3.2 In order to carry out an experiment with neutron diffraction on a crystal, the neutron wavelength must be of the same order as the lattice constant a. Find the neutron energy corresponding to $a = 10^{-10}$ m.

3.3 Express the probability flux density in terms of the phase of the wave function.

3.4 Show that probability flux density (3.18) is invariant with respect to global phase transformation, but not invariant with respect to a local phase transformation.

3.5 Find the vector operator of linear momentum.

3.6 Find the *average deviation* $\langle \delta q \rangle$ of a variable q from its expectation value $\langle q \rangle$.

3.7 Prove the relationship (3.42).

3.8 Find the minimal size of the hydrogen atom (its Bohr's radius) and the corresponding minimal energy from expression (3.44).

3.9 Consider a hypothetical universe where the electrostatic potential energy is, instead of Coulomb's law, determined by $U(r) = \kappa q_1 q_2 / r^3$.

  a) Using the same argument as one presented in Section 3.6, determine whether the QM indeterminacy of position and momentum can ensure the stability of atoms.

  b) For the electrostatic potential energy given by $U(r) = \kappa q_1 q_2 r^{-\alpha}$, with positive α, what is the critical value of α that could in principle produce stable atoms?

3.10 Imagine a fictitious universe in which the electrostatic interaction between the charges is described by Equation 3.46 with some $\nu \neq 1$. Find the minimal possible size of a hydrogen atom and its corresponding energy as a function of ν for $\nu < 2$.

3.11 Derive the expressions (3.56).

3.12 Separate the state (3.60) with the given set of four coefficients c_j, $j = 1, 2, 3, 4$.

References

1 Bayerlein, R. (2002) *Newton to Einstein: The Trail of Light: An Excursion to the Wave–Particle Duality and the Special Theory of Relativity*, Cambridge University Press, Cambridge, p. 170.

2 Davisson, C. and Germer, L.H. (1927) Diffraction of electrons by a crystal of nickel. *Phys. Rev.*, **30** (6), 705–740.

3 Davisson, C.J. (1928) Are electrons waves? *Franklin Inst. J.*, **205**, 597.

4 Hecht, E. (2002) *Optics*, 4th edn, Addison-Wesley, Reading, MA, p. 385.
5 Blokhintsev, D.I. (1964) *Principles of Quantum Mechanics*, Allyn & Bacon, Boston, MA.
6 Klein, A.G. and Werner, S.A. (1983) Neutron optics. *Rep. Prog. Phys.*, **46**, 259–335
7 Klepp, J., Pruner, C., Tomita, Y., Plonka-Spehr, C., Geltenbort, P., Ivanov, S., Manzin, G., Andersen, K.H., Kohlbrecher, J., Ellabban, M.A., and Fally, M. (2011) Diffraction of slow neutrons by holographic SiO_2 nanoparticle–polymer composite gratings. *Phys. Rev. A*, **84**, 013621.
8 Cronin, A.D., Schmiedmayer, J., and Pritchard, D.E. (2009) Optics and interferometry with atoms and molecules. *Rev. Mod. Phys.*, **81**, 1051–1129.
9 Krane, K. (1992) *Modern Physics*, John Wiley & Sons, Inc., New York.
10 Feynman, R., Leighton, R., and Sands, M. (1963–1964) *The Feynman Lectures on Physics*, vols. 1–3, Addison-Wesley, Reading, MA.
11 Frish, S.É. and Timoreva, A.V. (2007) *Course of General Physics* (in Russian), vols. 1–3, Lan Publishing House, Saint Petersburg.
12 Born, M. (1963) *Atomic Physics*, Blackie & Son, London.
13 Dirac, P. (1964) *The Principles of Quantum Mechanics*, Yeshiva University, New York.
14 Zommerfeld, A. (1952) *Mechanics*, Academic Press, New York, Section 44. Davydov, A. (1965) *Quantum Mechanics*, Pergamon Press/Addison-Wesley, Oxford/Reading, MA.
15 Merzbacher, E. (1998) *Quantum Mechanics*, 3rd edn, John Wiley & Sons, Inc., New York, pp. 257–270.
16 Born, M. and Wolf, E. (1980) *Principles of Optics*, Pergamon Press, Oxford.
17 Landau, L. and Lifshits, E. (1965) *Quantum Mechanics: Non-Relativistic Theory*, 2nd edn, Pergamon Press, Oxford.
18 Polchinski, J. (1998) *String Theory*, Cambridge University Press, Cambridge.
19 Greene, B. (2000) *The Elegant Universe: Superstrings, Hidden Dimensions, and the Quest for the Ultimate Theory*, W.W. Norton & Company, New York, NY.
20 Misner, C.W., Thorn, K.S., and Wheeler, J.A. (1973) *Gravitation*, Freeman & Co., San Francisco, CA.
21 Zeldovich, Ya.B. and Novikov, I.D. (2011) *Stars and Relativity*, Dover Publications, New York.
22 Hawking, S. (2010) *The Grand Design*, Bantam Books, New York.
23 Fayngold, M. (2008) *Special Relativity and How It Works*, Wiley-VCH Verlag GmbH, Weinheim, pp. 485–487.
24 Gell-Mann, M. (1964) A schematic model of baryons and mesons. *Phys. Lett.*, **8** (3), 214–215.
25 Zweig, G. (1964) *An SU3 Model for Strong Interaction Symmetry and Its Breaking*. CERN Report Nos. 8182/TH.401 and 8419/TH.412.
26 Halzen, F. and Martin, A.D. (1984) *Quarks & Leptons: An Introductory Course in Modern Particle Physics*, John Wiley & Sons, Inc., New York.
27 Nambu, Y. (1984) *Quarks: Frontiers in Elementary Particle Physics*, World Scientific, Singapore.
28 Gell-Mann, M. (1994) *The Quark and the Jaguar: Adventures in the Simple and the Complex*, Freeman & Co., New York.
29 Landau, L. and Lifshits, E. (2003) *Mechanics*, 3rd edn, Butterworth-Heinemann, Oxford.
30 Loudon, R. (2003) *The Quantum Theory of Light*, 3rd edn, Oxford University Press, Oxford.
31 Walls, D.F. and Milburn, G.J. (1994) *Quantum Optics*, Springer, Berlin.
32 Walls, D. (1983) Squeezed states of light. *Nature*, **306**, 141.
33 Bell, J.S. (1964) On the Einstein–Podolsky–Rosen paradox. *Physics*, **1**, 195–200.
34 Bell, J.S. (1966) On the problem of hidden variables in QM. *Rev. Mod. Phys.*, **38**, 447.

4
Playing with the Amplitudes

What we have learned in the previous chapters is already sufficient to describe many important QM phenomena. In this chapter, we apply the basic concepts of QM – the amplitudes and superposition of states – to consider how they work in some real situations. As a first step, we introduce convenient notations for QM states (the Dirac notations) and formulate the basic rules for operating with them.

4.1
Composition of Amplitudes

The original notation for a wave function is $\Psi(\mathbf{r}, t)$. The rationale for Dirac's notations [1] is the analogy between superposition of states

$$\Psi(x) = \sum_j c_j \psi_j(x) \tag{4.1}$$

and geometrical composition of vectors

$$\mathbf{e} = \sum_j a_j \hat{\mathbf{e}}_j. \tag{4.2}$$

If the set ψ_j is orthonormal and Ψ normalized, we have

$$c_j = \int \Psi(x)\psi_j^*(x)\mathrm{d}x \quad \text{and} \quad \sum_j |c_j|^2 = 1. \tag{4.3}$$

This is similar to

$$a_j = \hat{\mathbf{e}}\,\hat{\mathbf{e}}_j^* \quad \text{and} \quad \sum_j |a_j|^2 = 1 \tag{4.4}$$

for the orthonormal set $\mathbf{e}_j$ with normalized $\mathbf{e} = \hat{\mathbf{e}}$ in the *complex vector space*, in which we write the dot product in the same way as we do in (3.2): we take the product of one vector by complex conjugate of the other. This similarity shows that quantum states can be described as vectors in an abstract vector space – the *Hilbert space* [2]. Hence, the term *state vector* is frequently used in QM. In the case of an eigenstate we can as

Quantum Mechanics and Quantum Information: A Guide through the Quantum World,
First Edition. Moses Fayngold and Vadim Fayngold.
© 2013 Wiley-VCH Verlag GmbH & Co. KGaA. Published 2013 by Wiley-VCH Verlag GmbH & Co. KGaA.

well call this vector the *eigenvector*. The Hilbert space spanned by these vectors will henceforth be denoted as $\mathcal{H}$.

From a physical viewpoint, a coefficient c_j in (4.3) is the probability amplitude of finding state ψ_j in Ψ; from a mathematical viewpoint, it is the magnitude of projection of state vector Ψ onto an eigenvector ψ_j. This is the essence of the von Neumann *projection postulate* [2].

The most straightforward way to emphasize the outlined analogy would be to cap the state vectors as we frequently do with the unit vectors in algebra. But that could produce confusion with operators, whose symbols are also frequently capped. Instead of Ψ capped, Dirac suggested

$$\Psi \to |\Psi\rangle, \qquad \Psi^* \to \langle\Psi|, \qquad \psi_j \to |\psi_j\rangle. \tag{4.5}$$

Accordingly, instead of (4.1) we will have

$$|\Psi\rangle = \sum_j c_j |\psi_j\rangle. \tag{4.6}$$

In this notation, the amplitude c_j in (4.3) is

$$c_j = \langle\psi_j|\Psi\rangle. \tag{4.7}$$

The symbol $|\rangle$ here is called the *ket* – the second half of the word *bracket* for the whole symbol $\langle|\rangle$. Its mirror image $\langle|$ for the *conjugate state* is accordingly called the *bra* since it makes the first half of the bracket in (4.7). The $|\Psi\rangle$ and $\langle\psi_j|$ are the ket and bra vectors, respectively. The whole bracket (4.7) is just the *dot* (or *inner*) product of the two vectors. The reader can appreciate the elegance of simple expression (4.7) versus (4.3).

The dot product is a scalar quantity represented by a single (possibly complex) number. In order to satisfy this requirement, the bra in (4.7) must be more than just a complex conjugate of the corresponding ket. This is evident from the simple example of a two-state system, $j = 1, 2$. We can represent a state $|\Psi\rangle$ in some basis as a column matrix $|\Psi\rangle = \begin{pmatrix} c_1 \\ c_2 \end{pmatrix}$, with matrix elements being the corresponding probability amplitudes. For instance, we can take spin-up and spin-down states of an electron as the two basis states $|\psi_1\rangle$ and $|\psi_2\rangle$; let $|\psi_1\rangle$ stand for spin-up state. Then the spin measurement in this state will always find its spin up (the amplitude $c_1 = 1$) and never (the amplitude $c_2 = 0$) spin down. The pair of these amplitudes forms the matrix $|\psi_1\rangle = \begin{pmatrix} 1 \\ 0 \end{pmatrix}$. Similarly, the spin-down state can be represented as $|\psi_2\rangle = \begin{pmatrix} 0 \\ 1 \end{pmatrix}$. Then, as just mentioned, an arbitrary state with the nonzero amplitude of either outcome will be

$$|\Psi\rangle = c_1|\psi_1\rangle + c_2|\psi_2\rangle = c_1\begin{pmatrix} 1 \\ 0 \end{pmatrix} + c_2\begin{pmatrix} 0 \\ 1 \end{pmatrix} = \begin{pmatrix} c_1 \\ c_2 \end{pmatrix}.$$

The complex conjugate of $|\Psi\rangle$ multiplying some other state $|\Phi\rangle = \begin{pmatrix} b_1 \\ b_2 \end{pmatrix}$ would give

$$|\Psi\rangle^*|\Phi\rangle = \begin{pmatrix} c_1 \\ c_2 \end{pmatrix}^* \begin{pmatrix} b_1 \\ b_2 \end{pmatrix} = \begin{pmatrix} c_1^* & b_1 \\ c_2^* & b_1 \end{pmatrix}. \tag{4.8}$$

This is again a matrix rather than a single number. For an inner product as in (4.7) to produce a number, the bra must be defined as a complex conjugate *and* transpose of the matrix representing $|\psi_1\rangle$; that is,

$$|\Psi\rangle \to \langle\Psi| \equiv (c_1^* \ c_2^*). \tag{4.9}$$

Then we will get

$$\langle\Psi|\Phi\rangle = (c_1^* \ c_2^*) \begin{pmatrix} b_1 \\ b_2 \end{pmatrix} = b_1 c_1^* + b_2 c_2^*. \tag{4.10}$$

Now we have only one nonzero matrix element, which allows us to treat it as a single number, and this is precisely what we need. Thus, any bra is complex conjugate and transpose of the corresponding ket. The result of both operations together is called the *Hermitian conjugate* or just the *conjugate*. With this definition, bra *must be the first factor* in any inner product. Indeed, reversing their ordering would give

$$|\Phi\rangle\langle\Psi| = \begin{pmatrix} b_1 \\ b_2 \end{pmatrix} (c_1^* \ c_2^*) = \begin{pmatrix} b_1 c_1^* & b_1 c_2^* \\ b_2 c_1^* & b_2 c_2^* \end{pmatrix}. \tag{4.11}$$

This is a full-fledged matrix. In other words, the product in order (ket) × (bra), called the *outer product*, is an *operator* that generally cannot be reduced to a single number (see Section 6.6).

The presented examples illustrate another important property of bra and ket vectors. As seen from (4.8), any product of kets of the type $|\psi_j\rangle|\psi_k\rangle$ is a matrix. However, such products do not represent a physical state *in a vector space spanned by kets* $|\psi_j\rangle$. The reason is seen most clearly in case when $|\psi_j\rangle = |\mathbf{r}_j\rangle$ (the state of definite position $\mathbf{r} = \mathbf{r}_j$). The product $|\mathbf{r}_j\rangle|\mathbf{r}_k\rangle$ would represent a fictitious state with the particle entirely at $\mathbf{r}_j$ *and* entirely at $\mathbf{r}_k$ at the same time. But two distinct locations of a particle can coexist only as potentiality (*before* position measurement) and never as reality (after such measurement). Any position measurement finds the whole particle in one place and never in two places at once. No single measurement can show two different values of a measured observable. Therefore, such products of kets (or bras) have no physical meaning. On the other hand, they are totally legitimate when the multiplied kets (or bras) belong to *different* vector spaces. For instance, take two particles, one described by $|\psi_j\rangle_1$ and the other by $|\psi_k\rangle_2$. Each particle "lives" in its own $\mathcal{H}$-space. Denote them $\mathcal{H}_1$ and $\mathcal{H}_2$, respectively. Then the product $|\psi_j\rangle_1|\psi_k\rangle_2$ describes the whole pair with particle 1 in state j and particle 2 in state k. These legitimate products are denoted as $|\Psi_{jk}\rangle_{1,2} \equiv |\psi_j\rangle_1 \otimes |\psi_k\rangle_2$ (*direct*, or *tensor*, *products*). If j or k are numbering N distinct eigenvectors of the same operator, then $\mathcal{H}$-space of each particle has N dimensions, but the $\mathcal{H}$-space of the whole pair, which we denote as $\mathcal{H} \equiv \mathcal{H}_1 \otimes \mathcal{H}_2$, has N^2 dimensions according to the number of different tensor products $|\Psi_{jk}\rangle_{1,2}$.

One of the advantages of bra and ket symbols is their versatility. Let, for example, $\Psi(\mathbf{r}, t)$ be a wave (2.32) with definite momentum $\mathbf{p} = \mathbf{p}' = \hbar \mathbf{k}'$ (and $E = E'$). In our terminology, this is the *momentum eigenstate*. To each such eigenstate there corresponds a *unit vector* in $\mathcal{H}$. All these vectors are mutually orthogonal according to (3.9), and the whole set of them forms the corresponding basis. Altogether we have as many mutually orthogonal basis vectors as we have distinct eigenstates. When each such state is written as a function of $\mathbf{r}$, we say that it is given in the "r-representation." But the Dirac notation for this would be

$$\Psi_{\mathbf{p}'}(\mathbf{r}) \Rightarrow |\Psi_{\mathbf{p}'}\rangle, \quad \text{or just} \quad \Psi_{\mathbf{p}'}(\mathbf{r}) \Rightarrow |\mathbf{p}'\rangle, \tag{4.12}$$

and the whole expression reads "The particle in a state with $\mathbf{p} = \mathbf{p}'$." Such notation allows us to suppress information about representation when it is not necessary. Indeed, there is no mention of position $\mathbf{r}$ on the right side of (4.12). The symbol inside the ket stands for the variable of interest (in the given case, $\mathbf{p}$), and the ket itself denotes the *eigenstate* corresponding to eigenvalue $\mathbf{p} = \mathbf{p}'$. Therefore, it contains exactly the same amount of information about the system as the whole expression (2.32).

Thus, $|\Psi\rangle$ and $\Psi(x)$ are different ways of representing the same physical state: $|\Psi\rangle$ is a vector, whereas $\Psi(x)$ is a *number*, namely, the *projection of* $|\Psi\rangle$ *onto* $|x\rangle$ (physically, the amplitude for $|\Psi\rangle$ to collapse to $|x\rangle$ in an appropriate measurement). This number can be written in Dirac's notation as $\Psi(x) = \langle x|\Psi\rangle$. Combining it with (4.6) and (4.7), we can write

$$\Psi(x) = \langle x|\Psi\rangle = \sum_j c_j \langle x|\psi_j\rangle = \sum_j \langle x|\psi_j\rangle\langle\psi_j|\Psi\rangle. \tag{4.13}$$

Now let us see how the Dirac notations work in various situations.

The basis states are usually chosen so that they are the eigenstates of some relevant operator. For *any* two such states corresponding to different eigenvalues, we have a zero probability of finding in one state the eigenvalue of the other. If we consider an eigenstate $|\mathbf{p}'\rangle$, then the appropriate momentum measurement in this state will never give a result $\mathbf{p}''$ different from $\mathbf{p}'$, and a similar measurement in state $|\mathbf{p}''\rangle$ will never give the result $\mathbf{p}'$. In Dirac's notation, we have $\langle \mathbf{p}'|\mathbf{p}''\rangle = 0$. This holds true for any observable. Two eigenstates corresponding to eigenvalues α' and α'' of an observable α are represented by mutually orthogonal unit vectors $|\alpha'\rangle$ and $|\alpha''\rangle$ in $\mathcal{H}$, that is,

$$\langle \alpha'|\alpha''\rangle = \begin{cases} \delta_{\alpha'\alpha''}, & \alpha \text{ discrete,} \\ \delta(\alpha' - \alpha''), & \alpha \text{ continuous.} \end{cases} \tag{4.14}$$

This is the rehash of (3.9) and (3.10) in Dirac's notation.

If the two states $|\alpha\rangle, |\beta\rangle$ are not orthogonal, $\langle \alpha|\beta\rangle \neq 0$, the component of one state along the other state is nonzero. According to the projection postulate, such a component is the amplitude of finding the value α in state $|\beta\rangle$. This means that $|\beta\rangle$ is a superposition of two or more different $|\alpha\rangle$-eigenstates: $|\beta\rangle = \sum_\alpha c_\alpha |\alpha\rangle$, where $c_\alpha = \langle \alpha|\beta\rangle$. This allows us to write $|\beta\rangle$ in the form

$$|\beta\rangle = \sum_\alpha |\alpha\rangle c_\alpha = \sum_\alpha |\alpha\rangle\langle\alpha|\beta\rangle. \tag{4.15}$$

This holds for an arbitrary β; therefore,

$$\sum_\alpha |\alpha\rangle\langle\alpha| = I. \tag{4.16}$$

As we saw in example (4.11), an outer product $|\alpha\rangle\langle\alpha|$ is a matrix whose elements are generally all different. However, the *sum* of all $|\alpha\rangle\langle\alpha|$ is a diagonal matrix with all diagonal elements equal to 1 (the *identity operator* I). This result is also seen from (4.13).

Summary: For any two states the amplitude of finding in one state the eigenvalue of the other is determined by the inner product of the unit vectors representing these states.

In some cases, we may have a set of distinct eigenstates belonging to the same eigenvalue α. We call this phenomenon *degeneracy*. Such eigenstates may be nonorthogonal to each other, but we can always form a set of their superpositions, in which all new state vectors are mutually orthogonal. A known example of degeneracy is the case of different momentum eigenstates all belonging to the same energy eigenvalue $E = \mathbf{p}^2/2\mu$. In this case, the respective eigenfunctions are mutually orthogonal in $\mathcal{H}$, so we do not have to form a new orthonormal set out of them. But we must distinguish between different eigenstates with the same E by indicating, apart from E, also the polar and azimuthal angles (θ, φ) to specify the direction of respective momentum $\mathbf{p}$, for instance, $|E, \mathbf{p}\rangle = |E; \theta, \varphi\rangle = |\theta, \varphi\rangle_E$. Generally, we can denote a degenerate state as $|\alpha, \beta\rangle$ or $|\beta\rangle_\alpha$, where β is an additional variable similar to (θ, φ) in our example with energy and momentum eigenstates. Then condition (4.14) can be generalized for orthogonal degenerate eigenstates as

$$\langle \alpha, \beta | \alpha', \beta' \rangle = \begin{cases} \delta_{\alpha\alpha'}\delta_{\beta\beta'}, & \alpha, \beta \text{ discrete}, \\ \delta(\alpha - \alpha')\delta_{\beta\beta'}, & \alpha \text{ continuous}, \beta \text{ discrete}, \\ \delta_{\alpha\alpha'}\delta(\beta - \beta'), & \alpha \text{ discrete}, \beta \text{ continuous}, \\ \delta(\alpha - \alpha')\delta(\beta - \beta'), & \alpha, \beta \text{ continuous}. \end{cases} \tag{4.17}$$

The inner product of the bra and ket forming the entire bracket as in (4.7) admits different observables in the two parts of the bracket. Then the resulting expression may represent an amplitude of two events in succession. Thus, the expression $\langle q|s\rangle$ is the amplitude for a particle initially prepared in an eigenstate of observable s to be found later, after an appropriate measurement, in an eigenstate of observable q. This is sometimes referred to as the $(s \rightarrow q)$ *transition amplitude*. That is, the symbol on the right of the vertical bar determines the *initial* condition, and the symbol on the left describes the *final* state. The entire bracket must be read backward, from right to left.

This applies to cases like (2.32) as well. Namely, instead of talking about the amplitude of the particle in a state (2.32) to be found in the vicinity of position $\mathbf{r}$, we can talk about the amplitude of transition from the momentum eigenstate $|\mathbf{p}'\rangle$ to position eigenstate $|\mathbf{r}\rangle$ in an appropriate measurement. Accordingly, we can write (2.32) as $\Psi_{\mathbf{p}'}(\mathbf{r}) = \langle \mathbf{r}|\mathbf{p}'\rangle$, which is just a special case of (4.13).

4.2
Double Slit Revised I

Consider again the double-slit experiment. We analyze its more complicated version designed to answer the question: What happens when we attempt to "peep into" the process of formation of the diffraction pattern on the screen [3]? To this end, we introduce a source of light illuminating the electrons passing through a slit and two identical photon detectors that can record a single photon scattered by an electron (Figure 4.1).

Consider two amplitudes: A_1 for an electron from source S to land on the screen at point V having vertical coordinate y after having passed through slit 1, and A_2 to land at y via slit 2. They can be written as

$$A_1 = \langle y|1\rangle\langle 1|S\rangle, \qquad A_2 = \langle y|2\rangle\langle 2|S\rangle. \tag{4.18}$$

If both slits are open and no attempt is made to watch the electron's progress from S to y, the amplitude at y is the superposition

$$A = A_1 + A_2 = \langle y|1\rangle\langle 1|S\rangle + \langle y|2\rangle\langle 2|S\rangle. \tag{4.19}$$

This expression is totally equivalent to (1.18); the difference is only in notations. Accordingly, it describes the same diffraction pattern, shown in Figure 1.9.

Now we install the source of photons SP between the slits right behind the screen and two photon detectors such that detector D_1 looks exactly at slit 1, while D_2 looks

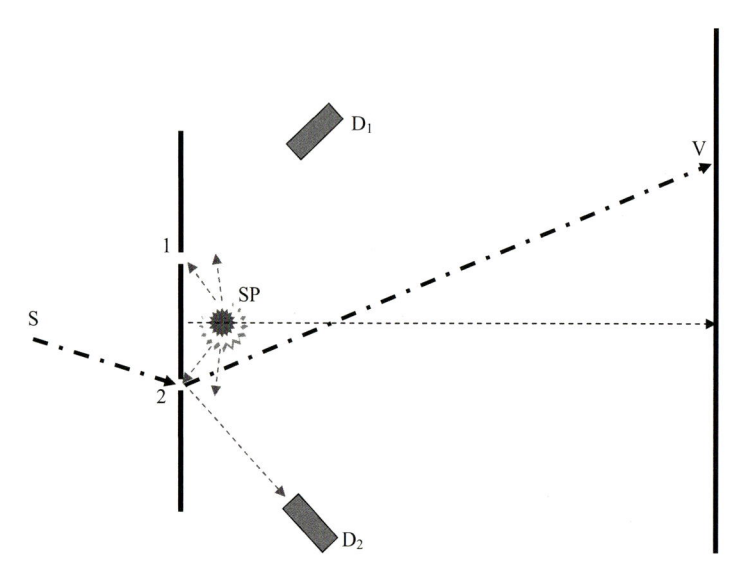

Figure 4.1 The double-slit experiment with the equipment to "peep in" to catch an electron passing through a specific slit. 1, 2: The two slits; S: the source of electrons; SP: source of photons; D_1 and D_2 : photon detectors; $S \rightarrow 2 \rightarrow V$: virtual electron path from S to V via slit 2, actualized by a photon scattered at the slit 2 and recorded in D_2.

at slit 2. How will this affect the outcome? The answer is that now, instead of one single event (an electron landing at y), we will have two distinct pairs of events:

1) an electron landing at y after having scattered a photon into D_1 (event (D_1,y)); and
2) an electron landing at y after having scattered a photon into D_2 (event (D_2,y)).

In other words, we can observe either electron at y and the photon at D_1 or the electron at y and the photon at D_2. The two pairs are distinguishable because the clicks of detector D_1 or D_2 are different macroscopic events. Each pair has its respective probability amplitude. In order to find them, we need to introduce the additional amplitudes describing the photon scattering by the electron. Generally, there will be four such amplitudes, since either detector can record the photon arrival from either slit. For instance, D_1 can "catch" a photon coming from the electron passing through slit 1, but it can as well record a photon from the electron passing through slit 2. The same is true for D_2.

One might ask "How can a photon from slit 2 enter D_1 if D_1 looks exactly at slit 1, not at slit 2?" This is a good question, and the answer to it would be "It can't," were the photon a classical particle. But while being recorded as a particle, it propagates as a wave, and if its wavelength is long enough, there is a nonzero amplitude for a photon scattered at slit 2 to get into D_1. Similarly, a photon scattered at slit 1 can be found in D_2. But if we pass only one electron and send only one photon at a time, then at each trial we will have *either* D_1 *or* D_2 clicking.

Let f_{11} stand for the amplitude of photon scattered to D_1 from the electron at slit 1, and f_{12} stand for scattering into the same detector from the electron at slit 2. Similarly, we introduce two amplitudes for scattering into D_2: f_{22} from slit 2, and f_{21} from slit 1. The four amplitudes form a matrix

$$\hat{f} = \begin{pmatrix} f_{11} & f_{12} \\ f_{21} & f_{22} \end{pmatrix}. \tag{4.20}$$

If the detectors are arranged symmetrically with respect to the central axis of the experiment as shown in Figure 4.1, the matrix (4.20) is symmetrical and has equal diagonal elements:

$$f_{11} = f_{22} \equiv f \quad \text{and} \quad f_{12} = f_{21} \equiv g. \tag{4.21}$$

Now we can write the amplitudes for our two pairs of events. The pair (D_1, y) can occur in two different ways: the electron lands at y after scattering the photon to D_1 from slit 1, or it lands at y after scattering it from slit 2. Using our notations (4.18) and (4.21), we can write this as

$$\langle D_1 | y \rangle = f A_1 + g A_2. \tag{4.22}$$

Similarly, for pair (D_2, y) we will have

$$\langle D_2 | y \rangle = g A_1 + f A_2. \tag{4.23}$$

Note that fA_1 and gA_2 are the amplitudes of the same event (the photon in D_1 and the electron at y) that can occur in two different but indistinguishable ways; therefore, we add these amplitudes to form $\langle D_1 | y \rangle$; similarly, fA_2 and gA_1 are the amplitudes of another event (the photon in D_2 and the electron at y), so they add to make $\langle D_2|y\rangle$. But $\langle D_1|y\rangle$ and $\langle D_2|y\rangle$ themselves, being the amplitudes of the two *distinguishable* events, *do not add*, so they *cannot* interfere. We cannot add such amplitudes even if we do not care which detector clicks. The Supreme Arbiter – Nature – takes care of that automatically even if we don't, since the events are objectively distinct. Therefore, the net probability of finding the electron at y is just the sum of probabilities:

$$\mathcal{P}(y) = \mathcal{P}_1(y) + \mathcal{P}_2(y) = |\langle D_1|y\rangle|^2 + |\langle D_2|y\rangle|^2. \tag{4.24}$$

In view of (4.23) and (4.24), this yields

$$\mathcal{P}(y) = |fA_1 + gA_2|^2 + |gA_1 + fA_2|^2$$
$$= (|f|^2 + |g|^2)|A_1|^2 + 4\operatorname{Re}(fg^*)\operatorname{Re}(A_1 A_2^*) + (|f|^2 + |g|^2)|A_2|^2. \tag{4.25}$$

Now we have a general expression embracing a wide variety of cases, and we can study them by playing with amplitudes f and g. Here we consider two extremes.

1) $f = g$. This corresponds to a situation in which each detector accepts equally efficiently the photons scattered from each slit. Clearly, in such a situation both detectors are as good as nonexistent: the click of D_1 only tells us about the photon's arrival, but it does not tell what we actually want to know – *where the photon came from*. The same holds for D_2. As mentioned above, this happens when the photons have a sufficiently long wavelength. In this case not even Nature knows which slit the given photon comes from. The classical notion of the electron landing at y after traveling along a single path fails. Both possible paths (traveling though slit 1 or through slit 2) are represented on the same footing in the resulting event y. In this case, instead of one actual history we have the result of two *virtual* histories, equally contributing to the final outcome. We then should first add their amplitudes in order to find the probability. And, indeed, setting $f = g$ in (4.25) reduces it to

$$\mathcal{P}(y) = 2\mathcal{P}_0|A_1 + A_2|^2, \qquad \mathcal{P}_0 \equiv |f|^2, \tag{4.26}$$

which describes the original double-slit diffraction without any detectors.

2) $g = 0$. This corresponds to the case of photon detection with ideal selective efficiency with respect to direction, when D_1 only records photons from slit 1, and D_2 does the same only for photons from slit 2. Accordingly, for each landing event it is known through which slit the electron came; the "travel histories" $\langle y|1\rangle\langle 1|S\rangle \Leftrightarrow \langle D_1|y\rangle$ and $\langle x|2\rangle\langle 2|S\rangle \Leftrightarrow \langle D_2|y\rangle$ are distinguishable. Virtual histories become actual histories, and the net probability is the sum of their individual probabilities, with no interference. Indeed, setting $g = 0$ in (4.25) gives

$$\mathcal{P}(y) = |f|^2(|A_1|^2 + |A_2|^2). \tag{4.27}$$

Each of the two terms on the right of (4.27) can also be obtained if we record only the corresponding *subset* of all landings with the respective detector clicking and ignore the rest of the landings. For instance, if we record *only* the events with detector D_1 clicking and plot the landing distribution for these events, the obtained graph will be exactly the same as if the slit 2 were closed. The same holds for landings with detector D_2 clicking. Thus, we can effectively "close" each slit by ignoring the electrons coming from it, which can be achieved by using the highly selective photon detection.

All intermediate cases between the above two extremes display intermediate distributions corresponding to partial interference between A_1 and A_2, which combine with different weights, depending on the relative magnitudes of f and g.

4.3
Double Slit Revised II

Thus, the direct attempt to catch an electron at a definite slit disturbs the process. If performed with maximal efficiency ($g = 0$), it totally destroys the interference pattern. But there seems to be a more sophisticated way to pinpoint the electron's path. What if we, instead of shooting at the electron with photons, use the conservation of momentum? The incident electron has momentum p along the z-direction. If the electron lands at distance y from the center of the screen, its trajectory must be deflected from the symmetry axis toward this point. Accordingly, it must acquire a transverse momentum $p_y \approx py/L$. Due to conservation of momentum, the first screen must recoil, acquiring the equal transverse momentum in the opposite y-direction. The value of p_y necessary for the electron to land at y from the first slit is different from the value needed to land there from the second slit. Accordingly, the respective recoils of the screen are different. An accurate measurement of the screen's recoil can identify the slit chosen by the electron. Since such measurement will not affect the electron itself, it must not affect the interference pattern either. Thus, the outlined procedure must provide us with the information about the electron path without destroying the interference.

This scheme will not work! Look back at Figure 1.8. In order to land at y from slit 1, the electron must acquire the transverse momentum $p_y^{(1)}$, and to do the same from slit 2, it needs the transverse momentum $p_y^{(2)}$. Consulting with Figure 1.8, we have

$$p_y^{(1)} = p\frac{y - (1/2)d}{L}, \quad p_y^{(2)} = p\frac{y + (1/2)d}{L}. \tag{4.28}$$

In each case, the screen will get the recoil with the respective opposite momentum. In order to distinguish between the two slits, we must measure the screen's momentum with an accuracy of at least half the difference between $p_y^{(1)}$ and $p_y^{(2)}$:

$$\Delta p_y \leq \frac{1}{2}\left(p_y^{(2)} - p_y^{(1)}\right) = p\frac{d}{2L} = \pi\frac{\hbar d}{\lambda L}. \tag{4.29}$$

But according to the indeterminacy principle, the screen's momentum measurement will introduce indeterminacy in its position Y:

$$\Delta Y \geq \frac{1}{2}\frac{\hbar}{\Delta p_y} = \frac{1}{2\pi}\frac{L}{d}\lambda. \tag{4.30}$$

And now look back at Equation 1.20. We see that the resulting indeterminacy in the slit's position (4.30) exceeds ($1/2\pi$) of the period of the diffraction pattern that would be observed without momentum measurement. This means that even the least allowed accuracy in momentum measurement will introduce noticeable indeterminacy in the observed picture on the second screen. If the accuracy of p_y measurement increases by a factor π, the bright fringes will totally overlap with the neighboring dark ones and vice versa. The whole pattern will be degraded to a uniform fuzzy spot over the whole screen.

More about various attempts to "peep in without disturbing" can be found in Ref. [4]. All have the same outcome: if an experimental arrangement does reveal the answer to "which slit?", it inevitably destroys the interference pattern.

4.4
Neutron Scattering in Crystals

Here we use the previously introduced mathematical tools to consider in more detail the neutron diffraction in a crystal mentioned in Section 3.2. Suppose we have a monochromatic beam of neutrons. As they pass though the crystal, they can scatter due to interaction with the nuclei of the crystalline lattice. The result of their scattering can be described as diffraction of neutron waves on a three-dimensional grating formed by the crystal lattice. One of the basic characteristics of this process is the diffraction angle θ (Figure 3.3 with neutrons instead of electrons). Focusing on this angle as an observable, we start with the question: "What is the amplitude of neutron scattering by an angle θ?"

Suppose we have a movable neutron detector collecting the data. Take the irradiated crystal as the origin. As we move the detector from one position to another, the number of detected neutrons changes; it is generally a function of two angles – θ (polar angle) and φ (azimuthal angle). To simplify the discussion, we will consider it as a function of θ alone. Denote this function for a scattering by a single nucleus as $f(\theta)$. In some cases, this individual amplitude may be the same for all θ. Such cases are known as *isotropic scattering*.

The amplitude of the neutron coming from the source S to detector D via a nucleus No. j is

$$A_j = \langle D|j\rangle f(\theta)\langle j|S\rangle. \tag{4.31}$$

Here $\langle j|S\rangle$ is the amplitude of finding the neutron from S at the nucleus No. j, $f(\theta)$ is the amplitude of scattering off this nucleus in direction θ (this amplitude is common for all nuclei), and $\langle D|j\rangle$ is the amplitude of reaching D from location j. Usually, the dependence $f(\theta)$ is much slower than that arising from interference of waves coming to D from different scattering centers j. Therefore, we will for the sake of simplicity drop the variable θ in f. Altogether we have N amplitudes $A_1, A_2, \ldots, A_j, \ldots, A_N$, where N is the total number of atoms in the crystal.

Next, we assume that the physical state of a nucleus remains the same after the interaction with the passing neutron. This is a valid assumption for sufficiently slow neutrons. In this case, there is no way to distinguish between various possibilities – various virtual paths that could be traveled by the neutron between the source and the detector. The paths are indistinguishable because, with nuclei unaffected in the process, there remains no "mark" of the scattering event that would indicate its location. Then, if we are looking for the net probability of the neutron to wind up at D, we again start with adding up the path amplitudes. The result is the superposition of the individual amplitudes much in the same way as in case of a photon passing through a grating. It can be written by analogy with diffraction on a grating as

$$A = \sum_{j=1}^{N} A_j = \sum_{j=1}^{N} \langle D|j\rangle f \langle j|S\rangle = f \sum_{j=1}^{N} \langle D|j\rangle \langle j|S\rangle. \tag{4.32}$$

Since reaching the detector from different locations j involves different path lengths, the incoming waves have different phases, so that when added, they produce, after taking the square modulus of the sum, characteristic interference terms:

$$I_{ES}(\theta) = \frac{d\mathcal{P}_{ES}(\theta)}{d\Omega} = \frac{|f|^2}{4\pi} \left| \sum_{j=1}^{N} \langle D|j\rangle f \langle j|S\rangle \right|^2$$

$$= \frac{|f|^2}{4\pi} \left\{ \sum_{j=1}^{N} |\langle D|j\rangle\langle j|S\rangle|^2 + 2\,\mathrm{Re} \sum\sum_{j\neq j'} \langle D|j\rangle\langle j|S\rangle\langle D^*|j'\rangle\langle j'|S^*\rangle \right\}. \tag{4.33}$$

Here $I_{ES}(\theta)$ is the counting rate for scattering without spin-flip (*elastic scattering*), and $d\mathcal{P}_{ES}(\theta)$ is the probability of such scattering into the element of solid angle $d\Omega(\theta) = 2\pi \sin\theta\, d\theta$ (assuming axial symmetry). We emphasize again that the angular dependence here comes not so much from $f(\theta)$ as from the interference of different terms in (4.32), which is described by the double sum in (4.33). Specifically, there may appear sharp diffraction spikes like those shown in Figure 4.2a. However, for some crystals the spikes are smoothed out, and there also appears a more or less uniform background, indicating the increased probability of uniform scattering in all directions. This can also be explained by general rules of composition of amplitudes if we take into account *spin interactions* between the neutron and nuclei.

The neutron spin is 1/2, and as will be seen later in Chapter 7, it has two possible projections on a chosen direction. If the nuclei of a given crystal are *spinless*, then the spin of an incident neutron has nothing to interact with, and the result will be pure diffraction with sharp maxima (Figure 4.2a). The same happens for the nuclei with spin if all of them are in the same spin state as the incident neutrons.

Consider now a crystal whose nuclei each have the same spin 1/2 as a single neutron, but they are all polarized opposite to the neutrons in the beam. For instance, all of them have their spins up, whereas the neutrons of the beam have their spins down. In this case, there is a chance of spin exchange (*inelastic scattering*): a neutron can undergo spin-flip when scattering from some nucleus, and the same happens to the nucleus as well. This is required by the conservation of angular momentum. And this can profoundly affect the observed pattern, since now the corresponding nucleus is "marked" by the change of its spin state, so that Nature has the means to distinguish the neutron path via this nucleus from all the rest. Accordingly, the probability with such events will be the sum of the individual probabilities.

Let us include this effect into our description of scattering. To this end, introduce the scattering amplitude with a spin-flip. Denote it as g. Then the spin-flip amplitude

4 Playing with the Amplitudes

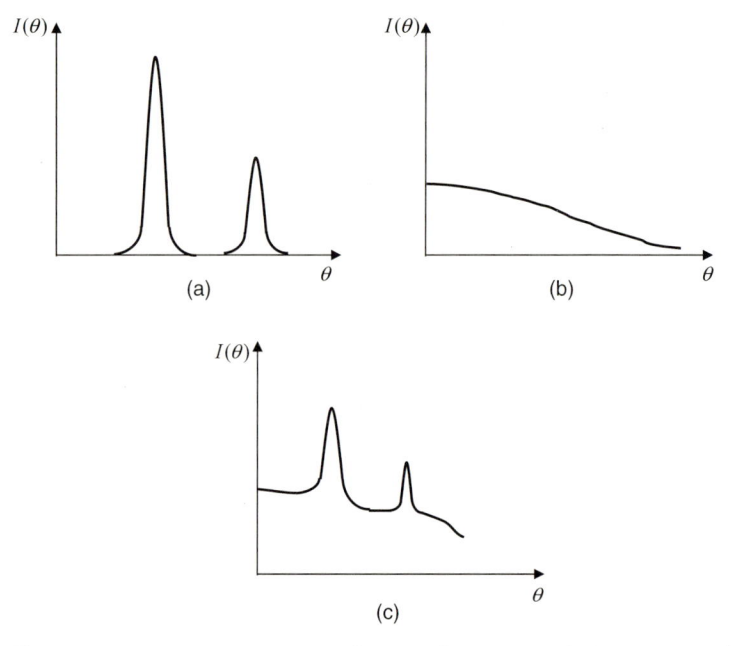

Figure 4.2 Neutron scattering rate as a function of scattering angle θ: (a) on a periodic system of nuclei with the zero spin; (b) on a system of nuclei causing neutron spin-flip; (c) observed scattering rate on system of nuclei with spin 1/2.

for a neutron scattering by a nucleus No. n will be

$$B_n = \langle D|n\rangle g \langle n|S\rangle, \tag{4.34}$$

and the corresponding probability is $|B_n|^2 = |g|^2 |\langle D|n\rangle\langle n|S\rangle|^2$. The second factor here only weakly depends on n, and all phases vanish after we take the square modulus. The net probability (per unit solid angle) of scattering with the spin-flip from *any* nucleus is the sum

$$I_{\text{in}}(\theta) = \frac{\mathrm{d}\mathcal{P}_{\text{in}}(\theta)}{\mathrm{d}\Omega} = \frac{|g|^2}{4\pi} \sum_{n=1}^{N} |\langle D|n\rangle\langle n|S\rangle|^2. \tag{4.35}$$

This sum describes a smooth probability distribution – there are no interference terms left.

If we *do not record* the neutron spin, then both types of scattering contribute, and the total probability is the sum of both expressions – (4.33) and (4.35). The resulting picture of scattering depends on the relative weight of these two contributions. If the spin interaction is so strong that the probability of spin-flip is close to 1, then the term (4.35) dominates, and we will observe the smooth picture close to that in Figure 4.2b. If there is a considerable chance for a neutron to pass through the crystal without spin-flip, then there will be noticeable contribution from elastic scattering (4.33), and the observed picture will be given by the composition of the first two graphs as shown in Figure 4.2c.

4.5
Bosonic and Fermionic States

Already at this stage, a clear understanding of the borderline between distinguishable and indistinguishable ways for an event to occur leads to predictions of some quantum effects in systems of *identical particles*, which are totally alien to our classical intuition.

Let us first consider collision of two different particles, say, a C nucleus and an α-particle (a He nucleus). Assume that both nuclei are in their respective ground states and the kinetic energy of collision is low, so that the collision results only in a change of their velocities but does not affect their internal states. This is *elastic scattering*.

The analysis of scattering of two particles is much simpler in the system of center of mass (i.e., in the *reference frame* (RF), where the center of mass of the particles is stationary (C_M frame), or, equivalently, the net momentum of the system is zero). In this RF, the individual momenta of both particles are equal in magnitude and opposite in direction at any stage of collision. So the only change that we can expect after the collision is the rotation of the line connecting the particles. The momentum magnitudes before and after elastic collision are the same. The angle of rotation is just the scattering angle θ. Geometrically, θ is the angle between initial and final directions of motion of either particle. But, since we describe the two particles as one system, we choose one direction along the initial motion as positive and measure their deflection angles with respect to this *common* direction (Figure 4.3). Therefore, if the scattering angle of one particle is θ, the scattering angle of the second particle

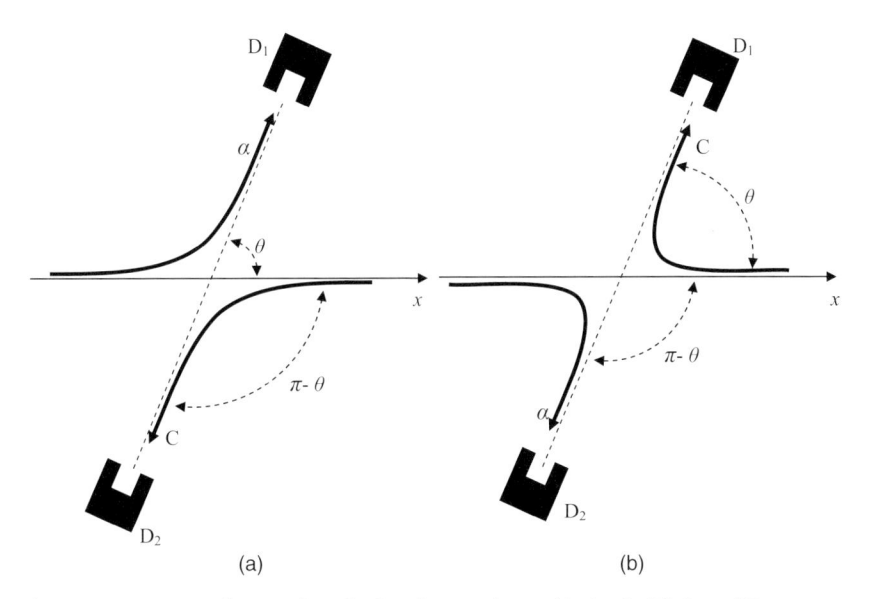

Figure 4.3 Scattering of α-particle and a C nucleus as observed in the C_M RF. D_1 and D_2 : detectors 1 and 2; α: ^{4}He nucleus; C: carbon (^{12}C) nucleus. (a) α is deflected to D_1 and C is deflected to D_2. (b) α is deflected to D_2 and C is deflected to D_1.

is $\pi - \theta$. Scattering at different θ occurs with different probability, so the natural question is to find the scattering amplitude f as a function of θ. For a complete description, we must also specify the azimuthal angle φ for the scattered particle. Then for its counterpart, such an angle will be $\varphi + \pi$. For a spherically symmetric interaction, the scattering amplitude depends only on θ but not on φ.

Of course, we cannot derive the function $f(\theta)$ without considering the forces between the particles, which are the long-range Coulomb force and short-range nuclear force. The Coulomb force is repulsive, since both nuclei have positive electric charge. Here we consider a low-energy collision, so the particles do not have enough energy to overcome the Coulomb repulsion and get close enough to engage in nuclear interaction. Therefore, in our case we can ignore the contribution of nuclear force. The scattering in Coulomb's field has been well known since Rutherford's time, and it is interchangeably called the *Coulomb* or *Rutherford scattering*.

A remarkable feature of this scattering is that in the case when the two particles can be distinguished from one another, QM gives the same result for the probability distribution $\mathcal{P}(\theta)$ as CM. But here we will be concerned with basic characteristics of $\mathcal{P}(\theta)$ (such as symmetry) rather than with its detailed form.

Consider the experimental setup shown in Figure 4.3. Two movable detectors D_1 and D_2 are separated by a fixed distance and look exactly at one another, since we know that in the C_M frame, the two particles are scattered in exactly opposite directions. Recording the counts at different orientations of line $D_1 D_2$ will give the scattering probability distribution $\mathcal{P}(\theta)$.

We consider two different scenarios. In the first one, D_1 and D_2 are designed so that each can detect only one fixed type of particles – for instance, D_1 can detect α-particles, and D_2 only C nuclei. Denote the amplitude of scattering at an angle θ as $f(\theta)$. From the viewpoint of QM, $f(\theta)$ is the angular part of the wave function describing the particles in the scattered state. The corresponding probability per unit solid angle is $\mathcal{P}(\theta) = |f(\theta)|^2$.

In the second scenario, each detector can respond to the arrival of either particle. Suppose we do not care which particle hits which detector. Now the only thing we know for certain is that if α is scattered at an angle θ then C is scattered at $\pi - \theta$, and vice versa (the scattering angles for the two particles are mutually supplementary [3]). Let $f(\theta)$ be the scattering amplitude for α. According to Figure 4.3, an angle θ takes α to D_1. Then C goes to D_2 at an angle $\pi - \theta$. But, since the events $(\alpha \to D_1)$ and $(C \to D_2)$ are parts of one combined event, the scattering amplitude for C in this event is also $f(\theta)$. The same argument shows that the amplitude for α going to D_2 and C going to D_1 is $f(\pi - \theta)$. Accordingly, the observed probability of scattering of *either* particle at an angle θ is, in this case,

$$\mathcal{P}(\theta) = |f(\theta)|^2 + |f(\pi - \theta)|^2. \tag{4.36}$$

We have the sum of two probabilities because we have two distinct pairs of events – α in D_1 and C in D_2, or C in D_1 and α in D_2 (Figure 4.3). Let us write this as

$$\begin{pmatrix} \text{Event 1:} \\ (\alpha \to D_1, C \to D_2) \end{pmatrix} \text{ or } \begin{pmatrix} \text{Event 2:} \\ (C \to D_1, \alpha \to D_2) \end{pmatrix}. \tag{4.37}$$

If the first event has the amplitude $f(\theta)$, then the amplitude of the second event is $f(\pi - \theta)$. Hence, the net probability of either event is the sum (4.36). One may wonder why the probabilities are added rather than the amplitudes, if the detectors record α and C with equal efficiency, that is, if they just click without distinguishing the particles.

The answer is that "being recorded with equal efficiency" and "being indistinguishable" are totally different things. The latter does not follow from the former. If a school teacher records equally accurately the high grade of a good student and failing grade of a poor student, it does not mean that the recorded grades (let alone the students earning them) are indistinguishable.

Thus, the prediction (4.37) is correct and is confirmed by numerous experiments.

But there is one special case when this prediction fails! It happens when in our pair of particles (α, C) we replace C with another α, so that now we have truly identical particles. In this case, assuming both α to be in the same state, the pairs of events in (4.37) become identical:

$$\begin{pmatrix} \text{Event 1 :} \\ (\alpha \to D_1, \alpha \to D_2) \end{pmatrix} \quad \text{or} \quad \begin{pmatrix} \text{Event 2 :} \\ (\alpha \to D_1, \alpha \to D_2) \end{pmatrix}. \tag{4.38}$$

With the internal states of both particles being the same, any attempt to mark the particles will fail. And the reason is the wave aspect of matter! Suppose we call the α-particle coming from the left α_1 and its counterpart coming from the right α_2. In CM, we could mark the particles by their initial positions and velocities and then watch their progress along their respective trajectories *without disturbing them*. Then, if α_1 gets deflected through an angle θ and is recorded in D_1, we know that the particle deflected at $\pi - \theta$ and recorded in D_2 is α_2. Thus, even identical particles could be distinguished due to some differences in their initial coordinates or momenta or both. We can utilize this difference and always differentiate one particle from another because of the continuity in the change of these parameters.

This approach does not work in QM! Watching which particle came from where involves observing the particles' position measurements, which must be especially accurate at the points of the closest approach. But due to position–momentum indeterminacy, this would disrupt the very same process that we want to study. To avoid disturbances, we must stop tracking the particles' motion. But then we cannot tell which particle, α_1 or α_2, showed up in the given detector. The indices 1 and 2 ascribed to them are purely nominal. In this case, the *amplitudes* add and interfere with one another. The two compound events in (4.38) now blend into one single event (both detectors record arrival of α), but this event can occur in two different ways: $\{\alpha$ from the left hits D_1, and α from the right hits $D_2\}$, or vice versa. Note that the "vice versa" is obtained just by swapping the particles. Therefore, denoting the *net* amplitude of the *pair* of events (4.38) as $\mathcal{F}(\theta)$, we have

$$\mathcal{F}(\theta) = f(\theta) + e^{i\delta} f(\pi - \theta). \tag{4.39}$$

Here δ is a phase shift between the two amplitudes. As we will learn later, for α-particles, $\delta = 0$. Therefore, the amplitudes of these compound events interfere constructively – they enter the superposition with the same sign:

$$\mathcal{F}(\theta) = f(\theta) + f(\pi - \theta), \tag{4.40}$$

and

$$\mathcal{P}(\theta) = |\mathcal{F}(\theta)|^2 = |f(\theta)|^2 + 2\operatorname{Re} f^*(\theta)f(\pi - \theta) + |f(\pi - \theta)|^2. \tag{4.41}$$

We see here the interference term that is absent in (4.36). The difference between (4.41) and (4.36) is maximal at $\theta = \pi/2$. In this case, (4.36) gives

$$\mathcal{P}\left(\frac{\pi}{2}\right) = 2|f(\pi/2)|^2, \tag{4.42}$$

whereas from (4.41) we get

$$\mathcal{P}\left(\frac{\pi}{2}\right) = |2f(\pi/2)|^2 = 4|f(\pi/2)|^2. \tag{4.43}$$

Thus, α-particles are scattered at 90° *at twice the rate* given by the classical equations. The difference between classically predicted and really observed scattering for arbitrary angle for the α-particles is illustrated in Figure 4.4.

All known experimental data show an interesting regularity: the observed probability $\mathcal{P}(\pi/2)$ exceeds the classical prediction by a factor of 2 for an entire class of identical particles – namely, for those with spin projection onto an arbitrary axis being $n\hbar$ with integer n. These *integer spin* particles are called *bosons* – after Satyendra Nath Bose, who first studied their statistics. The doubled scattering rate at 90° is the common feature of identical bosons in the *same states*.

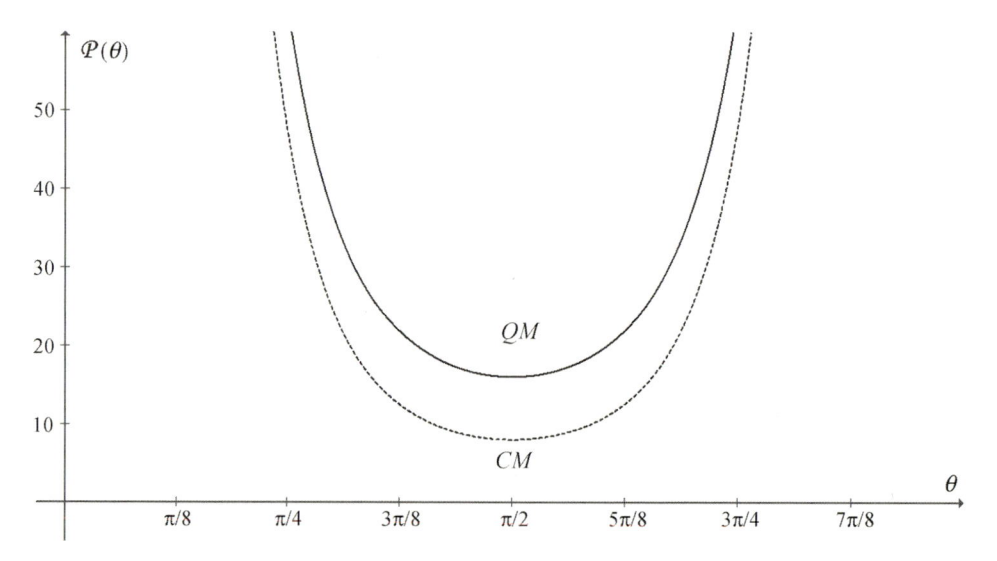

Figure 4.4 Coulomb scattering of two α-particles. CM: according to classical predictions; QM: according to quantum mechanics.

4.5 Bosonic and Fermionic States

Consider now another type of particles – those with half-integer spin. They are called *fermions* – after Enrico Fermi, who made the major contribution in the study of their properties. Electrons, protons, and neutrons are fermions.

Suppose we have two electrons. Assume again that their individual states do not change in the process of collision, which implies sufficiently low energy. Let $|\uparrow\rangle$ and $|\downarrow\rangle$ denote spin "up" and "down," respectively. Consider first the case of the opposite spins. Since we assume no spin exchange in the process, these spin states are "frozen." Therefore, the particles are distinguishable – we can use their different spin states as the natural marks. The particle carries its initial spin state as its "ID number," which is different from that of its partner.

Let the electron coming from the left be in state $|\uparrow\rangle$ and the electron from the right in state $|\downarrow\rangle$. Then we can measure the electron spin state of either partner after the collision and tell where the corresponding particle came from. For instance, if the spin of the electron recorded in D_1 is up, we know that it came from the left, and if it is down, we know that it came from the right. We do not even have to measure the spin of the particle recorded in D_1 – we know the result from records in D_1. The two outcomes are different pairs of events that are (or can be) objectively distinguished from one another. Using the same notations as in (4.37), we can write these events as

$$\begin{pmatrix} \text{Event 1}: \\ (|\uparrow\rangle \to D_1, |\downarrow\rangle \to D_2) \end{pmatrix} \text{ or } \begin{pmatrix} \text{Event 2}: \\ (|\downarrow\rangle \to D_1, |\uparrow\rangle \to D_2) \end{pmatrix}. \tag{4.44}$$

Since both events are different, each has its own probability. And, for the *net* angular distribution with *either* particle scattered at an angle θ, we obtain the same result (4.36) as in case of α and C.

Let now both spins be parallel, say, both point up. In this case, we have no natural marks that could identify the electrons – their "spin IDs" are the same, and they are totally indistinguishable. The corresponding table for pairs of events obtained from readings of our detectors becomes

$$\begin{pmatrix} \text{Event 1}: \\ (|\uparrow\rangle \to D_1, |\uparrow\rangle \to D_2) \end{pmatrix} \text{ or } \begin{pmatrix} \text{Event 2}: \\ (|\uparrow\rangle \to D_1, |\uparrow\rangle \to D_2) \end{pmatrix}. \tag{4.45}$$

Now there is no distinction between the events! Accordingly, the resulting probability is determined by superposition of the interfering amplitudes like that in (4.39). In this case, we find the feature opposite to that for bosons (Figure 4.5): when both particles are in the same state, their scattering rate is considerably less than that for distinguishable particles, and it is just zero at 90°! This can be explained if we assume that in expression (4.39) for indistinguishable fermions, the phase shift $\delta = \pi$, so the two amplitudes come with the opposite signs and interfere destructively:

$$\mathcal{F}(\theta) = f(\theta) - f(\pi - \theta), \qquad \mathcal{P}(\theta) = |f(\theta) - f(\pi - \theta)|^2. \tag{4.46}$$

As a result, $\mathcal{F}(\theta)$ vanishes at $\theta = 90°$. Therefore, the identical fermions in the same spin state never scatter at 90° in the C_M frame.

Note that in all considered cases the momenta of the particles were opposite. This is manifest in the fact that the particles wind up in different detectors on the opposite sides from the center. Therefore, the *total* states of the two particles in this setup are

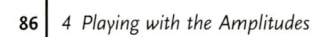

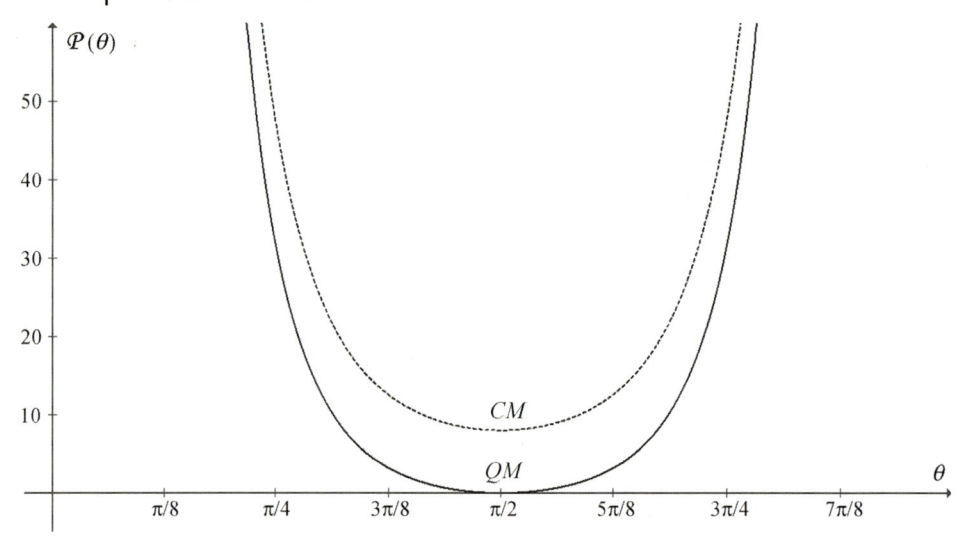

Figure 4.5 Coulomb scattering of two fermions in the same spin state. CM: according to classical predictions; QM: according to quantum mechanics.

not identical – rather, they are described by different complete sets of observables. For instance, if both have their spin up, their individual states are

$$|\Psi_1\rangle = |\uparrow, \mathbf{p}\rangle_1 = |\uparrow\rangle \otimes |\mathbf{p}\rangle_1, \quad |\Psi_2\rangle = |\uparrow, -\mathbf{p}\rangle_2 = |\uparrow\rangle \otimes |-\mathbf{p}\rangle_2. \quad (4.47)$$

In spherical coordinates, if momentum of the first particle is $\mathbf{p}_1 = (p, \theta, \varphi)$, then momentum of the second particle is $\mathbf{p}_2 = (p, \pi - \theta, \varphi + \pi)$. Therefore, in this kind of experiment we do not really need two detectors. If we leave only one of them, say, D_1, and find one particle there, we know, without any recording, that the second particle will be in D_2. Even if we remove both detectors, there still remains one important piece of information: we know that the momentum of either particle is opposite to that of its partner. Actually, this is another case of quantum entanglement! Indeed, for each given pair of angles (θ, φ) we can, without specifying particle type, write the combined state of the whole system in the form

$$|\Psi\rangle = |\theta, \varphi\rangle = |\mathbf{p}\rangle_1 \otimes |-\mathbf{p}\rangle_2 \pm |\mathbf{p}\rangle_2 \otimes |-\mathbf{p}\rangle_1 \quad (4.48)$$

(we drop the normalizing factor together with common spin states, which can be factored out). We must take "+" for bosons and "−" for fermions. But in either case, it is immediately seen that we have an entangled superposition similar to that discussed in Section 3.7. The only difference is that entanglement is with respect to momentum states rather than with respect to spin states.

Consider now the situation when both particles scatter into *the same momentum state*. This may happen, for example, when they both approach some other particle from different directions but get deflected in the same direction (Figure 4.6). This is a thought experiment that could in principle be performed with a specially selected target particle. Alternatively, it could be performed using two independent target

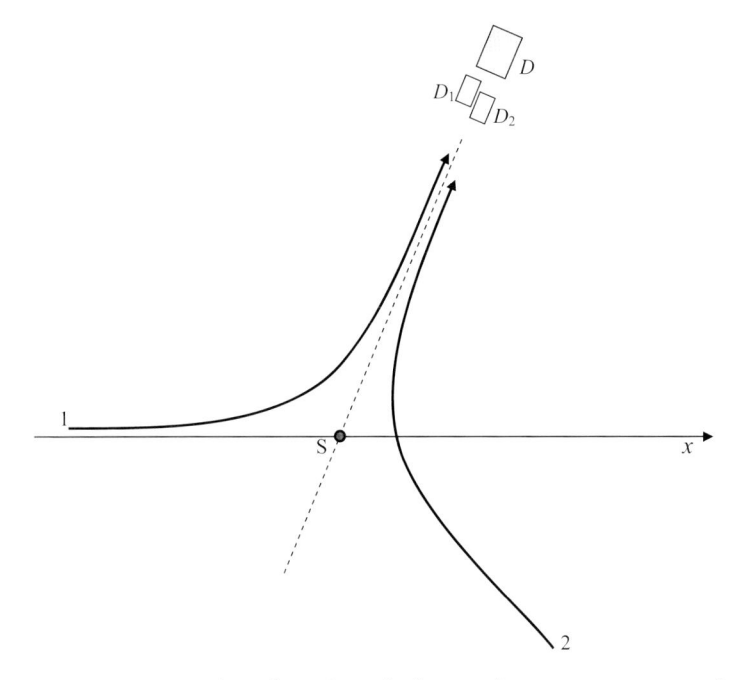

Figure 4.6 Scattering of two identical particles from an object S into one common final state.

particles [3]. In all such cases, the *complete set* specifying the final state would be the same for both scattered particles. Suppose that there are two detectors D_1 and D_2 in this path so close to each other that either particle can get into either detector. Let $\langle D_1|1\rangle$ be the amplitude of particle 1 getting into D_1, and $\langle D_2|2\rangle$ be the amplitude of particle 2 getting into D_2. Then the amplitude and the corresponding probability of both these events happening in one occasion are

$$\mathcal{F}_{1,2} = \langle D_2|2\rangle\langle D_1|1\rangle, \qquad \mathcal{P}_{1,2} = |\mathcal{F}_{1,2}|^2. \tag{4.49}$$

The amplitude and the corresponding probability of combined recording with the particles swapped are

$$\mathcal{F}_{2,1} = \langle D_2|1\rangle\langle D_1|2\rangle, \qquad \mathcal{P}_{2,1} = |\mathcal{F}_{2,1}|^2. \tag{4.50}$$

If the particles are distinguishable, the probability of concurrent clicks without specifying which particle is in which detector is

$$\tilde{\mathcal{P}} = \mathcal{P}_{1,2} + \mathcal{P}_{2,1}. \tag{4.51}$$

In CM, (4.51) is the only possible outcome. In QM, for identical particles in the same state we will have

$$\tilde{\mathcal{F}} = \mathcal{F}_{1,2} \pm \mathcal{F}_{2,1}, \qquad \tilde{\mathcal{P}} = |\mathcal{F}_{1,2} \pm \mathcal{F}_{2,1}|^2, \tag{4.52}$$

with "+" for bosons and "−" for fermions.

Let us now replace D_1 and D_2 with one detector D as shown in Figure 4.6. Then we will have from (4.49) and (4.50)

$$\mathcal{F}_{1,2} = \langle D|2\rangle\langle D|1\rangle = \langle D|1\rangle\langle D|2\rangle = \mathcal{F}_{2,1} \equiv \mathcal{F}, \qquad \mathcal{P}_{1,2} = \mathcal{P}_{2,1} = |\mathcal{F}|^2 \equiv \mathcal{P}, \tag{4.53}$$

and expression (4.51) for probability of capturing both particles in one state will give $\tilde{\mathcal{P}} = 2\mathcal{P}$. But from (4.52) there follows

$$\tilde{\mathcal{P}} = \begin{cases} 4\mathcal{P}, & \text{for bosons,} \\ 0, & \text{for fermions.} \end{cases} \tag{4.54}$$

The probability of finding two identical bosons in the same state is twice that of the classical particles. Somehow nature favors the bosons being in the same state. Their fundamental feature is the tendency to flock together. A spectacular demonstration of this tendency in collisions was observed in experiments with photons prepared in identical states [5,6]. Instead of a third particle shown in Figure 4.6, the researchers used a macroscopic scatterer – the beam splitter (BS) (Figure 4.7). If both photons are identical and come simultaneously, then out of the four possible outputs only two are realized in which both photons recede in the same direction. This phenomenon is known as the *HOM effect* (after Hong, Ou, and Mandel, who first observed it in 1987).

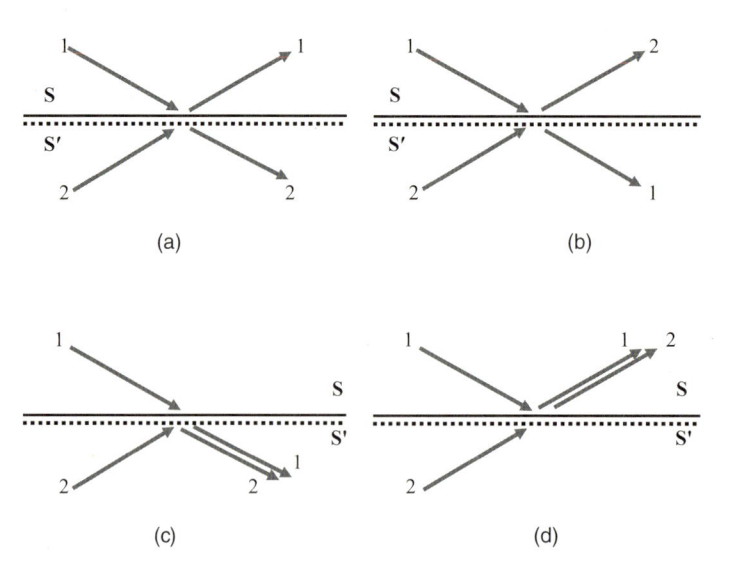

Figure 4.7 Schematic of light interaction with a BS considered as a two-sided surface. S and S' are opposite sides of the splitting surface of BS. If the BS is symmetric and neutral (not sensitive to photon polarization), there are four equally probable outcomes: (a) both photons reflected; (b) both transmitted; (c) photon 1 transmitted and photon 2 reflected; (d) photon 1 reflected and photon 2 transmitted. Each possible output is observed 25% of the time. However, if the polarization and frequency states of both incident photons are the same, *and* they come simultaneously, they always wind up in states (c) or (d), and never in states (a) or (b).

In contrast, the probability of finding two identical fermions in the same state is zero (the *Pauli exclusion principle*). For instance, two fermions in the same state never scatter in one direction.

We will say more about both types of particles in Chapters 21 and 22. But already at this stage we can see how the QM postulates cause dramatic departure from familiar classical behavior.

4.6
Path Integrals

The Dirac notation is frequently used for describing propagation of a particle between two locations in space. Such propagation can be described as the amplitude of finding the particle at a point $\mathbf{r}_2$ some time after its passing a point $\mathbf{r}_1$. The corresponding amplitude is conventionally denoted as $\langle \mathbf{r}_2 | \mathbf{r}_1 \rangle$, but *it has nothing to do with the inner product* $\langle \mathbf{r}_2 | \mathbf{r}_1 \rangle$ of eigenstates $|\mathbf{r}_1\rangle$ and $|\mathbf{r}_2\rangle$ as in (4.14). The latter refers to eigenvectors of $\hat{\mathbf{r}}$ and is always zero for $|\mathbf{r}_1\rangle \neq |\mathbf{r}_2\rangle$. The reader must distinguish between the two cases from the context.

For a free particle with definite energy, the transition amplitude is [3]

$$\langle \mathbf{r}_2 | \mathbf{r}_1 \rangle = \Phi \frac{e^{i k \cdot r_{12}}}{r_{12}}, \quad r_{12} \equiv |\mathbf{r}_2 - \mathbf{r}_1|, \tag{4.55}$$

where Φ is the numerical factor, and $k = p/\hbar = \sqrt{2\mu E}/\hbar$. The expression "with definite energy" already implies that $|\mathbf{r}_1\rangle$ in (4.55) cannot be regarded as a position eigenstate, which has energy indeterminacy $\Delta E = \infty$. Rather, it plays the role of a virtual stationary source of the secondary wave, as in the Huygens principle. This wave, unlike the plane wave (2.32), has the form of a *spherical wave* diverging from the "source" at $\mathbf{r}_1$.

In many problems, we have situations in which a particle on its way from s to q passes through one (or more) known location(s). Consider the simplest case, with only one such intermediate location (call it location 1). For instance, there may be an opaque screen between s and q with only one narrow slit in it, and we know both the position of the screen and the position of the slit on the screen, so the location 1 is exactly known. In this case, it is known with certainty that the particle does show up in this location before it lands at q, and therefore the amplitude of particle's propagation between s and q is the product of two amplitudes – one from s to 1 and the other from 1 to q:

$$\langle q | s \rangle_{\text{via 1}} = \langle q | 1 \rangle \langle 1 | s \rangle. \tag{4.56}$$

If the probability $\mathcal{P}$ of passing through the slit is less than 1 (e.g., due to particle's reflection from the edges), we take it into account by introducing the corresponding amplitude f of passing, $|f| = \mathcal{P}^{1/2}$, and (4.56) generalizes to

$$\langle q | s \rangle_{\text{via 1}} = \langle q | 1 \rangle f \langle 1 | s \rangle. \tag{4.57}$$

Expressions (4.56) and (4.57), like $\langle q | s \rangle$, are also written and should be read from right to left.

The next composition rule handles the cases when the particle can reach q from s by more than one way (e.g., the double-slit experiment). If the experimental setup allows one to actually see the particle in one of the ways, then the resulting probability of reaching q is just the sum of the individual probabilities. If the setup does not provide such a possibility (so that it is in principle impossible to distinguish between these ways), then the particle proceeds along all the available paths at once. In this case, we have a superposition of the corresponding amplitudes. For instance, if there is a screen between s and q with N slits in it (e.g., diffraction grating), so that there are N different but nondistinguishable paths from s to q, then

$$\langle q|s\rangle = \langle q|s\rangle_{\text{via }1} + \langle q|s\rangle_{\text{via }2} + \cdots + \langle q|s\rangle_{\text{via }N} = \sum_{n=1}^{N} \langle q|s\rangle_{\text{via }n}, \qquad (4.58)$$

or, in view of (4.56),

$$\langle q|s\rangle = \sum_{n=1}^{N} \langle q|n\rangle\langle n|s\rangle. \qquad (4.59)$$

Now we can consider more complicated cases [3,4]. Suppose that we have more than one screen between s and q, each containing a different number of slits (the succession of different diffraction gratings between s and q), and we want to find the amplitude of particle's arrival at point q on the observation screen Q (Figure 4.8). In this case, we have many paths, but the solution, albeit more involved technically, uses the same principle: the amplitude for each chosen path is the product of the corresponding consecutive amplitudes along this path; the *total* amplitude of particle's arrival at q is the sum of all such products.

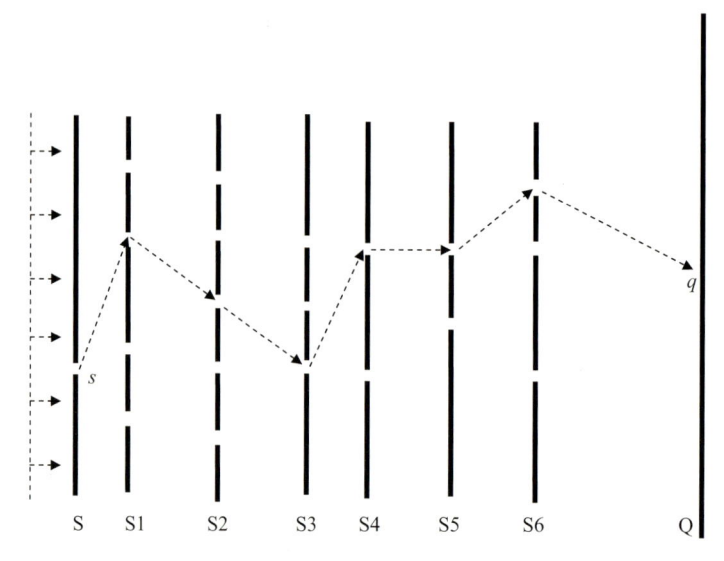

Figure 4.8 The multiple-screen-slit experiment with many virtual paths between s and q. One of the paths is shown by the dashed line. All available paths contribute (with different phase factors) to the net amplitude of arrival at q. S and Q are the initial and final screens.

Start with an expression for a *path amplitude*. Let M be the number of perforated screens between the initial (s) and final (q) points. Denote the first screen as S_1, the second one as S_2, and so on up to S_M. The slits in the first screen can be numbered as n_1. If the total number of slits in this screen is N_1, then n_1 ranges from 1 to N_1. Similarly, we denote as N_2 the total number of slits in screen 2, and number the individual slits in this screen as n_2. Then n_2 ranges from 1 to N_2. Finally, the number of slits in the last screen S_M will be denoted as N_M, and the individual slits in this screen as n_M, so that n_M ranges from 1 to N_M. All these conditions can be written as

$$n_1 = (1,2,3,\ldots,N_1),\ n_2 = (1,2,3,\ldots,N_2),\ \ldots,\ n_M = (1,2,3,\ldots,N_M). \tag{4.60}$$

Now write the expression for the amplitude of arrival at q by one of the paths. The particle from s can reach q by passing through slit number n_1 in the first screen, slit number n_2 in the second screen, and so on. Using rule (4.56), we have for this amplitude

$$\begin{aligned}\langle q|s\rangle_{\text{via } n_1,n_2,\ldots,n_M} &= \langle q|n_M\rangle\langle n_M|n_{M-1}\rangle\cdots\langle n_3|n_2\rangle\langle n_2|n_1\rangle\langle n_1|s\rangle \\ &= \langle q|n_M\rangle\left\{\prod_{j=1}^{M-1}\langle n_{j+1}|n_j\rangle\right\}\langle n_1|s\rangle.\end{aligned} \tag{4.61}$$

Next, we need to sum up all such products for different paths, that is, apply the rule for the net amplitude of an event that can occur in many indistinguishable ways. The total number N of different indistinguishable paths from s to q is $N = N_1 N_2 \cdots N_M = \prod_{j=1}^{M} N_j$. For instance, if we have five intermediate screens between s and q with the respective numbers of slits $N_1 = 13$, $N_2 = 7$, $N_3 = 6$, $N_4 = 8$, and $N_5 = 4$, the total number of possible paths between s and q will be the product $N = N_1 N_2 \cdots N_5 = \prod_{j=1}^{5} N_j = 17472$. So, generally it may be a huge number of terms. Summing (4.61) gives

$$\begin{aligned}\langle q|s\rangle &= \sum_{n_1=1}^{N_1}\sum_{n_2=1}^{N_2}\cdots\sum_{n_M=1}^{N_M}\langle q|s\rangle_{\text{via } n_1,n_2,\ldots,n_M} \\ &= \sum_{n_1=1}^{N_1}\sum_{n_2=1}^{N_2}\cdots\sum_{n_M=1}^{N_M}\langle q|n_M\rangle\left\{\prod_{j=1}^{M-1}\langle n_{j+1}|n_j\rangle\right\}\langle n_1|s\rangle.\end{aligned} \tag{4.62}$$

This expression can be considered as the embryo of Feynman's *path integrals*.

Box 4.1 Path integrals

The described procedure can be carried out in a slightly different way, considering each slit as a vertex on the particle's trajectory (a special case illustrating this is considered in Supplement to Chapter 4). We extend the above approach to trajectories with an arbitrary number of vertices and consider the limit when this number becomes infinite, thus bringing smooth paths into the picture [7-9].

The generalization is rather straightforward. We just reiterate the first step. Let $\mathbf{r}_0$ stand for position vector of P_0, and $\mathbf{r}_N$ for position vector of P. Then the amplitude (4.55) of transition from P_0 to P can be written as

$$\langle P|P_0\rangle = \langle \mathbf{r}_N|\mathbf{r}_0\rangle = \Phi \frac{e^{ikr_{0N}}}{r_{0N}}, \qquad (4.63)$$

where $r_{0N} = |\mathbf{r}_N - \mathbf{r}_0|$. Now insert $N-1$ additional points $\mathbf{r}_1, \mathbf{r}_2, \ldots, \mathbf{r}_j, \ldots, \mathbf{r}_{N-1}$ and consider them as the locations of the corresponding virtual events, each event being the passing of the particle through the respective point. The succession of these events generates a virtual trajectory. Any such trajectory is legitimate, albeit from the viewpoint of CM most of them would be completely crazy for a free particle. For the virtual trajectory determined as a set of given $\mathbf{r}_j$, we can think of transition $\mathbf{r}_0 \Rightarrow \mathbf{r}_N$ as a succession of N consecutive transitions $\mathbf{r}_0 \Rightarrow \mathbf{r}_1 \Rightarrow \cdots \Rightarrow \mathbf{r}_j \cdots \Rightarrow \mathbf{r}_N$. The corresponding amplitude is

$$\langle \mathbf{r}_N|\mathbf{r}_0\rangle_\alpha = \langle \mathbf{r}_N|\mathbf{r}_{N-1}\rangle \langle \mathbf{r}_{N-1}|\mathbf{r}_{N-2}\rangle \cdots \langle \mathbf{r}_{j+1}|\mathbf{r}_j\rangle \langle \mathbf{r}_j|\mathbf{r}_{j-1}\rangle \cdots \langle \mathbf{r}_2|\mathbf{r}_1\rangle \langle \mathbf{r}_1|\mathbf{r}_0\rangle. \qquad (4.64)$$

Here the subscript α labels a single path specified by the given set of $\mathbf{r}_j$, that is,

$$\alpha = (\mathbf{r}_1, \mathbf{r}_2, \ldots, \mathbf{r}_j, \ldots, \mathbf{r}_{N-1}). \qquad (4.65)$$

It can be considered as a vector in $\mathcal{H}$ (Problem 4.2.2). At $N \to \infty$ and $r_{j,j+1} = |\mathbf{r}_{j+1} - \mathbf{r}_j| \to 0$ for all j, the path becomes a continuous, albeit not necessarily differentiable, curve.

Applying (4.63) to each individual transition and taking their product, we obtain

$$\langle \mathbf{r}_N|\mathbf{r}_0\rangle_\alpha = \prod_{j=0}^{N-1} \Phi_j \frac{e^{ikr_{j,j+1}}}{r_{j,j+1}}, \qquad (4.66)$$

where $r_{j,j+1} = |\mathbf{r}_{j+1} - \mathbf{r}_j|$. For a continuous path, the product of exponents in (4.66) takes the form

$$\prod_{j=0}^{N-1} e^{ikr_{j,j+1}} \xrightarrow[r_{j,j+1} \to 0]{} e^{iS(\mathbf{r}_0, \mathbf{r}_N)}; \quad S(\mathbf{r}_0, \mathbf{r}_N) \equiv \int_{\mathbf{r}_0}^{\mathbf{r}_N} \mathbf{k} \cdot d\mathbf{r}. \qquad (4.67)$$

The total amplitude (4.63) is the superposition of amplitudes (4.66) with all possible α. Since parameter α is multidimensional and continuous, we have the multidimensional integral over α with $d\alpha \equiv \prod_j (d\mathbf{r}_j/r_{0N})$:

$$\langle \mathbf{r}_N|\mathbf{r}_0\rangle = \int \langle \mathbf{r}_N|\mathbf{r}_0\rangle_\alpha \, d\alpha = \int \prod_{j=0}^{N-1} \Phi_j \frac{e^{ikr_{j,j+1}}}{r_{j,j+1}} d\alpha. \qquad (4.68)$$

This expression determines the transition amplitude between two events in terms of all virtual paths connecting them. It is the essence of the theory of path integrals, which is one of the formulations of QM.

Summary: Expressions (4.63) and (4.64) describe different transitions. The former embraces all virtual paths; the latter specifies one actual path. By singling out such a path, we take a step toward the particle-like aspect of the process. But by making the path index α a variable, we take a step toward its wave aspect.

By making each possible path only a virtual entity with a certain amplitude attributed to it, we take the second step toward QM. By taking the sum of all these amplitudes, we pay full tribute to the wave aspect of the transition $P_0 \Rightarrow P$. All three steps together constitute integration over trajectories and encapsulate the essence of QM.

Problems

4.1 Describe qualitatively the basic requirements for the target particle in Figure 4.6 for it to be able to scatter two incoming neutrons in one common direction.

4.2 Find the number of available virtual paths connecting s and q in Figure 4.8.

4.3 What is the number of dimensions of the vector space spanned by α in Equation 4.64 if the corresponding trajectory is represented by (a) 1 vertex point; (b) 23 vertex points; (c) 10^{23} vertex points?

4.4 Consider the case in which all points $\mathbf{r}_0, \mathbf{r}_1, \ldots, \mathbf{r}_j, \ldots, \mathbf{r}_N$ in (4.64) are on the straight line $P_0 P$ and are equidistant with separation a, so that the net distance $P_0 P = R = N a$. Ignoring the normalizing factors, find the transition amplitude $\langle \mathbf{r}_N | \mathbf{r}_1 \rangle_\alpha$ for the process (4.64) and compare with (4.63). Explain the result.

(*Hint*: Consult Supplement to Chapter 4.)

References

1 Dirac, P. (1964) *The Principles of Quantum Mechanics*, Yeshiva University, New York.

2 von Neumann, J. (1955) *Mathematical Foundations of Quantum Mechanics*, Princeton University Press, Princeton, NJ.

3 Feynman, R., Leighton, R., and Sands, M. (1963–1964) *The Feynman Lectures on Physics*, vols. 1–3, Addison-Wesley, Reading, MA.

4 Aharonov, Y. and Rohrlich, V. (2005) *Quantum Paradoxes*, Wiley-VCH Verlag GmbH, Weinheim.

5 Hong, C.K., Ou, Z.Y., and Mandel, L. (1987) Measurement of subpicosecond time intervals between two photons by interference. *Phys. Rev. Lett.*, 59(18), 2044–2046.

6 Beugnon, J., Jones, M.P., Dingjan, J., Darquié, B., Messin, G., Browaeys, A., and Grangier, P. (2006) Quantum interference between two single photons emitted by independently trapped atoms. *Nature*, 440(7085), 779–782.

7 Feynman, R.P. and Hibbs, A.R. (1965) *Quantum Mechanics and Path Integrals*, McGraw-Hill, New York.

8 Goodman, J.W. (1968) Chapter 3, in *Introduction to Fourier Optics*, McGraw-Hill, San Francisco, CA.

9 Fikhtengol'ts, G.M. (1970) *Course of Differential and Integral Calculus* (in Russian), vol. 2, Nauka, Moscow.

5
Basic Features and Mathematical Structure of QM

5.1
Observables: the Domain of Classical and Quantum Mechanics

> Indeterminacy is the property of the very structure of matter, and location and momentum cannot even exist simultaneously as exactly measurable characteristics.
>
> *David Bohm*

The basic characteristics of the state of a classical particle are position **r** and momentum **p**. They determine its energy E and orbital angular momentum **L**. Thus, we have variables E, **r**, **p**, and **L**, describing the system. This summarizes the classical description.

In QM, we encounter new characteristics such as *spin* **S** and *parity* $\mathcal{P}$ (Section 5.4). In addition, the concept of EM potentials, considered subsidiary in CM, is much more important in QM (Chapter 16).

In CM, a complete description of the system specifies all relevant quantities at once. In QM, some combinations, for example, pairs (x, p_x) and (L_x, L_y), cannot be determined simultaneously. However, things may be different when these quantities form an expression; thus, exact values L_x and L_y cannot coexist, but L_x and $(L_x^2 + L_y^2 + L_z^2)$ can. Determining some quantities may depend on the situation: If the particle is free, definite $\boldsymbol{p}$ and E can coexist, but if there is a potential field, at least one component of **p** cannot be exactly determined even separately.

Various properties of physical objects are revealed in observations. We accordingly call them *observables*. This may look like just another fancy word for "quantities," but it actually has a deeper meaning, due to the fundamental role of observation in QM. Some interpretations of QM (e.g., one suggested by Bohm [1,2]) use *unobservable* properties – the so-called *hidden parameters* (or *hidden variables*), which were supposed to account for the probabilistic nature of quantum-mechanical predictions. The problem of hidden parameters is discussed in Chapters 23 and 24.

Since some pairs of observables cannot be exactly determined simultaneously (which, from the classical viewpoint, is necessary to fully describe a system), QM description is essentially different from the classical one: A *QM system generally needs fewer observables* for its complete description than would be needed in CM. A set of observables completely describing a system possesses the following properties:

1) All elements of the set can be accurately measured at one moment.
2) If this is done, no other independent observable can have a definite value in this state.

Such a set, ensuring the full description of a system, is called the *complete set*.

Box 5.1 Commutability

Quantities that can be known simultaneously are said to be *compatible*. Sometimes they are called *commutable* because their respective operators *commute* with each other. Otherwise, the corresponding operators do not commute; for instance, as we know from Section 3.5, operators x and $\hat{p}_x$ are noncommutable. In a complete set of observables, all their respective operators commute with each other, and adding any new observable that is not a combination of those already in the set will destroy this mutual commutability.

Box 5.2 "Complete set" of observables

Note the word *generally* in the phrase "a QM system generally needs fewer observables." In many cases, we can use more observables than in a complete set, but this will not yield more information. For example, if we measure x with infinite precision, then p_x will be totally unknown and will not show up in our minimal set. But measuring x less precisely would allow us to get some information about p_x at the cost of information about x. That would be another state with both x and p_x necessary for description, but both known only to a certain precision. Although nature restricts the amount of information about a system, it allows us to select the relevant aspects of the system's behavior; we may trade off information about position for information about momentum and vice versa.

Thus, "complete set" does not necessarily mean "the most informative"! Another set can be just as informative; however, it will contain more elements. The number of its observables can be reduced with information trade-off but without information loss.

Physically, "complete set" is a set of all simultaneously measurable characteristics of the system, neither of which is a combination of others, if we measure all quantities with infinite precision.

5.2
Quantum-Mechanical Operators

Now we can discuss the origin of operators in a more general way, which illustrates the operational structure of QM. Consider a state $\Psi(\mathbf{r})$. As we know, $|\Psi(\mathbf{r})|^2 = \rho(\mathbf{r}) = d\mathcal{P}(\mathbf{r})/d\mathbf{r}$ is the probability density of finding the particle at point $\mathbf{r}$. Due to the probabilistic nature of quantum states, the same measurement in the same state could find it at another location $\mathbf{r}'$, with probability $d\mathcal{P}(\mathbf{r}') = \rho(\mathbf{r}')d\mathbf{r} = |\Psi(\mathbf{r}')|^2 \, d\mathbf{r}$. Because of the variability of the measurement outcomes for the same observable, we characterize the state by the *mean* or *expectation value* (the average) of this observable. Unlike an individual outcome, which is generally unpredictable, the *average* of all outcomes is a predictable characteristic of the state. Let us derive a general rule determining it.

To find the mean value of an observable L in a state Ψ, we represent Ψ as a superposition of eigenstates of operator $\hat{L}(\mathbf{r})$:

$$\Psi(\mathbf{r}) = \sum_j a_j \psi_j(\mathbf{r}). \tag{5.1}$$

Here, $\mathbf{r}$ indicates representation (in the considered case – position), and the set ψ_j gives the basis. The L-measurement "throws the dice," selects one of the eigenstates ψ_j with its eigenvalue L_j, and completely obliterates the memory of all others.[1] We repeat the measurement, performing it on an *ensemble* of N identical systems all in the same state (the *pure ensemble*). In the limit $N \to \infty$, QM predicts the average precisely, even though individual values can be predicted only probabilistically. We denote this average as $\bar{L}$ or $\langle L \rangle$.

To find $\bar{L}$, we multiply each L_j by the number of times it occurs, and divide the sum of products by N. This is just the sum of products of each eigenvalue by its respective probability:

$$\bar{L} = \frac{N_1 L_1 + N_2 L_2 + \cdots + N_n L_n}{N} = \sum_j \mathcal{P}_j L_j = \sum_j |a_j|^2 L_j = \sum_j a_j^* L_j a_j,$$

$$\mathcal{P}_j \equiv \frac{N_j}{N} = |a_j|^2. \tag{5.2}$$

If L is a continuous variable, we have $L_j \to L$, $a_j \equiv a(L_j) \to a(L)$, and (5.2) takes the form

$$\bar{L} = \int L \, d\mathcal{P}(L) = \int L|a(L)|^2 \, dL = \int a^*(L) L a(L) dL. \tag{5.3}$$

In this approach, variable $\mathbf{r}$, as well as state $\Psi(\mathbf{r})$, is absent in both Equations 5.2 and 5.3 for $\bar{L}$. Any information about the original representation $\mathbf{r}$ is lost. Instead, both

---

1) The analogy with a dice is not very accurate, because unlike the dice case, the quantum probabilities are generally different for different outcomes, so the outcomes are not completely random. Nevertheless, we use the analogy to emphasize the unpredictability of a specific outcome in both cases.

expressions include explicit functions of L. This happens because in both expressions, the discrete set $a_j = a(L_j)$ or continuous set $a(L)$ completely determines $\Psi(\mathbf{r})$; we say that $a(L)$ gives the same state in the *L-representation*. Accordingly, $a(L)$ can be considered as the initial input instead of $\Psi(\mathbf{r})$. Expressing all relevant variables in terms of observable L represented by operator $\hat{L}$ is the most natural and simple way to find $\bar{L}$.

The algorithm works for any physically possible $\Psi(\mathbf{r})$, always returning the correct mean of a relevant observable for that $\Psi(\mathbf{r})$. Turning to the case when the observable in question is $\mathbf{r}$, we see that $\Psi(\mathbf{r})$ was from the very beginning given in the **r**-representation, so $\bar{\mathbf{r}}$ is automatically obtained by the same rule:

$$\bar{\mathbf{r}} = \int \mathbf{r}\rho(\mathbf{r})d\mathbf{r} = \int \mathbf{r}|\Psi(\mathbf{r})|^2 d\mathbf{r} = \int \Psi^*(\mathbf{r})\mathbf{r}\Psi(\mathbf{r})d\mathbf{r}. \tag{5.4}$$

But we can look for the mean of L in state $\Psi(\mathbf{r})$ *without switching* to the *L*-representation. Then the result (5.3) for $\bar{L}$ will be expressed in terms of $\Psi(\mathbf{r})$ and $\hat{L}(\mathbf{r})$, rather than $a(L)$ and L:

$$\bar{L} = \int \Psi^*(\mathbf{r})\hat{L}(\mathbf{r})\Psi(\mathbf{r})d\mathbf{r}. \tag{5.5}$$

Here, $\hat{L}(\mathbf{r})$ is the *L*-operator in **r**-representation. The diligent reader can try to derive this formula from (5.2) or (5.3) by expressing $a(L_j)$ in terms of $\Psi(\mathbf{r})$ and using the properties of the orthonormal set $\psi_j(\mathbf{r})$. For $\hat{L}(\mathbf{r}) \to \hat{p}(\mathbf{r})$, expression (5.5) gives the *mean momentum* in state $\Psi(\mathbf{r})$, and $\hat{p}(\mathbf{r})$ in the integrand is precisely the operator (3.24) (see Appendix C). The fact that using this operator in (5.5) determines the mean momentum is another reason to name it the *momentum operator*. This confirms expression (3.24) from a quite different perspective.

The operators representing physical observables must satisfy two important requirements.

First, they must be linear: Applying $\hat{L}$ to a superposition (5.1) must return the superposition

$$\hat{L}\Psi = \hat{L}\sum_j a_j\Psi_j = \sum_j a_j\hat{L}\Psi_j. \tag{5.6}$$

The corollary of this definition is $\hat{L}(a\Psi) = a\hat{L}\Psi$. Second, they must be self-conjugate, which means the following. As we have seen in Section 3.4, an expression for $\hat{L}$ may be complex. But in order for it to represent a *physical observable*, the whole expression on the right of (5.5) must be real. Indeed, $\bar{L}$ is the measurement results averaged over the ensemble. These results are real numbers.[2] Hence, all of the above-discussed mean values must be equal to their complex conjugates. This imposes a specific requirement on the operators representing observables: The

---

[2] This does not cancel complex quantities. Recall such characteristics as, for example, complex refraction index or complex impedance. But *they cannot be obtained in one measurement*. We must perform two measurements to find the real and the imaginary parts of a complex number, but we cannot imagine *one* measurement returning a complex number. An observable that can be defined in a single measurement is real.

complex conjugation of expression on the right in (5.5) must not change the expression. More generally, in any representation q, and for any two state functions $\Phi(q)$ and $\Psi(q)$, operator $\hat{L}(q)$ must satisfy the condition

$$\int \Phi^*(q)\hat{L}(q)\Psi(q)dq = \int \Psi(q)\hat{L}^*(q)\Phi^*(q)dq. \tag{5.7}$$

In the special case when $\Phi(q) = \Psi(q)$, this reduces to expression (5.5) for $\bar{L}$ in state Ψ. Therefore, the integral (5.7) must be a real number in this case. Taking the complex conjugate when $\Phi(q) = \Psi(q)$ is equivalent to replacing $\hat{L} \to \hat{L}^*$ and exchanging positions of Ψ^* and Ψ (complex conjugation for operators is defined in the same way as for complex numbers). Those $\hat{L}$ for which the integral (5.5) is insensitive to this operation are candidates to represent QM observables. The more strict selection requires the same insensitivity even for two arbitrary functions $\Phi(q)$ and $\Psi(q)$ as in (5.7). Operators satisfying this condition are called *self-adjoint, self-conjugate, Hermitian conjugate,* or just *Hermitian*. Only Hermitian operators are admitted to the privileged class of the observables' representatives.

Reversing the argument and applying (5.7) to the case $q \to \mathbf{r}$, $\Phi = \Psi$, one can easily check that the right-hand side of (5.5) is, indeed, real, if $\hat{L}$ is Hermitian. One can also prove that operator (3.24) is Hermitian in the class of the square-integrable functions. If we just take $\hat{p}(x) \equiv \hbar(\partial/\partial x)$ for this operator (which, on the face of it, looks like a more natural choice to represent real momentum), it will *not* be Hermitian, its average will not be real, and it therefore will not represent a physical observable. In order for some operators to represent real quantities, they must be imaginary! The reason for this is that the corresponding state functions are themselves complex, as is, for instance, the de Broglie wave.[3]

Thus, we have an algorithm (5.5) that takes a state function and returns the average $\bar{L}$ of observable L in this state. Naturally, the algorithm involves the operator $\hat{L}$ of this observable. If $\Psi(\mathbf{r})$ is itself an eigenstate $\psi_k(q)$ of $\hat{L}$, then getting $\bar{L}$ from (5.5) is quite straightforward: We simply take the eigenvalue L_k for it,

$$\hat{L}\psi_k(q) = L_k \psi_k(q), \qquad \bar{L} = L_k. \tag{5.8}$$

The operator just "pulls out" the expectation value (which equals L_k in this case) from an eigenstate. If $\hat{L}$ is applied to a superposition (5.1), the result, in view of (5.6) and (5.8), will be

$$\hat{L}\Psi(q) = \hat{L}\sum_j a_j \psi_j(q) = \sum_j a_j \hat{L}\psi_j(q) = \sum_j a_j L_j \psi_j(q). \tag{5.9}$$

Taking the inner product of $\Psi(q)$ and $\Phi(q) \equiv \hat{L}\Psi(q)$ (in that order) gives

$$\langle \Psi | \hat{L} \Psi \rangle = \int \Psi^*(q)\hat{L}(q)\Psi(q)\, dq = \int \left(\sum_j a_j \psi_j(q) \right)^* \left(\sum_k a_k L_k \psi_k(q) \right) dq$$
$$= \sum_{j,k} L_k a_j^* a_k \int \psi_j^*(q)\psi_k(q)dq = \sum_j L_j |a_j|^2 = \bar{L}. \tag{5.10}$$

---

3) This conclusion is not universal. For instance, the operator of an observable, when taken *in its own representation*, is just that observable.

(We have used here Equation 5.8 and orthonormality of eigenstates.) As defined in Section 3.4, the operator acts on the function to the right; the result is *then* multiplied by the function on the left. Thus, working "backward" and using linearity of $\hat{L}$ (5.6), we recover Equation 5.5 for $\overline{L}$.

Let us rewrite all these results in Dirac's notation:

$$|\Psi\rangle = \sum_j a_j |\psi_j\rangle, \qquad \hat{L}|\psi_k\rangle = L_k |\psi_k\rangle, \tag{5.11}$$

$$\hat{L}|\Psi\rangle = \hat{L}\sum_k a_k |\psi_k\rangle = \sum_k a_k \hat{L}|\psi_k\rangle = \sum_k a_k L_k |\psi_k\rangle, \tag{5.12}$$

$$\langle \Psi | \hat{L} \Psi \rangle = \left(\sum_j a_j^* \langle \psi_j | \right) \left(\sum_k a_k L_k |\psi_k\rangle \right) = \sum_{j,k} L_k a_j^* a_k \langle \psi_j | \psi_k \rangle = \sum_j L_j |a_j|^2$$
$$= \overline{L} \tag{5.13}$$

Here, variable q is conveniently suppressed, since the result depends only on state, *not* on its representation. Also note that although it is okay to write $\overline{L} = \langle \Psi | \hat{L} \Psi \rangle$, it is the accepted practice to put an additional vertical bar after the operator, writing the whole expression as

$$\overline{L} = \langle \Psi | \hat{L} | \Psi \rangle. \tag{5.14}$$

This has exactly the same meaning as (5.5) and (5.13); the operator is assumed to act on the ket vector to the right, and *then* the result is multiplied by the bra vector on the left.

Now we can summarize some basic properties of Hermitian operators.

1) They are all defined to produce the physically observed eigenvalues when acting on the corresponding eigenfunctions by rule (5.8).
2) The same definition guarantees that they accurately predict expectation values of their respective observables by rules (5.5) and (5.14).

5.3
Algebra of Operators

Here, we consider the question of how to deal with a combination of operators.

A combination of operators can be regarded as a single operator. When a sum of several operators acts on a function, the result is defined as the sum of results returned by the individual operators:

$$(\hat{A} + \hat{B} + \hat{C})\Psi = \hat{A}\Psi + \hat{B}\Psi + \hat{C}\Psi. \tag{5.15}$$

When a product of several operators acts on a function, the result is defined as follows: The operator nearest to the function acts on it, the result is fed to the next nearest operator, and so on, until the last operator returns its result:

$$\hat{A}\hat{B}\hat{C}\Psi = \hat{A}(\hat{B}\hat{C}\Psi) = \hat{A}(\hat{B}(\hat{C}\Psi)). \tag{5.16}$$

We can write powers of operators as $\hat{A}^2 = \hat{A}\hat{A}$, $\hat{A}^3 = \hat{A}\hat{A}\hat{A}$, and so on. With these rules, we can add, subtract, or multiply operators as usual, except that whenever multiplication is involved, the original order of multipliers must always be preserved. But, as is clear from the definition, for any given product of operators we can put parentheses anywhere, for example, $\hat{A}\hat{B}\hat{C} = \hat{A}(\hat{B}\hat{C}) = (\hat{A}\hat{B})\hat{C}$.

Since the product of two operators may depend on their order, the important characteristic of the given pair $\hat{A}$, $\hat{B}$ is the *difference* between $\hat{A}\hat{B}$ and $\hat{B}\hat{A}$. We call this difference the *commutator* and define it as

$$[\hat{A}, \hat{B}] \equiv \hat{A}\hat{B} - \hat{B}\hat{A}. \tag{5.17}$$

We say that $\hat{A}$ and $\hat{B}$ *commute* if $[\hat{A}, \hat{B}] = 0$, that is, their product does not depend on the order. Otherwise, the two operators are *noncommuting*. As we have seen from examples in Section 3.5, their respective observables are incompatible.

Note also that, as we mentioned in the previous section, a number can be considered as an operator whose action is multiplying a function by that number. It is easy to see that any *linear* operator commutes with any constant a. The word *linear* is necessary here; for instance, an operator $\hat{Q}$ that squares a function will not commute with a: $a\hat{Q}\Psi = a\Psi^2$ is not the same as $\hat{Q}(a\Psi) = (a\Psi)^2$. In QM we deal only with linear operators, so from now on we will usually skip the word *linear*, as linearity is always implied.

Using these results, we can prove the theorem: *if the two operators commute, they have common eigenfunctions.* Consider commuting operators $\hat{L}$ and $\hat{M}$. Let Ψ_L be an eigenfunction of $\hat{L}$ belonging to an eigenvalue L, so that

$$\hat{L}\Psi_L = L\Psi_L. \tag{5.18}$$

Now consider the expression $\hat{L}\hat{M}\Psi_L$. Since $\hat{L}$ and $\hat{M}$ commute, we have $\hat{L}\hat{M}\Psi_L = \hat{M}\hat{L}\Psi_L$. In view of (5.18) this converts into

$$\hat{L}\hat{M}\Psi_L = \hat{M}\hat{L}\Psi_L = \hat{M}L\Psi_L = L\hat{M}\Psi_L. \tag{5.19}$$

Denoting $\hat{M}\Psi_L \equiv \Phi_L$, we see that $\hat{L}\Phi_L = L\Phi_L$. This means that Φ_L is an eigenfunction of $\hat{L}$ belonging to the same eigenvalue L. Therefore, it can differ from Ψ_L only by a numerical factor. Denoting this factor as M, we have

$$\Phi \equiv \hat{M}\Psi_L = M\Psi_L. \tag{5.20}$$

But this is an equation for an eigenfunction of $\hat{M}$. Thus, Ψ_L, apart from being an eigenfunction of $\hat{L}$ with an eigenvalue L, turns out to be also an eigenfunction of $\hat{M}$ with eigenvalue M. This proves the theorem as well as compatibility of observables L and M.

Box 5.3 Functions of operators

We write a function of an operator just like the function of a number, for example, $\sqrt{\hat{A}}$, $2/(3\hat{A}^3 + 5)$, or $e^{\hat{A}}$. The meaning of such a function, however, is less intuitive. For example, if the operator takes the derivative of a function, it is not easy to explain what a square root of the operation of differentiation is! The answer is that we *define* the function of an operator by its power expansion around an arbitrary point, as we would do with the similar function of a variable. For example, we can write for a function $\sqrt{x}$

$$\sqrt{x} = 1 + \frac{1}{2}(x-1) - \frac{1}{2!}\frac{(x-1)^2}{4} + \cdots \quad \text{(Taylor expansion around } x = 1),$$

or for a function e^x

$$e^x = 1 + x + \frac{1}{2!}x^2 + \cdots \quad \text{(Taylor expansion around } x = 0), \text{ and so on.}$$

Similarly, we treat a function of an operator as if the operator in this function were a variable. Then the square root of an operator means

$$\sqrt{\hat{A}} = 1 + \frac{1}{2}(\hat{A}-1) - \frac{1}{2!}\frac{(\hat{A}-1)^2}{4} + \cdots. \tag{5.21}$$

The exponential function of an operator means

$$e^{\hat{A}} = 1 + \hat{A} + \frac{1}{2!}\hat{A}^2 + \cdots, \tag{5.22}$$

and so on. Any analytic function can be represented in terms of powers of the argument, and since power of an operator has been defined as a product of operators, *we can always write the meaningful expression for a function of any operator* if the corresponding function of a variable is analytic.

Based on this result, one can show that an operator always commutes with any analytic function of itself:

$$[\hat{A}, f(\hat{A})] = 0. \tag{5.23}$$

It is easy to establish the following properties of operators:

$$[\hat{A}, c\hat{B}] = [c\hat{A}, \hat{B}] = c[\hat{A}, \hat{B}], \quad [\hat{A}, \hat{B}] = -[\hat{B}, \hat{A}], \quad [\hat{A}+\hat{B}, \hat{C}] = [\hat{A}, \hat{C}] + [\hat{B}, \hat{C}],$$
$$[\hat{A}, \hat{B}+\hat{C}] = [\hat{A}, \hat{B}] + [\hat{A}, \hat{C}], \quad [\hat{A}\hat{B}, \hat{C}] = \hat{A}[\hat{B}, \hat{C}] + [\hat{A}, \hat{C}]\hat{B}, \quad [\hat{A}, \hat{B}\hat{C}] = \hat{B}[\hat{A}, \hat{C}] + [\hat{A}, \hat{B}]\hat{C}.$$
$$\tag{5.24}$$

5.4
Eigenvalues and Eigenstates

Imagine that you are a statistician studying the population of a certain country, or a coach of a basketball team. In the first case, you may be interested not only in the average longevity of the population but also in the individual longevities' spread around the average. In the second case, you will be interested not only in the average height of your players but, no less important, in its distribution among the players.

Similarly, if we measure an observable L in a system in state $\Psi(\mathbf{r})$, we may be interested not only in the average $\bar{L}$ but also in how the individual results are spread around $\bar{L}$. In all such cases, we are dealing with the individual deviations

$$\Delta L \equiv L - \bar{L} \tag{5.25}$$

and then considering their distribution. As we found in Section 3.5, the mean deviation $\overline{\Delta L} \equiv 0$ gives no information. The measure of the spread of L around $\bar{L}$ is given by the *mean-square deviation* or *variance* $\overline{(\Delta L)^2}$

$$var\{L\} \equiv \overline{(\Delta L)^2} \equiv \overline{(L - \bar{L})^2} = \overline{L^2} - (\bar{L})^2. \tag{5.26}$$

This generalizes (3.39) and (3.40) for an arbitrary observable. The (positive) square root of variance gives the *root-mean-square or standard deviation*:

$$\widehat{\Delta L} \equiv \sqrt{\overline{(\Delta L)^2}} = \sqrt{\overline{(L - \bar{L})^2}}. \tag{5.27}$$

As in (3.39), we use the arc sign " $\widehat{}$ " here to distinguish the *standard deviation* from an *individual deviation*. The standard deviation is a statistical characteristic of an *ensemble*, whereas a single deviation is the outcome of a *single measurement*. However, farther in the text we drop this sign – only to be consistent with the universally used notation. So there is always the risk of confusing the standard deviation with an individual deviation, and the reader has to get the true meaning of the symbol from the context.

Just as L is represented by its operator $\hat{L}$, so is an individual deviation ΔL, as well as its square. Their respective operators are

$$\Delta \hat{L} \equiv \hat{L} - \bar{L} \quad \text{and} \quad var\{L\} \equiv (\Delta \hat{L})^2 \equiv (\hat{L} - \bar{L})^2. \tag{5.28}$$

We leave it to the reader to show that for $\hat{L}$ Hermitian, $\Delta \hat{L}$ and $var\{L\}$ are also Hermitian.

By rule (5.5) we have

$$\overline{(\Delta L)^2} = \int \Psi^*(q)(\Delta \hat{L})^2 \Psi(q) dq. \tag{5.29}$$

It is important to distinguish between the classical and QM deviation. In classical physics, the deviation originates from natural experimental errors in a measurement. Such errors can be made much smaller than the measured value, so that standard deviation ΔL may be negligible.

The situation is fundamentally different in QM. Generally, QM variance cannot *in principle* be reduced below a certain level characteristic for a given state, even under ideal conditions with all experimental errors eliminated so that classical variance would be zero. There are, however, *special* states for any given observable, in which it has a sharply defined value and accordingly a zero variance, $\overline{(\Delta L)^2} = 0$. Since operator $\Delta \hat{L}$ is Hermitian, this condition leads to

$$\int \Psi^*(q) \Delta \hat{L}(\Delta \hat{L} \Psi(q)) dq = \int (\Delta \hat{L} \Psi(q))(\Delta \hat{L}^* \Psi^*(q)) dq = \int |\Delta \hat{L} \Psi(q)|^2 dq = 0. \tag{5.30}$$

It follows that in these states $\Delta \hat{L}\Psi(q) = 0$ or, in view of (5.28) and the fact that $\bar{L} = L$ when L is sharply defined,

$$\hat{L}\Psi_L(q) = L\Psi_L(q). \tag{5.31}$$

Here, we have labeled the corresponding state as $\Psi_L(q)$ to emphasize that it is a state with a definite value of L. We see that $\Psi_L(q)$ is an eigenfunction of $\hat{L}$, and L is its eigenvalue. If $q \to \mathbf{r}$, then $\hat{L}(q) \to \hat{L}(\mathbf{r})$ may be, like in (3.24), a differential operator acting on $\Psi(\mathbf{r})$, and $\Psi(\mathbf{r})$ is in this case a solution of a linear differential equation.

Now we will focus on possible representations of some operators and their eigenstates.

Let us start with eigenfunctions of the $\hat{\mathbf{r}}$-operator. Evidently, this operator in $\mathbf{r}$-representation is just $\mathbf{r}$. It is an example of the general (and somewhat tautological) statement: *an operator of any observable in its own representation is the observable itself*. Therefore, for position operator in position representation we have $\hat{\mathbf{r}} = \mathbf{r}$, and its action on any function of position is just multiplying it by $\mathbf{r}$, that is, $\hat{\mathbf{r}}\Psi(\mathbf{r}) = \mathbf{r}\Psi(\mathbf{r})$. But does this mean that *any* $\Psi(\mathbf{r})$ is an eigenfunction of $\hat{\mathbf{r}}$? An obvious (but wrong) answer would be: "This is obvious!" The argument is that if $\hat{\mathbf{r}} = \mathbf{r}$, then

$$\hat{\mathbf{r}}\Psi(\mathbf{r}) \equiv \mathbf{r}\Psi(\mathbf{r}) = \mathbf{r}\Psi(\mathbf{r}). \tag{5.32}$$

Here, the eigenvalue equation becomes an identity, implying that any $\Psi(\mathbf{r})$ would be an eigenfunction. This solution is wrong because an identity cannot be an eigenvalue equation. The factor $\mathbf{r}$ on both sides is a variable, whereas in a true eigenvalue equation, $\mathbf{r}$ on the right must be a *specific eigenvalue*, not a variable. To distinguish between them, denote the eigenvalue as $\mathbf{r}'$. Then the correct eigenvalue equation is

$$\mathbf{r}\Psi(\mathbf{r}) = \mathbf{r}'\Psi(\mathbf{r}). \tag{5.33}$$

The only (properly normalized) function satisfying this equation is the delta function

$$\Psi_{\mathbf{r}'}(\mathbf{r}) = \delta(\mathbf{r} - \mathbf{r}'). \tag{5.34}$$

The same argument shows that the *momentum* operator in its own representation is just multiplication by the momentum vector $\mathbf{p}$, that is, $\hat{\mathbf{p}}\Phi(\mathbf{p}) \equiv \mathbf{p}\Phi(\mathbf{p})$. An eigenstate belonging to an eigenvalue $\mathbf{p}'$ of this operator is

$$\Phi_{\mathbf{p}'}(\mathbf{p}) = \delta(\mathbf{p} - \mathbf{p}'). \tag{5.35}$$

This is just de Broglie wave in $\mathbf{p}$-representation! Generally, an eigenfunction $\psi_{q'}(q)$ of an operator in its own representation reduces to the δ-function, $\psi_{q'}(q) = \delta(q - q')$. If the observable's spectrum is discrete, then $\psi_{q_m}(q_n) = \delta_{mn}$.

Box 5.4 Mathematical symmetry and differential operators

We can take a look at the material of this and the previous sections about operators from a different angle, showing how the mathematical symmetry of an expression can reveal another dimension in the observable–operator relationship. Let us go back to Section 3.4 and drop the temporal factor from Equation 3.22 (or just consider it at $t = 0$). We will have

$$\Psi_k(x) = \Psi_0 e^{ikx}. \tag{5.36}$$

It is the eigenfunction of the momentum operator, belonging to the eigenvalue k (in units of $\hbar$). From the mathematical viewpoint, this function is totally symmetric with respect to both variables: k and x. What happens if we now consider it as a function of k, with x as a fixed parameter? Using the same terminology as before, we can say that the result will be a certain function in k-representation. Will this have a physical meaning and if yes, what is it? For example, can we now change the labels, write $\Phi_x(k) = \Phi_0 e^{ixk}$, and go on to write the position operator in k- representation "by analogy" as $\hat{x} = -i\hbar(\partial/\partial p)$? Alas, this analogy is misleading! The trouble with the above approach is that we indeed got a *certain* wave function in k-representation, but not the *same* function, and therefore operator written "by analogy" is also not the same as the position operator. In fact, it has exactly the wrong sign! To use symmetry correctly, you should make a switch $\Psi_k(x) \Leftrightarrow \Phi_x(k)$ via a Fourier transform, which gives

$$\Phi_x(k) = \Phi_0 e^{-ixk} \tag{5.37}$$

The original function (5.36) describes a state with a sharply defined momentum and totally undefined position. The *new* function (5.37) describes a state with sharply defined position x and totally undefined momentum. One can object that this cannot be true since, as we just found, the state with definite position $x' = x$ is described by the δ-function (5.34) (for one dimension). But this is not a valid objection since (5.34) describes a state in the x-representation, whereas now we are talking about *the same state in k-representation* (i.e., given *in the momentum space*). It is natural that by changing the representation of a state we may change the mathematical expression describing it. In addition, we can argue that the momentum operator in its own representation is just multiplication by $p = \hbar k$, and its eigenfunction is the δ-function (5.35). And this does not preclude the same operator from being a differential operator (3.26) in the x-representation, with the eigenfunction (5.36). Reversing this argument (and by symmetry with (3.26)), it is now a straightforward matter to show (Problem 5.8) that the position operator $\hat{x}$ in p-representation is the *differential operator* $\hat{x} = i\hbar(\partial/\partial p)$, and (5.37) is, indeed, its eigenfunction belonging to the eigenvalue x. In three dimensions, we have

$$\hat{\vec{r}} = i\hbar \vec{\nabla}_p \equiv i\hbar \left(\frac{\partial}{\partial p_x} \hat{\mathbf{x}} + \frac{\partial}{\partial p_y} \hat{\mathbf{y}} + \frac{\partial}{\partial p_z} \hat{\mathbf{z}} \right) \tag{5.38}$$

and

$$\Phi_r(\mathbf{k}) = \Phi_0 e^{-i\mathbf{r}\cdot\mathbf{k}}. \tag{5.39}$$

We know from Section 3.4 that operators $\hat{\mathbf{p}}$ and $\hat{H}$ generate translations in space and time, respectively. The form (5.39) of $\hat{\mathbf{r}}$ in **p**-representation shows that it generates translations in the momentum space. We will see later in Chapter 7 that the operator of angular momentum generates *spatial rotations*. All of these are continuous transformations. Now we will discuss briefly an operator associated with a totally different class of transformations that cannot be reduced to a succession of incremental variations. They can be called discrete transformations. We consider one of them – *space inversion*. It can be envisioned physically as the mirror reflection of a system under study, with coordinate axes fixed, or mirror reflection of coordinate axes, with the object fixed. We are dealing with this type of transformation every day. Each time when you look in the mirror you see your inverted self. An interesting thing about spatial reflections is that a complete inversion constitutes an independent type of transformation only in spaces of odd dimensions (see more details in Ref. [3]). Here, we restrict to inversion in one or three dimensions.

Consider an inversion in 3D:

$$x \to x' = -x, \quad y \to y' = -y, \quad z \to z' = -z, \quad \text{or} \quad \mathbf{r} \to \mathbf{r}' = -\mathbf{r}. \tag{5.40}$$

We can describe this either as a succession of three consecutive mirror reflections along the x, y, and z directions, respectively, or as converting a right triplet of the basis vectors to the left triplet.

Now we introduce the operator performing transformation (5.40). Call it the *parity operator* and denote as $\hat{\mathcal{P}}$. By definition, action of $\hat{\mathcal{P}}$ on a function given in coordinate representation inverts its coordinates:

$$\hat{\mathcal{P}}\Psi(\mathbf{r}) = \Psi(-\mathbf{r}). \tag{5.41}$$

Call $\Psi(\mathbf{r})$ an eigenfunction of $\hat{\mathcal{P}}$ corresponding to an eigenvalue $\mathcal{P}$, if it satisfies the equation

$$\hat{\mathcal{P}}\Psi(\mathbf{r}) = \mathcal{P}\Psi(\mathbf{r}) \quad \text{or} \quad \Psi(-\mathbf{r}) = \mathcal{P}\Psi(\mathbf{r}). \tag{5.42}$$

To determine these eigenfunctions and eigenvalues of $\hat{\mathcal{P}}$, apply $\hat{\mathcal{P}}$ again to (5.41):

$$\hat{\mathcal{P}}\Psi(-\mathbf{r}) = \hat{\mathcal{P}}^2\Psi(\mathbf{r}) = \mathcal{P}^2\Psi(\mathbf{r}). \tag{5.43}$$

But applying the parity operator twice is just the identical transformation, $\hat{\mathcal{P}}^2\Psi(\mathbf{r}) = \Psi(\mathbf{r})$. Therefore, (5.43) gives

$$\hat{\mathcal{P}}^2 = 1, \quad \mathcal{P} = \pm 1. \tag{5.44}$$

The parity operator has only two eigenvalues given by (5.44). We can thus write (5.42) in the form

$$\hat{\mathcal{P}}\Psi(\mathbf{r}) = \pm\Psi(\mathbf{r}). \tag{5.45}$$

According to this result, the eigenfunctions of $\hat{\mathcal{P}}$ come in two varieties: functions Ψ^+, which are not changed under inversion: $\hat{\mathcal{P}}\Psi^+ = \Psi^+$, and functions Ψ^-, which change sign under inversion: $\hat{\mathcal{P}}\Psi^- = -\Psi^-$. The functions Ψ^+ (known in mathematics as the *even functions*) describe the *even states*; the functions Ψ^- are the *odd functions* and describe the *odd states*. So we can now define parity as the physical property of being either in an odd state or in an even state.

5.5
Orthogonality of Eigenstates

We already know that the eigenstates corresponding to different eigenvalues of a Hermitian operator are mutually orthogonal. Here, we show that this result directly follows from hermiticity.

Consider an operator $\hat{L}$ with eigenvalues and eigenfunctions determined by Equation 5.31. Assume a discrete spectrum and write the equations for two eigenfunctions Ψ_m and Ψ_n:

$$\hat{L}\Psi_m = L_m \Psi_m \quad \text{and} \quad \hat{L}\Psi_n = L_n \Psi_n. \tag{5.46}$$

Take the complex conjugate of the first equation and multiply it by Ψ_n; then multiply the second equation by Ψ_m^*:

$$\Psi_n \hat{L}^* \Psi_m^* = L_m \Psi_n \Psi_m^* \quad \text{and} \quad \Psi_m^* \hat{L} \Psi_n = L_n \Psi_m^* \Psi_n. \tag{5.47}$$

Now, integrate both equations and take the difference:

$$\int \Psi_n \hat{L}^* \Psi_m^* \, dq - \int \Psi_m^* \hat{L} \Psi_n \, dq = (L_m - L_n) \int \Psi_m^* \Psi_n \, dq. \tag{5.48}$$

But according to the definition of a Hermitian operator, the two integrals on the left must be equal. Therefore, for $m \neq n$ we obtain

$$\int \Psi_m^*(q) \Psi_n(q) dq = 0, \quad m \neq n. \tag{5.49}$$

If $m = n$, then the integral on the right is multiplied by 0, and Equation 5.48 will be satisfied for any finite value of this integral. But we already know that in this case the integral's value is fixed by the normalization condition (3.25). Now we can combine both cases in one equation for the discrete spectrum:

$$\int \Psi_m^*(q) \Psi_n(q) dq = \delta_{mn}. \tag{5.50}$$

This result derives from the very general requirements to the state eigenfunctions and their operators: normalization and hermiticity. Therefore, it generalizes the result (3.8) onto an arbitrary discrete set of the eigenfunctions given in an arbitrary representation.

Applying the same procedure to eigenfunctions with a continuous spectrum gives

$$\int \Psi_L^*(q) \Psi_{L'}(q) dq = \delta(L - L'). \tag{5.51}$$

In many cases, we have more than one distinct eigenfunction of $\hat{L}$ belonging to the same eigenvalue L_m. A simple example is two eigenfunctions of the Hamiltonian with $U(x) = 0$ – de Broglie waves running in opposite directions – both belonging to one eigenvalue E, but with opposite momenta $p_1 = \sqrt{2\mu E}$ and $p_2 = -\sqrt{2\mu E}$. We say that the given energy level is *degenerate* and call the whole phenomenon *degeneracy*. We have to consider such a case.

Since we have now a *set of eigenstates* with one common eigenvalue L_n, we need a second index specifying a state:

$$\Psi_n(q) \to \Psi_{n,l}(q) = \{\Psi_{n1}(q), \Psi_{n2}(q), \ldots, \Psi_{n\nu}(q)\}, \tag{5.52}$$

where ν is the maximal possible l in this set (this maximum may be different for different sets, that is, it may depend on n). Equation 5.46 can be applied to two different eigenstates *within* the set (5.52):

$$\hat{L}\Psi_{n,l} = L_n \Psi_{n,l} \quad \text{and} \quad \hat{L}\Psi_{n,l'} = L_n \Psi_{n,l'}. \tag{5.53}$$

In this case, the same argument as before, but applied to two members of the set (5.52), leads to an equation similar to (5.48), but now it becomes the identity

$$\int \Psi_{n,l} \hat{L}^* \Psi_{n,l'}^* \, dq - \int \Psi_{n,l'}^* \hat{L}\Psi_{n,l} \, dq = 0 \cdot \int \Psi_{n,l'}^* \Psi_{n,l} \, dq, \tag{5.54}$$

and the hermiticity of $\hat{L}$ does not ensure orthogonality of $\Psi_{n,l}(q)$ and $\Psi_{n,l'}(q)$. Generally, different eigenstates belonging to the same eigenvalue of an operator may be nonorthogonal. But we can always form a new set of *mutually orthogonal* functions $\tilde{\Psi}_{n,l}(q)$ such that each element of the set remains an eigenfunction of $\hat{L}$ belonging to the same eigenvalue L_n. All we need for this is to utilize the superposition principle

$$\tilde{\Psi}_{n,1}(q) = \sum_{l=1}^{\nu} c_{1l} \Psi_{n,l}(q), \quad \tilde{\Psi}_{n,2}(q) = \sum_{l=1}^{\nu} c_{2l} \Psi_{n,l}(q), \ldots, \quad \tilde{\Psi}_{n,\nu}(q) = \sum_{l=1}^{\nu} c_{\nu l} \Psi_{n,l}(q)$$

(5.55)

and select the coefficients $c_{ll'}$ so that the set $\tilde{\Psi}_{n,l}(q)$ is orthonormal. The transformation (5.55) can be written in the matrix form

$$\begin{pmatrix} \tilde{\Psi}_{n,1}(q) \\ \tilde{\Psi}_{n,2}(q) \\ \vdots \\ \tilde{\Psi}_{n,\nu}(q) \end{pmatrix} = \begin{pmatrix} c_{11} & c_{12} & \cdots & c_{1\nu} \\ c_{21} & c_{22} & \cdots & c_{2\nu} \\ \vdots & \vdots & \ddots & \vdots \\ c_{\nu 1} & c_{\nu 2} & \cdots & c_{\nu\nu} \end{pmatrix} \begin{pmatrix} \Psi_{n,1}(q) \\ \Psi_{n,2}(q) \\ \vdots \\ \Psi_{n,\nu}(q) \end{pmatrix}. \tag{5.56}$$

The transformation matrix $\hat{C}$ converting the initial nonorthogonal set $\Psi_{n,l}(q)$ into a new orthogonal set $\tilde{\Psi}_{n,l}(q)$ is said to perform an orthogonalization. The problem then is to find such a matrix, which is always possible (see Problems 5.10 and 5.11).

The properties of quantum states as vectors in the Hilbert space are seen even more explicitly if we use the Dirac notations. Then the basic relations derived here will take the following forms:

- Definition of eigenfunctions of $\hat{L}$:

$$\hat{L}|\Psi_n\rangle = L_n|\Psi_n\rangle \quad \text{or} \quad \hat{L}|n\rangle = L_n|n\rangle. \tag{5.57}$$

- Orthonormality of the eigenvectors:

$$\langle \Psi_m | \Psi_n \rangle = \delta_{mn} \quad \text{or} \quad \langle m | n \rangle = \delta_{mn}. \tag{5.58}$$

- Superposition of eigenstates:

$$|\Psi\rangle = \sum_m a_m |\Psi_m\rangle \quad \text{or} \quad |\Psi\rangle = \sum_m a_m |m\rangle. \tag{5.59}$$

- The expansion coefficients:

$$a_n = \langle \Psi_n | \Psi \rangle \quad \text{or} \quad a_n = \langle n | \Psi \rangle. \tag{5.60}$$

With these notations, the geometrical interpretation of the state functions as vectors in the Hilbert space becomes almost self-obvious.

Finally, we will prove another important property. Write the superposition (5.59) in q-representation, $\Psi(q) = \sum_m a_m \Psi_m(q)$, and use expression (5.60) for the expansion coefficients. Returning to the Schrödinger notations and changing the ordering of summation and integration, we will get

$$\Psi(q) = \sum_m \left(\int \Psi(q') \Psi_m^*(q') dq' \right) \Psi_m(q) = \int \left(\sum_m \Psi_m(q) \Psi_m^*(q') \right) \Psi(q') dq'. \tag{5.61}$$

Finally, recall one of the basic properties of the δ-function, $\Psi(q) = \int \delta(q - q') \Psi(q') dq'$.

Comparing it with (5.61) gives

$$\sum_m \Psi_m(q) \Psi_m^*(q') = \delta(q - q'). \tag{5.62}$$

This can be considered as another definition of δ-function. Alternatively, it can be considered as an expansion of $\delta(q - q')$ over the eigenstates of $\hat{L}$. If we consider Ψ_m as functions of q, $\Psi_m = \Psi_m(q)$, then the expansion coefficients are $\Psi_m(q')$. If we consider Ψ_m as functions of q', that is, $\Psi_m = \Psi_m(q')$, then the expansion coefficients are $\Psi_m(q)$.

This rule can be extended onto the continuous spectrum, $\Psi_m(q) \to \Psi_L(q)$. In this case, (5.62) takes the form

$$\int \Psi_L(q) \Psi_L^*(q') dL = \delta(q - q'). \tag{5.63}$$

It is interesting to compare this with normalization condition (3.9) for states with a continuous spectrum, which we will rewrite here in the form

$$\int \Psi_L^*(q) \Psi_{L'}(q) dq = \delta(L - L'). \tag{5.64}$$

The comparison shows a symmetry revealing the double-faced nature of state $\Psi_L(q)$: On the one hand, it is a function of q, belonging to the eigenvalue L of the $\hat{L}$-operator; on the other hand, if we focus on a certain fixed q and regard L as a variable, it is the probability amplitude of finding the value L in the q-eigenstate of the corresponding $\hat{Q}$-operator.

As an example, consider again the de Broglie wave $\Psi_\mathbf{k}(\mathbf{r}) = (1/\sqrt{2\pi})e^{i\mathbf{k}\mathbf{r}}$. Here $\mathbf{k}$ plays the role of set L, and $\mathbf{r}$ plays the role of set q. Applying (5.63) to a pair of such functions with fixed $\mathbf{r}$ and $\mathbf{r}'$, respectively, we obtain

$$\int e^{i\mathbf{k}\mathbf{r}} e^{-i\mathbf{k}\mathbf{r}'} d\mathbf{k} = \int e^{i(\mathbf{r}-\mathbf{r}')\mathbf{k}} d\mathbf{k} = 2\pi\delta(\mathbf{r}-\mathbf{r}'). \tag{5.65}$$

Similarly, applying (5.64) to a pair of such functions with fixed $\mathbf{k}$ and $\mathbf{k}'$, respectively, we will have

$$\int e^{i\mathbf{k}\mathbf{r}} e^{-i\mathbf{k}'\mathbf{r}} d\mathbf{r} = \int e^{i(\mathbf{k}-\mathbf{k}')\mathbf{r}} d\mathbf{r} = 2\pi\delta(\mathbf{k}-\mathbf{k}'). \tag{5.66}$$

Either of the last two equations is one of the known definitions of the δ-function (see Appendix B). De Broglie wave, when considered as a function of $\mathbf{r}$, is an eigenfunction of the momentum operator for the vector eigenvalue $\mathbf{k}$. But if considered as a function of $\mathbf{k}$, it is the $\mathbf{k}$th probability amplitude (Fourier component) of the eigenfunction of the position operator. In other words, it is the Fourier transform of $\delta(\mathbf{r}-\mathbf{r}')$.

We will have more to say about various faces of a state function in the next chapter.

5.6
The Robertson–Schrödinger Relation

> "... one is always so much less clear the minute one tries to be complete."
> Marguerite Yourgenar, Alexis

We learned that momentum eigenstates have a totally undefined position and position eigenstates have a totally undefined momentum. But either type of eigenstate is just a very special case that results from a precise position or momentum measurement. In the corresponding eigenstates, we know everything about one observable and nothing about the other. Generally, however, a particle is not in an eigenstate of either of the two operators, and the general case represents a compromise between these two extremes.

The same problem arises with other pairs of observables represented by noncommuting Hermitian operators. Since such observables cannot be exactly known simultaneously, we need to determine the maximum possible efficiency of the information trade-off. The question is how precisely we can measure one observable given a required level of precision for its counterpart.

By convention, we take the standard deviation as the appropriate measure of the uncertainty in an observable:

$$\Delta A \equiv \sqrt{\overline{(A-\overline{A})^2}}. \tag{5.67}$$

5.6 The Robertson–Schrödinger Relation

Here, $\bar{A}$ and $var\{A\}$ are defined as usual:

$$\bar{A} = \int \psi^*(q)\hat{A}\psi(q)dq, \qquad var\{A\} \equiv \overline{(A - \bar{A})^2} = \int \psi^*(q)(\hat{A} - \bar{A})^2 \psi(q)dq. \tag{5.68}$$

We consider $\hat{A} - \bar{A} \equiv \hat{\delta}_A$ as an operator representing deviation $A - \bar{A}$; similarly, $\hat{\delta}_B \equiv \hat{B} - \bar{B}$ represents deviation of B from $\bar{B}$. If both operators $\hat{A}$ and $\hat{B}$ are Hermitian, so are their respective deviation operators – the property that we have used in (5.32). Therefore,

$$(\Delta A)^2 = \overline{(\hat{\delta}_A)^2} = \int \psi^*(\hat{\delta}_A)^2 \psi\, dq = \int \left|\hat{\delta}_A \psi\right|^2 dq (\Delta B)^2 = \overline{(\hat{\delta}_B)^2} = \int \psi^*(\hat{\delta}_B)^2 \psi\, dq = \int \left|(\hat{\delta}_B)\psi\right|^2 dq. \tag{5.69}$$

Denote the expressions in the integrands (5.69) as $\hat{\delta}_A \psi \equiv \Phi_A$ and $\hat{\delta}_B \psi \equiv \Phi_B$. We can consider Φ_A and Φ_B as the new derived functions related to ψ. Then we can use the Cauchy–Schwartz inequality [4–6]

$$\int |\Phi_A|^2 dx \cdot \int |\Phi_B|^2 dq \geq \left|\int \Phi_A^* \Phi_B\, dq\right|^2. \tag{5.70}$$

This is a generalization onto the $\mathcal{H}$-space of the known geometrical fact that the product of the lengths of two vectors Φ_A and Φ_B equals or exceeds their inner (dot) product $\Phi_A \cdot \Phi_B$. In Dirac's notation,

$$\langle \Phi_A | \Phi_A \rangle \langle \Phi_B | \Phi_B \rangle \geq |\langle \Phi_A | \Phi_B \rangle|^2. \tag{5.71}$$

($|\Phi_A\rangle$ and $|\Phi_B\rangle$ are not state vectors, so their norms are not necessarily 1.) Due to hermiticity of the deviation operators, the right-hand side of (5.70) can be written as

$$\int \Phi_A^* \Phi_B\, dq = \int \hat{\delta}_A \psi^* \hat{\delta}_B \psi\, dq = \int \psi^* \hat{\delta}_A \hat{\delta}_B \psi\, dq. \tag{5.72}$$

Next we write

$$\hat{\delta}_A \hat{\delta}_B = \frac{1}{2}\left(\hat{\delta}_A \hat{\delta}_B + \hat{\delta}_B \hat{\delta}_A\right) + \frac{1}{2}\left(\hat{\delta}_A \hat{\delta}_B - \hat{\delta}_B \hat{\delta}_A\right) = \hat{G} + \frac{1}{2}[\hat{A}, \hat{B}] \equiv \hat{G} + \frac{i}{2}\hat{C}, \tag{5.73}$$

where $\hat{G}$ denotes the first term in the sum, and $i\hat{C} \equiv [\hat{A}, \hat{B}]$. Put this into (5.71) and use the fact that $\hat{G}$ and $\hat{C}$ are Hermitian and their expectation values are real:

$$(\Delta A)^2 (\Delta B)^2 \geq \left|\int \psi^*\left(\hat{G} + \frac{i}{2}\hat{C}\right)\psi\, dq\right|^2 = \left|\bar{G} + \frac{i}{2}\bar{C}\right|^2 = \bar{G}^2 + \frac{1}{4}\bar{C}^2. \tag{5.74}$$

This is the *Robertson–Schrödinger relation* [7–10]. Dropping $\bar{G}$ simplifies it to

$$(\Delta A)^2 (\Delta B)^2 \geq \frac{1}{4}\bar{C}^2, \qquad \Delta A \Delta B \geq \frac{1}{2}|\bar{C}|. \tag{5.75}$$

Heisenberg's uncertainty principle follows from (5.75) when $\hat{C} = \hbar$, as in case (3.42).

5.7
The Wave Function and Measurements (Discussion)

Here, we summarize what we have learned in the form of two postulates and discuss them.

Postulate I. A state of a system in a fixed environment is described uniquely by a single-valued function $\Psi(q, t)$ of a relevant observable q (and time), which possesses the following properties:

a) It is generally a complex function.
b) It provides all the information available about the system within the framework of QM.
c) The square modulus $|\Psi(q, t)|^2$ determines the probability of finding the value q at a moment t.
d) This function satisfies a *linear homogeneous wave equation* (the *Schrödinger equation*).

This postulate has several important consequences.

Property (a) means a radical break with classical concepts, both mathematically and physically. Classically, the information about q is contained in one real number. In the quantum world, all available information is in a complex function determining the probabilities of various values of q.

Property (b) tells us that the goal of QM is to determine the wave function describing the system. With $\Psi(q, t)$ at hand, we know all there is to know about the system. This also implies that actually it is enough to find Ψ at some moment. The information about the system's future behavior is already present in Ψ; in other words, the time evolution of Ψ is uniquely determined (but only between the measurements, since a measurement can abruptly change the previous state in an unpredictable way).

Property (c) provides us with the probabilistic interpretation of the wave function. If a system is described by the function $\Psi(q)$, we know the probabilities for all values of q. But we cannot answer right away what the probabilities will be for the values of some other observable s. This is a tricky question, and it has two answers. If s is exactly measurable together with q, then, by property (b), it *must* be included in $\Psi(q)$ as an additional independent variable on the same footing as q to form a complete set, that is, $\Psi(q) = \Psi(q, s)$. This is why postulate I says that Ψ is a *function of the relevant observables*. If s is not exactly measurable together with q (is incompatible with q), we need to have the wave function in the form $\Phi(s)$, and it is always possible to find $\Phi(s)$ from $\Psi(q)$ (and vice versa), because by property (b), our $\Psi(q)$ provides *all* the available information about the system, and this includes information about probabilities for the values of s. Both forms – $\Psi(q)$ and $\Phi(s)$ – while being mathematically different, will hold the same amount of information about the system. If, for example, $\Psi(q)$ gives the probability 1 for $q = q'$ and 0 for all $q \neq q'$ (i.e., q is known exactly), then the corresponding $\Phi(s)$ will give the distributed[4] probability for all values of s (i.e., s is

---
4) As we know, in some cases this distribution may be uniform (constant for all s).

totally unknown). This is due to the indeterminacy relation between q and s that holds for any incompatible observables. Thus, as a consequence of properties (b) and (c), the wave function has many equivalent forms. We have said earlier that there is a one-to-one correspondence between the physical state and the wave function. This does not contradict our conclusion, because our $\Psi(q)$ and $\Phi(s)$ are different forms (aka different *representations*) of the same state. Indeed, if we go from $\Psi(q)$ to $\Phi(s)$ and then transform the latter back to $\Psi(q)$, we will again obtain our original function. This is the inevitable conclusion from the statements formulated so far, and mathematics will have to work out to produce this result; in the next chapter we will show that that is indeed the case.

Property (d) gives us a recipe for finding Ψ: We need to solve the Schrödinger equation for the system. According to (d), Ψ is determined up to an arbitrary complex factor $Ce^{i\delta}$ (with real C and δ): If the equation for Ψ is linear and homogeneous, then $Ce^{i\delta}\Psi$ with $\delta = $ const will also satisfy that equation. Two functions differing by only such a factor describe the same physical state and are therefore considered to be the same. According to (b), all these "different" solutions correspond to the same physical reality; either of them can be used as long as we are interested in relative probabilities of different events rather than absolute probabilities. But property (c) determines the magnitude C if we want to obtain *absolute* probabilities: We select C so that the sum of probabilities for all values of q equals 1; this still leaves us with an arbitrary phase factor $e^{i\delta}$.

Postulate II. If there is a chance $\mathcal{P}_1$ of finding the system in the state ψ_1 and a chance $\mathcal{P}_2$ of finding it in some other state ψ_2, then the *current* state of the system is given by $\Psi = a_1\psi_1 + a_2\psi_2$, where $\mathcal{P}_1 = |a_1|^2$, $\mathcal{P}_2 = |a_2|^2$. This can be generalized for any number of possible *different* states:[5]

$$\Psi = \sum_n a_n \psi_n, \quad \text{where } \mathcal{P}_n = |a_n|^2. \tag{5.76}$$

This is the *superposition principle*. The coefficients a_n are *probability amplitudes*, and the sum of their squares must be 1, as the system is certain to be found in *some* state:[6]

$$\sum_n \mathcal{P}_n = \sum_n |a_n|^2 = 1. \tag{5.77}$$

Thus, the square modulus of any coefficient can range from 0 (corresponding to the absence of the given state in superposition) to 1 (meaning that only this state results from an appropriate measurement).

Notice, however, that although it is perfectly okay to multiply the wave function by an arbitrary factor, in a superposition of different ψ_n such multiplication must be performed uniformly across the board!

Postulate II is consistent with property (d): If a linear equation has more than one different solution ψ_n, then a superposition (5.76) with arbitrary a_n is also a solution. If

---

5) By "different" we mean that none of them can be represented in terms of others.
6) We assume here that there is no *degeneracy* (see the discussion of degeneracy in Section 5.6).

all ψ_n satisfy the given boundary conditions and are properly normalized, they describe the set of the respective physical states and form the set of the basis functions.

The superposition principle is extremely important. If we know the possible states of the system, then we know the possible form of the system's wave function, although we still don't know the coefficients. If in addition we are given the amplitudes a_n, then we know the wave function exactly from (5.76). And vice versa, if we know Ψ, we know all the relevant probabilities.

It is now important to understand the relation between wave function and measurement. As said in property (c) of Postulate I, $|\Psi(q)|^2$ determines the probability for the system to have these values of the observables at the given moment of time. But what does it mean "to have a value"? In CM, having a value has nothing to do with any measurements. We may perform no measurements, but the system *does* possess some definite value of q – we just may not know it. But we are still sure that there is "objective reality." Even if the system is in conditions when the value *cannot* be measured in principle (e.g., particle in a dark box with no photons, so we cannot "see" its position), we assume that nature doesn't care about our experimental difficulties and still assigns the definite q to the system, so the value exists as a physical reality, just unknown to us. In QM, this is not the case. There is generally no "value that exists objectively." As mentioned in Section 3.9, the attempts to include the concepts of such values (hidden parameters) into the theory have failed; moreover, it was experimentally proved that they cannot be included into QM as we know it (see Chapter 24). So (at least until we discover some other QM) if we cannot *in principle* measure the definite value of an observable without changing the given physical state, this means the observable does not have a definite value in this state. Thus, when we know Ψ, we can find probabilities for each measurement result. But knowing the probability is not the same as knowing the actual value! Let us now see what this implies. Let's say we are interested in the value of some observable x. In the general case, we have the wave function; before the measurement, we do not (and cannot) know x because it does not have a definite value. Right after the measurement, we do know x (it now has a definite value), but the state (and wave function) is no longer the same. Indeed, there was a state where the system possessed no definite value of x, and now there is a state where x does possess a value. So, obviously, the new state is different from the old one, and it corresponds to a different wave function. The measurement of x took the system from the state with no definite x into a state with a definite x. We have no right to say that now, retrospectively, we see what the value of x really was back then; "back then," it was a different state with no such thing as x having some definite value (recall discussion in Section 3.9). So, measurement does not generally provide us with any new information about the original state of the system. In most cases, it creates some new state, realizing one of the possibilities (or potentialities) inherently present in the original state. Now, why do we say "in most cases"? Because we can conceive of an original state in which x is known (after all, this is exactly what we obtained after we measured x: Now, why not take this state and repeat the measurement?). Going from a state with known x to a state with known (and the same) x does not require the creation of a new state (unless we are waiting too long or do the measurement so clumsily that we change some other observable). An ideal measurement would find x without changing anything else.

According to this discussion, quantum measurements fall into two categories (in "measuring x," there can be two cases):

1) either x already has some value, or
2) it has no definite value.

In the first case, the goal is to get from nature the information it actually possesses. An ideal measurement would achieve the goal without changing anything, thus preserving the original state; a clumsy one will needlessly change the system but can still report some x close to the pre-existing one.

In the second case, we ask from nature something she herself doesn't know. Under these circumstances, the function of measurement is not to report information about x that lies ready waiting for us but to create a new state with some definite x. Accordingly, it will be described by a new wave function, which (if x is found *exactly*) will be a corresponding eigenfunction of the x-operator. The measurement must purchase certainty of x at the expense of certainty of some other observable(s). The ideal measurement will result in a fair trade-off: creation of new information about x, and loss of previous information about another variable q incompatible with x. The clumsy measurement may still produce some definite value of x but will needlessly increase indeterminacy in some other observables above the allowed limit, so the exchange will be unequal, with decrease of net information about the system.

In CM, the information about an observable is obtained directly. For instance, information about the position x of a particle is obtained from equation $F = \mu\ddot{x}$, and its solution $x = x(t)$ for given initial conditions yields x for any t. In QM, the information about an observable is obtained indirectly. Instead of an equation for $x = x(t)$, QM has an equation for $\Psi(x)$. Second (and worse), $\Psi(x)$ does not carry information about x such that knowing $\Psi(x)$ and finding $x(\Psi)$, you know x. Instead, it gives only the *probability* of finding x between x and $x + dx$. As mentioned before, this probability may happen to be 1 for some value and 0 for all other values, but that is just a very special case, and even that could later be spread over a wide range. So, for the same state of the system and, accordingly, the same wave function $\Psi(x, t)$, you can get different values of x from the identical measurements. Suppose you measure the position of a particle in the given physical state and find it to be within $x_1 - \Delta x < x < x_1 + \Delta x$. Then you *restore the original state* and repeat the measurement and this time you find $x_2 - \Delta x < x < x_2 + \Delta x$. Could it be the result of an experimental error? Yes, it may happen that x_1 and x_2 are really closer than the experimental range of error Δx. But now you repeat your measurement again from the initial state and this time get $x_3 - \Delta x < x < x_3 + \Delta x$ with x_3 beyond the range of error. Any attempt to detect some difference in the experimental conditions will be fruitless, since the conditions are identical. The only conclusion from these results is that nature is intrinsically probabilistic. In QM, unlike CM, fixed experimental conditions produce fixed $\Psi(x)$ but not necessarily fixed x. The *probability* of finding x within $x, x + \Delta x$ is the most we can get from $\Psi(x)$. Generally, the way such probability is extracted depends on how the wave function is represented. If it is given (or found) in the x-representation, you just take $|\Psi(x)|^2$. If it is in another representation, say, $\Phi(q)$, you first transform it into $\Psi(x)$.

We have also emphasized that $\Psi(\mathbf{r}, 0)$ uniquely determines the behavior of the system for all future moments $t > 0$ until another measurement is performed. We can always find $\Psi(\mathbf{r}, t)$ at future moments using equations of QM, which apply to the system. In some special cases when a system is bound to a limited region of space by fixed forces that do not depend explicitly on time (e.g., the electrons in a stable isolated atom), there exist states whose ψ-function does not change with time, except for the inessential phase factor $e^{i\delta(t)}$, whose explicit time dependence will be studied in Chapters 8 and 9. Such states are called the stationary states.

The phrase "in a fixed environment" means that the system is either totally isolated or interacts with other bodies that *do not depend* on the system's behavior (e.g., the system may be placed in some given external field, such as an electron in a hydrogen atom).

But in many situations, the environment changes with time. Or it is itself affected by the system we study. One could, for example, consider two electrons whose *total* momentum is known but not their individual momenta. Then the measurement of one momentum would obviously mean that we have found the other one as well, so we cannot regard one electron as an independent system and the other one as its fixed surroundings. No electron here can be described as situated in a fixed environment.

Another example was discussed in Section 3.7: a pair of electrons with the zero net spin. Then all we (or the electrons themselves!) know is that their spins point in opposite directions, without knowing the direction itself. This is another example of the reduced quantum-mechanical information as compared to the classical one. In classical description, we can *always* say that in such a system the first electron's spin points up, and the second electron's spin points down, or vice versa. According to QM, in the above situation we can only say that the respective spins are opposite, but their directions are not determined. If you chose the z-basis, it would be up–down or down–up. You could also choose another basis, say, the x-basis. Then it would be right–left or left–right, and so on. This is an entangled state considered in Section 3.7. In this state, neither electron separately can be characterized by its own wave function, because the actual measurement of its observable (spin in this case) will affect the observable of the second electron.

Problems

5.1 Derive the expression (5.5) for the average of operator $\hat{L}(\mathbf{r})$ in a state $\Psi(\mathbf{r})$.

(*Hint:* Start from (5.2) and express a_j in terms of $\Psi(\mathbf{r})$ from (5.1) using the basic properties of eigenstates $\psi_j(\mathbf{r})$; then use the hermiticity of $\hat{L}$.)

5.2 Prove that the momentum operator $\hat{\mathbf{p}}(\mathbf{r}) = -i\hbar\,\vec{\nabla}$ is Hermitian.

5.3 Show that any operator commutes with any analytic function of itself.

5.4 Prove relations (5.24) at the end of Section 5.3.

5.5 Express the mean square deviation of a variable in terms of its average and the average of its square

5.6 Prove that if $\hat{L}$ is Hermitian, then the operators $\Delta\hat{L}$ and $var\{L\}$ defined in (5.27) and (5.28) are also Hermitian.

5.7 Prove that the operator of a physical observable in its own representation reduces to this observable.

(*Hint:* Use the expression for the expectation value of an observable.)

5.8 Find the position operator in momentum representation.

5.9 Show that the complete inversion in 2D space can also be reproduced by rotation through $180°$ and thus does not constitute an independent transformation.

5.10 Consider a degenerate eigenvalue L_n of an operator $\hat{L}$, with two distinct eigenfunctions $\Psi_{n,1}(q)$ and $\Psi_{n,2}(q)$.

  a) Form two new functions $\tilde{\Psi}_{n,l}(q)$, $l = 1, 2$, by making linear superpositions of $\Psi_{n,1}(q)$ and $\Psi_{n,2}(q)$ and write the transformation $\Psi_{n,l}(q) \Rightarrow \tilde{\Psi}_{n,l}(q)$ in the matrix form.

  b) Show that the new functions $\tilde{\Psi}_{n,l}(q)$ remain the eigenfunctions of $\hat{L}$ belonging to the same eigenvalue L_n.

5.11 Assume that the functions $\Psi_{n,l}(q)$ in the previous problem are nonorthogonal, with given nonzero projection η onto each other. Find the elements of the transformation matrix $\hat{C}$ converting $\Psi_{n,l}(q)$ into $\tilde{\Psi}_{n,l}(q)$ such that $\tilde{\Psi}_{n,l}(q)$ are orthonormal.

5.12 Derive the normalization condition for a set of coefficients in expansion of an arbitrary state over the eigenstates of a Hermitian operator.

5.13 Given two Hermitian operators $\hat{A}$ and $\hat{B}$, find the properties of their product $\hat{C} \equiv \hat{A}\hat{B}$. Is $\hat{C}$ a Hermitian operator?

5.14 a) Consider the commutator $\hat{C} \equiv [\hat{A}, \hat{B}]$. Is $\hat{C}$ Hermitian?
  b) Consider the anticommutator $\hat{C}_+ \equiv \{\hat{A}, \hat{B}\}$. Is it Hermitian?

References

1 Bohm, D. (1952) A suggested interpretation of the quantum theory in terms of "hidden" variables. I. *Phys. Rev.*, **85**, 166–180.

2 Bohm, D. (1989) *Quantum Theory*, Dover Publications, New York.

3 Fayngold, M. (2008) *Special Relativity and How It Works*, Wiley-VCH Verlag GmbH, Weinheim, pp. 485–487.

4 Schwarz, H.A. (1988) Über ein Flächen kleinsten Flächeninhalts betreffendes Problem der Variationsrechnung (PDF). *Acta Soc. Sci. Fenn.*, **XV**, 318.

5 Solomentsev, E.D. (2001) Cauchy inequality, in *Encyclopedia of Mathematics* (ed. M. Hazewinkel), Springer, New York.

6 Steele, J.M. (2004) *The Cauchy–Schwarz Master Class*, Cambridge University Press, Cambridge.

7 Robertson, H.P. (1929) The uncertainty principle. *Phys. Rev.*, **43**, 163–164.

8 Schrödinger, E. (1930) Zum Heisenbergschen Unschärfeprinzip. *Sitzungsberichte der Preussischen Akademie der Wissenschaften. Physikalisch-Mathematische Klasse*, vol. 14, pp. 296–303.

9 Kim, Y.-H. C and Shih, Y. (1999) Experimental realization of Popper's experiment: violation of the uncertainty principle? *Found. Phys.*, **29**(12), 1849–1861.

10 Yosida, K. (1968) *Functional Analysis*, Springer, Berlin.

6
Representations and the Hilbert Space

6.1
Various Faces of a State Function

Here we consider some state functions in different physical situations.

First of all, we need to draw a clear distinction between a physical state and its representation.

For instance, a state described by (2.32) is the one with *definite momentum* $\mathbf{p} = \hbar\mathbf{k}$ – in our terminology, the *momentum eigenstate*, and we can represent it as the ket $|\mathbf{p}\rangle$ – a *basis vector* in $\mathcal{H}$. But (2.32) expresses it as a running wave in the *coordinate space*, that is, presents this state in the "**r**-representation." An appropriate Dirac notation for this is $\langle \mathbf{r}|\mathbf{p}\rangle$. The bracket $\langle \mathbf{r}|\mathbf{p}\rangle$, with $\mathbf{p}$ showing the basis and $\mathbf{r}$ indicating representation, is totally equivalent to (2.32); it is the *complex number*, the amplitude of finding the particle at $\mathbf{r}$ in a position measurement from state $|\mathbf{p}\rangle$.

Consider now *the same state* in the momentum representation. Accordingly, we suppress information about $\mathbf{r}$ and switch to variable $\mathbf{p}$. This will *not* change the state itself. But it reflects the change in the corresponding experimental program: instead of an assumed position measurement in state $|\mathbf{p}\rangle$, we now assume the (future!) momentum measurement in this state. Whereas the first program would destroy state $|\mathbf{p}\rangle$ to produce instead an eigenstate $|\mathbf{r}\rangle$ of the position operator, the second program must return the vector eigenvalue $\mathbf{p}$, just making it known to us. The difference in physical tools used to explore the state, and the corresponding difference in the measurement outcomes, is reflected in the difference seen in the mathematical expression describing the state. The state $|\mathbf{p}\rangle$ in $\mathbf{p}$-representation is described by the δ-function $\Phi_\mathbf{p}(\mathbf{p}') = \delta(\mathbf{p} - \mathbf{p}')$ (Equation 5.35) or, in Dirac's notation, $\langle \mathbf{p}'|\mathbf{p}\rangle$. Just like $\langle \mathbf{r}|\mathbf{p}\rangle$, it represents the *same* state (2.32): we have one state $|\mathbf{p}\rangle$ in two different representations. Equation 2.32 expresses $|\mathbf{p}\rangle$ in an "alien" $\mathbf{r}$-representation. Equation 5.35 gives it in its own representation.

We can choose some other observable (e.g., angular momentum $\mathbf{L}$) to be measured in state $|\mathbf{p}\rangle$. Expressing $|\mathbf{p}\rangle$ in terms of $\mathbf{L}$ is equivalent to expansion of the wave (2.32) over *spherical harmonics* (see Chapters 7 and 10). The expansion coefficients are the amplitudes of finding the respective values of **L**. The set of all these amplitudes will give $|\mathbf{p}\rangle$ in **L**-representation.

The same works for an arbitrary state $|\Psi\rangle$. Starting with the **r**-representation $|\Psi\rangle \to \Psi(\mathbf{r})$, we can express it as a Fourier expansion over momentum eigenstates

$$\Psi(\mathbf{r}) = (2\pi)^{-3/2} \int \Phi(\mathbf{p}) e^{(i/\hbar)\mathbf{p}\cdot\mathbf{r}} \, d\mathbf{p}, \quad \text{or} \quad |\Psi\rangle = \int \Phi(\mathbf{p})|\mathbf{p}\rangle \, d\mathbf{p}. \tag{6.1}$$

Then the expansion coefficients $\Phi(\mathbf{p}) = \langle \mathbf{p}|\Psi\rangle$ show us the "**p**-face" of state $\Psi(\mathbf{r})$. Alternatively, we can expand $\Psi(\mathbf{r})$ over spherical harmonics $Y_{l,m}(\theta,\varphi) = P_l^m(\theta)\Phi_m(\varphi)$ (eigenfunctions of $\hat{L}$) as

$$\Psi(\mathbf{r}) = \sum_{l,m} \Lambda(L_{l,m}) R_l(r) Y_{l,m}(\theta,\varphi), \quad \text{or} \quad |\Psi\rangle = \sum_{l,m} \Lambda(L_{l,m}) |\psi_{l,m}\rangle, \tag{6.2}$$

where $R_l(r)$ is the corresponding radial part in the expansion terms. Then the expansion coefficients $\Lambda(L_{l,m}) = \langle \psi_{l,m}|\Psi\rangle$ show us the "**L**-face" of the state. These three faces can be written as

$$|\Psi\rangle \to \begin{cases} \Psi(\mathbf{r}) = \langle \mathbf{r}|\Psi\rangle, & \text{r-face,} \\ \Phi(\mathbf{p}) = \langle \mathbf{p}|\Psi\rangle, & \text{p-face,} \\ \Lambda(L_{l,m}) = \langle \psi_{lm}|\Psi\rangle, & \text{L-face.} \end{cases} \tag{6.3}$$

They represent the *same* state when viewed from different perspectives. Geometrically, each perspective is a projection of $|\Psi\rangle$ onto an eigenvector of respective basis.

The terms *basis* and *representation* are frequently used interchangeably, but here we can see the subtle difference between them. A *basis state* as such is a pure vector given as a geometrical object without reference to any other vector and expressed as Dirac's ket or bra. *Representation* is a relation of this vector to vectors of the same or some other basis. This relation is given by von Neumann's projection: the inner product between the two vectors, expressed by the corresponding bracket. As we saw in the examples with Equations 2.32 and 5.25 for an eigenstate $|\mathbf{p}\rangle$, a basis state may itself be given in different representations, and vice versa, we may have different bases given in the same representation.

The capacity for a given state to have different representations is an exclusively QM property, originating from incompatibility of some characteristics of the state. Mathematically, the description of *two representations* of one state is given by *two functions of different variables*. The description of two *states* in the *same* representation is given by two *different functions of the same* variable. And two *different states* in *different representations* may have *mathematically identical expressions*, but they may depend on *different* observables (see Equations 5.34 and 5.35).

Summary. A quantum state turns out to be a multifaced entity, with all its faces equally legitimate. Each stores all the information necessary to construct all others. This is just a fancier way of saying that a state function stores all information about the system it describes, regardless of the chosen basis or representation.

6.2 Unitary Transformations

A state can be treated as a vector in $\mathcal{H}$ spanned by a certain basis. Because our choice of observable is arbitrary, there is nothing particular about this basis. Now we learn how to transfer from one basis to another.

Here is the precise formulation of the problem: Consider a state $\psi(x)$ and two incompatible observables q and s, *which are also incompatible with x*. We know all the eigenstates of each observable. Each of the corresponding sets of eigenstates is orthonormal. We can express $\psi(x)$ in terms of eigenstates $\chi_k(x)$ of q (the q-basis); alternatively, we can express it in terms of eigenstates $\phi_l(x)$ of s (the s-basis):

$$\psi(x) = \sum_{k=1}^{N} a_k \chi_k(x) \quad (q\text{-basis}) \tag{6.4a}$$

or

$$\psi(x) = \sum_{l=1}^{N} b_l \phi_l(x) \quad (s\text{-basis}). \tag{6.4b}$$

We require x, q, and s to be incompatible because otherwise (6.4a) and (6.4b) would *not* represent the *same* physical state. For instance, if s were a spin component of a particle, which can be measured simultaneously with x, then (6.4b) should be "upgraded" to $\psi(x, s) = \sum_{l=1}^{N} b_l \phi_l(x, s_l)$. Note also that the finite number N of eigenstates for q and s is not the typical case. This number can be infinite. Moreover, a set of eigenstates can be innumerable, like the set of momentum states. We assume finite N because it illustrates the subject in the simplest way.

The problem now is this: How, knowing only the set a_k, can we find the set b_l?

Let us switch to Dirac's notation $\psi(x) \to |\psi\rangle$, $\chi_k(x) \to |q_k\rangle$, and $\phi_l(x) \to |s_l\rangle$, so that Equation 6.4 takes the form

$$|\psi\rangle = \sum_{k=1}^{N} a_k |q_k\rangle \quad (q\text{-basis}) \tag{6.5a}$$

or

$$|\psi\rangle = \sum_{l=1}^{N} b_l |s_l\rangle \quad (s\text{-basis}). \tag{6.5b}$$

For discrete q and s considered here, $|\psi\rangle$ can be written as a column matrix:

$$|\psi\rangle = \mathbf{A} \equiv \begin{pmatrix} a_1 \\ \vdots \\ a_N \end{pmatrix} \tag{6.6a}$$

or

$$|\psi\rangle = \mathbf{B} \equiv \begin{pmatrix} b_1 \\ \vdots \\ b_N \end{pmatrix}. \tag{6.6b}$$

Both matrices (6.6a) and (6.6b) are legitimate representatives of state $|\psi\rangle$. With $|\psi\rangle$ given, we can find a_n and b_n directly as projections $\langle q_n|\psi\rangle$ and $\langle s_n|\psi\rangle$. But here we are interested in how, knowing *only* the face A, we can find face B? In other words, how is the set b_n expressed in terms of a_n? This means finding direct relation between matrices **A** and **B**.

Let us expand each $\chi_k(x)$ over $\phi_l(x)$:

$$\chi_k(x) = \sum_{l}^{N} T_{lk}\phi_l(x). \tag{6.7}$$

Putting this into (6.4a) and collecting terms with the same $\phi_l(x)$ gives

$$\psi(x) = \sum_{k=1}^{N} a_k \sum_{l=1}^{N} T_{lk}\phi_l(x) = \sum_{l=1}^{N}\left(\sum_{k=1}^{N} T_{lk}a_k\right)\phi_l(x). \tag{6.8}$$

Comparing with (6.4b), we find the sought-after relation:

$$b_l = \sum_{k=1}^{N} T_{lk}a_k. \tag{6.9}$$

The set T_{lk} forms the $N \times N$ *transformation matrix* $\hat{T}$ that takes us from q-basis to s-basis according to the formula

$$\begin{pmatrix} b_1 \\ b_2 \\ \vdots \\ b_N \end{pmatrix} = \hat{T}\begin{pmatrix} a_1 \\ a_2 \\ \vdots \\ a_N \end{pmatrix} = \begin{pmatrix} T_{11} & T_{12} & \cdots & T_{1N} \\ T_{21} & T_{22} & \cdots & T_{2N} \\ \vdots & \vdots & \ddots & \vdots \\ T_{N1} & T_{N2} & \cdots & T_{NN} \end{pmatrix}\begin{pmatrix} a_1 \\ a_2 \\ \vdots \\ a_N \end{pmatrix}. \tag{6.10}$$

Looking at an element T_{lk}, we see that it is the *l*th coefficient in the expansion (6.7) of the *k*th vector of the old basis relative to the new basis; that is,

$$T_{lk} = \langle\phi_l|\chi_k\rangle \equiv \langle s_l|q_k\rangle. \tag{6.11}$$

This is just the amplitude of finding the value $s = s_l$ in the s-measurement performed on state $|\chi_k\rangle$.

In a similar way we can derive the inverse transformation from the **B**- to the **A**-basis:

$$\begin{pmatrix} a_1 \\ a_2 \\ \vdots \\ a_n \end{pmatrix} = \tilde{T}^*\begin{pmatrix} b_1 \\ b_2 \\ \vdots \\ b_n \end{pmatrix} = \begin{pmatrix} T_{11}^* & T_{21}^* & \cdots & T_{n1}^* \\ T_{12}^* & T_{22}^* & \cdots & T_{n2}^* \\ \vdots & \vdots & \ddots & \vdots \\ T_{1n}^* & T_{2n}^* & \cdots & T_{nn}^* \end{pmatrix}\begin{pmatrix} b_1 \\ b_2 \\ \vdots \\ b_n \end{pmatrix}, \tag{6.12}$$

where $T_{kl}^* = \langle\phi_k|\chi_l\rangle$. Matrix element $\tilde{T}_{lk}^*$ is the amplitude of finding $q = q_k$ in the q-measurement on state $|\phi_l\rangle$. Matrix $\tilde{T}^*$ is called *conjugate* to T and denoted as $T^\dagger$. It is obtained by transposing $T \to \tilde{T}$ and taking its complex conjugate. The result is independent of the order of these two operations.

On the other hand, the *inverse transformation* is performed by taking the *inverse* T^{-1} of T, so that their product is the identity matrix $TT^{-1} = T^{-1}T = I$. It follows that T satisfies the *unitarity condition* $\tilde{T}^* = T^{-1}$. Such matrices and their corresponding transformations are called *unitary* because there is no change in the (unit) length of a transformed state vector.

Summarizing this part, we see that the switching between mutually exclusive perspectives showing two distinct faces of a single quantum state is, mathematically, a unitary transformation from one basis to another, which is just an appropriate rotation in $\mathcal{H}$. The elements of transformation matrix T are projections of the eigenvectors of one basis onto respective vectors of the other. The inverse transformation is performed by the matrix $T^\dagger$ conjugate to T. In this way, we can switch directly from the set of amplitudes $a(q)$ in the q-basis to the set $b(s)$ in the s-basis, and so on. All these alternative *forms of a wave function* inherit its physical meaning. They satisfy all of the properties formulated in Section 5.7. It will be shown later (in Chapter 8) that they also satisfy the Schrödinger equation in the corresponding basis.

If we consider the case in which all the states are given in the q- instead of the x-representation, we get a general proof that *eigenstates of a variable in its own representation are the Kronecker deltas for the discrete spectrum and delta functions for the continuous spectrum*. Also, the described approach can be extended to continuous matrices (see Supplement to Chapter 6).

Box 6.1 Matrices: concepts and definitions

For any matrix $T = \begin{pmatrix} T_{11} & T_{12} & \cdots & T_{1n} \\ T_{21} & T_{22} & \cdots & T_{2n} \\ \vdots & \vdots & \ddots & \vdots \\ T_{n1} & T_{n2} & \cdots & T_{nn} \end{pmatrix}$, we can define several related matrices. We formulate their definitions now and use them in this and subsequent sections.

Complex conjugate matrix T^* is defined as the matrix whose elements are complex conjugates of elements of T, that is, $(T^*)_{ij} = T^*_{ij}$.

The *transpose* of a matrix T is designated as $\tilde{T}$. It is the matrix obtained by interchanging the rows and columns of T. Thus, the ith row of T becomes the ith column of $\tilde{T}$ and the jth column of T becomes the jth row of $\tilde{T}$. The matrix elements of $\tilde{T}$ are given by $(\tilde{T})_{ij} = T_{ji}$. It is clear from these definitions that $(T^*)^* = T$ and $\tilde{\tilde{T}} = T$.

By performing the operations of transposition *and* taking the complex conjugate on matrix T (in any order), we obtain *adjoint*, or *conjugate*, matrix $T^\dagger$ (often pronounced "T dagger"). Thus,

$$T^\dagger \equiv \tilde{T}^* \quad \text{and} \quad (T^\dagger)_{ij} = T^*_{ji}. \tag{6.13}$$

In particular, if T is symmetric ($T_{ij} = T_{ji}$), then $T^\dagger = T^*$ and if T is real, then $T^\dagger = \tilde{T}$.

A matrix that is equal to its adjoint is called a *self-adjoint* or *Hermitian* matrix: $T^\dagger = T$.

One possibility, of course, is that T is both real and symmetric, but this is not the only way to make the matrix Hermitian: there are also complex matrices for which the complex conjugation will reverse the effect of transposition. It is easy to see that if every element T_{jk} of an $N \times N$ matrix is complex conjugate to the element T_{kj} symmetric relative to the diagonal (i.e., $T_{jk} = T_{kj}^*$), then T is Hermitian. This definition automatically makes the diagonal elements of a Hermitian matrix real.

The product of a square matrix T and its *inverse* T^{-1} is the identity matrix, I. That is, $T^{-1}T = TT^{-1} = I$ (it is easy to show that the result does not depend on order).

Another important type of matrix is the *orthogonal*. It is a square matrix for which if we take *any* two rows (we can also take the same row twice), multiply their first elements, their second elements, and so on, and sum all the products, our result will be 0 for different rows and 1 for the same row; and similarly for columns. Thus, an orthogonal matrix is defined by condition

$$\sum_{k=1}^{n} T_{ik}T_{jk} = \delta_{ij} \quad \text{or} \quad \sum_{k=1}^{n} T_{ki}T_{kj} = \delta_{ij} \tag{6.14}$$

(for continuous matrices, the integral replaces summation and Dirac's delta function replaces the Kronecker delta). It follows immediately that for orthogonal matrices, $T^{-1} = \tilde{T}$, because multiplying ith and jth rows of T is the same as multiplying ith row of T by jth column of $\tilde{T}$, which gives the element $(T\tilde{T})_{ij}$ of their matrix product; since it is Kronecker delta, $T\tilde{T}$ is the identity matrix, and thus $\tilde{T}$ is the inverse of T.

A *unitary* matrix is defined similarly to an orthogonal, except when we multiply two rows or columns, we multiply elements in one row (column) by the complex conjugates of the corresponding elements in the other row (column):

$$\sum_{k=1}^{n} T_{ik}^* T_{jk} = \delta_{ij}, \quad \sum_{k=1}^{n} T_{ki}^* T_{kj} = \delta_{ij}. \tag{6.15}$$

It follows that for unitary matrices, $T^{-1} = T^\dagger$ because now multiplying two rows of T is the same as multiplying the row of T by the column of $\tilde{T}^*$, which is the adjoint matrix of T; therefore, $TT^\dagger$ is the identity matrix and $T^\dagger$ is the inverse of T.

You can see now that the names "orthogonal" and "unitary" were invented by mathematicians, not quantum physicists. "Orthogonal matrix" in particular may seem misleading because there is no analogy with the definition of "orthogonal" in QM. In fact, it is the unitary matrix that correlates more accurately with the quantum-mechanical notion of orthogonal basis or orthogonal eigenstates. The "product" (6.14) of two rows or columns of an orthogonal matrix is actually more like the familiar dot product of two vectors in the real space. It is the "product" (6.15) of two rows or columns of a unitary matrix that resembles scalar products in Hilbert space. When we find the transformation matrix between two *orthonormal* bases formed from eigenstates, we will have found what mathematicians call a *unitary* matrix. This is a potential source of confusion, but it's too late to change the conventional terminology! You can find a more detailed description of the algebra of the transformation matrices in Appendix D.

> **Box 6.2 "Wave function" versus "representation"**
>
> As you can see from this discussion, the term "wave function" is something of a misnomer. Instead of an analytic function of some argument, as this term implicitly suggests, the concept actually implies a set of different functions of different arguments, where all functions provide an equivalent description of the system. In this picture, a function becomes just a viewpoint from which we look at a system. Hence, the word "representation" seems particularly appropriate.

6.3 Operators in the Matrix Form

Transformation of a wave function at the change of basis is accompanied by the corresponding transformation of operators. As we saw, the momentum operator is $\hat{\mathbf{p}} = -i\hbar \vec{\nabla}$ in position representation and $\hat{\mathbf{p}} = \mathbf{p}$ in its own representation. Similarly, the position operator is $\hat{\mathbf{r}} = \mathbf{r}$ in its own representation and $\mathbf{r} = i\hbar \vec{\nabla}_{\mathbf{p}}$ in momentum representation. But even when we remain in some fixed basis, the eigenstates of a variable can be represented basically in three different forms: one as an analytic expression, another as a ket (or bra), and the third as a matrix. We have used all three forms in description of a state. For instance, we described the electron spin-up eigenstate in the s_z-basis as a ket $|s_z^+\rangle = |\uparrow\rangle$ or as a column matrix $\begin{pmatrix} 1 \\ 0 \end{pmatrix}$. Alternatively, we could describe it as a bra $\langle s_z^+| = \langle\uparrow|$, which is a row matrix $(1 \quad 0)$. The corresponding analytic form could be $\delta_{0j}, j = 0, 1$, with 0, 1 standing for states + and −, respectively. Generally, once the basis is selected, we can write *any* relevant wave function as a vector in $\mathcal{H}$, and this vector can also be represented as a column (ket vector) or row (bra vector) matrix.

An operator acting on the matrix form of a wave function needs to be expressed as a matrix, too. As in (6.10), an operator $\hat{Q}$ acting on a wave function in matrix form converts one column matrix into another, and the same is true for a row matrix. The transforming algorithm requires that for a column (row) with N elements, $\hat{Q}$ must be represented by some $N \times N$ square matrix. It acts on the column matrix from the left or on the row matrix from the right.

Now we want to find the matrix form of $\hat{Q}$ for some observable q. So far we have considered operators in analytic form, mostly in a position representation in which many of them are differential operators like (3.24) and (3.30). Let us work in this representation (reduced, for the sake of simplicity, to one x-dimension). As usual, we define $\hat{Q}(x)$ by its action on its eigenstates $\chi_l(x)$ such that $\hat{Q}(x)\chi_l(x) = q_l\chi_l(x)$. To find the matrix form of $\hat{Q}$, we use the same approach as in finding the transformation matrix. Expand an arbitrary state $\psi(x)$ over eigenstates $\phi_k(x)$, $k = 1, 2, \ldots, N$, of observable s:

$$\psi(x) = \sum_{k=1}^{N} a_k \phi_k(x). \tag{6.16}$$

We are interested in the measurement of an observable q that is different from both s and x. Then operator $\hat{Q}$ of this observable gives us the expectation value of q according to (5.14), $\bar{q} = \langle \psi | \hat{Q} | \psi \rangle$. Putting here (6.16) gives

$$\bar{q} = \int \left(\sum_{k=1}^{N} a_k^* \phi_k^*(x) \right) \hat{Q}(x) \left(\sum_{l=1}^{N} a_l \phi_l(x) \right) dx = \sum_{k=1}^{N} \sum_{l=1}^{N} a_k^* a_l \langle \phi_k | \hat{Q} | \phi_l \rangle, \tag{6.17}$$

where

$$q_{kl} \equiv \langle \phi_k | \hat{Q} | \phi_l \rangle \equiv \int \phi_k^*(x) \hat{Q}(x) \phi_l(x) dx. \tag{6.18}$$

The elements q_{kl} form an $N \times N$ matrix

$$\hat{Q} = \begin{pmatrix} q_{11} & q_{12} & \cdots & q_{1N} \\ q_{21} & q_{22} & \cdots & q_{2N} \\ \vdots & \vdots & \ddots & \vdots \\ q_{N1} & q_{N2} & \cdots & q_{NN} \end{pmatrix}. \tag{6.19}$$

For each given basis, it defines $\hat{Q}$ as completely as does its analytic form $\hat{Q}(x)$. We treat the first index in q_{kl} as the row index and the second as the column index, thus following the standard convention for matrix element notation. We can write (6.17) as $\bar{q} = \sum_{k=1}^{N} \sum_{l=1}^{N} a_k^* q_{kl} a_l$. The form on the right is just the "matrix image" of (5.14) at $\hat{L} \to \hat{Q}$.

In view of hermiticity of $\hat{Q}$ we have from (6.18) $q_{lk} = q_{kl}^*$; that is, the complex conjugate of $\hat{Q}$ is equal to its transpose. Denoting the latter as $\tilde{Q}$, we have

$$\hat{Q}^* = \tilde{Q} \quad \text{or} \quad \hat{Q}^\dagger \equiv \tilde{\hat{Q}}^* = \hat{Q}. \tag{6.20}$$

This is the definition of a self-adjoint (or Hermitian) matrix. Again, this is what one should expect: a Hermitian operator is represented by Hermitian matrix. This immediately implies that the diagonal elements of $\hat{Q}$ are all real. No surprise – as seen from (6.18) at $k = l$, they are just the expectation values of $\hat{Q}$ in the respective eigenstates.

We can derive a known expression for $\bar{q}$, processing (6.17) backward and using the bras and kets there in the matrix form:

$$|\phi_1\rangle = \begin{pmatrix} 1 \\ 0 \\ \vdots \\ 0 \end{pmatrix}, \quad |\phi_2\rangle = \begin{pmatrix} 0 \\ 1 \\ \vdots \\ 0 \end{pmatrix}, \quad \ldots, \quad |\phi_N\rangle = \begin{pmatrix} 0 \\ 0 \\ \vdots \\ 1 \end{pmatrix}. \tag{6.21}$$

Each column matrix representing its ket has a simple physical meaning. In eigenstate 1, the observable s has with certainty (matrix element 1) the value

6.3 Operators in the Matrix Form

$s = s_1$ and never (zero matrix elements) any other values. In eigenstate 2, it will never (zeros) have the values other than s_2, which always (matrix element 1) shows in the s-measurement, and so on. If we now put all these kets into (6.16), we will obtain the state $\psi(x)$ as a column matrix (ket vector $|\psi\rangle$)

$$\psi(x) \to |\psi\rangle = a_1 \begin{pmatrix} 1 \\ 0 \\ \vdots \\ 0 \end{pmatrix} + a_2 \begin{pmatrix} 0 \\ 1 \\ \vdots \\ 0 \end{pmatrix} + \cdots + a_N \begin{pmatrix} 0 \\ 0 \\ \vdots \\ 1 \end{pmatrix} = \begin{pmatrix} a_1 \\ a_2 \\ \vdots \\ a_N \end{pmatrix}. \tag{6.22}$$

According to definitions in the previous section, the conjugate $|\psi\rangle^\dagger$ of $|\psi\rangle$ is the row matrix making the bra vector

$$|\psi\rangle^\dagger = (a_1^* \ a_2^* \ \cdots \ a_N^*) \equiv \langle\psi|. \tag{6.23}$$

Then (6.17) is the matrix form of $\bar{q} = \langle\psi|\hat{Q}|\psi\rangle$, with $\hat{Q}$ given by (6.18).

Since $\hat{Q}$ in (6.18) is formed from eigenstates of observable s, we say that it is given in the s-basis. Thus, as we "convert" the operator into a matrix form, the representation generally changes to that of the observable from whose eigenstates we form the basis. Since observable s was not specified, the result (6.18) is general. But the specific form of $\hat{Q}$ depends on basis. For instance, let us switch from the s-basis to the q-basis formed from the eigenfunctions $\chi_k(x)$. Then (6.18) becomes

$$q_{kl} \equiv \langle\chi_k|\hat{Q}|\chi_l\rangle \equiv \int \chi_k^*(x)\hat{Q}(x)\chi_l(x)dx. \tag{6.24}$$

But $\hat{Q}(x)\chi_l(x) = q_l\chi_l(x)$; therefore,

$$q_{kl} = q_l \langle\chi_k|\chi_l\rangle = q_l \delta_{kl}. \tag{6.25}$$

Thus, in the special (and the easiest) case when $\hat{Q}$ is defined in its own basis, the matrix is diagonal, with the corresponding elements being just the eigenvalues q_l:

$$\hat{Q}_{\text{in an alien basis}} = \begin{pmatrix} q_{11} & q_{12} & \cdots & q_{1N} \\ q_{21} & q_{22} & \cdots & q_{2N} \\ \vdots & \vdots & \ddots & \vdots \\ q_{N1} & q_{N2} & \cdots & q_{NN} \end{pmatrix} \xrightarrow{\text{in its own basis}} \begin{pmatrix} q_1 & 0 & \cdots & 0 \\ 0 & q_2 & \cdots & 0 \\ \vdots & \vdots & \ddots & \vdots \\ 0 & 0 & \cdots & q_N \end{pmatrix}. \tag{6.26}$$

Now the set of amplitudes a_l in (6.16) and (6.17) representing state $|\psi\rangle$ in the s-basis will be replaced by the set b_k expressing the same state in the q-basis. The expectation value from (6.17) becomes simply

$$\bar{q} = \sum_{k=1}^{N} \sum_{l=1}^{N} b_k^* b_l q_l \delta_{kl} = \sum_{k=1}^{N} b_k^* b_k q_k = \sum_{k=1}^{N} |b_k|^2 q_k, \tag{6.27}$$

just as we should expect.

Generally, we can write $\hat{Q}$ as a matrix in an arbitrary u-basis formed from eigenstates $\vartheta_i(x)$ of an observable u. We expand $\psi(x)$ over the new basis states, $\psi(x) = \sum_k b_k \vartheta_k(x)$, repeat the procedure leading to (6.17), and obtain $\bar{q} = \sum_k \sum_l b_k^* b_l q_{kl}$. Now the matrix q_{kl} gives $\hat{Q}$ in the u-basis:

$$q_{kl} \equiv \langle \vartheta_k | \hat{Q} | \vartheta_l \rangle \equiv \int \vartheta_k^*(x) \hat{Q}(x) \vartheta_l(x) dx. \tag{6.28}$$

Clearly, the elements q_{kl} in (6.28) are different from those in (6.18). So we must distinguish between the corresponding representations of the operator. In our example with two different bases – s and u – we may want to add the corresponding index indicating the basis, for instance, $q_{kl}^{(s)}$ and $q_{kl}^{(u)}$ for the matrix elements of the same operator in the s- and u-basis, respectively.

The natural question arises: How $q_{kl}^{(s)}$ and $q_{kl}^{(u)}$ are related to each other? In other words, how, knowing $q_{kl}^{(s)}$, do we find $q_{kl}^{(u)}$? The answer: Go back to the previous section and use operation (6.7) performing transformations between bases. With that in hand, express the eigenstates $\vartheta_i(x)$ in terms of $\phi_k(x)$ and put in (6.28). You will get the output for $q_{kl}^{(u)}$ in terms of $q_{kl}^{(s)}$ and the transformation matrix T_{ij}:

$$q_{kl}^{(u)} = \sum_i \sum_j T_{ik}^* T_{jl} q_{ij}^{(s)} \tag{6.29}$$

or, in the matrix form,

$$\hat{Q}^{(u)} = \hat{T}^+ \hat{Q}^{(s)} \hat{T}. \tag{6.30}$$

The last equation is known in Linear Algebra as the *similarity transformation* for matrices. One can easily obtain from this the inverse transformation from $\hat{Q}^{(u)}$ to $\hat{Q}^{(s)}$.

We see that conversion of $\hat{Q}$ from one basis to another is executed by the same matrix $\hat{T}$ that performs transformations between the bases themselves. All these transformations do not change the state $|\psi\rangle$ or operator $\hat{Q}$ as such. We have the same $\hat{Q}$ acting on the same $|\psi\rangle$. But $|\psi\rangle$ is now specified by the new set of amplitudes b_i given in the u-basis, and accordingly, the matrix elements q_{kl} are now different. Equations 6.25 and 6.26 illustrate one special case of this general rule.

The obtained results are trivially generalized to the case when $\hat{Q}$ is a vector function of a vector observable, say, $\hat{\mathbf{Q}} = \hat{\mathbf{Q}}(\mathbf{r}) = \hat{Q}_x(\mathbf{r})\hat{\mathbf{x}} + \hat{Q}_y(\mathbf{r})\hat{\mathbf{y}} + \hat{Q}_z(\mathbf{r})\hat{\mathbf{z}}$, where $\hat{\mathbf{x}}$, $\hat{\mathbf{y}}$, and $\hat{\mathbf{z}}$ are the unit vectors along axes x, y, and z, respectively. In this case, we will have each vector component represented by its own matrix, so $\hat{\mathbf{Q}}$ will be a vector sum of three distinct matrices.

We can now summarize our findings: A Hermitian operator can be represented by the corresponding Hermitian matrix. If the matrix form of an operator is given in the operator's own representation, it is a diagonal matrix whose diagonal elements are the eigenvalues of the operator. Otherwise, the matrix is not diagonal; its matrix elements are determined by (6.18) for each operator in each basis.

6.4
The Hilbert Space

Here we summarize some basic features of mathematical structure of QM in a more general (and thereby more abstract!) form, based on the concept of the Hilbert space $\mathcal{H}$. As we learned in Section 4.1, this concept originates from the analogy between superposition of quantum states in $\mathcal{H}$ and the geometric sum of vectors in the *physical vector space V*. Before going into details, we can capture the essence of the analogy in one statement: an arbitrary state $|\Psi\rangle = \sum c_n |\psi_n\rangle$ as a superposition of eigenstates in $\mathcal{H}$ is similar to an arbitrary vector $\mathbf{A} = \sum a_n \hat{\mathbf{e}}_n$ as a superposition of basis vectors in V. Here $|\psi_n\rangle$ are usually the eigenstates of a Hermitian operator $\hat{L}$ chosen as the basis in $\mathcal{H}$, and $\hat{\mathbf{e}}_n$ are the chosen basis vectors in V.

Once we know all a_n, we know $\mathbf{A}$. Similarly, once we choose the basis $|\psi_n\rangle$, the expansion coefficients c_n specify $|\Psi\rangle$. To emphasize this analogy, we represent both $\mathbf{A}$ and $|\Psi\rangle$ as a column matrix:

$$\mathbf{A} = \begin{pmatrix} a_1 \\ \vdots \\ a_n \\ \vdots \end{pmatrix}, \quad |\Psi\rangle = \begin{pmatrix} c_1 \\ \vdots \\ c_n \\ \vdots \end{pmatrix}. \tag{6.31}$$

In V-space, we usually use an orthonormal system of vectors $\hat{\mathbf{e}}_n$, for which $\hat{\mathbf{e}}_m \cdot \hat{\mathbf{e}}_n = \delta_{mn}$. Similarly, for the orthonormal set of eigenstates in $\mathcal{H}$ we have

$$\int \psi_m^*(q)\psi_n(q)dq \equiv \langle\psi_m|\psi_n\rangle \equiv \langle m|n\rangle = \delta_{mn}, \quad \text{discrete spectrum,} \tag{6.32}$$

$$\int \psi_\alpha^*(q)\psi_{\alpha'}(q)dq \equiv \langle\psi_\alpha|\psi_{\alpha'}\rangle \equiv \langle\alpha|\alpha'\rangle = \delta(\alpha - \alpha'), \quad \text{continuous spectrum.} \tag{6.33}$$

This is simply Equation 4.14 in some q-representation. Conditions (6.32) and (6.33) show that eigenstates $|\psi_n\rangle \equiv |n\rangle$ (or $|\psi_\alpha\rangle \equiv |\alpha\rangle$) of any Hermitian operator can indeed be interpreted as vectors of an orthonormal set. In the language of probabilities, for a particle in an eigenstate $|m\rangle$ the probability of finding it in this state is 1, and the probability of finding it in another eigenstate $|n\rangle$ is zero. Both these properties are expressed in (6.32). Similar properties are described by (6.33) for a continuous spectrum. So, just as the first column matrix in (6.31) gives $\mathbf{A}$ in the $\hat{\mathbf{e}}_n$ basis in V, the second one gives $|\Psi\rangle$ in the $|\psi_n\rangle$-basis in $\mathcal{H}$.

The scalar product of any two vectors $\mathbf{A}$ and $\mathbf{B}$ in V can be written in two ways:

$$\mathbf{A} \cdot \mathbf{B} = AB \cos\theta \tag{6.34a}$$

or

$$\mathbf{A} \cdot \mathbf{B} = \sum_n a_n b_n. \tag{6.34b}$$

Equation 6.34b uses a certain orthonormal basis with respect to which the components a_n and b_n have been determined, whereas (6.34a) uses only the properties of vectors themselves (their norms and relative orientation) without any reference to a coordinate system. We can also express the components a_n and b_n in terms of the basis vectors $\hat{e}_n$: $a_n = \mathbf{A} \cdot \hat{e}_n$ and $b_n = \mathbf{B} \cdot \hat{e}_n$, and apply Dirac's notation:

$$\hat{e}_n \to |n\rangle, \qquad \mathbf{A} \cdot \hat{e}_n \to \langle A|n\rangle, \qquad \hat{e}_n \cdot \mathbf{B} \to \langle n|B\rangle. \tag{6.35}$$

Then (6.34b) takes the form[1]

$$\mathbf{A} \cdot \mathbf{B} = \sum_n (\mathbf{A} \cdot \hat{e}_n)(\hat{e}_n \cdot \mathbf{B}) = \sum_n \langle A|n\rangle\langle n|B\rangle. \tag{6.36}$$

The same features show pairs of states in $\mathcal{H}$. Indeed, consider another normalized state function $|\Phi\rangle$ in the same basis as $|\Psi\rangle$, with amplitudes $d_m = \langle m|\Phi\rangle$:

$$|\Phi\rangle = \sum_m d_m |\psi_m\rangle \equiv \sum_m |m\rangle\langle m|\Phi\rangle. \tag{6.37}$$

Then we can define the inner product of the two states as the bracket $\langle \Psi|\Phi\rangle$ and write it in two ways:

$$|\langle \Psi|\Phi\rangle| = \left|\int \Psi^*(q)\Phi(q)dq\right| = |\cos\theta| \tag{6.38a}$$

or

$$\langle \Psi|\Phi\rangle = \sum_n c_n^* d_n. \tag{6.38b}$$

The result (6.38a) reflects the fact that for two normalized states $|\Psi\rangle$ and $|\Phi\rangle$, the absolute value of the integral (6.38) can be interpreted as $|\cos\theta| \leq 1$ (the Plancherel theorem [1]), so the two states are now 100% qualified to be called unit vectors in $\mathcal{H}$ with angle θ between them. Physically, (6.38) determines the amplitude for the particle in a state $|\Phi\rangle$ to be found in a state $|\Psi\rangle$, and vice versa. The result (6.38b) is obtained if we write $|\Phi\rangle$ and $|\Psi\rangle$ in (6.38a) in the $|\psi_m\rangle \equiv |m\rangle$-basis and apply condition (6.32). Both results together are quite similar to (6.34). Finally, multiplying (6.37) by the bra $\langle\Psi|$ from the left, we get the inner product of $|\Phi\rangle$ and $|\Psi\rangle$ in the form

$$\langle\Phi|\Psi\rangle = \sum_m \langle\Phi|m\rangle\langle m|\Psi\rangle. \tag{6.39}$$

This is mathematically identical to (6.36). Thus, the mathematical structures of $\mathcal{H}$-space and of V-space run along the same track! One important corollary is that $\sum |m\rangle\langle m| = I$ where summation is over all states of the chosen basis, and I is the identity matrix. When written in that form, this result is known as the *closure* (or *completeness*) relation. The above sum can always be sandwiched between the bra and the ket to assist us in finding the inner product.

Physically, the result (6.39) has a very straightforward interpretation: on the left, we have the transition amplitude between the states – the amplitude for a system initially in a state $|\Psi\rangle$ to be found, upon an appropriate measurement, in a state $|\Phi\rangle$. This is just a (generally, complex) number, which does not specify the possible ways

---

[1] Since the capped symbols for unit vectors are also used in many books (including this one) to denote *operators*, it may cause confusion. The reader should be careful to distinguish between the two cases from the context.

for transition $|\Psi\rangle \to |\Phi\rangle$. On the right, the same amplitude is given as a superposition of all possible ways from $|\Psi\rangle$ to $|\Phi\rangle$ via distinct intermediate states $|m\rangle$. There is an amplitude $\langle 1|\Psi\rangle$ for a system in state $|\Psi\rangle$ to be found in $|1\rangle$, and once in $|1\rangle$, there is an amplitude $\langle \Phi|1\rangle$ for finding it in $|\Phi\rangle$. So the amplitude of transition $|\Psi\rangle \to |\Phi\rangle$ via $|1\rangle$ is the product of the two, $\langle \Phi|1\rangle\langle 1|\Psi\rangle$. The same holds for transition via state $|2\rangle$, and so on. Thus, when all possibilities are open and no attempts are made to "peep into" and record any path, all of them are active and interfere with each other, so the result is determined by their superposition. It is precisely what we see on the right in (6.38). Thus, the abstract vectors in $\mathcal{H}$ reflect real physical effects. Recall also an example with the particle's propagation from one point to another via different virtual paths between them (Section 4.6).

All physically possible state vectors of a system reside in the corresponding $\mathcal{H}$-space. Each dimension of $\mathcal{H}$ corresponds to an eigenstate of the operator representing the variable of interest. As a system evolves, its state vector changes in direction while retaining magnitude. The tip of the vector traces out a curve on the spherical surface of unit radius. The difference from geometric vectors in V is that a state vector and the eigenstates representing it can be complex. Accordingly, the $\mathcal{H}$-space is a *complex vector space*. There is also some difference in normalization: $\sum a_n^2 = A^2$ for a vector in V and $\sum c_n^2 = 1$ for a vector in $\mathcal{H}$. This difference is only marginal and can be eliminated by normalizing **A** (dividing it by $|\mathbf{A}|$). Actually, there is no need even for this, since all transformations of interest in both spaces are rotations. In this case, all vectors **A** vary only in their directions; they have their tips sliding over the surface of radius $|\mathbf{A}|$. Once the radius is fixed, its specific numerical value is immaterial as far as mathematics is concerned.

The fact that the Hilbert space is generally complex is not a problem either. As mentioned, the concept of a vector space in algebra has long been generalized onto the complex vector space, in which both the unit vectors $\hat{\mathbf{e}}_j$ and projections b_j are complex numbers. A complex plane $z = x + yi$ can be considered from this viewpoint as a special case of a *one-dimensional* (1D) complex vector space with one axis real and another purely imaginary (for a modern view of the intimate ties between vectors and complex numbers see, for example, Refs [2,3]).

Below we consider some examples.

Start with the $\mathcal{H}$-space associated with spin variable. In the case of an electron, its spin component onto a chosen axis can have only two values. Accordingly, its spin state can be described by a vector in the 2D $\mathcal{H}$-space – one dimension for spin up, and the other for spin down. Note that the two spin components are two different spin projections onto *the same* axis in V, whereas the corresponding two states are represented by two *different* (mutually perpendicular) basis axes in $\mathcal{H}$. It may seem that a two-dimensional sheet of paper, or the surface of a flat board, will suffice to plot the state graphically as a vector with its tip on the unit circle. Denoting the eigenstate with spin up as $\mathbf{q}_1 = |\uparrow\rangle$ and its counterpart with spin down as $\mathbf{q}_2 = |\downarrow\rangle$, we can write the most general expression for an arbitrary state vector describing spin as a superposition

$$|\Psi\rangle = c_1\hat{\mathbf{q}}_1 + c_2\hat{\mathbf{q}}_2 = c_1|\uparrow\rangle + c_2|\downarrow\rangle, \tag{6.40}$$

with amplitudes c_1 and c_2 being projections of $|\Psi\rangle$ onto basis axes $\hat{\mathbf{q}}_1$ and $\hat{\mathbf{q}}_2$, respectively. But, since the $\mathcal{H}$-space is complex, these projections are, generally, complex numbers. As a result, the graphical representation of state vector (6.40) may be more subtle than just a vector with its tip on the unit circle. Two complex amplitudes are described by four real numbers. The normalization condition reduces this to three. And one of the two phase factors can be set equal to an arbitrary value for our convenience (due to global phase invariance!). This reduces the number of independent variables to two – still one more than the single angle needed to specify orientation of a unit vector in a plane. With two variables, it will be a unit vector in 3D space, whose orientation is determined by two parameters, for example, polar and azimuthal angles. We conclude that a state vector in 2D $\mathcal{H}$-space can be represented graphically by a 3-vector of unit length in real V-space. We will find the exact relationship between the two amplitudes and the corresponding angles in the next chapter.

What if we want to add also the *momentum* states for a more complete description of the electron? Then we need to expand $\mathcal{H}$ by adding a new dimension (a new basis vector linearly independent of all others) for each possible momentum. Since the number of distinct momentum states $|p\rangle$ of an electron is infinite, even if only in one direction, the resulting new $\mathcal{H}$-space is not just multidimensional; it has an infinite number of dimensions! Moreover, since the p-spectrum is continuous, all possible dimensions in $\mathcal{H}$ form an innumerable set! And the total dimensionality is twice innumerable (the number of dimensions $\mathcal{N} = 2 \cdot \infty$) due to two spin states for each p.

We might as well represent the electron's state in terms of position rather than momentum. Then, again, even in case of only one dimension (say, only the x-direction) in real space, each point of the electron's residence on the x-axis is a distinct physical state and should be represented by an independent eigenvector; so with spin included, we have again twice as many dimensions of the associated $\mathcal{H}$-space as the number of points on the line x!

An innumerable (continuous) set is infinitely more powerful than the set of all integers [4,5]. We don't even try to visualize the corresponding $\mathcal{H}$-space in its "wholeness." We can make only a few steps in this direction. We can imagine an eigenvector associated with a point $x = x_\alpha$ on the line as a unit vector "sticking out" from this point in some direction. But, since the directions for the unit vectors – each representing its respective point – must be not only different but also mutually orthogonal, our ordinary 3D space can only "accommodate" three such vectors, each corresponding to one out of three different arbitrarily chosen points on the line (Figure 6.1a). We can make it look more similar to the familiar 3D Cartesian coordinate system, with its triplet of unit basis vectors with common origin, if we bring all three vectors to one origin. In doing this, we do not lose any information about the locations of the points represented by these vectors, since each vector carries the corresponding label (Figure 6.1b).

For each new point on the line, we have to add a new direction perpendicular to all three already employed – this goes beyond our ability to depict things graphically. Besides the fact that the $\mathcal{H}$-space is complex, and so are its basis vectors, we cannot

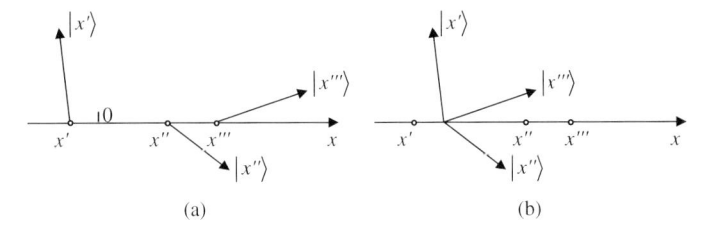

Figure 6.1 A subspace of the Hilbert space for three eigenfunctions of the position operator corresponding to the eigenvalues x', x'', and x'''. (a) Points $x = x'$, $x = x''$, and $x = x'''$ on the x-axis are different eigenvalues of the $\hat{x}$-operator. The corresponding eigenstates can be represented graphically by mutually perpendicular unit eigenvectors $|x'\rangle$, $|x''\rangle$, and $|x'''\rangle$ sticking out each from its respective point. (b) All three vectors brought to a common origin by parallel translations, which do not change the vectors and thereby the represented states.

visualize even one of them as a single directed segment representing an ordinary vector.[2]

But even though we cannot visualize multidimensional (and complex) vector spaces, we can understand them as abstract spaces with the same rules of operation as those in familiar spaces of one, two, and three dimensions.

If you are still in doubt how a line can, forming only a 1D space, give rise to an infinite-dimensional $\mathcal{H}$-space, ask this: What do we plot along a basis direction of our ordinary space, and along, say, a basis "x'-direction" of the corresponding $\mathcal{H}$-space?

Plotted along the basis x-direction in V-space is the geometrical projection of a position vector **A** onto the x-axis. Plotted along the x'-direction of the corresponding $\mathcal{H}$-space is the "projection" $c_{x'}$ of the state vector $|\Psi\rangle$, but, in addition, it has a physical meaning that makes a fundamental departure from classical intuition: the square modulus of this projection gives only the *probability* of finding the particle at $x = x'$ in a position measurement. In the case when the corresponding observable has a discrete spectrum, we can find the total probability by summing all these individual probabilities. Geometrically, the result is square of the length of the state vector (the Pythagorean theorem in multidimensional space). Physically, the result is the net probability of finding the system in any of its allowed states. Since the system can certainly be found in one of its allowed states, this net probability is, of course, equal to 1. Thus,

$$|a(n)|^2 \equiv \mathcal{P}(n) \leq 1, \qquad \sum_n |a(n)|^2 = |\langle \Psi | \Psi \rangle|^2 = 1. \qquad (6.41)$$

The state vector in this case turns out to have always the fixed length equal to 1. In other words, its tip cannot depart from the surface of multidimensional sphere of the unit radius. It can only slide over this sphere, "scanning" its surface.

---

2) Strictly speaking, the $\mathcal{H}$-space has been originally defined for square-integrable functions. Since the spectrum of eigenvalues of the position operator is continuous, its eigenfunctions are not square integrable in the ordinary sense. The normalization condition for them is given by (3.8) or the second Equation 6.33.

If the characteristic in question has a continuous spectrum (say, a space coordinate x), then we can only talk about some finite probability $d\mathcal{P}(x)$ of finding it within a small interval $(x, x+dx)$. By analogy with mechanics of a continuous medium, we define the corresponding *probability density* $w(x)$ as $w(x) \equiv d\mathcal{P}(x)/dx$.

So, the variable plotted along the "direction" x' is $c_{x'}$, whose modulus can have any value between 0 and ∞, depending on state Ψ. The variable plotted along some other "direction," say, x'', would be $c_{x''}$, also depending on state Ψ and generally with the same range of change as $c_{x'}$.[3] Again, keep in mind that here we consider, for instance, $c'_x = \Psi(x')$ as a variable depending *not* on x' (x' is fixed as an eigenvalue of the position operator) but on the physical state of the system. As an example, consider different possible solutions of the Schrödinger equation corresponding to different boundary conditions, or even the same solution of the time-dependent Schrödinger equation, but just evolving with time. In all these cases, the state vector $\Psi(x', t)$, evaluated for the same x' but at different t, goes through the corresponding $\mathcal{H}$-space, changing thereby its component $c_{x'}$ onto the x'-direction in $\mathcal{H}$. Physically this means changing the probability amplitude of finding the particle at $x = x'$.

Finally, we want to show another aspect of the $\mathcal{H}$-space by comparing it with the phase space used in classical statistical physics. For one particle moving in one direction, the corresponding *phase space* is 2D: one dimension representing x and the other representing p. The direction p_x on the phase diagram is perpendicular to the x-direction. Therefore, if we consider a QM state with definite p, it may be tempting to think that the corresponding direction must be "orthogonal" to the $\mathcal{H}$-space representing all the "x-eigenstates" considered above, and, if we want to include it into the picture, we must expand the previous $\mathcal{H}$-space by adding a new dimension perpendicular to all other preexisting directions. This argument seems especially compelling in view of the realization that each new point on the x-line requires a new dimension in the $\mathcal{H}$-space. It seems to be even more so for an additional variable different from x. But this time QM says that such an extension of our new intuition would be wrong! Indeed, according to the general relation (3.2), each p-eigenstate can be represented as an expansion over different "x-eigenstates":

$$\Psi_p(x) = \Psi_0 e^{ikx} = \Psi_0 \int e^{ikx'} \delta(x-x') dx' \tag{6.42}$$

or

$$|p\rangle = \frac{1}{\sqrt{2\pi}} \int e^{ikx'} |x'\rangle dx'. \tag{6.43}$$

---

[3] You may wonder why the amplitudes a are allowed here to go beyond $|a|=1$, if the state vector is to be unitary. The answer is that since the variable x is continuous, the corresponding expansion is an integral, not the sum, and $|\psi|^2$ is the probability density, rather than just a probability. The probability density may be arbitrarily large; hence the range of variable a_x may be infinite. If this variable were discrete ($a_x \to a_n$, with n integer), its modulus would range between 0 and 1.

The whole set of $\delta(x - x')$ consists of distinct eigenstates for different x', and $e^{ikx'}$ are the expansion coefficients. If a vector can be represented as a superposition of the eigenvectors of a given $\mathcal{H}$-space, this vector is itself an element of this space. The only thing that distinguishes it from all $|x\rangle$-eigenvectors used in the expansion is that this vector is not along any of $|x\rangle$ (if it were, it would be an x-eigenstate, not p-eigenstate). Thus, all states with definite p (the p-eigenstates) *also reside in the same $\mathcal{H}$-space formed by the x-eigenstates* (and vice versa)! The only difference between them is that the corresponding bases $(|x\rangle, |x'\rangle, \ldots)$ and $(|p\rangle, |p'\rangle, \ldots)$ are rotated with respect to each other, so that no p-eigenstate is along an x-eigenstate, and vice versa.

Summary. For any additional variable *compatible* with the x-variable, there should be the set of additional new dimensions in $\mathcal{H}$ for it to represent completely the state of the system. For any additional variable *incompatible* with the x-variable, there are no additional dimensions in $\mathcal{H}$. All the eigenstates of such a variable are already within the existing $\mathcal{H}$-space. This illustrates the fundamental difference between the *phase space* (p, x) formed by coordinate x and momentum p, on the one hand, and the $\mathcal{H}$-space formed by the *eigenstates* $|x'\rangle$ and/or *eigenstates* $|p'\rangle$, on the other. The phase space (p, x) is the 2D combination of independent (mutually perpendicular) 1D x-dimension + 1D p-dimension. The $\mathcal{H}$-space spanned by $|x'\rangle$ and by $|p'\rangle$ is *the same* for both sets. One set can be obtained from the other just by an appropriate rotation in the same space.

This is a special case of a general rule: suppose we have two different observables p and q. They form a 2D *phase space* (p, q), regardless of whether they are compatible or not. Let N_p denote the total number (which can be infinite!) of *eigenstates* of observable p, and let N_s be the total number of *eigenstates* of q. Then, if the observables are *compatible*, the corresponding two sets of their respective *eigenstates* form $(N_p N_s)$D $\mathcal{H}$-space. If they are incompatible like momentum and coordinate, then $N_p = N_s = N$, and the corresponding two sets of their eigenstates span the same common N-dimensional $\mathcal{H}$-space.

6.5
Operations in the Hilbert Space

It will be instructive to compare some operations on vectors in $\mathcal{H}$ and V spaces.

Start with V-space restricted to three dimensions. Vector addition is obvious and we skip it. Multiplication is of the two kinds – the *scalar* (dot) product and the *vector* (cross) product. We will focus here on the scalar product. The other term frequently used in QM for this operation is the *inner product*. From definition (6.34a) of the scalar product of two arbitrary vectors **A** and **B** making an angle θ with one another, we obtain the "multiplication table" for the system of three mutually orthogonal unit vectors $\hat{\mathbf{e}}_j$, $j = 1, 2, 3$. Namely, we see that all the products $T_{ij} \equiv \hat{\mathbf{e}}_i \cdot \hat{\mathbf{e}}_j = \delta_{ij}$ form a table

$$\hat{T} = \begin{pmatrix} 1 & 0 & 0 \\ 0 & 1 & 0 \\ 0 & 0 & 1 \end{pmatrix}. \tag{6.44}$$

6 Representations and the Hilbert Space

We say that the set T_{ij} forms a 3D second-rank tensor. We can use this to represent the scalar product (6.34a) in terms of components:

$$\mathbf{A} = \sum_i a_i \hat{\mathbf{e}}_i, \qquad \mathbf{B} = \sum_j b_j \hat{\mathbf{e}}_j. \tag{6.45}$$

Putting this in (6.34a) and using (6.44) gives $\mathbf{A} \cdot \mathbf{B} = \sum_i \sum_j \hat{\mathbf{e}}_i \cdot \hat{\mathbf{e}}_j a_i b_j = \sum_i a_i b_i$, that is, we recover (6.34b). Since $AB\cos\theta \le AB$, the right side satisfies the condition

$$\left(\sum_i a_i b_i\right)^2 \le \left(\sum_i a_i^2\right)\left(\sum_j b_j^2\right). \tag{6.46}$$

This is a special case of the Schwarz inequality [6–8]. If we normalize all vectors in V (make their magnitudes equal to 1), this condition takes the form $\sum_i a_i b_i \le 1$.

We could as well express $\hat{\mathbf{e}}_i$ themselves in the matrix form. For the 3D case, we have

$$\hat{\mathbf{e}}_1 = \begin{pmatrix} 1 \\ 0 \\ 0 \end{pmatrix}, \qquad \hat{\mathbf{e}}_2 = \begin{pmatrix} 0 \\ 1 \\ 0 \end{pmatrix}, \qquad \hat{\mathbf{e}}_3 = \begin{pmatrix} 0 \\ 0 \\ 1 \end{pmatrix}. \tag{6.47}$$

In the scalar product of two such matrices, the first must be transposed, for instance,

$$\hat{\mathbf{e}}_1 \cdot \hat{\mathbf{e}}_2 = (1 \ 0 \ 0)\begin{pmatrix} 0 \\ 1 \\ 0 \end{pmatrix} = 0, \qquad \hat{\mathbf{e}}_1 \cdot \hat{\mathbf{e}}_1 = (1 \ 0 \ 0)\begin{pmatrix} 1 \\ 0 \\ 0 \end{pmatrix} = 1.$$

$$\tag{6.48}$$

With these rules, it is easy to check that all products (6.48) satisfy $\hat{\mathbf{e}}_i \cdot \hat{\mathbf{e}}_j = \delta_{ij}$ and can accordingly be arranged into the table (6.44).

Using (6.47), we can represent $\mathbf{A}$ and $\mathbf{B}$ in the matrix form as

$$\mathbf{A} = a_1\begin{pmatrix} 1 \\ 0 \\ 0 \end{pmatrix} + a_2\begin{pmatrix} 0 \\ 1 \\ 0 \end{pmatrix} + a_3\begin{pmatrix} 0 \\ 0 \\ 1 \end{pmatrix} = \begin{pmatrix} a_1 \\ a_2 \\ a_3 \end{pmatrix},$$

$$\mathbf{B} = b_1\begin{pmatrix} 1 \\ 0 \\ 0 \end{pmatrix} + b_2\begin{pmatrix} 0 \\ 1 \\ 0 \end{pmatrix} + b_3\begin{pmatrix} 0 \\ 0 \\ 1 \end{pmatrix} = \begin{pmatrix} b_1 \\ b_2 \\ b_3 \end{pmatrix}. \tag{6.49}$$

Their dot product in this form is

$$\mathbf{A} \cdot \mathbf{B} = (a_1 \ a_2 \ a_3)\begin{pmatrix} b_1 \\ b_2 \\ b_3 \end{pmatrix} = a_1 b_1 + a_2 b_2 + a_3 b_3, \tag{6.50}$$

which again recovers (6.34b). (In a V-space with N dimensions, we would accordingly have N-dimensional matrices and N terms in the resulting sum of products in (6.50).)

Consider now the same operation in $\mathcal{H}$. In this case, the role of unit vectors $\hat{\mathbf{e}}_i$ is played by basis states $|e_j\rangle \equiv |j\rangle$, which are the eigenstates of a chosen operator $\hat{L}$.

Dropping the restriction to three dimensions, we can write their bras and kets in the matrix form

$$\langle j| = \begin{pmatrix} 0 & \cdots & \underset{\text{(jth column)}}{1} & \cdots & 0 & \cdots \end{pmatrix} \quad \text{and} \quad |j\rangle = \begin{pmatrix} 0 \\ \vdots \\ 1 \text{ (jth row)} \\ \vdots \\ 0 \\ \vdots \end{pmatrix}$$

(6.51)

(we assume the discrete spectrum here; for a continuous spectrum we would have continuous matrices, which cannot be represented by a table). All possible inner products $\langle j|i\rangle$ of the basis states satisfy $\langle j|i\rangle = \delta_{ij}$ and can be arranged into the identity matrix $\hat{I}$ similar to (6.44).

Any two state vectors $|\Psi\rangle$ and $|\Phi\rangle$ can be represented as the respective superpositions

$$|\Psi\rangle = \sum_i a_i |i\rangle, \qquad |\Phi\rangle = \sum_j b_j |j\rangle. \tag{6.52}$$

The coefficients a_i and b_j can be arranged into column matrices similar to (6.49). Now we can apply the same rules used in the V-space, in either vector or matrix form, to get the inner product $\langle \Phi|\Psi\rangle$. The bra of $|\Phi\rangle$ is $\langle\Phi| = \begin{pmatrix} b_1^* & b_2^* & \cdots & b_j^* & \cdots \end{pmatrix}$. It follows

$$\langle\Phi|\Psi\rangle = \begin{pmatrix} b_1^* & b_2^* & \cdots & b_i^* & \cdots \end{pmatrix} \begin{pmatrix} a_1 \\ \vdots \\ a_i \\ \vdots \end{pmatrix} = \sum_i a_i b_i^*, \tag{6.53}$$

in close analogy with (6.50).

Summary. The inner product of two state vectors in $\mathcal{H}$ is the projection of one vector onto the other. It can be expressed either directly in terms of vectors themselves (the left side of (6.53)) or in terms of their components (or projections) in some chosen basis (the right side of (6.53)). The direct expression $\langle\Phi|\Psi\rangle$ is more straightforward in saying that it is a probability amplitude of finding the state $|\Phi\rangle$ in state $|\Psi\rangle$; in other words, the amplitude of collapse, upon appropriate measurement, from $|\Psi\rangle$ to $|\Phi\rangle$. The alternative expression on the right indicates possible virtual paths for this transition in the chosen basis.

Below we consider some operations including sums and products of state vectors. To make further discussion more concrete, let us focus on position ($\mathbf{r}$) and momentum ($\mathbf{k} = \mathbf{p}/\hbar$) states. In this case, it will be convenient to switch back from (x_1, x_2, x_3), (k_1, k_2, k_3) to old notations (x, y, z), (k_x, k_y, k_z).

I) Start with the states $|x\rangle$ and $|x'\rangle$ and consider their direct product $|x\rangle \otimes |x'\rangle$. Formally, this is the probability amplitude of the combined event: finding the particle at x *and* finding it at x'. We know (recall Section 4.1) that this is impossible – such an event does not exist. Therefore, the product $|x\rangle \otimes |x'\rangle$ is an illegal operation – it does not reflect any experimental situation.

It may be informative to compare this with a legal operation – the superposition

$$a|x\rangle + a'|x'\rangle. \tag{6.54}$$

Classically, this would also be a meaningless expression since a classical object cannot be in two different places at once. Quantum mechanically, this is a mundane effect – superposition of two states with different positions. It is a state with coexisting *potentialities*, and as such it is a fundamental part of QM. The two possibilities represented in (6.54) cannot actualize together: the position measurement will find the particle either at x or at x'.

II) Now consider the product $|x\rangle \otimes |y\rangle$. It determines the probability amplitude of finding the particle at a point with coordinates (x, y). Both coordinates coexist as components of position vector $\mathbf{r} = x\hat{\mathbf{x}} + y\hat{\mathbf{y}}$ in V. In $\mathcal{H}$, the product $|x\rangle \otimes |y\rangle$ is just an eigenstate $|\mathbf{r}\rangle$ of position operator:

$$|\mathbf{r}\rangle = |x\rangle|y\rangle \equiv |x\rangle \otimes |y\rangle. \tag{6.55}$$

In position representation, this eigenstate is described by Equation (5.34) for 2D. Thus, (6.55) is a legitimate tensor product considered in Section 4.1.

III) The linear combination of states

$$|\psi\rangle = \alpha|x, 0\rangle + \beta|0, y\rangle = \alpha|x\rangle|0\rangle + \beta|0\rangle|y\rangle \tag{6.56}$$

is a superposition in $\mathcal{H}$, but it is *not* an eigenstate $|\mathbf{r}\rangle$ of position operator. Upon appropriate position measurement, it will collapse to one of the eigenstates $|x, 0\rangle$ or $|0, y\rangle$, but not to position $\mathbf{r}$. Even the vector expectation value $\langle \mathbf{r} \rangle \equiv \bar{\mathbf{r}}$ in state (6.56) (Problem 6.11) will be different from the position vector $\mathbf{r} = (x, y)$. The latter is determined exclusively by the chosen values of x and y, whereas the former depends also on the amplitudes α and β. The two kinds of states described by operations II and III are represented graphically in Figure 6.2.

IV) The same arguments apply to the momentum variables. The product $|k_x\rangle|k'_x\rangle$ is illegal. But the linear combination $b|k_x\rangle + b'|k'_x\rangle$ describes a known state – superposition of two de Broglie waves with different propagation numbers, running along the same x-direction. In the x-representation, we have

$$\psi(x) = b\,e^{ik_x x} + b'\,e^{ik'_x x}. \tag{6.57}$$

For a single particle in such state, both waves coexist as potentialities, one of which will be actualized by the momentum measurement. In a special case when $k'_x = -k_x$, we have the superposition of two oncoming waves of the same frequency, and if, in addition, $|b'| = |b|$, they form a standing wave.

V) On the other hand, the product

$$|k_x\rangle|k_y\rangle \equiv |k_x\rangle \otimes |k_y\rangle = |\mathbf{k}\rangle \tag{6.58}$$

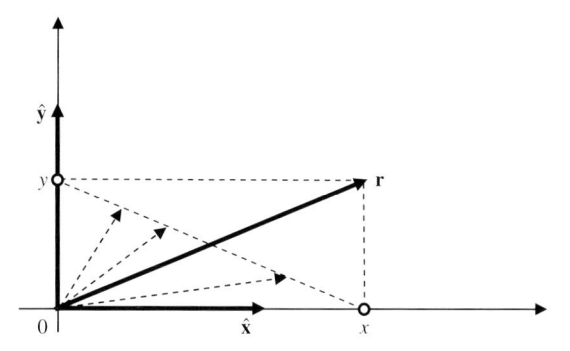

Figure 6.2 The tensor product $|\mathbf{r}\rangle = |x\rangle \otimes |y\rangle$ in $\mathcal{H}$ is an eigenfunction of position operator $\hat{\mathbf{r}}$. Its experimentally measured vector eigenvalue is represented graphically as composition $\mathbf{r} = x\hat{\mathbf{x}} + y\hat{\mathbf{y}}$ in V, with selected values of x, y indicated by the dots on the corresponding axes. The linear superposition $|\phi\rangle = \alpha|x\rangle|0\rangle + \beta|0\rangle|y\rangle$ is *not* an eigenfunction of $\hat{\mathbf{r}}$. Its vector expectation value $\langle \mathbf{r}\rangle \equiv \bar{\mathbf{r}}$ is a function of ratio α/β. All possible $\bar{\mathbf{r}}$ have their tips on the diagonal xy determined by the chosen values of x, y. Three of them are shown as the dashed vectors.

is a legitimate tensor product. It describes an eigenstate of the momentum operator with the vector eigenvalue

$$\mathbf{k} = k_x\hat{\mathbf{x}} + k_y\hat{\mathbf{y}}. \tag{6.59}$$

Such state exists because the *x*- and *y*-components of momentum are compatible observables (Figure 6.3). In position representation,

$$\psi_\mathbf{k}(\mathbf{r}) = \psi_0\, e^{i\mathbf{k}\mathbf{r}} = \psi_0\, e^{ik_x x}\, e^{ik_y y} = e^{i(k_x x + k_y y)}. \tag{6.60}$$

Since k_x and k_y are components of $\mathbf{k}$, they satisfy the *dispersion equation*

$$k_x^2 + k_y^2 = k^2 = \begin{cases} \dfrac{\omega^2}{c^2}, & \text{for a photon,} \\ \dfrac{2\mu E}{\hbar^2}, & \text{for a nonrelativistic particle.} \end{cases} \tag{6.61}$$

This means that the state (6.60) is an eigenstate of both the momentum and the energy operator.

VI) Finally, consider the combination

$$|\psi\rangle = b_1|k_1, 0\rangle + b_2|0, k_2\rangle. \tag{6.62}$$

In position representation,

$$\psi(\mathbf{r}) = \psi(x, y) = b_1\, e^{ik_1 x} + b_2\, e^{ik_2 y}. \tag{6.63}$$

There are no typos here! We deliberately changed (k_x, k_y) back to (k_1, k_2) in order to emphasize that now there are no connections between them – they are not components of a single propagation vector $\mathbf{k}$, but independent propagation numbers for the two waves running along *x* and *y*, respectively. For instance, in

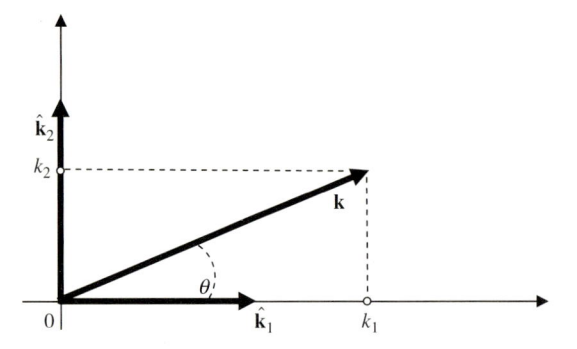

Figure 6.3 The tensor product of two momentum eigenstates $|k_1\rangle$ and $|k_2\rangle$ produces a new momentum *eigenstate* $|k\rangle$ in $\mathcal{H}$. Its vector eigenvalue **k** is represented graphically in V as a superposition $\mathbf{k} = k_1\hat{\mathbf{k}}_1 + k_2\hat{\mathbf{k}}_2$. This superposition of *observables in V* should never be confused with superposition of *states in* $\mathcal{H}$.

the case of a photon, we will have instead of (6.61)

$$k_1 = \frac{\omega_1}{c}, \quad k_2 = \frac{\omega_2}{c}. \tag{6.64}$$

Even in the special case when $\omega_1 = \omega_2 = \omega$, the components $k_1 = k_2$ will not satisfy the dispersion equation! Generally, (6.63) is not a monochromatic state, that is, not an eigenstate of the system's Hamiltonian. Nor is it a momentum eigenstate, because in each momentum measurement we will obtain one of the two mutually orthogonal momenta with respective magnitudes given by (6.64) instead of the single momentum (6.59) (Figure 6.4).

The photon in superposition (6.62) of such motions is totally different from the photon (6.58) of the same polarization with momentum **k**. The state (6.58) is a monochromatic plane wave moving along direction **k**. Its wavefronts are continuous flat planes perpendicular to **k** and propagating along it with phase velocity $\omega/|\mathbf{k}|$ (Figure 6.4a). The state (6.62) is a superposition of two different plane waves moving along the *x*- and *y*-directions, respectively. Even in a region where the individual waves overlap so that they can produce an interference pattern, their superposition does not produce a single plane wave moving along direction **k** (Figure 6.4b). The superposition does not have a single wavefront in the plane perpendicular to **k**; it forms a set of adjacent strips in this plane with local field amplitudes depending on position and with opposite phases in neighboring strips; each such set can be moving with a superluminal velocity (see a more detailed discussion in Ref. [9], Section 6.14).

The difference between the states (6.58) and (6.62) is manifest already at the stage of their preparation. State (6.58) can be prepared in the laboratory just by launching a photon from a laser with its barrel making an angle $\varphi = \arctan(k_y/k_x)$ with the *x*-axis. If we position three detectors D_1, D_2, and D_3 as shown in Figure 6.4, very far from each other in order to avoid the effects of

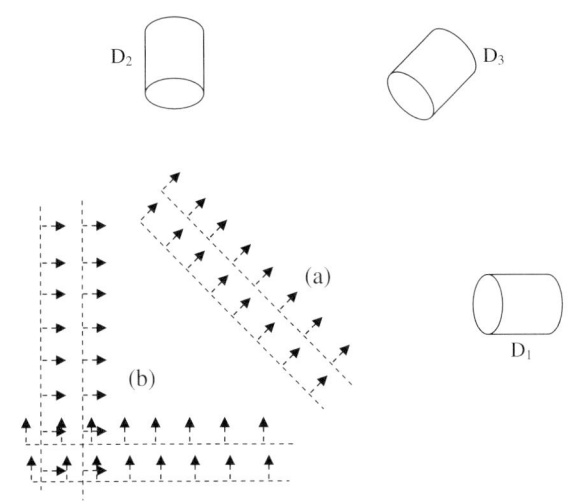

Figure 6.4 The difference in two states. (a) An eigenstate $|\mathbf{k}\rangle$ of momentum operator $\hat{\mathbf{k}} = \hat{k}_x \hat{\mathbf{x}} + \hat{k}_y \hat{\mathbf{y}}$ in V is a *tensor product* of states $|k_x\rangle$ and $|k_y\rangle$ in $\mathcal{H}$; it is *not* a superposition of these states. (b) A superposition of states $|k_x\rangle$ and $|k_y\rangle$ in $\mathcal{H}$ is *not* an eigenstate of operator $\hat{\mathbf{k}} = \hat{k}_x \hat{\mathbf{x}} + \hat{k}_y \hat{\mathbf{y}}$. Here $\hat{k}_x = -i(\partial/\partial x)$ and $\hat{k}_y = -i(\partial/\partial y)$ are operators, whereas $\hat{\mathbf{x}}$ and $\hat{\mathbf{y}}$ are the unit vectors.

diffraction spread of the initial laser beam, then the photon will always hit detector D_3 and never detector D_1 or D_2.

A state (6.62) for a single photon can be prepared by using the same source, but with an appropriate beam splitter (BS) added to it. In principle, it is possible to produce a single photon in a *nonmonochromatic* superposition of states (6.5.23), for instance, by passing it through a *moving* BS (Figure 6.5). But if we place the same detectors in the same positions very far from the beam splitter, as in Figure 6.4, the photon will be always found either in D_1 (with probability $|b_1|^2$) or in D_2 (with probability $|b_2|^2$) and never in D_3.[4]

Summary. In expression (6.59), k_x and k_y are compatible observables; they figure directly as coefficients at $\hat{\mathbf{e}}_1 = \hat{\mathbf{x}}$ and $\hat{\mathbf{e}}_2 = \hat{\mathbf{y}}$. In state (6.62), they are only mutually exclusive alternatives of an appropriate momentum measurement.

These comments show that frequently used term "superposition" in V may have totally different meaning from the one in $\mathcal{H}$. In the former, the superposed entities coexist before and after the measurement, and therefore can be measured simultaneously. In the latter, they coexist only as potentialities before the measurement, and collapse only to a single actual state in the process of measurement. Therefore, an

---

4) Strictly speaking, the described experimental setup using detectors constitutes *position* measurement rather than *momentum* measurement. Nevertheless, we employ it here simply because it is sufficient to illustrate the main point.

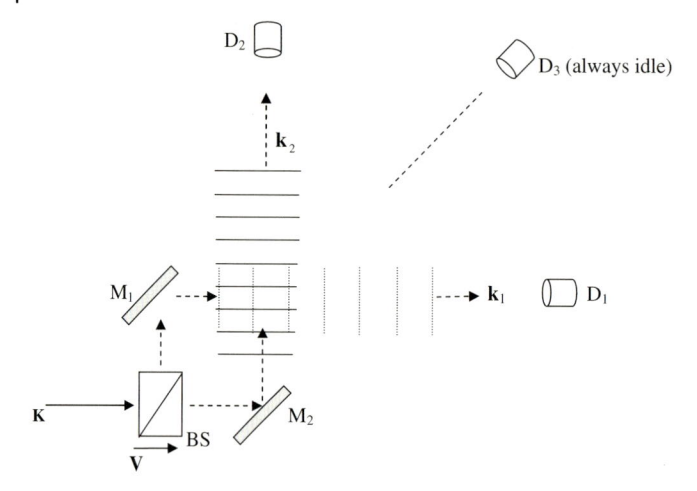

Figure 6.5 Schematic of an experiment transforming a photon state with a sharply defined momentum **K** into superposition of states with momenta k_1 and k_2. The initial state is the incident monochromatic wave with momentum **k**. It can be split into two waves with *different frequencies* if at least one of the transforming elements (beam splitter BS or the mirrors M_1 and M_2) is moving. In the case illustrated here, we have the BS moving horizontally with velocity **V** away from the source, and the mirrors are stationary. In this case, $|k_1| < |k_2| = |k|$ (notations k_1 and k_2 are introduced after reflection in the respective mirrors, in order to maintain the correspondence $(x, y) \leftrightarrow (1, 2)$). Here k_1 and k_2 are characteristics of a single photon state, and yet they are not components of any single momentum **k**. Accordingly, we will observe in each trial either detector D_1 or D_2 activated, but never D_3.

attempt to visualize QM superposition as the vector sum of the type (6.59), based on its *mathematical* similarity with (6.62), may be dangerously misleading. We have to recognize that trying to *visualize* QM phenomena using familiar classical images may be totally futile. In this respect, it would be safer to use the word "combination" or "composition" instead of "superposition" for a vector sum in V.

The remaining essential part of operations in $\mathcal{H}$ is outer product of states. We will discuss some of its features in the next section.

6.6
Nonorthogonal States

Up to now we were specifying a state relative to a basis formed by mutually orthogonal eigenvectors. Here we will discuss nonorthogonal states.

As mentioned in Section 5.4, such states can appear in case of degeneracy. Another possibility is to create them from *orthogonal* eigenstates belonging to different eigenvalues of some operator. We can always form from them two or more mutually nonorthogonal superpositions.

6.6 Nonorthogonal States

In this section, we will consider nonorthogonal states of a two-state system – the most simple but important case. We start with the familiar orthogonal "spin-up" and "spin-down" states and use them to construct, in obvious analogy with geometry of the real space, two nonorthogonal states:

$$|u_0\rangle = a|\uparrow\rangle + b|\downarrow\rangle = a\begin{pmatrix}1\\0\end{pmatrix} + b\begin{pmatrix}0\\1\end{pmatrix} = \begin{pmatrix}a\\b\end{pmatrix} \qquad (6.65)$$

and

$$|u_1\rangle = c|\uparrow\rangle + d|\downarrow\rangle = c\begin{pmatrix}1\\0\end{pmatrix} + d\begin{pmatrix}0\\1\end{pmatrix} = \begin{pmatrix}c\\d\end{pmatrix}. \qquad (6.66)$$

The amplitudes a, b, c, and d are subjected to normalization conditions $|a|^2 + |b|^2 = 1$ and $|c|^2 + |d|^2 = 1$, but otherwise they are arbitrary. If, in addition, they satisfied

$$ac^* + bd^* = 0, \qquad (6.67)$$

then the new states $|u_0\rangle$ and $|u_1\rangle$ would be also mutually orthogonal. But now we assume that (6.67) is *not* satisfied, so we know these states are nonorthogonal. Let us try to manipulate these states now.

Using the matrix representations (6.65) and (6.66), write first the bras $\langle u_0| = (a^*\ b^*)$ and $\langle u_1| = (c^*\ d^*)$.

Obviously, the new states remain normalized:

$$\langle u_0|u_0\rangle = (a^*\ b^*)\begin{pmatrix}a\\b\end{pmatrix} = a^*a + b^*b = 1, \qquad \langle u_1|u_1\rangle = (c^*\ d^*)\begin{pmatrix}c\\d\end{pmatrix}$$
$$= c^*c + d^*d = 1. \qquad (6.68)$$

Consider now the inner product of $|u_0\rangle$ and $|u_1\rangle$,

$$\langle u_0|u_1\rangle = (a^*\ b^*)\begin{pmatrix}c\\d\end{pmatrix} = a^*c + b^*d, \qquad \langle u_1|u_0\rangle = (c^*\ d^*)\begin{pmatrix}a\\b\end{pmatrix}$$
$$= ac^* + bd^*. \qquad (6.69)$$

The *absence* of condition (6.67) guarantees that the inner products, and thereby the projections of the states $|u_0\rangle$, $|u_1\rangle$ onto each other, are nonzero, so the states are indeed nonorthogonal.

Also, (6.69) is in compliance with the general rule $\langle u_1|u_0\rangle = \langle u_0|u_1\rangle^*$, as it should be.
Now consider the *outer products* $|u_0\rangle\langle u_0|$ and $|u_1\rangle\langle u_1|$:

$$|u_0\rangle\langle u_0| = \begin{pmatrix}a\\b\end{pmatrix}(a^*\ b^*) = \begin{pmatrix}aa^* & ab^*\\ a^*b & bb^*\end{pmatrix}, \qquad |u_1\rangle\langle u_1| = \begin{pmatrix}c\\d\end{pmatrix}(c^*\ d^*)$$
$$= \begin{pmatrix}cc^* & cd^*\\ c^*d & dd^*\end{pmatrix}. \qquad (6.70)$$

In accordance with the general theory, they are operators. Denote the first one as $\hat{P}_0$ and find the result of its action on our states:

$$\hat{P}_0|u_0\rangle \equiv |u_0\rangle\langle u_0||u_0\rangle \quad \text{and} \quad \hat{P}_0|u_1\rangle \equiv |u_0\rangle\langle u_0||u_1\rangle. \tag{6.71}$$

In view of the associative rule for multiplication, together with normalization (6.68), we immediately see that

$$\hat{P}_0|u_0\rangle = |u_0\rangle \quad \text{and} \quad \hat{P}_0|u_1\rangle = \langle u_0|u_1\rangle|u_0\rangle = \lambda|u_0\rangle,$$
$$\lambda \equiv \langle u_0|u_1\rangle = a^*c + b^*d. \tag{6.72}$$

The reader can check this by directly multiplying matrices (6.70) and (6.65), (6.66).

Thus, $|u_0\rangle$ is the eigenfunction of operator $\hat{P}_0$ with eigenvalue 1. In other words, $\hat{P}_0$ projects $|u_0\rangle$ onto itself. But $|u_1\rangle$ is *not* the eigenfunction of $\hat{P}_0$. Under the action of $\hat{P}_0$, it converts into $|u_0\rangle$ with an (generally complex) multiplying factor $\lambda \equiv \langle u_0|u_1\rangle$ equal to projection of $|u_1\rangle$ onto $|u_0\rangle$. We see that $\hat{P}_0$ is a *projection operator* with projection axis $|u_0\rangle$. In a similar way, we can introduce another projection operator $\hat{P}_1 \equiv |u_1\rangle\langle u_1|$ that projects both states onto $|u_1\rangle$.

Consider now the operator

$$\hat{\Pi}_0 \equiv 1 - \hat{P}_0 = \begin{pmatrix} 1 & 0 \\ 0 & 1 \end{pmatrix} - \begin{pmatrix} aa^* & ab^* \\ a^*b & bb^* \end{pmatrix} = \begin{pmatrix} bb^* & -ab^* \\ -a^*b & aa^* \end{pmatrix}. \tag{6.73}$$

Our intuition may tell us that if $\hat{P}_0$ projects the two states onto $|u_0\rangle$, then $\hat{\Pi}_0$ must project them onto direction orthogonal to $|u_0\rangle$. Let us check it:

$$\hat{\Pi}_0|u_0\rangle \equiv (1 - |u_0\rangle\langle u_0|)|u_0\rangle = |u_0\rangle - |u_0\rangle\langle u_0|u_0\rangle = 0 \cdot |u_0\rangle. \tag{6.74}$$

Thus, $|u_0\rangle$ is an eigenfunction of $\hat{\Pi}_0$ with the eigenvalue 0. Geometrically, 0 is a projection of a vector onto the direction perpendicular to it. Therefore, $\hat{\Pi}_0$ projects $|u_0\rangle$ onto direction orthogonal to $|u_0\rangle$. Applying $\hat{\Pi}_0$ to $|u_1\rangle$ gives

$$\hat{\Pi}_0|u_1\rangle = |u_1\rangle - |u_0\rangle\langle u_0||u_1\rangle = |u_1\rangle - \langle u_0|u_1\rangle|u_0\rangle. \tag{6.75}$$

Here $\langle u_0|u_1\rangle|u_0\rangle$ is projection of $|u_1\rangle$ onto $|u_0\rangle$; therefore, the whole expression on the right is the projection of $|u_1\rangle$ onto the direction perpendicular to $|u_0\rangle$ (Figure 6.6). Our intuition was correct!

In the same way, we can introduce an operator $\hat{\Pi}_1 \equiv 1 - \hat{P}_1$ and show that it projects $|u_0\rangle$ and $|u_1\rangle$ onto direction perpendicular to $|u_1\rangle$.[5]

An interesting (and important) question arises when we expand an *arbitrary* state $|\Psi\rangle$ over the complete set of nonorthogonal states $|u_j\rangle$. Namely, if we represent $|\Psi\rangle$ as a superposition of such states, what is the normalization condition for the superposition amplitudes? Of course, each state vector $|u_j\rangle$ of the new basis remains normalized – its length (its inner product with itself) remains 1, and the same is true for $|\Psi\rangle$. But now, even though the length of each state vector remains 1, its

---

5) Our notations for $\hat{\Pi}_0$ and $\hat{\Pi}_1$ may be different from those in the original publications; see, for example, Ref. [10].

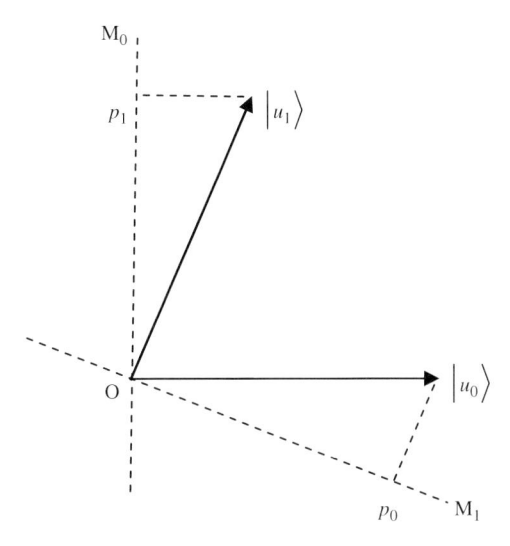

Figure 6.6 Projection operations $\hat{\Pi}_1$ and $\hat{\Pi}_2$ with a nonorthogonal basis in the 2D Hilbert space. $|u_0\rangle$ and $|u_1\rangle$: two nonorthogonal basis vectors; OM_0: direction orthogonal to $|u_0\rangle$; OM_1: direction orthogonal to $|u_1\rangle$; Op_0: projection of $|u_0\rangle$ onto OM_1; Op_1: projection of $|u_1\rangle$ onto OM_0.

projections will not satisfy (3.7). We want to find the general form of the normalization condition that would hold in a nonorthogonal basis.

Here, we will again restrict to two states. In this case, the analogy between the $\mathcal{H}$-space and V-space, albeit not as simple as for orthogonal bases, still remains pretty straightforward. Consider a regular plane with two reference directions x_1 and x_2 making an angle $\varphi \neq \pi/2$ with each other (Figure 6.7). The unit vectors $\mathbf{e}_1$ and $\mathbf{e}_2$ along these directions represent the two respective basis states. An arbitrary state can be represented by a vector $\mathbf{A}$. Geometrically, $\mathbf{A}$ is a vector sum of its *oblique* components $\tilde{a}_1$ and $\tilde{a}_2$:

$$\mathbf{A} = \tilde{a}_1 \mathbf{e}_1 + \tilde{a}_2 \mathbf{e}_2. \tag{6.76}$$

In terms of quantum states in $\mathcal{H}$, we have $\mathbf{A} \to |\Psi\rangle$, $\mathbf{e}_1 \to |u_1\rangle$, $\mathbf{e}_2 \to |u_2\rangle$, and (6.76) converts into

$$|\Psi\rangle = \tilde{a}_1 |u_1\rangle + \tilde{a}_2 |u_2\rangle. \tag{6.77}$$

This expression looks very familiar, like an ordinary superposition. But since the states $|u_1\rangle$ and $|u_2\rangle$ are nonorthogonal, the coefficients $\tilde{a}_1$ and $\tilde{a}_2$ are *not* the probability amplitudes! The latter are the *normal projections* a_1 and a_2 of $|\Psi\rangle$ onto $|u_1\rangle$ and $|u_2\rangle$, respectively, whereas $\tilde{a}_1$ and $\tilde{a}_2$ are the oblique projections. In physics, the normal projections are called *covariant* components of a vector, whereas oblique projections are called *contravariant* components (see, for example, Refs [11,12]). The latter are usually labeled with superscripts, but here we distinguish them by tilde sign in order to avoid confusion with powers of a.

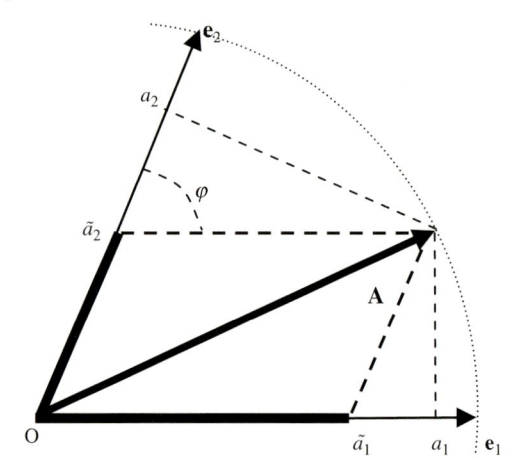

Figure 6.7 Contravariant $(\tilde{a}_1, \tilde{a}_2)$ and covariant (a_1, a_2) coordinates of a vector **A** in a nonrectangular coordinate system in 2D space. Contravariant components are oblique projections and covariant components are normal (orthogonal) projections of **A**.

According to the law of cosines, we have $A^2 = \tilde{a}_1^2 + 2\tilde{a}_1\tilde{a}_2 \cos\varphi + \tilde{a}_2^2$, and since **A** (or $|\Psi\rangle$) is normalized, this gives us the normalization condition for the contravariant components:

$$\tilde{a}_1^2 + 2\tilde{a}_1\tilde{a}_2 \cos\varphi + \tilde{a}_2^2 = 1. \tag{6.78}$$

Next, take into account that $\mathcal{H}$ is the complex vector space, so that $\tilde{a}_j^2 \to \tilde{a}_j\tilde{a}_j^*$ ($j = 1, 2$), $\cos\varphi \to \langle u_1|u_2\rangle$, and instead of (6.78) we must write

$$\tilde{a}_1\tilde{a}_1^* + \left(\tilde{a}_1\tilde{a}_2^* + \tilde{a}_2\tilde{a}_1^*\right)\langle u_1|u_2\rangle + \tilde{a}_2\tilde{a}_2^* = 1. \tag{6.79}$$

It obviously differs from the "normal" normalization condition by an extra term depending on φ in (6.78) or explicitly containing the dot product of the basis states in (6.79). But it still seems that all is not lost. Rewriting (6.78) as $1 = \tilde{a}_1(\tilde{a}_1 + \tilde{a}_2 \cos\varphi) + \tilde{a}_2(\tilde{a}_2 + \tilde{a}_1 \cos\varphi)$, we notice that the expressions in parentheses on the right are nothing else but the normal projections of **A** (Figure 6.7):

$$a_1 = \tilde{a}_1 + \tilde{a}_2 \cos\varphi, \qquad a_2 = \tilde{a}_2 + \tilde{a}_1 \cos\varphi. \tag{6.80}$$

Therefore, we can rewrite (6.79) as

$$a_1\tilde{a}_1 + a_2\tilde{a}_2 = 1. \tag{6.81}$$

It looks like we have recovered the familiar expression for normalization. But alas, this comes at the cost of using *two* types of vector components (co- and contravariant) in one expression. We are looking for the expression containing *only* the normal components a_1 and a_2 (orthogonal projections of **A**), since only they have the physical meaning of probability amplitudes. The sought-for expression is easy to obtain by solving Equation 6.80 for $\tilde{a}_1$ and $\tilde{a}_2$:

$$\tilde{a}_1 = \frac{a_1 - a_2 \cos\varphi}{\sin^2\varphi}, \qquad \tilde{a}_2 = \frac{-a_1 \cos\varphi + a_2}{\sin^2\varphi}. \tag{6.82}$$

Putting this into (6.81) yields

$$a_1^2 - 2a_1 a_2 \cos\varphi + a_2^2 - \sin^2\varphi. \tag{6.83}$$

In the language of the Hilbert space, we have $a_j^2 \to a_j a_j^*$, and (6.84) converts to

$$a_1 a_1^* - (a_1 a_2^* + a_2 a_1^*)\langle u_1|u_2\rangle + a_2 a_2^* = 1 - |\langle u_1|u_2\rangle|^2. \tag{6.84}$$

We obtained the normalization condition for a state vector in a nonorthogonal basis. In the limit $\varphi \to \pi/2$, this reduces to the known expression for an orthogonal basis.

For an arbitrary number of linearly independent states and respective dimensions in $\mathcal{H}$, the result can be written as

$$\sum_n |a_n|^2 + \sum_{m\neq n} a_m^* a_n \langle u_m|u_n\rangle + \frac{1}{2}\sum_{m\neq n}|\langle u_m|u_n\rangle|^2 = 1. \tag{6.85}$$

If we want to express this in some specific representation q, the result will be

$$\sum_n |a_n|^2 + \sum_{m\neq n} a_m^* a_n \int u_m^*(q)u_n(q)dq + \frac{1}{2}\sum_{m\neq n}\left|\int u_m^*(q)u_n(q)dq\right|^2 = 1. \tag{6.86}$$

It becomes increasingly cumbersome as we go to a higher number of dimensions (let alone the $\mathcal{H}$-space for continuous variables).

Problems

6.1 Show that the transpose of the product of two $N \times N$ matrices A and B is the product of their transposes taken in the reverse order, that is, $(\widehat{AB}) = \tilde{B}\tilde{A}$.

6.2 Show that $(AB)^\dagger = B^\dagger A^\dagger$.

6.3 Let operator $\hat{Q}$ represent observable q. Find the matrix form of $\hat{Q}$ in the q-basis in case when observable q is continuous (see Supplement to Section 6.2)

6.4 Find the matrix of momentum operator $\hat{\mathbf{p}}$ (a) in its own basis and (b) in the **r**-basis

6.5 Find the matrix of position operator $\hat{\mathbf{r}}$ (a) in its own basis and (b) in the momentum basis

6.6 Suppose you switch from the s-basis to the u-basis in the Hilbert space spanned by the eigenfunctions of observable s and observable u. Find the rule for expressing new matrix elements $q_{kl}^{(u)}$ of operator $\hat{Q}$ in terms of the old elements $q_{kl}^{(s)}$.

6.7 A quantum system is characterized by a *complete set* of observables (p, q, s). These observables have, respectively, N_p, N_q, and N_s eigenvalues. Find the dimensionality of the Hilbert space of this system.

6.8 Derive the general algorithm determining the dimensionality of the $\mathcal{H}$-space for a given quantum system if you know the number N of different observables forming the complete set and the number of different eigenvalues for each observable (assume all these numbers to be finite).

6.9 Find the expectation value of **r** in state (6.56). Compare it with the vector eigenvalue of state (6.55) given by the tensor product $|x\rangle \otimes |y\rangle$ and explain the results.

6.10 Find the condition under which the vector expectation value of **p** in state (6.60) is parallel to **r** in state (6.55)

a) graphically;
b) analytically.

6.11 Derive Equation 6.71 in matrix form by direct multiplication of the corresponding matrices and using the normalization condition.

6.12 Derive Equation 6.72

6.13 Find the matrix form of projection operator $|u_1\rangle\langle u_1|$, where $|u_1\rangle$ is a superposition (6.66) of two mutually orthogonal states. Show that it projects both nonorthogonal states (6.65) and (6.66) onto $|u_1\rangle$.

6.14 Find the explicit matrix form for an operator $\hat{\Pi}_1 \equiv 1 - \hat{P}_1$ and show that it projects states of the 2D Hilbert space onto the direction perpendicular to $|u_1\rangle$.

References

1 Yosida, K. (1968) *Functional Analysis*, Springer, Berlin.
2 Hestenes, D. (1971) Vectors, spinors and complex numbers in classical and quantum physics. *Am. J. Phys.*, **39**, 1013–1028.
3 Hestenes, D. (2003) Oersted Medal Lecture 2002: Reforming the mathematical language of physics. *Am. J. Phys.*, **71** (2), 104–121.
4 Hausdorf, F. (1991) *Set Theory*, 4th edn, Chelsea Publishing Company, New York.
5 Vilenkin, N.Ya. (1968) *Stories About Sets*, Academic Press, New York.
6 Arfken, G.B. and Weber, H.J. (1995) *Mathematical Methods for Physicists*, Academic Press, New York.
7 Solomentsev, E.D. (2001) Cauchy inequality, in *Encyclopedia of Mathematics* (ed. M. Hazewinkel), Springer, Berlin.
8 Steele, J.M. (2004) *The Cauchy–Schwarz Master Class*, Cambridge University Press, Cambridge.
9 Fayngold, M. (2002) *Special Relativity and Motions Faster than Light*, Wiley-VCH Verlag GmbH, Weinheim.
10 Bennett, C.H. (1992) Quantum cryptography using any two nonorthogonal states. *Phys. Rev. Lett.*, **68** (21), 3121–3124.
11 Landau, L. and Lifshits, E. (1981) *The Field Theory*, Butterworth-Heinemann, Oxford.
12 Fayngold, M. (2008) *Special Relativity and How It Works*, Wiley-VCH Verlag GmbH, Weinheim, pp. 485–487.

7
Angular Momentum

This chapter is devoted to one of the most important (and most fascinating) members of the family of observables – angular momentum. It comes in two varieties – *orbital* angular momentum and *intrinsic* angular momentum (*spin*). The first describes motion of a particle or system of particles relative to a certain point taken as the origin. The second is an irreducible characteristic of some particles, determining important features of their behavior.

7.1
Orbital and Spin Angular Momenta

In CM, both varieties of angular momentum have common origin. Orbital angular momentum $\mathbf{L} = \mathbf{r} \times \mathbf{p}$ arises when a body with nonzero $\mathbf{p}$ moves around (or just passes by!) a point. Spin angular momentum $\mathbf{S}$ arises from the body's rotation about an axis passing through its center of mass. For a finite-sized object such as a spinning top, we find $\mathbf{S}$ by adding up the individual angular momenta $\mathbf{L}_i$ of each point of the object. If the spinning object also has orbital motion (e.g., the Earth is spinning about its axis while also orbiting the Sun), we find its *total* angular momentum $\mathbf{J}$ as the vector sum

$$\mathbf{J} = \mathbf{L} + \mathbf{S}. \tag{7.1}$$

Here $\mathbf{L}$ and $\mathbf{S}$ are not conceptually different. They both arise from rotation, that is, from the change of angular coordinates describing their position and orientation. That's why the addition (7.1) goes "smoothly," without causing any mathematical or conceptual difficulties. In many situations, we don't care what type of angular momentum we are dealing with. Then we can introduce another level of abstraction:

$$\mathbf{J} = \mathbf{J}_1 + \mathbf{J}_2. \tag{7.2}$$

Here $\mathbf{J}_1$ and $\mathbf{J}_2$ are "generic" angular momenta, which can be spin, or orbital, or some combination of the two. Equation 7.2 applies to a single object participating in two rotational motions as in (7.1), or to a two-body system with total angular momentum $\mathbf{J}$ and individual net angular momenta $\mathbf{J}_1$ and $\mathbf{J}_2$, respectively.

What are the properties of the net angular momentum $\mathbf{J}$ of a system? The answer to this question is not trivial, even in CM. If the system is isolated and there is no

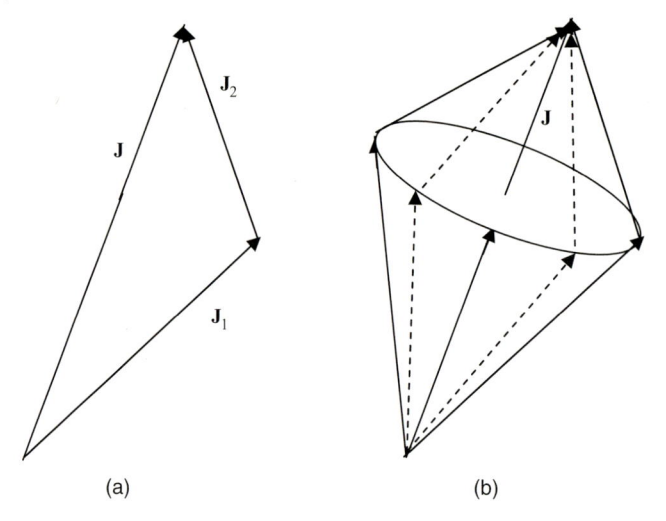

Figure 7.1 The classical net angular momentum of two particles: (a) in the absence of interaction; (b) in the presence of a weak interaction causing precession.

interaction between spin and orbital motion, each term in (7.1) is conserved, and we have the sum of two *fixed* vectors. A weak interaction between them still leaves, in the first approximation, the *magnitudes* of J_1 and J_2 intact. Vector **J** is conserved, but not the orientations of J_1 and J_2. This can be visualized as a synchronous rotation of J_1 and J_2 (precession) about **J** (Figure 7.1). A known example is slow precession of the Earth's rotational axis and its orbit (with perturbations from the Moon and other planets neglected [1,2]). But for any moment of time **J** is completely defined by (7.2).

Strictly speaking, the statement that **L** and **S** are essentially similar quantities is not exactly true, even in CM. One major difference emerges when we think of a point particle. We can ascribe to such a particle an angular momentum **L** (the particle can still orbit around a given center!), but we get into serious conceptual trouble when we try to ascribe to it a spin angular momentum **S**. It is meaningless to speak of a point spinning around itself. Upon some thought, one *could* possibly justify extending the notion of spin to a point particle. One could say, "We get spin of a point particle if we compress a spinning top into a single point using only radial forces." But this can only be accepted as an ersatz model approximating an object with negligible size. One simple argument is sufficient to show the deadly flaw in such model: in the process of infinite compression, the rotation rate of the top would increase until its equatorial velocity runs into the light barrier. Therefore, in CM one has to consider a point particle as spinless. An alternative would be to define spin of a point particle as a quantity that *behaves* as an ordinary spin (e.g., adding to **L** as in Equation 7.1), yet has nothing to do with the ordinary rotation. That approach was never taken in CM because there was no need for it. There had been much controversy about the electron size until the view of a point particle was eventually endorsed by modern physics. Until the *line splitting* in the spectrum of alkali atoms

7.2 The Eigenstates and Eigenvalues of $\hat{L}$

was detected, and the idea of electron spin was put forth to explain the results of the *Stern–Gerlach experiment*, there seemed to be no need to assign spin to point particles, so it was left to QM to take up the latter approach.

In QM, spin **S** is "promoted" to a status equal with **L**. There are no requirements on the size of a "spinning" object, and there is an express prohibition against visualizing an elementary particle as a small spinning top. Instead of following this misleading model, QM *defines* spin as a quantity whose operator $\hat{S}$ has the same commutation properties as the operator $\hat{L}$. This opens up a straightforward way to define the generic angular momentum: We derive commutation properties of $\hat{L}$, then *postulate* the same properties for $\hat{S}$, and finally, consider an operator of the net angular momentum $\hat{J}$, defined according to (7.1), and determine what happens when we add two angular momenta together. But before we embark further, we must warn the reader that finding the eigenfunctions of $\hat{L}^2$ involves rather extensive mathematics (see Appendix E), which can be skipped during the first reading. In many problems, knowing the explicit form of these eigenfunctions is unnecessary. At the same time, while learning how to *write* these eigenstates is tricky, learning how to *manipulate* them correctly is easier! This seems to be one of those cases when the result matters more than the method of its derivation. This is why the next section will be devoted to the final results while their derivation can be postponed for later.

7.2
The Eigenstates and Eigenvalues of $\hat{L}$

We construct the operator for the orbital angular momentum by using the already known procedure: taking the classical definition $\mathbf{L} = \mathbf{r} \times \mathbf{p}$ and replacing each observable there with its operator. In position representation $\hat{\mathbf{r}} = \mathbf{r}$ and $\hat{\mathbf{p}} = -i\hbar \vec{\nabla}$, we obtain

$$\hat{\mathbf{L}} = \hat{\mathbf{r}} \times \hat{\mathbf{p}} = -i\hbar (\mathbf{r} \times \vec{\nabla}). \tag{7.3}$$

This is a vector operator with components

$$\begin{aligned}
\hat{L}_x &= y\hat{p}_z - z\hat{p}_y = -i\hbar \left(y\frac{\partial}{\partial z} - z\frac{\partial}{\partial y} \right), \\
\hat{L}_y &= z\hat{p}_x - x\hat{p}_z = -i\hbar \left(z\frac{\partial}{\partial x} - x\frac{\partial}{\partial z} \right), \\
\hat{L}_z &= x\hat{p}_y - y\hat{p}_x = -i\hbar \left(x\frac{\partial}{\partial y} - y\frac{\partial}{\partial x} \right).
\end{aligned} \tag{7.4}$$

The next interesting question is the relation between the components of **L** and its magnitude L. Let us write the expression for $\hat{L}^2$:

$$\hat{L}^2 = \hat{L}_x^2 + \hat{L}_y^2 + \hat{L}_z^2. \tag{7.5}$$

Putting here expressions (7.5) will give $\hat{L}^2$ in *Cartesian coordinates*. The eigenvalues of this operator will determine the allowed magnitudes L. So our next task is to solve the eigenvalue problem

$$\hat{L}^2|\Psi\rangle = L^2|\Psi\rangle. \tag{7.6}$$

However, solving it in Cartesian coordinates will be next to impossible. The best strategy is to first rewrite (7.4) and (7.5) in spherical coordinates (this is the most natural representation, since angular momentum is associated with orbiting most conveniently described in angular variables). So we write $x = r\sin\theta\cos\varphi$, $y = r\sin\theta\sin\varphi$, and $z = r\cos\theta$ and leave it as an exercise for the reader to show that

$$\hat{L}_x = i\hbar\left(\sin\varphi\frac{\partial}{\partial\theta} + \cot\theta\cos\varphi\frac{\partial}{\partial\varphi}\right),$$

$$\hat{L}_y = -i\hbar\left(\cos\varphi\frac{\partial}{\partial\theta} - \cot\theta\sin\varphi\frac{\partial}{\partial\varphi}\right),$$

$$\hat{L}_z = -i\hbar\frac{\partial}{\partial\varphi}.$$

We can also express $\hat{L}^2$ in terms of the spherical coordinates:

$$\hat{L}^2 = -\hbar^2\left[\frac{1}{\sin\theta}\frac{\partial}{\partial\theta}\left(\sin\theta\frac{\partial}{\partial\theta}\right) + \frac{1}{\sin^2\theta}\frac{\partial^2}{\partial\varphi^2}\right].$$

This is the known angular part of the Laplacian operator.

Then we can solve the problem in variables r, θ, and φ, which is done in Appendix E. Here we present the final result. The eigenfunctions of (7.6) are known as *spherical harmonics* and written as $Y_{l,m}(\theta,\varphi)$. Each one is defined by two integers l and m, where for any given l the second number m can take values $m = -l, -l+1, \ldots, l$. The set of all spherical harmonics forms an orthogonal basis. With this solution, Equation 7.6 takes the form

$$\hat{L}^2 Y_{l,m}(\theta,\varphi) = L^2 Y_{l,m}(\theta,\varphi) \tag{7.7}$$

and the corresponding eigenvalues are

$$L_l^2 = l(l+1)\hbar^2, \quad l = 0, 1, 2, \ldots. \tag{7.8}$$

Suppose our system is in an eigenstate $Y_{l,m}(\theta,\varphi)$ so that L_l^2 is known. What can we say about L_x, L_y, and L_z? Solving Equation 7.7 answers this question as well. Namely, $Y_{l,m}(\theta,\varphi)$ turns out to be also the eigenfunction of $\hat{L}_z$:

$$\hat{L}_z Y_{l,m}(\theta,\varphi) = L_z Y_{l,m}(\theta,\varphi), \tag{7.9}$$

with eigenvalue

$$L_z = m\hbar, \quad m = -l, -l+1, \ldots, l. \tag{7.10}$$

Why did we single out the L_z-operator, and what happened to the other two components? There are two ways to answer these questions. The first has to do with the method used to solve the problem. In the Cartesian system, all three coordinate axes are treated the same way. But this "equality" does not hold in the spherical system. When switching to r, θ, and φ, we had to choose an arbitrary axis, labeled as z, relative to which the *polar* angle θ is measured. The other two axes then define the plane in which the *azimuthal* angle φ "lives." This arrangement distinguishes the z-axis from the other two. In a spherical system, the z-axis is "more equal" than others. The deep physical reason for this is that rotation singles out one direction (rotational axis) from all the rest. Thus, we can write Equation 7.9 for $\hat{L}_z$ but not for $\hat{L}_x$ (or $\hat{L}_y$). Or, more accurately, we *can* do that, but then we would have to redirect the z-axis of the spherical system along the x-axis (or y-axis) of the Cartesian system. The spherical harmonic $\Upsilon_{l,m}(\theta, \varphi)$ is only compatible with *one* component of **L** – the one corresponding to the axis relative to which the polar angle θ was defined!

The other answer has to do with commutation rules for components of $\hat{\mathbf{L}}$. A reader with a good physical intuition should anticipate this answer in advance. The very fact that in order to solve the problem we had to use spherical coordinates and thereby single out the z-axis already provides a sufficient hint. As we will see later, operator $\hat{L}_z$ does not commute with either $\hat{L}_x$ or $\hat{L}_y$. We can measure the value of L^2 and L_z simultaneously, but once we know L_z, we cannot know L_x and L_y. These two components will remain indeterminate.

Another important fact to keep in mind: *only* spherical harmonics $\Upsilon_{l,m}(\theta, \varphi)$ are *common* eigenfunctions of $\hat{\mathbf{L}}^2$ and $\hat{L}_z$. A superposition of harmonics $\sum c_m \Upsilon_{l,m}(\theta, \varphi)$ with different possible m is an eigenfunction of $\hat{\mathbf{L}}^2$ but not an eigenfunction of $\hat{L}_z$.

Finally, we must say a few words about notations. In Dirac's notation, an eigenstate is the corresponding eigenvalue inside the ket. To be "scrupulous," we should represent $\Upsilon_{l,m}$ in (7.7) and (7.9) as $|l(l+1)\hbar^2, m\hbar\rangle$. This form would make it explicitly clear that the given eigenstate contains a pair of eigenvalues, $l(l+1)\hbar^2$ and $m\hbar$ for operators $\hat{\mathbf{L}}^2$ and $\hat{L}_z$, respectively. But the eigenket so written is a little bulky. In such cases, we use a shorthand notation. For angular momentum, we represent eigenstates just by the pair of numbers l and m inside the ket. Then Equations 7.7 and 7.9 take (in units of $\hbar$) the form

$$\hat{\mathbf{L}}^2|l, m\rangle = l(l+1)|l, m\rangle, \qquad \hat{L}_z|l, m\rangle = m|l, m\rangle. \tag{7.11}$$

One only has to keep in mind that l and m in (7.11) are not themselves eigenvalues but merely serve as *labels* for the respective eigenvalues. These are often referred to as *quantum numbers*. The name *orbital quantum number* for l reflects its relation to orbital motion. The number m is called *magnetic quantum number* because interaction with magnetic field was used in early experiments to study the properties of angular momentum (see also Box 7.5).

What is the meaning of restrictions (7.10) on m? If we ignore all observables other than **L**, there is only one single state with orbital momentum $l = 0$ (no surprise – there is only one way in which a particle *does not orbit* around the center). In general, condition (7.10) admits $N_l = 2l + 1$ different states for any given l. This multiplicity of allowed states is by itself not a big surprise either, because a fixed magnitude of **L** admits motions in any orbital plane: in CM we have a *continuous set* of such planes with different orientations of **L**. In contrast, Equation 7.10 leaves only $2l + 1$ possibilities for a given l, which is an entirely QM phenomenon.

7.3
Operator $\hat{\mathbf{L}}$ and Its Commutation Properties

We are now going to prove that components of $\hat{\mathbf{L}}$ do not commute with each other but commute with $\hat{\mathbf{L}}^2$, as was mentioned in the previous section. For that purpose, it will be convenient to go back to the Cartesian system.

First, we find out how components of $\hat{\mathbf{L}}$ commute with $\hat{\mathbf{r}}$ and $\hat{\mathbf{p}}$. We need this intermediate step because $\hat{\mathbf{L}}$ is expressed in terms of $\hat{\mathbf{r}}$ and $\hat{\mathbf{p}}$ anyway. So let's combine any component of $\mathbf{r}$ and any component of $\mathbf{L}$. For instance, are the observables x and L_y compatible or not? The answer is given by

$$[x, \hat{L}_y] = x(z\hat{p}_x - x\hat{p}_z) - (z\hat{p}_x - x\hat{p}_z)x = z[x, \hat{p}_x] = i\hbar z. \tag{7.12}$$

The corresponding operators do not commute, and therefore the answer is negative. On the other hand, a similar treatment of x and $\hat{L}_x$ gives $[x, \hat{L}_x] = 0$, so there are no restrictions on their simultaneous measurement. Following this procedure, we can find the set of all commutation rules for $\mathbf{r}$ and $\hat{\mathbf{L}}$ (Problem 7.1):

$$\begin{aligned} &[x, \hat{L}_x] = 0, & &[x, \hat{L}_y] = i\hbar z, & &[x, \hat{L}_z] = -i\hbar y, \\ &[y, \hat{L}_x] = -i\hbar z, & &[y, \hat{L}_y] = 0, & &[y, \hat{L}_z] = i\hbar x, \\ &[z, \hat{L}_x] = i\hbar y, & &[z, \hat{L}_y] = -i\hbar x, & &[z, \hat{L}_z] = 0. \end{aligned} \tag{7.13}$$

Box 7.1 Permutations

As we see from (7.13), if two different indices appear inside the commutator, then $i\hbar$ times the last remaining index appears on the right, with the sign given by one simple rule: if the three indices, when read from left to right, form an *even* permutation of x, y, z, the sign is positive, and if they form an *odd permutation*, the sign is negative. The phrase "even/odd permutation" means a reshuffling of x, y, z where the swapping of two indices occurs an even/odd number of times; for example, both y, x, z and x, z, y are odd permutations of x, y, z, whereas y, z, x is an even permutation. If the indices inside the commutator are the same, the result is zero. All this can be written in one expression if we use the tensor notation. Let x_1, x_2, and x_3 stand for x, y, and z, respectively. Introduce the symbols i, j, and k, each of which takes on the values 1, 2, or 3. Then, (7.13) can be written as

$$[x_i, \hat{L}_j] = -i\hbar e_{ijk} x_k. \tag{7.14}$$

Here e_{ijk} is antisymmetric third-rank unit tensor, known as the *Levi–Civita* symbol, or just the permutation symbol. It has two basic properties: first, the element e_{123} is set to be 1, that is, $e_{123} = 1$; second, its components change sign at the swapping of any two indices i, j, or k, that is, $e_{jik} = -e_{ijk} = e_{ikj}$.

It follows that all elements with equal indices must be zero since according to the definition, we have for elements with $i = j$: $e_{iij} = -e_{iij}$, which can be satisfied only if $e_{iij} = 0$. Thus, only the elements with all three indices i, j, and k different will survive, and for these elements we have, for instance, $e_{132} = -e_{123} = -1$, and so on.[1] As an exercise, one can play with this tensor and show (Problem 7.2) that components of any cross product $\mathbf{C} = \mathbf{A} \times \mathbf{B}$ can also be written as

$$C_i = e_{ijk} A_j B_k. \tag{7.15}$$

Similar commutation relations can be found (Problem 7.3) for the components of $\hat{\mathbf{p}}$ and $\hat{\mathbf{L}}$. They look exactly the same as (7.13), with x, y, z replaced by $\hat{p}_x$, $\hat{p}_y$, $\hat{p}_z$. Or, using the compact notation from Box 7.1, we can write

$$\left[\hat{p}_i, \hat{L}_j\right] = -i\hbar e_{ijk} \hat{p}_k. \tag{7.16}$$

Summarizing this part, we see that any chosen component of **L** is only compatible with the *same* component of **r** and **p** and incompatible with the other two components. Thus, a particle with definite L_z does not have definite x- and y-components of its position and momentum vectors.

Equations 7.14 and 7.16 set the stage for deriving the commutation relations between components of angular momentum itself. One can now show (Problem 7.4) that

$$\left[\hat{L}_x, \hat{L}_y\right] = i\hbar \hat{L}_z, \quad \left[\hat{L}_y, \hat{L}_z\right] = i\hbar \hat{L}_x, \quad \left[\hat{L}_z, \hat{L}_x\right] = i\hbar \hat{L}_y, \tag{7.17}$$

or, in the tensor notation,

$$\left[\hat{L}_i, \hat{L}_j\right] = i\hbar e_{ijk} \hat{L}_k. \tag{7.18}$$

We see the remarkable thing – incompatibility between components of **L**. In this respect, **L** is totally different from **r** and **p**. All three components of **r** can be measured simultaneously. The same is true for **p**. In this respect, each observable, **r** and **p**, is a true vector, which can be represented by a directed segment. In contrast, the **L**-components do not exist simultaneously! Only one can be defined at a time. An exact measurement of another component will destroy information about the first one. So there is an irreducible indeterminacy in any two components of **L**. Therefore, **L** is,

---

1) In tensor algebra (see, for example, Ref. [3]), summation is assumed over the indexes occurring twice in an expression. This is superfluous here because when we sum over γ, only one term is nonzero – the one with γ different from both α and β (assuming that $\alpha \neq \beta$; otherwise all three terms are zero since $e_{\alpha\beta\gamma} = 0$). Therefore, it is acceptable to just check that $\alpha \neq \beta$ and then write the term with the remaining index γ.

Box 7.2 Angular momentum and rotations

We can also justify the result (7.18) by the following intuitive argument based on an example from the classical theory of rotations [4]. Denote a rotation about the x_i-axis as $\hat{R}_i$. Now consider two rotations $\hat{R}_1$ and $\hat{R}_2$, which are rotations about the axes x and y, respectively. Then the two rotations in sequence can be written as $\hat{R}_2\hat{R}_1$. The corresponding expression can be called the group product of the two rotation operations and should be read from right to left: we apply first the operation $\hat{R}_1$, and then the operation $\hat{R}_2$. The same two operations in the reverse order can be written as $\hat{R}_1\hat{R}_2$. This is another group product: now we apply first $\hat{R}_2$ and then $\hat{R}_1$. As is well known, the group of rotations is not commutative. Only in the special case of rotations about *the same* axis will the result be independent of the order: rotating an object about axis 1 through an angle χ and then through an angle φ will produce the same effect as rotating in the reverse order: $\hat{R}_1(\varphi)\hat{R}_1(\chi) = R_1(\varphi + \chi) = \hat{R}_1(\chi)\hat{R}_1(\varphi)$. But generally, in cases involving rotations about different axes, the product of two rotations depends on their order, so $\hat{R}_1\hat{R}_2 \neq \hat{R}_2\hat{R}_1$. As an example, take a figure in the xy-plane and rotate it, say, counterclockwise through 90° first in the xz-plane, and then in the same sense and through the same angle in the yz-plane. The operation can be written as $\hat{R}_1\hat{R}_2$ (Figure 7.2a). In another trial, do the same rotations of the figure first in the yz-plane, and then in the zx-plane (succession $\hat{R}_2\hat{R}_1$, with the result shown in Figure 7.2b). The results are different. Based on these examples and using the fact that the rotation operator $\hat{R}_i$ is closely related to the corresponding component $\hat{L}_i$ of the orbital momentum operator [3], we could have predicted at once that operators $\hat{L}_i$ and $\hat{L}_j$ with $i \neq j$ do not commute.

---

strictly speaking, not a vector, but a more complex entity, which can be roughly visualized as a conical surface rather than a geometrical segment (Figure 7.3).

The next interesting question is the relation between the components of **L** and its magnitude. Using (7.18), we get (Problem 7.5)

$$\left[\hat{L}^2, \hat{L}_i\right] = 0 \quad \text{for any } i. \tag{7.19}$$

This means that, even though all three *components* of the angular momentum vector cannot be known at the same time, the *magnitude* of this vector can nevertheless have a definite value, together with the exact value of *one* out of the three components. It could be any one of the three components, although the chosen component is frequently labeled as z by default. We have now justified the result mentioned in the previous section.

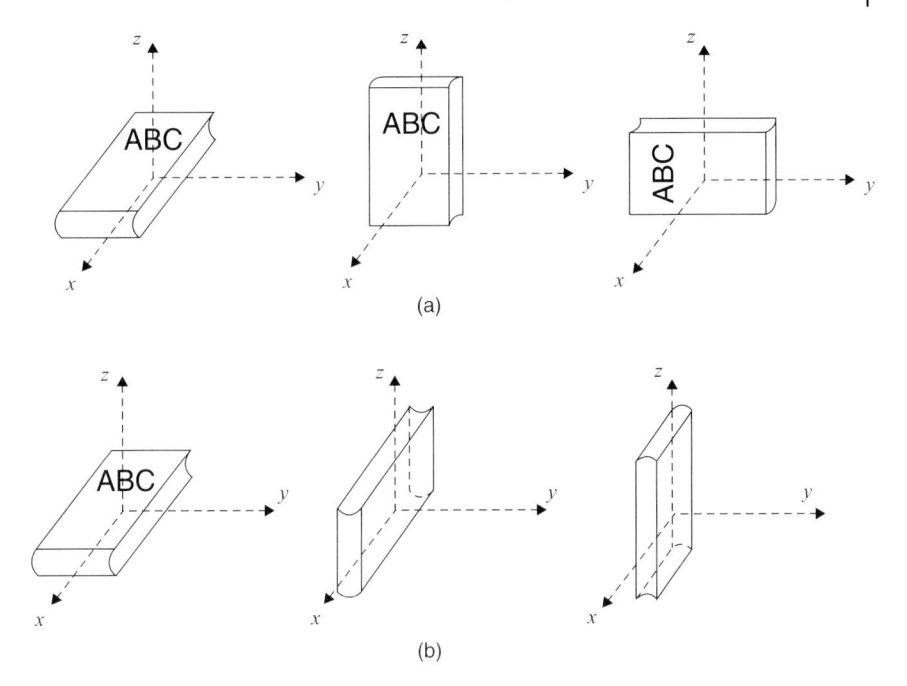

Figure 7.2 Rotation operations.

Box 7.3 Uncertainty relations for angular momentum components

The measure of the incompatibility between components of **L** is determined by the commutation relations between them. Combining the Robertson–Schrödinger theorem with (7.17), we have

$$\Delta L_x \Delta L_y \geq \frac{1}{2}\left|\overline{[\hat{L}_x, \hat{L}_y]}\right| = \frac{1}{2}\hbar\left|\overline{\hat{L}_z}\right| = \frac{1}{2}\hbar\left|\overline{L_z}\right|. \quad (7.20)$$

Similarly,

$$\Delta L_x \Delta L_z \geq \frac{1}{2}\hbar\left|\overline{L_y}\right|, \quad \Delta L_y \Delta L_z \geq \frac{1}{2}\hbar\left|\overline{L_x}\right|. \quad (7.21)$$

If the i-component of **L** is measured precisely, then the system is in an eigenstate of the $\hat{L}_i$ operator with an eigenvalue L_i, and we have $\overline{L}_i = L_i$. We will have more to say about this later in Chapter 13.

But already at this stage we can make a certain prediction about the expectation values of some components in certain states. Suppose that the particle is in an eigenstate of $\hat{L}_z$ with an eigenvalue L_z. In this case $\Delta L_z = 0$, and so must be the product $\Delta L_x \Delta L_z$ or $\Delta L_y \Delta L_z$. On the other hand, (7.21) gives, for instance,

$(1/2)\hbar|\overline{L_y}|$ as the lowest possible boundary for $\Delta L_x \Delta L_z$ in this state. Both statements can be consistent with each other only if $|\overline{L_y}| = 0$ in the considered state. We conclude that in a state with definite L_z, the expectation value of L_x and L_y is zero. We will see below that this conclusion is, indeed, satisfied with an even stronger constraint: the possible components perpendicular to z form an axially symmetric distribution around z, which automatically makes for the zero average as seen in Figure 7.3.

There is, however, an obvious exception from the rules (7.20) and (7.21): a state with the zero **L**. In this state, all three components are exactly zeros. Nature allows all three components to exist together in one state when all three are reduced to nonexistence!

Box 7.4 Visualizing L and its components

The result (7.19) can be visualized in the following way. Start with the classical vector **L** and represent it geometrically with respect to a triplet of Cartesian axes x, y, z (Figure 7.3a). It is specified either by its three (x, y, z) components (Cartesian coordinates) or by its magnitude L and two angles θ and φ (spherical coordinates). All three components are defined. Now rotate the initial vector about the z-axis (Figure 7.3b). We obtain a continuous set of vectors with $0 \leq \varphi \leq \pi$, forming a conical surface with z as its symmetry axis and the open angle 2θ (as we know from geometry, any specific vector **L'** on this surface is one of its generatrices). In Cartesian coordinates, all vectors **L'** have the same z-component but differ from one another by their x- and/or y-components. Either of these components forms a continuous set within the same range

$$-L\sin\theta \leq L_x \leq L\sin\theta, \quad -L\sin\theta \leq L_y \leq L\sin\theta \tag{7.22}$$

but is subject to the additional condition

$$L_x^2 + L_y^2 = L^2 - L_z^2 = L^2 \sin^2\theta, \tag{7.23}$$

so that neither of them can be chosen independently of the other.

But if we satisfy all these classical conditions and choose any *single* **L'**, we return to the classical notion of a vector, and this will be inconsistent with the quantum rules (7.20) and (7.21). The only way to comply with (7.20) and (7.21) is to go beyond the notion of **L** as a vector and introduce instead a totally new concept of a physical characteristic represented geometrically by a conical surface (*conoid*) as in Figure 7.3b. Rather crudely, we can imagine an object originally prepared as a rod as in Figure 7.3a, and then smeared out so that its substance is distributed over the surface as in Figure 7.3b. If this distribution is uniform, the system is in a state with definite L and L_z. And the components L_x and L_y become fuzzy, with dramatically reduced information sufficient to determine only the ranges (7.22)

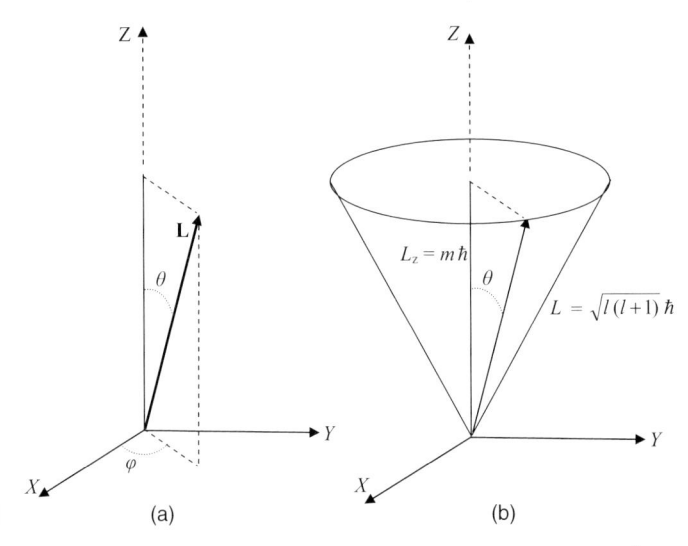

Figure 7.3 (a) The classical vector of the angular momentum. (b) The conical surface representing a *quantum state* with the definite z-component of angular momentum. In this state, only the sum of the squares of L_x and L_y can be determined: $L_x^2 + L_y^2 = [l(l+1) - m^2]\hbar^2$, but not L_x and L_y separately. The state can be considered as an entangled superposition of all possible x- and y-components of **L**, such that $L_x^2 + L_y^2 = [l(l+1) - m^2]\hbar^2$.

instead of two exact numbers. (Actual situations require more accurate illustrations, so we must consider this only as an approximation!)

According to Figure 7.3b, there is an infinite continuous set of different orbital planes corresponding to vectors with the same L and θ and differing only in φ. This determines the set of different **L** with the same z-component – the case we have just considered. But we can also have different **L** with the same magnitude but different z-components. Based on our result, according to which an angular momentum with definite L_z can be represented by the corresponding conoid in Figure 7.3b, we can represent L-states with different L_z by conoids with different open angles. However, according to quantization rules (7.8) and (7.10), the allowed magnitudes of L form a discrete set, and for each possible L only a discrete set of such angles is allowed. Here is where the QM predictions differ from classical ones.

The discussed features explain Equation 7.8, which may initially seem a little strange. Why, indeed, does it require that $L = \sqrt{l(l+1)}\hbar$ instead of just $L = l\hbar$? The answer is that the latter would be inconsistent with the indeterminacy principle. For instance, in a state with $|m| = l$, the corresponding conoid would degenerate into a classical vector along the z-axis, with components $L_z = l\hbar$ and $L_y = L_z = 0$. All three components would be defined, but this is allowed only when all three are zero, that is, $l = 0$. Any $L = l\hbar \neq 0$ would contradict (7.20) and (7.21). Hence condition (7.8).

Let us now introduce two new operators, defined as follows:

$$\hat{L}_+ = \hat{L}_x + i\hat{L}_y, \qquad \hat{L}_- = \hat{L}_x - i\hat{L}_y, \tag{7.24}$$

or, in spherical coordinates,

$$\hat{L}_\pm = \hbar e^{\pm i\varphi}\left(\pm\frac{\partial}{\partial\theta} + i\cot\theta\,\frac{\partial}{\partial\varphi}\right). \tag{7.25}$$

These operators are of special interest because of the way they act on the eigenstates. Let us first establish the commutation rules for them. One can easily prove (Problem 7.6) that

$$\left[\hat{L}^2, \hat{L}_+\right] = 0, \qquad \left[\hat{L}^2, \hat{L}_-\right] = 0,$$

$$\left[\hat{L}_+, \hat{L}_-\right] = 2\hbar \hat{L}_z, \tag{7.26}$$

$$\left[\hat{L}_z, \hat{L}_+\right] = \hbar \hat{L}_+, \qquad \left[\hat{L}_z, \hat{L}_-\right] = -\hbar \hat{L}_-.$$

Suppose now that operator $\hat{L}_+$ acts on an eigenket $|l, m\rangle$. We are going to prove that the result is a different eigenket, $|l, m+1\rangle$. Look at the following two equations that make use of (7.26):

$$\begin{cases} \hat{L}^2 \hat{L}_+ |l, m\rangle = \hat{L}_+ \hat{L}^2 |l, m\rangle = l(l+1)\hbar^2 \hat{L}_+ |l, m\rangle, \\ \hat{L}_z \hat{L}_+ |l, m\rangle = (\hbar \hat{L}_+ + \hat{L}_+ \hat{L}_z)|l, m\rangle = \hbar(m+1)\hat{L}_+|l, m\rangle. \end{cases} \tag{7.27}$$

Clearly, $\hat{L}_+|l, m\rangle$ is an eigenstate of $\hat{L}^2$ and $\hat{L}_z$ with eigenvalues $l(l+1)\hbar^2$ and $(m+1)\hbar$; therefore, in our shorthand notation it must be written as $|l, m+1\rangle$. In a similar manner, we can show that $\hat{L}_-|l, m\rangle$ produces the eigenket $|l, m-1\rangle$. So the effect of $\hat{L}_+$ ($\hat{L}_-$) is to take an eigenstate into a new eigenstate with quantum number m increased (decreased) by 1. For this reason, $\hat{L}_+$ and $\hat{L}_-$ are known as *raising* and *lowering* operators or simply as *ladder operators*. This is the first example of operators of this important class.

There remains to find a constant multiplier, which may be introduced by a ladder operator into the resulting state. We cannot tell its value from Equation 7.27. So, to account for this possibility we now have to write the ladder property of $\hat{L}_+$ and $\hat{L}_-$ in the general form

$$\hat{L}_+|l, m\rangle = c_+|l, m+1\rangle, \qquad \hat{L}_-|l, m\rangle = c_-|l, m-1\rangle. \tag{7.28}$$

To find the complex constants c_+, we compose the expression $\langle l, m|\hat{L}_-\hat{L}_+|l, m\rangle$ and note that $\langle l, m|\hat{L}_- = (\hat{L}_+|l, m\rangle)^+ = (c_+|l, m+1\rangle)^+$, since the raising and lowering operators are Hermitian adjoints of each other. Multiplying by

$\hat{L}_+|l, m\rangle = c_+|l, m+1\rangle$, we get simply $|c_+|^2$. As a result,

$$|c_+|^2 = \langle l, m|\hat{L}_-\hat{L}_+|l, m\rangle = \langle l, m|(\hat{L}_x - i\hat{L}_y)(\hat{L}_x + i\hat{L}_y)|l, m\rangle$$
$$= \langle l, m|\left(\hat{L}_x^2 + i\hat{L}_x\hat{L}_y - i\hat{L}_y\hat{L}_x + \hat{L}_y^2\right)|l, m\rangle.$$

After applying the commutation rule $[\hat{L}_x, \hat{L}_y] = i\hbar\hat{L}_z$ to the right side, we have

$$|c_+|^2 = \langle l, m|\left(\hat{L}_x^2 - \hbar\hat{L}_z + \hat{L}_y^2\right)|l, m\rangle = \langle l, m|\left(\hat{L}^2 - L_z^2 - \hbar\hat{L}_z\right)|l, m\rangle$$
$$= (l(l+1) - m(m+1))\hbar^2. \tag{7.29}$$

In a similar manner, by composing the expression $\langle l, m|\hat{L}_+\hat{L}_-|l, m\rangle$, we can show that

$$|c_-|^2 = \langle l, m|\left(\hat{L}_x^2 + \hbar\hat{L}_z + \hat{L}_y^2\right)|l, m\rangle = \langle l, m|\left(\hat{L}^2 - L_z^2 + \hbar\hat{L}_z\right)|l, m\rangle$$
$$= (l(l+1) - m(m-1))\hbar^2. \tag{7.30}$$

Therefore, Equation 7.28 can be written as

$$\begin{cases} \hat{L}_+|l, m\rangle = \hbar\sqrt{l(l+1) - m(m+1)}|l, m+1\rangle, \\ \hat{L}_-|l, m\rangle = \hbar\sqrt{l(l+1) - m(m-1)}|l, m-1\rangle. \end{cases} \tag{7.31}$$

This completes the discussion of commutation rules for *orbital* angular momentum. The question arises: Can one write similar equations for *spin* $\hat{S}$ and for general angular momentum $\hat{J}$? We said earlier that $\hat{S}$ is defined as an operator whose components obey the same commutation rules as the components of $\hat{L}$. With all that, the spin operator is different from (7.3) – it is not a function of either position $\mathbf{r}$ or momentum $\mathbf{p}$. Accordingly, it must commute with their respective operators, as well as with $\hat{L}$ itself. Thus, in contrast with (7.14) and (7.16) we have

$$[\hat{x}_\alpha, \hat{S}_\beta] = [\hat{p}_\alpha, \hat{S}_\beta] = 0. \tag{7.32}$$

Physically, this means that particle's spin can coexist with either of observables $\mathbf{r}$ and $\mathbf{p}$. Also, the same components of $\mathbf{S}$ and $\mathbf{L}$ must commute:

$$[\hat{S}_\alpha, \hat{L}_\alpha] = 0. \tag{7.33}$$

But the commutation relations between the spin components themselves are identical to (7.18) and (7.19).

We can also introduce the spin ladder operators $\hat{S}_+ = \hat{S}_x + i\hat{S}_y$ and $\hat{S}_- = \hat{S}_x - i\hat{S}_y$ similar to (7.24) and derive for them the commutation rules similar to (7.26).

There remains only one distinction with regard to eigenvalues of $\hat{S}^2$ versus $\hat{L}^2$. The reader remembers that the latter are determined by (7.8) with integer l. However, the requirements that $L^2 = l(l+1)\hbar^2$ and that l be an integer do not follow from any commutation rules discussed thus far. They follow from the solution of the eigenvalue problem for $\hat{L}^2$, which was written in terms of position and momentum operators. This was also the origin of the condition $L_z = m\hbar$ with the restriction $m = -l, -l+1, \ldots, l$. But we now want to sever the connection between operator eigenvalues and any specific physical models of rotation! Therefore, in order to make a transition to the general case, we must now revisit Equation 7.11 and rederive these rules for the case of spin, this time using the commutation properties of $\hat{\mathbf{S}}$ exclusively.

Since $\hat{S}^2$ and $\hat{S}_z$ commute, we can postulate the existence of common eigenstates for these two operators. However, this time we will not make any assumptions about their eigenvalues. We will denote them by symbols a and b and write the common eigenstates as $|a, b\rangle$. In other words,

$$\hat{S}^2|a, b\rangle = a|a, b\rangle, \qquad \hat{S}_z|a, b\rangle = b|a, b\rangle. \tag{7.34}$$

Application of the raising operator to these eigenstates leads to an equation similar to (7.27):

$$\begin{cases} \hat{S}^2 \hat{S}_+|a, b\rangle = \hat{S}_+ \hat{S}^2|a, b\rangle = a\hat{S}_+|a, b\rangle, \\ \hat{S}_z \hat{S}_+|a, b\rangle = \left([\hat{S}_z, \hat{S}_+] + \hat{S}_+ \hat{L}_z\right)|a, b\rangle = \left(\hbar \hat{S}_+ + \hat{S}_+ \hat{S}_z\right)|a, b\rangle = (b + \hbar)\hat{S}_+|a, b\rangle. \end{cases} \tag{7.35}$$

A similar equation for the lowering operator results in the term $(b - \hbar)\hat{S}_-|a, b\rangle$ on the right. Introducing the constant multipliers as before, we get

$$\hat{S}_+|a, b\rangle = c_+|a, b + \hbar\rangle, \qquad \hat{S}_-|a, b\rangle = c_-|a, b - \hbar\rangle. \tag{7.36}$$

We can now establish the restrictions on a and b using the following argument. It is clear that the z-component of spin cannot be greater than the total magnitude of the spin "vector". In other words, we must require that

$$-\sqrt{a} \leq b \leq \sqrt{a}. \tag{7.37}$$

Therefore, no eigenstates can exist where b goes over the bound (7.37). At the same time, (7.36) allows us to keep generating new eigenstates forever by applying the ladder operators repeatedly. The only way to escape this contradiction is to assume that after some maximum (minimum) value, the raising (lowering) operator will annihilate the state, thus terminating the eigenvalue series:

$$\hat{S}_+|a, b_{max}\rangle = 0, \qquad \hat{S}_-|a, b_{min}\rangle = 0. \tag{7.38}$$

If we apply the lowering operator to the first equation of (7.38) and the raising operator to the second equation, we should again get zeros. But then, by the same reasoning that helped us derive Equations 7.29–30, we have

$$\begin{cases} \hat{S}_-\hat{S}_+|a, b_{max}\rangle = \left(\hat{S}_x^2 + \hat{S}_y^2 - \hbar \hat{S}_z\right)|a, b_{max}\rangle = \left(\hat{S}^2 - \hat{S}_z^2 - \hbar \hat{S}_z\right)|a, b_{max}\rangle = 0, \\ \hat{S}_-\hat{S}_+|a, b_{min}\rangle = \left(\hat{S}_x^2 + \hat{S}_y^2 + \hbar \hat{S}_z\right)|a, b_{min}\rangle = \left(\hat{S}^2 - \hat{S}_z^2 + \hbar \hat{S}_z\right)|a, b_{min}\rangle = 0. \end{cases} \tag{7.39}$$

It follows that

$$\begin{cases} a - b_{max}^2 - \hbar b_{max} = 0, \\ a - b_{min}^2 + \hbar b_{min} = 0, \end{cases} \tag{7.40}$$

or simply
$$b_{max}^2 + \hbar b_{max} = b_{min}^2 - \hbar b_{min} = a. \tag{7.41}$$

This equation is satisfied when $b_{min} = -b_{max}$. Thus, the allowed values of b are distributed symmetrically about zero. As evidenced by (7.36), these values change in increments of $\hbar$, so we have

$$b = -b_{max}, -b_{max} + \hbar, \ldots, b_{max} - \hbar, b_{max}. \tag{7.42}$$

Let $2s$ be the number of terms in this sequence, meaning that s can be integer or half-integer. Then

$$b_{max} = s\hbar \tag{7.43}$$

and, from the first equation of (7.40),

$$a = b_{max}(b_{max} + \hbar) = s\hbar(s\hbar + \hbar) = s(s+1)\hbar^2. \tag{7.44}$$

Thus, (7.34) goes over to

$$\begin{aligned}\hat{S}^2|s(s+1)\hbar^2, m\hbar\rangle &= s(s+1)\hbar^2|s(s+1)\hbar^2, m\hbar\rangle, \\ \hat{S}_z|s(s+1)\hbar^2, m\hbar\rangle &= m\hbar|s(s+1)\hbar^2, m\hbar\rangle,\end{aligned} \tag{7.45}$$

where

$$m = -s, -s+1, \ldots, s-1, s, \tag{7.46}$$

and we recover the results obtained previously for orbital angular momentum. Introducing the same shorthand notation as before, we get the spin analogue of Equation 7.11:

$$\begin{aligned}\hat{S}^2|s, m\rangle &= s(s+1)\hbar^2|s, m\rangle, \\ \hat{S}_z|s, m\rangle &= m\hbar|s, m\rangle\end{aligned} \tag{7.47}$$

It may seem by looking at (7.47) that there never was any real need to solve the eigenvalue equations for $\hat{L}^2$ and L_z in spherical coordinates when we could obtain the same result following the same straightforward procedure that we just used for $\hat{S}^2$ and S_z. However, that effort was not entirely wasted. The result (7.47) establishes the general rule that s can take integer *or* half-integer values. That rule holds for l as well as it does for s. However, if we impose on top of the commutation rules an additional requirement that the process must be described by a direct physical model (a "real" rotation, involving an actual change of position and momentum coordinates), this requirement puts an additional restriction on the quantum number. That's why l in (7.11) takes integer values only. This is something that we could not have predicted from (7.47). It only becomes evident after we go through all steps of the solution for $\hat{L}^2$, L_z as outlined in Appendix E.

7 Angular Momentum

It remains to introduce the last level of abstraction. Let $\hat{\mathbf{J}}$ denote the generic angular momentum that could involve any combination of orbital and spin angular momenta of a given system (which is generally multiparticle). Then, by the same reasoning that led us to (7.47),

$$\hat{J}^2|j,m\rangle = j(j+1)\hbar^2|j,m\rangle, \quad j = \begin{cases} 0,1,2,3,\ldots \\ 1/2, 3/2, 5/2, \ldots \end{cases} \quad (7.48)$$
$$\hat{J}_z|j,m\rangle = m\hbar|j,m\rangle, \quad m = -j,\ldots,j.$$

Any additional restrictions on j and m would follow from the specific conditions of a given problem, detailing the type (orbital versus spin) and the quantum numbers associated with individual angular momenta $\hat{\mathbf{J}}_1, \hat{\mathbf{J}}_2, \ldots, \hat{\mathbf{J}}_n$ comprising the total angular momentum $\hat{\mathbf{J}}$.

7.4
Spin as an Intrinsic Angular Momentum

We have seen that spin of an elementary particle must be described by a quantum number that can take values $s = 0, 1/2, 1, 3/2, 2, 5/2, \ldots$, all consistent with the commutation rules for angular momentum. However, this fact by itself does not tell us which one of the possible values we should take. Two questions arise. First, can we change s by varying experimental conditions (we would say, by making the particle spin faster or slower if such intuitive models were applicable), or is s fixed? Second, if it is fixed, then given an elementary particle, how can we tell this value?

The first question is easy to answer. It is well established that the spin quantum number is an *intrinsic* property of elementary particles and it cannot be changed by the experimenter.

Box 7.5 Experiments involving spin quantum number s

The earliest observations of electron spin (albeit not properly explained at that time) were made in spectroscopy experiments with alkali atoms [5]. These atoms (Li, Na, K, Rb, Cs, Fr) have one valence electron in the outer shell, which makes them convenient objects for study if one is interested in spin properties of individual electrons. Upon a first glance, their spectra conform to the simple model discussed in Chapter 3. But upon a closer look we see some differences. For example, in the spectrum of atomic sodium, there are two very close yellow lines with wavelengths $\lambda_1 = 5895.93$ Å and $\lambda_2 = 5889.95$ Å, known as the Na doublet. The lines are associated with optical transitions 2p → 1s from two very close energy levels onto the ground level (Figure 7.4). This fact shows that the 2p level ($n = 2$, $l = 1$) in the sodium atom is split into two close levels. This line splitting has received a special name: *fine structure*.

```
2p ─────────                    ═════════ }2p


                    │              │ │
                    │              │ │
                    │              │ │
                    │              │ │
                    ▼              ▼ ▼
1s ─────────                    ───────── 1s
    (a)                            (b)
```

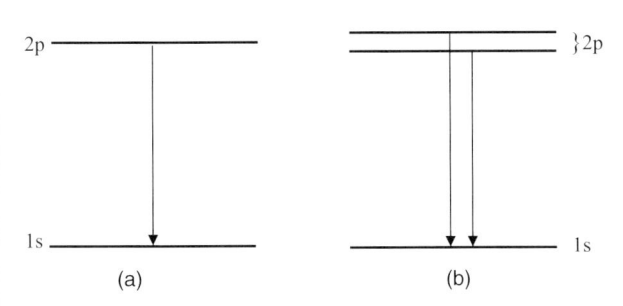

Figure 7.4 The energy level 2p and the corresponding splitting of the spectral line in the absence of an external magnetic field: (a) the degenerated level expected in the absence of the electron spin; (b) the splitting of the state 2p due to the spin interaction with the atomic magnetic field (spin–orbit interaction).

As we now know, this effect is caused by the interaction between the orbital and spin angular momenta. The electron whose orbital motion is characterized by $l = 1$ can have $m_l = 0, \pm 1$ (we used the subscript in order to distinguish the orbital number m_l from the spin quantum number m_s). As we will see in Chapter 10, the corresponding three states are degenerate (they have the same energy, at least in the absence of an external field). Accordingly, only one spectral line would be expected. However, the electron also possesses spin. As we know from electrodynamics, a spinning charged object produces a circular current that generates a magnetic field. Such a field is characterized by the corresponding *magnetic moment* $\mathcal{M}$, which is proportional to the angular momentum of rotation. Therefore, a classical model would predict the relation $\mathcal{M} = g\mathbf{S}$, where the proportionality coefficient g is known as the *gyromagnetic ratio* and will be discussed in detail at the end of Chapter 10.[2] Even though classical rotation analogies fail for spin, the above relation turns out to be far more general than the initial assumption, and applies also in QM, including point-like charged particles such as the electron (except that the value of g is about two times greater than the one predicted by electrodynamics). What is essential is that $\mathbf{S}$ and $\mathcal{M}$ are rigidly coupled to each other and we can infer the existence and properties of $\mathbf{S}$ by looking at magnetic interactions of the particle with its environment. In our case, there is no *external* field. However, in all states with nonzero l and m the bound electron produces a circular electric current that in turn produces magnetic field (when it comes to *orbital* motion, it is alright to apply classical models). Therefore, even in the absence of an

---

2) If spin could be modeled as classical rotation, one would expect the gyromagnetic ratio to be $g = -q_e/2\mu_e \equiv -\mathcal{M}_B/\hbar$, where μ_e is electron mass, q_e is the absolute value of electron charge, and $\mathcal{M}_B \equiv q_e\hbar/2\mu_e$ is Bohr's magneton. The rotation analogy fails and the actual QM formula reads $g = -g_e q_e/2\mu_e \equiv -g_e \mathcal{M}_B/\hbar$ where the constant $g_e = 2$ is called the electron spin *g-factor*. If one takes into account QED (quantum electrodynamics) effects, its value must be adjusted to $g_e = 2.002319$.

external field, the electron (which is now conceptualized as a small magnet $\mathcal{M}$) is moving in the "atmosphere" of magnetic field **B** produced by its own orbital motion. Thus, it acquires an additional energy depending on the orientation of $\mathcal{M} = g\mathbf{s}$ relative to **B**. This effect is called *spin–orbit coupling*. Since there are two possible orientations (parallel or antiparallel to the field), there are two values of this additional contribution, and thus the initial single energy level splits into two levels. By measuring the split, one can obtain the value of $\mathcal{M}$ and check the proportionality relation $\mathcal{M} = g\mathbf{s}$.

A very straightforward evidence of spin and its coupling with magnetic moment was obtained in the Einstein–de Haas experiment [6–9]. Imagine a ferromagnetic cylinder suspended on a string so that it can easily rotate around the vertical axis along the string. The whole setup is put into a vertical magnetic field, so that the cylinder is magnetized along the field direction. The net magnetic moment of the cylinder in this state originates from the individual magnetic moments of the outer atomic electrons all lined up in one direction parallel to the field. Suppose now that we reverse the field direction. This will cause the magnetic moment of each electron to flip over and thus the whole cylinder will get remagnetized in the opposite direction. Because the magnetic moment is rigidly coupled with the spin through $\mathcal{M} = g\mathbf{s}$, its flipping over is accompanied by the same action of spin. The spin contribution to the net angular momentum of the cylinder changes sign. The *net* angular momentum of the cylinder was initially zero and must remain zero due to its conservation. Therefore, there must appear a compensating *orbital* momentum directed opposite to the net spin. This can be crudely modeled as a crowd of dancers on the initially stationary circular stage that can rotate about an axle. When all dancers at once change direction of spinning from clockwise to counter-clockwise, the whole stage starts rotating clockwise around the axle. In the real experiment, the whole cylinder must start rotating about the vertical axis! In this process, a microscopic effect is magnified to an observable macro-scopic scale, and the corresponding information gets transported onto the macroscopic level.

The amount of acquired rotation and thereby the net spin of the system can be easily determined by measuring the twist of the string. In this way, Einstein and de Haas were able to not only confirm the existence of electron spin but also measure its magnitude.

Another early observation was the Stern–Gerlach experiment [10–15], in which the beam of Ag atoms (in a state with $l = 0$) was passed through the region with an inhomogeneous magnetic field perpendicular to the beam. The naive expectation is that in the absence of the orbital angular momentum, the atoms will have no magnetic moment and therefore will not interact with the external field. The atomic beam must pass through the field region without deflection. But the observation showed deflection (Figure 7.5), indicating that the atoms had nonzero magnetic moment. The only possible explanation was to attribute this moment to the electron as such and therefore to assume that the electron

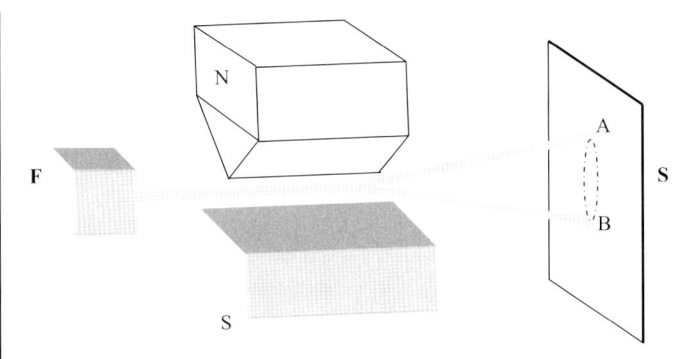

Figure 7.5 Schematic of the Stern–Gerlach experiment. F: furnace with collimator producing a directed beam of atoms with spin $S = \hbar/2$ due to an unpaired valence electron; $\binom{N}{S}$: north and south poles of a magnet producing a nonhomogeneous magnetic field; A, B: observed landing points on the screen indicating the splitting of the initial beam into two subbeams; shaded oval between A and B: classically expected area of atoms' landings on the screen.

must have a nonzero spin according to $\mathcal{M} = g\mathbf{s}$.[3] Furthermore, the observed deflections, instead of forming a continuous range, had only two values, one along the magnetic field gradient and the other in the opposite direction. This means that the initial beam splits into two parts. This looks similar to the splitting of the 2p level producing spectral Na doublet in the first example, doesn't it? If we take the gradient direction to be the z-axis, we can say this indicates the electron spin is quantized (the so-called *spatial quantization*), with only two possibilities for its z-component. Numerical data from the experiment show that $s_z = \pm(1/2)\hbar$, which can be reconciled with (7.47) if we assume that $s = 1/2$ and $m = \pm 1/2$.

The difference between the last two eigenvalues is $\hbar$ – exactly the same as the difference between two closest eigenvalues (2.40) of the $\hat{L}_z$-operator for the case of *orbital* angular momentum. Now we see another example of generalization – Equation 2.40 turns out to be more general than the case with quantization of the de Broglie waves from which it was derived. There are no "waves" associated with the electron spin, and yet it is quantized in the same way as is the *orbital* angular momentum.

---

3) The observed magnetic moment cannot be associated with the atomic nucleus, since the gyromagnetic ratio g for the proton is about 2000 times smaller than for the electron. Accordingly, magnetic moment of the nucleus (the Ag nucleus has 47 protons) is too small to produce the observed deflections. Also, electrons in the inner shells balance each other neatly, so their spins do not produce any noticeable effect either. One could ask, why then should we use Ag atoms at all instead of experimenting with single electrons? The answer is that the electron's wave packet spreads out too fast, so we would not be able to measure the deflection with sufficient accuracy. A heavy atom like Ag removes this difficulty.

The second question (about specific value of spin for a given particle) has no easy answer. A complete theory of spin is studied in relativistic QM (which is another way of saying that spin is a relativistic phenomenon!) and is outside the scope of this text. If we restrict ourselves to the *nonrelativistic* QM, we really have no choice but to cite experimental values and take them for granted. Therefore, we must simply state the known facts without explanation (see Box 7.6).

Box 7.6 Spin quantum number s and elementary particles

The electrons, the muon, the tau particle (tauon), all three flavors of neutrino (electron, muon, and tau neutrino), and their respective antiparticles have $s = 1/2$. These 12 particles are called leptons (their defining property is that they are not subject to the strong interaction and can only interact via gravitation, electromagnetic, or weak interactions). In addition, all six quarks (*up, down, charm, strange, top,* and *bottom*) and their antiquarks also have $s = 1/2$. So altogether we have 24 spin-1/2 elementary particles. Collectively, they are known as *Fermi particles* or *fermions*. The word "fermion" is actually meant to refer to *any* particles with half-integer spin. However, *elementary* particles with spin-3/2 or higher have not been observed so far.

Particles having integer spin are called *Bose particles* or *bosons*. We can distinguish three cases corresponding to $s = 0$, $s = 1$, and $s = 2$. The particle with $s = 0$ is the elusive Higgs boson, whose existence has long been predicted by the *standard model* but is still awaiting a decisive confirmation as of 2013. The case $s = 1$ is represented by the photon, the three W and Z bosons (together known as the *weak* bosons, they are the *exchange particles* underlying the weak force – just as the photon is the exchange particle for the electromagnetic force), and the eight independent types of gluons (exchange particles for the strong force) known in quantum chromodynamics; 12 particles altogether. Finally, the hypothetical (as of today, it has not been directly observed) exchange particle for the gravity force – the graviton – represents the case $s = 2$. No elementary particles with higher spin values are known today.[4]

Then there are the *composite* particles – hadrons and atomic nuclei. Skipping a discussion of the nuclei as there are too many varieties of them, we dwell on the hadrons. They are the strongly interacting particles composed of quarks (which are themselves surrounded by a cloud of gluons!). The two-quark combinations are called *mesons*, and three-quark combinations, *baryons*. As one can imagine, combining two quarks with spin-1/2 can yield a total spin $s = 0$ or $s = 1$, depending on whether we add or subtract spins (more about that later), so

---

[4] There are some other hypothetical particles predicted by various supersymmetric theories that we have not mentioned. Most of them would also have spin-0 or spin-1/2. Standing apart is the hypothetical *gravitino* particle, which will have $s = 3/2$ when and if discovered.

the mesons are bosons. Examples of mesons include pions ($s = 0$), kaons ($s = 0$), and the J/ψ meson ($s = 0$). Combining three quarks can yield a total spin $s = 1/2$ or $s = 3/2$, so the baryons are fermions. Baryons themselves are divided into two categories: nucleons (protons and neutrons) and hyperons. Both the proton and the neutron have $s = 1/2$. Some hyperons also have $s = 1/2$ (examples include the Λ^0, Σ^+, Σ^0, and Σ^- hyperons) while others (Ω^- and the so-called *sigma resonances* Σ^{*+}, Σ^{*0}, and Σ^{*-}) have $s = 3/2$.

Below we will focus on the spin of an elementary particle, primarily the electron. Since $s = 1/2$ for the electron, we have only two possible (s, s_z) eigenstates: $|1/2, 1/2\rangle$ and $|1/2, -1/2\rangle$. If we measure the square of spin angular momentum, the first equation of (7.47) yields $S^2 = (3/4)\hbar^2$ and the spin has magnitude $S = \sqrt{s(s+1)}\hbar = (\sqrt{3}/2)\hbar$. If we measure the projection of spin on an arbitrarily chosen z-axis, the only allowed possibilities are $S_z = \pm(1/2)\hbar$.

Before we move on, let us emphasize again that it would be totally meaningless to try visualizing the electron spin as its actual "spinning" on its axis. We already mentioned two reasons for this: the electron being a point particle (though it can be formally attributed an *effective radius* $r_e = e^2/\mu_e c^2 \approx 10^{-15}$ m [16], its actual size is not manifest in any known phenomena, including high-energy collisions), and the relativistic light barrier (if we were to assume the finite radius $r_e \approx 10^{-15}$ m, the rate of rotation necessary to produce $S = (\sqrt{3}/2)\hbar$ would cause the equatorial rotational velocity to exceed the speed of light). The third reason is that classical rotation about a fixed axis would produce a fixed angular momentum represented by a geometric arrow with all three components defined. But we already know that this is not the case with QM angular momentum. This last point deserves more attention. In CM, we can treat angular momentum (whether orbital or spin) as a vector because in the classical limit its components are much greater than $\hbar$. Accordingly, each of them is much greater than the corresponding indeterminacy. The condition for this is $J_i \gg \hbar$, $i = 1, 2, 3$. (Here symbol J stands for "generic" angular momentum, which can be either orbital or spin, or a sum of both. We use this notation when the type of angular momentum is not specified. In this case, instead of l or s we use symbol j, which as a generic quantum number for the square of angular momentum is now allowed to be integer or half-integer.) In terms of quantum numbers j and m, the classical limit can be written as $j, m \gg 1$. This condition can be satisfied for the orbital quantum number l but not for the spin quantum number s of any known elementary particle. We can in principle imagine an electron orbiting the Sun and having huge orbital momentum $L = \sqrt{l(l+1)}\hbar$. But when we measure its spin, we will always get the same value $S = (\sqrt{3}/2)\hbar$. There can be no extension into the classical realm in this case. That is another reason for saying that spin is exclusively a quantum-mechanical phenomenon.

As the projection of spin onto any direction can have only two values, operators $\hat{S}_x$, $\hat{S}_y$, and $\hat{S}_z$ can be represented by 2×2 matrices. The eigenvalues of these matrices can be $S = (1/2)\hbar$ (spin projection is positive) or $-(1/2)\hbar$ (spin projection is negative). It is therefore convenient to isolate the $-(1/2)\hbar$ factor by writing

$$\hat{\mathbf{S}} = \frac{1}{2}\hbar\hat{\boldsymbol{\sigma}}. \tag{7.49}$$

In this case, the problem reduces to finding three matrices $\hat{\sigma}_\gamma$ (where $\gamma = 1, 2, 3$ stands for x, y, z, respectively) whose eigenvalues are 1 and -1. It follows that for any γ, the eigenvalues of the *squares* $\hat{\sigma}_\gamma^2$ are all equal to 1. This condition is satisfied by the identity matrix

$$\hat{\sigma}_\gamma^2 = \begin{pmatrix} 1 & 0 \\ 0 & 1 \end{pmatrix}, \quad \gamma = 1, 2, 3. \tag{7.50}$$

Since the identity matrix is invariant (the same in any representation), the matrices $\hat{\sigma}_\gamma^2$ have the same form (7.50) regardless of the chosen basis.

Let us work in the σ_z-basis $\{|\uparrow\rangle, |\downarrow\rangle\}$ formed by the two eigenstates of $\hat{\sigma}_z$. The z-direction corresponds to $\gamma = 3$, so the matrix $\hat{\sigma}_3$ will be in its own representation. A matrix operator in its own representation is diagonal, with diagonal elements being its eigenvalues. Therefore, in this representation we have

$$\hat{\sigma}_3 = \begin{pmatrix} 1 & 0 \\ 0 & -1 \end{pmatrix}. \tag{7.51}$$

It remains to find the explicit form of $\hat{\sigma}_1$ and $\hat{\sigma}_2$. They can be found from the commutative relations between different spin components. As mentioned, they are the same as those for the orbital angular momentum and when expressed in terms of σ_γ, will have the form

$$\hat{\sigma}_1\hat{\sigma}_2 - \hat{\sigma}_2\hat{\sigma}_1 = 2i\hat{\sigma}_3, \quad \hat{\sigma}_2\hat{\sigma}_3 - \hat{\sigma}_3\hat{\sigma}_2 = 2i\hat{\sigma}_1, \quad \hat{\sigma}_3\hat{\sigma}_1 - \hat{\sigma}_1\hat{\sigma}_3 = 2i\hat{\sigma}_2. \tag{7.52}$$

Since the commutators (7.15) are not zeros, there are indeterminacy relations between the variances of the different components. So an eigenstate of σ_z leaves the other two components undefined. This means that in the σ_3-basis the matrices $\hat{\sigma}_1$ and $\hat{\sigma}_2$ are nondiagonal:

$$\hat{\sigma}_1 = \begin{pmatrix} a_{11} & a_{12} \\ a_{21} & a_{22} \end{pmatrix}, \quad \hat{\sigma}_2 = \begin{pmatrix} b_{11} & b_{12} \\ b_{21} & b_{22} \end{pmatrix}. \tag{7.53}$$

We put this into (7.50), which holds in any representation. After simple algebra, we obtain

$$a_{12}(a_{11} + a_{22}) = 0, \quad a_{11}^2 + a_{12}a_{21} = a_{22}^2 + a_{12}a_{21} = 1, \tag{7.54}$$

and similar relations for the elements of matrix $\hat{\sigma}_2$. Since the nondiagonal elements of matrices (7.53) are nonzero, the first condition (7.54) requires that the traces of these matrices be zero. Therefore, denoting $a_{11} = -a_{22} \equiv a$ and $b_{11} = -b_{22} \equiv b$, we have

$$\hat{\sigma}_1 = \begin{pmatrix} a & a_{12} \\ a_{21} & -a \end{pmatrix}, \quad \hat{\sigma}_2 = \begin{pmatrix} b & b_{12} \\ b_{21} & -b \end{pmatrix}. \tag{7.55}$$

7.4 Spin as an Intrinsic Angular Momentum

Now we put this result into the last two equations of (7.52). After the same simple algebra, we obtain

$$a = b = 0, \quad b_{12} = -ia_{12}, \quad b_{21} = ia_{21}. \tag{7.56}$$

It follows then from the second equation of (7.54) that

$$a_{12}a_{21} = 1. \tag{7.57}$$

Since the spin operators must be Hermitian, the two nondiagonal elements must have the same magnitude, in which case there follows $|a_{12}| = |a_{21}| = 1$, and we can write

$$a_{12} = e^{i\delta}, \quad a_{21} = e^{-i\delta}, \tag{7.58}$$

where δ is an arbitrary real number. Without loss of generality, we can set $\delta = 0$, so that $a_{12} = a_{21} = 1$. Combining this with the last two equations (7.58), we finally obtain

$$\hat{\sigma}_1 = \begin{pmatrix} 0 & 1 \\ 1 & 0 \end{pmatrix}, \quad \hat{\sigma}_2 = i \begin{pmatrix} 0 & -1 \\ 1 & 0 \end{pmatrix}, \quad \hat{\sigma}_3 = \begin{pmatrix} 1 & 0 \\ 0 & -1 \end{pmatrix}. \tag{7.59}$$

This matrix representation of components of the spin operator in the s_z-basis is known as the *Pauli matrices*. Note that all three matrices are traceless.

Exercise 7.1

Find the anticommutators for operators (7.59). An *anticommutator* of two operators $\hat{A}$ and $\hat{B}$ is defined as

$$\{\hat{A}, \hat{B}\} \equiv \hat{A}\hat{B} + \hat{B}\hat{A}. \tag{7.60}$$

Solution:
Using (7.22), we find

$$\hat{\sigma}_1\hat{\sigma}_2 = -\hat{\sigma}_2\hat{\sigma}_1 (= i\hat{\sigma}_3), \quad \hat{\sigma}_2\hat{\sigma}_3 = -\hat{\sigma}_3\hat{\sigma}_2 (= i\hat{\sigma}_1), \quad \hat{\sigma}_3\hat{\sigma}_1 = -\hat{\sigma}_1\hat{\sigma}_3 (= i\hat{\sigma}_2). \tag{7.61}$$

It follows

$$\{\hat{\sigma}_1, \hat{\sigma}_2\} = \{\hat{\sigma}_2, \hat{\sigma}_3\} = \{\hat{\sigma}_3, \hat{\sigma}_1\} = 0. \tag{7.62}$$

Now we want to find the *eigenstates* of these operators. Let us start with the eigenstates of $\hat{\sigma}_3$. Denote them as $|\uparrow\rangle$ for spin-up state ($\sigma_3 = 1$) and as $|\downarrow\rangle$ for spin-down state ($\sigma_3 = -1$):

$$\hat{\sigma}_3|\uparrow\rangle = |\uparrow\rangle \quad \text{and} \quad \hat{\sigma}_3|\downarrow\rangle = -|\downarrow\rangle, \tag{7.63}$$

or, using expression (7.59) for $\hat{\sigma}_3$,

$$\begin{pmatrix} 1 & 0 \\ 0 & -1 \end{pmatrix}|\uparrow\rangle = |\uparrow\rangle \quad \text{and} \quad \begin{pmatrix} 1 & 0 \\ 0 & -1 \end{pmatrix}|\downarrow\rangle = -|\downarrow\rangle. \tag{7.64}$$

It is immediately seen that the eigenstates satisfying these equations are column matrices:

$$|\uparrow\rangle = \begin{pmatrix} 1 \\ 0 \end{pmatrix}, \quad |\downarrow\rangle = \begin{pmatrix} 0 \\ 1 \end{pmatrix}. \tag{7.65}$$

The meaning of the matrix elements here is obvious: they represent the probability amplitudes of finding the spin pointing up or down. In the spin-up eigenstate (the left matrix in (7.65)), we will certainly (amplitude 1, top matrix element) find the electron spin pointing up, and never (zero amplitude, the bottom element) find it pointing down. In the spin-down eigenstate (the right matrix), we will have the opposite distribution of amplitudes.

Now we find the eigenstates of $\hat{\sigma}_1$ and $\hat{\sigma}_2$. Denote an eigenstate of $\hat{\sigma}_1$ as $|\chi\rangle = \begin{pmatrix} \chi_1 \\ \chi_2 \end{pmatrix}$ and let λ stand for the corresponding eigenvalue. Similarly, $|\Upsilon\rangle = \begin{pmatrix} \Upsilon_1 \\ \Upsilon_2 \end{pmatrix}$ and μ will be an eigenstate and the corresponding eigenvalue of $\hat{\sigma}_2$. Then we have two eigenvalue equations

$$\begin{pmatrix} 0 & 1 \\ 1 & 0 \end{pmatrix} \begin{pmatrix} \chi_1 \\ \chi_2 \end{pmatrix} = \lambda \begin{pmatrix} \chi_1 \\ \chi_2 \end{pmatrix}, \quad \begin{pmatrix} 0 & -i \\ i & 0 \end{pmatrix} \begin{pmatrix} \Upsilon_1 \\ \Upsilon_2 \end{pmatrix} = \mu \begin{pmatrix} \Upsilon_1 \\ \Upsilon_2 \end{pmatrix}. \tag{7.66}$$

A nonzero solution exists for the first equation of (7.31) if

$$\lambda^2 = 1, \quad \lambda = \pm 1. \tag{7.67}$$

For $\lambda = 1$ we have $\chi_2 = \chi_1$, and for $\lambda = -1$ we have $\chi_2 = -\chi_1$. Together with the normalization condition, these eigenstates (up to an arbitrary phase factor) can be written as

$$\left|\underset{x}{\rightarrow}\right\rangle \equiv |\chi^+\rangle = \frac{1}{\sqrt{2}} \begin{pmatrix} 1 \\ 1 \end{pmatrix}, \quad \left|\underset{x}{\leftarrow}\right\rangle \equiv |\chi^-\rangle = \frac{1}{\sqrt{2}} \begin{pmatrix} 1 \\ -1 \end{pmatrix}. \tag{7.68}$$

In a similar way (Problem 7.9), we find the eigenvalues and respective eigenfunctions of $\hat{\sigma}_2$:

$$\left|\underset{y}{\rightarrow}\right\rangle \equiv |\Upsilon^+\rangle = \frac{1}{\sqrt{2}} \begin{pmatrix} 1 \\ i \end{pmatrix}, \quad \left|\underset{y}{\leftarrow}\right\rangle \equiv |\Upsilon^-\rangle = \frac{1}{\sqrt{2}} \begin{pmatrix} 1 \\ -i \end{pmatrix}. \tag{7.69}$$

Note that each of these states can also be represented as an equally weighted superposition of the "up" and "down" states (Problem 7.10):

$$\left|\underset{x}{\rightarrow}\right\rangle = \frac{1}{\sqrt{2}}(|\uparrow\rangle + |\downarrow\rangle), \quad \left|\underset{x}{\leftarrow}\right\rangle = \frac{1}{\sqrt{2}}(|\uparrow\rangle - |\downarrow\rangle) \tag{7.70}$$

and

$$\left|\underset{y}{\rightarrow}\right\rangle = \frac{1}{\sqrt{2}}(|\uparrow\rangle + i|\downarrow\rangle), \quad \left|\underset{y}{\leftarrow}\right\rangle = \frac{1}{\sqrt{2}}(|\uparrow\rangle - i|\downarrow\rangle). \tag{7.71}$$

This is a special case of the general rule (6.9) for transformation between different sets of mutually orthogonal basis states.

7.4 Spin as an Intrinsic Angular Momentum

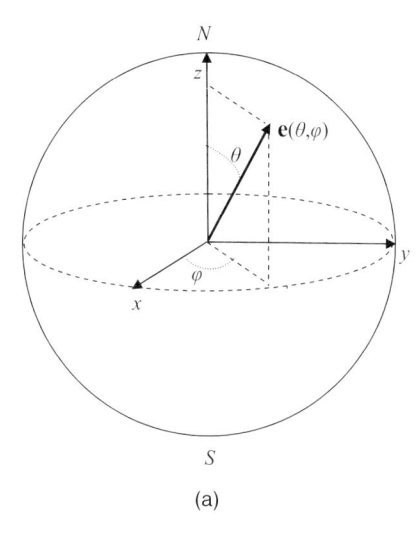

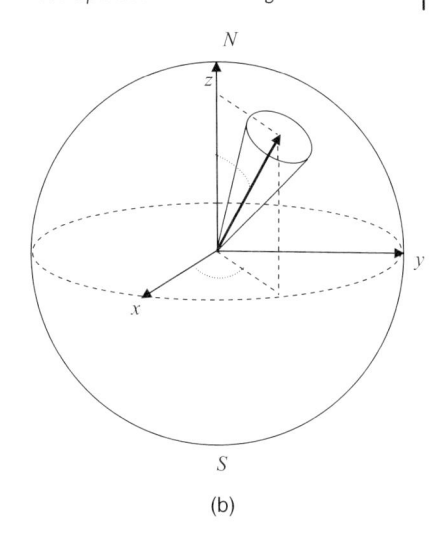

Figure 7.6 The Bloch (or Poincaré) sphere. It provides a convenient graphical representation of the spin states. (a) For a spin-1/2, its various possible states are represented by points (θ, φ) on the sphere, where θ and φ are the polar and azimuthal angles, respectively. (b) A more detailed representation of a spin state on the Bloch sphere (not to scale). For a spin-1/2, the vector $\mathbf{e}(\theta, \varphi)$ actually represents the symmetry axis of the conoid corresponding to the given state.

All these results have a simple physical meaning. No matter whether we try a σ_x-measurement, a σ_y-measurement, or a σ_z-measurement, or choose some other axis, we will always find the projection of σ on the chosen axis equal to $+1$ or -1. The amplitudes of these two possible outcomes depend on two things: the current state and the axis chosen for the measurement. Let us say we were given a "spin-right" state $|\rightarrow\rangle_x$: a σ_x-measurement will definitely yield $+1$. Then the projection of σ onto any other axis is indeterminate. The indeterminacy is maximal for the axes (y and z, for example) perpendicular to x, so when we make a σ_y- or σ_z-measurement both outcomes ± 1 will have the same 50% probability.

The most important result of this part is that we have *the same pair of spin eigenvalues* $\sigma_\gamma = \pm 1$, for any γ, that is, for any of the three directions x, y, and z. This means that these two values make the only allowed pair of values of the spin component along an *arbitrary* direction $\hat{\mathbf{e}}$.

Electron spin states admit a graphical representation, using the *Bloch sphere* of unit radius (Figure 7.6). Any two antipodal points on this sphere correspond to mutually orthogonal eigenstates of a two-state quantum system. In optics, the Bloch sphere (also known as the *Poincaré sphere*, see Section 24.6) is used to represent different types of polarization states.

It is convenient to choose the polar axis of the Bloch sphere along the z-direction. Then the state $|\uparrow\rangle$ can be represented by the unit vector from the center pointing at the north pole of the sphere. The state $|\downarrow\rangle$ will be represented by a vector pointing to the south pole.

Note that the two states in question are distinct eigenstates of an operator, and as such they are represented by *mutually orthogonal* vectors in $\mathcal{H}$. But in the Bloch sphere (which is in V-space!) they are represented by two unit vectors that are *antiparallel*. In this respect, the Bloch sphere directly represents *observable eigenvalues* of the spin operator, and only indirectly – its eigenstates. Such a situation is similar to that of the particle position being on the x-axis: as we found in Section 6.5, actual physical positions (the different points on one axis) are represented by different mutually orthogonal unit vectors in $\mathcal{H}$.

The unit vector on the Bloch sphere represents *the vector expectation value of the spin operator in the corresponding state*. As we already know, the "spin vector" is not really a vector. It is represented by a conical surface as in Figure 7.3. So when we refer to the unit vector on Bloch sphere as a "spin vector" or just "spin," actually it is the direction along the *symmetry axis* of the corresponding conical surface (Figure 7.6). For $s = 1/2$, the direction of the symmetry axis defined by spherical coordinates (θ, φ) uniquely specifies the spin state. On the other hand, the normalized unit radius of the sphere is *not* the magnitude of angular momentum! The latter for an electron spin is (in units of $\hbar$) $\sqrt{s(s+1)} = \sqrt{3}/2 \neq 1$.

Let us now find a general expression for an *arbitrary* spin-1/2 state $|S\rangle$. In this case, $|S\rangle$ is represented by a vector $\mathbf{e}(\theta, \varphi)$ with arbitrary θ and φ. As before, we will use the s_z-basis, with the polar axis of the Bloch sphere taken as the z-axis. Using notations $|\uparrow\rangle$ and $|\downarrow\rangle$ for the "up" and "down" states, we can write $|S\rangle$ as a superposition:

$$|S\rangle = \tilde{c}_1 |\uparrow\rangle + \tilde{c}_2 |\downarrow\rangle = \tilde{c}_1 \begin{pmatrix} 1 \\ 0 \end{pmatrix} + \tilde{c}_2 \begin{pmatrix} 0 \\ 1 \end{pmatrix} = \begin{pmatrix} \tilde{c}_1 \\ \tilde{c}_2 \end{pmatrix}. \tag{7.72}$$

Here $\tilde{c}_1$ and $\tilde{c}_2$ are the amplitudes of finding the electron in a spin-up or spin-down state, respectively. The column matrix on the right provides the corresponding matrix representation of state $|S\rangle$. For the normalized state,

$$|\tilde{c}_1|^2 + |\tilde{c}_2|^2 = \mathcal{P}(\uparrow) + \mathcal{P}(\downarrow) = 1. \tag{7.73}$$

Before doing any mathematics, let us play the game of an intelligent guess. We know that the eigenstates $|\uparrow\rangle$ and $|\downarrow\rangle$, which are separated by the angle $\theta = \pi$ on the Bloch sphere, are separated by half of this angle in $\mathcal{H}$. The amplitudes $\tilde{c}_1$ and $\tilde{c}_2$ in (7.72) are projections of $|S\rangle$ onto the eigenvectors $|\uparrow\rangle$ and $|\downarrow\rangle$ in $\mathcal{H}$. Therefore, we expect that variable θ on the Bloch sphere will be mapped onto $\theta/2$ in $\mathcal{H}$. In other words, $\tilde{c}_1$ and $\tilde{c}_2$ must be functions of $\theta/2$.

Now let us do the mathematics. We start by writing the amplitudes as $\tilde{c}_1 = c_1 e^{i\varphi_1}$ and $\tilde{c}_2 = c_2 e^{i\varphi_2}$ with normalization condition $c_1^2 + c_2^2 = 1$. The z-component of $|S\rangle$ is the projection of $\mathbf{e}(\theta, \varphi)$ onto the z-axis, so it is just $\cos\theta$. On the other hand, it is the expectation value of the σ_z-component in the state (7.37). Applying the general expression (5.2) for the expectation value to our case, we have

$$\langle \sigma_z \rangle = 1 c_1^2 + (-1) c_2^2 = c_1^2 - c_2^2 = \cos\theta. \tag{7.74}$$

The system of two simultaneous equations $c_1^2 + c_2^2 = 1$ and $c_1^2 - c_2^2 = \cos\theta$ is easily solved:

$$c_1 = \cos\left(\frac{\theta}{2}\right), \quad c_2 = \sin\left(\frac{\theta}{2}\right). \tag{7.75}$$

7.4 Spin as an Intrinsic Angular Momentum

Using the fact that a quantum state is defined up to an arbitrary phase factor, we can set $\phi_1 = 0$ and $\phi_2 = \varphi$, with φ being the azimuthal angle on the Bloch sphere. Then the amplitudes in (7.72) can be written as

$$\tilde{c}_1 = \cos\left(\frac{\theta}{2}\right), \qquad \tilde{c}_2 = \sin\left(\frac{\theta}{2}\right)e^{i\varphi}. \qquad (7.76)$$

Putting this into (7.72) and using (7.65), we obtain a general spin state in the matrix form in the S_z-representation:

$$|S(\theta,\varphi)\rangle = \cos\left(\frac{\theta}{2}\right)|\uparrow\rangle + \sin\left(\frac{\theta}{2}\right)e^{i\varphi}|\downarrow\rangle = \begin{pmatrix} \cos(\theta/2) \\ \sin(\theta/2)e^{i\varphi} \end{pmatrix}. \qquad (7.77)$$

As we expected, the amplitudes are proportional to cos and sin functions of $\theta/2$ rather than θ. This is why spin-1/2 is sometimes referred to as "half-vector." The presence of the azimuthal angle φ is necessary for the represented spin states to span all directions in V.

The cases $\theta = 0$ and $\theta = \pi$ correspond to states $|\uparrow\rangle$ and $|\downarrow\rangle$, respectively, recovering their matrix form (7.65). Another interesting case is $\theta = \pi/2$, when the spin spans the equatorial plane. For any spin state represented by a vector in this plane, the indeterminacy of its z-component is maximal: in a σ_z-measurement, it will be equally probable to obtain $\sigma_z = 1$ or $\sigma_z = -1$ (the state vector will jump from the equatorial plane to the polar axis pointing north or south with equal probability). The corresponding amplitudes have a common modulus $|\tilde{c}_1| = |\tilde{c}_2| = 1/\sqrt{2}$. And this is precisely what we have from (7.75) and (7.77) at $\theta = \pi/2$.

The direction of the spin vector in the equatorial plane is determined by φ. For example, $\varphi = 0$ gives us "spin-right" state $|\rightarrow\rangle_x$, so $|S\rangle \to |\chi^+\rangle$. Indeed, setting in (7.77) $\theta = \pi/2$ and $\varphi = 0$, we recover $|\chi^+\rangle$ of Equation 7.68. The $\varphi = \pi$ gives us "spin-left" state $|\leftarrow\rangle_x$, so $|S\rangle \to |\chi^-\rangle$, recovering the second Equation 7.68. In a similar way, one can check that at $\varphi = \pm\pi/2$ (positive or negative y-direction, respectively) we recover $|\Upsilon^\pm\rangle$ of Equation 7.69.

Equation 7.72 gives the spin state $|S\rangle$ in the S_z-representation, with spin-up and spin-down states forming the basis. But this is only one out of the many possible representations of state $|S\rangle$ (one out of various possible faces the state can show us, depending on which component we want to "see," that is, to measure). Imagine that instead of asking about the probability amplitudes for S_z, we are now interested in the amplitudes for S_y. In other words, we want to know the chances of finding the electron spin pointing right or left along the y-axis if we perform the appropriate S_y-measurement. In this case, we choose the "spin-right" $|\rightarrow\rangle_y$ and "spin-left" $|\leftarrow\rangle_y$ states as our basis. The two S_y states *in their own representation* can be written in the matrix form in a way analogous to (7.72):

$$\left|\rightarrow\right\rangle_y = \begin{pmatrix} 1 \\ 0 \end{pmatrix} \quad \text{and} \quad \left|\leftarrow\right\rangle_y = \begin{pmatrix} 0 \\ 1 \end{pmatrix}, \qquad (7.78)$$

but the matrix elements here stand for the spin amplitudes along the y-axis rather than z-axis. In state $\left|\rightarrow\right\rangle_y$, according to its definition, we will certainly (the first matrix element is 1) find $S_y = \hbar/2$ and never (the second matrix element is 0) find $S_y = -\hbar/2$. And contrariwise, in state $\left|\leftarrow\right\rangle_y$ we will always get $S_y = -\hbar/2$ and never $S_y = \hbar/2$. A physical question then arises: Suppose we have the *actual* state $|S\rangle$ initially given in the S_z-representation, but we want to perform an S_y-measurement; then what are the chances of finding $S_y = \hbar/2$ and $S_y = -\hbar/2$? In other words, how are these chances expressed in terms of the given $\tilde{c}_1$ and $\tilde{c}_2$?

Let $\tilde{b}_1$ and $\tilde{b}_2$ be the probability amplitudes of spin-right and spin-left states, respectively. Then $|S\rangle$ can be written in terms of these new amplitudes as

$$|S\rangle = \tilde{b}_1 \left|\rightarrow\right\rangle_y + \tilde{b}_2 \left|\leftarrow\right\rangle_y = \tilde{b}_1 \begin{pmatrix} 1 \\ 0 \end{pmatrix} + \tilde{b}_2 \begin{pmatrix} 0 \\ 1 \end{pmatrix} = \begin{pmatrix} \tilde{b}_1 \\ \tilde{b}_2 \end{pmatrix}. \tag{7.79}$$

The state is now written in the s_y-representation. The corresponding probabilities are

$$\mathcal{P}\left(\rightarrow_y\right) = |\tilde{b}_1|^2, \quad \mathcal{P}\left(\leftarrow_y\right) = |\tilde{b}_2|^2, \quad \mathcal{P}\left(\rightarrow_y\right) + \mathcal{P}\left(\leftarrow_y\right) = 1. \tag{7.80}$$

So now we have two different representations (7.72) and (7.79) of *the same* state and can combine both in one equation:

$$|S\rangle = \tilde{c}_1 |\uparrow\rangle + \tilde{c}_2 |\downarrow\rangle = \tilde{b}_1 \left|\rightarrow\right\rangle_y + \tilde{b}_2 \left|\leftarrow\right\rangle_y. \tag{7.81}$$

Looked at from one side (the S_z-representation) it shows us one facet of this state, whose mathematical expression is given by the column matrix (7.72). Looked at from another side (the S_y-representation) it shows its other facet, represented by matrix (7.79). Since both are different representations of the same entity – state $|S\rangle$ – there must be some correlation between them. Let us find this correlation, that is, express one matrix in terms of the other. Using our previous result (7.36), putting it into (7.81), and comparing the coefficients, we get

$$\tilde{b}_1 = \frac{1}{\sqrt{2}}(\tilde{c}_1 - i\tilde{c}_2), \quad \tilde{b}_2 = \frac{1}{\sqrt{2}}(\tilde{c}_1 + i\tilde{c}_2). \tag{7.82}$$

The probabilities are easily found:

$$\mathcal{P}\left(\rightarrow_y\right) = |\tilde{b}_1|^2 = \frac{1}{2}(\tilde{c}_1 - i\tilde{c}_2)(\tilde{c}_1^* + i\tilde{c}_2^*) = \frac{1}{2} - \operatorname{Im}\tilde{c}_1\tilde{c}_2^*,$$
$$\mathcal{P}\left(\leftarrow_y\right) = |\tilde{b}_2|^2 = \frac{1}{2}(\tilde{c}_1 + i\tilde{c}_2)(\tilde{c}_1^* - i\tilde{c}_2^*) = \frac{1}{2} + \operatorname{Im}\tilde{c}_1\tilde{c}_2^*. \tag{7.83}$$

The last step makes use of the identity $c - c^* = 2i\operatorname{Im}(c)$. We got the answer to our question: The probability distribution for S_y is generally *asymmetric* even when it is symmetric for S_z-measurements. The latter symmetry holds when $\theta = \pi/2$, that is, when $\mathbf{e}(\theta, \varphi)$ lies in the equatorial plane of the Bloch sphere. On the other hand, when the $\tilde{c}$-amplitudes are such that $\operatorname{Im}\tilde{c}_1\tilde{c}_2^* = 0$, we have a 50–50 chance of finding spin right or spin left in S_y-measurements. Consulting with (7.76) and (7.77), we see that condition $\operatorname{Im} \tilde{c}_1 \tilde{c}_2^* = 0$ is met at $\varphi = 0, \pi$. Geometrically, this happens when $\mathbf{e}(\theta, \varphi)$ lies in the xz-plane. The two cases $|S\rangle = |\uparrow\rangle$ and $|S\rangle = |\downarrow\rangle$, where φ is undefined, also fall into the same category – they correspond to the situation when one of the amplitudes $\tilde{c}$ is zero so that $\operatorname{Im} \tilde{c}_1 \tilde{c}_2^* = 0$. These are the cases of extreme S_z-asymmetry, when state $|S\rangle$ always produces one outcome and never produces the other, but we have instead the maximum (50–50) symmetry for S_y-measurements.

Exercise 7.2

Using the general expression $\bar{\sigma} = \langle S|\hat{\sigma}|S\rangle$ for the mean value of the spin operator in an arbitrary state $|S\rangle$, find (a) $\bar{\sigma}_z$ and $\bar{\sigma}_x$ when $|S\rangle = |\uparrow\rangle$ and (b) the same when $|S\rangle = |\rightarrow\rangle$.

Solution:

Representing $|S\rangle$ as $\begin{pmatrix} \tilde{c}_1 \\ \tilde{c}_2 \end{pmatrix}$, we can write the spin operator in the matrix form:

$$\bar{\sigma} = (\tilde{c}_1^* \ \tilde{c}_2^*) \begin{pmatrix} \sigma_{11} & \sigma_{12} \\ \sigma_{21} & \sigma_{22} \end{pmatrix} \begin{pmatrix} \tilde{c}_1 \\ \tilde{c}_2 \end{pmatrix} = \tilde{c}_1^* \sigma_{11} \tilde{c}_1 + \tilde{c}_1^* \sigma_{12} \tilde{c}_2 + \tilde{c}_2^* \sigma_{21} \tilde{c}_1 + \tilde{c}_2^* \sigma_{22} \tilde{c}_2. \quad (7.84)$$

In case (a) we expect $\bar{\sigma}_z \equiv \bar{\sigma}_3 = 1$, since $|\uparrow\rangle$ is an eigenstate of the $\hat{\sigma}_z$-operator corresponding to spin up. And indeed, from the third Pauli matrix (7.59), we have $\sigma_{11} = -\sigma_{22} = 1$, $\sigma_{12} = \sigma_{21} = 0$ and $\tilde{c}_1 = 1$, $\tilde{c}_2 = 0$, so Equation 7.58 yields $\bar{\sigma}_z = 1$. The same state $|S\rangle$ should also give us $\bar{\sigma}_x = 0$, since the spin-up state is an *equally weighted* superposition of the spin-right and spin-left states. The same is seen from Figure 7.2: the *average* horizontal projection of generatrices forming a conical surface with the vertical symmetry axis is zero. Formally, we now apply the $\hat{\sigma}_x$ Pauli matrix instead of $\hat{\sigma}_z$. Its matrix elements are $\sigma_{11} = \sigma_{22} = 0$ and $\sigma_{12} = \sigma_{21} = 1$. Plugging them into (7.58) with the same amplitudes $\tilde{c}_1 = 1$ and $\tilde{c}_2 = 0$ gives $\bar{\sigma}_x = 0$ as expected. The case (b) is solved trivially by analogy with (a) and we get $\bar{\sigma}_z = 0$ and $\bar{\sigma}_x = 1$.

With the operators of all three spin components known, we can do two more things.

I. First, we can write the expression for the *vector spin operator* as we do for the operator $\hat{\mathbf{p}}$ in Equations 3.27 and 3.29: similar to

$$\hat{\mathbf{p}} = \hat{p}_x \mathbf{x} + \hat{p}_y \mathbf{y} + \hat{p}_z \mathbf{z} \equiv \sum_j \hat{p}_j \mathbf{x}_j, \quad j = 1, 2, 3, \quad (7.85)$$

for the linear momentum, we have

$$\hat{\sigma} = \hat{\sigma}_x \mathbf{x} + \hat{\sigma}_y \mathbf{y} + \hat{\sigma}_z \mathbf{z} \equiv \sum_j \hat{\sigma}_j \mathbf{x}_j \quad (7.86)$$

for the vector spin operator. Here $\mathbf{x}_j$ indicate *unit vectors*, but we dropped the caps in order to avoid confusion with *operators* $\hat{p}_j$ and $\hat{\sigma}_j$. Note that despite the mathematical similarity between Equations 7.85 and 7.86, there is a fundamental physical difference between them: different components of $\hat{\mathbf{p}}$ commute, $[p_i, p_j] = 0$, whereas different components of $\hat{\boldsymbol{\sigma}}$ do not. Therefore, in contrast with $\hat{\mathbf{p}}$ having a set of eigenstates $|\psi_{\mathbf{p}}\rangle$ according to $\hat{\mathbf{p}}|\psi_{\mathbf{p}}\rangle = \mathbf{p}|\psi_{\mathbf{p}}\rangle$, the operator $\hat{\boldsymbol{\sigma}}$ *does not have* the vector eigenvalues or the corresponding eigenstates! We *cannot* write a vector equation $\hat{\boldsymbol{\sigma}}|\psi_{\boldsymbol{\sigma}}\rangle = \boldsymbol{\sigma}|\psi_{\boldsymbol{\sigma}}\rangle$ because a definite eigenvalue for one component of $\hat{\boldsymbol{\sigma}}$ would automatically "kill" the two others by making them indefinite. Similarly, determining an eigenstate for one component of $\hat{\boldsymbol{\sigma}}$ excludes the simultaneous existence of such states for two others. This means the absence of a *vector eigenvalue* $\boldsymbol{\sigma} = \sigma_x \mathbf{x} + \sigma_y \mathbf{y} + \sigma_z \mathbf{z}$. Even though the angular momentum *operator* $\hat{\boldsymbol{\sigma}}$ is treated as a vector, the represented *observable* $\boldsymbol{\sigma}$ is not a vector quantity (see again Figure 7.6). As in case of **L**, QM defines the *magnitude* of $\boldsymbol{\sigma}$ and *only one* (e.g., σ_z) out of its three components. The magnitude of $\boldsymbol{\sigma}$ is determined from $\hat{\boldsymbol{\sigma}}^2 = \sum_j \hat{\sigma}_j^2$, which, in view of (7.50), reduces to a pure number

$$\hat{\boldsymbol{\sigma}}^2 = \sum_j \hat{\sigma}_j^2 = 3 = \boldsymbol{\sigma}^2. \tag{7.87}$$

This number determines σ according to

$$\sigma = |\boldsymbol{\sigma}| = \sqrt{\boldsymbol{\sigma}^2} = \sqrt{3}. \tag{7.88}$$

The last result can be checked by writing the magnitude of electron spin in an analogous form $S = |\mathbf{S}| = \sqrt{\mathbf{S}^2} = \sqrt{s(s+1)\hbar^2} = (\sqrt{3}/2)\hbar$ and recalling that $\mathbf{S} = (\hbar/2)/\boldsymbol{\sigma}$.

The same results can be obtained from another argument, more sophisticated but accordingly more instructive, since it shows explicitly the absence of a common eigenstate for *all three* components of $\hat{\boldsymbol{\sigma}}$. Let us use the Pauli matrices to write (7.88) in the $\hat{\sigma}_z \equiv \hat{\sigma}_3$ representation:

$$\hat{\boldsymbol{\sigma}} = \begin{pmatrix} 0 & 1 \\ 1 & 0 \end{pmatrix} \mathbf{x} + \begin{pmatrix} 0 & -i \\ i & 0 \end{pmatrix} \mathbf{y} + \begin{pmatrix} 1 & 0 \\ 0 & -1 \end{pmatrix} \mathbf{z} = \begin{pmatrix} \mathbf{z} & \mathbf{x} - i\mathbf{y} \\ \mathbf{x} + i\mathbf{y} & -\mathbf{z} \end{pmatrix}. \tag{7.89}$$

The expression on the right represents $\hat{\boldsymbol{\sigma}}$ as a single matrix, whose elements are unit vectors or their compositions. Suppose for a moment that observable $\boldsymbol{\sigma}$ represented by $\hat{\boldsymbol{\sigma}}$ does exist as a geometrical vector and therefore has *all* its three components defined. Denote them as $\sigma_x, \sigma_y, \sigma_z$. Then $\boldsymbol{\sigma}$ must be obtained as a vector eigenvalue with these components specified from the equation $\hat{\boldsymbol{\sigma}}|S\rangle = \boldsymbol{\sigma}|S\rangle$, that is,

$$\begin{pmatrix} \mathbf{z} & \mathbf{x} - i\mathbf{y} \\ \mathbf{x} + i\mathbf{y} & -\mathbf{z} \end{pmatrix} \begin{pmatrix} \tilde{c}_1 \\ \tilde{c}_2 \end{pmatrix} = (\sigma_x \mathbf{x} + \sigma_y \mathbf{y} + \sigma_z \mathbf{z}) \begin{pmatrix} \tilde{c}_1 \\ \tilde{c}_2 \end{pmatrix}. \tag{7.90}$$

The reader can solve this by using the same approach as is done for an "ordinary" eigenvalue problem, setting up the characteristic equation with the determinant

$$\mathcal{D} = [\sigma_x \mathbf{x} + \sigma_y \mathbf{y} + (\sigma_z - 1)\mathbf{z}][\sigma_x \mathbf{x} + \sigma_y \mathbf{y} + (\sigma_z + 1)\mathbf{z}] - 2. \tag{7.91}$$

A nontrivial solution will exist if $\mathcal{D} = 0$. This condition gives the equation for $\boldsymbol{\sigma}$:

$$\sigma_x^2 + \sigma_y^2 + \sigma_z^2 \equiv \sigma^2 = 3. \tag{7.92}$$

It confirms the result (7.88). But at the same time, it is immediately evident that (7.92), being the solution of Equation 7.90 for vector eigenvalues of $\hat{\boldsymbol{\sigma}}$, determines only the *magnitude* of $\boldsymbol{\sigma}$, but not its components. Even if one of them (say, σ_z) is defined, Equation 7.92 contains no additional hints necessary to find the other two, so it fails to define $\boldsymbol{\sigma}$ as a vector. Accordingly, the corresponding eigenstates of $\hat{\boldsymbol{\sigma}}$ remain undefined too. Such a situation seems unusual, but it is natural for a vector operator with noncommuting components. In the given case, it even provides us with another (and very simple!) way of computing the degree of indeterminacy in the corresponding components.

Exercise 7.3

Find the indeterminacy in σ_x and σ_y for a state with $\sigma_z = 1$.

Solution:

For such state, we have from (7.92): $\sigma_x^2 + \sigma_y^2 = 2$. It follows

$$\langle \sigma_x^2 + \sigma_y^2 \rangle = \langle \sigma_x^2 \rangle + \langle \sigma_y^2 \rangle = 2. \quad (7.93)$$

Since the x- and y-directions are equally represented here (both make an angle $\theta = \pi/2$ with z), we have $\langle \sigma_x^2 \rangle = \langle \sigma_y^2 \rangle = 1$, and, in view of the symmetry, $\langle \sigma_x \rangle = \langle \sigma_y \rangle = 0$. This gives $\langle \Delta\sigma_x^2 \rangle = \langle \Delta\sigma_y^2 \rangle = 1$, so that $\Delta\sigma_x \Delta\sigma_y = 1$ and, accordingly,

$$\Delta S_x \Delta S_y = \frac{1}{4}\hbar^2. \quad (7.94)$$

This is consistent with the general expression (7.20) for $L_z = (1/2)\hbar$. Note that here we have calculated the *exact* product $\Delta S_x \Delta S_y$ for the given state, and it turned out to be equal to its lowest possible value allowed by the indeterminacy principle.

Summary. Both observables – L and s – are represented by their respective *vector operators*, but are themselves *not* geometrical vectors. Either operator is a powerful mathematical tool for calculating relevant physical characteristics, as illustrated in part II.

II. Knowing all three components of the spin operator allows us to find the operator for the spin component onto an *arbitrary* direction given by a unit vector **e**. It is just the *projection* of $\hat{\boldsymbol{\sigma}}$ onto **e**: $\hat{\boldsymbol{\sigma}}_e = \hat{\boldsymbol{\sigma}} \cdot \mathbf{e}$. Specifying components of **e** in terms of spherical coordinates $r = 1$, θ, φ, we have $e_x = \sin\theta\cos\varphi$, $e_y = \sin\theta\sin\varphi$, and $e_z = \cos\theta$, and the dot product takes the form

$$\hat{\sigma}_e(\theta, \varphi) \equiv \sigma_x e_x + \sigma_y e_y + \sigma_z e_z$$

$$= \begin{pmatrix} 0 & 1 \\ 1 & 0 \end{pmatrix} \sin\theta\cos\varphi + \begin{pmatrix} 0 & -i \\ i & 0 \end{pmatrix} \sin\theta\sin\varphi + \begin{pmatrix} 1 & 0 \\ 0 & -1 \end{pmatrix} \cos\theta$$

$$= \begin{pmatrix} \cos\theta & \sin\theta\, e^{-i\varphi} \\ \sin\theta\, e^{i\varphi} & -\cos\theta \end{pmatrix}. \quad (7.95)$$

The obtained matrix satisfies the general rule (7.50) (prove it!). Also, it is easy to see that it reduces to (7.59) when **e** is coincident with one of the coordinate axes. The case **e** = **z** corresponds to $\theta = 0$ and yields the $\hat{\sigma}_z$ matrix. Similarly, by directing **e** along x and y we recover the $\hat{\sigma}_x$ and $\hat{\sigma}_y$ matrices.

Unlike the vector operator $\hat{\boldsymbol{\sigma}}$, operator $\hat{\sigma}_e \equiv \hat{\boldsymbol{\sigma}} \cdot \mathbf{e}$, being the *dot product* of two vectors, represents a scalar quantity – the set of possible spin projections onto **e**. Mathematically, $\hat{\boldsymbol{\sigma}}$ is represented by matrix (7.89) whose elements are vectors, whereas $\hat{\sigma}_e$ is represented by matrix (7.95) whose elements are numbers. In contrast to $\hat{\boldsymbol{\sigma}}$ that does not have any eigenfunctions, $\hat{\sigma}_e$ has eigenvalues σ_e and eigenstates $|S\rangle = |S_e\rangle$.

The eigenvalues are easy to predict. The fact that $\sigma_{e_i} = \pm 1$ for all three i strongly suggests that possible spin projections onto *any* direction are $\pm (1/2)\hbar$; hence, we should expect $\sigma_e = \pm 1$ regardless of the direction of **e**. (We can even strengthen this statement: Since we can choose the z-axis arbitrarily and still get $\sigma_z = \pm 1$ in all measurements, it is clear that the electron simply does not have any other results in store for us and will always give us either 1 or -1; we did not even need to mention the other two axes.) As for the eigenstates $|S_e\rangle$, we already have them from (7.77). Let us now derive σ_e and $|S_e\rangle$ from a slightly different perspective, by solving the eigenvalue problem

$$\hat{\sigma}_e |S_e\rangle = \sigma_e |S_e\rangle. \tag{7.96}$$

Representing $|S_e\rangle$ in the matrix form as $|S_e\rangle = \begin{pmatrix} C_1 \\ C_2 \end{pmatrix}$, we have

$$\begin{pmatrix} \cos\theta & \sin\theta\, e^{-i\varphi} \\ \sin\theta\, e^{i\varphi} & -\cos\theta \end{pmatrix} \begin{pmatrix} C_1 \\ C_2 \end{pmatrix} = \sigma_e \begin{pmatrix} C_1 \\ C_2 \end{pmatrix}. \tag{7.97}$$

The resulting characteristic equation has determinant

$$\mathcal{D} = (\cos\theta - \sigma_e)(-\cos\theta - \sigma_e) - \sin^2\theta \tag{7.98}$$

and setting $\mathcal{D} = 0$ gives us exactly the two eigenvalues we expected:

$$\sigma_e^2 = 1, \qquad \sigma_e = \pm 1. \tag{7.99}$$

For the eigenvalue $\sigma_e = +1$, we get by solving the two simultaneous equations (7.97)

$$C_2^+ = \frac{1 - \cos\theta}{\sin\theta} e^{i\varphi} C_1^+ = \tan\left(\frac{\theta}{2}\right) e^{i\varphi} C_1^+. \tag{7.100}$$

Together with the normalization condition $|C_1^+|^2 + |C_2^+|^2 = [1 + \tan^2(\theta/2)]|C_1^+|^2 = 1$, this gives

$$C_1^+ = \cos\left(\frac{\theta}{2}\right) e^{i\delta}, \qquad C_2^+ = \sin\left(\frac{\theta}{2}\right) e^{i(\delta+\varphi)}, \tag{7.101}$$

7.4 Spin as an Intrinsic Angular Momentum

where δ is an arbitrary phase angle. We can simplify the result by making use of a permitted operation – multiplying the solution by an arbitrary common phase factor. Choosing this factor as $e^{-i\delta}$ (which is equivalent to just setting $\delta = 0$ in (7.101)), we recover the already familiar result (7.77) for a state represented by $\mathbf{e}(\theta, \varphi)$ on the Bloch sphere:

$$C_1^+ = \cos\left(\frac{\theta}{2}\right), \qquad C_2^+ = \sin\left(\frac{\theta}{2}\right)e^{i\varphi}. \tag{7.102}$$

Thus,

$$|S_e^+\rangle = \begin{pmatrix} \cos(\theta/2) \\ \sin(\theta/2)e^{i\varphi} \end{pmatrix} \tag{7.103}$$

In a similar way, we find the second eigenfunction corresponding to $\sigma_e = -1$:

$$C_1^- = \sin\left(\frac{\theta}{2}\right), \qquad C_2^- = -\cos\left(\frac{\theta}{2}\right)e^{i\varphi}, \tag{7.104}$$

so that

$$|S_e^-\rangle = \begin{pmatrix} \sin(\theta/2) \\ -\cos(\theta/2)e^{i\varphi} \end{pmatrix}. \tag{7.105}$$

This can be considered as a state represented by the vector $-\mathbf{e}(\theta, \varphi)$ antiparallel to $\mathbf{e}(\theta, \varphi)$. It can be obtained directly from (7.103) by replacing $\theta \to \pi - \theta$ and $\varphi \to \varphi + \pi$.

The obtained results have a simple physical meaning. The two found eigenvalues (7.99) and the respective eigenstates (7.103) and (7.105) describe two possible spin states represented by vector $\mathbf{e}$. Since the eigenstates (7.103) and (7.105) are written in the σ_z-representation, their matrix elements determine the probabilities for the outcomes of the future s_z-measurements made from these eigenstates. The top element in the column matrix $|\chi_e\rangle = \begin{pmatrix} C_1 \\ C_2 \end{pmatrix}$ determines the probability of finding the z-component of spin being "up" when measured from state $|\chi_e\rangle$. The lower element determines the probability of finding spin down in such measurement. As we noted before, the probability distribution between these two outcomes is symmetric (50–50) at $\theta = \pi/2$, that is, when $\mathbf{e}$ lies in the equatorial plane of the Bloch sphere.

Exercise 7.4

a) Find the average $\langle \sigma_e \rangle$ of observable σ_e and the corresponding variance $\Delta \sigma_e$ in the state $|\uparrow\rangle$.
b) Find the average and the variance of the same observable in the state $|\rightarrow\rangle_x$.

7 Angular Momentum

Solution:

a) On a qualitative level, we expect that the *average* σ_e in state $|\uparrow\rangle$ must be equal to the average of $\sigma_z \equiv \sigma_3$ in state $|S_e\rangle$: both are determined by projection of **z** onto the **e**-direction, or vice versa. Since the angle between **e** and **z** is θ, we expect that $\langle\sigma_e\rangle = \cos\theta$. Quantitatively, we obtain this result by multiplying matrices:

$$\langle\sigma_e\rangle = \langle\uparrow|\hat{\sigma}_e(\theta,\varphi)|\uparrow\rangle = (1\ 0)\begin{pmatrix}\cos\theta & \sin\theta\,e^{-i\varphi} \\ \sin\theta\,e^{i\varphi} & -\cos\theta\end{pmatrix}\begin{pmatrix}1\\0\end{pmatrix} = \cos\theta. \quad (7.106)$$

Using (7.59) and (7.77), one can show that the same result is given by $\langle S_e|\hat{\sigma}_3|S_e\rangle$. For the standard deviation $\Delta\sigma_e$, we first determine $\langle\Delta\sigma_e^2\rangle = \langle\sigma_e^2\rangle - \langle\sigma_e\rangle^2$. But $\langle\sigma_e^2\rangle = \langle\hat{\sigma}_e^2\rangle = 1$; therefore, in view of (7.106), we obtain $\langle\Delta\sigma_e^2\rangle = \sin^2\theta$ and

$$\Delta\sigma_e = |\sin\theta|. \quad (7.107)$$

b) The solution is the same as in (a), with the distinction that now, instead of $|\uparrow\rangle$, we use the state function (7.68): $|\chi^+\rangle = (1/\sqrt{2})\begin{pmatrix}1\\1\end{pmatrix}$. The results are

$$\langle x|\sigma_e|x\rangle = \sin\theta\cos\varphi = \cos\alpha, \qquad \Delta\sigma_e = \sqrt{1-\sin^2\theta\cos^2\varphi} = |\sin\alpha|. \quad (7.108)$$

Here α is the angle between **e** and **x**, so that projection **e** on the x-axis is $e_x = \cos\alpha$, which is also expressed as $\sin\theta\cos\varphi$ in spherical coordinates. Naturally, these results are the same for all **e** on the conical surface with angle α around **x** determined by the first equation of (7.108). This surface must not be confused with a conoid representing a single spin state $|S_e\rangle$, shown in Figure 7.6 The conoid in Figure 7.6 is formed around one chosen direction **e** and is characterized by a fixed angle η such that

$$\cos\eta = \frac{m}{l} = \frac{1}{\sqrt{3}}. \quad (7.109)$$

Consider now, apart from $|S_e\rangle$, another arbitrary state $|S_{e'}^+\rangle = \begin{pmatrix}\cos(\theta'/2)\\\sin(\theta'/2)e^{i\varphi'}\end{pmatrix}$. This is just an eigenfunction of operator $\hat{\sigma}_{e'}$ for $e' = (\theta',\varphi')$. The question arises, "What is the average of observable σ_e measured from state $|S_{e'}^+\rangle$?" Applying the same rules as in Exercise 7.4, one can show (Problem 7.11) that

$$\langle S_{e'}^+|\hat{\sigma}_e|S_{e'}^+\rangle = \cos\theta\cos\theta' + \sin\theta\sin\theta'\cos(\varphi-\varphi'). \quad (7.110)$$

But the expression on the right is just $\cos\delta'$, where δ' is the angle between **e** and **e'**. Thus,

$$\langle S_{e'}^+|\hat{\sigma}_e|S_{e'}^+\rangle = \cos\delta'. \quad (7.111)$$

Also, applying the same rules, we can find the variance $\Delta\sigma_e$ under measurements from the state $|S_{e'}^+\rangle$ (Problem 7.12):

$$\Delta\sigma_e = \sin\delta'. \quad (7.112)$$

In view of isotropy of space, the same result will hold for measurements of spin component $\sigma_{e'}$ from state $|S_e^+\rangle$. If we now introduce one more direction $\mathbf{e}''$ making an angle δ'' with $\mathbf{e}$, we can write for measurements from state $|S_e^+\rangle$:

$$\Delta\sigma_{e'} = \sin\delta', \qquad \Delta\sigma_{e''} = \sin\delta''. \tag{7.113}$$

The product of the corresponding variances is

$$\Delta\sigma_{e'}\Delta\sigma_{e''} = \sin\delta' \sin\delta'' \tag{7.114}$$

These results generalize the indeterminacy relations between spin components σ_x, σ_y, and σ_z of a particle to the case of *arbitrarily* oriented components (note also that any mention of the choice of basis has been removed from (7.114)). In particular, when both δ' and δ'' approach $\pi/2$ (both directions $\mathbf{e}'$ and $\mathbf{e}''$ become perpendicular to $\mathbf{e}$), the product of indeterminacies reaches its maximum possible value

$$\Delta\sigma_{e'}\Delta\sigma_{e''} = 1 \tag{7.115}$$

(which is at the same time the minimum possible value for mutually orthogonal components!).

In terms of the spin itself, this is consistent with the familiar result $\Delta S_x \Delta S_y = (1/4)\hbar^2$.

7.5
Angular Momentum of a Compound System

Suppose we have a compound system where the total $\mathbf{J}$ consists only of two parts $\mathbf{J}_1$ and $\mathbf{J}_2$. The question is, which of the six quantities J^2, J_z, J_1^2, J_{1z}, J_2^2, and J_{2z} can we determine simultaneously?

Suppose first that we have measured J_1^2 and J_2^2. This is possible because the respective operators commute (each operator acts on the respective part of the combined wave function). Since J_1^2 is compatible with J_{1z}, and J_2^2 with J_{2z}, we can also measure these two z-components. Once again, J_{1z} does not conflict with J_{2z} because their operators refer to different parts of the system. The trouble starts when we want to know the values of J^2 and J_z. In contrast with geometrical vectors $\mathbf{v}_1$ and $\mathbf{v}_2$ uniquely defining the vector sum $\mathbf{v} = \mathbf{v}_1 + \mathbf{v}_2$, the individual angular momenta $\mathbf{J}_1$ and $\mathbf{J}_2$ generally do not define $\mathbf{J}$ uniquely. That is to say, equation

$$\hat{\mathbf{J}} = \hat{\mathbf{J}}_1 + \hat{\mathbf{J}}_2 \tag{7.116}$$

holds for the *operators* but not for the vector *observables* $\mathbf{J}$, $\mathbf{J}_1$, and $\mathbf{J}_2$. We are only allowed to make a simultaneous measurement of four observables out the six. Such a measurement will leave the compound system in a state $|j_1, j_2, m_1, m_2\rangle$, which is a common eigenstate for the operators $\hat{J}_1^2$, $\hat{J}_2^2$, $\hat{J}_{1z}$, and $\hat{J}_{2z}$. These states are mutually orthogonal and thus form an appropriate basis. When a compound system is written in this way, we say its state is given in the *uncoupled representation*.

Alternatively, we could decide to measure J^2 and J_z. This measurement is compatible with the knowledge of J_1^2 and J_2^2; however, we lose information about J_{1z} and J_{2z}. The system is now in state $|j_1, j_2, j, m\rangle$, which is a common eigenstate for the operators $\hat{J}_1^2, \hat{J}_2^2, \hat{J}^2$, and $\hat{J}_z$. Once again, such states form an orthonormal basis. A system defined in this way is said to be in a *coupled representation*.

To take an example, suppose we have a system of two electron spins. We can write the compound system in an uncoupled representation $|s_1, s_2, m_1, m_2\rangle$ or in a coupled representation $|s_1, s_2, s, m\rangle$. Since the spin quantum number is an inherent property of the electron, we always have $s_1 = s_2 = 1/2$, so we can drop these two numbers and shorten the kets, writing them as $|m_1, m_2\rangle$ and $|s, m\rangle$. Then each basis will consist of four eigenkets:

Uncoupled : Coupled :

$$\{|1/2, 1/2\rangle, |1/2, -1/2\rangle, |-1/2, -1/2\rangle, |-1/2, -1/2\rangle\} \quad \{|1, 1\rangle, |1, 0\rangle, |1, -1\rangle, |0, 0\rangle\}$$
(7.117)

Let us now find the number of eigenstates for the general case (7.116). Since $m_1 = -j_1, \ldots, j_1$ and $m_2 = -j_2, \ldots, j_2$, there will be $2j_1 + 1$ allowed values for m_1 and $2j_2 + 1$ allowed values for m_2, so the uncoupled basis will consist of $(2j_1 + 1)(2j_2 + 1)$ eigenstates. A change of basis should not change the dimensionality of the Hilbert space, so $(2j_1 + 1)(2j_2 + 1)$ is also the number of eigenstates in the coupled basis.

We can now ask the next question: How do we switch from one basis to the other? For example, we may have the system written in the coupled basis and want to write it in the uncoupled basis. To that end, we must expand each coupled eigenket as a sum of uncoupled eigenkets:

$$|j_1, j_2, j, m\rangle = \sum_{m_1} \sum_{m_2} |j_1, j_2, m_1, m_2\rangle \langle j_1, j_2, m_1, m_2 | j_1, j_2, j, m\rangle. \quad (7.118)$$

The expansion coefficients appearing in (7.118) are known as *Clebsch–Gordan coefficients*:

$$C_{j_1, j_2, j, m}(m_1, m_2) = \langle j_1, j_2, m_1, m_2 | j_1, j_2, j, m\rangle. \quad (7.119)$$

Let us establish several important properties. First of all, Clebsch–Gordan coefficients vanish unless $m = m_1 + m_2$. To prove that, use the operator identity

$$\hat{J}_z = \hat{J}_{1z} + \hat{J}_{2z}. \quad (7.120)$$

The expression

$$\langle j_1, j_2, m_1, m_2 | \hat{J}_z - \hat{J}_{1z} - \hat{J}_{2z} | j_1, j_2, j, m\rangle = 0 \quad (7.121)$$

must hold due to (7.120), but applying $\hat{J}_z$ to the coupled eigenket on the right and $\hat{J}_{1z}$ and $\hat{J}_{2z}$ to the uncoupled eigenket on the left, we get

$$(m - m_1 - m_2)\hbar \langle j_1, j_2, m_1, m_2 | j_1, j_2, j, m\rangle = 0, \quad (7.122)$$

and if the quantity in the parentheses is nonzero, it means the Clebsch–Gordan coefficient must vanish. This theorem has a simple physical meaning: when we know J_z, we cannot predict the individual z-components J_{1z} and J_{2z} but we do know that if we measure them, they will add up to J_z.

To see how this theorem works in practice, suppose a two-electron system has $S_z = 0$. This can be achieved in two ways: $|\uparrow\downarrow\rangle$ and $|\downarrow\uparrow\rangle$, where we used vertical arrows to denote "spin-up" and "spin-down" states. Both ways are possible, so we must find two Clebsch–Gordan coefficients, $C_{1/2,1/2,s,0}(1/2, -1/2)$ and $C_{1/2,1/2,s,0}(-1/2, 1/2)$. If the system has $S_z = \hbar$, then $m = 1$, and we know immediately that $m_1 = m_2 = 1/2$. The only possible state is $|\uparrow\uparrow\rangle$, and we have only one Clebsch–Gordan coefficient that must be equal to 1 because of the normalization requirement. If $S_z = -\hbar$, then the state is $|\downarrow\downarrow\rangle$ and the only nonzero coefficient is again 1.

The second property is that Clebsch–Gordan coefficients vanish unless

$$|j_1 - j_2| \leq j \leq j_1 + j_2. \tag{7.123}$$

To prove this, simply notice that the number of allowed pairs (m_1, m_2) must be equal to the number of allowed pairs (j, m), but the equality holds if we assume the upper and lower bounds on j imposed by (7.123):

$$(2j_1 + 1)(2j_2 + 1) = \sum_{j=|j_1-j_2|}^{j_1+j_2} (2j + 1). \tag{7.124}$$

The right-hand side is an arithmetic progression, which readily reduces to the product on the left. So if we know the angular momentum quantum numbers of individual parts, Equation 7.123 immediately gives us a set of possible values for the quantum number of the whole system.

Finally, Clebsch–Gordan coefficients are defined up to an arbitrary phase factor, but we can choose this phase so that all coefficients will be real numbers:

$$\langle j_1, j_2, m_1, m_2 | j_1, j_2, j, m \rangle = \langle j_1, j_2, j, m | j_1, j_2, m_1, m_2 \rangle. \tag{7.125}$$

Let us now go back to the expansion (7.118) and apply the ladder operators $\hat{J}_+ = (\hat{J}_1)_+ + (\hat{J}_2)_+$ and $\hat{J}_- = (\hat{J}_1)_- + (\hat{J}_2)_-$ to both sides:

$$\begin{cases} \hbar\sqrt{j(j+1) - m(m+1)}|j_1, j_2, j, m+1\rangle \\ \quad = \sum_{m_1}\sum_{m_2} \Big(\hbar\sqrt{j_1(j_1+1) - m_1(m_1+1)}|j_1, j_2, m_1+1, m_2\rangle \\ \quad\quad + \hbar\sqrt{j_2(j_2+1) - m_2(m_2+1)}|j_1, j_2, m_1, m_2+1\rangle\Big) C_{j_1, j_2, j, m}(m_1, m_2), \\ \hbar\sqrt{j(j+1) - m(m-1)}|j_1, j_2, j, m-1\rangle \\ \quad = \sum_{m_1}\sum_{m_2} \Big(\hbar\sqrt{j_1(j_1+1) - m_1(m_1-1)}|j_1, j_2, m_1-1, m_2\rangle \\ \quad\quad + \hbar\sqrt{j_2(j_2+1) - m_2(m_2-1)}|j_1, j_2, m_1, m_2-1\rangle\Big) C_{j_1, j_2, j, m}(m_1, m_2). \end{cases}$$

Multiplying by $\langle j_1, j_2, m_1, m_2 |$ and using orthogonality, we get the following recurrence relations:

$$\begin{cases} \sqrt{j(j+1) - m(m+1)} C_{j_1, j_2, j, m+1}(m_1, m_2) \\ = \left(\sqrt{j_1(j_1+1) - (m_1-1)m_1} \, C_{j_1, j_2, j, m}(m_1 - 1, m_2) \right. \\ \left. + \sqrt{j_2(j_2+1) - (m_2-1)m_2} \, C_{j_1, j_2, j, m}(m_1, m_2 - 1) \right), \\ \sqrt{j(j+1) - m(m-1)} C_{j_1, j_2, j, m-1}(m_1, m_2) \\ = \left(\sqrt{j_1(j_1+1) - (m_1+1)m_1} \, C_{j_1, j_2, j, m}(m_1 + 1, m_2) \right. \\ \left. + \sqrt{j_2(j_2+1) - (m_2+1)m_2} \, C_{j_1, j_2, j, m}(m_1, m_2 + 1) \right), \end{cases} \quad (7.126)$$

which allow us to derive Clebsch–Gordon coefficients by moving from point to point on the (m_1, m_2)-plane. Of course, condition (7.122) implies that coefficients on the left vanish, unless

$$m_1 + m_2 = m \pm 1. \quad (7.127)$$

This fact, in conjunction with the normalization requirement

$$\sum_{m_1, m_2} |C_{j_1, j_2, j, m}(m_1, m_2)|^2 = 1, \quad (7.128)$$

helps determine all coefficients almost uniquely.

We are now able to complete the example (7.117). Starting with the first coupled eigenket, which we said is equal to $|s = 1, m = 1\rangle = |\uparrow\uparrow\rangle$, and applying the lowering operator to it:

$$\hat{S}_-|s = 1, m = 1\rangle = ((\hat{S}_1)_- + (\hat{S}_2)_-)|\uparrow\uparrow\rangle.$$

After simple algebra, this reduces to

$$\sqrt{2}|s = 1, m = 0\rangle = |\downarrow\uparrow\rangle + |\uparrow\downarrow\rangle,$$

which determines the second coupled eigenket. The next application of the lowering operator gives us the already familiar result $|s = 1, m = -1\rangle = |\downarrow\downarrow\rangle$. It remains to determine the last eigenket. Since the coupled basis must be orthogonal, let us demand that this eigenket be orthogonal to the first three. We write it as

$$|s = 0, m = 0\rangle = A|\uparrow\uparrow\rangle + B|\uparrow\downarrow\rangle + C|\downarrow\uparrow\rangle + D|\downarrow\downarrow\rangle$$

and determine the constants A, B, C, and D from this requirement. Orthogonality with the first and the third eigenkets immediately implies $A = D = 0$. Orthogonality with the second eigenket tells us that $((\langle\downarrow\uparrow| + \langle\uparrow\downarrow|)(B|\uparrow\downarrow\rangle + C|\downarrow\uparrow\rangle)) = 0$ or $B = -C = 1/\sqrt{2}$. The four coupled spin states are determined as

$$\begin{cases} |s = 1, m = 1\rangle = |\uparrow\uparrow\rangle, \\ |s = 1, m = 0\rangle = \dfrac{1}{\sqrt{2}}(|\uparrow\downarrow\rangle + |\downarrow\uparrow\rangle), \\ |s = 1, m = -1\rangle = |\downarrow\downarrow\rangle, \\ |s = 0, m = 0\rangle = \dfrac{1}{\sqrt{2}}(|\uparrow\downarrow\rangle - |\downarrow\uparrow\rangle). \end{cases} \quad (7.129)$$

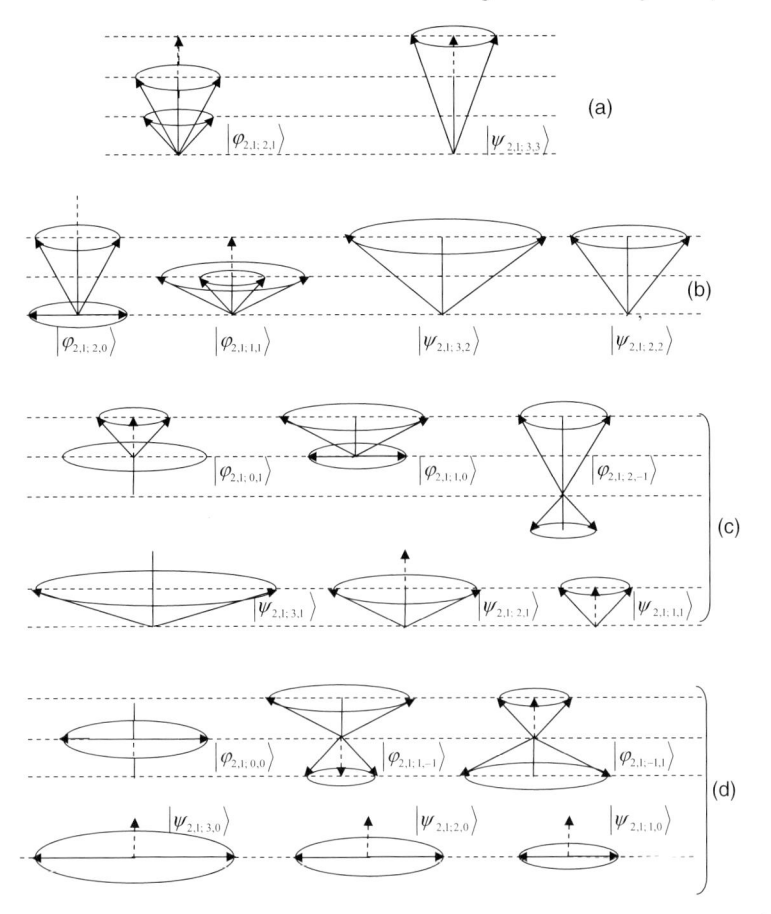

Figure 7.7 Angular momentum of a compound system in two different representations, $\left|\varphi_{j_1,j_2;m_1,m_2}\right\rangle$ and $\left|\psi_{j_1,j_2;j,m}\right\rangle$ for $j_1 = 2$ and $j_2 = 1$: (a) $m = 3$; (b) $m = 2$; (c) $m = 1$; (d) $m = 0$.

The first three states correspond to $s = 1$ and are known collectively as the spin *triplet*. The last state, which corresponds to $s = 0$, is called the *singlet*. This state plays an important role in many applications such as entanglement and production of EPR pairs.

Figure 7.7 shows a slightly more complicated case – the uncoupled and coupled eigenstates corresponding to the addition of angular momenta $j_1 = 2$ and $j_2 = 1$.

Box 7.7 Coupled versus uncoupled representation

As evidenced by (7.124), the total number of eigenstates does not depend on the choice of basis. This comes as no surprise, since either way, we are working in the same $\mathcal{H}$-space that can be spanned by a coupled basis with the same ease as it can be spanned by the uncoupled basis. The two bases are just rotated with respect to each other. The relationship between the two bases seems very simple and elegant in the framework of the abstract geometry of the $\mathcal{H}$-space. At the same time, it has a very rich physical meaning. Recall Section 6.4, in which the rotation of two bases – $|x\rangle$ and $|p_x\rangle$ with respect to each other in one common $\mathcal{H}$-space – reflects the noncommutability of the $\hat{x}$ and $\hat{p}_x$ operators and indeterminacy relationship between the corresponding observables. Similarly, rotation of the two bases considered here reflects noncommutability of different components of the angular momentum operators and indeterminacy relationship between the corresponding observables. But here, in view of the compound nature of the considered system, it also reveals another stunning feature of QM: it turns out that indeterminacy extends onto the relationship between the parts and the whole. *Maximum knowledge about all parts of a system does not necessarily imply maximum knowledge about the whole system, and vice versa.* It goes even farther than that: in fact, as seen in the described case, maximal possible information about both parts of a system (two individual angular momenta **L** and **S**) is generally *inconsistent* with maximal information about the system as a whole (its total angular momentum **J**). The uncoupled eigenstates imply the best possible knowledge about the *individual* angular momenta of the system. And it is precisely because of this that we lose information about the *net angular momentum*: as shown by the Clebsch–Gordan expansion, we can only make probabilistic predictions for the values of j and m.

Similarly, having maximal possible information about the whole system precludes us from maximizing our information about its parts. In this case, knowledge of j and m comes at the cost of knowledge of m_1, m_2, which are "excluded" from the corresponding quadruplet. Altogether, we see a certain complementarity between the whole and its parts [17], similar to the wave–particle dualism in description of a single object.

7.6
Spherical Harmonics

It is now time to revisit the eigenfunctions of orbital angular momentum introduced in (7.7) and (7.9). The following equation gives the explicit form of these functions:

$$Y_{l,m}(\theta, \varphi) = \sqrt{\frac{(2l+1)(l-|m|)!}{4\pi(l+|m|)!}} P_l^{|m|}(\cos\theta) e^{im\varphi}, \quad |m| = 0, 1, 2, \ldots, l-1, l.$$

(7.130)

Here $P_l^{|m|}(\cos\theta)$ is another special function, called the *associated Legendre polynomial* [18,19], and the exponent $\Phi_m(\varphi) = e^{im\varphi}$ is the eigenfunction of the operator $\hat{L}_z$ (in spherical coordinates, $\hat{L}_z = \partial(-i\hbar)/\partial\varphi$ and it is trivial to show that $\hat{L}_z Y_{l,m}(\theta, \varphi) = m\hbar Y_{l,m}(\theta, \varphi)$). Here we have the reason why l could not be half-integer: if it were, we would have to allow for half-integer m, and a complete rotation by 360° would make $e^{im\varphi}$ change sign, so the wave function would not be single-valued.

Below we list the spherical harmonics for $l = 0$, $l = 1$, and $l = 2$:

$$Y_{0,0}(\theta, \varphi) = \frac{1}{2}\sqrt{\frac{1}{\pi}},$$

$$Y_{1,-1}(\theta, \varphi) = \sqrt{\frac{3}{8\pi}}\sin\theta\, e^{-i\varphi}, \qquad Y_{1,0}(\theta, \varphi) = \frac{1}{2}\sqrt{\frac{3}{\pi}}\cos\theta,$$

$$Y_{1,1}(\theta, \varphi) = -\sqrt{\frac{3}{8\pi}}\sin\theta\, e^{i\varphi},$$

$$Y_{2,-2}(\theta, \varphi) = \frac{1}{4}\sqrt{\frac{15}{2\pi}}\sin^2\theta\, e^{-2i\varphi}, \qquad Y_{2,2}(\theta, \varphi) = \frac{1}{4}\sqrt{\frac{15}{2\pi}}\sin^2\theta\, e^{2i\varphi},$$

$$Y_{2,-1}(\theta, \varphi) = \sqrt{\frac{15}{8\pi}}\sin\theta\cos\theta\, e^{-i\varphi}, \qquad Y_{2,0}(\theta, \varphi) = \frac{1}{4}\sqrt{\frac{5}{\pi}}(3\cos^2\theta - 1),$$

$$Y_{2,1}(\theta, \varphi) = -\sqrt{\frac{15}{8\pi}}\sin\theta\cos\theta\, e^{i\varphi}.$$

Of course, we do not mention all of this in order to bring up the subject of spherical harmonics now, when the spin theory is already derived and it seems these functions should be next to useless at this point. The reason is much deeper: we want to be able to derive *transition probabilities* between different eigenstates of angular momentum. That is, we are given *some* operator $\hat{M}$ that shows up in the Hamiltonian for the interaction between atomic electrons and an external field, and we want to find matrix elements $\langle j', m'|\hat{M}|j, m\rangle$. If we are lucky, $\hat{M}$ will be a scalar operator and our task will be relatively easy. But in the majority of problems $\hat{M}$ will be a vector operator, $\hat{\mathbf{M}} = (\hat{M}_x, \hat{M}_y, \hat{M}_z)$. Sometimes it will be a second-rank tensor with elements $\hat{M}_{ij}$. Finally, in a small percentage of problems we will be dealing with third-rank tensor operators $\hat{M}_{ijk}$ or tensor operators of even higher rank.

The standard definition says that a tensor is a set of elements $T_{i,j,k,\dots}$ (the number of subscripts is the rank of the tensor; subscripts will usually be allowed take values 1, 2, 3 as long as we working with a 3D Cartesian space) such that the elements transform under rotation of coordinates in the same way as products of the corresponding components of a vector. Vector components M_i, $i = 1, 2, 3$, would transform according to the simple rule

$$M_i \Rightarrow \sum_{i'} R_{ii'} M_{i'}, \tag{7.131}$$

where $R_{ii'}$ is a 3 × 3 rotation matrix. By analogy, tensor components $T_{ijk\dots}$ transform according to

$$T_{ijk\dots} \Rightarrow \sum_{i'} R_{ii'} R_{jj'} R_{kk'} \cdots T_{i'j'k'\dots}. \tag{7.132}$$

By that definition, a scalar quantity can be described as a zero-rank tensor, an ordinary 3D vector as a first-rank tensor, a 3 × 3 matrix as a second-rank tensor, a three-dimensional 3 × 3 × 3 matrix as a third-rank tensor, and so on.

However, the above definition describes just one type of tensors – the *Cartesian tensors*. Another type of tensors that are of special interest to us are the so-called *irreducible spherical tensors* (ISTs). We can loosely define such a tensor as a set of elements ("tensor components") that change under rotations in the same way as spherical harmonics.

But then we must be careful with this definition so as to avoid possible confusion. The table of spherical harmonics is infinite: we have one harmonic for $l = 0$, three harmonics for $l = 1$, five harmonics for $l = 2$, and so on. So, how many elements should be contained in the tensor?

To refine the above definition, we say that an IST of rank l is a set of $2l + 1$ elements that transform under rotations as spherical harmonics $Y_{l,m}(\theta, \varphi)$. So, only one set of harmonics (the "l-sequence") is involved in the definition of an IST of a given rank. This has an interesting consequence: we can establish a one-to-one correspondence between an lth rank Cartesian tensor and lth rank IST only for $l = 0$ and $l = 1$, but not for the larger l. For example, if $l = 2$, a Cartesian tensor has nine elements while an IST has only five elements, so no one-to-one mapping can be established between tensor elements! It should be possible, however, to represent the Cartesian tensor as some linear combination of *different* ISTs. The last point can be illustrated by the following simple example. Suppose we are given a second-rank antisymmetric tensor A_{ij}, where $A_{ij} = -A_{ji}$, whose elements are arranged into a 3×3 matrix. Then if we record the values of three elements to the upper right of the main diagonal, we can immediately reconstruct this tensor on demand. The elements on the lower left are *dependent* on the first three since the latter can be obtained from the former by a change of sign, and the diagonal elements are zeros and hence also known. So even though we have a total of nine elements, only three of them are *independent* and the other six follow from them, thus for all practical purposes we can think of A_{ij} as consisting of three elements only. If these three elements transform under rotations like spherical harmonics $l = 1$ and $m = 0, \pm 1$, we call A_{ij} an IST. Next, imagine a second-rank symmetric tensor $B_{ij} = B_{ji}$ whose diagonal elements add up to 0. This time we need to know five independent elements to reconstruct the tensor, and we call the B_{ij} an IST if they transform under rotations like spherical harmonics $l = 2$ and $m = 0, \pm 1, \pm 2$. Finally, imagine a second-rank tensor C_{ij} whose diagonal elements are the same and all nondiagonal elements are zero. It has only one independent element that will remain invariant under rotation (recall the property of the identity matrix!). This corresponds to the case of a single spherical harmonic $l = 0$ and $m = 0$, which is a constant number and thus invariant under rotations, and so C_{ij} is again a spherical tensor.

Imagine now that we composed a fourth Cartesian tensor as some linear combination of the first three, for example, $D_{ij} = A_{ij} + B_{ij} + C_{ij}$. In spherical tensor language, D_{ij} is called *reducible* because it can be decomposed into simpler parts A_{ij}, B_{ij}, and C_{ij}, which correspond to different sequences of spherical harmonics ($l = 1$, $l = 2$, $l = 0$). Tensors A_{ij}, B_{ij}, and C_{ij}, on the other hand, cannot be decomposed in this manner, which explains the origin of the word *irreducible* in their name. A second-rank Cartesian tensor D_{ij} is now expanded as a sum of ISTs of ranks 0, 1, and 2.

We must now agree on notation. If T is an IST of rank l, we will be using the symbol $T_{l,m}$ to denote the element of this tensor that is like the spherical harmonic $Y_{l,m}(\theta, \varphi)$.

7.6 Spherical Harmonics

Our next task is to find out how to express an arbitrary Cartesian vector or tensor in terms of ISTs. This is where the knowledge of spherical harmonics (or at least the first few sequences) comes in handy. First of all, we notice that spherical and Cartesian coordinates are related as follows:

$$\cos\theta = \frac{z}{r}, \qquad \sin\theta = \frac{\sqrt{x^2+y^2}}{r}, \qquad \sin\theta\, e^{i\varphi} = \frac{x+iy}{r}. \tag{7.133}$$

Then we can rewrite our spherical harmonics in terms of Cartesian coordinates:

$$Y_{1,-1} = \sqrt{\frac{3}{4\pi}}\frac{x-iy}{r\sqrt{2}}, \qquad Y_{1,0} = \sqrt{\frac{3}{4\pi}}\frac{z}{r}, \qquad Y_{1,-1} = -\sqrt{\frac{3}{4\pi}}\frac{x+iy}{r\sqrt{2}},$$

$$Y_{2,-2} = \sqrt{\frac{15}{32\pi}}\frac{(x-iy)^2}{r^2}, \qquad Y_{2,2} = \sqrt{\frac{15}{32\pi}}\frac{(x+iy)^2}{r^2},$$

$$Y_{2,-1} = 2\sqrt{\frac{15}{32\pi}}\frac{x-iy}{r^2}z, \qquad Y_{2,0} = \frac{1}{\sqrt{3}}\sqrt{\frac{15}{32\pi}}\left(3\frac{z^2}{r^2}-1\right), \qquad Y_{2,1} = -2\sqrt{\frac{15}{32\pi}}\frac{x+iy}{r^2}z.$$

Without loss of generality, we can set $r=1$ and also drop the common coefficient for each l-sequence:

$$Y_{1,-1} = \frac{x-iy}{\sqrt{2}}, \qquad Y_{1,0} = z, \qquad Y_{1,-1} = -\frac{x+iy}{\sqrt{2}}, \tag{7.134}$$

$$Y_{2,-2} = (x-iy)^2, \qquad Y_{2,-1} = 2(x-iy)z, \qquad Y_{2,0} = \frac{1}{\sqrt{3}}(3z^2-1),$$
$$Y_{2,1} = -2(x+iy)z, \qquad Y_{2,2} = (x+iy)^2. \tag{7.135}$$

Now we do the following trick. Variables x, y, and z in these equations can be thought of as Cartesian components of vector **r**. By analogy, we can apply the same equations to components of some arbitrary vector **U**. Quantities defined in this way will again behave like spherical harmonics under rotations. Using the new notation $U_{l,m}$ to denote the tensor elements obtained this way, we have

$$U_{1,-1} = \frac{U_x - iU_y}{\sqrt{2}}, \qquad U_{1,0} = U_z, \qquad U_{1,1} = -\frac{U_x + iU_y}{\sqrt{2}}. \tag{7.136}$$

Using the same procedure, we can determine the tensor elements for higher l, for example:

$$U_{2,-2} = (U_x - iU_y)^2, \qquad U_{2,-1} = 2(U_x - iU_y)U_z,$$
$$U_{2,0} = \frac{3U_z^2 - 1}{\sqrt{3}}, \qquad U_{2,1} = -2(U_x + iU_y)U_z, \qquad U_{2,2} = (U_x + iU_y)^2.$$

The quantity $(U_{1,-1}, U_{1,0}, U_{1,1})$ defined by Equation 7.136 is called a *spherical vector*. It defines an alternative way in which we can think of a Cartesian vector $\mathbf{U} = (U_x, U_y, U_z)$ and for all practical purposes is equivalent to writing the vector in a new representation. Indeed, it is easy to check that **U** is uniquely determined by the components appearing in (7.136):

$$U_x = \frac{U_{1,-1} - U_{1,1}}{\sqrt{2}}, \quad U_y = -\frac{U_{1,-1} + U_{1,1}}{i\sqrt{2}}, \quad U_z = U_{1,0}. \quad (7.137)$$

The last equation guarantees that we can always perform a reverse operation, taking spherical tensor T back to the familiar Cartesian form.

Suppose we are given two vectors, $\mathbf{U} = (U_{1,-1}, U_{1,0}, U_{1,1})$ and $\mathbf{V} = (V_{1,-1}, V_{1,0}, V_{1,1})$, written in the spherical form. How do we compose their dot product? Since $\mathbf{U} \cdot \mathbf{V}$ is a scalar, its value should be invariant, so we can just rewrite the familiar Cartesian dot product in the new representation:

$$\begin{aligned}\mathbf{U} \cdot \mathbf{V} &= U_x V_x + U_y V_y + U_z V_z \\ &= \frac{(U_{1,-1} - U_{1,1})(V_{1,-1} - V_{1,1})}{2} - \frac{(U_{1,-1} + U_{1,1})(V_{1,-1} + V_{1,1})}{2} + U_{1,0} V_{1,0} \\ &= U_{1,0} V_{1,0} - U_{1,-1} V_{1,1} - U_{1,1} V_{1,-1}.\end{aligned} \quad (7.138)$$

As one might expect, the above equations will still hold if we replace vector components by their operators. We can therefore define an irreducible spherical tensor operator $\hat{T}_{l,m}$ whose elements are operators that transform under rotations like spherical harmonics $Y_{l,m}$. It is possible to show (a detailed proof, not given here, involves applying infinitesimal rotation operators to tensor components $\hat{T}_{l,m}$) that the following commutation relations hold:

$$\begin{cases} [\hat{J}_z, \hat{T}_{l,m}] = m\hbar \hat{T}_{l,m}, \\ [\hat{J}_\pm, \hat{T}_{l,m}] = \sqrt{l(l+1) - m(m \pm 1)}\hbar\, \hat{T}_{l,m\pm 1}. \end{cases} \quad (7.139)$$

Let us look carefully at the last equation. Here the reader who followed the discussion of ladder operators attentively may have a *déjà vu* moment. Indeed, these formulas look exactly is if $\hat{T}_{l,m}$ were not an operator but an eigenstate $|l, m\rangle$ of the angular momentum. It seems that we have hit upon something profound!

We now state a theorem that plays a crucial role in QM. In a basis composed of angular momentum eigenstates, matrix elements $\langle j, m | \hat{T}_{j_2, m_2} | j_1, m_1 \rangle$ of an irreducible spherical tensor operator $\hat{T}_{j_2, m_2}$ are proportional to Clebsch–Gordan coefficients that one would obtain by expanding a coupled eigenket $|j_1, j_2, j, m\rangle$ in terms of uncoupled eigenkets $|j_1 j_2, m_1, m_2\rangle$:

$$\langle j, m | \hat{T}_{j_2, m_2} | j_1, m_1 \rangle = C_{j_1 j_2 j, m}(m_1, m_2) \frac{\langle j \| \hat{T}_{j_2} \| j_1 \rangle}{\sqrt{2j_1 + 1}}. \quad (7.140)$$

Here the term denoted as $\langle j \| \hat{T}_{j_2} \| j_1 \rangle$ is the proportionality factor that depends only on $j_1, j_2,$ and j. It is sometimes called the *reduced* matrix element since it does not contain a dependence on the three quantum numbers $m_1, m_2,$ and m, unlike the original matrix element. Notice the double bar notation in (7.140). Don't let this notation frighten you. No special mathematical operation is implied here. The reduced matrix element is just a number, and the double bar is nothing more than yet another convention. Moreover, one does not calculate $\langle j\|\hat{T}_{j_2}\|j_1\rangle$. Instead, one finds its value (for a given triplet j_1, j_2, j) using just one of the matrix elements on the left whose value either is known from experiment or has been estimated from some theoretical considerations. Since all Clebsch–Gordan coefficients are known, the reduced matrix element follows immediately. Then the entire sequence of matrix elements for these j_1, j_2, j follows from (7.140).

Equation 7.140 is the *Wigner–Eckart theorem*. Its proof follows once we use the second commutation rule (7.139) to write

$$\langle j, m | [\hat{J}_\pm, \hat{T}_{j_2, m_2}] | j_1, m_1 \rangle = \sqrt{l(l+1) - m(m \pm 1)}\hbar \langle j, m | \hat{T}_{l, m \pm 1} | j_1, m_1 \rangle$$

and transform the left-hand side as follows:

$$\langle j, m | (\hat{J}_\pm \hat{T}_{j_2, m_2} - \hat{T}_{j_2, m_2} \hat{J}_\pm) | j_1, m_1 \rangle = \langle j, m \mp 1 | \sqrt{j(j+1) - m(m \mp 1)}\hbar \hat{T}_{j_2, m_2} | j_1, m_1 \rangle$$
$$- \langle j, m | \hat{T}_{j_2, m_2} \sqrt{j(j+1) - m(m \pm 1)}\hbar | j_1, m_1 \pm 1 \rangle.$$

Equating the right-hand sides, we obtain equations having exactly the same form as the recursive relations (7.126) for the sequence of Clebsch–Gordan coefficients corresponding to a given triplet j_1, j_2, j. Since the factors in front of inner products are the same as in (7.126), it follows that the inner products themselves must be equal to the respective coefficients times some arbitrary multiplier common for all terms in the sequence. Calling this multiplier (times the factor $1/\sqrt{2j_1 + 1}$ introduced for convenience) the reduced matrix element, we complete the proof.

The theorem essentially tells us to treat the expression $\hat{T}_{j_2, m_2} | j_1, m_1 \rangle$ as addition of angular momenta (j_1, m_1) and (j_2, m_2). This works despite the fact that operator $\hat{T}_{j_2, m_2}$, generally speaking, has nothing to do with orbital motion or spin! It is sufficient that $\hat{T}_{j_2, m_2}$ is an element numbered m_2 of an irreducible spherical tensor operator of rank j_2. And since the conditions

$$\begin{cases} |j_1 - j_2| \leq j \leq j_1 + j_2 \\ m = m_1 + m_2 \end{cases} \tag{7.141}$$

must be met in order for Clebsch–Gordan coefficients to be nonzero, (7.141) give us the *selection rules* that we need in order to eliminate the vanishing matrix elements.

To illustrate the usefulness of this theorem, suppose we are asked to calculate matrix element $\langle l = 1, m = 0 | \hat{x} | l = 0, m = 0 \rangle$ of a spinless hydrogen atom. Since we are told to neglect spin, l will be the entire angular momentum.[5] Here $\hat{x}$ is the position operator, which of course is totally unrelated to angular momentum. Without the Wigner–Eckart

---

[5] Of course, a spinless atom is an "unphysical" situation, but ignoring spin does not cause any contradictions here because $\hat{x}$ is a coordinate operator that does not act on spin.

theorem we would be forced to write the wave functions for the ket and the bra explicitly, sandwich the operator between these functions, and then integrate. Instead, we have a simple and elegant shortcut. In the language of spherical tensors, $x = r_x = (r_{1,-1} - r_{1,1})/\sqrt{2}$ according to (7.137), and so we have a difference of two matrix elements, $\langle l=1, m=0|(\hat{r}_{1,-1}/\sqrt{2} - \hat{r}_{1,1}/\sqrt{2})|l=0, m=0\rangle$. Both elements are zero because of the m-selection rule (7.141,) and the atom will never undergo transition from state $|l=0, m=0\rangle$ to state $|l=1, m=0\rangle$ under the influence of a Hamiltonian term proportional to $\hat{x}$. What if we were given a different operator, $3\hat{z}^2 - \hat{r}^2$? We notice that $3z^2 - r^2$ is proportional to $Y_{2,0}$ (see Equation 7.135) and thus we have a $\hat{T}_{2,0}$ irreducible spherical tensor operator. The problem reduces to that of combining angular momentum states $|l_1=0, m_1=0\rangle$ and $|l_2=2, m_2=0\rangle$ and checking whether the sum is consistent with the coupled eigenket $|l=1, m=0\rangle$. This time the m-selection rule is satisfied, but the l-selection rule is violated and the answer is again zero.

A very important special case of the Wigner–Eckart theorem takes place when $j = j_1$ and $j_2 = 1$. This condition is clearly consistent with the selection rule since $j_1 - 1 \le j \le j_1 + 1$ is obviously true. Also, $j_2 = 1$ automatically implies that $\hat{T}_{j_2,m_2}$ is a first-rank tensor, that is, vector operator. Let it be denoted as $\hat{T}_{1,m_2} \equiv \hat{V}_{m_2}$, where $\mathbf{V} = (V_{-1}, V_0, V_1)$ is a spherical vector, which we would previously write $\mathbf{V} = (V_{1,-1}, V_{1,0}, V_{1,1})$, but have now dropped the first subscript to make the notation more compact. Equation 7.140 takes the simplified form

$$\langle j, m|\hat{V}_{m_2}|j, m_1\rangle = C_{j,1,j,m}(m_1, m_2) \frac{\langle j\|\hat{V}\|j\rangle}{\sqrt{2j+1}}.$$

In this case, the so-called *projection theorem* holds:

$$\langle j, m|\hat{V}_{m_2}|j, m_1\rangle = \frac{\langle j, m_1|\hat{\mathbf{J}}\cdot\hat{\mathbf{V}}|j, m_1\rangle}{\hbar^2 j(j+1)} \langle jm|\hat{J}_{m_2}|jm_1\rangle, \tag{7.142}$$

where $\hat{\mathbf{J}} = (\hat{J}_{-1}, \hat{J}_0, \hat{J}_1)$ is the angular momentum operator written in the same spherical vector notation. For a proof, apply (7.140) to both sides:

$$C_{j,1,j,m}(m_1, m_2)\frac{\langle j\|\hat{V}\|j\rangle}{\sqrt{2j+1}} = \frac{\langle j, m_1|\hat{\mathbf{J}}\cdot\hat{\mathbf{V}}|j, m_1\rangle}{\hbar^2 j(j+1)} C_{j,1,j,m}(m_1, m_2)\frac{\langle j\|\hat{J}\|j\rangle}{\sqrt{2j+1}}.$$

Canceling the like terms, we get

$$\langle j\|\hat{V}\|j\rangle = \frac{\langle j, m_1|\hat{\mathbf{J}}\cdot\hat{\mathbf{V}}|j, m_1\rangle}{\hbar^2 j(j+1)} \langle j\|\hat{J}\|j\rangle. \tag{7.143}$$

We will now show that (7.143) is always true. Write the dot product according to (7.138), $\hat{\mathbf{J}}\cdot\hat{\mathbf{V}} = \hat{J}_0\hat{V}_0 - \hat{J}_1\hat{V}_{-1} - \hat{J}_{-1}\hat{V}_1$, and note that $\hat{J}_0 = \hat{J}_z$, $\hat{J}_1 = -(\hat{J}_x + i\hat{J}_y)/\sqrt{2} \equiv -\hat{J}_+/\sqrt{2}$, and $\hat{J}_{-1} = (\hat{J}_x - i\hat{J}_y)/\sqrt{2} \equiv \hat{J}_-/\sqrt{2}$.

Then the matrix element transforms as

$$\langle j, m_1 | \hat{\mathbf{J}} \cdot \hat{\mathbf{V}} | j, m_1 \rangle = \langle j, m_1 | \hat{J}_z \hat{V}_0 | j, m_1 \rangle + \langle j, m_1 | \frac{\hat{J}_+ \hat{V}_{-1}}{\sqrt{2}} | j, m_1 \rangle - \langle j, m_1 | \frac{\hat{J}_- \hat{V}_1}{\sqrt{2}} | j, m_1 \rangle$$

$$= m_1 \hbar \langle j, m_1 | \hat{V}_0 | j, m_1 \rangle + \sqrt{j(j+1) - m_1(m_1 - 1)} \hbar \langle j, m_1 - 1 | \frac{\hat{V}_{-1}}{\sqrt{2}} | j, m_1 \rangle$$

$$- \sqrt{j(j+1) - m_1(m_1 + 1)} \hbar \langle j, m_1 + 1 | \frac{\hat{V}_1}{\sqrt{2}} | j, m_1 \rangle$$

$$= \left\{ m_1 \hbar C_{j,1,j,m_1}(m_1, 0) + \frac{\sqrt{j(j+1) - m_1(m_1 - 1)} \hbar}{\sqrt{2}} C_{j,1,j,m_1-1}(m_1, -1) \right.$$

$$\left. - \frac{\sqrt{j(j+1) - m_1(m_1 + 1)} \hbar}{\sqrt{2}} C_{j,1,j,m_1+1}(m_1, 1) \right\} \frac{\langle j \| \hat{V} \| j \rangle}{\sqrt{2j+1}}.$$

In the last step, we simply applied the Wigner–Eckart theorem to each matrix element. Now, the expression inside the curly bracket is just a number that might depend on the parameters j and m_1 only. Upon second thought we realize that it cannot depend on m_1. If it did, $\langle jm_1 | \hat{\mathbf{J}} \cdot \hat{\mathbf{V}} | jm_1 \rangle$ would depend on the direction of spin, but $\hat{\mathbf{J}} \cdot \hat{\mathbf{V}}$ is a scalar operator, so its matrix elements should be invariant. So the curly bracket is only a function of j. Call it $\lambda(j)$ and write the last result as $\langle j, m_1 | \hat{\mathbf{J}} \cdot \hat{\mathbf{V}} | j, m_1 \rangle = \lambda(j) (\langle j \| \hat{V} \| j \rangle / \sqrt{2j+1})$. Also note that we did not specify $\hat{\mathbf{V}}$ in advance, so $\hat{\mathbf{V}}$ can be chosen arbitrarily. If we choose $\hat{\mathbf{V}} = \hat{\mathbf{J}}$, then $\langle j, m_1 | \hat{J}^2 | j, m_1 \rangle = j(j+1)\hbar^2 = \lambda(j)(\langle j \| \hat{J} \| j \rangle / \sqrt{2j+1})$. Dividing the equations, we get $\langle j, m_1 | \hat{\mathbf{J}} \cdot \hat{\mathbf{V}} | j, m_1 \rangle / j(j+1)\hbar^2 = \langle j \| \hat{V} \| j \rangle / \langle j \| \hat{J} \| j \rangle$, which is equivalent to (7.143).

The projection theorem is useful for calculating when we want to calculate transition amplitudes between states characterized by the same j and differing only in their magnetic quantum number m. In case when the magnetic quantum numbers are also the same, the theorem helps us find expectation values. Suppose we want to find the expectation value of $\langle \hat{L}_z \rangle$ in some state $|l, s, j, m\rangle$. Since $\hat{L}_z$ is a spherical vector operator, $\hat{L}_z \equiv \hat{L}_0$, the projection theorem applies:

$$\langle \hat{L}_z \rangle = \langle l, s, j, m | \hat{L}_z | l, s, j, m \rangle = \frac{\langle l, s, j, m | \hat{\mathbf{J}} \cdot \hat{\mathbf{L}} | l, s, j, m \rangle}{\hbar^2 j(j+1)} \langle j, m | \hat{J}_z | j, m \rangle \quad (7.144)$$

and the problem has been restated in terms of the total angular momentum operators. The dot product can be written in Cartesian representation by $\hat{\mathbf{J}} = \hat{\mathbf{L}} + \hat{\mathbf{S}}$ as $\hat{\mathbf{J}} - \hat{\mathbf{L}} = \hat{\mathbf{S}}$ and squaring both sides, and the expectation value is easily obtained from the resulting expression:

$$\langle \hat{L}_z \rangle = \frac{\langle l, s, j, m | \left(\hat{J}^2 + \hat{L}^2 - \hat{S}^2 \right) / 2 | l, s, j, m \rangle}{\hbar^2 j(j+1)} \langle j, m | \hat{J}_z | j, m \rangle \quad (7.145)$$

$$= \frac{1}{2} \frac{j(j+1) + l(l+1) - s(s+1)}{j(j+1)} m\hbar.$$

Problems

7.1 Derive commutation rules (7.13).

7.2 Derive the expression (7.14) for the cross product of two vectors in tensor notation.

7.3 Find the commutation relations between components of the vector operators of linear and angular momentum, that is, find $[p_i, L_j]$.

7.4 Find the commutation relations between the different components of the angular momentum operator.

7.5 Prove that all three components of the angular momentum are compatible with its magnitude, even though they are not compatible with one another.

7.6 Derive the commutation rules (7.26).

7.7 Find the indeterminacy relations between (a) L_x and L_z and (b) L_y and L_z. Apply them to a state with definite L_z. Explain the result using the cone model shown in Figure 7.3.

7.8 Consider a person of mass 70 kg sitting on a rotating chair of mass 20 kg in front of a desktop. Suppose that in describing rotation about the vertical axis, the whole system can be approximated by a uniform cylinder with effective radius 0.5 m. If the total angular momentum of the system is equal to one electron spin (i.e., $L_z = (1/2)\hbar$), how long will it take for the system to make one complete rotation?

7.9 Find the eigenfunctions and eigenvalues of the $\hat{S}_y$ operator in the $\hat{S}_z$-basis for a spin-$(1/2)\hbar$ particle.

7.10 Derive Equations 7.70 and 7.71.

7.11 Derive Equation 7.110.

7.12 Derive Equation 7.112.

7.13 Suppose that we try to visualize an electron as a classical solid sphere with uniform distribution of charge and mass. In order to have a nonzero angular momentum, the sphere must be spinning. If the effective radius of this sphere is $r_e = e^2/\mu_e c^2 \approx 10^{-15}$ m, then

  a) What must be the electron angular velocity corresponding to its spin $S = \hbar/2$?

  b) What must be the linear velocity of the equatorial points?

7.14 For a spin-1/2 particle, express spin-up and spin-down state in terms of spin-right and spin-left eigenstates of $\hat{S}_x$, $\hat{S}_y$.

7.15 Prove that $(\hat{\sigma}_e(\theta, \varphi))^2 = 1$.

7.16 Prove that $\langle \uparrow | \sigma_e | \uparrow \rangle = \langle \chi^+ | \sigma_z | \chi^+ \rangle$.

7.17 Consider an object consisting of two parts characterized each by its individual angular momentum $\mathbf{J}_1$ and $\mathbf{J}_2$, respectively. Suppose that their magnitudes are measured and found to be (in units $\hbar$) $|\mathbf{J}_1| \equiv J_1 = \sqrt{l_1(l_1+1)}$ and $|\mathbf{J}_2| \equiv J_2 = \sqrt{l_2(l_2+1)}$, respectively. Find the number of different states with the same pair of quantum numbers (l_1, l_2). Using the same notations as in (7.121), apply your result to only one particle (e.g., an atomic electron) possessing both orbital (**L**) and spin (**S**) angular momentum.

7.18 Consider the addition of angular momenta $\mathbf{J}_1 + \mathbf{J}_2$. The resulting state can be written as $|\varphi\rangle = |j_1, j_2, m_1, m_2\rangle$ or as $|\psi\rangle = |j_1, j_2, j, m\rangle$. In the case $m_1 = j_1$ and $m_2 = j_2$, we have one-to-one correspondence between the $|\varphi\rangle$ and $|\psi\rangle$ states: $|j_1, j_2, m_1 = j_1, m_2 = j_2\rangle \Leftrightarrow |j_1, j_2, j = j_1 + j_2, m = j\rangle$.

Both sides are just two equivalent representations of *the same* physical state. But there are *two* distinct uncoupled eigenstates with $m = m_1 + m_2 = j_1 + j_2 - 1$ and, consequently, there are two distinct coupled states with this m, so this time we have the correspondence between the *pairs* of states:

$$\begin{cases} |\varphi_1\rangle = |j_1, j_2, m_1 = j_1, m_2 = j_2 - 1\rangle \\ |\varphi_2\rangle = |j_1, j_2, m_1 = j_1 - 1, m_2 = j_2\rangle \end{cases}$$
$$\Leftrightarrow \begin{cases} |\psi_1\rangle = |j_1, j_2, j = j_1 + j_2, m = j - 1\rangle \\ |\psi_2\rangle = |j_1, j_2, j = j_1 + j_2 - 1, m = j\rangle \end{cases}.$$

Suppose each uncoupled eigenstate was written as a superposition:

$$|\varphi_1\rangle = a_{11}|\psi_1\rangle + a_{12}|\psi_2\rangle, \qquad |\varphi_2\rangle = a_{21}|\psi_1\rangle + a_{22}|\psi_2\rangle.$$

a) Given the transformation matrix formed by coefficients a_{ij}, write the inverse transformation expressing the coupled states in terms of uncoupled states.
b) Consider the case $m = m_1 + m_2 = j_1 + j_2 - 2$. Write the resulting eigenstates in uncoupled and coupled representations, and the corresponding transition matrix.
c) What is the maximum number of eigenstates that we can get as we keep decreasing m?
d) What happens when m becomes less than $|j_1 - j_2|$?

7.19 Derive the components of $\hat{\mathbf{L}}$ in spherical coordinates.

References

1 Zommerfeld, A. (1952) *Mechanics*, Academic Press, New York, Section 44. Davydov, A. (1965) *Quantum Mechanics*, Pergamon Press/Addison-Wesley, Oxford/Reading, MA.
2 Walls, D. (1983) Squeezed states of light. *Nature*, **306**, 141.
3 Landau, L. and Lifshits, E. (1965) *Quantum Mechanics: Non-Relativistic Theory*, 2nd edn, Pergamon Press, Oxford.
4 Sakurai, J.J. (1994) *Modern Quantum Mechanics*, Addison-Wesley, Reading, MA, p. 153.
5 Herzberg, G. (1944) *Atomic Spectra and Atomic Structure*, 2nd edn, Dover Publications, New York.
6 Richardson, O.W. (1908) A mechanical effect accompanying magnetization. *Phys. Rev. (Ser. I)*, **26** (3), 248–253.
7 Einstein, A. and de Haas, W.J. (1915) Experimenteller Nachweis der Ampereschen Molekularströme. *Dtsch. Phys. Ges. Verhandl.*, **17**, 152–170.
8 Einstein, A. and de Haas, W.J. (1915) Experimental proof of the existence of Ampère's molecular currents (in English). *Koninklijke Akademie van Wetenschappen te Amsterdam. Proceedings*, vol. 18, pp. 696–711.
9 Ya Frenkel, V. (1979) On the history of the Einstein–de Haas effect. *Sov. Phys. Usp.*, **22** (7), 580–587.

10 Stern, O. (1921) Ein Weg zur experimentellen Pruefung der Richtungsquantelung im Magnetfeld. *Z. Phys.*, **7**, 249–253.
11 Gerlach, W. and Stern, O. (1922) Das magnetische Moment des Silberatoms. *Z. Phys.*, **9**, 353–355.
12 Tomonaga, S.I. (1997) *The Story of Spin*, University of Chicago Press, Chicago, IL, p. 35.
13 Weinert, F. (1995) Wrong theory–right experiment: the significance of the Stern–Gerlach experiments. *Stud. Hist. Philos. Mod. Phys.*, **26**, 75–86.
14 Phipps, T.E. and Taylor, J.B. (1927) The magnetic moment of the hydrogen atom. *Phys. Rev.*, **29**(2), 309–320.
15 Scully, M.O., Lamb, W.E., and Barut, A. (1987) On the theory of the Stern–Gerlach apparatus. *Found. Phys.*, **17**(6), 575–583.
16 Landau, L. and Lifshits, E. (1981) *The Field Theory*, Butterworth-Heinemann, Oxford.
17 Vidick, Th. and Wehner, S. (2011) Does ignorance of the whole imply ignorance of the parts? Large violations of noncontextuality in quantum theory. *Phys. Rev. Lett.*, **107**, 3.
18 Hildebrand, F.B. (1976) *Advanced Calculus for Applications*, Prentice-Hall, Englewood Cliffs, NJ, p. 161.
19 Pipes, L.A. (1958) *Applied Mathematics for Engineers and Physicists*, McGraw-Hill, New York, pp. 364–371.

8
The Schrödinger Equation

8.1
The Schrödinger Equation

We already noted that a QM system changes continuously between measurements but undergoes a quantum jump with generally unpredictable outcome during a measurement.

In this chapter we will study the first type, in which a system evolves continuously. This evolution is described by a wave equation. Its important special case – the Schrödinger equation – is known to us from Chapter 3. Here we describe it in more detail.

We will work in position representation and denote the initial and final state functions as $\Psi(\mathbf{r}, t_0)$ and $\Psi(\mathbf{r}, t)$, respectively. Suppose that there are no measurements between these moments. Then the system's evolution between t_0 and t is deterministic; that is, $\Psi(\mathbf{r}, t)$ is determined by $\Psi(\mathbf{r}, t_0)$. This requires $\Psi(\mathbf{r}, t)$ to be not only continuous but also analytic, so it can be expanded in Taylor series:

$$\Psi(\mathbf{r}, t) = \Psi(\mathbf{r}, t_0) + \left(\frac{\partial \Psi(\mathbf{r}, t)}{\partial t}\right)_{t=t_0} (t - t_0) + \frac{1}{2}\left(\frac{\partial^2 \Psi(\mathbf{r}, t)}{\partial t^2}\right)_{t=t_0} (t - t_0)^2 + \cdots. \tag{8.1}$$

For sufficiently small $|t - t_0|$, we can retain in (8.1) only the first two terms.

The experimental evidence shows that knowing $\Psi(\mathbf{r}, t_0)$ for slow ($v \ll c$) particles is sufficient to determine $\Psi(\mathbf{r}, t)$. This means that the first derivative at $t = t_0$ must be completely determined by $\Psi(\mathbf{r}, t_0)$; that is, it can be obtained by applying a certain operator $\hat{Q}(\mathbf{r}, t)$ to $\Psi(\mathbf{r}, t_0)$. Since the initial moment is arbitrary, this requirement holds for any moment t:

$$\frac{\partial \Psi(\mathbf{r}, t)}{\partial t} = \hat{Q}(\mathbf{r}, t)\Psi(\mathbf{r}, t). \tag{8.2}$$

In other words, $\Psi(\mathbf{r}, t)$ must be a solution of the first-order differential equation with respect to time. All higher order time derivatives necessary for determining $\Psi(\mathbf{r}, t)$ by (8.1) at any t can be obtained by reiteration of (8.2).

Quantum Mechanics and Quantum Information: A Guide through the Quantum World,
First Edition. Moses Fayngold and Vadim Fayngold.
© 2013 Wiley-VCH Verlag GmbH & Co. KGaA. Published 2013 by Wiley-VCH Verlag GmbH & Co. KGaA.

The simplest way to determine the operator $\hat{Q}(\mathbf{r}, t)$ is just to examine the simplest wave function $\Psi(\mathbf{r}, t) = \Psi_0 \, e^{(i/\hbar)(\mathbf{p} \cdot \mathbf{r} - Et)}$ known to us. We have for it

$$\frac{\partial \Psi(\mathbf{r}, t)}{\partial t} = -\frac{i}{\hbar} E \Psi(\mathbf{r}, t) = -\frac{i}{\hbar} \frac{p_x^2 + p_y^2 + p_z^2}{2\mu} \Psi(\mathbf{r}, t). \tag{8.3}$$

But

$$-\frac{i}{2\mu\hbar}\left(p_x^2 + p_y^2 + p_z^2\right)\Psi(\mathbf{r}, t) = \frac{i\hbar}{2\mu}\left(\frac{\partial^2}{\partial x^2} + \frac{\partial^2}{\partial y^2} + \frac{\partial^2}{\partial z^2}\right)\Psi(\mathbf{r}, t) = \frac{i\hbar}{2\mu}\nabla^2\Psi(\mathbf{r}, t). \tag{8.4}$$

Comparing this with (8.2) gives

$$\hat{Q}(\mathbf{r}, t) = -\frac{i}{\hbar}\left[-\frac{\hbar^2}{2\mu}\nabla^2\right]. \tag{8.5}$$

We see that $\hat{Q}$ is a differential operator that in the given case does not depend on either t or $\mathbf{r}$; this makes perfect sense for a free particle. Further, we recognize in the brackets the operator $\hat{K}$ of kinetic energy, so $\hat{Q} = (-i/\hbar)\hat{K}$. Now we only need to generalize this result to the cases when a particle is subject to external forces. To this end, we make a natural assumption that the action of forces can be described by adding to $\hat{K}$ an operator representing the corresponding potential energy. Since the latter is a function of $\mathbf{r}$ and can vary with time, this operator is generally position and time dependent. And since the position operator in its own representation is simply the variable $\mathbf{r}$, we have in this case just potential energy function $U(\mathbf{r}, t)$ to be added to $\hat{K}$ to get $\hat{Q}(\mathbf{r}, t) = -(i/\hbar)[\hat{K} + U(\mathbf{r}, t)]$. The expression in the brackets is the Hamiltonian operator $\hat{H}$ of the total energy of a particle:

$$\hat{H}(\mathbf{r}, t) = \hat{K} + U(\mathbf{r}, t) = \frac{\hat{\mathbf{p}}^2}{2\mu} + U(\mathbf{r}, t). \tag{8.6}$$

Using this notation, we can finally write the sought for equation for Ψ in the form

$$i\hbar \frac{\partial \Psi(\mathbf{r}, t)}{\partial t} = \hat{H}(\mathbf{r}, t)\Psi(\mathbf{r}, t). \tag{8.7}$$

We have recovered the Schrödinger equation (3.10). Note that it does not derive directly from basic principles. The above derivation is a compelling but not a 100% rigorous argument to justify its form. So it is better to consider it as an additional postulate of nonrelativistic QM.

8.2
State Function and the Continuity Equation

In most cases, the state represented by a state function $\Psi(\mathbf{r}, t)$ changes with time. For instance, the wave packet considered in Section 3.3 can move and change its shape. Some regions of its probability cloud $|\Psi(\mathbf{r}, t)|^2$ may grow denser due to

8.2 State Function and the Continuity Equation

redistribution of probability, that is, at the expense of some other regions. This is accompanied by a flow of probability between the regions. The problem arises to determine this flow.

Just as $|\Psi(\mathbf{r},t)|^2$ determines the probability density, the probability flux is described by another quadratic form of Ψ. This form satisfies the continuity equation similar to that for the mass and flow density in fluid dynamics, or charge and current density in EM.

Consider the Schrödinger equation and its complex conjugate

$$i\hbar \frac{\partial \Psi}{\partial t} = \hat{H}\Psi \tag{8.8a}$$

and

$$-i\hbar \frac{\partial \Psi^*}{\partial t} = \hat{H}\Psi^*. \tag{8.8b}$$

Now multiply (8.8a) through from the left by Ψ^* and (8.8b) by Ψ:

$$i\hbar \Psi^* \frac{\partial \Psi}{\partial t} = \Psi^* \hat{H}\Psi, \qquad -i\hbar \Psi \frac{\partial \Psi^*}{\partial t} = \Psi \hat{H}\Psi^*. \tag{8.9}$$

Take the difference:

$$i\hbar \left(\Psi^* \frac{\partial \Psi}{\partial t} + \Psi \frac{\partial \Psi^*}{\partial t} \right) = \Psi^* \hat{H}\Psi - \Psi \hat{H}\Psi^*. \tag{8.10}$$

This simplifies to

$$\frac{\partial (\Psi^* \Psi)}{\partial t} = -\frac{i\hbar}{2\mu} \vec{\nabla} \cdot \left(\Psi \vec{\nabla} \Psi^* - \Psi^* \vec{\nabla} \Psi^* \right). \tag{8.11}$$

Denote

$$\rho \equiv \Psi^* \Psi \quad \text{and} \quad \mathbf{j} \equiv \frac{i\hbar}{2\mu} \left(\Psi \vec{\nabla} \Psi^* - \Psi^* \vec{\nabla} \Psi \right) = -\frac{\hbar}{\mu} \text{Im} \left(\Psi \vec{\nabla} \Psi^* \right). \tag{8.12}$$

Then (8.11) can be written as

$$\frac{\partial \rho}{\partial t} = -\vec{\nabla} \cdot \mathbf{j}. \tag{8.13}$$

This is a continuity equation for a flowing entity with a local density ρ and the corresponding flux density $\mathbf{j}$. We interpret $\mathbf{j}$ defined by (8.12) as the probability flux density. Later we will generalize (8.12) to include the case of the charged particles in an external magnetic field.

Integrating Equation 8.13 over a volume V, we will have on the left the rate of change of probability $\mathcal{P}(V)$ of finding the particle within this volume; this rate equals the net flux of vector $\mathbf{j}$ through the surface of the volume (Problem 8.1). If there is no flux ($\mathbf{j} = 0$) or the net flux through the surface is zero (the inflow equals the outflow), then $\mathcal{P}(V)$ does not change. If it does change, for example, decreases, this is accompanied by the corresponding increase in some other place, so that the net probability is conserved. And the mechanism of this is exactly the same as in the fluid dynamics: the probability leaks out from the place where its density decreases and flows into a place where its density increases. This expresses the *conservation of probability* for a stable particle. In this respect, the distribution $\rho(\mathbf{r})$ behaves as a continuous fluid, whose net mass is also conserved. The value $|\mathbf{j}|$ is the measure of the probability flow rate (similar to flow density of air in the wind or current density in a conductor). A component of $\mathbf{j}$, say, j_x, at a given location is the probability per unit time for a particle to cross a unit area (at this location) perpendicular to $\hat{\mathbf{x}}$. And under the usual sign convention, the crossing in the $\hat{\mathbf{x}}$-direction is considered as positive flux, and in the opposite direction as negative flux. In the case when we have a *closed* surface including some volume V, another convention is used: at each point of the surface, the local flux is considered as positive when it is directed out of the volume; otherwise, it is negative. In the first case, it describes the particle quitting the volume, in the second case entering it. If we take the surface integral of $\mathbf{j}$ and find it positive, we know that, on the average, the particle is quitting the enclosed region, so the probability of finding it there decreases with time. Otherwise, the particle, on the average, enters the region, and the probability of finding it there increases with time. All these results follow from Equation 8.13.

If the boundary enclosing V recedes to infinity so that V includes *all space*, the *net* flux must be zero, for a very simple reason: No matter what happens, at any moment of time the particle is certain to be found somewhere, so the net probability is always 1, and therefore the net flux through the infinitely remote boundary must be zero. A nonzero net flux in this case would mean that the particle is created in space or disappears in it, depending on the sign of the flux. Thus, the requirement of the zero net flux through the closed but infinitely remote surface can be considered a kind of the boundary conditions, corresponding to conservation of particles. It is generally *not equivalent* to a requirement that the *local* flux density must vanish at infinity. The de Broglie wave does not vanish anywhere, and yet its *net* flux through any closed surface is zero because positive and negative contributions exactly balance each other.

The described conditions refer to stable nonrelativistic particles. They show the most important feature of the probability distribution: it can flow and is a continuous function of position and time, which satisfies the same equation as does the material fluid.

Generally (recall Section 3.10), particles can decay or convert into other particles in an interaction, especially in high-energy collisions. Some cases of particle's decay will be considered in Chapters 11 and 19. The detailed description of the corresponding phenomena is the subject of quantum field theory and lies beyond the scope of this book.

Box 8.1 Visualizing probability density

We must be very careful when interpreting the continuity equation as an indication that the particle with undefined position is "smeared out" over the extended region of space. If this were literally true, we could "chip off " a certain fraction of the particle the same way as we would take a teaspoon of water from a filled cup. If we attempted to do this, we would find either the whole particle in one place or no particle. Such an attempt is equivalent to taking a position measurement, which interrupts the continuous flow described above. We will postpone the detailed discussion of what happens to the whole extended probability cloud when we find the whole particle at a certain point or within a small volume. Here we will only say that in some situations (specifically, between the measurements) the particle *does* behave *as if* it were spread out, as prescribed by the Schrödinger equation, and we can find the correct values of all its relevant characteristics in the corresponding state. But on the other hand, we should not try to envision this as some actual substance filling the volume of space, because the $|\Psi(\mathbf{r},t)|^2$ is the measure of *probability* density, not mass density or charge density. We can calculate the latter characteristics as the products $\mu|\Psi(\mathbf{r},t)|^2$ and $q|\Psi(\mathbf{r},t)|^2$, respectively, but even so they would only have the statistical meaning as the corresponding *local* averages for a set of measurements in the same state $\Psi(\mathbf{r},t)$. And we must also keep in mind the argument in Section 3.2 about the meaning of the wave function and probability flow for a *system* of N particles. In this case, we will have Equation 8.8 and expressions (8.12) and (8.13) for a function $\Psi(\mathbf{r}_1, \mathbf{r}_2, \ldots, \mathbf{r}_N, t)$ in an abstract multidimensional space.

8.3
Separation of Temporal and Spatial Variables: Stationary States

Consider the case when the system is in a stationary environment. Then the potential energy of the particle does not depend on t, that is, $U(\mathbf{r},t) \to U(\mathbf{r})$, and the state is described by the Schrödinger equation with time-independent Hamiltonian. What can we say in such case about state Ψ? Will it be stationary as the environment is, or will it evolve? The answer is: It depends. In order to see why and how, let us find the *general solution* of Equation 8.8a with time-independent $U(\mathbf{r})$. Such a solution can be found by a well-known method — separation of variables [1,2]. Namely, we seek the solution as a product of two functions — one depending only on $\mathbf{r}$ and the other only on t:

$$\Psi(\mathbf{r},t) = \psi(\mathbf{r})\chi(t). \tag{8.14}$$

Putting this into (8.8a) gives

$$i\hbar\psi(\mathbf{r})\frac{\partial\chi(t)}{\partial t} = \chi(t)\hat{H}(\mathbf{r})\psi(\mathbf{r}). \tag{8.15}$$

Since $\Psi(\mathbf{r},t) \neq 0$, we can divide this through by $\Psi(\mathbf{r},t)$ and get

$$i\hbar\frac{\dot\chi(t)}{\chi(t)} = \frac{\hat{H}(\mathbf{r})\psi(\mathbf{r})}{\psi(\mathbf{r})} \tag{8.16}$$

(the dot on $\chi(t)$ stands for the time derivative). The left-hand side of the resulting equation depends exclusively on t, whereas the right-hand side depends only on $\mathbf{r}$.

Since both variables are independent, the equation can only hold if either side is a constant. Denote this constant as E. Then (8.16) splits into two equations – one for the temporal part and the other for the spatial part:

$$i\hbar \frac{\partial \chi(t)}{\partial t} = E\chi(t) \tag{8.17a}$$

and

$$\hat{H}\psi_E(\mathbf{r}) = E\psi_E(\mathbf{r}). \tag{8.17b}$$

We are already familiar with them! They are characteristic equations for the eigenfunctions of the energy operator (3.33) and of the Hamiltonian, respectively. *Both operators produce the same result in the set of solutions of the Schrödinger equation.*

Now let us go to solutions. The first equation gives

$$\chi_E(t) = \chi_0\, e^{-i(E/\hbar)t}, \tag{8.18}$$

where χ_0 is a constant. This solution is just the temporal factor in the de Broglie wave.

As to the solutions $\psi_E(\mathbf{r})$, they are determined by $U(\mathbf{r})$ and the boundary conditions.

It is easy to see that the product

$$\Psi_E(\mathbf{r}, t) = \psi_E(\mathbf{r})\chi_E(t) = \psi_E(\mathbf{r})\, e^{-i(E/\hbar)t} \tag{8.19}$$

satisfies either of Equations 8.17 and gives, according to (8.14), the *complete solution* corresponding to the eigenvalue E. Here the constant χ_0 figuring in (8.18) is absorbed by $\psi_E(\mathbf{r})$ and must be determined from the normalization condition. The eigenvalues E are the possible values of the energy of the system.

Suppose we have a particle in one such state with definite E. Its probability density $|\Psi_E(\mathbf{r}, t)|^2$, as well as the probability of finding a certain value of any other observable, will not depend on t. So, even though the eigenfunction (8.19) *is time dependent,* the *physical state* described by it is a *stationary state*. It does not evolve.

A definite energy makes the state unaware of the flow of time. Does this make sense? Recall the time–energy relationship (3.39). In a state with $\Delta E \to 0$, the "longevity" $\Delta t \to \infty$. If your energy were *exactly* defined, you would stay in your current state forever, but with little if any advantage. While becoming formally immortal, you would freeze in time, with your heartbeat stopped and your thoughts fixed, unable to tell the difference between today and 160 million years BC. *Your time* will stop flowing.

Thus, any observable change of a physical system is unavoidably linked with an indeterminacy of its energy. How can this be the case if the solution (8.19) describes a state with fixed energy? The answer is in the mathematics of differential equations: first, Equation 8.16 has more than one solution; and second, if $\Psi_E(\mathbf{r}, t)$ and $\Psi_{E'}(\mathbf{r}, t)$ are such solutions corresponding to eigenvalues E and E', respectively, then their weighted sum $\Psi(\mathbf{r}, t) = \alpha \Psi_E(\mathbf{r}, t) + \beta \Psi_{E'}(\mathbf{r}, t)$ is also a solution. This is just a special case of the superposition principle. Applying this principle, we can construct the most general solution of the Schrödinger equation from special solutions (8.19) used as building blocks:

$$\Psi(\mathbf{r},t) = S_E \Phi(E)\psi'_E(\mathbf{r})e^{-i(E/\hbar)t}. \quad (8.20)$$

Here the symbol S_E denotes summation over discrete energy eigenvalues and integration over a continuous spectrum of eigenvalues. Evidently, the general solution is *not an eigenstate of* $\hat{H}(\mathbf{r})$ but a superposition of its different eigenstates $\Psi_E(\mathbf{r})$ with *time-dependent* amplitudes $\Phi(E)e^{-i(E/\hbar)t}$. Accordingly, it describes a state with indeterminate energy, and this state is time dependent due to the interference between different terms in (8.20). For instance, the probability distribution will be

$$\rho(\mathbf{r},t) = |\Psi(\mathbf{r},t)|^2$$
$$= S_E |\Phi(E)\psi_E(\mathbf{r})|^2 + 2 S_{E>} S_{E'} |\Phi(E)\Phi(E')\psi_E(\mathbf{r})\psi_{E'}(\mathbf{r})| \cos\left(\frac{E-E'}{\hbar}t + \varphi_{E,E'}\right), \quad (8.21)$$

where $\varphi_{E,E'}$ are the phase angles. This is an explicit function of time.

Alternatively, we can regard (8.20) as the expansion over the *time-dependent* eigenfunctions (8.19) forming an orthonormal set, with the *time-independent* amplitudes $\Phi(E)$. The latter are determined by the initial conditions, for instance, by the shape of $\Psi(\mathbf{r},t)$ at $t=0$. If $\Psi(\mathbf{r},0)$ is known, the coefficients $\Phi(E)$ are

$$\Phi(E) = \int \Psi(\mathbf{r},0)\psi'^*_E(\mathbf{r})d\mathbf{r}. \quad (8.22)$$

We see that a system in a stationary environment can be stationary or not depending on whether it is in an energy eigenstate or in a superposition of such states with *different* eigenvalues of $\hat{H}$. In the latter case, the system evolves *despite* the fact that its environment is stationary. As we will see later, this evolution can be fairly complicated.

8.4
The Helmholtz Equation and Dispersion Equation for a Free Particle

For a free particle, the Schrödinger equation reduces to

$$i\hbar \frac{\partial \Psi}{\partial t} = -\frac{\hbar^2}{2\mu}\nabla^2 \Psi. \quad (8.23)$$

Select a stationary state with a fixed energy value E, so that

$$\Psi(\mathbf{r},t) = \psi(\mathbf{r})e^{-i\omega t}, \quad \omega = \frac{E}{\hbar}. \quad (8.24)$$

Plugging this into (8.23) gives, according to the already known procedure the equation for $\psi(\mathbf{r})$,

$$\nabla^2 \psi(\mathbf{r}) + k^2 \psi(\mathbf{r}) = 0, \quad k^2 = \frac{p^2}{\hbar^2} = \frac{2\mu E}{\hbar^2}. \quad (8.25)$$

This rehash of (8.4) is the *Helmholtz equation*, widely used in many areas of physics dealing with waves. No surprise it made its way into QM dealing with matter waves. But a matter wave in QM is associated with a single particle. The simplest solution of the Helmholtz equation is the spatial part of the de Broglie wave $\psi(\mathbf{r}) = \psi_0\, e^{i\mathbf{kr}}$. Combining this with (8.24) recovers the complete expression $\Psi(\mathbf{r},t) = \psi_0 e^{i(\mathbf{kr}-\omega t)}$ for this wave including the time dependence. (Remember, this kind of time dependence through an exponential factor with real ω does not involve any time dependence of relevant observables.) Its specific feature is the relationship between the wave vector $\mathbf{k}$ and frequency ω. This relationship follows directly from the second equation (8.25) and gives

$$k^2 = k_x^2 + k_y^2 + k_z^2 = \frac{2\mu\omega}{\hbar}, \tag{8.26}$$

where frequency ω is defined in (8.24). This is the *dispersion equation*, which relates the wave vector and frequency of the de Broglie wave in nonrelativistic QM.

In the relativistic case, the wave equation is different from (8.23,) and accordingly, the dispersion equation for $\omega = E/\hbar$ and $\mathbf{k}$ is different from (8.26):

$$k^2 = k_x^2 + k_y^2 + k_z^2 = \frac{1}{(\hbar c)^2}(E^2 - \mu_0^2 c^4) = \left[\frac{\omega^2}{c^2} - \left(\frac{\mu_0 c}{\hbar}\right)^2\right]. \tag{8.27}$$

Here μ_0 is the *rest mass* of the particle and $E \geq \mu_0 c^2$ is its relativistic energy, which includes also the energy stored by the stationary particle (*rest energy*) [3]. Solving (8.27) for E (or ω), we find

$$E = \sqrt{\mu_0^2 c^4 + (\hbar c)^2 k^2} \quad \text{or} \quad \omega = \pm\sqrt{\lambda_c^{-2} + k^2}, \quad \lambda_c \equiv \frac{\hbar}{\mu_0 c}, \tag{8.28}$$

where λ_c is the *Compton wavelength* of a particle with the rest mass μ_0. In the nonrelativistic approximation, we expand (8.28) into the Taylor series

$$E(k) = \mu_0 c^2 + \frac{\hbar^2 k^2}{2\mu_0} + \cdots, \quad \omega(k) = \frac{\mu_0 c^2}{\hbar} + \frac{\hbar k^2}{2\mu_0} + \cdots, \tag{8.29}$$

take the *rest energy* as the reference point on the energy scale ($E - \mu_0 c^2 \to E$), and obtain $\omega = \hbar k^2/2\mu_0$, which is just Equation 8.26. Conversely, in the ultrarelativistic case the rest mass in (8.27) can be neglected. In particular, for a particle with the zero rest mass (a photon), this reduces to familiar $|\mathbf{k}| = \omega/c$.

The de Broglie wave $\psi(\mathbf{r}) = \psi_0 e^{i\mathbf{kr}}$ is only a special solution of the Helmholtz equation, describing a wave propagating along a direction $\mathbf{k}/k$. A wave with *the same magnitude of* $\mathbf{k}$ but propagating in another direction would be as good a solution, albeit it would describe a different physical state. Thus, we have an infinite set of distinct states all with the same energy. In the language of QM, $\psi(\mathbf{r}) = \psi_0 e^{i\mathbf{kr}}$ is an eigenfunction of $\hat{\mathbf{p}}$ corresponding to a *vector eigenvalue* $\mathbf{p}$, that is, to a triplet (p_x, p_y, p_z) satisfying (8.25) in the nonrelativistic case. But it is only one out of continuous set of different eigenfunctions belonging to the same eigenvalue (8.25). This is an example of *degeneracy* discussed in Chapter 5 – the coexistence of *different* eigenstates all with the *same* energy. We will study some cases in more detail in Section 8.6.

8.5
Separation of Spatial Variables and the Radial Schrödinger Equation

In Section 8.3, we separated the temporal and spatial variables in the Schrödinger equation.

Here we will use a similar approach for further separation of variables; now it will be different spatial variables in the case of a spherically symmetric potential $U(\mathbf{r}) \to U(r)$, which is a function of one variable r – the distance from the source (origin) of the field. Two different points on a sphere centered at the origin have the same potential; hence, the equipotential surfaces are concentric spheres. A well-known example is the Coulomb potential energy $U(r) = e^2/r$.

The most straightforward approach to find the eigenstates of such a system is to write the Schrödinger equation in spherical coordinates and represent the solution as the product of two functions – one being $R(r)$, depending only on r, and the other, $\Upsilon(\theta, \varphi)$, depending on angles θ and φ.

Here we consider the general features of solutions in spherically symmetric potentials that fall off to zero at infinity but are allowed to have a singularity at the origin. The singularity must be less than of the second order. In other words, it is allowed to go to infinity when $r \to 0$, but more slowly than r^{-2}, so that

$$\lim_{r \to 0} r^2 U(r) = 0. \tag{8.30}$$

Following the outlined procedure, we express $\hat{K} = -(\hbar^2/2\mu)\nabla^2$ in spherical coordinates (see, for example, Ref. [4]):

$$\hat{K}(r, \theta, \varphi) = -\frac{\hbar^2}{2\mu}\left\{\frac{1}{r^2}\frac{\partial}{\partial r}\left(r^2 \frac{\partial}{\partial r}\right) + \frac{1}{r^2}\left[\frac{1}{\sin\theta}\frac{\partial}{\partial \theta}\left(\sin\theta \frac{\partial}{\partial \theta}\right) + \frac{1}{\sin^2\theta}\frac{\partial^2}{\partial \varphi^2}\right]\right\}. \tag{8.31}$$

In the square brackets, we recognize the angular momentum operator (Appendix E, Equation E.1). This is precisely what one should expect, since the classical kinetic energy is the sum of radial and angular parts, namely,

$$K(r, \theta, \varphi) = \frac{p_r^2}{2\mu} + \frac{L^2}{2\mu r^2}. \tag{8.32}$$

The first term here is the energy associated with the radial component of momentum, and the second one is associated with its orbital component responsible for the angular momentum L. Accordingly, we can identify the first term on the right in

(8.32) as the operator of radial kinetic energy [4]

$$\hat{K}_r = -\frac{\hbar^2}{2\mu} r^{-2} \frac{\partial}{\partial r}\left(r^2 \frac{\partial}{\partial r}\right) \qquad (8.33)$$

(based on this expression, one can find the expression for the radial component $\hat{p}_r$ of the linear momentum operator, such that $\hat{K}_r = \hat{p}_r^2/2\mu$). Note that the angular momentum L in (8.32) is a conserving quantity (the integral of motion) in a spherically symmetric field.

Thus, the stationary Schrödinger equation can be written as

$$\left[\hat{K}_r(r) + \frac{\hat{L}^2(\theta,\varphi)}{2\mu r^2} + U(r)\right]\Psi = E\Psi. \qquad (8.34)$$

The observables L and E are compatible because their operators commute. Accordingly, they must have common eigenfunctions. Therefore, Ψ must also be an eigenfunction of $\hat{L}^2$; that is, apart from (8.34), it must satisfy the equation $\hat{L}^2\Psi = L^2\Psi$. We know from Sections 7.2 and 7.3 that $L^2 = \hbar^2 l(l+1)$, so we must have

$$\hat{L}^2\Psi = \hbar^2 l(l+1)\Psi. \qquad (8.35)$$

Operator $\hat{L}^2$ acts only on the angular variables in Ψ. This means that the radial dependence can be factored out; that is, Ψ can be represented as a product of two functions – one depending on r and the other on θ and φ (recall Sections 7.2 and 7.3):

$$\Psi(r,\theta,\varphi) = R(r)Y_{l,m}(\theta,\varphi). \qquad (8.36)$$

Here m is the magnetic quantum number labeling the z-component of L (recall that $Y_{l,m}$ is itself an eigenfunction of two operators, $\hat{L}^2$ and $\hat{L}_z$).

Thus, the separation of variables in a spherically symmetric field is automatically executed by the property (8.36). Now, putting (8.36) into (8.34), using (8.35), and dropping the angular factor $Y_{l,m}(\theta,\varphi)$ from both sides leads to the equation for the radial function $R(r)$:

$$\left[\hat{K}_r(r) + \frac{\hbar^2 l(l+1)}{2\mu r^2} + U(r)\right]R(r) = ER(r). \qquad (8.37)$$

We will call it the radial Schrödinger equation. It determines the set of eigenvalues E. We see that these eigenvalues can depend on L through quantum number l, but not on L_z, which is absent in (8.37). This reflects the fact that the particle's energy is independent of direction of its angular momentum vector in a spherically symmetric field. Altogether, since the angular part of the state function is determined by numbers l and m, the whole function in a spherically symmetric field is completely determined by the values E, l, and m. In other words, energy E, angular momentum L, and its projection L_z onto a chosen axis form together a complete set of observables needed for the full description of particle's motion.

Look for the solution of (8.37) in the form

$$R(r) = \frac{G(r)}{r}. \tag{8.38}$$

Putting this into (8.37) and using (8.33) gives after some algebra (Problem 8.7) the equation for $G(r)$:

$$-\frac{\hbar^2}{2\mu}\frac{d^2 G}{dr^2} + \frac{\hbar^2}{2\mu}\frac{l(l+1)}{r^2}G + U(r)G = EG. \tag{8.39}$$

Formally, it is identical with equation for 1D motion in the field with the effective potential energy

$$\tilde{U}_l(r) = U(r) + \frac{\hbar^2}{2\mu}\frac{l(l+1)}{r^2}, \tag{8.40}$$

where the additional term $\hbar^2 l(l+1)/2\mu r^2$ is due to the angular momentum. This term is called sometimes the *centrifugal energy* [4]. Thus, the study of motion in a field with spherical symmetry can be reduced to the problem of motion in one dimension restricted to positive values of r. Accordingly, the normalization condition for $R(r)$ takes the form

$$\int_0^\infty |R(r)|^2 r^2 \, dr = \int_0^\infty |G(r)|^2 \, dr = 1. \tag{8.41}$$

8.6
Superposition of Degenerate States

We found (Equation 8.17b) that a stationary state is the eigenstate of the Hamiltonian corresponding to an eigenvalue E. Here E may belong to a discrete spectrum ($E = E_j$ with integer $j \geq 0$) or to continuous spectrum (e.g., $E = E(k)$). The corresponding time-dependent state function is described by (8.19).

As was emphasized in Section 8.3, an arbitrary state (a *general* solution of the time-dependent Schrödinger equation) can be expressed as a superposition (8.20) of eigenstates (8.19). The important point here is that each complete eigenstate (8.19), being the function of position *and* time, is itself the solution of the *time-dependent* Schrödinger equation as well, but it is its *special* solution. The expectation value of any observable in the corresponding eigenstate is time independent. And yet the time-dependent exponential factors in (8.20) are absolutely necessary since they produce time-dependent interference terms that determine the evolution of the general state $\Psi(\mathbf{r}, t)$. The important feature of such state is the possibility it affords for finding, in an appropriate measurement, any one of the allowed values E for which $\Phi(E) \neq 0$. This means that the energy in state (8.20) is indeterminate.

But we may have a *restricted superposition* in which *all* different stationary states belong to *the same E* (the degenerate states). The resulting state has definite energy and remains stationary. The simple example is the relationship (8.25) for a free particle, admitting an infinite set of different momentum states with the same energy. Here we consider it in more detail.

In the 1D case, the same energy corresponds to momentum states $|p\rangle$ and $|-p\rangle$ or, in the x-representation, $|p\rangle \to e^{i(kx-\omega t)}$ and $|-p\rangle \to e^{-i(kx+\omega t)}$. The spatial factors here are the corresponding eigenfunctions of $\hat{p}_x$, and at the same time each of them is the eigenfunction of the free-particle Hamiltonian with the same E. Both waves may coexist in a superposition

$$\Psi_E(x,t) = \Phi_1 e^{i(kx-\omega t)} + \Phi_2 e^{-i(kx+\omega t)} = \left(\Phi_1 e^{ikx} + \Phi_2 e^{-ikx}\right) e^{-i\omega t}. \quad (8.42)$$

This expression, in which the eigenfunction $e^{-i\omega t}$ of $\hat{E}$ can be factored out, describes a stationary state with energy $E = \hbar\omega$. But its momentum is indefinite. Depending on the amplitudes Φ_1 and Φ_2, it may have different expectation values $\langle p \rangle$ corresponding to different flux densities j (Problem 8.8). In case $|\Phi_1| = |\Phi_2|$, we have $\langle p \rangle = 0$, $j = 0$. The energy does not flow anywhere in this state – we have a standing wave.

These results can be generalized onto two and three dimensions. Superposition of states with the same $|\mathbf{p}|$ will now involve a continuous set of waves propagating in various directions. Mathematically, it comprises the Fourier expansions of $\Psi_E(\mathbf{r},t)$ over monochromatic plane waves. In the case when the expansion coefficients $\Phi(\mathbf{k}) = $ const, we have an axially symmetric standing wave in 2D and a spherically symmetric standing wave in 3D. They are standing (the states are stationary) as long as the participating eigenstates are restricted to $|\mathbf{k}| = $ const.

Let us form a stationary state from various p-eigenstates in 3D. We can do it in the following way. Take a plane wave with $\mathbf{k} = \mathbf{p}/\hbar$, but write its amplitude as $\Phi(k\mathbf{n})$ instead of Φ_j. Here $\mathbf{n}$ is a unit vector with variable direction. Accordingly, $k\mathbf{n}$ is a variable propagation vector with a fixed magnitude. When we use the set of all corresponding waves, their amplitudes $\Phi_j \to \Phi(k\mathbf{n})$ form a continuous function of $\mathbf{n}$, and superposition (8.20) is an integral with $\omega = E/\hbar$ and E determined from (8.25). For k fixed, the time dependency factors out and the integral can be written as

$$\Psi_E(\mathbf{r},t) = e^{-i(E/\hbar)t} \int \Phi(k\mathbf{n}) e^{i\mathbf{k}\mathbf{r}}\, d\mathbf{k}. \quad (8.43)$$

This form explicitly shows that such state is, indeed, a stationary state with fixed energy.

The integral (8.43) must be taken over the sphere of fixed radius k, so it is a surface integral. It is encountered in the problem of elastic scattering, in which a plane monochromatic wave incident on a scattering center is partially converted into a system of diverging waves of the same frequency. Such a system can be described either as a continuous set of plane waves passing the center in various directions or, equivalently, as a diverging spherical wave with direction-dependent amplitude. We will deal with this problem in Chapter 19.

Now we will show that this set does indeed form one spherical wave. Let all constituent plain waves have *the same amplitude regardless of direction*, that is,

$\Phi(\mathbf{kn}) \to \Phi_0 = $ const. Then the corresponding system is spherically symmetric and can be described by coordinates k, θ, and φ (we have k instead of r because we are using the **p**-basis and are accordingly integrating in the *momentum space* rather than in coordinate space!). Take the direction of chosen **r** (due to spherical symmetry, it is immaterial which one!) as the z-axis, so that the variable θ will be the polar angle between **r** and **k**. With all these notations, (8.43) becomes a rather simple surface integral with the area element $d\mathbf{k} \to da = k^2 \sin\theta \, d\theta \, d\varphi$:

$$\Psi_E(\mathbf{r}, t) = \Phi_0 \, e^{-i(E/\hbar)t} \int e^{ikr\cos\theta} k^2 \sin\theta \, d\theta \, d\varphi. \tag{8.44}$$

The polar angle θ ranges between 0 and π. The integration is straightforward and yields

$$\Psi_E(\mathbf{r}, t) = -4\pi k \Phi_0 \frac{\sin kr}{r} e^{-i(E/\hbar)t}. \tag{8.45}$$

The result (8.45) is a typical expression for a spherical wave, whose unmistakable signature is the radial distance r in the denominator. An interesting feature of this result is that the wave here is neither diverging ($\sim \exp(ikr)$) nor converging ($\sim \exp(-ikr)$) but is an equally weighted superposition of the two (a spherical standing wave). This is a natural consequence of condition $\Phi(\mathbf{kn}) = $ const assigning to each **k** in (8.43) its counterpart $-\mathbf{k}$ with the same amplitude. So by making all amplitudes equal, we get a standing wave in each radial direction. No surprise that the sum of all standing plane waves with the same amplitude produces a standing spherical wave.

A superposition of states with *different* energies would correspond to the most general expression (8.20) for the state function. Applying this to our current problem, we would have to include the states with different magnitudes of k, and accordingly to extend the integration in (8.43) onto the radial k-dimension, but now the temporal factor depending on $E(k)$ must also be included in integration! Even if we keep restricting to only one dimension, the features of the resulting state may be so weird that we will devote to this case a few separate sections in Chapter 11.

Some of the results obtained here may hold for systems with $U(\mathbf{r}) \neq 0$. For instance, in the Coulomb field, the total energy

$$E = K + U = \frac{p_r^2}{2\mu} + \frac{L^2}{2\mu r^2} + U(r) \tag{8.46}$$

can be distributed in many different ways between K and U, and K, in turn, being the sum of energies of the radial and orbital motion relative to the center, can be distributed differently between these two motions. Therefore, there may exist many distinct physical states, all corresponding to the same net energy E. So we can expect degeneracy in this case as well.

Summary. Superposition of stationary states is also stationary if it includes only degenerate states with the same E. Superposition of stationary states is nonstationary if it includes the states with different E. The energy of the resulting state in this case is indeterminate.

8.7
Phase Velocity and Group Velocity

Here we consider the *general solution* of the time-dependent Schrödinger equation for a free particle. Let us start with the particle having definite momentum. Assume that the corresponding de Broglie wave propagates along the *x*-direction. This motion can be visualized as the graph of the cosine or sine function (being, respectively, the real or imaginary part of the wave) sliding along *x*. For each moment t_0, this graph can be considered as an instantaneous snapshot of the wave, and its shape at a later moment t will be obtained by parallel shifting of the graph. If at the moment t_0 we focus on a certain point x_0 of the graph, then at a later moment t this point, while retaining its position on the curve, will have another coordinate x. So the curve will slide the distance $x - x_0$ during the time interval $t - t_0$, and its speed is

$$u = \frac{x - x_0}{t - t_0}. \tag{8.47}$$

Since the chosen point presents a fixed mark on the curve, its coordinate x at moment t must correspond to the same phase ϕ as coordinate x_0 at moment t_0:

$$\phi = kx - \omega t = kx_0 - \omega t_0. \tag{8.48}$$

In other words, u is defined as velocity of a point on the wave for which the phase ϕ remains constant. Therefore, the wave velocity is frequently referred to as the phase velocity.

Combining Equations 8.47 and 8.48 yields the expression for phase velocity in terms of k and ω:

$$u = \frac{\omega}{k}. \tag{8.49}$$

We know only one case when the phase velocity is the same for waves of all frequencies: light waves in a vacuum.[1] These all propagate with the same speed c, which makes c a universal constant of nature. In the language of QM, all photons move through space with the universal speed regardless of their energy. But this harmony of all frequencies (or all energies) breaks down in a medium. The reason is that the speed of a light wave in a medium is a certain trade-off between its tendency to move with speed c and the medium's response to the EM field of the wave. This response is different for different frequencies, and as a result, the speed of light in a medium becomes frequency dependent. It seems natural to expect the same for de Broglie waves: for any particle other than a photon (or graviton) its phase velocity may be frequency dependent when in a medium, but frequency independent in vacuum.

Remarkably, the second part of this expectation is wrong! The reason is rather deep and comes from fundamental laws of relativity. According to these laws, the speed of light in vacuum is an invariant characteristic; that is, it stays the same in

---

1) The same must hold for gravitational waves (and their quanta – gravitons) as well, but the latter have not been directly observed at the time of writing this book.

any reference frame. If you board a spaceship moving in the same direction as a laser pulse, then no matter how fast the ship moves, the measured speed of the pulse relative to the ship remains c. This can be rephrased by saying that a photon cannot be brought to rest – it does not exist at rest. A nonexisting object cannot have any mass. Therefore, the rest mass of the photon is zero. This is a fundamental difference between a photon and other known particles that *can* be brought to rest and, accordingly, have a rest mass. And this results in a dramatic difference in propagation characteristics: for a particle with $\mu_0 \neq 0$, its phase velocity in vacuum is a function of μ_0. Indeed, combining (8.49) with de Broglie's postulate (2.9) yields the phase velocity in terms of particle's energy and momentum:

$$u = \frac{\omega}{k} = \frac{E}{p}. \tag{8.50}$$

In view of (8.28), we obtain

$$u = \frac{\sqrt{\mu_0^2 c^4 + p^2 c^2}}{p} = c\sqrt{1 + \frac{\mu_0^2 c^2}{p^2}}. \tag{8.51}$$

Using $p = \hbar k$, we can also express u in terms of k or the wavelength λ:

$$u = c\sqrt{1 + \frac{\mu_0^2 c^2}{\hbar^2 k^2}} = c\sqrt{1 + 4\pi^2 \frac{\lambda_c^2}{\lambda^2}}, \quad \lambda_c \equiv \frac{\hbar}{\mu_0 c}. \tag{8.52}$$

We see a few interesting things here.

First, through the rest mass μ_0, dependence on p and thereby on λ crawls in! This is how u becomes a function of λ, $u \to u(\lambda)$, and thereby, automatically, $u \to u(\omega)$ (Problem 8.10).

Second, we see that regardless of λ, the particle's phase velocity $u \geq c$! It even becomes infinite when $p \to 0$. In other words, de Broglie wave of a resting particle ($v = 0$) propagates with infinite speed! These results, however, do not constitute any contradiction with relativity since the phase velocity is *not* velocity of a signal [3].

Third, the property $u \geq c$ implies that the phase velocity of the particle's de Broglie wave is something quite different from particle's actual velocity, which is always $v < c$. More accurately, u and v must be inversely related since, as seen from (8.52), u increases when v decreases.

And finally, the expression (8.51) or (8.52) approaches c for any wavelength when $\mu_0 \to 0$, so any massless particle must have the same universal speed c in vacuum regardless of its energy.

Let us now find the direct relationship between u and v. To this end, we put in (8.51) $p = \mu v$ and then take account of the fact that coefficient μ (*relativistic mass*) multiplying v for a moving particle is related to μ_0 by

$$\mu = \frac{\mu_0}{\sqrt{1 - \beta^2}}, \quad \text{so that} \quad p = \frac{\mu_0 v}{\sqrt{1 - \beta^2}}, \quad \beta \equiv \frac{v}{c}. \tag{8.53}$$

Combining this with (8.51) gives after simple algebra the universal relation between u and v, which is independent of μ_0 and thus holds for *any* object:

$$u = \frac{c^2}{v}, \quad \text{or} \quad uv = c^2. \tag{8.54}$$

With all this information, we are better prepared to resume our discussion of time evolution of quantum state for a free particle.

As far as the particle is in a state with *definite* momentum (and thereby definite energy and frequency), its wave function is just the de Broglie wave. According to our definition of evolution, such a wave propagates but does not evolve, which seems quite natural in the absence of external forces. But according to QM, a particle can be in a state with *undefined* momentum, that is, in a superposition of eigenstates with different momenta and hence different phase velocities. Such superposition (the wave packet) produces a shape that can, apart from propagating along x, also change with time. This reason for the change can be visualized if we imagine a few sinusoids with different wavelengths and consider their sum. This sum will be represented by a graph (called the envelope) with a certain shape. If all the individual sinusoids have the same u, then their envelope will follow the motion of all its constituents without changing its shape, like a group of cars moving with the same speed along a highway: the size of the group, as well as the distances between the individual cars, remains fixed. But if the phase velocities of individual monochromatic waves are different, they will slide with respect to one another. In this case, the resulting shape will change. Imagine again the group of cars, this time with different speeds. This comparison, while being very crude, conveys the basic feature: if the individual speeds of the cars are different, the shape of the group will change; for instance, if initially all the cars were close to each other (the group was short), at a later moment they will be farther apart (the group spreads out along the way). The same happens with the envelope of different sinusoids if $u = u(\lambda)$. If initially the envelope was very narrow, resembling a spike, then in most cases it will eventually spread along the direction of motion. This phenomenon is called the *spread of a wave packet*. The state undergoes evolution.

Another way to see the origin of such evolution is to use analytic approach. In a state with indeterminate momentum, we have the superposition

$$\Psi(x, t) = \int \Phi(k) e^{i(kx - \omega t)} \, dk. \tag{8.55}$$

The coefficients $\Phi(k)$ describe the same state in the p-basis. Since the momentum of the particle is uniquely related to its energy, the continuous set $\Phi(k)$ forms the energy (and thereby the frequency) spectrum of the particle in state $\Psi(x, t)$. Note that the probabilities $|\Phi(k)|^2$ (with or without the temporal factors $e^{-i\omega(k)t}$) are time independent. Therefore, the *spectrum* of the state is a fixed characteristic – it does not evolve. Nevertheless, if we switch from p- to x-representation, we will see a different picture. To see the difference, use (8.50) and write the shape (8.55) as

$$\Psi(x,t) = \int \Phi(k) e^{ik(x-ut)}\, dk. \tag{8.56}$$

Consider first the case with $u = \text{const}$. Then, regardless of the form of $\Phi(k)$, the shape of $\Psi(x,t)$ will be described by

$$\Psi(x,t) = \Psi(x - ut), \tag{8.57}$$

and $\Psi(x,t)$ depends only on combination $x - ut$. Such dependence describes propagation of a waveform with velocity u without any distortion. The state does not evolve. Such a situation is characteristic for the photons. A laser pulse in vacuum retains its initial shape.

Consider now the general case, when $u = u(k)$. Then the integral (8.56) cannot produce the result (8.57). The state $\Psi(x,t)$ will depend on x and t separately rather than in combination $(x - ut)$. Accordingly, the resulting shape, apart from moving in space, will also change in time. It will evolve. In this (most general!) case, we have a rather unusual situation: the same state is "frozen" in the momentum space but evolves in the coordinate space.

The reason for this becomes almost self-evident if we rewrite (8.55) in the form

$$\Psi(x,t) = \int \left\{ \Phi(k) e^{-i\omega(k)t} \right\} e^{ikx}\, dk. \tag{8.58}$$

If $\omega(k) = uk$ with $u = \text{const}$, we recover the result (8.57) for a group with a fixed shape, moving with the speed u. But we know that this can happen only for the photons or gravitons. Generally, (8.58) does not sum to a wave with a fixed shape; the corresponding packet will be both moving *and* evolving.

Now we can discuss both aspects – motion of the packet and its reshaping – in more detail. We start with the motion aspect. Since the packet changes its form, defining the velocity of its motion is not trivial. Let us take as a guide the following argument. Suppose we observe a moving cloud that changes its shape and mass distribution. We can define the velocity of the whole system as the velocity of its center of mass. In many cases (but *not* always!), the center of mass is close to the densest part of the cloud. Then we can take the velocity of this part as the velocity of the whole cloud. Since in this analogy the value $|\Psi(x,t)|^2$ plays the role of the mass density, we can *define* the velocity of the packet as the velocity of maximum of function $|\Psi(x,t)|^2$.[2] The velocity of the packet thus defined is called the *group velocity*. We denote it as v_g.

---

[2] The general situation may be more subtle. As mentioned in the text, the *center of mass* is not necessarily coincident with location of *maximal density*. For instance, the center of mass of a bagel or a doughnut is at its geometric center, which is an empty space.

8 The Schrödinger Equation

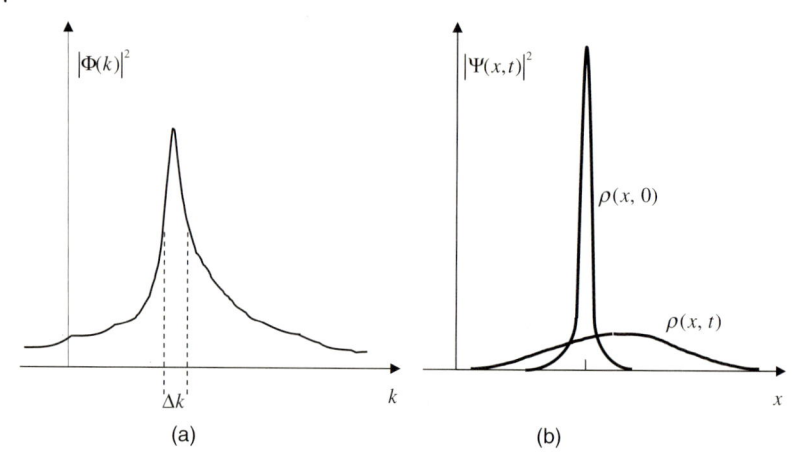

Figure 8.1 Spectrum and spread of a wave packet. (a) A possible shape of a wave spectrum in the momentum space. This is not to be confused with the shape (waveform) of the packet in the configuration space shown in (b). (b) An instantaneous probability distribution $\rho(x, t)$ of a wave packet at the initial moment $\rho(x, 0)$ and at a later moment $\rho(x, t)$. The packet generally moves as a whole and changes its shape. But its spectrum shown in (a) *does not change*.

Since we found that the *phase velocity* of the de Broglie wave of the particle is something totally different from the particle's actual velocity v, we turn to a possibility that maybe the *group* velocity will represent v. In other words, we expect that for a particle in vacuum, $v_g = v$.

Let us consider a group of waves whose spectrum $\mathcal{P}(k) = |\Phi(k)|^2$ peaks sharply within a narrow region around a certain value $k = k_0$ and falls off quickly outside this region. Such a group can be characterized by an effective spectral half-width Δk – the distance from k_0 at which the value of $\mathcal{P}(k)$ is one-half of the maximum $\mathcal{P}(k_0)$ (Figure 8.1a). We assume that only the wave amplitudes within the region

$$k_0 - \Delta k \leq k \leq k_0 + \Delta k \tag{8.59}$$

contribute to the packet and neglect contributions from the waves outside this region. Then (8.10) can be written as

$$\Psi(x, t) = \int_{k_0-\Delta k}^{k_0+\Delta k} \Phi(k) e^{i(kx-\omega t)}\, dk. \tag{8.60}$$

Here ω is the function of k determined by Equation 8.29. For a narrow region (8.59), we can represent $\omega(k)$ as

$$\omega(k) \approx \omega(k_0) + \left.\frac{d\omega}{dk}\right|_{k_0}(k - k_0). \tag{8.61}$$

Denote $k - k_0 = \xi$ and take ξ as a new variable. Assume that $\Phi(k)$ is a slowly varying function of k within the region (8.59), so that we can take there $\Phi(k) \cong \Phi(k_0)$. Then we obtain

$$\Psi(x,t) = \Phi(k_0) e^{i(k_0 x - \omega_0 t)} \int_{-\Delta k}^{\Delta k} e^{i\left(x - (d\omega/dk)|_{k_0} t\right)\xi} \, d\xi. \tag{8.62}$$

This is easily integrated and gives

$$\Psi(x,t) = \Psi_0(x,t) e^{i(k_0 x - \omega_0 t)}, \tag{8.63}$$

where

$$\Psi_0(x,t) \equiv 2\Phi(k_0) \frac{\sin\left\{\left(x - (d\omega/dk)|_{k_0} t\right) \Delta k\right\}}{\left(x - (d\omega/dk)|_{k_0} t\right)}. \tag{8.64}$$

When Δk is sufficiently small, $\Psi_0(x,t)$ is a slowly varying function of x and t. Therefore, it can be regarded as slowly varying amplitude (and was denoted as such in (8.63)) of the wave $\exp(k_0 x - \omega_0 t)$. Now, by the above definition, in order to find the group velocity, we only need to find the point x at which the amplitude $\Psi_0(x,t)$ is maximal. This point can be considered as the center of the packet, and its velocity will be the group velocity. It is easy to see that (8.64) is maximal at

$$x - \frac{d\omega}{dk}\bigg|_{k_0} t = 0 \quad \text{or} \quad x = \frac{d\omega}{dk}\bigg|_{k_0} t. \tag{8.65}$$

It follows that the group velocity is

$$v_g = \frac{x}{t} = \frac{d\omega}{dk}\bigg|_{k_0} \quad \text{or} \quad v_g = \frac{dE}{dp}\bigg|_{p_0}. \tag{8.66}$$

Note that it is different from the phase velocity (8.50), which is just the ratio of ω and k, or E and p. Next, we go to (8.28) and find

$$\frac{dE}{dp} = \frac{p}{E} c^2. \tag{8.67}$$

But, according to the most famous equation in physics, $E/c^2 = \mu$, where μ is the relativistic mass of the particle, determined by Equation 8.53. Therefore, dropping the index "0" in p_0, we finally get

$$v_g = \frac{p}{\mu} = v. \tag{8.68}$$

This is precisely what we had expected!

Thus, it can be shown that the group velocity of a wave packet *in vacuum* can indeed be equal to the actual velocity of a particle represented by this group. This fact was a strong additional argument in favor of the de Broglie waves.

On the other hand, if we turn to the second aspect – the evolution of a wave packet – it appears to contradict the de Broglie hypothesis. First of all, a real group may have a broad k-spectrum, for which the concept of group velocity is poorly defined [5–7]. We cannot talk of definite velocity of a particle with wildly indeterminate momentum. Further, even a spectrally narrow group of the de Broglie waves, as mentioned above, changes its shape, and eventually it spreads over the whole coordinate space (Figure 8.1b). As we will see later, the character and rate of this change dramatically depend on the initial shape of the packet, since this shape determines the spectrum according to (8.55). In addition, the wave packet can under certain conditions even split into two or more parts due to splitting of its monochromatic constituents. The classical particles do not behave this way. Therefore, all attempts to show that a particle really *is* an appropriately constructed group of the matter waves failed. The de Broglie postulate was firmly established only after Born came up with his probabilistic interpretation, according to which the wave function is, by itself, not an observable characteristic.

8.8
de Broglie's Waves Revised

An attentive reader of the previous section may have noticed what appears to be an inconsistency in the mathematical structure of QM. We can spot the contradiction between the correct relativistic description of the phase velocity and its non-relativistic limit.

Indeed, the relativistic relation (8.54) between the physical velocity v of a particle and the phase velocity u of its de Broglie wave must hold for *any* speed, since relativity reigns over all the domain $0 \leq v \leq c$ (as well as for all speeds beyond, $u > c$; see, for example, Ref. [3]). Let us check whether the expression (8.54) for the phase velocity is consistent with the *relativistic* (or Lorentz) transformations; that is, it gives the same result in all inertial reference frames. To this end, consider another reference frame K′ moving parallel to **k** with a speed V relative to initial frame K. According to relativistic rules, in this frame the particle's velocity and the phase velocity of its de Broglie wave are given by [3]

$$v' = \frac{v + V}{1 + vV/c^2}, \quad u' = \frac{u + V}{1 + uV/c^2}. \tag{8.69}$$

Their product is

$$u'v' = \frac{uv + (u + v)V + V^2}{1 + (u + V)V/c^2 + uvV^2/c^4}. \tag{8.70}$$

Setting here $uv = c^2$, which holds in frame K, immediately gives $u'v' = c^2 = uv$. Thus, the relation (8.54) is Lorentz invariant.

But on the other hand, for slow motions $v \ll c$, Newtonian mechanics is the extremely accurate approximation. We found that the definition of group velocity of a narrow-spectrum wave packet gives the *actual velocity* of the corresponding particle in both relativistic and nonrelativistic cases. It seems therefore that the same approach to u and v in the nonrelativistic domain must give for them the result very close to the correct formula (8.54). For the nonrelativistic case $v \ll c$, the formula predicts that $u \to \infty$. This by itself is no concern, since the Newtonian mechanics does not impose any restrictions on velocities (formally, it follows from SR if we set $c \to \infty$), and in any event, velocity u cannot be utilized for signaling [3].

Let us check it. Take the same universal equation (8.50), but now put into it the *nonrelativistic* expression (8.4) for the energy:

$$u = \frac{\omega}{k} = \frac{E}{p} = \frac{1}{2}v. \tag{8.71}$$

This is totally different from what we had expected! Instead of being overwhelmingly greater than v, the phase velocity calculated by the rules of nonrelativistic QM is only $(1/2)v$.

But the situation is even worse than this. Formula (8.71) not only contradicts (8.54), but is not consistent even with Galilean transformations. According to the latter, in frame K' the speeds u and v are

$$u' = u + V, \qquad v' = v + V. \tag{8.72}$$

In view of (8.71), this gives for u':

$$u' = \frac{1}{2}v + V. \tag{8.73}$$

But on the other hand, because of the equivalence of all inertial reference frames, the phase velocity in frame K' must be

$$u' = \frac{\omega'}{k'} = \frac{E'}{p'} = \frac{1}{2}v', \tag{8.74}$$

or, in view of (8.72),

$$u' = \frac{1}{2}(v + V). \tag{8.75}$$

This differs from (8.73). The nonrelativistic calculations of u by two different methods give different results.

We face a chilling revelation: the de Broglie relationship for a free particle, while being totally consistent with the Lorentz transformations, is not consistent with Galilean transformations when we use the nonrelativistic expressions for the energy and momentum. Therefore, there should be no such thing as nonrelativistic QM!

Two questions arise: (1) Is the discovered inconsistency the manifestation of a fundamental flaw in the whole mathematical structure of QM? And if so, then (2) why is there a self-consistent nonrelativistic theory of QM?

The answer to the first question is no. Part of the answer to the second question is that the phase velocity of the de Broglie wave is not a directly observable characteristic since we cannot directly measure the de Broglie wave itself. In an actual experiment, we can observe the *group* velocity. And its definition (8.66) works equally well and gives the same result regardless of whether we use a relativistic or a nonrelativistic expression for E. Indeed, we found at the end of the previous section that if, instead of $E = \sqrt{\mu_0^2 c^4 + p^2 c^2}$, we put into (8.66) $E = p^2/2\mu_0$, we will get the same result $v_g = v$. And the physical velocity v is what we can actually measure, for instance, by measuring momentum and mass of the particle, or by measuring the average time between its emission from the source and arrival to a detector a known distance away. If the theory provides consistent relationships between the observable characteristics, it is consistent.

The second part of the answer comes from examining the origin of the nonrelativistic expression $E = p^2/2\mu_0$. As already mentioned, this expression leaves off the first term $E_0 = \mu_0 c^2$ in (8.29). This term describes the energy associated with the rest mass of the particle. In the nonrelativistic domain, the second term is much smaller than the first. Why do we discard the bigger term? The answer: because it is a constant, and we know that energy in the nonrelativistic theory can be defined up to an additive constant. We routinely use this, for instance, in the case of gravitational potential energy. Even though the rest mass stores the huge amount of energy associated with it, this amount is not manifest in any transformations or processes of interest in Newtonian mechanics. Why carry this useless baggage? So, it is disregarded and what is left (and called "energy" in Newtonian mechanics) is actually the truncated energy – a varying but small fraction of its full amount. Therefore, in many cases the transition from SR to Newtonian mechanics uses this truncation, although it is not always explicitly stated.

The truncation of E and ω leads unavoidably to truncation of u. Indeed, we have

$$u = \frac{E}{p} = \frac{\sqrt{\mu_0^2 c^4 + p^2 c^2}}{p} = \frac{\mu_0 c^2}{p} + \frac{p}{2\mu_0} + \cdots. \tag{8.76}$$

We see immediately that the truncation is okay for observed quantities such as E or v (which is described by the *group velocity* (8.67)) since the discarded part of E is constant and v is related only to the remaining part – the kinetic energy (formally, we just have a constant term discarded when taking the derivative in (8.66)). The same can be said about ω if it is defined as $E/\hbar$ rather than considered an independent variable. If we accept such a definition, then ω in the de Broglie wave becomes merely a substitute for E, and measuring the latter would be also measuring the former.

But in case of the *phase velocity*, the truncation leads to the above-mentioned inconsistency, and Equation 8.76 clearly shows the reason: in this case, the discarded part (the first term on the right) is not a constant – it depends on p. So strictly speaking, as far as the phase velocity is concerned, the truncation is an illegal procedure. And yet nonrelativistic QM is a vibrant theory providing an accurate description of the huge amount of phenomena and possessing an astounding predictive power! The reason for this is the probabilistic nature of the wave function. Since we cannot measure $\Psi(x,t)$ itself, the truncated terms drop from the equations describing measurable characteristics, and they do it in three different ways. First, expression (8.66) for v_g is insensitive to truncation because it contains dE rather than E itself, and the differential of a function, as was just mentioned, is insensitive to any constant added to or subtracted from it. Second, in cases when the subtracted part is a variable as in (8.76), the corresponding physical characteristic like phase velocity is not a physical observable. Third, in cases when a particle is in a superposition of states with different k and thereby *different E* (wave packet), the truncated part is the *same* for all energy states. Indeed, take the general expression (8.55) for a wave packet and put there the Taylor expansion (8.29) for ω. Retaining the first two terms, we get

$$\Psi(x,t) = \int \Phi(k) e^{i\left(kx - (\mu_0 c^2/\hbar)t - (\hbar k^2/2\mu_0)t\right)} dk = e^{-i(E_0/\hbar)t} \int \Phi(k) e^{i\left(kx - (\hbar k^2/2\mu_0)t\right)} dk. \tag{8.77}$$

Here the term associated with the rest energy $E_0 = \mu_0 c^2$ is the same for all Fourier components and is factored out. Truncation of E will just remove the exponential coefficient before the integral in (8.77). But this is of no consequence, since $\Psi(x,t)$ is defined only up to an arbitrary phase factor $e^{i\alpha}$ with real α, so $\Psi(x,t)$ and $e^{i\alpha}\Psi(x,t)$ describe the same state.

Note that this happy outcome is only due to the fact that the discarded part of the energy is the same for all possible energy states since it is the rest energy of the same particle. But what if we have a system of two or more different particles?

Consider the case of two particles with different rest masses, for example, a system of He and C nuclei discussed in Section 4.5. Denote the rest energies of He and C as E_0^{He} and E_0^{C}, respectively. Then the individual wave functions of He and C can be written as

$$\tilde{\Psi}_{He}(\mathbf{x},t) = e^{-i(E_0^{He}/\hbar)t}\Psi_{He}(\mathbf{r},t) \quad \text{and} \quad \tilde{\Psi}_C(\mathbf{x},t) = e^{-i(E_0^C/\hbar)t}\Psi_C(\mathbf{r},t). \tag{8.78}$$

Here $\Psi_{He}(\mathbf{r},t)$ can be a wave packet described by an integral similar to (8.77), and the same for $\Psi_C(\mathbf{r},t)$. But then the finding, say, He at $\mathbf{r}$ and C at $\mathbf{r}'$ is a *combined* event $(\mathbf{r},\mathbf{r}')$ (i.e., a pair of distinct events – He at $\mathbf{r}$ and C at $\mathbf{r}'$) and its probability amplitude is the *product* of the individual amplitudes:

$$\begin{aligned}\tilde{\Psi}_{He,C}(\mathbf{r},\mathbf{r}';t) &= e^{-i(E_0^{He}/\hbar)t}e^{-i(E_0^C/\hbar)t}\Psi_{He}(\mathbf{r},t)\Psi_C(\mathbf{r}',t) \\ &= e^{-i(E/\hbar)t}\Psi_{He}(\mathbf{r},t)\Psi_C(\mathbf{r}',t).\end{aligned} \tag{8.79}$$

Here $E_0 \equiv E_0^{He} + E_0^C$ is the sum of the rest masses of the two nuclei (it is generally *not* the same as the rest mass of the whole system [3]!). The probability amplitude of an "inverse" combined event $(\mathbf{r}', \mathbf{r})$ of finding He at $\mathbf{r}'$ and C at $\mathbf{r}$ is obtained from (8.79) by swapping primed and unprimed variables:

$$\tilde{\Psi}_{He,C}(\mathbf{r}', \mathbf{r}; t) = e^{-i(E_0/\hbar)t} \Psi_{He}(\mathbf{r}', t) \Psi_C(\mathbf{r}, t). \tag{8.80}$$

The probability of finding one particle at $\mathbf{r}$ and the other at $\mathbf{r}'$ without specifying which particle is found where is the sum of the probabilities found from (8.79) and (8.80:)

$$\mathcal{P}_{He,C}(\mathbf{r}', \mathbf{r}; t) = |\Psi_{He}(\mathbf{r}, t)\Psi_C(\mathbf{r}', t)|^2 + |\Psi_{He}(\mathbf{r}', t)\Psi_C(\mathbf{r}, t)|^2. \tag{8.81}$$

The rest energy of the system drops from the result.

Thus, if we use only the kinetic energies of the two nuclei in the nonrelativistic approximation, this does not lead to any changes in either predicted experimental outcomes or their theoretical description. The result, again, turns out to be independent of the "discarded" part of energy.

Finally, consider the case of two *indistinguishable* particles, say, two He in the same state. Then there is an *interference of the amplitudes* of the type (8.79) or (8.80), but now the subscripts (say, 1 and 2) must only label the coordinate state of the respective particle:

$$\tilde{\Psi}(\mathbf{r}', \mathbf{r}; t) = e^{-i(E_0/\hbar)t} (\Psi_1(\mathbf{r}', t)\Psi_2(\mathbf{r}, t) \pm \Psi_1(\mathbf{r}, t)\Psi_2(\mathbf{r}', t)). \tag{8.82}$$

Here "+" stands for the bosons and "−" for the fermions. But again, the term associated with the discarded part of E factors out and will drop from the observable outcomes. This is consistent with the fact that the time-dependent wave function (or the probability amplitude $\Psi(\mathbf{r}, t)$, for that matter) is, at least in the nonrelativistic domain, a complex number.

8.9
The Schrödinger Equation in an Arbitrary Basis

Up to now, we have used the Schrödinger equation in position representation, that is, both the state function and Hamiltonian were represented as functions of $\mathbf{r}$. Here we will rewrite the equation in a more general form, which does not specify the basis.

Start with the familiar form

$$i\hbar \frac{\partial \Psi(\mathbf{r}, t)}{\partial t} = \hat{H}(\mathbf{r}, t) \Psi(\mathbf{r}, t). \tag{8.83}$$

Let us write the state $\Psi(\mathbf{r}, t)$ in some arbitrary basis, forgetting about representations. Assume that the basis is a discrete set of eigenstates $|\psi_n\rangle$ of some Hermitian operator $\hat{Q}(\mathbf{r})$. Then $\Psi(\mathbf{r},t) \to |\Psi\rangle = \sum_n c_n(t)|\psi_n\rangle$. The set of amplitudes $c_n(t)$ describes state $\Psi(\mathbf{r},t)$ in the $\hat{Q}$-basis. Now put it into (8.83)

$$i\hbar \sum_n \frac{\partial c_n(t)}{\partial t} |\psi_n\rangle = \sum_n c_n(t) \hat{H} |\psi_n\rangle \tag{8.84}$$

and project the result onto $|\psi_m\rangle$:

$$i\hbar \sum_n \frac{\partial c_n(t)}{\partial t} \langle \psi_m|\psi_n\rangle = \sum_n c_n(t) \langle \psi_m|\hat{H}|\psi_n\rangle. \tag{8.85}$$

We can use the shorthand $|n\rangle$ for $|\psi_n\rangle$, and rewrite (8.85) as

$$i\hbar \sum_n \frac{\partial c_n(t)}{\partial t} \langle m|n\rangle = \sum_n \langle m|\hat{H}|n\rangle c_n(t). \tag{8.86}$$

Here we recognize in $\langle m|\hat{H}|n\rangle$ the matrix element H_{mn} of the Hamiltonian in the $|n\rangle$-basis. Recalling that $\langle m|n\rangle = \delta_{mn}$ we have

$$i\hbar \frac{\partial c_m(t)}{\partial t} = \sum_n H_{mn} c_n(t), \quad m = 1, 2, 3, \ldots. \tag{8.87}$$

This is the Schrödinger equation in an arbitrary basis $|n\rangle$. Actually, it is a system of equations for the amplitudes $c_n(t)$. In some sources, this system is called the *Hamiltonian equations*. Altogether we have a system of simultaneous linear differential equations of the first order, in which all amplitudes c_n are coupled with each other. The rate of change of each amplitude c_m stands in some proportion not only to c_m, but also to all other amplitudes c_n. The proportionality coefficients are the corresponding matrix elements of the Hamiltonian.

We can write the Hamiltonian system for a *continuum* of basis states as well. Denote them as $|\nu\rangle$, where ν is a continuous parameter, and the amplitudes $c_n(t)$ will become $c_\nu(t)$. Then, instead of the sum $|\Psi\rangle = \sum_n c_n(t)|\psi_n\rangle$ we will have the integral $|\Psi\rangle = \int c_\nu(t)|\nu\rangle d\nu$.

The same procedure as before, using the projection operation and the rule $\langle \nu|\nu'\rangle = \delta(\nu - \nu')$, will give us the equations for $|c_\nu(t)\rangle$:

$$i\hbar \frac{\partial c_\nu(t)}{\partial t} = \int H_{\nu\nu'} c_{\nu'}(t) d\nu'. \tag{8.88}$$

Here $H_{\nu\nu'} = \langle \nu|\hat{H}|\nu'\rangle$ is a continuous matrix, which cannot be represented in a table form.

The formal solution of Equation 8.87 can be obtained by known algorithm of linear algebra (see Supplement to Chapter 8 (Section 8.9). Here we will restrict to the special case of solution in the *E*-basis.

Box 8.2 Solving system (8.87) in the energy basis

We can illustrate the above analysis and simplify our system by choosing a special (and most natural) basis: the set of eigenstates of the Hamiltonian itself. We will not write any transformations at this stage but just assume from the very beginning that operator $\hat{Q}$ is the Hamiltonian, so that all $|n\rangle$ are the eigenstates of $\hat{H}$. According to this assumption, we now have

$$\hat{H}|n\rangle = E_n|n\rangle, \tag{8.89}$$

where E_n is the corresponding energy eigenvalue. In this case, the matrix H_{mn} is diagonalized (it is now in its own basis), so that

$$H_{mn} = \langle m|\hat{H}|n\rangle = E_n\langle m|n\rangle = E_n\delta_{mn}. \tag{8.90}$$

The system (8.87) simplifies to

$$i\hbar \frac{\partial c_m(t)}{\partial t} = E_m c_m(t) \tag{8.91}$$

with immediate solution

$$c_m(t) = c_m(0) e^{-i(E_m/\hbar)t}. \tag{8.92}$$

The coefficients $c_m(0)$ must be determined, as usual, from the initial conditions.

Box 8.3 Relativistic wave equations

After a course in non-relativistic QM, a student may get the impression that *all* QM is essentially described by the Schrödinger equation. Nothing can be farther from the truth! When we wrote the Hamiltonian operator, we used the Newtonian formula $E = p^2/2\mu_0 + U$ for energy. Had we used instead the more accurate relativistic mass-energy relation $E^2 = p^2c^2 + \mu_0^2 c^4$ and replaced observables with their operators, we would arrive after elementary algebra at the following relativistic wave equation:

$$\Box^2 \Psi(\mathbf{r}, t) = \frac{\mu_0^2 c^2}{\hbar^2} \Psi(\mathbf{r}, t), \quad \Box^2 \equiv \nabla^2 - \frac{1}{c^2}\frac{\partial^2}{\partial t^2} \tag{8.93}$$

This is the *Klein-Gordon equation*. It can be used for *massless* (zero rest mass) and *massive* particles alike; in the latter case it reduces to the Schrödinger equation in the non-relativistic limit. Ironically, it was Schrödinger who derived (8.93) first in 1925, some months before O. Klein and W. Gordon, and it would certainly be named after him had he published it at the time. What caused Schrödinger to abandon this equation? When he applied it to the Hydrogen atom, he was hoping to get not only the correct Bohr energy levels (for that purpose, Equation 8.6 worked just as fine) but also something that (8.6) could not provide - the correct *fine structure* pattern. Because (8.93) included relativistic effects, that hope was well justified. But he was in for a disappointment: the solution that he got did not match the available data for Hydrogen. Realizing that he could not explain the fine structure, Schrödinger abandoned Equation (8.93) and adopted the cruder non-relativistic version, which at least did not promise more than it could deliver.

We now know what caused this failure. Equation (8.93) does not take into account the effects of spin and therefore cannot describe atomic electrons. It only works for *spinless* particles. Examples of those include composite spin-0 bosons: pions, kaons, and J/ψ mesons. The Higgs boson (spin-0), when/if finally discovered, should also obey the Klein-Gordon equation. One can also apply (8.93) when the spin exists but plays no role. One such example is the *free* electron; in that case, both Klein-Gordon and Schrödinger equations give the same solution $\Psi(\mathbf{r}, t) = e^{i(\mathbf{k}\cdot\mathbf{r} - \omega t)}$ for the spatial part of the wave function.

For massive spin-1/2 particles, one must instead use the equation obtained in 1928 by Dirac:

$$i\hbar\frac{\partial \Psi}{\partial t} = (\beta\mu_0 c^2 - i\hbar c\boldsymbol{\alpha}\cdot\nabla)\Psi, \qquad \alpha_j \equiv \begin{pmatrix} 0 & \sigma_j \\ \sigma_j & 0 \end{pmatrix}, \beta \equiv \begin{pmatrix} I & 0 \\ 0 & -I \end{pmatrix} \qquad (8.94)$$

The *Dirac equation* makes use of 4×4 matrices $\alpha_1, \alpha_2, \alpha_3$ and β. In the definition of these matrices on the right of (8.94), each entry stands for a 2×2 matrix: I is the identity matrix, 0 is the matrix composed of zeros, σ_j are three Pauli spin matrices with $j = 1, 2, 3$ corresponding to x, y, z. In this equation Ψ itself must also be regarded as a 4-component column matrix. Applying (8.94) to a Hydrogen atom, one can correctly describe the spin-orbit interaction and recover the results mentioned in Chapter 7: the fine structure splitting and the value 2 for the electron spin g-factor.

The Klein-Gordon and Dirac equations are the most "popular" ones, but there are others as well: the *Proca equation* for massive spin-1 particles, the *Rarita-Schwinger equation* for massive spin-3/2 particles, and the *Weyl equation* for massless spin-1/2 particles. The case of massless spin-1 particles is covered in any Electromagnetism course: they are described by the Maxwell equations. As for the massless spin-2 particles, they are described by *Einstein field equations* - these hypothetical particles are the gravitons, and the equations describing them form the General Theory of Relativity.

Most generally, there are two classes of correct relativistic equations: equations for bosons (integer spin), and equations for fermions (half-integer spin). The solutions of relativistic equations transform under the general Lorentz transformation (including spatial rotations) as tensors whose rank is equal to the spin quantum number. For instance, the solutions for spin-0 bosons transform as a zero rank tensor, that is, they are either scalar or *pseudo-scalar* functions. The solutions describing spin-1 particles (e.g. photons) transform as a first rank tensor, and accordingly they are vector functions. On the other hand, fermionic solutions transform as half-integer rank tensors. Such tensors are called *spinors*. The wave functions of spin-1/2 fermions (electrons, protons, neutrons, neutrinos, and their antiparticles) are spinors of rank 1/2, which transform as a tensor of rank 1/2. For this reason, they are sometimes even called "half-vectors". The solution Ψ of the Dirac equation is a *four-component Dirac spinor*.

Problems

8.1 Show that the rate of change of probability of finding a particle within a fixed volume V is equal to the probability flux through the surface enclosing the volume.

8.2 Find the probability flux density in the following stationary states:
  a) The de Broglie wave $\psi = \psi_0 e^{i\mathbf{k}\mathbf{r}}$.
  b) A superposition of the two oncoming waves $\psi = \psi_1 e^{i\mathbf{k}\mathbf{r}} + \psi_2 e^{-i\mathbf{k}\mathbf{r}}$ with the amplitudes ψ_1 and ψ_2.
  c) Consider the special case in (b) when $|\psi_1| = |\psi_2|$.
  d) The superposition of the two crossed waves $\psi = \psi_1 e^{ik_x x} + \psi_2 e^{ik_y y}$.

8.3 Find the probability flux density
  a) For each of two different stationary states $\Psi_j(x,t) = A_j \cos k_j x \, e^{i\omega_j t}$, $j = 1, 2$.
  b) For a superposition of these states: $\Psi(x,t) = (1/\sqrt{2})(\Psi_1(x,t) + \Psi_2(x,t))$.

8.4 Prove that the superposition (8.20) is a solution of the Schrödinger equation.

8.5 Show that the de Broglie wave is the common eigenfunction of three different operators – energy, momentum, and the Hamiltonian operator for a free space.

8.6 Find the expression for operator of radial component of linear momentum in the r-representation.

(*Hint*: The square of this operator must be equal, up to numerical coefficient, to expression (8.33).)

8.7 Derive Equation 8.39.

8.8 Find $\langle p \rangle$, probability density, and the probability flux density in the state (8.42).

8.9 For a free particle, derive the expression describing its *spherically symmetric* state with definite energy E.

8.10 Find the phase velocity of a free relativistic particle as a function of its energy.

References

1 Arfken, G.B. and Weber, H.J. (1995) *Mathematical Methods for Physicists*, Academic Press, New York.

2 Hildebrand, F.B. (1976) *Advanced Calculus for Applications*, Prentice-Hall, Englewood Cliffs, NJ, p. 161.

3 Fayngold, M. (2008) *Special Relativity and How It Works*, Wiley-VCH Verlag GmbH, Weinheim, pp. 485–487.

4 Landau, L. and Lifshits, E. (1965) *Quantum Mechanics: Non-Relativistic Theory*, 2nd edn, Pergamon Press, Oxford.

5 Brillouin, L. (1960) *Wave Propagation and Group Velocity*, Academic Press, New York, pp. 1–83, 113–137.

6 Milonni, P.W. (2005) Chapters 2 and 3, in *Fast Light, Slow Light, and Left-Handed Light*, Institute of Physics Publishing, Bristol, UK/Philadelphia, PA.

7 Gehring, G., Schweinsberg, A., Barsi, C., Kostinski, N., and Boyd, R. (2006) Observation of a backward pulse propagation through a medium with a negative group velocity. *Science*, **312** (5775), 895–897.

9
Applications to Simple Systems: One Dimension

The problems considered in this chapter are idealized models of real situations. But their solutions retain typical features of more realistic quantum systems. Therefore, they, being much simpler, give a more straightforward illustration of fundamental concepts, which otherwise may be obscured under the heaps of formulas.

9.1
A Quasi-Free Particle

The title of this section reflects a certain ambiguity of the case considered in it. In this case, the particle is absolutely free ($U(\mathbf{r}) = 0$) within a certain domain of space, but is absolutely forbidden to move beyond this domain. Physically, this means that there is an infinite force on the particle at the boundary. This force is directed to the interior of the region and is such that an infinite work is required for the particle to escape. Accordingly, the particle's potential energy outside the region would be infinite. It is therefore impossible for the particle to cross the boundary, and it is doomed to remain forever trapped inside. But once within, it can travel without any restrictions.

A typical example is a photon within a cavity with reflecting walls (*optical resonator*). In most simple case of one dimension, this would be the *Fabry–Pérot interferometer* [1,2]. We will consider the corresponding 1D problem.

Let the motion be restricted to a region of size a. Taking the origin at the left edge of this region, we can write the potential as

$$U(x) = \begin{cases} \infty, & x < 0, \\ 0, & 0 < x < a, \\ \infty, & x > a. \end{cases} \quad (9.1)$$

The potential described by this expression is called the *potential box*. The problem now is to find the eigenstates for the corresponding Hamiltonian.

In this problem, the potential (9.1) imposes the boundary conditions that do not exist for a truly free particle. In view of (9.1), it is certain that the trapped particle will never be found on the outside. Therefore, the state function must satisfy the condition $\Psi(x) = 0$ outside the box:

$$\Psi(x) = 0 \quad \text{for} \quad x \leq 0 \quad \text{and} \quad \text{for} \quad x \geq a. \tag{9.2}$$

Condition (9.2) is required by the Schrödinger equation to eliminate the otherwise infinite term $U\psi$ from outside the box. We included the boundary in the ban as well, because Ψ is required to be continuous, and thus it cannot immediately jump from the zero when outside to nonzero on the boundary. The requirements (9.2) constitute the boundary conditions for this problem. They produce severe restrictions on the possible energies of the particle, allowing only a discrete set of them (discrete spectrum).

We need to find a nonzero solution of the Schrödinger equation, under conditions (9.2). Separating the variables as in Section 8.3, we have Equations 8.17 for the temporal and spatial parts, respectively. The temporal factor is given by (8.18). For the spatial part, we have inside the box

$$-\frac{\hbar^2}{2\mu}\frac{d^2}{dx^2}\psi(x) = E\psi(x). \tag{9.3}$$

The most general solution of (9.3) is a superposition

$$\psi(x) = C_1 e^{ikx} + C_2 e^{-ikx}, \quad 0 \leq x \leq a, \tag{9.4}$$

with the wave number $k = \sqrt{2\mu E}/\hbar$ (Equation 8.42) at $t = 0$. Now, apply the boundary conditions (9.2). The first of them requires

$$\psi(0) = C_1 + C_2 = 0, \quad \text{that is,} \quad C_1 = -C_2 = C. \tag{9.5}$$

It follows

$$\psi(x) = C\left(e^{ikx} - e^{-ikx}\right) = \tilde{C}\sin kx, \quad \tilde{C} = 2iC. \tag{9.6}$$

The second condition

$$\psi(a) = \tilde{C}\sin ak = 0 \tag{9.7}$$

selects out of all k a discrete set k_n for which it can be satisfied:

$$k \to k_n = n\frac{\pi}{a}, \quad n = 1, 2, \ldots. \tag{9.8}$$

The value $n = 0$ is excluded since it would give the trivial solution $\psi(x) = 0$. Thinking in terms of the de Broglie waves, condition (9.8) selects only those for which the integral number of half-wavelengths fit into the box. The corresponding energy eigenvalues are

$$E_n = \frac{\hbar^2 k_n^2}{2\mu} = \frac{\pi^2 \hbar^2}{2\mu a^2} n^2, \quad n = 1, 2, 3, \ldots. \tag{9.9}$$

All eigenvalues are positive – the energy is measured from the bottom of the box. But all possible states are bound, and accordingly, the spectrum is discrete.

Now it only remains to normalize the solution (9.7). Using (3.5) with the state function (9.6) within the box of size a, one can find (Problem 9.1) that $|\tilde{C}| = \sqrt{2/a}$. And since any additional phase factor $e^{i\alpha}$ with a real α is immaterial, we can set $\alpha = 0$. The final result for the whole solution including the temporal factor is

$$\Psi_n(x,t) = \begin{cases} 0, & x < 0, \\ \sqrt{\dfrac{2}{a}} \sin(k_n x) e^{i(E_n/\hbar)t}, & 0 < x < a, \\ 0, & x > a, \end{cases} \quad (9.10)$$

with E_n determined by (9.9). The first three eigenfunctions are shown in Figure 9.1.

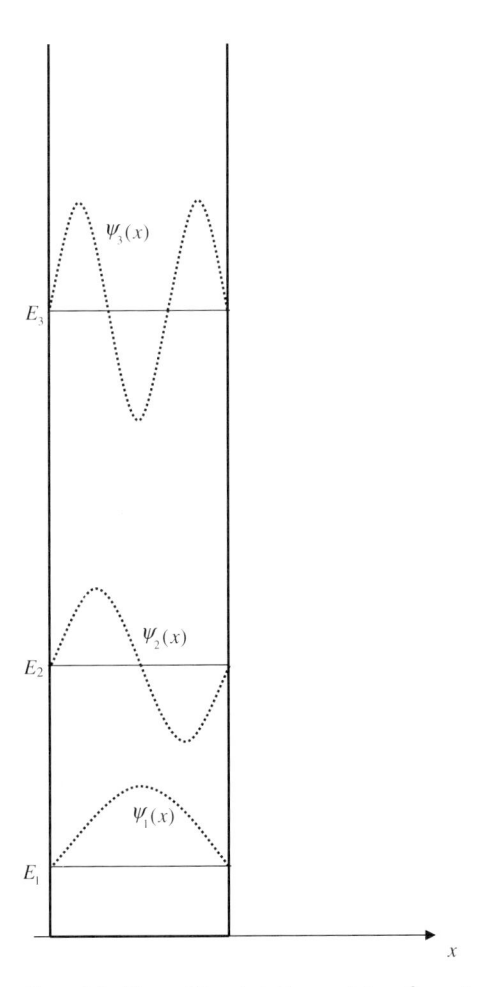

Figure 9.1 Three different stationary states of a particle in a box. $\psi_1(x)$: the ground state $E = E_1$ ($n = 1$); $\psi_2(x)$: the first excited state $E = E_2$ ($n = 2$); $\psi_3(x)$: the next excited state $E = E_3$ ($n = 3$).

As one can check by direct inspection, the parity of energy eigenstates $|n\rangle$ in a box is opposite to the parity of the labeling number n. (The parity is not immediately seen from the analytic expression because we chose the *edge* of the box as the origin.)

It is interesting to note some analogy between these solutions and the corresponding shapes for a vibrating string of length a with the fixed ends. But we should not overestimate its significance, since the quantities involved have different physical meaning. Expression (9.10) gives the *probability* amplitude of finding the particle at a point x in a state with fixed energy (9.9). The corresponding classical expression gives the *actual amplitude of vibration* of a point x on the string. The classical harmonics are equidistant on the frequency scale, but the box energy eigenvalues are not. And there are no vibrations in any of the eigenstates (9.10), whereas the classical string with definite nonzero energy is vibrating.

Now we can write the general solution for this problem as a superposition of eigenstates (9.10):

$$\Psi(x,t) = \sum_n C_n \Psi_n(x,t) = \sum_n C_n \psi_n(x) e^{i(E_n/\hbar)t}, \qquad (9.11)$$

with the set C_n satisfying the normalization condition, $\sum_n |C_n|^2 = 1$. The complete set of coefficients C_n can be considered as the set of adjustable parameters, and having an infinite number of them, we can represent *any* function within the box in terms of $\psi_n(x)$. One of the most weird functions known in physics is Dirac's δ-function. We can represent even it, if we manage to appropriately choose the expansion coefficients. To this end, we set in (9.11) $t=0$ and $\Psi(x,0) = \delta(x-x')$, with x' being a chosen point within the box. Then

$$\sum_n C_n \psi_n(x) = \delta(x-x'). \qquad (9.12)$$

Multiplying (9.12) through by $\psi_m^*(x)$, integrating over x, and using the fact that $\psi_n(x)$ form an orthonormal set, we obtain

$$C_m = \int \delta(x-x') \psi_m^*(x) dx = \psi_m^*(x'). \qquad (9.13)$$

Putting this back into (9.11) gives the equivalent of already familiar relationship (5.63):

$$\sum_n \psi_n(x) \psi_n^*(x') = \delta(x-x'). \qquad (9.14)$$

Here we obtained it differently as an exercise.

Applying (9.13) to our box and using (9.10), we have

$$C_m = \sqrt{\frac{2}{a}} \sin k_m x' \qquad (9.15)$$

and

$$\Psi(x,0) = \delta(x-x') = \frac{2}{a}\sum_{n=1}^{\infty} \sin(k_n x)\sin(k_n x'), \quad 0 \le x, x' \le a. \tag{9.16}$$

Considering this as an initial waveform at $t = 0$, we can write the expression for this state at an arbitrary t, according to (9.11):

$$\Psi(x,t) = \frac{2}{a}\sum_{n=1}^{\infty} \sin(k_n x)\sin(k_n x') e^{i(\hbar k_n^2/2\mu)t}, \quad 0 \le x, x' \le a. \tag{9.17}$$

It is easy to see that expression (9.17) satisfies the temporal Schrödinger equation with the Hamiltonian of potential box (9.1) (Problem 9.2). Therefore, it describes the evolution of the initial "δ-spike" predicted by the nonrelativistic QM.

Box 9.1 Force on the walls

The obtained results have an interesting implication: the particle trapped in a box exerts a force on its walls. This follows from a simple observation that if we decrease the size of the box by decreasing a, it will increase the particle's energy. According to (9.9), the energy change corresponding to the change of box's size by da is

$$dE_n(a) = -\frac{\pi^2 \hbar^2 n^2}{\mu a^3} da, \quad n = 1, 2, 3, \ldots. \tag{9.18}$$

Therefore, an equal amount of work has to be done by an external agent pushing on the box, and therefore a certain force F must be exerted for pushing the walls inward: $dW = -F\,da = dE_n$. This means that the equal force is exerted on the wall by the particle from the inside:

$$F_n(a) = -\frac{dE_n(a)}{da} = \frac{\pi^2 \hbar^2}{\mu a^3} n^2. \tag{9.19}$$

The force increases unboundedly at $n \to \infty$ and/or $a \to 0$; otherwise, it is finite. One may wonder how this result is consistent with the initial statement that the particle is subject to an infinite force on the surface of the walls *regardless* of its state and the width of the box. The answer is that such statement was about particle *being* at the surface, but the particle in an eigenstate (9.10) has the zero probability of being found at the boundary. So the infinite force at the edges is multiplied by the zero probability density there. The product $0 \times \infty$ can be evaluated by a limiting procedure and may be zero, infinite, or finite, depending on specific conditions describing it [3]. In the considered case, the conditions lead to the finite result (9.19).

9.2
Potential Threshold

Consider a particle in a field with potential

$$U(x) = U\Theta(x), \quad \Theta(x) = \begin{cases} 0, & x < 0, \\ 1, & x \geq 0. \end{cases} \quad (9.20)$$

Here $\Theta(x)$ is the *Heaviside step function* that takes on only two values in two adjacent regions [4,5]. Below we assume that $U = \text{const} > 0$ (Figure 9.2). How would a quantum particle behave in such potential field?

The behavior of a classical particle in this situation would critically depend on its energy E. If $E > U$, then the particle incident from the left will enter the region $x > 0$ and move farther. It will only be slowed down at the threshold due to increase of its potential energy at the cost of its kinetic energy. If $E < U$, the particle will

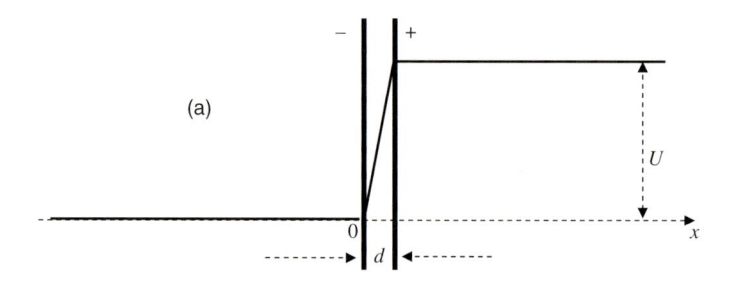

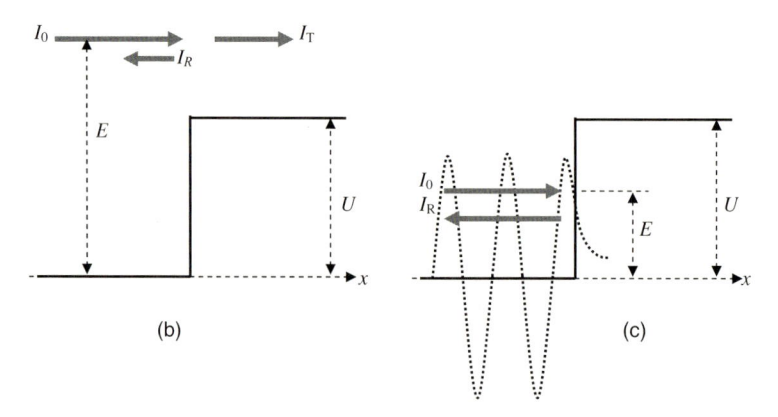

Figure 9.2 (a) Potential step produced by a charged parallel-plate capacitor; its simplified "rectangular" shape used in the text is obtained in the limit $d \to 0$ at fixed U. (b) $E > U$: The incident wave is partially reflected and partially transmitted (the amplitude splitting). (c) $E < U$: There is no transmitted wave; all the incident particles bounce back (total reflection). The net flux in the x-direction is zero. But there still is a nonzero probability amplitude of finding the particle on the right side of the interface. If there were an additional component of motion *parallel to the interface*, this amplitude would slide along the interface clinging to it, forming the *evanescent wave* known in optics.

9.2 Potential Threshold

bounce back. The point $x = 0$ where it turns back is called the *turning point*. In scientific language, the particle will be transmitted through the threshold if $E > U$ and reflected if $E < U$.

QM predicts much more complex behavior displaying all possible gradations between the two outlined extremes. For complete description, we need to find the eigenfunctions of the Hamiltonian with potential (9.20). Before doing it, consider an optical analogy: monochromatic light wave incident on the interface between vacuum and a transparent medium with refraction index n. The wave is partially transmitted and partially reflected, with the corresponding amplitudes depending on ω and n. Such splitting of the wave amplitude occurring over all wavefront is the example of the *amplitude splitting* [1] as opposed to the *wavefront splitting* caused by the slits in an opaque screen. At low light intensity, the particle aspect becomes prominent and we say that the transmission and reflection *probability amplitudes* are functions of n and the photon's energy $E = \hbar\omega$.

This analogy enables us to expect that at any $E > U$, the de Broglie wave of a particle will split into two waves associated with transmitted and reflected parts, respectively. And even at $E < U$ there will be a certain probability of leaking through the interface. Both outcomes are totally impossible for a classical particle.

Now let us check if our intuition is true. We need to solve the time-independent Schrödinger equation

$$-\frac{\hbar^2}{2\mu}\frac{d^2\Psi}{dx^2} + U\Psi = E\Psi, \tag{9.21}$$

with $U(x)$ determined by (9.20). Since the potential is constant in each semispace, the equation can be easily solved if we consider it separately for the left and right regions and then match the solutions on the interface. We have

$$\frac{d^2\Psi_L}{dx^2} = -k^2\Psi_L, \quad k \equiv \frac{\sqrt{2\mu E}}{\hbar}, \quad \text{and} \quad \frac{d^2\Psi_R}{dx^2} = k^2 n^2 \Psi_R, \quad n \equiv \sqrt{1 - \frac{U}{E}}, \tag{9.22}$$

where the subscripts L and R stand for the left and right semispace, respectively (note that $U > 0$ corresponds to $n < 1$). The solutions are

$$\Psi_L(x) = Ae^{ikx} + Be^{-ikx} \quad \text{and} \quad \Psi_R(x) = \tilde{A}e^{iknx} + \tilde{B}e^{-iknx}. \tag{9.23}$$

Each solution is a superposition of the two oncoming waves. But the propagation numbers of the waves in different semispaces are different.

The solutions (9.23) reflect the most general situation, when there are incident waves from the two opposite directions. But we are interested in a special case usually encountered in practice, when we have only one source. This corresponds to our initial assumption that we have only one wave incident, say, from the left. We can, without loss of generality, normalize the amplitude of this wave to $A = 1$. And since there is no wave incident from the right, we set $\tilde{B} = 0$. It is worthwhile to redenote the specific subset of solutions so defined: $\Psi_L \to \Psi_I$ (the left semispace will be the incidence side) and $\Psi_R \to \Psi_T$ (the right semispace will be the transmission side). This takes us to

$$\Psi_I(x) = e^{ikx} + Be^{-ikx} \quad \text{and} \quad \Psi_T(x) = \tilde{A}e^{iknx}. \tag{9.24}$$

But this is still a couple of two disjoint solutions. We get one solution describing a single particle by "stitching" or *matching* them on the interface (such matching amounts to applying the boundary conditions). The first condition is that $\Psi(x)$ must be continuous. This gives

$$\Psi_I(0) = \Psi_T(0). \tag{9.25}$$

The second condition requires the continuity of the first spatial derivative of $\Psi(x)$. This requirement follows from the Schrödinger equation. Applying it to our case, integrate Equation 9.21 over a narrow region $-\delta \leq x \leq \delta$ including the point $x = 0$:

$$\left.\frac{d\Psi_T}{dx}\right|_\delta - \left.\frac{d\Psi_I}{dx}\right|_{-\delta} = -\frac{2\mu}{\hbar^2} \int_{-\delta}^{\delta} (E - U)dx. \tag{9.26}$$

The integrand on the right is discontinuous at $x = 0$ but it is integrable since the discontinuity is finite (we can consider it as the Stieltjes integral [6–9]). At $\delta \to 0$ the integral vanishes, and we have

$$\left.\frac{d\Psi_I}{dx}\right|_{x=0} = \left.\frac{d\Psi_T}{dx}\right|_{x=0}. \tag{9.27}$$

This is the second boundary condition. Applying these conditions to our two functions (9.24) gives

$$1 + B = \tilde{A} \quad \text{and} \quad 1 - B = n\tilde{A}. \tag{9.28}$$

It follows

$$\tilde{A} = \frac{2}{1+n}, \quad B = \frac{1-n}{1+n}. \tag{9.29}$$

With the amplitudes determined by (9.29), the two separate functions (9.24) become a single state function describing a particle with energy E in potential field (9.20). We can write it as

$$\Psi(x) = \begin{cases} e^{ikx} + Be^{-ikx}, & x < 0, \\ \tilde{A}e^{iknx}, & x > 0. \end{cases} \tag{9.30}$$

The difference from (9.24) is that now the amplitudes B and $\tilde{A}$ are correlated and specified by (9.29).

Equation 9.30 confirms our qualitative prediction. At $x < 0$, we have a superposition of the incident and reflected waves. The latter exists even when $E > U$. And at $x > 0$ we have a nonzero wave even when $E < U$. This is purely wave-like behavior. On the other hand, if our source emits only one particle at a time, and we perform the position measurement, for instance by putting detectors everywhere, only one of them will fire in each trial. This is particle-like behavior. But the *probability* of finding the particle at a certain location, for example, in the left or the right semispace, is uniquely determined by the amplitudes (9.29). This is quantum-mechanical behavior, embracing both extremes.

Now determine the probability flux for our solution. Assuming first that $E > U$ and applying definition (3.17) gives

$$j = \frac{\hbar k}{\mu} \begin{cases} (1 - |B|^2), & x < 0 \quad \text{(a)}, \\ n|\tilde{A}|^2, & x > 0 \quad \text{(b)}, \end{cases} \tag{9.31}$$

with B and $\tilde{A}$ determined by (9.30). The first term in Equation 9.31a gives the flux j_0 associated with the incident wave, and the second term determines the flux j_R produced by the reflected wave. Equation 9.31b gives the flux j_T due to the transmitted wave. The ratios

$$R \equiv \frac{j_R}{j_0} \quad \text{and} \quad T \equiv \frac{j_T}{j_0} \tag{9.32}$$

are the coefficients of reflection and transmission, respectively [1,2]. Using (9.31) and (9.29) gives

$$R = |B|^2 = \left(\frac{1-n}{1+n}\right)^2, \quad T = n|\tilde{A}|^2 = \frac{4n}{(1+n)^2}. \tag{9.33}$$

It is immediately seen that

$$R + T = 1. \tag{9.34}$$

In optics, this is just conservation of energy: at normal incidence and in the absence of absorption, the intensities in reflected and transmitted waves sum to that in the incident wave. In QM, this is the conservation of probability: the incident particle is with certainty either reflected or transmitted, so the sum of probabilities of the two outcomes is 1.

So far we assumed that $E > U$. What happens if $E < U$? The answer to this is also encoded in the solution of the Schrödinger equation. According to (9.22), $E < U$ corresponds to imaginary n:

$$n \to i\nu, \quad \nu \equiv \sqrt{\frac{U}{E} - 1} > 0, \quad E < U. \tag{9.35}$$

This immediately kills the two *running* waves in the right semispace and converts Ψ_R in (9.23) into superposition of two real exponentials – one decaying and one increasing as we go farther to the right. Now we discard the increasing one, but for a different reason than before: not because of the absence of a source of particles shooting from the right but because $\Psi(x)$ exponentially increasing without any bounds does not correspond to any physical situation and violates the requirement that Ψ be finite at infinity. Thus, there remains, as before, only one term in the right semispace, although now it does not describe any flow away from the interface. But it is nonzero, which means that there still is a nonzero probability of finding the particle in the right semispace, albeit exponentially decreasing with the distance. This is a fundamental difference from the classical particle that cannot progress beyond the turning point. But it is a known effect in wave optics – the emergence of an *evanescent wave* in the process of total internal reflection [1,2].

Since there is no running wave in the right semispace, there is no flux there. The reader can show this, applying again the definition of flux to a function with real exponent. Once we know that there is no flux from the interface on the right side, we know that the coefficient of transmission is zero. Then Equation 9.34 tells us that the incident wave must be entirely reflected ($R = 1$). And indeed, (9.29) shows that the reflection amplitude B in this case is

$$B = \frac{1 - i\nu}{1 + i\nu}. \qquad (9.36)$$

Its modulus is equal to 1, and it differs from the incident amplitude only by a phase factor. Therefore, the coefficient of reflection is 1. We have the *standing wave* in the left semispace now. This is quite natural: the *net* flux is zero on both sides.

Looking ahead, it is important to note that the considered problem can be viewed from another perspective. Namely, it is a very special case (1D limit) of the process of scattering that we will consider in Chapter 19. From the viewpoint of scattering, the reflected wave in (9.24) is the wave *emitted by the interface* toward the source (the backscattered wave). The transmitted wave is the result of superposition of the incident wave and the wave *emitted by the interface* in the forward direction (forward-scattered wave).

Finally, a brief note on the case when $U < 0$ (the particle comes from the region with higher potential toward the region with lower potential). The results can be obtained from already derived equations just by changing sign in U. This would also be equivalent to the case $E > U > 0$ but with the incident particles coming from the right.

9.3
Tunneling through a Potential Barrier

Quantum tunneling is a remarkable phenomenon, in which a particle passes through a region totally forbidden classically. Here is a simple example. Imagine that you are driving a car with engine off. On a horizontal straight road and without friction you could still keep on moving at a constant speed. But if you go uphill, your car will gain potential energy at the cost of its kinetic energy, and will accordingly slow down. If the hill is high enough, the car will eventually stop at some point on the slope, and your motion will reverse to rolling back. The hill becomes an insurmountable barrier, and the point where the motion turns back is the *turning point*. Graphically, it is the intersection of the line representing the mechanical energy E of your car and the curve representing its potential energy $U(x)$ (Figure 9.3). We say that this curve represents a potential barrier.

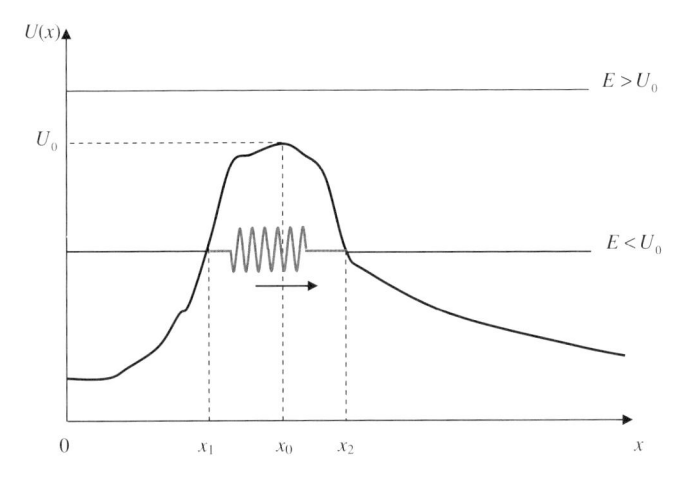

Figure 9.3 The graph of potential energy $U(x)$ for an arbitrary barrier. (i) $E < U(x_0) \equiv U_0$: The region $x_1 < x < x_2$ is classically inaccessible; according to CM, the barrier cannot be passed; x_1 and x_2 are the turning points for a particle approaching the barrier from the outside, left or right, respectively. (ii) $E > U_0$: The barrier can and will be passed. (iii) The particle is entirely within the barrier ($\Delta x < x_2 - x_1$). According to QM, this automatically changes particle's energy from a sharply defined value into an energy band, which gives a nonzero probability of finding, in an appropriate measurement, the value $E > U_0$. This, in turn, gives the corresponding probability for the particle to pass the barrier. Actually, the shown *position* of the wave packet representing the particle is not essential for this conclusion.

In CM, particles with E exceeding the top U_0 of the barrier will merely slow down as they pass, only to regain their kinetic energy on the other side, but for those with $E < U_0$, the barrier cannot be passed. Such a particle is reflected at the turning point.

For a one-top barrier, at $E < U_0$, there are two turning points, x_1 and x_2 – each one for a particle approaching from the corresponding side. The region $x_1 < x < x_2$ is inaccessible – we have there $E < U(x)$, so the particle would have negative kinetic energy and, accordingly, imaginary velocity.

QM changes all that. An incident particle with definite momentum has totally indeterminate position. Therefore, introduction of a barrier can only change probability for the particle to be found on this or that side of the barrier; but it cannot restrict the particle's motion to only one side, unless the barrier is infinitely high or has an infinitely wide top above E (the latter would effectively convert the barrier into a potential step considered in the previous section). If the particle's momentum is not sharply defined, its wave function will form a packet of a finite width representing the indeterminacy Δx in its position. Now we can ask: What happens if, after a position measurement, the particle (its wave packet) is found entirely within the barrier (as in Figure 9.3), where it cannot be classically? We might expect that the particle here must have the imaginary velocities. Does this constitute a paradox? The answer is no, for two reasons. First, the position indeterminacy

$\Delta x \leq |x_2 - x_1|$ automatically introduces, through $\Delta p \geq \hbar/2\Delta x$, the corresponding indeterminacy into its energy:[1]

$$E \to E \pm \frac{1}{2}\Delta E. \tag{9.37}$$

Accordingly, there appears a chance to find in the particle's spectrum an energy exceeding U_0; this is sufficient for the particle to get a chance to pass over the barrier "legally," with its velocity remaining a real quantity. Second, the wave aspect of matter *allows* an imaginary velocity and momentum, because this would only mean the imaginary wave number k, and the latter is well known in the wave theory. An imaginary part of k indicates attenuation (or amplification – depending on sign of this part) of the wave. A purely imaginary k converts the running wave into a stationary one that only oscillates with its amplitude being an exponential function of position. This is precisely what we get in Equation 9.30 at $E < U$ and accordingly $n \to iv$.

Thus, a particle with definite $E < U_0$ can pass through a barrier without invoking any contradictions because at no time is it entirely within the barrier. A particle entirely within the barrier and with its *average* energy $\bar{E} < U_0$ can pass through without any contradictions because now its energy spectrum has a part above U_0. By this trade-off Nature avoids the paradox. This allows one to represent graphically the passing with $E < U_0$ as if there were a tunnel cut through the barrier. Hence, the (very appropriate) name "tunneling" for this effect. Of course, with dissipation negligible, the barrier is not affected by the passage of a particle. But, since such passage is allowed (although not guarantied!), it is *as if* there were some virtual tunnel in the barrier. In the everyday language, when the possibility of tunneling is actualized, it is as if a solid object banged into a concrete wall, and instead of bouncing back, emerged on the opposite side, both the object and the wall remaining intact. A description of such an effect can be found in Lewis Carroll's immortal story "Through the Looking Glass" [10].

We want to stress again that quantum-mechanical *possibility* of tunneling is intimately linked with the *impossibility* to predict an individual outcome. In CM, the situation is strictly deterministic: if you know *exactly* the initial conditions, you can predict the individual outcome with certainty; if $E > U_0$, the particle will pass; otherwise, it will not. In contrast, QM gives *any* particle (even with $E \ll U_0$) a chance to pass – but this is only a chance. This is what we mean when we say that QM determinism (which gives a far more accurate description of Nature!) is "less determined" than classical one. It has nothing to do with elimination of determinism, which some claim is inherent in QM. The probability P_T of tunneling is an

---

1) We must draw a subtle distinction here from a stationary state with sharply defined energy in an external potential field. In such state, even though the total energy E is determined exactly, its constituent parts – the particle's potential and kinetic energy (and thereby its momentum) – are undetermined (this is reminiscent of a system with its *net* spin component exactly known (Section 7.5), while the individual spins of its constituents are not determined). The position measurement destroys the initial stationary state and introduces the indeterminacy (9.37) into the *total* energy E.

objective characteristic of a given system, which can be accurately calculated and tested. If a barrier is tall and/or wide and $E \ll U_0$, then $\mathcal{P}_T \ll 1$. If the barrier is thin and/or low, and $E \approx U_0$, then $\mathcal{P}_T \approx 1$. If the quantum-mechanical solution for a given system gives, say, $\mathcal{P}_T \approx 0.43$, then in a large pure ensemble, 43% of the particles will tunnel.

Now we turn to quantitative description of the phenomenon. We use a simplified model – a rectangular barrier (Figure 9.4a):

$$U(x) = \begin{cases} 0, & x < 0, \\ U, & 0 < x < a, \\ 0, & x > a. \end{cases} \quad (9.38)$$

This will allow us to utilize some results already obtained in the previous section. Now we find the energy eigenstates by solving the Schrödinger equation separately in three regions defined in (9.38) and "stitching" the solutions on the boundaries. We specify the solutions assuming, as before, only one incident wave from the left and nothing incident from the right. Denote the solutions on the incidence and transmission sides of the barrier as Ψ_I and Ψ_T, respectively, and the solution within the barrier as Ψ_M. Then using the same notations as in Section 9.2, we have

$$\Psi_I(x) = e^{ikx} + Be^{-ikx}, \quad \Psi_M(x) = \alpha e^{iknx} + \beta e^{-iknx}, \quad \Psi_T(x) = \tilde{A}e^{ikx}. \quad (9.39)$$

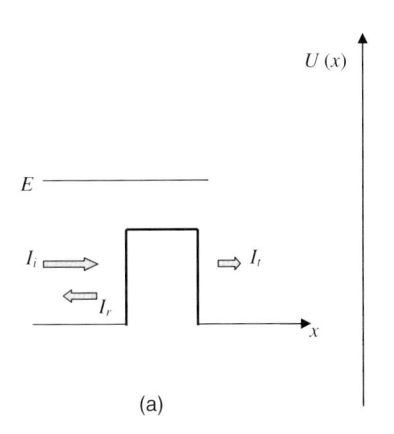

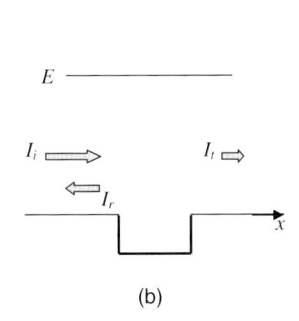

Figure 9.4 Rectangular barrier and well. (a) A particle with energy $E > U_{max}$, which classically would pass above the barrier, actually has a chance to be reflected from it. If the energy is less than U_{max} (this case is not shown here), the particle has a chance to pass ("tunnel") through the barrier. (b) If, instead of the barrier, we have a potential well, there is still a nonzero probability for the particle to be reflected from it. Arrows represent intensities of the incident (I_i), reflected (I_r), and transmitted (I_t) flux, respectively (recall that we can speak about flux even in the case of only one particle). If there is no absorption in the region of barrier or well, we have $I_r + I_t = I_i$.

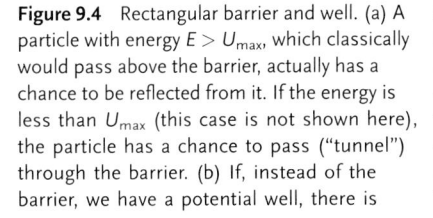

Next, we must match these solutions at the boundaries. Now we have two boundaries and, accordingly, four conditions:

$$\Psi_I(0) = \Psi_M(0), \qquad \Psi_M(a) = \Psi_T(a),$$
$$\Psi'_I(0) = \Psi'_M(0), \qquad \Psi'_M(a) = \Psi'_T(a) \tag{9.40}$$

(here the primes stand for d/dx). Applied to (9.39), this gives four equations:

$$1 + B = \alpha + \beta, \qquad \alpha e^{iakn} + \beta e^{-iakn} = \tilde{A} e^{iak},$$
$$1 - B = n(\alpha - \beta), \qquad n(\alpha e^{iakn} - \beta e^{-iakn}) = \tilde{A} e^{iak}. \tag{9.41}$$

After some algebra (Problem 9.10), we find the solutions

$$B = \frac{2i(1-n^2)\sin(akn)}{(1-n)^2 e^{iakn} - (1+n)^2 e^{-iakn}}, \qquad \tilde{A} = -\frac{4n e^{-iak}}{(1-n)^2 e^{iakn} - (1+n)^2 e^{-iakn}},$$

$$\alpha = -\frac{2(1+n) e^{-iakn}}{(1-n)^2 e^{iakn} - (1+n)^2 e^{-iakn}}, \qquad \beta = -\frac{2(1-n) e^{iakn}}{(1-n)^2 e^{iakn} - (1+n)^2 e^{-iakn}}. \tag{9.42}$$

As before, we consider two cases.

1) $E > U$. In this case, n is real, and using the same procedure and notations as in the previous section, we obtain the reflection and transmission coefficients

$$R = \frac{\gamma^2 \sin^2 akn}{1 + \gamma^2 \sin^2 akn}, \qquad T = \frac{1}{1 + \gamma^2 \sin^2 akn}, \qquad \gamma^2 \equiv \frac{1}{4}(n - n^{-1})^2. \tag{9.43}$$

Keep in mind that n and γ here are themselves functions of k. But, regardless of the nature of these functions, the transmission coefficient is now explicitly shown to be generally less than 1, and accordingly, we have a nonzero reflection. This confirms our qualitative predictions that a particle can be reflected even when $E > U$.

2) $E < U$. In this case, we have $n = i\nu$ with positive ν given by (9.35), and both terms of $\Psi_M(x)$ in (9.39) become the evanescent waves! The same procedure now gives

$$R = \frac{\tilde{\gamma}^2 \operatorname{sh}^2 ak\nu}{1 + \tilde{\gamma}^2 \operatorname{sh}^2 ak\nu}, \qquad T = \frac{1}{1 + \tilde{\gamma}^2 \operatorname{sh}^2 ak\nu}, \qquad \tilde{\gamma}^2 \equiv \frac{1}{4}(\nu + \nu^{-1})^2. \tag{9.44}$$

In both cases, we have $R + T = 1$. The results confirm our qualitative predictions: we can have reflection from the barrier at $E > U$ and transmission through it at $E < U$.

If $ak\nu \gg 1$, we have $\text{sh}(ak\nu) \approx (1/2)e^{ak\nu} \gg 1$. In this case, Equation 9.44 simplifies to

$$R \approx 1 - T_0(E)e^{-2ak\nu}, \qquad T \approx T_0(E)e^{-2ak\nu}, \qquad T_0(E) = \tilde{\gamma}^{-2} \approx 4\frac{E}{U}. \tag{9.45}$$

As in case with the potential threshold, we can look at the whole process from the viewpoint of scattering. We can consider the barrier as a 1D scatterer. The wave reflected from the barrier is the backscattered wave. The wave transmitted (tunneled) through the barrier is a superposition of the incident and forward-scattered wave. The latter can be considered as emitted by (originated in) the barrier itself under the influence of the incident wave. The same analogy with scattering can be drawn and the similar results obtained if we replace the barrier with the well.

Summary. The incidence of a particle onto a potential step with $U > E$ is analogous to the total internal reflection of a wave in acoustics and optics. Total as it is, it does not occur exactly at the interface between the two mediums. It is preceded by partial penetration of the incident wave into the second medium, with the exponentially decaying amplitude; if **k** has the component *along* the interface, the penetrated part is "crawling" along the surface (evanescent wave). If the second medium is a layer of finite width, the penetrated part partially leaks out onto the other side of the layer (frustrated internal reflection), thus causing transmission.

Similarly, the *quantum particle* incident onto potential threshold has a nonzero amplitude of penetrating into the reflecting region. If this region has a finite width a (a barrier!), the particle has a nonzero chance to reach the opposite face and to leak out into space on the other side (tunneling!), where it can again propagate freely. This effect has, by symmetry, its "antipode": just as a particle with $E < U$ can tunnel through the barrier, the same particle with $E > U$ may reflect from it. Moreover, even if we replace the barrier with a rectangular *potential well*, there may be a chance for the particle to get reflected from it (Figure 9.4b) – a totally crazy situation from the classical viewpoint!

One experimental manifestation of tunneling will be considered in the next section.

9.4
Cold Emission

Here we consider an important application of the theory developed in the previous section. To this end, we note first that the expression $ak\nu$ determining the probability of tunneling in Equation 9.45 has a simple geometrical interpretation

$$ak\nu = \frac{a\sqrt{2\mu(U-E)}}{\hbar} \equiv \frac{a\tilde{p}}{\hbar} \equiv \frac{\tilde{S}}{\hbar}, \tag{9.46}$$

where $\tilde{p}$ and $\tilde{S}$ can be defined as a *would be* (or *virtual*) classical momentum and action, respectively, along the virtual path within the barrier. For a narrow barrier

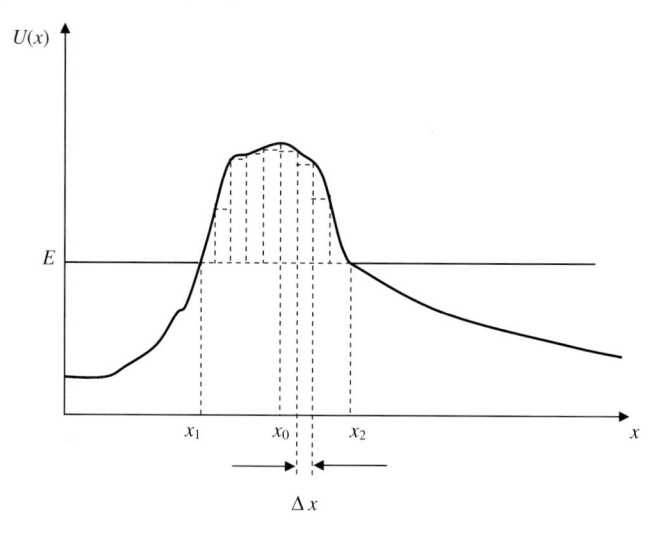

Figure 9.5 The graph of potential energy $U(x)$ for a barrier with an arbitrary shape. The transmission probability is approximated by the area above the energy level.

$a \to \Delta a = \Delta x$, we would have $kv\Delta x = \Delta \tilde{S}/\hbar$ in (9.45). This suggests that tunneling through an arbitrarily shaped barrier $U \to U(x)$ can be considered as a succession of transmissions through a row of N adjacent narrow rectangular barriers of variable height $U(x_j)$ (Figure 9.5). The resulting amplitude can then be approximated as the product of the individual amplitudes:

$$\tilde{T} = \prod_{j=1}^{N} \tilde{T}_j \approx \prod_{j=1}^{N} \tilde{T}_{0j}\, e^{-(1/\hbar)\sum_{j=1}^{N}\Delta\tilde{S}_j}. \tag{9.47}$$

If $U(x)$ does not change too rapidly within the considered region, then the product of pre-exponential factors can be replaced with sufficient accuracy with an effective single factor $\tilde{T}_{\text{Ef}}(E)$ depending on $U(x)$ and E. In the limit $\Delta x \to dx$, the sum in the exponent becomes the integral over the region between the two turning points:

$$\sum_{j=1}^{N} \Delta\tilde{S}_j \to \int_{x_1(E)}^{x_2(E)} d\tilde{S}(x) = \int_{x_1(E)}^{x_2(E)} \sqrt{2\mu(U(x)-E)}\, dx = \tilde{S}(E) \tag{9.48}$$

and the net transmission probability can be written as

$$\tilde{T}(E) \approx \tilde{T}_{\text{Ef}}(E)e^{-\tilde{S}(E)/\hbar}. \tag{9.49}$$

Now we apply this result to an important experimental situation. Consider a conducting slab whose surface we want to investigate. The slab is so thick that we may not bother about the opposite face and can consider the slab as filling all semispace to the left of the surface of interest, where we place the origin. The free electrons within the slab occupy all allowed energy levels forming practically continuous energy band up to the so-called *Fermi level E* (thermal motion neglected). The electrons are trapped within the slab because E is less than the threshold U_0. The minimal additional energy ΔW (the work function) necessary for an electron to break loose from the slab in one-dimensional motion is given by $\Delta W = U_0 - E$.

Suppose now that instead of pumping in this energy, we apply sufficiently high positive potential with electric field $\mathcal{E}$ to the slab, so the latter becomes a cathode. This changes the potential so that instead of $U = \text{const} = U_0$ we now have outside the slab

$$U \to U(x) = U_0 - q\mathcal{E}x, \quad x > 0, \tag{9.50}$$

where q is the electron charge. The potential threshold now is converted into triangular potential barrier (Figure 9.6), and the electrons get a chance to tunnel outside through it *without acquiring any additional energy,* for example, without any illumination or heating of the slab. Minuscule as this chance may be for an individual electron, it may, in view of a huge number of them, result in the emergence of an experimentally measurable electron emission. This effect is known as *cold emission*. Do not confuse it with the photoemission effect (PEE) discussed in Section 2.3. In the PEE, there is no applied field, and the electrons get necessary energy from the incident photons. In the cold emission there are no photons, but there is applied field and electrons get out by tunneling [11,12].

We can describe the cold emission quantitatively within our simplified model. We need to take an integral (9.48) for potential (9.50):

$$\tilde{S}(E) = \int_{x_1(E)}^{x_2(E)} \sqrt{2\mu(W - q\mathcal{E}x)}\, dx = -\frac{2}{3}\frac{\sqrt{2\mu}}{q\mathcal{E}}(W - q\mathcal{E}x)^{3/2}\Big|_{x_1(E)}^{x_2(E)}. \tag{9.51}$$

The two turning points restricting the integration region are determined as intersections of the horizontal E with the slopes of the barrier. The first one is just $x_1 = 0$, and the second is determined from equation $U(x) - E = W - q\mathcal{E}x = 0$. Thus, we get

$$x_1 = 0, \quad x_2(E) = \frac{U_0 - E}{q\mathcal{E}}. \tag{9.52}$$

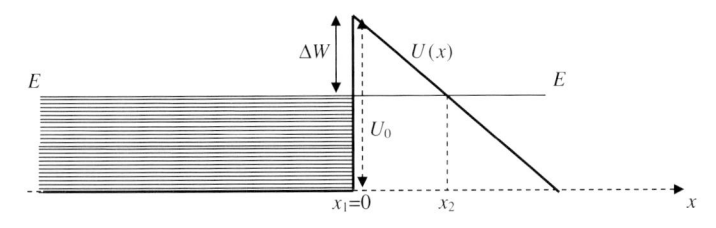

Figure 9.6 Triangular potential barrier produced by an appropriate electric field perpendicular to the surface of a conducting slab. The electrons can break loose from the slab by tunneling through the barrier.

Putting this into (9.51) gives

$$\tilde{S}(E) = \frac{2}{3} \frac{\sqrt{2\mu}}{q\mathcal{E}} (U_0 - E)^{3/2}. \tag{9.53}$$

Finally, we obtain the transmission coefficient as a function of E and $\mathcal{E}$:

$$T(E, \mathcal{E}) = T_{\text{Ef}}(E) e^{-\mathcal{E}_0/\mathcal{E}}, \qquad \mathcal{E}_0 \equiv \frac{2}{3} \frac{\sqrt{2\mu}}{q} (U_0 - E)^{3/2}. \tag{9.54}$$

A similar expression can be written for the emission current.

For a better comparison with the experiment, we need to take into account that in the original theory E stands for the motion in one dimension – along the x-direction, whereas the actual electron motion within the bulk of the slab is three-dimensional. Taking the bottom of the valence band as the zero potential, we find from $E = \langle p_x^2 + p_y^2 + p_z^2 \rangle / 2\mu$ that the averaged energy of one-dimensional motion entering our original equations is only one-third of the actual electron energy. If we want to express the result in terms of the actual energy, we must change $E \to (1/3)E$ in our equations. The obtained results are in perfect agreement with the experiment [13–15].

9.5
Potential Well

As already mentioned, another simple model for a particle in a potential field is a rectangular potential well. In contrast with the potential box that is infinitely deep, the depth U of the well is finite (Figure 9.4b). In this case, we take the reference level for potential energy at the top of the well. Figure 9.4b suggests the possibility of states with $E < 0$ (the bound states).

A classical particle has continuous energy spectrum apart from $E > 0$, also in the region $-U \leq E \leq 0$. Quantum-mechanical picture is different from the classical one in two respects: first, the energy spectrum in the negative range is discrete, and second, a particle with $E > 0$ has a chance to get reflected from the well.

There is a well-known optical analogue of both effects. A transparent material with index of refraction $n > 1$ (e.g., glass) imitates a potential well. But light partially reflects from a glass plate. Due to the wave aspect of matter, any particle can, in principle, do the same. On the other hand, the same glass plate can work as a waveguide admitting a discrete set of modes for the trapped photons and allowing them to propagate only along the plate.

Our problem now is to describe both effects quantitatively, that is, to find the allowed energies for $E < 0$ and the reflection coefficient for $E > 0$. We will start with the second problem because its solution can be obtained directly from the results of Section 9.3; the mathematics is the same in both cases. The particle's state is described by the same equation (9.39), and the solutions will have the same physical meaning: Ψ_I is the superposition of the incident and reflected waves; Ψ_M is the solution within the well, describing the result of multiple reflections on the two

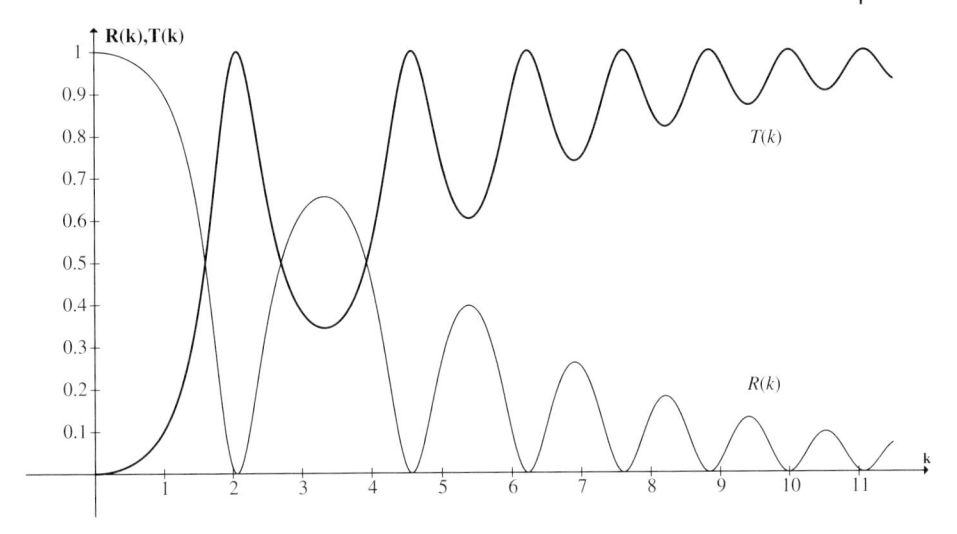

Figure 9.7 The transmission and reflection coefficients as functions of k for rectangular potential well.

interfaces; and Ψ_T is the transmitted wave. The reflection and transmission coefficients are given by (9.43). The only difference from (9.43) is that now (9.22) gives $n > 1$.

Figure 9.7 shows the graphs of functions (9.43). *In both cases, $T(k) = 1$ for certain values of k*, and accordingly, there is no reflection at these values. This is similar to passing of light of some specific wavelengths through a layer of plasma (barrier) or glass plate (well) of thickness a, when the light reflected from an inner face returns to it after the inner reflection from the opposite face with the phase shift $2\pi N$ with an integer N. The mathematical condition for it is, as seen from (9.43),

$$akn = \pi N \quad \text{or} \quad a = N\frac{\lambda}{2n}. \tag{9.55}$$

This means that an integer number N of half-wavelengths $\lambda(n) \equiv \lambda/n$ fits into the plate. In this case, all consecutive transmitted amplitudes interfere constructively producing the maximal possible net result equal to the incident amplitude. The layer or plate is 100% transparent for the corresponding wavelengths.

Exercise 9.1

Exactly the same argument applies to the secondary waves contributing to reflection – they also interfere constructively under condition (9.55), but in this case it leads to the opposite result: there is no reflection. This is consistent with conservation of energy (or probability, for that matter): surely, there must be no reflection when $T(k) = 1$. But how is it consistent with constructive interference of all waves returning to the source?

Solution:

The answer is that the described interference is completely constructive only for all *secondary* waves, that is, the waves having traversed the layer (plate) $2, 4, \ldots, 2N, \ldots$ times. But they all interfere destructively with the *primary* reflected wave that did not enter the medium. As is well known [1,2], a wave undergoes the phase shift $= \pi$ when reflected from the surface of a medium with higher n, and therefore all secondary reflected waves under condition (9.55) are in phase with each other but out of phase with the primary reflected wave, and the net result is zero. There is no reflection – complete passing through. The same holds for the barrier! In this case $n < 1$, and the primary reflected wave does not change its phase, but the internally reflected waves do, and those sent back to the source have undergone an odd number of reflections and are all out of phase with the primary. All those transmitted have undergone an even number of reflections, and thus are all in phase. Barrier/well is totally transparent (100% transmitting) for the corresponding wavelengths.

The transmission peaks observed under these conditions are called the *resonance transmission maxima*. The corresponding values of E are the *resonance energies*. Sometimes they are also called the virtual energy levels. The reader should remain alert to this effect, since it is the "embryo" of a more general effect of the resonance scattering in 3D problems (Chapter 19). Altogether, we see that major contributions to the reflection and transmission amplitudes B and $\tilde{A}$ (Problem 9.11) come from multiple reflections of the wave within the barrier or well.

All considered cases of particle's encounters with the well (or barrier) are *asymmetric* in the sense that only one-sided incidence (from the left *or* from the right) was assumed. In such situations, the "asymmetric" position of the origin (at an edge of the well/barrier rather than in the middle) seems to be the natural (albeit not mandatory) choice. Generally, however, each side may be the incidence and transmission side at once, if there is a source of particles on either side of barrier/well. Then we may have for each E a superposition of de Broglie waves propagating in both directions on either side. If the amplitudes of such superposition on each side are equally weighted, we have $\bar{p} = 0$. Such a state can also be an eigenstate (even or odd) of the parity operator (Problem 9.14). In such cases, it may be convenient to choose the origin in the middle of the well/barrier.

Let us now turn to the second part of our problem – determining of the bound states ($E < 0$). In this case there are no incidences, and the notion of refraction index n is irrelevant. Therefore, we change notations as follows:

$$k \to i\kappa, \quad \kappa \equiv \sqrt{2\mu|E|}/\hbar, \quad kn \to \nu \equiv \sqrt{2\mu(U - |E|)}/\hbar,$$
$$\kappa^2 + \nu^2 = \nu_0^2, \quad \nu_0 \equiv \frac{\sqrt{2\mu U}}{\hbar} \tag{9.56}$$

and return to Ψ_L and Ψ_R for solutions on the left and right from the well, respectively. Expressions (9.39) take the form

$$\Psi_L = B e^{\kappa x}, \quad \Psi_M = \alpha e^{i\nu x} + \beta e^{-i\nu x}, \quad \Psi_R = \tilde{A} e^{-\kappa x} \tag{9.57}$$

(the first term in Ψ_L is rejected because the corresponding function diverges at $x \to -\infty$). Note that ν here is the wave number corresponding to energy $E' = U - |E|$ measured from the bottom of the well, so that $E' = \hbar^2\nu^2/2\mu$. Now it is more convenient to shift the origin to the middle of the well so that the potential can be written as

$$U(x) = \begin{cases} 0, & x < -a/2, \\ -U, & -a/2 < x < a/2, \\ 0, & x > a/2. \end{cases} \quad (9.58)$$

The new choice of the origin, although not essential for the solution, has the advantage of making more explicit the role of parity mentioned in Chapter 5. The Hamiltonian in our problem is symmetric with respect to the center of the well; with the origin at the center, this symmetry is expressed as the invariance under inversion $x \to -x$. Therefore, with the same boundary conditions on both sides, the Hamiltonian eigenstates are also the eigenstates of the parity operator $\hat{P}$. They must be either even or odd under inversion. Thus, we set the amplitudes in (9.57) to satisfy the condition

$$\begin{array}{ll} \alpha = \beta, & \tilde{A} = B \quad \text{(even states)}, \\ \alpha = -\beta, & \tilde{A} = -B \quad \text{(odd states)} \end{array} \quad (9.59)$$

Utilizing this property simplifies the solution. It is sufficient to find the solution, say in the region $0 \le x < \infty$. The solution on the other side of the origin will be its extension – either even or odd depending on parity. Also, depending on parity, we must have either $\Psi_M(0) = 0$ or $\Psi'_M(0) = 0$ (Figure 9.1 illustrates this feature for the case of potential box). Then, the stationary solution to the right of the well ($x > a/2$) is given by the third equation of (9.57). Inside the well ($0 \le x \le a/2$), using (9.59), we will have (dropping factor 2)

$$\psi(x) = \begin{cases} \alpha \cos \nu x, & \hat{P} \text{ even}, \\ \beta \sin \nu x, & \hat{P} \text{ odd}. \end{cases} \quad (9.60)$$

For states of even parity, the continuity requirements give

$$\begin{cases} \alpha \cos\left(\frac{1}{2}a\nu\right) = \tilde{A} e^{-(1/2)a\kappa}, \\ \alpha\nu \sin\left(\frac{1}{2}a\nu\right) = \tilde{A}\kappa e^{-(1/2)a\kappa}. \end{cases} \quad (9.61)$$

Taking the ratio of these equations and using (9.56), we obtain

$$\tan\left(\frac{1}{2}a\nu\right) = \sqrt{\frac{\nu_0^2}{\nu^2} - 1} \quad (9.62)$$

with ν within the range $0 \le \nu \le \nu_0$. Equation 9.62 selects the energy eigenvalues $E_j = (\hbar^2 \nu_j^2/2\mu) - U$ for which there exist the corresponding eigenstates of even

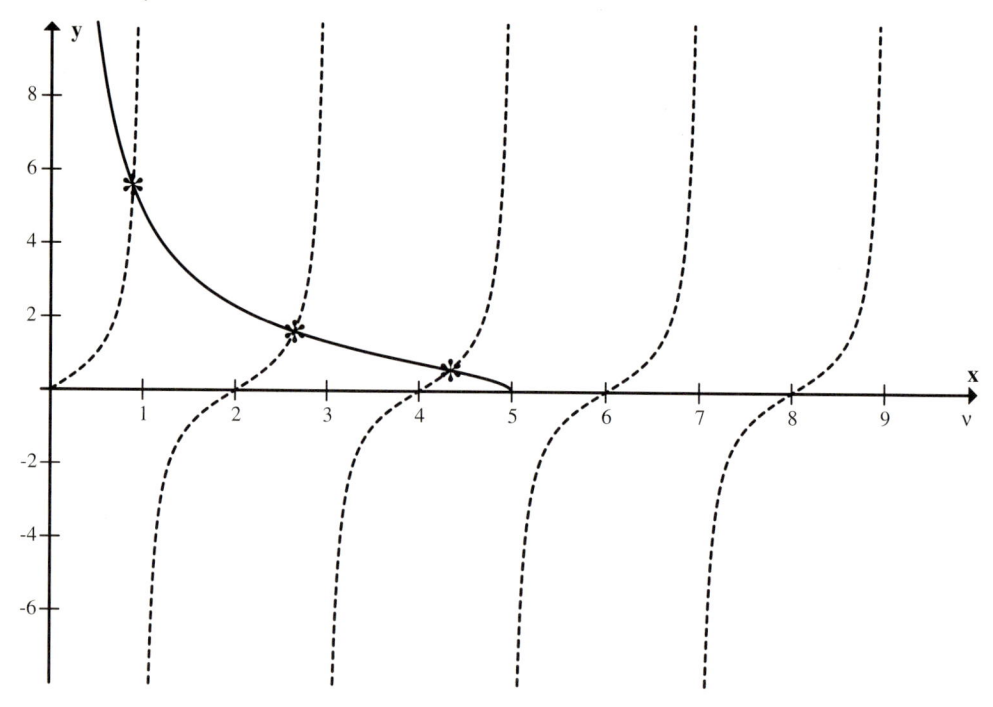

Figure 9.8 The graphical solution of equation for the bound stationary states within a rectangular well.

parity. It is an example of the *transcendental equations* and can be solved graphically (Figure 9.8). Since the tangent is periodic function, $\tan \xi = \tan(\xi + j\pi), j = 1, 2, \ldots$, we can generally have more than one solutions ν_j. And in any event, as is evident from Figure 9.8, there always exists at least one solution ν_1 with the energy eigenvalue below the top of the well. Thus, the symmetric 1D well, no matter how shallow and/or narrow, always has at least one bound state to accommodate the trapped particle (and this turns out to be the even state). The maximal number N of allowed even eigenstates is determined by the value ν_0 and can be found from the same figure as equal to

$$N = \mathrm{mod}\left(\frac{a\nu_0}{2\pi}\right), \tag{9.63}$$

where mod (x) is the integer part of x. We see that the number of allowed states increases unboundedly with the depth of the well. This is consistent with the results of Section 9.1.

Putting one of the solutions $\nu = \nu_j$ back into (9.61), we can find the ratio of α and $\tilde{A}$ for the corresponding state j:

$$\frac{\tilde{A}_j}{\alpha_j} = \frac{\nu_j}{\nu_0} e^{(1/2)a\sqrt{\nu_0^2 - \nu_j^2}}. \tag{9.64}$$

Both amplitudes are finally determined by the normalization condition (Problems 9.15 and 9.16).

In a similar way, we derive the solutions for the odd states. In this case, the continuity requirements lead to the equations

$$\begin{cases} \beta \sin\left(\frac{1}{2}a\nu\right) = \tilde{A} e^{-(1/2)a\kappa}, \\ \beta\nu \cos\left(\frac{1}{2}a\nu\right) = -\tilde{A}\kappa e^{-(1/2)a\kappa}. \end{cases} \quad (9.65)$$

Their ratio gives the condition selecting the set of energy eigenvalues for the odd states:

$$\cot\left(\frac{1}{2}a\nu\right) = -\sqrt{\frac{\nu_0^2}{\nu^2} - 1}. \quad (9.66)$$

This is the equation of the same type as (9.62), but now the allowed energy eigenvalues will be numbered by integers $2j$. In both cases, the found energy eigenvalues relative to the bottom of the well reduce to the familiar result (9.9) in the limit $U \to \infty$, when the well converts into a box with the impenetrable walls (Problem 9.18).

Now we can fully answer the question why the *bound* states in a potential (9.55) are also the eigenstates of $\hat{P}$, while the states of the same Hamiltonian with $E > 0$ are not.

The answer is that the solutions are determined, apart from the Hamiltonian, also by the boundary conditions. The Hamiltonian eigenstates for $E > 0$ had been found under the asymmetric boundary conditions. In contrast, the eigenstates for $E < 0$ are subject to the same condition on either side of the well. This selects only those solutions that are also the eigenstates of $\hat{P}$. The lack of symmetry in the boundary conditions for $E > 0$ leaves in the wagon the nonsymmetric states as well. Generally, we could also impose symmetric boundary conditions for the nonlocalized states. The resulting solutions would then be the eigenfunctions of $\hat{P}$. But we were describing typical experimental situations when the well is "illuminated" from one side only. The asymmetry of such situation is naturally reflected in the asymmetry of the corresponding solutions.

9.6
Quantum Oscillator

The system we consider in this section provides some additional insights into relationship between QM and its classical limit. Therefore, we start with brief review of the classical oscillator. It is described as a particle in the potential field $U(x) = (1/2) kx^2$ generating the restoring force $F = -kx$, where k is the spring constant. The classical equation of motion $\ddot{x}(t) = F/\mu = -(k/\mu)x(t)$ under initial condition $x(0) = 0$ yields

$$x(t) = a \sin \omega_0 t, \qquad v(t) = \dot{x}(t) = a\omega_0 \cos \omega_0 t. \quad (9.67)$$

It describes periodic motion with frequency $\omega_0 = \sqrt{k/\mu}$. The oscillation amplitude a is determined by the particle's energy

$$E = K + U = \frac{p^2}{2\mu} + \frac{1}{2}kx^2. \tag{9.68}$$

Using (9.67) gives $E = (1/2)\mu\omega_0^2 a^2$, so that

$$a = \sqrt{\frac{2E}{\mu\omega_0^2}}. \tag{9.69}$$

Here the energy is a continuous variable that can take any positive value starting from $E = 0$ (particle resting at the origin).

Consider now QM description. We want to solve the stationary Schrödinger equation

$$\hat{H}\psi = E\psi \quad \text{with} \quad \hat{H} = -\frac{\hbar^2}{2\mu}\frac{d^2}{dx^2} + \frac{1}{2}\mu\omega_0^2 x^2. \tag{9.70}$$

Introduce two new variables

$$\xi = \frac{x}{x_0} \quad \text{with} \quad x_0 \equiv \sqrt{\frac{\hbar}{\mu\omega_0}}, \quad \text{and} \quad \eta = \frac{2E}{\hbar\omega_0}. \tag{9.71}$$

Then (9.70) simplifies to

$$\frac{d^2\psi}{d\xi^2} + (\eta - \xi^2)\psi = 0, \tag{9.72}$$

with ψ being now the function of ξ.

Just as for the bound states considered in the previous sections, Equation 9.72 has the finite continuous solutions only for a discrete set $\eta_n = 2n + 1$, $n = 0, 1, 2, \ldots$ (Appendix F). In view of (9.71), this converts into

$$E \to E_n = \left(n + \frac{1}{2}\right)\hbar\omega_0. \tag{9.73}$$

Instead of continuous energy spectrum for a classical oscillator, we now have the discrete spectrum. The allowed energies form the set of equidistant levels starting from the lowest possible energy $E_0 = (1/2)\hbar\omega_0$. The corresponding eigenfunctions are given by (Appendix F)

$$\psi_n(\xi) = e^{-(1/2)\xi^2} H_n(\xi), \quad H_n(\xi) = \frac{(-1)^n}{\sqrt{2^n n! \sqrt{\pi}}} e^{\xi^2} \frac{d^n}{d\xi^n} e^{-\xi^2}, \tag{9.74}$$

where $H_n(\xi)$ are the Hermite polynomials of the nth order [16]. The coefficient in it is chosen so as to satisfy the normalization condition. The structure of this solution gives the algorithm for determining all $\psi_n(\xi)$:

$$\psi_n(\xi) = \frac{(-1)^n}{\sqrt{2^n n! \sqrt{\pi}}} e^{(1/2)\xi^2} \frac{d^n}{d\xi^n} e^{-\xi^2}. \tag{9.75}$$

At $n = 0$, this gives the wave function $\psi_0(\xi)$ of the ground state.

Expressions (9.74) and (9.75) also suggest an alternative way to obtain all eigenfunctions from $\psi_0(\xi)$ by writing the recurrent relations between $\psi_n(\xi)$ and $\psi_{n+1}(\xi)$ (Problem 9.19):

$$\psi_{n+1}(\xi) = \frac{1}{\sqrt{2(n+1)}} \left[\xi \psi_n(\xi) - \frac{d\psi_n(\xi)}{d\xi} \right]. \tag{9.76}$$

Returning to the initial variable x, and noticing that $d/dx = \hat{p}/(-i\hbar)$, we can rewrite this as

$$|\psi_{n+1}\rangle = \frac{1}{\sqrt{n+1}} \frac{1}{\sqrt{2\hbar\omega}} \left(\sqrt{k}x - \frac{i}{\sqrt{\mu}} \hat{p} \right) |\psi_n\rangle. \tag{9.77}$$

We obtained a very interesting result: any eigenstate $|\psi_{n+1}\rangle$ can be obtained from the previous eigenstate $|\psi_n\rangle$ by applying to it a specific combination of position and momentum operators. The fact that this combination works throughout all ladder of eigenstates makes it universal and worthy of special notation. So let us introduce

$$a^\dagger \equiv \frac{1}{\sqrt{2\mu\hbar\omega}} (\mu\omega x - i\hat{p}). \tag{9.78}$$

Then (9.77) can be written as

$$a^\dagger |\psi_n\rangle = \sqrt{n+1} |\psi_{n+1}\rangle. \tag{9.79}$$

The result (9.83) suggests that any eigenstate $|\psi_n\rangle$ can be obtained from $|\psi_0\rangle$ by successive application of operator $a^\dagger$:

$$|\psi_n\rangle = \frac{1}{\sqrt{n!}} (a^\dagger)^n |\psi_0\rangle. \tag{9.80}$$

Here we write explicitly the first three eigenstates ($n = 0, 1, 2$)

$$\psi_0(\xi) = \frac{1}{\sqrt[4]{\pi}} e^{-(1/2)\xi^2}, \quad \psi_1(\xi) = \frac{\sqrt{2}}{\sqrt[4]{\pi}} e^{-(1/2)\xi^2} \xi, \quad \psi_2(\xi) = \frac{1}{\sqrt[4]{4\pi}} e^{-(1/2)\xi^2} (2\xi^2 - 1). \tag{9.81}$$

They are shown in Figure 9.9. For comparison, we also show the three vibration states (fundamental and first two harmonics) for a classical string with the fixed ends. It is easy to see that the oscillator's eigenstates are even or odd functions of ξ depending on n. Thus, the number n determines the state's parity. We also see that $\psi_0(\xi)$ ($n = 0$) is nonzero everywhere. The function $\psi_1(\xi)$ has a node at $x = 0$; $\psi_2(\xi)$ has two nodes. This is manifestation of a general rule: the quantum number n is equal to the number of nodes of the corresponding eigenfunction *within* (edges excluded!) its range. If you look back at the potential box (Section 9.1), you will see the similar rule in action there. Recall also a certain (but restricted) analogy with the classical states of a vibrating string indicated in that section. Figure 9.9 shows that this rule holds for a classical standing wave as well.

Let us now compare the probabilities in the classical and QM description of oscillator. Consider, for instance, the level E_2. Classically, the particle with energy E_2 can move only within the available region $-a_2 \leq x \leq a_2$, with $-a_2$ and a_2 being the turning points. The distance a_2 is the amplitude (9.69) of the classical oscillations

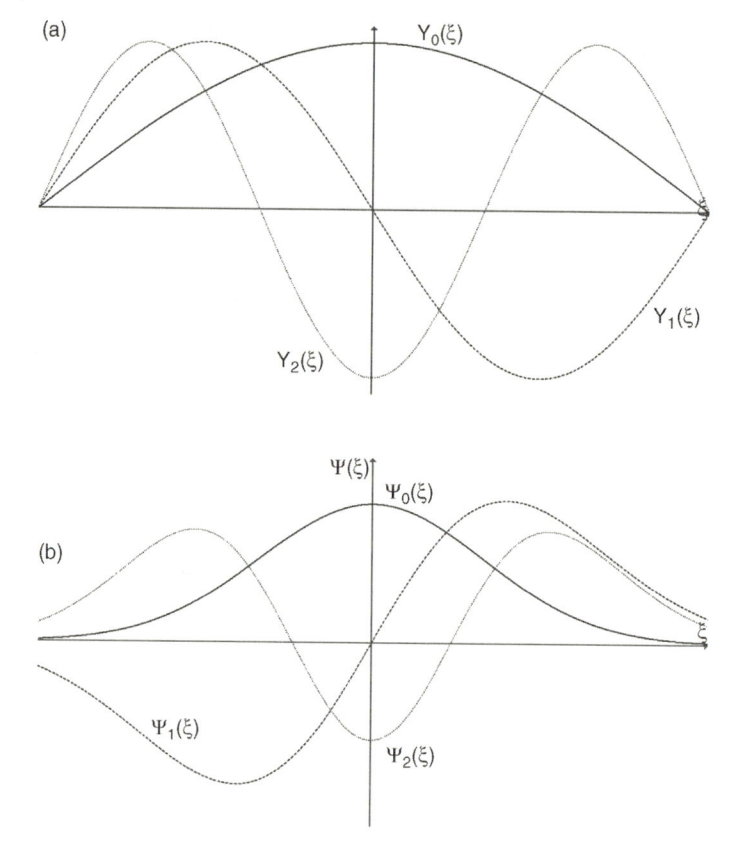

Figure 9.9 The first three eigenstates of quantum oscillator (b) versus the fundamental mode and the first two harmonics of classical vibrations on a sting (a).

corresponding to $E = E_2$. The classical probability of finding the particle with this energy beyond the region $-a_2 \leq x \leq a_2$ is strictly zero. Let us evaluate this probability also *within* this region and compare it with the QM probability for the same energy. The classical probability $d\mathcal{P}_{CM}(x)$ of finding the oscillating particle within the small interval $(x, x + dx)$ is proportional to the time dt it takes for the particle to travel the distance dx in this interval:

$$d\mathcal{P}_{CM}(x) = \frac{dt}{T} = \frac{1}{T}\frac{dx}{v(x)}. \tag{9.82}$$

Here $T = 2\pi/\omega_0$ is the period of oscillations. From (9.67) we have $v(x) = a\omega_0\sqrt{1 - (x^2/a^2)}$, so that

$$d\mathcal{P}_{CM}(x) = \frac{1}{2\pi a}\frac{dx}{\sqrt{1 - (x^2/a^2)}}, \quad -a \leq x \leq a. \tag{9.83}$$

This probability density $d\mathcal{P}_{CM}(x)/dx$ for $n = 2$ is shown in Figure 9.10. It is minimal at the center, where the particle's speed is the highest, so it passes this region very fast,

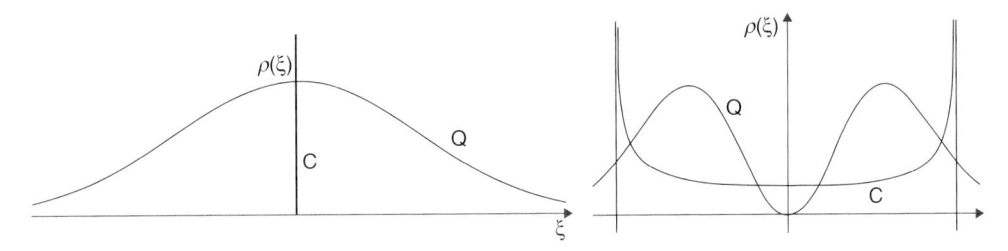

Figure 9.10 The probability distribution for a quantum oscillator (Q) and a classical particle (C): (a) the ground state ($n = 0$); (b) the first excited state ($n = 1$).

which makes it the least likely to be found here. On the other hand, it diverges at the boundaries – a very natural result for the turning points, where the particle's instantaneous velocity (*not* acceleration!) is zero.

Compare this with the QM probabilities for the same energy, shown in the figure. Like the classical probability, it is minimal (is strictly zero) at the center. It peaks near the classical turning points (Problem 9.19) but without any singularities. And the most important distinction is its being nonzero beyond the turning points! There is a nonzero probability of finding the particle in the region that is classically forbidden in the given state. We have again the same result as those found earlier for a potential well or in the tunneling problem. We already know that this does not constitute any contradiction in QM, since the kinetic and potential energies are not compatible observables.

The difference between quantum-mechanical and classical pictures is especially dramatic for the state with the minimal energy. As mentioned before, classically the minimal energy is zero and corresponds to the particle at rest at the equilibrium position. According to (9.69), the span of oscillations in this case is $a \to 0$, and Equation 9.83 gives for $dP_{CM}(x)/dx$ the result similar to $\delta(x)$: we have infinite probability density for finding the particle at $x = 0$ and zero for finding it elsewhere. But QM says that, first, the minimal energy is $E_0 = (1/2)\hbar\omega_0 \neq 0$, and second, the probability density is a continuous function determined by the first equation of (9.81). Both distributions are shown in Figure 9.10.

The residual energy $E_0 = (1/2)\hbar\omega_0$ cannot be taken away from the quantum oscillator. The existence of such energy is typical for bound systems (recall the hydrogen atom or particle in a potential well) and is a consequence of the quantum-mechanical indeterminacy. It is easy to show (Problem 9.20) that the averages $\langle p \rangle$ and $\langle x \rangle$ in any of the bound states (9.81) are zeros, and according to (3.40), the indeterminacy principle in this case reduces to

$$\langle p^2 \rangle \langle x^2 \rangle \geq \frac{1}{4}\hbar^2. \tag{9.84}$$

On the other hand, the average energy in a stationary state is equal to the corresponding eigenvalue E_n and can be written as

$$\langle E \rangle = \frac{\langle p^2 \rangle}{2\mu} + \frac{1}{2}\mu\omega_0^2 \langle x^2 \rangle. \tag{9.85}$$

Combining this with (9.84) gives

$$\langle E \rangle \geq \frac{\langle p^2 \rangle}{2\mu} + \frac{\mu \omega_0^2 \hbar^2}{8 \langle p^2 \rangle}. \tag{9.86}$$

This result describes the same situation as in Section 3.5: any decrease of the potential energy results in the increase of kinetic energy and vice versa. Minimal possible value of energy in the last expression considered as a function of $\langle p^2 \rangle$ is $E_{\min} = E_0 = (1/2)\hbar\omega_0$.

9.7
Oscillator in the E-Representation

As we know, the Hamiltonian in the E-representation is a diagonal matrix $H_{mn} = E_n \delta_{mn}$, and its eigenfunctions are the column matrices $|\psi_m\rangle = \delta_{mn}$. For an oscillator,

$$\hat{H} = \begin{pmatrix} \frac{1}{2}\hbar\omega_0 & 0 & 0 & 0 & \cdots \\ 0 & \frac{3}{2}\hbar\omega_0 & 0 & 0 & \cdots \\ 0 & 0 & \frac{5}{2}\hbar\omega_0 & 0 & \cdots \\ \vdots & \vdots & \vdots & \vdots & \ddots \end{pmatrix}, \quad |\psi_1\rangle = \begin{pmatrix} 1 \\ 0 \\ 0 \\ \vdots \end{pmatrix},$$

$$|\psi_2\rangle = \begin{pmatrix} 0 \\ 1 \\ 0 \\ \vdots \end{pmatrix}, \quad |\psi_3\rangle = \begin{pmatrix} 0 \\ 0 \\ 1 \\ \vdots \end{pmatrix}. \tag{9.87}$$

The most general state is a superposition of basis states (9.87):

$$|\Psi\rangle = \sum_m c_n |\psi_n\rangle, \quad \sum_n |c_n|^2 = 1 \tag{9.88}$$

or, in the matrix form,

$$|\Psi\rangle = c_1 \begin{pmatrix} 1 \\ 0 \\ 0 \\ \vdots \end{pmatrix} + c_2 \begin{pmatrix} 0 \\ 1 \\ 0 \\ \vdots \end{pmatrix} + c_3 \begin{pmatrix} 0 \\ 0 \\ 1 \\ \vdots \end{pmatrix} + \cdots = \begin{pmatrix} c_1 \\ c_2 \\ c_3 \\ \vdots \end{pmatrix}. \tag{9.89}$$

Here the c-amplitudes include the corresponding temporal factors (the *Schrödinger representation*): $c_n \to c_n(t) = c_n e^{i\omega_n t}$, $\omega_n = (n+1/2)\omega_0$ (recall also Section 8.9). Similarly, applying the general rules of Chapters 5 and 6 to coordinate x, and using

Equation F.22, we can find its matrix elements in the E-representation:

$$x_{mn} = \langle m|x|n\rangle = \frac{x_0}{\sqrt{2}}\left(\sqrt{n}\delta_{m,n-1} + \sqrt{n+1}\delta_{m,n+1}\right) \quad (9.90)$$

or

$$\hat{x} = \frac{1}{2}x_0 \begin{pmatrix} 0 & 1 & 0 & 0 & 0 & \cdots \\ 1 & 0 & \sqrt{2} & 0 & 0 & \cdots \\ 0 & \sqrt{2} & 0 & \sqrt{3} & 0 & \cdots \\ 0 & 0 & \sqrt{3} & 0 & \sqrt{4} & \cdots \\ \vdots & \vdots & \vdots & \vdots & \vdots & \ddots \end{pmatrix}. \quad (9.91)$$

Alternatively to the Schrödinger representation, we can include the temporal factors into matrix elements of the x-operator (the *Heisenberg representation*):

$$x_{mn} \rightarrow x_{mn}(t) = x_{mn}\,e^{i\omega_{mn}t}, \quad \omega_{mn} = (\omega_m - \omega_n) = (m-n)\omega_0. \quad (9.92)$$

Expressions (9.90)–(9.92) provide additional insights into the light–matter interactions. As we will see in Section 18.3, a matrix element x_{mn} determines the amplitude for quantum transition between states $|m\rangle$ and $|n\rangle$ for an oscillator in the radiation field. The structure of matrix (9.91) shows that (unless radiation is too intense) only transitions between two *neighboring* states are possible, with absorption or emission of a photon of energy $\hbar\omega_0$. Therefore, we can consider a state $|n\rangle$ as a result of absorbing n equal quanta of energy starting from state $|0\rangle$. This suggests a possibility of a *totally different interpretation of state $|n\rangle$ as a pure ensemble of n identical particles (not necessarily photons!)* with energy $\hbar\omega_0$ each. The irreducible energy $(1/2)\hbar\omega_0$ corresponding to a state $|0\rangle$ (no particles, vacuum state) can be attributed to the energy of vacuum fluctuations. We will explore this venue and its implications in Chapter 22.

Further, Equations 9.89–9.91 can be used for calculations of oscillator's characteristics in an arbitrary state $|\Psi\rangle$ given by (9.88) and (9.89). Since state $|\Psi\rangle$ is nonstationary, we can expect that $\rho(\mathbf{r},t)$, as well as the average position and momentum of the oscillator, will change with time. Regardless of representation (Schrödinger or Heisenberg), we get the same result when using expression (6.21) to find the average x-coordinate:

$$\langle x(t)\rangle = \sum_{m,n} c_m^*(t)x_{mn}c_n(t) = \sum_{m,n} c_m^* x_{mn}(t)c_n. \quad (9.93)$$

In view of (9.90)–(9.92), it is immediately seen that the last expression results in $\langle x(t)\rangle$ being the harmonic function of time with frequency ω_0:

$$\langle x(t)\rangle = a\sin(\omega_0 t + \varphi), \quad (9.94)$$

where a is the amplitude and φ is the phase of oscillations. Their exact values are determined by the amplitudes c_n.

A simple example will help visualize this transition from a set of basis states (9.87) with "frozen" probability distributions obtained from (9.81) and shown in Figure 9.10 to a vibrant wiggling cloud already resembling a familiar classical oscillator. Consider a superposition of two eigenstates $|\psi_1\rangle$ and $|\psi_2\rangle$ with the amplitudes $c_1 = c_2 = 1/\sqrt{2}$. Using (9.81) and (9.87), we have

$$\Psi(\xi, t) = \frac{1}{\sqrt{2}}\left(\psi_1(\xi)e^{-(E_1/\hbar)t} + \psi_2(\xi)e^{-(E_2/\hbar)t}\right)$$

$$= \pi^{-(1/4)} e^{-(1/2)\xi^2} \left[\xi e^{-i(3/2)\omega_0 t} + \left(\xi^2 - \frac{1}{2}\right)e^{-i(5/2)\omega_0 t}\right]. \quad (9.95)$$

The corresponding probability density is now a periodic function of time:

$$\rho(\xi, t) = \frac{1}{\sqrt{\pi}} e^{-\xi^2} \left[\xi^4 + \frac{1}{4} + \xi(2\xi^2 - 1)\cos\omega_0 t\right]. \quad (9.96)$$

The important thing about this expression is that the time-dependent term is an odd function of ξ and thus brings in the contributions of the opposite signs to the left and right semispace (Figure 9.11). Both these contributions themselves periodically change sign with the frequency ω_0. The density of the probability cloud oscillates in such a way that the local densities are phase shifted by 180° in the left and right sides of the cloud: when one on the left increases, the one on the right decreases and vice versa. As a result, the center of mass of the whole cloud is oscillating with frequency ω_0. In superposition of any two states other than neighboring states, the

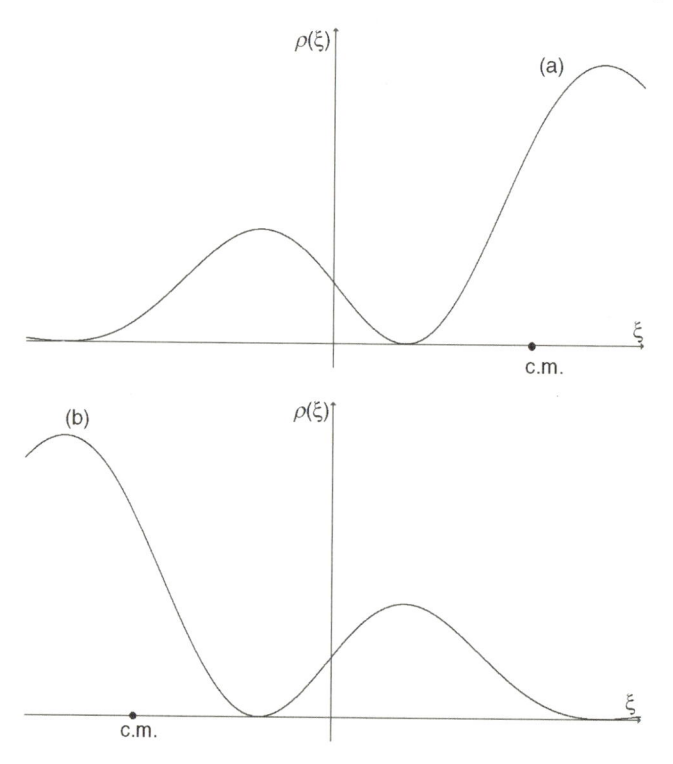

Figure 9.11 Nonstationary quantum oscillator; (a) and (b) represent $\rho(x, t)$ at two different moments of time separated by half of the period. Point c.m. indicates position of the center of mass.

9.8
The Origin of Energy Bands

Here we make a step to considering particle's motion in a periodic potential field. Such field represents real conditions encountered, for instance, in a perfect crystal. As always, we consider a simplified model: an unbounded (the field is truly periodic) 1D crystal. What are the basic features of particle's motion in such a field?

We can predict at least some of these features from the results of the previous sections. Let us start with a single rectangular potential well. We know that there is always at least one discrete bound energy level $E = -E_0$ in such a well. Suppose that this level is occupied. What happens if we bring in another identical well from infinity, so that now we have two wells sufficiently close to each other? The resulting potential function is shown in Figure 9.12. The second well, when isolated, can also accommodate at least one particle with the same energy. But when the wells are brought sufficiently close to each other, we have a new system – a pair of wells described by one Hamiltonian. We can consider this system as two valleys separated by a potential barrier in the middle. Since a barrier of finite size is penetrable, the wells are no longer isolated. This brings profound difference in the energy spectrum.

Suppose we have two identical particles both in the same state, and yesterday one particle was put into the left well and the other into the right one. Then when we look into the wells today and see one particle in each, we cannot tell whether the particle, say, in the left well is the same one that was put into it yesterday. The particles could

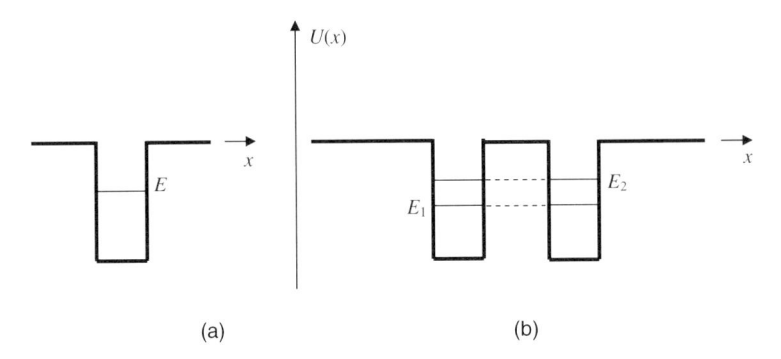

Figure 9.12 Double well. (a) An isolated rectangular potential well with one discrete energy state. Another such well infinitely far away can accommodate exactly the same single state. (b) The two wells are brought close together. Now they form one system, and instead of the initial single state with energy E, the system accommodates two states with energies E_1 and E_2, respectively. Each of these states *belongs to both wells*.

have exchanged their wells by tunneling. And there is no way – absolutely no way – to say whether this really happened or not.

It is also possible to find both particles in one of the wells, whereas the other one will be empty. But what if the particles are fermions? In this case, the only way they can be found in one common well is for them to have different energies. But this contradicts the initial condition ascribing the same energy state $E = -E_0$ to each well. The contradiction can only be resolved if the initial discrete level characterizing an isolated well splits into two levels E_1 and E_2 when the wells are coupled (Figure 9.12b). Each of the split levels belongs to both wells. The reason for splitting can be seen from the following argument. The appearance of another well close to the previous one can be considered as a perturbation. We will see later (in Chapter 17) that even a small perturbation shifts the initial level of a system. In our case, the perturbation converts the original single well into a double well with more than doubled effective width. Such a well can admit twice as many bound states as the original well. This becomes nearly obvious in the limiting case when the initial well was a potential box of width a, and we ask what would happen to level 1 if another box of the same width is brought close together. Suppose "close together" means so close that the remaining barrier between them becomes infinitesimally thin, making the tunneling probability close to 1. Then we can ignore the barrier and consider the result as just doubling the width of the original box (Figure 9.13). Consulting Equation 9.9, we see that in this case the first level will go down to a quarter of its original value, and the second level will lower down

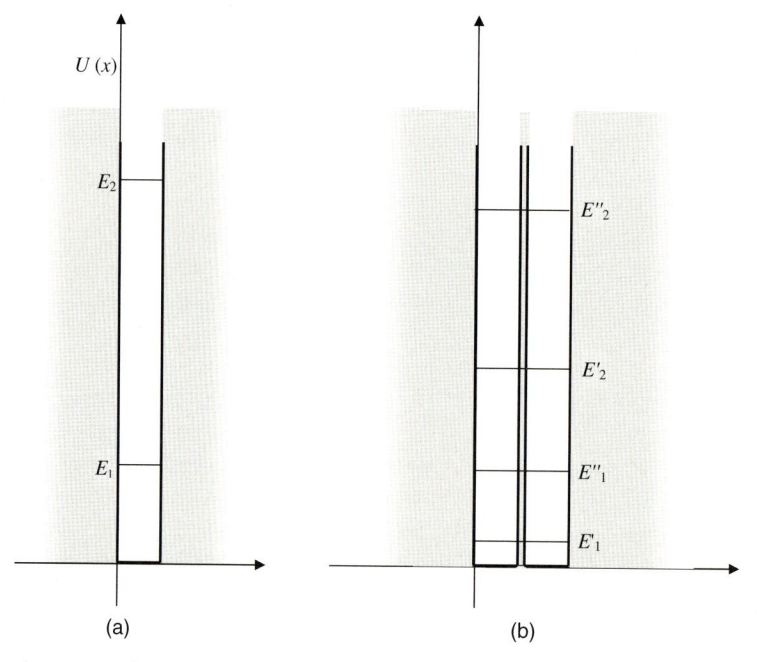

Figure 9.13 Illustration of the mechanism for the energy level splitting in a system of wells.

to the initial value of the first one. Now we will have two levels instead of one within the same stretch of the energy scale. And this holds for any n. The energy range $(0, E_n)$ of the original box will now contain $2n$ levels. The result is as if each initial level E_n shifts down and splits into two new ones. Focusing on a single original level of interest, we can write, without specifying n,

$$E_n \to \begin{cases} E_n + \Delta E_{n1} = E_{n1}, \\ E_n + \Delta E_{n2} = E_{n2}. \end{cases} \quad (9.97)$$

As must be evident from the above argument, the average $(1/2)(E_{n1} + E_{n2}) < E_n$.

The result (9.97) is the Nature's way to secure two *different* energy states for two fermions that were initially in one spin state and one energy state, but in two separate wells. The position of a well was an additional observable that, together with energy and spin, formed a complete set. But after the wells are coupled, position becomes incompatible with energy (a particle cannot have a definite energy *and* permanently reside in one definite well). The complete set shrinks down to two variables (energy and spin), and energy level has to split in order to save unitary (i.e., to ensure the coexistence of the two fermions in the new system without violating the Pauli principle).

This argument qualitatively explains the necessity for splitting within the rules of QM.

Reiterating it, we expect that in a system of J identical wells, a single level E_n of a separate well splits into J sublevels

$$E_n \to E_{n1}, E_{n2}, \ldots E_{nj}, \ldots, E_{nJ}, \quad (9.98)$$

so that there emerges a bundle of close discrete sublevels. But for a bound state the width of the bundle is restricted from above by the depth U of the wells, so that the sublevels get closer together with increase of J and produce the continuous energy band of a finite width in the limit when $J \to \infty$. In this limit, we will need, instead of j, a *continuous* parameter, to numerate different values of energy within the band. Such a parameter might be a variable k representing the average momentum of the particle (called *quasimomentum*), so that

$$E_n \to (E_{n1}, E_{n2}, \ldots, E_{nj}, \ldots) \to E_n(k). \quad (9.99)$$

Here n is numbering the levels in an isolated well, and j labels the wells. The relevance of quasimomentum k is connected with another specific feature of the structure: due to the possibility of successive tunneling, it can support the particle's *one-way* motion with a nonzero average momentum. The term "quasi" reflects the fact that the variable k is *not* an eigenvalue of the momentum operator whose eigenstates describe a *free* particle. And yet this new variable describes a one-way motion of a particle in the structure, which is similar to the one-way motion of a free particle. The one-way motion is impossible in a system of *finite* number of wells, whose bound stationary states have only zero average momentum. The contrast illustrates the role of the boundary conditions, which are different in both cases: we have the eigenfunctions vanishing at infinity and accordingly square integrable for discrete spectrum of a system of J wells, and eigenfunctions not vanishing at infinity and not square integrable for a periodic structure.

The reader familiar with the set theory [17,18] can ask: It is easy to understand that a discrete set of new levels (9.98) becomes infinite when $J \to \infty$. But how come such *discrete*, even if infinite, set becomes a *continuous* band (9.99)? In the language of the set theory, how does a numerable set become innumerable even though the set of wells is *always* numerable? The question appears purely mathematical, but it has a deep physical meaning. Because of the presence of other wells, the lifetime of a particle on the original level of a separate well becomes finite: the particle can now escape from its well by tunneling into other wells. In the limit $J \to \infty$, it can tunnel out *without returning*. As a result, the original level not only splits, but also spreads into a narrow zone. Such spreading occurs only when the "exodus" from the initial well becomes permanent, much in the same way as the process of electron "decay" from the trapped state in a slab (cold emission) discussed in Section 9.4. We will see in Chapter 11 that possibility of decay makes the corresponding energy level indeterminate. Thus, when the number of wells becomes infinite, it affects the particle's initial level E_0 in a twofold way: first, the level splits into an infinite number of sublevels as in (9.98) with $J \to \infty$; and second, each sublevel "smears out" into a subzone due to possibility of permanent escape from a given well. The subzones overlap to form a continuous energy band (9.99).

Consider now a few initial levels in an isolated well, say, $n-1$, n, and $n+1$. According to (9.99), in a periodic array of wells the levels spread into the corresponding bands $E_{n-1}(k)$, $E_n(k)$, and $E_{n+1}(k)$. The resulting picture of energy bands and gaps depends on parameters of the structure. For instance, in succession of widely separated narrow wells, their mutual influence on each other is relatively small, and the levels E_n will spread into narrow bands separated by wide gaps. In the opposite case of very close wells, we can expect broad bands separated by a very narrow energy gaps, or even overlapped bands.

Thus, already simple qualitative analysis yields some predictions of basic features of particle's states in a periodic structure.

9.9
Periodic Structures

Now we consider the behavior of a particle in periodic potential more rigorously. We will determine quantitatively the energy bands described in the previous section. The system of such bands is frequently referred to as the band or zone structure. We can use the term "energy spectrum" as well.

To find the spectrum, we need to find the stationary states of a particle. Once they are determined, we can also find their superposition, for example, describe the motion of a wave packet in a periodic field.

We need to find the eigenstates of a Hamiltonian with $U(x)$ being a periodic function of x with a period a:

$$\hat{H} = -\frac{\hbar^2}{2\mu}\frac{d^2}{dx^2} + U(x), \quad U(x) = U(x+a). \tag{9.100}$$

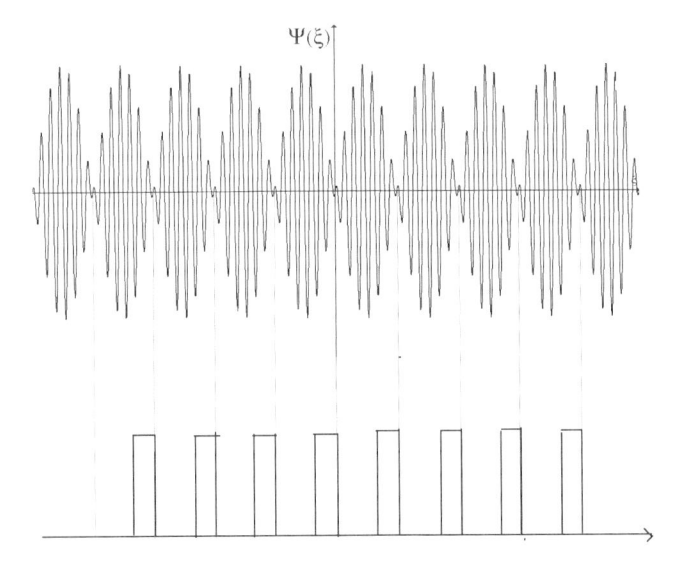

Figure 9.14 Kronig–Penney potential and the corresponding Bloch function.

The characteristic equation for such states is known as *Hill's equation* [19]. Already at this stage we can spot out an interesting property of its solutions. Consider an eigenstate $\Psi_{nk}(x)$ of (9.100) belonging to an eigenvalue $E_n(k)$ in an nth band. What happens to it under translation by distance a along the x-axis? In view of periodicity, Equation 9.100 looks after translation exactly the same as before. Therefore, its solution at $x + a$ can differ from $\Psi_{nk}(x)$ only by a phase factor. This can be satisfied if $\Psi_{nk}(x)$ is a product of periodic function $Q_{nk}(x)$ and a phase factor with real phase:

$$\Psi_{nk}(x) = e^{ikx} Q_{nk}(x), \quad Q_{nk}(x) = Q_{nk}(x + a). \tag{9.101}$$

Here k is the *quasimomentum* that describes the particle's motion along the x-direction.[2] From the mathematical viewpoint, the eigenfunction $\Psi_{nk}(x)$ is a plane wave e^{ikx} periodically modulated in step with periodicity of potential (Figure 9.14). It is called the *Bloch function* [20]. The property (9.101) of solutions of Hill's equation is known as the *Floquet theorem* [19].

The periodic factor $Q_{nk}(x)$ in an eigenstate (9.101) must be a *complex* function of x. Indeed, applying expression (3.17) for the probability flux density to a state (9.101) with real $Q_{nk}(x)$, we obtain $j = (\hbar k/\mu)\rho(x)$ with $\rho(x) = Q_{nk}^2(x)$. For real $Q_{nk}(x)$, the flux would be a periodic function of x. But this, in view of the continuity equation (3.16), would contradict the condition that state $\Psi_{nk}(x)$ is stationary. This condition can be satisfied only if the "amplitude" $Q_{nk}(x)$ is itself a complex

---

2) For a free particle, the wave vector **k** is sometimes loosely called momentum because it is rigidly connected to momentum by $\mathbf{p} = \hbar \mathbf{k}$. The term *quasimomentum* for **k** refers to a particle in a medium.

function (Problems 9.29 and 9.30). Writing it as $Q_{nk}(x) = Q^0_{nk}(x)e^{i\varphi(x)}$ with real $Q^0_{nk}(x)$ and $\varphi(x)$, and applying (3.17), we obtain

$$j = \frac{\hbar}{\mu}\left(Q^0_{nk}(x)\right)^2\left(k + \frac{d\varphi(x)}{dx}\right). \tag{9.102}$$

Now **j**, being the product of two functions of x, can be constant if $Q^0_{nk}(x)$ and $\varphi(x)$ are subject to the restrictive condition $\left(Q^0_{nk}(x)\right)^2(k + d\varphi(x)/dx) = \text{const}$.

A general quantitative treatment of the whole problem involves the solution of the Hill equation for an arbitrary periodic potential [19–21]. Here we illustrate basic features of the solution using a simple model of a *rectangular* periodic potential formed as an above-discussed infinite sequence of identical rectangular barriers and wells (the Kronig–Penney potential, Figure 9.14). Coincidentally, such a model describes an important optical device – a system of alternating layers with different index of refraction. In this model, it is convenient to find a solution separately in each region – one in a well and one in the adjacent barrier – and match the found solutions at the boundary between them.

The solution in region I (a well, $0 < x < a_1$) is

$$\Psi_I = A_1 e^{iq_1 x} + B_1 e^{-iq_1 x}, \quad q_1 \equiv \sqrt{2\mu(E - U_1)}/\hbar. \tag{9.103a}$$

And in region II (a barrier, $-a_2 < x < 0$) we have

$$\Psi_{II} = A_2 e^{iq_2 x} + B_2 e^{-iq_2 x}, \quad q_2 \equiv \sqrt{2\mu(E - U_2)}/\hbar. \tag{9.103b}$$

In the region of barrier III, the solution, according to (9.101), must differ from (9.103b) only by a phase factor e^{iak}:

$$\Psi_{III}(x) = e^{iak}\Psi_{II}(x-a) = e^{iak}\left(A_2 e^{iq_2(x-a)} + B_2 e^{-iq_2(x-a)}\right). \tag{9.104}$$

The continuity conditions for Ψ and Ψ' at $x = 0$ and $x = a_1$

$$\Psi_I(0) = \Psi_{II}(0), \quad \Psi'_I(0) = \Psi'_{II}(0), \quad \Psi_I(a_1) = \Psi_{III}(a_1), \\ \Psi'_I(a_1) = \Psi'_{III}(a_1) \tag{9.105}$$

give four equations for A_1, B_1, A_2, and B_2:

$$A_1 + B_1 = A_2 + B_2, \quad A_1 e^{ia_1 q_1} + B_1 e^{-ia_1 q_1} = e^{iak}\left(A_2 e^{-ia_2 q_2} + B_2 e^{ia_2 q_2}\right),$$

$$q_1(A_1 - B_1) = q_2(A_2 - B_2), \quad q_1\left(A_1 e^{ia_1 q_1} - B_1 e^{-ia_1 q_1}\right) = q_2 e^{iak}\left(A_2 e^{-ia_2 q_2} - B_2 e^{ia_2 q_2}\right).$$

$$\tag{9.106}$$

We can simplify this system by solving the two left equations for A_2 and B_2 and plugging the result into the equations on the right to get after some algebra (Problem 9.31)

$$\begin{pmatrix} \left(1+\frac{q_1}{q_2}\right)\left[e^{ia_1 q_1} - e^{i(ak-a_2 q_2)}\right] & \left(1-\frac{q_1}{q_2}\right)\left[e^{-ia_1 q_1} - e^{i(ak-a_2 q_2)}\right] \\ \left(1-\frac{q_1}{q_2}\right)\left[e^{ia_1 q_1} - e^{i(ak+a_2 q_2)}\right] & \left(1+\frac{q_1}{q_2}\right)\left[e^{-ia_1 q_1} - e^{i(ak+a_2 q_2)}\right] \end{pmatrix}\begin{pmatrix} A_1 \\ B_1 \end{pmatrix} = 0$$

$$\tag{9.107}$$

(we wrote the system in the matrix form). The system has a nonzero solution only if its determinant is zero. This requirement yields (Problem 9.32)

$$\cos ak = \cos a_1 q_1 \cos a_2 q_2 - \frac{1}{2}\left(\frac{q_1}{q_2} + \frac{q_2}{q_1}\right) \sin a_1 q_1 \sin a_2 q_2. \tag{9.108}$$

At $E < U_2$, the value q_2 becomes imaginary and Equation 9.108 takes the form

$$\cos ak = \cos a_1 q_1 \operatorname{ch} a_2 |q_2| - \frac{1}{2}\left(\frac{q_1}{|q_2|} - \frac{|q_2|}{q_1}\right) \sin a_1 q_1 \operatorname{sh} a_2 |q_2| \tag{9.109}$$

In view of (9.103a) and (9.103b), the transcendental equations (9.108) and (9.109) are the dispersion equations between energy E and quasimomentum k. They give for each k generally an infinite number of solutions $E_n(k)$ determining the corresponding energy bands. For each solution, we can determine the corresponding set of amplitudes A and B using Equations 9.106 and 9.107.

Figure 9.15 shows a few solutions $E_n(k)$. We see that in each band the energy $E_n(k)$ is a periodic function of k with the period $2\pi/a$:

$$E_n(k) = E_n(k + 2lk_a), \quad k_a \equiv \frac{\pi}{a}, \quad l = \text{integer}. \tag{9.110}$$

The set of points $k_l = 2lk_a$ on the k-axis is referred to as the 1D *reciprocal lattice* [21–23]. It can be considered as an imprint onto k-space of the periodic potential in the coordinate space.

In view of the symmetry of a stationary state with respect to direction of the one-way motion of a particle across the system, the function (9.110) must also be symmetric:

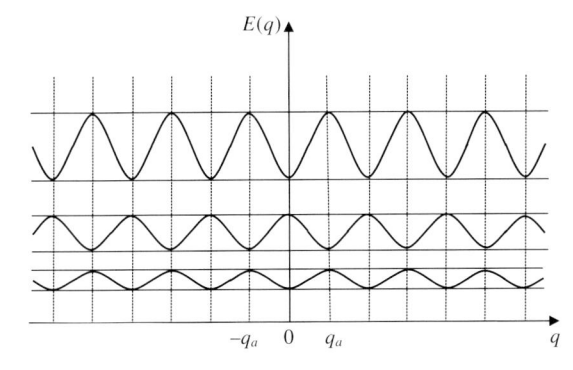

$$E_n(-k) = E_n(k). \tag{9.111}$$

Figure 9.15 The zone structure of a particle's energy spectrum in a periodic potential. The region $-q_a < q \leq q_a$ is the first Brillouin zone. The energy spectrum within this region is given by a multivalued function $E(q)$, and the corresponding representation is called the *reduced zone scheme*. Considering each brunch $E_j(q)$ as a separate periodic function of q within the same allowed energy band is a *repeated zone* or *periodic zone* scheme.

Being a periodic and an even function of k, the energy in each band can be expanded into Fourier series over cosine functions:

$$E_n(k) = \sum_{l=0}^{\infty} E_{nl} \cos(lak). \qquad (9.112)$$

The expansion coefficients E_{nl} depend only on the form of the potential energy $U(x)$.

The multiplicity of $E_n(k)$, together with periodicity (9.110), means that neither E nor k are *single-valued* functions of one another. The periodicity (9.110) illustrates the fundamental difference between quasimomentum k and momentum k of a free particle, for which $E(k) = \hbar^2 k^2 / 2\mu$. An immediate implication of this difference is that, as far as the energy is concerned, the quasimomentum, in contrast to ordinary momentum k, can be restricted to domain

$$-k_a \leq k < k_a, \qquad (9.113)$$

since the change of k within this domain gives all range of energy in each band. Indeed, consider an arbitrary value k' beyond this domain and associated energy $E_n(k')$. Then there is a value k *within* this domain, which differs from k' by a multiple integer of k_a, for instance, $k' = k + 2lk_a$. Due to periodicity (9.110), the value k' will be associated with the same energy (and other observable characteristics of the particle) as the value k, and therefore it will label *the same* physical state. In this respect, all physical characteristics of the particle can be represented by values of k within the range (9.113). This range forms what we call the *first Brillouin zone* in the k-space.

This property of the system can also be understood from a slightly different perspective.

The periodic function $Q_{nk}(x)$ can be expanded into the Fourier series, so that the whole state (9.101) can be written as

$$\Psi_{kn}(x) = \frac{1}{\sqrt{2\pi}} \sum_{l=-\infty}^{\infty} \Phi_{nl}(k + 2k_a l) e^{i(k+2k_a l)x}. \qquad (9.114)$$

Here $\Phi_n(k + 2k_a l)$ are the Fourier coefficients of function $Q_{nk}(x)$ (Problem 9.33). Physically, the energy eigenstate $\Psi_{kn}(x)$ is a superposition of states with different momenta

$$p \to p_l = \hbar(k + 2k_a l) \qquad (9.115)$$

with integer l. In other words, a state with definite energy here is a state with indeterminate momentum, as is always the case in a nonzero external field. The specific feature of our case is that this indeterminacy is "dispersed" in the momentum space – each given k produces discrete spectrum of equidistant values of p given by (9.115) (we will study the dispersed indeterminacy in Section 13.7). Once the solution is known, specifying a certain k within domain (9.113) automatically determines all other values of momentum outside this domain and the corresponding amplitudes $\Phi_n(k + 2k_a l)$, which are necessary for complete description of the system.

We can indicate a few other important properties of the system described by Equations 9.110 and 9.111.

I) As mentioned above, these equations generally have an infinite number of distinct solutions $E_n(k)$, which means an infinite set of distinct states characterized by *the same k* within the region (9.113), and as we already know, each nth branch of multivalued solution $E_n(k)$ is a periodic function of k with the period $2k_a$. In view of this, there are two possibilities:
   a) The neighboring regions of different branches, say, $E_n(k)$ and $E_{n\mp1}(k)$, overlap. In this case, the spectrum is continuously spread over a large range of E.
   b) These regions do not overlap. In this case, the spectrum has forbidden zones or gaps. A particle with energy within a forbidden gap cannot penetrate the structure.

   The regions of forbidden energies are determined by the conditions

$$\left|\cos a_1 q_1 \cos a_2 q_2 - \frac{1}{2}\left(\frac{q_1}{q_2}+\frac{q_2}{q_1}\right)\sin a_1 q_1 \sin a_2 q_2\right| > 1, \quad E > U_2$$
(9.116a)

or

$$\left|\cos a_1 q_1 \operatorname{ch} a_2|q_2| - \frac{1}{2}\left(\frac{q_1}{|q_2|}-\frac{|q_2|}{q_1}\right)\sin a_1 q_1 \operatorname{sh} a_2|q_2|\right| > 1, \quad E < U_2.$$
(9.116b)

II) As follows from (9.103b) and (9.109), the lowest possible value $E = U_1$ is forbidden (Problem 9.34). By continuity, the energies sufficiently close to U_1 are forbidden as well. In other words, the band spectrum starts with a forbidden zone at the bottom.

III) The width of allowed energy bands decreases together with the band number n, and has the least value at $n = 1$. This property can be understood intuitively in the framework of perturbations: the deeper the energy level in a separate well, the less it is perturbed by the presence of the neighboring well. Hence, the width of the zone originating from the lowest level is the least possible for a given system.

IV) It follows from condition (9.116a) and (9.116b) that the energy gaps can exist not only for $E < U_2$ but also for $E > U_2$. This result may appear somewhat puzzling. We can understand the allowed energy bands in the region $E < U_2$ as the result of tunneling through the barriers between wells, which converts into coordinated tunneling through an infinite system of barriers. We can also understand the gaps between the allowed bands in this region as the remainder of the corresponding gaps between the original discrete levels in a separate well. But why some energies *above the barriers* are banned even though *all* such energies are allowed for a finite number J of barriers? Since there are no forbidden energies above the barriers for a finite J, they seem to come from nowhere in the limit $J \to \infty$.

The answer to this puzzle is that *just as energy bands are generated due to tunneling, the energy gaps are generated due to reflection*. A system of infinite number of barriers can cause the *total reflection* for certain energies not only below but also above the barriers (recall that even a single barrier or even a well can reflect a particle no matter how energetic). The reflection becomes total for some energies in case of an infinite J.

The origin of *total* reflection can be understood in terms of interference and superposition of states. The net reflected wave is a superposition of waves reflected from individual barriers. Suppose that for a certain E the wave reflected from a barrier differs in phase from the wave reflected from the previous barrier by a multiple integer of 2π. Then these two waves as well as the waves reflected from all other barriers interfere constructively, thus producing maximal possible amplitude of reflection. Total reflection means that a particle with a corresponding energy cannot propagate across the system or cannot penetrate deep into it when incident from the outside. Such reflection from a layered system or from a system of parallel atomic planes in an ideal crystal (the *Bragg reflection*) is a well-known effect in the X-ray diffraction [21].

V) An unbounded periodic field accommodates the stationary states with nonzero momentum:

$$\Psi_{E_n}(x,t) = \Psi_{kn}(x)e^{-i(E_n(k)/\hbar)t} = Q_{kn}(x)e^{i[kx - (E_n(k)/\hbar)t]}. \tag{9.117}$$

A superposition of these states with k within a range $k - \Delta k \le k \le k + \Delta k$ makes a group (wave packet) similar to that of a free particle:

$$\tilde{\Psi}_n(x,t) = \int_{k-\Delta k}^{k+\Delta k} \Phi(k) Q_{kn}(x) e^{i[kx - (E_n(k)/\hbar)t]} \, dk \tag{9.118}$$

(we consider the simplest case when such packet is formed from the energy eigenstates belonging to one zone). Its only difference from the wave packet of a free particle is the way energy depends on k. The same argument as in Section 8.7 shows that the packet propagates with the group velocity

$$v_{ng} = \frac{1}{\hbar} \frac{dE(k)}{dk}. \tag{9.119}$$

The average quasimomentum associated with this motion is $\bar{p} = \mu v_{ng} = (\mu/\hbar)(dE_n(k)/dk)$ or, in view of (9.112),

$$\bar{p} = -a\frac{\mu}{\hbar} \sum_{l=0}^{\infty} l E_{nl} \sin(lak). \tag{9.120}$$

We immediately see that on the edges of the energy band, where $k = mk_a$, the group velocity, as well as the average momentum, is zero. The solution in such cases is the *modulated standing wave*. It is quite consistent with property IV, according to which the energy gaps are the regions of Bragg reflection, and the edge of an energy band is also the edge of the adjacent energy gap. The obtained result holds even for arbitrarily large m, that is, very far from the center of the first Brillouin zone in the quasimomentum space. This is another illustration of the fact that restricting to the first Brillouin zone is sufficient for description of all possible states of the particle in a periodic structure.

VI) Another interesting feature of the system is that the particle in periodic structure acquires an *effective mass* μ^* different from μ [20,24,25]. We can see it if we attempt to describe the particle's energy by the same equation as in free space. As seen from Figure 9.14, this can be done when the energy is sufficiently close to the edge of the corresponding allowed band. In this case, we can approximate $E_n(k)$ as

$$E_n(k) \approx \left\{ E_n^+ + \frac{1}{2} \frac{d^2 E_n}{dk^2} \bigg|_{k=0} k^2, \quad \text{in the middle of the first Brillouin zone,} \right. \tag{9.121a}$$

$$E_n(k) \approx \left\{ E_n^\pm + \frac{1}{2} \frac{d^2 E_n}{dk^2} \bigg|_{k=k_a} (k - k_a)^2, \quad \text{close to the border of the first Brillouin zone.} \right. \tag{9.121b}$$

Here $E_n^\pm$ is the top or the bottom, respectively, of the nth band. Comparing this with $E(k) = (\hbar^2/2\mu)k^2$ for a free particle, we see that we can attribute to the particle in a periodic structure an effective mass determined as

$$\frac{1}{\mu^*} = \frac{1}{\hbar^2} \frac{d^2 E_n}{dk^2} \bigg|_{k=0 \text{ or } k_a}. \tag{9.122}$$

The second-order derivative in (9.122) is a function of E, and thus is the effective mass. It is positive when the particle's energy is close to the bottom of the corresponding band. When the energy is close to the top of the band, the effective mass is negative!

As a transition to the next chapter, we complete the described treatment by generalizing it onto three dimensions. We will need to increase the number of variables: $x, y, z = x_1, x_2, x_3$ instead of x and $k_x, k_y, k_z = k_1, k_2, k_3$ instead of k. This will complicate some of the results. For instance, energy in a 3D crystal will be a function of $\mathbf{k} = (k_1, k_2, k_3)$, and generally it may depend differently on each component. The Brillouin zones will be not necessarily rectangular, they may be rather complicated polygons – see, for example, Ref. [22]. The Taylor expansion (9.121a) relative to the center of the first Brillouin zone will now read

$$E_n(\mathbf{k}) \approx E_n^\mp + \frac{1}{2} \sum_{i,j} \frac{\partial^2 E_n}{\partial k_i \partial k_j} \bigg|_{\mathbf{k}=0} k_i k_j, \quad i, j = 1, 2, 3 \tag{9.123}$$

(we assume that the energy band has its minimum or maximum at the zone center). In contrast to a free particle, expression (9.123) does not generally have spherical symmetry in the **k**-space. Physically, this means mass dependence on the direction of travel; mathematically, this is represented by mass being a second-rank tensor in 3D space (an *effective mass tensor* [24,25]). Comparison with (9.122) suggests that (9.123) can also be written as

$$E_n(\mathbf{k}) \approx E_n^{\mp} + \frac{\hbar^2}{2}\sum_{i,j}\left(\frac{1}{\mu^*}\right)_{ij} k_i k_j, \quad i,j=1,2,3, \quad \text{with} \quad \left(\frac{1}{\mu^*}\right)_{ij} \equiv \frac{1}{\hbar^2}\frac{\partial^2 E_n}{\partial k_i \partial k_j}\bigg|_{k=0}.$$
(9.124)

In most cases, $E_n(\mathbf{k})$ has minimum (or maximum) at the center of the first Brillouin zone. Then the elements of the effective mass tensor are represented by the diagonal matrix $(1/\mu^*)_{ii} \equiv (1/\hbar^2)(\partial^2 E_n/\partial k_i^2)_{k=0}$. Mass dependence on direction should not come as a big surprise when the particle is in an anisotropic environment. In relativity, even a free particle singles out the direction of its own motion in space. The space as such remains isotropic, but in terms of the particle's response to a force, it becomes "effectively anisotropic" when the particle is in motion. As a result, the particle's mass, defined by its reaction to a force [26], is different for a force acting along direction of motion and a force acting perpendicular to this direction [27–29]. In the first case, the corresponding mass is frequently referred to as *longitudinal mass*, and in the second case as the *transverse mass*. But the mechanism of emergence of the direction-dependent effective mass in QM is fundamentally different: it is the wave aspect of matter, which brings in the different response of the wave to different conditions imposed by the lattice onto its motion in different directions. The anisotropy of the lattice is the objectively existing fact irrespective of the particle's state. Accordingly, the particle's response to a force (acceleration) depends not only on magnitude of the force, but also on its direction, *even though the particle may be initially at rest*. Hence, the anisotropy of the particle's effective mass: it is now a tensor quantity.

Apart from direction, the effective mass of the electron depends on the energy band number and on the type of the crystal lattice. It may vary over a broad range of values, in most cases between $0.01\,\mu$ and $10\,\mu$.

The quantum-mechanical description of particle's motion in a periodic structure is of paramount importance in solid-state physics, and particularly, in the physics of metals. It gives full explanation practically for all observed phenomena in this area. For instance, the result about motion of a wave packet shows that the electron in an ideal crystal must move with constant average momentum, that is, without experiencing any resistance. The actually observed resistance is due to some local impurities and thermal motion producing fluctuating irregularities of the crystal lattice. Electrons scatter on these irregularities, and this affects their collective directed motion. With applied voltage turned off, such motion becomes increasingly less organized and eventually converts into heat with no collective charge transfer. Purifying the crystal and lowering its temperature decreases the resistance, in agreement with the general theory. The resistivity of an ideal metallic crystal must vanish at zero temperature. This prediction of QM contradicts CM, according to which there is always a nonzero resistance to the electron motion through a crystal, regardless of the crystal purity and temperature. The observed facts indisputably confirm QM to a highest accuracy achievable in the experiment. In particular, they show that resistivity of a conductor with perfect lattice is indeed becoming vanishingly small as the temperature approaches zero.[3]

Problems

9.1 Find the normalizing factor for an nth eigenstate of a particle in the potential box.

9.2 Show that expression (9.17) for the evolving "δ-spike" locked within a potential box is the solution of the corresponding Schrödinger equation.

9.3 Find the expression for the pushing force from a trapped particle in state (9.10).

9.4 Express the Heaviside step function in terms of the δ-function.

9.5 For a particle encountering a potential threshold, we require, in addition to continuity of its wave function, also continuity of its first derivative at the interface (Equation 9.27). Why is the similar condition not imposed for a particle in a box?

9.6 Find the probability flux when the particle with energy E is in a superposition of two de Broglie waves with amplitudes A and B, respectively, running in the opposite directions.

9.7 Prove the conservation of probability expressed in Equation 9.34.

9.8 Show that a function of the type $\psi(x) = \alpha e^{-kvx}$ with real k and v does not produce any flux.

9.9 Show that the coefficient of reflection of a particle incident on the potential threshold U is equal to 1 when the particle's energy $E < U$.

9.10 Solve the system (9.41).

9.11 Consider the reflection from the potential well as the result of the following multistaged process: an incident wave partially reflects from the first interface at $x = a$ interacting with it as with the potential step. The wave entering the well is partially reflected at the opposite face; upon returning to the first interface, it is partially transmitted and partially reflected back into the well, and so on. Calculate the amplitude of the net reflected wave as the result of interference of the first such wave and all consecutive transmitted waves returned to and passing through the first interface after the reflection from the second one. Show that the result amounts to R in (9.43).

9.12 Do the same with all the amplitudes transmitted in the forward direction and show that their sum amounts to T in (9.43).

---

3) This must not be confused with *superconductivity*, which is a totally different phenomenon including pairing of electrons into couples with the zero net spin (*Cooper pairs*), formation of the superfluidic ensemble of bosons and sudden disappearance of resistance at a finite temperature [30,31]. This is another purely quantum-mechanical effect totally alien to classical physics.

9.13 Explain why the reflection coefficient from a well of finite thickness (or its optical analogue – the corresponding glass plate) does not reach the value of 1.

(*Hint*: Recall that the external reflection from the plate with $n > 1$ is accompanied by the phase shift $\Delta\varphi = \pi$.)

9.14 Show that an eigenstate of parity operator can only have the zero expectation value of momentum.

9.15 Find the ratio a/A of "inner" and "outer" amplitudes for a particle bound in a potential well described in this section, in terms of the eigenvalues ν_j:

a) for the even states;
b) for the odd states.

9.16 For both types of the bound states in the well, determine the solutions completely, using the normalization condition.

9.17 Let k_n (n being an integer) denote one of the discrete sets of k at which the resonance transmission is observed. Define the half-width δk of the corresponding virtual level as the quantity such that for $k = k_n + \delta k$ the transmission coefficient is $T = 1/2$. Find δk in terms of k, particle's mass μ, and parameters of the well (a and U).

9.18 Show that Equations 9.62 and 9.66, with the reference point shifted to the bottom of the well, will give the known result (9.9) for a particle in a box in the limit $U \to \infty$.

9.19 Derive recurrent relation (9.76) for the eigenstates of harmonic oscillator.

9.20 Using (9.74), find

a) the peaks of the probability distribution for a quantum oscillator of mass μ in states with $n = 1, 2, 3$;
b) do the same for a classical oscillator with the same mass and energies corresponding to the same states and compare the results.

9.21 Prove that the expectation values of x and p are zeros in any bound stationary state with symmetric probability distribution.

9.22 Find the average energy distribution between kinetic and potential energies for the lowest state of quantum oscillator with characteristic frequency ω_0. Find it in two different ways:

a) from the general expressions for the average value of an observable in a given quantum state;
b) by minimizing the right-hand side of expression (9.85) and comparing the contributions of both terms.

9.23 Find the average kinetic and potential energies in the classical limit:

a) using the classical equations (9.67) and finding the averages U and K in (9.68) over the period;
b) using the classical probability (9.83);
c) compare your results with those of the previous problem.

9.24 Using the indeterminacy relation, prove that the energy of a quantum oscillator cannot be made less than the irreducible minimum $(1/2)\hbar\omega_0$.

9.25 Consider a superposition $\Psi(x,t) = c_1\psi_1(x,t) + c_2\psi_2(x,t)$ similar to (9.95) but with (a) $c_1 = -c_2 = 1/\sqrt{2}$ and (b) $c_1 = -ic_2 = 1/\sqrt{2}$. For each case, find the corresponding probability distribution and the average $\langle x(t)\rangle$.

9.26 Do the same with a superposition $\Psi(x,t) = c_1\psi_1(x,t) + c_3\psi_3(x,t)$. Explain the results.

9.27 Write down the system of equations (boundary conditions) determining the bound eigenstates and their eigenvalues for a particle trapped by a system of two identical rectangular wells shown in Figure 9.12b.

9.28 For additional point, solve this system.

9.29 Show that if a periodic function $Q_{nq}(x)$ in (9.101) is real, it cannot describe a stationary state in a nonzero periodic potential.

(*Hint*: Consider the expression for probability flux density and recall that it must be a constant in a stationary state.)

9.30 Find the relationship between the real and imaginary parts of $Q_{nq}(x)$ necessary for the flux to be constant.

9.31 Derive the system (9.107).

9.32 Derive Equation 9.108.

9.33 Without specifying explicit form of functions $k_1(q)$ and $k_2(q)$ determined by the dispersion equations (9.108) and (9.109) and the corresponding amplitudes A and B, find the amplitudes of the Fourier expansion (9.118) of an eigenstate (9.101) in terms of k_1 and k_2 and the amplitudes A and B.

9.34 Using Equation 9.103b and condition (9.109), show that the value $E = U_1$ is forbidden.

References

1 Hecht, E. (2002) *Optics*, 4th edn, Addison-Wesley, Reading, MA, p. 385.
2 Born, M. and Wolf, E. (1999) *Principles of Optics*, 7th edn, Cambridge University Press, Cambridge.
3 Fikhtengol'ts, G.M. (1970) *Course of Differential and Integral Calculus* (in Russian), vol. 2, Nauka, Moscow.
4 Arfken, G.B. and Weber, H.J. (1995) *Mathematical Methods for Physicists*, Academic Press, New York.
5 Weisstein, E.W. *Heaviside Step Function*. Available at http://mathworld.wolfram.com/HeavisideStepFunction.html.
6 Hildebrandt, T.H. (1938) Definitions of Stieltjes integrals of the Riemann type. *Am. Math. Mo.*, **45** (5), 265–278.
7 Pollard, H. (1920) The Stieltjes integral and its generalizations. *Q. J. Pure Appl. Math.*, **19**.
8 Shilov, G.E. and Gurevich, B.L. (1978) *Integral, Measure, and Derivative: A Unified Approach*, Dover Publications, New York.
9 Stroock, D.W. (1998) *A Concise Introduction to the Theory of Integration*, 3rd edn, Birkhauser, Boston, MA.
10 Carroll, L. (1999) *Through the Looking Glass*, Dover Publications, New York.
11 Schottky, W. (1923) Uber kalte und warme Elektronenentladungen. *Z. Phys. A*, **14** (63), 63–106.
12 Millikan, R.A. and Eyring, C.F. (1926) Laws governing the pulling of electrons out of metals under intense electrical fields. *Phys. Rev.*, **27**, 51.
13 Fowler, R.H. and Nordheim, D.L. (1928) Electron emission in intense electric fields. *Proc. R. Soc. Lond.*, **119** (781), 173–181..
14 Kleint, C. (1993) On the early history of field emission including attempts of tunneling spectroscopy. *Prog. Surf. Sci.*, **42** (1–4), 101–115.
15 Kleint, C. (2004) Comments and references relating to early work in field electron emission. *Surf. Interface Anal.*, **36** (56), 387–390.

16 Courant, R. and Hilbert, D. (2004) *Methods of Mathematical Physics*, vols. **1–2**, Wiley-VCH Verlag GmbH, Weinheim.
17 Hausdorf, F. (1991) *Set Theory*, 4th edn, Chelsea Publishing Company, New York.
18 Vilenkin, Ya. N. (1968) *Stories about Sets*, Academic Press, New York.
19 Magnus, W. and Winkler, S. (1979) *Hill's Equation*, Dover Publications, New York.
20 Blokhintsev, D.I. (1964) *Principles of Quantum Mechanics*, Allyn & Bacon, Boston, MA.
21 James, R.W. (1965) *The Optical Principles of Diffraction of X-Rays*, Cornell University Press, Ithaca, NY.
22 Ziman, J.M. (1972) *Principles of the Theory of Solids*, Cambridge University Press, Cambridge.
23 Merzbacher, E. (1998) *Quantum Mechanics*, 3rd edn, John Wiley & Sons, Inc., New York.
24 Pekar, S. (1946) The method of effective electron mass in crystals. *Zh. Eksp. Teor. Fiz.*, **16**, 933.
25 Pastori Parravicini, G. and Franco Bassani, G. (1975) *Electronic States and Optical Transitions in Solids*, Pergamon Press, Oxford.
26 Mermin, N.D. (2005) *It's About Time: Understanding Einstein's Relativity*, Princeton University Press, Princeton, NJ.
27 Landau, L. and Lifshits, E. (1981) *The Field Theory*, Butterworth-Heinemann, Oxford.
28 Fayngold, M. (2008) *Special Relativity and How it Works*, Wiley-VCH Verlag GmbH, Weinheim, pp. 485–487.
29 Tolman, R. (1969) *Relativity, Thermodynamics, and Cosmology*, Clarendon Press, Oxford.
30 Feynman, R., Leighton, R., and Sands, M. (1963–1964) *The Feynman Lectures on Physics*, vols. **1–3**, Addison-Wesley, Reading, MA.
31 Kittel, C. (1996) *Introduction to Solid State Physics*, 7th edn, John Wiley & Sons, Inc., New York.

10
Three-Dimensional Systems

10.1
A Particle in a 3D Box

Consider a quasi-free particle ($U = 0$ within a certain 3D region A and $U = \infty$ beyond A). Such a particle can be described by solutions of the Helmholtz equation (8.25) in A in terms of the momentum eigenstates

$$\Psi_{\mathbf{k}}(\mathbf{r}) = \Phi(\mathbf{k})e^{i\mathbf{k}\cdot\mathbf{r}} = \Phi(\mathbf{k})e^{i(k_1 x_1 + k_2 x_2 + k_3 x_3)} = \Phi(\mathbf{k})e^{ikr\cos\theta}. \tag{10.1}$$

The simplest case of A is the interior of a rectangular box with the sides a_1, a_2, and a_3 – a 3D extension of a box studied in Section 9.1. Let us call it the 3D box. In this case, it is convenient to use Cartesian coordinates x_1, x_2, and x_3 standing for x, y, and z, respectively, with the origin at one of the vertices and the axes along the respective edges. A solution in A will be determined by conditions

$$\Psi_{\mathbf{k}}(\mathbf{r}) = 0 \quad \text{at} \quad \begin{cases} x_1 = 0, & a_1, \\ x_2 = 0, & a_2, \\ x_3 = 0, & a_3. \end{cases} \tag{10.2}$$

Now note that the first form of (10.1) naturally splits into product of three functions, each depending only on its respective variable: $\Psi_{\mathbf{k}}(\mathbf{r}) = \prod_j \Phi_j\, e^{ik_j x_j}$, $j = 1, 2, 3$. This suggests looking for solution as a product of three functions $\Psi_{\mathbf{k}}(\mathbf{r}) = \prod_j \psi_j(x_j)$ with yet unknown $\psi_j(x_j)$. Putting this into (8.25) leads to

$$\frac{\psi_1''(x_1)}{\psi_1(x_1)} + \frac{\psi_2''(x_2)}{\psi_2(x_2)} + \frac{\psi_3''(x_3)}{\psi_3(x_3)} = -k^2 \tag{10.3}$$

(here double prime in each term stands for the second-order derivative with respect to the corresponding variable). Since each term is a function of a separate variable independent of others, they can sum up to a constant only if each of them is a constant as well. Denoting such constant for term number j as $-k_j^2$, we have

$$\psi_j''(x_j) = -k_j^2 \psi_j(x_j), \tag{10.4}$$

provided that $k_1^2 + k_2^2 + k_3^2 = k^2$. This condition is just the dispersion equation (8.26) for a particle with definite energy $E = \hbar\omega = \hbar^2 k^2/2\mu$. The general solution of (10.4)

Quantum Mechanics and Quantum Information: A Guide through the Quantum World,
First Edition. Moses Fayngold and Vadim Fayngold.
© 2013 Wiley-VCH Verlag GmbH & Co. KGaA. Published 2013 by Wiley-VCH Verlag GmbH & Co. KGaA.

is $\psi_j(x_j) = A_j\, e^{ik_j x_j} + B_j\, e^{-ik_j x_j}$, and the boundary condition (10.2) at $x_j = 0$ immediately selects $B_j = -A_j$, so that the solution takes the familiar form for a particle in a 1D box: $\psi_j(x_j) = A_j \sin k_j x_j$. The same condition at $x_j = a_j$ gives also familiar $k_j = n_j \pi / a_j$ with $n_j = 1, 2, \ldots$. Combining all these results, we obtain the set of eigenstates for a 3D box:

$$\psi(x_1, x_2, x_3) = \Phi \sin\left(\frac{\pi n_1}{a_1} x_1\right) \sin\left(\frac{\pi n_2}{a_2} x_2\right) \sin\left(\frac{\pi n_3}{a_3} x_3\right) \tag{10.5}$$

(with Φ as a normalizing factor), and the corresponding eigenvalues

$$E_{n_1 n_2 n_3} = \frac{\pi^2 \hbar^2}{2\mu}\left(\frac{n_1^2}{a_1^2} + \frac{n_2^2}{a_2^2} + \frac{n_3^2}{a_3^2}\right). \tag{10.6}$$

Any possible state here is determined by three quantum numbers n_1, n_2, and n_3.

10.2
A Free Particle in 3D (Spherical Coordinates)

Let us now get back to a completely free particle. If we want to use the corresponding states later for studying, say, scattering problems, the spherical coordinates would be more convenient. In this case, we choose the second form in (10.1), where θ is the angle between **k** and **r**. Choosing the z-axis along certain **r** (or **k**, depending on convenience) will make θ the polar angle of our coordinate system. Usually such axis is singled out by some experimental condition. For instance, in a scattering problem it is the direction of incidence. Note that even though the Hamiltonian with $U = 0$ is spherically symmetric, a plain wave (10.1) is not. However, as we saw in Section 8.6, a set of such waves with a fixed $|\mathbf{k}| = k$ can combine to form a spherically symmetric solution (8.45) describing a spherical standing wave. We had obtained this solution by *imposing*, apart from $k = \text{const}$, also the condition $\Phi(k\mathbf{n}) \to \Phi_0 = \text{const}$, which ensures the spherical symmetry. Now recall from Chapter 7 that such symmetry is observed only in states with $l = 0$, when the probability cloud is not spinning. Any spinning ($l \ne 0$) singles out the corresponding rotational axis, which breaks the spherical symmetry. In addition, it changes the radial dependence in such a way that maximum is observed at a certain radial distance from the origin. This makes sense because in order for a classical particle to have a nonzero **L**, it must orbit the center at a certain finite distance. In QM, this corresponds to the radial probability distribution peaking at a certain $r = r_l \ne 0$. Thus, any state of a free particle of energy E is a solution of the Helmholtz equation (8.25) and can be represented as a 3D Fourier integral of the type (8.43) (a superposition of plane waves with $k = \sqrt{2\mu E}/\hbar$).

Conversely, any eigenstate of the *linear* momentum (a plane wave solution of the Helmholtz equation) can be represented as a superposition of eigenstates of *angular* momentum, which are solutions of the same equation in spherical coordinates. Such reciprocity reflects the fact that $\hat{L}$- and $\hat{p}$-operators do not commute (Equation 7.16), so that each eigenstate of one operator can be a superposition of eigenstates of the other.

In the rest of this section, we will show how it works. Our problem now will consist of two parts. First, we must solve the Helmholtz equation in spherical coordinates, and second, we must expand a plane wave (10.1) over these solutions, and for this we will have to find the expansion coefficients.

The first part has been partially solved in Section 8.5, where the Schrödinger equation for stationary states was split into radial and angular parts. The solutions of the angular part are already familiar spherical harmonics. It remains now to find the radial factors. This takes us back to the radial Schrödinger equations (8.37) and (8.39) with $U(r) = 0$. The equation then reduces to

$$G_l'' + \left[k^2 - \frac{l(l+1)}{r^2}\right] G_l = 0. \tag{10.7}$$

Here primes stand for derivatives with respect to r, and subscript l shows that solution $G_l(r)$ depends on l.

Consider first the case $l = 0$. The solution in this case is trivial:

$$G_0(r) = B_1 e^{ikr} + B_2 e^{-ikr}, \quad R_0(r) = \frac{G_0(r)}{r} = \frac{B_1 e^{ikr} + B_2 e^{-ikr}}{r}. \tag{10.8}$$

The physically acceptable solution must be finite at $r \to 0$, which selects $B_2 = -B_1$.

Therefore, $G_0(r) = B \sin kr$, and we recover the radial part of (8.36). We obtained in different way already familiar result that spherically symmetric superposition (8.43) and (8.44) of plane waves describes the state with $l = 0$.

For an arbitrary l, the normalized solution is (Appendix G)

$$R_l(r) = 2(-1)^l \left(\frac{r}{k}\right)^l \left(\frac{1}{r}\frac{d}{dr}\right)^l \frac{\sin kr}{r}. \tag{10.9}$$

Radial functions $R_l(r)$ can be expressed in terms of the Bessel functions $J_{l+1/2}(kr)$ of half-integer order, or, alternatively, of the *Bessel spherical functions* $I_l(kr)$ [1]:

$$R_l(r) = \sqrt{2\pi\frac{k}{r}} J_{l+1/2}(kr) = 2k I_l(kr), \quad \text{where} \quad I_l(kr) \equiv \sqrt{\frac{\pi}{2kr}} J_{l+1/2}(kr). \tag{10.10}$$

For future applications, it is essential to know the asymptotic expressions of the radial functions for a free particle at $r \to 0$, as well as at $r \to \infty$. At small r, we expand $\sin kr$ in the Taylor series and retain only the term giving after all differentiations the lowest power of r:

$$\left(\frac{1}{r}\frac{d}{dr}\right)^l \frac{\sin kr}{r} \approx (-1)^l \left(\frac{1}{r}\frac{d}{dr}\right)^l \frac{(kr)^{2l+1}}{(2l+1)!r} = (-1)^l \frac{k^{2l+1}}{(2l+1)!!}. \tag{10.11}$$

Combining this with (10.9), we have in the vicinity of the origin

$$R_l(r) \approx k^{l+1} r^l. \tag{10.12}$$

In the second domain ($r \to \infty$), we retain only the term $(d/dr)^l \sin kr$: it falls off most slowly at large r, since each differentiation of $\sin \xi$ produces just $\sin(\xi - \pi/2)$. As a result, we get

$$R_l(r) \xrightarrow[r \to \infty]{} \frac{2}{r} \sin\left(kr - l\frac{\pi}{2}\right). \tag{10.13}$$

At $l = 0$, this reduces, as it should, to the radial part of (8.45).

Returning to the exact radial function (10.9) and combining it with the spherical function (8.36) for given l,[1] we obtain the solution for a free particle in a state with definite E (given k) and $\mathbf{L}$ (given l and m). The angular part is independent of k, whereas the radial function contains k as a parameter. We explicitly indicate this dependence here by providing Ψ and R with subscript k:

$$\Psi_{klm}(r, \theta, \varphi) = R_{kl}(r) Y_{lm}(\theta, \varphi) = 2k I_l(kr) Y_{lm}. \tag{10.14}$$

This completes our first task – to find the *energy eigenstates* with *definite l and m* for a free particle. Now we are in a position to expand a plane wave over these eigenstates. To simplify the expansion, direct the z-axis along $\mathbf{k}$, so that our plane wave takes the form $e^{ikz} = e^{ikr\cos\theta}$. This function is axially symmetric with respect to the z-axis; therefore, its expansion will contain only states independent of φ, that is, states with $m = 0$:

$$e^{ikr\cos\theta} = \sum_{l=0}^{\infty} \Phi_l R_{kl}(r) P_l(\cos\theta). \tag{10.15}$$

The expansion coefficients can be found using the standard procedure (see, for example, Ref. [1]) or by expanding both sides of (10.15) over powers $(r\cos\theta)^n$ and comparing coefficients (Problem 10.6). The result is

$$\Phi_l = \frac{l + 1/2}{k} i^l. \tag{10.16}$$

Using this and (10.13), we find asymptotic form of (10.15) far from the origin:

$$e^{ikr\cos\theta} \approx \frac{1}{kr} \sum_{l=0}^{\infty} (2l + 1) i^l P_l(\cos\theta) \sin\left(kr - l\frac{\pi}{2}\right). \tag{10.17}$$

Each term in this expansion contains the ratio $\sin(kr - l\pi/2)/r$, which is an intrinsic property of a spherical standing wave. The zeroth term ($l = 0$) is $\sin kr/kr$, which is, up to an independent factor, equal to (8.45). The terms with higher l are phase shifted with respect to the zeroth term and, in addition, are *not* spherically symmetric: as mentioned before, the lack of spherical symmetry is characteristic of states with $l > 0$.

Equations 10.15–10.17 complete the problem outlined in the beginning of this section.

---

1) The reader must be cautious using the term "spherical." We follow the conventional use of this term (spherical functions, spherical harmonics, etc.) according to which "spherical" is not necessarily spherically symmetric!

10.2.1
Discussion

Dealing with spherical *standing* waves provokes a question: Is it possible to use instead purely *converging* or *diverging* waves, that is, spherical analogues of the plane waves, with *flowing* energy? The answer is yes. A spherical standing wave is a superposition of diverging and converging waves, just as a linear standing wave is a superposition of oppositely running plane waves. In the simplest case $l = 0$, we have

$$\frac{\sin kr}{r} = 2i\frac{1}{r}\left(e^{ikr} - e^{-ikr}\right). \qquad (10.18)$$

Each term on the right is a possible (albeit not finite at $r \to 0$!) solution of the radial Schrödinger equation (8.37) with $l = 0$, $U(r) = 0$, and we can label it by conventional notation R used for the radial functions: $R_{k0}^{\pm} = e^{\pm ikr}/r$ (for simplicity, we drop normalizing factor here). For higher l, using the same argument as before, we can obtain the solution of the radial equation in the form

$$R_{kl}^{\pm} = (-1)^l \left(\frac{r}{k}\right)^l \left(\frac{d}{r\,dr}\right)^l \frac{e^{\pm ikr}}{r} \quad \text{or} \quad R_{kl}^{\pm}(r) = \pm i\sqrt{\frac{\pi k}{2r}} H_{l+1/2}^{1,2}(kr), \qquad (10.19)$$

where $H_{l+1/2}^{1,2}(kr)$ are the *Hankel functions* [1] of the first and the second kind for diverging and converging waves, respectively. Due to the factor i in the exponent in (10.19), each successive differentiation returns $\pm i = e^{\pm i(\pi/2)}$ as a coefficient in front of the exponent, which is equivalent to adding the term $\pm i(\pi/2)$ in the exponent itself. Therefore, using the same argument as we did for the standing wave solution, we get a similar asymptotic expression for diverging/converging wave solution:

$$R_{kl}^{\pm}(r) \approx \frac{1}{r} e^{\pm i(kr - l(\pi/2))}, \quad r \to \infty. \qquad (10.20)$$

10.3
Some Properties of Solutions in Spherically Symmetric Potential

Here we derive one general property of unbound states ($E > 0$) in potentials that fall off to zero at infinity. Consider the asymptotic behavior of solution at $r \to \infty$. Then the potential energy at infinity can be set equal to zero, $U(\infty) = 0$, as in case of the Coulomb potential. Accordingly, we must distinguish two cases, $E > 0$ and $E < 0$.

Under the above assumptions, both terms of the effective potential energy (8.40) in Equation 8.39 can be neglected sufficiently far from the origin if $U(r)$ falls off faster than r^{-1} (for the Coulomb potential, the situation is a little more subtle [2,3], but this does not affect the general conclusions below). As a result, Equation 8.39 far from the center reduces to (10.7) at $l = 0$ with solution (10.8), where $k \equiv \sqrt{2\mu E}/\hbar$ and $r \to \infty$. This solution critically depends on the sign of E. For $E > 0$, the state is unbound and has a continuous spectrum; for $E < 0$, we have $k \to i\kappa \equiv i\sqrt{2\mu|E|}/\hbar$, the state is bound and has a discrete spectrum. Indeed, in view of (8.38) we can write the asymptotic solutions as

$$R(r) \approx \begin{cases} B_1 \dfrac{e^{ikr}}{r} + B_2 \dfrac{e^{-ikr}}{r}, & E > 0 \quad \text{(a)}, \\ B_1 \dfrac{e^{-\kappa r}}{r} + B_2 \dfrac{e^{\kappa r}}{r}, & E < 0 \quad \text{(b)}. \end{cases} \qquad (10.21)$$

The first solution describes a superposition of diverging and converging spherical waves. Such states are unbounded and correspond to particle's scattering in the given potential, when the particle approaches the center from infinity and recedes back to infinity. For a definite energy E the state is stationary, and the probability flux of the particle in the incoming state must equal the flux in its outgoing state. This implies $|B_1| = |B_2| \equiv B$, and (10.21)a) can, up to an inessential phase factor, be written as a standing spherical wave

$$R_k(r) = B \frac{\sin(kr + \eta)}{r} e^{i\delta}, \quad E > 0. \qquad (10.22)$$

This generalizes formula (10.13) for a free particle. The scattering due to the potential field brings in a nonzero phase shift η. We will find more about scattering in Chapter 19.

The details of solution for $E < 0$ can be found in Supplement to Section 10.3.

10.4
Spherical Potential Well

As a specific example of a 3D system with nonzero spherically symmetric potential, consider a simplified model of a spherical potential well:

$$U(r) = \begin{cases} -U, & r < a, \\ 0, & r > a. \end{cases} \qquad (10.23)$$

The corresponding Hamiltonian generally has, apart from the spectrum of positive eigenvalues, also a discrete set of negative eigenvalues (bound states). Consider first the range $E > 0$. The radial Schrödinger equation with potential (10.23) is

$$G''(r) + \left[q^2 - \frac{l(l+1)}{r^2}\right] G = 0, \quad q^2 = \frac{2\mu(E+U)}{\hbar^2}, \quad r < a, \qquad (10.24a)$$

and

$$G''(r) + \left[k^2 - \frac{l(l+1)}{r^2}\right] G = 0, \quad k^2 = \frac{2\mu E}{\hbar^2}, \quad r > a. \qquad (10.24b)$$

Based on our results in Section 10.2 (Equation 10.10), we can write the solution for the radial function $R(r) = G(r)/r$ in the region within the well as

$$R_{ql}(r) = A \sqrt{\frac{q}{r}} J_{l+1/2}(qr), \quad r < a. \qquad (10.25)$$

10.4 Spherical Potential Well

Our naïve impulse might be to write the similar expression for region $r > a$, with the only change $A \to B$ and $q \to k$. But this would be wrong! Recall that the Bessel functions of all orders originate from $\sin kr/r$ by its consecutive differentiation according to (10.9). But this procedure was derived for $U = \text{const}$ everywhere. In our case, U makes a jump at $r = a$, and as a result, the radial Schrödinger equation is to be solved separately for the inner and outer regions. And the outer solution for $r > a$ cannot be subjected to the boundary condition $G_{kl}(r) = 0$ at $r = 0$. Therefore, its amplitudes B_1 and B_2 do not have to satisfy the requirement $B_2 = -B_1$, and we return to the general form (10.21). If we need the solution describing a stationary state, it must still be a standing wave, but the sufficient condition for this is $|B_1| = |B_2|$, which is much less restrictive than $B_1 = -B_2$. With $|B_1| = |B_2| \equiv B$, and focusing on the case $l = 0$, we can write the two amplitudes in (10.8) as $Be^{i\eta}$ and $Be^{-i\eta}$ with real B and η. This takes us back to already familiar expression (10.22), $R_{k0}(r) = B\sin(kr + \eta)/r$. The phase shift η here is due entirely to nonzero potential in the inner region. This is how the additional shift in (10.22) comes about!

Starting from this, one can build up the states with $l > 0$ applying the same rule (10.9) with the distinction that now the "generating function" $R_{k0}(r)$ will be (10.22) instead of (8.45), and the resulting radial functions $R_{kl}(r)$ will no longer be the Bessel functions, although they will be closely related to them.

But here, to illustrate the general principles, it will be sufficient to stick with the case $l = 0$. Denoting the solutions in the inner (I) and outer (II) regions as $R_{kl}^{(I)}$ and $R_{kl}^{(II)}$, respectively, we have

$$R_{k0}^{(I)}(r) = A\frac{\sin qr}{r}, \quad r < a, \quad \text{and} \quad R_{k0}^{(II)}(r) = B\frac{\sin(kr + \eta)}{r}, \quad r > a. \tag{10.26}$$

Now we must stitch them together at the well's boundary by imposing the continuity conditions

$$R_{k0}^{(I)}(a) = R_{k0}^{(II)}(a) \quad \text{and} \quad \left.\frac{dR_{k0}^{(I)}(r)}{dr}\right|_a = \left.\frac{dR_{k0}^{(II)}(r)}{dr}\right|_a. \tag{10.27}$$

Applying this to expressions (10.26) yields

$$\begin{aligned} A\sin aq &= B\sin(ak + \eta), \\ A(aq\cos aq - \sin aq) &= B(ak\cos(ak + \eta) - \sin(ak + \eta)). \end{aligned} \tag{10.28}$$

It follows from the first equation that

$$B = A\frac{\sin aq}{\sin(ak + \eta)}. \tag{10.29}$$

The ratio of the two equations in (10.28) determines η:

$$q\cot aq = k\cot(ak + \eta) \quad \text{or} \quad \eta = \text{arccot}\left(\frac{q}{k}\cot aq\right) - ak. \tag{10.30}$$

As mentioned in Section 10.1, the phase shift η turns out to be a function of k whose form depends on $U(r)$. Altogether we have a spherical standing wave with parameters determined by Equations 10.29 and 10.30. We can also consider such state as a superposition of converging and diverging spherical waves.

Now we turn to the bound states ($E < 0$). For these states, Equations 10.24a and 10.24b give

$$k = i\frac{\sqrt{2\mu|E|}}{\hbar} = i\kappa, \qquad q = \frac{\sqrt{2\mu(U - |E|)}}{\hbar} = \sqrt{\frac{2\mu U}{\hbar^2} - \kappa^2}. \qquad (10.31)$$

The solution in region I leads to the same result as in the first equation of (10.26). But in region II we now have a superposition of exponentially increasing and decreasing waves, and only the latter can be selected to describe a physical situation, so we are left with $R_{k0}^{(II)}(r) = B e^{-\kappa r}/r$. With this radial function, the two conditions (10.27) at $r = a$ give

$$A \sin aq = B e^{-a\kappa}, \qquad (10.32a)$$

$$A(aq \cos aq - \sin aq) = -B(a\kappa + 1)e^{-a\kappa}. \qquad (10.32b)$$

It follows that

$$B = A e^{a\kappa} \sin aq \quad \text{and} \quad aq \cot aq = -a\kappa. \qquad (10.33)$$

Combining this with (10.31) gives

$$\sin aq = \pm \frac{\hbar}{\sqrt{2\mu a^2 U}} aq. \qquad (10.34)$$

This equation selects allowed values of q (keep in mind that we must select only those of them for which $\cot aq < 0$, in view of (10.33)). The set of q found from (10.34) determines, through (10.31), the discrete set of allowed energies.

Denote

$$aq \equiv \xi \quad \text{and} \quad \frac{\hbar}{\sqrt{2\mu a^2 U}} \equiv \alpha, \qquad (10.35)$$

so that Equation 10.34 takes the form $\alpha \xi = \pm \sin \xi$. The solutions of this equation are given by the intersection points between the lines $y(\xi) = \alpha \xi$ and $y(\xi) = \pm \sin \xi$ (Figure 10.1). The intersections for which $\cot aq < 0$ are within the ranges $(\pi/2, \pi)$, $(3\pi/2, 2\pi)$, ... The points within the ranges $(0, \pi/2)$, $(\pi, 3\pi/2)$, ... must be discarded. Then we can see that for sufficiently large $\alpha \geq 1$ there are no intersections satisfying all physical conditions formulated above. The first intersection has the coordinates $\xi = \pi/2$, $y = \alpha \xi = 1$. In view of (10.31) and (10.35), this corresponds to

$$E = 0, \qquad U = U_{\min} = \frac{\pi^2 \hbar^2}{8\mu a^2}. \qquad (10.36)$$

In other words, the first bound state with $E = 0$ appears when the depth of the well is given by (10.36). We see that for a spherical well to be able to trap a particle of mass μ, its depth and size must satisfy the condition $a^2 U \geq \pi^2 \hbar^2/8\mu$. If the well is so shallow or small that this condition is not satisfied, then it is doomed to remain empty – it cannot accommodate a particle of given mass.

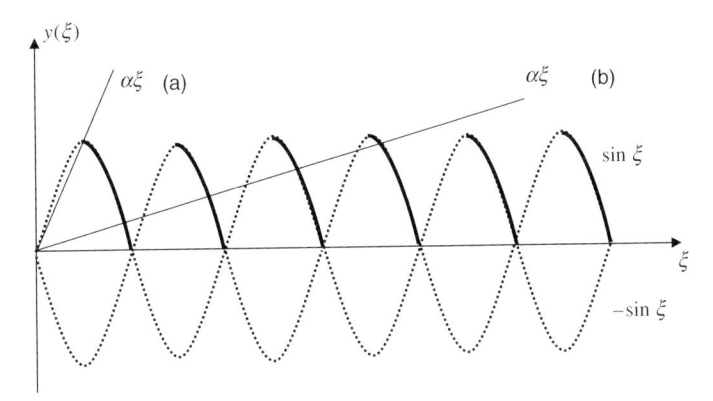

Figure 10.1 Graphical solution of Equation 10.34. The solutions are represented by intersection points of the curves on the left and right sides of this equation. Only those intersections count for which $\cot a q < 0$. The corresponding segments of curves $\pm \sin a q$ are shown by solid lines. At sufficiently small ξ, there are no intersections (no bound states in the well). (a) The critical value of α at which there appears first intersection – the emergence of the first bound state ($E_1 = 0$) at the critical size of the well. (b) A value of α at which the well can harbor four bound states.

Could the states with $l > 0$ be more easily accommodated by the well? The answer is no. Higher l bring in the centrifugal energy $l(l+1)\hbar^2/2\mu r^2$, which gives a positive contribution to E. Therefore, the eigenstates with $l \geq 1$ would have higher energy eigenvalues, and if the well cannot bind a state with $l = 0$, it would be even less so for states with higher l. So the critical condition (10.36) for trapping is general for 3D, and its existence is the fundamental difference from 1D wells.

10.5
States in the Coulomb Field and a Hydrogen Atom

Here we will study in more detail the states in the Coulomb potential. This is a case of paramount importance since it is the first step in our understanding of atoms. It describes the motion of an electron in a hydrogen atom, as well as in a helium ion (He^+), doubly ionized lithium atom (Li^{2+}), and so on. It also gives a good description of atoms of alkali metals such as Li, Na, Ka, and so on. The reason is that in these atoms a single valence electron is far outside the inner shell of other electrons tightly enveloping the nucleus and partially screening its field. As a result, the outer electron finds itself in a field only slightly different from the Coulomb field of the unit charge. Here we will restrict mostly to the electron bound states with $E < 0$. Our problem will be to find the corresponding eigenstates and the energy eigenvalues in these cases.

Strictly speaking, we have here a two-body problem, in which the constituting particles – the electron and respective nucleus – both move around their common center of mass. Their *relative* motion can be accurately described by introducing the

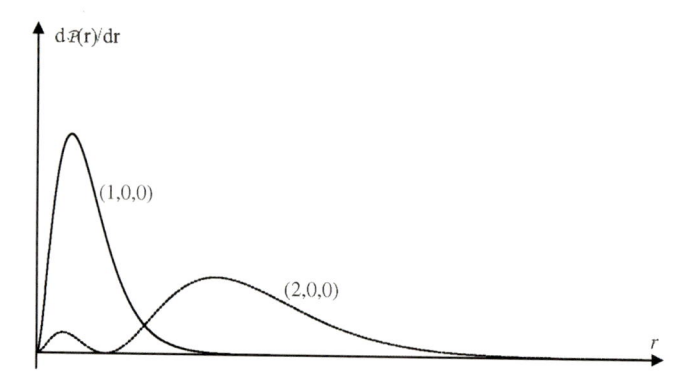

Figure 10.2 Radial probability distributions for the first two bound states in the Coulomb field.

reduced mass of the system $\mu^* \equiv M\mu/(M+\mu)$, where M and μ are the masses of the nucleus and electron, respectively. But since $M \gg \mu$, we can set $\mu^* = \mu$ with practically no loss of accuracy.

The radial functions found in Appendix H allow us to recover the energy eigenvalues (2.40) and write the complete expression for bound states in the Coulomb field as the product of the radial and angular solutions of the Schrödinger equation: $\Psi_{nlm}(r, \theta, \varphi) = R_{nl}(r) Y_{lm}(\theta, \varphi)$. The full solution is determined by the quantum numbers n, l, and m. As an example, we write here the normalized states for $Z = 1$, $n = 1, 2$, and $l = 0$:

$$\Psi_{100} = \frac{1}{\sqrt{\pi}} (a_0)^{-3/2} e^{-r/a_0}, \qquad \Psi_{200} = \frac{1}{4\sqrt{2\pi}} (a_0)^{-3/2} \left(2 - \frac{r}{a_0}\right) e^{-r/2a_0}.$$

(10.37)

The corresponding probability distributions are shown in Figure 10.2.

We will see that found solutions account for all basic features of a hydrogen atom. The atom's energy is determined by expression (H.12), which depends only on $n \equiv n_r + l + 1$. If n is given, then the orbital quantum number l can take on the values $l = 0, 1, \ldots, n-1$. And the magnetic quantum number, as we know from Section 7.1, for each given l can take on any of the values $m = 0, \pm 1, \pm 2, \ldots, \pm l$. Thus, the same energy eigenvalue E_n can correspond to different eigenstates due to multiplicity of the angular momentum states. We can evaluate the number of distinct states belonging to a given E_n by noticing that according to (7.10), for each l we have $2l+1$ possible states differing by number m. But l itself ranges from 0 to $n-1$. Therefore, the total number of different states with the same energy E_n is

$$\sum_{0}^{n-1} (2l+1) = n^2.$$

(10.38)

The succession of energy levels of a hydrogen atom is shown in Figure 10.3. In contrast to the spherical well, which can accommodate only a finite number of

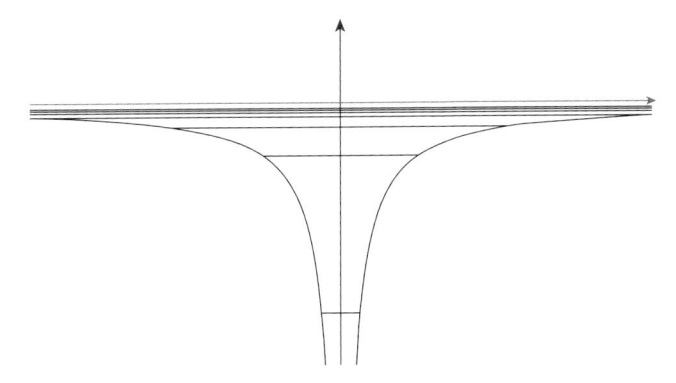

Figure 10.3

bound states or no such states at all, here we have an infinite number of energy levels and associated states. And in contrast to the quantum oscillator, for which the levels are equidistant, here they are getting ever closer to one another as we climb up the energy scale. Keep in mind that going up this scale in (2.40) means $E_n \xrightarrow[r \to \infty]{} 0$ since all E_n are negative. In the limit $n \to \infty$, the levels get infinitesimally close to one another, changing into continuous spectrum (electron freed from bondage, atom ionized) at $E \geq 0$. The minimal energy needed to ionize the atom from the ground level E_1 is called *ionization energy*, and is usually denoted as I. From (2.40) we get $I_{\text{Hy}} = -E_1 = 13.56\,\text{eV}$. The energy absorbed by the atom can come from the environment in the form of electromagnetic radiation (recall Sections 1.1 and 2.4). What happens if the incoming photon has the energy less than I_{Hy}? The answer depends on radiation frequency. If the frequency is such that the photon energy does not match any of the values $\Delta E_{1n} = E_n - E_1$, the atom usually remains unaffected in its ground state. Otherwise, the photon can be absorbed, in which case the electron will be transferred to a higher allowed level E_n (excited state). Usually such states are short-lived, with a typical average lifetime of about $10^{-12}\,\text{s}$. The state eventually decays back to the ground state, and the released energy is radiated in the form of a photon with the same frequency as before. This is the spontaneous emission discussed in the introductory chapters. But this is not the only one possible outcome. If n was high enough, the excited electron can "drop down" to the ground state in a few successive steps, via some intermediate energy levels. In this case, it will radiate a few photons with the corresponding frequencies, whose sum is equal to the frequency of the initially absorbed photon. Alternatively, instead of falling down onto the ground level, the excited electron can be excited again to even a higher level. If there are other photons whizzing around, and some have energy that matches one of the values $\Delta E_{nn'} = E_{n'} - E_n$ with $n' > n$, then there is a chance that the atom will absorb one of such photons while still on level n, in which case electron will be transferred to level n'. Thus, the atom can gain the energy $\Delta E_{nn'} = E_{n'} - E_n$ for some n and n' with $n' > n$; the radiation field will lose equal energy in the form of the absorbed photon (the number of photons will decrease by 1). So generally, the atom can interact with radiation through quantized energy exchange according to quantum rule $\hbar\omega_{nn'} = \Delta E_{nn'} = E_{n'} - E_n$. Putting here expression (2.40) for the energy values for a hydrogen atom, we have

$$\omega_{nn'} = \frac{\mu e^4}{2\hbar^3}\left(\frac{1}{n^2} - \frac{1}{n'^2}\right), \quad n' > n. \tag{10.39}$$

This expression describes a set of absorbed or emitted frequencies.

We have naturally come to observed structure of the absorption or emission spectrum as a set forming a matrix in Equation 1.16. The intensities for each transition will be considered in Chapter 18, where we will study light interaction with atoms in more detail.

The set of frequencies corresponding to optical transitions from different excited levels $n' > n$ onto the same lower level n forms a *spectral series*. For a hydrogen atom, all transitions to the ground level ($n = 1$) form the *Lyman series*. They follow from Equation 10.39 at $n = 1$, $n' = 2, 3, \ldots$. All corresponding lines are in the UV range of spectrum.

The transitions onto level $n = 2$ from the higher levels form the visible range of spectrum. The set of the corresponding spectral lines is called the *Balmer series* after the name of the first scientist who had formulated the rule (10.39) for the case $n = 2$ from analysis of the experimental data.

Let us now turn to a more detailed description of the quantum states $\Psi_{nlm}(r, \theta, \varphi)$ given by Equations 8.36 and H.16. The numbers n, l, and m, indicating physical characteristics E, L^2, and L_z, uniquely define the corresponding state $\Psi_{nlm}(r, \theta, \varphi)$ and therefore they form the complete set of observables. The number of variables in the set is 3, as one would expect for a system (atomic electron) with three degrees of freedom describing electron's motion in an atom.[2] Knowing these three numbers, we know $\Psi_{nlm}(r, \theta, \varphi)$, which gives us all relevant information. In particular, we can find the probability distribution

$$d\mathcal{P}_{nlm} = |\Psi_{nlm}(r, \theta, \varphi)|^2 r^2\, dr\, d\Omega = R_{nl}^2(r) r^2\, dr |Y_{lm}(\theta, \varphi)|^2\, d\Omega, \tag{10.40}$$

where $d\Omega = \sin\theta\, d\theta\, d\varphi$ is the element of solid angle. This equation allows us to consider separately radial and angular dependences. We can integrate (10.40) over the angles. Then, since the spherical harmonics $Y_{lm}(\theta, \varphi)$ here are normalized to 1, the result is the full probability $d\mathcal{P}_{nl}(r)$ of finding the electron somewhere between the two spheres of radii r and $r + dr$:

$$d\mathcal{P}_{nl}(r) = R_{nl}^2(r) r^2\, dr. \tag{10.41}$$

The factor r^2 comes from the 3D geometry: the probability of finding the electron within a shell of radius r is proportional to its volume $4\pi r^2\, dr$.

The distributions $R_{nl}^2(r) r^2$ in Figure 10.2 illustrate some basic features of one-electron states in an atom. The curves are labeled by the corresponding quantum numbers n and l, and the normalized radial coordinate r/a is plotted along the horizontal axis. As seen from the graphs, the number $n_r = n - l - 1$ first introduced

---

[2] We do not consider here the additional degrees of freedom corresponding to the electron spin and to the internal states (spin, etc.) of the nucleus.

in Appendix H determines the number of nodes in the radial function $R_{nl}(r)$. Note that we actually have here a nodular sphere rather than a nodular point, since r is the radius of the corresponding sphere. Thus, n_r has the meaning of number of nodes along an arbitrary radial direction, and accordingly it is called the *radial quantum number*.

The analysis of radial distributions sheds more light on the physical meaning of the Bohr radius a introduced in (2.41) and (3.45). Consider the distribution for the ground state $n = 1$. In this state, l can take only one value $l = 0$. The corresponding radial function, according to (H.16), is

$$R_{10}(\zeta) = N_{10}\, e^{-(1/2)\zeta} L_1^1(\zeta) = N_0\, e^{-r/a}, \tag{10.42}$$

and the radial probability distribution is

$$\rho_1(r) = \frac{d\mathcal{P}_1}{dr} = N_0^2 r^2\, e^{-2r/a}. \tag{10.43}$$

This function has a maximum at $r = a$. In other words, it is most likely to find the electron at a distance $r = a$ from the proton. Recalling (3.45), we can say that $a = \hbar^2/\mu e^2$ is the "preferred" distance at which we will find the electron most frequently in position measurement. In terms of Bohr's initial version of quantum theory with E and L quantized but the selected states still remaining the classical orbits, the quantity a is just the radius of the smallest possible orbit. Hence, its name – the Bohr radius. But, whereas in the Bohr theory there was a nonzero angular momentum associated even with the lowest state, in reality, the orbital momentum in this state is zero! The electron does not have to orbit the attracting center to avoid the "fall" onto it – the fall is precluded by the indeterminacy (recall Section 3.5). Therefore, the term "the Bohr radius" by no means must be taken literary as the radius of a classical orbit. Classically, the probability of finding the electron with $n = 1$ would be nonzero only at the surface of the sphere with radius a. In the language of the probability density, it could be described as $\rho(r) \sim \delta(r - a)$. With a distribution that sharp, it becomes clear why the *classical* description of a simple dynamic system does not need the concept of probability!

According to QM, the probability of finding the bound electron is nonzero in all space. This makes the term "probability distribution" intuitively clear.

Note that in considered states, the electron's kinetic energy $K(p)$ and potential energy $U(\mathbf{r})$ do not have definite values since, in contrast with L and L_z, neither $\mathbf{p}$ nor $\mathbf{r}$ have definite values; however, the *net* energy $E = \langle K(p) + U(\mathbf{r}) \rangle = E_n$ in each such state is sharply defined. This is analogous to situation with angular momentum $L^2 = L_x^2 + L_y^2 + L_z^2$, when the two components, say, L_x^2 and L_y^2, do not have definite values (since L_x and L_y do not have them), but their sum (together with L_z^2) is sharply defined.

But the difference between classical and QM description is not a complete chasm. There remains connection between them: the *maximum* of the QM distribution is at the classical radius. Therefore, from the viewpoint of current QM, it would be more appropriate to term a just as the *effective size* of the atom in its ground state.

For the states with higher n, we can find similarly that $a_n = an^2$. The effective size of the atom swells as it gets more excited. This seems natural since the higher n, the closer is the atom to its ionized state at which the electron breaks loose and can be found arbitrary far. The excited states with $n \gg 1$ exhibit many interesting features [4–8]. The atom in such states is called the *Rydberg atom*, and the states are called the *Rydberg states*.

We can now combine the outlined picture with the angular distributions derived in Section 7.1 for the eigenstates of the $\hat{L}^2$-operator. In all these states, we have the axially symmetric probability distribution – it does not depend on φ. In other words, the form of the resulting electron cloud has the symmetry of a body of revolution about the z-axis (the axis with definite components of L). The θ-dependence of the few corresponding probability clouds is roughly shown in Figure 3.1, where the polar system is used, so that the value of $\mathcal{P}_{lm}(\theta)$ is plotted along a direction specified by θ. For a state with $l = 0$ (called the s-state), the distribution has no angular dependence, so the electron cloud is spherically symmetric.

Clearly, there is no single classical orbit that could even approximately be associated with any of the atomic states. It is especially true for the s-states ($l = 0$). This was one of the difficulties of the Bohr old theory, but we know now that the lowest state does not need for its stability any orbital motion – it is maintained by the QM indeterminacy.

Consider now a state with $l = 1$ (the p-state). This state has three "substates" with $m = 0, \pm 1$, respectively. The corresponding angular distributions are given by the functions $P_1^1(\cos\theta)$ and $P_1^0(\cos\theta)$. From (E.36) we have

$$\mathcal{P}_{1,\pm 1}(\theta) = \frac{3}{8\pi}\sin^2\theta, \qquad \mathcal{P}_{1,0}(\theta) = \frac{3}{4\pi}\cos^2\theta. \tag{10.44}$$

The first of these expressions, as well as Figure 3.1, shows that, in contrast with classical description, we have a nonzero probability of finding an electron not only in the "equatorial" plane $\theta = \pi/2$, but also outside it. The correspondence with the classical picture is that the probability is maximal on this plane. The origin of the "angular spreading" of probability becomes more clear if we recall that the magnitude of **L** is $L_l = \hbar\sqrt{l(l+1)}$, whereas its z-component with $|m| = l$ is only $L_z = l\hbar$. Geometrically, this means that the symmetry axis of a virtual classical orbit is not along the z-axis but forms an angle $\alpha = \arccos(L_z/L) = \arccos\sqrt{l/l+1}$ with it. In other words, the angular momentum vector makes a nonzero angle α with the z-axis. Accordingly, the orbital plane, which is perpendicular to this vector, makes the same angle with the plane $z = 0$. But the angular momentum vector in QM cannot be represented by a directed arrow since in the state with definite L_z its x- and y-components are indeterminate. Therefore, it can be visualized as a continuous set of generatrices along the conical surface with z being its symmetry axis. Accordingly, instead of one orbital plane we have a continuous set of such planes, all tilted to the equatorial plane by angle α. For each orbital plane which passes, at a given φ, above the equatorial plane, there exists its symmetric counterpart below this plane. For a sufficiently small α, the maximal probability is between the orbital planes, right on the equatorial plane.

For the state $(1, 0)$ the angular momentum vector is perpendicular to z-axis, and the corresponding orbital plane contains this axis. But again, we have a continuous set of such vectors, this time all lying in the (x, y)-plane, and the continuous set of the corresponding orbital planes, all containing the z-axis. Therefore, in such state the probability of finding an electron bulges toward the z-axis, being maximal on it.

In a similar way, we can describe the shape of the electron configuration in states with higher quantum numbers l and m. Generally, the equation $P_l^m(\cos\theta) = 0$ has $s = l - |m|$ real solutions $\theta_1, \theta_2, \ldots, \theta_s$. These angles determine the corresponding nodal conical surfaces. The azimuthal part of the state function, $\Phi(\varphi) = e^{im\varphi}$, does not have the nodes, but its real or imaginary parts *taken separately* have m of them. In space, they form m nodal planes all passing through the z-axis. This is similar to the succession of m de Broglie waves "wrapped around" the symmetry axis; as mentioned in Chapter 1, this analogy was one of the motivations for de Broglie's groundbreaking work.

It is also interesting to note that the system of the nodal surfaces of the function $\Psi_{nlm}(r, \theta, \varphi)$ (set of n_r spheres, s conical surfaces, and $|m|$ planes) has the same geometry as the nodal surfaces of a vibrating solid sphere. In this respect, $\Psi_{nlm}(r, \theta, \varphi)$ are similar to the functions describing such vibrations. This situation repeats the one in Section 9.5, where the eigenfunctions of a particle in a box look similar to the functions describing vibrations of a string with the fixed edges.

10.6
Atomic Currents

Here we consider some experimentally observed properties of atomic states with $l \neq 0$. In a typical experiment, applied magnetic field **B** that singles out direction of z as a reference is much weaker than the internal atomic fields. Therefore, in the first approximation, it does not affect the basic characteristics of the electron state.

Even though QM has done away with the classical picture of a point-like particle in orbital motion, we still expect some circular motion associated with atomic electron – something that we have loosely named as the "rotation of probability cloud" around its center. For $m = 0$, one could try to visualize it as a doughnut-shaped cloud rotating around its symmetry axis *perpendicular to z*. Since the cloud is electrically charged (it represents an electron!), one could think of it as a circular current loop with the symmetry axis, say, along the y-direction. A current loop produces magnetic field and is characterized by magnetic moment along its symmetry axis. Thus, one could expect the existence of magnetic moment along the y-direction due to electron motion in a state with $l \neq 0$, $m = 0$. Well, this expectation is wrong because it does not go far enough along QM route: it does not take into account the QM indeterminacy for components of **L**. The assumption that $\mathbf{L} \upuparrows \hat{y}$ does not reflect this indeterminacy!

Using the conoid representation of **L** (Section 7.1), we can evaluate the product of indeterminacies $\Delta L_x \Delta L_y$. In the state $m = 0$, it reaches, according to Figure 7.1, its maximum value. This value can be estimated as $\Delta L_x \Delta L_y = (1/2)l(l+1)\hbar^2$, which means that either of the two components ranges within

$$-\sqrt{\frac{1}{2}l(l+1)}\hbar \leq L_x, L_y \leq \sqrt{\frac{1}{2}l(l+1)}\hbar. \tag{10.45}$$

In other words, **L** can have, with equal probability, any direction in the (x, y)-plane, and accordingly, we have a continuous set of orbital planes all containing the z-axis. The electron probability cloud is spread over *each* such plane, which also means over *all* of them. We already discussed such situation in the previous section.

For any orbit with the symmetry axis along some $\hat{n}$ in the (x, y)-plane, there exists its counterpart with symmetry axis along $-\hat{n}$ in this plane. The probability clouds representing these two orbits rotate in the opposite senses. Equivalently, we can say that the corresponding de Broglie waves flow in two opposite directions at once. The net flow is zero, and with it the net electric current and accordingly net magnetic field and magnetic moment are zero. Thus, a more careful QM treatment allows us to predict, without any calculations, that in a state with $m = 0$ the atom is neutral not only electrically but also magnetically (intrinsic magnetism of the electron and the nucleus not included!). This is another apparent paradox in QM: a state with nonzero angular momentum but zero projection onto a chosen direction does not show any experimental evidence of rotation.

Keep in mind that a state with definite *nonzero* value of a square of a certain vector quantity **A** may still have the zero *expectation value* of this quantity. A known example is a particle with definite nonzero energy E in a free space. Due to $E = p^2/2\mu$, this is also a state with definite nonzero p^2. However, some of these states may be characterized by $\langle \mathbf{p} \rangle = 0$, so that the particle is not moving anywhere (standing wave!). Similarly, in a state with $l \neq 0$, but $m = 0$, the particle has $\mathbf{L}^2 \neq 0$ and yet is not orbiting around anything. More accurately, its probability cloud is (on the average!) not rotating around anything. The words "on the average" in parenthesis are very important. If, in a state with $l \neq 0$, $m = 0$, we measure **L** in another basis (say, L_x-component), we may find $L_x \neq 0$. However, this will be already a different state, and, in addition, the statistical average of such measurements in a pure ensemble will be $\langle L_x \rangle = 0$.

Summary. In a "dumbbell-like" state $m = 0$ shown in Figure 3.1, there is no rotation. If, however, $m \neq 0$, then the virtual orbits with the same m have different but not opposite orientations, so there remains a certain irreducible circulation around the z-axis. Accordingly, we can expect a nonzero magnetic moment associated with such a state.

Now we describe this effect quantitatively. We use expressions (3.17) and (3.18) for probability current density.[3] Multiplied by the electron charge, it will give us the electric current density in a state $\Psi_{nlm}(\mathbf{r})$. Since we express $\Psi_{nlm}(\mathbf{r})$ in

---

[3] We will see in Chapter 16 that this expression changes in the presence of magnetic field, but in our case such change is negligibly small.

spherical coordinates, we must use the del operator in the same coordinates as well:

$$\nabla_r = \frac{\partial}{\partial r}, \quad \nabla_\theta = \frac{\partial}{r\,\partial\theta}, \quad \nabla_\varphi = \frac{\partial}{r\sin\theta\,\partial\varphi}. \tag{10.46}$$

Then, recalling that $\Psi_{nlm}(\mathbf{r}) = R_{nl}(r) P_l^m(\cos\theta) e^{im\varphi}$ with the first two factors being the real functions, we immediately see from Equation 3.18 that $j_r = j_\theta = 0$. There are no radial or "meridional" currents. The atom in a stationary state does not shrink or expand. Nor are there any flows from the "north" to the "south" pole or vice versa. The only possibility is the azimuthal current along the φ-coordinate lines. Since $\Phi(\varphi) \sim e^{im\varphi}$, its magnitude and direction ("west" or "east") depend on m. If $m = 0$, Equation 3.18 gives $j_\varphi = 0$. Thus, we now have a rigorous proof that there are no atomic currents in a state with $m = 0$. If, however, $m \neq 0$, we have a nonzero current in the "east–west" direction or vice versa, depending on the sign of m:

$$j_\varphi = -\hbar \frac{q_e}{\mu_e} \frac{\rho(r,\theta)}{r\sin\theta} m, \quad \rho = |\Psi|^2. \tag{10.47}$$

We can roughly visualize the atom in such state as rotating about the z-axis, producing a microscopic current loop.

Now we can find the net magnetic moment produced by this current. Consider a loop corresponding to a circular "filament" of radius $r\sin\theta$ and cross-sectional area $d\sigma = r\,d\theta\,dr$. The current in this loop is $dI = j_\varphi\,d\sigma$, and its contribution to the net magnetic moment is

$$d\mathcal{M} = \pi(r\sin\theta)^2\,dI = \pi(r\sin\theta)^2 j_\varphi\,d\sigma. \tag{10.48}$$

Combining this with (10.47) and integrating gives

$$\mathcal{M} = \int d\mathcal{M} = -\frac{1}{2}\frac{q_e}{\mu_e} m \int |\Psi_{nlm}(r,\theta,\varphi)|^2 2\pi r\sin\theta\,d\sigma. \tag{10.49}$$

But $2\pi r\sin\theta\,d\sigma$ is the volume of the filament loop and therefore the integral on the right is just the volume integral of $|\Psi_{nlm}|^2$, which gives 1 by normalization condition. We are left with

$$\mathcal{M} = -\frac{1}{2}\frac{q_e}{\mu_e}\hbar m = -m\mathcal{M}_B. \tag{10.50}$$

Here

$$\mathcal{M}_B \equiv \frac{1}{2}\frac{q_e}{\mu_e}\hbar \tag{10.51}$$

is the atomic unit of magnetic moment, called the *Bohr magneton*. Thus, the atom in a stationary state n, l, m with $m \neq 0$ is a microscopic magnet. Its magnetic moment is quantized in step with its orbital angular momentum along the z-direction. The "vectors" $\langle \mathcal{M} \rangle$ and $\langle \mathbf{L} \rangle$ are pointing in the opposite directions due to the " $-$ " sign in Equation 10.50 (this sign can be included into q_e to reflect the negative charge of the electron).

We can express $\mathcal{M}$ in terms of L (by writing $m\hbar = L_z$ in (10.50)) to show intimate connection between magnetic moment and angular momentum:

$$\mathcal{M}_z = -\frac{1}{2}\frac{q_e}{\mu_e} L_z. \tag{10.52}$$

Since the choice of reference direction is arbitrary (rotational invariance of space), the relation (10.52) would hold for any other component as well, which allows us to drop the z-subscript. The ratio

$$\frac{\mathcal{M}}{L} = -\frac{\mathcal{M}_B}{\hbar} = -\frac{1}{2}\frac{q_e}{\mu_e} \tag{10.53}$$

is called the *gyromagnetic* ratio. (We encountered *spin* gyromagnetic ratio earlier in Box 7.5; it turns out to be twice the *orbital* ratio (10.53) because we cannot model spin magnetic moment by a current loop). In view of the rotational invariance, this ratio is a universal property of the orbital motion of an atomic electron.

Summary. The magnetic properties of known particles are intimately connected with their electric and mechanical properties, specifically with their charge and angular momentum. The strength of the atomic magnet is determined by the quantum number m. This gives another explanation why m is called the *magnetic* quantum number.

10.7
Periodic Table

We have seen in Chapters 1–3 how CM fails and QM succeeds in explaining the existence of atoms. Now we can go much farther and explain the way the set of chemical elements is arranged, forming the *periodic system*. The many-electron atoms are extremely complex systems. And yet QM gives a natural explanation of their basic features and chemical characteristics. This explanation is based on two fundamental facts: (1) E- and L-quantization and (2) the Pauli exclusion principle. These two facts explain periodicity in chemical characteristics and behavior of chemical elements as we go to heavier and heavier atoms. An important breakthrough was the realization that actual factor determining the structure and chemical properties of atoms is the *atomic number N* rather than atomic weight A. The atomic number is determined by the number of protons in the nucleus, and therefore by the number of electrons needed to make the atom electrically neutral (this number is usually denoted as N in spectroscopy and as Z in many physical texts, so we can use both symbols interchangeably). Of course, each electron in the atom is affected by the presence of its partners. But even a highly simplified model considering only attraction of electrons toward the nucleus and neglecting the electron interactions between themselves is generally sufficient to understand the basic features. Therefore, in what follows we will use such an idealized model.

We can get the basic idea of how quantum rules work in chemistry if we make a few steps up the periodic table and in each step consider a ground state of the corresponding atom. No new information is needed – we must just apply the two rules formulated above and use the "magic" of quantum numbers.

The first step – hydrogen – is just a rehash of what we know. The electron in the ground state occupies the lowest level with $n = 1$. The quantum numbers l and m in this state are both zero, so the state is described by the triplet (1, 0, 0). The corresponding probability distribution is spherically symmetric and is known as the s-state. The probability density for $n = 0$ peaks at the center, and we have the probability cloud shown in Figure 3.1a.[4] We can denote this state as 1 s (one electron in the s-state). Even though there is only one state of orbital motion at $n = 1$, the cloud can still accommodate one more electron with the opposite spin. This explains why the hydrogen atom is chemically active – it has a tendency to take an electron from some other atom close by if that electron is weakly bound to its original host.

Now go to He atom ($N \equiv Z = 2$). It has two electrons. The lowest available ground state has only one place (the state $n = 0$ is not degenerate). It can accommodate two electrons only if they are in different spin states. So now the cloud shown in Figure 3.1a is formed by two electrons with opposite spins. In this case, they more than just tolerate – they welcome each other's presence. The opposite electron spins produce opposite magnetic moments that attract each other. This diminishes the effect of Coulomb repulsion. And, with strong attraction of both electrons toward the nucleus, it is not easy to separate them. According to Equations 2.40 and 2.41, the absolute value of energy would be proportional to Z^2 and effective radius of the electron cloud would be inversely proportional to Z^2 *for a single electron* in the field of He nucleus. The electrons are closer to the center, and it would take more energy to shift one of them a little farther away from the other, let alone separating them. The ionization energy necessary to remove one electron from the He atom is 24.6 eV – nearly twice that of the hydrogen atom. On the other hand, there is no way for a third electron to get close, since its spin state would be the same as that of one of the occupants. The first "electron shell" associated with level $n = 1$ is filled. As a result, the He electrons do not form any chemical bonds. He is a chemically inert monatomic gas. The spectroscopic notation for the corresponding electron structure is $1s^2$ (two electrons both in 1s state).

The monovalent hydrogen and inert He make the first period of the periodic table. The third element of the table (Li) has three electrons. Since the first shell is already filled, the best the third electron can do to lower its energy is to occupy the state with $n = 2$ (the second electron shell). This state is $n^2 = 4$-degenerate with respect to angular momentum (one s-state and three p-states with $l = 1$). Each of these states can accommodate two electrons with the opposite spins, so altogether there are eight vacancies to form the second electron shell. Already from this fact we can predict the second period of the table as the group of eight elements. And this is what had been known long before the onset of QM but could not be explained. Li is the first element of the second group. The third electron in Li occupies the second shell, whose effective radius is $n^2 = 4$ times greater than the radius of the first shell. Therefore, the third electron feels the

---

4) We must distinguish between the *radial distribution* in Figure 10.2 (probability to find an electron within a thin spherical shell of radius r) and *probability density* $\rho(\mathbf{r})$ of finding the electron in the vicinity of point **r**, shown in Figure 3.1.

Coulomb field of the Li nucleus partially screened by the first shell. Since it is the only occupant of the second shell, it can choose either the s-state with $l = 0$ or any of three p-states with $l = 1$. In a purely Coulomb field, all these choices would have equal chances because all of them would be associated with the same energy. But the field in the Li atom is not purely Coulomb field, and the degeneracy is lifted. The energies of the s- and p-states are slightly different, and electron has an opportunity to select the state with smaller energy. This is the s-state, because its probability distribution has, as mentioned before, its peak at the center. So for an electron in the s-state the chance to be found close to the center is higher than that for a p-electron. But closer to the center means stronger Coulomb attraction corresponding to $Z = 3$, whereas farther from it means weaker (and partially screened) Coulomb field corresponding to $Z = 1$. The former option is associated with lower energy, so the electron is in the 2s state. But even so, the energy of this state is of course much higher than that of the ground state (the ionization of Li would require only 5.4 eV). Therefore, Li is very active – totally opposite to He. In the periodic table, it is in the same column as hydrogen. The formula describing electron configuration in Li reads $1s^2 2s$: two electrons in 1s state and one electron in 2s state.

Adding, one by one, a new proton (together with neutron) to the nucleus and electron to the atomic shell, we will obtain successively the Be atom, B atom, and then C, N, O, F, and Ne. In F, we have the second shell being one electron short of completion. Its electron configuration is $1s^2 2s^2 2p^5$. An extra electron from the environment, for example, a valence electron in another atom close by, is attracted by the F shell since filling it could lower the net energy of both atoms. This makes F a very aggressive chemical element, a member of halogen family. In Ne, all eight states of the second shell are occupied. Its electron configuration is $1s^2 2s^2 2p^6$ (two s-electrons in the first shell, and two s-electrons and six p-electrons in the second shell). The exclusion principle does not allow to add a new electron to the filled shell. On the other hand, since the shell is complete and all magnetic moments are compensated, it is difficult to deform or break it by removing one or more electrons. We have again an inert monatomic gas that does not participate in any chemical reactions and does not form any compositions. This completes the second period of the periodic system. Any new electrons required for the nuclei with more protons than in Ne can be added starting only with state with $n = 3$, thus forming the third shell. The number of states in this shell is determined from Equation 10.40 as $2n^2 = 18$. Filling here the s- and p-states leads to a group from Na to Ar. This makes the third period of the table, similar to the period from Li to Ne. The next element after Ar is K. Starting from K, the correspondence between the simplified model we use and actual succession of chemical elements becomes less straightforward, due to increasing contribution of electron interactions. Our original model suggests that the valence electron of K must be in the d-state of the third energy level $(n, l) = (3, 2)$. In reality, this electron is in the s-state of the shell no. 4, where it resides leaving the shell no. 3 incomplete. This means that the state $(n, l) = (4, 0)$ in K has lower energy than the state $(n, l) = (3, 2)$. But even in such cases, the system of all possible electron states described by quantum numbers n, l, and m plus the spin variables remains as the basis of electron configuration, up to the heaviest elements of the periodic table (the reader can find more details in Refs [2,9]).

Summary. QM explains not only the existence of atoms, but also all their observed properties. Atoms turn out to be systems with complex electron configurations, exhibiting 3D layered structure (electronic shells). In particular, all chemical properties of atoms are explained by behavior of their external (valence) electrons. The properties of solutions of the Schrödinger equation for a hydrogen atom (quantization of energy and angular momentum), together with the Pauli exclusion principle, are sufficient for understanding the behavior and basic chemical characteristics of all known atoms.

Problems

10.1 In the beginning of Section 10.1 we said that the de Broglie waves (10.1) can serve only as building blocks for a solution describing a physical situation for a quasi-free particle. Explain why a *single* de Broglie wave will not work in this case?

10.2 Find the eigenstates of a particle in a rectangular potential box with the sides a_1, a_2, and a_3. Formulate the conditions under which the energy eigenstates can be degenerate.

10.3 Find the normalizing factor for an eigenstate (n_1, n_2, n_3) of a particle in a 3D potential box.

10.4 Determine the force exerted by a particle in a 3D box on its walls.

10.5 In the previous problem, find the pressure exerted by the particle on the walls. Is this pressure isotropic?

10.6 Derive expression (10.16) for the coefficients in the expansion of a plane wave over eigenstates of angular momentum

10.7 Find the eigenstates of a 3D quantum oscillator with potential energy $U(\mathbf{r}) = (1/2)k(x^2 + y^2 + z^2)$

   a) using Cartesian coordinates;
   b) using spherical coordinates

10.8 For the first solution (10.37), find

   a) the average radial distance of a bound particle from the center;
   b) the variance of its radial distance from the center.

10.9 For the principal quantum number n, find the possible values of l (orbital quantum number) and n_r (number of terms in the corresponding Laguerre polynomial) for an electron in the Coulomb field.

10.10 Using the known atomic constants, evaluate from Equation 10.39 the wavelengths of the first three lines ($n' = 2, 3, 4$) in the Lyman series.

10.11 Calculate the frequencies and wavelengths for the same transitions in hydrogen-like ions He^+ and Li^{2+}.

10.12 a) Find the frequencies for transitions between different Rydberg states, $\omega_{nn'} = \Delta E_{nn'}/\hbar$, as a function of n, assuming $n' - n \ll n$.

b) Find the fractional difference $\Delta a_{nn'}/a \equiv (a_{n'} - a_n)/a$ of the atomic sizes in such states as a function of n, under the same assumption as in (a).

10.13 Derive the expression for atomic current in a one-electron atom in a state with a nonzero magnetic quantum number m.

References

1 Arfken, G.B. and Weber, H.J. (1995) *Mathematical Methods for Physicists*, Academic Press, New York.

2 Blokhintsev, D.I. (1964) *Principles of Quantum Mechanics*, Allyn & Bacon, Boston, MA.

3 Landau, L. and Lifshits, E. (1965) *Quantum Mechanics: Non-Relativistic Theory*, 2nd edn, Pergamon Press, Oxford.

4 Herzberg, G. (1937) *Atomic Spectra and Atomic Structure*, Prentice-Hall, Englewood Cliffs, NJ.

5 Karplus, M. and Porter, R.N. (1970) *Atoms & Molecules*, Benjamin Cummings & Co., Inc., Menlo Park, CA.

6 Gallagher, T.F. (1994) *Rydberg Atoms*, Cambridge University Press, Cambridge.

7 Metcalf Research Group (2004) *Rydberg Atom Optics*, Stony Brook University.

8 Murray-Krezan, J. (2008) The classical dynamics of Rydberg Stark atoms in momentum space. *Am. J. Phys.*, **76** (11), 1007–1011.

9 Feynman, R., Leighton, R., and Sands, M. (1963–1964) *The Feynman Lectures on Physics*, vols. **1–3**, Addison-Wesley, Reading, MA.

11
Evolution of Quantum States

Up to now, we were mostly concerned with stationary states. But they are only a special case of quantum phenomena. This chapter is devoted to a much more general case of evolving states. We will focus primarily on continuous evolution described by the Schrödinger equation.

11.1
The Time Evolution Operator

The Schrödinger equation gives an algorithm expressing a state $\Psi(\mathbf{r}, t)$ in terms of the initial state $\Psi(\mathbf{r}, 0)$, provided the Hamiltonian of a system is known. For the purposes of this section, we shall rewrite it as

$$\frac{\partial}{\partial t}\Psi = \frac{\hat{H}}{i\hbar}\Psi. \tag{11.1}$$

Restrict to an important class of problems with time-independent $\hat{H}$ and differentiate (11.1) with respect to time:

$$\frac{\partial^2}{\partial t^2}\Psi = \frac{\hat{H}}{i\hbar}\frac{\partial}{\partial t}\Psi. \tag{11.2}$$

Expressing the time derivative on the right in terms of $\hat{H}\Psi$, we get

$$\frac{\partial^2}{\partial t^2}\Psi = \left(\frac{\hat{H}}{i\hbar}\right)^2 \Psi. \tag{11.3}$$

Iterating this process gives

$$\frac{\partial^n}{\partial t^n}\Psi = \left(\frac{\hat{H}}{i\hbar}\right)^n \Psi, \quad n = 0, 1, 2, \ldots. \tag{11.4}$$

Now use the Taylor expansion to write $\Psi(\mathbf{r}, t)$ at an arbitrary moment in terms of the initial state $\Psi(\mathbf{r}, 0)$:

$$\Psi(\mathbf{r}, t) = \sum_n \frac{1}{n!}\left.\frac{\partial^n \Psi(\mathbf{r}, t)}{\partial t^n}\right|_{t=0} t^n = \sum_n \frac{1}{n!}\frac{\partial^n}{\partial t^n}\Psi(\mathbf{r}, 0)t^n. \tag{11.5}$$

Quantum Mechanics and Quantum Information: A Guide through the Quantum World,
First Edition. Moses Fayngold and Vadim Fayngold.
© 2013 Wiley-VCH Verlag GmbH & Co. KGaA. Published 2013 by Wiley-VCH Verlag GmbH & Co. KGaA.

Though the Schrödinger equation itself only provides the first time derivative, the corollary (11.4) that follows from this equation gives us all we need:

$$\Psi(\mathbf{r}, t) = \left(\sum_n \frac{1}{n!} \left(\frac{\hat{H}}{i\hbar} \right)^n t^n \right) \Psi(\mathbf{r}, 0). \tag{11.6}$$

Treating the sum of powers of an *operator* $\hat{H}t/i\hbar$ in the same way as we would treat the sum of powers of a *variable*, we can write the whole expression in parentheses as an exponential function of the operator:

$$\sum_n \frac{1}{n!} \left(\frac{\hat{H}}{i\hbar} \right)^n t^n \equiv e^{\hat{H}t/i\hbar}. \tag{11.7}$$

Then (11.6) becomes

$$\Psi(\mathbf{r}, t) = e^{\hat{H}t/i\hbar} \Psi(\mathbf{r}, 0) \equiv \hat{U}(\hat{H}, t) \Psi(\mathbf{r}, 0), \tag{11.8}$$

where we have introduced

$$\hat{U}(\hat{H}, t) \equiv e^{-i\hat{H}t/\hbar}. \tag{11.9}$$

We see that $\Psi(\mathbf{r}, t)$ emerges by acting with operator $\hat{U}(\hat{H}, t)$ on the initial state $\Psi(\mathbf{r}, 0)$. It is therefore natural to call $\hat{U}(\hat{H}, t)$ the *time evolution operator*.

To see how it works, look at the simplest case when $\Psi(\mathbf{r}, 0)$ is an eigenfunction of $\hat{H}$:

$$\Psi(\mathbf{r}, 0) = \psi_a(\mathbf{r}), \qquad \hat{H}\psi_a(\mathbf{r}) = E_a \psi_a(\mathbf{r}). \tag{11.8}$$

In this case, the total state function oscillates with frequency $\omega_a = E_a/\hbar$:

$$\Psi(\mathbf{r}, t) = \psi_a(\mathbf{r}) e^{-i\omega_a t} = \Psi(\mathbf{r}, 0) e^{-i\omega_a t}. \tag{11.9}$$

The time evolution operator will produce the same result:

$$\hat{U}(\hat{H}, t)\psi_a(\mathbf{r}) = \left(\sum_n \frac{1}{n!} \left(\frac{\hat{H}}{i\hbar} \right)^n t^n \right) \psi_a(\mathbf{r}) = \sum_n \frac{1}{n!} \left(\frac{E_a}{i\hbar} \right)^n t^n \psi_a(\mathbf{r})$$
$$= \psi_a(\mathbf{r}) e^{-i(E_a/\hbar)t}. \tag{11.10}$$

Now let $\Psi(\mathbf{r}, 0)$ be a superposition $\Psi(\mathbf{r}, 0) = \sum_a c_a \psi_a(\mathbf{r})$. Since $\hat{U}(\hat{H}, t)$ is linear, it will act on each term in (11.10) independently and $\Psi(\mathbf{r}, t)$ will be the weighted sum of the individual results:

$$\Psi(\mathbf{r}, t) = \sum_a c_a \psi_a(\mathbf{r}) e^{-i\omega_a t}. \tag{11.11}$$

This is precisely the expected evolution of such a state in a fixed environment. The result shows that the time evolution operator does not change the *probabilities* of eigenstates but it changes *the phases* of their respective amplitudes according to (11.11). Therefore, it can dramatically change the future shape of a state. We will discuss this result in more detail in Sections 11.5–11.8 for the case of continuous energy spectrum.

Looking again at (11.10), we see that an energy eigenvector is also an eigenvector of $\hat{U}(\hat{H}, t)$. But its eigenvalue is a complex exponential, which means $\hat{U}(\hat{H}, t)$ is *not* Hermitian. This does not contradict anything since hermiticity is a feature of operators representing *observables*, and $\hat{U}(\hat{H}, t)$ does not belong to this class.

The matrix form of $\hat{U}$ depends on the basis used. In the energy basis, we obtain from (11.10):

$$U_{\alpha\beta} = e^{-i\omega_\beta t} \begin{cases} \delta_{\alpha\beta}, & \alpha, \beta \text{ integers,} \\ \delta(\alpha - \beta), & \alpha, \beta \text{ continuous.} \end{cases} \quad (11.12)$$

The $\hat{U}$-matrix in this representation is diagonal, and it is immediately seen that the diagonal elements (11.12) are complex numbers. The elements of the Hermitian adjoint $\hat{U}^+$ differ from those of $\hat{U}$ only in the sign in the exponent. It follows that $\hat{U}^+\hat{U} = 1$ and $\hat{U}$ is a unitary operator. This is what one should expect, since evolution must preserve probability. Operationally, the evolution of a state is a succession of unitary transformations from one moment to the next.

What if instead of the energy basis we used another basis, composed of eigenstates $|q_n\rangle$ of some operator $\hat{Q}$? To make this case truly different, we shall require that $\hat{Q}$ does not commute with $\hat{H}$. Then expanding the initial and final state functions over $|q_n\rangle$ we see that $\hat{U}$ transforms a matrix column of coefficients $a_n(0)$ into matrix column of $a_n(t)$. Indeed, we have

$$|\Psi(t)\rangle = \sum_n a_n(t)|q_n\rangle, \qquad |\Psi(0)\rangle = \sum_n a_n(0)|q_n\rangle. \quad (11.13)$$

In view of (11.8), the two expansions are related as $\sum_n a_n(t)|q_n\rangle = \hat{U}(t)\sum_n a_n(0)|q_n\rangle$. Projecting this onto $|q_m\rangle$ gives

$$\sum_n a_n(t)\langle q_m|q_n\rangle = \sum_n a_n(0)\langle q_m|\hat{U}(t)|q_n\rangle \quad (11.14a)$$

or, in view of orthonormality of the Q-basis,

$$a_m(t) = \sum_n U_{mn}(t) a_n(0), \quad (11.14b)$$

where $U_{mn}(t)$ are the matrix elements of $\hat{U}(t)$ in this basis. To elucidate their physical meaning, consider a special case when the system was prepared in an eigenstate of $\hat{Q}$ with eigenvalue q_l. In this case $a_n(0) = \delta_{ln}$, and (11.14b) reduce to

$$a_m(t) = U_{ml}(t). \quad (11.15)$$

But amplitude $a_m(t)$ determines the probability $\mathcal{P}_{ln}(t) = |a_m(t)|^2$ of obtaining q_m in a Q-measurement. So for a system prepared at $t = 0$ in an eigenstate $|q_l\rangle$ there appears after time t a nonzero probability to be in an eigenstate $|q_m\rangle$. So $|U_{ml}(t)|^2$ is equal to the probability of transition $|q_l\rangle \Rightarrow |q_m\rangle$ after the system was allowed to evolve for time t. Elements of the time evolution matrix in Q-basis acquire a clear physical meaning of transition amplitudes between the corresponding eigenstates of this basis.

Note that eigenstates $|q_n\rangle$ themselves remain unchanged. It is the state $\Psi(t)$ that evolves with time – from $|\Psi(0)\rangle = |q_l\rangle$ to $|\Psi(t)\rangle = \sum_n a_n(t)|q_n\rangle = \sum_n U_{ln}(t)|q_n\rangle$

having nonzero projections $U_{ln}(t)$ onto $|q_n\rangle$ at a time $t > 0$. We can regard this as another aspect of the motion of the state vector: as time goes on, its components change continuously. This is another facet of the phenomenon discussed in Sections 8.3 and 8.7, when we realized that probability density $\rho(\mathbf{r})$ of finding the particle at point $\mathbf{r}$ changes as the wave packet evolves with time: $\rho(\mathbf{r}, 0) \rightarrow \rho(\mathbf{r}, t)$. In particular, as we saw in Figure 8.3, there appears a nonzero probability of finding the particle at location $\mathbf{r}$ far beyond the narrow region around $\mathbf{r}'$ that was its initial "residence."

Box 11.1 Probability amplitudes: time dependence

Note that the transition amplitudes are generally time dependent, and accordingly, the state vector is always "wandering" over the Hilbert space even in a stationary environment. This can look as a contradiction since the stationary environment means a time-independent Hamiltonian with definite eigenvectors. If we take the set of these vectors as the basis and the amplitudes c_α in (11.11) are constants, then the state vector appears to be stationary in this basis. A vector stationary in one basis is stationary in any other basis (unless the basis itself is rotating, which is not supposed here). Accordingly, the transition probabilities must be time independent if the Hamiltonian is time independent! How come then that we obtain the *time-dependent* transition amplitudes in this case?

The answer to this is that we must again draw a sharp distinction between *probability* and *probability amplitude*. The former may be stationary even when the latter is not. When we said "the amplitudes c_α in (11.11) are constants," we overlooked the temporal factors $e^{-i\omega_\alpha t}$. They cannot be attributed to the basis states $\psi_\alpha(\mathbf{r})$ since we decreed that the latter should not rotate or oscillate. Then they must go to c_α making the complete amplitudes time dependent. In some basis, *the probabilities may be fixed while the amplitudes oscillate*. It is these oscillations that result in probabilities being functions of time *with respect to some other basis*. This is why the spectrum of a particle may be fixed, say, in the momentum space (recall Sections 8.3 and 8.7) but evolve in the configuration space. The meaning of the word "spectrum" is "spectrum of eigenvalues," and it is different in different bases. It is the energy or momentum spectrum in the momentum space, and coordinate spectrum in the configuration space. But as we turn our attention from spectrum (i.e., from $|\Phi(\mathbf{k})|^2$) to the amplitudes themselves, they are time dependent in any basis! Or, if you choose the Heisenberg picture, the corresponding *operators* are time dependent. In any event, this leads to evolution of the state vector. As you think of it, there is another, even simpler analogy: the vector evolves, but its norm does not. This is another manifestation of the fact that the Hilbert vector space is a complex space. In this space, a state may have a fixed probability for each axis and yet evolve with time due to oscillations of its phases. In a real vector space, the fixed probabilities would automatically involve fixed projections and thus the state vector just frozen in space.

This is another reason why a quantum state must be described by a complex function!

11.2
Evolution of Operators

Our next task is to find out how an arbitrary observable changes with time. In CM, the evolution rate of observable L is given by the time derivative dL/dt. What would be the quantum analogue of this derivative? Let us start with a certain state $|\Psi(0)\rangle$ that evolves with time to become $|\Psi(t)\rangle$ at a moment t. Our first, "classical" impulse is to measure L at this moment, repeat the measurement at a later moment $t' = t + \Delta t$, and see by how much the result has changed. But then we realize that generally in either measurement there is a whole set of possible eigenvalues that L could take. The eigenvalue L measured at time t and eigenvalue L' measured at time $t' = t + \Delta t$ in a *different sample* that was prepared identically to the first (the original sample will be "spoiled" by the first measurement!) do not have to be close at all. It would be totally meaningless to seek a quantum analogue of the derivative by comparing individual values obtained in a pair of consecutive measurements. The only quantum characteristic that behaves like the classical derivative is the rate of change of the *expectation value* of L. So we will be looking for the quantity

$$\frac{d\langle L \rangle}{dt} \equiv \frac{\langle L \rangle_{t+dt} - \langle L \rangle_t}{dt}, \tag{11.16}$$

where angle brackets indicate *ensemble average*. This quantity defines the time evolution of the corresponding dynamic variable.

Generally, the change of expectation value consists of two different contributions: one from the possible explicit time dependence of the *observable itself* (which changes its eigenvalues), and the other from the time evolution of the *state* (which changes the probability for each eigenvalue). A case in point: the potential energy of an electron inside a capacitor may change directly due to the change of $U(\mathbf{r}, t)$ when charging the capacitor, and also due to the evolution of the electron state. On the other hand, the variance $\Delta x(t)$ characterizing the width of a spreading wave packet depends on time even when the operator $\Delta \hat{x} = x - \bar{x}$ is time independent. In this case, Δx changes only because of the change of state. Based on this principle, we shall differentiate the average, applying the product rule:

$$\frac{d\bar{L}}{dt} = \frac{d}{dt}\langle \Psi | \hat{L} | \Psi \rangle = \left\langle \frac{\partial \Psi}{\partial t} \Big| \hat{L} \Psi \right\rangle + \left\langle \Psi \Big| \frac{\partial \hat{L}}{\partial t} \Psi \right\rangle + \left\langle \Psi \Big| \hat{L} \frac{\partial \Psi}{\partial t} \right\rangle. \tag{11.17}$$

The middle term on the right is the expectation value of $\partial \hat{L}/\partial t$. Next make the replacement $|\partial \Psi/\partial t\rangle = -(i/\hbar)|\hat{H}\Psi\rangle$ in the last term, and $\langle \partial \Psi/\partial t| = (i/\hbar)\langle \Psi \hat{H}^\dagger| = (i/\hbar)\langle \Psi|\hat{H}$ in the first one (using the hermiticity property $\hat{H}^\dagger = \hat{H}$). Then

$$\frac{d\bar{L}}{dt} = \frac{i}{\hbar}\langle \Psi|\hat{H}\hat{L}\Psi \rangle + \left\langle \frac{\partial \hat{L}}{\partial t} \right\rangle - \frac{i}{\hbar}\langle \Psi|\hat{L}\hat{H}\Psi \rangle = \left\langle \frac{\partial \hat{L}}{\partial t} \right\rangle + \frac{i}{\hbar}\langle [\hat{H}, \hat{L}] \rangle. \tag{11.18}$$

Here $[\hat{H}, \hat{L}]$ is the commutator of $\hat{H}$ and $\hat{L}$. The operator $(i/\hbar)[\hat{H}, \hat{L}]$ is called the *Poisson quantum bracket*, by analogy with the *classical* expression for dL/dt in terms of the Hamiltonian of the system [1]. Time rate of change of the observable is now expressed in terms of its operator and the system's Hamiltonian in a quite general way, without any reference to a problem-specific basis or state.

One would naturally like to introduce an *operator* representing $d\langle L\rangle/dt$. Such an operator would enable us to determine $\langle L\rangle_{t+dt}$ from $\langle L\rangle_t$, and generally, determine $\langle L\rangle_t$ at an arbitrary moment from $\langle L\rangle_0$ much in the same way as reconstructing $\Psi(t)$ from $\Psi(0)$. The result (11.18) suggests the natural *definition* of the time derivative operator as

$$\frac{d\hat{L}}{dt} \equiv \frac{\partial \hat{L}}{\partial t} + \frac{i}{\hbar}[\hat{H},\hat{L}], \qquad (11.19)$$

so that time derivative of the expectation value is equal to the expectation value of the time derivative operator: $d\langle L\rangle/dt = \langle d\hat{L}/dt\rangle$. As seen from (11.19), if $\hat{L}$ has no explicit dependence on time and commutes with the Hamiltonian, then $d\bar{L}/dt = 0$ and $\bar{L}$ is time independent. The corresponding observable L is then the *integral of motion*.

Let us apply these results to the coordinate and momentum of a particle. Their operators are time independent; therefore, (12.4) reduces to

$$\frac{d\hat{\mathbf{r}}}{dt} = \frac{i}{\hbar}[\hat{H},\hat{\mathbf{r}}], \qquad \frac{d\hat{\mathbf{p}}}{dt} = \frac{i}{\hbar}[\hat{H},\hat{\mathbf{p}}]. \qquad (11.20)$$

These operator equations are analogous to the classical Hamilton equations

$$\dot{q} = \frac{\partial H}{\partial p}, \qquad \dot{p} = -\frac{\partial H}{\partial q} \qquad (11.21)$$

for the generalized coordinates and momenta q and p [2]. When generalized coordinates are identified with the "ordinary" ones, the first equation relates velocity to momentum as $\mathbf{v} = \mathbf{p}/\mu$, and the second equation relates the change of momentum to the gradient of potential, that is, formulates Newton's second law.

The quantum equations (11.20) have a similar meaning. Indeed, using the known expressions for $\hat{\mathbf{r}}$, $\hat{\mathbf{p}}$, and $\hat{H}$, we obtain from (11.20):

$$\frac{d\hat{\mathbf{r}}}{dt} = \frac{\hat{\mathbf{p}}}{\mu}, \qquad \frac{d\hat{\mathbf{p}}}{dt} = -\vec{\nabla}U. \qquad (11.22)$$

In terms of averages, this amounts to

$$\frac{d}{dt}\langle\mathbf{r}\rangle = \frac{1}{\mu}\langle\mathbf{p}\rangle, \qquad \frac{d}{dt}\langle\mathbf{p}\rangle = -\left\langle\vec{\nabla}U\right\rangle. \qquad (11.23)$$

These relations are known as the *Ehrenfest theorem* [1,3]. Time derivative of the first equation of (11.23) combined with the second one gives

$$\mu\frac{d^2\langle\mathbf{r}\rangle}{dt^2} = -\left\langle\vec{\nabla}U\right\rangle = \langle\mathbf{F}\rangle. \qquad (11.24)$$

This is QM formulation of Newton's second law. One example of time-dependent $\langle x\rangle$ must be familiar to the reader from Section 9.7 and Figure 9.11.

A careful look at these results shows the intimate relation between conservation laws and invariant characteristics of the system's Hamiltonian. Thus, according to the second equation of (11.23), the net momentum of a system is conserved (is integral of motion) only if $U = $ const, that is, the system is invariant under displacement. Similarly, if it is invariant with respect to rotations, then its net angular momentum is integral of motion, and the energy conservation follows from the invariance with respect to temporal translations. These deep connections between the invariance and conservation laws are known as *Noether's theorem* [4–6].

A system and the Hamiltonian representing it can also be invariant with respect to discrete transformations. One of them (spatial inversion) and its related characteristic (*parity*) were briefly discussed in Sections 3.4 and 5.4.[1] According to the general theory, if the Hamiltonian of a system is invariant under inversion, we can expect the conservation of parity. An example is a system in a spherically symmetric field. The corresponding Hamiltonian is invariant under rotations *and* complete inversion with respect to the center of symmetry. In all such cases, the Hamiltonian commutes with $\hat{P}$, $[\hat{H}, \hat{P}] = 0$, and this leads to conservation of parity.

For a long time, parity conservation was thought to be a universal law. Now we know that there are some interactions that are not invariant with respect to inversions [7]. They are called the *weak interactions*. These interactions are responsible for β-decays considered in Section 3.6. Parity is not conserved in such processes.

There is a more general characteristic – the product of parity and charge, which is invariant under combined inversion in the V-space and in the "charge space." Simply put, the system's characteristics are invariant under combination of mirror reflection and replacing all electric charges by their opposite (CP invariance [8,9]). Then it was discovered that there is a class of processes that are not invariant even under this transformation. They were not invariant under time inversion either. However, combined invariance under CP transformation *and* time inversion (CPT) held for these processes. It seemed that finally a type of symmetry was discovered that was truly universal in the world of elementary particles and their transformations. But more recent evidence suggested that even CPT symmetry could break in certain antineutrino experiments. We shall not comment further on this because the conclusions are still tentative and there is intense ongoing research in this field.

11.3
Spreading of a Gaussian Packet

One example of state evolution is spatial spreading of a wave packet. This example is all the more important because it shows the appearance of time dependence in the indeterminacy relation $\Delta x \Delta p \geq \hbar/2$.

---

1) There are other important integrals of motion: charge, baryon number, lepton number, color, and strangeness. They are similar to parity in the sense that they are also discrete. All these characteristics have one common feature: in contrast to classical integrals of motion, which are continuous characteristics depending on the initial conditions, the quantum integrals of motion can be discrete characteristics of a particle, independent of the initial conditions.

The typical textbook example is a normalized Gaussian packet $\psi(x)$ centered about x_0. The Gaussian distribution is a popular case study because it is the only one whose Fourier transform is also Gaussian, and one of the few that can minimize the product of the corresponding indeterminacies. The latter are reciprocals of each other: if a waveform has standard deviation σ in the k-space, its spatial standard deviation is $1/\sigma$, and vice versa. The corresponding waveforms are

$$f(k) = f_0\, e^{-(k-k_0)^2/2\sigma^2}, \qquad f_0 = \frac{\pi^{-1/4}}{\sqrt{\sigma}}, \tag{11.25}$$

$$\psi(x) = \sigma f_0\, e^{-(1/2)\sigma^2(x-x_0)^2}\, e^{ik_0(x-x_0)}. \tag{11.26}$$

The parameter k_0 determines the average momentum of the packet, and x_0 is its *initial* position in the x-space. These are not essential characteristics, because we can always use a reference frame in which $k_0 = 0$ and choose its origin so that $x_0 = 0$. In this frame, the center of mass of the packet is stationary ($\bar{p} = 0$), and its shape remains symmetric with respect to the origin. This will simplify the calculations while retaining all the basic features of the phenomenon.

The wave packet for a free particle is a superposition of infinitely long sinusoidal wave trains. Each wave train is defined by its own wave number k and therefore it is in a definite energy state. So $\Psi(x,t)$ for any t can be determined by superposition of eigenstates $e^{ikx-\omega t}$ with the k-depending amplitudes (11.25) and (in nonrelativistic case) frequencies $\omega(k) = \hbar k^2/2\mu$:

$$\Psi(x,t) = \frac{1}{\sqrt{2\pi}}\int f(k) e^{i(kx-(\hbar k^2/2\mu)t)}\, dk = \frac{f_0}{\sqrt{2\pi}}\int e^{-k^2/2\sigma^2}\, e^{i(kx-(\hbar k^2/2\mu)t)}\, dk. \tag{11.27}$$

Here $f(k)$ is given by (11.25) with $k_0 = 0$. The integration gives the result

$$\Psi(x,t) = \frac{\sigma f_0}{\sqrt{1+i(\hbar/\mu)\sigma^2 t}}\, e^{-x^2/2(1+(\hbar^2/\mu^2)\sigma^4 t^2)}\, e^{i\varphi(x,t)}. \tag{11.28}$$

Here φ is the additional "phase shift":

$$\varphi(x,t) = \frac{\hbar}{2\mu}\sigma^4 t\, \frac{x^2}{1+(\hbar^2/\mu^2)\sigma^4 t^2}. \tag{11.29}$$

At $t = 0$, the function $\Psi(x,t)$ reduces, as it should, to (11.26) with $k_0 = x_0 = 0$. At arbitrary t, the shape of the packet is described by the amplitude and the first exponent of expression (11.28). The packet remains Gaussian, but its width and height change with time: the packet spreads and its height decreases. On the other hand, its shape in k-space (its spectrum) remains constant.

One can show that $\langle k^2 \rangle = \sigma^2/2$, so $\Delta k = \sigma/\sqrt{2}$. In the x-space, by our choice of the origin in the comoving frame, $\langle x \rangle = 0$ at any t. Further, denoting $1+i(\hbar/\mu)\sigma^2 t \equiv \Omega$, we have

$$\langle x^2 \rangle = \int_{-\infty}^{\infty} x^2 |\Psi(x,t)|^2\, dx = \frac{f_0^2 \sigma^2}{|\Omega|}\int_{-\infty}^{\infty} x^2\, e^{-(\sigma^2/|\Omega|^2)x^2}\, dx = \frac{1}{2}\frac{|\Omega|^2}{\sigma^2} \tag{11.30}$$

and

$$\Delta x = \frac{1}{\sqrt{2}}\frac{|\Omega|}{\sigma} = \frac{1}{\sqrt{2}\sigma}\sqrt{1+\frac{\hbar^2}{\mu^2}\sigma^4 t^2}. \tag{11.31}$$

As mentioned in the beginning, this has interesting implications in the indeterminacy relation for Δk and Δx. Indeed, we see that the product $\Delta k \Delta x$ turns out to be time dependent:

$$\Delta k \Delta x = \frac{1}{2}|\Omega| = \frac{1}{2}\sqrt{1+\frac{\hbar^2}{\mu^2}\sigma^4 t^2}. \tag{11.32}$$

At $t=0$, this reduces to $\Delta k \Delta x = 1/2$, attaining its minimal possible value. Thus, even the Gaussian distribution, most famous for its property to minimize the product of indeterminacies, only does so at one moment of time; the rest of eternity the product remains greater than the minimal possible and unboundedly increases with $|t| \to \infty$.

11.4
The B-Factor and Evolution of an Arbitrary State

Here we develop a new technique for determining evolution of the waveforms. It will work also in cases that cannot be treated by the time evolution operator.

The term "evolution," when applied to a wave packet in a free space, means change of its shape, rather than its motion as a whole. As emphasized in the previous section, the latter can always be eliminated by boarding a reference frame comoving with the center of mass of the packet. And vice versa, once we know the evolution of a packet with stationary center of mass, we can easily find its evolution in any other reference frame by performing transformation of function $\Psi(\mathbf{r}, t)$ to that frame. Therefore, we will again restrict, with no loss of generality, to the packets with stationary center of mass. Otherwise, there are no restrictions on the initial state.

First, consider a 1D case with an arbitrary state $\Psi(x, t)$. Its spectrum $\Phi(k)$ is determined by the initial shape $\Psi(x, 0)$:

$$\Psi(x,0) = \frac{1}{\sqrt{2\pi}}\int \Phi(k)e^{ikx}\,dk, \qquad \Phi(k) = \frac{1}{\sqrt{2\pi}}\int \Psi(x,0)e^{-ikx}\,dx. \tag{11.33}$$

Then the shape at any t is

$$\Psi(x,t) = \frac{1}{\sqrt{2\pi}}\int \Phi(k)e^{i[kx-\omega(k)t]}\,dk. \tag{11.34}$$

With the use of (11.33), this gives

$$\begin{aligned}\Psi(x,t) &= \frac{1}{2\pi}\int\left\{\int \Psi(x',0)e^{-ikx'}\,dx'\right\}e^{i[kx-\omega(k)t]}\,dk \\ &= \frac{1}{2\pi}\int dx'\,\Psi(x',0)\int e^{i[k(x-x')-\omega(k)t]}\,dk \equiv \int B(x-x',t)\Psi(x',0)\,dx',\end{aligned} \tag{11.35}$$

where

$$B(\Delta x', t) \equiv \frac{1}{2\pi} \int e^{-i[\omega(k)t - k\Delta x']} dk, \quad \Delta x' \equiv x - x'. \tag{11.36}$$

This can be easily generalized to the 3D case:

$$\Psi(\mathbf{r}, t) = \hat{\Gamma}(\mathbf{r}, t)\Psi(\mathbf{r}, 0) \equiv \int B(\mathbf{r} - \mathbf{r}', t)\Psi(\mathbf{r}', 0) d\mathbf{r}', \tag{11.35a}$$

where

$$B(\Delta \mathbf{r}', t) \equiv \frac{1}{2\pi} \int e^{-i[\omega(k)t - k\Delta \mathbf{r}']} d\mathbf{k}, \quad \Delta \mathbf{r}' \equiv \mathbf{r} - \mathbf{r}', \tag{11.36a}$$

and $d\mathbf{k} = d\mathbf{p}/\hbar^3$ is the volume element in the 3D momentum space.

Representing $\Psi(\mathbf{r}, t)$ in the form (11.35) or (11.35a) allows one to factor out the initial shape (an input) in the integrand. The evolution factor $B(\Delta \mathbf{r}', t)$ can be considered as the kernel of the integral operator

$$\hat{\Gamma}(x, t) \equiv \int B(x - x', t) \ldots dx' \quad (1D) \quad \text{or} \quad \hat{\Gamma}(\mathbf{r}, t) \equiv \int B(\mathbf{r} - \mathbf{r}', t) \ldots d\mathbf{r}'. \tag{11.37}$$

This operator converts the input $\Psi(\mathbf{r}, 0)$ into the final shape (output) $\Psi(\mathbf{r}, t)$ according to (11.35) and (11.35a). It is completely determined by the function $\omega(k)$ obtained from the dispersion equation (8.28). Until this function is specified, (11.36) (or (11.36a) in 3D) can be considered as the *general definition* of $B(\Delta x', t)$.

On the face of it, the B-factor looks like the *Green function* $G(\xi, \xi')$, but they are totally different. The Green function of operator $\hat{L}(\xi)$ is a special solution of an inhomogeneous differential equation $\hat{L}(\xi) G(\xi - \xi') = \delta(\xi - \xi')$. It acts on the source $f(\xi)$ of equation $\hat{L}(\xi) F(\xi) = f(\xi)$ and produces the corresponding special solution $F(\xi)$ [10]. The conversion works by integrating over the same argument ξ', so $G(\xi, \xi')$ acts as the kernel of the integral operator $\hat{L}^{-1}(\xi) \equiv \int G(\xi, \xi') \ldots d\xi'$ (you can find more details in Ref. [11]). The B-factor is the kernel of an integral operator (11.37) that takes the initial wave packet (input) to that of a later moment (output).

Thus, the Green function converts the source $f(\xi)$ of an inhomogeneous differential equation into a *different* function $F(\xi)$ (special solution of this equation) through integration over the *same* variable ξ (which can be the time variable). The B-factor takes a *general* solution $\Psi(x, 0)$ of a *homogeneous* (e.g., Schrödinger's) equation to the *same* solution $\Psi(x, t)$ at a later moment t through integration over the variable x other than t.

Also, the $\hat{\Gamma}(\mathbf{r}, t)$-operator with the B-factor is different from the time evolution operator $\hat{U}(\hat{H}, t)$ ((11.8) and (11.9)). The *result* of application of $\hat{\Gamma}(\mathbf{r}, t)$ is the same as one obtained with the $\hat{U}(\hat{H}, t)$, but the *algorithm* is totally different. The $\hat{U}(\hat{H}, t)$ converts the initial state into a final one by a specific translation through time, without integration over spatial coordinates. Its disadvantage is the necessity for numerous reiterated applications of the Hamiltonian to $\Psi(\mathbf{r}, 0)$ and summing up the results. And as is evident from its structure including the Laplacian $\vec{\nabla}^2$, it is not applicable to any waveforms with a discontinuity. In contrast, the $\hat{\Gamma}(\mathbf{r}, t)$-operator with the B-factor determines the final result by performing only one integration that tolerates even such outrageous discontinuities as that of δ-function.

Equations 11.35 and 11.36 or 11.35a and 11.36a provide us with the general rules for evaluation of a state of a free particle at an arbitrary moment of time.

Let us now find the explicit expression for the B-factor in the nonrelativistic QM. In this case $\omega(k) = \hbar k^2/2\mu$, and (11.36) takes the form

$$B(\Delta x', t) = \frac{1}{2\pi} e^{i(\Delta x'^2/2a)} \int_{-\infty}^{\infty} e^{-(1/2)ia(k-(\Delta x'/a))^2} dk, \quad a \equiv \frac{\hbar}{\mu} t. \tag{11.38}$$

The integral in (11.38) is an improper integral that can be taken by the corresponding technique ("the generalized summation") [12,13]. Alternatively, it can be evaluated as the limit

$$\int_{-\infty}^{\infty} e^{-(1/2)ia(k-(\Delta x'/a))^2} dk = 2\sqrt{\frac{2}{a}} \int_0^{\infty} e^{-i\xi^2} d\xi = 2\sqrt{\frac{\pi}{a}} \lim_{\kappa \to \infty} [C(\kappa) - iS(\kappa)],$$

where

$$C(\kappa) \equiv \sqrt{\frac{2}{\pi}} \int_0^{\kappa} \cos \xi^2 \, d\xi \quad \text{and} \quad S(\kappa) \equiv \sqrt{\frac{2}{\pi}} \int_0^{\kappa} \sin \xi^2 \, d\xi \tag{11.39}$$

are the Fresnel integrals [13]. The result is

$$\int_{-\infty}^{\infty} e^{-(1/2)ia(k-(\Delta x'/a))^2} dk = \sqrt{\frac{2\pi}{ia}}. \tag{11.40}$$

Putting this into (11.38) gives

$$B(\Delta x', t) = \sqrt{\frac{\mu}{2\pi i \hbar t}} e^{i(\mu \Delta x'^2/2\hbar t)}, \tag{11.41}$$

and (11.35) takes the form

$$\Psi(x, t) = \sqrt{\frac{\mu}{2\pi i \hbar t}} \int \Psi(x', 0) e^{i\left(\mu(x-x')^2/2\hbar t\right)} dx'. \tag{11.42}$$

For the 3D case, evaluation of $B(\Delta \mathbf{r}', t)$ requires writing (11.38) separately for each variable k_x, k_y, and k_z, integrating each the same way as before, and then taking the product of the results, with the outcome

$$B(\Delta \mathbf{r}', t) = \left(\frac{\mu}{2\pi i \hbar t}\right)^{3/2} e^{i(\mu \Delta r'^2/2\hbar t)}. \tag{11.43}$$

The difference from 1D case is only in power of the coefficient, and instead of (11.42) we will have

$$\Psi(\mathbf{r}, t) = \left(\frac{\mu}{2\pi i \hbar t}\right)^{3/2} \int \Psi(\mathbf{r}', 0) e^{i\left(\mu(\mathbf{r}-\mathbf{r}')^2/2\hbar t\right)} d\mathbf{r}'. \tag{11.44}$$

This expression allows us to find the shape of a *nonrelativistic* wave packet at an arbitrary t if we know its shape at $t = 0$.

In relativistic domain, we must use for $\omega(k)$ the relativistic formula $\omega^2 = c^2(\kappa_c^2 + k^2)$, $\kappa_c \equiv \lambda_c^{-1} \equiv \mu_0 c/\hbar$. Already at this stage we see the fundamental difference from nonrelativistic situation – the existence of two branches of the dispersion function known as positive and negative frequency solutions $\omega(k) = \pm c\sqrt{\kappa_c^2 + k^2}$. This equation assigns to any k two different ω with opposite signs. They represent two different eigenvalues (and corresponding eigenfunctions) of relativistic Hamiltonian, which give rise to particle–antiparticle interpretation [14] of the corresponding solutions. According to the completeness and closure relations [12], *all* eigenfunctions (and accordingly, all corresponding eigenvalues) are necessary for the expansion of an arbitrary quantum state $\Psi(\mathbf{r}, t)$. Therefore, with the relativistic expression for $\omega(k)$, the evolution factor (11.36) takes the form

$$B(\Delta x', t) \equiv \frac{1}{\sqrt{2}} \frac{1}{2\pi} \int \left\{ e^{i\left[k\Delta x' - ct\sqrt{\kappa_c^2 + k^2}\right]} + e^{i\left[k\Delta x' + ct\sqrt{\kappa_c^2 + k^2}\right]} \right\} dk$$

$$= \frac{1}{\pi\sqrt{2}} \int e^{ik\Delta x'} \cos\left(ct\sqrt{\kappa_c^2 + k^2}\right) dk, \quad \Delta x \equiv x - x'. \tag{11.45}$$

Here the additional factor $1/\sqrt{2}$ on the right takes care of the doubling of eigenstates and preserves normalization.

For our purposes, it is sufficient to consider the special case of a zero-mass particle (photon or graviton), for which $\kappa_c = 0$. In this case, (11.45) simplifies to

$$B(\Delta x', t) \equiv \frac{1}{\sqrt{8}} \frac{1}{\pi} \int \left\{ e^{ik(\Delta x' - ct)} + e^{ik(\Delta x' + ct)} \right\} dk$$

$$= \frac{1}{\sqrt{2}} [\delta(x - x' - ct) + \delta(x - x' + ct)]. \tag{11.46}$$

In the next two sections, we apply both nonrelativistic and relativistic expressions to some important cases of quantum evolution. We obtain from them two dramatically different versions of evolution depending on the specific form of the function $\omega(k)$.

11.5
The Fraudulent Life of an "Illegal" Spike

> For some solutions to the Schrödinger equation, the integral is infinite; in that case no multiplicative factor is going to make it 1.
>
> D. Griffiths

11.5 The Fraudulent Life of an "Illegal" Spike

Now we apply the algorithm developed in the previous section to a wave packet whose initial shape in the 1D case is a "spike" represented by δ-function:

$$\Psi(x,0) = \Psi_{x'}(x) = \delta(x - x'). \tag{11.47}$$

This is an eigenstate of the position operator and it satisfies the normalization condition (3.9) for states with continuous spectrum.[2] In our case, this condition takes the form

$$\int \delta(x - x')\delta(x - x'')dx = \delta(x' - x''). \tag{11.48}$$

The δ-function has somewhat ambiguous status in QM. First, it is not square integrable, but this "misdemeanor" is shared by all functions satisfying condition (3.9). Second, it is outrageously discontinuous, and in this respect, it violates one of the basic rules for a state function. But on the other hand, it constitutes a necessary element in mathematical structure of QM and is universally (see, for example, Refs [1,15–17]) used as such, albeit not always interpreted as a state function. Apart from being an eigenfunction of $\hat{x}$, it meets the most fundamental criteria for a state function: it stores all the relevant information about the system and (if treated correctly!) accurately predicts its evolution. Therefore, the peculiarities of δ-function do not make it an illegal entity in the framework of QM. In many respects, it can be treated on the same footing as a regular function. The expression $|\Psi_a(\mathbf{r},t)|^2$ correctly represents the probability density for this state everywhere except for the point of singularity [17]. Note that the most respected wave function – the de Broglie wave – represents the uniform probability distribution only up to an unspecified factor and is not square integrable, either. In view of all this, we call the spike (11.47) "illegal" with the quotation mark.

Now we consider two different versions of its time evolution. First, we apply to it the rules of nonrelativistic QM. The analysis will show a severe hidden pitfall that makes nonrelativistic QM unqualified to treat this problem, even when the spike is at rest! Then we will treat the same spike relativistically, and in this case will obtain the predictions consistent with known behavior of the ultrashort laser pulses.

1) *Nonrelativistic treatment.* Consider an eigenstate of $\hat{x}$ with eigenvalue $x = x_0$, that is, $\Psi_{x_0}(x,0) = \delta(x - x_0)$. Applying Equations 11.41 gives

$$\Psi_{x_0}(x,t) = \sqrt{\frac{\mu}{2\pi i \hbar t}} \int \delta(x' - x_0) e^{i\left(\mu(x-x')^2/2\hbar t\right)} dx'$$

$$= \sqrt{\frac{\mu}{2\pi i \hbar t}} e^{i\left(\mu(x-x_0)^2/2\hbar t\right)}, \quad t \neq 0. \tag{11.49}$$

---

2) Dimensionality of $\Psi(x)$ determined by (11.47) is m^{-1} instead of required m$^{-1/2}$ ("m" stands for meter).

The combined expression for the modulus $|\Psi_{x_0}(x,t)|$ at *any* moment including $t = 0$ is

$$|\Psi_{x_0}(x,t)| = \begin{cases} \delta(x-x_0), & t = 0 \quad (a), \\ \sqrt{\dfrac{\mu}{2\pi\hbar|t|}}, & t \neq 0 \quad (b). \end{cases} \quad (11.50)$$

We have a function of time, with a singularity at $|t| \to 0$. For $x = x_0$, this singularity is what one should expect, since it must recover the δ-function at $t = 0$. But we see the same feature at all other locations as well, since (11.50b) is independent of x. The function (11.49) oscillates with frequency

$$\Omega = \Omega(x,t) = \frac{\mu(x-x_0)^2}{2\hbar t^2}, \quad (11.51)$$

which is itself a function of x and t and increases unboundedly as $|t| \to 0$. A quantity oscillating with infinite frequency around the zero point is effectively equal to zero. This again would be precisely what one would expect, since $\Psi(x,t)$ must be zero at $t = 0$ for all $x \neq x_0$. But this argument does not work for the modulus $|\Psi(x,t)|$. According to this result, the original probability distribution $|\Psi(x,t)|^2$ has spread out instantly right after the "creation" or "release" of the spike, to fill uniformly the whole space. Conversely, if it is uniformly spread *before* $t = 0$, the corresponding distribution will be converging to $x = x_0$, first continuously, and then abruptly to produce the spike there at $t = 0$. The instantaneous spreading or convergence (having *nothing to do with collapse* at position measurement!) means that the standard deviation of the particle's momentum, and thereby of its velocity, is infinite. This is unusual, but does not contradict the logical structure of *nonrelativistic* physics, which does not impose any restrictions on velocity.

The described evolution is shown in Figure 11.1. Note that according to definition of the δ-function, the spike itself is integrable, but *not square integrable*. As to $|\Psi(x,t)|$ at $t \neq 0$, it is not integrable even in the first power. This by itself does not mean any inconsistency because the theory does not impose any conditions on the *integrability* of $|\Psi(x,t)|$. The fact that $|\Psi(x,t)|$, and accordingly $|\Psi(x,t)|^2$, is a function of time will not violate unitarity for two reasons: first, the

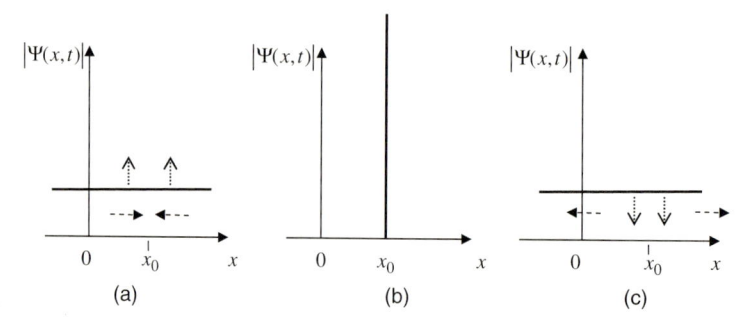

Figure 11.1 Evolution of delta spike (nonrelativistic treatment): (a) $t < 0$; (b) $t = 0$; (c) $t > 0$. The dotted vertical arrows indicate the direction of change of $|\Psi(x,t)|$ with time. It increases before the zero moment and decreases after it. The dashed horizontal arrows indicate the local probability flux.

amplitude of a *non-square-integrable* function is defined only up to an arbitrary factor; second, and more important, the corresponding state, as will be seen below, satisfies the local continuity equation.

The whole picture is, of course, exotic, but it is logically consistent. One can show that the state (11.50) satisfies the Schrödinger equation for all $|t| \neq 0$. Applying condition (3.9) to an evolved state (11.49) and to a similar state with x-eigenvalue x_1 gives

$$\int \Psi^*_{x_0}(x,t) \Psi_{x_1}(x,t)\, dx = \frac{\mu}{2\pi\hbar t} \int e^{i[\mu[(x-x_1)^2 - (x-x_0)^2]/2\hbar t]}\, dx. \quad (11.52)$$

The integration on the right yields

$$\int \Psi^*_{x_0}(x,t) \Psi_{x_1}(x,t)\, dx = e^{-i[\mu(x_0^2 - x_1^2)/2\hbar t]} \delta(x_0 - x_1), \quad t \neq 0. \quad (11.53)$$

This expression is consistent with (11.48).

The function $\rho(x,t) = |\Psi(x,t)|^2$ describes (as mentioned above, up to an undefined constant) the probability distribution

$$\rho(x,t) = \frac{\mu}{2\pi\hbar|t|}, \quad t \neq 0. \quad (11.54)$$

Expression (11.54) goes to infinity at $t \to 0$ *everywhere*, rather than exclusively at $x = x_0$. This, again, reflects the non-square integrability of the initial state.

Consider now the time derivative of ρ:

$$\frac{\partial \rho}{\partial t} = \mp \frac{\mu}{2\pi\hbar t^2}. \quad (11.55)$$

Here "−" stands for $t > 0$ (after the release of the δ-spike), and "+" stands for $t < 0$ (before the formation of the spike). At any point in space the probability density decreases with time *after* the initial moment and increases *before* this moment. And the rate of this change is independent of x. Together with spatial uniformity of ρ, this seems to contradict the local probability conservation. But that would be a wrong conclusion. The probability flux is determined not only by ρ, but also by the phase $\varphi(x)$ in the exponent of Ψ, since the gradient of phase represents the effective velocity of the wave. In state (11.49), the phase is $\varphi = \mu(x - x_0)^2/2\hbar t$, which makes the flow rate proportional to $|x - x_0|$. Indeed, applying the basic definition (8.12) of the flux, $\mathbf{j} = -(\hbar/\mu)\mathrm{Im}(\Psi \vec{\nabla} \Psi^*)$, to our case (11.49), we obtain

$$j = \pm \frac{\mu}{2\pi\hbar t^2}(x - x_0) = \rho(x - x_0)/t \quad (11.56)$$

with the "+" sign for $t > 0$ and the "−" sign for $t < 0$.

Thus, even though the density is uniform, the *flux density* is not. It is a function of position despite the uniformity of ρ. The farther away from the initial spike, the greater the flux! On the face of it, this appears to be nonsense, but as you think more, it looks reasonable: the flux originates from the single source at $x = x_0$, and whatever reaches the farther locations, must have greater instantaneous velocity, as is evident from (11.56).

As seen from (11.56), after the initial moment the *direction* of flux is to the right for all $x > x_0$ and to the left for all $x < x_0$. In other words, it is away from the point x_0: the spike is spreading and leveling out. Before the initial moment, the directions are the opposite – the flux is toward the spike (actually producing it!).

With the results (11.55) and (11.56), it is easy to see that the evolved spike satisfies the local continuity equation for one dimension:

$$\frac{\partial \rho}{\partial t} = -\frac{\partial j}{\partial x}. \tag{11.57}$$

All these results are trivially generalized to the 3D case, giving the expressions

$$\Psi(\mathbf{r}, t) = \left(\frac{\mu}{2\pi i \hbar t}\right)^{3/2} e^{i(\mu(\mathbf{r}-\mathbf{r}_0)^2/2\hbar t)}, \qquad \rho(x, t) = \left(\frac{\mu}{2\pi \hbar |t|}\right)^3, \qquad t \neq 0 \tag{11.58}$$

and

$$|\mathbf{j}| = \left(\frac{\mu}{2\pi \hbar |t|}\right)^3 \frac{1}{t} |\mathbf{r} - \mathbf{r}_0|, \qquad \frac{\partial \rho}{\partial t} + \vec{\nabla} \cdot \mathbf{j} = 0. \tag{11.59}$$

Now we have a spherical wave converging to the singular point $\mathbf{r}_0$ before and diverging from it after the zero moment, and its intensity changes faster with time than in the 1D case. It satisfies both the Schrödinger and continuity equations everywhere except for $\mathbf{r}_0$. But its amplitude does not fall off far from the center, as is the case with a "normal" spherical wave. And other aspects of its behavior are weird – the instantaneous convergence and spread, the uniform spatial distribution at all t except for $t = 0$, and the flux rate increasing with distance from $\mathbf{r}_0$. All these results, with all their formal consistency, contradict observations! Indeed, we see that at $t = 0$ the particle was certainly at $\mathbf{r} = \mathbf{r}_0$ and nowhere else. If we know *exactly* the position $\mathbf{r}_0$ of the particle at $t = 0$, then we are certain that the particle is *absent elsewhere*. If at any other moment of time no matter how close to $t = 0$, there appears a *nonzero* chance to find the particle elsewhere no matter how far from $\mathbf{r}_0$ (maybe, in another galaxy!), that would mean the instant arrival of the edge of the initial wave packet. The advance of the edge constitutes a signal transfer [11,18]. Therefore, the obtained equations describe an infinitely fast (although uncontrolled) signal. But superluminal signals contradict relativistic causality.

Thus, nonrelativistic treatment of the δ-spike predicts the results inconsistent with observations. This forces us to turn to the relativistic treatment of the same problem.

2) *Relativistic treatment (real life of a spike)*. For our purposes, it is sufficient to consider the special case of a zero-mass particle (photon or graviton), for which $\kappa_c = 0$, and we assume that it is sharply localized at $t = 0$. Actually, one can sharply localize only electric or magnetic "footprint" of a photon, and only momentarily [19,20]. So the considered spike will be only an idealized model chosen for the sake of simplicity. Applying Equation 11.46 to our case gives

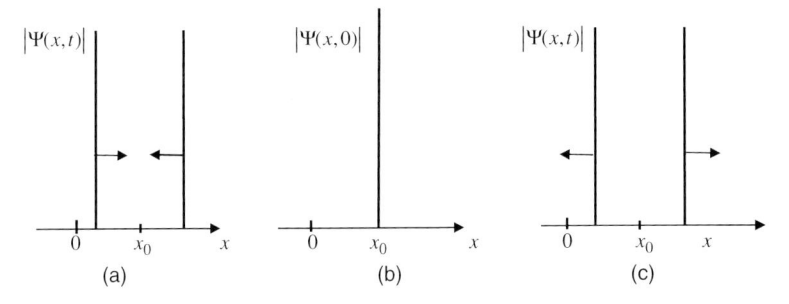

Figure 11.2 Evolution of delta spike (relativistic treatment): (a) $t < 0$; (b) $t = 0$; (c) $t > 0$.

$$\Psi(x,t) = \int B(x-x',t)\delta(x'-x_0)dx'$$
$$= \frac{1}{\sqrt{2}}[\delta(x-x_0-ct) + \delta(x-x_0+ct)]. \tag{11.60}$$

This state represents two δ-spikes moving with the speed c in two opposite directions:

$$x(t) = \begin{cases} x_0 + ct, \\ x_0 - ct. \end{cases} \tag{11.61}$$

They move toward each other at $t < 0$ and away from one another at $t > 0$ (Figure 11.2). At the moment $t = 0$, they merge together at $x = x_0$ to produce one common spike satisfying condition (11.48). Focusing on the future, we say that the initial spike splits into two twins, which together represent their "parent" existing at the moment $t = 0$. Focusing on the past, we say that the two approaching spikes represent the prehistory of the fleeting state (11.47), with $x' = x_0$. Altogether, we have again the time evolution of state (11.47). But this evolution, in contrast to the one predicted by nonrelativistic QM, is quite natural and consistent with all known evidence. Especially important result is that now the violent discontinuity, which is characteristic of δ-function (and which makes the spike fit for signaling!), propagates with the invariant speed c, in accord with relativistic causality.

Relativistic QM triumphs where nonrelativistic QM fails. And it fails despite the fact that the center of mass of the whole system was stationary. What then was the cause of failure? To find the answer to this question, we need to go to the next two sections.

11.6
Jinnee Out of the Box

Now we take for the initial state $\Psi(x,0)$ a "decent" function

$$\Psi_n(x,0) = \psi_n(x) = \begin{cases} 0, & x < 0, \\ \sqrt{\frac{2}{a}}\sin k_n x, & 0 \leq x \leq a, \quad k_n \equiv n\frac{\pi}{a}, \quad n = 1, 2, \ldots, \\ 0, & x > a. \end{cases} \tag{11.62}$$

This is our "old friend" – a quasi-free particle from Section 9.1. It is a typical textbook example of a normalized solution of the Schrödinger equation, with the energy spectrum $E_n = \hbar^2 k_n^2/2\mu_0$. As all nonrelativistic expressions, it leaves out the particle's rest energy $E_0 = \mu_0 c^2$. If a is large enough so that $E_n \ll \mu_0 c^2$ for $1 \leq n \leq N$, the set of functions (11.62) with eigenvalues E_n gives an accurate description of the corresponding physical states and the related effects. This is also true for any superposition of such states as long as the particle remains trapped in the box.

But what happens if we suddenly open the box and make the particle free? Suppose the particle was initially in one of eigenstates satisfying condition $E_n \ll \mu_0 c^2$. Then we can consider (11.62) as the initial state for the Schrödinger equation with the new (now force-free) Hamiltonian. The state will start spreading, and the question is: Will the nonrelativistic QM describe the spread as correctly as it describes the initial state?

To learn the answer, let us apply the algorithm with $\hat{\Gamma}$-operator ((11.37) and (11.43)) to state (11.62):

$$\Psi(x,t) = \sqrt{\frac{2}{\pi i a}} \tau \int_0^a e^{i\tau^2(x-x')^2} \sin k_n x' \, dx', \quad \tau \equiv \tau(t) \equiv \sqrt{\frac{\mu_0}{2\hbar t}}. \quad (11.63)$$

Integration gives a rather bulky expression in terms of the Fresnel integrals [13]:

$$\Psi(x,t) = -\frac{1}{2}\sqrt{\frac{i}{a}} e^{-i(k_n^2/4\tau^2)} \{[C(p) + iS(p) - C(q) - iS(q)]e^{ik_n x}$$

$$+ [C(r) + iS(r) - C(s) - iS(s)]e^{-ik_n x}\}. \quad (11.64)$$

Here

$$p \equiv \frac{k_n}{2\tau} - \tau(x-a), \quad q \equiv \frac{k_n}{2\tau} - \tau x, \quad r \equiv \frac{k_n}{2\tau} + \tau(x-a), \quad s \equiv \frac{k_n}{2\tau} + \tau x. \quad (11.65)$$

At any $t > 0$, expression (11.64) is nonzero for $x > a$ or $x < 0$. At $t \to \infty$, we have $\tau \to 0$ and (11.64) approaches zero everywhere, which is a natural outcome of the total spread of the initial waveform.

The nonzero values of (11.64) for all x show instantaneous spread, so already at this stage we see violation of relativistic causality. But it may be instructive to look at this in more detail. So let us turn to the *initial* stage of this process. At sufficiently small t, the τ is so large that the first terms on the right of (11.65) are negligible. The corresponding conditions for this are

$$\frac{k_n}{2\tau} \ll \tau|x-a|, \quad \frac{k_n}{2\tau} \ll \tau|x| \quad \text{or} \quad \frac{\hbar k_n}{\mu} t \ll |x|, |x-a|. \quad (11.66)$$

These conditions exclude the areas around the respective initial edges. The later in time, the farther from the edges we must be in order for the following simplified expressions to be accurate (note that $\hbar k_n/\mu \equiv v_n$ is the group velocity corresponding to $k = k_n$). With conditions (11.66) satisfied, (11.64) reduces to

$$\Psi(x,t) \approx i\sqrt{\frac{i}{a}} e^{-i(k_n^2/4\tau^2)} \{C(\tau(x-a)) + iS(\tau(x-a)) - C(\tau x) - iS(\tau x)\} \sin k_n x. \tag{11.67}$$

Using asymptotic behavior of the Fresnel integrals

$$C(z) \underset{z\to\infty}{\approx} \frac{1}{2} + \frac{\sin z^2}{\sqrt{2\pi z}}, \qquad S(z) \underset{z\to\infty}{\approx} \frac{1}{2} - \frac{\cos z^2}{\sqrt{2\pi z}}, \tag{11.68}$$

one can bring (11.67) to the form

$$\Psi(x,t) \approx \frac{1}{\tau}\sqrt{\frac{i}{2\pi a}} e^{-i(k_n^2/4\tau^2)} \left(\frac{e^{i(\tau(x-a))^2}}{x-a} - \frac{e^{i(\tau x)^2}}{x}\right) \sin k_n x. \tag{11.69}$$

Finally, using (11.63), we obtain

$$\rho(x,t) = |\Psi(x,t)|^2 \approx 2\hbar t \frac{a^2 + 4x(x-a)\sin^2[(\mu_0/2\hbar t)a(x-(a/2))]}{\pi a \mu_0 x^2 (x-a)^2} \sin^2 k_n x. \tag{11.70}$$

The result is symmetric with respect to the center of the box, $x = a/2$, and expressions (11.69) and (11.70), as they should, approach zero at $t \to 0$.

The similar treatment within the region $0 < x < a$ gives, after rather lengthy calculation,

$$\rho(x,t) \approx \frac{2}{a}\sin^2 k_n x \left\{1 + \frac{x\left[\sin(\tau(a-x))^2 - \cos(\tau(a-x))^2\right] + (a-x)\left[\sin(\tau x)^2 - \cos(\tau x)^2\right]}{\sqrt{2\pi}\tau x(a-x)}\right\}. \tag{11.71}$$

At $t \to 0$, it reduces to the initial distribution (11.62).

According to (11.66) and (11.70), $\rho(x)$ outside the box first increases from zero to a small but nonzero value, and then decreases. The corresponding distributions are shown in Figure 11.3. As seen from the figure, we have nonzero $\rho(x)$ *arbitrarily far from the origin* immediately after the release. In other words, the nonrelativistic treatment of initially nonrelativistic state predicts that the probability cloud will spread instantaneously beyond its initial area $0 \le x \le a$. This means an infinite speed of both edges of the form (11.62) and thereby an infinite signal velocity.[3]

Now we consider the same problem using the *relativistic* expression for $\omega(k)$. To simplify, we use as before, an ultrarelativistic limit $\omega(k) = \pm ck$, and the B-factor reduces to (11.46). Physically, this may correspond to a photon within a Fabry–Pérot resonator of length a (actually, in this case (11.62) represents only electric "footprint" of a photon [19,20] and must therefore have the additional normalizing

---

3) We have two such signals associated with motion of the edges – one traveling to the right and the other to the left from the packet's center of mass, which itself remains stationary.

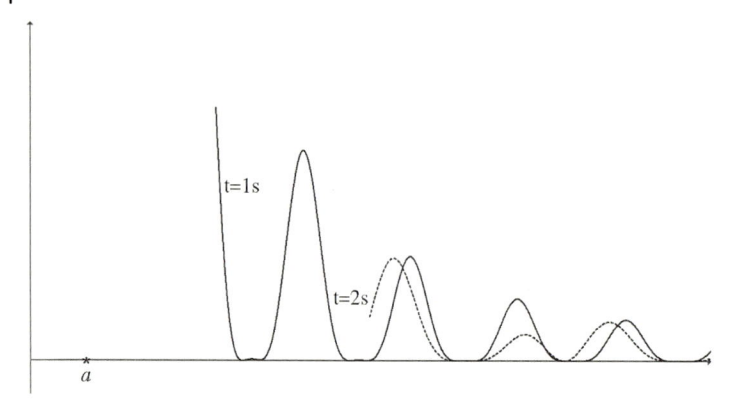

Figure 11.3 Evolution of the "jinnee" after release (nonrelativistic treatment). Shown are two snapshots at two different moments of time on the right side of the initial box.

factor $1/\sqrt{2}$). It is also important to realize that the conventional notion of the photon as a "massless" particle does not apply to its eigenstates within a resonator. In any such state, the photon has a nonzero rest mass $\mu_0^{(n)} = \hbar\omega_n/c^2 = \pi(\hbar/ac)n$ [11], which depends on quantum number n.[4] If the resonator's mirrors are now suddenly made transparent, the initial photon state becomes free and its evolution will be described by (11.35) with the evolution factor (11.46). Applying this to (11.62) gives

$$\Psi_n(x,t) = \sqrt{\frac{1}{a}} \left\{ \begin{bmatrix} 0, & x+ct<0 \\ \sin k_n(x+ct), & 0 \leq x+ct \leq a \\ 0, & x+ct>a \end{bmatrix} + \begin{bmatrix} 0, & x-ct<0 \\ \sin k_n(x-ct), & 0 \leq x-ct \leq a \\ 0, & x-ct>a \end{bmatrix} \right\}.$$

(11.72)

The initial state splits into two identical waveforms propagating *without change of shape* in two opposite directions corresponding to "+" or "−" signs in $\omega(k) = \pm ck$ (Figure 11.4). Similar result is obtained for the magnetic component of the photon state. For each component, both wave packets move with the speed of light, thus satisfying relativistic causality and all known observations of behavior of laser pulses.

With the use of the more general evolution factor (11.45) corresponding to a particle with $\mu_0 \neq 0$, the shape of the split parts will change with time, but the luminal speed of the edges will not be affected [18].

It is important to realize that the packet *as a whole* remains stationary and *in this respect* remains a nonrelativistic object even after the release from the box. Also, it would be totally wrong to interpret its split into two receding parts as the emergence of two photons – one flying to the right and the other flying to the left. We had only one

4) In all problems involving evolution of a wave packet, one must draw a sharp distinction between two different rest masses. One rest mass $\mu_0^{(n)}$ is that of the initial (stationary) state $|n\rangle$ (e.g., when trapped in a box) or that of the resulting state after release. The other rest mass μ_0 corresponds to a *free* particle with *definite momentum*. The B-factor is determined by the latter rest mass.

11.7 Inadequacy of Nonrelativistic Approximation in Description of Evolving Discontinuous States

(a)

(b)

(c)

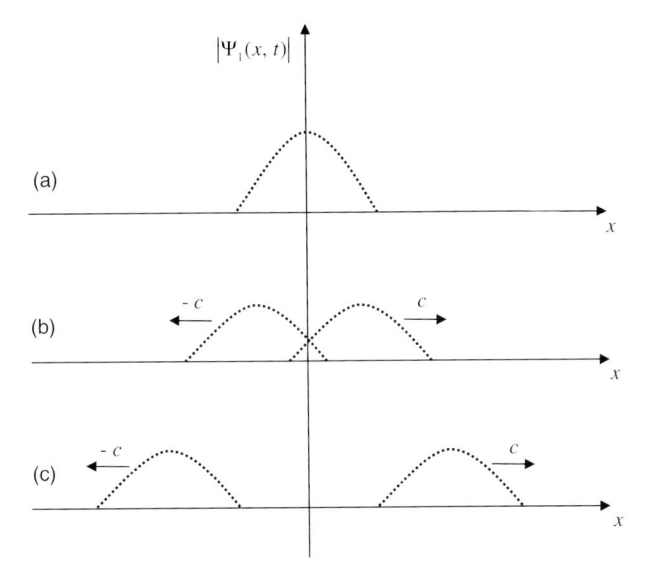

Figure 11.4 Evolution of the "jinnee" from potential box according to relativistic QM. Before the release, the jinnee was in the first eigenstate of the old Hamiltonian. After the release, it is in a superposition of the eigenstates of the new Hamiltonian. The edges propagate with the speed of light. For a massless particle, both parts of the whole wave packet are receding from each other with the same speed, and their shapes do not change. (a) $t = 0$; (b) $0 < t < a/2c$; (c) $t > a/2c$.

photon to begin with! And already in the beginning it was in a superposition of two states of motion in two opposite directions. This superposition persists after release with the only distinction that now it is unimpeded by the mirrors, so the oppositely moving parts separate from each other. We still have only one (now free!) photon, and it still has a nonzero rest mass, because its center of mass remains at rest! Nor is this photon split into two halves – we know that if we place two photon detectors on left and right, each time only one of them will click, recording the whole photon. This is just another miracle for classical world and mundane thing in quantum world: one photon in two places at once that are moving apart with the speed of light, but we find it in one place and in one piece in a measurement.

In contrast with the results (11.70) of nonrelativistic treatment, now we have the results consistent with observations. The same question then arises: Why does the nonrelativistic prediction contradict reality? The answer is discussed in the next section.

11.7
Inadequacy of Nonrelativistic Approximation in Description of Evolving Discontinuous States

Here we summarize the results obtained for two objects – a spike and a "jinnee." In both cases, the nonrelativistic description of evolution fails even though the center of mass of the object remains at rest. The general feature in both cases was the existence of edges. This feature is *necessary* for a form to be used for signaling, and the maximal signal velocity is c. Already the existence of this scaling parameter for *any* waveform with an edge shows that *time evolution of such a form cannot be described using nonrelativistic dispersion equation.*

11.7.1
Discussion

Formally, inadequacy of nonrelativistic QM for treating evolving discontinuous states can be seen from the following argument. According to the *Sommerfeld–Brillouin theorem* [18], the maximal possible speed u_s of a *signal* associated with motion of the edge is given by a *high-frequency limit* of phase velocity $u(\omega)$ in the corresponding medium: $u_s = \lim_{\omega \to \infty} u(\omega)$. The nonrelativistic dispersion equation $\omega(k) = \hbar k^2/2\mu_0$ for a free particle of mass μ_0 gives

$$u(\omega) = \frac{\omega}{k} = \frac{\hbar k}{2\mu_0} = \sqrt{\frac{\hbar\omega}{2\mu_0}}, \quad u_s = \lim_{\omega \to \infty} u(\omega) = \infty, \quad (11.73)$$

and this is precisely what we got applying this equation, contrary to an overwhelming amount of scientific evidence.[5] In contrast, the *relativistic* dispersion equation $\omega(k) = \pm c\sqrt{\kappa_c^2 + k^2}$, $\kappa_c \equiv \mu_0 c/\hbar$, gives

$$u(\omega) = \frac{\omega}{k} = \omega\left(\frac{\omega^2}{c^2} - \kappa_c^2\right)^{-1/2}, \quad u_s = c. \quad (11.74)$$

So all the difference comes from the nonrelativistic expression (11.73) used for $\omega(k)$. This may seem strange because after all the nonrelativistic approximation follows from relativistic equations in the corresponding limit. The answer to this is that the nonrelativistic approximation for $\omega(k)$ in the considered case *cannot in principle* be done accurately. Such approximation is acceptable only within the range

$$\omega \approx \frac{\hbar k^2}{2\mu_0} \ll \frac{\mu_0 c^2}{\hbar}, \quad \text{that is,} \quad |k| \ll \kappa_c. \quad (11.75)$$

But when we perform integration (11.36) in the momentum space to calculate the time evolution of a state, we go beyond the landmark (11.75), and the used approximation for $\omega(k)$ loses its validity. In this case, given the presence of an edge that is always associated with pronounced high-frequency Fourier components, it is absolutely imperative to use the relativistic expression for $\omega(k)$ in all range of spectrum.

---

5) In 1990s, claims were made of observed *superluminal signal* velocities in quantum tunneling [21,22], but they were shown to be unsubstantiated [23–27]. Similar claims were made quite recently [28] about observation of superluminal neutrinos. The corresponding experiments (interesting as they are), their results, and especially their interpretation need thorough scrutiny [29,30].

Thus, the reason for the wrong predictions of nonrelativistic QM in this case is the incompatibility of the two mutually excluding requirements: on the one hand, the use of *all range* of Fourier spectrum of such state is absolutely necessary for the accurate representation of the state; on the other hand, this automatically renders the nonrelativistic QM unfit for the correct handling of the problem.

The δ-spike in Section 11.5 just shows the inadequacy of nonrelativistic QM in the most simple way. But taking the "jinnee," which is initially a sample object in nonrelativistic QM, leads to the same result. This shows that the problem was not in the δ-spike as such. The nonrelativistic predictions for the δ-spike are wrong not because the spike is illegal (it is not), but because applying the nonrelativistic expression for $\omega(k)$ to such a case is illegal. In all considered cases, nonrelativistic QM is inadequate for describing correctly the time evolution of states with a discontinuous waveform.[6]

The relationship between the shape of a form, its bandwidth, and its ability to transmit a signal still remains a sensitive and sometimes even controversial issue [22–35] in both classical and quantum theory of signaling. Therefore, elucidating possible pitfalls in description of evolution of relevant states may also help to gain deeper insights into the whole problem of signal transfer.

11.8
Quasi-Stationary States

Now we turn back to a barrier studied in Section 9.3 and ask what happens if we add to it another barrier of the same height and width (Figure 11.5)? The resulting changes are of twofold nature:

1) For an incident particle with some definite energy E, the second barrier introduces relatively minor difference – it will only change the R and T coefficients.
2) A far more dramatic change is a possibility to "prepare" the particle trapped between the barriers (Figure 11.5) at the initial moment $t = 0$, so that its wave function at this moment is nonzero only in the region between the barriers:

$$\Psi(x, 0) = \begin{cases} \psi_0(x), & -a \leq x \leq a, \\ 0, & |x| > a. \end{cases} \quad (11.76)$$

In this case, we may observe a new phenomenon: under certain conditions the particle can remain trapped there for a long time before it leaks outside due to tunneling. Now we must be more accurate in the use of language. The expression "being trapped" when there is a chance of leaking out makes sense when the

---

6) In relativistic quantum field theory, an initially localized state also spreads instantaneously, but this is due to the stochastic contribution of vacuum fluctuations – spontaneous birth and annihilation of virtual particles [21,31–34]. This effect cannot be used for signaling. Without such contribution, the effective rate of the spreading reduces to (11.74).

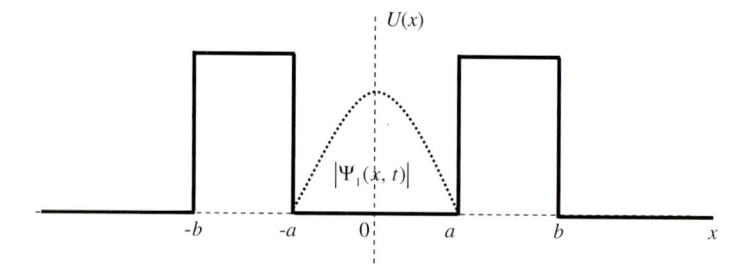

Figure 11.5 Initial state of a particle "prepared" between the two barriers. According to CM, the particle would remain trapped inside. According to QM, the particle will not stay between the barriers forever. Eventually, it will leak out. Its energy in this state cannot be sharply defined.

initial state satisfies one more condition apart from (11.76). Namely, the chance of leaking out must be so small that the average time τ spent by the particle between the barriers by far exceeds the average period T of its motion there. On the average, a particle must make many bounces off the barriers – many rounds back and forth between them before it manages to escape.

The region between the barriers can roughly be considered as a potential well, and the particle in a state (11.76) has position indeterminacy $\Delta x \leq 2a$. The material of Section 9.3 suggests that the high average longevity between the barriers may happen only if the expectation value of the particle's energy is less than the height of the barriers, $\langle E \rangle < U$. Assuming the (anti)symmetry of the trapped state, which implies $\langle p \rangle = 0$, one can show that this condition can be satisfied simultaneously with $\Delta x \leq 2a$ if

$$a^2 U > \frac{\hbar^2}{8\mu}. \tag{11.77}$$

The solutions of the Schrödinger equation for our system can be found in two different but equivalent ways. The first way can be called traditional: forming a nonstationary solution as a superposition of stationary states in the same way as in the previous sections. The only difference is that now we will have a superposition of eigenstates $\Psi_E(x, t) = \psi_E(x)\exp(-iEt/\hbar)$ of the Hamiltonian with two barriers, rather than of the eigenstates $\exp(kx - \omega t)$ of a free particle. This approach gives direct analytic expression for the evolved state and shows that the studied case is another version of spreading of a wave packet, possibly impeded by the barriers. On the other hand, it relies on the knowledge of the initial form, and can give wrong result in the nonrelativistic approximation if the form has edges as in (11.76). Also, it does not show directly the average lifetime in the trapped state.

These two problems are addressed in the second approach. Its basic idea is to consider the evolving state *formally* as a stationary state and treating it as such. This can be done consistently but leads to the result that the energy eigenvalues must be complex!

Thus, we will try to represent a trapped state as an energy eigenfunction, as usual, in the form

$$\Psi(x,t) = \psi(x)e^{-(E/\hbar)t}. \tag{11.78}$$

Then, taking the origin at the center of the system and restricting to the semispace $x > 0$, we have for the spatial factor

$$\psi(x) = \begin{cases} Ae^{ikx} + Be^{-ikx}, & 0 \le x \le a, \\ \alpha e^{qx} + \beta e^{-qx}, & a \le x \le b, \\ \mathcal{A}e^{ikx}, & x > b. \end{cases} \tag{11.79}$$

Here, as in (9.35),

$$k = \frac{\sqrt{2\mu E}}{\hbar}, \quad q \equiv kv \equiv k\sqrt{\frac{U}{E} - 1}, \tag{11.80}$$

and we consider the energies within the range $0 \le E \le U$. Since $\bar{p} = 0$, there must be also $|A| = |B|$. On the other hand, the absence of the second term in the third of Equations 11.79 reflects the absence of the incoming flux. Note that if we set in (11.79) $\alpha = \beta = \mathcal{A} = 0$ in an attempt to satisfy the initial condition of $\psi(x)$ being zero outside the barriers, then it will remain zero at all times because the state is formally stationary.

Thus, we have a wave with a finite amplitude moving away to the right of the system. Similarly, we have one wave moving away on the left of the system. Altogether we have two fluxes – one to the right and one to the left of the system – describing the probability flow away from the center. This comes at the expense of probability in the region within and between the barriers. Using the continuity equation (3.16) and the property $j(b) = -j(-b)$ reflecting the symmetry of the system, we can evaluate the rate of change of probability in the region $-b \le x \le b$ in terms of the net flux through the outer faces of the barriers:

$$\frac{\partial}{\partial t} \int_{-b}^{b} \Psi^*(x,t)\Psi(x,t)dx = -2j(b) < 0. \tag{11.81}$$

This means that the described state is only formally stationary! But on the other hand, if the eigenvalue E in expression (11.78) for $\Psi(x,t)$ were *real*, any temporal dependence in the integrand in (11.81) would vanish, and the time derivative would be zero. Therefore, real E in the considered situation is inconsistent with conservation of probability. The only way to reconcile them is to realize that a nonstationary state cannot be described by a single *and* real energy eigenvalue. We have either a superposition of many eigenstates with different real E, as studied in Sections 9.3–9.6, or we can still describe such state by a solution with one definite E, but this E will be complex! Therefore, we must write

$$E \to \tilde{E} \equiv E - \frac{1}{2}i\Gamma. \tag{11.82}$$

Then the time-independent Schrödinger equation takes the form (for the 3D case)

$$\left(-\frac{\hbar^2}{2\mu}\nabla^2 + U(\mathbf{r})\right)\psi(\mathbf{r}) = \left(E - \frac{1}{2}i\Gamma\right)\psi(\mathbf{r}). \tag{11.83}$$

There arise a few questions.

1) How does expression (11.82) sit together with the theorem that the Hermitian operators have only real eigenvalues?

   The answer to this is that the hermiticity of an operator depends not only on the operator itself but also on a set of functions on which it acts. The change (11.82) automatically transforms eigenfunctions $\psi_j(\mathbf{r})$ in such a way that *the same* $\hat{H}$ with real $U(\mathbf{r})$ becomes non-Hermitian in the new set. Indeed,

$$\int \psi_\alpha^*(\mathbf{r})\hat{H}(\mathbf{r})\psi_\beta(\mathbf{r})d\tau = \left(E_\beta - i\frac{1}{2}\Gamma_\beta\right)\int \psi_\alpha^*(\mathbf{r})\psi_\beta(\mathbf{r})d\tau,$$

$$\int \psi_\beta(\mathbf{r})\hat{H}^*(\mathbf{r})\psi_\alpha^*(\mathbf{r})d\tau = \left(E_\alpha + i\frac{1}{2}\Gamma_\alpha\right)\int \psi_\alpha^*(\mathbf{r})\psi_\beta(\mathbf{r})d\tau. \tag{11.84}$$

   We see that the condition of hermiticity is not satisfied even in case $\psi_\beta = \psi_\alpha$. And, once we have $\hat{H}^+ \neq \hat{H}$, we cannot require the orthogonality of its eigenfunctions, either.

2) What is the physical meaning of a complex energy eigenvalue and, particularly, of its imaginary part $\Gamma/2$?

   This can be understood if we form the expression for the probability density in (11.81):

$$\rho(x,t) = |\Psi(x,t)|^2 = |\psi(x)|^2\,e^{-(\Gamma/\hbar)t} \equiv \rho(x,0)e^{-(\Gamma/\hbar)t} \tag{11.85}$$

and evaluate the quantity

$$\mathcal{P}_a(t) = \mathcal{P}_a(0)e^{-\gamma t}, \quad \gamma \equiv \frac{\Gamma}{\hbar}, \quad \mathcal{P}_a(0) = \int_{-a}^{a}\rho(x,0)dx. \tag{11.86}$$

Here $\mathcal{P}_a(0)$ is the *initial* probability of finding the particle between the barriers. For other moments, the probability $\mathcal{P}_a(t)$ will be different from $\mathcal{P}_a(0)$, exponentially decreasing with time. We describe this by saying that the state is decaying. The constant γ in the exponent gives the decay rate and is accordingly called the *decay constant*. Thus, the *imaginary term in (11.83) is directly connected with the observable decay rate of a state*. We can also rewrite (11.86) as

$$\mathcal{P}_a(t) = \mathcal{P}_a(0)e^{-(t/\tau)}, \quad \tau \equiv \frac{\hbar}{\Gamma}. \tag{11.87}$$

Here τ is the lifetime of the state. According to the time–energy indeterminacy (3.36), the time τ is associated with the minimal possible indeterminacy δE in energy (the energy-level width) of the state, $\tau = (1/2)\hbar/\delta E$. Comparing this with (11.87) shows that

$$\frac{1}{2}\Gamma = \delta E. \tag{11.88}$$

Thus, the imaginary term in complex energy is directly related to the width of the quasi-stationary energy level.

3) How does the complex energy eigenvalue affect the values of k and q?

It follows from (11.80) that they also become complex:

$$k \to \tilde{k} = k - i\chi, \qquad q \to \tilde{q} = q - i\varsigma. \tag{11.89}$$

In view of (11.80), this gives

$$E = \frac{\hbar^2}{2\mu}(k^2 - \chi^2), \qquad \frac{1}{2}\Gamma = \frac{\hbar^2}{\mu}k\chi. \tag{11.90}$$

Conversely, solving this for k and χ when E and Γ are known yields

$$k = \frac{1}{\hbar}\sqrt{\mu\left(\sqrt{E^2 + \frac{1}{4}\Gamma^2} + E\right)}, \qquad \chi = \frac{1}{\hbar}\sqrt{\mu\left(\sqrt{E^2 + \frac{1}{4}\Gamma^2} - E\right)}. \tag{11.91}$$

The similar expressions can be obtained for q and ς.

4) What is the physical meaning of complex $\tilde{k}$ and $\tilde{q}$?

This question will be explained in the process of solution of the whole problem. Since the Hamiltonian here is invariant under inversion $x \to -x$, it has eigenstates with definite parity – either even or odd functions. We will, for definiteness, restrict to the odd solutions, that is, set $B = -A$ in (11.79). Then we can write (11.79) in the form explicitly showing the odd parity of the state (as well as the complex values of $\tilde{k}$ and $\tilde{q}$):

$$\psi(x) = \begin{cases} A\sin\tilde{k}x, & 0 \leq x \leq a, \\ \alpha e^{\tilde{q}x} + \beta e^{-\tilde{q}x}, & a \leq x \leq b, \\ \mathcal{A}e^{i\tilde{k}x}, & x > b. \end{cases} \tag{11.92}$$

(For the left semispace we will have the same expression, with the only difference being in signs of amplitudes α, β, and $\mathcal{A}$.) The boundary conditions at a and b are

$$A\sin a\tilde{k} = \alpha e^{a\tilde{q}} + \beta e^{-a\tilde{q}}, \qquad \alpha e^{b\tilde{q}} + \beta e^{-b\tilde{q}} = \mathcal{A}e^{ib\tilde{k}}, \\ \tilde{k}A\cos a\tilde{k} = \tilde{q}(\alpha e^{a\tilde{q}} - \beta e^{-a\tilde{q}}), \quad \tilde{q}(\alpha e^{b\tilde{q}} - \beta e^{-b\tilde{q}}) = i\tilde{k}\mathcal{A}e^{ib\tilde{k}}. \tag{11.93}$$

It follows that

$$\tan a\tilde{k} = \frac{\tilde{k}}{\tilde{q}}\frac{\alpha e^{a\tilde{q}} + \beta e^{-a\tilde{q}}}{\alpha e^{a\tilde{q}} - \beta e^{-a\tilde{q}}}, \qquad i\frac{\tilde{k}}{\tilde{q}} = \frac{\alpha e^{b\tilde{q}} - \beta e^{-b\tilde{q}}}{\alpha e^{b\tilde{q}} + \beta e^{-b\tilde{q}}}. \tag{11.94}$$

This gives the system of two simultaneous equations for α and β:

$$\left(\tan a\tilde{k} - \frac{\tilde{k}}{\tilde{q}}\right)e^{a\tilde{q}}\alpha = \left(\tan a\tilde{k} + \frac{\tilde{k}}{\tilde{q}}\right)e^{-a\tilde{q}}\beta, \\ (i\tilde{k} - \tilde{q})e^{b\tilde{q}}\alpha = -(i\tilde{k} + \tilde{q})e^{-b\tilde{q}}\beta. \tag{11.95}$$

For a nonzero solution, the system's determinant must be zero:

$$(i\tilde{k} - \tilde{q})\left(\tan a\tilde{k} + \frac{\tilde{k}}{\tilde{q}}\right)e^{\tilde{q}D} + (i\tilde{k} + \tilde{q})\left(\tan a\tilde{k} - \frac{\tilde{k}}{\tilde{q}}\right)e^{-\tilde{q}D} = 0. \qquad (11.96)$$

Here $D \equiv b - a$ is the width of the barrier. As was the case in Section 9.9, we have a transcendental equation for determining $\tilde{k}$. But the fundamental difference from (9.108) and (9.109) is that the unknown in (11.96) is a *complex* variable (11.89), so that actually we now have two equations for two unknowns – k and χ.

What specifies the "chosen" energies E_j at which the states are quasi-stationary? Since the leakage from the region between the barriers is associated with tunneling, we can expect that quasi-stationary states are possible for sufficiently wide or high barriers. In this case, it is difficult for the particle to "feel" the outer side of the barrier. The barriers effectively act as if each one of them was the potential threshold, and the spacing between them becomes a potential well. Then the discrete set of allowed energies for such a well can be considered as an embryo of the set E_j.

The mathematical condition for such a situation is low transmittance of the barrier. We know (Equation 9.45) that small transmission coefficient corresponds to a large qD. In this region, in the zero-order approximation, we can neglect the term with $e^{-\tilde{q}D}$ in Equation 11.96, thus reducing it to

$$\left(\tan a\tilde{k} + \frac{\tilde{k}}{\tilde{q}}\right) = 0. \qquad (11.97)$$

Except for some difference in notations, this is identical to Equation 9.62 for the E-eigenvalues in a potential well of depth U and width $2a$. Apart from the graphical solution obtained in Section 9.5, we can in the given case evaluate the solution of Equation 11.97 analytically if we first neglect Γ, which we can do in the considered approximation. Then we can go to the next approximation in which we assume Γ_j small but nonzero. The method of consecutive approximations will be studied in more detail in Chapter 17, and here we just present the result of the first approximation:

$$\Gamma_j = 2\frac{\hbar^2}{\mu}k_j\chi_j = 2\frac{(E_j(U - E_j))^{3/2}}{U^2(1 + aq_j)}e^{-2q_jD}, \qquad q_j \equiv \frac{\sqrt{2\mu(U - E_j)}}{\hbar}. \qquad (11.98)$$

Since $\hbar k/\mu \simeq v$ is the velocity of the particle inside the well, and $k \approx (2a)^{-1}$, the last equation, in view of (11.80), can be written as

$$\Gamma_j \approx 2\frac{\hbar}{T_j}e^{-(2/\hbar)\sqrt{2\mu(U-E_j)}D} = \hbar\gamma_j e^{-(2/\hbar)\sqrt{2\mu(U-E_j)}D}, \qquad T_j = \frac{4a}{v_j}. \qquad (11.99)$$

This expression has a simple physical interpretation. T_j is the period of particle's motion (its "oscillations") in the region $(-a, a)$ between the barriers, and its doubled reciprocal is the collision frequency γ_j with the inner sides of the barriers in the given state j. This frequency can be crudely envisioned as the oscillation frequency of the classical particle between the two barriers. The exponential factor, according to (9.45), determines the transmission coefficient T in the given approximation.

Therefore, the right-hand side of the equation is, up to the factor $\hbar$, the product of transmissivity and collision frequency (the number of "attempts" to get through per unit time). This product determines the total transmission probability per unit time.

In terms of experimentally observed characteristics, the imaginary part of $\tilde{E}$ in (11.82) has a twofold physical interpretation. On the one hand, it is the width of the corresponding energy level E_j, and on the other hand it is, up to $\hbar$, the probability per unit time for the particle to leak out. The level width and the probability per unit time to tunnel out (the decay rate) are intimately linked characteristics of a quasi-stationary state.

11.8.1
Discussion

Now we are in a position to turn to question No. 4 in the beginning of the section, about the physical meaning of the imaginary part χ_j of the wave number. We can get the answer immediately by plugging the complex wave vector into the third equation in (11.92) for the de Broglie wave of the particle outside the barriers:

$$\psi(x) = A\, e^{\chi_j x} e^{ik_j x}. \tag{11.100}$$

This describes the outgoing wave whose amplitude exponentially increases with distance! The imaginary part of $\tilde{k}$ describes the rate of this increase. The similar situation is well known in optics. Usually a wave is attenuated in an absorbing medium, or amplified in a medium with inversed population (e.g., in a laser). The difference with the current situation is that here the de Broglie wave from decaying source is propagating in vacuum – there is nothing to amplify it, and in contrast with the laser, we do not see any restrictions here on the traveled distance. Therefore, no matter how small the χ_j may be, sufficiently far from the origin the amplitude becomes arbitrarily large. This is definitely a nonphysical result, and the question arises: What does it mean? The answer is that this is not the whole result but only part of it. We forgot the temporal factor

$$e^{-i(\tilde{E}/\hbar)t} = e^{-i\left[\left(E_j - (1/2)i\Gamma_j\right)/\hbar\right]t} = e^{-(1/2)(\Gamma_j/\hbar)t}\, e^{-i\omega_j t}, \quad \omega_j = E_j/\hbar. \tag{11.101}$$

This factor describes exponential decay of the state (Equation 11.87)! To get the complete result, we must substitute it into the whole expression for the eigenstate. This gives

$$\Psi(x,t) = \psi(x)e^{-i(\tilde{E}/\hbar)t} = A\, e^{\chi_j x - (1/2)(\Gamma_j/\hbar)t}\, e^{i(k_j x - \omega_j t)} = A_j(x,t)e^{i(k_j x - \omega_j t)}. \tag{11.102}$$

The amplitude of the outgoing wave

$$A_j(x,t) \equiv A\, e^{\chi_j x - (1/2)(\Gamma_j/\hbar)t} \tag{11.103}$$

is a function of distance *and* time. It increases with distance because at large distances we observe the particle that left its initial residence long ago, when the probability (11.86) was accordingly higher. But by the *current moment* t this probability has exponentially dropped from that initially high value, and this is reflected by the temporal term in the exponent (11.103). Therefore, if the theory gives a consistent description of the process, both terms in this exponent must cancel each other in the case when x and t are associated with the propagation of a fixed particle. *In this case*, x and t (which are generally *independent* variables!) are related by $x = vt$. Putting this into (11.103) and using (11.90) gives the result that along the particle's "path" we have

$$\mathcal{A}_j(x,t) \Rightarrow \mathcal{A}\, e^{(v-(\hbar/\mu)k)\chi_j t}. \tag{11.104}$$

But $\hbar k/\mu$ is the group velocity v (this result is the same in relativistic and nonrelativistic mechanics!). We immediately see that both terms in the exponent do indeed cancel each other out, leaving the amplitude constant.

A possible objection against this argument is that we are not mandated to follow the motion of a fixed wave packet, and can instead consider (11.104) at a fixed moment of time, say, at $t = 0$, with x as a single *independent* variable. Physically, this means that instead of considering one particle at different moments of its life, we switch to an ensemble of particles whose respective state functions are all evaluated at one moment but at different locations. This would take us back to the case (11.101) with exponentially increasing spatial factor as the only actor in the play. It does blow up at $x \to \infty$, and for already indicated reason: a greater x corresponds to a particle that left the origin earlier, when $\mathcal{P}_a(t)$ was higher.

Such argument is technically correct, but it shows again that we cannot turn away from time in our discussion. No matter how close the *quasi*-stationary state is to being stationary, it evolves, and the time cannot be ignored. Therefore, if one insists that we can take arbitrarily large distances within the same mathematical model, that is, with function (11.100) remaining valid for all positive x, one automatically introduces infinitely remote past with function (11.101) remaining valid for all negative t. Both functions are merely different sides of the same coin – the full function $\Psi(x,t)$.

The only way to avoid the divergence of the state function at infinity *at fixed t* is to recognize that the used model of the quasi-stationary state determined for all x and t is an idealization. When we define the amplitudes A, $\mathcal{A}$, α, and β of the state at a moment taken as $t = 0$, we must consider this moment as the beginning of time. This automatically restricts the possible values of x for each given $t > 0$ by condition

$$0 \leq x \leq v(E)t. \tag{11.105}$$

In other words, the possible spatial range of the evolving state at a moment t is naturally restricted by the distance the emitted particle can travel from the source by this moment.

11.9
3D Barrier and Quasi-Stationary States

Consider now a more realistic case of a 3D region enclosed by a spherical potential barrier of height U, with inner and outer radii a and b, respectively, and a particle with $0 < \bar{E} < U$ initially locked within the barrier. The state is described by an appropriate solution of the radial Schrödinger equation. As in the case of a 3D potential well (Section 10.3), we will restrict to states with $l = 0$. Then the needed solution is, asymptotically, diverging spherical wave:

$$\Psi(r, \theta, \varphi) \Rightarrow \Psi(r) = \Phi \frac{e^{ikr}}{r}. \tag{11.106}$$

This solution is nonstationary since it is not balanced by an incoming wave, so we have only diverging flux

$$j_r = -\frac{i\hbar}{\mu} \operatorname{Im}\left(\Psi \frac{\partial \Psi^*}{\partial r}\right) = |\Phi|^2 \frac{\hbar k}{\mu r^2}. \tag{11.107}$$

Therefore, integrating the continuity equation (8.13) over a volume V within a spherical surface A around the origin and applying the Gauss theorem, we find

$$\frac{d}{dt}\int_V \Psi^* \Psi \, d\tau = -\oint_A j_r \, da = -4\pi |\Phi|^2 \frac{\hbar k}{\mu}. \tag{11.108}$$

This tells us that the probability of finding the particle within V decreases with time (the minus sign on the right). The basic characteristics of the state are the same as in the 1D case in the previous section. The main difference is that we now have a complex propagation vector in each radial direction:

$$\tilde{k}_j = k_j - i\chi_j = k_j - \frac{1}{2}i\frac{\mu}{\hbar^2 k_j}\Gamma_j \tag{11.109}$$

with Γ_j determined by (11.98). The corresponding radial functions differ from (11.79) only by the factor r^{-1}:

$$\psi_j(r) \approx \frac{\Phi}{r}\begin{cases} e^{\chi_j r} e^{ik_j r} - e^{-\chi_j r} e^{-ik_j r}, & 0 \leq r \leq a \\ \alpha e^{q_j r} e^{-i\xi r} + \beta e^{-q_j r} e^{i\xi r}, & a \leq r \leq b, \\ A e^{\chi_j r} e^{ik_j r}, & r > b \end{cases} \tag{11.110}$$

and the formal quasi-stationary solution is

$$\Psi(r, t) = \psi_j(r) e^{-(1/2)(\Gamma_j/\hbar)t} e^{-i(E_j/\hbar)t}. \tag{11.111}$$

The radial function here exponentially increases with r, whereas the temporal function exponentially decreases with t.

The temporal dependence (11.111) already suggests the law of decay similar to (11.87) in the 1D case. As in that case, assume that the particle is entirely within the barrier at $t = 0$. Then its initial radial function is equal to zero at $r > a$:

$$\Psi(r, 0) = \begin{cases} \psi_j(r), & r < a, \\ 0, & r > a. \end{cases} \tag{11.112}$$

As we found in the previous section, the corresponding evolution can formally be described as an "everlasting" state (11.78) with complex energy. Actually, we have a nonstationary state "created" at $t = 0$. According to the general theory, it can be represented as a superposition of eigenstates $\psi_E(r)$ with various E:

$$\Psi(r,0) = \int_0^\infty C(E)\psi_E(r)dE, \quad t=0, \tag{11.113}$$

and

$$\Psi(r,t) = \int_0^\infty C(E)\psi_E(r)e^{-i(E/\hbar)t}dE, \quad t>0. \tag{11.114}$$

(The requirement $\langle E \rangle < U$ for the *average* energy of a trapped state does not preclude the existence of individual *stationary* states with arbitrarily high energies in (11.113) and (11.114).)

The expansion coefficients are determined from the initial state

$$C(E) = \int \Psi(r,0)\psi_E^*(r)dr, \tag{11.115}$$

and

$$|C(E)|^2 dE = w(E)dE = d\mathcal{P}(E) \tag{11.116}$$

gives the probability of finding energy in the vicinity of E in the evolving state (11.114).

The natural question arises – how this probability depends on E? It turns out that the decay law (11.87) determines the explicit energy dependence $w(E) \equiv d\mathcal{P}(E)/dE$. To see it, let us first determine the amplitude $\Upsilon(t)$ of finding the particle still residing within its initial domain at a moment $t > 0$. This is just the amplitude with which the initial state $|\Psi(0)\rangle$ is represented in the current state $|\Psi(t)\rangle$. As we know, such an amplitude is given by the inner product $\Upsilon(t) = \langle \Psi(t)|\Psi(0)\rangle$. In position representation,

$$\Upsilon(t) = \int \Psi^*(r,0)\Psi(r,t)dr. \tag{11.117}$$

Putting here (11.113) and (11.114) and changing the ordering of integration gives

$$\Upsilon(t) = \int C(E)C^*(E')e^{-i(E/\hbar)t}\left\{\int \psi_E(r)\psi_{E'}^*(r)dr\right\}dE\,dE'. \tag{11.118}$$

In view of orthogonality of eigenstates $\psi_E(r)$ (3.10) and of (11.115), this reduces to

$$\Upsilon(t) = \int w(E)e^{-i(E/\hbar)t}dE. \tag{11.119}$$

The value $|\Upsilon(t)|^2 = \mathcal{P}(t)$ gives the sought-for law of decay of the initial state. The form of this law is thus determined by the energy distribution $w(E)$ in the initial state. This result is known as the Krylov–Fock theorem [1,36,37].

Let us evaluate $\Upsilon(t)$ for a typical situation when $\Psi(r,t)$ is a quasi-stationary state (11.111). We know that the wave function of such state exponentially increases with r at $t > 0$. This is not exact radial function of quasi-stationary state when considered for all r and t, as we have found in the previous section. Next, we can choose $\Psi(r,0)$ in the form (11.112). As noted above, this it is not an exact choice either, since the quasi-stationary solution in the form (11.110) and (11.111) at no time reduces to (11.112). So it is only an approximation, but it is sufficiently accurate. The function (11.112) describes the actual probability cloud in the initial state; its "temporal extension" $\Psi(r, t)$ is, as we know, sufficiently accurate if $r < vt$. But we can ignore the behavior of $\Psi(r, t)$ outside the barrier since we have there $\Psi^*(r, 0)\Psi(r, t) = 0$ due to (11.112). Thus, with $\Psi(r, t)$ chosen in the form (11.111), the integral (11.116) reduces to

$$\Upsilon(t) = \left\{ \int_0^a |\Psi(r,0)|^2 \, dr \right\} e^{-(1/2)(\Gamma_j/\hbar)t} \, e^{-i(E_j/\hbar)t} = e^{-(1/2)(\Gamma_j/\hbar)t} \, e^{-i(E_j/\hbar)t} \tag{11.120}$$

(the function $\Psi(r, 0)$ is normalized to 1). We have found the decay law

$$|\Upsilon(t)|^2 = \mathcal{P}(t) = e^{-(\Gamma_j/\hbar)t}. \tag{11.121}$$

It is the well-known law of the exponential decay: the probability of finding the particle still lingering in the initial state by a moment t exponentially decreases with t. The value $\gamma \equiv \Gamma_j/\hbar$ is the decay constant.

Now we can use the obtained result to determine $C(E)$ by combining (11.120) with (11.119):

$$|\Upsilon(t)| = \left| \int |C(E)|^2 \, e^{-i(E_j/\hbar)t} \, dE \right| = e^{-(1/2)(\Gamma_j/\hbar)t}. \tag{11.122}$$

This is satisfied when (Problem 11.14)

$$|C(E)|^2 = \frac{d\mathcal{P}(E)}{dE} = \frac{1}{2\pi} \frac{\Gamma_j}{\left(E - E_j\right)^2 + (1/4)\Gamma_j^2}. \tag{11.123}$$

We obtained the well-known *dispersion formula* [36,37] for the energy distribution in a nonstationary state. If you represent a nonstationary state of the type (11.122) as a superposition of stationary states, they will be there with probabilities (11.123). The system may have more than one quasi-stationary states ($j = 1, 2, \ldots$), in which case the probability distribution will be described by the sum of expressions (11.123) with different E_j and Γ_j (Figure 11.6). The quantity $(1/2)\Gamma_j$ determines the respective maximum width (the width of the corresponding energy level E_j) as well as the average lifetime $\tau_j = \hbar/\Gamma_j$ of the particle in its initial state on the level E_j.

The states (11.122) with energy distribution (11.123) are called the *Gamow states*. We will consider some examples of such states in the next section.

11.10
The Theory of Particle Decay

The QM theory of quasi-stationary states provides a natural explanation of radioactive decay [38]. The most known example of this phenomenon is the α-decay.

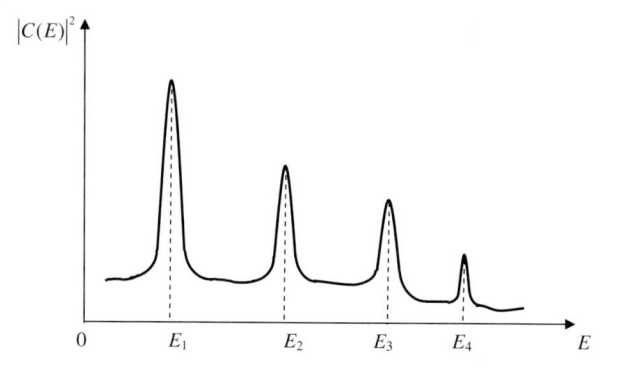

Figure 11.6 Probability distribution of finding a particle with a certain energy in one of quasi-stationary states (graphical representation of the dispersion formula).

Many radioactive nuclei decay by emitting α-particles. An α-particle (He nucleus) is a very stable system of 2p and 2n held together by short-ranged ($r \leq 10^{-15}$ m) but very strong forces of nuclear attraction. This attraction, as was found in the Rutherford experiments, is two to three orders of magnitude stronger than the Coulomb force at the same distance, but it falls off exponentially with distance. The graph of potential energy of an α-particle (or of a proton) in combined electric and nuclear field is shown in Figure 11.7. Nuclear force is independent of particle's charge, and acts between p and n as well. It holds protons and neutrons in a nucleus tightly packed together within a very small region.

The extreme stability of an α-particle implies that it can retain its individuality within a nucleus, constituting one of its building blocks. This conclusion is not necessary. The α-particle can as well be formed from two protons and two neutrons shortly before quitting the mother nucleus (U^{239} or U^{235}). But once formed, it can remain as a single whole and interact with the remaining part of the nucleus. In either case, we can consider it as a single particle during the whole process.

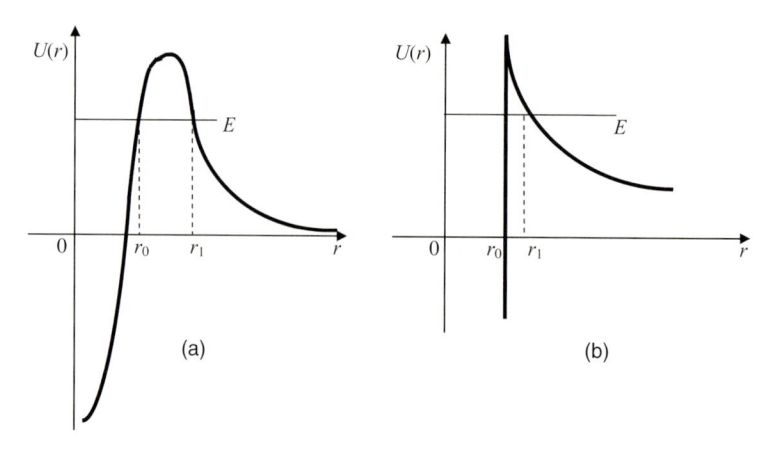

Figure 11.7 The combined field of nuclear and Coulomb potential for an α-particle in a heavy nucleus.

As seen from Figure 11.7a, the combined field of nuclear and electric force produces a 3D potential well surrounded by potential barrier. The α-particle before its emission is in the extremely narrow but deep well within the barrier. The range of action of nuclear forces is so short and they fall off with the distance so rapidly that we can with sufficient accuracy model the resulting potential as a very deep and narrow well with vertical inner wall (Figure 11.7b).

In his experiments, Rutherford bombarded the U-nuclei with α-particles from Th. Their measured initial kinetic energy was about $K \simeq 1.3 \times 10^{-12}$ J. The α-particles with such energy can overcome the Coulomb repulsion of U-nucleus and approach its center up to the turning point r_1 determined from $K = 2Ze^2/r$, where e^2, as usual, is a shorthand for $q_p^2/4\pi\varepsilon_0$. Putting here $Z = 92$ for U and the initial value of K gives $r_1 = 3 \times 10^{-14}$ m. The experiment showed that even at such close proximity to the center, the resulting angular distribution of scattered particles was exactly the same as the one due to purely Coulomb field of the "naked" U nucleus. The initial energy, high as it was, still failed to bring α-particles closer to the center where they would encounter the nuclear potential, which could change their angular distribution. In other words, the closest approach r_1 was still greater than r_0. This confirms the validity of the model shown in Figure 11.7b. We conclude that α-particles bound within the well by nuclear forces must reside there at $r < r_1$. Reversing the argument, we can say that α-particles once emitted from the nucleus must be subject to electric force alone starting from about the same distance r_1. (Actual electric potential would involve $Z' = Z - 2 = 90-$ the remaining charge of the nuclear core after the initial nucleus splits into two, U → R + He. This produces only minor correction $\simeq 2\%$ in our calculations.) Thus, the *classical* physics predicts the "exodus" of α-particles from U nucleus as emission from the region $r \leq r_0$ with energy no less than $U(r_0) = U(r_1)$. Accordingly, far from the nucleus it would have kinetic energy significantly higher than $K = U(r_1) = 1.3 \times 10^{-12}$ J.

This prediction contradicted the observations. In addition, it is very unlikely that an α-particle with energy matching or even exceeding the top of the barrier could have been firmly trapped by it before emission. A hypothesis was proposed by T. Kelvin that the average energy of an α-particle inside the nucleus is significantly lower, but there are many other nucleons there in a state of random motion similar to that of molecules in boiling water. Then, according to laws of statistical mechanics, one of the particles receives sometimes a "kick" and gets energy sufficient to escape from the well. This means that the energy of the escaped particles must exceed the value of 1.3×10^{-12} J. But the actual measurements showed less than half of this value.

QM gave a simple solution of that puzzle: the particle within the nucleus has an energy less than the top of the barrier (otherwise it could not exist there so long), but tunneling allows it to eventually leak through with this low energy. The quantitative theory of radioactive decay based on the concept of quasi-stationary states was developed by G. Gamow [38]. It describes correctly all observed features of the phenomenon and, in particular, very strong dependence of the decay constant on energy of emitted particles.

As mentioned in Section 11.8, a particle before escaping from its confinement can be considered as "oscillating" between the barriers with the effective group velocity v_g. The same argument applies to α-particle in our problem. Then the decay constant can be found as

$$\gamma = \frac{v_{g,j}}{2r_0} T = f_j T, \qquad (11.124)$$

where $f_j = v_{g,j}/2r_0 = 1/\tau_j$ is the frequency of oscillations with the period τ_j and T is the barrier transmissivity for a state j (recall the comments to Equation 11.99). Since the barrier's shape in Figure 11.7 is not rectangular, we have

$$T_j = T_0 \, e^{-(2/\hbar) \int_{r_0}^{r_1} \sqrt{2\mu[U(r) - E_j]} \, dr}. \qquad (11.125)$$

For the case shown in Figure 11.7, r_1 is determined as the turning point from the equation $U(r) = 2Z'^2 e^2/r_1 = E_j$, where $E_j = K_j$ is the particle's kinetic energy after decay. The integral I with this $U(r)$ can be taken by denoting $\xi = r/r_1$ and then introducing a new variable η according to $\xi = \cos^2 \eta$. The result is (Problem 11.17)

$$I = 2Z' e^2 \sqrt{\frac{2\mu}{E_j}} \left(\arccos \sqrt{\xi_0} - \sqrt{\xi_0(1-\xi_0)} \right), \quad \xi_0 \equiv \frac{r_0}{r_1} = \frac{r_0 E_j}{2Z' e^2}. \qquad (11.126)$$

Since $\xi_0 < 1$, we can expand this expression into the Taylor series and retain only first two terms:

$$I \approx \pi Z' e^2 \sqrt{\frac{2\mu}{E_j}} - 2e\sqrt{\mu r_0 Z'}. \qquad (11.127)$$

Putting this into (11.125) and combining with (11.124), we obtain the following expression for γ:

$$\gamma = \frac{\hbar T_0}{2\mu r_0^2} e^{-(2/\hbar)\left(\pi Z' e^2 \sqrt{2\mu/E_j} - 2e\sqrt{\mu r_0 Z'}\right)}. \qquad (11.128)$$

According to this expression, γ must exhibit very strong dependence on E_j, r_0, and Z'. Such dependence was discovered experimentally by Geiger and Nuttall [39,40] in 1911, long before G. Gamow formulated his theory of α-decay. We know from numerous experimental data that γ varies depending on these parameters in a very wide range between 10^6 s^{-1} and 10^{-18} s^{-1}. This corresponds to the average lifetimes ranging between 10^{-6} s and 10^{18} s $\simeq 3 \times 10^{10}$ years. The major contribution into variation of γ comes from variation in E_j and the least one from variation of r_0, which is approximately the same for all quasi-stationary states and all nuclei. If we reverse the argument and determine r_0 for different nuclei from (11.128) using experimental values for γ, E_j, and Z, we find all r_0 within a narrow region 5×10^{-15} m $\leq r_0 \leq 9 \times 10^{-15}$ m. QM gives very accurate predictions of all these features in the radioactive decay.

11.11
Particle–Antiparticle Oscillations

> If there is a place where we could verify the basic principles of QM – superposition of amplitudes, it is right here.
> *Richard Feynman*

> There isn't anything truly beautiful without a touch of strangeness.
> *Francis Bacon*

We know that for a state to change in time it must be in a superposition of distinct stationary states with different energies. The simplest case is superposition of two such states

$$|\Psi(t)\rangle = \psi_1 e^{i\omega_1 t} + \psi_2 e^{i\omega_1 t}, \qquad (11.129)$$

where, as usual, $\omega = E/\hbar$, and we are interested only in the time dependence of a state.

The corresponding probability is

$$W(t) = \left| \psi_1 e^{i\omega_1 t} + \psi_2 e^{i\omega_1 t} \right|^2. \qquad (11.130)$$

Assuming both amplitudes to be real, we obtain

$$W(t) = w_1 + w_2 + 2\sqrt{w_1 w_2}\cos(\omega_1 - \omega_2)t, \quad w_{1,2} \equiv \psi_{1,2}^2. \qquad (11.131)$$

The case (11.131), strictly speaking, is not really an evolution, since we have just periodic reiterations of the same state. Actual evolution requires superposition of more than two states, and their energy levels must not be equidistant.

But let us look out of the box. What happens if we consider a superposition of two *quasi-stationary* states? Recall that we can formally treat a quasi-stationary state as stationary, but with complex energy. So let us consider $\omega \to \omega + i\gamma$, $\gamma \equiv \Gamma/2\hbar$, and put this into (11.129) and (11.130). We will obtain

$$W(t) = w_1 e^{-2\gamma_1 t} + w_2 e^{-2\gamma_2 t} + 2\sqrt{w_1 w_2}\, e^{-(\gamma_1 + \gamma_2)t} \cos(\omega_1 - \omega_2)t. \qquad (11.132)$$

Now we see oscillations combined with already familiar exponential decay of the state.

This turns out to be a description of a fascinating quantum-mechanical effect that could be called *particle–antiparticle oscillations*. Here we will consider the case of oscillation between being K-meson and anti-K-meson.

K-mesons (or just kaons) come in two "brands" – one charged (K^+) and one neutral (K^0). And this pair has its anti-pair, K^- and $\tilde{K}^0$. In some respects, this is similar to familiar case with nucleons: we have a proton and neutron (p, n) and their antiparticles – antiproton and antineutron ($\tilde{p}, \tilde{n}$). Both – proton and neutron – have *baryon charge* $B = +1$, whereas their counterparts have each the baryon charge $B = -1$. Therefore, n and ñ are different (albeit symmetric to one another) particles, even though they are both electrically neutral. Their gyromagnetic ratios also have opposite signs. But kaons, being of meson family, do not have baryonic charge. What then distinguishes K^0 from $\tilde{K}^0$? It is a specific characteristic of some elementary particles discovered by Gell-Mann and Nishijima and called *strangeness* [41,42]. A K^0 has the strangeness $S = 1$, whereas $\tilde{K}^0$ has strangeness $S = -1$. So, *they are different particles*. For instance, $\tilde{K}^0$ bumping into a proton can produce Λ^0-particle *(lambda hyperon)* according to

$$\tilde{K}^0 + p \to \Lambda^0 + \pi^+, \tag{11.133a}$$

whereas for K^0, having opposite strangeness, the corresponding reaction will be

$$K^0 + p^- \to \tilde{\Lambda}^0 + \pi^-. \tag{11.133b}$$

Strangeness is conserved in each reaction because $\tilde{\Lambda}^0$ has $S = 1$ and Λ^0 has $S = -1$. The $\tilde{K}^0$ cannot substitute K^0 in reaction (11.133b), and vice versa – this would violate conservation of strangeness [43].

Since strangeness is an integral of motion, there seems to be no way for K^0 to transform into $\tilde{K}^0$ and vice versa. However, there is one strange thing about strangeness: it is a "weak" integral of motion – it is conserved in strong interactions, but not in weak interactions. And this brings in one of the strangest phenomena that we are going to discuss.

Due to the "weakness" of strangeness, its value may not be manifest in the weak interactions. Therefore, K^0 and $\tilde{K}^0$, different as they are, can behave in the same way in such interactions. For instance, K^0 can decay into a pair of charged π-mesons (π^+, π^-):

$$K^0 \to \pi^+ + \pi^-. \tag{11.134}$$

Actually, the mere existence of neutral kaons was discovered from studying the products of their decay. This is a very slow (by the standards of micro-world) process with the average lifetime $\tau \sim 2 \times 10^{-10}$ s. For comparison, the typical lifetime in strong interactions (responsible, among other things, for nuclear force) is $\tau \sim 10^{-23}$ s – about 13 orders of magnitude shorter than that in process (11.134)!

Note that strangeness does not conserve in this process (π-mesons do not have strangeness), so we have $S = +1$ on the left (before the process) and $S = 0$ on the right (after the process). Due to symmetry, $\tilde{K}^0$ is entitled to the same violation – it undergoes exactly the same decay

$$\tilde{K}^0 \to \pi^+ + \pi^-. \tag{11.135}$$

And strangeness does not conserve here either: we have $S = -1$ on the left of (11.135) and $S = 0$ on the right. In both cases, the strangeness disappears after decay, and two different particles (or two different initial states) may have identical final states.

This has implications of fundamental importance. We can envision it by considering the time reverse of either process:

$$\pi^+ + \pi^- \to K^0 \text{ or } \tilde{K}^0. \tag{11.136}$$

If (11.134) and (11.135) are equally possible, either of the reversed processes (11.136) is also possible. Then it is possibly a two-staged combination of the direct and reversed process: for instance, $K^0 \to \pi^+ + \pi^-$ followed by conversion of the emerged (π^+, π^-) pair back into K^0 or into $\tilde{K}^0$. Restricting to K^0 at the initial stage, we can write this as

$$K^0 \to \pi^+ + \pi^- \to K^0 \text{ or } \tilde{K}^0. \tag{11.137}$$

In the first case, the initial K^0 comes back to life after having decayed into pair (π^+, π^-) and then being rebuilt from them; in the second case, it does the same but reincarnates into its counterpart with the opposite S – nothing horrible: the strangeness does not have to conserve in the weak process. This means that there must be an amplitude of spontaneous transition $K^0 \to \tilde{K}^0$ via decay and reassembling from its products. Similarly, there is the same privilege – to decay and then resurrect or to decay and reincarnate as its antiparticle, for $\tilde{K}^0$. The pair (π^+, π^-) in such processes is a virtual pair. Denoting as $\hat{G}$ the term in Hamiltonian responsible for weak interactions, we can write for the respective transition amplitudes

$$\left\langle \tilde{K}^0 | \hat{G} | K^0 \right\rangle = \left\langle K^0 | \hat{G} | K^0 \right\rangle = \left\langle K^0 | \hat{G} | \tilde{K}^0 \right\rangle = \left\langle \tilde{K}^0 | \hat{G} | \tilde{K}^0 \right\rangle \equiv A. \tag{11.138}$$

The amplitudes of all four possible transitions are equal to each other and denoted as A. Since it is this amplitude that gives rise to the imaginary part in the energy of K^0, it must be a complex number, with the real part responsible for complete transitions "there and back" (11.137) and the imaginary part responsible for pure decays (11.134) and (11.135) without resurrection or reincarnation. The existence of an imaginary part seems to contradict the rule that the amplitudes of direct and reverse processes must be complex conjugate of one another; that is, in our case,

$$\left\langle \tilde{K}^0 | \hat{G} | K^0 \right\rangle = \left\langle K^0 | \hat{G} | \tilde{K}^0 \right\rangle^*. \tag{11.139}$$

This may be consistent with (11.138) only if A is real. However, this argument does not work here since unitarity is not conserved in decays, and accordingly, the rule (11.139) is not obeyed. Thus, A can (and must, in order to include the cases of irreversible decays!) be complex without contradicting anything.

The transitions $K^0 \rightleftarrows \tilde{K}^0$ suggest that we can treat K^0 and $\tilde{K}^0$ as two distinct states of one object rather than two different objects. Denote these states as $|K^0\rangle$ and $|\tilde{K}^0\rangle$. The object can switch from one state to another due to weak interaction. If we choose $|K^0\rangle$ and $|\tilde{K}^0\rangle$ as two basis states, then any possible state $|K\rangle$ of neutral kaon can be expressed as superposition

$$|K\rangle = c|K^0\rangle + \tilde{c}|\tilde{K}^0\rangle. \tag{11.140}$$

As always, the amplitudes c and $\tilde{c}$ are projections of $|K\rangle$ onto respective states: $c = \langle K^0|K\rangle$ and $\tilde{c} = \langle \tilde{K}^0|K\rangle$. If the states $|K^0\rangle$ and $|\tilde{K}^0\rangle$ were not influenced by the interaction with virtual (π^+, π^-) field, we would have

$$i\hbar \frac{dc}{dt} = E_0 c, \qquad i\hbar \frac{d\tilde{c}}{dt} = E_0 \tilde{c}. \tag{11.141}$$

(This is just the Schrödinger equation in its own basis, recall Section 8.9.) But due to existence of these interactions, there is a transition amplitude A that affects the rate of change of c and $\tilde{c}$. Therefore, an accurate description must include this amplitude:

$$i\hbar \frac{dc}{dt} = E_0 c + Ac + A\tilde{c}, \qquad i\hbar \frac{d\tilde{c}}{dt} = E_0 \tilde{c} + A\tilde{c} + Ac. \tag{11.142}$$

The first term on the right of both equations, when acting alone, would produce a stationary state. The second term describing the amplitude of transition of K^0 to *itself* would only slightly shift the stationary level E_0. But the third term describes transition between the states $|K^0\rangle \rightleftarrows |\tilde{K}^0\rangle$ via intermediate decay. Accordingly, Equations 11.142 become coupled.

We can simplify the system (11.142) by taking the sum and difference:

$$i\hbar \frac{d}{dt}(c + \tilde{c}) = (E_0 + 2A)(c + \tilde{c}), \qquad i\hbar \frac{d}{dt}(c - \tilde{c}) = E_0(c - \tilde{c}). \tag{11.143}$$

Denote

$$\frac{1}{\sqrt{2}}(c + \tilde{c}) = c_1, \qquad \frac{1}{\sqrt{2}}(c - \tilde{c}) = c_2. \tag{11.144}$$

Then

$$i\hbar \frac{dc_1}{dt} = (E_0 + 2A)c_1, \qquad i\hbar \frac{dc_2}{dt} = E_0 c_2. \tag{11.145}$$

(The new coefficients can be considered as representing the same general state $|K\rangle$ in another basis $|K_1\rangle, |K_2\rangle$. Mathematically, they are projections of $|K\rangle$ onto the respective states of this basis.) The equations are now decoupled and easy to solve:

$$c_1(t) = c_1(0)e^{-i[(E_0+2A)/\hbar]t}, \qquad c_2(t) = c_2(0)e^{-i(E_0/\hbar)t}. \tag{11.146}$$

Dropping the common factor $e^{-(E_0/\hbar)t}$ (which is equivalent to taking E_0 as the reference value for measuring E), we obtain

$$c_1(t) = c_1(0)e^{-i(2A/\hbar)t}, \qquad c_2(t) = c_2(0). \tag{11.147}$$

We have noted already that A is the complex number; therefore, the corresponding state is nonstationary. In order to explicitly express this fact, we set

$$2A = E - i\Gamma. \tag{11.148}$$

Then

$$c_1(t) = c_1(0)e^{-\gamma t}e^{-i\omega t}, \qquad c_2(t) = c_2(0), \qquad \omega \equiv \frac{E}{\hbar}. \tag{11.149}$$

Returning back to c and $\tilde{c}$, we obtain using (11.144)

$$c(t) = \frac{1}{\sqrt{2}}[c_1(t) + c_2(t)] = \frac{1}{\sqrt{2}}[c_1(0)e^{-\gamma t}e^{-i\omega t} + c_2(0)],$$
$$\tilde{c}(t) = \frac{1}{\sqrt{2}}[c_1(t) - c_2(t)] = \frac{1}{\sqrt{2}}[c_1(0)e^{-\gamma t}e^{-i\omega t} - c_2(0)]. \quad (11.150)$$

Coefficients $c_1(0)$ and $c_2(0)$ can be determined from the initial conditions.

Suppose that initially a kaon was born in a nuclear collision as K^0, that is, in state $|K^0\rangle$. This means that $c(0) = 1$ and $\tilde{c}(0) = 0$. Putting this into (11.150) at $t = 0$ gives

$$1 = \frac{1}{\sqrt{2}}[c_1(0) + c_2(0)], \qquad 0 = \frac{1}{\sqrt{2}}[c_1(0) - c_2(0)], \qquad c_1(0) = c_2(0) = \frac{1}{\sqrt{2}}. \quad (11.151)$$

Therefore,

$$c(t) = \frac{1}{2}\left(e^{-\gamma t}e^{-i\omega t} + 1\right), \qquad \tilde{c}(t) = \frac{1}{2}\left(e^{-\gamma t}e^{-i\omega t} - 1\right). \quad (11.152)$$

The corresponding probabilities are

$$|c(t)|^2 = \frac{1}{4}\left(1 + 2e^{-\gamma t}\cos\omega t + e^{-2\gamma t}\right),$$
$$|\tilde{c}(t)|^2 = \frac{1}{4}\left(1 - 2e^{-\gamma t}\cos\omega t + e^{-2\gamma t}\right). \quad (11.153)$$

A strange and totally counterintuitive result! Initially (at $t = 0$) we have $c(0) = 1$ and $\tilde{c}(0) = 0$. At this moment, we are dealing with a pure K^0 characterized by strangeness number $S = 1$. Then, as time evolves, there appears a chance to find instead of K^0 its antiparticle $\tilde{K}^0$ characterized by the strangeness number $S = -1$. This means conversion of one isolated particle into another (Figure 11.8). The strangeness number swaps periodically in this process. More accurately, the *expectation value* $\langle S(t)\rangle$ of S oscillates continuously with time between the two extremes $S = 1$ and $S = -1$, which are the eigenvalues of S. It is only these two values that are actually observed experimentally when we perform a test and "look in" – say, study the photographs of the traces left by a particle or its decay products in a bubble chamber. Then the analysis of the results tells us that each specific time we have either

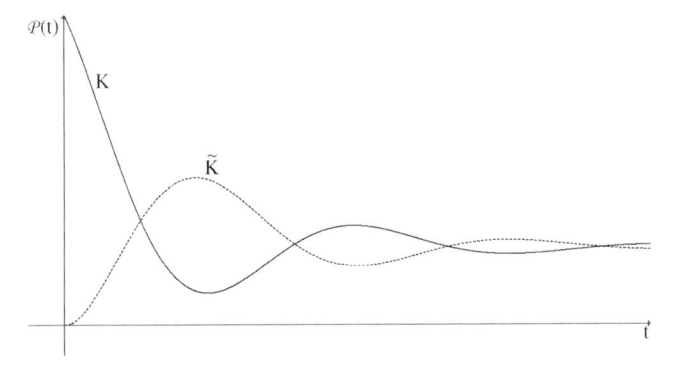

Figure 11.8 K–$\tilde{K}$ oscillations.

K^0-meson or $\tilde{K}^0$-meson. Equation 11.153 gives us only the *probability* of observing this or that outcome at any given t. How strong and how frequently these probabilities oscillate cannot be inferred from the general principles alone. This information can be found from the experiment.

Usually, kaons move with a speed close to the speed of light. Then the curves in Figure 11.8 can represent the probability change along the path of a given kaon.

The total probability of finding the initial kaon in either K^0 or $\tilde{K}^0$ state at a moment t is

$$W(t) = |c(t)|^2 + |\tilde{c}(t)|^2 = \frac{1}{2}(1 + e^{-2\gamma t}). \tag{11.154}$$

It is a decreasing function of time – no surprise, we are dealing with an unstable object, so eventually both possible states of this object are vacated due to decay. As mentioned before, unitarity of state $|K\rangle$ is not conserved once the decays are involved! What was born as K^0-meson will half of the time disappear in about 2×10^{-10} s (which corresponds to an average path length about 6 cm), giving birth to some other particles, primarily a pair of π-mesons, which themselves will, in turn, convert into other particles. But it also can, with the oscillating probability, transform into its antiparticle, and then back to itself. Unlike an ordinary quantum state showing its different faces when watched from different sides, the kaon periodically shows itself as two opposite entities in stationary environment. One of the strangest phenomena ever observed!

Another strange feature of this phenomenon is the unusual statistical finish of the initial state: its probability $W(t)$ exponentially decreases down to 0.5, in sharp contrast with the familiar decay of the Gamow states for which $\mathcal{P}(t) \sim e^{-\gamma t} \to 0$ as $t \to \infty$. In the given case, we have a nonzero probability of finding an unstable particle still in its initial state no matter how long after its birth, because there is always a chance for it to reassemble from its own decay products!

All this is a spectacular demonstration of predictive power of QM. The described effect was first predicted theoretically [41,42] on the basis of purely logical argument without any information on the inner structure of K-mesons. It was only a few years later that the predicted effect was discovered in the experiments with high-energy particles (see more detailed description in Refs [17,43]). And the experimental data give us the originally missing information, for example, about the specific numerical values of E and Γ in Equation 11.148. They showed that the ratio $\Gamma/E \approx 1.04$. From this information, one can calculate the mass difference of the "normal" (stationary) states $|K_1\rangle$ and $|K_2\rangle$. It turned out to be about $\Delta\mu \approx 3.5 \times 10^{-6}$ eV $\approx 6 \times 10^{-12}\mu_e$ (six trillionth of the electron mass). This is the least difference between the masses of two particles known today.

Note that the described process is not something totally exotic. There is another similar effect known in the particle physics – the *neutrino oscillations* [44–48]. In this process, the different types of light particles (leptons), for example, muon neutrino (ν_μ) and electron neutrino (ν_e), also transform periodically from one to another, which produces interesting observable effects involving solar neutrinos.

11.11.1
Discussion

It may be instructive to note the analogy between the discussed phenomenon and the "normal oscillations" known in CM. Here we can give a specific "classical" example – two coupled oscillators (two masses on springs or two pendulums, connected by a weak spring).

We consider the case of equal masses $\mu_1 = \mu_2 \equiv \mu$ on identical springs with spring constant k. This is close to our QM system with the same mass of K^0 and $\tilde{K}^0$. Equal spring constants reflect the fact that weak interaction is the same for both kaons. The two oscillators are coupled by an additional weak spring with a spring constant $k_c < k$.

Suppose that initially the first oscillator is "excited" (e.g., somebody or something pushed it hard to the left, compressing its spring). After being released, the mass starts vibrating around its equilibrium position with the frequency $\omega = \sqrt{k/\mu}$. In the absence of coupling, these vibrations would last forever (assuming no dissipation!), leaving the second mass intact. The initial energy input to mass I would remain concentrated on this mass.

Coupling, even if very weak, changes all that. The intermediate spring gets periodically deformed and accordingly is gently but persistently "nagging" the second mass with the frequency close to ω. Any oscillator under even very weak but closely tuned external undulation will eventually start vibrating with large amplitude. The necessary energy will, of course, come from mass I. Thus, each cycle of mass I will produce an energy flux through the coupling spring to mass II. After a certain time, all input energy E that was initially on mass I will have flown to mass II. So mass I will now be nearly idle, whereas mass II will be at full swing. The masses swapped their roles. After that, the energy flow will reverse, until restoring the initial status quo. Then the process will reiterate. In the language of QM, mass I is analogous to K^0 and mass II is analogous to $\tilde{K}^0$. Therefore, we will use the same notation for them, denoting the first one as K^0 and the second one as $\tilde{K}^0$. The periodic energy exchange between the oscillating masses is analogous to periodic conversion (kaon oscillations) of one particle into the other and back. The state of each mass is nonstationary, and so is the state of each kaon. Therefore, we can say that when mass I is in its full swing, this models kaon in state K^0, and when the second mass is in its full swing, it would be the initial kaon in state $\tilde{K}^0$.

But we know that there are also two *stationary* states in the K^0–$\tilde{K}^0$ system, which we have denoted as $|K_1\rangle$ and $|K_2\rangle$. They may have slightly different eigenvalues of energy, but these eigenvalues are sharply defined, which makes both states stationary. The oscillations of experimentally observed K^0 and $\tilde{K}^0$ states can be explained as a result of superposition of two stationary eigenstates. In this picture, what is the classical analogue of the *stationary states* $|K_1\rangle$ and $|K_2\rangle$? In other words, what are the two *normal modes* in our system of two classical oscillators?

There are two such states. One can be created when *both* masses are pushed *simultaneously* in one direction by two equal external forces. Both start vibrating with their natural frequency ω without affecting each other since they vibrate in phase and the distance between them remains fixed. The coupling spring under these conditions is as good as nonexistent. It remains relaxed and therefore does not perturb the vibration of either mass. The vibration state of either mass remains stationary – each vibrates with the same constant frequency and amplitude.

The second stationary state is created if the two masses are hit *simultaneously* by instant, equal and *oppositely* directed forces, say, mass I is pushed to the right, and mass II is at the same moment pushed with equal strength to the left. In other words, the initial action must be *symmetric* with respect to the center of the system. In this case, both masses will start vibrating again synchronously, but now out of phase. This will affect the coupling spring, subjecting it to periodic deformations. The reaction of this spring will be opposing the action of masses. When the coupling spring is being compressed by the two masses, it will repel them, and when stretched, it will pull them back to each other. The result will be as if each mass were under the action of a single effective spring with a higher spring constant and accordingly oscillating with a smaller amplitude but higher frequency $\omega' = \omega + \delta\omega$. But the vibrations of each mass will be, again, stationary.

Thus, we have two distinct eigenstates. Their common feature is the absence of energy flow between parts of the system, so the vibration state of each part is stationary.

Apart from the described two stationary (eigen)states, we have a continuous spectrum of various nonstationary states. Out of them, there are two "maximally nonstationary" states, with only one out of two masses initially hit by a force. If that was mass I, we have the model of K^0 born in a high-energy collision, and if it was mass II, we have the model of $\tilde{K}^0$ born in a high-energy collision. But which of them was born is not really important since they start converting into one another anyway.

A thoughtful reader may ask: "OK, this may be rather convincing, but how can we in the framework of this picture explain the fact that each stationary state (or normal mode) has a single frequency? After all, transformations between the bases can be carried out both ways. It is quite clear that any nonstationary state has its energy indeterminate since it is a superposition of two (or more) stationary states with different energies. But by the same token, we can take two appropriately chosen nonstationary states as a basis. Then each stationary state can be considered as a superposition of at least two such nonstationary states. How can this produce a single-energy state if each of the superposed states is nonstationary?"

This is a good question. And the answer is: in most cases it cannot, but sometimes it can. Generally, the same superposition that produces nonstationary states from stationary ones can do the opposite job. The only condition for this is the appropriate choice of the basis states, and appropriate choice of the amplitudes in their superposition. We can always find such basis and such amplitudes that extra energy eigenvalues in superposed nonstationary states cancel each other, leaving a state with only one energy eigenvalue. This is precisely what happens when we combine the amplitudes c and $\tilde{c}$ in Equation 11.144 in such a way that their combinations c_1 and c_2 become decoupled. The same holds in the presented classical analogy. Each of the considered normal modes can be represented as an appropriate superposition of nonmonochromatic modes.

Altogether, we have a very close classical analogue of our QM process of kaon (or neutrino) oscillations. It only remains to include dissipation into our mechanical system to model general decrease of probability in (11.154) due to decay into (π^+, π^-) pairs.

After reading this, one might think that after all we contradict ourselves when we say that described QM effect is counterintuitive. "What is so counterintuitive here? Quite the contrary, it seems pretty clear now. The observed particle oscillations are just QM version of classical coupled oscillations."

This seems true, but a second thought shows two subtle distinctions. First, no mechanical model can explain why the decreasing net probability in (11.154) *stops halfway* between 1 and 0. Any dissipation in CM would eventually stop all oscillations completely. We do not know anything better than dissipation to model the decay into (π^+, π^-) pairs, but no dissipation can model a possibility of inverse process – getting reassembled from such pair after the decay. So, the mechanical analogy, while being useful, works only partially.

Second, even this partial analogy works only when we start thinking about *one free particle* in terms of *oscillations of a certain system* like coupled pendulums or masses on a spring, which are not free. Such oscillations are already an embryo of a wave. It is still counterintuitive if we insist on describing a *single* particle named kaon just as a free particle (with no springs attached!). And if we want to grasp, at least partially, the mystery of its changing appearances, we can only do it in terms of *oscillations or waves* and their interference, not in terms of classical particles. This is another spectacular illustration that our "particle" is not exactly a particle, nor is it a pure wave. It is something totally new, which could, perhaps, be more accurately described as an excited state of a quantum field. Our limited intuition, if restricted to mutually excluding concepts of a wave only or a particle only, cannot capture the real QM nature of the world in its entirety.

11.12
A Watched Pot Never Boils (Quantum Zeno Effect)

The *watched pot rule* is encapsulated in the proverb "a watched pot never boils" and refers to the fact that under certain conditions, continuous observation of a quantum system can freeze its evolution. This peculiar quantum-mechanical behavior is also known as the *quantum Zeno effect*, named so after the ancient Greek philosopher Zeno who argued that an arrow in flight cannot move since at any moment it has a sharply defined position.

To derive the rule, suppose we take a look at an evolving quantum system at some moment t when its state is

$$|\Psi(t)\rangle = \hat{U}(\hat{H}, t)|\Psi(0)\rangle = e^{-(i/\hbar)\hat{H}t}|\Psi(0)\rangle. \tag{11.155}$$

The amplitude $\langle 0|t\rangle$ of finding it at its *initial* state at this moment is given by

$$\langle 0|t\rangle = \langle \Psi(0)|\Psi(t)\rangle = \langle \Psi(0)|\hat{U}(\hat{H}, t)|\Psi(0)\rangle. \tag{11.156}$$

For a sufficiently small t, we can expand $\hat{U}(\hat{H}, t)$ into a Taylor series and retain only the terms up to the second power of t:

$$\langle 0|t\rangle \approx \langle \Psi(0)|\left(1 - i\frac{\hat{H}}{\hbar}t - \frac{1}{2}\left(\frac{\hat{H}}{\hbar}\right)^2 t^2\right)|\Psi(0)\rangle \approx 1 - \frac{i}{\hbar}\langle E\rangle t - \frac{1}{2\hbar^2}\langle E^2\rangle t^2. \quad (11.157)$$

Here we did not indicate that the expectation values are taken at the initial moment because in an isolated system with time-independent Hamiltonian, expectation values of any function of energy are also time independent.

The corresponding probability is (we again ignore the higher order terms)

$$\mathcal{P}(0,t) = |\langle 0|t\rangle|^2 = 1 - \frac{1}{\hbar^2}\left(\langle E^2\rangle - \langle E\rangle^2\right)t^2. \quad (11.158)$$

But the difference in parentheses is $\langle \Delta E^2\rangle$ in state $|\Psi\rangle$. Therefore,

$$\mathcal{P}(0,t) = 1 - \frac{\Delta E^2}{\hbar^2}t^2, \quad t \ll \frac{\hbar}{\Delta E}. \quad (11.159)$$

In a stationary state, $\Delta E = 0$ and $\mathcal{P}(0,t)$ is equal to 1 for any t just as it should. But generally, this probability depends on time. Naturally, it is close to 1 at small t.

Now imagine a curious child who takes a look at the system very soon after the zero moment. According to (11.159), he/she almost certainly finds the system in the initial state. Starting from the moment t_1 of this first observation, the system again evolves according to the rule (11.155) in which t_1 now plays the role of the initial moment; that is, we must set $\mathcal{P}(0,t_1) = 1$. At the next moment, the child takes another look, again (with overwhelming probability) throwing the system back to its initial state. The process can reiterate for a very long time but in the end the child still finds the system in its initial state. In the limit, when we measure the system's state every Δt second with $\Delta t \to 0$ (i.e., watch continuously) we stop the evolution of the system.

This effect is noteworthy for the following reason. In the majority of situations where we are interested in the preservation of the original state, interaction with surrounding objects is seen as the enemy that always tries to perturb the system. The watched pot rule shows how the same parasitic effect can, counterintuitively, become an ally by helping to freeze the system's evolution. One elegant application of this rule follows from optics. When incident light of intensity I_0 is polarized at an angle θ to the transmission axis of a polarizing filter (henceforth "polarizer"), only a fraction with intensity $I(\theta) = I_0 \cos^2\theta$ will pass (Malus' law [49]). This fraction is now polarized along the transmission axis. If we have only one incident photon at a time, we must speak about probabilities instead of intensities: $\mathcal{P}_0 = 1$ instead of I_0 and $\mathcal{P}(\theta)$ (transmission probability for the photon) instead of $I(\theta)$. Then Malus' law reduces to $\mathcal{P}(\theta) = \cos^2\theta$. The photon certainly passes at $\theta = 0$, and never passes at $\theta = \pi/2$. If θ is arbitrary and the photon happens to pass, it comes out polarized along the transmission axis, irrespective of the initial value of θ. It follows that if we put a second polarizing filter in the photon's path and orient its transmission axis

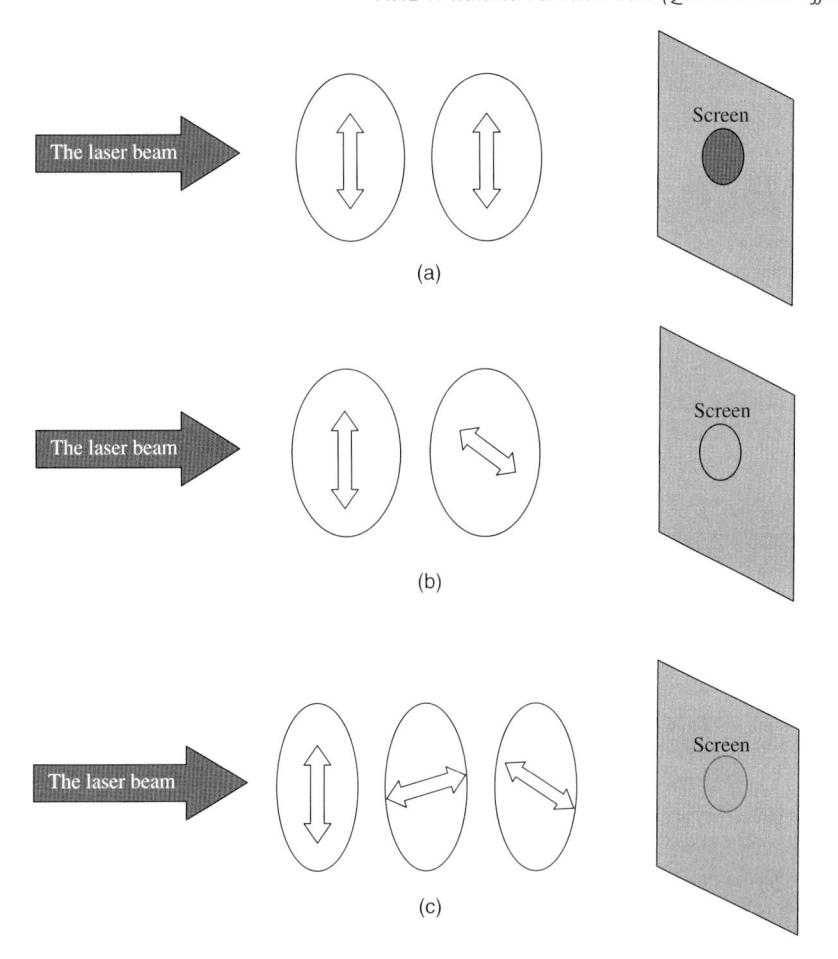

Figure 11.9 Action of a pair of polarizing filters. (a) With parallel transmission axes. The component of the incident light polarized parallel to these axes will pass without attenuation. The pair is fully transparent for a photon linearly polarized along the axes. (b) With mutually perpendicular transmission axes. Such pair (crossed filters) totally blocks any incident light. (c) The same pair with an intermediate polarizer in between. Such a polarizer partially restores the transmission.

vertically (Figure 11.9a), all photons that passed through the first filter will definitely pass through the second one. If the transmission axes are perpendicular to each other (Figure 11.9b), no photon will ever pass through the pair, regardless of its initial polarization state. The pair (frequently referred to as *crossed filters*) is absolutely opaque to light even though each filter separately may be ideally transparent for the light with appropriate polarization.

Consider now a linearly polarized photon propagating in a transparent gyrating medium (e.g., sugar water). As it travels in such medium, its polarization vector undergoes rotation proportional to the distance traveled [49]. (Do not confuse this

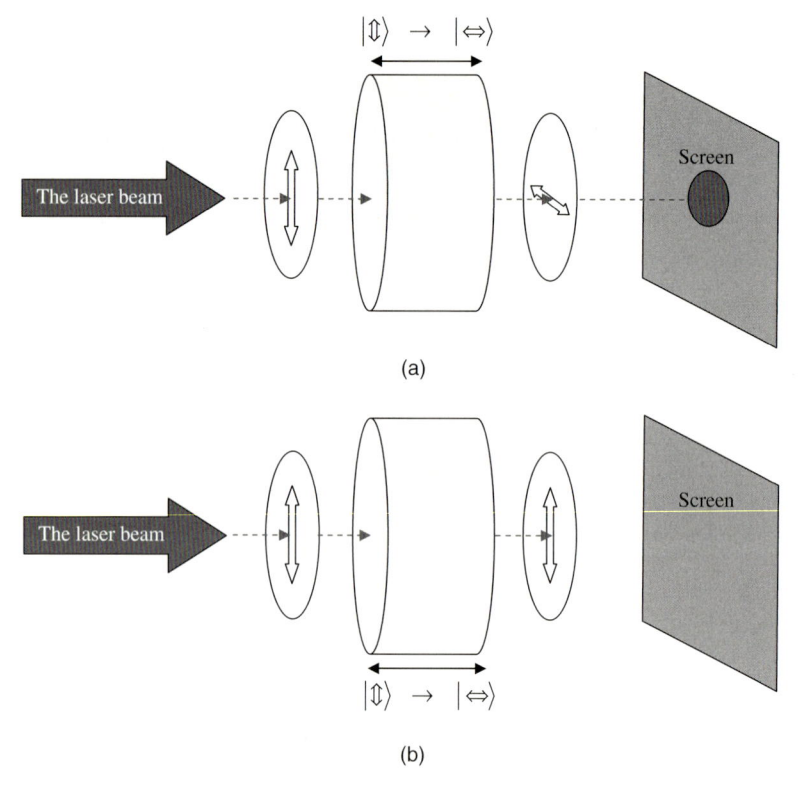

Figure 11.10 Pair of polarizing filters with a gyrating slab inside. The slab turns linear polarization of the incoming light through 90°. (a) The filters form a crossed pair. The whole system is transparent for the incident vertically polarized light. (b) The transmission axes of the filters are parallel. The pair is totally opaque.

process with circular polarization! Here the polarization *remains linear* at each point even in gyrating medium, whereas in a circularly polarized state the polarization vector is rotating at each point even in the absence of any medium.) Suppose a medium layer turns polarization by 90°. A vertically polarized photon entering the layer will exit it horizontally polarized, which can be easily checked with a polarizing filter.

Now put such a slab between the two filters (Figure 11.10). Suppose the axis of the front filter is oriented vertically. Any photon that enters the slab will now be vertically polarized. As it propagates, its polarization state gradually changes to horizontal. The second filter with a horizontal axis that is placed behind the slab will now be fully transparent for the photons. If initial polarization is vertical, all incident light will pass successfully through the pair of crossed filters, indicating the presence of a gyrating medium. Alternatively, if we make the transmission axes parallel, the *absence* of transmission will be the signature of the gyrating slab. Either arrangement gives us the same information: there is something between the filters that rotates the state through a certain angle (90° in our case). This is quantum evolution of the polarization state.

11.12 A Watched Pot Never Boils (Quantum Zeno Effect)

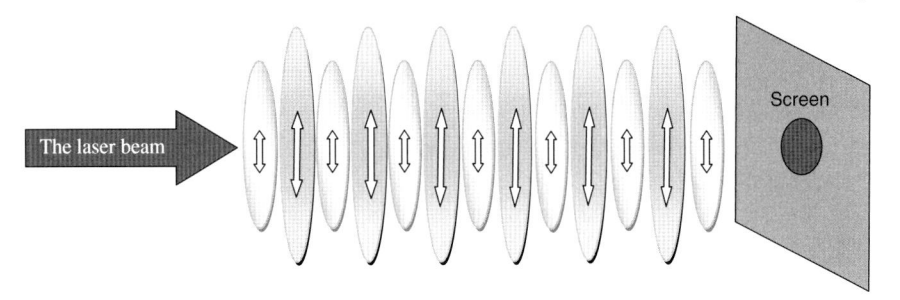

Figure 11.11 Quantum Zeno effect in a gyrating medium (dark gray) interspersed by polarizing filters (light gray) with parallel transmission axes. The system of sufficiently dense stack of such filters freezes the polarization vector of passing light at one fixed direction parallel to the common direction of their axes.

Is it possible to watch this evolution from moment to moment in real time, instead of just detecting its final result? Suppose we divide the slab equally into N parallel layers and put vertical filters in between [50] as shown in Figure 11.11. Each layer will turn polarization by

$$\Delta\theta = \frac{\pi}{2N}. \quad (11.160)$$

Now we have to distinguish the pair of outer polarizers sandwiching the system from the interim polarizers within the gyrating slab. To this end, let us write Polarizers of the pair with capital P.

Without gyrating medium, a photon that passed through the first Polarizer would not even notice the presence of other polarizers inside. However, as the first layer turns its polarization by $\Delta\theta$, its chance to pass through the next polarizer is $\mathcal{P}_1(\Delta\theta) = \cos^2 \Delta\theta$.

If the photon does pass, it is now again polarized vertically. As it faces the next gyrating layer with a vertical polarizer behind it, it finds itself in exactly the same situation as in the first cycle. The process reiterates periodically. The probability to pass polarizer No. j (under the same initial condition as before) is given by the same equation as for No. 1, with the single distinction – subscript j numbering the cycles:

$$\mathcal{P}_j(\Delta\theta) = \cos^2 \Delta\theta, \quad j = 1, 2, \ldots, N. \quad (11.161)$$

Each transmission/absorption event tells us the exact polarization state of the photon at the corresponding layer, so we have information about the photon's state throughout all its path. In other words, our setup watches the photon's evolution continuously.

The probability of passing all layers and emerging safely from Polarizer 2 is the product of all individual probabilities (11.161):

$$\tilde{\mathcal{P}}(N) = \prod_{j=1}^{N} \mathcal{P}_j(\Delta\theta) = \prod_{j=1}^{N} \cos^2 \Delta\theta = \left(\cos^2 \Delta\theta\right)^N. \quad (11.162)$$

In view of (11.160), this can be written as

$$\tilde{P}(N) = \left(\cos^2 \frac{\pi}{2N}\right)^N \underset{N \gg 1}{\Rightarrow} \left(1 - \left(\frac{\pi}{2N}\right)^2\right)^N = \left[1 - \left(\frac{\pi}{2N}\right)\right]^N \left[1 + \left(\frac{\pi}{2N}\right)\right]^N. \tag{11.163}$$

In the limit $N \to \infty$ (continuous watching), we have $\tilde{P}(N) \to e^{-\pi/2} e^{\pi/2} = 1$ and the photon gets transmitted 100% of the time. This means that polarization remains vertical despite the gyrating environment. An attempt to watch the gyration using the set of polarizers as detectors freezes the process at its initial state. The watched pot never boils!

11.13
A Watched Pot Boils Faster

We will now show a counterexample to the "watched pot rule" when watching *accelerates* the system's evolution [51]. Imagine a series of polarizers where orientations of any two neighboring polarizers differ by a small angle $\Delta\theta$. Such a stack of polarizers acts as a good model of the gyrating medium of Section 11.12. But if we try to *watch* the instantaneous photon polarization now, the outcome will be quite different.

Generally, if the pair of polarization states $\{|\,\|\,\rangle, |\perp\rangle\}$ (i.e., states parallel and perpendicular to the transmission axis) is taken as the basis, an arbitrary polarization forming angle θ with the transmission axis is described by the superposition

$$|\theta\rangle = \cos\theta |\,\|\,\rangle + \sin\theta |\perp\rangle. \tag{11.164}$$

Now let the kets $|\leftrightarrow\rangle$ and $|\updownarrow\rangle$ denote the horizontal and vertical polarization states, respectively (we will use this same notation for a photon and also for a polarizer having a horizontal or vertical transmission axis). Since these kets are orthogonal, we have $\langle\leftrightarrow|\updownarrow\rangle = 0$: a vertically polarized photon cannot pass through a horizontal filter and vice versa. However, photon $|\updownarrow\rangle$ has a nonzero amplitude $A(\theta)$ of passing through the filter oriented along $\hat{\mathbf{e}} = |\theta\rangle$:

$$A(\theta) = \langle\theta|\updownarrow\rangle = \cos\theta. \tag{11.165}$$

The same expression (up to an inconsequential phase factor) will give the amplitude for a photon in state $|\hat{\mathbf{e}}\rangle = |\theta\rangle$ to pass through a vertical filter. In either case, transmission probability for the photon is $\cos^2\theta$. Transmission indicates that the photon has "collapsed" from the initial state $|\theta\rangle$ to the basis state $|\,\|\,\rangle$; if it chooses the other basis state $|\perp\rangle$, it gets absorbed and we end up with the zero-photon or *vacuum state* $|0\rangle$. The probability of this outcome is $\sin^2\theta$.

If the incident light is *unpolarized* (i.e., one can find there with equal probability any state $|\theta\rangle$ with arbitrary θ), then our filter will pass only those photons that choose to collapse to state $|\,\|\,\rangle$. The transmitted beam will now contain only one state $|\,\|\,\rangle$ (this is why such a filter is called polarizer). But of course, this "purification" of light comes at a price: the intensity of the "purified" beam is lower. For unpolarized light (initial states are completely random!), exactly one half of the incident intensity gets transmitted.

What happens if we try to watch the evolution of photon's polarization in such a medium? We will describe the whole process as a succession of steps.

Step 1. Direct vertically polarized light onto a polarizer in the state $|\updownarrow\rangle$, and add another filter in the same state. We will have 100% transmission.

Step 2. Rotate the second filter. The resulting change of brightness on the screen is described by the familiar equation $\mathcal{P}(\theta) = \cos^2 \theta$, with θ now being the angle between two polarizers. At $\theta = \pi/2$ (crossed polarizers), the screen becomes totally dark: a single-photon state $|\updownarrow\rangle$ incident upon a pair of crossed polarizers ends up in the vacuum state, $|\updownarrow\rangle \Rightarrow |0\rangle$. This absorption is a 100% predictable outcome and in this sense it is as deterministic as any classical evolution. But on the other hand, it is a discontinuous process and therefore it is not what we would call "evolution."

Step 3. Let us now insert the third polarizer between the first two (Figure 11.9c). What happens then? Common wisdom dictates that three polarizers must be more opaque than two. But if we rotate the third polarizer so that its transmission axis points anywhere between the first two ($\theta \neq 0; \theta \neq \pi/2$), there appears an illuminated spot on the screen, whose brightness changes with rotation angle – a rather unusual effect from the purely corpuscular viewpoint, but well known in classical optics. The intermediate polarizer will absorb a fraction of the incident beam, but the transmitted part will emerge polarized at the angle θ and therefore it will find the last polarizer partially transparent.

If $A_1(\theta)$ is the amplitude for a photon emerging from the first ("vertical") polarizer to pass through the inserted polarizer, and $A_2(\theta)$ is the amplitude for the photon emerging from the inserted polarizer to pass through the second ("horizontal") polarizer, the resulting amplitude for the photon to emerge from the apparatus alive and reach the screen is (Problem 11.15.2)

$$A(\theta) = A_1(\theta) A_2(\theta) = \langle \leftrightarrow | \theta \rangle \langle \theta | \updownarrow \rangle = \frac{1}{2} \sin 2\theta. \qquad (11.166)$$

The maximum brightness of the spot (1/4 of I_0) is observed for $\theta = 45°$ and QM thus explains this classical phenomenon.

Step 4. Now it is natural to ask whether it is possible, by adding more polarizers, to amplify the brightness of the illuminated spot on the screen to 100% of the original brightness. The answer is yes. We use a set of N polarizers, $N \geq 3$, whose directions will partition the 90° angle equally into $N-1$ smaller angles $\Delta\theta$ (Figure 11.12). It is easy to show that such a set for $N > 3$ is more transparent for initially vertically polarized light than the set of three polarizers. Physically, this is due to the fact that at small $\Delta\theta$ the amplitude of transmission is close to 1, while the amplitude of absorption is very small.

11 Evolution of Quantum States

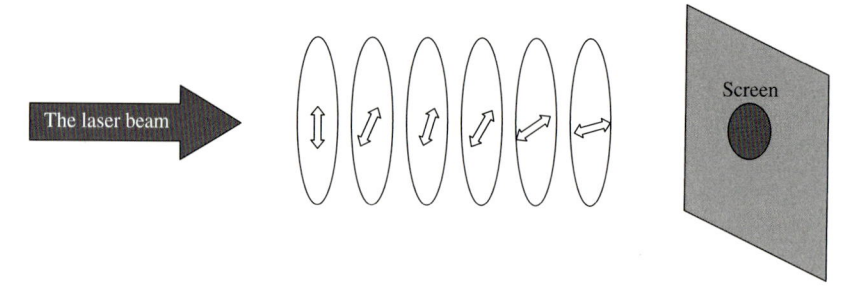

Figure 11.12 The stack of polarizers imitating a gyrating medium. The vertical polarization of the incident beam is being turned as the light passes through the system.

Step 5. We now find the probability of passing through N polarizers. For the sake of generality, we specify the final rotation by the angle θ_N (dropping the requirement that $\theta_N = \pi/2$). Let N polarizers partition this angle into $N-1$ small rotations $\Delta\theta_j$, so that $\theta_N = \sum_{j=1}^{N-1} \Delta\theta_j$. For a photon that just left polarizer j in state $|\theta_j\rangle$ and now enters polarizer $j+1$, the transmission amplitude is $\langle\theta_{j+1}|\theta_j\rangle = \cos(\Delta\theta_j)$. The amplitude of passing through all N polarizers oriented successively from $|\updownarrow\rangle$ to $|\theta_N\rangle$ is the product of individual transmission amplitudes at each polarizer:

$$A(\theta_N) = \langle\theta_N|\theta_{N-1}\rangle \cdots \langle\theta_2|\theta_1\rangle\langle\theta_1|\updownarrow\rangle = \prod_{j=1}^{N-1} \cos(\Delta\theta_j). \tag{11.167}$$

Suppose $\Delta\theta_1 = \Delta\theta_2 = \cdots \equiv \Delta\theta$. Then

$$A(\theta_N) = (\cos\Delta\theta)^N = \left(\cos\left(\frac{\theta_N}{N}\right)\right)^N. \tag{11.168}$$

For $N \gg 1$, the angle $\Delta\theta = \theta_N/N$ is small enough to approximate $\cos(\Delta\theta) \approx 1 - (\Delta\theta)^2/2$. Squaring the resulting amplitude $A(\theta_N) \approx (1 - \theta_N^2/2N^2)^N \approx e^{-\theta_N^2/2N}$, we get transmission probability

$$\mathcal{P}(\theta_N) \approx e^{-\theta_N^2/N}. \tag{11.169}$$

At $N \to \infty$, the system becomes fully transparent for the initial state $|\updownarrow\rangle$ and converts it into state $|\theta_N\rangle$ at the end (e.g., if $\theta_N = \pi/2$, the photon comes out horizontally polarized). That's why the set of $N \to \infty$ polarizers (Figure 11.12) is a suitable model of a transparent gyrating medium. Though the physical mechanism of rotating the polarization vector is quite different from that in a true gyrating medium, the final result will be the same, provided initial polarization is aligned with the first polarizer. In both cases, polarization state evolves smoothly with time, rotating through an angle proportional to the traversed distance. The smoothness of this process in the limit $N \to \infty$ is the physical instance of a well-known mathematical fact: any continuous function can be derived from certain discontinuous functions in the limit of infinite number of infinitely small jumps. Photon polarization $|\theta(x)\rangle$ thus becomes a continuous function of position x within the stack of polarizers. Or we can write it is as a time-dependent state $|\theta(t)\rangle$, where

$t = x/u$ and u is the group velocity of this photon within the stack. A photon prepared in state $|\updownarrow\rangle$ will evolve as $|\theta(t)\rangle = e^{i\hat{H}(\updownarrow,\theta)t}|\updownarrow\rangle$, where $\hat{H}(\updownarrow,\theta)$ is the Hamiltonian for the stack of polarizers (Figure 11.12) in the limit of $N \to \infty$.

Just as in case of gyrating medium, this process has nothing to do with circular polarization! The electric field is oscillating, not rotating, at any fixed point of space. For a single photon, polarization is linear at each point, and rotation only accompanies the photon's progress from one point to another.

Even though the described stack of polarizers (call it "Set 1") acts as a gyrating medium, this similarity only holds when the initial polarization aligned with the first polarizer. If there is an angular mismatch Θ between them, the transmitted light is attenuated by $\cos^2 \Theta$. And the final polarization state at the exit will still be $|\theta\rangle$, whereas it would be $|\Theta + \theta\rangle$ for a "true" gyrating medium.

One would think that at least for $\Theta = 0$, watching the photon's evolution should replicate the results of the previous section: a set of vertical polarizers (call it "Set 2") interspersed with our artificial "medium" will do the same job of freezing the evolution. This turns out not to be the case.

Step 6. Prepare Set 1 such that $\theta = \theta_N = \pi/2$, place Set 2 in front of Set 1, and illuminate the system with vertically polarized light as shown in Figure 11.13a. The beam will pass through without attenuation. In terms of photons, each photon passes through with certainty, but with its polarization turned through 90°.

Step 7. Now place Set 2 *behind* Set 1 (Figure 11.13b). The illuminated spot on the screen disappears.

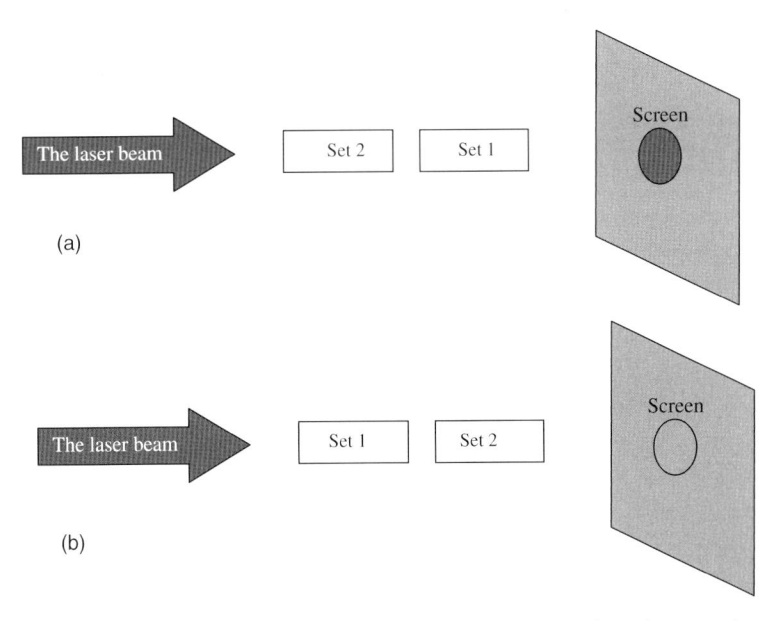

Figure 11.13 Two sets of polarizers. Set 1 imitates gyrator turning the polarization through 90°. Set 2 is a stack of polarizers with parallel transmission axes. Here all of them are also parallel to the axis of the first polarizer in Set 1.

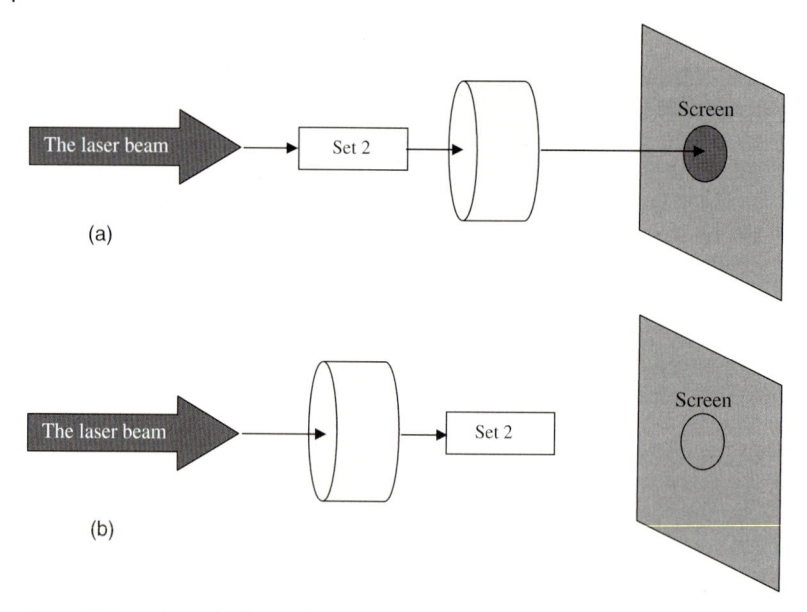

Figure 11.14 The stack of vertical polarizers and the gyrating slab in two different orders.

Steps 8 and 9. Repeat the two previous steps, but now using the *actual* gyrating slab instead of polarizers in Set 1. If the slab rotates the polarization through 90°, we will observe exactly the same results! The light will pass when Set 2 is placed before the slab, and will be blocked when Set 2 is placed after the slab (Figure 11.14). In either case, the pair of sets that are transparent for the initial state $|\updownarrow\rangle$ when acting separately or in the order 2–1 is opaque in the order 1–2.

All this seems to justify our expectations that using Set 2 as a measuring device will freeze the gyration in Set 1 in the same way as it did with actual gyrating medium.

These expectations are wrong! To see this, let us complete our study.

Step 10. The measuring device (Set 2) is inserted into the "medium" (Set 1) so that individual polarizers of Set 2 are interspersed uniformly among individual polarizers of Set 1 (Figure 11.15). In contrast with the results of the previous section, now the illuminated spot on the screen disappears.

The reason is very simple: the angle between a pair of neighboring elements (one from Set 1 and the other from Set 2) increases as we look at pairs farther down the stack. Consider nth pair of polarizers. Right after the photon emerges from a $|\updownarrow\rangle$ polarizer, it faces polarizer $|\theta_n\rangle$ oriented at $\hat{\mathbf{e}} = \theta_n$, which it passes with probability $\cos^2 \theta_n$. Then it encounters the next $|\updownarrow\rangle$ polarizer and passes with the same probability $\cos^2 \theta_n$, after which the process reiterates for $n + 1$, and so on. So the nth pair transmits the proton with probability $\cos^4 \theta_n$ and absorbs it with probability $1 - \cos^4 \theta_n$. Imagine that each separate element of our sets is a fancy device capable of detecting even a single photon absorption (or a lack thereof).

11.13 A Watched Pot Boils Faster

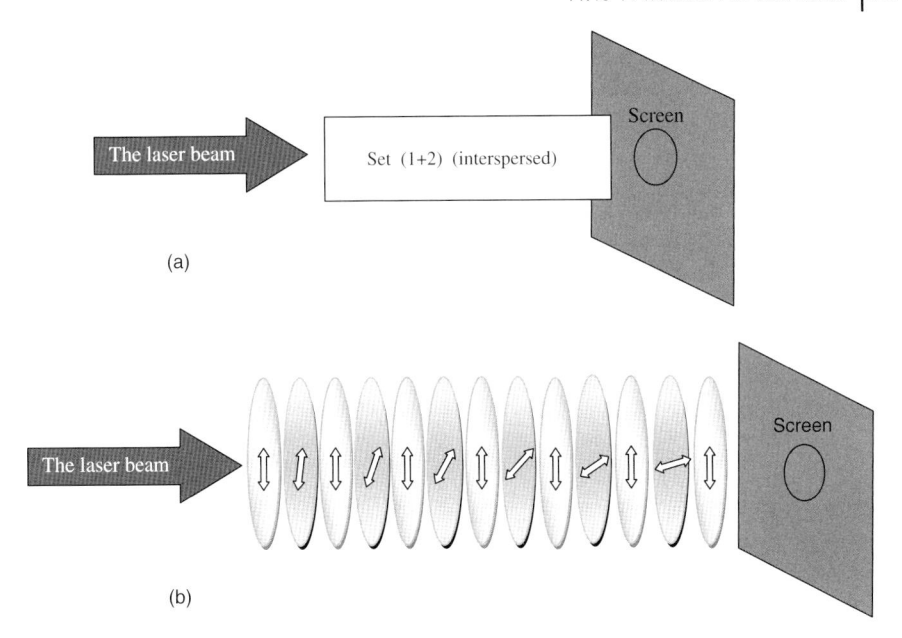

(a) The laser beam → Set (1+2) (interspersed) → Screen

(b) The laser beam → [stack of polarizers] → Screen

Figure 11.15 Sets 1 and 2 interspersed. Set 1 (dark gray) imitates action of a gyrating medium; Set 2 is used for watching the evolution of the polarized state in its motion through the system, in the same way as in Figure 11.11.

We are then able to track the state of polarization for each step j as the photon makes its way through the stack, so our program to watch the system continuously succeeds at large N. But now the probability of absorption increases as we keep watching! Sooner or later the observation will kill the state whose evolution we wanted to observe!

This is where the physical difference from the true gyrating medium steps into the picture. Up till now this difference was masked by the similarity of the outcome. But now the difference is unmasked. In a true gyrating medium (G), the inserted polarizers would merely freeze the gyration but the photon itself would be preserved because G is transparent *for any polarization state*. In our artificial "gyrator," there appears an ever-greater chance of photon absorption because of the growing misalignment between the transmission axes of neighboring filters.

As a result, the photon's evolution is severely affected in two respects. First, it breaks up into a series of abrupt jumps between the states $|\theta = 0\rangle$ and states $|\theta_n\rangle$ with $\theta_n = (n/N)\theta_N$ and the size of a jump being a linear function of n (Figure 11.16a). Second, the survival probability $\mathcal{P}(n)$ for a watched photon falls off very rapidly with n:

$$\mathcal{P}(n) = \prod_{j=1}^{n} \cos^4(j\Delta\theta) = \prod_{j=1}^{n} \cos^4\left(\frac{j}{N}\theta_N\right). \tag{11.170}$$

In the case when $\theta_N = \pi/2$, the photon won't make it at all. But even for smaller angles, the diligent reader can study a few numerical examples to see that a

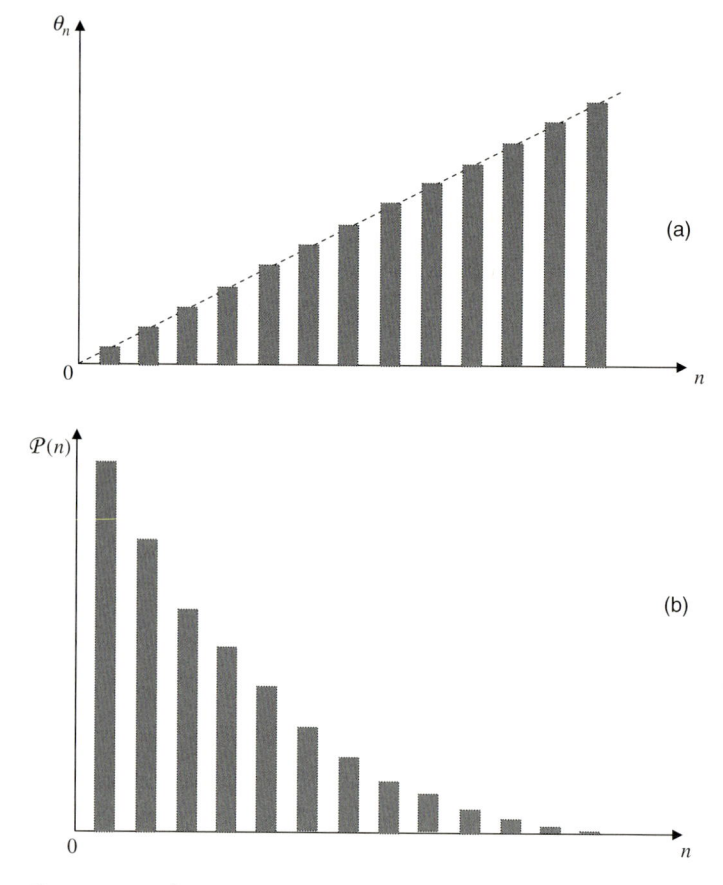

Figure 11.16 Jerky evolution of a watched polarization state in a system of polarizers. (a) Evolution of observable θ. At a step No. n, it is jumping between 0 (state $|\updownarrow\rangle$) and θ_n (state $|\theta_n\rangle$). The dashed line is its predicted evolution in continuous gyrating medium. (b) The photon survival probability as a function of n in Set $(1+2)$ of interspersed polarizers, with Set 1 imitating the gyrating medium, and Set 2 serving as a measuring device.

watched photon has essentially zero chance to survive for long (i.e., to reach polarizer pair $n \to N$) if only one takes $N \gg 1$ (Figure 11.16b). Continuous ($N \to \infty$) detection of instantaneous polarization kills the photon with certainty. A watched pot not just boils faster – it rapidly evaporates!

Problems

11.1 Find the Hermitian adjoint of the time evolution operator $\hat{U}(t)$.
11.2 Prove that $\hat{U}(t)$ is unitary.

(*Hint*: Consider it in the energy representation.)

11.3 Prove that if $\hat{L} = \hat{A}\hat{B}$, then $d\hat{L}/dt$ is determined by the same product rule as for regular variables.

11.4 Show that complete inversion in a 2D space can also be reproduced by rotation through 180°, and thus does not constitute an independent transformation.

11.5 Show that the wave function in (11.25) is normalized.

11.6 In Equation 11.25, take the Fourier transform of $f(k)$ and find position representation $\psi(x)$ of the wave function: (a) at $t = 0$; (b) at an arbitrary t.

11.7 Find the expectation values of k and k^2 for a Gaussian wave packet (11.25).

11.8 Find the spread of an "illegal" δ-spike in Section 11.5 as a function of time: (a) as described by nonrelativistic QM; (b) as described by relativistic QM.

11.9 (a) Show that the *evolved* spike considered in Section 11.5 satisfies the Schrödinger equation; (b) do the same for a 3D spike.

11.10 Evaluate the spread of a "jinnee" from Section 11.6 as a function of time: (a) in the framework of nonrelativistic QM; (b) in the framework of relativistic QM.

11.11 Derive Equation 11.67 from (11.64).

11.12 Find the rest mass of a photon in a Fabry–Pérot resonator of length 0.4 m with ideally reflecting mirrors: (a) in an eigenstate with $n = 137$; (b) in an eigenstate with $n = 3 \times 10^4$.

11.13 In the previous problem, find the rest mass of the photon in an equally weighted superposition $|\Psi\rangle = (1/\sqrt{N})\sum_{n=1}^{N} |\psi_n\rangle$ of harmonic eigenstates for $N = 3 \times 10^4$.

11.14 Prove that the dispersion formula (11.123) satisfies Equation 11.122.

(*Hint*: Use integration in the complex energy plane and apply the theorem of residues.)

11.15 Find the average lifetime of a particle in the Gamow state with decay constant γ.

11.16 Find the indeterminacy Δt of the moment of decay of the Gamow state.

11.17 Derive expression (11.126) determining transmissivity of a 3D barrier in the α-decay problem.

11.18 (a) Write an arbitrary state $|K\rangle$ of a neutral K-meson in the $\{|K_1\rangle, |K_2\rangle\}$ basis. (b) Write the basis states $|K_1\rangle$ and $|K_2\rangle$ in the $\{|K^0\rangle, |\bar{K}^0\rangle\}$ basis, and vice versa.

11.19 Find the expectation value of strangeness $\langle S \rangle$ as a function of time for a neutral K-meson initially born in the state with $S = +1$ (K^0-state).

11.20 Prove that the intensity of natural (unpolarized) light passing through a linear polarizer is one half of its initial intensity.

11.21 Derive Equation 11.166.

11.22 Find the law of decrease of probability for an initially vertically polarized photon to pass through the system of interspersed polarizers described in Section 11.13.

11.23 Express the solution of the previous problem in terms of x and Δx, where x is the distance traveled within the stack and Δx is the thickness of one pair of filters.

References

1 Blokhintsev, D.I. (1964) *Principles of Quantum Mechanics*, Allyn & Bacon, Boston, MA.
2 Walls, D. (1983) Squeezed states of light. *Nature*, **306**, 141.
3 Sen, D., Das, S.K., Basu, A.N., and Sengupta, S. (2001) Significance of Ehrenfest theorem in quantum–classical relationship. *Curr. Sci.*, **80** (4), 25.
4 Noether, E. (1918) *Invariante Varlationsprobleme*, Nechr. d. Konig. Gesellsch. d. Wiss. zu Gottingen, Math-Phys. Klasse, pp. 235–257. English translation: Travel, M.A. (1971) Invariant variation problem. *Transport Theory Stat. Phys.*, **1** (3), 183–207.
5 Byers, N. (1999) Noether's discovery of the deep connection between symmetries and conservation laws. *Proceedings of Israel Mathematical Conference*, vol. 12.
6 Byers, N. (1996) *History of Original Ideas and Basics Discoveries in Particle Physics* (eds H.B. Newman and T. Ypsilantis), Plenum Press, New York.
7 Khriplovich, I.B. (1991) *Parity Nonconservation in Atomic Phenomena*, Gordon and Breach Science Publishers, Philadelphia, PA.
8 Halzen, F. and Martin, A.D. (1984) *Quarks & Leptons: An Introductory Course in Modern Particle Physics*, John Wiley & Sons, Inc., New York.
9 Nambu, Y. (1984) *Quarks: Frontiers in Elementary Particle Physics*, World Scientific, Singapore.
10 Challis, L. and Sheard, F. (2003) The Green of Green functions. *Phys. Today*, **56** (12), 41–46.
11 Fayngold, M. (2008) *Special Relativity and How it Works*, Wiley-VCH Verlag GmbH, Weinheim, pp. 485–487.
12 Courant, R. and Hilbert, D. (2004) *Methods of Mathematical Physics*, vols. 1–2, Wiley-VCH Verlag GmbH, Weinheim.
13 Gradshteyn, I. and Ryzhik, I. (2007) *Table of Integrals, Series, and Products*, 7th edn, Academic Press/Elsevier, Burlington, MA, pp. 887–890.
14 Dirac, P. (1964) *The Principles of Quantum Mechanics*, Yeshiva University, New York.
15 Landau, L. and Lifshits, E. (1965) *Quantum Mechanics: Non-Relativistic Theory*, 2nd edn, Pergamon Press, Oxford.
16 Park, D. (1992) *Introduction to the Quantum Theory*, McGraw-Hill, New York, pp. 31–32.
17 Griffiths, D. (1994) *Introduction to Quantum Mechanics*, Prentice-Hall, Englewood Cliffs, NJ, pp. 11, 54–57.
18 Brillouin, L. (1960) *Wave Propagation and Group Velocity*, Academic Press, New York, pp. 1–83, 113–137.
19 Paierls, R. (1979) *Surprises in Theoretical Physics*, Princeton University Press, Princeton, NJ, Section 1.3.
20 Bialynicki-Birula, I. and Bialynicki-Birula, Z. (2009) Why photons cannot be sharply localized. *Phys. Rev. A*, **79** (3), 2112–2120.
21 Hegerfeldt, G.C. (2001) *Particle localization and the notion of Einstein causality*. arXiv: quant-ph/0109044v1.
22 Nimtz, G. (1999) Evanescent modes are not necessarily Einstein causal. *Eur. Phys. J. B*, **7**, 523–525.
23 Azbel', M.Ya. (1994) Superluminal velocity, tunneling traversal time and causality. *Solid State Commun.*, **91** (6), 439–441.
24 Diener, G. (1996) Superluminal group velocities and information transfer. *Phys. Lett. A*, **223** (5), 327–331.
25 Chiao, R.Y., Kwiat, P.G., and Steinberg, A.M. (2000) Faster than light? *Sci. Am.*, (Special Issue), 98–106.
26 Mitchell, M.W. and Chiao, R.Y. (1998) Causality and negative group delays in a simple bandpass amplifier. *Am. J. Phys.*, **66**, 14–19.
27 Bigelow, M., Lepeshkin, N., Shin, H., and Boyd, R. (2006) Propagation of a smooth and discontinuous pulses through materials with very large or very small group velocities. *J. Phys.: Condens. Matter*, **18** (11), 3117–3126.
28 Adam, T. et al. (The OPERA Collaboration) (2011) *Measurement of the neutrino velocity with the OPERA detector in the CNGS beam*. arXiv:1109.4897v1 [hep-ex].
29 Fayngold, M. (2011) *On the report of discovery of superluminal neutrinos*. arXiv.org/abs/1109.5743.

30 Cohen, A.G. and Glashow, Sh.L. (2011) *New constraints on neutrino velocities.* arXiv.org/hep/-ph/1109.6562.
31 Hegerfeldt, G.C. (1994) Causality problems for Fermi's two-atom system. *Phys. Rev. Lett.*, **72**, 596.
32 Buchholz, D. and Yngvason, J. (1994) There are no causality problems for Fermi's two-atom system. *Phys. Rev. Lett.*, **73**, 613.
33 Labarbara, A. and Passante, R. (1994) Causality and spatial correlations of the relativistic scalar field in the presence of a static source. *Phys. Lett. A*, **206** (1), 1–6.
34 Milonni, P.W., James, D.F.V., and Fearn, H. (1995) Photodetection and causality in quantum optics. *Phys. Rev. A*, **52**, 1525.
35 Heitmann, W. and Nimtz, G. (1994) On causality proofs of superluminal barrier traversal of frequency band limited wave packets. *Phys. Lett. A*, **196** (3–4), 154–158.
36 Krylov, N.S. and Fock, V.A. (1947) On two main interpretations of the uncertainty relation for energy and time. *Sov. J. Exp. Theor. Phys. (JETP)*, **17**, 93–107.
37 Davydov, A.V. (2003) On the duration of the absorption and emission of nuclear γ-rays. *Phys. Atom. Nuclei*, **66** (12), 2113–2116 (translated from *Yadernaya Fizika*).
38 Gamow, G. (1928) Zur Quantentheorie des Atomkernes (On the quantum theory of the atomic nucleus). *Z. Phys.*, **51**, 204–212.
39 Geiger, H. and Nuttall, J.M. (1911) The ranges of the α particles from various radioactive substances and a relation between range and period of transformation. *Philos. Mag. Ser. 6*, **22** (130), 613–621.
40 Geiger, H. and Nuttall, J.M. (1912) The ranges of α particles from uranium. *Philos. Mag. Ser. 6*, **23** (135), 439–445.
41 Nishijima, K. (1955) Charge independence theory of V particles. *Prog. Theor. Phys.*, **13**, 285.
42 Gell-Mann, M. (1964) A schematic model of baryons and mesons. *Phys. Lett.*, **8** (3), 214–215.
43 Feynman, R., Leighton, R., and Sands, M. (1963–1964) *The Feynman Lectures on Physics*, vols. 1–3, Addison-Wesley, Reading, MA.
44 Maki, Z., Nakagawa, M., and Sakata, S. (1962) Remarks on the unified model of elementary particles. *Prog. Theor. Phys.*, **28**, 870.
45 Pontecorvo, B. (1968) Neutrino experiments and the problem of conservation of leptonic charge. *Sov. Phys. JETP*, **26**, 984.
46 Davis, R., Jr., Harmer, D.S., and Hoffman, K.C. (1968) Search for neutrinos from the sun. *Phys. Rev. Lett.*, **20**, 1205.
47 Gribov, V. and Pontecorvo, B. (1969) Neutrino astronomy and lepton charge. *Phys. Lett. B*, **28**, 493.
48 Cohen, A.G., Glashow, S.L., and Ligeti, Z. (2009) Disentangling neutrino oscillations. *Phys. Lett. B*, **678**, 191.
49 Hecht, E. (2002) *Optics*, 4th edn, Addison-Wesley, Reading, MA, p. 385.
50 Kwiat, P., Weinfurter, H., and Zeilinger, A. (1996) Quantum seeing in the dark. *Sci. Am.*, **275**, 72–78.
51 Milonni, P.W. (2000) Quantum decay: a watched pot boils quicker. *Nature*, **405**, 525–526.

12
Quantum Ensembles

12.1
Pure Ensembles

Solving the Schrödinger equation to find the state function of a system and extract all information from it constitutes the *direct problem* of QM. But the corresponding Hamiltonian is not always known. In such cases, we try to determine possible states of the system from the available experimental results. This constitutes the so-called *inverse problem* of QM. There is a special subset of the inverse problems mostly encountered in high-energy physics. It is to determine the interaction forces between two particles and/or the structure of one of them from experimental data about their scattering on one another at different energies of collision. This information can be derived from the angular distribution of scattered particles, changes of their spin, energy, and so on.

Let us consider some important aspects of such experiments. Since the connection between the state function and the measurement outcomes is generally only probabilistic, the result of a single measurement of a certain observable does not tell us much about it. For instance, if we have one photon and direct it onto a double-slit screen, its landing position on the second screen does not tell us the photon's energy. After such landing, the photon is lost, so we cannot use it to replay the game. If, however, we have a large number of such photons all in one state, the result (e.g., the distribution of their landing positions) can reveal this state. If this distribution forms a well-resolved succession of bright and dark fringes, we know that the photons were in a state with definite wavelength λ. Knowing the geometry of the apparatus (the width of the slits, the distance between them, and the distance between the screens), we can determine λ from Equation 1.20 and λ determines the common momentum and energy of each photon. And vice versa, if we know λ but do not know, say, the distance between the slits, we can find it (inverse problem!) from the same experiment. In either case, in order to carry out a large number ($N \gg 1$) of identical measurements, we need a large number of independent identical particles. We can carry out the experiment with one particle at a time. If we want to save time, we can use many of them simultaneously. If their mutual interactions are negligible, the net result will be the same.

A set in which all particles are *in the same state* ψ is called the *pure ensemble*. There are no restrictions on state ψ – it is not necessarily a state with definite momentum as in the above example with the photons. The state ψ may be a superposition of different momentum eigenstates (wave packet) or any other superposition. What is essential is that this state ψ is common for all particles of the ensemble. Then all probabilities and expectation values obtained from ψ refer to measurements in such an ensemble.

12.2
Mixtures

A pure ensemble is only a very special case of quantum ensembles. Now turn to a more general case: a set of different particles, or identical particles but in different states.

Consider a system of N identical particles. Suppose N_1 particles are all in a state ψ_1, N_2 particles are all in a state ψ_2, and so on, so that N_n particles are all in a state ψ_n and $N_1 + N_2 + \cdots + N_n = N$. Then there is a certain probability $\mathcal{P}_1$ of finding the particle in a state ψ_1, probability $\mathcal{P}_2$ of finding it in a state ψ_2, and so on, where $\mathcal{P}_j$ is determined as

$$\mathcal{P}_j = \frac{N_j}{N}, \quad j = 1, 2, \ldots, n, \quad \text{and} \quad \sum_{j=1}^{n} \mathcal{P}_j = 1. \tag{12.1}$$

In other words, the set of N particles consists of n subsets labeled as subset 1, subset 2, and so on, each subset in the corresponding common state ψ_j. But the latter condition defines a pure ensemble! Therefore, we have n distinct pure ensembles, each characterized by its own state ψ_j. When considered as one combined system, they form a *mixed ensemble* or just a *mixture*.

One example of a mixture must be familiar to the reader. It is a container with gas with only one kind of molecules. But the molecules may be in different states and are accordingly described by different state functions. For instance, they may differ in energy. When in thermal equilibrium, the probability of finding a molecule in a state with energy E_j is given by the Boltzmann distribution (2.3).

Whereas a pure ensemble is described by one state function ψ common for all its particles, a mixture is described by a set of distinct states ψ_j together with the corresponding probabilities $\mathcal{P}_j$ with which the states are represented in the mixture.

As noted in the previous section, a single state function ψ characterizing a pure ensemble may itself be a *superposition* of eigenfunctions ψ_j. Suppose we have both a pure ensemble with N particles all in one state ψ

$$\psi = \sum_{j=1}^{n} A_j \psi_j, \quad A_j = a_j \, e^{i\alpha_j} \tag{12.2}$$

and the mixture of N such particles in various eigenstates ψ_j with the weights (12.1). The difference between the two is that in pure ensemble *each* particle is in *the same superposition* (12.2), whereas in the mixture each particle is in *some single eigenstate*

ψ_j, but particles from two different subsets are in different eigenstates, and each subset is characterized by the corresponding probability $\mathcal{P}_j$. In other words, a mixture needs two sets for its characterization:

$$\begin{cases} \psi_1, \psi_2, \ldots, \psi_j, \ldots, \psi_n, \\ \mathcal{P}_1, \mathcal{P}_2, \ldots, \mathcal{P}_j, \ldots, \mathcal{P}_n. \end{cases} \quad (12.3)$$

The difference between pure ensemble in state (12.2) and mixture (12.3) remains even if we adjust the amplitudes A_j so that

$$|A_j|^2 \equiv a_j^2 = \mathcal{P}_j. \quad (12.4)$$

To illustrate the difference quantitatively, perform a position measurement (or calculate the probability of finding the particle in a certain location). For a particle of pure ensemble we will have, using (12.2),

$$\rho_{\text{Pure}}(\mathbf{r}') = |\psi(\mathbf{r}')|^2 = \sum_{j=1}^{n} \mathcal{P}_j \Big|\psi_j(\mathbf{r}')\Big|^2 + \sum\sum_{j\neq k} \sqrt{\mathcal{P}_j \mathcal{P}_k} e^{i(\alpha_j - \alpha_k)} \psi_j(\mathbf{r}') \psi_k^*(\mathbf{r}'). \quad (12.5)$$

For the corresponding mixture we have probability $\mathcal{P}_j$ that the particle found at $\mathbf{r}'$ was from subset j, that is, was in the state ψ_j. For this state we have $\rho_j(\mathbf{r}') = |\psi_j(\mathbf{r}')|^2$, so the total probability of collapsing to $\mathbf{r}'$ from state j is the product $\mathcal{P}_j |\psi_j(\mathbf{r}')|^2$. The net probability density of finding *any* particle of the mixture at $\mathbf{r}'$ is

$$\rho_{\text{Mixture}}(\mathbf{r}') = \sum_{j=1}^{n} \mathcal{P}_j \Big|\psi_j(\mathbf{r}')\Big|^2. \quad (12.6)$$

Comparison with (12.5) shows the difference. For a pure ensemble we have the interference terms, which are absent in the case of mixture. This is analogous to difference between coherent and incoherent light. The coherent light intensity at a point $\mathbf{r}'$ is determined by the squared sum of the different wave amplitudes at this point, similar to (12.5); otherwise, it is just the sum of individual intensities similar to (12.6). Thus, the coherent light is a pure ensemble of photons; the incoherent light is a mixture of photons.

The analysis of measurements reveals another important feature. Suppose we start with a pure ensemble of particles all in state (12.2), with functions ψ_j being the eigenstates of an observable q. If we measure q for each particle of the ensemble, we will obtain the value q_1 with probability $\mathcal{P}_1$, that is, in $\mathcal{P}_1 \cdot 100\%$ of all cases, the value q_2 in $\mathcal{P}_2 \cdot 100\%$ of all cases, and so on. Each measurement collapses superposition (12.2) to some single eigenstate – the one obtained as a measurement result. In the end, the set of N particles initially all in one state (12.2) (pure ensemble!) will be converted into n different subsets: N_1 particles in state 1, N_2 particles in state 2, and so on, up to N_n particles in state n. In other words, the original pure set will be sorted out into n *different* pure subsets. But such a system is a mixture described by (12.3). Thus, a pure ensemble can be converted into mixture in the process of measurement.

What if we are interested in some observable f other than q and incompatible with it?

In this case, f is indeterminate in a state ψ_j, but we can find its average in this state

$$\langle f \rangle_j = \int \psi_j^*(\mathbf{r}) \hat{f}(\mathbf{r}) \psi_j(\mathbf{r}) d\mathbf{r}. \tag{12.7}$$

Keep in mind that $\langle f \rangle_j$ is *not* an eigenvalue of $\hat{f}$, but the average in subensemble No. j. Since the state ψ_j itself is represented with probability $\mathcal{P}_j$ in the given mixture, the net average will be the weighted sum of the results (12.7) over all subensembles:

$$\langle f \rangle = \sum_{j=1}^{n} \mathcal{P}_j \langle f \rangle_j. \tag{12.8}$$

We leave it as a problem for the reader to find the average of f in the pure ensemble characterized by (12.2) and (12.4), and compare the result with (12.8).

Not every collection of particles forms a pure or mixed ensemble as defined above. The electrons in an atom are neither a pure ensemble nor a mixture, because one cannot ascribe a wave function to either electron. Only the whole system of them can be described accurately by the corresponding wave function. Atomic electrons are said to be *coupled*: the state of any chosen electron depends on states of its partners, but the latter, in turn, depend on the state of the chosen electron. In all atoms other than hydrogen, the picture of a single electron as existing in a *fixed* environment is far less accurate than in the case of hydrogen atom. Other examples of this are systems of entangled particles. Coupled or entangled systems form an important special class of ensembles. However, we show later that they can be successfully described by the same formalism that applies to mixtures.

12.3
The Density Operator

The *density matrix* introduced by John von Neumann [1] plays the same role in mixed ensembles as the wave function in pure ensembles. We will now try to motivate the density matrix by considering a mixture composed of normalized states $|\psi_1\rangle, \ldots, |\psi_n\rangle$ with weights $\mathcal{P}_1, \ldots, \mathcal{P}_n$, so that $\sum_i \mathcal{P}_i = 1$. Suppose we want to know the average of some observable f. If we pick an electron in the state $|\psi_i\rangle$, then f in that state will have the average $\langle f \rangle_i = \langle \psi_i | \hat{f} | \psi_i \rangle$. But we only get state $|\psi_i\rangle$ with probability $\mathcal{P}_i$. Therefore, the *ensemble average* $\langle f \rangle$ obtained in a large number of experiments is the sum (12.8) of all weighted QM averages $\langle f \rangle_i$:

$$\langle f \rangle = \sum_i \mathcal{P}_i \langle f \rangle_i = \sum_i \mathcal{P}_i \langle \psi_i | \hat{f} | \psi_i \rangle. \tag{12.9}$$

Suppose our states $|\psi_i\rangle$ are written in the orthonormal $\{|b_1\rangle, \ldots, |b_n\rangle\}$ basis as $|\psi_i\rangle = \sum_l c_{il} |b_l\rangle = \sum_l |b_l\rangle \langle b_l | \psi_i \rangle$. Here $c_{il} = \langle b_l | \psi_i \rangle$ is the amplitude of finding the eigenvalue b_l in the pure subensemble No. i.[1]) We then find $\hat{f}$ sandwiched between the two identity operators, $I = \sum_k |b_k\rangle\langle b_k|$ and $I = \sum_l |b_l\rangle\langle b_l|$, with $\{|b_k\rangle\}$ and $\{|b_l\rangle\}$ representing *the same* basis, but labeled by different indices:

$$\langle f \rangle = \sum_{i,k,l} \mathcal{P}_i \langle \psi_i | b_k \rangle \langle b_k | \hat{f} | b_l \rangle \langle b_l | \psi_i \rangle. \tag{12.10}$$

In terms of the b-basis, the middle factor $\langle b_k | \hat{L} | b_l \rangle$ is the matrix element f_{kl}.[2]) To the right and to the left of f_{kl} are two inner products; they are just numbers that can be shuffled without changing the result. Therefore, we can put the right inner product before the left one. Such reordering takes f_{kl} to the end of the expression. This will bring (12.10) to the form $\langle f \rangle = \sum_{i,k,l} \mathcal{P}_i \langle b_l | \psi_i \rangle \langle \psi_i | b_k \rangle f_{kl}$. After the final regrouping of terms, we can write

$$\langle f \rangle = \sum_{k,l} \langle b_l | \left(\sum_i \mathcal{P}_i |\psi_i\rangle\langle\psi_i| \right) |b_k\rangle f_{kl} \tag{12.11}$$

and interpret the part in the parentheses as a new operator

$$\hat{\rho} \equiv \sum_i \mathcal{P}_i |\psi_i\rangle\langle\psi_i|. \tag{12.12}$$

This is von Neumann's *density operator*. Its matrix elements in the chosen basis are

$$\rho_{lk} = \langle b_l | \hat{\rho} | b_k \rangle = \sum_i \langle b_l | \psi_i \rangle \mathcal{P}_i \langle \psi_i | b_k \rangle \equiv \sum_i \mathcal{P}_i c_{il}^* c_{ik}. \tag{12.13}$$

The diagonal matrix elements

$$\rho_{kk} = \langle b_k | \hat{\rho} | b_k \rangle = \sum_i \mathcal{P}_i |c_{ik}|^2 \tag{12.14}$$

have a clear physical meaning: ρ_{kk} is the probability of finding the eigenvalue $b = b_k$ in the whole mixture. Since a matrix is just a specific "face" of the operator, the terms *density operator* and *density matrix* have essentially the same meaning and are used interchangeably; the exact form in each case is always clear from the context. The origin of the term "density" becomes clear if we take b as position variable **r**, write (12.13) as

$$\rho_{\mathbf{rr}'} = \sum_i \langle \mathbf{r} | \psi_i \rangle \mathcal{P}_i \langle \psi_i | \mathbf{r}' \rangle \equiv \sum_i \mathcal{P}_i \psi_i^*(\mathbf{r}') \psi_i(\mathbf{r})$$

(a continuous matrix of the type discussed in Section 6.3), and consider a *diagonal* element $\rho_{\mathbf{rr}} = \sum_i \mathcal{P}_i |\psi_i(\mathbf{r})|^2 = \mathcal{P}(\mathbf{r})$. The result $\mathcal{P}(\mathbf{r})$ is the probability density of

---

1) Note that we *do not* require states $|\psi_i\rangle$ to be orthogonal. This is natural because one can prepare an ensemble consisting of partially overlapping states $\langle \psi_i | \psi_j \rangle \neq 0$. Also, the number of states $|\psi_i\rangle$ could be less than the dimensionality of the $\mathcal{H}$-space, so they do not necessarily form a complete basis. In contrast, the basis $\{|b_1\rangle, \ldots, |b_n\rangle\}$ in which these pure states are written should of course be orthonormal.

2) As already mentioned, the default basis $\{|b_k\rangle\}$ is generally not the same as the eigenbasis of $\hat{f}$. In other words, f_{kl} is not necessarily diagonal.

finding a particle at **r** in the mixed ensemble. So the density matrix is generalization of probability density defined for pure ensembles onto mixtures. To show this in a slightly different way, consider first a pure ensemble with only one state $|\psi\rangle$ present. Then the density matrix reduces to $\hat{\rho} = |\psi\rangle\langle\psi|$. Representing $|\psi\rangle$ as the column matrix in $|b\rangle$-basis gives

$$\hat{\rho} = |\psi\rangle\langle\psi| = \begin{pmatrix} c_1 \\ \vdots \\ c_n \end{pmatrix} (c_1^* \quad \cdots \quad c_n^*) = \begin{pmatrix} |c_1|^2 & \cdots & c_1 c_n^* \\ \vdots & \ddots & \vdots \\ c_n c_1^* & \cdots & |c_n|^2 \end{pmatrix}. \quad (12.15)$$

Here jth diagonal element is the probability of finding $b = b_j$ (i.e., $|\psi\rangle$ collapsing into one of basis eigenstates $|b_j\rangle$), and their sum (Tr $\hat{\rho}$) is equal to 1. The generic density matrix $\hat{\rho} = \sum_i \mathcal{P}_i \hat{\rho}_i$ is thus composed as a weighted sum of such special cases. Each of the above probabilities $|c_i|^2$ is then assigned its weight $\mathcal{P}_i$ of the corresponding pure state in the given mixture. This explains the physical meaning of diagonal elements of $\hat{\rho}$: they are the total probabilities of measurement outcomes, taking into account the composition of the given mixture. Each element ρ_{ii} is equal to the probability that a system picked randomly from the assembly will be found in ith eigenstate of the observable b.

With the introduction of the density matrix, the expectation value of f will now take a very simple form:

$$\langle f \rangle = \sum_{k,l} \rho_{lk} f_{kl} \quad (12.16a)$$

or

$$\langle f \rangle = \text{Tr}(\hat{\rho}\hat{f}). \quad (12.16b)$$

The second form (12.16b) becomes evident if we note that with l fixed and k changing, the summation in (12.16a) will traverse lth row of the $\hat{\rho}$-matrix and lth column of the $\hat{f}$-matrix; in other words, we get the (l, l) element of $\hat{\rho}\hat{f}$-matrix. Then, letting l vary, we get the sum of elements along the main diagonal, which is the trace of $\hat{\rho}\hat{f}$.

Thus, if we know $\hat{\rho}$, we can find the average of any observable by taking the trace (12.16b). Of course, both factors in $\hat{\rho}\hat{f}$ must be taken in the same basis.

Note that the order of factors in (12.16b) is not important – we may as well write $\langle f \rangle = \text{Tr}(\hat{f}\hat{\rho})$. The equivalence of the two forms is based on an important property of traces (see Box 12.1):

$$\langle \psi | \varphi \rangle = \text{Tr}(|\varphi\rangle\langle\psi|). \quad (12.17)$$

Using this, we get

$$\text{Tr}(\hat{f}\hat{\rho}) = \text{Tr}\left(\hat{f} \sum_i \mathcal{P}_i |\psi_i\rangle\langle\psi_i|\right) = \sum_i \mathcal{P}_i \text{Tr}\left(\hat{f}|\psi_i\rangle\langle\psi_i|\right) = \sum_i \mathcal{P}_i \langle\psi_i|\hat{f}|\psi_i\rangle$$
$$= \sum_i \mathcal{P}_i \langle f_i \rangle = \langle f \rangle,$$

which completes the proof.

Apart from being simple and elegant, expression (12.16) is invariant with respect to transformation of basis. Since the used basis $\{b\}$ was arbitrary, the final form (12.16) for the mean value would be the same in any other basis. This reflects the known property of the trace – it is invariant relative to unitary transformations.

Consider a simple example showing how the scheme works. Let S denote the spin, and suppose we have an ensemble of electrons with 25% of them in state $|S_{z+}\rangle$ ($S_z = \hbar/2$), 50% in state $|S_{x+}\rangle$ ($S_x = \hbar/2$), and 25% in state $|S_{x-}\rangle$ ($S_x = -\hbar/2$). What is the average of S_z? We must already have sufficient quantum intuition to predict that $\langle S_z \rangle = 0.25\hbar/2 = \hbar/8$. Indeed, our mixture contains a pure spin-up state with $\langle S_z \rangle = \hbar/2$ and horizontal spin states with $\langle S_z \rangle = 0$. With 25% of electron spins pointing up and 75% pointing in the horizontal (right or left) direction, we can see the answer in advance! Now let us get it formally with the density matrix. Using the basis $\{|\uparrow\rangle, |\downarrow\rangle\}$, we can describe the given ensemble as

$$|\psi_1\rangle = \begin{pmatrix} 1 \\ 0 \end{pmatrix}, \quad |\psi_2\rangle = \frac{1}{\sqrt{2}}\begin{pmatrix} 1 \\ 1 \end{pmatrix}, \quad |\psi_3\rangle = \frac{1}{\sqrt{2}}\begin{pmatrix} -1 \\ 1 \end{pmatrix}, \quad P_1 = 0.25,$$

$$P_2 = 0.5, \quad P_3 = 0.25. \tag{12.18}$$

According to (12.12), the density matrix for this ensemble is

$$\hat{\rho} = 0.25 \begin{pmatrix} 1 \\ 0 \end{pmatrix}(1\ 0) + \frac{0.5}{\sqrt{2}}\begin{pmatrix} 1 \\ 1 \end{pmatrix}\frac{1}{\sqrt{2}}(1\ 1) + \frac{0.25}{\sqrt{2}}\begin{pmatrix} -1 \\ 1 \end{pmatrix}\frac{1}{\sqrt{2}}(-1\ 1)$$

$$= \frac{1}{8}\begin{pmatrix} 5 & 1 \\ 1 & 3 \end{pmatrix} \tag{12.19}$$

and with the Pauli matrix we easily find

$$\langle S_z \rangle = \text{Tr}(\hat{\rho}\hat{S}_z) = \text{Tr}\left(\frac{1}{8}\begin{pmatrix} 5 & 1 \\ 1 & 3 \end{pmatrix}\frac{\hbar}{2}\begin{pmatrix} 1 & 0 \\ 0 & -1 \end{pmatrix}\right) = \frac{\hbar}{16}\text{Tr}\begin{pmatrix} 5 & -1 \\ 1 & -3 \end{pmatrix} = \frac{\hbar}{8}. \tag{12.20}$$

We realize that $\hat{\rho}$-operator stores all possible information about a mixture, as does a state function in case of a pure ensemble. In other words, $\hat{\rho}$ plays the same role for a mixture as $|\psi\rangle$ for a pure ensemble. Calculating the averages for any observable s and probabilities of all possible measurement outcomes can be done for a much broader range of conditions by using the density matrix. In fact, some physicists have become so accustomed to the language of density matrices that they even consider the wave function notation a *mauvais ton*! Given a system in state $|\psi\rangle$, they will say "the system in state $\hat{\rho} = |\psi\rangle\langle\psi|$," using the word "state" generally to refer to *any* ensemble, whether pure or mixed. While most papers won't go to such extremes, the reader should be aware of the above notation.

We will now formulate five important properties of density matrices.

1) $\text{Tr}\,\hat{\rho} = 1$. This property is just the statement that all probabilities add to 1.
2) For a pure ensemble, $\text{Tr}(\hat{\rho}^2) = 1$; conversely, if $\text{Tr}(\hat{\rho}^2) = 1$, then the ensemble is pure.

This follows from writing $\hat{\rho}^2 = (|\psi\rangle\langle\psi|)(|\psi\rangle\langle\psi|) = |\psi\rangle\langle\psi| = \hat{\rho}$ and using property 1. An operator having the property $\hat{\rho}^2 = \hat{\rho}$ is called *idempotent*. We have just established the idempotence of the density operator for a pure state. Obviously, such an operator satisfies $\hat{\rho}^n = \hat{\rho}$ for any integer $n > 0$. For a mixture, we have

$$\hat{\rho}^2 = \left(\sum_i \mathcal{P}_i |\psi_i\rangle\langle\psi_i|\right)\left(\mathcal{P}_j \sum_j |\psi_j\rangle\langle\psi_j|\right) = \sum_{i,j} \mathcal{P}_i \mathcal{P}_j |\psi_i\rangle\langle\psi_i|\psi_j\rangle\langle\psi_j|$$

$$= \sum_i \mathcal{P}_i^2 |\psi_i\rangle\langle\psi_i|,$$

and therefore $\text{Tr}(\hat{\rho}^2) = \sum_i \mathcal{P}_i^2 < 1$.

3) $\hat{\rho}$ is Hermitian. This can be established from the form of the pure state density matrix, which is clearly self-adjoint; the weighted sum of self-adjoint matrices will be self-adjoint as well.

4) $\hat{\rho}$ is *positive semidefinite*, that is, $\langle\varphi|\hat{\rho}|\varphi\rangle \geq 0$ for any nonzero state $|\varphi\rangle$ (when $\langle\varphi|\hat{\rho}|\varphi\rangle$ is strictly positive, the corresponding operator is called *positive definite*).

This follows directly from the definition $\langle\varphi|\hat{\rho}|\varphi\rangle \equiv \sum_i \mathcal{P}_i \langle\varphi|\psi_i\rangle\langle\psi_i|\varphi\rangle$, and the sum is nonnegative since we are adding nonnegative numbers weighted by nonnegative probabilities. Note that this proof only uses the fact that square of a complex number is nonnegative.[3] So the last requirement is just another way of saying that we can't have negative probabilities. (On a side note, since $\mathcal{P}_i$ are not the "quantum" probabilities of Born but the ordinary "statistical" probabilities, the positive semidefiniteness is not even a feature of QM, but rather a requirement that follows from the ordinary logic.)

5) All eigenvalues of $\hat{\rho}$ are nonnegative. This is a corollary of property 4. The hermiticity property implies that ρ is diagonalizable, so there must exist a *spectral decomposition* $\rho = U\Lambda U^{-1}$, where Λ is the diagonal matrix whose elements are the eigenvalues λ_i of ρ and U is a unitary matrix whose columns are the orthogonalized eigenvectors of ρ. Then property 4 guarantees that $\langle\varphi|\hat{\rho}|\varphi\rangle = \langle\varphi|U\Lambda U^{-1}|\varphi\rangle \geq 0$ for any choice of $|\varphi\rangle$. Denote $U^{-1}|\varphi\rangle \equiv |\xi\rangle$ to rewrite the last condition as $\langle\varphi|\hat{\rho}|\varphi\rangle = \langle\xi|\Lambda|\xi\rangle \geq 0$. Now Λ is positive semidefinite for any choice of $|\xi\rangle$, which is only possible if λ_i are nonnegative.

The above theorem is used frequently in applications. For example, suppose you study some mixture and find that the elements of its density matrix depend on some variable parameter τ. Then the eigenvalues of $\hat{\rho}$ will also depend on this parameter. By requiring that all eigenvalues $\lambda_i = \lambda_i(\tau)$ be nonnegative, you can find the allowed interval of values for τ. Then if somebody asks you to explain your reasoning, you will simply say that τ must be confined to that interval in order to exclude negative probabilities – an incontrovertible argument as conceptually simple as proofs get! We'll see examples of this technique used later in Chapters 24 and 25.

---

[3] Of course, $|\varphi\rangle$ should be defined in the *same* Hilbert space as $|\psi_i\rangle$; otherwise the inner products above will have no physical significance. For example, if $\hat{\rho}$ was defined in the spin space, then we mean by $|\varphi\rangle$ some arbitrary spin state and not, say, energy state of an oscillator.

Let $\hat{\rho}$ have a set of eigenvalues λ_i corresponding to eigenstates $|\lambda_i\rangle$ forming an orthogonal basis $\{|\lambda_i\rangle\}$ (with degenerate states orthogonalized). Then, acting with $\hat{\rho}$ on an arbitrary state and applying the closure relation $\sum_i |\lambda_i\rangle\langle\lambda_i| = I$, we can write

$$\hat{\rho}|\varphi\rangle = \hat{\rho}\left(\sum_i |\lambda_i\rangle\langle\lambda_i|\right)|\varphi\rangle = \sum_i \lambda_i |\lambda_i\rangle\langle\lambda_i|\varphi\rangle. \tag{12.21}$$

The resulting identity

$$\hat{\rho} = \sum_i \lambda_i |\lambda_i\rangle\langle\lambda_i|, \tag{12.22}$$

known as *spectral decomposition theorem* (SDT) [2], holds for any Hermitian operator.

The question arises: Given a density matrix, how can we find all possible mixed ensembles that would generate this matrix? To answer this question, we are now going to introduce a very useful mathematical trick: we will absorb the weights into the wave functions themselves:

$$\hat{\rho} = \sum_i \mathcal{P}_i |\psi_i\rangle\langle\psi_i| = \sum_i |\tilde{\psi}_i\rangle\langle\tilde{\psi}_i|, \quad |\tilde{\psi}_i\rangle = \sqrt{\mathcal{P}_i}|\psi_i\rangle. \tag{12.23}$$

Since the original states $|\psi_i\rangle$ had a unit norm, the new functions $|\tilde{\psi}_i\rangle$ will have norms less than unity. Matrix $\hat{\rho}$ is thus associated with the (generally) nonorthogonal, unnormalized set $|\tilde{\psi}_i\rangle$ defined in (12.23). But it is now expressed in a more simple form.

Next, we will show that it can be written in terms of another set $|\tilde{\phi}_i\rangle$ as $\hat{\rho} = \sum_i |\tilde{\phi}_i\rangle\langle\tilde{\phi}_i|$ if and only if the two generating sets are obtained one from the other by a unitary transformation, $|\tilde{\phi}_i\rangle = \sum_j u_{ij}|\tilde{\psi}_j\rangle$. The proof consists of two parts. First, if we compose a new set $\{|\tilde{\phi}_i\rangle\}$ according to the above transformation, then both sets will generate the same ρ:

$$\hat{\rho}' = \sum_i |\tilde{\phi}_i\rangle\langle\tilde{\phi}_i| = \sum_{i,k,l} u_{ik} u_{il}^* |\tilde{\psi}_k\rangle\langle\tilde{\psi}_l| = \sum_{k,l}\left(\sum_i u_{ik} u_{il}^*\right)|\tilde{\psi}_k\rangle\langle\tilde{\psi}_l|$$

$$= \sum_{k,l} \delta_{kl}|\tilde{\psi}_k\rangle\langle\tilde{\psi}_l| = \sum_k |\tilde{\psi}_k\rangle\langle\tilde{\psi}_k| = \hat{\rho}$$

(the sum in parentheses is equal to δ_{kl} because matrix U is unitary). Therefore, we have a *unitary freedom* in choosing the ensemble: since we can choose any U we want, there is an infinite variety of possible ensembles $\{|\tilde{\phi}_i\rangle\}$.

Conversely, suppose that both sets – $\{|\tilde{\psi}_i\rangle\}$ and $\{|\tilde{\phi}_i\rangle\}$ – result in the same $\hat{\rho}$. Then they must be connected by a unitary matrix. Since $\hat{\rho}$ is Hermitian, we can use the spectral decomposition theorem to write it in terms of some orthonormal basis $\{|k\rangle\}$: $\rho = \sum_k \lambda_k |k\rangle\langle k| = \sum_k |\tilde{k}\rangle\langle\tilde{k}|$. Then each $|\tilde{\psi}_i\rangle$ must be expressible as a linear combination of states $\{|\tilde{k}\rangle\}$.

Proof: If it is not, then $|\tilde{\psi}_i\rangle$ must have a component orthogonal to all $\{|\tilde{k}\rangle\}$. Call that component $|\chi\rangle$. Then $\langle\chi|\tilde{\psi}_i\rangle \neq 0$. But on the other hand, we have $\langle\chi|\rho|\chi\rangle = \langle\chi|\left(\sum_k |\tilde{k}\rangle\langle\tilde{k}|\right)|\chi\rangle = 0 = \langle\chi|\left(\sum_i |\tilde{\psi}_i\rangle\langle\tilde{\psi}_i|\right)|\chi\rangle = \sum_i |\langle\chi|\tilde{\psi}_i\rangle|^2$, which implies that $\langle\chi|\tilde{\psi}_i\rangle = 0$, contradicting the initial assumption. If $|\tilde{\psi}_i\rangle$ is a linear combination, $|\tilde{\psi}_i\rangle = \sum_k c_{ik}|\tilde{k}\rangle$, then

$$\rho = \sum_i |\tilde{\psi}_i\rangle\langle\tilde{\psi}_i| = \sum_{i,k,l} c_{ik} c_{il}^* |\tilde{k}\rangle\langle\tilde{l}| = \sum_{k,l} \left(\sum_i c_{ik} c_{il}^*\right) |\tilde{k}\rangle\langle\tilde{l}| = \sum_k |\tilde{k}\rangle\langle\tilde{k}|.$$

The last equation will hold as long as the sum in parentheses equals δ_{kl}, but this means that matrix C is unitary.[4] In the same way, one proves that $|\tilde{\beta}_i\rangle = \sum_k d_{ik}|\tilde{k}\rangle$ for some unitary matrix D. Then matrix $U = CD^+$ will transform the set $\{|\tilde{\beta}_i\rangle\}$ into $\{|\tilde{\alpha}_i\rangle\}$ and since the C and D are unitary matrices, U must be unitary as well, completing the proof.

The statement proved above is sometimes called *the theorem about unitary freedom in the ensemble for density matrices*. This theorem gives us an algorithm for constructing other ensembles for a given density matrix. If you start with $\{|\psi_i\rangle, \mathcal{P}_i\}$, you will only need to compose the new set $\{|\tilde{\psi}_i\rangle\}$, write it in the column matrix form, multiply by an arbitrary unitary matrix U, look at the resulting set $\{|\tilde{\phi}_i\rangle\}$, and extract the new weights by requiring that $|\tilde{\phi}_i\rangle = \sqrt{\mathcal{P}_i}|\phi_i\rangle$, where $|\phi_i\rangle$ is normalized. Since an infinite variety of possible ensembles can be constructed in this way, the knowledge of $\hat{\rho}$ does not tell us which ensemble was used. Consequently, the formation of a mixed ensemble is generally *irreversible*: once we scramble a set of pure states together to form a mixture and forget the original states $|\psi_i\rangle$ and their weights $\mathcal{P}_i$, we can no longer reconstruct the original "recipe" by just looking at the representing matrix $\hat{\rho}$ (see also Box 12.2). In particular, if you encode a message in the weights $\{\mathcal{P}_i\}$ of selected pure states $\{|\psi_i\rangle\}$, scramble these states together to form a mixed ensemble, and send it to a recipient, your addressee will not be able to read this message. This fact has extremely important implications for quantum communication theory as we will see later on.

Box 12.1 Trace of the outer product is equal to the inner product!

Let us take two wave functions, $|\varphi\rangle = \begin{pmatrix} c_1 \\ \vdots \\ c_n \end{pmatrix}$ and $|\psi\rangle = \begin{pmatrix} d_1 \\ \vdots \\ d_n \end{pmatrix}$, and compose their outer and inner products:

---

[4] There may be an additional step involved in the proof because C is not necessarily a square matrix. However, in such cases we can always append the missing rows or columns to C to make it a unitary square matrix.

$$|\varphi\rangle\langle\psi| = \begin{pmatrix} c_1 \\ \vdots \\ c_n \end{pmatrix} \begin{pmatrix} d_1^* & \cdots & d_n^* \end{pmatrix} = \begin{pmatrix} c_1 d_1^* \\ \vdots \\ c_n d_n^* \end{pmatrix},$$

$$\langle\psi|\varphi\rangle = \begin{pmatrix} d_1^* & \cdots & d_n^* \end{pmatrix} \begin{pmatrix} c_1 \\ \vdots \\ c_n \end{pmatrix} = c_1 d_1^* + \cdots + c_n d_n^*$$

(in the outer product, we wrote only the diagonal elements because they are the ones that define the trace). One can see immediately that the second expression is simply the trace of the first — just as stated in the title of this box. (Note that there is only *one* way in which one can compose an inner product from the given ket and bra pair.)

Box 12.2 Loss of information

To illustrate the erasure of information about the ensemble, consider the example (12.18). The density matrix $\hat{\rho} = \frac{1}{8}\begin{pmatrix} 5 & 1 \\ 1 & 3 \end{pmatrix}$ could equally well have been constructed from the ensemble $\{|\phi_1\rangle = (1/\sqrt{5})(2|\uparrow\rangle + |\downarrow\rangle), |\phi_2\rangle = i|\downarrow\rangle, |\phi_3\rangle = -(i/\sqrt{2})(|\uparrow\rangle - |\downarrow\rangle); \mathcal{P}_1 = 5/8, \mathcal{P}_2 = 1/8, \mathcal{P}_3 = 1/4$
In general, one can construct a given $\hat{\rho}$ in infinitely many ways!

Sometimes you may see a density matrix that is already written in the ket–bra form, for instance, $\hat{\rho} = (3/4)|\uparrow\rangle\langle\uparrow| + (1/4)|\downarrow\rangle\langle\downarrow|$. This clearly suggests one possible original ensemble: $\{|\psi_1\rangle = |\uparrow\rangle, |\psi_2\rangle = |\downarrow\rangle; \mathcal{P}_1 = 3/4, \mathcal{P}_2 = 1/4\}$. This form is very self-intuitive, following from the very definition of $\hat{\rho}$. But again, this is just one possible solution. We could as well have obtained the same ρ from $\{|\phi_1\rangle = (\sqrt{3}|\uparrow\rangle + \langle\downarrow|)/2, |\phi_2\rangle = (\sqrt{3}|\uparrow\rangle - \langle\downarrow|)/2; \mathcal{P}_1 = \mathcal{P}_2 = 0.5\}$.
There is no "preferred" way to deduce the form of the ensemble from a given $\hat{\rho}$.

By solving the eigenvalue problem for $\hat{\rho}$, we can immediately obtain one possible ensemble $\{|\psi_i\rangle, \mathcal{P}_i\}$. Namely, if the characteristic equation for $\hat{\rho}$ produces a set of eigenvalues λ_i corresponding to eigenstates $|\lambda_i\rangle$, then $\sum_i \lambda_i |\lambda_i\rangle\langle\lambda_i| = \hat{\rho}$ is a legitimate outer product form of $\hat{\rho}$ (because the density operator defined this way yields the correct eigenvalues and eigenvectors, $\hat{\rho}|\lambda_i\rangle = \lambda_i|\lambda_i\rangle$), so it only remains to set $\mathcal{P}_i$ equal to λ_i. The last example also illustrates the following interesting observation: when it comes to guessing the composition of a mixed ensemble, eigenvalues and eigenvectors of a density matrix do not have any special significance! Since $\rho|\uparrow\rangle = (3/4)|\uparrow\rangle$ and $\rho|\downarrow\rangle = (1/4)|\downarrow\rangle$, our "most intuitive" mixed ensemble was composed of the eigenstates and eigenvalues of this density matrix. But this is just *one* possible composition of the ensemble, which is as good as any other (other possibilities can then be obtained using the unitary freedom theorem). We still don't know which ensemble actually led to the given density matrix. Therefore, we *cannot* interpret the obtained eigenvalues of ρ as probabilities $\mathcal{P}_i$ or as eigenvalues of the constituent states $|\psi_i\rangle\langle\psi_i|$. To discourage this line of thinking once and for all, note that the number of eigenvalues thus obtained need not even be the same as the number of pure states in the mixture. For example, the two eigenvalues[5] $\lambda_{1,2} = (4 \pm \sqrt{2})/8$ of the density operator $\hat{\rho} = \frac{1}{8}\begin{pmatrix} 5 & 1 \\ 1 & 3 \end{pmatrix}$ from the problem ((12.16)–(12.18) have no relation whatsoever to the weights $\mathcal{P}_1 = 1/4$, $\mathcal{P}_2 = 1/2$, and $\mathcal{P}_3 = 1/4$, and the two respective eigenvectors to the pure states $|\psi_1\rangle\langle\psi_1|$, $|\psi_2\rangle\langle\psi_2|$, and $|\psi_3\rangle\langle\psi_3|$. Likewise, if the density operator is given in the outer product form,[6] then the corresponding weights suggested by this form are *not* the eigenvalues of $\hat{\rho}$.

12.4
Time Evolution of the Density Operator

Finally, we want to consider the time dependence of $\hat{\rho}$. A state function satisfies the Schrödinger equation and evolves with time. We can therefore expect that the density operator also satisfies a certain temporal equation, all the more so that a mixture itself is evolving with time. The density operator must be time dependent to reflect this evolution.

Since time was not explicitly present in the previous equations, we can assume that all of them refer to an initial moment $t = 0$. Then we can find the expression for $\hat{\rho}_{mn}(t)$ by applying its definition (12.13) to an arbitrary $t \neq 0$ in much the same way as we did in Chapter 11 to find evolution of a wave packet released at the initial moment:

$$\rho_{mn}(t) \equiv \sum_i \mathcal{P}_i c_{in}^*(t) c_{im}(t). \tag{12.24}$$

Differentiate this with respect to time:

$$\frac{\partial \rho_{mn}(t)}{\partial t} \equiv \sum_i \mathcal{P}_i \frac{\partial c_{in}^*(t)}{\partial t} c_{im}(t) + \sum_i \mathcal{P}_i c_{in}^*(t) \frac{\partial c_{im}(t)}{\partial t}. \tag{12.25}$$

But the time derivatives for each $c_{im}(t)$ and $c_{im}^*(t)$ can be found from the Schrödinger equation (or its conjugate, respectively) in the corresponding representation

---

[5] The Hilbert space here is the two-dimensional spin space of the electron; hence, only two eigenvalues are obtained from the eigenvalue equation.

[6] The trouble with this form is that it introduces ambiguity: Was it only intended as a statement about the ultimate matrix form of the density operator (since there are infinitely many outer product forms that lead to the same matrix!), or is it supposed to be a statement about the actual composition of the mixture?

(recall Section 8.9!):

$$i\hbar \frac{\partial c_{im}(t)}{\partial t} = \sum_n H_{mn} c_{in}(t) \quad \text{and} \quad -i\hbar \frac{\partial c_{im}^*(t)}{\partial t} = \sum_n H_{mn}^* c_{in}^*(t). \tag{12.26}$$

Putting this into (12.25) and again using (12.24) yields

$$\frac{\partial \rho_{mn}}{\partial t} = \frac{1}{i\hbar} \sum_n [H_{mn'} \rho_{n'n} - \rho_{mn'} H_{n'n}] \tag{12.27}$$

or, in the operator form,

$$\frac{\partial \hat{\rho}}{\partial t} = -[\hat{H}, \hat{\rho}], \tag{12.28}$$

where $[\hat{H}, \hat{\rho}]$ is the Poisson quantum bracket. Thus, knowing the Hamiltonian of the system and $\hat{\rho}(0)$, we can determine $\hat{\rho}(t)$ for an arbitrary t using Equation 12.28.

What happens if we interrupt continuous time evolution by performing a measurement of observable b in the mixture? Each such measurement collapses the corresponding state from superposition $|\psi_j\rangle = \sum_n c_{jn} |b_n\rangle$ to only one eigenstate $|b_m\rangle$. In this process, each *pure* subset of the mixture characterized by a certain fixed a_i is eventually converted into a number of different pure subsets characterized each by b_m with different m. As a result, the original mixture converts into a totally different mixture in which new pure states $|b_m\rangle$ are represented with probabilities $\mathcal{P}(b_m)$. If $|\psi_j\rangle$ were eigenstates corresponding to eigenvalues a_j, the mixture changes from the one sorted out with respect to observable a to one sorted out with respect to a complementary observable b. The corresponding elements of the new density matrix will be $\rho'_{mn} = \sum_j \mathcal{P}_j(b_m)_j c_{jn}^* c_{jm}$, where c_{jm} is the amplitude of finding b_m in a state $|\psi_j\rangle$.

In classical statistics, an ensemble of independent particles (the Gibbs ensemble) is characterized by probability density $D(p, q)$ of finding a particle with momentum p and coordinate q, respectively (recall the concept of the phase space in Section 3.6). According to the Liouville theorem, this density is constant; that is,

$$\frac{dD}{dt} = \frac{\partial D}{\partial t} + [H, D] = 0, \tag{12.29}$$

where $[H, D]$ is the classical Poisson bracket discussed in Section 11.2. Equation 12.29 says that

$$\frac{\partial D}{\partial t} = -[H, D]. \tag{12.30}$$

We see the striking similarity with (12.28), which turns out to be the quantum analogue of (12.30). In this respect, a quantum mixture is similar to a classical Gibbs ensemble (a system of subsets of particles with all particles within one subset being in one state). The difference is that in the classical limit we can talk about p and q having simultaneously definite values, whereas in quantum physics we can only talk about either q or p alone having a definite value. Accordingly, we have, say, $\rho_{qq'}$ instead of $D(p, q)$. But both obey essentially the same equations. The function $D(p, q)$ is the classical probability density. This analogy endorses again the name *density operator* for $\rho_{qq'}$.

12.5
Composite Systems

Let us now generalize the density matrix formalism to include ensembles of entangled systems. To choose a case study that illustrates all important features of composite systems, we'll consider a system of two electrons labeled as A and B, and focus on their spin states. If both electrons are mutually independent, each of them lives in its own 2D subspace. Write these subspaces as $\mathcal{H}_A = \{|\uparrow\rangle_A, |\downarrow\rangle_A\}$ and $\mathcal{H}_B = \{|\uparrow\rangle_B, |\downarrow\rangle_B\}$. A measurement on one electron would be carried out in that electron's space, leaving the other one unaffected. The whole system has four independent degrees of freedom and hence will be described by a 4 × 4 matrix. A convenient way to refer to the corresponding $\mathcal{H}$-space is $\mathcal{H} = \mathcal{H}_A \otimes \mathcal{H}_B$ – a notation that indicates the composition of this space in a self-intuitive way. Obviously, any vector in that space can be represented in terms of vectors in the subspaces $\mathcal{H}_A$ and $\mathcal{H}_B$. To keep the same consistent notation, we are going to use $\otimes$ when combining those "subvectors." We have already encountered the tensor product symbol $\otimes$ in Chapter 4. It is time now to discuss its physical significance and its use in the matrix algebra.

A system state where spin A points up and spin B points down will be constructed as $|\uparrow\rangle_A \otimes |\downarrow\rangle_B$. It is easily seen that the basis

$$\mathcal{H} = \mathcal{H}_A \otimes \mathcal{H}_B : \quad \{|\uparrow\rangle_A \otimes |\uparrow\rangle_B, \; |\uparrow\rangle_A \otimes |\downarrow\rangle_B, \; |\downarrow\rangle_A \otimes |\uparrow\rangle_B, \; |\downarrow\rangle_A \otimes |\downarrow\rangle_B\} \quad (12.31)$$

composed of four such product states is orthonormal and spans all the $\mathcal{H}$-space. Any possible spin state of pair (A, B) can be expressed as superposition of basis states (12.31), so $\mathcal{H} = \mathcal{H}_A \otimes \mathcal{H}_B$ is the right space to describe a composite system. A tensor product of "subvectors" $|\psi_A\rangle = c_1|\uparrow\rangle_A + c_2|\downarrow\rangle_A$ and $|\psi_B\rangle = d_1|\uparrow\rangle_B + d_2|\downarrow\rangle_B$ will be a vector in $\mathcal{H}$:

$$\begin{aligned}|\psi_A\rangle \otimes |\psi_B\rangle &= (c_1|\uparrow\rangle_A + c_2|\downarrow\rangle_A) \otimes (d_1|\uparrow\rangle_B + d_2|\downarrow\rangle_B) \\ &= c_1 d_1 |\uparrow\rangle_A \otimes |\uparrow\rangle_B + c_1 d_2 |\uparrow\rangle_A \otimes |\downarrow\rangle_B + c_2 d_1 |\downarrow\rangle_A \otimes |\uparrow\rangle_B + c_2 d_2 |\downarrow\rangle_A \otimes |\downarrow\rangle_B. \end{aligned} \quad (12.32)$$

The role of coefficients is the same as usual: $|c_1 d_1|^2$ is the probability for the system to collapse into product state $|\uparrow\rangle_A \otimes |\uparrow\rangle_B$, and so on.

If there is some correlation between the electrons, we cannot measure one of them without affecting the other. Then any measurement we can think of must be regarded as a measurement of the entire system. A familiar example is the spin singlet state $s = m = 0$ already discussed in Sections 3.7 and 7.5 (see Equation 7.129). Write this state as

$$|0,0\rangle = \frac{1}{\sqrt{2}} \left(|\uparrow\rangle_A \otimes |\downarrow\rangle_B - |\downarrow\rangle_A \otimes |\uparrow\rangle_B \right). \tag{12.33}$$

Now the electrons are maximally entangled, and (12.33) cannot be written as a product (12.32) of two separate kets. If the system (12.33) collapses into a "reduced" product state $|\uparrow\rangle_A \otimes |\downarrow\rangle_B$ or $|\downarrow\rangle_A \otimes |\uparrow\rangle_B$, with each electron independent of its "colleague" and "living" in its respective subspace, we will again have a mixture.

Now we want to see how *all possible states* can be described by the density matrix. Start with separable state (12.32) in the matrix form

$$|\psi_A\rangle \otimes |\psi_B\rangle = \begin{pmatrix} c_1 \\ c_2 \end{pmatrix} \otimes \begin{pmatrix} d_1 \\ d_2 \end{pmatrix} = \begin{pmatrix} c_1 d_1 \\ c_1 d_2 \\ c_2 d_1 \\ c_2 d_2 \end{pmatrix} \Rightarrow \begin{pmatrix} c_1 \begin{pmatrix} d_1 \\ d_2 \end{pmatrix} \\ c_2 \begin{pmatrix} d_1 \\ d_2 \end{pmatrix} \end{pmatrix}. \tag{12.34}$$

The rightmost matrix gives us an easy way to represent $|\psi_A\rangle \otimes |\psi_B\rangle$. Let us adopt the rule that eigenstates in the basis should always be ordered in such a way as to make the form (12.34) valid. To illustrate the role of ordering, write the $\mathcal{H}$-space as $\mathcal{H} = \mathcal{H}_B \otimes \mathcal{H}_A$. Then Equation 12.32 modifies to

$$|\psi_B\rangle \otimes |\psi_A\rangle = (d_1|\uparrow\rangle_B + d_2|\downarrow\rangle_B) \otimes (c_1|\uparrow\rangle_A + c_2|\downarrow\rangle_A)$$
$$= d_1 c_1 |\uparrow\rangle_B \otimes |\uparrow\rangle_A + d_1 c_2 |\uparrow\rangle_B \otimes |\downarrow\rangle_A + d_2 c_1 |\downarrow\rangle_B \otimes |\uparrow\rangle_A + d_2 c_2 |\downarrow\rangle_B \otimes |\downarrow\rangle_A \tag{12.35}$$

or, in the matrix form,

$$|\psi_B\rangle \otimes |\psi_A\rangle = \begin{pmatrix} d_1 \\ d_2 \end{pmatrix} \otimes \begin{pmatrix} c_1 \\ c_2 \end{pmatrix} = \begin{pmatrix} d_1 c_1 \\ d_1 c_2 \\ d_2 c_1 \\ d_2 c_2 \end{pmatrix} \Rightarrow \begin{pmatrix} d_1 \begin{pmatrix} c_1 \\ c_2 \end{pmatrix} \\ d_2 \begin{pmatrix} c_1 \\ c_2 \end{pmatrix} \end{pmatrix}. \tag{12.36}$$

This is the same rule for constructing the system state in the column matrix form. Note, however, that the basis implied by (12.35) and (12.36) is

$$\mathcal{H} = \mathcal{H}_B \otimes \mathcal{H}_A : \quad \{|\uparrow\rangle_B \otimes |\uparrow\rangle_A, |\uparrow\rangle_B \otimes |\downarrow\rangle_A, |\downarrow\rangle_B \otimes |\uparrow\rangle_A, |\downarrow\rangle_B \otimes |\downarrow\rangle_A\}. \tag{12.37}$$

Compare this with (12.31). As far as algebra is concerned, the change is only nominal: we just switched the labels A and B. But as physicists, we may notice an important subtlety: while $|\uparrow\rangle_B \otimes |\uparrow\rangle_A$ is the same state as $|\uparrow\rangle_A \otimes |\uparrow\rangle_B$ (both spins point up!), we cannot make the same claim about $|\uparrow\rangle_B \otimes |\downarrow\rangle_A$ and $|\uparrow\rangle_A \otimes |\downarrow\rangle_B$. Those two states can be physically distinguished. Imagine that before the measurement the electrons were taken apart and put in separate boxes. This, if carried out properly, does not affect any prior entanglement between the electrons. But now, sealing the boxes, putting labels A and B on them, and doing the measurement, we can tell which spin points up and which one points down! So the bases (12.31) and (12.37) are *not* exactly the same. The *second* and *third* eigenstates in (12.31) are physically identical to the *third* and *second* eigenstates in (12.37), respectively. In other words, one basis is obtained from the other by swapping the two middle terms. This illustrates an important feature of composite $\mathcal{H}$-spaces: the order of terms in the basis depends on the order in which subspaces $\mathcal{H}_A$ and $\mathcal{H}_B$ have been multiplied! To avoid mistakes, remember the algorithm used in (12.31) and (12.37): notice which subspace is on the left, look up the order of terms in its basis, and combine the *first* term consecutively with *all* terms from the second subspace's basis, then reiterate this for the second term, and so on.

Using this rule, we can also define operators on the $\mathcal{H}$-space of a composite system. Let $\mathcal{M}_A$ be an operator acting in $\mathcal{H}_A$ and $\mathcal{M}_B$ an operator in $\mathcal{H}_B$. The action of the operators on their respective subspaces will be defined by $\mathcal{M}_A |\psi\rangle_A = |\psi'\rangle_A$ and $\mathcal{M}_B |\psi\rangle_B = |\psi'\rangle_B$, or, in the matrix form,

$$\begin{pmatrix} \mathcal{M}_{11} & \mathcal{M}_{12} \\ \mathcal{M}_{21} & \mathcal{M}_{22} \end{pmatrix}_A \begin{pmatrix} c_1 \\ c_2 \end{pmatrix} = \begin{pmatrix} c'_1 \\ c'_2 \end{pmatrix}, \quad \begin{pmatrix} \mathcal{M}_{11} & \mathcal{M}_{12} \\ \mathcal{M}_{21} & \mathcal{M}_{22} \end{pmatrix}_B \begin{pmatrix} d_1 \\ d_2 \end{pmatrix} = \begin{pmatrix} d'_1 \\ d'_2 \end{pmatrix}. \tag{12.38}$$

A simultaneous action of these operators on state $|\psi\rangle_A \otimes |\psi\rangle_B$ defined in $\mathcal{H} = \mathcal{H}_A \otimes \mathcal{H}_B$ should then simply append primes to all elements in the rightmost matrix (12.34). This can be achieved by defining the operator on a composite system as

$$\mathcal{M}_A \otimes \mathcal{M}_B = \begin{pmatrix} (\mathcal{M}_{11})_A \begin{pmatrix} \mathcal{M}_{11} & \mathcal{M}_{12} \\ \mathcal{M}_{21} & \mathcal{M}_{22} \end{pmatrix}_B & (\mathcal{M}_{12})_A \begin{pmatrix} \mathcal{M}_{11} & \mathcal{M}_{12} \\ \mathcal{M}_{21} & \mathcal{M}_{22} \end{pmatrix}_B \\ (\mathcal{M}_{21})_A \begin{pmatrix} \mathcal{M}_{11} & \mathcal{M}_{12} \\ \mathcal{M}_{21} & \mathcal{M}_{22} \end{pmatrix}_B & (\mathcal{M}_{22})_A \begin{pmatrix} \mathcal{M}_{11} & \mathcal{M}_{12} \\ \mathcal{M}_{21} & \mathcal{M}_{22} \end{pmatrix}_B \end{pmatrix}. \tag{12.39}$$

The reader can check using simple algebra and (12.38) that this definition works:

$$\mathcal{M}_A \otimes \mathcal{M}_B \begin{pmatrix} c_1 \begin{pmatrix} d_1 \\ d_2 \end{pmatrix} \\ c_2 \begin{pmatrix} d_1 \\ d_2 \end{pmatrix} \end{pmatrix} = \begin{pmatrix} (\mathcal{M}_{11})_A c_1 d'_1 + (\mathcal{M}_{12})_A c_2 d'_1 \\ (\mathcal{M}_{11})_A c_1 d'_2 + (\mathcal{M}_{12})_A c_2 d'_2 \\ (\mathcal{M}_{21})_A c_1 d'_1 + (\mathcal{M}_{22})_A c_2 d'_1 \\ (\mathcal{M}_{21})_A c_1 d'_2 + (\mathcal{M}_{22})_A c_2 d'_2 \end{pmatrix} = \begin{pmatrix} c'_1 d'_1 \\ c'_1 d'_2 \\ c'_2 d'_1 \\ c'_2 d'_2 \end{pmatrix}$$

$$= \begin{pmatrix} c'_1 \begin{pmatrix} d'_1 \\ d'_2 \end{pmatrix} \\ c'_2 \begin{pmatrix} d'_1 \\ d'_2 \end{pmatrix} \end{pmatrix}. \tag{12.40}$$

Equations 12.34 and 12.39 expose the mathematical meaning of the $\otimes$ symbol. In matrix algebra, the *direct* (or *Kronecker*) *product* of two matrices A and B of arbitrary dimension is defined as

$$A \otimes B = \begin{pmatrix} A_{11} & \cdots & A_{1n} \\ \vdots & \ddots & \vdots \\ A_{m1} & \cdots & A_{mn} \end{pmatrix} \otimes \begin{pmatrix} B_{11} & \cdots & B_{1l} \\ \vdots & \ddots & \vdots \\ B_{k1} & \cdots & B_{kl} \end{pmatrix}$$

$$= \begin{pmatrix} A_{11}B & \cdots & A_{1n}B \\ \vdots & \ddots & \vdots \\ A_{m1}B & \cdots & A_{mn}B \end{pmatrix}, \qquad (12.41)$$

where symbol B on the right stands for the entire $k \times l$ matrix B. Thus, the product of two states or of two operators defined by (12.34) or (12.39) is given by the direct product[7] of the respective matrices. This is why we took special care to order the terms in the basis the way we did![8]

One important case of $\mathcal{M}_A \otimes \mathcal{M}_B$ is the operator $\mathcal{M}_A \otimes I$. It is easy to check that it acts only in $\mathcal{H}_A$ while leaving $\mathcal{H}_B$ alone. Operator $I \otimes \mathcal{M}_B$ is defined analogously.

Formally, we can apply the direct product formula (12.41) to the definition of $\mathcal{H} = \mathcal{H}_A \otimes \mathcal{H}_B$. Since defining the basis of a Hilbert space is equivalent to defining the space itself, we can write the following mnemonic rule giving the correct order of terms for the basis of $\mathcal{H}$:

$$\mathcal{H}_A \otimes \mathcal{H}_B \Rightarrow \left\{\begin{array}{c} |\uparrow\rangle_A \\ |\downarrow\rangle_A \end{array}\right\} \otimes \left\{\begin{array}{c} |\uparrow\rangle_B \\ |\downarrow\rangle_B \end{array}\right\} \Rightarrow \left\{\begin{array}{c} |\uparrow\rangle_A \otimes \left\{\begin{array}{c} |\uparrow\rangle_B \\ |\downarrow\rangle_B \end{array}\right\} \\ |\downarrow\rangle_A \otimes \left\{\begin{array}{c} |\uparrow\rangle_B \\ |\downarrow\rangle_B \end{array}\right\} \end{array}\right\} \Rightarrow \begin{pmatrix} |\uparrow\rangle_A \otimes |\uparrow\rangle_B \\ |\uparrow\rangle_A \otimes |\downarrow\rangle_B \\ |\downarrow\rangle_A \otimes |\uparrow\rangle_B \\ |\downarrow\rangle_A \otimes |\downarrow\rangle_B \end{pmatrix}$$

$$\Rightarrow \mathcal{H}. \qquad (12.42)$$

(We have used curly brackets to emphasize that this is a "matrix" of eigenkets, *not* the matrix of complex numbers. With that reservation, we could just as easily have written the same rule in the row matrix form.)

We have said earlier that $\mathcal{M}_A$ and $\mathcal{M}_B$ are only defined on their respective $\mathcal{H}$-spaces. The reader can ask: But what about entanglement? Isn't it true that by acting on electron A, operator $\mathcal{M}_A$ will automatically affect electron B? And if it

---

[7] In the literature, the terms "tensor product" and "direct product" are used interchangeably. Some authors will say "tensor product" when they want to keep focus on the underlying physics (i.e., an arrangement where each operator acts on its own subspace of $\mathcal{H}$) and "direct product" when they want to emphasize the underlying matrix formula (12.41). Overall, the term "tensor product" seems to be taking over. This is in a sense unfortunate because one might prefer to reserve the term for the actual tensor operators (we have encountered them briefly in Chapter 7). Of course, broadly speaking, all QM entities – scalars, vectors, and matrices – are also tensors (this includes spinors, which are treated as tensors of half-integer rank).

[8] Another related concept is the direct sum, defined as $A \oplus B = \begin{pmatrix} A & O \\ O & B \end{pmatrix}$, where symbols A and B stand for the respective matrices and O is a matrix of zeros having the appropriate dimensions. Note that $A \oplus B \neq B \oplus A$.

affects B, shouldn't we say that $\mathcal{M}_A$ also has some "$\mathcal{M}_B$ aspect" to it? The first statement is true, and the second is false. The "$\mathcal{M}_B$ aspect" would exist if electron B had its separate wave function $|\psi\rangle_B$ and if $\mathcal{M}_A$ would then change that function by manipulating coefficients d_1 and d_2. But as evidenced by (12.40), in this case the action of $\mathcal{M}_A$ amounts strictly to transforming the pair of coefficients c_1 and c_2 to c'_1 and c'_2. When there is entanglement, there is no independent state $|\psi\rangle_B$ and hence no coefficients d_1 and d_2 to be transformed. When operator $\mathcal{M}_A$ affects electron B, it does so by changing the left terms in the product states (12.31). For example, let's see what $\mathcal{M}_A$ does to the product state $(1/\sqrt{2})|\uparrow\rangle_A \otimes |\downarrow\rangle_B$. As coefficients $c_1 = 1$ and $c_2 = 0$ characterizing the term $|\uparrow\rangle_A$ change to c'_1 and c'_2, the product state becomes $(1/\sqrt{2})(c'_1|\uparrow\rangle_A + c'_2|\downarrow\rangle_A) \otimes |\downarrow\rangle_B$. The second product state transforms similarly to $-(1/\sqrt{2})(c''_1|\uparrow\rangle + c''_2|\downarrow\rangle_A) \otimes |\uparrow\rangle_B$. The result is a new system state,

$$\mathcal{M}_A|0,0\rangle = \frac{1}{\sqrt{2}}(-c''_1|\uparrow\rangle_A \otimes |\uparrow\rangle_B + c'_1|\uparrow\rangle_A \otimes |\downarrow\rangle_B + c''_2|\downarrow\rangle_A \otimes |\uparrow\rangle_B + c'_2|\downarrow\rangle_A \otimes |\downarrow\rangle_B),$$

in which the odds of finding electron B with its spin pointing in a given direction are not the same as before. But our operator $\mathcal{M}_A$ has never changed the d-coefficients pertaining to kets labeled by B! In fact, if the singlet $|s=0, m=0\rangle$ contained just one product state, the statistics of electron B would remain exactly the same.

To complete the direct product formalism, we must now define density matrices on $\mathcal{H} = \mathcal{H}_A \otimes \mathcal{H}_B$. Let ρ^A and ρ^B be density matrices defined, respectively, on $\mathcal{H}_A$ and $\mathcal{H}_B$ according to

$$\rho^A = |\psi_A\rangle\langle\psi_A| = \begin{pmatrix} c_1 \\ c_2 \end{pmatrix}(c_1^* \ c_2^*) = \begin{pmatrix} |c_1|^2 & c_1 c_2^* \\ c_1^* c_2 & |c_2|^2 \end{pmatrix},$$

$$\rho^B = |\psi_B\rangle\langle\psi_B| = \begin{pmatrix} d_1 \\ d_2 \end{pmatrix}(d_1^* \ d_2^*) = \begin{pmatrix} |d_1|^2 & d_1 d_2^* \\ d_1^* d_2 & |d_2|^2 \end{pmatrix}.$$

Then it is natural to form the density matrix of the composite system as

$$\rho^{AB} \equiv \rho^A \otimes \rho^B = |\psi_A\rangle\langle\psi_A| \otimes |\psi_B\rangle\langle\psi_B|. \tag{12.43}$$

To convince yourself that this definition is logically self-consistent, just note that its matrix form

$$\rho^{AB} = \begin{pmatrix} |c_1|^2 & c_1 c_2^* \\ c_1^* c_2 & |c_2|^2 \end{pmatrix} \otimes \begin{pmatrix} |d_1|^2 & d_1 d_2^* \\ d_1^* d_2 & |d_2|^2 \end{pmatrix} = \begin{pmatrix} |c_1|^2|d_1|^2 & |c_1|^2 d_1 d_2^* & c_1 c_2^*|d_1|^2 & c_1 c_2^* d_1 d_2^* \\ |c_1|^2 d_1^* d_2 & |c_1|^2|d_2|^2 & c_1 c_2^* d_1^* d_2 & c_1 c_2^*|d_2|^2 \\ c_1^* c_2|d_1|^2 & c_1^* c_2 d_1 d_2^* & |c_2|^2|d_1|^2 & |c_2|^2 d_1 d_2^* \\ c_1^* c_2 d_1^* d_2 & c_1^* c_2|d_2|^2 & |c_2|^2 d_1^* d_2 & |c_2|^2|d_2|^2 \end{pmatrix}$$

$$\tag{12.44}$$

is identical to the density matrix obtained from the column vector on the right of (12.34).

Let us now ask the following question: Given the matrix ρ^{AB}, how does one recover ρ^A and ρ^B? One look at the 4×4 matrix (12.44) should be sufficient to convince you that the solution involves traces. If we subdivide this matrix into four 2×2 matrices, then the trace of the upper left matrix is equal to $|c_1|^2 (|d_1|^2 + |d_2|^2) = |c_1|^2 \operatorname{Tr} \rho^B = |c_1|^2$, which is just the element in the upper left corner of ρ^A. Similarly, taking the traces of the three remaining matrices, we recover the other three elements of ρ^A. Note that in each case the element gets multiplied by the trace of matrix ρ^B, which is equal to 1.

The procedure for recovering ρ^B will be similar. Just imagine the square formed by the first and third rows and the first and third columns, and the corner elements will give you the first 2×2 matrix. Shifting the outline of that imaginary square one position to the right, one position down, or one position down and to the right, you will obtain the other three matrices. Arranging them in a self-intuitive order and replacing matrices with their traces, you will obtain ρ^B:

$$\rho^A = \begin{pmatrix} \operatorname{Tr}\begin{pmatrix} \rho^{AB}_{11} & \rho^{AB}_{12} \\ \rho^{AB}_{21} & \rho^{AB}_{22} \end{pmatrix} & \operatorname{Tr}\begin{pmatrix} \rho^{AB}_{13} & \rho^{AB}_{14} \\ \rho^{AB}_{23} & \rho^{AB}_{24} \end{pmatrix} \\ \operatorname{Tr}\begin{pmatrix} \rho^{AB}_{31} & \rho^{AB}_{32} \\ \rho^{AB}_{41} & \rho^{AB}_{42} \end{pmatrix} & \operatorname{Tr}\begin{pmatrix} \rho^{AB}_{33} & \rho^{AB}_{34} \\ \rho^{AB}_{43} & \rho^{AB}_{44} \end{pmatrix} \end{pmatrix}, \quad \rho^B = \begin{pmatrix} \operatorname{Tr}\begin{pmatrix} \rho^{AB}_{11} & \rho^{AB}_{13} \\ \rho^{AB}_{31} & \rho^{AB}_{33} \end{pmatrix} & \operatorname{Tr}\begin{pmatrix} \rho^{AB}_{12} & \rho^{AB}_{14} \\ \rho^{AB}_{32} & \rho^{AB}_{34} \end{pmatrix} \\ \operatorname{Tr}\begin{pmatrix} \rho^{AB}_{21} & \rho^{AB}_{23} \\ \rho^{AB}_{41} & \rho^{AB}_{43} \end{pmatrix} & \operatorname{Tr}\begin{pmatrix} \rho^{AB}_{22} & \rho^{AB}_{24} \\ \rho^{AB}_{42} & \rho^{AB}_{44} \end{pmatrix} \end{pmatrix}.$$

The above procedure on the composite system's matrix ρ^{AB} that yields ρ^A (or ρ^B) is called taking the *partial trace* over system B (or system A). It is denoted by the corresponding subscript after the Tr symbol:

$$\rho^A = \operatorname{Tr}_B \rho^{AB}, \qquad \rho^B = \operatorname{Tr}_A \rho^{AB}. \tag{12.45}$$

The word "partial" refers to the fact we only reduce a large matrix to a smaller matrix (a "completed" tracing operation is supposed to reduce a matrix to one single number). In the process of taking $\operatorname{Tr}_B \rho^{AB}$, the probability amplitudes d_i describing system B[9] vanish (get "traced out"), so only amplitudes c_i find their way into the resulting matrix. Similarly, $\operatorname{Tr}_A \rho^{AB}$ traces out amplitudes pertaining to system A, leaving only those pertaining to system B.

Strictly speaking, the above procedure did not yield ρ^A, but rather $\rho^A \operatorname{Tr} \rho^B$. In declaring the result identical to ρ^A, we have relied on $\operatorname{Tr} \rho^B$ being equal to unity. The latter fact can be seen from trace rule (12.16): $\operatorname{Tr} \rho^B = \operatorname{Tr}(|\psi_B\rangle\langle\psi_B|) = \langle\psi_B|\psi_B\rangle = 1$. With this in mind and retaining $\operatorname{Tr}(|\psi_B\rangle\langle\psi_B|)$ in the equation for the reason that will be made clear shortly, we can now formally define partial trace as follows:

$$\operatorname{Tr}_B \rho^{AB} = \operatorname{Tr}_B(|\psi_A\rangle\langle\psi_A| \otimes |\psi_B\rangle\langle\psi_B|) = (|\psi_A\rangle\langle\psi_A|)\operatorname{Tr}(|\psi_B\rangle\langle\psi_B|), \tag{12.46}$$

and a similar equation for the other partial trace.

Although in the above derivation both ρ^A and ρ^B were assumed to be pure states, the derivation will remain substantially the same if ρ^A and ρ^B are mixtures. We would just write the last equation multiple times – once for each term – and then

---

9) We can already substitute the word "system" for "electron," which was of course only used for illustrative purposes. All concepts and equations in this section can be generalized for arbitrary systems A and B (which may involve higher dimensional Hilbert spaces) in a self-obvious way.

multiply the terms by their respective weights and add the equations together to obtain the final result in the form (12.45).

So far, it was assumed that both A and B possess their own wave functions, so the composite system is in the product state: $|\psi_{AB}\rangle = |\psi_A\rangle \otimes |\psi_B\rangle$ and thus $\rho^{AB} = \rho^A \otimes \rho^B$. This is not true for entangled systems such as the singlet (12.33). What happens if we apply the above procedure to such a system? While ρ^{AB} can no longer be written as a direct product of two matrices labeled by A and B, it always can be (and usually is) written as a sum of outer products $|a_j\rangle\langle a_j| \otimes |b_k\rangle\langle b_l|$, where $|a_i\rangle, |a_j\rangle$ and $|b_k\rangle, |b_l\rangle$ are pairs of eigenstates belonging to $\mathcal{H}_A$ and $\mathcal{H}_B$, respectively, and where, generally speaking, $i \neq j$ and $k \neq l$. For example, we have for the singlet (12.33)

$$\rho^{AB} = \frac{1}{2}(|\uparrow\rangle_A \otimes |\downarrow\rangle_B - |\downarrow\rangle_A \otimes |\uparrow\rangle_B)(\langle\uparrow|_A \otimes \langle\downarrow|_B - \langle\downarrow|_A \otimes \langle\uparrow|_B)$$

$$= \frac{1}{2}(|\uparrow\rangle_A\langle\uparrow|_A \otimes |\downarrow\rangle_B\langle\downarrow|_B - |\uparrow\rangle_A\langle\downarrow|_A \otimes |\downarrow\rangle_B\langle\uparrow|_B - |\downarrow\rangle_A\langle\uparrow|_A \otimes |\uparrow\rangle_B\langle\downarrow|_B + |\downarrow\rangle_A\langle\downarrow|_A \otimes |\uparrow\rangle_B\langle\uparrow|_B).$$

If we perform the same partial trace procedure as before – split the larger matrix into smaller matrices, take their traces, and use them to compose the final matrix – the result obtained should clearly be linear in the inputs; in other words, we may as well take the partial trace of each summand and add the partial traces:

$$\mathrm{Tr}_B\,\rho^{AB} = \mathrm{Tr}_B\left(\frac{1}{2}|\uparrow\rangle_A\langle\uparrow|_A \otimes |\downarrow\rangle_B\langle\downarrow|_B\right) - \mathrm{Tr}_B\left(\frac{1}{2}|\uparrow\rangle_A\langle\downarrow|_A \otimes |\downarrow\rangle_B\langle\uparrow|_B\right)$$

$$- \mathrm{Tr}_B\left(\frac{1}{2}|\downarrow\rangle_A\langle\uparrow|_A \otimes |\uparrow\rangle_B\langle\downarrow|_B\right) + \mathrm{Tr}_B\left(\frac{1}{2}|\downarrow\rangle_A\langle\downarrow|_A \otimes |\uparrow\rangle_B\langle\uparrow|_B\right).$$

The problem has been reduced to that of finding the terms $\mathrm{Tr}_B|a_j\rangle\langle a_j| \otimes |b_k\rangle\langle b_l|$. To make it as general as possible, we can now drop the requirement that the terms in this product be eigenstates. (This is reasonable because we might as well have written ρ^{AB} in terms of some other $|a_i\rangle, |a_j\rangle$ and $|b_k\rangle, |b_l\rangle$). It is sufficient for our purposes that they are some *arbitrary* states belonging to $\mathcal{H}_A$ and $\mathcal{H}_B$, respectively.

Let us write both outer products in the matrix form and compose the direct product:

$$|a_i\rangle\langle a_j| = \begin{pmatrix} c_1 \\ c_2 \end{pmatrix}\begin{pmatrix} d_1^* & d_2^* \end{pmatrix}, \quad |b_k\rangle\langle b_l| = \begin{pmatrix} e_1 \\ e_2 \end{pmatrix}\begin{pmatrix} f_1^* & f_2^* \end{pmatrix},$$

$$\begin{pmatrix} c_1 d_1^* & c_1 d_2^* \\ c_2 d_1^* & c_2 d_2^* \end{pmatrix} \otimes \begin{pmatrix} e_1 f_1^* & e_1 f_2^* \\ e_2 f_1^* & e_2 f_2^* \end{pmatrix} = \begin{pmatrix} c_1 d_1^* e_1 f_1^* & c_1 d_1^* e_1 f_2^* & c_1 d_2^* e_1 f_1^* & c_1 d_2^* e_1 f_2^* \\ c_1 d_1^* e_2 f_1^* & c_1 d_1^* e_2 f_2^* & c_1 d_2^* e_2 f_1^* & c_1 d_2^* e_2 f_2^* \\ c_2 d_1^* e_1 f_1^* & c_2 d_1^* e_1 f_2^* & c_2 d_2^* e_1 f_1^* & c_2 d_2^* e_1 f_2^* \\ c_2 d_1^* e_2 f_1^* & c_2 d_1^* e_2 f_2^* & c_2 d_2^* e_2 f_1^* & c_2 d_2^* e_2 f_2^* \end{pmatrix}.$$

The same procedure that we used in connection with (12.44) will now yield the matrix $|a_i\rangle\langle a_j|$ multiplied by the number $e_1 f_1^* + e_2 f_2^*$, which is just the trace of

$|b_k\rangle\langle b_l|$. So we have

$$\text{Tr}_B(|a_j\rangle\langle a_j| \otimes |b_k\rangle\langle b_l|) = |a_j\rangle\langle a_j|\text{Tr}(|b_k\rangle\langle b_l|). \tag{12.47}$$

This formula is more general than (12.46) and in fact is commonly used (in conjunction with the linearity of QM operators) as the *definition* of the partial trace.

It should now be clear why we wanted to retain the trace on the right side of (12.46). In the more general case (12.47), this trace is not equal to unity:

$$\text{Tr}(|b_k\rangle\langle b_l|) \equiv \langle b_l|b_k\rangle \neq 1. \tag{12.48}$$

This should not surprise us because the outer product $|b_k\rangle\langle b_l|$ is not a density matrix. The role of the factor $\langle b_l|b_k\rangle$ (which represents the overlap between states $|b_k\rangle$ and $|b_l\rangle$) is more subtle: it is the weighting coefficient with which matrix $|a_j\rangle\langle a_j|$ will appear in the final result after summation. Since this is a constant number, it is clear that our algorithm does in fact remove all references to system B, which is just what we expect from the partial trace. Thus, even as we broadened the scope of the partial trace definition from (12.46), which was perfectly good for composite systems in a product state, to (12.47), which is applicable to *all* composite systems, including entangled ones, we did not lose the physical meaning established earlier. The effect of Tr_B is to trace out all amplitudes pertaining to system B, thereby giving us the statistics of measurement outcomes for system A.

We now have the tool we need to solve the electron singlet. Out of the four traces, two are ones and two are zeros: $_B\langle\downarrow\,|\downarrow\rangle_B = {}_B\langle\uparrow\,|\uparrow\rangle_B = 1$ and $_B\langle\downarrow\,|\uparrow\rangle_B = {}_B\langle\uparrow\,|\downarrow\rangle_B = 0$. Therefore,

$$\text{Tr}_B\,\rho^{AB} = \frac{1}{2}|\uparrow\rangle_A\langle\uparrow|_A + \frac{1}{2}|\downarrow\rangle_A\langle\downarrow|_A = \frac{I}{2}. \tag{12.49}$$

A similar calculation for the other partial trace yields the same result:

$$\text{Tr}_A\,\rho^{AB} = \frac{I}{2} \tag{12.50}$$

The trace of $I/2$ is 1, so we obtained a legitimate density matrix after the traces of summands added to unity as expected.

We are now in a position to describe the statistics of experimental outcomes. Namely, a measurement of either electron will be equally likely to find it in spin-up or spin-down state. Since such a measurement will break the entanglement (an electron is no longer entangled when it knows its spin state), the obtained matrices describe the two mixed[10] ensembles that *will* emerge after the disentangling of the electrons. The future tense here is of extreme importance. It would be wrong to interpret these results as telling us that there are two mixed states $\rho^A = (1/2)I$ and $\rho^B = (1/2)I$ existing *currently* within their subspaces of composite space ρ^{AB}. In fact, such separate states do not exist while entanglement is still present. That is why a student who naively multiplies $I/2 \otimes I/2$ will *not* get the composite state ρ^{AB} that we started with (check it!).

The matrices $\rho^A = \text{Tr}_B\,\rho^{AB}$ and $\rho^B = \text{Tr}_A\,\rho^{AB}$ obtained by taking the respective partial traces have the special name *reduced density matrices*. In the absence of

---

10) They are mixed because $\text{Tr}\left((\rho^A)^2\right) = \text{Tr}\left((\rho^B)^2\right) = (1/2) < 1$.

entanglement, when the composite system is described by a product state $\rho^{AB} = \rho^A \otimes \rho^B$, they coincide with the density matrices ρ^A and ρ^B appearing in this product state:

$$\rho^{AB} = \left(\mathrm{Tr}_B\, \rho^{AB}\right) \otimes \left(\mathrm{Tr}_A\, \rho^{AB}\right). \tag{12.51}$$

This formula is no longer true when systems A and B are entangled so that ρ^{AB} cannot be written as a direct product. The hallmark of entangled systems is that

$$\rho^{AB} \neq \left(\mathrm{Tr}_B\, \rho^{AB}\right) \otimes \left(\mathrm{Tr}_A\, \rho^{AB}\right). \tag{12.52}$$

So if we are only given the two reduced density matrices, we don't have sufficient knowledge to recover the composite state. We must then write ρ^{AB} in as general form as possible, as a superposition similar to (12.31), including all possible superposition terms, and then do our best to establish relations between coefficients based on general physical grounds. We'll see an example of this approach in Chapter 24, where rotational invariance will be used to relate the coefficients and establish an upper bound on the quality of cloning.

Problems

12.1 Consider a pure ensemble in a state described by superposition (12.2) of eigenstates of an observable q. Find the average of another observable M incompatible with q.

12.2 Compare the result of the previous problem with the average of M for a mixture satisfying the condition (12.4).

12.3 A certain mixture contains the eigenstates $\psi_j(x)$ of operator $\hat{M}$ with weights $\mathcal{P}_j$. Suppose we choose instead of $\psi_j(x)$ another set – the eigenstates $\phi_m(x)$ of operator $\hat{Q}$. Since, by assumption, both sets are the functions of x, we can expand each $\psi_a(x)$ over $\phi_m(x)$: $\psi_j(x) = \sum_m b_{jm} \phi(x)$. Find the average $\langle M \rangle$ in terms of $\mathcal{P}_j$ and b_{jm}.

12.4 Consider two operators: $\hat{A}$ with the set of eigenfunctions $\psi_a(x)$ and $\hat{B}$ with the set of eigenfunctions $\phi_m(x)$. It is known that each of functions $\psi_a(x)$ can be expanded into series over $\phi_m(x)$ (i.e., can be represented as a superposition of eigenstates $\phi_m(x)$), and vice versa. Prove that operators $\hat{A}$ and $\hat{B}$ do not commute.

12.5 Prove that the composition of a mixed ensemble can be recovered from its density matrix if all states in the ensemble are orthonormal.

12.6 Derive Equation 12.27.

References

1 von Neumann, J. (1955) *Mathematical Foundations of Quantum Mechanics*, Princeton University Press, Princeton, NJ.

2 Nielsen, M.A. and Chuang, I.L. (2000) *Quantum Computation and Quantum Information*, Cambridge University Press, Cambridge.

13
Indeterminacy Revisited

We have learned in Chapter 9 that indeterminacy, even for an isolated system, may be a function of time. We know that indeterminacy may exist for observables other than position and momentum. We added new members, such as number N of particles in an ensemble, to our compendium of observables. All this will allow us to gain a deeper insight into the nature of indeterminacy and study it in more detail.

13.1
Indeterminacy Under Scrutiny

Suppose a moving particle with the initial momentum **p** passes through a very narrow slit of width a. This constitutes a position measurement, yielding coordinate y with uncertainty $\Delta y \approx a$. As it squeezes through, the particle effectively finds itself within a potential box of width a. The existence of such box is especially evident if the screen is not absorbing and has finite thickness. The new momentum state has a certain spread in its p_y component around $p_y = 0$. As usual, we place a screen behind the slit at a distance L and record the landing position y. One asks, did the particle acquire a definite transverse component of momentum $p_y \approx p(y/L)$? If yes, then we can measure this component with any desired accuracy by taking $y, L \gg a$ until Δp_y becomes small enough that $\Delta p_y \Delta y \approx \Delta p_y a \ll \hbar$, violating the indeterminacy principle.

Where is the flaw in this reasoning? Our error was ascribing to the particle a classical path making an angle $\theta = \arctan(y/L)$ with the initial direction. Actually, de Broglie waves will be running in *all* directions from the slit. The slit here acts as the source of diverging waves in analogy with the Huygens principle of classical optics. Depending on the length of the slit, these waves will be close to spherical or cylindrical and they can always be represented as a superposition of de Broglie waves. In other words, the particle recedes from the slit *in all available directions at once*. To each direction there corresponds a separate p_y with its own probability amplitude. The distribution of p_y has standard deviation Δp_y and if we calculate it we will find that it exceeds $\hbar/a$.

The second error in the argument is a direct result of the first. Since the particle is not taking any single classical trajectory, connecting the initial and final points does not amount to a momentum measurement. When the particle hits the screen, its wave function collapses from the infinite superposition of possible landing points to a singularity (δ-function) corresponding to the actual landing point.[1] This is position measurement, which is incompatible with that of momentum! A real momentum measurement must be set up differently. For instance, one might look at the y-momentum transferred by the particle to the screen. Then one would know momentum, but at the cost of creating indeterminacy in position y.

There were other attempts to use the above thought experiment for refutation of the indeterminacy principle. The corresponding argument is as follows. The conversion of the incident plane wave on one side of the screen to the diverging wave on the other side can be formally considered as a result of a certain interaction between the particle and the edge of the slit during the brief time interval of passing through the slit. Since the experimental setup is fixed, this interaction is time independent and therefore it cannot change the particle's energy. This is a well-known and indisputable fact. Any diffraction experiment with light, while changing its angular distribution, will not change its frequency. So the slit can change direction but not magnitude of particle's momentum (recall the dispersion equation from Section 8.4). The maximum possible projection p_y is obtained for the maximum deflection of the particle, $\theta = \pi/2$, when $|p_y| = p$. Therefore, the range of possible p_y is restricted to the region $-p \leq p_y \leq p$ (Figure 13.1) and accordingly, the corresponding indeterminacy cannot exceed $2p$. This conclusion is totally independent of the slit width a and accordingly of the particle's *position indeterminacy* $\Delta y \cong a$ when passing through the slit. We can in principle narrow the slit down to zero. Then on the one hand we have $\Delta p_y \leq 2p$ and on the other hand $\Delta y \to 0$. By reducing the magnitude of momentum (e.g., illuminating the slit with monochromatic infrared light) or by fixing p and reducing the slit width, we can make $\Delta p_y \Delta y \to 0$.

The fallacy of this argument is in the "obvious" statement that p_y cannot exceed p. Actually, it can and does. We can see it from two complementary perspectives.

First, the diffraction pattern on the screen is determined to a high accuracy by the Fourier transform of the so-called *aperture function*, which is the truncated "patch" of the incident wavefront intercepted by the slit [1–3]. This transform contains waves with *all* p_y, that is, $-\infty < p_y < \infty$, *regardless* of the width a. Moreover, the narrower the slit, the more pronounced are its Fourier components with high p_y (recall

---

1) We can still think of such collapse as a two-staged process: collapse to a single plane wave directed to y, which immediately collapses to a single point at y. Both stages occur instantaneously at the moment of collision with the second screen; therefore, even if we admit the fleeting existence of the first stage, it does not constitute the transverse momentum measurement at the much earlier moment of passing through the first screen. Moreover, there is an exchange of transverse momentum between the particle and the first screen: if the particle lands on the second screen at y, say, above the symmetry axis, the first screen bounces by the vanishingly small distance $\delta y = (\mu/M) y$ below this axis (M is the mass of the screen). But again, this can, in principle, be observed only *after* the particle makes its mark at y on the second screen. Before that, the particle and the screen are in the entangled superposition of states with all possible $(\delta y, y)$ satisfying condition that the center of mass of the system particle + screen remains fixed.

13.1 Indeterminacy Under Scrutiny

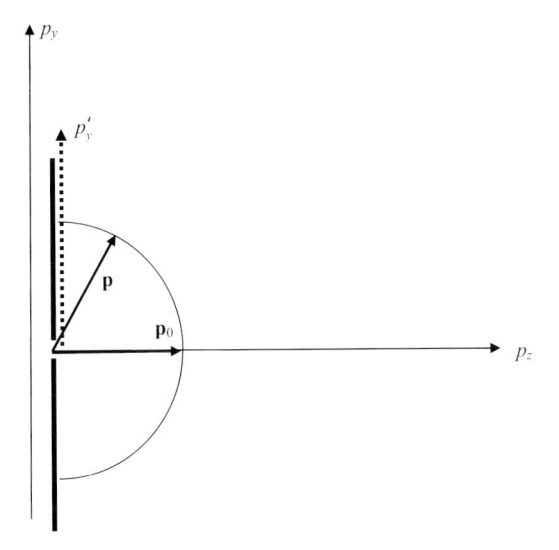

Figure 13.1 One-slit experiment in the momentum space. **p** is the momentum vector of one of the plane waves propagating from the slit (one of the Fourier components of the slit-diffracted wave); all these waves have the same magnitude $|\mathbf{p}| = |\mathbf{p}_0|$ of momentum (the whole set is monochromatic) since all of them "originate" from the plane monochromatic incident wave with momentum $\mathbf{p}_0$. Therefore, according to naïve expectations, neither p_y, nor p_z-component in any such wave can exceed $|\mathbf{p}|$. In reality, there exist waves with arbitrarily large $p'_y > |\mathbf{p}_0|$. One of them is shown as dotted vector. This does not contradict dispersion equation (8.28) because such waves have imaginary p_z-component. They are "crawling" strictly along the screen and their amplitude is exponentially decreasing function of z. They are similar to the evanescent waves known in optics.

Section 11.2). But this is a rather formal argument, and, in addition, one wonders what happened to the conservation of the magnitude p.

In order to address the latter concern, we turn to the second perspective: recall that the magnitude of p is determined by the dispersion equation (8.26) or (8.27), which in our case (i.e. for the photons) reads $p^2 = p_y^2 + p_z^2 = \hbar^2\omega^2/c^2$. Here ω is the frequency and u is the phase velocity of the corresponding de Broglie wave. Also recall that in QM, and generally in the theory of waves, the momentum space is a *complex* vector space! In our case, this becomes manifest precisely for waves deflected by 90°. When the values $p_y > p$ are required for such waves by the Fourier expansion of the aperture function, this can be achieved without changing p if p_z becomes imaginary:

$$p_z = \sqrt{p^2 - p_y^2}\bigg|_{p_y > p} = i\eta, \quad \eta \equiv |p_z|. \tag{13.1}$$

And if we plug an imaginary p_z into the expression for the de Broglie wave deflected by 90°, we will obtain

$$\Psi(\mathbf{r}) = \Psi_0 e^{-\kappa_z z} e^{ik_y y}, \quad \kappa_z = \frac{\eta}{\hbar}. \tag{13.2}$$

Here $\Psi(\mathbf{r})$ is a wave running parallel to the screen with amplitude exponentially decaying in the z-direction. This is quite a natural result for a wave that is deflected by 90° and therefore cannot reach the second screen to leave its mark there.

This kind of waves are well known in optics as *evanescent* waves [1,2] and, as a result of de Broglie relations, they made their way into QM to ensure, among other things, that the indeterminacy principle holds. If you calculate Δp_y either from the aperture function (the y-representation) or from its Fourier transform (p_y-representation), you will find no violation of the indeterminacy relation $\Delta p_y \Delta y \geq \hbar/2$ (Problems 13.2–13.4).

One important conclusion is that observable k_z in the incident beam is affected by the slit. For a *pure ensemble* of incident particles with definite but unknown k_z, the analysis of the resulting diffraction pattern will tell us the wavelength and hence the value of k_z, but only in the retrospect, since the state of each particle will be changed after passing through the slit, let alone hitting the screen.

13.2
The Heisenberg Inequality Revised

As we learned in Section 3.6, the indeterminacy principle not only explains stability of atoms, but also correctly predicts some of their parameters, such as the size and energy spectrum of a hydrogen atom. However, we want to caution the reader of the possible pitfalls in this venue. Our calculations for the electron ground state produced a numerically correct result because we had a "well-behaved" probability distribution – with only one "peak" and no "valleys." Generally, we have to be very careful when doing such estimations. When a particle is confined within a region of size a, the common rule of thumb is to estimate indeterminacy in Δp as the ratio $\hbar/a$, regardless of the details of particle's state within this region. This is generally wrong, and the actual indeterminacy in p may greatly exceed $\hbar/a$. Consider for illustration a particle in a potential box of width a. The normalized wave function of eigenstate $|n\rangle$ is $\psi_n(x) = \sqrt{\pi/a}\sin((\pi n/a)x)$. Since all probability distributions in these states are symmetric about the center of the box, $x = a/2$, we have $\langle p \rangle = 0$ and $\langle x \rangle = a/2$ so that $\langle \Delta p^2 \rangle = \langle p^2 \rangle$ and $\langle \Delta x^2 \rangle = \langle x^2 \rangle - (a/2)^2$ (Figure 9.1). Let us use this to find the actual uncertainties Δp, Δx, and finally $\Delta p \Delta x$:

$$\langle \Delta p^2 \rangle \equiv \langle \hat{p}^2 \rangle = \frac{\pi}{a} \int_0^a \sin\left(\frac{\pi n}{a}x\right)\left(-i\hbar\frac{\partial}{\partial x}\right)^2 \sin\left(\frac{\pi n}{a}x\right)dx = \frac{\pi^2 \hbar^2}{a^2}n^2, \quad (13.3)$$

$$\langle \Delta x^2 \rangle = \langle x^2 \rangle - \langle x \rangle^2 = \frac{2}{a}\int_0^a x^2 \sin^2\frac{\pi}{a}nx\,dx - \left(\frac{a}{2}\right)^2 = \frac{a^2}{12}\left[1 - \frac{6}{\pi^2 n^2}\right], \quad (13.4)$$

$$\Delta p \Delta x = \left(\frac{\pi \hbar}{a}n\right)\left(\frac{a}{2\sqrt{3}}\sqrt{1 - \frac{6}{\pi^2 n^2}}\right) = \frac{\pi \hbar}{2\sqrt{3}}\sqrt{n^2 - \frac{6}{\pi^2}}. \quad (13.5)$$

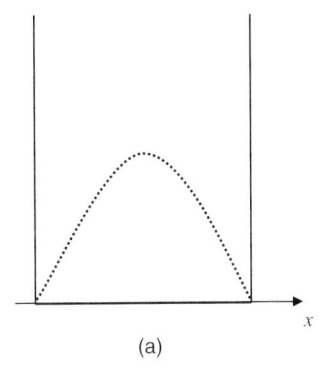

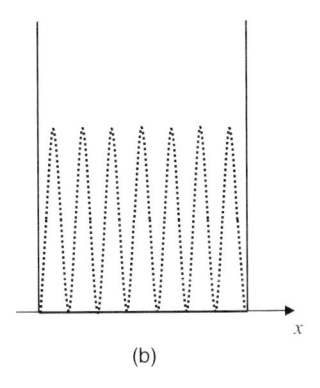

Figure 13.2 Two different stationary states of a particle in a box: (a) the ground state $E = E_1$; (b) the excited state $E = E_n$ (with $n = 7$). In both states, the position probability distribution spans (albeit not uniformly) the whole width of the box, but the corresponding momentum indeterminacy in (b) is about seven times greater than that in (a).

Now let us read these equations. Only for the lowest state $n = 1$ is the product of uncertainties close to $\hbar/2$. (Actually, even this product still slightly exceeds $\hbar/2$, because the Ψ function in (9.10) is not Gaussian.) For large n, the uncertainty in momentum, and accordingly, the product $\Delta p \Delta x$, increases unboundedly.

The second and more important feature is the emergence of "fine structure" for large n. It is only in the lowest state $n = 1$ that the probability distribution $\rho(x) = |\psi(x)|^2$ is monotone increasing in the left half of the box and monotone decreasing in its right half. As n grows, the distribution starts exhibiting a succession of peaks and valleys (nodes) within the box (Figure 13.2). We have $\rho(x) = 0$ at the very bottom (in the middle) of each valley and accordingly cannot find the particle there. By indicating the troughs of probability within the box, the excited state gives us more features of the particle's spatial distribution than the ground state. As n grows, spatial distribution becomes increasingly detailed. This gain in "x-information" comes at the expense of p-information! Each time we switch to a higher n, Δp increases (in proportion to n as $n \to \infty$).

If you want a less formal and a more "physical" explanation, consider this: in a succession of peaks and valleys, position indeterminacy is effectively reduced only by a factor of 2; if we add up the effective widths of the "hills" within which we *can* find the particle, we will get roughly $a/2$. Actual indeterminacy is even greater than $a/2$ because there still exists a small chance to find the particle in the vicinity of a trough (in fact, only a discrete set of points – the nodes – are excluded). By all means, this should not significantly change the indeterminacy of particle's location. But as it comes to estimating the corresponding p-indeterminacy, Nature counts differently. Nature tells us to disregard the width of the whole box and focus instead on the width of a single peak.

The effective width of one peak in the nth state is about $\Delta x_n \cong a/n$. Nature takes this width a/n rather than just a as a measure for the position indeterminacy, *when* we want to trade it for the corresponding momentum indeterminacy. Then the estimation for Δp is $\Delta p \geq \hbar/\Delta x_n = \hbar n/a$, in accordance with Equation 13.3. That is, instead of rigorously calculating Δp in (13.3), we could have estimated it simply by pretending that position uncertainty is only a/n rather than a. For highly excited states (large n), the product of uncertainties is thus much greater than Planck's constant!

One might argue that this is not exactly a fair bargain. After all, knowledge of the positions of the peaks does not tell us which peak the particle will choose to collapse in. This remains totally undetermined. The exclusion of n valleys (actually, just n points!) reduces the indeterminacy in x roughly only by half, and not by the factor of n. Why then should we pay more than the double price when it comes to the indeterminacy in p? But Nature does not give any discounts when the valleys become narrower! It keeps charging the same amount for each of the n valleys. As a result, the total price we must pay increases with n.

Box 13.1 The role of short wavelengths

Actually, this is a direct consequence of the basic postulates of QM. The moment we accepted the concept of matter waves and probabilistic nature of their amplitudes, we let the jinnee out of the bottle. If the particle's probability distribution has a fine structure (e.g., narrow peaks and valleys), and we describe it as a superposition of the de Broglie waves, we need in this superposition a range of short waves, which are absolutely necessary to shape out the fine details of the wave packet. And the finer the details, the shorter the waves needed to build them up. According to the mathematical properties of Fourier integrals, the contribution of the short waves becomes more and more dominating for the packets with sharp peaks, edges, or discontinuities (we encountered this on another occasion in Sections 11.3 and 11.5). According to QM, short waves mean big momenta! It should therefore come as no surprise that it is not so much the overall width of particle's spatial distribution, as its *fine structure*, that determines corresponding actual distribution in the momentum space. The reader will see a few other dramatic illustrations of this phenomenon in the problems at the end of this chapter.

13.3
The Indeterminacy of Angular Momentum

We have learned in Section 7.1 that a common eigenstate of $\hat{L}_z$ and $\hat{L}^2$ can be visualized by plotting a vector of length $L = \hbar\sqrt{l(l+1)}$ with the z-component equal to $m\hbar$, and then rotating it about the z-axis (Figure 7.1). We will get a conical surface. Classically, the angular momentum **L** would at each moment be represented by a single fixed generatrix of this surface. This generatrix would be specified by corresponding azimuthal angle φ, or, equivalently, by one of the components L_x or L_y. But in QM, neither of these two components has a definite value in a state with definite L_z. This means that the azimuthal angle φ is *not* determined either; that is, the actual state is represented geometrically by *the whole cone*. A state with definite L_z can be described as an equally weighted continuous entangled superposition of all states with the corresponding L_x and L_y.

We can now find the lowest limit of the indeterminacy in L_x and L_y. Applying the Robertson–Schrödinger relations (5.75) to our case yields

$$\Delta L_x \Delta L_y \geq \frac{1}{2}|\langle[\hat{L}_x,\hat{L}_y]\rangle| = \frac{1}{2}\hbar|\langle\hat{L}_z\rangle| = \frac{1}{2}\hbar|m\hbar| = \frac{|m|\hbar^2}{2}. \tag{13.6}$$

The right-hand side gives us only the lowest possible limit for the product of indeterminacies. On the other hand, we can find the *actual* amount of indeterminacy in L_x and L_y from the "conoidal" representation in Figure 7.1. Indeed, since $\langle L_x \rangle$ and $\langle L_y \rangle$ are both zero due to the axial symmetry of the system, we have

$$\Delta L_x \equiv \sqrt{\langle L_x^2\rangle - \langle L_x\rangle^2} = \sqrt{\langle L_x^2\rangle}, \qquad \Delta L_y = \sqrt{\langle L_y^2\rangle - \langle L_y\rangle^2} = \sqrt{\langle L_y^2\rangle}. \tag{13.7}$$

Averaging the operator equation $L_x^2 + L_y^2 = L^2 - L_z^2$ gives

$$\langle L_x^2\rangle + \langle L_y^2\rangle = \langle L^2\rangle - \langle L_z^2\rangle = \hbar^2[l(l+1) - m^2]. \tag{13.8}$$

The axial symmetry leads to $\langle L_x^2\rangle = \langle L_y^2\rangle$, so in view of (13.7), we finally obtain

$$\Delta L_x = \Delta L_y = \frac{\hbar}{\sqrt{2}}\sqrt{l(l+1) - m^2}, \qquad \Delta L_x \Delta L_y = \frac{1}{2}\hbar^2(l(l+1) - m^2). \tag{13.9}$$

We see that angular momentum presents a unique case with two incompatible observables having equal indeterminacy. The expression on the right is smallest when $|m|$ is maximal. In other words, at a given l, the maximal information about L_x and L_y is stored in the states with $m = \pm l$. The same is seen from Figure 7.3: the greater the $|m|$, the narrower the conical surface and the smaller its base representing the allowed values for L_x and L_y. Putting this maximal $|m| = l$ in (13.9) gives $(\Delta L_x \Delta L_y)_{\min} = (1/2)l\hbar^2$. This coincides with the right-hand side of Equation 13.6 for $m = \pm l$. The product of indeterminacies in this case reaches its smallest possible value allowed by Nature.

In the opposite case $m = 0$, the right-hand side of (13.6) becomes zero. One might naively expect that in this state Nature will admit definite values for L_x and L_y. But the uncertainty principle is an *inequality*, not the equation! In fact, according to (13.9), the product of indeterminacies in the $m = 0$ state is $(\Delta L_x \Delta L_y)_{\max} = (1/2)l(l+1)\hbar^2$.

Instead of shrinking to zero, it blows up to its maximum possible value for the given l.

Exercise 13.1

For eigenstates of L_z, compare Equation 13.6 giving the lowest possible value for $\Delta L_x \Delta L_y$ with Equation 13.9 giving its actual value. Find those m for which both give the same result.

Solution:
The requirement $[l(l+1) - m^2]\hbar^2 = |m|\hbar^2$ gives a quadratic equation for $|m|$, with solutions

$$|m| = \frac{1}{2}(-1 \pm (2l+1)). \tag{13.10}$$

Leaving the positive solution for $|m|$, we recover the already familiar result that the product $\Delta L_x \Delta L_z$ is minimized when $|m|$ is maximum: $|m| = l$.

Another indeterminacy associated with angular momentum is $\Delta L_z \Delta \varphi \geq (1/2)\hbar$ (Equation 3.40), and it is also far from trivial. For instance, in eigenstates with $L_z = m\hbar$ we have $\Delta L_z = 0$ and (3.40) requires in this case $\Delta \varphi \to \infty$. But on the other hand, φ is a periodic variable, that is, all its values beyond the range $0 \leq \varphi \leq 2\pi$ are redundant, and therefore the physically reasonable value for $\Delta \varphi$ must be close to π. This apparent contradiction will be addressed in the next section.

13.4
The Robertson–Schrödinger Relation Revised

We want now to see how the Robertson–Schrödinger relation applies in the most general case, not based on any assumptions about the distribution averages. To this end, we will revise its derivation in Section 5.7, now keeping in mind the possibility of periodic boundary conditions.

Consider two noncommuting Hermitian operators $\hat{A}$ and $\hat{B}$, which in a state Ψ have expectation values $\langle A \rangle$ and $\langle B \rangle$. Following Section 5.6, introduce two related (and likewise Hermitian!) operators $\hat{\delta}_A \equiv \hat{A} - \langle A \rangle$ and $\hat{\delta}_B \equiv \hat{B} - \langle B \rangle$, which represent deviations of A and B from their averages. Due to the hermiticity of these "deviation operators," we can write for the variances

$$\langle \Delta A^2 \rangle = \langle \Psi | \hat{\delta}_A^2 \Psi \rangle = \langle \Psi | \hat{\delta}_A \hat{\delta}_A \Psi \rangle = \langle \hat{\delta}_A \Psi | \hat{\delta}_A \Psi \rangle,$$
$$\langle \Delta B^2 \rangle = \langle \Psi | \hat{\delta}_B^2 \Psi \rangle = \langle \Psi | \hat{\delta}_B \hat{\delta}_B \Psi \rangle = \langle \hat{\delta}_B \Psi | \hat{\delta}_B \Psi \rangle. \tag{13.11}$$

The expressions appearing in the kets can be considered as the new derived functions $\hat{\delta}_A \Psi \equiv \Phi_A$ and $\hat{\delta}_B \Psi \equiv \Phi_B$. With these notations, we can now write

13.4 The Robertson–Schrödinger Relation Revised

the variances as the squares of corresponding vectors (Φ_A or Φ_B, respectively) in $\mathcal{H}$:

$$\langle \Delta A^2 \rangle = \langle \Phi_A | \Phi_A \rangle = |\Phi_A|^2, \qquad \langle \Delta B^2 \rangle = \langle \Phi_B | \Phi_B \rangle = |\Phi_B|^2. \tag{13.12}$$

And from this, we can write the Cauchy–Schwarz inequality – the result (6.68):

$$\langle \Delta A^2 \rangle \langle \Delta B^2 \rangle \geq |\langle \Phi_A | \Phi_B \rangle|^2. \tag{13.13}$$

We now again use the hermiticity of $\hat{\delta}_A$ and $\hat{\delta}_B$ to rewrite the inner product as $\langle \Phi_A | \Phi_B \rangle = \langle \hat{\delta}_A \Psi | \hat{\delta}_B \Psi \rangle = \langle \Psi | \hat{\delta}_A \hat{\delta}_B \Psi \rangle$, and (13.13) takes the form

$$\langle \Delta A^2 \rangle \cdot \langle \Delta B^2 \rangle \geq \left| \langle \Psi | \hat{\delta}_A \hat{\delta}_B \Psi \rangle \right|^2. \tag{13.14}$$

At this stage, relation (13.14) presents the most accurate and general statement of the indeterminacy principle. It explicitly shows, among other things, that the lowest possible limit for the product of the uncertainties depends not only on the kind of observables in question, but also on *physical state* of the system (state function Ψ). Therefore, it is not necessarily a fixed quantity (as $\hbar/2$ in Heisenberg's inequality), but rather a *variable* depending on Ψ. This variable, as we have seen in (13.6) and will see again below, can take on any value *including* zero!

Unfortunately, this important point has been lost in the *final form* (5.75) of the Robertson–Schrödinger inequality. To see how it happened, consider again the operators introduced in Section 5.6:

$$\hat{G} \equiv \frac{1}{2}\{\hat{\delta}_A, \hat{\delta}_B\}, \qquad \hat{C} \equiv \frac{1}{i}[\hat{\delta}_A, \hat{\delta}_B] = \frac{1}{i}[\hat{A}, \hat{B}]. \tag{13.15}$$

We have $\hat{\delta}_A \hat{\delta}_B = \hat{G} + i\hat{C}/2$. Putting this into (13.14) and noticing that due to hermiticity of *all* the operators involved,[2] their averages are real numbers, we finally get

$$\langle \Delta A^2 \rangle \cdot \langle \Delta B^2 \rangle \geq \left| \left\langle \Psi \left| \left(\hat{G} + \frac{i}{2} \hat{C} \right) \Psi \right\rangle \right|^2 = \left| \langle G \rangle + \frac{i}{2} \langle C \rangle \right|^2 = \langle G \rangle^2 + \frac{1}{4} \langle C \rangle^2 \geq \frac{1}{4} \langle C \rangle^2, \tag{13.16}$$

$$\Delta A \Delta B \geq \frac{1}{2} |\langle C \rangle|. \tag{13.17}$$

Note that in this last form the generality of (13.14) has been lost, first, by dropping the term $\langle G \rangle^2$, and second (and more importantly), by claiming that the averages $\langle G \rangle$ and $\langle C \rangle$ are real numbers. As we look at the right-hand side of (13.16), we see that it can never be zero, except for an exotic special case when $\langle C \rangle$ and $\langle G \rangle$ are *both* zero. We do not even know whether such a case can be realized in a physical state (Problem 13.9). The possibility of having zero on the right has thus been lost!

---

2) As a general rule, caution must be exercised when claiming hermiticity of some *combination* of Hermitian operators. In the current example, it is straightforward to see that $\hat{G}$ and $\hat{C}$ are both Hermitian operators. That statement would not be true if we had skipped the coefficient $1/i$ in the definition of $\hat{C}$.

13 Indeterminacy Revisited

To appreciate the importance of this change, take a pair of operators in spherical coordinates,

$$\hat{A} = \phi, \qquad \hat{B} = \hat{L}_z = -i\hbar \frac{\partial}{\partial \phi}. \tag{13.18}$$

Consider first the general form (13.14). It claims that $\langle \Delta \varphi^2 \rangle \cdot \langle \Delta L_z^2 \rangle \geq \left| \langle \Psi | \hat{\delta}_A \hat{\delta}_B \Psi \rangle \right|^2$. Is this claim true, for example, when Ψ is an eigenstate of $\hat{L}_z$? Since $\Delta \varphi$ is finite and ΔL_z is zero, the left side is zero and (13.14), if correct, *must* give us zero on the right. This is indeed the case: $\hat{\delta}_B \Psi = (\hat{L}_z - \langle \hat{L}_z \rangle) \Psi = (\hat{L}_z - m\hbar) \Psi = 0$ and the right side becomes zero as it should.

Consider now the *truncated* form (13.17) of the Robertson–Schrödinger inequality. This time the claim is that $\Delta \varphi \Delta L_z \geq (1/2)|\langle C \rangle|$. The left side is zero for the same reason as before. But the right side, as the reader can check, is $\hbar/2 \neq 0$. The inequality fails!

Obviously, some subtle error has been made in deriving (13.17). Where did the error originate? The answer is that for a *periodic variable* like φ, the assumption of hermiticity of operator $\hat{L}_z$ "for all functions involved," which we used in deriving (13.17), breaks down. This operator remains Hermitian in a *subspace* of the Hilbert space comprised of its eigenfunctions $\Psi_m(\varphi)$ and their superpositions

$$\Phi(\phi) = \sum_{m=-l}^{l} c_m \Psi_m(\phi). \tag{13.19}$$

All these functions are periodic, because all Ψ_m are exponents $\exp(im\varphi)$ with integer m. But, incredible as it may seem, the same operator is *not* Hermitian in a broader function space containing the derived functions such as

$$\tilde{\Phi}(\varphi) \equiv \varphi \Phi(\varphi). \tag{13.20}$$

The reason for this is that the function $\tilde{\Phi}(\varphi)$ is *not* periodic. As a result, Equation 13.17 does not follow from (13.14) in this case. The derivation ((13.14)–(13.17)) breaks down at the last stage because, as we have mentioned above, at least one of the averages $\langle G \rangle$ and $\langle C \rangle$ is not a real number.

One might argue that the function of the type (13.20) is (or can be made) periodic if we recall that mathematically the allowed range for variable φ is the same as for x, that is, $-\infty < \varphi < \infty$. The only difference is that, unlike x, any two values of φ separated by a multiple of 2π correspond to the same spatial orientation of the system; that is, $\phi = 2\pi$ describes the same situation as $\phi = 0$. To make the essence of this argument maximally clear, consider the most simple case of a system in an eigenstate $m = 0$, that is, $\Phi(\varphi) = 1/\sqrt{2\pi}$. Then $\tilde{\Phi}(\varphi) \sim \varphi$, and following the above argument that φ and $\varphi + 2\pi$ actually denote the same thing, we can make it periodic by reiterating it in each new cycle of φ (Figure 13.3).

This argument does not work. It is clearly seen from Figure 13.3 that the function $\tilde{\Phi}(\varphi)$, artificially broken into reiterating identical pieces, is not continuous and

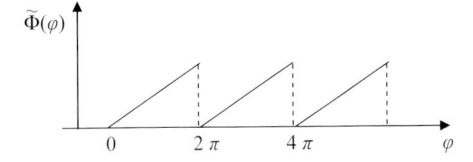

Figure 13.3 Periodic variable.

accordingly does not correspond to a possible physical situation.[3] It does not belong to a set of "decent" residents of the $\mathcal{H}$-space. In mathematical terms, the sum on the right in Equation 13.19 will converge to $\tilde{\Phi}(\varphi)$ only for infinite number of integers m. This would require an infinite angular momentum, which is obviously not a physical situation.

As a result, operator $\hat{B}$ as defined in (13.18), while being Hermitian in the family of functions $\Psi(\varphi)$ or $\Phi(\varphi)$, is *not* Hermitian in family of functions $\tilde{\Phi}(\varphi)$. Indeed,

$$\int_0^{2\pi} \tilde{\Phi}^* \hat{B} \tilde{\Phi} \, d\varphi = -i\hbar \int_0^{2\pi} \varphi \Phi^* \frac{\partial \Phi}{\partial \varphi} \, d\varphi = \cdots,$$

and using the fact that $\hat{B}^* = -\hat{B}$, the last integral can be transformed as

$$\cdots = -i\hbar \left[\int_0^{2\pi} \frac{\partial}{\partial \varphi}(\varphi \Phi^* \Phi) d\varphi - \int_0^{2\pi} \Phi \frac{\partial}{\partial \varphi}(\varphi \Phi^*) d\varphi \right] = -i\hbar(\varphi \Phi^* \Phi)|_0^{2\pi} + \int_0^{2\pi} (\hat{B}\tilde{\Phi})^* \Phi \, d\varphi.$$

Now, if $\tilde{\Phi} \equiv \varphi \Phi$ were periodic (as is Φ), then the first term on the right would be zero, and we would have

$$\int_0^{2\pi} \tilde{\Phi}^* \hat{B} \tilde{\Phi} \, d\varphi = \int_0^{2\pi} (B\tilde{\Phi})^* \tilde{\Phi} \, d\varphi, \qquad (13.21)$$

that is, operator $\hat{B}$ would be Hermitian. But this function is not periodic; therefore,

$$\int_0^{2\pi} \tilde{\Phi}^* \hat{B} \tilde{\Phi} \, d\varphi = -i\hbar 2\pi Q + \int_0^{2\pi} (B\tilde{\Phi})^* \tilde{\Phi} \, d\varphi, \quad Q \equiv |\Phi(0)|^2, \qquad (13.22)$$

and we see that $\hat{B}$ is *not* Hermitian in the space of functions including $\tilde{\Phi}(\phi)$ [4–6].

In view of this result, we must recast the Robertson–Schrödinger inequality into a form different from (13.16) and (13.17), using manipulations that are not based on any hermiticity assumptions for the operators involved. Writing the product $\hat{\delta}_A \hat{\delta}_B$ explicitly in terms of $\hat{A}$ and $\hat{B}$ and putting the result into (13.14), we get instead of (13.16) a more general inequality, which can be considered as a natural

---

3) We know only one allowed discontinuous function – the δ-function, considered in Chapter 11.

generalization of the familiar identity for the variance:

$$\Delta A \Delta B \geq |\langle \Psi|(\hat{A} - \langle \hat{A} \rangle)(\hat{B} - \langle \hat{B} \rangle)|\Psi \rangle| = |\langle AB \rangle - \langle A \rangle \langle B \rangle|. \tag{13.23}$$

It reduces to familiar $\langle \Delta A^2 \rangle \equiv \langle A^2 \rangle - \langle A \rangle^2$ when $\hat{A} = \hat{B}$. Note that while the averages $\langle A \rangle$ and $\langle B \rangle$ remain real, the average $\langle AB \rangle$ here is not necessarily real, because the expression for this average contains a nonperiodic function $\tilde{\Phi} = \varphi \Psi^*$. It is interesting, then, to evaluate the real and imaginary parts of the average product for the operators defined in (13.18). We have

$$2\operatorname{Re}(\langle AB \rangle) = \langle AB \rangle + \langle AB^* \rangle = -i\hbar \int_0^{2\pi} \phi \left(\Psi^* \frac{\partial \Psi}{\partial \phi} - \Psi \frac{\partial \Psi^*}{\partial \phi} \right) d\phi = -2\hbar \int_0^{2\pi} \phi \operatorname{Im} \left(\Psi \frac{\partial \Psi^*}{\partial \phi} \right) d\phi,$$

$$2i\operatorname{Im}(\langle AB \rangle) = \langle AB \rangle - \langle AB^* \rangle = -i\hbar \int_0^{2\pi} \phi \left(\Psi^* \frac{\partial \Psi}{\partial \phi} + \Psi \frac{\partial \Psi^*}{\partial \phi} \right) d\phi = -i\hbar \int_0^{2\pi} \phi \frac{\partial}{\partial \phi} (\Psi^* \Psi) d\phi.$$

The last integral can be easily taken by parts, yielding

$$\operatorname{Im}(\langle AB \rangle) = -\frac{1}{2}\hbar(2\pi Q - 1). \tag{13.24}$$

Therefore,

$$\langle AB \rangle = -\hbar \left[\int_0^{2\pi} \phi \operatorname{Im} \left(\Psi \frac{\partial \Psi^*}{\partial \phi} \right) d\phi + \frac{1}{2}i(2\pi Q - 1) \right]. \tag{13.25}$$

Putting this into (13.23), we finally get

$$\Delta A \Delta B \geq \left| -\hbar \left[\int_0^{2\pi} \varphi \operatorname{Im} \left(\Psi \frac{\partial \Psi^*}{\partial \varphi} \right) d\varphi + \frac{1}{2}i(2\pi Q - 1) \right] - \langle A \rangle \langle B \rangle \right|. \tag{13.26}$$

It is now easy to show that, for instance, if Ψ is an eigenstate $\Psi = \Psi_m$, then the right-hand side of (13.26) shrinks down to zero, as it should (Problem 13.5.2). In this case, the generalized Robertson–Schrödinger relationship consistently describes the already known situation, when the product of uncertainties can be equal to zero, but this only happens beyond the strictly defined conditions of hermiticity.

13.5
The N–φ Indeterminacy

In a typical classical wave, all relevant quantities (wavelength λ, frequency f, amplitude A, and the initial phase ϕ) are fully determinate. In particular, ϕ can always be measured precisely. That was the case in the original (classical) Young's double-slit experiment where one could in principle determine the phases ϕ_1 and ϕ_2 of both partial waves emerging from their respective slits.

But when a *single photon* passes through a screen with two slits, we cannot have a direct measurement of even the phase difference $\Delta\phi = \phi_2 - \phi_1$, let alone ϕ_1 and ϕ_2. We can only *calculate* $\Delta\phi$ as the difference between the optical paths provided that we know the wavelength. We could also measure it from the *statistics* of landing points (i.e., local intensities in the interference pattern) if we use expression (1.18) for the corresponding state under the typical conditions (incident plane monochromatic wave propagating along the optical axis, large distance between the two screens, etc.):

$$\Psi(r,t) = A\left(e^{i(kr_1 - \omega t)} + e^{i(kr_2 - \omega t)}\right). \tag{13.27}$$

(Here we ignore the radial dependence of waves diverging from the respective slits.) In this case, the probability distribution on the second screen is

$$I(\mathbf{r}) = A^2(1 + \cos\Delta\phi), \quad \Delta\phi = 2\pi\frac{d\sin\theta}{\lambda}. \tag{13.28}$$

But this expression depends only on $\Delta\phi$, not on ϕ_1 and ϕ_2 separately. Only the phase difference figures in the experimentally measurable picture. The *absolute* phase remains arbitrary. This agrees with the fact that quantum state functions are defined up to an exponential factor $e^{i\phi}$ with arbitrary ϕ.

But even the phase difference $\Delta\phi$ cannot be determined from this experiment if performed with a *single* photon! Instead, we need a pure ensemble of $N \gg 1$ photons. Similar reasoning will apply if we attempt to determine the phase from a direct measurement of the oscillating quantity. In a classical EM wave, one could accurately measure **E** (or **B**) and thereby know the phase. But this plan fails in the case of a quantum wave of one photon. And the difference, again, is in the number N. We see that the classical limit of QM (a monochromatic EM wave) can be described as the flux of a huge number ($N \gg 1$) of photons all in one state.

So what is lost when we take a *coherent* (all photons are in the same state!) beam and start removing photons from the beam until only a single photon is left? To see the answer, let us first ask: What do we *gain* from this procedure? Careful thought leads to the answer: We gain information about the number of particles! When we were boasting about knowing *everything* about the classical wave, we forgot to mention one "trifle": How many photons does it contain? This omission may appear totally inconsequential in the classical picture, and even nonsensical: after all, a classical wave was not supposed to be comprised of corpuscular objects to begin with! But once we quantized that wave, the question of how many quanta it contains becomes relevant. But the answer is not trivial. In a wave that admits classical approximation, this number must be infinite, $N \to \infty$. Under this condition, the addition or removal of one quantum will not change any measurable characteristics, so amplification or attenuation of the beam can be considered as a continuous rather than discrete process.

The reader of this book understands that infinity is an idealization of anything very big, so we can say in more practical terms that the number N should be much greater than the increments ΔN in which it changes. We can even have $\Delta N \gg 1$ as long as $N \gg \Delta N$.

But this immediately opens up another perspective of the problem: we can describe the classical wave in terms of quantum indeterminacy. Namely, in a classical wave anything relevant to waves, phase included, is totally defined.

However, the number of quanta (or just number of particles) is totally undefined because any classical measurement will determine this number up to $\Delta N \gg 1$. We had not noticed this feature before the onset of QM because property N, as just mentioned, seemed totally irrelevant to waves. The phase ϕ was determinate and the wave behaved classically. Thus, with λ, f, A, and ϕ sharply defined, we have complete information about the system in its wave aspect, but no information about its corpuscular aspect as we can't tell how many particles are there. When we switch to one particle only, we have $\Delta N \to 0$, and we gain maximal information about N. But with this, we lose knowledge of the classical wave aspect – the value of ϕ.

This answers our original question. Instead of exact phase ϕ with $\Delta \phi = 0$, we now have an arbitrary phase, $\Delta \phi \to \infty$. We cannot have both $\Delta N \ll 1$ and $\Delta \phi \ll 1$ at once. There is a lower boundary on the product of these two indeterminacies. The wave–particle duality, in addition to introducing restrictions such as $\Delta p \Delta x \geq (1/2)\hbar$, $\Delta E \Delta t \geq (1/2)\hbar$, $\Delta L_x \Delta L_y \geq (1/2)\hbar L_z$, and so on for a single particle, also requires that

$$\Delta N \Delta \phi \geq 2\pi \qquad (13.29)$$

must hold for a pure ensemble of particles. The indeterminacy ΔN here is defined as the number of particles such that its addition to or subtraction from the existing number N of such particles in the system does not produce any noticeable change in its properties.

13.6
Dispersed Indeterminacy

Here we consider an unusual case. We have seen before that indeterminacy allows us to localize a particle in its phase space no better than within a cell of minimum possible area $(1/2)\hbar$. The phase space is quantized (granulated). The remarkable thing here is that indeterminacy restricts only the minimal size of granules, but not their shape. Its leniency in this respect goes as far as to allow a single granule to be stretched into a line (recall the extremely squeezed states in Section 3.5). Here we look at even more interesting situations when a single granule is split into separate smaller patches (Figure 13.4). We can say that the corresponding state is represented by a *dispersed* granule.

On the face of it, this appears to contradict the principal claim of the indeterminacy principle. After all, if a granule can be separated into a number of splinters, then their sizes can be made arbitrarily small! The subtle point here is that the daughter granules are not independent entities. Even when separated as in Figure 13.4, they remain parts of a single (albeit dispersed) granule. Their combined area is still no less than $(1/2)\hbar$. In other words, neither of them can exist alone without all the rest – they form *one irreducible set*. Even more importantly, all of them together represent the state of a *single* particle. These states display even more bizarre departure from classical physics than the fuzziness of dynamic characteristics of a particle that we considered up to now.

As an illustration, consider an extreme case when an "indivisible" cell of area $(1/2)\hbar$ is dispersed into a periodic array of point-like dots over the whole phase

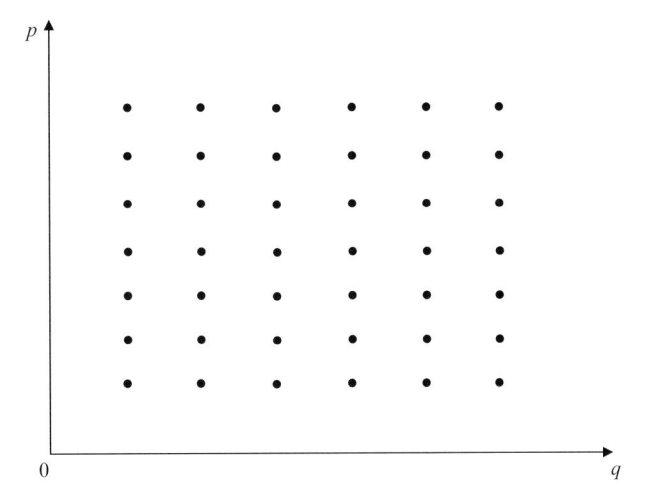

Figure 13.4 Dispersed indeterminacy in the phase space. The whole set of dots here belongs to one state. The minimal net area of all dots exceeds $(1/2)\hbar$ even if the area of each dot is vanishingly small. Even when each dot shrinks to a point, indeterminacy Δq becomes arbitrarily large, $\Delta q \to \infty$ and the product $\Delta p \Delta q \to \infty$ if the number of dots $N \to \infty$.

space as shown in Figure 13.4. To be specific, suppose q and p in this figure stand for position and momentum, respectively. As we see, the position and momentum of each dot are known exactly! And yet the overall indeterminacy is this case is maximal! We have here $\Delta p \Delta x \to \infty$ since we do not (and cannot!) know which dot the particle is in (more accurately, which dot represents the particle's coordinates p and x). The particle itself does not know it either: it is in a superposition of all possibilities – residing in all dots at once. This indeterminacy lasts until an appropriate measurement will change the state so that all dots will instantly merge together into a single granule.

A state described by such a set of dots seems to have a purely academic interest, but in fact we can find it in a textbook example: diffraction through a grating. This is a version of diffraction with wavefront splitting [1] – actually, just the extension of the double-slit experiment for a large number of slits. Let the light intensity be low so that we are dealing with one incident photon at a time. Provided there are no attempts to watch the photon in the process, it takes all available virtual paths and then interferes with itself. An idealized model features $N \to \infty$ infinitely narrow slits with two neighbors separated by a gap d. The photon's state function as it passes through the grating is

$$\psi(x) = \sum_n \delta(x - nd), \quad |n| = 0, 1, 2, \ldots. \tag{13.30}$$

Each slit ($x_n = nd$) is represented by the respective column of dots indexed by n in the phase diagram in Figure 13.4.

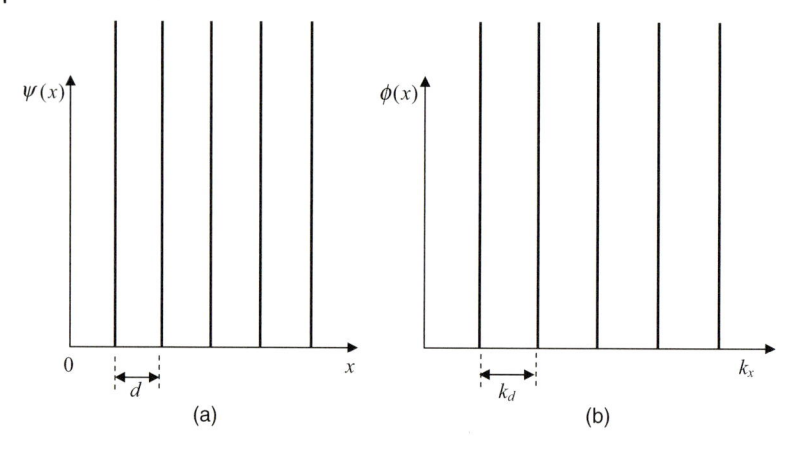

Figure 13.5 The aperture function of a grating (the comb function) in a multiple-interference experiment: (a) in configuration space; (b) in momentum space.

Following optics, we call part(s) of the wavefront passing through the slit(s) the "aperture function." The aperture function (13.30) is represented graphically in Figure 13.5a. Its Fourier transform is

$$\phi(k_x) = \int \psi(x) e^{-ik_x x} \, dx = \sum_n e^{-ik_x nd}. \tag{13.31}$$

There exists an identity [7]

$$\sum_n e^{-in\xi} = 2\pi \sum_n \delta(\xi + 2\pi n). \tag{13.32}$$

In our case $\xi = k_x d$, so that

$$\phi(k_x) = \phi_0 \sum_n \delta(k_x - k_d n), \quad k_d \equiv \frac{2\pi}{d}. \tag{13.33}$$

Thus, all possible k_x form a discrete set of equidistant values $k_x^{(n)} = nk_d$. Each value corresponding to a particular n is represented by nth row of dots in the phase diagram.

The series of δ-functions on the right in Equation 13.32 is known as the *Dirac comb function*, a term made self-intuitive by its graphical representation in Figure 13.5a and b (it is also known as *Shah function*, *sampling symbol*, or *replicating symbol* [7,8]). We see that a comb function in the configuration space (Figure 13.5a) converts into a comb function in the momentum space (Figure 13.5b), and vice versa. We read sometimes that the Gaussian function is the only one whose Fourier transform is also Gaussian. Figure 13.5 shows another quantum entity with two different but similarly looking "faces."

Each slit $x_n = nd$ takes its part in the formation of *all* waves with $k_x^{(n)} = k_d n$, and conversely, every such wave is generated by the entire succession of slits. Using

(13.33), we can write the diffracted wave immediately after the grating ($z = 0$) as

$$\psi(x) = \int \phi(k_x) e^{ik_x x} \, dk_x = \phi_0 \sum_n e^{ik_d nx}. \tag{13.34}$$

The complete expression for this wave in the entire semispace $z \geq 0$ above the grating is

$$\Psi(\mathbf{r}, t) = \phi_0 \sum_n e^{i\left(k_x^{(n)} x + k_z^{(n)} z\right)} e^{-i\omega t}. \tag{13.35}$$

Here

$$k_z^{(n)} = \pm \sqrt{\frac{2\mu E}{\hbar^2} - k_d^2 n^2} \quad \text{or} \quad k_z^{(n)} = \pm \sqrt{\frac{\omega^2}{c^2} - k_d^2 n^2} \tag{13.36}$$

for a nonrelativistic particle or the photon, respectively. (We assume that the light is incident from below on a horizontal grating.) In both cases, we must select the "+" sign in order for our solution to describe waves *receding* from the grating along the positive z. Dropping in (13.35) the temporal factor and setting $z = 0$, we recover (13.34).

Box 13.2 Evanescent de Broglie waves

An attentive reader will notice a pitfall in the above argument similar to that discussed in Section 13.1 (Equations 13.1–13.2). Here we will discuss a few other aspects of the same effect and in more detail. The existence of arbitrarily high n in (13.33) implies the possibility of arbitrarily high transverse momenta p_x (i.e., momenta along the screen) in a diffracted state. If a measurement gives a certain definite $p_x^{(n)}$ for a diffracted particle, then due to conservation of momentum of the whole system, the grating must acquire the equal and opposite momentum $-p_x^{(n)}$. This implies that before the measurement, the particle and the grating must be in an entangled superposition of the type considered in Section 3.7. So, we can include the quantum states of the grating for a more accurate description of the process. To this end, we introduce the corresponding momentum states of the grating, denoting them just as $|g(n)\rangle$. The states of the particle in the same notations can be written as $|p(n)\rangle$. Then the entangled state of the system (particle + grating) will be

$$|\Psi(p, g)\rangle = \sum_n c_n |p(n)\rangle |g(-n)\rangle. \tag{13.37}$$

According to (13.36), the increase of k_x along the grating can come only at the cost of k_z. The energy increase due to k_x will be compensated for by decrease of energy associated with k_z, while the particle's net energy E remains constant. At sufficiently high n, the square root in (13.36) becomes imaginary, $k_z^{(n)} \to i\kappa_z^{(n)}$. This converts the corresponding plane wave into an evanescent wave similar to that known in optics [1,2].

$$\psi_n(\mathbf{r}) = \phi_0 \, e^{i\left(k_d nx + k_z^{(n)} z\right)} \Rightarrow \phi_0 \, e^{-\kappa_z^{(n)} z} \, e^{ik_d nx}. \tag{13.38}$$

In this domain ($k_d n > \sqrt{2\mu E}/\hbar, \omega/c$), Equation 13.36 gives

$$\kappa_z^{(n)} = \sqrt{k_d^2 n^2 - \frac{2\mu E}{\hbar^2}} \quad \text{or} \quad \kappa_z^{(n)} = \sqrt{k_d^2 n^2 - \frac{\omega^2}{c^2}}. \tag{13.39}$$

Instead of departing from the grating, the evanescent wave component of No. n propagates clinging to it, with a speed $u_n = \omega d/2\pi n$. The word "clinging" reflects the fact that the wave's amplitude is maximal at the grating ($z = 0$) and exponentially decreases with z.

But the exponential decrease of its amplitude along direction perpendicular to propagation makes the evanescent wave fundamentally different from the de Broglie wave of the same frequency (energy). It remains an eigenfunction of the momentum operator, but does not have a *real* vector eigenvalue (its z component is imaginary). Its wavelength and phase velocity may be much less than those in the initial incident wave. Actual momentum measurement can be performed only on a free particle. For a p_x-measurement, this corresponds to the wave that has "slid off" the screen. But any such wave restores its original phase velocity and wavelength [10].

Problems

13.1 Consider the aperture function $\psi(y) = \psi_0$ within the slit of width a and $\psi(y) = 0$ outside the slit in the plane of the first screen (Figure 13.1). Take this function as the state function of the particle in the plane of the screen. What can you say about the y-derivative of this function at the edges of the slit region?

13.2 For the function given in the previous problem, find the standard deviations Δy and Δp_y. What can you say about their product?
(*Hint*: When calculating Δp_y, pay special attention to the edges of the slit region.)

13.3 Suppose the screen in Section 13.1 is made of a superconducting material and is illuminated by a normal incident light beam. If polarization of photons is aligned with the slit, their electric field component must be zero at the edges of the slit. In this case, a more realistic expression for the electric component of the photon's wave function may be the one given by Equation 9.10. For this case, calculate Δy and Δp_y and find their product.

13.4 Do the same as in the previous problem, but now in p_y-representation.

13.5 Find the Fourier transform of nth energy eigenstate for a particle in a box.

13.6 Evaluate the product of indeterminacies for $\Delta L_x \Delta L_z \equiv Z(m)$ as a function of m using the Robertson–Schrödinger *complete* equation (5.74).

13.7 Prove (or disprove) that the averages of two operators $\hat{G}$ and $\hat{C}$ defined in (13.15) cannot both be zero.

13.8 Show that in an eigenstate, the right-hand side of (13.26) is zero.

13.9 Express the relationship (13.26) in terms of the probability amplitudes if operators $\hat{A}$ and $\hat{B}$ are defined as in (13.18), and the state is a superposition (13.19).

13.10 Prove that if the probability amplitudes in Problem 13.9 are real and symmetric ($c_m = c_m^* = c_{-m}$), so that $\bar{A} = \pi$ and $\bar{B} = 0$, then the product of uncertainties can be made arbitrarily small when $c_0 \to 1$.

13.11 Find the indeterminacy in φ for the normalized wave function $\psi_m(\varphi) = (1/\sqrt{2\pi})e^{im\varphi}$.

References

1 Hecht, E. (2002) *Optics*, 4th edn, Addison-Wesley, Reading MA, p. 385.
2 Born, M. and Wolf, E. (1999) *Principles of Optics*, 7th edn, Cambridge University Press, Cambridge.
3 Louisell, W. (1963) Amplitude and phase uncertainty relations. *Phys. Lett.*, **7**, 60.
4 Judge, D. (1964) On the uncertainty relation for angle variables. *Nuovo Cimento*, **31**, 332.
5 Carruthers, P. and Nieto, N. (1968) Phase and angle variables in quantum mechanics. *Rev. Mod. Phys.*, **40**, 411.
6 Roy, C.L. and Sannigrahi, A.B. (1979) Uncertainty relation between angular momentum and angle variable. *Am. J. Phys.*, **47**(11), 965–967.
7 Bracewell, R.N. (1999) The sampling or replicating symbol, in *The Fourier Transform and Its Applications*, 3rd edn, McGraw-Hill, New York, pp. 77–79, 85.
8 Córdoba, A. (1989) Dirac combs. *Lett. Math. Phys.*, **17** (3), 191–196.
9 von Neumann, J. (1927) Thermodynamik quantenmechanischer Grossen. *Gott. Nachr.*, 273–291 (in German).
10 Fayngold, M. (2008) *Special Relativity and How it Works*, Wiley-VCH, Weinheim.

14
Quantum Mechanics and Classical Mechanics

In this chapter, we study the relationship between quantum and classical aspects in our description of the world. We will see that this relationship is far from being trivial.

14.1
Relationship between Quantum and Classical Mechanics

As emphasized in Chapter 1, one immanent element of quantum physics is the Planck constant $\hbar$ representing the minimum possible (albeit not necessarily square-shaped) "pixel" of the phase space. In particular, it provides a criterion as to whether we should choose quantum or classical description. If physical properties of the system are such that $\hbar$ can be neglected, then classical description may be sufficiently accurate. Since $\hbar$ has the dimensionality of action, the best comparison to consider is the one between $\hbar$ and the actual action S. Accordingly, the initial form of the Planck constant $h = 2\pi\hbar$ is frequently referred to as the minimal possible amount of action per cycle in any periodic motion. If

$$S \gg \hbar, \tag{14.1}$$

the object is, in most cases, classical. Otherwise, we need QM for its description.

Why do we say "in most cases," rather than "always"? The answer is, from the mathematical viewpoint, condition (14.1) is necessary but not sufficient for a system to behave classically. When this condition is not met, system's behavior is definitely nonclassical. But when it is met, this does not by itself guarantee classicality: phenomena such as superfluidity and superconductivity occur in *macroscopic* systems characterized by huge classical action $S \gg \hbar$ and yet display purely quantum behavior.

But in this chapter, we deal with single particles only and the criterion (14.1) will be definitive.

Let us first consider some internal characteristics describing an object as such (e.g., its mass or size) rather than its motion as a whole. We will also focus on such characteristics as indeterminacy of object's position or momentum because they define the spreading of the corresponding wave packet (purely quantum behavior),

Quantum Mechanics and Quantum Information: A Guide through the Quantum World,
First Edition. Moses Fayngold and Vadim Fayngold.
© 2013 Wiley-VCH Verlag GmbH & Co. KGaA. Published 2013 by Wiley-VCH Verlag GmbH & Co. KGaA.

and try to find the borderline in these properties that separate them from the classical behavior.

As an illustration, use the expression (11.31) for the spreading of a Gaussian wave packet. With Δ standing for the initial indeterminacy, and $\Delta(t)$ for indeterminacy at a later moment t, we have

$$\Delta(t) = \Delta\sqrt{1 + \frac{1}{2}\frac{\hbar^2}{\mu^2 \Delta^4}t^2}. \tag{14.2}$$

In this expression, the property having the dimensionality of action is

$$S(t) \equiv \mu \Delta^2 / t. \tag{14.3}$$

Note that it is *inversely* proportional to time – a very unusual feature if we compare this with the "conventional" action describing classical motion along a trajectory.

Making the substitution, we can rewrite (14.2) as

$$\Delta(t) = \Delta\sqrt{1 + \frac{1}{2}\frac{\hbar^2}{S^2(t)}}. \tag{14.4}$$

The condition for classicality in terms of μ and Δ is

$$\mu \Delta^2 \gg \hbar t. \tag{14.5}$$

We see immediately that it is satisfied when the particle is sufficiently massive and its "size" (the initial width of the packet) is sufficiently large. Under these conditions, the "action term" $\hbar^2/2S^2(t)$ can be neglected and the size of the system remains constant – just what we expect from a classical system.

Let our system be the hydrogen atom. Suppose the atom was prepared in its ground state ($E = E_0$), but with indeterminacy of its position equal to the atomic size. In other words, $\Delta \approx 2a_0$, where a_0 is the Bohr radius. We emphasize that atomic position indeterminacy Δ has nothing to do with position indeterminacy of the atomic electron! The latter is an intrinsic characteristic of the electron's stationary state, and it defines the Bohr radius a_0. The former is related to the indeterminacy of kinetic energy ΔE of the atom as a whole. One can show (Problem 14.1) that energy indeterminacy ΔE associated with position indeterminacy $\Delta \approx 2a_0$ is much less than $|E_0|$, so it will not perturb the internal state of the atom. Accordingly, the initial numerical approximation of Δ in terms of a_0 should not cause any confusion. The Bohr radius remains constant because the internal structure of the atom remains constant, whereas Δ evolves with time, and its being numerically close to $2a_0$ at the beginning is merely the initial condition. In fact, using the table constants for $\hbar$, a_0, and mass μ of a hydrogen atom, we see that 1 s later, the size of the atomic wave packet will be $\Delta(t = 1\,\text{s}) \approx 5 \times 10^{12}\Delta \approx 500$ m. The probability cloud describing indeterminacy of the atom's position will expand by more than 12 orders of magnitude in 1 s. It is true that 1 s, being a very short time for us, is a huge time for an atom if we compare it with characteristic atomic time $\tau \sim 10^{-9}$ s (the typical lifetime of an excited state). But even within this much shorter timescale, we would have $\Delta(\tau) \approx 5 \times 10^3 \Delta \approx 5 \times 10^{-7}$ m. This may seem a very small value, but it is more than three orders of magnitude larger than the initial size, and this change occurs in just 10^{-9} s. This is QM behavior.

Let us now take something more massive: a speck of dust with mass $\mu = 1.7 \times 10^{-9}$ kg and size $a = 10^{-4}$ m. A particle of this size is barely visible by the naked eye. And let the initial position indeterminacy Δ of its center of mass be 10 times the diameter of a hydrogen atom, that is, $\Delta \approx 20\, a_0$. For a single hydrogen atom in its ground state, the position described by Δ would be very poorly determined, but for a bound ensemble of $\simeq 10^{18}$ of them in the speck, it is determined with sniper's accuracy. Now formula (14.2) gives us $\Delta(t = 1\,\text{s}) \approx (1 + 2.5 \times 10^{-13})\Delta$. This time there was practically no change. The initial indeterminacy increased by about $\delta\Delta \approx 2.5 \times 10^{-22}\Delta = 2.5 \times 10^{-22}$ m. It is safe to say that the size of a freely evolving speck remains essentially constant. Even if we increase the time from 1 to 10^{12} s (about 30 000 years), the change will still be barely noticeable. You can plug the mass of a coin or a rock into the formula (14.2), only to see that any reasonable initial indeterminacy will remain the same for all practical purposes even if we wait for the lifetime of the Universe. This is classical behavior.

Now we are prepared for the next question. The example with the atom obviously represents the QM, and the one with the speck of dust – the CM domain. Where is the borderline between the two? We can see that mass alone, or size alone, or even both at once are not enough to determine the borderline. If we take a look at the atom in the first example 10^{-14} s after its preparation, we will not see a noticeable change in Δ. On the other hand, if we take a look at the speck 1 million years from the initial moment (keeping it well isolated in between!), its position will already be fuzzy, with indeterminacy well above its size. It turns out that the atom's behavior is classical for the first $t < 10^{-14}$ s and the speck's behavior is quantum-mechanical after $t > 1$ million years: time is also an actor in the play unless we restrict it to "practically meaningful" values. Mathematical structure of the theory shows that the borderline between quantum and classical can itself be shifting with time: an isolated object with classical features at one moment may exhibit quantum features at another moment, and vice versa. Strictly speaking, we should take into account all three characteristics – mass, size, and time, which takes us back to the formal condition (14.1). But in practical terms, the numerical estimates we just made should teach us proper respect for the practicalities, in particular, for time- and spatial scales we live in. Not only are the times involved too long for macroscopic objects, but even more importantly we must make sure not to freeze the expected quantum evolution by watching! And "watching" here has a more general meaning than just a chance look of a curious conscious observer. It means *any* interactions of the wave packet with its environment, including random fluctuations that may cause *decoherence* and the collapse of the packet to its initial state. And the more massive the object, the harder it is to keep it truly isolated. That answers the question as to why we never see quantum evolution of objects even as small as a speck of dust.

14.2
QM and Optics

Our next question is about the *motion* of the center of mass of a packet in the quasi-classical approximation, rather than about the *size* of the packet. How does the criterion (14.1) apply in this case?

Start with the simplest possible case: one-dimensional de Broglie wave in free space. Note that the exponent in $\Psi(x,t) = \Psi_0\, e^{i(kx-\omega t)}$ is proportional to the classical action:

$$px - Et = S(x,t), \tag{14.6}$$

so $\Psi(x,t)$ can be written as

$$\Psi(x,t) = \Psi_0\, e^{iS(x,t)/\hbar}. \tag{14.7}$$

Now suppose that instead of free space the particle encounters a potential barrier. We could still approximate the motion by a single monochromatic wave as long as the particle energy is far above the top of the barrier. But we know that even in this case there may appear a reflected wave; moreover, such wave may appear even if we have a potential well instead of the barrier. In all such cases, the quantum state of a single particle is a superposition of incident and reflected waves. But classically, there must be no reflection, and the particle must always proceed in its original direction. This classical behavior must be ensured under condition (14.1). What does this condition require from the potential function $U(x)$? Let us apply it to the spatial part of action (14.6). Since we still approximate the actual state by $\Psi_0\, e^{i(kx-\omega t)}$ even in the region Δx where the potential $U(x)$ is changing, we have from Equation 14.1 for the action in this region:

$$S(x) \approx p\Delta x \gg \hbar \quad \text{or} \quad k\Delta x \gg 1, \quad \text{that is,} \quad \lambda \ll 2\pi\Delta x. \tag{14.8}$$

In other words, the particle's de Broglie wavelength should be much shorter than the characteristic distance over which the potential energy undergoes a noticeable change. Using $\Delta U \approx (dU/dx)\Delta x$, we can also rewrite (14.8) as

$$\frac{dU}{dx} \ll 2\pi \frac{\Delta U}{\lambda}. \tag{14.9}$$

The rate of change of the potential must be sufficiently low.

Since $|\Delta U| \leq |U|$, the last condition can be written in terms of the so-called logarithmic derivative:

$$\frac{|U'|}{|U|} = \frac{d}{dx}\ln|U(x)| \ll k \tag{14.10}$$

(here we assume U to be measured in units U_0 taken at some reference point x_0). For a rectangular potential barrier or well considered in Chapter 9, conditions (14.8)–(14.10) cannot be met no matter how high the kinetic energy of the particle is. There will always be a small but nonzero chance for the particle to be reflected from the obstacle, which classically is totally passable.

If the potential also changes with time, then similar reasoning applies for the temporal part of the action and leads to the condition $T \ll 2\pi \Delta t$. This means that period T of the wave has to be much shorter than the characteristic time of potential change Δt, or, for each fixed location,

$$\frac{d}{dt} \ln|U(x,t)| \ll \omega. \tag{14.11}$$

The rate of change of the potential should be much less than the frequency of the wave.

These conditions are familiar in the ray optics. We know that if the refractive index of a transparent medium is a sufficiently slow function of position (and/or time), such a medium will not be strongly reflective and the propagation of light can be described in terms of continuous bundles of rays, which is equivalent to the propagation of a wave packet with negligible spread and without splitting.

To summarize, if the particle's kinetic energy is much higher than its potential energy and the wavelength is much shorter than the characteristic range of potential change, the particle's motion is close to classical. The same holds when the frequency is much higher than the time rate of change of the potential. This is analogous to the transition from wave optics to geometric (ray) optics when the wavelength of light approaches zero.

14.3
The Quasi-Classical State Function

When we described the particle in a potential field by a monochromatic wave, that was a rather crude picture, since neither $\mathbf{k}$ nor ω have exact values in such environment. Even if potential depends only on coordinates but not on time, which enables stationary states with definite energy (and thereby definite ω), each of those states will be a superposition of de Broglie waves with various $\mathbf{k}$ within a certain range $\Delta \mathbf{k}$. Classically, the particle's momentum will be a function of position.

One asks, what would be the quasi-classical description of the corresponding motion? And how can we actually apply the criterion (14.1) to the system's state function?

The classical action for one particle in a stationary environment can be written as

$$S(\mathbf{r},t) = \int_0^{\mathbf{r}} \mathbf{p}(\mathbf{r}')d\mathbf{r}' - Et. \tag{14.12}$$

Then the answer to both questions is to seek the solution to the Schrödinger equation in the form

$$\Psi(\mathbf{r},t) = e^{iS(\mathbf{r},t)/\hbar}, \tag{14.13}$$

where $S(\mathbf{r},t)$, in the limit $\hbar \to 0$, would be the classical action (14.12). The normalizing factor is absorbed in $S(\mathbf{r},t)$ (recall that action is determined up to an arbitrary additive constant). The choice of the form (14.13) for the wave function can be considered as the natural extension of the de Broglie wave, which is a solution to the Schrödinger equation for the free space, with $S(\mathbf{r}, t) \to \mathbf{p} \cdot \mathbf{r} - \omega t$ being the classical action in this case.

Let us now find the QM expression for $S(\mathbf{r}, t)$. Putting (14.13) into the stationary Schrödinger equation $-\hbar^2 \nabla^2 \Psi = 2\mu(E - U)\Psi$, we obtain the equation for $S(\mathbf{r}, t)$:

$$-i\hbar \vec{\nabla}^2 S + (\vec{\nabla} S)^2 = 2\mu(E - U). \tag{14.14}$$

For a stationary state, we can consider the action in (14.14) with the temporal term dropped, that is, as a function of coordinates only, $S(\mathbf{r}, t) \to S(\mathbf{r})$. We can also restrict to one dimension without losing any important details. Then (14.14) simplifies to

$$-i\hbar S'' + S'^2 = 2\mu(E - U). \tag{14.15}$$

This is an exact equation, but is generally difficult to solve. Since we now study systems under condition (14.1), the idea is to look for the approximate solution as an expansion over the powers of $\hbar$:

$$S = S_0 - i\hbar S_1 + (-i\hbar)^2 S_2 + \cdots. \tag{14.16}$$

In the zeroth approximation, drop all terms with $\hbar$ in (14.15) and (14.16), to get equation for S_0:

$$S_0'^2 = 2\mu(E - U(x)). \tag{14.17}$$

This can be immediately solved:

$$S_0(x) = \pm \int p(x) dx, \quad p(x) = \sqrt{2\mu(E - U(x))}. \tag{14.18}$$

The integrand here is just the classical momentum of the particle – a quite natural result for an approximation that assumes a negligible $\hbar$.

As seen from (14.15), for the used approximation to be legitimate, there must hold

$$\left| \frac{S''}{S'^2} \right| \hbar \ll 1 \quad \text{or} \quad \left| \frac{d}{dx}\left(\frac{\hbar}{S'} \right) \right| \ll 1. \tag{14.19}$$

Since, according to (14.18) in this approximation, $S' = p$, the condition (14.19) can also be written in three equivalent forms:

$$\left| \frac{d\lambda}{dx} \right| \ll 2\pi, \quad \frac{\mu |f|}{p^3} \hbar \ll 1, \quad \text{or} \quad \left| \frac{d}{dx}\left(\frac{\hbar}{p(x)} \right) \right| \ll 1. \tag{14.20}$$

Here $p(x)$ is determined by (14.18), and $f = -U'$ is the classical force. The first form shows that the wavelength must change very slowly with distance. This is equivalent to the condition (14.10) obtained in the previous section in a quite different way. Both conditions can be regarded as a version of the indeterminacy principle: wavelength cannot be a definite function of x because momentum cannot be a definite function of x, and therefore (14.18) can make sense only if $U(x)$ (and thereby λ) changes sufficiently slowly with x.

The second form shows that the approximation fails for sufficiently small momenta, and particularly, in the vicinity of the turning points $x = x_j$, where $U(x_j) = E$. As we know, on one side of such a point we have $U > E$, so this region is classically inaccessible, and the classical particle would stop at $x = x_j$ for a moment before turning back. In quantum terms, the corresponding wavelength goes to infinity, $\lambda \to \infty$, instead of being small as required by (14.8).

Strictly speaking, the condition (14.19) is necessary but not sufficient for this approximation, since it is obtained by dropping the higher order derivative term in Equation 14.15. Therefore, there may be cases when the condition is met and yet the approximation fails [1], for instance, when $S(x)$ has a term close to but not exactly linear in x over a large range of x. In some cases, this can be remedied in the first approximation where we retain all terms of the first order in $\hbar$. In this approximation, we obtain

$$S_0' S_1' = -\frac{1}{2} S_0''. \tag{14.21}$$

It follows that

$$S_1'(x) = -\frac{S_0''(x)}{2 S_0'(x)} = -\frac{p'(x)}{2p(x)}. \tag{14.22}$$

Integration gives

$$S_1(x) = -\frac{1}{2} \ln p(x), \tag{14.23}$$

so that

$$S(x) \approx \pm \int p(x) dx + \frac{i\hbar}{2} \ln p(x). \tag{14.24}$$

In view of (14.13), the wave function in this approximation is

$$\Psi(x) \approx \frac{C_1}{\sqrt{p(x)}} e^{(i/\hbar) \int p(x) dx} + \frac{C_2}{\sqrt{p(x)}} e^{-(i/\hbar) \int p(x) dx}. \tag{14.25}$$

Accordingly, the probability of finding the particle in the vicinity of point x is

$$d\mathcal{P}(x) = |\Psi(x)|^2 dx \sim p^{-1}(x) dx. \tag{14.26}$$

This reflects a feature of the classical motion – the corresponding probability is proportional to the time dt it takes for a classical particle to travel the distance dx. This time is $dt = dx/v(x) \sim dx/p(x)$ – in accordance with (14.26). And again, we see that the quasi-classical expression (14.25) diverges at the turning points. Thus, the first approximation, while generally improving the accuracy, does not help in the vicinity of such points. Because (14.25) diverges, the next approximations cannot be rigorously applied at these points either, and therefore the corresponding regions should be considered separately. We will do that in the next section.

14.4
The WKB Approximation

We will now try to find an exact solution in the vicinity of the turning points. Let $x = a$ be one such point. Then the exact value $\Psi(a)$ can be considered as a boundary condition for matching the approximate values of $\Psi(x)$ found in quasi-classical approximation sufficiently far from a on the left and on the right of a. The corresponding procedure is known as the WKB (Wentzel–Kramers–Brillouin) approximation [2–4].

Suppose that $U(x) > E$ for all $x > a$ (Figure 14.1). In this region, the first term of (14.24) takes the form of an exponentially decaying function

$$\Psi(x) = \frac{A}{2\sqrt{|p(x)|}} e^{-(1/\hbar)\int_a^x |p(x')|dx'}. \tag{14.27}$$

The second term must be dropped because it increases exponentially with x, so (14.27) remains the sole actor in the play. We have to match this with expression (14.24) representing the state function to the left of a. To this end, one can use the *exact* solution of Equation 14.15 in the vicinity of a, where approximations (14.24) and (14.27) both fail.

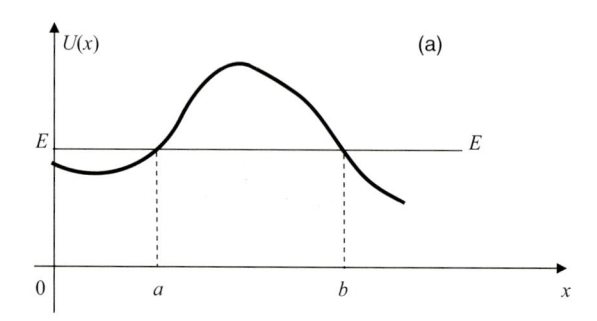

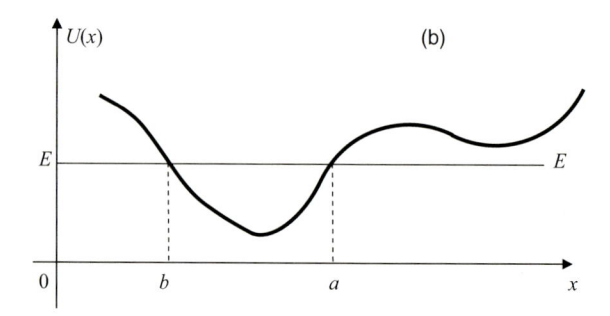

Figure 14.1 The potential function and the turning points $x = a, b$. The WKB approximation breaks down in the vicinity of these points. (a) Potential barrier; (b) potential well.

We will use here another option that does not require an exact solution [5]. Namely, we will consider $\Psi(x)$ as a function of the complex variable x and perform a transition from one side of a to another side along a path going *around a* and sufficiently far from it in the plane of complex x so that the condition (14.22) is satisfied at all points of this path [1]. This can be done successfully if the potential does not change too rapidly in the vicinity of a so that we can represent $E - U(x)$ and $p(x)$ there as

$$E - U(x) \approx f_a(x-a), \qquad p(x) = \sqrt{2\mu f_a(x-a)}. \qquad (14.28)$$

Here $f_a = -U'(a) < 0$ is the classical force at a. Then we have

$$\int_a^x p(x')dx' = \frac{2}{3}\sqrt{2\mu f_0}(x-a)^{3/2}, \qquad (14.29)$$

and the expression (14.27) to the right of a becomes

$$\Psi(x) \underset{x>a}{\approx} \frac{1}{2} A \big(2\mu|f_a|(x-a)\big)^{-1/4} e^{-(2/3\hbar)\sqrt{2\mu|f_a|}(x-a)^{3/2}}. \qquad (14.30)$$

The corresponding expression to the left of a becomes

$$\Psi(x) \underset{x<a}{\approx} \big(2\mu|f_a|(a-x)\big)^{-1/4} \left[A_1\, e^{i(2/3\hbar)\sqrt{2\mu|f_a|}(a-x)^{3/2}} + A_2\, e^{-i(1/3\hbar)\sqrt{2\mu|f_a|}(a-x)^{3/2}} \right]. \qquad (14.31)$$

Let us trace the change of this function as we go around point a counterclockwise along the semicircle of radius β in the upper semiplane of complex x. On this semicircle,

$$x - a = \beta e^{i\varphi}, \qquad 0 \le \varphi \le \pi. \qquad (14.32)$$

The radius β is such that at $x \ge a + \beta$ the approximation (14.30) is sufficiently accurate, and the same is true for the approximation (14.31) at $x \le a - \beta$. If we now put (14.32) into (14.30) and go around a from $x = a + \beta (\varphi = 0)$ to $x = a - \beta (\varphi = \pi)$ as indicated above, we get

$$\Psi(x) \underset{x<a}{\approx} \frac{1}{2} \frac{A e^{-i(\pi/4)}}{(2\mu|f_a|\beta)^{1/4}} e^{-(2i/3\hbar)\sqrt{2\mu|f_a|}\beta^{3/2}}. \qquad (14.33)$$

Now we are to the left of a where $\Psi(x)$ is approximated by (14.31), and comparison shows that (14.33) is identical to the second term of (14.31) with the coefficient

$$A_2 = \frac{1}{2} A e^{-i(\pi/4)}. \qquad (14.34)$$

The above procedure leaves the first coefficient in (14.31) undetermined (except for its absolute value). This has a simple explanation. If we reverse the procedure, going back from $x = a - \beta$ to $x = a + \beta$ along the same semicircle, then the first term acquires the exponentially decaying factor, which will make it negligible as compared with the second term. As a result, the first term will be "lost" by the end of this transition. We leave it to the reader to show (Problem 14.3) that the amplitude A_1 of this term can be found if we repeat the direct procedure, but now along the semicircle in the *lower* part of the complex plane x. In this way, we will find

$$A_1 = -\frac{1}{2} A e^{i(\pi/4)}. \tag{14.35}$$

Thus, to the exponentially decreasing function (14.27) on the right of the turning point there corresponds the trigonometric function on the left:

$$\Psi(x) = \begin{cases} \dfrac{A}{2\sqrt{|p(x)|}} e^{-(1/\hbar) \int_a^x |p(x')| dx'}, & E < U(x), \quad x \geq a + \beta, \\[1em] \dfrac{A}{\sqrt{p(x)}} \sin\left(\dfrac{1}{\hbar} \int_x^a p(x') dx' + \dfrac{\pi}{4}\right), & E > U(x), \quad x \leq a - \beta. \end{cases} \tag{14.36}$$

A similar argument works for the case when $U(x) > E$ to the left of the turning point $x = b$ and $U(x) < E$ to the right of that point. In this case, denoting as B the amplitude of the exponentially decreasing function to the left of b, we obtain

$$\Psi(x) = \begin{cases} \dfrac{B}{2\sqrt{|p(x)|}} e^{-(1/\hbar) \int_x^b |p(x')| dx'}, & E < U(x), \quad x \leq b - \gamma, \\[1em] \dfrac{A}{\sqrt{p(x)}} \sin\left(\dfrac{1}{\hbar} \int_b^x p(x') dx' + \dfrac{\pi}{4}\right), & E > U(x), \quad x \geq b + \gamma. \end{cases} \tag{14.37}$$

The parameter γ here plays the same role as β in (14.36): it is the radius of the circle around b in the complex x-plane such that quasi-classical approximation is sufficiently accurate outside this circle. Combining both results, we can calculate quasi-stationary states of the particle in a potential well where classical motion would be confined in the region between the two turning points. This problem will be the topic of the next section.

14.5
The Bohr–Sommerfeld Quantization Rules

The results (14.36) and (14.37) of the previous section allow us to describe the motion of a *bound* particle in the quasi-classical approximation. Suppose the particle is bound within region II between two classically inaccessible regions I and III (Figure 14.2). In the WKB approximation, motion in these regions is described by different analytical expressions. Denote them as Ψ_I, Ψ_II, and Ψ_III, respectively. Using the results (14.36) and (14.37), we can write the expressions for classically inaccessible regions as

$$\Psi_\text{I}(x) = \frac{A}{2\sqrt{|p(x)|}} e^{-(1/\hbar) \int_x^a |p(x')| dx'}, \quad E < U(x), \quad x \leq a - \alpha \tag{14.38}$$

14.5 The Bohr–Sommerfeld Quantization Rules

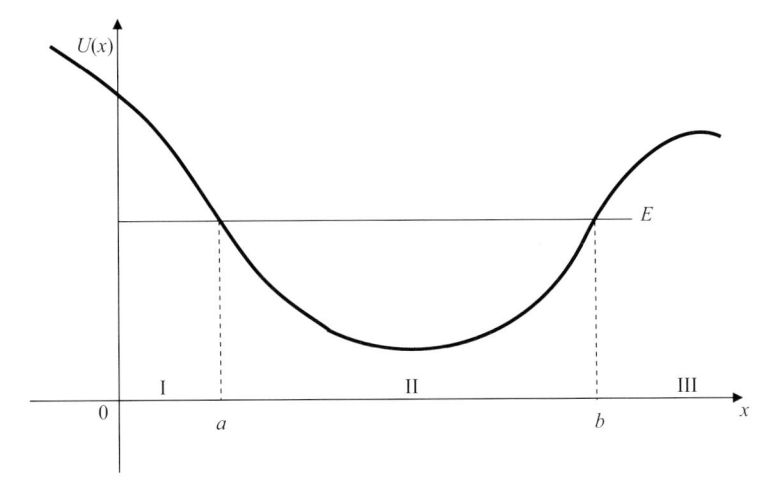

Figure 14.2 The potential function for a particle in a "valley" II between turning points $x = a, b$. The regions I ($x < a$) and III ($x > b$) are classically inaccessible.

and

$$\Psi_{III}(x) = \frac{B}{2\sqrt{|p(x)|}} e^{-(1/\hbar)\int_b^x |p(x')|dx'}, \quad E < U(x), \quad x \geq b + \beta. \tag{14.39}$$

As for the region $a < x < b$, we have two different expressions:

$$\Psi_{II}^A(x) = \frac{A}{\sqrt{p(x)}} \sin \Phi_A(x), \quad \Phi_A(x) \equiv \frac{1}{\hbar}\int_a^x p(x')dx' + \frac{\pi}{4}, \quad x \geq a + \alpha \tag{14.40}$$

and

$$\Psi_{II}^B(x) = \frac{B}{\sqrt{p(x)}} \sin \Phi_B(x), \quad \Phi_B(x) \equiv \frac{1}{\hbar}\int_x^b p(x')dx' + \frac{\pi}{4}, \quad x \leq b - \beta. \tag{14.41}$$

Clearly, both expressions are equivalent since they describe the same state. Therefore, they must be equal to each other:

$$\Psi_{II}^A(x) = \Psi_{II}^B(x) = \Psi_{II}(x). \tag{14.42}$$

We know that the energy of a bound particle is quantized, and quantization results from a certain condition imposed on the corresponding state function. For two expressions (14.40) and (14.41), we note that the sum of their phases

$$\Phi_A(x) + \Phi_B(x) = \frac{1}{\hbar}\int_a^b p(x)dx + \frac{\pi}{2} = \Phi(E) \tag{14.43}$$

is a constant depending on the particle's energy and geometry of the problem (e.g., the distance $\Delta = b - a$ and the shape of $U(x)$ in the "valley"), but not on x. This immediately tells us that the condition determining the energy spectrum will be a condition imposed on $\Phi(E)$. Both expressions (14.42) are equal if

$$A = (-1)^n B \quad \text{and} \quad \Phi(E) = (n+1)\pi, \quad n = 1, 2, \ldots. \tag{14.44}$$

Putting here the expression (14.43) for $\Phi(E)$ gives

$$\frac{1}{\hbar}\int_a^b p(x)\,dx = \left(n + \frac{1}{2}\right)\pi. \tag{14.45}$$

We can also introduce the phase integral over the *whole cycle* of classical motion back and forth between a and b:

$$\oint p(x)\,dx = 2\int_a^b p(x)\,dx. \tag{14.46}$$

Then condition (14.45) becomes

$$\oint p(x)\,dx = 2\pi\hbar\left(n + \frac{1}{2}\right). \tag{14.47}$$

This formula determines the allowed energies of a bound particle in the quasi-classical approximation. It corresponds to the Bohr–Sommerfeld quantization condition [6,7] in the old quantum theory. Indeed, according to (14.45), the phase of the sine function in region II changes by $(n + 1/2)\pi$ between the turning points a and b, which means that n determines the number of nodes of the state function in this region. We can also say that it is the number of standing waves within this region. Since the quasi-classical approximation is accurate when the wavelength is small relative to the characteristic size of the system, n must satisfy the condition $n \gg 1$, which is equivalent to (14.20). In terms of energy, the WKB approximation accurately describes states with high energy levels.

The reader may recall Equation 3.62, where the rule (14.47) expresses quantization of states in terms of the *phase space*. From this viewpoint, the integral (14.47) is the area within the classical phase trajectory of the bound particle, and $2\pi\hbar$ is the least possible area (the quantum of the phase space). Therefore, n can be interpreted as the number of indivisible cells for an eigenstate with $E = E_n$.

The obtained results show that h is the conversion factor between the classical action for the given path and the corresponding phase shift *in radians* for the quantum amplitude (14.13) of that path. If one measures the phase shift in full cycles of 2π radians, the conversion factor goes from h to $\hbar$. This is another aspect of the Planck constant.

The quantization rule (14.47) indicates the general feature of energy distribution in the spectrum. Denote $p_n(x) = \sqrt{2\mu(E_n - U(x))}$. Then the energy difference ΔE between two neighboring levels E_n and E_{n+1} can easily be found from (14.47):

$$\oint p_{n+1}(x)\,dx - \oint p_n(x)\,dx = \oint [p_{n+1}(x) - p_n(x)]\,dx = 2\pi\hbar. \tag{14.48}$$

Since for high n we have $\Delta E \ll E_n$, $\Delta p(x) \equiv p_{n+1}(x) - p_n(x) \ll p_n(x)$, there follows

$$\Delta p(x) \approx \frac{\mathrm{d}p(x)}{\mathrm{d}E}\Delta E = \frac{\Delta E}{v(x)}, \qquad (14.49)$$

where $v(x) = \mathrm{d}E/\mathrm{d}p(x)$ is the classical velocity of a particle with energy E and momentum $p(x)$. Therefore, the integral in (14.49) can be evaluated as

$$\oint [p_{n+1}(x) - p_n(x)]\,\mathrm{d}x \approx \Delta E \oint \frac{\mathrm{d}x}{v(x)} = \Delta E \cdot T. \qquad (14.50)$$

Here $T = 2\pi/\omega$ is the classical period of motion and ω is the corresponding frequency.

The condition (14.47) then amounts to

$$\Delta E = \frac{2\pi}{T}\hbar = \hbar\omega. \qquad (14.51)$$

We recover the formula that was first introduced by Bohr. Energy levels turn out to be equidistant, separated by the same energy gap $\hbar\omega$ that does not depend on n. Conversely, the transition frequency between neighboring energy levels coincides with the classical frequency ω, and the frequencies corresponding to transitions $n \rightleftarrows n'$ are multiples of ω. These results hold for the quasi-classical region of the spectrum, $n \gg 1$.

There is one model already familiar to the reader, which is exactly solvable and gives the result (14.51) that is exact and true for *all* n. This is the harmonic oscillator with potential energy $U(x) = (1/2)\mu\omega_0^2 x^2$. This potential is symmetric with respect to $x = 0$; therefore, the corresponding turning points in Figure 14.2 must be related as $b = -a$. Expressing the momentum in terms of a and ω_0, putting it into (14.47), and performing integration, one recovers (Problem 14.6) the already familiar result (9.76).

Problems

14.1 Evaluate the indeterminacy in the kinetic energy of a hydrogen atom whose *position* is determined up to the Bohr radius.

14.2 Derive the conditions (14.20).

14.3 A particle is placed into a potential field, with the classically forbidden region to the right of the turning point $x = a$. Find the relation between the oscillating function in the classically accessible region and the exponentially decreasing function in the forbidden region by considering the transition between the two regions along a circle of an appropriate radius centered at a.

14.4 Suppose you want to normalize the quasi-classical state function (14.25) for a bound state in a potential well shown in Figure 14.2. To this end, you need to take the integral $\int |\Psi(x)|^2\,\mathrm{d}x$ over all three regions I, II, and III. Evaluate by the order of magnitude the fractional contribution of the integrals over regions I and III in terms of parameters γ and β.

14.5 Find the normalized amplitude of a quasi-classical state with the quantum number $n \gg 1$.

(*Hint*: Use the result of the previous problem and the fact that position-dependent amplitude $(p(x))^{-1/2}$ in region II is a slowly changing function of x as compared to the corresponding sine function describing the standing wave for a given n.)

14.6 Find the quantization rule for a quantum harmonic oscillator in the framework of the WKB approximation.

References

1 Landau, L. and Lifshits, E. (1965) *Quantum Mechanics: Non-Relativistic Theory*, 2nd edn, Pergamon Press, Oxford.

2 Wentzel, G. (1926) Eine Verallgemeinerung der Quantenbedingungen für die Zwecke der Wellenmechanik. *Z. Phys.*, **38** (6–7), 518–529.

3 Kramers, H.A. (1925) Wellenmechanik und halbzählige uantisierung. *Z. Phys.*, **39** (10–11), 828–840.

4 Brillouin, L. (1926) La mécanique ondulatoire de Schrödinger: une méthode générale de resolution par approximations successive. *C. R. Acad. Sci.*, **183**, 24–26.

5 Young, L.A. and Uhlenbeck, G.E. (1930) On the Wentzel–Brillouin–Kramers approximate solution of the wave equation. *Phys. Rev.*, **36**, 1154–1167.

6 ter Haar, D. (1967) *The Old Quantum Theory*, Pergamon Press, Oxford, p. 206.

7 Sommerfeld, A. (1919) *Atombau und Spektrallinien*, Friedrich Vieweg & Sohn, Braunschweig.

15
Two-State Systems

15.1
Double Potential Well

Here we turn again to the particle bound by a couple of identical wells considered in Section 9.8. This time we will be interested in the way the state can evolve from the initial condition of a particle prepared in one of the wells, say, in well I.

Before any analytical treatment, let us make a few qualitative predictions. The probability cloud has a natural tendency to spread over the whole available volume. Suppose the initial preparation (at $t = 0$) placed the particle in well I. This could happen if someone simply brought it from the outside and dropped it into this well. Alternatively, the particle might have been spread over the two wells, being in an eigenstate of $\hat{H}$ (thus it had a definite energy and parity and accordingly indefinite position) until a position measurement just before $t = 0$ found it in well I. Such a measurement must have disturbed the pre-existing state, giving rise to indeterminacy $\Delta E \neq 0$ (it would also destroy parity). Therefore, by virtue of being inside well I, the particle must be in a nonstationary state with energy spread ΔE. It will immediately start spreading to well II via tunneling, but the spreading will not reduce indeterminacy ΔE.

First of all, let us get the idealized case out of the way. An ideal measurement would prepare the particle in a position eigenstate $\Psi(x,0) = \delta(x - x_1)$ (with x_1 within well I) giving us maximum possible information about position, but at the cost of making the energy totally indeterminate, $\Delta E \to \infty$. The huge energy boost received by the particle during this measurement would raise $\langle E \rangle$ far above the wells, making the wells irrelevant and the particle essentially free. The post-measurement state would spread rapidly over the whole space as described in Sections 11.5 and 11.6, and this process will be irreversible since there is nothing beyond the wells to reflect the particle back to the origin.

We now turn to the realistic (i.e., nonideal) position measurement where ΔE is comparable with the depth of the wells, so we can expect significant reflection from the walls. This will slow down the spreading and may also cause some undulations in its rate. But the main effect will be a gradual decrease of $\mathcal{P}_1$ (probability of being found in well I) accompanied by increase of $\mathcal{P}_2$ (probability of being found in well II). After $\mathcal{P}_2$ reaches its maximum, probability will start "flowing" back to well I. Thus, both probabilities will oscillate in such a way that one is maximum when the other is minimum, and vice versa. Aside from this, the net probability $\mathcal{P} = \mathcal{P}_1 + \mathcal{P}_2$ of being found in any of two wells may gradually decrease due to the probability leakage into the outer region from both wells.

We will focus on a situation when the position measurement is performed with accuracy $\Delta x \sim a$, where a is the width of the well. In this case, we know which well the particle is in, but we do not know its exact location in that well. The corresponding indeterminacy in particle's energy is $\Delta E \approx \Delta(p^2/2\mu) \approx \hbar^2/2\mu a^2$. We also assume that each well is deep enough to keep the particle in such a state strongly bound. This means that $|\overline{E}| \gg \Delta E \approx \hbar^2/2\mu a^2$ (the work necessary to set the trapped particle free is much greater than the energy available to it due to indeterminacy ΔE). This will rule out the leakage of probability into the outer world, and the only remaining effect will be the interplay of $\mathcal{P}_1$ and $\mathcal{P}_2$ starting, say, with $\mathcal{P}_1 = 1$ and $\mathcal{P}_2 = 0$ at $t = 0$.

The particle might as well have been initially found (or prepared) in well II. Then the evolution of the state would start from this well. In either case, we can describe it in terms of two coupled oscillators from Section 11.10. Using that analogy, we can associate the particle prepared in well I with a mass on the spring that receives an initial impulse from an external force. Let's call it mass 1 (and accordingly mass 2 for a particle prepared in well II). Then we expect the particle's behavior similar to that of a coupled pendulum or oscillator. Initially located with certainty in well I, it will later have an increasing likelihood to be found in well II. After a certain time determined by the particle's average energy and transmittance of the interim barrier, all probability will have flown into well II. At this moment, a measurement would certainly find the particle there. Then the flow will turn back until the particle returns to well I and the new cycle begins.

We remember that for coupled classical oscillators, one other type of oscillations is possible: normal vibrations, when both of them oscillate with constant amplitude without energy exchange. One normal mode occurs when both masses are pushed simultaneously with equal force in the same direction, and the other mode occurs when they are pushed in the opposite directions. The corresponding normal frequencies are slightly different by the frequency close to the frequency of the coupling spring.

The QM counterparts of these two normal vibration modes are the two *stationary* states, that is, eigenstates of the system's Hamiltonian. According to our conditions, their energies are pretty close to each other. The difference between them depends on transmittance of the barrier, just as the difference in two normal frequencies depends on coupling force provided by the connecting spring (i.e., on the spring constant). In each stationary state, there is no probability flux anywhere, the probability cloud has a fixed shape and is distributed evenly between the wells, so the particle does not have a definite location until an appropriate position measurement is performed.

The mathematical description of both problems – the coupled classical oscillators, on the one hand, and two adjacent wells, on the other hand – is essentially the same. There is, however, a fundamental difference in the nature of the described entities: *oscillation amplitudes* in the classical problem, and *probabilities* associated with the two wells in QM. The former can be watched continuously for one and the same sample; information about the latter is provided by a set of successive measurements on a pure ensemble.

After this prelude, we turn to the quantitative description. Within the accuracy outlined above, we can consider a system having only two distinct states: state $|1\rangle$ (particle in well I) and state $|2\rangle$ (particle in well II). Then an arbitrary state in the chosen basis is

$$|\Psi\rangle = c_1(t)|1\rangle + c_2(t)|2\rangle. \tag{15.1}$$

The problem is to find the amplitudes $c_1(t)$ and $c_2(t)$ and the corresponding probabilities $\mathcal{P}_1(t)$ and $\mathcal{P}_2(t)$. This will be an opportunity to see how the general scheme from Section 8.9 works in a specific situation – for $N = 2$ states. The Schrödinger equation in the given basis for $N = 2$ results in a system of two equations:

$$i\hbar \frac{dc_1(t)}{dt} = H_{11} c_1(t) + H_{12} c_2(t),$$
$$i\hbar \frac{dc_1(t)}{dt} = H_{21} c_1(t) + H_{22} c_2(t). \tag{15.2}$$

First, we seek the *stationary* solutions (eigenstates of the Hamiltonian, analogous to the normal vibrations in CM). In other words, we start with the case when $|\Psi\rangle$ is an energy eigenstate with yet unknown eigenvalue E. Then the solution must be of the form

$$c_1(t) = C_1 e^{-i(E/\hbar)t}, \qquad c_2(t) = C_2 e^{-i(E/\hbar)t}. \tag{15.3}$$

There follows a system of two algebraic equations for C_1 and C_2 (or one matrix equation)

$$\begin{pmatrix} H_{11} - E & H_{12} \\ H_{21} & H_{22} - E \end{pmatrix} \begin{pmatrix} C_1 \\ C_2 \end{pmatrix} = 0. \tag{15.4}$$

In order for it to have a nonzero solution, its determinant must be zero:

$$\mathrm{Det}\begin{pmatrix} H_{11} - E & H_{12} \\ H_{21} & H_{22} - E \end{pmatrix} = (H_{11} - E)(H_{22} - E) - H_{12} H_{21} = 0. \tag{15.5}$$

This gives a quadratic equation for E:

$$E^2 - (H_{11} + H_{22})E + H_{11} H_{22} - |H_{12}|^2 = 0, \tag{15.6}$$

with solutions

$$E_{1,2} = \frac{1}{2}\left(H_{11} + H_{22} \pm \sqrt{(H_{11} - H_{22})^2 + 4|H_{12}|^2}\right). \tag{15.7}$$

In our case, the Hamiltonian is symmetric with respect to the midpoint between the wells (see Figure 9.1); therefore, its diagonal elements in the $|n\rangle$-basis are the same for both states (for the same reason, we expect $H_{12} = H_{21} = A$, where A must be a real number). Therefore, introducing notations

$$H_{11} = H_{22} = E_0, \qquad |H_{12}| = A, \tag{15.8}$$

we obtain from (15.7) two energy eigenvalues

$$E_{1,2} = E_0 \pm A. \tag{15.9}$$

(It is precisely the eigenvalues shown graphically in Figure 9.12b). Thus, the system (15.4) splits into two systems – one for each eigenvalue:

$$[E_0 - (E_0 + A)]C_1^{(E_1)} - AC_2^{(E_1)} = 0 \quad \Rightarrow \quad -A\left[C_1^{(E_1)} + C_2^{(E_1)}\right] = 0, \tag{15.10}$$

$$[E_0 - (E_0 - A)]C_1^{(E_2)} - AC_2^{(E_2)} = 0 \quad \Rightarrow \quad A\left[C_1^{(E_2)} - C_2^{(E_2)}\right] = 0. \tag{15.11}$$

Together with normalization condition, this gives

$$C_1^{(E_1)} = -C_2^{(E_1)} = \frac{1}{\sqrt{2}}, \qquad C_1^{(E_2)} = C_2^{(E_2)} = \frac{1}{\sqrt{2}} \tag{15.12}$$

and

$$c_1^{(E_1)}(t) = \frac{1}{\sqrt{2}} e^{-i(E_1/\hbar)t}, \qquad c_2^{(E_1)}(t) = -\frac{1}{\sqrt{2}} e^{-i(E_1/\hbar)t}, \tag{15.13}$$

$$c_1^{(E_2)}(t) = \frac{1}{\sqrt{2}} e^{-i(E_2/\hbar)t}, \qquad c_2^{(E_2)}(t) = \frac{1}{\sqrt{2}} e^{-i(E_2/\hbar)t}. \tag{15.14}$$

Thus, we obtain two special solutions for a state (15.1) – the eigenstates $|\Psi_{E_1}\rangle \equiv |E_1\rangle$ and $|\Psi_{E_2}\rangle \equiv |E_2\rangle$. Using the found eigenvalues (15.9), we can write them as

$$|E_1\rangle = \frac{1}{\sqrt{2}} e^{-i[(E_0+A)/\hbar]t}(|1\rangle - |2\rangle), \qquad |E_2\rangle = \frac{1}{\sqrt{2}} e^{-i[(E_0-A)/\hbar]t}(|1\rangle + |2\rangle). \tag{15.15}$$

As seen from these expressions, each of the obtained functions is a stationary state. In each of them, the probabilities $\mathcal{P}_1$ and $\mathcal{P}_2$ are constants (no surprise for a stationary state!). They are equal to each other, indicating a 50–50 chance to find the particle in either basis state. No surprise again: it is only at this distribution of probabilities that the state of the given system can remain stationary. The particle's probability cloud is "smeared" over the available region so that each well gets an equal share. To use the coupled oscillator analogy, these eigenstates correspond to the two normal vibrations.

An arbitrary state $|\tilde{\Psi}\rangle$ is constructed as a superposition of the two eigenstates. Using (15.13)–(15.15) as "building blocks," we get after simple algebra (Problem 15.1)

$$|\tilde{\Psi}\rangle = a_1|E_1\rangle + a_2|E_2\rangle = (\tilde{c}_1(t)|1\rangle + \tilde{c}_2(t)|2\rangle)e^{-i\omega_0 t}, \tag{15.16}$$

where

$$\tilde{c}_1(t) = a_1 e^{-i\omega_A t} + a_2 e^{i\omega_A t}, \qquad \tilde{c}_2(t) = -a_1 e^{-i\omega_A t} + a_2 e^{i\omega_A t} \tag{15.17}$$

and the factor $1/\sqrt{2}$ is absorbed into the coefficients a_1 and a_2. The frequencies are

$$\omega_0 \equiv \frac{E_0}{\hbar}, \qquad \omega_A \equiv \frac{A}{\hbar}. \tag{15.18}$$

The state $|\tilde{\Psi}\rangle$ is nonstationary, and the amplitudes $\tilde{c}_1(t)$ and $\tilde{c}_2(t)$ describe the corresponding nonstationary probabilities of finding the particle in one of the respective wells. All we need now is to find these probabilities by specifying a_1 and a_2 from the initial conditions. According to our assumption (particle initially in well I), we can write

$$P_1(0) = |\tilde{c}_1(0)|^2 = 1, \qquad P_2(0) = |\tilde{c}_2(0)|^2 = 0 \tag{15.19}$$

or, in view of (15.17),

$$\tilde{c}_1(0) = a_1 + a_2 = 1, \qquad \tilde{c}_2(0) = -a_1 + a_2 = 0. \tag{15.20}$$

It follows $a_1 = a_2 = 1/2$, so that

$$\tilde{c}_1(t) = \cos\omega_A t, \qquad \tilde{c}_2(t) = i\sin\omega_A t \tag{15.21}$$

and

$$P_1(t) = \cos^2\omega_A t, \qquad P_2(t) = \sin^2\omega_A t. \tag{15.22}$$

Both probabilities change periodically with time, but their sum is 1.

15.2
The Ammonium Molecule

The apparently abstract problem of the double potential well is a good model of a real object – the ammonium molecule, NH_3. As seen from its chemical formula, it consists of a nitrogen (N) atom bound together with three hydrogen (H) atoms. The N tries to appropriate electrons from the three H atoms, and partially succeeds. The result is that these electrons do, on the average, get closer to N and farther away from their respective H atoms. In the quantum picture, the H-electron clouds "gravitate" toward N, leaving their H atoms partially naked. This results in a redistribution of charges in the (overall neutral) molecule: there is an effective negative charge $-Q_{eff}$ on N and positive charge Q_{eff} on the H atoms. There accordingly appears attraction between N and H atoms, which holds them together.[1] In the equilibrium configuration, the three H atoms form an equilateral triangle. This triangle defines a plane. If one draws a "symmetry axis," that is, a line normal to the plane and passing through the center of the triangle (Figure 15.1), there will be two symmetric equilibrium positions for N – one on either side of the plane. One of them can be treated as the center of potential well I, and the other as the center of well II. The

---

1) This is only a very crude description of chemical bonding between the atoms in a molecule. The bonding does not amount to electrical forces only. The latter alone cannot keep a system in a stable equilibrium (the Earnshaw theorem [1,2]).

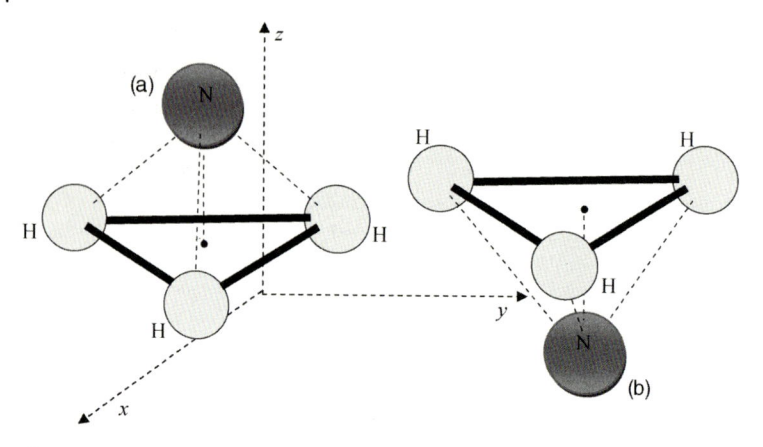

Figure 15.1 Two possible states of NH$_3$ molecule. (a) and (b) are the two equilibrium positions of N. Their locations (states!) are permanent, but their occupancy is not, due to tunneling.

plane itself plays the role of the potential barrier between the wells because classically it would take additional energy for N to squeeze through it, pushing apart the three H atoms guarding the passage. But N can accomplish this trick by tunneling. So we have yet another double-well system, and if we neglect all other degrees of freedom (translational motion, rotations, vibrations, etc.) to focus on the tunneling of N through the plane, this system will have only two states. All results of the previous section apply here. Let us denote the plane by the symbol H$_3$, and the symmetry axis by the symbol x. The Hamiltonian describing the motion of N along x has two distinct eigenstates $|E_1\rangle$ and $|E_2\rangle$ in which N does not belong to either well but exists in both at once. As in the last section, we will have the "well states" $|1\rangle$ and $|2\rangle$. These are *not* eigenstates of $\hat{H}$ (just as before, they are described in terms of position eigenstates, with all the reservations concerning ideal position measurements). As states with a well-defined position, separated by a penetrable barrier, they are nonstationary in the sense that neither of them can be used by N as a permanent residence. The residency periodically "flows over" from one place to the other, so if we initially see the N on one side of H$_3$, then a little later there will appear a small chance to find it on the other side. We again have the quantum analogue of a classical oscillator, moving back and forth between two extremes.

According to the general rules, we can express each of the states $|1\rangle$ and $|2\rangle$ in terms of the stationary states. In these terms, by forcing N to occupy with certainty one of the states $|1\rangle$ or $|2\rangle$, we give rise to a superposition of two stationary states $|E_1\rangle$ and $|E_2\rangle$, which produces specific "beats" by combining two different frequencies.

Quantitatively, the "tug of war" between $|1\rangle$ and $|2\rangle$ is described by the equations of the previous section. But NH$_3$ is a real physical system, not an abstract entity! It is thus natural to wonder about the behavior of its mass and charge distributions. The N atom, apart from being rather massive (about five times the combined mass of the three H atoms), also carries the charge $-Q_{\text{eff}}$. Therefore, oscillations $|1\rangle \leftrightarrow |2\rangle$ are accompanied by mass and charge transfer through the H$_3$-plane. Of course, the

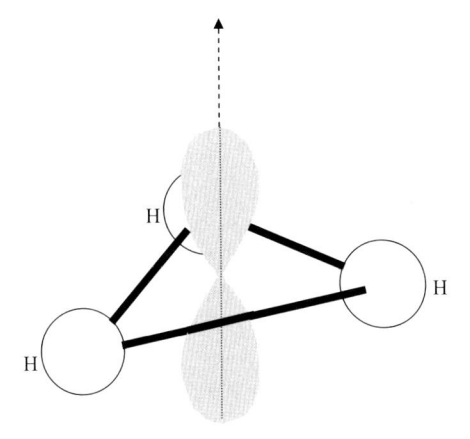

Figure 15.2 A stationary state of NH$_3$ molecule. In contrast with states in Figure 15.1, here we cannot label by N the top and bottom "semi-clouds" of the nitrogen atom, because there is only one N in the NH$_3$ molecule.

center of mass of an isolated NH$_3$ molecule remains stationary, but configuration of the molecule changes from one shape to its mirror image and then back. But since the "N" part is more massive than the "H$_3$" part, it is accordingly less mobile, and it would be more accurate to say that it is the H$_3$-plane that oscillates back and forth about N.

Let us visualize both pairs of states in the position representation (i.e., expressing a state function as the spatial probability amplitude of finding N at a given location) [3]. Consider the stationary states $|E_1\rangle$ and $|E_2\rangle$ first. In these states, there is no probability flux. This is consistent with a probability distribution symmetric relative to the H$_3$-plane (Figure 15.2). At any moment, there is a 50–50 chance to find N on either side of the plane. For each side, expectation value d_0 for the distance between N and the plane is the same. Since N carries an effective charge, its probability distribution is matched by a charge distribution having the same functional dependence on position. In view of the symmetry of this distribution, there is no dipole moment in either of the states $|E_1\rangle$ and $|E_2\rangle$.

Now we turn to the states $|1\rangle$ and $|2\rangle$. Being nonstationary, they can offer only a short-term (but periodically renewed!) residence to a particle, and are best represented graphically at the moments of such residence (Figure 15.1a and b). At such moments, the charge distribution is maximally asymmetric. And since this charge is negative, while the H$_3$-plane carries effective positive charge, we have a dipole moment $\mathbf{p}$[2] perpendicular to H$_3$. As the probability distribution, together with the charge, flows through the separating barrier toward the second residence and then back, it changes gradually as an oscillating function of time, $\mathbf{p} = \mathbf{p}(t)$.

---

2) Since there is no mention of momentum in this section, we hope that the conventional notation **p** for dipole moment will not cause any confusion with momentum, which is usually denoted by the same symbol.

This graphical picture helps us visualize some features of the whole phenomenon. The NH_3 molecule prepared in one of the two *energy eigenstates* is not polarized, has the plane of symmetry (H_3), and remains in that state as long as it is not disturbed. The molecule prepared in one of the two *position eigenstates* has asymmetric charge distribution and forms a vibrating dipole. In this respect, it behaves as a microscopic antenna. Therefore, it must, as any antenna does, radiate EM waves! We can evaluate their frequency in two slightly different but equivalent ways. First, we can use the fact that either of the states $|1\rangle$ and $|2\rangle$ is an equally weighted superposition of eigenstates $|E_1\rangle$ and $|E_2\rangle$. Therefore, we can expect, in accordance with the Bohr rules, radiation resulting from quantum transitions between the corresponding states with the frequency

$$\omega = \omega_{12} = \frac{E_2 - E_1}{\hbar}. \tag{15.23}$$

Second, we can consider radiation originating from the oscillating charge distribution. The above-mentioned dipole moment $p(t) = |\mathbf{p}(t)|$ associated with it stands in proportion to Q_{eff} and $P_j(t)$ with $j = 1$ or 2, so we can write

$$p(t) = Q_{\text{eff}} D_0 \mathcal{P}_j(t) = Q_{\text{eff}} D_0 \begin{cases} \cos^2(\omega_A t) = \frac{1}{2}(1 + \cos 2\omega_A t), \\ \sin^2(\omega_A t) = \frac{1}{2}(1 - \cos 2\omega_A t). \end{cases} \tag{15.24}$$

Here $D_0 = 2d_0$ is the effective distance between two extreme positions of N, that is, the spatial separation between states $|1\rangle$ and $|2\rangle$. We also assumed that Q_{eff} remains constant, which is, of course, only an approximation.

The frequency of radiation is equal to the oscillation frequency, so in view of (15.19)

$$\omega = 2\omega_A = 2\frac{A}{\hbar}. \tag{15.25}$$

This is consistent with (15.23) since, in view of (15.10), $\omega_{12} = 2\omega_A$.

Alternatively, instead of behaving like the dipole (15.24) vibrating between the states $|1\rangle$ and $|2\rangle$ and radiating with frequency (15.25), the molecule may be in one of the energy eigenstates $|E_1\rangle$ and $|E_2\rangle$ with the zero dipole moment. The frequency (15.23) of the optical transition $|E_1\rangle \Leftrightarrow |E_2\rangle$ between energy eigenstates is the same as vibration frequency. Therefore, if the molecule is subject to the *incident* radiation with the same frequency, it may undergo a resonant optical transition. Depending on the initial state, it may absorb a photon, which will take it from level E_1 to the higher level E_2, or emit a photon (in the act of spontaneous or stimulated emission), which will take it from E_2 to E_1. For the NH_3 molecule, it was found experimentally that $E_2 - E_1 = 10^{-4}$ eV. This corresponds to frequency $f = \omega/2\pi = 24\,000$ MHz and wavelength is $\lambda = 1.25$ cm, which is in the microwave range of the spectrum. From these data, using formula (15.25), we can determine parameter A, which is very difficult to calculate theoretically.

Now we can consider a more complex experimental situation, when NH_3 is subject to external electric field $\mathcal{E}$, normal to the H_3-plane. This will lead to an additional term in the system's Hamiltonian. How will this affect the system? As long as the field is not strong enough to cause noticeable changes in the electron configuration of the molecule, the corresponding term will be relatively small and can be treated as a small perturbation. Accordingly, the above-described states of the molecule will, in first approximation, remain the same.

Under these conditions, the additional term will be determined only by the pre-existing intrinsic dipole moment of the molecule. The dipole moment **p** in an external electric field $\mathcal{E}$ has potential energy $U = \mathbf{p} \cdot \mathcal{E}$. In the case when both vectors are parallel,

$$U = \mathbf{p} \cdot \mathcal{E} = p\mathcal{E}. \tag{15.26}$$

(An equivalent way to see it is that for charge q and potential $V = \mathcal{E}x$ the additional potential energy is $U = qV = q\mathcal{E}x$. But qx is the dipole moment p of the molecule; therefore, both expressions are equivalent.) So the response to the field depends on p. But we found that the dipole moment of the molecule is different in different states:

$$p = \begin{cases} 0, & \text{states}\, |E_1\rangle \text{ and } |E_2\rangle, \\ D_0 Q_{\text{eff}}\, P(t), & \text{states}\, |1\rangle \text{ and } |2\rangle. \end{cases} \tag{15.27}$$

Accordingly, within the approximations made, the effect of the field is different for energy eigenstates, on the one hand, and for position eigenstates, on the other hand. The molecule in an energy eigenstate will not feel the applied electric field and so will not respond to it. The molecule in a position eigenstate will acquire an additional energy that depends on the dipole moment of that state. The expectation values $H_{11} = H_{22} = E_0$ of energy in states $|1\rangle$ and $|2\rangle$, which were the same for both states in the absence of $\mathcal{E}$, will now change to

$$H_{11} = E_0 + p\mathcal{E}, \qquad H_{22} = E_0 - p\mathcal{E}. \tag{15.28}$$

On the other hand, under the assumption that the applied electric field does not distort the molecular configuration, it will not change the amplitude of transition between states $|1\rangle$ and $|2\rangle$. Therefore, we can set it to be the same as in (15.8), that is, $H_{12} = H_{21} = A$.

15.3
Qubits Introduced

In computer science, information is measured in bits. One *classical* bit (C-bit) is the amount of information stored in a two-state system that can be either in state 1 or in state 0. (For now, we shall assume the a priori probability of either state is 50%; in Chapter 26 we will refine this definition by introducing classical Shannon information.) The simplest example of a classical two-state system is the tossed coin (heads or tails correspond to bit values 1 and 0). Old computers stored C-bits in radio lamps; modern computers store them in tiny ($\simeq 20$ nm) transistors placed on the surface of a silicon microchip. Further reduction of transistor size will take us into the quantum territory. A quantum computer would store the state of a bit in a two-level quantum system such as the ammonium molecule. Such a computer will possess quantum (rather than classical) properties that form the subject matter of quantum information science.

The best way to learn the main features of quantum information is by using a representative case study: a particle on a Bloch sphere. With this archetypical two-level system, one employs the two eigenstates of spin to store bit values 1 and 0. Most of the current research focuses on just two particles: the electron and the photon. In case of the electron, we can associate bit value 1 with spin-up state $|\uparrow\rangle$ and bit value 0 with spin-down state $|\downarrow\rangle$. For the photon, we can associate 1 with horizontal polarization $|\leftrightarrow\rangle$ and 0 with vertical polarization $|\updownarrow\rangle$. One could in principle use other spin-1/2 particles as well, except that nuclear spin states or those of individual protons or neutrons are much harder to measure (recall Chapter 7), and particles such as muons are much harder to produce. Since such particles would not add any new features or offer any practical advantage over the electron, we are going to focus on electrons and photons exclusively without any loss of generality.

D. Mermin suggested the term "quantum bit" or Q-bit for the generic two-state quantum system whose orthogonal states $|1\rangle$ and $|0\rangle$ spanning the $\mathcal{H}$ space are associated with the corresponding bit values. Thus, we have

$$|1\rangle = \begin{bmatrix} |\leftrightarrow\rangle, & \text{photon,} \\ |\uparrow\rangle, & \text{electron,} \end{bmatrix} \qquad |0\rangle = \begin{bmatrix} |\updownarrow\rangle, & \text{photon,} \\ |\downarrow\rangle, & \text{electron.} \end{bmatrix}$$

The orthogonal eigenbasis $\{|1\rangle, |0\rangle\}$ is the most natural choice of the measurement basis for the Q-bit. It is often called *computational basis*, a term that is quite self-intuitive in our computer context.

Below we list the four important properties of Q-bits:

1) A Q-bit can be in an arbitrary superposition of 0 *and* 1:

$$\text{Q-bit} \quad \Rightarrow \quad |\psi\rangle = a|0\rangle + b|1\rangle, \qquad |a|^2 + |b|^2 = 1. \tag{15.29}$$

Thus, a Q-bit can in principle *store* an infinite amount of information in its complex amplitudes a and b because it can be prepared in an infinite number of ways. While the two states of a C-bit can be represented by a pair of antipodal points on the Bloch sphere (recall Figure 7.4), the possible Q-bit states will cover its entire surface.

2) It seems that Q-bits are infinitely more powerful than C-bits when it comes to storage of data. But when it comes to the *retrieving* of data stored, one realizes that regardless of the way in which the Q-bit was prepared, any spin/polarization measurement one uses to read off the data will be limited in scope to just *two* possible outcomes. Thus, neighboring points on the Bloch sphere cannot be distinguished reliably.

3) Q-bits can get entangled with each other, making the composite system and its behavior much richer than a comparable system of C-bits. An entangled Q-bit

does not have its own wave function and must be represented by a reduced density matrix – a concept unheard of for a C-bit.

4) While a classical computation just flips C-bits between 0 and 1, a quantum computation subjects the Q-bits to a much more general unitary transformation that is specified by the time evolution of the system. In other words, the amplitudes are generally functions of time: $a = a(t)$ and $b = b(t)$. In case of entanglement, the unitary transformation is imposed on the state amplitudes of the composite system, rather than on amplitudes of the individual states that do not exist in this case.

Mermin's term Q-bit did not survive in the literature for long. After a while, the majority of authors chose to spell it the way it is pronounced. To keep up with this convention, we are going to use term *qubit* from now on.

Most often one has to work not with individual qubits, but with composite systems of qubits. Usually, qubits are regarded as distinguishable particles, that is, one can label them and tell a state $|0_A\rangle \otimes |1_B\rangle$ from a state $|0_B\rangle \otimes |1_A\rangle$. Generally, a qubit pair A and B must be described by a density matrix ρ^{AB}, and then each qubit must be described by a reduced density matrix. The composite state can be pure, $\rho^{AB} = |\psi^{AB}\rangle\langle\psi^{AB}|$. However, even then individual qubits in that state can be entangled.

It turns out that we can derive a quantitative measure of entanglement between two qubits. The pure state of a composite system can be written in the computational basis as

$$|\psi^{AB}\rangle = a|0\rangle \otimes |0\rangle + b|0\rangle \otimes |1\rangle + c|1\rangle \otimes |0\rangle + d|1\rangle \otimes |1\rangle. \tag{15.30}$$

At the same time, we can define a new ket with complex-conjugated coefficients

$$|\psi^{AB}\rangle^* = a^*|0\rangle \otimes |0\rangle + b^*|0\rangle \otimes |1\rangle + c^*|1\rangle \otimes |0\rangle + d^*|1\rangle \otimes |1\rangle. \tag{15.31}$$

Acting on this ket with the operator $\sigma_y \otimes \sigma_y$ where the first Pauli operator acts on the left ket in a tensor product and the second on the right ket, we obtain another state

$$|\tilde{\psi}^{AB}\rangle = \sigma_y \otimes \sigma_y |\psi^{AB}\rangle^*. \tag{15.32}$$

We leave is as a problem for the reader to prove that

$$\left|\langle\tilde{\psi}^{AB}|\psi^{AB}\rangle\right| = 2|ad - bc| \equiv C(\psi^{AB}). \tag{15.33}$$

The quantity $C(\psi^{AB})$ is known as *concurrence*. To see how it can serve as a measure of the degree of entanglement, suppose the state can be written as a product $|\psi^{AB}\rangle = (c_1|0\rangle + c_2|1\rangle) \otimes (d_1|0\rangle + d_2|1\rangle)$. Then $a = c_1 d_1$, $b = c_1 d_2$, $c = c_2 d_1$, $d = c_2 d_2$, and $C(\psi^{AB}) = 0$, indicating zero entanglement. On the other hand, consider a fully entangled state such as electron singlet $\psi^{AB} = (1/\sqrt{2})(|0\rangle \otimes |1\rangle - |1\rangle \otimes |0\rangle)$. Then we have $a = d = 0$, $b = c = -1/\sqrt{2}$, and $C(\psi^{AB}) = 1$. Thus, the degree of entanglement can range from 0 to 1:

$$0 \leq C(\psi^{AB}) \leq 1. \tag{15.34}$$

Of course, we could have written $C(\psi^{AB}) \equiv 2|ad - bc|$ from the very beginning instead of deriving it as the absolute value of an inner product. However, we used the more circuitous path with the view of generalization for the case of density matrices. Indeed, expressing the pure states (15.30) and (15.31) as density matrices, $\rho^{AB} = |\psi^{AB}\rangle\langle\psi^{AB}|$ and $\tilde{\rho}^{AB} = |\tilde{\psi}^{AB}\rangle\langle\tilde{\psi}^{AB}|$, we can rewrite the previous result as follows:

$$C^2(\psi^{AB}) \equiv \langle\psi^{AB}|\tilde{\psi}^{AB}\rangle\langle\tilde{\psi}^{AB}|\psi^{AB}\rangle = \mathrm{Tr}\Big(|\psi^{AB}\rangle\langle\tilde{\psi}^{AB}|\psi^{AB}\rangle\langle\tilde{\psi}^{AB}|\Big) = \mathrm{Tr}(\rho^{AB}\tilde{\rho}^{AB}).$$

Then, by analogy, we have for the mixed state $C(\rho^{AB}) \equiv \sqrt{\mathrm{Tr}(\rho^{AB}\tilde{\rho}^{AB})}$. Since density matrices are positive semidefinite, and the Hilbert space of the qubit pair is four-dimensional, the product under the root is guaranteed to have four positive eigenvalues. Let us denote the *square roots* of these eigenvalues by $\lambda_1, \lambda_2, \lambda_3,$ and λ_4, arranging them so that the first number is the largest. A theorem by Wootters, which we shall state here without proof, says that the square root of the above trace is equal either to $\lambda_1 - \lambda_2 - \lambda_3 - \lambda_4$ or to zero if that expression is negative:

$$C(\rho^{AB}) \equiv \sqrt{\mathrm{Tr}(\rho^{AB}\tilde{\rho}^{AB})} = \max(0, \lambda_1 - \lambda_2 - \lambda_3 - \lambda_4). \tag{15.35}$$

Equation 15.35 defines concurrence of a mixed system.

Problem

15.1 Derive the expressions (15.16) and (15.17).

References

1 Griffiths, D. (1999) *Introduction to Electrodynamics*, 3rd edn, Addison-Wesley, Reading, MA.
2 Jackson, J. (1999) *Classical Electrodynamics*, 3rd edn, John Wiley & Sons, Inc., New York.
3 Feynman, R., Leighton, R., and Sands, M. (1963–1964) *The Feynman Lectures on Physics*, vols. 1–3, Addison-Wesley, Reading, MA.

16
Charge in Magnetic Field

16.1
A Charged Particle in EM Field

Up to now we studied situations with a particle subject to forces depending (classically) only on its position $\mathbf{r}$. Such forces can be described by a potential function $U(\mathbf{r})$ such that $\mathbf{f} = -\vec{\nabla} U$. Here we consider nondissipative forces depending on particle's velocity. Such forces emerge when a charged particle is moving in the magnetic field $\mathbf{B}$. In this case, the particle is subject to the *Lorentz force* $\mathbf{f}_M = q\mathbf{v} \times \mathbf{B}$. If there is also an electric field $\mathcal{E}$, we have

$$\mathbf{f} = q(\mathcal{E} + \mathbf{v} \times \mathbf{B}) = \mathbf{f}_{El} + \mathbf{f}_{Mgn}. \tag{16.1}$$

The magnetic force $\mathbf{f}_{Mgn}$ is perpendicular to both $\mathbf{v}$ and $\mathbf{B}$. The net force in this equation is also called sometimes the Lorentz force. In the absence of $\mathcal{E}$, the resulting motion is well-studied circular motion around the direction of $\mathbf{B}$.

In classical theory, an arbitrary EM field with external sources (charges and currents) in vacuum is described by a system of coupled differential equations (Maxwell's equations [1,2]):

$$\vec{\nabla} \cdot \mathcal{E} = \frac{\rho}{\varepsilon_0}, \quad \vec{\nabla} \times \mathcal{E} + \frac{\partial \mathbf{B}}{\partial t} = 0, \quad \vec{\nabla} \times \mathbf{B} - \frac{1}{c^2} \frac{\partial \mathcal{E}}{\partial t} = \mu_0 \mathbf{j}, \quad \vec{\nabla} \cdot \mathbf{B} = 0, \tag{16.2}$$

where ρ and $\mathbf{j}$ are the charge and current density, respectively, and ε_0 and μ_0 are the permittivity and permeability of free space, respectively.

It is often convenient to describe the field by the 4-vector of potential $A = (V, \mathbf{A})$ whose "temporal" component is the familiar electrical potential V, and three spatial components form the magnetic vector potential $\mathbf{A}$. The physical observables $\mathcal{E}$ and $\mathbf{B}$ represent forces acting on a test charge, and are determined through V and $\mathbf{A}$ as

$$\mathcal{E} = -\vec{\nabla} V - \frac{\partial \mathbf{A}}{\partial t}, \quad \mathbf{B} = \vec{\nabla} \times \mathbf{A}. \tag{16.3}$$

The 4-vector A is not uniquely defined. As seen from (16.3), $\mathcal{E}$ and $\mathbf{B}$ remain unaffected if we change A according to

$$\mathbf{A} \to \mathbf{A}' = \mathbf{A} + \vec{\nabla}\Lambda, \quad V \to V' = V - \frac{\partial \Lambda}{\partial t}, \tag{16.4}$$

where $\Lambda(\mathbf{r}, t)$ is an arbitrary differentiable function of $\mathbf{r}$ and t. In particular, we can subject $\mathbf{A}$ to the *Lorentz condition*

$$\vec{\nabla} \cdot \mathbf{A} + \frac{1}{c}\frac{\partial V}{\partial t} = 0 \tag{16.5}$$

without affecting the EM field, if Λ satisfies the D'Alembert equation (or the wave equation [3,4]). Transformation (16.4) is called the *gauge transformation*, and the invariance of physical observables and equations determining them with respect to the gauge transformation is called the *gauge invariance*.

If we now combine (16.3) with the Maxwell equations and then subject $\mathbf{A}$ to condition (16.5), the equations take the form

$$\nabla^2 V - \frac{1}{c^2}\frac{\partial^2 V}{\partial t^2} = -\frac{\rho}{\varepsilon_0}, \quad \nabla^2 \mathbf{A} - \frac{1}{c^2}\frac{\partial^2 \mathbf{A}}{\partial t^2} = -\mu_0 \mathbf{j}. \tag{16.6}$$

The advantage of this form is that it contains four uncoupled equations (a scalar equation for V and a vector equation for $\mathbf{A}$) whereas (16.2) is a system of eight coupled equations. But this comes at a price – Equations 16.6 are of the second order with respect to time. Nevertheless, these equations are usually easier to solve than the set (16.2).

Another reason why the potentials V and $\mathbf{A}$ are important is that the Hamiltonian simply depends on them. In nonrelativistic CM, the motion of a particle can be described by the Hamilton equations $\dot{\mathbf{p}} = -\vec{\nabla} H$ and $\dot{\mathbf{r}} = \vec{\nabla}_p H$. Here the dots denote time derivative, and the subscript $\mathbf{p}$ in the second equation indicates the gradient with respect to the components of momentum $\mathbf{p}$ (gradient in the momentum space). The corresponding Hamiltonian in the presence of EM field is

$$H = \frac{1}{2\mu}(\mathbf{p} - q\mathbf{A})^2 + qV. \tag{16.7}$$

If we plug this into the Hamilton equations, the result will be the Newton equations $\mu \ddot{\mathbf{r}} = q(\mathcal{E} + \mathbf{v} \times \mathbf{B})$ for a particle under the Lorentz force (16.1).

The quantum-mechanical Hamiltonian originates from (16.7) if we substitute all the observables there with their respective operators. In position representation, this gives

$$\hat{H} = \frac{1}{2\mu}(\hat{\mathbf{p}} - q\mathbf{A})^2 + qV \equiv \frac{1}{2\mu}\left(i\hbar\vec{\nabla} + q\mathbf{A}\right)^2 + qV. \tag{16.8}$$

The corresponding Schrödinger equation for a charged spinless particle in the EM field is

$$i\hbar \frac{\partial \Psi}{\partial t} = \left[\frac{(\hat{\mathbf{p}} - q\mathbf{A})^2}{2\mu} + qV\right]\Psi. \tag{16.9}$$

We leave it to the reader (Problem 16.2) to show that this can be brought to the form

$$i\hbar \frac{\partial \Psi}{\partial t} = \left[\frac{1}{2\mu} \left(\hat{p}^2 - q(\hat{p} \cdot A + A \cdot \hat{p}) + q^2 A^2 \right) + qV \right] \Psi \tag{16.10}$$

or

$$i\hbar \frac{\partial \Psi}{\partial t} = \left[\frac{1}{2\mu} \left(\hat{p}^2 - 2 A \cdot \hat{p} + i q \hbar \vec{\nabla} \cdot A + q^2 A^2 \right) + qV \right] \Psi. \tag{16.11}$$

If the particle has also a nonzero spin, then its Hamiltonian acquires additional term $-\mathcal{M} \mathbf{B}$ describing the interaction of the corresponding magnetic moment $\mathcal{M}$ with the external magnetic field. We will consider such case in Section 16.5.

Since the Hamiltonian explicitly contains the 4-vector A, it does change under the gauge transformation (16.4). But the *equations of motion* in CM as well as the Schrödinger equation in QM must remain unaffected. The latter can be the case only if Ψ is subjected to the appropriate *phase transformation* of the type (3.18) concurrently with (16.4). To see the link between the two transformations, it will be instructive to obtain the same results from another perspective [3]. As we know from Section 4.7, the amplitude for a particle to pass from point $\mathbf{r}_1$ to point $\mathbf{r}_2$ along a virtual path connecting both points is determined by the corresponding path integral

$$\langle \mathbf{r}_2 | \mathbf{r}_1 \rangle = \frac{\Psi_0}{r_{12}} e^{i[S(\mathbf{r}_1, \mathbf{r}_2)/\hbar]}, \quad r_{12} \equiv |\mathbf{r}_2 - \mathbf{r}_1|, \quad S(\mathbf{r}_1, \mathbf{r}_2) \equiv \int_{\mathbf{r}_1}^{\mathbf{r}_2} \mathbf{k} \cdot d\mathbf{r}. \tag{16.12}$$

The presence of $\mathbf{A}$ brings in the additional phase factor

$$\langle \mathbf{r}_2 | \mathbf{r}_1 \rangle_{l,A} = \langle \mathbf{r}_2 | \mathbf{r}_1 \rangle_l \exp\left\{ i \frac{q}{\hbar} \int_{\mathbf{r}_1}^{\mathbf{r}_2} \mathbf{A} \cdot d\mathbf{r} \right\} = \frac{\Psi_0}{r_{12}} \exp\left\{ \frac{i}{\hbar} \int_{\mathbf{r}_1}^{\mathbf{r}_2} (\mathbf{p} + q\mathbf{A}) \cdot d\mathbf{r} \right\}. \tag{16.13}$$

This is equivalent to changing $-i\hbar \vec{\nabla} \to -i\hbar \vec{\nabla} - q\mathbf{A}$. Thus, the additional phase in (16.13) automatically involves the additional term in (16.4). This gives us another illustration of how the local phase transformation and gauge transformation are intimately connected, at least for charged particles. If we perform the gauge transformation (16.4) and simultaneously the phase transformation according to (3.15) with $\tilde{\Lambda} \equiv (q/\hbar)\Lambda(\mathbf{r}, t)$, then both transformations together will take us back to the initial Schrödinger equation (16.9) for Ψ.

16.2
The Continuity Equation in EM Field

How does the local phase transformation affect the continuity equation? It is easy to see that expression (3.2) for $\rho(\mathbf{r})$ remains the same since any additional exponent of the type $e^{i(q/\hbar)\Lambda(\mathbf{r},t)}$ with real Λ will be canceled. However, the flux density $\mathbf{j}(\mathbf{r})$ described by $\Psi \vec{\nabla} \Psi^*$ is more sensitive and remains invariant only under a global phase transformation. The local transformation (3.15) will change it (Problem 16.5). In order to conserve the flux, transformation (3.15) must be carried out concurrently with the corresponding gauge transformation (16.4). This allows us to obtain the expression for $\mathbf{j}(\mathbf{r})$ in the same way as in Section 8.2, but now we must use the Hamiltonian (16.8) instead of (8.6). Thus, we write $\partial \rho / \partial t = \Psi^*(\partial \Psi / \partial t) + \Psi(\partial \Psi^* / \partial t)$, and replace the time derivatives using (16.9):

$$\frac{\partial \rho}{\partial t} = \frac{1}{i\hbar} \left\{ \Psi^* \left[\frac{(\hat{\mathbf{p}} - q\mathbf{A})^2}{2\mu} + qV \right] \Psi - \Psi \left[\frac{(\hat{\mathbf{p}}^* - q\mathbf{A})^2}{2\mu} + qV \right] \Psi^* \right\}. \tag{16.14}$$

This equation can, after some tedious algebra, be brought to the form (8.13) with

$$\mathbf{j} = \frac{1}{\mu} \{ \mathrm{Re}\, \Psi[(\hat{\mathbf{p}} - q\mathbf{A})\Psi]^* \} = -\frac{1}{\mu} \left\{ \hbar\, \mathrm{Im}\, \Psi \vec{\nabla} \Psi^* + q\Psi \mathbf{A} \Psi^* \right\}. \tag{16.15}$$

As in Equation 8.13, vector $\mathbf{j}$ is the probability flux density. It is different from (8.12) by the additional term that describes the impact of magnetic field. In the absence of the field, it reduces to (8.12). As in case (8.13), Equation 16.14 describes conservation of probability. If a state Ψ is evolving in time, then any change of density ρ at a point $\mathbf{r}$ is associated with the corresponding probability flux toward or away from $\mathbf{r}$. The additional term $q\Psi\mathbf{A}\Psi^* = q\rho\mathbf{A}$ in Equation 16.15 ensures that the probability flux will remain the same under transformation (16.4) when carried out concurrently with the corresponding local phase transformation (3.18).

Let us now focus on a more detailed physical interpretation of this term.

Any flux density can be represented as the product $\mathbf{j} = \rho \mathbf{v}$ (the density of the flowing entity times its velocity). In our case, the density is $\rho = \Psi\Psi^*$. To simplify the discussion, consider a special case when Ψ is an eigenstate of $\hat{\mathbf{p}}$. In this case, $(\hat{\mathbf{p}}\Psi)^* = \mathbf{p}\Psi^*$, and (16.15) reduces to

$$\mathbf{j}_\mathbf{p} = \frac{\mathbf{p} - q\mathbf{A}}{\mu} \rho. \tag{16.16}$$

If the coefficient multiplying ρ in the expression $\mathbf{j} = \rho \mathbf{v}$ is interpreted as velocity, then it follows that

$$\mathbf{p} - q\mathbf{A} = \mu \mathbf{v}. \tag{16.17}$$

In magnetic field, it is the combination (16.17), rather than just $\mathbf{p}/\mu$, that determines the average velocity. In other words, instead of $\mathbf{p} = \mu\mathbf{v}$ we now have $\mathbf{p} = \mu\mathbf{v} + q\mathbf{A}$. It looks like the magnetic vector potential $\mathbf{A}$ gives additional contribution $q\mathbf{A}$ to the momentum $\mathbf{p}$ of a charged particle. The resulting net momentum consists of two parts. The "traditional" part $\mu\mathbf{v} \equiv \mathbf{p}_\mathrm{v}$ is sometimes called the $\mu\mathbf{v}$-momentum or *kinematic momentum*. The second part

$$q\mathbf{A} \equiv \mathbf{p}_\mathbf{A} \tag{16.18}$$

can be called the A-momentum. The total

$$\mathbf{p} = \mathbf{p}_\mathrm{v} + \mathbf{p}_\mathbf{A} \tag{16.19}$$

is called in some sources a *generalized momentum* or *dynamic momentum* [3]. Thus, $-i\hbar \vec{\nabla}$ in the presence of magnetic field is operator of dynamic momentum $\mathbf{p}$.

Exercise 16.1

Express the kinematic momentum $\mathbf{p}_v = \mu\mathbf{v}$ of a classical particle moving in a circle in a homogeneous magnetic field $\mathbf{B}$ in terms of magnetic vector potential $\mathbf{A}$.

Solution:
Equating the centripetal force to the Lorentz force gives $qvB = \mu v^2/r$ or $qB = \mu v/r = p_v/r$. It follows that $p_v = qBr$. Since the motion of the charged particle is circular, it is convenient to use cylindrical coordinates with the z-axis along $\mathbf{B}$ and polar angle φ, with the lines of vector field $\mathbf{A}$ circling around the rotational axis. In this case, from $\mathbf{B} = \vec{\nabla} \times \mathbf{A}$ we have, for $\mathbf{B} = $ const,

$$B = \frac{2\pi r A(r)}{\pi r^2} = 2\frac{A(r)}{r}. \tag{16.20}$$

This gives

$$p_v = 2qA(r). \tag{16.21}$$

16.3
Origin of the A-Momentum

According to (16.3), the "magnetic" part of the electric field (due to change of $\mathbf{A}$) is

$$\mathcal{E} = -\frac{\partial \mathbf{A}}{\partial t}. \tag{16.22}$$

How is it incorporated into QM?

The state function Ψ of a particle, together with its Hamiltonian, determines its own rate of change $\partial\Psi/\partial t$ through the Schrödinger equation (16.9). According to this equation, even a very rapid change of $\mathbf{A}$ immediately affects only $\partial\Psi/\partial t$, not Ψ itself. The latter will feel the change and accordingly respond only after some delay.

With this in mind, consider the following situation [3]. Suppose we have a long solenoid and a stationary charged particle outside (Figure 16.1). Turn on the current at a certain moment, say, $t = 0$. Almost immediately, there emerges magnetic field within the solenoid and the corresponding magnetic vector potential $\mathbf{A}$ circling around the symmetry axis z (in order for this to happen "almost immediately," the ratio L/R of the solenoid's inductance to resistance must be very small; see the theory of $L–R$ circuits [3,5]).

Note that $\mathbf{A}$ jumps into existence outside the solenoid as well, although it does not create any magnetic field there (there is no magnetic field outside a tightly wound and sufficiently long solenoid, and the fringe fields around its end points are so far that they can be neglected within the region of our interest). Thus, we have no field ($\mathbf{A} = 0$) before the initial moment; right after this moment, we have $\mathbf{A} \neq 0$ everywhere, and

$$\mathbf{B} = \vec{\nabla} \times \mathbf{A} = \begin{cases} 0, & \text{outside the solenoid,} \\ \text{const} \neq 0, & \text{within the solenoid.} \end{cases} \tag{16.23}$$

16 Charge in Magnetic Field

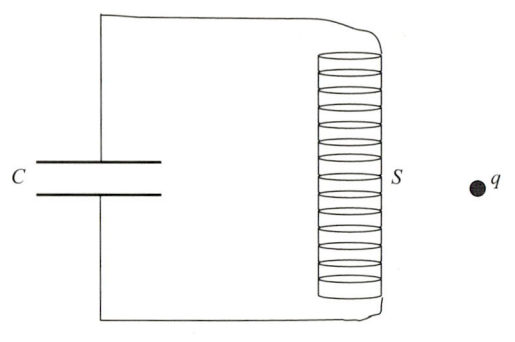

Figure 16.1 The origin of A-momentum. A changing magnetic field inside of solenoid S produces A-field *outside* as well. The changing **A**, in turn, generates electric field that acts on the charged particle outside.

But, as mentioned above, Ψ (and therefore its gradient $\vec{\nabla}\Psi$) will not immediately respond. Since the operator $-i\hbar\nabla$ represents the generalized or dynamic momentum $\mathbf{P} = \mathbf{p}_v + \mathbf{p}_A$ of the particle, this momentum remains essentially unchanged, either. But this does not pertain to its constituents $\mathbf{p}_v$ and $\mathbf{p}_A$. Indeed, the A-momentum is directly related to **A** by Equation 16.18 and therefore jumps up together with **A** at $t = 0$. And since the sum (16.19) remains constant at this moment, the instantaneous change of $\mathbf{p}_A$ must be compensated for by the immediate and opposite change of $\mathbf{p}_v$. The velocity of the particle must jump together with **A**. What physical process is responsible for it?

Well, it is the change of **A**. According to (16.22), the rapid change of **A** is associated with a huge electric field that accelerates the charge so rapidly that it picks up the nonzero velocity simultaneously with the change of **A**. And this velocity is just right to compensate for the change in $\mathbf{p}_A$ and ensure the constancy of **P**. Indeed, we have, in view of (16.22),

$$\mathbf{v} = \int_0^{\Delta t} \mathbf{a}\, dt = \int_0^{\Delta t} \frac{q\mathbf{E}}{\mu} dt = -\frac{q}{\mu}\mathbf{A}, \tag{16.24}$$

so that

$$\mu\mathbf{v} = \mathbf{p}_v = -q\mathbf{A} = -\mathbf{p}_A, \tag{16.25}$$

and $\mathbf{P} = \mathbf{p}_v + \mathbf{p}_A$ remains zero if the charge was initially at rest. It is due to *its conservation at sudden change of* **A** that the dynamic momentum is such an important characteristic of a charged particle in both CM and QM. Another important conclusion from this argument is that not only A-momentum, but also the v-momentum can be expressed as an explicit function of **A**.

The mathematical structure of QM allows us to derive the obtained results in a formal way. Namely, we can derive Equation 16.17 by determining the velocity operator as time derivative of **r** and then determining the latter as the Poisson quantum bracket by applying Equations 11.21 and 11.23:

$$\hat{\mathbf{v}} = \frac{d\hat{\mathbf{r}}}{dt} = \frac{d\mathbf{r}}{dt} = \frac{1}{i\hbar}(\mathbf{r}\hat{H} - \hat{H}\mathbf{r}). \tag{16.26}$$

We know that in the absence of magnetic field ($\mathbf{A} = 0$) this gives $\hat{\mathbf{v}} = \hat{\mathbf{p}}/\mu$. In the presence of magnetic field,

$$\frac{d\mathbf{r}}{dt} = \frac{1}{2i\mu\hbar}\left[\mathbf{r}(\hat{\mathbf{p}} - q\mathbf{A})^2 - (\hat{\mathbf{p}} - q\mathbf{A})^2\mathbf{r}\right], \tag{16.27}$$

and the calculation of the commutator is more involved. If the reader takes the pain to do it (Problem 16.23), he/she will obtain $d\mathbf{r}/dt = (1/\mu)(\hat{\mathbf{p}} - q\mathbf{A})$, which is identical with (16.17).

16.4
Charge in Magnetic Field

Once we found the general expressions for the Hamiltonian of a charged particle in the EM field, we can describe its behavior. We need to solve the corresponding Schrödinger equation (16.9). Since we are now interested in those aspects of particle's behavior that are related to magnetic field alone, we will drop the term with V. We also assume that magnetic field $\mathbf{B}$ is homogeneous. Choosing the z-axis along $\mathbf{B}$, we have

$$B_z = B, \qquad B_x = B_y = 0, \qquad V = 0. \tag{16.28}$$

Classically, the motion would be described by Newton's law with the Lorentz force $\mathbf{f}_M = q\mathbf{v} \times \mathbf{B}$ or, in coordinates,

$$\ddot{x} = \omega\dot{y}, \qquad \ddot{y} = \omega\dot{x}, \qquad \ddot{z} = 0, \qquad \omega \equiv \frac{q}{\mu}B. \tag{16.29}$$

The well-known general solution of these equations describes the particle's motion in a helix with the symmetry axis along $\mathbf{B}$. This is a combination of a uniform motion along $\mathbf{B}$ and a circular motion in the plane perpendicular to $\mathbf{B}$ (the xy-plane). The center of the corresponding circle, its radius, and the pitch of the helix are determined by the initial position vector $\mathbf{r}_0$ and the initial momentum $\mathbf{p}_0 = \mu\mathbf{v}_0$ of the particle. The rotational frequency ω is often called the *cyclotron frequency* since it is observed in special kind of accelerators – *cyclotrons* – used in high-energy physics.

Already at this point, we can envision some basic features of the corresponding QM solution. Sliding along the z-axis, as any free motion with definite p_z, must be described by the de Broglie wave $\psi(z) = \psi_0 e^{ik_z z}$ with $k_z = p_z/\hbar$. This trivial component of motion will not interest us and we will drop it from our equations or just set $k_z = 0$. The circular motion with frequency ω in the xy-plane can be represented as a superposition of two harmonic oscillations with the same frequency and amplitude, one along x and one along y, but phase shifted by 90° with respect to one another. Each oscillation separately could be described as the motion in the corresponding parabolic potential studied in Section 9.6. Their QM superposition would be described as the motion in 2D with the *effective potential*

$$U(x,y) = \frac{1}{2}\mu\omega^2(x^2 + y^2) \equiv \frac{1}{2}\mu\omega^2 s^2. \tag{16.30}$$

Therefore, it may seem that instead of solving the problem for motion in magnetic field **B**, we could just use already familiar solution for quantum oscillator, generalized for two dimensions. But the described analogy is deceptive, and in many respects the potential (16.30) is different from magnetic field. The linear (1D) oscillations of a charged particle, which are possible in the field (16.30), are absolutely excluded in any magnetic field. Generally, the potential (16.30) admits combination of *independent* motions along x- and y-directions with different amplitudes and with an *arbitrary* phase shift between them. That would produce elliptical orbits, so the circular motion is only a very special case in this field. In contrast, the charge motion (in the xy-plane) in homogeneous magnetic field can be *only circular*.

Therefore, we only mention about invoking the effective potential (16.30) as a tempting possibility, and turn to description of the effect in magnetic field. This will also allow us to demonstrate the gauge invariance at work.

The magnetic field (16.28) can be described by a vector potential **A** with components

$$A_x = -\frac{1}{2}By, \qquad A_y = \frac{1}{2}Bx, \qquad A_z = 0. \tag{16.31}$$

Taking $\vec{\nabla} \times \mathbf{A}$ yields the given field **B**. The advantage of this choice of **A** is that it is axially symmetric as is the magnetic field itself, and therefore using it must give us the expected axially symmetric solution. The disadvantage is that the solution involves two nonzero components and is accordingly much more complicated. We can utilize the gauge invariance to select another form of **A**, having only one nonzero component. So we perform the gauge transformation (16.4) $\mathbf{A} \to \mathbf{A}' = \mathbf{A} + \vec{\nabla}\Lambda$, and the appropriate choice of Λ in our case will be

$$\Lambda = -\frac{1}{2}Bxy. \tag{16.32}$$

With this Λ, (16.4) gives

$$A'_x = -By, \qquad A'_y = A'_z = 0. \tag{16.33}$$

Taking curl $\vec{\nabla} \times \mathbf{A}'$ gives the same magnetic field (16.28). Therefore, we can use **A**′ as confidently as **A** but with much less hassle in computations. Putting **A**′ into Equation 16.9, we obtain

$$\left(-\frac{\hbar^2}{2\mu}\nabla^2 - i\frac{q\hbar}{\mu}By\frac{\partial}{\partial x} + \frac{q^2}{2\mu}B^2y^2\right)\Psi(x,y) = E\Psi(x,y). \tag{16.34}$$

Look for solution in the form

$$\Psi(x,y) = e^{ik_x x}\Psi(y), \tag{16.35}$$

with an arbitrary k_x. This expression is not axially symmetric: instead of expected circular motion in the xy-plane, we have a plane wave propagating along the x-direction and modulated by some function of y. But let us postpone the judgment until we obtain the exact solution.

16.4 Charge in Magnetic Field

Putting (16.35) into (16.34) gives the equation for $\Psi(y)$:

$$\left(-\frac{\hbar^2}{2\mu}\frac{d^2}{dy^2} + \frac{q\hbar}{\mu}Bk_x y + \frac{q^2}{2\mu}B^2 y^2\right)\Psi(y) = \left(E - \frac{\hbar^2 k_x^2}{2\mu}\right)\Psi(y). \qquad (16.36)$$

Change the variable $y \to y'$ so that

$$y' = y - y_0, \qquad y_0 \equiv -\frac{\hbar k_x}{qB}. \qquad (16.37)$$

Then (16.36) simplifies to

$$\left(-\frac{\hbar^2}{2\mu}\frac{d^2}{dy'^2} + \frac{1}{2}\mu\omega^2 y'^2\right)\Psi(y') = E\Psi(y'), \qquad (16.38)$$

with ω determined in (16.29). We recognize the equation for a quantum oscillator with known solutions

$$\Psi_n(y') = e^{-(1/2)\xi^2} H_n(\xi), \qquad E_n = \hbar\omega\left(n + \frac{1}{2}\right), \qquad (16.39)$$

where

$$\xi = \sqrt{\frac{\mu\omega}{\hbar}}y' = \sqrt{\frac{\mu\omega}{\hbar}}(y - y_0). \qquad (16.40)$$

The total function $\Psi(x, y)$ is

$$\Psi(x, y) = \Psi_{nk_x}(x, y) = e^{-(1/2)\xi^2} H_n(\xi) e^{ik_x x}. \qquad (16.41)$$

This solution does indeed give both the function $\Psi_n(y')$ in (16.39) and the allowed energies (called the *Landau levels*) – the same as those for a quantum oscillator. This is no surprise in view of our introductory comments.

The energy (16.39) can be expressed in terms of magnetic moment $M = gL$ of the circular current in magnetic field. We just note that $\hbar$ is the quantum of angular momentum L. Then, using (10.51), we can write $\hbar\omega = (\hbar q/\mu)B = gLB = M \cdot \mathbf{B}$ and

$$E_n = MB\left(n + \frac{1}{2}\right). \qquad (16.42)$$

Thus, we got the energy quantization for a charged particle in magnetic field. The found eigenstates form the basis for periodic motion with the cyclotron frequency. This is all very well, but what about the *x*-component of motion? On the face of it, we have, as mentioned above, just uniform propagation of a modulated plane wave having nothing to do with oscillations around a fixed center. But this apparent contradiction arises because we naively used the equation $v_x = p_x/\mu$, which holds only in the absence of **B**. At $\mathbf{B} \neq 0$, it is replaced by Equation 16.17 with generalized momentum. Applying this to our case and using (16.33), we obtain

$$v_x = \frac{\hbar k_x}{\mu} + \omega y = \omega(y - y_0). \qquad (16.43)$$

We see that the x-component of particle's velocity depends on y-coordinate! There is already a correlation between motions along two perpendicular directions. But there is more to it! As we found in Section 9.7, the classical oscillations emerge as periodic changes of the relevant average characteristics in certain nonstationary states. In particular, we found (Equation 9.94) that $\langle y(t)\rangle = a \sin \omega t$, where a is an effective oscillation amplitude. Applying this result to (16.43), we can write for the expectation value of v_x at a certain moment t:

$$\langle v_x(t)\rangle = \left\langle \frac{dx}{dt}\right\rangle = \omega(\langle y(t)\rangle - y_0) = \omega(a \sin \omega t - y_0). \tag{16.44}$$

The first term on the right (oscillating y-coordinate), determining v_x-oscillations on the left, is a unique signature of circular motion! The second term, as is clearly seen from the structure of the above expressions, corresponds to the classical y-coordinate of the center of rotation. However, a *homogeneous* field **B** does not single out any point in space, so the value of y_0 in (16.37) is determined by our choice of k_x. But the corresponding expression (16.41) is only a special solution. Since the energy E_n in (16.39) does not depend on k_x, we have infinite degeneracy – a continuous set of distinct states (16.41) with different k_x, all belonging to the same E_n. Therefore, any superposition of such states with different k_x will also be a possible eigenstate with the same energy. Thus, the general solution for the Hamiltonian with vector potential (16.33) can be written as

$$\tilde{\Psi}_n(x,y) = \int c(k_x) e^{-(1/2)\xi^2} H_n(\xi) e^{ik_x x} dk_x. \tag{16.45}$$

In particular, $c(k_x)$ can be chosen so that $\tilde{\Psi}(x,y)$ will describe a state with definite classical x-coordinate of the center of rotation, defined as $x_0 \equiv x + (\hbar k_y/qB)$. Here both terms making x_0 are *variables*, but the corresponding operator $\hat{x}_0 \equiv x + (\hbar \hat{k}_y/qB)$ commutes with $\hat{H}$, so x_0 is the conserving quantity. However, the observables x_0 and y_0 are incompatible because their operators do not commute. Therefore, we can measure one of them, but not both. In contrast with the classical situation, the center of rotation in magnetic field cannot be completely defined! This explains the lack of axial symmetry in states (16.41) and (16.45).

16.5
Spin Precession

Another example of intimate connection between the phase and gauge transformations is spin precession of a charged particle in magnetic field. We start with its classical description. A particle with magnetic moment $\mathcal{M}$ in a magnetic field **B** acquires energy $U_\mathcal{M} = -\mathcal{M} \cdot \mathbf{B} = -\mathcal{M} B \cos\theta$, where θ is the angle between $\mathcal{M}$ and **B**. We can interpret $U_\mathcal{M}$ as the potential energy associated with θ, but θ here specifies particle's orientation, not its position in space. Angle θ determines the torque on the particle

$$\boldsymbol{\tau} = \mathcal{M} \times \mathbf{B} = \mathcal{M} B \sin\theta \, \hat{\mathbf{n}}, \tag{16.46}$$

where $\hat{\mathbf{n}}$ is the unit vector perpendicular to both $\mathcal{M}$ and $\mathbf{B}$ and forming the right triplet with them. If the field is nonhomogeneous, there is also a net magnetic force on the particle

$$\mathbf{F} = -\vec{\nabla} U_\mathcal{M}(\theta) = -\vec{\nabla}(\mathcal{M} \cdot \mathbf{B}). \tag{16.47}$$

It vanishes when $\mathbf{B} = $ const.

Here we will focus only on the effects associated with the torque, which exists even in a homogeneous field. According to (16.46), if $\mathcal{M} \| \mathbf{B}$, there is no torque. But if $\mathcal{M}$ and $\mathbf{B}$ are misaligned, there appears torque tending to line up $\mathcal{M}$ with $\mathbf{B}$ (Figure 16.2). One might think that $\mathcal{M}$ in this case would just start swinging about $\mathbf{B}$ as a pendulum swings about the vertical in the gravitational field. But magnetic moment is associated with angular momentum, $\mathcal{M} = g\mathbf{L}$. A nonzero $\mathbf{L}$ changes the particle's response to torque: instead of swings, there will be precession. Torque affects the angular momentum $\mathbf{L}$ according to

$$\boldsymbol{\tau} = \frac{d\mathbf{L}}{dt}, \tag{16.48}$$

which is the rotational analogue of $\mathbf{F} = d\mathbf{p}/dt$ for translational motion. Since, according to (16.46), $\boldsymbol{\tau} \perp \mathcal{M}, \mathbf{L}, \mathbf{B}$, the tip of $\mathbf{L}$ is moving in direction perpendicular to $\mathbf{B}$, which means rotation (precession) of $\mathbf{L}$ around $\mathbf{B}$ (Figure 16.2). Let us estimate its period. Using (16.48) and Figure 16.2, we have

$$T = \frac{2\pi L \sin\theta}{dL/dt}. \tag{16.49}$$

In view of (16.48) and (16.46), this gives

$$T = \frac{2\pi L}{\mathcal{M} B}, \quad \omega = \frac{2\pi}{T} = \frac{\mathcal{M}}{L} B = 2\frac{q}{\mu} B. \tag{16.50}$$

The result is intuitively clear: it is natural that stronger magnetic field would cause more rapid precession. On the other hand, ω does not depend on θ. Change of

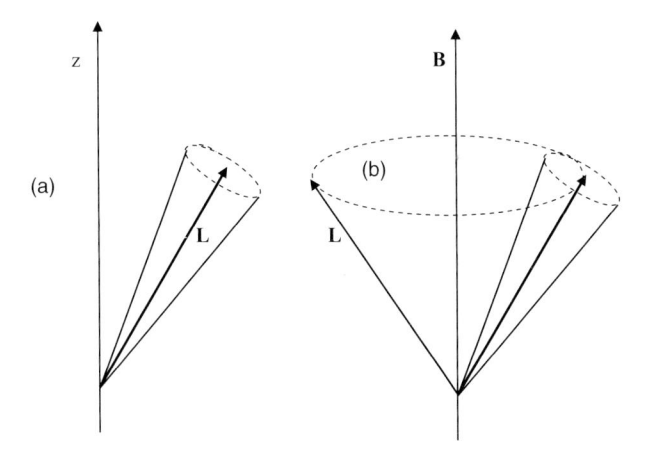

Figure 16.2 The net angular momentum **L** of a charged particle: (a) in the absence of magnetic field; (b) in the presence of magnetic field **B** (precession).

orientation of a bar magnet relative to **B** will not affect its precession rate. But the effect itself is most noticeable when $\theta = 90°$ and the tip of $\mathcal{M}$ traces out the largest possible circle.

Now we turn to the QM description. As we know, any change of a system in time involves superposition of stationary states with different energies. Similarly, we expect that spin precession originates from superposition of two different eigenstates of the system's Hamiltonian. Since we are now interested only in precession, we focus only on the part of $\hat{H}$ that represents magnetic interaction:

$$\hat{H}_\mathcal{M} = U_\mathcal{M} = -\mathcal{M}\mathbf{B} = -g\mathbf{sB} \tag{16.51}$$

(when dealing with spin, we switch to its conventional notation $\mathbf{L} \to \mathbf{s} = (1/2)\hbar\boldsymbol{\sigma}$).

Let us take the direction along **B** as a basis. Then, for $s = 1/2$, there are two basis states:

$$s_z = \frac{1}{2}\hbar, \quad |\uparrow\rangle = \begin{pmatrix} 1 \\ 0 \end{pmatrix}, \quad \text{and} \quad s_z = -\frac{1}{2}\hbar, \quad |\downarrow\rangle = \begin{pmatrix} 0 \\ 1 \end{pmatrix}. \tag{16.52}$$

For a negatively charged particle (an electron or negative muon), the magnetic moment is opposite to **s**, so in spin-up state the magnetic moment points down, and vice versa. The corresponding eigenstates are

$$|\psi_1\rangle = |\uparrow\rangle e^{-i\omega t}, \quad |\psi_2\rangle = |\downarrow\rangle e^{i\omega t}, \quad \omega = \frac{\mathcal{M}B}{\hbar}, \tag{16.53}$$

with the respective energy eigenvalues $E_M^{1,2} = \mp\hbar\omega$. Each state separately, being an eigenstate of $\hat{H}$, is stationary. All its observable characteristics do not depend on time.

Consider now a state most favorable for observing precession – when $\mathbf{s} \perp \mathbf{B}$ (i.e., $m = 0$). For $s = 1/2$ such a state is not an eigenstate of $\hat{H}$, since a half-integer spin does not have integer eigenvalues of s_z. The only way the state with $\mathbf{s} \perp \mathbf{B}$ can be realized is through an equally weighted superposition of the two states (16.53). Another way to see it is to notice that spin-up and spin-down states are both needed to form a state with the zero *expectation value* $\langle s_z \rangle = 0$, which we want to describe.

We are coming to an interesting point: so far, we have not learned the rules governing the behavior of spin 1/2 in an external magnetic field. This ignorance, however, does not preclude us from correct description of the process because the knowledge of the eigenstates (16.53) is sufficient for describing their superposition. Once we understood this point, we can generalize it to states with arbitrary θ (Problem 16.10) and even to particles with spin higher than 1/2, but here we will stick with the case at hand. All we need is to take the superposition

$$|\Psi(t)\rangle = c_1|\uparrow\rangle e^{-i\omega t} + c_2|\downarrow\rangle e^{i\omega t}. \tag{16.54}$$

Since we chose the case with $\mathcal{M} \perp \mathbf{B}$ (i.e., the expectation value $\langle s_z \rangle = 0$), we know that there must be $|c_1| = |c_2| = 1/\sqrt{2}$. Let us choose such an axis in plane perpendicular to **B** that with respect to this axis, both amplitudes have the same phase, and since the value of the common phase factor is immaterial, we can set it to zero. Call this axis x. Thus, if it is known that at some initial moment the x-component of particle's spin is $s_x = 1/2$ (in $\hbar$ units), then, denoting this state as $|+\rangle$, we can express it in terms of up and down states as

$$|\Psi(0)\rangle = \frac{1}{\sqrt{2}}(|\uparrow\rangle + |\downarrow\rangle) = |+\rangle. \tag{16.55}$$

And at a later moment,

$$|\Psi(t)\rangle = \frac{1}{\sqrt{2}}\left(|\uparrow\rangle e^{-i\omega t} + |\downarrow\rangle e^{i\omega t}\right). \tag{16.56}$$

Then the amplitude of finding the particle's spin along $+x$ at a moment t is given by projection of $|\Psi(t)\rangle$ onto $|+\rangle$: $c_x(t) = \langle +|\Psi(t)\rangle$. Using (16.55), making the inner products, and taking into account that basis $|\uparrow\rangle, |\downarrow\rangle$ is orthonormal, we obtain

$$c_x(t) = \frac{1}{2}\left(e^{i\omega t} + e^{-i\omega t}\right) = \cos \omega t. \tag{16.57}$$

The probability of finding the spin pointing along $+x$ is

$$\mathcal{P}_{+x}(t) = \cos^2 \omega t. \tag{16.58}$$

The amplitude $c_x(t)$ oscillates with the period $T = 2\pi/\omega = 2\pi s/MB$ determined by (16.50). But the probability oscillates with half of this period, and accordingly, its oscillation frequency is 2ω. In accord with our assumption that initially the spin is with certainty along the $+x$ direction, we have $\mathcal{P}_{+x}(0) = 1$. For the chance to find the spin pointing along some other direction making an angle φ (or $-\varphi$, depending on sense of rotation) with $+x$ under the same initial condition, the result will be phase shifted by φ:

$$\mathcal{P}_{\pm\varphi}(t) = \cos^2(\omega t \mp \varphi). \tag{16.59}$$

Generally, we have a uniform rotation of spin in the xy-plane about the z-axis.

The described precession is difficult to observe directly. But it can be observed indirectly by monitoring some other effects connected with precession. One of such effects is the decay of muon (μ^-) in a magnetic field. A muon belongs to the same family of particles as an electron, and in many respects it can be considered just as a more massive, that is, excited and accordingly unstable, state of an electron. It decays into an electron and a pair of neutrinos: $\mu^- \to e^- + \nu + \tilde{\nu}$. An essential feature of this decay is that the electron is emitted primarily in the direction opposite to the spin direction of μ^- [3]. So detecting the electrons emitted in a certain direction, we know the spin direction of the μ^- just before the decay. If we let a beam of polarized muons with the known initial spin state into a region with a uniform magnetic field with $\mathbf{B} \perp \mathbf{s}_\mu$, the spin will start precessing. The precession frequency is determined by (16.50) with $\mu_\mu \simeq 206\mu_e$. The probability of finding the muon's spin pointing in a certain direction will periodically change with time. Accordingly, the same will happen with the preferable direction of the emitted electrons. Therefore, the detection rate of the electrons arriving at a certain detector placed in a fixed position must periodically change between maximum and minimum (which is close to zero) with the period given by (16.50) for muons. Such periodic variation will be an experimental evidence of the spin precession and we could measure its frequency by measuring the period of probability oscillations. Such an experiment is technically difficult since it requires all the muons to be initially polarized in the same state, say, $+x$. In other words, we must prepare a pure ensemble of muons, at least with respect to the spin variable. But it is executable. It has been carried out and confirmed theoretical predictions. Now the described effect is used as a tool in studies of many effects in magnetic field [6–9].

Box 16.1 Spinors

In QM, the angle θ in a spin eigenstate cannot approach zero because of quantization of both $|\mathbf{L}|$ and L_z. Therefore, θ is also quantized, with allowed values given by $\cos\theta = m/\sqrt{l(l+1)}$. Albeit this equation was derived for the orbital momentum, it describes the properties of spin as well. We must just change the notation $l \to s$ to get $\cos\theta = m/\sqrt{s(s+1)}$. For spin, the minimal possible value of θ is determined by condition $m = s$. It is easy to check that for $s = 1/2$ we have two possible angles, $\theta_{1,2} = 30°, 150°$. Either of them is pretty far from being aligned with the z-axis. All we could say is that the spin in this state is represented by a conical surface with vertex at the origin, symmetry axis z, and an open angle of 60°.

Generally (recall Chapter 7), with a certain direction as a basis, the spin eigenstates are characterized each by the corresponding conical surface with the *symmetry axis* along this direction, *not* the spin itself in this direction. In QM, angular momentum in general, and spin in particular, is not a vector as it is understood in geometry and in CM. The professional term for it is *spinor*.

Box 16.2 Precession and the superposition of $\hat{s}_z$-eigenstates

The above point provides another illustration of the difference between the spin eigenstates and their superpositions. Consider, for example, an eigenstate of $\hat{s}_z$-operator. It is also the eigenstate of the Hamiltonian, and as such, is a stationary state. There is nothing changing in such a state, all physical characteristics remain constant, just as they are constant in an eigenstate of $\hat{p}$-operator. There is nothing moving in an eigenstate of $\hat{p}$ (except for the probability flux that is not directly observed!) and, similarly, there is no precession in an eigenstate of $\hat{s}_z$, even though the particle may be in a magnetic field parallel to z. For precession to exist, QM *demands* superposition of the corresponding states. The superposition of distinct eigenstates of $\hat{s}_z$ is fundamentally different from each constituent eigenstate taken separately. In particular, in the presence of magnetic field, different eigenstates have different energy, so their superposition is not a stationary state. This is a necessary condition for precession. Note that while the eigenstates of $\hat{s}_z$ form a restricted discrete set ($2s + 1$ distinct states), their possible superpositions form a continuous set of states. Accordingly, the possible expectation values of θ form a continuous set.

16.6
The Aharonov–Bohm Effect

One of the most spectacular manifestations of the wave aspect of matter is a version of the double-slit experiment in the presence of magnetic field **B**. Running slightly ahead of schedule, we tell here the outcome right upstart: the magnetic field affects the interference pattern – the pattern undergoes a displacement depending on **B**, particle's momentum, and geometry of the experiment. One may think "What is so spectacular here? The shift of the pattern is an obvious effect: the magnetic field acts on moving electrons with the Lorentz force. If the electrons move horizontally from the slits to the second screen and **B** is vertical, the Lorentz force will be horizontal and parallel to the screen – precisely along the direction of the pattern shift. Since the velocities of all the electrons are practically the same, and assuming **B** homogeneous, the Lorentz force is also the same for all of them, so the landing places for all electrons will be shifted by the same amount – hence a uniform shift of the whole pattern."

This argument by itself is correct, but in the version of the experiment we are going to discuss, there is no magnetic field in the region of motion of electrons! Whatever field there is in this experiment, it is not uniform – it is squeezed into a very thin cylindrical region. This can be achieved by placing a very long and thin, tightly wound solenoid between the slits (Figure 16.3). The word "long" means that the solenoid's ends are so far away that we may not bother about the fringe fields; the word "thin" means that the diameter of the solenoid is much less than the distance d between the slits. Under these conditions, the magnetic field outside the solenoid is practically zero, no matter how strong it may be inside. So magnetic field does exist when we turn on the current, or place a magnetized iron whisker instead of solenoid, but it exists in the extremely narrow region, where there are no electrons, and is zero

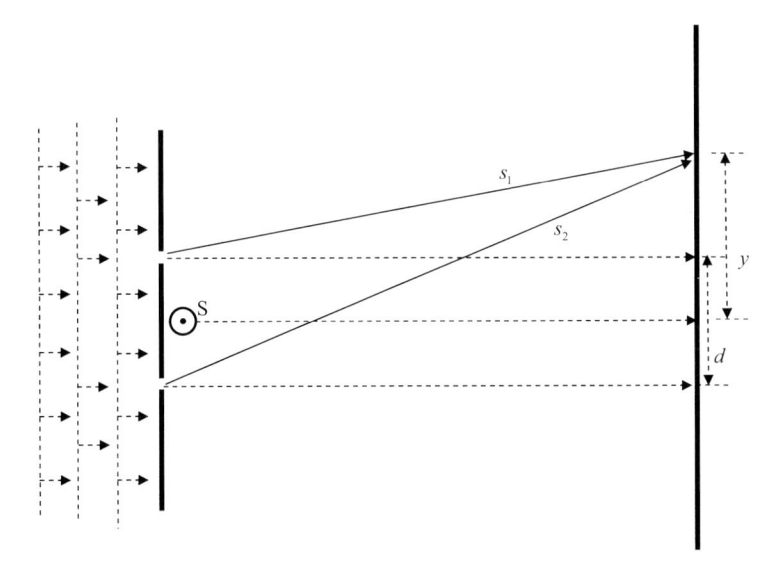

Figure 16.3 The double-slit experiment with magnetic field inside of a long narrow solenoid S.

in the region where we have moving electrons. So there is no additional force on the electrons in this experiment, and yet we observe the shift of the pattern. On the other hand, there would be no shift without the field, for instance, if we remove the magnetized whisker or turn off the current in the solenoid. Thus, the field *is* necessary to produce the shift, but it does this in some distant region far from the place where it is on. It looks like QM has restored the mystical *action at a distance*, which was so much hated by Newton and other great creators of classical physics. How can we account for this strange result?

There is no way to do it in classical physics, but we can, indeed, easily do it in QM. The pattern on the screen originates from interference of each single electron with itself. And this interference results from the phase difference between any two virtual paths – one through slit I and the other through slit II – to a given landing point. And the phase along the path is determined, apart from the path's geometry, also by *potential function* of the fields acting in some region even if all the paths are beyond this region. We emphasized in Sections 16.1–16.3 the distinction between a *field* producing a local force on a particle and the field-generating *potential* that can extend even in a field-free region, and that it is the potential, not the field, that determines the Hamiltonian. Now we came to a situation where this feature is manifest in its full fledge. In the case of EM field, the phase is determined by a 4-vector $A = (V, \mathbf{A})$. The actual observables – electric and magnetic fields – are determined from $(V, \mathbf{A})$ by Equation 16.3.

As a starting example consider a narrow parallel-plate capacitor with a strong electric field inside. We know that there is no field outside the capacitor and therefore it may seem that the capacitor cannot affect the particles' motion in the outer region. But the field inside the capacitor will produce the potential difference between the plates. Some people would prefer to say that, conversely, it is the potential difference between the plates that produces the field within. This also makes sense. In any event, the potential on one side of the capacitor will be higher than that on the other side (Figure 9.2). Accordingly, the kinetic energy and momentum of a charged particle moving *along the plates* in the outer region will depend on which side of the capacitor it is moving. Suppose we take the potential on the left side as a reference point, so $U = 0$ in the left semispace. On the right, we will have $U = qV$, where V is the potential difference across the plates. As a result, the momentum of the particle with a virtual path in the left region will remain equal to its initial momentum $p = \sqrt{2\mu E}$; for a virtual path on the right, the momentum will be $p' = \sqrt{2\mu(E - U)}$ (assuming this route is passable, that is, $E > U$). If such a capacitor is placed between the slits instead of the solenoid in the double-slit experiment (Figure 16.4), there will appear an additional phase difference between the paths leading to the landing point from different slits:

$$\Delta\varphi = \varphi - \varphi' = (k - k')s = \frac{\sqrt{2\mu}}{\hbar}\left(\sqrt{E} - \sqrt{E - U}\right)s, \qquad (16.60)$$

where s is the length of the plates along the path. For $U \ll E$, this gives

$$\Delta\varphi \approx \frac{1}{2}\varphi\frac{U}{E}, \qquad (16.61)$$

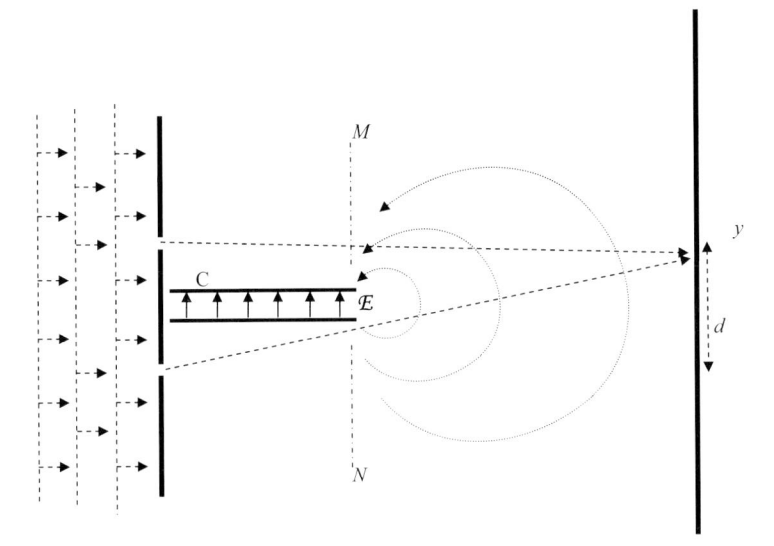

Figure 16.4 The double-slit experiment with electric field inside of a large narrow parallel-plate capacitor C. The interior of the capacitor is practically inaccessible for the electrons in the given setup. In the region to the left of MN, the capacitor does not affect *directly* the electrons' motion, but it produces the phase shift between the interfering virtual paths. This shift by itself would cause the corresponding shift of the diffraction pattern on the screen. However, it is counterbalanced by the opposite shift due to the *direct effect* of the fringe field (dotted lines) to the right of MN. Such direct effect (actual physical force on the electrons) can be eliminated in an experiment with *magnetic field*, which is characterized by different topology: magnetic potential is a *vector* field **A** with closed field lines and without equipotential surfaces. This allows the experimental arrangement shown in Figure 16.3, in which the fringe magnetic field **B** is practically absent in all region of electrons' motion. As a result, the magnetic potential **A** is the sole actor in the play and produces the observable shift of the diffraction pattern.

where $\varphi = ks$ is the phase along *the same stretch* of either path in the absence of capacitor (see more details in Ref. [10]). Another way to see it is by thinking of U as the change in the *refractive index* of a medium [11]: the whole thing is analogous to the accumulation of phase difference between the two paths with different refractive index for a photon in, say, Mach–Zehnder interferometer. As a result of this phase shift, the point on the second screen on the symmetry axis of the optical device (right in the middle of the screen between the slits), instead of being the center of the bright fringe, may have lower brightness or even be the center of a dark fringe. The *additional* phase shift due to difference of refractive index for the two paths is essentially the same for all landing points. Therefore, it would cause the uniform shift of the whole interference pattern.

Returning to the language of potential U, one can show (Problem 16.13) that for an electron traveling all the distance s along the capacitor, the accumulated phase difference between the two paths would cause the shift on the screen by

$$\delta y = \frac{1}{2}\frac{U}{E}\frac{L}{d}s, \tag{16.62}$$

where L is the distance between the screens. We see the embryo of another miracle of the quantum world: the constant potential does not produce any force on the electron, and yet it produces the phase difference between the paths separated by the capacitor. Since the electron takes both of them, and they are at different potentials, it feels the *potential difference* between the two paths! And the potential difference in this situation is equivalent to an effective force. The same old secret that cannot be explained for a classical particle, but can be for a wave spread over a region. This is the way in which the potential can affect the interference pattern without producing any *field* along the path.

The attentive reader may notice that expression (16.62) was referred to as a *would-be* shift of the pattern on the screen, and the phase difference (16.60) and (16.61) accumulated along the segment s is only an embryo of the miracle. Why? Because there is a fringe field *directly* acting on the electron after it passes the capacitor (Figure 16.4), and this action produces the additional phase shift between the paths opposite to the one accumulated before. As a result, when the paths meet at a landing point, only the phase difference due to difference in geometric path lengths survives, and the actual diffraction pattern remains practically the same as in the absence of capacitor. The fringe electric field kills the embryo of the effect we want to describe. The only reason we still included this example in our description is that starting with the electric potential difference may be easier for the first reading than starting with magnetic vector potential $\mathbf{A}$. Also, it may provoke the reader to look deeper into the nature of the phenomenon.

Now we can apply the same argument to the vector potential $\mathbf{A}$. According to (16.7), it affects the Hamiltonian through the electron's momentum rather than through kinetic energy as in (16.60). And the geometry (we would even say, topology) of the $\mathbf{A} \leftrightarrow \mathbf{B}$ relationship is such that allows us to remove the fringe fields from the region of the electron motion. Moreover, we can get rid of these fields altogether by extending the solenoid and connecting its ends together far behind the experimental setup. Otherwise, the mechanism is the same: the presence of magnetic field in one region of space can affect the probability amplitude of a charged particle in another region now totally *free of the field*. There is no local action of a force causing acceleration. Instead, we have the phase shift along a path (another example of nonlocality!) caused by the vector potential:

$$\varphi(S) = \frac{q}{\hbar} \int_S \mathbf{A} \, d\mathbf{s}. \tag{16.63}$$

Here the integral is taken along the path S of interest. This is all we need in order to describe quantitatively the effect of the solenoid on the interference pattern. We just apply (16.63) to both paths leading to a landing point from different slits (Figure 16.3) and then the difference between the results will give us the shifting effect. We have

$$\varphi(S_1) = \frac{q}{\hbar} \int_{S_1} \mathbf{A} \, d\mathbf{s}, \quad \varphi(S_2) = \frac{q}{\hbar} \int_{S_2} \mathbf{A} \, d\mathbf{s}. \tag{16.64}$$

The difference is

$$\Delta\varphi(S_1, S_2) = \varphi(S_1) - \varphi(S_2) = \frac{q}{\hbar} \int_{S_1} \mathbf{A}\,d\mathbf{s} - \frac{q}{\hbar} \int_{S_2} \mathbf{A}\,d\mathbf{s}. \tag{16.65}$$

But this is the same as taking one integral along the *closed loop L* embracing both paths:

$$\Delta\varphi(S_1, S_2) = \frac{q}{\hbar} \oint_L \mathbf{A}\,d\mathbf{s}. \tag{16.66}$$

(Here we, in order to complete the two paths to the closed loop, have added the short stretch running directly from one slit to the other along the screen. This does not add any noticeable contribution to the result.)

This can be expressed in terms of **B** as well, if we apply the Stokes theorem known in vector calculus [12] and relating the line integral of a vector along a closed loop to the curl of this vector producing the flux through the loop. Using this theorem and recalling that $\vec{\nabla} \times \mathbf{A} = \mathbf{B}$, we have

$$\Delta\varphi(S_1, S_2) = \frac{q}{\hbar} \iint_A \mathbf{B}\,d\mathbf{a} = \frac{q}{\hbar} \Phi_B, \tag{16.67}$$

where A is the area enclosed by the loop and Φ_B is the magnetic field flux through A. In our geometry, all contribution to Φ_B comes from cross-sectional area Ω of the solenoid, and (16.67) is expressed in terms of experimental characteristics as $\Delta\varphi = (q/\hbar)\Omega B$. Together with the shift due to purely geometric path difference, these expressions determine the resulting interference pattern that is shifted relative to the pattern without field. It is easy to see that both $\mathbf{B}$ and $\Delta\varphi(S_1, S_2)$ are invariant under the gauge transformation (16.4) together with appropriate local phase transformation (3.15).

16.6.1
Discussion

Both expressions (16.66) and (16.67) are equally legitimate representations of the phase shift caused by the presence of magnetic field. This may produce an impression that **B** retains its status of something real whereas **A** is merely a mathematical construction producing the same result in calculations. But keep in mind that the classical idea of the actual field is based on its local action on a particle at a given point. In our case, the field is zero in all regions available to the particle, so it is impossible to explain its influence in terms of local action. On the other hand, there is a field **A** circulating *around the solenoid* – in the region of particle's motion. Therefore, **A** is *at least* as important as **B**. The field **B** does affect the experimental results, but it cannot do it directly. Rather, it does this indirectly – through its plenipotentiary and omnipresent representative **A**, and **A** does it non-locally, without exerting any force – in the way unavailable for **B**. The classical image of the world can be described (although less conveniently) without V and **A**; the real quantum world cannot. According to the classical picture, the only way to learn whether there is a current through a solenoid is to launch a charged particle *into* it. According to QM, the presence of field **B** inside the solenoid can be detected by moving around it at an arbitrary large distance, utilizing the field **A**.

But we must also see the other side of the coin: classically, the only way to do it is to "peep in" (do a probe), which can be done using only one charged particle. In QM, we do not have to scrutinize the tiny interior, but in exchange, we need a pure ensemble of particles: since the world is probabilistic, the landing of a single particle will not be informative! Only in the extremely rare cases when it lands exactly at the center of a "dark fringe" of the field-free pattern could it tell us about the presence of some magnetic field in the solenoid, but a single event in micro-world is never 100% reliable. Apart from this, it would tell nothing about the field strength. For this, we need a pattern produced by a pure ensemble. And even in this case, the quantitative result will be true up to a multiple integer of the minimal field value shifting the pattern by one fringe. If we need the *full* information, we have to perform the measurement with at least two pure ensembles differing from each other by the electrons' energy.

With all that, the mere fact that the potentials V and **A** act in a field-free region and affect the outcome of the experiments shows that they are as essential characteristics of a process as the fields $\mathcal{E}$ and **B**. This fact was known from the onset of QM. It is the potentials, not fields that figure in the Schrödinger equation. And yet the inertia of thought inherited from classical physics and enhanced by the fact that potentials are not uniquely defined had prevented the researchers from seeing full importance of potentials. W. Ehrenberg and R. Siday were the first who did it in 1949 [13], but the effect was named after Aharonov and Bohm [14,15] who suggested the experiment discussed here. Now there is an extensive literature on this phenomenon (see, for example, Refs [16–18]).

16.7
The Zeeman Effect

So far we have considered a single charged particle in magnetic field. Now we extend our query onto an atom with a single valence electron. It may be hydrogen or a hydrogen-like atom (e.g., Na or K). What happens when such an atom is in a magnetic field **B**?

Take the z-axis along **B** and express **B** in terms of magnetic vector potential **A** using Equation 16.31. Putting it into the Hamiltonian, we obtain the Schrödinger equation for stationary states in the form

$$\left(\hat{H}_0 - i\frac{q\hbar}{2\mu} B \left(x\frac{\partial}{\partial y} - y\frac{\partial}{\partial x} \right) + \frac{q^2}{2\mu} B^2 (x^2 + y^2) + \frac{q\hbar}{2\mu} (\sigma_z B) \right) \Psi = E\Psi.$$

(16.68)

Here $\hat{H}_0$ is the Hamiltonian in the absence of **B**, the next two terms represent the effect of **B** on the orbital motion of the electron, and the last term describes

16.7 The Zeeman Effect

interaction between **B** and the electron spin magnetic moment $\mathcal{M}_s$. In Section 16.4, the field **B** was the only actor in the play; therefore, both terms – one linear and one quadratic with respect to **B** – were considered on the same footing. The situation here is different since the electric field within the atom in $\hat{H}_0$ is dominating, so the term with B^2 can be neglected in the first approximation. In the remaining term, the operator $-i\hbar(x(\partial/\partial y) - y(\partial/\partial x)) = -i\hbar(\partial/\partial\varphi) = \hat{L}_z$ is just the z-component of **L**. Therefore, the remaining part of the equation is

$$\left(\hat{H}_0 + \frac{Bq}{2\mu}(\hat{L}_z + \hbar\hat{\sigma}_z)\right)\Psi = E\Psi. \tag{16.69}$$

Take the basis in which the spin matrix is diagonal (i.e., consider σ_z in its own basis). Then

$$\hat{\sigma}_z\Psi = \begin{pmatrix} 1 & 0 \\ 0 & -1 \end{pmatrix}\begin{pmatrix} \Psi_1 \\ \Psi_2 \end{pmatrix} = \begin{pmatrix} \Psi_1 \\ -\Psi_2 \end{pmatrix}, \tag{16.70}$$

and (16.69) splits into two uncoupled equations

$$\left(\hat{H}_0 + \frac{Bq}{2\mu}(\hat{L}_z + \hbar)\right)\Psi_1 = E\Psi_1, \quad \left(\hat{H}_0 + \frac{Bq}{2\mu}(\hat{L}_z - \hbar)\right)\Psi_2 = E\Psi_2. \tag{16.71}$$

In the absence of **B**, this reduces to equation for a hydrogen atom with known solution

$$\Psi_1^0 = \begin{pmatrix} \psi_{nlm} \\ 0 \end{pmatrix}, \quad \sigma_z = 1, \quad \Psi_2^0 = \begin{pmatrix} 0 \\ \psi_{nlm} \end{pmatrix}, \quad s_z = -1 \tag{16.72}$$

and $E = E_{nl}^0$ independent of m.

It is immediately seen that the same solutions will satisfy Equation 16.71 with **B**, but their energy eigenvalues will be different from those in (16.72). Indeed, the states (16.72) are the eigenfunctions of $\hat{L}_z$ with the eigenvalues $m\hbar$; therefore, if we put them into (16.71), the result will be the same states with the eigenvalues

$$\tilde{E}'_{nlm} = E_{nl}^0 + \frac{q\hbar}{2\mu}B(m+1), \quad \sigma_z = 1,$$

$$\tilde{E}''_{nlm} = E_{nl}^0 + \frac{q\hbar}{2\mu}B(m-1), \quad \sigma_z = -1. \tag{16.73}$$

Thus, the solutions are the same as in the absence of **B**: as far as we can neglect B^2, the magnetic field does not deform the atom. But the energy level splits into sublevels (m-degeneracy is eliminated) due to interaction between **B** and the *total* magnetic moment $\mathcal{M} = \mathcal{M}_L + \mathcal{M}_s$ due to both orbital angular momentum of the electron and its spin. The interaction energy depends on orientation of $\mathcal{M}$ with respect to **B**. And direction of $\mathcal{M}$ is determined by that of angular momentum according to $\mathcal{M} = g\mathbf{J}$. Hence, the dependence of splitting on m and σ. For instance, the s-term ($m = 0$) splits into two levels – one is shifted down by amount $(q\hbar/2\mu)B$ for spin down, and the other goes up by the same amount for spin up. Note that the splitting of the s-term occurs only due to the electron spin. This is precisely the splitting into two sublevels that was first observed in the Stern–Gerlach experiments. For the *p*-term without spin, there would be three split sublevels corresponding to $m = 0, 1, -1$, respectively. With the spin, we have five sublevels (three for spin up and three for spin down, but two of them are merged since for them shift in one direction due to the *m*-value is compensated for by equal shift in the opposite direction due to the σ-value). So we have to distinguish between the two kinds of shifts.

The splitting of the energy levels in the magnetic field increases the number of optical transitions and accordingly the number of observed spectral lines. The splitting of spectral lines due to the external magnetic field is called the *simple Zeeman effect*.

Problems

16.1 Derive the general expression for the Lorentz force from the Hamilton equations.

16.2 Using the rules for multiplying operators, derive the forms (16.10) and (16.11) of the Schrödinger equation for a particle in the EM field.

16.3 Show that combination of the gauge and phase transformations with the same scalar function $\Lambda(\mathbf{r}, t)$ leaves the Schrödinger equation invariant.

16.4 Derive the continuity equation and the expression for probability flux from the Schrödinger equation for a particle in magnetic field.

16.5 Show that a local phase transformation without the corresponding gauge transformation would change the probability flux density, which is an observable characteristic of the state Ψ.

16.6 Show that the phase transformation (3.15) together with the gauge transformation (16.4) will not affect the flux density.

16.7 Find the velocity operator in magnetic field using the expression for the time derivative of **r** with the corresponding Hamiltonian.

16.8 Find the precession frequencies for an electron, a muon, and a proton, respectively, in a magnetic field of 0.3 T.

16.9 A neutron has the same spin as an electron. Will it precess in a magnetic field? Explain your answer.

16.10 How would you make a QM description of spin precession with an arbitrary angle θ between **s** and **B** (16.54)?

(*Hint*: Consider the eigenstates (16.53) and try to express θ in terms of the corresponding superposition amplitudes.)

16.11 Suppose you have a charged particle with a spin higher than 1/2. The particle's magnetic moment makes an angle θ with an external magnetic field **B**. Find the necessary condition on superposition amplitudes for the QM description of particle's precession.

16.12 Consider the conical surface representing a spin-1/2 state $|\psi\rangle = a^+|+\rangle + a^-|-\rangle$, where $|+\rangle$ and $|-\rangle$ are the two eigenstates of $\hat{s}$ in the basis along an arbitrary axis, and a^+ and a^- are $1/\sqrt{3}$ and $-i\sqrt{2/3}$, respectively. What is the average $\bar{\theta}$ for the open angle of the cone?

16.13 Find the amount of shift of the diffraction pattern in the double-slit experiment with a narrow parallel-plate capacitor between the slits as shown in Figure 16.4, assuming the distance between the plates is much smaller than distance d between the slits.

16.14 Suppose we perform the Aharonov–Bohm version of a double-slit experiment using electrons with kinetic energy $K = 4.8 \times 10^{-3}$ eV. The distance between the slits is $d = 3 \times 10^{-5}$ cm and the distance between the two screens is $L = 10$ cm (this is a thought experiment). For a solenoid with cross-sectional area $\sigma \ll d^2$ between the slits, find the magnetic field that would shift the interference pattern by one fringe.

References

1. Griffiths, D. (1999) *Introduction to Electrodynamics*, 3rd edn, Addison-Wesley, Reading, MA.
2. Jackson, J. (1999) *Classical Electrodynamics*, 3rd edn, John Wiley & Sons, Inc., New York.
3. Feynman, R., Leighton, R., and Sands, M. (1963–1964) *The Feynman Lectures on Physics*, vols. 1–3, Addison-Wesley, Reading, MA.
4. Frish, S.É. and Timoreva, A.V. (2007) *Course of General Physics* (in Russian), vols. 1–3, Lan Publishing House, Saint Petersburg.
5. Griffiths, D. (1994) *Introduction to Quantum Mechanics*, Prentice-Hall, Englewood Cliffs, NJ, pp. 11, 54–57.
6. Uemura, Y.J., Kossler, W.J., Yu, X.H., and Kempton, J.R. (1987) Antiferromagnetism of La_2CuO_{4-y} studied by muon-spin rotation. *Phys. Rev. Lett.*, **59**, 1045–1048.
7. Cox, S.F.J. (1987) Implanted muon studies in condensed matter science. *J. Phys. C*, **20**, 3187–3319.
8. Dalmas de Reotier, P. and Yaouanc, A. (1997) Muon spin rotation and relaxation in magnetic materials. *J. Phys.: Condens. Matter*, **9**, 9113–9166.
9. Blundell, S.J. (1999) Spin-polarized muons in condensed matter physics. *Contemp. Phys.*, **40**, 175–192.
10. Aharonov, Y. and Rohrlich, V. (2005) *Quantum Paradoxes*, Wiley-VCH Verlag GmbH, Weinheim.
11. Hecht, E. (2002) *Optics*, 4th edn, Addison-Wesley, Reading, MA, p. 385.
12. Hildebrand, F.B. (1976) *Advanced Calculus for Applications*, Prentice-Hall, Englewood Cliffs, NJ, p. 161.
13. Ehrenberg, W. and Siday, R. (1949) The refractive index in electron optics and the principles of dynamics. *Proc. Phys. Soc. B*, **62**, 8–21.
14. Aharonov, Y. and Bohm, D. (1959) Significance of electromagnetic potentials in quantum theory. *Phys. Rev.*, **115**, 485–491.
15. Aharonov, Y. and Bohm, D. (1961) Further considerations on electromagnetic potentials in the quantum theory. *Phys. Rev.*, **123**, 1511–1524.
16. Peshkin, M. and Tonomura, A. (1989) *The Aharonov–Bohm Effect*, Springer, Berlin.
17. Batelaan, A. and Tonomura, A.A. (2009) The Aharonov–Bohm effects: variations on a subtle theme. *Phys. Today*, **62**, 38–43.
18. Sjöqvist, E. (2002) Locality and topology in the molecular Aharonov–Bohm effect. *Phys. Rev. Lett.*, **89** (21), 210401.

17
Perturbations

Examples studied in Chapters 9 and 10 allowed an analytical solution of the corresponding eigenvalue problem. Most of such cases are merely idealized models of real systems. In the majority of real-life problems, such simple analytical solutions do not exist. We then have two choices. One is to attempt the *numerical* solution of a problem. Its advantage is high accuracy afforded by modern computers. Its disadvantage is the difficulty of retrieving the general features of system's behavior from the heaps of numerical data.

The second choice is to seek an approximate analytical solution of the problem. It may be not quite as accurate as the numerical solution, but in exchange it allows direct physical interpretation of the result. This chapter is devoted to the second approach.

17.1
Stationary Perturbation Theory

We shall seek an approximate solution for the case when the system under study is not very different from some simpler and, hopefully, well-studied system with a known exact solution. Denote the Hamiltonian of the known system as $\hat{H}_0$ and suppose the actual system we study is described by a slightly different Hamiltonian $\hat{H}$:

$$\hat{H} = \hat{H}_0 + W. \tag{17.1}$$

The additional term W is called *perturbation* to the original Hamiltonian $\hat{H}_0$. The method that follows applies when W is small enough to cause only a minor change in the eigenfunctions and eigenvalues of $\hat{H}_0$. What is the meaning of this requirement when $\hat{H}_0$ contains some potential energy term $U(\mathbf{r})$? On the face of it, before actually attempting this method, we must convince ourselves that $W \ll U$ is true at *every* point $\mathbf{r}$ where the wave function exists:

$$\xi \equiv \frac{|W(\mathbf{r})|}{|U(\mathbf{r})|} \ll 1. \tag{17.2}$$

But in practice, the requirement is less stringent than it appears. As an example, consider a hydrogen atom in a constant electric field $\mathcal{E}$ that is much weaker than the

Quantum Mechanics and Quantum Information: A Guide through the Quantum World,
First Edition. Moses Fayngold and Vadim Fayngold.
© 2013 Wiley-VCH Verlag GmbH & Co. KGaA. Published 2013 by Wiley-VCH Verlag GmbH & Co. KGaA.

Coulomb field. This condition cannot be satisfied for the whole range of the Coulomb field, which approaches zero at large distances from the nucleus. But it can be satisfied within a certain atomic region harboring a few states with sufficiently low quantum numbers n (Problem 17.1). Another example is a constant external magnetic field, which must be weaker than the field produced by atomic currents. In these examples, the perturbation method is good enough at describing what really matters – the ground state and the next several states that occur with reasonably high probabilities.

So let us suppose we have the right perturbation W and suppose further that both $\hat{H}_0$ and W are time independent. Our task, obviously, is to solve $\hat{H}|\Psi_n\rangle = E_n|\Psi_n\rangle$. The first trivial case occurs when the term W is strictly zero:

$$\hat{H}_0|\Psi_n^{(0)}\rangle = E_n^{(0)}|\Psi_n^{(0)}\rangle, \quad n = 1, 2, \ldots . \tag{17.3}$$

This equation (we assume a discrete spectrum here and use the superscript (0) to remind ourselves that we are dealing with the unperturbed system) will not give us any trouble: By the initial assumption, it describes a relatively simple system, so all eigenstates $|\Psi_n^{(0)}\rangle$ and eigenvalues $E_n^{(0)}$ are presumably well known. The normalized set of $|\Psi_n^{(0)}\rangle$ forms a basis that spans the $\mathcal{H}$-space. Here, we make another assumption that the form of the $\mathcal{H}$-space will not change just because we introduced the perturbation.[1] Then every "real" solution $|\Psi_n\rangle$ will also live in the same space, so it can be written as some superposition $|\Psi_n\rangle = \sum_m c_{nm}|\Psi_m^{(0)}\rangle$ (note that we have omitted the *time-dependent* factor $e^{-i(E_n/\hbar)t}$). Our program then is to find the set of coefficients c_{nm} for each eigenstate $|\Psi_n\rangle$ of the perturbed Hamiltonian. Once we determine new eigenstates $|\Psi_n\rangle$, we will also find new energy eigenvalues E_n and thereby will know the corresponding corrections to the unperturbed eigenvalues $E_n^{(0)}$.

After the self-explanatory algebra and introducing matrix elements W_{mn} of the perturbation Hamiltonian W in the $|\Psi_n^0\rangle$ basis, we obtain

$$(\hat{H}_0 + W)\left(\sum_n c_n|\Psi_n^{(0)}\rangle\right) = E_n\left(\sum_n c_n|\Psi_n^{(0)}\rangle\right), \quad \langle\Psi_m^{(0)}|\sum_n c_n W|\Psi_n^{(0)}\rangle$$

$$= \langle\Psi_m^{(0)}|\sum_n c_n(E_n - \hat{H}_0)|\Psi_n^{(0)}\rangle, \quad \sum_n c_n W_{mn} = c_m(E_m - E_m^{(0)}).$$

The idea of the perturbation method is to seek both $|\Psi_n\rangle$ and E_n in the form of a series, as a sum of successive and ever more "fine-tuned" corrections:

$$|\Psi_n\rangle = |\Psi_n^{(0)}\rangle + |\Psi_n^{(1)}\rangle + |\Psi_n^{(2)}\rangle + \cdots, \quad E_n = E_n^{(0)} + E_n^{(1)} + E_n^{(2)} + \cdots . \tag{17.4}$$

The order of the term in the expansion is given by the superscript symbol. Here, the zero-order correction is simply the solution of the unperturbed system.

---

1) This sounds trivial given how small W is, but it does require some attention from the reader. For instance, we may solve the electron in the spinless approximation and then treat spin interaction with the field as a perturbation. Then spinors, superfluous as they were in the initial problem, must be absorbed into the basis eigenstates $|\Psi_n^0\rangle$ before we can claim that $\mathcal{H}$ has "remained the same."

The first-order correction will give us a rough estimate of the *actual* energy levels and eigenstates, and will suffice as a working approximation in many problems. If we need better accuracy, we should include the higher order correction terms. Practice shows that for the majority of problems, there is little need to go beyond the second-order correction.

Here, $|\Psi_n^{(1)}/\Psi_n^{(0)}|$ and $|E_n^{(1)}/E_n^{(0)}|$ are of the same order as ξ in (17.2), $|\Psi_n^{(2)}/\Psi_n^{(0)}|$ and $|E_n^{(2)}/E_n^{(0)}|$ are of the same order as ξ^2, and so on. Disregarding the higher order corrections, we write

$$(\hat{H}_0 + W)\left(|\Psi_n^{(0)}\rangle + |\Psi_n^{(1)}\rangle + |\Psi_n^{(2)}\rangle\right) = \left(E_n^{(0)} + E_n^{(1)} + E_n^{(2)}\right)\left(|\Psi_n^{(0)}\rangle + |\Psi_n^{(1)}\rangle + |\Psi_n^{(2)}\rangle\right). \quad (17.5)$$

Now we start consecutive steps of the procedure.

Zeroth approximation

In this approximation, we set $W = 0$. Accordingly, $\xi = 0$ and all additional terms in $|\Psi_n\rangle$ and E_n will disappear. Then, (17.5) reduces to $\hat{H}_0|\Psi_n^{(0)}\rangle = E_n^{(0)}|\Psi_n^{(0)}\rangle$. This gives the set of known eigenstates $|\Psi_n^{(0)}\rangle$ and the corresponding eigenvalues $E_n^{(0)}$, where $n = 1, 2, \ldots$.

First approximation

Focus on one of the eigenstates, say, $|\Psi_n^{(0)}\rangle$, and ask how it will be affected by W. To look only for the corrections linear with respect to ξ, drop all terms with subscript 2 and higher and also drop the products $W|\Psi_n^{(1)}\rangle$, $E_n^{(1)}|\Psi_n^{(1)}\rangle$ since they are proportional to ξ^2. Also, drop both zero-order terms since they just cancel each other. We are left with

$$\left(\hat{H}_0 - E_n^{(0)}\right)|\Psi_n^{(1)}\rangle = \left(E_n^{(1)} - W\right)|\Psi_n^{(0)}\rangle. \quad (17.6)$$

By the same reasoning as before, $|\Psi_n^{(1)}\rangle$ is a denizen of the $\mathcal{H}$-space and therefore expands in unperturbed eigenstates:

$$|\Psi_n^{(1)}\rangle = \sum_l c_{nl}^{(1)}|\Psi_l^{(0)}\rangle. \quad (17.7)$$

The additional superscript in the coefficients $c_{nl}^{(1)}$ is a label reminding us that this is a set of coefficients for the *first* correction. Putting that into the last formula, and multiplying both sides by a bra, we get the first correction to the energy level n:

$$\langle\Psi_m^{(0)}|\sum_l c_{nl}^{(1)}\left(E_l^{(0)} - E_n^{(0)}\right)|\Psi_l^{(0)}\rangle = \langle\Psi_m^{(0)}|\left(E_n^{(1)} - W\right)|\Psi_n^{(0)}\rangle, \quad (17.8)$$

$$c_{nm}^{(1)}\left(E_m^{(0)} - E_n^{(0)}\right) = E_n^{(1)}\delta_{mn} - W_{mn}, \quad (17.9)$$

$$E_n^{(1)} = W_{nn}. \quad (17.10)$$

The last step is derived from the special case $m = n$. The energy correction for unperturbed state $|\Psi_n^{(0)}\rangle$ turns out to be equal to the expectation value of perturbation Hamiltonian W in that state – quite a believable result! The case $m \neq n$ yields the amplitudes $c_{nm}^{(1)}$, from which there follows the first-order correction:

$$c_{nm}^{(1)} = \frac{W_{mn}}{E_n^{(0)} - E_m^{(0)}}, \quad m \neq n, \tag{17.11}$$

$$|\Psi_n^{(1)}\rangle = \sum_{m \neq n} \frac{W_{mn}}{E_n^{(0)} - E_m^{(0)}} |\Psi_m^{(0)}\rangle + c_{nn}^{(1)} |\Psi_n^{(0)}\rangle. \tag{17.12}$$

17.1.1
Discussion

There is one remarkable feature in the above solution. Note that the algebra gives us all the amplitudes $c_{nm}^{(1)}$ except for $c_{nn}^{(1)}$, which corresponds to the case $m = n$. The solutions in Equation 17.11 apply only for $m \neq n$. This result has a simple explanation. Let us focus on an arbitrary eigenstate No. k. In order for the function $|\Psi_k\rangle \simeq |\Psi_k^{(0)}\rangle + |\Psi_k^{(1)}\rangle$ to describe a real system, it must be normalized, as is $|\Psi_k^0\rangle$. Since we haven't enforced normalization yet, it is only natural that our equations left the value of $c_{kk}^{(1)}$ adjustable, to give us the freedom to regulate the norm of the state vector. Let us try to determine $c_{kk}^{(1)}$ by writing the matrix column of the state to be normalized (we will only write the three relevant rows as the rest is self-evident):

$$|\Psi_k\rangle \simeq |\Psi_k^{(0)}\rangle + |\Psi_k^{(1)}\rangle = \begin{pmatrix} 0 \\ 1 \\ 0 \end{pmatrix} + \begin{pmatrix} c_{k,k-1}^{(1)} \\ c_{kk}^{(1)}(?) \\ c_{k,k+1}^{(1)} \end{pmatrix} = \begin{pmatrix} c_{k,k-1}^{(1)} \\ 1 + c_{kk}^{(1)} \\ c_{k,k+1}^{(1)} \end{pmatrix}, \tag{17.13}$$

with yet unknown $c_{kk}^{(1)}$. For the last matrix on the right-hand side, the normalization condition reads as follows:

$$\sum_{j \neq k} |c_{kj}^{(1)}|^2 + |1 + c_{kk}^{(1)}|^2 = 1. \tag{17.14}$$

There follows

$$2 \operatorname{Re} c_{kk}^{(1)} + |c_{kk}^{(1)}|^2 = -\sum_{j \neq k} |c_{kj}^{(1)}|^2. \tag{17.15}$$

But the second term on the left-hand side, as well as the sum on the right-hand side, must be neglected in the first approximation, so we are left with $\operatorname{Re} c_{kk}^{(1)} = 0$. The imaginary part remains undetermined, but in any event, we can write for it $|\operatorname{Im} c_{kk}^{(1)}| \ll 1$. Therefore, we can neglect $c_{kk}^{(1)}$ in the first approximation: $c_{kk}^{(1)} = 0$, or conclude that it must be taken to the set of the second-order corrections.

This result has a very simple $\mathcal{H}$-space interpretation. Figure 17.1 shows a 2D subspace of $\mathcal{H}$ featuring two states with different quantum numbers k, m. The two "rectilinear" basis vectors in the figure represent the unperturbed eigenstates $|\Psi_k^{(0)}\rangle$, $|\Psi_m^{(0)}\rangle$. The two tilted basis vectors represent the perturbed eigenstates $|\Psi_k\rangle, |\Psi_m\rangle$. The difference between vectors $|\Psi_k^{(0)}\rangle$ and $|\Psi_k\rangle$ determines the sought-for correction $|\Psi_k^{(1)}\rangle$. At small perturbations (small tilt angle), the projection $c_{km}^{(1)}$ of $|\Psi_k^{(1)}\rangle$ onto $|\Psi_m^{(0)}\rangle$ is of the same order as $|\Psi_k^{(1)}\rangle$ itself (i.e., of the order ξ). But its projection $c_{kk}^{(1)}$ onto $|\Psi_k^{(0)}\rangle$ is of the second order ($\sim \xi^2$) and is therefore negligible in the first approximation (mathematically, $c_{kk}^{(1)} \sim \xi^2$). Or, to state it in simple terms, in order to

17.1 Stationary Perturbation Theory

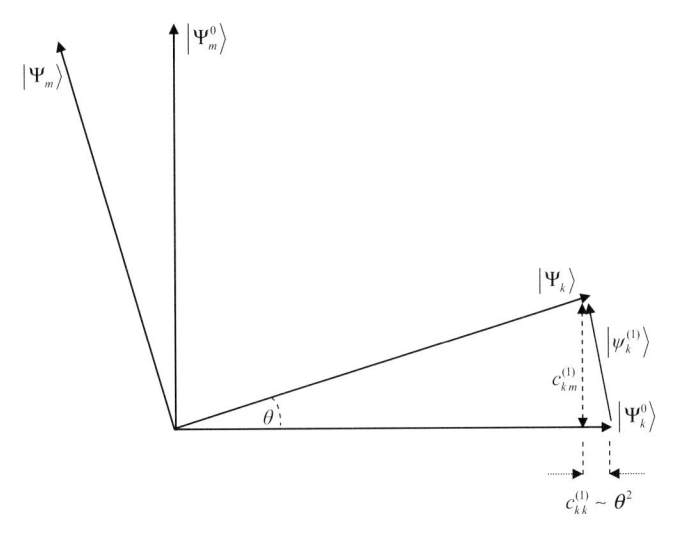

Figure 17.1 A crude visual illustration of stationary perturbations in the Hilbert space. The mutually orthogonal unit eigenvectors $|\Psi_k^0\rangle$, $|\Psi_m^0\rangle$ represent the pair of eigenstates of unperturbed Hamiltonian $\hat{H}_0$. Rotation through a small angle θ takes them to the pair $|\Psi_k\rangle$, $|\Psi_m\rangle$ of eigenstates with the same quantum numbers of Hamiltonian $\hat{H}$. It is immediately seen that θ is proportional to perturbation W when both are small. The vector $|\psi_k^{(1)}\rangle$ is the first-order correction to $|\Psi_k^0\rangle$; when added to $|\Psi_k^0\rangle$, it produces $|\Psi_k\rangle$, but by itself it is not an eigenvector of either $\hat{H}_0$ or $\hat{H}$. The coefficients $c_{km}^{(1)}$, $c_{kk}^{(1)}$ are its projections onto $|\Psi_m^0\rangle$ and $|\Psi_k^0\rangle$, respectively. The latter projection $c_{kk}^{(1)} \approx c_{km}^{(1)}\theta \sim \theta^2$ is proportional to θ^2 and in the first approximation vanishes at small θ.

preserve the norm of $|\Psi_k^{(0)}\rangle$, the small correction must be essentially perpendicular to $|\Psi_k^{(0)}\rangle$, so its projection $c_{kk}^{(1)}$ onto $|\Psi_k^{(0)}\rangle$ is zero. This is why many textbooks introduce the constraint $c_{kk}^{(1)} = 0$ in an equivalent form, $\langle\Psi_k^{(0)}|\Psi_k^{(1)}\rangle = 0$. Based on the same reasoning, we can state the constraint $c_{kk}^{(s)} = \langle\Psi_k^{(0)}|\Psi_k^{(s)}\rangle = 0$ for a correction of the arbitrary order s.

Box 17.1 First-order correction $|\Psi_k^{(1)}\rangle$ and the normalization requirement

As we see from Figure 17.1, the magnitude of the first-order correction is much less than 1. One might think that this does not look right, because a state function with a nonunitary norm would contradict basic postulates of QM. Indeed, both $|\Psi_k^{(0)}\rangle$ and $|\Psi_k^{(0)}\rangle + |\Psi_k^{(1)}\rangle$ are normalized (unitary) solutions of the Schrödinger equation. The difference ($|\Psi_k^{(1)}\rangle$ in our case) between two solutions of any linear differential equation is also a solution. Why is it not normalized?

The flaw of this argument is that it treats $|\Psi_k^{(0)}\rangle$ and $|\Psi_k^{(0)}\rangle + |\Psi_k^{(1)}\rangle$ as solutions of *the same* equation. But they are eigenfunctions of two different Hamiltonians and accordingly, they are solutions to different equations. This becomes self-evident if we recall that the perturbed and unperturbed kth eigenstates have different

> eigenvalues, $E_k \neq E_k^{(0)}$. The difference $|\Psi_k^{(1)}\rangle$ between these solutions *is not itself an eigenfunction of any Hamiltonian* and not a solution to any Schrödinger equation. Whereas $|\Psi_k^{(0)}\rangle$ and $|\Psi_k^{(0)}\rangle + |\Psi_k^{(1)}\rangle$ are both normalized eigenkets, we cannot make the same claim about $|\Psi_k^{(1)}\rangle$ or any higher order corrections. Accordingly, the normalization condition does not apply to it.

Second approximation:

Write down the obtained results, put them back into (17.8), and retain all terms proportional to ξ^2. Discard the higher order terms and also drop the zero- and first-order terms since they were already shown to cancel each other. This means that we keep the terms with $|\Psi_n^{(2)}\rangle$, $E_n^{(2)}$, $E_n^{(1)}|\Psi_n^{(1)}\rangle$, $W|\Psi_n^{(1)}\rangle$, but drop the products $W|\psi_n^{(2)}\rangle$, $E_n^{(1)}|\Psi_n^{(2)}\rangle$, $E_n^{(2)}|\Psi_n^{(1)}\rangle$. By the same reasoning as before, we get the following sequence of steps:

$$\begin{aligned}
\left(\hat{H}_0 - E_n^{(0)}\right)|\Psi_n^{(2)}\rangle &= \left(E_n^{(1)} - W\right)|\Psi_n^{(1)}\rangle + E_n^{(2)}|\Psi_n^{(0)}\rangle, \\
\langle\Psi_k^{(0)}|\left(\hat{H}_0 - E_n^{(0)}\right)|\Psi_n^{(2)}\rangle &= \langle\Psi_k^{(0)}|(W_{nn} - W)|\Psi_n^{(1)}\rangle + E_n^{(2)}\delta_{kn}, \\
\langle\Psi_k^{(0)}|\sum_l c_{nl}^{(2)}\left(E_l^{(0)} - E_n^{(0)}\right)|\Psi_l^{(0)}\rangle &= \langle\Psi_k^{(0)}|(W_{nn} - W)|\Psi_n^{(1)}\rangle + E_n^{(2)}\delta_{kn}, \\
c_{nk}^{(2)}\left(E_k^{(0)} - E_n^{(0)}\right) &= \langle\Psi_k^{(0)}|(W_{nn} - W)|\Psi_n^{(1)}\rangle + E_n^{(2)}\delta_{kn}.
\end{aligned} \qquad (17.16)$$

The second energy correction follows from the case $k = n$:

$$E_n^{(2)} = \langle\Psi_n^{(0)}|(W - W_{nn})|\Psi_n^{(1)}\rangle = \sum_{m \neq n}\langle\Psi_n^{(0)}|(W - W_{nn})\frac{W_{mn}}{E_n^{(0)} - E_m^{(0)}}|\Psi_m^{(0)}\rangle = \sum_{m \neq n}\frac{|W_{nm}|^2}{E_n^{(0)} - E_m^{(0)}}. \qquad (17.17)$$

Next we find coefficients $c_{nk}^{(2)}$. To this end, we choose $k \neq n$:

$$\begin{aligned}
c_{nk}^{(2)}\left(E_n^{(0)} - E_k^{(0)}\right) &= \langle\Psi_k^{(0)}|(W - W_{nn})|\Psi_n^{(1)}\rangle = \langle\Psi_k^{(0)}|(W - W_{nn})\left(\sum_{m \neq n}\frac{W_{mn}}{E_n^{(0)} - E_m^{(0)}}|\Psi_m^{(0)}\rangle\right) \\
&= \sum_{m \neq n}\frac{W_{mn}W_{km}}{E_n^{(0)} - E_m^{(0)}} - \sum_{m \neq n}\frac{W_{mn}W_{nn}\delta_{km}}{E_n^{(0)} - E_m^{(0)}} = \sum_{m \neq n}\frac{W_{km}W_{mn}}{E_n^{(0)} - E_m^{(0)}} - \frac{W_{nn}W_{kn}}{E_n^{(0)} - E_k^{(0)}}.
\end{aligned} \qquad (17.18)$$

With the knowledge of $c_{nk}^{(2)}$ for $k \neq n$ and setting the last coefficient $c_{nn}^{(2)}$ to zero as mentioned above, we write the second correction to the eigenstate:

$$|\Psi_n^{(2)}\rangle = \sum_{k \neq n} c_{nk}^{(2)}|\Psi_k^{(0)}\rangle = \sum_{k \neq n}\left(-\frac{W_{nn}W_{kn}}{\left(E_n^{(0)} - E_k^{(0)}\right)^2} + \sum_{m \neq n}\frac{W_{km}W_{mn}}{\left(E_n^{(0)} - E_m^{(0)}\right)\left(E_n^{(0)} - E_k^{(0)}\right)}\right)|\Psi_k^{(0)}\rangle. \qquad (17.19)$$

This procedure can be reiterated, giving more and more accurate approximations.

Exercise 17.1

Write the eigenstates and eigenenergies in the first-order approximation for the case of a continuous energy spectrum.

Solution:

Labeling the eigenstates in the continuous part of the spectrum by a variable ν, we gereneralize formula (17.10) as follows:

$$|\Psi_k^{(1)}\rangle = \sum_n c_{kn}^{(1)}|\Psi_n^{(0)}\rangle + \int_\nu c_{k\nu}^{(1)}|\Psi_\nu^{(0)}\rangle d\nu. \tag{17.20}$$

The coefficients $c_{k\nu}^{(1)}$ derive in the same way as for the discrete spectrum and can be written as

$$c_{k\nu}^{(1)} = \frac{W_{k\nu}}{E_k^{(0)} - E_\nu^{(0)}}. \tag{17.21}$$

Here $W_{k\nu}$ is the matrix element of W for the states $|\Psi_k^{(0)}\rangle$ and $|\Psi_\nu^{(0)}\rangle$, and E_ν is an energy eigenvalue in the continuous region.

Box 17.2 Interpretation of amplitudes $c_{kn}^{(1)}$

It is worthwhile to comment on two other features of the obtained results. One of them is the physical interpretation of the amplitudes $c_{kn}^{(1)}$. Imagine we prepared the system in an eigenstate $|\Psi_k^{(0)}\rangle$ of $\hat{H}_0$. Then we apply perturbation W very slowly and keep it constant long enough for the system to readjust to the new conditions. This readjustment may involve, among other things, the change of state, $|\Psi_k^{(0)}\rangle \to |\Psi_k\rangle$, so the system is now in the kth eigenstate of $\hat{H}$ but not of $\hat{H}_0$. Then we "switch off" the perturbation and measure the system's energy. The removal of W restores the Hamiltonian $\hat{H}_0$ and the system state $|\Psi_k\rangle$ now partially overlaps with states of the restored eigenbasis: $\langle\Psi_n^{(0)}|\Psi_k\rangle \simeq \langle\Psi_n^{(0)}|\Psi_k^{(0)} + \Psi_k^{(1)}\rangle = \langle\Psi_n^{(0)}|\Psi_k^{(1)}\rangle = c_{kn}^{(1)}$. Accordingly, there is now a finite probability for the system to jump into eigenstate $|\Psi_n^{(0)}\rangle$. The ultimate result of the experiment is the transition $|\Psi_k^{(0)}\rangle \to |\Psi_n^{(0)}\rangle$ caused by the intermediate perturbation W, and thus coefficients $c_{kn}^{(1)}$ have the meaning of transition amplitudes. We have seen that they are proportional to the (nondiagonal!) matrix elements W_{kn}. Therefore, conversely, such matrix elements determine the amplitudes of transition between the corresponding states. The transition amplitude $c_{kn}^{(1)}$ is also inversely proportional to the energy gap between the initial and final states. We can express this by saying that it is harder for a system to make quantum jumps across a wide gap. But on the other hand, we must be very careful when the system has closely spaced energy levels: If the gap is too narrow, we no longer have

$$\left|\frac{W_{kn}}{E_k^{(0)} - E_n^{(0)}}\right| \ll 1, \tag{17.22}$$

which is necessary for the whole approximation!

The next interesting observation comes from rewriting the second correction in terms of the first. Substituting for W_{mn} and W_{nk} their respective first-order expressions, we get

$$c_{km}^{(2)} = \sum_{n \neq k} \frac{\left(E_n^{(0)} - E_m^{(0)}\right)}{\left(E_k^{(0)} - E_m^{(0)}\right)} c_{kn}^{(1)} c_{nm}^{(1)} - \frac{W_{kk} W_{mk}}{\left(E_k^{(0)} - E_m^{(0)}\right)^2}, \quad m \neq k. \tag{17.23}$$

It tells us that in the second approximation, the transition amplitude is the weighted sum of products of amplitudes found in the first approximation. Each of the latter amplitudes describes a *direct* transition between the two eigenstates. It depends only on properties of these states themselves – their energies and the corresponding nondiagonal matrix element between them. But the product of two such amplitudes with one common index, say, $c_{kn}^{(1)} c_{nm}^{(1)}$, describes an *indirect* transition $k \to m$ via n, that is, $k \to n \to m$. We can call it *two-stage* transition. Therefore, in the second approximation, we see the Pandora box opened. One and the same transition from k to m can now involve *any* of the remaining eigenstates! The system can jump from k to n and then from n to m or from k to $n+1$ and then from $n+1$ to m, and so on. All such possibilities are present in (17.23). Based on this observation, we can predict that the third approximation will describe *three-stage* transitions of the type $k \to l \to n \to m$ with different l, n, and so on. If we depict this graphically as the trajectories in the energy space (Figure 17.2), it will be analogous to the path integral in the configuration space.

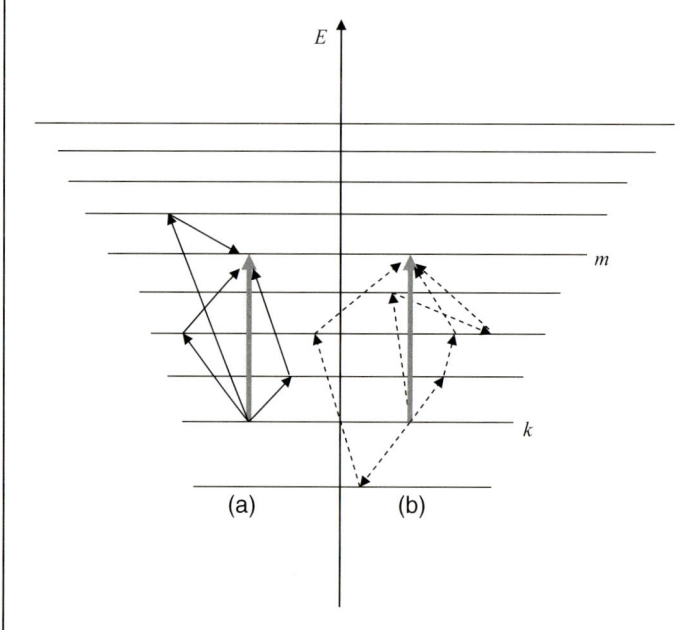

Figure 17.2 Representation of transition amplitudes as path integrals in the energy space. (a) The two-step transitions (solid lines). (b) The three-step transitions (dashed lines).

17.2
Asymptotic Perturbations

As noted in Box 17.2, the analytical approximations described there may fail for some states with closely spaced energy levels. For instance, the difference between two neighboring energy eigenvalues for a bound electron in the Coulomb field is

$$E_n^{(0)} - E_{n\pm 1}^{(0)} = E_1^{(0)}\left(\frac{1}{(n\pm 1)^2} - \frac{1}{n^2}\right) = \frac{\mp 2n - 1}{n^2(n\pm 1)^2} E_1^{(0)}, \quad n > 1, \qquad (17.24)$$

where E_1^0 is the energy of the ground state (the Rydberg constant). At small n, the gap between neighboring energy levels is wide enough ($E_n^{(0)} - E_{n\pm 1}^{(0)} \gg W_{n,n\pm 1}$) and our theory works flawlessly. But for the highly excited states (the Rydberg states), the gap approaches zero as n^{-3}, and the theory will fail to find the right corrections for the high energy levels.

There is one interesting special situation when the gaps are wide enough and yet the perturbation theory cannot be applied. This happens when W, while being small within a certain region, dramatically changes the field beyond this region, especially at very large distances from it. Consider a hydrogen atom inside of a parallel plate capacitor. The Hamiltonian for the atomic electron will now contain two potential fields, as shown in Figure 17.3. The net field is now asymmetric. It bounds the

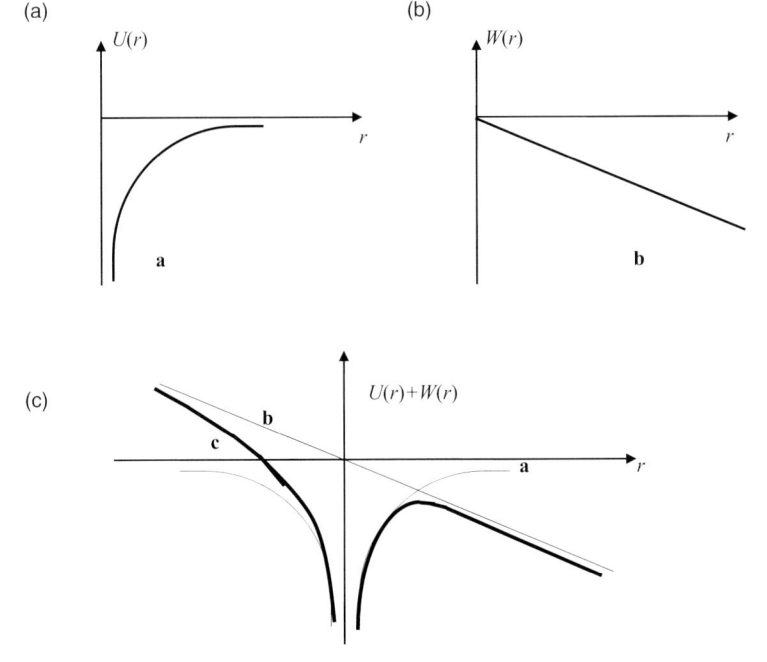

Figure 17.3 (a) Coulomb potential. (b) Linear potential of a homogeneous external field. (c) Combined potential along the direction of the external field.

electron on one side (Figure 17.3a) even more efficiently than the atomic field alone. But on the opposite side (Figure 17.3b), the result is disastrous for the mere existence of the atom or at least for its stability. The electron is now separated from the outer world only by the formed potential barrier, which allows tunneling to take place. The region sufficiently close to the origin is effectively a potential well. The distance between capacitor plates is so much greater than the atomic size that we can describe its field as extending to infinity. So the entire semispace behind the barrier becomes, in principle, available to this electron. This produces a twofold outcome. First, the electron gets a chance to break loose from the atom. Second, once the ban on traveling to infinity is lifted, the energy spectrum becomes continuous. Energies that were forbidden in the pure Coulomb field are now allowed! All this happens no matter how weak the applied external field is. It may be much smaller than the average Coulomb field within the atomic region and thereby treatable by the perturbation theory. But it is huge with respect to this field at large distances (as we say, asymptotically) and accordingly changes asymptotic characteristics of the whole system. We can call such cases *asymptotic perturbations*. Their basic feature is that asymptotically, the original field U and additional field W swap their roles: The former can be considered as a small perturbation of the latter.

With all that, we can still calculate corrections to the unperturbed functions $|\Psi_n^{(0)}\rangle$ and eigenvalues $E_n^{(0)}$ using the perturbative approach within the atomic region. The obtained eigenstates $|\Psi_n\rangle$ and their eigenvalues E_n will be approximately accurate within the atomic region. But the question now arises as to their physical meaning. What distinguishes E_n from other energies of the newly formed continuous spectrum, which cannot be calculated by this approach and yet are now totally legitimate? What is the point of calculating energy levels close to the unperturbed set $E_n^{(0)}$ if *any other* level E is also allowed? The answer to this question is that *other energies* are allowed in a different way. The energies $E_k = E_k^{(0)} + E_k^{(1)}$ are not only allowed but also *encouraged*, whereas all the rest are merely tolerated. Then our question can be restated in the following way: What is the criterion that distinguishes the "privileged" states from the ones that are formally allowed but "unwelcome?"

The criterion is the behavior of respective wave functions. The privileged states $|\Psi_n\rangle$ calculated with the regular perturbation theory are distinguished by the large values of $\langle r|\Psi_n\rangle = \Psi_n(r)$ within the potential well (where the "native" field is strong) and small values outside it (where the asymptotic perturbation is strong). The wave functions of "unwelcome" states have the opposite behavior. In the first case, we can say that the electron resides, most probably, within the well (i.e., inside of the atom); in the second case, it is most probably in the outer world, far from the atom.

The first case can be prepared in two different ways. We can irradiate the well (initially containing the nucleus only) by electrons with energy close to E_n. In the case shown in Figure 17.3, such electrons will be coming from the right (where the potential is lowest). This involves the incident wave coming from infinity and therefore it is under the jurisdiction of the scattering theory, to be discussed in detail in Chapter 19. What we need to know for now is this: If the wave function solution indicates a high probability for a particle to be found near the center, then scattered particles must spend a long time there before leaving in the form of an outgoing wave. This is the so-called *resonant scattering* characterized by a large cross-sectional area. The corresponding state can be stationary (the energy can be exactly defined) if the outgoing waves are balanced by the incoming waves from a permanent source and at any moment there is no net flux toward or away from the center.

The second way involves a certain moment of time that can be singled out from all other moments and considered as the origin of time. At this moment, the particle is "created" inside the well or brought into it "by hand" and left there to its own devices. There are no incident waves, but there appear outgoing waves because of the tunneling through the barrier. Eventually, the entire particle will "flow out" of the well. Such situation falls under the jurisdiction of the theory of radioactive decay. The state cannot be stationary in principle because the probability changes with time and the initial state of residing within the well cannot be created with a sharply defined energy. The stationary perturbation theory in such case can be valid and sufficiently accurate only within a finite time interval close to the average lifetime of the particle within the well. In reality, this lifetime may be very long and it increases with the decrease of W. The corresponding nonstationary state may then be called quasi-stationary, and it belongs to the family of states considered in Sections 11.9–11.11.

Let us apply these ideas to the problem at hand. Suppose the external field is much weaker than the Coulomb field in the region $r \leq a_0 N^2$, where N is some integer and a_0 is the Bohr radius (recall that $a_0 N^2$ is the effective size of a hydrogen atom in Nth eigenstate). Then the perturbation theory is applicable for all $n \leq N$ and we can determine the appropriate corrections to the wave functions and to the energy levels. However, the result of this calculation will hold only on the timescale of τ (i.e., the lifetime determined by the transmissivity of the formed potential barrier), and the corrections $E_n^{(1)}$ will tell us nothing about energy indeterminacy ΔE_n in the corresponding states. Generally, the former can be much larger than the latter. The lifetime may be very long, but it will be finite even for the ground state. This enables us to predict that the atom placed in an external electric field will eventually lose all its electrons. Such loss of atomic electrons has been observed experimentally in sufficiently strong fields, and also in weak fields but for high energy levels (i.e., the Rydberg states). This effect is called *autoionization*.

17.3
Perturbations and Degeneracy

Suppose now that energy level $E_n^{(0)}$ is s-fold degenerate, so it is shared by s different eigenstates $|\Psi_{n\alpha}^{(0)}\rangle$, where $\alpha = 1, 2, 3, \ldots, s$, and try to find the first-order corrections in this case. We immediately run into trouble.

The first and most obvious difficulty is that our theory requires that all eigenstates be orthogonal. This is generally not true for the degenerate eigenstates. Fortunately, there is an easy remedy: We can always construct linear combinations of eigenstates in the set to make it orthogonal. Let us assume this has been done, and from now on, eigenstates in $\{|\Psi_{na}^{(0)}\rangle\}$ will be orthogonal to each other and to all eigenstates outside that set.

But then there is another problem. As far as the other nondegenerate eigenstates are concerned, the first-order correction formula will work as before. The only change will be that s terms in the sum will have equal denominators. But for the eigenstates *within* the degenerate set, the same formula now reads

$$|\Psi_{na}^{(1)}\rangle = \sum_{m\neq n} \frac{W_{mn}}{E_n^{(0)} - E_m^{(0)}} |\Psi_m^{(0)}\rangle + \sum_{\beta\neq a} \frac{W_{n\beta,ma}}{E_{na}^{(0)} - E_{n\beta}^{(0)}} |\Psi_{n\beta}^{(0)}\rangle. \tag{17.25}$$

Denominators in the second sum are all zeros! The frontal attack got us nowhere and it is clear that we must try to find a way around this sudden obstacle. Obviously, as long as we work with our set $\{|\Psi_{na}^{(0)}\rangle\}$, we will just keep getting singularity in the formula. But recall that the formula itself was derived from (17.9)! This exposes the cause of our trouble: We have implicitly assumed matrix element W_{mn} to be nonzero, and we then extended that assumption to the off-diagonal element $W_{n\beta,ma}$. Naturally then, as the energy difference was set to zero, the amplitude $c_{n\beta,ma}^{(1)}$ exploded.

Of course, for an *arbitrary* set $\{|\Psi_{na}^{(0)}\rangle\}$, we have every reason to expect $W_{n\beta,ma} \neq 0$. After all, this set was composed from eigenstates of $\hat{H}_0$, not from eigenstates of W. The perturbation W will generally *not* be diagonal in this basis. But we have the unitary freedom in the choice of $\{|\Psi_{na}^{(0)}\rangle\}$. Recall that a set can be orthogonalized in an infinite variety of ways. Alternatively, consider a rotation (i.e., unitary transformation) of the orthogonal basis to convince yourself that a suitable basis can always be found as long as W commutes with $\hat{H}_0$.

Our task then is to switch bases, $\{|\Psi_{na}^{(0)}\rangle\} \Rightarrow \{|\tilde{\Psi}_{na}^{(0)}\rangle\}$, to secure the condition $W_{n\beta,ma} = 0$ for $\beta \neq a$. That is to say, we must diagonalize the operator W by following the familiar procedure: Write it as a matrix in the original basis, find the s solutions of the secular equation $\det|W - \lambda I| = 0$, and construct a new basis from the obtained eigenvectors.

What is achieved by this procedure? First of all, as singularities vanish, the perturbation theory becomes usable again and we can start thinking about the second and higher approximations. Next, *after* we switch to the new basis, Equation 17.10 tells us the first-order corrections to the degenerate levels: They are just equal to the diagonal elements of the perturbation, $E_{na}^{(1)} = W_{na,na}$, and we remember from the operator theory that these diagonal elements are nothing else but the solutions λ_a ($a = 1, \ldots, s$) to the secular equation, lined up along the main diagonal. Finally, what about corrections to the eigenstates? On the face of it, we have another problem here because by selecting the diagonal basis, we left $c_{n\beta,ma}^{(1)}$ undefined. But there is a good reason for it. The eigenvectors of the diagonal basis, those that we found while solving the secular equation, are the energy eigenstates in the zeroth *and* in the first approximation. We solved not part of the problem, but the entire problem! This is easy to see when one considers that

$$(\hat{H}_0 + W)|\tilde{\Psi}_{na}^{(0)}\rangle = \hat{H}_0|\tilde{\Psi}_{na}^{(0)}\rangle + W|\tilde{\Psi}_{na}^{(0)}\rangle = E_n|\tilde{\Psi}_{na}^{(0)}\rangle + \lambda_a|\tilde{\Psi}_{na}^{(0)}\rangle. \tag{17.26}$$

In other words, the eigenstates we found are just the ones that will make the energies shift by the right amounts. There is no need to make any adjustments! We finally have this simple result:

$$E_{n\alpha}^{(1)} = \lambda_\alpha, \quad |\tilde{\Psi}_{n\alpha}^{(1)}\rangle = 0, \quad |\tilde{\Psi}_{n\alpha}\rangle \equiv |\tilde{\Psi}_{n\alpha}^{(0)}\rangle + |\tilde{\Psi}_{n\alpha}^{(1)}\rangle = |\tilde{\Psi}_{n\alpha}^{(0)}\rangle. \tag{17.27}$$

So to conclude, in the first-order approximation, the perturbation acts on nondegenerate eigenstates by changing both the energies and the eigenstates themselves. It acts on degenerate states by changing their energy only. The states themselves are left intact. They remain degenerate with respect to $\hat{H}_0$, but they are no longer degenerate with respect to the full Hamiltonian $\hat{H} = \hat{H}_0 + W$. In the presence of the perturbation, they will just correspond to different eigenenergies than before. To put it in a nutshell, the effect of W is to *lift the degeneracy*.

This completes the first-order perturbation theory. However, it may so happen that some of the solutions λ_α will be equal, and therefore *some* energy levels will remain degenerate even after the above procedure. In such cases, we say that degeneracy was *partially lifted*. Then, if we try the second-order approximation, we will encounter the same problem as before until we look at the derivation of the second-order correction formulas and realize that corrections to these degenerate states and energies should be zero, since W is already diagonal in the degenerate eigenbasis. It may be possible to lift the degeneracy further if in addition to the main perturbation W there is yet another, smaller perturbation term $W' \ll W$. Then, following the procedure outlined above and restricting to the first-order correction, we will focus on the new degenerate subset and diagonalize it with respect to W'. If some of the solutions are still the same, look at the next term, $W'' \ll W' \ll W$, and continue in this manner until degeneracy is fully lifted.

As an illustration, let us solve a degenerate two-state system (i.e., $s = 2$). Start with an arbitrary basis $\{|\Psi_1^{(0)}\rangle, |\Psi_2^{(0)}\rangle\}$ and diagonalize it with respect to W by solving the secular equation:

$$\det \begin{pmatrix} W_{11} - \lambda, & W_{12}, \\ W_{21}, & W_{22} - \lambda, \end{pmatrix} = 0. \tag{17.28}$$

It has two solutions, $\lambda_{1,2} \equiv E_{1,2}^{(1)} = \frac{1}{2}(T \pm \Delta E)$, where we denoted $T \equiv W_{11} + W_{22}$ and $\Delta E \equiv \sqrt{(W_{11} - W_{22})^2 + 4|W_{12}|^2}$. Thus, the perturbed Hamiltonian has two distinct eigenvalues, $E_1 = E^0 + E_1^{(1)}$ and $E_2 = E^0 + E_2^{(1)}$. The notation ΔE has a simple physical meaning: It is (one half of) the energy gap between the split energy levels. The symbol T is the trace of the perturbation matrix, which determines the shift of the initial energy level.

Solving the eigenvalue equation for the two found energies and normalizing the results, we find (after setting $\Delta W \equiv W_{11} - W_{22}$) the eigenstates of the diagonal basis:

$$|\tilde{\Psi}_1^{(0)}\rangle = \sqrt{\frac{W_{12}}{2|W_{12}|}}\left(1 + \frac{\Delta W}{\Delta E}\right)|\Psi_1^{(0)}\rangle + \sqrt{\frac{W_{21}}{2|W_{21}|}}\left(1 - \frac{\Delta W}{\Delta E}\right)|\Psi_1^{(0)}\rangle, \quad (17.29)$$

$$|\tilde{\Psi}_2^{(0)}\rangle = \sqrt{\frac{W_{12}}{2|W_{12}|}}\left(1 - \frac{\Delta W}{\Delta E}\right)|\Psi_1^{(0)}\rangle - \sqrt{\frac{W_{12}}{2|W_{12}|}}\left(1 + \frac{\Delta W}{\Delta E}\right)|\Psi_2^{(0)}\rangle. \quad (17.30)$$

17.4
Symmetry, Degeneracy, and Perturbations

The previous results raise a question: What specifically "lifts" degeneracy when a perturbation is applied? Will *any* perturbation do the trick or only some specific class of them?

In order to see the answer, it is good to ask another question: What causes the degeneracy itself? Knowing its cause, we will see the mechanism of its elimination.

Let us turn to the best-known case of degeneracy – that of eigenenergies of a free particle. A free particle admits an infinite number of distinct states with the same magnitude but different directions of linear momentum $\mathbf{p} = \hbar\mathbf{k}$. Change of direction leaves the energy the same. This is reflected in the dispersion equation (8.26) and results in degeneracy of the infinite order. The source of it is the same in both CM and QM: the isotropy of space. All directions in space are equivalent, hence energy independence of the direction of motion. This, in turn, is reflected in the invariance of the Hamiltonian with respect to rotations. If we write the Hamiltonian of a free particle in a certain basis $(\hat{\mathbf{x}}, \hat{\mathbf{y}}, \hat{\mathbf{z}})$ and then switch to a rotated basis $(\hat{\mathbf{x}}', \hat{\mathbf{y}}', \hat{\mathbf{z}}')$, it will not affect the Hamiltonian:

$$\hat{H}'(\hat{\mathbf{x}}', \hat{\mathbf{y}}', \hat{\mathbf{z}}') = -\frac{\hbar^2}{2\mu}\left(\frac{\partial^2}{\partial x'^2} + \frac{\partial^2}{\partial y'^2} + \frac{\partial^2}{\partial z'^2}\right) = -\frac{\hbar^2}{2\mu}\left(\frac{\partial^2}{\partial x^2} + \frac{\partial^2}{\partial y^2} + \frac{\partial^2}{\partial z^2}\right) = \hat{H}(\hat{\mathbf{x}}, \hat{\mathbf{y}}, \hat{\mathbf{z}}).$$

We can rotate the system instead of the basis – the result will be the same.

Another well-known example is the degeneracy of electron states in the Coulomb field. Ignoring spin, a state with *definite angular momentum* **L** (i.e., with a given quantum number l) is $(2l + 1)$-fold degenerate. In this case, the cause of degeneracy is also clearly seen from the classical analogy for the Coulomb (or gravitational) field: All bound states with a given size and shape of the elliptical orbit have the same magnitude $L = |\mathbf{L}|$ regardless of the orientation of the orbital plane in space. (This property would be observed in *any* field possessing spherical symmetry.) Once L is determined, we can create an infinite variety of degenerate states by changing the direction of **L**. As in the first example, this is due to the rotational invariance – this time because of the spherical symmetry of the potential field. Rotation of a system through an arbitrary angle, about an arbitrary axis passing through the center of the field, does not affect its Hamiltonian either. This again gives rise to a number of distinct states with different quantum numbers m but the same l and the same energy. In both cases, degeneracy is intimately linked with the symmetry of a system. The only difference is that in case of angular momentum, the order of

17.4 Symmetry, Degeneracy, and Perturbations

degeneracy is different in CM and QM. Classically, it is infinite because all possible orientations are allowed – they form a continuous set. Quantum-mechanically, there is only a finite number of possible configurations because of the indeterminacy and quantization of states. But the origin of the phenomenon is the same in both cases: Degeneracy of states arises from the symmetry. Once we understood this, we can answer the original question: A degenerate energy level can be split if the perturbation is such that it breaks the initial symmetry of the system.

Consider an illustration showing how degeneracy arises from a symmetry and is lifted by breaking that symmetry. We have a 2D oscillator when two potentials $U_1(x) = (1/2)k_1x^2$ and $U_2(y) = (1/2)k_2y^2$ cause vibrations along the x- and y-directions. In particular, if $k_1 = k_2 = k$, the net potential depends only on the distance s from the z-axis:

$$U(x,y) = U_1(x) + U_2(y) = \frac{1}{2}k(x^2+y^2) = \frac{1}{2}ks^2. \tag{17.31}$$

Accordingly, the Hamiltonian is invariant under rotations in the x-, y-plane, in other words, the system possesses axial symmetry. Therefore, we expect degeneracy, and want to see how it originates.

Write the Schrödinger equation and separate variables using $\Psi(x,y) = \psi_1(x)\psi_2(y)$ and $E = E_1 + E_2$:

$$\left[-\frac{\hbar^2}{2\mu}\left(\frac{\partial^2}{\partial x^2} + \frac{\partial^2}{\partial y^2}\right) + \frac{1}{2}k(x^2+y^2)\right]\Psi(x,y) = E\Psi(x,y). \tag{17.32}$$

The solutions are already familiar to us from Section 9.6:

$$E_1 = \hbar\omega_0\left(n_1 + \frac{1}{2}\right), \quad E_2 = \hbar\omega_0\left(n_2 + \frac{1}{2}\right); \quad \omega_0 \equiv \sqrt{\frac{k}{\mu}}, \quad n_1, n_2 = 0, 1, 2, \ldots,$$
$$\psi_1(x) = \psi_{n_1}(x), \psi_2(x) = \psi_{n_2}(x).$$

The complete solution $\Psi_{n_1 n_2}(x,y) = \psi_{n_1}(x)\psi_{n_2}(y)$ has energy $E_{n_1 n_2} = \hbar\omega_0(n_1 + n_2 + 1)$.

It is immediately seen that different pairs (n_1, n_2) with the same sum $n = (n_1 + n_2)$ determine different states with the same energy $E_{n_1 n_2} = \hbar\omega_0(n+1)$. For instance, states $(n_1, n_2) = (3, 6)$ and $(n'_1, n'_2) = (7, 2)$ have the same energy $E = 10\hbar\omega_0$. Therefore, especially from the viewpoint of spectroscopy dealing with the observed energies of a system, it is more convenient to describe such a state by a pair (n_1, n) rather than by (n_1, n_2). The number n here can be called the principal quantum number for an oscillator with axially symmetric Hamiltonian, and the oscillator's energy in terms of n will be

$$E_n = \hbar\omega_0(n+1), \quad n = 0, 1, \ldots. \tag{17.33}$$

Energy can be characterized by a single quantum number n even though the system is two dimensional. To each n, there correspond $n+1$ degenerate eigenstates differing from each other by quantum number n_1.

Suppose now that there is a perturbation imposed along the y-axis:

$$W(y) = \frac{1}{2}\kappa y^2, \quad \kappa \ll k, \quad \text{so that} \quad W(y) \ll U_2(y), \quad U(x, y). \tag{17.34}$$

Note that $W(y)$ is not an asymptotic perturbation because it does not change the asymptotic behavior of the system; and yet, no matter how small it may be, it breaks the axial symmetry and therefore kills the degeneracy. Denoting $k + \kappa \equiv \chi$, we now have

$$\left[-\frac{\hbar^2}{2\mu}\left(\frac{\partial^2}{\partial x^2} + \frac{\partial^2}{\partial y^2}\right) + \frac{1}{2}kx^2 + \frac{1}{2}\chi y^2\right]\Psi(x,y) = E\Psi(x,y). \tag{17.35}$$

The net potential energy is no longer the function of s alone due to an additional term $E_2^{(0)} \to E_1^{(0)}$. But the new Schrödinger equation can still be solved exactly by the same separation of variables as before. We obtain the same expression (17.36) for the eigenstates, but different expression for the eigenvalues:

$$E_{n_1 n_2} = \hbar\omega_0\left(n_1 + \frac{1}{2}\right) + \hbar\Omega\left(n_2 + \frac{1}{2}\right), \quad \Omega \equiv \sqrt{\frac{\chi}{\mu}}. \tag{17.36}$$

We can again introduce the principal quantum number $n = n_1 + n_2$ to write

$$E_{n,n_1} = \hbar\omega_0\left(n_1 + \frac{1}{2}\right) + \hbar\Omega\left(n - n_1 + \frac{1}{2}\right). \tag{17.37}$$

For each n we again have $n + 1$ eigenstates corresponding to all possible n_1, but their energies are no longer the same. The unperturbed level E_n now splits into $n + 1$ different levels. The degeneracy is gone together with the symmetry.

17.5
The Stark Effect

The Stark effect [1–3] consists in the splitting of atomic energy levels when the atom is placed in a constant external electric field. We already know that the action of this field on the atom is of twofold nature: It changes the asymptotic behavior, and it breaks the spherical symmetry. In this section, we elaborate on the second aspect. For the sake of simplicity, we shall consider only the hydrogen atom, whose unperturbed eigenstates are well known to us.

As we remember from Chapter 7, the unperturbed Hamiltonian $\hat{H}_0$ gives rise to eigenstates $|n, l, m\rangle$. We found then that $l = 0, 1, \ldots, n - 1$ and $m = -l, \ldots, l$ and a simple summation shows that there is a total of n^2 distinct eigenstates for each n. Thus, energy level E_n is (n^2)-fold degenerate.

To lift this degeneracy, let us apply a constant field $\mathcal{E}$ directed along the z-axis. This gives rise to the perturbation Hamiltonian:

$$W = -q\mathbf{r} \cdot \mathcal{E} = e\mathcal{E}z \tag{17.38}$$

(where electron charge e was taken as a positive number).

Let us start with the first-order correction to the energy levels E_n ($n = 1, 2, 3, \ldots$). The ground level $n = 1$ is nondegenerate, so the perturbation method applies immediately and since operator z connects states with opposite parities only, the correction is zero:

$$E_1^{(1)} = W_{11} = \langle 1,0,0|W|1,0,0\rangle = \langle 1,0,0|e\mathcal{E}\cdot\mathbf{r}|1,0,0\rangle = e\mathcal{E}\langle 1,0,0|z|1,0,0\rangle = 0. \quad (17.39)$$

We say therefore that there is no *linear Stark effect* for the ground state, because the first-order correction was linear with respect to $\mathcal{E}$; the higher order corrections will contain higher powers of $\mathcal{E}$.

Now proceed to the first excited state $n = 2$. There is fourfold degeneracy, so we should diagonalize W in the degenerate basis, which we will write as

$$\left\{|\Psi_1^{(0)}\rangle, |\Psi_2^{(0)}\rangle, |\Psi_3^{(0)}\rangle, |\Psi_4^{(0)}\rangle\right\} \equiv \{|200\rangle, |210\rangle, |211\rangle, |21-1\rangle\}. \quad (17.40)$$

To make notations more compact, we dropped the commas between the quantum numbers here and below (e.g. one should read $|21-1\rangle$ as $|2,1,-1\rangle$). Crossing out the matrix elements that are equal to zero, we obtain

$$W = \begin{pmatrix} \cancel{\langle 200|W|200\rangle} & \langle 200|W|210\rangle & \cancel{\langle 200|W|211\rangle} & \cancel{\langle 200|W|21-1\rangle} \\ \langle 210|W|200\rangle & \cancel{\langle 210|W|210\rangle} & \cancel{\langle 210|W|211\rangle} & \cancel{\langle 210|W|21-1\rangle} \\ \cancel{\langle 211|W|200\rangle} & \cancel{\langle 211|W|210\rangle} & \cancel{\langle 211|W|211\rangle} & \cancel{\langle 211|W|21-1\rangle} \\ \cancel{\langle 21-1|W|200\rangle} & \cancel{\langle 21-1|W|210\rangle} & \cancel{\langle 21-1|W|211\rangle} & \cancel{\langle 21-1|W|21-1\rangle} \end{pmatrix}. \quad (17.41)$$

The justification for the above step is as follows. To facilitate references to matrix elements, we now restore the "default" notation, specifying elements by their row and column numbers. All diagonal elements W_{kk} are zeros since the integrands are odd functions of z. The same is true for W_{34}, W_{43}. The elements W_{13}, W_{14}, W_{23}, W_{24} are zeros because of the factor $e^{\pm i\varphi}$ appearing in the wave functions (see Appendix E) whose integral over the period is zero. After straightforward algebra, the eigenvalue equation gives us four solutions:[2]

$$\begin{pmatrix} 0 & W_{12} & 0 & 0 \\ W_{21} & 0 & 0 & 0 \\ 0 & 0 & 0 & 0 \\ 0 & 0 & 0 & 0 \end{pmatrix} \begin{pmatrix} c_1 \\ c_2 \\ c_3 \\ c_4 \end{pmatrix} = \lambda \begin{pmatrix} c_1 \\ c_2 \\ c_3 \\ c_4 \end{pmatrix} \Rightarrow \begin{cases} \lambda_{1,2} \equiv E_{1,2}^{(1)} = \pm|W_{12}|, \\ \lambda_3 \equiv E_3^{(1)} = 0, \quad \lambda_4 \equiv E_4^{(1)} = 0. \end{cases}$$

$$(17.42)$$

One degenerate level will be raised by the perturbation, another will be lowered by the same amount, and two other levels will remain unchanged and retain their degeneracy. The corresponding energy levels are shown in Figure 17.4.

---

[2] When all elements outside of a certain submatrix are zeros, it is sufficient to solve the eigenvalue problem for that submatrix and set the remaining eigenvalues and coefficients to zero.

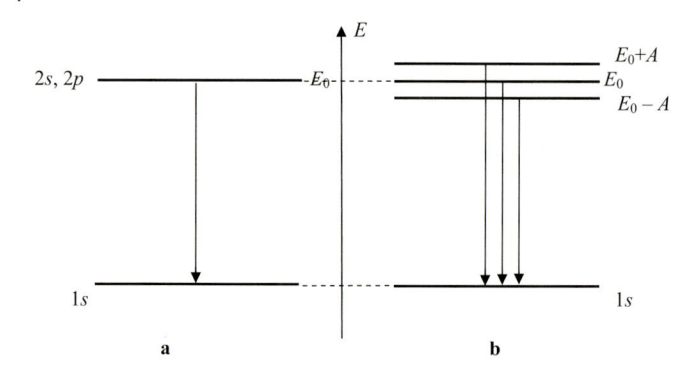

Figure 17.4 The Stark effect. (a) The degenerated level $n=2$ in the hydrogen atom. (b) Splitting of the level 2s, 2p due to the external electric field.

It remains to find the eigenstates corresponding to these eigenvalues. Solving the above equation and normalizing the resulting amplitude, we obtain the following:

For $E_1^{(1)} = |W_{12}|$: $c_1 = \dfrac{1}{\sqrt{2}}$, $c_2 = -\dfrac{1}{\sqrt{2}}$, $c_3 = 0$, $c_4 = 0$.

For $E_2^{(1)} = -|W_{12}|$: $c_1 = \dfrac{1}{\sqrt{2}}$, $c_2 = \dfrac{1}{\sqrt{2}}$, $c_3 = 0$, $c_4 = 0$.

For $E_3^{(1)}$ and $E_4^{(1)}$: $c_1 = c_2 = 0$ and $c_{3,4} \neq 0$. The amplitudes c_3, c_4 remain unspecified in the latter case (except for the normalization condition). There remains partial degeneracy of the second order.

With these results, we now compose the new degenerate basis from these four eigenstates:

$$|\tilde{\Psi}_1^{(0)}\rangle = \frac{1}{\sqrt{2}}\left(|\Psi_1^{(0)}\rangle + |\Psi_2^{(0)}\rangle\right) = \frac{1}{\sqrt{2}}(|200\rangle - |210\rangle). \tag{17.43}$$

$$|\tilde{\Psi}_2^{(0)}\rangle = \frac{1}{\sqrt{2}}\left(|\Psi_1^{(0)}\rangle - |\Psi_2^{(0)}\rangle\right) = \frac{1}{\sqrt{2}}(|200\rangle + |210\rangle). \tag{17.44}$$

$$|\tilde{\Psi}_3^{(0)}\rangle = c_3|\Psi_3^{(0)}\rangle + c_4|\Psi_4^{(0)}\rangle = c_3|211\rangle + c_4|21-1\rangle. \tag{17.45}$$

$$|\tilde{\Psi}_{3,4}^{(0)}\rangle = d_3|\Psi_3^{(0)}\rangle + d_4|\Psi_4^{(0)}\rangle = d_3|211\rangle + d_4|21-1\rangle. \tag{17.46}$$

We have switched from c's to d's in the last equation to distinguish the fourth eigenstate from the third. These coefficients should be chosen so as to make all four states mutually orthogonal. In particular, one can choose $c_3 = 1$, $c_4 = 0$ and $d_3 = 0$, $d_4 = 1$ to leave the last two states the same as they were in the old basis! Since we have established that they will not respond to W anyway, it only makes sense to leave them alone. The perturbation will thus be diagonalized in the basis

$$\left\{|\tilde{\Psi}_1^{(0)}\rangle, |\tilde{\Psi}_2^{(0)}\rangle, |\tilde{\Psi}_3^{(0)}\rangle, |\tilde{\Psi}_4^{(0)}\rangle\right\} \equiv \left\{\frac{|200\rangle - |210\rangle}{\sqrt{2}}, \frac{|200\rangle + |210\rangle}{\sqrt{2}}, |211\rangle, |21-1\rangle\right\}. \tag{17.47}$$

The splitting of the degenerate level $E_2^{(0)}$ into three levels enables optical transitions that could not be observed in the unperturbed atom. The transition $E_2^{(0)} \to E_1^{(0)}$ produces the first line in the Lyman UV series. When the field is turned on, the line spawns off two other lines, which move farther and farther apart in proportion to the applied field.[3]

One may ask: What is the reason that residual degeneracy still remains in the system? The answer is that perturbation W breaks the symmetry only partially. The electron cloud in the electric field is no longer spherically symmetric, but it remains axially symmetric – it is not affected by rotations about z.

One can obtain the nonvanishing matrix elements W_{12} and W_{21} by direct integration (Problem 17.6) that involves wave functions $\Psi_1^{(0)}$ and $\Psi_2^{(0)}$. The result is

$$|W_{12}| = |W_{21}| = W = e\mathcal{E}\left|\int \Psi_1^{(0)} z \Psi_2^{(0)} d\mathbf{r}\right| = 3ea_0\mathcal{E}. \tag{17.48}$$

This implies that we can interpret W_{12} and W_{21} as the potential energies of a dipole with the magnitude of the dipole moment $p = 3ea_0$ in the external field $\mathcal{E}$. Then eigenstates possessing these energies acquire a simple physical interpretation: In these eigenstates, the atom is *polarized even in the absence of the external field!* In state $|\tilde{\Psi}_1^{(0)}\rangle$, it has a *negative* average dipole moment $\langle p_z \rangle = -3ea_0$. When the field is applied, the atom acquires the additional potential energy $\mathcal{E}_1^{(1)} = -\langle p_z \rangle \mathcal{E} = 3ea_0 \mathcal{E}$. In state $|\tilde{\Psi}_2^{(0)}\rangle$, it is polarized in the *positive* z-direction, and accordingly acquires additional energy $\mathcal{E}_2^{(1)} = -3a_0 q\mathcal{E}$. But in states $|\tilde{\Psi}_3^{(0)}\rangle$ and $|\tilde{\Psi}_4^{(0)}\rangle$, the dipole moment is zero, so these states do not change under a weak perturbation. This is similar to the polarized states of the NH_3 molecule, with the difference that this time the polarized states are stationary (there is no possibility for tunneling!).

Note that the linear Stark effect (i.e., energy level shift is proportional to E and appears already in the first approximation) is only observed because energy levels with different l are degenerate. Accordingly, the new eigenstate written as a superposition of the old ones will be polarized only if it includes states with different l. This is precisely the case in superpositions (17.43) and (17.44) involving states with $l = 0, 1$. One can show (Problem 17.9) that any superposition restricted only to states with different m but the same l produces a state with a zero dipole moment.

17.6
Time-Dependent Perturbations

The perturbations considered before were constant in time. But the majority of real-life situations are not static. A typical practical problem might be, for example, what will happen to an atom prepared initially in one of its excited states after it interacts for a while with incident light. We would write $\mathcal{E}(t) = \mathcal{E}\cos(\omega t)$ for the electric field in the EM wave and then the perturbation that appeared in the Stark effect would become a function of time, $W(t) = ez\mathcal{E}(t)$. This perturbation would start when the front of the EM wave reaches the atom and stop at some later moment τ after the

---
[3] This phenomenon was first observed by Stark for a spectral line in the Balmer series (it corresponds to transitions from the higher levels to the level $n = 2$).

entire wave train has passed by. Another example is magnetic resonance, when atomic spin interacts with a time-dependent magnetic field. In all such cases, it is convenient to start counting the time from the moment $W(t)$ is "turned on." Then,

$$W = \begin{cases} 0, & t < 0, \\ W(\mathbf{r}, t), & 0 < t < \tau, \\ 0, & t > \tau. \end{cases} \tag{17.49}$$

Parenthetically, we shall always assume that $W(\mathbf{r}, t)$ turns to zero at the endpoints of the interval. This is done in order to avoid discontinuity at the endpoints.

Time-dependent perturbations are different from time-independent ones in that after turning off, the perturbation at $t = \tau$ will *not* restore the initial state of the system even though the Hamiltonian is again $\hat{H}_0$. The reason is that the system evolves while $W(\mathbf{r}, t)$ is active, so the moment $t = \tau$ will generally find it in a different state.

The observable of interest in applications is usually energy, and the problem is to find the probabilities of transitions $E_k \rightleftarrows E_n$ between different energy levels k, n of the system under a given perturbation $W(\mathbf{r}, t)$. It is important to remember that when subjected to a time-dependent perturbation, the system *does not* have a definite value of energy. Only at $t < 0$ and $t > \tau$ will energy be an integral of motion that could have a definite value. Therefore, our question makes sense only when we ask about the situation *after* the perturbation is turned off. That is, if we are told that at some $0 < t < \tau$ the amplitude of eigenstate $|\Psi_n\rangle$ reaches 1, it means that *if* we turned off the perturbation now, *then* an energy measurement would certainly result in an outcome E_n.

Now we use the approach outlined in the previous sections. Write the Schrödinger equation and expand $|\Psi\rangle$ over eigenstates $|\Psi_n\rangle$ of the unperturbed Hamiltonian:

$$i\hbar \frac{\partial}{\partial t} \left(\sum_n c_n(t) e^{-i\omega_n t} |\Psi_n\rangle \right) = (\hat{H}_0 + W(\mathbf{r}, t)) \left(\sum_n c_n(t) e^{-i\omega_n t} |\Psi_n\rangle \right). \tag{17.50}$$

We want to highlight the fact that exponential temporal factors $e^{-i\omega_n t}$ (where $\omega_n \equiv E_n/\hbar$) that were inconsequential in the time-independent case must now be included explicitly! As to the expansion amplitudes $c_n(t)$, they are now also functions of time. This seems natural since these amplitudes must depend on the perturbation.

We now differentiate on the left, cancel out equal terms on both sides keeping in mind that $\hat{H}_0 |\Psi_n\rangle = E_n |\Psi_n\rangle$, and, finally, project the equation onto $\langle e^{-i\omega_m t} \Psi_m|$:

$$\sum_n \left(i\hbar \dot{c}_n(t) e^{-i\omega_n t} + \cancel{\hbar \omega_n c_n(t) e^{-i\omega_n t}} \right) |\Psi_n\rangle$$
$$= \sum_n c_n(t) e^{-i\omega_n t} (\cancel{E_n |\Psi_n\rangle} + W(\mathbf{r}, t) |\Psi_n\rangle),$$

$$\langle \Psi_m | e^{i\omega_m t} \sum_n i\hbar \dot{c}_n(t) e^{-i\omega_n t} |\Psi_n\rangle = \langle \Psi_m | e^{i\omega_m t} \sum_n c_n(t) e^{-i\omega_n t} W(\mathbf{r}, t) |\Psi_n\rangle. \tag{17.51}$$

In the last step, we have introduced the *transition frequency* defined as

$$\omega_{mn} \equiv \omega_m - \omega_n = \frac{E_m - E_n}{\hbar} = -\omega_{nm}. \tag{17.52}$$

At this stage, we have yet to establish the relation between ω_{mn} and the perturbation. All we can say about ω_{mn} right now is that it is the frequency of the photon that would be emitted/absorbed in a transition $m \rightleftarrows n$.

The matrix element of the perturbation $W_{mn}(t)$ is now a function of time. This function has the same behavior as $W(\mathbf{r}, t)$, that is, it is zero beyond the domain $0 \le t \le \tau$. The set of equations (17.51) for different m define a system of coupled differential equations that amount to a single matrix equation:

$$i\hbar \begin{pmatrix} \dot{c}_1 \\ \dot{c}_2 \\ \dot{c}_3 \\ \vdots \end{pmatrix} = \begin{pmatrix} W_{11}(t)e^{i\omega_{11}t} & W_{12}(t)e^{i\omega_{12}t} & W_{13}(t)e^{i\omega_{13}t} \\ W_{21}(t)e^{i\omega_{21}t} & W_{22}(t)e^{i\omega_{22}t} & W_{23}(t)e^{i\omega_{23}t} \\ W_{31}(t)e^{i\omega_{31}t} & W_{32}(t)e^{i\omega_{32}t} & W_{33}(t)e^{i\omega_{33}t} \\ \vdots & \vdots & \vdots & \end{pmatrix} \begin{pmatrix} c_1 \\ c_2 \\ c_3 \\ \vdots \end{pmatrix} \qquad (17.53)$$

If we manage to solve this system, we will know all amplitudes $c_n(t)$ and thus know the probability of finding the system in any given eigenstate $|\Psi_n\rangle$ as a function of t; by setting $t = \tau$, we will know the situation right after the perturbation. As is the case with any differential equation, mathematics requires that we provide the initial condition. Of special practical interest is the case when at $t = 0$ the system was in some energy eigenstate $|\Psi_k\rangle$. Then the solution will give us transition probabilities $k \to n$ for all n.

One example for which (17.53) has a simple solution is a two-level system subject to a perturbation such that this equation reduces to

$$i\hbar \begin{pmatrix} \dot{c}_1 \\ \dot{c}_2 \end{pmatrix} = \begin{pmatrix} 0 & \gamma e^{i\omega t} \\ \gamma e^{-i\omega t} & 0 \end{pmatrix} \begin{pmatrix} c_1 \\ c_2 \end{pmatrix}. \qquad (17.54)$$

In other words, we have taken $W_{11}(t) = W_{22}(t) = 0$, $W_{12}(t) = W_{21}(t) = \gamma$, $\omega_{12} = \omega$. This special case represents a sinusoidal perturbation that oscillates with frequency ω. If the system was initially in the first eigenstate, that is, $c_1(0) = 1$, $c_2(0) = 0$, one obtains

$$c_2(t) = \frac{\gamma/\hbar}{\Omega} \sin \Omega t, \quad c_1(t) = \sqrt{1 - c_2^2(t)}, \qquad (17.55)$$

where Ω is the so-called generalized *Rabi frequency* defined as

$$\Omega = \sqrt{\left(\frac{\gamma}{\hbar}\right)^2 + \left(\frac{\omega - \omega_{21}}{2}\right)^2}. \qquad (17.56)$$

The minimum value $\gamma/\hbar$ of the Rabi frequency is attained at the so-called *resonance condition* $\omega = \omega_{21}$ (frequency of the sinusoidal perturbation is tuned to the transition frequency). The resonance condition allows $c_1(t)$ and $c_2(t)$ to oscillate with amplitude 1, so if we wait for $\tau = k\pi/\Omega$, $k = 0, 1, 2, \ldots$, we will find the system in its initial state with certainty. The quantity $\Delta \equiv \omega - \omega_{21}$ has a special name: It is called *detuning* and is a measure of how far the perturbation is out of tune with the transition frequency. Away from resonance, that amplitude will be lower and there will always be a chance to find the system in either eigenstate. Very far from resonance ($|\omega - \omega_{21}| \gg \gamma/\hbar$), probability of the second state becomes negligible and the net result is the same as if there were no perturbation. This is analogous to the behavior of the classical driven oscillator; one often uses the term *driven two-level system*.

The reason this example is so important to us is perturbation (17.54) successfully models the interaction between an atom and an incident light. This model is *semiclassical*: It treats the atom as a quantum system, but describes light as a classical EM wave. The *Rabi oscillations* described by (17.55) give rise to the *emission–absorption cycle*. Increase of probability $|c_2(t)|^2$ in the first half of the cycle describes absorption of photons from the incident field (the atom is now more likely to be found in the excited state); decrease of $|c_2(t)|^2$ in the second half of the cycle describes stimulated emission that tends to return the atom to the ground state. (A more rigorous treatment of the same problem is provided by the *Jaynes–Cummings model*, which considers the EM field as a flux of photons and is thus fully quantum-mechanical.) It is worth noting that if we dropped the exponential function in (17.54), we would recover the case (15.2), with $H_{11} = H_{22} = 0$ and $H_{12} = H_{21} = \gamma$. Indeed, dropping the exponential function means setting $\omega \to 0$ (so that $e^{\pm i\omega t} \simeq 1$ for any finite t), thus effectively removing the perturbation, which takes us back to the case of a particle oscillating between two wells.

Sadly, the well-behaved perturbation (17.54) is an exception rather than the rule. Generally, Equation (17.53) has no analytical solution. Then, we attempt to solve it using successive approximations as follows. First, express the initial condition (system prepared in state $|\Psi_k\rangle$) as $c_m(0) = \delta_{km}$. In the zero-order approximation, assume

$$c_m(t) \approx c_m(0) = \delta_{km} \qquad (17.57)$$

(in other words, with perturbation ignored, the initial state would never change) and apply that approximation to the *right-hand* side of (17.53), while leaving amplitudes on the *left-hand* side as they are – as some unknown functions of time. This results in the simplified equation:

$$i\hbar \dot{c}_m(t) \simeq W_{mk}(t) e^{i\omega_{mk} t}. \qquad (17.58)$$

Second, integrate this equation to get the first-order approximation:

$$c_m(t) \approx \delta_{km} - \frac{i}{\hbar} \int_0^t W_{mk}(t') e^{i\omega_{mk} t'} dt'. \qquad (17.59)$$

Apply it again to the right-hand side of (17.53), keeping the left-hand side unchanged. This gives

$$i\hbar \dot{c}_m(t) = \sum_n W_{mn}(t) e^{i\omega_{mn} t} c_n(t). \qquad (17.60)$$

Now integrate this equation again, substituting (after a trivial change of index) expression (17.59) for $c_n(t'')$:

17.6 Time-Dependent Perturbations

$$c_m(t) \approx \delta_{km} - \frac{i}{\hbar}\sum_n \int_0^t W_{mn}(t')e^{i\omega_{mn}t'}\left(\delta_{kn} - \frac{i}{\hbar}\int_0^t W_{nk}(t'')e^{i\omega_{nk}t''}dt''\right)dt'$$

$$= \delta_{km} - \frac{i}{\hbar}\int_0^t W_{mk}(t')e^{i\omega_{mk}t'}dt' + \left(-\frac{i}{\hbar}\right)^2 \sum_n \iint_0^{t\,t} W_{mn}(t')e^{i\omega_{mn}t'} W_{nk}(t'')e^{i\omega_{nk}t''}dt''dt'.$$

$$(17.61)$$

This procedure can be extended if we want to obtain more accurate approximations (obtaining the so-called *Dyson series*), but usually the first, and sometimes second, approximations are sufficient.

Note that the Dyson series makes use of *all* moments of time in the integration procedure, but measurement statistics after the perturbation is obtained by evaluating $c_m(t)$ at $t > \tau$. Also, since for all t outside the interval $(0, \tau)$ the perturbation is zero, we can formally extend the integration to the entire time axis. Then, the first-order approximation at the moment τ will be written as

$$c_{km}(\tau) = \delta_{km} - \frac{i}{\hbar}\int_{-\infty}^{\infty} W_{mk}(t)e^{i\omega_{mk}t}dt. \qquad (17.62)$$

By focusing on one specific moment $t = \tau$ we have eliminated time dependence from, the amplitudes $c_{km}(\tau)$ become time-independent coefficients; interval τ during which the perturbation is finite figures as a parameter in (17.62). Now we are going to explain our reason for extending the integration from $-\infty$ to ∞. Let us ask ourselves, what is the Fourier transform of the function $W_{mk}(t)$. Writing the Fourier integrals, we see[4] that

$$W_{mk}(t) = \frac{1}{\sqrt{2\pi}}\int \mathcal{W}_{mk}(\omega_{mk})e^{-i\omega_{mk}t}d\omega_{mk},$$

$$\mathcal{W}_{mk}(\omega_{mk}) = \frac{1}{\sqrt{2\pi}}\int W_{mk}(t)e^{i\omega_{mk}t}dt, \qquad (17.63)$$

the Fourier transform $\mathcal{W}_{mk}(\omega_{mk})$ of the matrix element $W_{mk}(t)$ for the transition $(k \to m)$ is nothing else but the integral appearing on the right-hand side of (17.62). Therefore,

$$c_{km}(\tau) = \delta_{km} - \frac{\sqrt{2\pi}i}{\hbar}\mathcal{W}_{mk}(\omega_{mk}), \quad t > \tau. \qquad (17.64)$$

Assuming $k \neq m$ (we are obviously interested in transitions between *different* levels!), the transition amplitude is nonzero only if the corresponding Fourier transform of the perturbation element is nonzero. The first equation in (17.63) describes expansion of $W_{mk}(t)$ over all possible frequencies ω_{mk} (note again that in this equation, ω_{mk} is a variable of integration!), which form the frequency spectrum of $W_{mk}(t)$. To say that $\mathcal{W}(\omega_{mk}) \neq 0$ for some *specific* ω_{mk} is to say that

---

[4] To make it easier to read these integrals, you can drop the subscripts mk everywhere in (17.63), convince yourself that $\mathcal{W}(\omega_{mk})$ is the Fourier transform of $W(t)$, and then reinstate the subscripts.

this specific frequency appears in the Fourier expansion of $W_{mk}(t)$. The transition $k \to m$ can only take place if the frequency spectrum of the perturbation element $W_{mk}(t)$ contains the transition frequency $\omega = \omega_{mk}$! In other words, transitions have a resonant nature. This justifies Bohr's hypothesis (2.10) that a quantum system is modeled by a set of oscillators, which respond to radiation tuned to their respective proper frequencies. Conversely, under a perturbation, only those oscillators get activated whose frequencies are equal to those in the external field.

To take an example, suppose the perturbation is nearly monochromatic within the time interval $0 \le t \le \tau$ (we had to say "nearly" because a fully monochromatic function would extend from $-\infty$ to ∞) and turns to zero at the endpoints:

$$W_{km}(t) = W_{km} \sin \omega t, \quad 0 \le t \le \tau. \tag{17.65}$$

Then, ω has to satisfy the boundary condition $\sin(\omega \tau) = 0$; in other words, $\omega = \pi j/\tau \equiv \omega_j$, where $j = 1, 2, \ldots$ is the number of half-cycles within the time interval. (Think of the particle-in-a-box analogy from Section 9.1 where an integer number of half-waves had to fit within the box to ensure the continuity of the wave function). The perturbation (17.62) could be a laser pulse from a Fabri–Perrot resonator whose mirrors were suddenly made transparent. An atom placed in the way of this pulse will feel the perturbation during the time interval $\tau = a/c$. The first-order approximation (17.59) gives (Problem 17.10)

$$c_{km} \approx \delta_{km} + \frac{\sqrt{2\pi}}{\hbar} W_{km} \frac{(\omega_j \cos \omega_j \tau - i\omega_{mk} \sin \omega_j \tau) e^{i\omega_{mk}\tau} - \omega_{mk}}{\omega_{mk}^2 - \omega_j^2}. \tag{17.66}$$

We are interested in *transitions*, when the particle quits its initial state, that is, $m \ne k$. For all such cases, the last expression, in view of the condition $\sin(\omega_j \tau) = 0$, reduces to

$$c_{km} \approx \frac{\sqrt{2\pi}}{\hbar} W_{km} \frac{(-1)^j \omega_j e^{i\omega_{mk}\tau} - \omega_{mk}}{\omega_j^2 - \omega_{mk}^2}. \tag{17.67}$$

Generally, this result gives a sufficiently good description of the phenomenon. But when we want to use in the vicinity of resonance, $\omega_j \approx \omega_{mk}$, we run into trouble as the transition amplitude diverges:

$$c_{km} \approx \sqrt{\frac{\pi}{2}} \frac{W_{km}}{\hbar} \frac{(-1)^j e^{i\omega_{mk}\tau} - 1}{\omega_j - \omega_{mk}}. \tag{17.68}$$

This is analogous to the problem we encountered with stationary perturbations in the degenerate case when the obtained amplitude also diverged at $E_k^{(0)} \to E_m^{(0)}$. We see that the assumption of sharply defined energy levels is not very consistent with cyclic perturbations. In the vicinity of resonance, we must use a more realistic model of quasi-stationary states and/or the exponentially decreasing perturbation in the spirit of Gamov's theory. This will be done in Chapter 18.

Problems

17.1 Consider a hydrogen atom inside a parallel plate capacitor with 1 cm separation between the plates, under a constant voltage 180 V. Estimate the set of quantum numbers n for which the external electric field E will not significantly disturb energy eigenstates.

17.2 Write the expression for the eigenfunction $|\Psi_k(\mathbf{r})\rangle$ of the perturbed Hamiltonian (a) in the first approximation and (b) in the second approximation.

17.3 Consider a two-state system. Apply Equation 17.5 to this case and show that it describes two Hilbert spaces rotated with respect to each other by a small angle θ.

17.4 In the previous problem, express the rotation angle in terms of perturbation W in the first approximation.

17.5 Evaluate the correction to an nth energy level of a hydrogen atom in a homogeneous electric field $\mathcal{E}$ within the region $r \leq a_0 N^2$ using the perturbation theory (a) in the first approximation and (b) in the second approximation.

17.6 Derive the expression (17.48).

17.7 In spectroscopic experiments, a meaningful characteristic is the ratio of the *energy* level shift to the energy of unperturbed transition. Find this ratio in terms of the atomic characteristics and applied field for the optical transition $E_2^{(0)} \to E_1^{(0)}$ in a hydrogen atom.

17.8 Identify the matrix elements in (17.41) that must vanish because of the Wigner–Eckart theorem.

17.9 Prove that in *any* eigenstate of a degenerate energy level $\mathcal{E}_{nl}^{(0)}$ of a hydrogen atom, the expectation value of the dipole moment is zero (the word *any* here means not only states with definite m, but also all their possible superpositions).

17.10 Calculate the Fourier transform and the transition amplitude in case of perturbation (17.65).

References

1 Condon, E.U. and Shortley, G.H. (1935) *The Theory of Atomic Spectra*, Cambridge University Press.

2 Friedrich, H. (1990) *Theoretical Atomic Physics*, Springer, Berlin.

3 Kroto, H.W. (1992) *Molecular Rotation Spectra*, Dover, New York.

18
Light–Matter Interactions

In this chapter, we apply the perturbation theory to the interactions between atoms and incident light. In this treatment, the perturbation $W(\mathbf{r}, t)$ will represent potential energy of an electron in the field of the incident EM wave.

18.1
Optical Transitions

We will address here the question asked at the end of Chapter 17: What happens when $\omega \to \omega_{km}$, that is, the perturbation frequency approaches the effective frequency of transition $k \rightleftarrows m$?

If the question is put in this way, it is physically meaningless. The phrase "perturbation frequency" implies that there is a single frequency ω exactly defined, while the word "approaches" implies that an optical transition is characterized by a single definite frequency ω_{km}. This is *mathematically* possible, but then the amplitude $c_{km}^{(1)}$ in (17.66) diverges at $\omega \to \omega_{km}$. This means that the obtained result, although describing correctly the tendency of the transition amplitude to increase at $\omega \to \omega_{km}$, does not hold in the immediate vicinity of the resonant frequency ω_{km}. The first approximation algorithm is restricted by the condition $c_{km}^{(1)} \ll 1$. This defines the range of frequencies near resonance that cannot be treated by the straightforward use of the algorithm with a *definite* perturbation frequency (Problem 18.1). Note that the divergence of the result at the resonant frequency is obtained even when the condition (17.2) for small perturbations is satisfied! So this condition by itself is necessary, but not sufficient for getting the correct result. What is the cause of this discrepancy? To get a hint, recall Equation (2.27) for an oscillator driven by a resonant force. The solution there also diverges at resonance in the absence of friction, no matter how small the driving force (see what happens with (2.26) at $\omega = \omega_0$. if you take $\gamma \to 0$!). In this limit, (2.26) must be replaced with a nonstationary solution, as described in Appendix A. As shown here, even in the absence of friction, there exist two types of solution that are finite for all finite t. The solution of the second type is a system's response to a driving force with a complex frequency, which actually describes slow exponential decrease of the force. Such decrease is similar to what we observe in the decay of a quasi-stationary state in QM.

This classical analogy seems purely formal, but eventually it suggests the answer: It is not the failure of the theory, but an indication that we did not take into account the real properties of the excited states. Recall Section 11.10 on radioactive decay. Each excited state $|\Psi_k\rangle$ of an *unperturbed* atom is actually a quasi-stationary state with a certain average lifetime τ_k and, due to the indeterminacy $\Delta E_k \geq \hbar/\tau_k$, it cannot be characterized by an exactly definite energy level E_k. Rather, it is characterized by a central energy E_k and the uncertainty in energy (width ΔE_k of the level E_k). The latter can be formally described by an additional parameter Γ in the complex energy plane. In our case we are interested in the partial width Γ_{km} of the energy level corresponding to a specific transition $k \to m$. This transition is accurately described by attributing to it, apart from the resonant frequency (2.10), the corresponding partial width Γ_{km}. Accordingly, the transition frequency can be formally described by a complex number

$$\omega_{km} \to \omega'_{km} = \omega_{km} + i\gamma_{km}, \quad \gamma_{km} = \frac{\Gamma_{km}}{\hbar}. \tag{18.1}$$

If we put this phenomenological parameter into (17.66), we get

$$c_{km}^{(1)}(\omega) \sim \frac{1}{\omega_{km} - \omega - i\gamma_{km}}. \tag{18.2}$$

The corresponding transition probability will be

$$\mathcal{P}_{km}^{(1)}(\omega) \sim \frac{1}{(\omega_{km} - \omega)^2 + \gamma^2}. \tag{18.3}$$

We could obtain the same result if, instead of describing the atomic states as quasi-stationary Gamov's states, we consider the perturbing radiation field as quasimonochromatic, of the type

$$E(t) = \begin{cases} 0, & t \leq 0, \\ E_0 e^{-(t/\tau)} e^{i\omega t}, & t > 0. \end{cases} \tag{18.4}$$

This would be similar to Equations A.15, A.18. Such form of perturbation field is quite natural. After all, the field irradiating our atom had been radiated before by some other atom in the process of its optical decay from level m to a lower level k. A nonstationary excited state is the Gamov state. Therefore, an optical emission, as any other decay, obeys the typical decay law (11.85) and (11.111) for the Gamov states. No surprise that expressions (18.3) and (18.4) for a natural physical perturbation satisfy this law.

Summary. The perturbation theory, if applied consistently, correctly describes the resonant nature of quantum transitions of atomic states under the EM radiation. Therefore, it can be used for description of the light–matter interaction near resonance.

18.2
Dipole Radiation

We are now equipped for outlining the basic features of light interaction with a hydrogen-like atom. We assume the incident light to be (quasi)monochromatic and linearly polarized. The electric field of a typical light wave (in the sunlight) is about

10^{-8} of the average Coulomb's field acting on the atomic electron. The effect of magnetic field is even less (in the EM wave $|\mathbf{B}| = |\mathcal{E}|/c$; so in the Lorentz force (16.2), the average magnetic force on the atomic electron in the EM wave is the small fraction of electric force, $f_{\text{magn}} = (v/c) f_{\text{el}}$). Under these conditions, we can consider the incident light as a perturbation. Another simplifying factor is that, unless we consider an ultrahard radiation, the light wavelength is much larger than the size of the atom. Therefore, we can neglect the spatial variation of the electric field of an EM wave within the atomic region, and consider it only as a function of time:

$$\mathcal{E}(t) = \mathbf{e}\mathcal{E}_0(t) e^{i\omega t}, \tag{18.5}$$

with a slowly varying amplitude $\mathcal{E}_0(t)$ and the unit vector of polarization $\mathbf{e}$. The corresponding perturbation can be written as

$$W(\mathbf{r}, t) = -q\mathbf{r} \cdot \mathcal{E}(t) = -\mathbf{d}(\mathbf{r}) \cdot \mathcal{E}(t), \tag{18.6}$$

where $\mathbf{d}(\mathbf{r}) \equiv q\mathbf{r}$ is the classical dipole moment for location $\mathbf{r}$. We can consider (18.6) either as potential energy of the electron in a spatially uniform electric field $\mathcal{E}(t)$ or as potential energy of the corresponding dipole $\mathbf{d}(\mathbf{r})$ in this field.

Now we can apply the results of perturbation theory. According to (17.61), the probability of an optical transition $n \rightleftarrows n'$ is determined by the Fourier transform $F(\omega') \equiv W_{nn'}(\omega')$, $\omega' = \omega_{nn'}$, of the matrix element of perturbation:

$$P_{nn'} = \frac{4\pi^2}{\hbar^2} |W_{nn'}(\omega_{nn'})|^2. \tag{18.7}$$

The latter is given by

$$W_{nn'}(t) = -q\langle n'|\mathbf{r} \cdot \mathcal{E}(t)|n\rangle = -q\mathcal{E}_0(t) e^{i\omega t} \langle n'|\mathbf{r} \cdot \mathbf{e}|n\rangle. \tag{18.8}$$

Since $\mathbf{e}$ is constant unit vector, this reduces to

$$W_{nn'}(t) = -q\langle n'|\mathbf{r} \cdot \mathcal{E}(t)|n\rangle = -q\mathcal{E}_0(t) e^{i\omega t} \mathbf{e} \cdot \mathbf{r}_{nn'}. \tag{18.9}$$

In other words, the result depends on the dot product of polarization vector and the matrix element of position operator:

$$\mathbf{r}_{nn'} = x_{nn'}\mathbf{x} + y_{nn'}\mathbf{y} + z_{nn'}\mathbf{z}, \tag{18.10}$$

with $\mathbf{x}, \mathbf{y}, \mathbf{z}$ being the corresponding unit vectors.

Thus, the spatial and temporal dependencies are represented each by their respective separate factor, and the result is determined by the Fourier component $\mathcal{E}(\omega_{nn'})$ of the electric field:

$$\mathcal{E}(\omega_{nn'}) = \frac{1}{\sqrt{2\pi}} \int_0^\infty \mathcal{E}_0(t) e^{i\omega t} e^{-i\omega_{nn'} t} dt \tag{18.11}$$

(recall that the acting field is nonzero only for $t > 0$). The resulting transition probability can be written as

$$P_{nn'} = 2\pi \frac{q^2}{\hbar^2} |\mathcal{E}(\omega_{nn'})|^2 |\mathbf{e} \cdot \mathbf{r}_{nn'}|^2. \tag{18.12}$$

We can rewrite it in the form more suitable for practical applications. Note that

$$|\mathcal{E}(\omega)|^2 = \frac{1}{c}\frac{dE}{d\omega} \qquad (18.13)$$

is the amount of energy E per unit frequency range, which has flown through the unit cross-sectional area perpendicular to the direction of propagation of light. We can write this amount as

$$\frac{dE(\omega)}{d\omega} = c\rho(\omega)T, \qquad (18.14)$$

where $\rho(\omega)$ is the spectral energy density (1.2) and $T > \tau$ is the "waiting time" exceeding the time of action of the incident field. Since both these times are usually indefinite, a more meaningful characteristic related to real observations is the transition probability per unit time:

$$w_{nn'} \equiv \frac{\mathcal{P}_{nn'}(\omega_{nn'})}{T}. \qquad (18.15)$$

Combining all these equations give

$$w_{nn'} = 2\pi \frac{q^2}{\hbar^2}\rho(\omega_{nn'})|\mathbf{e}\cdot\mathbf{r}_{nn'}|^2. \qquad (18.16)$$

We can also write this in terms of the photon number $N(\omega_{nn'})$ as

$$w_{nn'} = 2\pi\hbar\omega_{nn'}\frac{q^2}{\hbar^2}N(\omega_{nn'})|\mathbf{e}\cdot\mathbf{r}_{nn'}|^2. \qquad (18.17)$$

Here, $N(\omega_{nn'})$ is the number of the respective photons per unit frequency, all in one state of the same frequency, momentum, and polarization. Thus, the transition probability $w_{nn'}$ is completely determined by polarization of light, its spectral energy density, and the matrix elements of position operator $\mathbf{r}$.

Let us focus on one such element $\mathbf{r}_{nn'}$ with specified n and n' and consider the case when the radiation field (18.5) can be approximated by a plane wave with propagation direction $\hat{\mathbf{k}}$. The polarization vector $\mathbf{e}$ is perpendicular to $\hat{\mathbf{k}}$. It can lie in the plane containing $\mathbf{r}_{nn'}$ and $\hat{\mathbf{k}}$, in which case we denote it as $\mathbf{e}_1$, or it can be perpendicular to this plane (and thereby to $\mathbf{r}_{nn'}$), in which case we denote it as $\mathbf{e}_2$ (Figure 18.1).

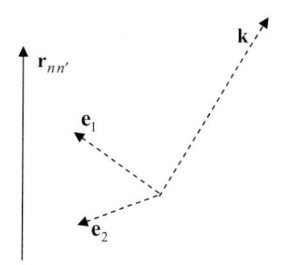

Figure 18.1 Relationship between direction of light wave, its polarization, and the vector matrix element corresponding to transition $n \rightleftarrows n'$.

Generally, it can be a superposition of both, but we do not need to consider the general case: According to (18.16), the atom would not respond to radiation polarized perpendicular to $\mathbf{r}_{nn'}$. And, since the electromagnetic interactions are time reversible, the atom cannot radiate a photon polarized perpendicular to the vector matrix element $\mathbf{d}_{nn'} = q\mathbf{r}_{nn'}$ of the corresponding transition. So for each dipole moment $\mathbf{d}_{nn'} = q\mathbf{r}_{nn'}$, the interaction occurs only with photons polarized in the plane containing this moment. For all such cases, denoting the angle between $\mathbf{d}_{nn'}$ and propagation direction $\hat{\mathbf{k}}$ of the corresponding photons as $\theta_{nn'}$, we can write the probability of absorption or stimulated emission as

$$w_{nn'}(\hat{\mathbf{k}}) = \frac{2\pi}{\hbar^2} \rho(\omega_{nn'}) |\mathbf{d}_{nn'}|^2 \sin^2 \theta_{nn'}. \tag{18.18}$$

This probability is axially symmetric with respect to $\mathbf{d}_{nn'}$, so we can integrate it over the azimuthal angle φ to get

$$w_{nn'}(\theta_{nn'}) = \frac{4\pi^2}{\hbar^2} \rho(\omega_{nn'}) |\mathbf{d}_{nn'}|^2 \sin^2 \theta_{nn'}. \tag{18.19}$$

According to (18.19), the emission probability is zero along the dipole axis and maximal in the equatorial plane of the dipole. All this is similar to classical dipole radiation.

Integrating (18.19) over $\theta_{nn'}$ gives

$$w_{nn'}(\omega_{nn'}) = \frac{16\pi^2}{3\hbar^2} \rho(\omega_{nn'}) |\mathbf{d}_{nn'}|^2 = \frac{16\pi^2 \omega_{nn'}}{3\hbar} N(\omega_{nn'}) |\mathbf{d}_{nn'}|^2. \tag{18.20}$$

These probabilities are determined for transitions caused by the external radiation. Therefore, they describe absorption and stimulated emission of light. Their ratio to $N(\omega_{nn'})$ will give us the transition probability per photon, which is nothing else but the Einstein coefficients for absorption and stimulated emission. We see that these coefficients are equal to each other, as was initially inferred from the conditions of thermal equilibrium.

18.3
Selection Rules

The results of the two preceding sections may produce an impression that transitions between any two eigenstates are possible. This is not so. Many direct transitions are forbidden, at least in the first-order perturbation theory. This becomes clear if we take a closer look at the matrix elements determining the transition amplitudes. In the light–matter interactions, these are the matrix elements in the basis of eigenstates of the unperturbed Hamiltonian. It turns out that some of the matrix elements $d_{nn'} = 0$. In such cases, the corresponding optical transition does not occur even when the light frequency is tuned to the transition frequency $\omega_{nn'}$. Therefore, the condition $d_{nn'} = 0$ acts as a selection rule for the corresponding transition. Note that this refers only to optical transitions or to any other transitions under perturbations whose matrix elements are proportional to $d_{nn'}$. It may happen that some transition $n \rightleftarrows n'$ that is impossible under illumination may occur, say, in the process of collision with a massive particle.

We consider separately selection rules for optical transitions of an oscillator and of a valence electron in a hydrogen-like atom.

18.3.1
Oscillator

For a 1D oscillator, we need to consider only those components of **r** that are parallel to oscillation axis. The corresponding dipole moment can be written as $d_{nn'} = qx_{nn'}$.

But we have already determined $x_{nn'}$ in Section 9.7 and found (Equations 9.91 and 9.92) that all its elements are zeros except for those with

$$n' = n \pm 1. \tag{18.21}$$

This gives us the selection rule for transitions associated with emission or absorption of light: Only those that satisfy (18.21) are possible. The corresponding frequencies are

$$\omega_{nn'} = \omega_0(n - n') = \pm\omega_0. \tag{18.22}$$

Thus, optical transitions are possible only between the neighboring energy levels. An oscillator with frequency ω_0 can only emit or absorb light of the same frequency.

Keep in mind that this result holds only for sufficiently weak perturbations and for the wavelengths λ by far exceeding the effective size of the system, that is, the distance between the two symmetric maxima of the oscillator's effective amplitude $a_n = \sqrt{(\hbar/\mu\omega_0)(n+1/2)}$. Thus, the found selection rule works for the wavelengths

$$\lambda \gg a_n = \sqrt{\frac{\hbar}{\mu\omega_0}\left(n + \frac{1}{2}\right)}. \tag{18.23}$$

Note also that parabolic potential considered in Section 9.6 is only an idealized model. Some real interactions can be approximated by such potential only within a restricted range, depending on the properties of a given system. Oscillations with large a_n exceeding this distance are nonharmonic, which may by itself break the derived selection rule.

18.3.2
Hydrogen-Like Atom

This case involves all three dimensions, and according to (18.16), the transition probability is determined by the dot product $\mathbf{e} \cdot \mathbf{r}_{nn'}$ where **e** is the unit vector along the polarization direction. Here, the selection rules emerge due to the angular momentum associated with each state. The angular momentum brings in, in addition to n, two quantum numbers l and m. Therefore, the matrix elements

$\mathbf{r}_{nn'}$ must be specified more completely as $\mathbf{r}_{nlm,\,n'l'm'}$. Accordingly, Equation 18.16, as well as other equations involving interatomic optical transitions, must be written as

$$W_{nn'} = 2\pi \frac{q^2}{\hbar^2} \rho(\omega_{nn'}) |\mathbf{e} \cdot \mathbf{r}_{nlm,n'l'm'}|^2. \tag{18.24}$$

Since the atomic eigenstates are expressed in spherical coordinates

$$\Psi_{nlm}(x,y,z) \rightarrow \Psi_{nlm}(r,\theta,\varphi) = R_{nl}(r) P_l^m(\cos\theta) e^{im\varphi}, \tag{18.25}$$

the position vector operator must also be expressed in terms of spherical coordinates: $x = r\sin\theta\cos\varphi$, $y = r\sin\theta\sin\varphi$, $z = r\cos\theta$. It is also convenient to consider, instead of x, y, their combinations

$$\alpha = x + iy = r\sin\theta e^{i\varphi}, \quad \beta = x - iy = r\sin\theta e^{-i\varphi}. \tag{18.26}$$

The corresponding matrix elements are

$$\alpha_{nlm,n'l'm'} = r_{nl,n'l'} \int_0^\pi P_l^m(\cos\theta) P_{l'}^{m'}(\cos\theta) \sin^2\theta\, d\theta \int_0^{2\pi} e^{i(m-m'+1)\varphi} d\varphi, \tag{18.27}$$

$$\beta_{nlm,n'l'm'} = r_{nl,n'l'} \int_0^\pi P_l^m(\cos\theta) P_{l'}^{m'}(\cos\theta) \sin^2\theta\, d\theta \int_0^{2\pi} e^{i(m-m'-1)\varphi} d\varphi, \tag{18.28}$$

$$z_{nlm,n'l'm'} = r_{nl,n'l'} \int_0^\pi P_l^m(\cos\theta) P_{l'}^{m'}(\cos\theta) \sin\theta\cos\theta\, d\theta \int_0^{2\pi} e^{i(m-m')\varphi} d\varphi. \tag{18.29}$$

Here, $r_{nl,n'l'}$ is the matrix element of radial coordinate:

$$r_{nl,n'l'} = \int_0^\infty R_n(r) r R_{n'}(r) r^3 dr, \tag{18.30}$$

which is the common factor in all three equations. It is important to distinguish the polar angle θ in these equations from the angle between $\mathbf{e}$ and $\mathbf{r}_{nlm,n'l'm'}$ in the dot product in Equation 18.16. The former is the integration variable determined by the choice of the z-axis; the latter is the set of angles determined by polarization direction of the incident light and matrix elements $\mathbf{r}_{nlm,n'l'm'}$.

A brief look at the last integrals in Equations 18.27–18.29 tells us that only matrix elements with $m' = m$ or with $m' = m \pm 1$ are nonzero. This is the first selection rule: Only the transitions preserving m, or changing it by 1, are allowed. To specify which case realizes in a specific situation, let us complete the calculations. Denote

$$\int_0^\pi P_l^m(\cos\theta) P_{l'}^{m'}(\cos\theta) \sin^2\theta\, d\theta \equiv P_{ll'}^{mm'}, \tag{18.31}$$

$$\int_0^\pi P_l^m(\cos\theta) P_{l'}^{m'}(\cos\theta) \sin\theta\cos\theta\, d\theta \equiv \tilde{P}_{ll'}^{mm'}. \tag{18.32}$$

Then, Equations 18.27–18.29 take the following form, respectively:

$$\alpha_{nlm,n'l'm'} = r_{nl,n'l'} \mathcal{P}_{ll'}^{mm'} \delta_{m,m'-1}, \tag{18.33}$$

$$\beta_{nlm,n'l'm'} = r_{nl,n'l'} \mathcal{P}_{ll'}^{mm'} \delta_{m,m'+1}, \tag{18.34}$$

$$z_{nlm,n'l'm'} = r_{nl,n'l'} \tilde{\mathcal{P}}_{ll'}^{mm'} \delta_{m,m'}. \tag{18.35}$$

Thus, the matrix elements for α are nonzero only for $m' = m + 1$, and β has nonzero matrix elements only for $m' = m - 1$. As to the z-component of $\mathbf{r}_{nlm,n'l'm'}$, only transitions preserving m are allowed. The *magnetic quantum number remains the same* in optical transitions associated with the z-component of $\mathbf{r}_{nlm,n'l'm'}$.

The integrals (18.31) and (18.32) allow us to find the selection rules also for the quantum number l. For the integral $\tilde{\mathcal{P}}_{ll'}^{mm'}$ in $z_{nlm,n'l'm'}$, we need to consider only the elements with $m = m'$. Setting $m = m'$ in (18.32) and denoting $\cos\theta \equiv \xi$ gives

$$\tilde{\mathcal{P}}_{ll'}^{mm} \equiv \int_0^\pi P_l^m(\xi) P_{l'}^m(\xi) \xi d\xi. \tag{18.36}$$

This integral is nonzero only for $l' = l \pm 1$ [1]. Thus, transitions represented by $z_{nlm,n'l'm'}$ preserve the magnetic quantum number and change the orbital quantum number by 1. The same result follows for the integral $\mathcal{P}_{ll'}^{mm'}$ at $m' = m \pm 1$. So the selection rule for the orbital quantum number is $l' = l \pm 1$. The optical transitions for dipole radiation are possible only between the neighboring states with respect to l. This result explains the known experimental facts in optical spectroscopy.

Problems

18.1 Find the range of frequencies around the resonant frequency ω_{km} in the light interaction with atoms, for which the probability of the corresponding optical transition cannot be correctly described by the Bohr model considering the atomic states as stationary and EM field as a monochromatic wave.

18.2 Find the transition probability in the first approximation of perturbation theory when the time-dependent perturbation starts at a certain moment and then is exponentially attenuated with time.

18.3 Evaluate the wavelength corresponding to an optical transition $|\Upsilon_{2,1}\rangle \to |\Upsilon_{1,0}\rangle$ between the angular momentum eigenstates of a chlorine molecule. Use the model of the molecule as a "dumbbell" with point masses connected by a massless rod of length $a = 0.8 \times 10^{-10}$ m. Assume that this length does not change in transition. The atomic weight of chlorine is 35.45.

Reference

1 Blokhintsev, D.I. (1964) *Principles of Quantum Mechanics*, Allyn & Bacon, Boston.

19
Scattering

19.1
QM Description of Scattering

Scattering (collision of particles) is, even in the simplest case, a two-body problem. Its treatment involves coordinates r_1 and r_2 of the first and second particles, respectively. But we can simplify mathematical description of the process by using instead two other variables: position vector **R** of the center of mass of the system, and position vector **r** of one particle with respect to the other. The direction of **r** depends on which particle we choose as its origin. Usually it is associated with the more massive particle; if the particles are identical, the choice is arbitrary.

In the new variables, the problem splits into two independent problems: one describing the motion of the whole system with mass $M = \mu_1 + \mu_2$, and the other describing the motion of one particle [with the reduced mass $\mu = (\mu_1\mu_2)/(\mu_1 + \mu_2)$] relative to the other [1]. The first of these problems is trivial since the center of mass of an isolated system is either stationary or moving uniformly. Thus, we are left with the motion of only one particle in the given interaction field.

The particle chosen as the origin remains, by definition, stationary and is usually referred to as the target; the other particle is called the oncoming, bombarding, or incident particle.

In the experiments, the target is irradiated by incident particles with approximately the same initial energy and momentum. Classically, the process can be described as flux of particles, all moving with the same initial velocity toward the target and deflected by it.

In the following, we will restrict to the case of interaction depending only on distance between the particles. The scattering in this case is axially symmetric – the deflection θ is independent of the azimuthal angle φ. In a given potential, it depends only on how close the incident particle would approach the target in the absence of interaction between them (Figure 19.1). This distance (which also determines classical angular momentum of the incident particle) is called the impact parameter. We will denote it as s.

In most cases, the interaction increases as the particles get closer to one another. In such cases, the particles with smaller s are deflected stronger (will have greater θ) than particles with larger s (Figure 19.1). At $s \to \infty$, we have $\theta \to 0$. In some exotic potentials of attraction [e.g., $U(r) \sim -r^{-\alpha}, \alpha \geq 2$] at sufficiently small s, there

Quantum Mechanics and Quantum Information: A Guide through the Quantum World,
First Edition. Moses Fayngold and Vadim Fayngold.
© 2013 Wiley-VCH Verlag GmbH & Co. KGaA. Published 2013 by Wiley-VCH Verlag GmbH & Co. KGaA.

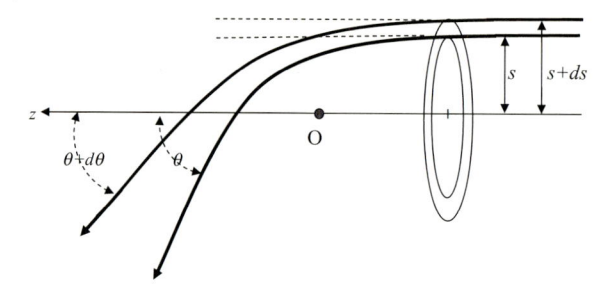

Figure 19.1 Scattering of a classical particle. Connection between the impact parameter s and the scattering angle θ. The change $d\theta$ is negative (a greater impact parameter corresponds to a smaller deflection).

appears the so-called *spiral scattering* [2–6], when the incident particle spirals toward and then away from the target before getting ultimately scattered. In such cases, the scattering angle may be zero (forward scattering – no deflection) also for some finite values of s. Generally, θ is a function of s, whose form depends on the interaction potential, and the scattering problem in classical physics is to find $\theta(s)$.

In real experiments it is difficult to control s, and we are dealing with an ensemble of particles with all possible s within the region $0 \leq s \leq s_{max}$, where s_{max} depends on the source of incident particles (specifically, on its aperture). And what we measure is the number of particles dN scattered per unit time within the interval $(\theta, \theta + d\theta)$. In space, for axially symmetric scattering, this interval is associated with the solid angle $d\Omega(\theta) = 2\pi \sin\theta d\theta$, and we actually measure $dN(\theta)$ – the number of particles scattered within this solid angle. As seen from Figure 19.1, $dN(\theta)$ is proportional to the number of incident particles passing through cross-sectional area $d\sigma = 2\pi s ds$. This number, however, not only depends on area $d\sigma$ but also stands in proportion to the flux intensity I. Therefore, the meaningful characteristic of scattering is the ratio dN/I, which has a dimensionality of area and is numerically equal to

$$\frac{dN(\theta)}{I} = d\sigma(\theta) = 2\pi s(\theta)\, ds(\theta). \tag{19.1}$$

This characteristic is called the *differential cross section of scattering*. Geometrically, it is a fractional area (19.1) in the incident beam, which must be crossed by the incident particles to be scattered within $(\theta, \theta + d\theta)$, that is, within the solid angle $d\Omega(\theta)$. It follows that the integral

$$\int_0^\pi d\sigma(\theta) = \int 2\pi s\, ds = \sigma \tag{19.2}$$

gives what we call the *total cross section of scattering*. As seen from its definition, $\sigma = \infty$ for any $U(r) \neq 0$ at an arbitrary r, since in all such cases the particle even with arbitrarily large s will have possibly very small but still nonzero deflection. And these small deflections will give the overwhelming contribution to σ. In other words, the quantity defined as

$$|f(\theta)|^2 \equiv \frac{d\sigma(\theta)}{d\Omega(\theta)} \tag{19.3}$$

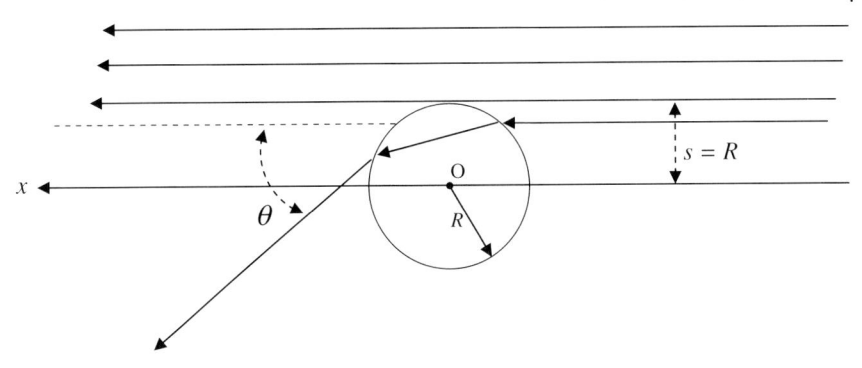

Figure 19.2 Scattering of a classical particle on a spherical potential well: O – scattering center; θ – the scattering angle.

diverges at $\theta \to 0$. This means, among other things, that in order to have the complete description of scattering in the *classical* domain, the aperture must be arbitrarily large.

The exception is the case when the potential is nonzero only within a finite region of space, for instance, a spherical well or barrier of radius R considered in Chapter 10. In such cases, as seen directly from geometry (Figure 19.2), we have total cross section $\sigma = \pi R^2$. Then the aperture can be finite, but not less than R. The reader can also find the *differential* cross section for this case using the appropriate boundary condition at $r = R$.

Combining Equations 19.1 and 19.3 gives

$$\frac{dN(\theta)}{I} = d\sigma(\theta) = |f(\theta)|^2 d\Omega(\theta). \tag{19.4}$$

The total cross section in terms of $f(\theta)$ is

$$\sigma = \int_0^\pi |f(\theta)|^2 d\Omega. \tag{19.5}$$

Equation 19.4 shows that the function $f(\theta)$ is experimentally measurable characteristic of scattering. It is called the *scattering amplitude*. It can be evaluated from the experimental data on $dN(\theta)$ within the range of scattering angles. As seen from its definition, it has dimensionality of length. This definition holds in QM as well.

Generally, if the potential is known, the classical differential cross section or, equivalently, the scattering amplitude $f(\theta)$ can be found by determining θ for each given s, which requires knowledge of the interaction potential [1].[1)] If the potential is not known, the above equations can be used to determine it from the experimental data. The determining of unknown potential from the experiments on scattering constitutes the so-called *inverse problem* in both classical and quantum physics.

---

1) In case of the above-mentioned spiral scattering, more than one s correspond to the same θ. In such cases, the relationship ($s \leftrightarrow \theta$) becomes more complicated [2–6].

Now turn to the QM description. In this domain, we must distinguish between two possible approaches. One of them represents an incident particle as a wave packet. Accordingly, the scattering state is nonstationary. We will use another approach, considering scattering as a state with definite energy E, which means that it is a stationary state. Since the scattered particle is unbounded, its energy is positive. The concept of trajectories, even the fuzzy ones that can figure in the first approach, does not apply in this case. Instead, since the particles in the incident beam have definite momentum, the corresponding state must be represented by a monochromatic plane wave even for a single particle. But such a wave alone only describes a free particle without any scattering. In our case, the particle is moving in a potential field. Therefore, the exact state function can be represented by the plane wave only asymptotically – far from the center, where the potential can be neglected. Also, the particle gets scattered. Experimentally, the scattering is manifest in the flux diverging from the center. Therefore, the correct description must, apart from the plane wave, also include a diverging spherical wave e^{ikr}/r. But the flux density in the scattered wave is generally different for different θ, so its amplitude must be a function of θ. The diverging wave, albeit spherical, is not necessarily spherically symmetric! Therefore, we have to introduce the angle-dependent amplitude:

$$\frac{e^{ikr}}{r} \Rightarrow f(\theta)\frac{e^{ikr}}{r}. \tag{19.6}$$

The factor $f(\theta)$ stores all information about the scattering process in the given field. Like its classical counterpart, it is called the scattering amplitude, has dimensionality of length, and its specific form is determined by potential $U(r)$.

Thus, the state of the particle in the given field is represented, far from the center, by a superposition of the incident and scattered waves:

$$\psi(r) \approx e^{ikz} + f(\theta)\frac{e^{ikr}}{r}, \tag{19.7}$$

where $k = \sqrt{2\mu E}/\hbar$ (for a relativistic particle, we would have $k = \sqrt{E^2 - \mu_0^2 c^4}/\hbar c$). The same value of k in both waves reflects the fact that the process is stationary. The particle's energy does not change during its interaction with the scattering center. This kind of scattering is called *elastic*, by analogy with elastic collisions in CM.

On the face of it, Equation 19.7 may appear to be total nonsense because it combines two apparently incompatible terms. Apart from the plane wave describing a passing particle, we have a diverging wave that seems to describe some additional particles emerging from the center. As we know from Section 10.2, the plane wave itself is a superposition of angular momentum eigenstates with all possible l. According to Equations 10.13 and 10.20, each radial coefficient $R_{kl}(r)$ is (asymptotically!) an equally weighted superposition of two spherical waves – one diverging and one converging, forming a stationary standing wave. The whole system of waves is perfectly balanced so that there is no net flux through any closed surface. The plane wave alone is already a stationary state. But in (19.7) there appears an additional diverging wave not balanced by any converging counterpart and therefore it must produce a nonzero flux from the center. This seems to imply some *source* at the origin, adding a new particle to each one in the passing wave.

This argument is wrong because the factor $f(\theta)$ in the diverging wave is the *amplitude*, not probability! It would produce a nonzero diverging flux *only when acting alone*. But when added to the balanced system of standing waves, it may interfere with it constructively at some θ and destructively at some other θ, so that the *net* total flux through any closed surface remains zero. And this is precisely what happens! We will see later that $f(\theta)$ causes only *redistribution* of the local flux density that does not change the net result. Specifically, at $\theta \to 0$ (forward scattering), the scattered wave interferes destructively with the incident wave, forming what is called the "shadow region" with lower (and sometimes even zero) average intensity. This is in line even with our classical intuition: The net flux in the forward direction is weakened by scattering, which deflects some of the particles from their initial path to other routes. There appears flux in these other directions, but it comes at the cost of decreased flux in the forward direction. In some cases of the so-called *planar scattering* (e.g., when the target is a monoatomic plane), the destructive interference between incident and forward-scattered waves may be so strong that there remains no transmitted wave, and all incident waves get entirely reflected [7].

We can also look at it in terms of what we know from the EM theory: A passing plane EM wave (in our case, it would describe a single incident photon), interacting with a charge (target) at the center, brings it to an oscillatory motion of the same frequency. The oscillating charge behaves like a microantenna, emitting its own wave with the same frequency as that of the incident wave. This already explains the equal magnitudes of k in both terms of (19.7). It is important to realize that the secondary wave from the source does not mean the emission of the second photon in addition to the initial one. That would contradict the conservation of energy: Initially the source was at rest, so the energy needed for its oscillation with resulting secondary radiation can only come from the incident wave. In the language of photons, energy coming from the incident photon does not mean its frequency decrease! The photon's energy $\hbar\omega$ is an indivisible quantum, its splitting would mean splitting of the photon itself into two or more particles. In the described process, we observe the initial frequency unaltered. The distinct photon states here are the unscattered state (incident plane wave – call it the *I*-state) and scattered state (diverging spherical wave, call it the *S*-state). In stationary scattering, each photon is in superposition of these two states. The photon interferes with itself, and "the energy requisition" from the *I*-state means that in the presence of an external field $U(r)$, the particle will not always be found in this state. It may be found in the *S*-state instead. Thus, the ultimate "energy requisition" for creating *S*-state from the *I*-state in elastic scattering is "all or nothing," but never partial – already known dramatic difference between QM and classical physics.

Another analogy would be that of a beam splitter, in which the incident wave $|i\rangle$ splits into transmitted wave $|t\rangle$ and reflected wave $|r\rangle$. Dropping, for the sake of simplicity, the phase factors, we can write this in terms of the amplitudes as a transition:

$$|i\rangle \Rightarrow \frac{1}{\sqrt{2}} (|r\rangle + |t\rangle). \tag{19.8}$$

The subtle (but important!) point here is that the transmitted wave itself is the superposition of incident and the forward-scattered waves [8]. Denoting the latter as $|f(0)\rangle$, we can write

$$|t\rangle = (|i\rangle + |f(0)\rangle). \tag{19.9}$$

The reason we put the terms on the right-hand side in parentheses is that the two states there, although being different (they definitely have different origin and different phases), are not distinguishable by any known experimental procedure: The phase difference between these states does not carry any information identifying the state, and all their other characteristics that are measurable are the same for both states. If you recorded the photon in a detector down the path of the $|t\rangle$-wave, there is no way to tell whether this photon was in the $|i\rangle$ state or in the $|f(0)\rangle$ state. In any event, the amplitude of the transmitted wave is (for a symmetric beam splitter) by a factor $\sqrt{2}$ less than the incident amplitude, which means that the forward-scattered wave interferes destructively with the incident wave.

This result is the same as for a point charge under an incident monochromatic EM wave. Its secondary radiation is phase shifted by 180° with respect to the incident one, thus weakening the net wave in the forward direction. So all considered examples confirm the general rule – that the interference between the incident and scattered waves is always destructive in the forward direction. And all this, taking so much space and time in words, is stored in one Equation 19.7.

Exercise 19.1

Suppose we want to describe the photon's interaction with a beam splitter in terms of scattering. This means that we must consider the beam splitter as a scatterer in the first place. Its interaction with the incident photon is not spherically symmetric, and accordingly, the resulting state does not possess the axial symmetry. Also, the scatterer is not point-like. It is a macroscopic object whose size is much larger than the photon's wavelength. The scattered state can be roughly represented by two beams – one in the forward direction and the other in transverse direction. We can restrict to the plane containing both split beams. If the two split beams make 90°, their directions correspond to $\theta = 0$ and $\theta = \pi/2$. In the first approximation, we can represent the scattered wave by expression

$$f(\theta) = f(0)\delta(\theta) + f\left(\frac{\pi}{2}\right)\delta\left(\theta - \frac{\pi}{2}\right). \tag{19.10}$$

Estimate the values of $f(0)$ and $f(\pi/2)$ for a symmetric beam splitter.

Solution:

In a symmetric beam splitter, the initial intensity is distributed equally (50/50) between the two split beams. Therefore, the amplitude of each split beam is $1/\sqrt{2}$. This immediately gives the answer for the reflected beam. In terms of scattering, this is the beam scattered at $\theta = \pi/2$. Thus, we have $f(\pi/2) = 1/\sqrt{2}$. As to the transmitted beam, we have, in view of (19.9), $1 + f(0) = 1/\sqrt{2}$ (superposition of the incident wave and the forward-scattered wave produces the transmitted wave with amplitude $1/\sqrt{2}$). It follows

$$f(0) = -\left(1 - \frac{1}{\sqrt{2}}\right). \tag{19.11}$$

The amplitude $f(0)$ is negative. This means that in the given experimental setup, the forward-scattered wave is totally out of phase with the incident wave. They interfere destructively so that the result (transmitted wave) is always weaker than the incident wave. This is a specific example illustrating the basic feature of the forward scattering that we discussed in comments to Equation 19.7. You will find more about this effect in the next two sections.

19.2
Stationary Scattering

We found in Section 10.2 (Equation 10.22) that the field $U(r)$ vanishing at infinity causes the phase shift η in the asymptotic expression of radial function. Equation 10.22 describes only a special case $l = 0$, so that $\eta = \eta_0$, but the same effect works for states with higher l as well, each state being phase shifted by the corresponding amount η_l:

$$\frac{1}{r}\sin\left(kr - l\frac{\pi}{2}\right) \Rightarrow \frac{1}{r}\sin\left(kr - l\frac{\pi}{2} + \eta_l\right). \tag{19.12}$$

The phase shifts η_l and the emergence of scattered spherical wave $f(\theta)$ are the two sides of the same coin. Both of them result from interaction with the scatterer; hence, there must be a correlation between them. We want to find this correlation, that is, to find an equation determining $f(\theta)$ through the set of η_l [9–11].

The most general form of the stationary and axially symmetric solution of the radial Schrödinger equation with positive energy is a superposition of all eigenstates with this energy:

$$\Psi(r, \theta) = \sum_l B_l P_l(\cos\theta) R_{kl}(r). \tag{19.13}$$

Just as in case of the de Broglie wave expansion (10.15) and (10.17), superposition (19.13) has only the states with $m = 0$. Using (10.17) and combining with (19.12), we obtain its asymptotic form:

$$\Psi(r, \theta) \approx \frac{2}{r} \sum_l B_l P_l(\cos\theta) \sin\left(kr - l\frac{\pi}{2} + \eta_l\right)$$

$$= \frac{1}{ir} \sum_l B_l P_l(\cos\theta) e^{-i\eta_l} \left[e^{2i\eta_l} e^{(ikr - l(\pi/2))} - e^{-(ikr - l(\pi/2))}\right]. \tag{19.14}$$

With appropriately adjusted expansion coefficients B_l, the superposition (19.14) will describe the state (19.7) of scattering process:

$$\frac{1}{ir}\sum_l B_l P_l(\cos\theta)\left[i^{-l}e^{i(kr+\eta_l)} - i^l e^{-i(kr+\eta_l)}\right] = e^{ikr\cos\theta} + f(\theta)\frac{e^{ikr}}{r}. \tag{19.15}$$

Now we isolate the scattering wave and use the expansion (10.15) of the de Broglie wave. The result is

$$\frac{f(\theta)e^{ikr}}{r} = \frac{1}{ir}\left\{\sum_l\left[B_l i^{-l}e^{i\eta_l} - \frac{2l+1}{2k}\right]P_l(\cos\theta)e^{ikr}\right.$$

$$\left. + \sum_l i^l\left[-B_l e^{-i\eta_l} + i^l\frac{2l+1}{2k}\right]P_l(\cos\theta)e^{-ikr}\right\}. \tag{19.16}$$

This equation can hold only if the right-hand side does not contain converging wave. This condition determines the coefficients B_l:

$$B_l = i^l\frac{2l+1}{2k}e^{i\eta_l}. \tag{19.17}$$

With these B_l, (19.16) gives

$$f(\theta) = \frac{1}{2ik}\sum_l (2l+1)\left(e^{2i\eta_l} - 1\right)P_l(\cos\theta). \tag{19.18}$$

Thus, we killed two hares with one stroke: We found the expansion of state (19.7) over angular momentum eigenstates, and found expression for the scattering amplitude in terms of the phase shifts η_l. We see that this expression is a (rather complex!) function of θ.

In many applications (see Section 19.3), the $\exp(2i\eta_l)$ is denoted as S_l, that is, we introduce

$$S_l \equiv e^{2i\eta_l}. \tag{19.19}$$

Then, (19.18) takes the form

$$f(\theta) = \frac{1}{2ik}\sum_l (2l+1)(S_l - 1)P_l(\cos\theta). \tag{19.20}$$

These results allow us to express the total cross section σ in terms of η_l as well. Putting (19.20) into (19.5) gives

$$\sigma = 2\pi\int_0^\pi \left|\frac{1}{2ik}\sum_l (2l+1)(S_l - 1)P_l(\cos\theta)\right|^2 \sin\theta\, d\theta. \tag{19.21}$$

Since $P_l(\cos\theta)$ form orthogonal set and are normalized to $2/(2l+1)$ [10], the calculation yields

$$\sigma = \frac{\pi}{k^2}\sum_l (2l+1)|S_l - 1|^2 = \frac{4\pi}{k^2}\sum_l (2l+1)\sin^2\eta_l. \tag{19.22}$$

Thus, σ is the sum $\sigma = \sum_l \sigma_l$ of terms:

$$\sigma_l = \frac{4\pi}{k^2}(2l+1)\sin^2\eta_l. \qquad (19.23)$$

Each such term can be interpreted as the effective cross section for a particle in the corresponding eigenstate l. Since each such state is only a part of the whole superposition, it is sometimes called the *partial wave*, and the corresponding value σ_l is called the *partial cross section*. In parallel with CM, we can envision the partial scattering of order l as associated with impact parameter s_l such that the corresponding angular momentum relative to the center is given by $L_l = s_l p = \sqrt{l(l+1)}\hbar$, that is,

$$s_l = \frac{\sqrt{l(l+1)}}{k} = 2\pi\sqrt{l(l+1)}\lambda. \qquad (19.24)$$

But in QM, definite l does not correspond to a definite s, so do not take this analogy too literary. The more accurate statement would be noticing that a radial function $R_{kl}(r)$ peaks close to $r = s_l$, which means that for a particle in state l, it is most probable to be found at a distance s_l from the center (Figure 19.3).

Now we want to consider the relationship between $f(0)$ and σ. Take the amplitude (19.18) in the limit $\theta \to 0$ (forward scattering). All Legendre polynomials are normalized so that $P_l(1) = 1$. Therefore,

$$f(0) = \frac{1}{2ik}\sum_l (2l+1)\left(e^{2i\eta_l} - 1\right). \qquad (19.25)$$

This looks very similar to series (19.22) for σ. Now take the imaginary part,

$$\mathrm{Im}f(0) = \frac{1}{2k}\sum_l (2l+1)(1 - \cos 2\eta_l) = \frac{1}{k}\sum_l (2l+1)\sin^2\eta_l. \qquad (19.26)$$

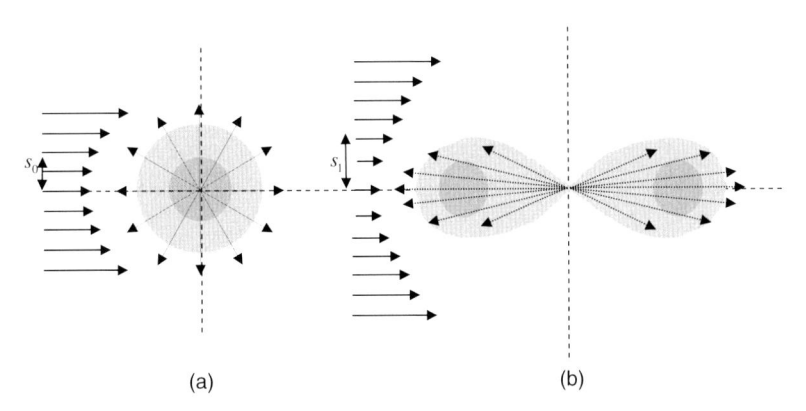

Figure 19.3 (a) Partial scattering with $l = 0$ (s-scattering). (b) Partial scattering with $l = 1$ (p-scattering).

Comparing with (19.22) gives

$$\text{Im} f(0) = \frac{k}{4\pi}\sigma. \qquad (19.27)$$

This is the first example of the *optical theorem* that relates the amplitude of the forward scattering and the total cross section in a given field.

The form (19.27) of this theorem, while being very general, is not universal. It was obtained for a scatterer with negligible size, that is, the corresponding scattered wave can be considered as a spherical wave diverging from a point-like object. In the absence of such condition, the relationship between $f(0)$ and σ may differ from (19.27). For instance, the BS considered at the end of Section 19.1 is much greater than the wavelength of incident light, and the corresponding scattered state consists of two approximately plane waves rather than one spherical wave. Accordingly, expression (19.11) for $f(0)$ does not contain imaginary part at all, so that someone familiar only with the optical theorem in the form (19.27) could conclude that $\sigma = 0$, which is equivalent to nonexistence of BS!

The cause of possible confusion here is that the optical theorem for a BS has another form that can be approximated by an expression for the planar scattering [7]:

$$\text{Re} f(0) = -\tilde{\Phi}|f(0)|^2. \qquad (19.28)$$

Here, the total probability of scattering $\mathcal{P}_s = 2\tilde{\Phi}|f(0)|^2$ plays the role of (dimensionless) total cross section and is expressed in terms of the *real part* of $f(0)$. The amplitude $f(0)$ itself is also dimensionless because in contrast to (19.16), all the scattered waves here are plane and do not form a spherical wave with radial distance in the denominator.

19.3
Scattering Matrix and the Optical Theorem

Consider again a state (19.14). We found that for such state the coefficients B_l are given by (19.17). Therefore, we can write the asymptotic expression for this state as

$$\Psi(r,\theta) \approx \frac{1}{2ikr}\sum_l (2l+1)P_l(\cos\theta)i^l\left[-e^{-i(kr-l(\pi/2))} + e^{2i\eta_l}e^{i(kr-l(\pi/2))}\right]. \qquad (19.29)$$

Select, for the sake of simplicity, only one partial wave with a certain l and write it as

$$\Psi_l(r,\theta) \approx -i^l\frac{(2l+1)}{2ik}P_l(\cos\theta)\left[\frac{e^{-i(kr-l(\pi/2))}}{r} - S_l\frac{e^{i(kr-l(\pi/2))}}{r}\right], \qquad (19.30)$$

with $S_l = e^{2i\eta_l}$ defined in (19.19). Written in this form, the expression for partial wave admits simple interpretation. We have a superposition of converging and diverging spherical waves with definite l. The phase shift η_l in the lth diverging wave differs from 0 and thereby quantity S_l differs from 1 only in the presence of potential field causing scattering. So the action of the field changes the amplitude of the outgoing partial wave No. l by a factor S_l. At $S_l = 1$ (no field!), *we have no scattering.* The outgoing waves would be merely a continuation of the incoming waves, collectively producing only the plane wave (10.17). The presence of $U(r) \neq 0$ is reflected mathematically in the emergence of factors S_l multiplying the corresponding amplitudes. According to (19.30), these factors take the set of "passive" outgoing waves to the set of the "active" (modified by scattering) outgoing waves. Let us denote the first set as $|I^+\rangle_0$ and arrange them in a column matrix with element No. l being $|I_l^+\rangle_0$. Then the second set $|I^+\rangle$ will form the column matrix with elements $|I_l^+\rangle = S_l |I_l^+\rangle_0$. The second matrix can be easily obtained from $|I^+\rangle_0$ if we arrange the factors S_l into diagonal matrix $\hat{S}$ and take the product $\hat{S}|I^+\rangle_0$:

$$|I^+\rangle = \hat{S}|I^+\rangle_0 \quad \text{or} \quad \begin{pmatrix} |I_0^+\rangle \\ |I_1^+\rangle \\ \vdots \end{pmatrix} = \begin{pmatrix} S_{00} & 0 & 0 & \cdots \\ 0 & S_{11} & 0 & \cdots \\ \vdots & \vdots & \vdots & \vdots \end{pmatrix} \begin{pmatrix} |I_0^+\rangle_0 \\ |I_1^+\rangle_0 \\ \vdots \end{pmatrix}. \tag{19.31}$$

This product describes the result of scattering in the given potential $U(\mathbf{r})$. Hence, the name of $\hat{S}$ – the *scattering matrix* or the *scattering operator*. In the considered case (stationary state, angular momentum basis, with $m \neq 0$ excluded), the matrix is diagonal, that is,

$$S_{ll'}(k) = e^{2i\eta_l(k)} \delta_{ll'}. \tag{19.32}$$

Here, we have explicitly indicated the phase shift $\eta_l = \eta_l(k)$ as a function of k. Generally, the scattering matrix $\hat{S}$ can be defined as an operator converting outgoing waves of a free particle into outgoing waves of the particle subjected to potential field $U(\mathbf{r})$. The phase shifts $\eta_l(k)$ determining the matrix elements of $\hat{S}$ are in turn determined by this field.

There is a slightly different way to formulate this basic definition of the scattering matrix. Namely, we can consider the phase factors $\exp(il\pi/2)$ and $\exp(-il\pi/2)$, respectively, as included in the radial exponents of converging and diverging partial waves in (19.29) and (19.30). Then, the amplitudes of converging and diverging waves would be $(1, -1)$ for a free particle and $(1, -S_l)$ for a particle in the field $U(\mathbf{r})$. In this approach, we could say that $\hat{S}$ matrix converts the amplitudes of the input state (converging waves) into the amplitudes of the output state (diverging waves) for a particle in the given field.

In cases we considered so far, k and $\eta_l(k)$ were real, and accordingly, S_l considered as operator is unitary. But it can be generalized to a much wider spectrum of states, including states with negative energy. We know that such states are bound, there is no scattering, and yet they can be described in terms of scattering matrix. But in this case it will not be unitary. This can be seen from the following simple argument.

For negative energies, $E < U(\infty)$, we have imaginary $k = -i\kappa$, $\kappa > 0$, and the partial wave (19.30) becomes

$$\Psi_l(r,\theta) \approx \frac{1}{2\kappa r} (2l+1) P_l e^{il(\pi/2)} \left[e^{il(\pi/2)} e^{-\kappa r} - e^{2i\eta_l} e^{-il(\pi/2)} e^{\kappa r} \right]. \tag{19.33}$$

Instead of two equally weighted radial waves in the brackets in (19.30), we now have two functions – one exponentially decreasing and the other exponentially increasing.

The increasing one is physically unacceptable. Therefore, we require its amplitude to be zero:

$$S_l(k) \equiv e^{2i\eta_l(k)} = 0, \quad k = -i\kappa. \tag{19.34}$$

Nothing is left of the original unitarity of $S_l(k)$!

The condition (19.34) can be satisfied only for a certain discrete set of allowed energies $E = E_n$, $n = 1, 2, 3, \ldots$, and corresponding $\kappa_n = \sqrt{2\mu|E_n|}/\hbar$. The phase shift at these values of E must be positively imaginary and infinite:

$$\eta_l(\kappa_n) = +i\infty. \tag{19.35}$$

In other words, $\eta_l(k)$ must generally be a complex function of k, where k can be considered as a complex variable as well:

$$k \to \tilde{k} = k + i\kappa, \quad \eta_l(k) \to \tilde{\eta}_l(\tilde{k}). \tag{19.36}$$

In the considered case, $\tilde{\eta}_l(\tilde{k})$ has poles at the values of $\tilde{k}$ selected by Equation 19.34, that is, on the imaginary axis $i\kappa$ of the complex $\tilde{k}$-plane. The matrix $S_l(\tilde{k})$ considered as a function of complex $\tilde{k}$ has zeros at these points.

Thus, we come at a far broader picture including positive energies (unbound states, real k) and negative energies (bound states, imaginary k) just as special cases of one problem. Add to this the complex energies that emerge when we consider decays formally as stationary states, and that will automatically bring to life generalization (19.36) that we had first introduced in Section 11.9 (Equation 11.109). We obtain a powerful algorithm for studying arbitrary states, considering E and k as complex variables, with the scattering matrix as a major tool – a function of these variables determining all physical observables in a wide range of possible conditions.

In particular, the scattering matrix allows us to generalize the previous results onto the nonelastic scattering. As mentioned, nonelastic scattering is associated with energy transitions of the scattered particle (as well as with possible transitions between its internal states). It turns out that this process can be described (in some respects!) similar to the just mentioned problem of decay (quasi-stationary states). Namely, we can consider the nonelastic scattering of a particle as its absorption and subsequent emission in another energy state. This is equivalent to its decay from the initial state into the final state with different energy. In other words, each nonelastic process removes the particle from its initial energy state, thus decreasing the amplitude to find it in this state. As a result, some eigenvalues of $S_l(k)$ for the elastically scattered wave are less than 1 – the matrix $S_l(k)$ is not unitary! The corresponding phase shift must be complex:

$$\begin{aligned}\eta_l(k) \to \tilde{\eta}_l(k) &= \eta_l(k) + i\gamma_l(k), \\ S_l(k) \to \tilde{S}_l(k) &= e^{2i\tilde{\eta}_l(k)} = e^{-2\gamma_l(k)}e^{2i\eta_l(k)}, \quad |\tilde{S}_l(k)| < 1.\end{aligned} \tag{19.37}$$

The difference from the decay theory is that here we can still consider E and k as real numbers: The scattered particle may be taken from the state with *definite* E to some other states due to its interaction with the external source at the center. This is different from its decay caused by intrinsic instability, which would make its energy innately indeterminate.

The partial cross section for elastic scattering is, according to (19.22),

$$\sigma_l^e = \frac{\pi}{k^2}(2l+1)|1 - \tilde{S}_l|^2. \tag{19.38}$$

But now we have added the superscript "e" to indicate that this is cross section of the elastic scattering as opposed to the inelastic one. Sometimes these two possibilities are referred to as scattering through the elastic or inelastic channel, respectively.

Putting into (19.38), the generalized expression (19.37) for $\tilde{S}_l$ yields

$$\sigma_l^e = \frac{\pi}{k^2}(2l+1)\left[\left(1 - e^{-2\gamma_l(k)}\right)^2 + 4\sin^2\eta_l(k)\right]. \tag{19.39}$$

At $\gamma_l(k) \to 0$, this reduces to (19.22). Now we want to find the partial section of *inelastic* scattering. Denote it as σ_l^{in}. According to its definition, it must be equal to the difference between the incoming partial flux (with amplitude 1) through a closed surface around the center and the outgoing flux of elastically scattered particles (with amplitude S_l). This difference is just the net flux, which is now nonzero because inelastic scattering removes some of the particles from the initial eigenstate of energy or some intrinsic variable. We can find it by choosing the surface as a sphere centered at the origin and integrating the known expression for the flux density over this surface:

$$J_l = \frac{i\hbar}{\mu}\int \text{Im}\,\Psi_l(r,\theta)\frac{\partial}{\partial r}\Psi_l^*(r,\theta)da, \quad da = r^2 d\Omega = 2\pi r^2 \sin\theta\, d\theta. \tag{19.40}$$

We are interested only in the radial flux and accordingly use here only the radial component of the operator (expressed in spherical coordinates), that is, $\nabla_r = \partial/\partial r$.

Putting here $\Psi_l(r,\theta)$ from (19.30) and performing integration gives (Problem 19.4)

$$J_l = \frac{\pi\hbar}{\mu k}(2l+1)\left(1 - |S_l|^2\right). \tag{19.41}$$

Its ratio to the flux density in the incident wave gives the sought-for partial cross section for inelastic scattering:

$$\sigma_l^{in} = \frac{\pi}{k^2}(2l+1)\left(1 - |S_l^e|^2\right). \tag{19.42}$$

Once we know the partial cross sections, we can find the respective total cross sections for each kind of scattering:

$$\sigma^e = \frac{\pi}{k^2}\sum_l (2l+1)|1 - S_l|^2 \quad \text{and} \quad \sigma^{in} = \frac{\pi}{k^2}\sum_l (2l+1)\left(1 - |S_l|^2\right). \tag{19.43}$$

Finally, adding these two gives the *total* (both elastic and inelastic) cross section:

$$\tilde{\sigma} \equiv \sigma^e + \sigma^{in} = \frac{2\pi}{k^2}\sum_l (2l+1)(1 - \text{Re}\,S_l). \tag{19.44}$$

Now let us go back to (19.25) and write its imaginary part as

$$\mathrm{Im} f(0) = \frac{1}{2k} \sum_l (2l+1)(1 - \mathrm{Re}\, S_l). \tag{19.45}$$

Comparing with (19.44), we obtain

$$\mathrm{Im} f(0) = \frac{k}{4\pi} \tilde{\sigma}. \tag{19.46}$$

This is already a familiar optical theorem, but now it is formulated for a much more general case including nonelastic scattering. The theorem holds in this case as well, and generally, it relates $\mathrm{Im} f(0)$ to the *total* cross section of all possible processes, rather than only to elastic cross section. In Sections 19.4 and 19.5, we consider a couple of examples illustrating the general theory outlined here.

19.4
Diffraction Scattering

Here, we consider a situation analogous to the known optical problem of the wave diffraction on a totally absorbing sphere of given radius R. Hence, the term "diffraction scattering" in the corresponding QM problem.

In optics, an absorbing medium is characterized by a complex refractive index, $n \to \tilde{n} = n + i\nu$ [12]. For a totally absorbing medium, the imaginary part becomes overwhelmingly dominating over real part so that the latter can be neglected.

In QM, the potential field $U(\mathbf{r})$ plays the same role as the refractive index in optics. In particular, the absorbing medium can be described by a complex potential $U \to \tilde{U} = U + iV$ with positive V. For a very strong absorption modeling a blackbody object in Chapter 1, we can leave only the imaginary part. Thus, the optical properties of the absolutely absorbing sphere will be described by a potential:

$$\tilde{U}(r) = \begin{cases} iV, & r < R, \\ 0, & r > R. \end{cases} \tag{19.47}$$

Consider the particle scattering on such sphere when the particle wavelength is much smaller than R, that is,

$$\lambda \ll R \quad \text{or} \quad k \gg \frac{2\pi}{R}, \quad E \gg \frac{4\pi^2 \hbar^2}{2\mu R^2}. \tag{19.48}$$

In terms of the impact parameters and partial cross sections, this condition means that we expect total absorption for all $s < R$. In terms of l, this means that the states with $l < l_R$ will be totally absorbed. Here, l_R is determined by $\sqrt{l_R(l_R+1)}\hbar = R\sqrt{2\mu E}$. For large $l_R \gg 1$, this simplifies to

$$l_R \approx kR (\gg 2\pi). \tag{19.49}$$

But total absorption means the absence of the corresponding outgoing partial waves. In the language of scattering matrix, the corresponding matrix elements are zeros,

$$S_l(k) = 0, \quad \text{for all } l \leq l_R. \tag{19.50}$$

On the other hand, for all $s > R$, that is, for all $l > l_R$, the particle does not seem to notice any potential at all (is effectively in a free space), and the diverging partial waves must remain approximately the same as for a free particle. Therefore, in this region, we can, to a high accuracy, set $S_l(k) = 1$ for all $l > l_R$. Altogether, a totally absorbing sphere of radius R is described by the scattering matrix:

$$S_l(k) = \begin{cases} 0, & l \leq l_R, \\ 1, & l > l_R. \end{cases} \tag{19.51}$$

So only the states with l satisfying condition (19.50) get scattered. This sounds as a total nonsense, because precisely for such states, according to the definition of $S_l(k)$, there are no diverging waves. How can we talk of any scattering in the absence of diverging waves describing it?

This is only an apparent paradox. It reflects a possible misconception that once scattering is associated with a diverging spherical wave in solution (19.14), then any diverging wave describes scattering. But the latter does not follow from the former. As seen from (19.15) and (19.16), any partial *scattered* wave is the *difference* between the *actual* diverging wave and diverging wave in the absence of $U(\mathbf{r})$. Recall that the plane wave describing a free particle is superposition of converging and diverging spherical waves (Equations 10.15, 10.17). If the actual diverging wave is zero, there still is a nonzero scattered wave. It is just opposite in sign to the corresponding diverging wave in the absence of potential. Consider another example. We know that any standing spherical wave (which is not associated with any scattering!) is itself an equally weighted superposition of diverging and converging waves. Thus, we have here a system including diverging wave and yet not associated with scattering. Now ask what happens when the two opposing amplitudes in the standing wave become, for whatever reason, not equally weighted? In this case, the wave is no longer standing! There will appear an inward or outward radial flux depending on which amplitude is now dominating. But such flux is already equivalent to scattering: Even if it is converging, that would mean an *inelastic scattering* associated with absorption of the incident particle by the center. This is precisely the case at hand: For a certain subset of l, we have the zero matrix elements, which means that no outgoing waves could counterbalance the incoming ones. This, in turn, means *total* absorption of particles with the respective l from the incident beam.

So we can try to evaluate the corresponding cross section by summing the partial cross sections for inelastic scattering. We go back to Equation 19.14 and set there $S_l = 0$ for all $l \leq l_R$ and $S_l = 1$ for all $l > l_R$. But the moment we try to do it, we notice another, may be even more embarrassing thing: Any inelastic cross section in (19.43) comes in pair with its elastic counterpart (Equation 19.38!), and according to (19.43), they must be equal to each other in the case $S_l = 0$ considered here. So to each event of absorption, there is equally probable event of elastic scattering of an incident particle.

We explained how the absence of diverging waves can be consistent with the presence of scattering by invoking the inelastic scattering. But it turns out that eventually the inelastic scattering brings in by hand its elastic partner anyway, without the corresponding diverging waves!

The explanation of this effect is a little more subtle, but is entirely in the spirit of QM.

Note that a partial cross section of elastic scattering is determined not just by the diverging wave alone but also by its difference from the partial amplitude in the incident wave. Therefore, even when $S_l(k) = 0$, so there is no diverging wave, there still is a nonzero partial cross section of elastic scattering! In particular, if we observe the particle in the region of geometric shadow of the sphere, it can be only elastically scattered particle. And indeed in a similar optical experiment, as we know, the distant screen behind an opaque obstacle has a bright illuminated spot (known as the *Poisson spot*) right at the center of the shadow region [12,13], where there could not be any light according to corpuscular theory. We can visualize this effect as diffraction of the incident wave – its deviation toward the symmetry axis when passing by the edge of the absorbing sphere (Figure 19.4). Note that this deviation has nothing to do with any forces, which, as already emphasized, are absent outside the sphere. But it has to do with the *Huygens construction* known in optics and explaining the deviation as a purely wave phenomenon, according to which "passing by" is not so passive after all!

All this rather lengthy comment is mathematically described by Equation 19.51.

With this in hand, we go back to Equation 19.43, set there $S_l = 0$, and truncate summation up to $l = l_R$. Both cross sections turn out to be equal:

$$\sigma^{\text{in}} = \sigma^{\text{e}} = \frac{\pi}{k^2} \sum_{l=0}^{l_R} (2l+1) = \frac{\pi}{k^2}(l_R + 1)^2 \approx \pi \left(\frac{l_R}{k}\right)^2. \tag{19.52}$$

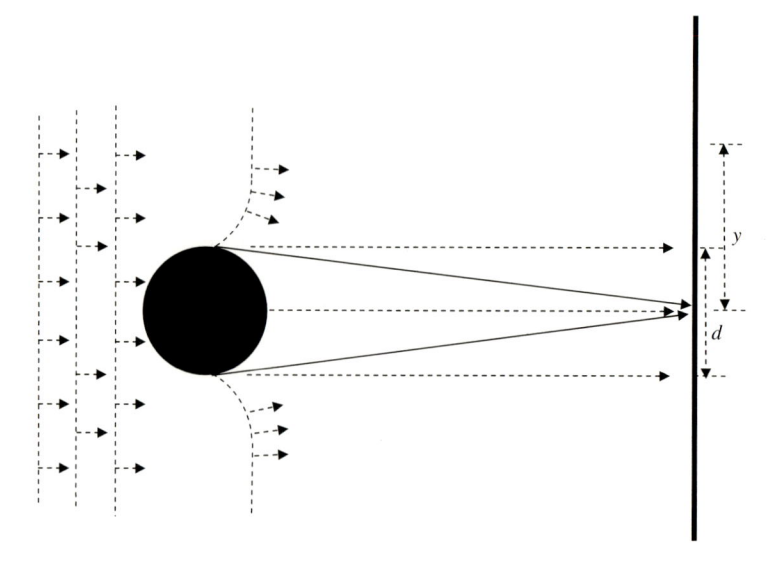

Figure 19.4 Huygens construction, illustrating the wave diffraction at the edge of an obstacle (not to scale). It leads to the appearance of the bright area (the Poisson spot) at the center of the geometrical shadow, which is not accessible to direct incident wave.

Each sum, in view of (19.49), is equal to πR^2. And the total cross section is $2\pi R^2$, that is, twice the area of classical shadow cast by the sphere, despite the fact that the summation was only up to $l = l_R$! This is another illustration of the fact that the scattering in a state with angular momentum L is not the same as classical scattering of the particle with definite impact parameter $s = L/p$. It involves much more fuzzy area, because the corresponding radial function has, as emphasized in Section 19.2 (see Figure 19.4 and comment to Equation 19.24), only its maximal value at $s = L/p$, and otherwise is spread over the whole space.

QM allows us to find more than that. We can extract much more information by determining *angular distribution* of elastically scattered particles, that is, *differential cross section* of elastic scattering from Equation 19.20. In our case, this equation takes the following form:

$$f(\theta) = \frac{1}{2ik} \sum_{l=0}^{l_R} (2l+1) P_l(\cos\theta). \tag{19.53}$$

For sufficiently large l_R and small angles, we can use approximation $P_l(\cos\theta) \approx J_0(l\theta)$, where J_0 is the Bessel function of the zero order [14]. In this range of variables,

$$f(\theta) \approx \frac{1}{2ik} \sum_{l=0}^{l_R} (2l+1) J_0(l\theta). \tag{19.54}$$

This sum can, in turn, be approximated by an integral [14]:

$$f(\theta) \approx \frac{1}{ik} \int_0^{kR} J_0(\theta l) l\, dl = -i\frac{R}{\theta} J_1(kR\theta). \tag{19.55}$$

Finally, we get

$$\frac{d\sigma^e}{d\Omega} = |f(\theta)|^2 = R^2 \frac{J_1^2(kR\theta)}{\theta^2}. \tag{19.56}$$

This, again, describes familiar diffraction pattern for light scattering on a large circular opaque screen. It has a sharp maximum at the center and secondary weaker maxima around the symmetry axis. And the similar picture is observed in quite different situation – in experiments on neutron scattering on heavy nuclei or π-meson–proton scattering. With all differences, the common angular distribution is recorded under two conditions common for these experiments and the considered optical diffraction: There must be strong interaction with the target leading to strong inelastic scattering, and the de Broglie wavelength of the incident particles must be much less than the effective size of the target. For instance, an incident neutron with its wavelength much less than the radius of the nucleus can be absorbed by the nucleus. The nucleus acts as a "blackbody" for the neutron. Similarly, a high-energy π-meson strongly interacts with nucleons, when its wavelength is so small (less than $\sim 10^{-15}$ m) that it allows it to get "in touch" with a nucleon. In such cases, there is a high probability to create new particles at the cost of the π-meson itself and/or its initial energy. This is an inelastic collision equivalent to absorption of the incident particle. In all these cases, we have 50/50 ratio of particles undergoing elastic or inelastic scattering, respectively. And those that scatter elastically show the angular distribution described by Equation 19.56.

19.5
Resonant Scattering

Consider a situation when the target has a set of allowed energies and the corresponding transition frequencies, and the incident particle has energy tuned to some of them. Such situation is an example of the known phenomenon of resonance. Its basic features can be illustrated in the case of a low-energy incident particle with the wavelength much greater than the effective size R of the target: $\lambda \gg R$, that is, $Z \equiv kR \ll 1$. Under this condition, the dominating contribution to scattering comes from the s-state. Therefore, we can, to a sufficient accuracy, represent the total wave function by the partial wave with $l = 0$:

$$\Psi_0(r) = R_0(r) = \frac{e^{-ikr}}{r} - S_0 \frac{e^{ikr}}{r}, \qquad (19.57)$$

where S_0 is the element of the scattering matrix S_l for $l = 0$. Since the interaction is negligible at $r > R$, we can again approximate the potential by a discontinuous function of the type (10.23) – such that $U = 0$ at distances r exceeding the characteristic distance R.

And, similarly to (10.26, 10.27), we look for a solution $G_0(r) \equiv r\Psi_0(r)$ satisfying the boundary conditions according to which $G_0(r)$ and its derivative $G'_0(r) = dG_0(r)/dr$ must be continuous at $r = R$. The problem is that now we do not know the potential in the inner region $r < R$ and accordingly do not know the corresponding solution there. So we do not know to what we should match the outer function (19.57) and its derivative. We want to get some information about the inner region from the scattering data, not vice versa!

We get around this difficulty by using the so-called phenomenological approach – formulating the most general relations that follow from the known conditions without going into details. In our case, the only thing we know is the outer function and continuity condition, so let us first just evaluate $G_0(r)$ and $G'_0(r)$ from (19.57) at $r = R$:

$$G_0(R) = e^{-ikR} - S_0 e^{ikR}, \quad G'_0(R) = -ik(e^{-ikR} + S_0 e^{ikR}). \qquad (19.58)$$

We do not know the equated values, but we know that whatever they are, if $G_0(r)$ and $G'_0(r)$ are both continuous, so is their ratio. So, as a second step, take the ratio $G'_0(r)/G_0(r)|_{r=R}$. It is just the derivative of $\ln G_0(r)$ and therefore is frequently referred to as the logarithmic derivative. In our case, this derivative evaluated "from the outside" according to (19.58) can be written as

$$\frac{G'_0(R)}{G_0(R)} R = -iZ \frac{e^{-iZ} + S_0 e^{iZ}}{e^{-iZ} - S_0 e^{iZ}}, \qquad (19.59)$$

with $Z(E) \equiv kR \ll 1$. Since we do not know the value of ratio (19.59), let us just denote it $f(E)$, where $E = \hbar^2 k^2/2\mu$ is the energy of the incident particle. Thus, we have

$$-iZ \frac{e^{-iZ} + S_0 e^{iZ}}{e^{-iZ} - S_0 e^{iZ}} = f(E). \tag{19.60}$$

The form of function $f(E)$ depends on the intrinsic properties of the targeted system (atom, nucleus, etc.). Whatever it is, let us pretend that we know it and solve Equation 19.60 for $S_0 = S_0(E)$ to express it in terms of $f(E)$:

$$S_0(E) = -e^{2iZ} \frac{Z - if(E)}{Z + if(E)}. \tag{19.61}$$

Generally, $f(E)$ is a complex function of E, so let us show this explicitly by writing

$$f(E) = g(E) + ih(E), \tag{19.62}$$

with real $g(E)$ and $h(E)$. So far all this is of little help, being just algebraic manipulations with formulas. Let us see if it makes sense under some additional physical assumptions. Our assumption will be that the known resonances in the photon–atom interaction postulated by Bohr and then confirmed in experiments are only a special case of a general rule. According to this rule, if an energy of the incident particle is tuned to some possible transition of the target between its different states with energies E_m and E_n, respectively, that is, $E = E_r \equiv E_n - E_m$ ($E_n > E_m$), then interaction between the particle and the target enormously increases. Experimentally, this must be manifest in the corresponding increase of the cross section of scattering. Accordingly, we must expect the sharp peaks of the scattering matrix at energies $E \approx E_r$. These energies are called the *resonant energies*, and the corresponding scattering is called the *resonant scattering*.

Let us see if this approach works. We need to make one additional assumption (which is justified by the final result): That real part $g(E)$ in (19.62) is zero at resonant energy, that is, $g(E_r) = 0$. Then, in the vicinity of E_r, the function $g(E)$ can be approximated as

$$g(E) \approx g(E_r) + g'(E_r)(E - E_r) = g'(E_r)(E - E_r), \tag{19.63}$$

where $g'(E_r)$ is the shorthand for $dG(E)/dE|_{E=E_r}$. The function $h(E)$ remains nonzero, so in the vicinity of E_r it can be replaced with $h = h(E_r)$. Thus, the whole function $f(E)$, unknown to us as it is, in the vicinity of resonance energy can be represented by

$$f(E) \approx g'(E_r)(E - E_r) + ih. \tag{19.64}$$

Putting this into (19.61) gives

$$S_0(E) \underset{E \approx E_r}{\approx} -e^{-2iZ} \frac{[(Z-h)/g'] - i(E - E_r)}{[(Z+h)/g'] + i(E - E_r)}. \tag{19.65}$$

Now denote

$$2\frac{Z}{g'} \equiv \Gamma_e, \quad 2\frac{h}{g'} \equiv \Gamma_r. \tag{19.66}$$

Then,

$$S_0 = e^{2iZ}\frac{E - E_r - (1/2)i(\Gamma_e - \Gamma_r)}{E - E_r + (1/2)i\Gamma}, \tag{19.67}$$

where $\Gamma \equiv \Gamma_e + \Gamma_r$. Finally, express the experimentally observed partial cross sections in terms of the introduced parameters Γ_e, Γ_r:

$$\sigma^{in} = \frac{\pi}{k^2}\left(1 - |S_0|^2\right) = \frac{\pi}{k^2}\frac{\Gamma_e\Gamma_r}{(E - E_r)^2 + (1/4)\Gamma^2} \tag{19.68}$$

and

$$\sigma^e = \frac{\pi}{k^2}|1 - S_0|^2 = \frac{\pi}{k^2}\left|\frac{\Gamma_e}{(E - E_r) + (1/2)i\Gamma} - 2e^{-iZ}\sin Z\right|^2. \tag{19.69}$$

Their sum gives the total cross section for the s-scattering:

$$\sigma = \sigma^{in} + \sigma^e = \frac{2\pi}{k^2}(1 - \mathrm{Re}\, S_0) = \frac{2\pi}{k^2}\frac{\Gamma_e((1/2)\Gamma \cos 2Z - (E - E_r)\sin 2Z)}{(E - E_r)^2 + (1/4)\Gamma^2}. \tag{19.70}$$

These are typical expressions describing resonance. They give us elastic and inelastic cross sections as functions of energy. We see that all three functions peak sharply at $E = E_r$. The quantity Γ is total *resonance half-width*. The term "half-width" reflects the fact that the corresponding cross section drops down to the half of its maximal value when $|E - E_r| = (1/2)\Gamma$. The quantities Γ_e and Γ_r are half-widths of the elastic and inelastic scattering, respectively. The amplitude $1 - S_0$ of the elastic scattering is the sum of two interfering terms: The term $\Gamma_e/[(E - E_r) + (1/2)i\Gamma]$ is responsible for the sharp maximum of $\sigma^e(E)$ at $E = E_r$ and is associated with *elastic resonant scattering*; the second term $2e^{-iZ}\sin Z$, in which out of all characteristics of the target only its effective size R is represented, is responsible for a much smaller contribution called *potential scattering*. This part of scattering is independent of the internal structure of the target, except for its size.

The potential scattering is noticeable only relatively far from exact resonance, that is, at $|E - E_r| \gg \Gamma$. At $|E - E_r| < \Gamma$, the resonant scattering becomes overwhelmingly dominating, and the potential scattering, in view of $Z \equiv kR \ll 1$, can be neglected.

The elastic and total cross sections can then be written as

$$\sigma^e = \frac{\pi}{k^2}\frac{\Gamma_e^2}{(E - E_r)^2 + (1/4)\Gamma^2}, \quad \sigma = \frac{\pi}{k^2}\frac{\Gamma\Gamma_e}{(E - E_r)^2 + (1/4)\Gamma^2}. \tag{19.71}$$

Their ratio is, as it should, equal to

$$\frac{\sigma^e}{\sigma} = \frac{\Gamma_e}{\Gamma}. \tag{19.72}$$

Thus, the phenomenological approach allows us to derive the expressions that correctly describe the scattering near resonance even when we have little or no information about the properties of the target. We do not know the parameters Γ_e, Γ_r since we do not know the function $f(E)$; and yet we have the correct expressions for the measurable observables – cross sections of scattering – as functions of these parameters. Hence, these parameters and thereby some properties of the target can be determined from the experimental data. Such procedure is part of what we call the inverse problem of scattering: extracting information about the interaction potential from experimentally measured characteristics.

Equations 19.68 and 19.71 were first obtained by Breit and Wigner [15,16]. They are similar to the formulas known in optics for the light scattering and absorption (a form of the inelastic scattering) near a resonant spectral line.

The reader can also notice similarity with Equation 11.123 describing the energy distribution of an excited state of a nucleus in the theory of radioactive decay. This similarity reflects an essential analogy in the physical nature of both effects. An incident particle undergoing scattering can be thought of as first absorbed by the target that becomes unstable like an unstable radioactive nucleus capable of decay. Then the target reemits the particle (decays!). This model is accurate even for purely elastic scattering, which corresponds to reemitting the particle in the same energy state as it had before. The same can be said about light interaction with atoms. Each scattered photon can be thought of as first absorbed by the atom and then reemitted with the same or different frequency. During the corresponding collision time, the excited atom is essentially a radioactive system doomed to decay.

Resonant scattering, in contrast with the resonance in mechanical vibrations, is a purely quantum-mechanical effect showing the wave aspect of matter. As seen from Equation 19.71, the cross section at resonance can take on huge values $\sigma \sim 4\pi/k^2 = \lambda^2/\pi(\Gamma^e \sim \Gamma)$, which, at $kR \ll 1$, by far exceed geometric cross section πR^2 of the target. Thus, in scattering of slow neutrons on nuclei Xe_{54}^{135}, they undergo strong resonant absorption whose cross section is about 10^5 of geometric area of a Xe nucleus. This resonance is very important in functioning of nuclear reactors [9].

19.6
The Born Approximation

Here, we will study the case when $f(\theta)$ can be evaluated with a sufficient accuracy directly from potential $V(r)$, without recurring to the phase shifts. Consider again scattering as a stationary state with $E > 0$. We need to solve the stationary Schrödinger equation:

$$\nabla^2 \Psi - V(r)\Psi = -k^2 \Psi, \tag{19.73}$$

where we again assume the field to be spherically symmetric and denote, as usual, $k^2 = 2\mu E/\hbar^2$, $V(r) \equiv 2\mu U(r)/\hbar^2$. The approach described below is much more straightforward, but it is restricted to sufficiently high energies.

We look for the solution in the familiar form, which is, far from the center, a superposition of the incident plane wave and an additional wave $u(\mathbf{r})$ caused by the potential $V(r)$:

$$\Psi \approx e^{ikz} + u(\mathbf{r}). \tag{19.74}$$

The additional wave then can be nothing else but the scattered wave. The problem is to find $u(\mathbf{r})$ and see if it has the familiar form:

$$u(\mathbf{r}) = f(\theta)\frac{e^{ikr}}{r}, \tag{19.75}$$

in which case the factor $f(\theta)$ would give us the sought-for scattering amplitude (Figure 19.5).

If the interaction is sufficiently weak, we can expect the accordingly weak scattering with relatively small $f(\theta)$. This suggests that mathematically we could treat weak interaction of incident particle with the target as a small perturbation and accordingly find the second term in (19.74) as a small correction to the plane wave. The question arises what is the criterion for the interaction to be considered as "weak." The form of Equation 19.73 suggests the answer: The interaction potential must be small with respect to kinetic energy, that is, at all r there must be

$$E \gg U(r) \quad \text{or} \quad V(r) \ll k^2. \tag{19.76}$$

This condition will definitely be satisfied for sufficiently high E and for potentials that are finite in all space. In cases when $U(r)$ has a pole at the origin, we must require that $E \gg |\langle U(r)\rangle|$. Under these conditions, we can consider $u(\mathbf{r})$ as a small correction necessary for Ψ to satisfy Equation 19.73. Note that $u(\mathbf{r})$ depends on position vector $\mathbf{r}$, rather than on radial distance alone, because it is a function of θ as well.

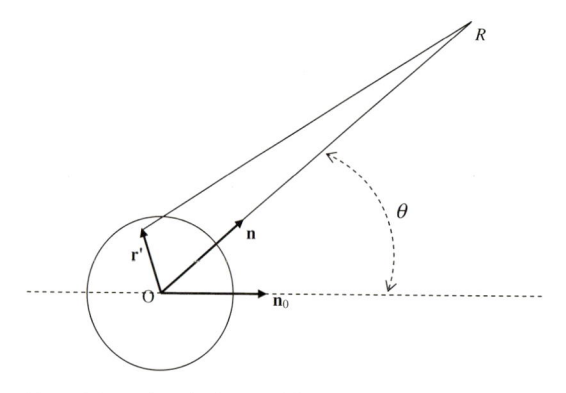

Figure 19.5 The calculation of the scattering amplitude $f(\theta)$ in the Born approximation. O – scattering center; θ – the scattering angle.

The operation we perform now is a special case of the perturbation theory. Put (19.74) into (19.73) and regroup the terms:

$$(\nabla^2 + k^2)e^{ikz} + \nabla^2 u(\mathbf{r}) + k^2 u(\mathbf{r}) = V(r)e^{ikz} + V(r)u(\mathbf{r}). \tag{19.77}$$

The first term on the left-hand side vanishes since the plane wave is a solution of the Helmholtz equation (8.25). The second term on the right-hand side can be neglected as a product of two small factors. We are left with

$$(\nabla^2 + k^2)u(\mathbf{r}) = Ve^{i\mathbf{k}\mathbf{r}}, \quad \mathbf{k}\mathbf{r} = kr\cos\theta = kz. \tag{19.78}$$

This is the Helmholtz *inhomogeneous* equation. We need to find its solution that would have asymptotic form (19.74).

The equation of the type (19.78) and its solutions are well known in electrodynamics. It emerges from the D-Alambert inhomogeneous equation for potential $\Phi(\mathbf{r}, t)$ (recall Equation 16.8):

$$\left(\nabla^2 - \frac{1}{c^2}\frac{\partial^2}{\partial t^2}\right)\Phi(\mathbf{r}, t) = -\frac{\rho(\mathbf{r}, t)}{\varepsilon_0}. \tag{19.79}$$

The solution of this equation is (see, for example, Ref. [17])

$$\Phi(\mathbf{r}, t) = \frac{1}{4\pi\varepsilon_0}\int\frac{\rho(\mathbf{r}', t - |\mathbf{r} - \mathbf{r}'|/c)}{|\mathbf{r} - \mathbf{r}'|}d\mathbf{r}'. \tag{19.80}$$

It just represents $\Phi(\mathbf{r}, t)$ as a superposition of incremental potentials from each volume element $d\mathbf{r}'$ that contained amount of charge $\rho(\mathbf{r}', t')d\mathbf{r}'$ at a retarded moment $t' = t - (|\mathbf{r} - \mathbf{r}'|/c)$ such that radiation from that element reaches observation point $\mathbf{r}$ at the moment t.

Suppose that the whole charge distribution oscillates with frequency ω, that is, $\rho(\mathbf{r}, t) = \rho_0(\mathbf{r})e^{-i\omega t}$. Then, $\Phi(\mathbf{r}, \theta)$ will oscillate with the same frequency:

$$\Phi(\mathbf{r}, t) = \frac{1}{4\pi\varepsilon_0}\int\frac{\rho_0(\mathbf{r}')e^{-i\omega(t-(|\mathbf{r}-\mathbf{r}'|/c))}}{|\mathbf{r} - \mathbf{r}'|}d\mathbf{r}' = \Phi_0(\mathbf{r})e^{-i\omega t}, \tag{19.81}$$

where

$$\Phi_0(\mathbf{r}) = \frac{1}{4\pi\varepsilon_0}\int\frac{\rho_0(\mathbf{r}')e^{ik|\mathbf{r}-\mathbf{r}'|}}{|\mathbf{r} - \mathbf{r}'|}d\mathbf{r}', \quad k = \frac{\omega}{c}. \tag{19.82}$$

If we put this into Equation 19.79 and drop from both sides the common factor $e^{-i\omega t}$, the result will be the equation for $\Phi_0(\mathbf{r})$:

$$(\nabla^2 + k^2)\Phi_0(\mathbf{r}) = -\frac{\rho_0(\mathbf{r})}{\varepsilon_0}, \tag{19.83}$$

with solution (19.82).

The form of Equation 19.78 for $u(\mathbf{r})$ is identical to that of Equation 19.83. Therefore, the solution for $u(\mathbf{r})$ will be obtained from (19.83) if we replace there $[\rho_0(\mathbf{r})]/\varepsilon_0$ with $V(r)e^{i\mathbf{k}\mathbf{r}}$:

$$u(\mathbf{r}) = -\frac{1}{4\pi}\int \frac{V(\mathbf{r}')e^{i\mathbf{k}\mathbf{r}'}e^{ik|\mathbf{r}-\mathbf{r}'|}}{|\mathbf{r}-\mathbf{r}'|}d\mathbf{r}'. \tag{19.84}$$

We are interested in the asymptotic form of the solution far from the origin. As mentioned above, the potential is assumed to fall off with distance sufficiently fast so that integration region is effectively finite and we consider $|\mathbf{r}| \gg |\mathbf{r}'|$. In this case,

$$|\mathbf{r}-\mathbf{r}'| \approx r - \hat{\mathbf{n}}\mathbf{r}', \quad \hat{\mathbf{n}} \equiv \frac{\mathbf{r}}{r}. \tag{19.85}$$

In the denominator of (19.84), the term $\hat{\mathbf{n}}\mathbf{r}'$ can be neglected with respect to r, but in the oscillating exponent it must be retained. Therefore,

$$u(\mathbf{r}) = -\frac{e^{ikr}}{4\pi r}\int V(\mathbf{r}')e^{ik(\hat{\mathbf{n}}-\hat{\mathbf{n}}_0)\mathbf{r}'}d\mathbf{r}', \tag{19.86}$$

where $\hat{\mathbf{n}}_0 = \mathbf{k}/k = \hat{\mathbf{z}}$ is the unit vector along the z-axis. We see that $u(r)$ has the expected form (19.75) of diverging spherical wave multiplied by the factor depending on the scattering angle. Comparing with (19.75) gives

$$f(\theta) = -\frac{1}{4\pi}\int V(\mathbf{r}')e^{ik(\hat{\mathbf{n}}_0-\hat{\mathbf{n}})\mathbf{r}'}d\mathbf{r}'. \tag{19.87}$$

We can rewrite the term on the right-hand side in a way explicitly showing that it is a function of θ. Indeed, denoting $k(\hat{\mathbf{n}}_0-\hat{\mathbf{n}}) \equiv \mathbf{q}$, we have $|\mathbf{q}| = 2k\sin(\theta/2)$. In terms of q, the integral in (19.87) takes the form

$$f(\theta) = -\frac{1}{4\pi}\int V(\mathbf{r}')e^{i\mathbf{q}\mathbf{r}'}d\mathbf{r}'. \tag{19.88}$$

It shows that $f(\theta)$ is (up to a numerical factor) just the Fourier transform $\mathcal{F}(\mathbf{q})$ of function $V(\mathbf{r})$, that is,

$$f(\theta) = \frac{1}{2\sqrt{2\pi}}\mathcal{F}(\mathbf{q}) = \frac{1}{2\sqrt{2\pi}}\mathcal{F}\left(2\hat{\mathbf{q}}k\sin\frac{\theta}{2}\right), \tag{19.89}$$

where $\hat{\mathbf{q}}$ is the unit vector along $\mathbf{q}$. The obtained result is known as the *Born approximation* [9,18].

Problems

19.1 Write the Schrödinger equation for two particles with mass μ_1 and μ_2, respectively, assuming that their interaction is described by potential depending only on the distance between them. Then introduce new coordinates $\mathbf{r} = \mathbf{r}_2 - \mathbf{r}_1$ and $\mathbf{R} = (\mu_1\mathbf{r}_1 + \mu_2\mathbf{r}_2)/\mu_1 + \mu_2$ (position vector of center of mass). Express the Schrödinger equation in terms of $\mathbf{r}$ and $\mathbf{R}$.

19.2 In classical approximation, find the differential cross section and function $f(\theta)$ in Equations 19.1–19.4

a) for a spherical potential well of depth U and radius R, and
b) for a spherical barrier of the same radius and height U.

(*Hint*: Determine first the corresponding boundary conditions for momentum.)

19.3 For each case in the previous problem, find the total cross section by integrating the function $|f(\theta)|^2$. Compare your results with direct geometric picture in Figure 19.2.

19.4 Use expression (19.40) for the net flux to find the inelastic partial cross section in the spherically symmetric potential field.

19.5 What happens if we change sign of V in mathematical formulation of scattering problem (Equation 19.73)? What physical situation would it correspond to? Explain your answer.

References

1. Walls, D. (1983) Squeezed states of light. *Nature*, **306**, 141.
2. Ford, K.W., Hill, D.L., Wakano, M., and Wheeler, J.A. (1959) Quantum effects near a barrier maximum. *Ann. Phys.*, **7**, 239.
3. Ford, K.W. and Wheeler, J.A. (1959) Semiclassical description of scattering. *Ann. Phys.*, **7**, 259–286.
4. Ford, K.W. and Wheeler, J.A. (1959) Application of semiclassical scattering analysis. *Ann. Phys.*, **7**, 287.
5. Fayngold, M. (1966) On a possible γ-radiation mechanism in reactions with heavy ions. *Sov. J. Nucl. Phys.*, **3** (4), 458–460.
6. Fayngold, M. (1968) Some features of spiral scattering. *Proc. Acad. Sci. USSR*, **6** (in Russian).
7. Fayngold, M. (1985) Resonant scattering and suppression of inelastic channels in a two-dimensional crystal. *Sov. Phys. JETP*, **62** (2), 408–414.
8. Feynman, R., Leighton, R., and Sands, M. (1963) *The Feynman Lectures on Physics*, vols **1–3**, Addison-Wesley, Reading, MA, p. 64.
9. Blokhintsev, D.I. (1964) *Principles of Quantum Mechanics*, Allyn & Bacon, Boston.
10. Landau, L. and Lifshits, E. (1965) *Quantum Mechanics: Non-Relativistic Theory*, 2nd edn, Pergamon Press.
11. Faxen, H. and Holtsmark, J. (1927) *Zeits. f. Phys.*, **45**, 307.
12. Hecht, E. (2002) *Optics*, 4th edn, Addison-Wesley, p. 385.
13. Born, M. and Wolf, E. (1999) *Principles of Optics: Electromagnetic Theory of Propagation, Interference and Diffraction of Light*, CUP Archive.
14. Whittaker, V.T. and Watson, G.H. (1996) *A Course of Modern Analysis*, Cambridge University Press.
15. Breit, G. and Wigner, E. (1936) Capture of slow neutrons. *Phys. Rev.*, **49**, 519.
16. Breit, G. (1959) Theory of resonance reactions. *Handbuch der Phy.*, **41**, 1.
17. Jackson, J. (1999) *Classical Electrodynamics*, 3rd edn, John Wiley & Sons, Inc., New York.
18. Mott, N. and Messi, G. (1965) *The Theory of Atomic Collisions*, 3rd edn, Oxford University Press.

20
Submissive Quantum Mechanics

This chapter is devoted to a new approach and the new kind of problems in QM – actually, to a new area of QM developed relatively recently [1–5]. The new area shows QM from a totally different perspective opening a breathtaking possibility of monitoring, producing, and operating new quantum states and manipulating some parameters of the known states. This allows one to transform gradually one quantum system into another. Such possibility is especially important in the rapidly developing physics of nanostructures, and may open a new era – an era of quantum engineering.

We will present only basic ideas on the qualitative level. We will describe the potential transformations that change any chosen spectral parameter, but keep all others unperturbed. Similarly, we will discuss some ways of changing evolution of a quantum system. Altogether, the considered examples illustrate the methods of the theory of spectral, scattering, and decay control.

20.1
The Inverse Problem

> It is easy to make a cake from a recipe; but can we write down recipe if we are given a cake?
>
> R. Feynman

Traditionally, most of quantum mechanics deals with the problem: Given the quantum system (e.g., the potential describing interactions between the parts of the system and its interaction with the environment), determine the wave function and hence the properties of the studied object. In other words, given the interaction potential, we can find the energy spectrum of a system, and if in addition we know the initial conditions, we can predict the evolution of the system. Mathematically, this requires the solution of the corresponding wave equation, which constitutes the so-called direct problem.

In many applications we face the opposite situation: We may have experimental data about some physical characteristics of a system and want to extract from these data the information about the system's structure, namely, how the parts of this system interact with each other and with the environment. For instance, by bombarding a nucleus with protons or electrons, we may measure the scattering cross sections and thereby some characteristics of its energy eigenstates, and we can extract from these data the nature and behavior of force binding the protons and neutrons in this nucleus. In other words, we recover the interaction potential (the most essential part of the Hamiltonian operator!) from some known eigenvalues of this operator. From the mathematical viewpoint, this would be similar to solving a wave equation in reverse – Instead of finding eigenstates and eigenvalues of a given Hamiltonian (direct problem), we now want to find the Hamiltonian (the interaction potential) from its eigenvalues determined, for instance, from the scattering data. The latter kind of problem was appropriately named the inverse problem (IP).

Mathematical formulation of IP in QM appeared in 1951 in Refs [6,7]. Its development has largely extended the domain of possible applications (see, for example, [8–10]) and opened a new dimension in our view of QM. This achievement gives us a far broader vision of QM than before. To quote from Ref. [1] (regarding the above-mentioned experiments with collisions and scattering), "There are no tools like our eyes with microscopes for examining details of such small objects as atomic and nuclear systems. For this we need the 'illumination' with such ultra-short waves which cannot be operated as the visible light. It appears that the formalism of IP replaces our eyes here. This is a wonderful present of mathematicians, which for a long time was not properly realized and used by physicists."

Figuratively, paraphrasing the authors of [1], we can say that QM, like the Moon, was known only from one side (direct problem), and the new development of the IP opens to us its unknown opposite side.

IP was traditionally understood as reconstruction of interaction potential $U(\mathbf{r})$ from experimental data. Now we can take a broader view of IP: in addition to finding previously unknown $U(\mathbf{r})$ from the experiments, we can use IP algorithms for transforming a known $U(\mathbf{r})$ to construct new quantum systems with new properties. IP gives analytical expressions for the transformed potentials and wave functions corresponding to the changes of selected spectral parameters S. This opens to us infinite manifolds of exactly solvable models (ESMs). In the 1D case, these ESMs form the complete set.

Some of the new possible quantum systems discovered by the IP approach have really unusual features, such as coexistence of confinement and propagating waves, possibility of discrete scattering states, and unusual (non-Gamow) decay states.

The emerging broad area of finding and, in some cases, possible construction of new quantum systems with unusual properties can be characterized as *quantum design*. We cannot give sufficiently detailed description of this new domain and refer the reader to some original publications [2–10]. Here, we will restrict to a few specific examples of quantum design that are described in Section 20.2.

20.2
Playing with Quantum States

> Quantum mechanics is after all the code of laws for connections between potentials V and observables S (spectral and scattering data).
>
> *Zakhariev and Chabanov*

Our intuition based on experience with solving direct problems tells us that each given potential determines its unique set of the corresponding eigenstates and energy eigenvalues. Most physicists were convinced that any thinkable variation of potentials V changes all spectral characteristics (S). For instance, one would expect that any deformation of potential box by "warping" its bottom inevitably affects all set of the corresponding eigenvalues. That intuition turned out to be wrong. Extensive studies of the IP showed that one can select a single energy level or arbitrary group of them and vary them with sniper's accuracy while keeping the rest of the set intact by appropriately perturbing the potential function. This opens the exciting possibility of changing selected physical properties of a system. So far, such possibility has not been mentioned in the majority of up-to-date manuals. But it arises naturally in the framework of the IP.

Here, we consider the simple qualitative rules for lowering or raising an arbitrary individual energy level E.

The most simple case of a potential box in one dimension will be sufficient to illustrate the basic idea. The variations of the potential in this case are especially visual against the background of the flat bottom of the initial box. Later we will frequently refer to variations $W(x)$ of potential as perturbations in order to maintain terminology used in other chapters. But perturbations considered here (just as asymptotic perturbations in Chapter 17) are not necessarily small relative to the initial potential.

Let us start with the question what happens if we lift the bottom of the box by an amount W. In other words, we uniformly increase the potential function in the Hamiltonian (9.1) by $W = $ const. How will this affect the corresponding eigenvalues? The answer is obvious: The whole set of allowed energies will be shifted up on the energy scale by the same amount, because the described perturbation is equivalent to just lifting the whole box. Similarly, if we uniformly decrease the potential by W, the system of energy levels will be uniformly lowered by the same amount. Both results are manifestations of invariance of the system with respect to a uniform gauge transformation. A uniform shift of all energy levels does not produce any change in the physical behavior of a system, since all physically observable effects are associated with *energy difference* between the levels, rather than with absolute values of levels themselves. It is easy to give a formal proof (Problem 20.1) that eigenstates obtained from the Schrödinger equation will not change and their respective eigenvalues will all undergo the same shift if we add a constant to the potential $U(x)$, and the amount of the shift will be equal to the added constant.

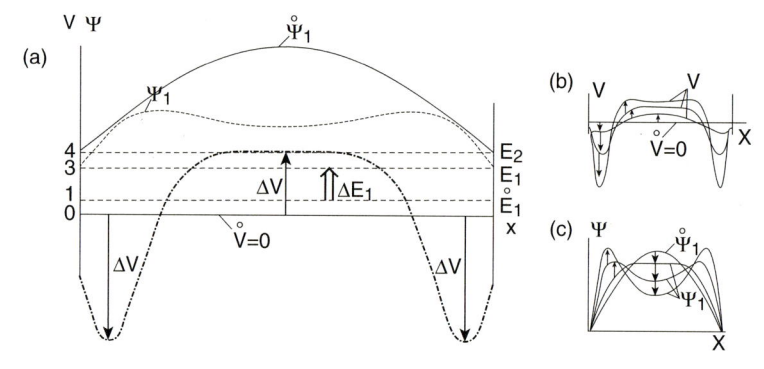

Figure 20.1 (a) Deformation (dashed–dotted line) of bottom of the potential box ($V(x) = 0$; $0 < x \lesssim a$) that raises the energy of the ground state $\mathring{E}_1 \to E_1$ by ΔE_1. At the "sensitive" center of the wave function Ψ_1, the barrier in V (shown by the bold black arrow) pushes E_1 upward. The wells at the edges (with bold downward arrows), where the Ψ_1 has low sensitivity, weakly influence E_1 but compensate the barrier action on all other levels $E_{n>1}$, leaving them immovable due to their higher sensibility at walls. The dashed horizontal lines depict energy levels and the dashed curves show deformed wave function Ψ_1 for the two values of energy shift. This is another example of energy level control for state with only bump, which will allow one to understand a more general case of shifting arbitrary energy levels in arbitrary potentials. (b and c) Change of the potential $V(x)$ and the wave function Ψ_1 for $\Delta E = 1, 2, 2.5$. (Adapted from Ref. [1] with permission.)

But the same shift for all energy levels is not what we want here. We want to shift a single selected energy level E_n (one element of spectrum) while keeping all others at their previous positions. The only remaining possibility to do this is to change the potential nonuniformly. In graphical terms, we must "warp" the bottom of the box. To see how, let us first consider the ground state $\psi_1(x)$, which has the simplest wave function, with only one "bump" (Figure 20.1a and b). The ground state is most sensitive to potential perturbations in the central part of the box, where the probability of finding the particle is maximal. For the same reason, it is practically insensitive to local perturbations close to the walls of the box: At these locations, the function $\psi_1(x)$ and accordingly the probability to find the particle is close to zero (recall that all energy eigenfunctions for the potential box are zero at the walls due to the boundary conditions). This gives us a hint that a *local* perturbation within a certain restricted area can play a double role: It may affect the chosen state function differently depending on location of the corresponding "warp"; and once the chosen location of the warp is fixed, it may affect the different states differently because they have different probabilities for the particle to be found at this location. This allows us to play with the states selectively, utilizing these two "degrees of freedom" (the value of the warp and its location).

Let us see how it works. Suppose we deform the bottom of the box by pushing up only its central part. This will definitely move up the energy level E_1, since the particle in the ground state is most sensitive to changes in the middle of the box, where its probability distribution has a peak. But the effect will be far less pronounced for the second level, since the corresponding state function $\psi_2(x)$

has a node in the middle of the box (Figure 9.2) and the probability to find the particle close to the center is small in this state. Accordingly, the response (upward shift of the second energy level E_2) will be smaller. This is already a step toward our goal. But our goal is to have no shift at all for the second level. Can we achieve this? The answer is yes. To this end, we must compensate for the rise of the second level by pulling it down, and this can be done if we add two more local perturbations: lower the two peripheral parts of the bottom of the box (Figure 20.1b). This will only slightly affect the ground level E_1 since the probability to find the particle close to the walls is small in state $\psi_1(x)$. In contrast, the probability peaks on both sides of the center closer to the walls for the second state $\psi_2(x)$ (Figure 9.2). Therefore, the two additional local warps of the bottom – one to the left and the other to the right of the center – will be largely unnoticed by the particle in state $\psi_1(x)$, but will be immediately "felt" by the same particle when in state $\psi_2(x)$. This means that an appropriate deformation of the bottom of the box – raising its central part and lowering the two symmetric peripheral parts – will do the job. It will raise the ground level, while leaving the second level intact. One can ask – how about the third level? In the state $\psi_3(x)$, there are three probability peaks, with one of them being at the center and therefore sensitive to the raising of the central part of the bottom. But it also has two additional peaks closer to the periphery, where the bottom of the box is warped down and "pulls" the energy level down. The positions of these peaks are closer to the walls than those in state $\psi_2(x)$, so that we can make both levels – with $n = 2$ and $n = 3$– immobile by making additional small wiggles in the deformed bottom of the box. Taking this into account, higher states will require additional wiggles in the perturbation $W(x)$, but this is achievable, at least in principle [1].

The similar procedure can be applied for pulling the selected level E_1 down by a desired amount. All we need is to change the sign in the variations of potential used in the first case (Figure 20.2).

Thus, it is possible to change the potential in such a way that only one energy level is affected – shifted up or down by a desired amount, while all the rest of them remain intact.

On the face of it, such a possibility may seem improbable, since the corresponding single perturbation function $W(x)$ performing such operation must satisfy an infinite number of conditions. Namely, while changing the energy level 1, it must leave intact the levels 2, 3, . . . , and so on. The number of levels in the box is infinite, and so is the number of conditions. Recall however that each analytical function is uniquely defined by an infinite number of parameters, for example, by the values of all its derivatives at a certain point, and this number is also infinite within the region of function's analyticity. Thus, we have an infinite number of adjustable parameters, which allows us to satisfy an infinite number of conditions. There is another way to see it, using the language of the Hilbert space: Since $W(x)$ is defined within the same region as the eigenstates $\psi_n(x)$, we can represent it as a superposition:

$$W(x) = \sum_n c_n \psi_n(x). \tag{20.1}$$

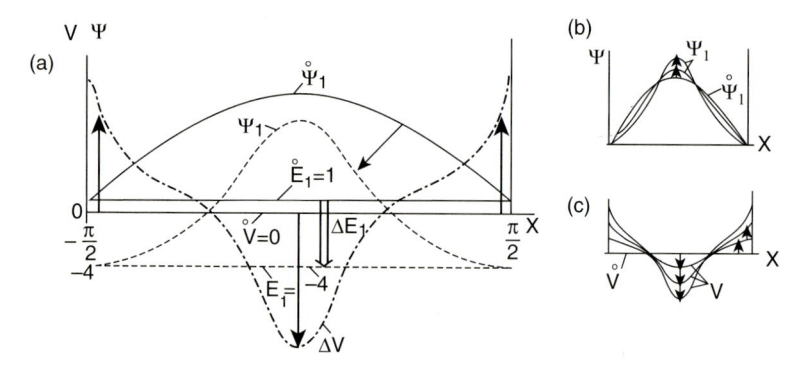

Figure 20.2 (a) The potential perturbation of the bottom of a potential box, which lowers only one ground state level $\overset{\circ}{E}_1 = 1 \to E_1 = -4$ (shown by the broad arrow ΔE_1). The bold black arrow directed downward points at the center of the box (dashed–dotted line $\Delta V(x)$ acting on the most sensitive, central region of the wave function to shift the level downward. The analogous arrows are directed upward near the walls of the initial box, where the function Ψ_1 has knots and where it is the least sensitive to potential changes. These arrows point at the barriers needed to keep fixed all other energy levels $E_{n>1}$ with $\Psi_{n>1}$ more sensitive to these barriers. This picture demonstrates qualitative features of universal, elementary transformation for *only one bump* of a wave function. It allows one to understand the rule of the wave transformations of arbitrary states with many bumps and in *arbitrary* potentials. (b) Evolution of the Ψ_1 when $\Delta E = 0, -1, -3$. (c) Evolution of the potential when $\Delta E = -1, -2, -3$. (Adapted from Ref. [1] with permission.)

The expansion coefficients c_n in this case are *not* the probability amplitudes, since $W(x)$ is not a physical state, and is not normalized. But from mathematical viewpoint, it can be considered as an *x*-projection of a vector, although not a unitary one, in the Hilbert space. We see that $W(x)$ is uniquely specified by the set of coefficients c_n. Again, we have an infinite number of adjustable parameters necessary and sufficient to satisfy the infinite number of conditions.

It might be tempting to think that if we managed to keep unchanged all eigenvalues E_n above E_1, then the corresponding eigenstates $|n\rangle$, $n = 2, 3, \ldots$, must also have unchanged eigenfunctions $\psi_n(x)$, $n = 2, 3, \ldots$. Such a conclusion would be wrong. Any perturbation $W(x)$ (except for $W = \text{const}$) changes the Hamiltonian of the system and thereby its eigenfunctions. This holds for *all* eigenstates regardless of whether their respective eigenvalues are changed by perturbation or not. In other words, not only all $\psi_n(x)$ with $n \geq 2$ will undergo change, but also the state $\psi_1(x)$ itself (Problem 20.2). This can be seen from the expression for energy eigenvalue:

$$E_n = \int \psi_n^*(x) \hat{H}(x) \psi_n(x) dx. \tag{20.2}$$

If we change the Hamiltonian by adding $W(x)$, we *must* also change $\psi_n(x)$ to keep E_n unchanged. The described algorithm can change in a desired way a single selected eigenvalue or selected subset of all eigenvalues; but it inevitably changes *all* eigenstates.

As mentioned above, the described variations $W(x)$ are not necessarily small. In fact, we can, for instance, push the ground level so far upward that it will get

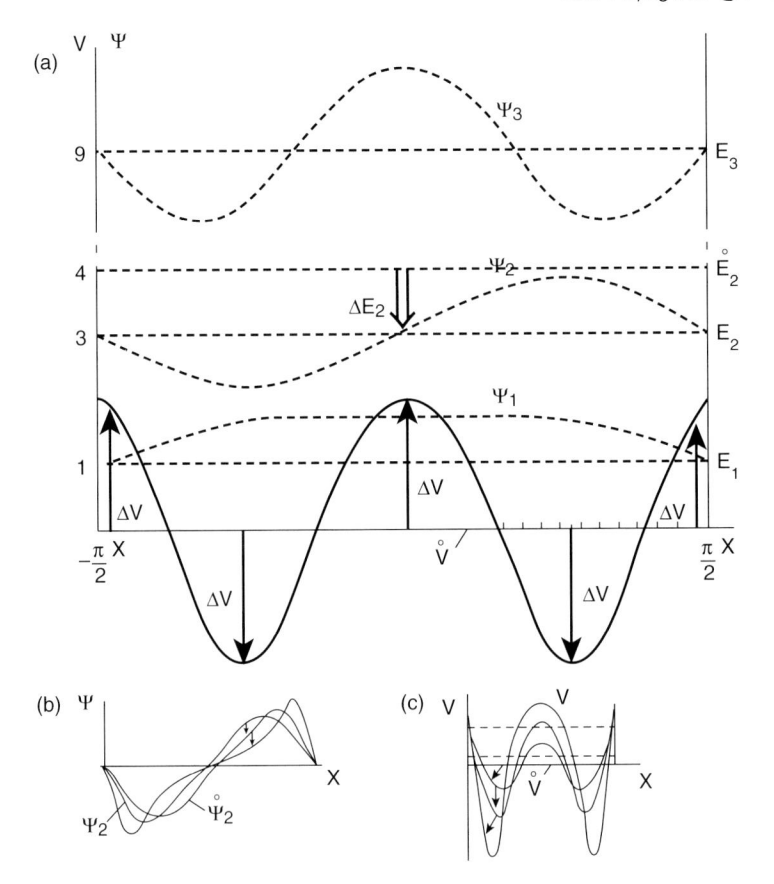

Figure 20.3 (a) Shift of the second level E_2 alone downward is performed by two potential wells in the region of two bumps of the initial wave function $\Psi_2(x)$ and by three barriers near its knots. These barriers are needed to keep all other levels $E_{n\neq 2}$ at their previous positions. The arrows point to changes of the energy E_2, potential $\Delta V(x)$, and function $\Psi_2(x)$. (b) Changes of function $\Psi_2(x)$ as E_2 approaches the first level. (c) Changes of the potential. (Adapted from Ref. [1] with permission.)

arbitrarily close to the second level. All we need for this is to first appropriately increase the "bottom bulge" at the central part of the box – this will push all the levels higher. Then we need to deepen the peripheral valleys to compensate for the excessive push of the central bulge. This will increase the pull down on all the levels higher than the ground level. So there are no restrictions on manipulating a single selected level, and this unrestricted manipulation can be performed without affecting all the rest.

What happens if the ground level is pushed up so high that it merges with the second energy level? The answer depends on the behavior of the two respective eigenstates. Let us denote the two changed eigenstates in question as

$$\psi_1(x) \to \psi_1(x) + \delta\psi_1(x) \equiv \tilde{\psi}_1(x) \quad \text{and}$$
$$\psi_2(x) \to \psi_2(x) + \delta\psi_2(x) \equiv \tilde{\psi}_2(x). \tag{20.3}$$

If the changes are such that the resulting new eigenfunctions turn out to be equal to each other, $\tilde{\psi}_1(x) = \tilde{\psi}_2(x)$, we will have two initially different states completely merged into one. A single new state instead of the initial two is equivalent to removing one state from the initial set, since the total number of distinct states has been reduced by 1. One could think that this is not the big deal if the total number of states is infinite. But even so, it is a remarkable phenomenon in its own right, showing that we can change the systems given by Nature and know the result of such change in advance.

If the new eigenfunctions remain different, $\tilde{\psi}_1(x) \neq \tilde{\psi}_2(x)$, we will have two different eigenstates with the same eigenvalue. This would be perhaps even more interesting result, meaning that we can create a degenerate state from the two nondegenerate states.

The list of possibilities can be extended further. For instance, the same algorithm for shifting of levels in a box (states with the finite energy differences that increase as we go to the higher energies) can be applied to the control of spectral zone positioning in periodic systems [9]. For this, the corresponding potential perturbation must be periodically continued from the given interval. As we know from Sections 9.8 and 9.9, each energy level of the discrete spectrum of a system converts into an energy band (allowed spectral zone) after the periodic reiteration of the potential. The shifting of one such level before reiteration is followed by the shifting of the corresponding zone generated from it after reiteration. In this way, we can join the two neighboring allowed zones, which means the disappearance of one spectral gap (forbidden zone) between them. Thus, it is in principle possible to apply the algorithms of discrete spectrum control to manipulate the states in periodic structures.

The described algorithm can be generalized to E-shifts in the complex energy plane. This can transform the initial stationary state into decaying one! But surprisingly, the resulting states can be square-integrable, in contrast with the Gamov's decaying states [10]. The reader can find many other interesting effects, including manipulations with decaying states, in Ref. [1]. We will discuss monitoring the decaying states from slightly different perspective in Section 20.3.

20.3
Playing with Evolution: Discussion

Here, we discuss another problem – how to play with *evolution* of states rather than with states as such. We consider some states as initially given and having a known natural way to evolve, and then figure out how this way could be changed under an external influence or our clever intervention.

Our first example will be a radioactive atom. Here, the word "radioactive" will mean an atom ready to decay by emitting a photon from an excited state. We know that the so-called stationary states deserve their name only for the ground state of a system. All the rest are nonstationary, and each of them is characterized by its respective average lifetime τ or, equivalently, by the "energy level width" Γ (determining the corresponding decay constant $\gamma = \Gamma/\hbar$).

Consider one such level excited from the ground state. We will, for the sake of simplicity, focus on this excited level alone, ignoring the rest of them. In this respect, we will deal again with a two-state system – one ground state and the other excited state. Let ω be the frequency of the corresponding optical transition, and $\Delta\omega$ the corresponding linewidth. We assume that $\Delta\omega \ll \omega$. Suppose that the atom has been prepared in the excited state (by being hit in a collision or by having absorbed an appropriate photon etc.). Then, we can describe its natural evolution by the law of radioactive decay discussed in Chapter 11.

We already know some ways to change the behavior of the excited atom. One of them is to illuminate it by a pure ensemble of photons all in one state with frequency ω tuned to the atomic transition frequency. This will encourage the atom to stimulated emission, which can have much higher rate than normal (spontaneous) decay. Accordingly, the state's lifetime will dramatically decrease relative to its natural value, due to an expedite emission of the photon. The emitted photon will have exactly the same frequency as the incident ones. If you give it a careful thought, this is a truly remarkable phenomenon. It seems to contradict the indeterminacy relationship for time and energy. Indeed, the reduced lifetime must lead to the increased linewidth according to

$$\Delta\omega \Delta t \geq 1 \quad \text{and} \quad \Delta t \sim \tau. \tag{20.4}$$

Here, Δt is indeterminacy in the moment of emission, and for the decaying states it is numerically close to τ, so (20.4) can be written as

$$\tau \Delta\omega \geq 1. \tag{20.5}$$

However, in the discussed situation, we have the decrease of τ without the corresponding increase of $\Delta\omega$, which seems to contradict (20.5). Indeed, according to (18.17), increasing the flux intensity (the photon number N) increases the emission rate, which in turn reduces the atom's lifetime in the excited state. So τ can in principle be made arbitrarily small for an atom in sufficiently intense laser beam. And this is not accompanied by increase of $\Delta\omega$. Moreover, the amplified component of the laser beam can be more monochromatic than the input photon(s) since the atom prefers to radiate at the frequency of *spectral maximum* of the given transition. As a result, the product $\tau\Delta\omega$ can be made arbitrarily small, thus blatantly violating the sacred rule (20.5)!

But this is only an apparent contradiction, because the reduction of τ in stimulated emission is only an apparent reduction. The true nature of this characteristic is much more subtle, and we need a closer look at the process to get a deeper insight into it.

We can formulate the change in our view of τ in the following way. In the flux of photons forcing the atom to the immediate emission, the atom does acquiesce and emits the required photon ahead of its average schedule. This photon is identical to those that provoked the emission, and its release is what we call the stimulated emission. But this does not preclude the atom from following its natural tendency to emit a photon on its own (probabilistic!) schedule – may be only once in a long while. And this is what it does – once in a long while: After being returned many times to the excited state by absorption, it does eventually (with unchanged τ!) emit the photon spontaneously, with the unchanged $\Delta\omega$, with indeterminate phase and generally not to the bunch of photons flying by, but in some random direction. We usually consider this as a nuisance preventing us from amplifying the initial light signal with 100% accuracy. Such goal would mean an *ideal multiple cloning* of each photon from the input signal. This goal is unachievable precisely because we cannot violate relationship (20.5) – the spontaneous emission remains as an irreducible background decreed by this relationship. It cannot be eliminated in the process of cloning.

This example illustrates the subtlety of relationship (20.5). Its interpretation is relatively straightforward for a spontaneous emission of an *isolated* atom. It is less straightforward when the atom is exposed to a flux of other photons. In this case, thorough scrutiny is required in order to discern spontaneous emission that remains unaffected, albeit less noticeable on the background of increased stimulated emission. Actually, we have to introduce two different lifetimes: τ determined with respect to spontaneous emission and τ_{st} determined with respect to stimulated emission. And both of them are characteristics of the same excited state. One of them is intrinsic and the other depends also on the photon population nearby.[1] We can also consider the *average* of the two and denote it as $\bar{\tau}$.

Thus, stimulated emission can be characterized by its own time indeterminacy. In a *pulse* laser, a huge amount of atoms in the preliminary prepared excited state cooperatively radiate in a highly coordinated way with dramatically decreased *average* lifetime $\bar{\tau}$ and accordingly *increased* $\Delta\omega$. In a *monochromatic* laser, $\bar{\tau}$ and Δt are no longer equal to each other! The average lifetime $\bar{\tau}$ decreases with respect to τ, but the time indeterminacy increases because of multiple reiteration of the emission–absorption process. In a way, the time indeterminacy Δt is now analogous to dispersed indeterminacy Δx in a grating described in Section 13.7, when Δx can be arbitrarily large despite the fact that each individual slit is very narrow. So $\bar{\tau}$ and Δt are now different characteristics of the phenomenon, and we need to return to the original indeterminacy relation (20.4) if we want to estimate $\Delta\omega$. In this process, Δt becomes equal to the *total emission time* T_L of the laser, $\Delta t \approx T_L$. This time may last hours, during which period each atom radiates at a highly increased rate, performing the trips between the excited and the ground states much more frequently than it does when left to its own devices. Billions of billions of photons can be radiated during the time T_L. But since they are all totally identical and the conditions preclude from measuring the emission moment of each of them, this moment becomes undefined up to within the total period of working of the laser. In other words, the indistinguishability of the photons makes the emission time of each single photon practically undefined. One could argue that we could determine this time by recording the photon's arrival time in a detector. Knowing the detector's distance from laser would allow us to find the moment of emission. But this argument does not work. First, the time of arrival is not so easy to define, to begin

---

[1] We will see later that even the intrinsic characteristic can be changed by changing the distant environment.

with [11–13]. Second and more important, there is no way to tell whether the recorded photon is the one actually emitted by the atom within the laser or the one that passed through the laser. Third, there is no way to tell how many rounds the detected photon had made within the laser before exiting it. This situation is analogous to the multiple beam interference, when the detection of a photon at a certain position on the screen does not tell from which slit in the diffraction grating the photon arrived at this position. We conclude again that in this respect, the discussed time indeterminacy is another example of the dispersed indeterminacy considered in Section 13.7. In this analogy, τ_{St} determines the average period of the emission–absorption process and plays the role of the distance d between the neighboring slits in the grating. The net indeterminacy Δt is close to the duration of the whole process and plays the role of the total size of the grating.

We can also look at this from a different angle. Even if we focus only on a single atom A and a single photon, their interaction during the whole process can be considered as an entangled superposition of the atom in its excited state $|\tilde{A}\rangle$ in the field of N other photons and the same atom in its ground state $|A\rangle$ in the field of $N+1$ photons (N previous plus the emitted one):

$$|\Psi(A, N)\rangle = c_N |\tilde{A}\rangle |N\rangle + c_{N+1} |A\rangle |N+1\rangle. \qquad (20.6)$$

Here, c_N, c_{N+1} are the superposition amplitudes, which can be functions of time. Generally, both of them are nonzero for each given moment. On this simplified level, we come to the same conclusion as before: The moment of emission is totally undetermined within the duration of the whole process of interaction of the atom with the EM field. Thus, paradoxically, and totally counterintuitive, the increased emission rate does not necessarily decrease Δt. Hence the quasimonochromatic radiation with decreased $\Delta \omega$ in total compliance with (20.4).

Another change in the process of decay is in the direction of emission. In the natural (spontaneous) emission, the direction of the emitted photon is largely arbitrary. For instance, if the transition $n \to m$ is characterized by a vector matrix element $\mathbf{p}_{mn}$ of the effective dipole moment, the possible directions of emitted photon are described by a vector function $\mathbf{n}(\theta, \varphi)$, where θ is the polar angle and φ is the azimuthal angle with respect to $\mathbf{p}_{mn}$. The dipole radiation is axially symmetric about $\mathbf{p}_{mn}$, so that for each given θ there is an equal chance to find the photon emitted in any direction φ. In contrast, within a working laser, the direction of stimulated emission is to a high degree predetermined by the direction of incident flux (recall Section 18.2 and Figure 18.1). So when the excited atom is subjected to a collimated flux of identical photons of the resonant frequency ω, the flux affects not only the emission rate but also the preferable direction of emission.

It is important to note that all these possibilities to change the evolution are not restricted to the photon emission alone. Consider, for instance, again the situation with the α-decay discussed in Chapter 11. This process is extremely slow. But we can, at least in principle, boost it by using the same approach as used in lasers. We can illuminate the sample of U_{92}^{235} by a beam of α-particles with the same energy as that of an α-particle emitted in the process of decay. Since α-particles are bosons, the presence of other α-particles must increase the emission probability. This would reduce the corresponding lifetime of the nucleus – without any change in its internal structure. In principle, one could even envision a *nuclear alpha laser* based on this effect. In such a laser, a sample of U irradiated by a collimated beam of α-particles, would act as a system with inverse population, and we could observe a stimulated emission of α-particles, amplifying the initial beam.

Of course, this is only an idea, and there is a long way to its technical realization.

There are many technical difficulties that today seem unsurpassable. For instance, the amplification rate is much less than in case of photons. It is very difficult to obtain a sufficiently intensive beam of α-particles. And it seems extremely difficult, if not impossible, to keep all α-particles staying in the beam – such beam would immediately disperse due to the electric repulsion forces between its particles.

But these problems are only technical – the effect itself is not forbidden by laws of Nature as we know them today. What is not forbidden is allowed. What is allowed may happen sooner or later.

Let us now get back to photons and reverse the argument. If it is possible to stimulate the emission, is it possible to suppress it as well? The answer to this question is yes [14]. In the above-considered case, the way to enhance evolution was to enhance the chances of photon emission. Similarly, the way to suppress evolution is to suppress the chance of photon emission. In order to see how this can be done, consider another way to affect the photon emission, without hitting the excited atom with a photon beam.

We can place an excited atom into a cavity with reflecting walls.[2] The cavity can dramatically change the process of emission. The effect can be best described (and understood) in the language of standing waves (harmonics) in a cavity or the cavity modes. The shape of the cavity may be favorable for formation of standing waves with given frequency emitted by the atom. We call such a cavity an optical resonator. Consider such resonator tuned to a transition frequency $\omega = \omega_{mn}$ of a given atom. Let it be a spherical resonator of radius R with the excited atom at its center. The reflecting walls will keep the emitted photon well inside. But in order for the emission to happen, the photon's wave function must be a solution of the wave equation with corresponding boundary conditions. For the ideally reflected walls, these conditions require the wave function to have a nodal surface at the wall, that is,

$$\Psi(r,\theta,\varphi) = \mathcal{R}(r)Y(\theta,\varphi) \underset{r=R}{=} 0. \tag{20.7}$$

The result will be a set of standing waves with the simplest radial function of the form

$$\mathcal{R}(r) = \frac{\sin k_R^{(j)} r}{r}, \quad k_R^{(j)} = \frac{\pi}{R} j, \quad j = 1, 2, \ldots. \tag{20.8}$$

---

[2] A cavity harboring an atom constitutes the potential well. Therefore, the atom's Hamiltonian will change by an additional term $W(\mathbf{r})$ describing this well. This is equivalent to changing the potential considered in Section 20.2. The atomic eigenstates and their energy eigenvalues will also accordingly change. But if the cavity greatly exceeds the size of the atom, such change is negligibly small.

Combining this with $k_R^{(j)} = \omega_{mn}/c = 2\pi/\lambda_{mn}$ gives the "tuning condition" for R:

$$R \Rightarrow R_j^{(mn)} = \pi \frac{c}{\omega_{mn}} j = \frac{\lambda_{mn}}{2} j. \tag{20.9}$$

If this condition is satisfied, the photon's wave function is in the list of allowed modes, and its emission is allowed. But the emission rate will now be a function of time, depending also on size of the cavity. The average lifetime of the excited atom in the tuned cavity will also change. The whole process becomes so subtle that sometimes we cannot even say whether the emission is spontaneous or stimulated!

Somehow we managed to change the intrinsic property of the atom that determines the evolution of its excited state. And we did it without any invasion into the atom itself. Changing the boundary condition for the photon wave function far away from the atom is sufficient to do the job.

How does the atom know about the new conditions affecting its evolution, when the affecting factor is so far from the atom? In principle, the atom can feel the presence of the cavity through its "electron cloud," which extends arbitrarily far from the nucleus. But the probability density of this cloud is exponentially small beyond the Bohr radius, and the cavity wall, while being reflecting for the photon, might be highly penetrable for an electron. So the electron is of little, if any, help. Then what mysterious agent changes the atom's way of photon emission? It is the photon itself! The whole picture may be very complicated, and here we only outline qualitatively how it works. Due to indeterminacy of the emission time, at any moment after the creation of the excited atomic state and before the actual detection of the emitted photon, the system is in an entangled superposition of the two states: the excited atom and no photons around, and the atom in the ground state and the emitted photon:

$$|\Psi(A,\gamma)\rangle = c_0(R,t)|\tilde{A}\rangle|0\rangle + c_1(R,t)|A\rangle|1\rangle. \tag{20.10}$$

This is a special case of (20.6) when there is only one photon, and now we explicitly showed the dependence of the superposition amplitudes on R and t. It is the virtual photon in the second term of the superposition that feels the shape of the cavity! Actually, it does not feel it due to the boundary condition (20.7), but this is precisely what it needs in order to materialize in its full fledge. It eventually does it, if j in (20.9) is sufficiently large. One can easily find the necessary low bound for j in terms of ω_{mn} and τ (Problem 20.6). Then the same boundary condition (total reflection!) converts the photon wave diverging from the atom into one converging to it. This brings back the possibility of its absorption by the atom with subsequent reemission. The process can reiterate, in which case $c_0(R,t)$ and $c_1(R,t)$ are periodic functions of time with the period $T_R = 2(R/c)$ and phase shifted by $\pi/2$ with respect to each other (the latter is due to normalization condition that is reinstated for a system within a cavity). Thus, the decay of a Gamov-type state is converted into succession of decays and self-assembles.

Another interesting case is associated with low j. In this case, the leading edge of the photon's wave packet after reflection converges back onto atom before the trailing edge "gets released." Such situation can be described best as a quasi-stationary state with a standing wave (20.8). The system is suspended in a limbo between emission and absorption, and in this case the question whether the emission is spontaneous or stimulated loses its meaning. From observer's viewpoint, this is neither emission nor absorption, the system's evolution is slowed down nearly to a stop, and at any moment there is a 50/50 chance to find either the excited atom with no photon around or the atom in ground state with the quasi-free photon in the cavity. Such system is even farther away from the Gamov state than in the previous case.

Now, what if the cavity's radius is such that the cavity is *detuned* from the photon's frequency? This means that the photon's spherical wave originating from the center would have the crest at the wall:

$$R \Rightarrow R_j^{(mn)} = \pi \frac{c}{\omega_{mn}} \left(j + \frac{1}{2}\right) = \frac{\lambda_{mn}}{2}\left(j + \frac{1}{2}\right). \tag{20.11}$$

Such a cavity, no matter how big, would give no room for a stationary (monochromatic) photon state with frequency $\omega = \omega_{mn}$ [14,15]. For the Gamov states, which are quasi-stationary, the cavity gives no room if

$$R_j^{(mn)} = \frac{\lambda_{mn}}{2}\left(j + \frac{1}{2}\right) < \frac{1}{2}\mathcal{L}^{(mn)}, \tag{20.12}$$

where $\mathcal{L}^{(mn)}$ is the effective length of the photon wave packet for the given transition. Since $\mathcal{L}^{(mn)} = c\tau$, this gives the condition for j:

$$R_j^{(mn)} = \frac{\lambda_{mn}}{2}\left(j + \frac{1}{2}\right) < \frac{1}{2}c\tau \quad \text{or} \quad j + \frac{1}{2} < \frac{1}{2\pi}\omega_{mn}\tau. \tag{20.13}$$

Under condition (20.11) with j satisfying (20.13), the cavity cannot accommodate the photon. Physically, this means the following. Suppose we prepared an atom in the excited state. The atom is poised to release a photon, which would inevitably happen in a free space, although with a moment of emission predictable only probabilistically. Put this atom at the center of just described inhospitable cavity. Then the atom will remain forever in its excited state! The frequency of a would-be photon after its would-be emission is now not in the list of the allowed frequencies for this cavity. And the photon seems to know it *before* it gets a chance to materialize and grope the walls or measure the size of the cavity, that is, before it is created. As a result, it remains in the vacuum state $|0\rangle$ (is not created).

Thus, we can freeze and even stop the natural process of the light emission or of radioactive decay! This situation is in some respects similar to suppression of inelastic channels in the planar resonant scattering [16] mentioned in Section 19.2. Even though the conditions and physical mechanism here are quite different from those considered in Ref. [16], the final outcome is the same.

The described inhibition of spontaneous emission also reminds of the process of stopping the evolution of polarization in the experiments with optical activity described in Section 11.13. This suggests that we can look at the whole process from the viewpoint of the "watched pot" rule. According to this rule, continuous watch of the atom must freeze its evolution. In the case of an excited atom, the continuous watch will preclude its decay.

The question is how can we watch it continuously? By surrounding it with a conducting spherical surface satisfying condition (20.13)! High conductivity ensures maximal interaction strength between the photon and material of the wall. The photon cannot penetrate a superconductor because its energy would be immediately absorbed. It is a known optical phenomenon – high reflectivity is associated with high absorbance [17]. Condition (20.13) (wave crest at the cavity surface), if realized, would guarantee the maximal possible probability of absorption. Any act of absorption can be detected because it leaves a trace in the form of the local disturbance that can, in principle, be registered [16]. And *we are not required* to put any detectors over the cavity wall – the mere *possibility of detection* is sufficient, because the surface with described properties would inform Nature of any photon reaching this surface. In this respect, the surface is a perfect observer. Thus, the cavity with properties (20.13) amounts to watching the process of the photon emission anywhere and at any time. And precisely in this case, the photon emission will stop, and the atom's excited state, no matter how unstable, now would freeze forever. This can be considered as another manifestation of the rule about a pot that never boils when watched. In this situation, the impossibility to have the EM wave crest on an ideally conducting wall, and the watched pot rule can be considered as two sides of the same coin.

All considered examples are varieties of a beautiful play in which the photon plays hide and seek with the atom and with itself. And we can use it to play with evolution of quantum states on our own, the way we want. We can slow it down, or just stop it, and in some cases, conversely, we can accelerate it as described in Section 11.14. We can make the Nature's pots boil faster, slower, or freeze them.

We can ask again – how it works? For instance, what specific physical agent stops the evolution? And again, that would be an attempt to get the right answer to a wrong question. As we have emphasized before, there is no such agent, no local influence on the atom, the atomic structure is not affected by the cavity. The question is of the same nature as "How does an electron in the double-slit experiment know about the places on the screen where it cannot land?"

The answer to both questions comes from the wave aspect of matter, specifically, from superposition of states and indeterminacy. In case of an electron, the position indeterminacy associated with its wave aspect makes it feel the configuration of both slits; the information received from the slits gets imprinted into the split electron waves, and the electron reads it by interfering with itself. In case of the photon, the time indeterminacy arising from the quasi-stationary character of the state makes the distinction between the past and the future fuzzy and enables it to collect information from some future moments of its history. In more prosaic language, the information about distant walls of the cavity is imprinted into the amplitudes of superposition (20.10) and the system reads these imprints by interfering with itself. In the absence of the walls, both superposition amplitudes are nonzero, which results in the "regular" evolution of the Gamov state. With the reflecting spherical surface around, satisfying condition (20.13), the second term in superposition (20.10) vanishes, and the remaining term immediately freezes – it cannot change in solo due to unitarity. The whole effect can be considered as another manifestation of quantum nonlocality.

Problems

20.1 Prove that adding a constant to a potential in the Schrödinger equation will shift all the energy levels of the corresponding system by the same amount equal to the added constant.

20.2 Prove that any perturbation $W(x)$ not equal to a constant changes all state functions, that is, causes transition $\psi_n(x) \to \tilde{\psi}_n(x) = \psi_n(x) + u_n(x)$ with $u_n(x) \neq 0$

   a) for any state whose eigenvalue E_k remains intact;
   b) for a state $\psi_k(x)$ with changed eigenvalue $E_k \to \tilde{E}_k = E_k + \varepsilon_k$.

20.3 Find the expectation value of perturbation $W(x)$

   a) for each of the initial unperturbed states $\psi_n(x)$ with unaffected eigenvalue E_n;
   b) for an initial unperturbed state $\psi_k(x)$ with the changed eigenvalue $E_k \to \tilde{E}_k = E_k + \varepsilon_k$.

20.4 In the previous problem, find the expectation value of $W(x)$

   a) for each of the *perturbed* states $\tilde{\psi}_n(x) = \psi_n(x) + u_n(x)$ with unaffected eigenvalue E_n;
   b) for a *perturbed* state $\tilde{\psi}_k(x) = \psi_k(x) + u_k(x)$ with the changed eigenvalue $E_k \to \tilde{E}_k = E_k + \varepsilon_k$. Compare your answer with that in the previous problem and explain the results.

20.5 Consider an electron in a hydrogen atom. Suppose that the atom is put at the center of a spherical cavity of radius R whose walls are impenetrable for the electron.

   a) Sketch the potential function in the electron's Hamiltonian.
   b) Sketch the radial function of the ground state in the vicinity of R and compare it with the case without cavity.

20.6 Suppose you have an initially excited two-state atom at the center of a spherical cavity tuned to the atomic transition frequency. Estimate the minimal size of the cavity necessary for the onset of the emission–absorption cycles.

References

1 Zakhariev, B.N. and Chabanov, V.M. (2007) *Submissive Quantum Mechanics: New Status of the Theory in Inverse Problem Approach*, Nova Science Publishers Inc., New York.

2 Chabanov, V.M. and Zakhariev, B.N. (1999) Coexistence of confinement and propagating waves: a quantum paradox. *Phys. Lett. A*, **255**, 123–128.

3 Chabanov, V.M. and Zakhariev, B.N., (2006) Unprecedented possibility of discrete scattering states and their control. *Int. J. Mod. Phys.*, **20** (28), 4779–4784.

4 Chabanov, V.M., Zakhariev, B.N., and Sofianos, S.A. (1997) Universal elementary constituents of potential transformations. *Ann. Phys.*, **6**, 136.

5 Chabanov, V.M. and Zakhariev, B.N. (2006) Surprises of quantum wave motion through potentials. *Concept Phys.*, **3**, 169.

6 Gelfand, I.M. and Levitan, B.M. (1951) On the determination of a differential equation from its spectral function. *Sov. J. Dokl. Acad. Nauk USSR*, **77**; 557–60

7 Marchenko, V.A. (1955) On reconstruction of the potential energy from phases of the scattered waves. *Sov. Dokl. Acad. Nauk USSR*, **104**, 695–698.

8 Agranovich, V.M. and Marchenko, V.A. (1960) *The Inverse Problem in Scattering Theory*, Kharkov University, Kharkov (English edition: Gordon and Breach, New York, 1963).

9 Chabanov, V.M. and Zakhariev, B.N. (2004) To the theory of waves in periodic structures in inverse problem approach. *Int. Mod. Phys.*, **18** (30), 3941–3957.

10 Chabanov, V.M. and Zakhariev, B.N. (2001) Unusual (non-Gamow) decay states. *Inverse Probl.*, **17**, 1–11.

11 Chiao, R.Y., Kwiat, P.G., and Steinberg, A.M. (2000) Faster than light? *Sci. Am.*, Special Issue, 98–106.

12 Leavens, C.R. (1998) Time of arrival in quantum and Bohmian mechanics. *Phys. Rev. A*, **58**, 840–847.

13 Anastopoulos, C. and Savvidou, N. (2006) Time-of-arrival probabilities and quantum measurements. *J. Math. Phys.*, **47**, 122106.

14 Kleppner, D. and Haroche, S. (1989) Cavity Quantum Electrodynamics. *Phys. Today*, **42**, 24.

15 Gabrielse, G. and Dehmelt, H. (1985) Observation of inhibited spontaneous emission. *Phys. Rev. Lett.*, **55** (1), 67–70.

16 Fayngold, M. (1985) Resonant scattering and suppression of inelastic channels in a two-dimensional crystal. *Sov. Phys. JETP*, **62** (2), 408–414.

17 Feynman, R., Leighton, R., and Sands, M. (19664) *The Feynman Lectures on Physics*, vols **1–3**, Addison-Wesley, Reading, MA.

21
Quantum Statistics

21.1
Bosons and Fermions: The Exclusion Principle

> To describe systems of identical particles, one can begin with enumerated distinguishable particles and then remove their distinguishability.
>
> *Jurgen Audretsch*

Suppose we prepare two boxes, labeled 1 and 2, separated by an infinitely high potential barrier, and we put particle A into the first box and an identical particle B into the second box. The composite state of this particle pair is described by the product

$$\Psi_{AB}(q_1, q_2) = \psi_A(q_1)\psi_B(q_2). \tag{21.1}$$

Here, q_1, q_2 refer to coordinates of the corresponding particles. An hour later we open box 1, and sure enough, see one particle there. Since the barrier is impenetrable, we can be sure the particle we found is the same particle that was put there an hour ago. The same holds for box 2. This means we can always track either particle. In other words, the particles are *distinguishable*. In classical statistical mechanics, all objects are thought to be distinguishable, at least in principle, and they obey Maxwell–Boltzmann statistics. We should expect analogous behavior for distinguishable quantum particles. For example, the average occupancy numbers of energy levels would depend on the energy as $\langle n_i \rangle \sim e^{-\hbar\omega/k_B T}$ as long as particles remain trackable.

Next, we reduce the height of the barrier. There appears a small transition amplitude between the boxes. This time we can only *suppose* that the particle discovered in box 1 is most likely A, but there is a small chance that each particle has tunneled into the opposite box and we are in fact looking at particle B. A physical particle carries no label except for its quantum numbers, and if they are the same for A and B, all we can say is that our detector has registered *a* particle, but not *the* particle. Since all detector readings are the same in either event, both possibilities must describe the *same* physical state. We must then write this state as a superposition (introducing a relative phase φ that is yet to be determined):

$$\Psi_{AB}(q_1, q_2) = \alpha\psi_A(q_1)\psi_B(q_2) + \beta e^{i\varphi}\psi_B(q_1)\psi_A(q_2). \tag{21.2}$$

Quantum Mechanics and Quantum Information: A Guide through the Quantum World,
First Edition. Moses Fayngold and Vadim Fayngold.
© 2013 Wiley-VCH Verlag GmbH & Co. KGaA. Published 2013 by Wiley-VCH Verlag GmbH & Co. KGaA.

One common although potentially misleading picture of this state is the two wave packets with very small spatial overlapping. The reason this picture should be viewed skeptically is that once we said that packets overlap, we cannot really talk of "two wave functions." The only wave function that is present there is that of the composite system $\Psi_{AB}(q_1, q_2)$. The form of Equation 21.2 is already familiar to us and it looks very much like an entangled state. Since it is not a product state and cannot be factorized, the particles are correlated. Then it only makes sense to talk about the *joint* amplitude $\Psi_{AB}(q_1, q_2)$ of detecting two particles at locations q_1 and q_2. Despite its resemblance to entangled states, we *do not* call such a state entangled, because we did not prepare our particle pair in a state with, say, zero net spin in order to cause correlated behavior. Whatever correlations will be found will be due to basic quantum properties that are common for all multiparticle systems. We, on our part, have washed our hands by preparing the particles independent of each other and suppressing any possible interactions between them in the preparation stage.

With all these reservations, the picture of overlapping wave packets is a good way to visualize the problem and it will not cause any substantial errors as long as $\alpha \gg \beta$ so that we can still act as if the particles were fully distinguishable. The upshot: we won't have to write a superposition such as (21.2) for every two electrons in the Universe. We only need to worry about those pairs for which $\alpha \simeq \beta$ so that our picture will show substantial overlapping.

Box 21.1 Thermal de Broglie wavelength

One asks, how can we tell when the crude model (21.1) is good enough, and when we must use the more refined model (21.2). One easy criterion is this: In an ideal gas (i.e., in the absence of interparticle forces) of N particles of mass μ, held in volume V at temperature T, in a thermal equilibrium, we may neglect correlations if the average distance between particles is greater than the average de Broglie wavelength of a particle.

The latter is derived in Statistical Mechanics as

$$\lambda_{th} = \begin{cases} \dfrac{h}{\sqrt{2\pi\mu k_B T}}, & \mu \neq 0, \\[1em] \dfrac{ch}{2\pi^{1/3} k_B T}, & \mu = 0, \end{cases}$$

where c is the speed of light and k_B is the Boltzmann constant and is called *thermal de Broglie wavelength*. (As a side note, derivation of this result takes more than applying $\lambda = h/p = h/\sqrt{(2\mu E)}$ where $E = (3/2)k_B T$. One must consider the statistics of Fermi gas instead, or if one wants a good classical approximation, Maxwell speed distribution of gas molecules will yield the correct formula, up to a factor of 2). The average interparticle distance is $d = (V/N)^{1/3}$ and then the system can be treated as a Maxwell–Boltzmann gas as long as $d \gg \lambda_{th}$.

Now we take down the barrier, allowing the wave packets in the picture to overlap completely. This time when we find a particle in box 1, there is no way of knowing that we are still looking at A. Indeed, if A decided to go to box 2 while B went into box 1, our detector would not notice any difference. And, with the barrier removed, there is nothing to prevent such journeys between the boxes; in fact, there is no longer any reason for A to prefer box 1 over box 2. We must conclude that when *all states in the Hilbert space of A are also equally accessible to B* and vice versa, there is a 50% probability that we are looking at the system with the particles swapped. Our task then is to write the function (21.2) of two variables for this case.

We will now take the general case and allow q_1, q_2 to have a continuous range. Furthermore, we no longer think of them in terms of position, but rather make them stand for the complete sets of generalized coordinates for the corresponding particles. The two terms in (21.2) represent the cases $(q_1 = 1, q_2 = 2)$ and $(q_2 = 1, q_1 = 2)$. These two points are symmetric relative to the bisector $q_1 = q_2$ of the first quadrant in the (q_1, q_2) plane. We are now going to claim that the complete function $\Psi_{AB}(q_1, q_2)$ must have the same magnitude at points (q_1, q_2) and (q_2, q_1). If this were not true, two possibilities – (A at q_1 and B at q_2) and (B at q_1 and A at q_2) would have different likelihood; this cannot be possible in a setup when the entire state space is equally accessible to both particles, with no particle being in a privileged situation with respect to its colleague. Mathematically, this requirement reads

$$\Psi_{AB}(q_1, q_2) = \Psi_{AB}(q_2, q_1), \tag{21.3}$$

where we have adapted the convention that arguments to the function are entered in the same order (coordinate of A goes first, and coordinate of B goes second). Equivalently, we could have left parenthesis the same and changed the order of terms on the right-hand side:

$$\Psi_{AB}(q_1, q_2) = \Psi_{BA}(q_1, q_2). \tag{21.4}$$

That would amount to swapping particle labels, A⇄B. If we use the latter form, Equation 21.2 leads us to

$$\alpha \psi_A(q_1) \psi_B(q_2) + \beta e^{i\varphi} \psi_B(q_1) \psi_A(q_2) = \alpha \psi_B(q_1) \psi_A(q_2) + \beta e^{i\varphi} \psi_A(q_1) \psi_B(q_2) \tag{21.5}$$

This means $\alpha = \beta e^{i\varphi}$, in other words, the amplitudes of two equivalent detection events are equal, up to a phase factor.

Mathematics also tells us that we can restore the original function by switching the labels again. Formally, we can introduce an *exchange operator* $\hat{\mathcal{E}}$ whose defining property is to swap the labels in the complex input function, making no other changes to its *magnitude*. Also, applying this operator twice should be equivalent to just leaving the function unchanged in both magnitude *and* phase (that last requirement is necessary in order for the function to remain single-valued). This means that $\hat{\mathcal{E}}^2$ is the identity operator, $\hat{\mathcal{E}}^2 = I$, with the single eigenvalue 1, and *any function is its eigenfunction*; for operator $\hat{\mathcal{E}}$ itself, the equivalent statement is *generally* false. But the symmetry requirements (21.3 and 21.4) then say that the corresponding composite wave function of the type (21.2) *must be an eigenfunction of* $\hat{\mathcal{E}}$. However, an operator $\hat{\mathcal{E}}$ that satisfies these requirements could also introduce a constant phase factor $e^{i\phi}$:

$$\hat{\mathcal{E}}\Psi_{AB}(q_1,q_2) = e^{i\phi}\Psi_{BA}(q_1,q_2), \tag{21.6}$$

$$\hat{\mathcal{E}}^2\Psi_{AB}(q_1,q_2) = e^{2i\phi}\Psi_{AB}(q_1,q_2). \tag{21.7}$$

The phase factor would have no effect on the *magnitude* of the wave function; however, Equation 21.7 clearly shows that there must hold

$$e^{2i\phi} = 1 \Rightarrow \{\phi = 0 \quad \text{or} \quad \phi = \pi\}. \tag{21.8}$$

In other words, $\hat{\mathcal{E}}$ has eigenvalues 1 or -1. The corresponding eigenfunctions are, respectively, symmetric (they are not affected by the exchange operator) and antisymmetric (exchange operator reverses their sign). In this respect, operator $\hat{\mathcal{E}}$ in q-space acts similar to the inversion or parity operator in **r**-space (see Figure 23.1).

Thus, all particles fall into two categories. Particles with "antisymmetric" behavior ($\phi = \pi$) are the Fermi particles (*fermions*), and particles with "symmetric" behavior ($\phi = \pi$) are the Bose particles (*bosons*). As mentioned in Section 4.5, the particle type

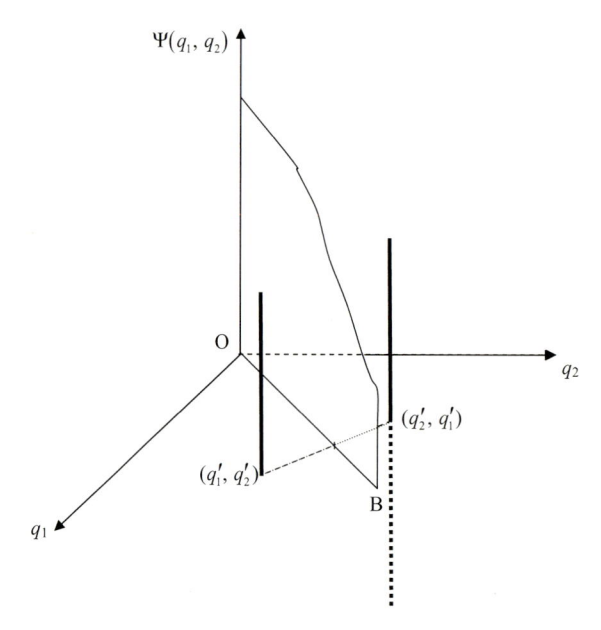

Figure 21.1 To the definition of exchange operation. An eigenfunction of the exchange operator $\hat{\mathcal{E}}(q_1, q_2)$, while having an arbitrary shape along bisector OB, must be symmetric or antisymmetric relative to the plane passing through OB and the vertical axis. Shown in the figure is the exchange operation at an arbitrarily chosen point (q'_1, q'_2) of the plane $q_1 O q_2$. The pair of solid lines (| |) at points (q'_1, q'_2) and (q'_2, q'_1) represents the symmetric function $\Psi(q_1, q_2)$ at this point; the pair $\begin{pmatrix} | \\ \vdots \end{pmatrix}$ represents the antisymmetric function.

(boson or fermion) is determined by its spin. This fact, known as *spin-statistics theorem*, was derived during 1939–1940 from relativistic considerations by M. Fierz and then by W. Pauli. Here we can only state the rule without derivation (the reader can find more details in Ref. [1]). All particles with half-integer spin are fermions (e. g., electrons, protons, neutrons, muons, neutrinos, and their antiparticles). The particles with integer spin are bosons (e.g., photons, π and K mesons, and their antiparticles). In particular, a spinless particle ($s = 0$) is also a boson.

It is straightforward, by requiring that the state (21.2) be an eigenstate of the exchange operator and setting $\alpha = \beta = 1/\sqrt{2}$, to see that phase φ in that equation should be identified with ϕ. Then, using the "+" and "−" subscripts for bosons and fermions,[1]

$$\Psi_+(q_1, q_2) = \frac{\psi_A(q_1)\psi_B(q_2) + \psi_B(q_1)\psi_A(q_2)}{\sqrt{2}},$$
$$\Psi_-(q_1, q_2) = \frac{\psi_A(q_1)\psi_B(q_2) - \psi_B(q_1)\psi_A(q_2)}{\sqrt{2}}.$$
(21.9)

Of course, to say that Ψ_+, Ψ_- are symmetric or antisymmetric is not to say that they must be composed of "building blocks" such as $\psi_A\psi_B$ and $\psi_B\psi_A$. This could only be said when the particles were separated by an impenetrable barrier. However, there is a good physical reason to consider these products even after the barrier was removed. In the absence of potential $V(q_1, \ldots, q_N)$ that depends on coordinates of more than one particle, the Schrödinger equation for the system will have a solution in the form

$$\Psi(q_1, q_2, \ldots, q_N) = \psi_1(q_1)\psi_2(q_2) \cdots \psi_n(q_n).$$
(21.10)

This is the so-called *Hartree product*. A permutation, defined as the Hartree product with some of its particle indices exchanged, is obviously also the solution. If satisfying the Schrödinger equation were the only requirement for the system, our analysis could stop right there. However, products $\psi_1(q_1)\psi_2(q_2) \cdots \psi_N(q_N)$ and their permutations are generally *not* eigenstates of $\hat{E}$ and therefore are not legitimate physical states. One obvious exception occurs when the states are equal, $\psi_1 = \psi_2 = \cdots = \psi_N$. But in all other cases, after establishing that $\psi_1, \psi_2, \ldots, \psi_N$ would be valid one-particle solutions that also satisfy initial conditions, we must next check whether their product is symmetric under all possible exchanges $\hat{\mathcal{E}}(q_i, q_j)$ where (i, j) run over all possible pairs of indices. If complete exchange symmetry is lacking, we must then run the symmetrization procedure that generalizes (21.9) for the N-particle case.[2] Since any superposition of solutions to the Schrödinger equation is also a solution, this will give us a legitimate N-particle state.

---

1) It is essential to remember that this procedure is only intended to derive the *form* of the wave function of the composite state. Its only goal was to show why this wave function should be symmetric or antisymmetric. The final notation for states (21.9) will be different. In particular, the labels must be removed (see also Boxes 21.2 and 21.3).
2) As a side note, any function of two variables has a *canonical decomposition* in terms of symmetric and antisymmetric functions: $\Psi(q_1, q_2) = (1/\sqrt{2})(\Psi_+(q_1, q_2) + \Psi_-(q_1, q_2))$.

The question then is how to symmetrize an arbitrary Hartree product. Here and below, we are going to label the particles as if they were distinguishable and then remove the labels. For fermions, relabeling A, B as 1, 2, it is convenient to write (21.9) in the form[3] $\Psi_-(q_1, q_2) = \frac{1}{\sqrt{2}} \det \begin{pmatrix} \psi_1(q_1) & \psi_2(q_1) \\ \psi_1(q_2) & \psi_2(q_2) \end{pmatrix}$. Although the advantages of introducing it are not obvious for a two-particle system, they become clear as we generalize it for systems with $N > 2$ particles. The antisymmetric expression for N particles is elegantly written as

$$\Psi_-(q_1, q_2, \ldots, q_N) = \frac{1}{\sqrt{N!}} \det \begin{pmatrix} \psi_1(q_1) \psi_2(q_1) \cdots \psi_N(q_1) \\ \psi_1(q_2) \psi_2(q_2) \cdots \psi_N(q_2) \\ \vdots \\ \psi_1(q_N) \psi_2(q_N) \cdots \psi_N(q_N) \end{pmatrix}. \quad (21.11)$$

The right-hand side of this equation is called the *Slater determinant*. We can justify this empirical formula as follows. For a system of N identical fermions, the wave function must be a superposition of all products $\Psi(q_1, \ldots, q_N) = \psi_1(q_1) \cdots \psi_N(q_N)$ obtained from one another by exchange operations $\hat{\mathcal{E}}(i,j)$ on any pair of indices i, j. Every such operation amounts to an index permutation with a change of sign. The number of terms in the sum is the number of distinct permutations that is equal to $N!$, so we need the normalizing coefficient $1/\sqrt{N!}$. By the well-known linear algebra theorem,[4] the sum of all possible permutations, with the sign determined by whether the particular permutation was obtained by an even or odd number of index swaps, produces the determinant (21.11).

To take an example, let us construct an antisymmetric function for the product $\Psi(q_1, q_2, q_3) = \psi_1(q_1) \psi_2(q_2) \psi_3(q_3)$. There are $3! = 6$ ways to rearrange three indices, so the new function must contain six terms: $\Psi_-(q_1, q_2, q_3) = (1/\sqrt{6}) \{\psi_1(q_1)\psi_2(q_2)\psi_3(q_3) - \psi_1(q_1)\psi_3(q_2)\psi_2(q_3) + \psi_2(q_1)\psi_3(q_2)\psi_1(q_3) - \psi_2(q_1)\psi_1(q_2)\psi_3(q_3) + \psi_3(q_1)\psi_1(q_2)\psi_2(q_3) - \psi_3(q_1)\psi_2(q_2)\psi_1(q_3)\}$.
The reader can check that this expression agrees with (21.11).

Consider two examples for a system of two spin 1/2 particles. To simplify, assume that our set q can be restricted to a single variable: spin projection σ_z. Suppose the spin wave functions ψ_1 and ψ_2 are different (particles are prepared in different spin states). We immediately see that the amplitude of two particles collapsing into the

---

3) It is important to emphasize that our matrix was two-dimensional only because we considered the case $N = 2$ (two particles), *not* because their spin projection is double-valued. We would still have a 2D matrix if the particles were π-mesons, whose spin projection onto an axis has three discrete values (zero included), or kaons (spin 3/2 particles). It is the number of particles, not the domain of variable q that determines the number of *selected* values of q (one value for each particle) and thereby dimensionality of the matrix.

4) This theorem states that if $\varepsilon_{ik\cdots m}$ is 1 or -1 depending on whether the number of "disorders" in a permutation is even or odd (where "disorder" means that a greater index precedes a smaller index), then $\det \begin{pmatrix} a_{11} & a_{12} \\ a_{21} & a_{22} \\ \vdots & \end{pmatrix} = \sum_{(ik\cdots m)} \varepsilon_{ik\cdots m} a_{1i} a_{2k} \cdots a_{nm}$.

same spin state (spin-up, say) is zero because then both variables have the same value $q_1 = q_2 = 1$ and expression (21.11) vanishes. The same holds for $q_1 = q_2 = -1$. Even more generally, the answer is zero for all $q_1 = q_2$. This result is evident from Figure 21.1 for the antisymmetric case (for fermions, $\psi(q_1,q_2) = 0$ on line OB). Linear algebra gives us the same conclusion: When all q's are the same, the rows in the Slater determinant become equal, and the determinant vanishes.

Suppose now that $q_1 \neq q_2$, but instead we prepared the particles in the same spin state $\psi_1 = \psi_2$. (That is, $\psi_1 = \psi_2$ is not an eigenstate as otherwise q_1, q_2 would be equal. One example is the spin-left state, $\psi_1 = \psi_2 = (1/\sqrt{2})(|\uparrow\rangle - i|\downarrow\rangle)$, that can collapse into either spin-up or spin-down state, so a pair of results $1,-1$ is possible.) If in the previous case we tried to make electrons *collapse* into the same eigenstate of q, this time we are trying to *prepare* the same state without enforcing the future measurement outcome. Still, the verdict does not change: This cannot be done, this time because two *columns* in the Slater determinant become equal to each other. We came by a different route to the same conclusion: Two fermions cannot occupy the same state. This is the famous *Pauli exclusion principle*. Physically, it corresponds to the amplitudes of equivalent detection events (A at q_1 and B at q_2) and (B at q_1 and A at q_2) canceling each other in destructive interference when $q_1 = q_2$.

21.1.1
Discussion

Actually, q may have arbitrary range and may be either discrete (e.g., spin or angular momentum) or continuous (e.g., linear momentum or coordinate), so (q_1, q_2) could be merely a discrete *subset* of selected values. Also, the used basis may contain an infinite, and even continuous, set of eigenvectors. For instance, we decide to measure spins of both electrons in a He atom, neglecting their interaction and *regardless* of their energy levels. In this case, (q_1, q_2) will stand for spin only, and we will use energy eigenstates as the basis. If the atom's state is nonstationary, then many energy eigenstates will contribute to the measured result. Let n_1 and n_2 be the principal quantum numbers for the first and the second electrons. Then for each pair (q_1, q_2), we will have instead of a single matrix a *set* of distinct matrices – a separate matrix $\begin{pmatrix} \psi_{n_1}(q_1) \psi_{n_2}(q_1) \\ \psi_{n_1}(q_2) \psi_{n_2}(q_2) \end{pmatrix}$ for each pair (n_1, n_2). Then we must take the determinant of each matrix and the general state of the system will be a weighted superposition of all of them: $\Psi_{1,2}(q_1, q_2) = \sum_{n_1 > n_2} c(n_1, n_2) \det \begin{pmatrix} \psi_{n_1}(q_1) \psi_{n_2}(q_1) \\ \psi_{n_1}(q_2) \psi_{n_2}(q_2) \end{pmatrix}$.

It is easy to show [2] that $c(n_1, n_2) = -c(n_2, n_1)$. This is an extension of the exchange antisymmetry requirement onto the c-amplitudes – after all, the set $c(n_1, n_2)$ is the same state function in another basis.

Next, we turn to the symmetric (bosonic) states. These states can again be symmetrized using the "antideterminant" – a variant of Slater determinant that takes all products with the "+" sign. There is one subtle point: if for fermions no two states could be the same, for bosons this is not only possible but even likely. Therefore, we must exempt such states from the symmetrization procedure, and only consider exchanges between bosons in different states: Otherwise, for every pair of bosons sharing the same state, $\psi_i = \psi_j$, we would get in our "antideterminant" two equal terms $(1/\sqrt{N!})\psi_1 \cdots \psi_i(q_l) \cdots \psi_j(q_m) \cdots \psi_N$ and $(1/\sqrt{N!})\psi_1 \cdots \psi_j(q_l) \cdots \psi_i(q_m) \cdots \psi_N$, both referring to the same physical state (the difference is no longer in the ordering of indices, but merely in the order of multipliers!). So we should drop the redundant terms resulting from permutations $i \rightleftarrows j$. That does *not* make any meaningful change to the wave function because it effectively amounts to a division by 2, which it then taken care of during normalization. Indeed, because the number of summands is no longer $N!$ but rather $(N!)/2$, normalization constant must be made $\sqrt{2}$ times larger. If three bosons in the antideterminant shared the same state $\psi_i = \psi_j = \psi_k$, every exchange of indices $i \rightleftarrows j \rightleftarrows k$ would give us six identical terms, and eliminating redundancy, we are left with only $(N!)/6$ summands in the expression. Extending this reasoning to the most general case of N bosons, out of which $N_1, N_2, N_3, \ldots$ are the numbers of bosons sharing the same states, and normalizing the state accordingly, we get

$$\Psi_+(q_1, q_2, q_3) = \left(\frac{N_1! N_2! N_3! \cdots}{N!}\right)^{1/2} \sum \psi_1(q_1) \psi_2(q_2) \cdots \psi_N(q_N), \quad (21.12)$$

where summation is carried out over the permutations obtained by the exchange operation between *different* bosonic states. Consider, for instance, the antisymmetric state $\Psi_-(q_1, q_2, q_3)$ discussed above and suppose the particles are now bosons. This only requires changing each "−" to "+". But suppose further that $\psi_1 = \psi_2$. Then, permutations $\psi_1\psi_2\psi_3 \leftrightarrow \psi_2\psi_1\psi_3$, $\psi_1\psi_3\psi_2 \leftrightarrow \psi_2\psi_3\psi_1$, and $\psi_3\psi_1\psi_2 \leftrightarrow \psi_3\psi_2\psi_1$ are not counted because they amount to exchange of coordinates between states ψ_2 and ψ_2. As a result, only three terms out of six should remain, and the normalizing coefficient therefore changes from $1/\sqrt{3!}$ to $\sqrt{2!}/\sqrt{3!}$:

$$\begin{aligned}\Psi_+(q_1, q_2, q_3) = \frac{1}{\sqrt{3}} \{ & \psi_1(q_1)\psi_2(q_2)\psi_3(q_3) + \psi_1(q_1)\psi_3(q_2)\psi_2(q_3) \\ & + \psi_2(q_1)\psi_3(q_2)\psi_1(q_3) + \psi_2(q_1)\psi_1(q_2)\psi_3(q_3) \\ & + \psi_3(q_1)\psi_1(q_2)\psi_2(q_3) + \psi_3(q_1)\psi_2(q_2)\psi_1(q_3) \}.\end{aligned}$$

Up till now we did not dwell much on the physical meaning of the determinant. Moreover, the whole procedure outlined above looked somewhat self-contradictory. Why did we say that particles are indistinguishable, only to assign fictitious labels to them and then swap these labels? There is a simple rationale behind this procedure. Because we cannot tell one particle from the other, we cannot prepare a system with photon A polarized horizontally and photon B polarized diagonally, and be sure that A and B did not exchange identities. The purpose of the determinant is to account for that possibility by giving us a completely symmetric initial preparation with respect to exchanges. At this point, we would like to make an important remark. Having learned that there is, in reality, no "photon A" or "photon B," some readers may be tempted to go into the opposite extreme and deny the very possibility of enumerating the photons. This is the wrong way of thinking. Just as the impossibility of tracking the photon's trajectory in a double-slit experiment did not stop us from talking about "slit 1" and "slit 2," or even more to the point, about the optical path difference between "path 1 and path 2," so we have the right to talk about "horizontal A, diagonal B" and "horizontal B, diagonal A" with the understanding that these are just possible "paths" that our system (photon pair) can take rather than pieces of paper glued to each photon. By writing the symmetrized state, we simply add (or subtract) the amplitudes associated with these two "paths."[5] Just keep in mind that different "paths" should represent *orthogonal* possibilities for the same reason we want to keep the slits separate with no overlapping between them. That is to say, we want to work with orthogonal Hartree products, where the word "orthogonal" means ones that would have a zero inner product if the particles could in fact be labeled. After that we can drop the labels, merge "states" that are the indistinguishable into one state, and, last but not least, renormalize the result (see Box 21.2).

Box 21.2 Normalization of the determinant

When introducing the normalization constant $1/\sqrt{N!}$ in the Slater determinant and in its "positive sign" variety, we tacitly assumed that all terms will be mutually orthogonal. This issue deserves more attention. Generally speaking, $\Psi(q_1, q_2, \ldots, q_N)$ need not be orthogonal to its permutation $\hat{E}(i,j)\Psi(q_1, q_2, \ldots, q_N)$. As an illustration of this, consider a pair of photons. If one is polarized horizontally and the other vertically, the product $|H_A\rangle|V_B\rangle$ is orthogonal to $\hat{E}(A,B)|H_A\rangle|V_B\rangle \equiv |H_B\rangle|V_A\rangle$. If both are polarized horizontally, the state and its permutation are "parallel," in fact, they are identical ($|H_A\rangle|H_B\rangle = |H_B\rangle|H_A\rangle$) even if we label the photons. In the intermediate case where one is polarized horizontally and the other diagonally, $|H_A\rangle|D_B\rangle$ and $|H_B\rangle|D_A\rangle$ are neither orthogonal nor parallel, but exactly in between: their inner product is $\langle H_A|D_A\rangle\langle D_B|H_B\rangle = 1/2$. If one plugs $|H_A\rangle|D_B\rangle$ directly into (21.12), the resulting expression $(1/\sqrt{2})(|H_A\rangle|D_B\rangle + |H_B\rangle|D_A\rangle)$ will have the norm $\sqrt{3/2}$. The correct strategy is to rewrite the state in a "well-behaved" form and then symmetrize each term separately: $|H_A\rangle|D_B\rangle \Rightarrow \Psi_+ = |H_A\rangle\left(\frac{1}{\sqrt{2}}|H_B\rangle + \frac{1}{\sqrt{2}}|V_B\rangle\right) = \frac{1}{\sqrt{2}}|H_A\rangle|H_B\rangle + \frac{1}{\sqrt{2}}|H_A\rangle|V_B\rangle$. Symmetrization yields $\Psi_+ = (1/\sqrt{2})|H_A\rangle|H_B\rangle + (1/\sqrt{2})[(|H_A\rangle|V_B\rangle + |H_B\rangle|V_A\rangle)/\sqrt{2}]$. This way we avoid the difficulty with normalization. Simply put, the original preparation of the system must be decomposed into orthogonal preparations.

Let us now run ahead of ourselves and discuss the meaning of this expression. The normalized state we got is clearly an intermediate result because it assumes that photons can be labeled. But we already know that, in reality, the last two terms are indistinguishable and therefore must be identified with the *same* state. Again,

---

[5] Equivalently, one can say that the symmetrized state describes a *mixture* of all possible preparations of the particle pair. This will be an equal mixture in case of fully overlapping wave packets. If the overlapping is partial (distinguishability increases), different preparations must be assigned different weights, and finally, when the overlapping disappears, the mixture goes over to a pure state.

this requires more than just removing the labels and adding identical terms. If we stopped at that, our answer would be $\Psi_+ = (1/\sqrt{2})|H\rangle|H\rangle + |H\rangle|V\rangle$ (now the last term means having one horizontal and one vertical photon *without* distinguishing the order). The norm is no longer 1. But that's only natural: Once we allow identical (i.e., nonorthogonal) summands, it changes normalization. Note that this time we cannot just remove the "redundant" $(1/2)|H\rangle|V\rangle$ as we did earlier. The difference is the presence of the term $(1/\sqrt{2})|H\rangle|H\rangle$ in the sum. We cannot arbitrarily change *some* amplitudes and leave the others intact! That would mean changing the interference pattern. The only way to proceed is to renormalize the whole expression: $|\Psi\rangle = \left(\sqrt{1/3}\right)|H\rangle|H\rangle + \left(\sqrt{2/3}\right)|H\rangle|V\rangle$. We arrive at a remarkable result. Two-thirds of the time the diagonal photon will decide to join its comrade in the horizontal state, and only one-third of the time it will choose to stay alone in the vertical state. In a single-photon experiment, both choices would have been equally likely.

Our primary goal of course is to predict the behavior and probabilities of outcomes for a system of *N* identical particles. The Slater determinant method has nothing to say about it. Its only use is to obtain possible symmetrized states that such a system *could* occupy. The initial Hartree product must either come from the solution of the Schrödinger equation or be conjectured based on physical intuition, or finally, be predetermined by the way in which the system and/or the measuring apparatus was prepared. Equation 21.11 is not going to tell us whether this or that symmetrized state will in fact be occupied by the system and with what probability. Moreover, the basis one employs to solve the Schrödinger equation and obtain the Hartree product will generally not be the right basis for studying the behavior of symmetrized solutions (see Boxes 21.3 and 21.4).

In particular, one natural choice of basis for bosons is the basis composed of *number states*, also called *Fock states*.[6] These states are written as $|n\rangle$ where *n* is the number of photons sharing the given state or, more generally, as $|n_1, n_2, n_3, \ldots\rangle$ where n_i means the number of photons in the *i*th *optical mode*, and all possible orthogonal[7] (i.e., mutually exclusive) modes are enumerated within the ket. The "mode" is defined by specifying the three wave numbers k_x, k_y, k_z and spin (which means polarization in case of photons). One often has a system of bosons whose wave numbers are known to be the same, in which case it is enough to define a mode by polarization only.

---

6) Another useful basis for many fields is a basis composed of *coherent* states, which have indefinite photon number and indefinite phase, but keep the product of respective uncertainties down to a minimum. They can always be expanded in terms of the number states. A basis-formed coherent states have some unusual features. For example, these are eigenstates of a non-Hermitian operator.
7) The modes are orthogonal if the respective states defined by these modes would be orthogonal for a single boson. Enumerating the modes becomes tricky when one deals with eigenstates of continuous variables such as $\mathbf{k} = (k_x, k_y, k_z)$. In that case one has to write the number state as $|n_\mathbf{K}\rangle$ or $|n_{1,\mathbf{K}}, n_{2,\mathbf{K}}\rangle$ to account for both polarizations, and let $\mathbf{k}$ take a continuous range of values.

Box 21.3 State space for indistinguishable particles

If the particles were distinguishable, we could construct the Hilbert space as usual, as the product of one-particle $\mathcal{H}$-spaces. For instance, the four-dimensional polarization space $\mathcal{H}_{AB}^4 = \mathcal{H}_A^2 \otimes \mathcal{H}_B^2$ of two identical photons would be spanned by four product states, $\mathcal{H}_{AB}^4 = \{|H_A\rangle \otimes |H_B\rangle, |H_A\rangle \otimes |V_B\rangle, |V_A\rangle \otimes |H_B\rangle, |V_A\rangle \otimes |V_B\rangle\}$. The photon pair would then be in some superposition of these four eigenstates. Now imagine we remove the labels. The first and the fourth states in the basis will be unaffected, but the second and the third now mean the same physical state! Working in $\mathcal{H}_{AB}^4$ is impossible: that would be like having a Cartesian space with two x-axes. Since the number of orthogonal states is now three, we need a 3D subspace of $\mathcal{H}_{AB}^4$. However, it shouldn't be an arbitrary subspace, but one that will be unaffected by the removal of the labels.

To that end, we introduce two subspaces of $\mathcal{H}_{AB}^4$, which can formally be defined as *symmetric* and *antisymmetric* tensor products (superscripts (S) and (A) in the expression below) of one-particle Hilbert spaces. These products are constructed as follows: $\mathcal{H}_{AB}^{(S,A)} \equiv \mathcal{H}_A^2 \odot \mathcal{H}_B^2 = \mathcal{H}_A^2 \otimes \mathcal{H}_B^2 \pm \mathcal{H}_B^2 \otimes \mathcal{H}_A^2$, with subsequent normalization of the resulting "unit vectors." The "+" sign corresponds to symmetric and "−" to antisymmetric products. The new symbol $\odot$ is intended to distinguish this operation from taking the ordinary tensor product of two spaces (an alternative symbol $\vee$ is also common in the literature).

This operation gives us a *symmetric subspace* and an *antisymmetric subspace*: $\mathcal{H}_{AB}^{(S)} = \{|H_A\rangle \otimes |H_B\rangle, |V_A\rangle \otimes |V_B\rangle, 1/\sqrt{2}(|H_A\rangle \otimes |V_B\rangle + |V_A\rangle \otimes |H_B\rangle)\}$, $\mathcal{H}_{AB}^{(A)} = \{1/\sqrt{2}(|H_A\rangle \otimes |V_B\rangle - |V_A\rangle \otimes |H_B\rangle)\}$. Pairs of identical bosons live in the symmetric subspace, and pairs of fermions in the antisymmetric one. Note that the former is spanned by three eigenstates, the latter is spanned by a single eigenstate in this example. Physically, this happened because the two other "unit vectors" $|H_A\rangle \otimes |H_B\rangle$ and $|V_A\rangle \otimes |V_B\rangle$ are impossible for fermions. Sometimes the basis vectors of $\mathcal{H}_{AB}^{(S,A)}$ are called *configurations*. Then every legitimate two-particle state can be written as a linear combination of configurations. We leave it to the reader to generalize these results for systems of more than two particles.

If the actual number of photons is unknown, one has to work in the *Fock space*, which is formed as a direct sum of symmetric subspaces for different photon numbers that range from 0 to infinity.

This space $\mathcal{F}$ can be written as $\mathcal{F} \equiv \mathbb{C} \oplus \mathcal{H} \oplus (\mathcal{H} \odot \mathcal{H}) \oplus (\mathcal{H} \odot \mathcal{H} \odot \mathcal{H}) \oplus \ldots \equiv \oplus_{n=0}^{\infty} \mathcal{H}^{\odot n}$. The first term in this summation is $\mathbb{C}$, the vector space for the *vacuum state* with no particles (photon number 0). The right-hand side offers a compact way of writing the Fock space, where $\oplus_{n=0}^{\infty}$ is the direct sum equivalent of the familiar summation symbol $\Sigma_{n=0}^{0}$, and $\odot n$ denotes nth power for the case when "multiplication" is accomplished by $\odot$.

The fact that the $\mathcal{H}$-space of a system of identical particles is so different from ordinary $\mathcal{H}$-spaces is one formal reason we don't call such particles entangled. Entangled systems are usually defined to live in ordinary product spaces such as $\mathcal{H}_{AB}^4$.

> **Box 21.4 Notation for multiparticle states**
>
> One common catchphrase in physics texts is that one should never label identical particles. That is true in the sense that *after* the appropriate Hilbert space is constructed, one should never follow the distinguishable particle model. It would be absurd to describe a pair consisting of a horizontally and a vertically polarized photon "by analogy" as $1/\sqrt{2}(|H\rangle \otimes |V\rangle + |V\rangle \otimes |H\rangle)$. Both summands here represent one and the same state! If you insist for some reason on writing it in this form, you should at least replace $1/\sqrt{2}$ with $1/2$, and that would basically state that $x = x/2 + x/2$. Having said that, if you recall how we constructed the Hilbert space, you will realize that particle labels cannot be banished from the conceptual picture. After all, the building blocks for that space did come from the tensor products of distinguishable particle states!
>
> To avoid confusion, imagine that we assigned labels to the particles, constructed the state space according to the instructions above, and then removed the labels and forgot about them forever. The question as to what building blocks we used will then have only historical importance. Now that we have a new basis to work with, writing its unit vectors in the old form as $|H\rangle \otimes |H\rangle, |H\rangle \otimes |V\rangle$, and so on, even with labels removed, would be conceptually inconsistent because that would imply that our particles *still* can exist, at least in principle, in their separate H-spaces. To address this difficulty, one can now rewrite the corresponding states in a modified notation: $\mathcal{H}_{AB}^{(S)} = \{|HH\rangle, |VV\rangle, |HV\rangle\}$, $\mathcal{H}_{AB}^{(A)} = \{|HV\rangle\}$. This helps enforce the indistinguishability of particles: thus, $|HH\rangle$ should be read "two horizontally polarized photons" and $|HV\rangle$ should be read "one horizontally and one vertically polarized photon." Though this notation is common, it is still imperfect: We may also have written $|VH\rangle$ instead of $|HV\rangle$. It is more expedient then to introduce the form $|n_1, n_2\rangle$ with the understanding that n_1 will be the number of horizontally polarized photons in the system and n_2 will be the number of vertically polarized photons. In this new notation, the system is said to have a *horizontal mode* and a *vertical mode*, with each mode having its own occupation number (n_1 or n_2, respectively). Bosonic states written in this form are called number states. Obviously, this way we can totally avoid meaningless statements about *which* photons are polarized in a given way. In this optimal notation, we have $\mathcal{H}_{AB}^{(S)} = \{|2,0\rangle, |0,2\rangle, |1,1\rangle\}$ and $\mathcal{H}_{AB}^{(A)} = \{|1,1\rangle\}$. We repeat here that the first basis contains bosonic states, and the second contains fermionic states, so the similarity in notation should not cause any confusion.
>
> There arises an inevitable question: Should one be writing the number state in terms of single-particle states? For example, in the spirit of the above procedure, one would have
>
> $|2,0\rangle = |1,0\rangle \otimes |1,0\rangle = |H\rangle \otimes |H\rangle,$
> $|1,1\rangle = (1/\sqrt{2})(|1,0\rangle \otimes |0,1\rangle + |0,1\rangle \otimes |1,0\rangle) = (1/\sqrt{2})(|H\rangle \otimes |V\rangle + |V\rangle \otimes |H\rangle),$
> $|0,2\rangle = |0,1\rangle \otimes |0,1\rangle = |V\rangle \otimes |V\rangle.$

21.1 Bosons and Fermions: The Exclusion Principle

The answer to this question is twofold. On the one hand, this is exactly what we would do (implicitly, at the very least) to answer questions such as what is the inner product $\langle 2, 0 | 1, 1 \rangle$ (the answer is, predictably, zero). Indeed, the validity of this procedure is guaranteed by linear algebra: any vector belonging to a 3D subspace of a 4D vector space can always be written as a superposition of the four basis vectors. But there is an alternative view that one can take. Note that the second expression already takes us back to the notion of distinguishable particles, or otherwise the expression in parenthesis would be tautological. If a physicist is reluctant to tolerate this conceptual ambiguity, he must take the view that though our number states $|2, 0\rangle, |0, 2\rangle, |1, 1\rangle$ were *derived* from $\mathcal{H}_{AB}^4$ vectors, they are not *identical* to these vectors. Purely formally, when we write $|H\rangle \otimes |H\rangle$, we are making an implicit claim that it is a 4D vector, whereas the physics of a photon pair is such that this vector should be described as 3D. In a nutshell, one faces this highly abstract but nevertheless strong objection: *isomorphism* between two sets of vectors does not imply *identity*. For this reason, it is advisable to avoid writing tensor products that we just wrote even though the algebra, if done correctly, should work out all right.

Consider now the following question. A system consists of two identical photons. We "seed" the Slater determinant with the Hartree product $|\psi_1\rangle \otimes |\psi_2\rangle$, where $|\psi_1\rangle$ and $|\psi_2\rangle$ are one-photon polarization states. We ask: What are the chances that a measurement in the $\{|R\rangle, |L\rangle\}$ basis will find both photons R-polarized, both photons L-polarized, and finally, one photon R-polarized and the other L-polarized?

To keep the numbers simple, assume that $|\langle R|\psi_1\rangle| = |\langle L|\psi_1\rangle| = |\langle R|\psi_2\rangle| = |\langle L|\psi_2\rangle| = a$. This will be the case if $|\psi_1\rangle$ and $|\psi_2\rangle$ represent linear or diagonal polarizations. For example, let's make both photons horizontally polarized: $|\psi_1\rangle = |\psi_2\rangle = |\leftrightarrow\rangle \equiv |H\rangle$. It is convenient to start the problem by assuming that the photons are distinguishable by some parameter other than polarization. In this case, we can track both of them from the beginning to the end. There is no interference of amplitudes. Each photon "minds its own business," choosing "right" or "left" polarization with the same probability a^2. We should simply square the amplitudes and add the resulting probabilities:

$$\mathcal{P}_{1,2}(R, R) = |\langle R|H\rangle\langle R|H\rangle|^2 = |a|^2|a|^2 = 1/4,$$
$$\mathcal{P}_{1,2}(L, L) = |\langle L|H\rangle\langle L|H\rangle|^2 = |a|^2|a|^2 = 1/4,$$
$$\mathcal{P}_{1,2}(R, L \text{ or } L, R) = |\langle R|H\rangle\langle L|H\rangle|^2 + |\langle R|H\rangle\langle L|H\rangle|^2 = 2|a|^2|a|^2 = 1/2.$$

Similar equations would also hold for distinguishable fermions, except that we would have to replace "polarization states" with "spin states." Our pair of distinguishable particles shows exactly the same statistics that we would observe when tossing two coins (just substitute "heads and "tails" for "R" and "L").

Now make the photons fully indistinguishable. Then interference can be observed: We must follow the principle "add, then square." Since amplitudes can interfere constructively or destructively, it is necessary to take into account the relative phases. Since the Hartree product $|H\rangle \otimes |H\rangle$ is already symmetric, symmetrizing it with (21.12) is not necessary. We thus have the initial state $|2, 0\rangle$ (two photons in the horizontal mode and zero photon in the vertical mode). The final state could be either $|2, 0\rangle$, $|0, 2\rangle$, or $|1, 1\rangle$ (remember that it's written in the circular basis). Recalling how these configurations were constructed from one-particle eigenstates,

$$\left\{ |H\rangle \otimes |H\rangle, |V\rangle \otimes |V\rangle, \frac{|H\rangle \otimes |V\rangle + |V\rangle \otimes |H\rangle}{\sqrt{2}} \right\}$$

$$\Rightarrow \{|HH\rangle, |VV\rangle, |HV\rangle\} \stackrel{\text{linear}}{\equiv} \{|2,0\rangle, |0,2\rangle, |1,1\rangle\},$$

$$\left\{ |R\rangle \otimes |R\rangle, |L\rangle \otimes |L\rangle, \frac{|R\rangle \otimes |L\rangle + |L\rangle \otimes |R\rangle}{\sqrt{2}} \right\}$$

$$\Rightarrow \{|RR\rangle, |LL\rangle, |RL\rangle\} \stackrel{\text{circular}}{\equiv} \{|2,0\rangle, |0,2\rangle, |1,1\rangle\},$$

and projecting the initial state onto each possible final state, we get

$$_{\text{circular}}\langle 2, 0|2, 0\rangle_{\text{linear}} = \langle R|H\rangle\langle R|H\rangle = 1/2,$$

$$_{\text{circular}}\langle 0, 2|2, 0\rangle_{\text{linear}} = \langle L|H\rangle\langle L|H\rangle = 1/2, \quad _{\text{circular}}\langle 1, 1|2, 0\rangle_{\text{linear}} = 1/\sqrt{2}.$$

We recover the familiar statistics of the two-coin toss: 1/4 of trials will result in a pair of R-polarized photons, 1/4 of trials in a pair of L-polarized photons, and half of the time the photons will emerge oppositely polarized (the same statistics would be obtained if we chose instead $|\psi_1\rangle = |\psi_2\rangle = |\uparrow\rangle \equiv |V\rangle$). But suppose we now try the third possibility: prepare a pair of photons with opposite linear polarizations, in other words, take the initial state $|1, 1\rangle$. Projecting it onto the same eigenstates as before and rerunning the calculations, we notice something unusual:

$$_{\text{circular}}\langle 2,0|1,1\rangle_{\text{linear}} = \frac{i}{\sqrt{2}}, \quad _{\text{circular}}\langle 0,2|1,1\rangle_{\text{linear}} = -\frac{i}{\sqrt{2}}, \quad _{\text{circular}}\langle 1,1|1,1\rangle_{\text{linear}} = 1/\sqrt{2}.$$

The first two probabilities are equal to 1/2, but the last one is zero! In other words, we will never see two photons emerge with opposite circular polarizations, but instead will always see them emerge in the same state. This radical departure from the familiar statistics of the coin pair is explained by the destructive interference where two amplitudes cancel each other. There exists a familiar physical analogy. The reader who remembers Chapter 4 must have already realized that we are looking here at a modified version of the HOM experiment! Indeed, in this experiment, it is essential that the photons arrive at the beam splitter simultaneously, to give their wave packets an opportunity to overlap. Only then can we treat them as an indistinguishable pair. If this condition is not met, one registers a dip in the frequency of coincident counts (i. e., where the photons emerge together from the same port). This so-called HOM dip can be used to measure the time delay between photons arrivals, thus providing a measure of how far the photons are from being indistinguishable.

Let us now ask a question that on the face of it looks exactly the same. Suppose we put two photons into a box and then partitioned the box equally with a wall. What is the probability to find both photons to the left of the wall, both photons to the right, and finally, one photon to the left and one to the right? The answer is quite unexpected: All three outcomes are equiprobable! When we put the photon pair into the box, we did not specify its state, so at the input we have an equal mixture of $|2,0\rangle$, $|0,2\rangle$, and $|1,1\rangle$ and the three respective probability distributions $(1/4, 1/4, 1/2)$, $(1/4, 1/4, 1/2)$, $(1/2, 1/2, 0)$ must be given the same weight $1/3$. Weighing, we obtain equal probabilities: $(1/3)[(1/4, 1/4, 1/2) + (1/4, 1/4, 1/2) + (1/2, 1/2, 0)] = (1/3, 1/3, 1/3)$.

As we saw in Box 21.2, once the symmetrized state is written and we start removing the labels, it turns out that amplitudes of composite bosonic states in the symmetric subspace behave in a rather peculiar way. Some of the amplitudes get amplified, particularly those of states that are represented by a large number of Hartree products (21.12). Conceptually, this is easy to understand: These states occur more frequently because there are more ways in which they can be created. Each extra path leading to these states diverts some of the total probability much in the same way as each extra resistor connected in parallel diverts some of the current from the other resistors. The difference is that amplitudes can interfere not only constructively but also destructively. Here, we will focus on the following important case of constructive interference. Look again at the symmetrized function (21.12). This bosonic system exhibits behavior opposite to that of fermions. If a boson has a strong individual preference for a certain state, that preference will be stronger still for a pair of bosons. This results in "group thinking" for bosons: They will tend to bunch together. In particular, if we have N bosons with clusters of N_1 bosons in state 1, N_2 in state 2, and so on, then, all $(N!)/(N_1!N_2!N_3!\cdots) \equiv \tilde{N}$ permutations appearing in (21.12) become equal when $q_1 = q_2 = \cdots = q_N \equiv q$. Denote their common value as $\psi_1(q)\psi_2(q)\cdots\psi_N(q) \equiv \Psi(q)$. The net probability for all N bosons to collapse to the same point q is equal to

$$\mathcal{P}(N,q) = \left| \frac{1}{\sqrt{\tilde{N}!}} \underbrace{(\Psi(q) + \cdots + \Psi(q))}_{\tilde{N}!} \right|^2 = \tilde{N}!|\Psi(q)|^2. \qquad (21.13)$$

Now imagine that we added a new boson. For the sake of simplicity, we ignore the question of how exactly this was achieved because this can be quite tricky. (Loosely speaking, one would need some process such as stimulated emission with postselection to coordinate the arrival of the new boson with the state of the existing bosons. One doesn't just put a new boson into the box with other bosons.) Suppose it was prepared in some new state ψ_{N+1} different from the states of boson clusters $N_1, N_2, N_3, \ldots$. [However, ensure that with the arrival of the newcomer, the Hartree products remain mutually orthogonal (see Box 21.2)]. Then the factorial increases by a factor of $N+1$ and a new factor $|\psi_{N+1}|^2$ is introduced into the expression:

$$\mathcal{P}(N+1,q) = (N+1)\tilde{N}!|\Psi(q)\psi_{N+1}|^2 = (N+1)|\psi_{N+1}|^2\mathcal{P}(N,q). \qquad (21.14)$$

Based on classical logic, we should have expected a different formula, $\mathcal{P}(N+1,q) = |\psi_{N+1}|^2\mathcal{P}(N,q)$. This would be the case if the new boson made its own decision independently, paying no attention to what its colleagues are doing. But this is not how things work in QM. Bosons possess a herd instinct: An additional boson has an $(N+1)$ times greater chance to occupy a state that already contains N such bosons[8] than it would have if the bosons were distinguishable.

On the other hand, if the new boson is prepared in the same state as the bosons in some cluster N_i, that chance only increases by $(N+1)/(N_i+1)$. In particular, if we add another boson to a state already shared by all N bosons (i.e., $\psi_1 = \psi_2 = \cdots = \psi_N = \psi_{N+1}$), then we have $N_i = N$ and there is no probability enhancement. Such a system behaves as if all $N+1$ bosons were distinguishable. We can see now why two photons entering the *same* port of the HOM apparatus will behave independently of each other.

On a final note, it could seem that for large N, the probability (21.14) increases unboundedly. However, we have seen already that after adding the amplitudes of indistinguishable outcomes, our next step is to renormalize the wave function, requiring that the probabilities of all possible outcomes add up to 1. With this, we can now reinterpret the equal distribution of probabilities in the box-partitioning experiment as follows. From the symmetric setup of the experiment, there is no reason for either photon to expect that its colleague should prefer one compartment over the other; therefore, both possibilities should be assigned *a priori* equal weights. Suppose the right compartment is chosen. The remaining photon is now two times more likely to choose the right compartment containing one photon than the left compartment containing zero photon. Then it chooses the right compartment with probability 2/3 and hence the chance for two photons to occupy the right compartment is $(1/2)(2/3) = 1/3$. By the same reasoning, there is 1/3 chance to find both photons in the left compartment, and then a 1/3 chance remains for the possibility of splitting the pair.

21.2
Planck and Einstein Again

Let us now apply the obtained results to some real physical process. We will again consider the BBR, but from a quite different perspective. Instead of introducing the Einstein coefficients "by hand," we will now derive them together with the Planck formula from basic principles.

We note that radiation field is about photons, and their spatial concentration (in case of thermodynamic equilibrium) is determined by the rates of their absorption and emission by atoms. An emitted photon leaves the atom and joins whatever photons may have been already present in the respective mode of the surrounding EM field; an absorbed photon leaves the EM field and disappears within the atom. Apply to the emission the fundamental rule we have learned in Section 21.1. The probability of photon emission into a certain state $|q\rangle$ increases by a factor of $(N+1)$ if this state already harbors N photons. In the language of amplitudes, this means the increase of the emission amplitude by a factor of $\sqrt{N+1}$.

---

[8] By saying "already contains" rather than "has a chance to contain" we did not change the condition of the problem: just apply (21.14) when $\mathcal{P}(N) = 1$.

The transition of a photon to state $|q\rangle$ from an initial state $|\chi\rangle$ is described by the inner product $\langle q|\chi\rangle$ that satisfies the condition $\langle q|\chi\rangle = \langle \chi|q\rangle^*$. The addition of a new photon to an EM mode already containing N photons (i.e., emission) can be considered as a transition between number states $|N\rangle$ and $|N+1\rangle$. The corresponding amplitude is

$$\langle q|\chi\rangle \Rightarrow \langle N+1|N\rangle = \sqrt{N+1}\,a, \tag{21.15}$$

where a is the transition amplitude in the absence of other photons. And conversely, the amplitude of the time-reversed transition from $|N+1\rangle$ to $|N\rangle$ (i.e., absorption), which decreases the photon number from $N+1$ to N, will be given by

$$\langle \chi|q\rangle \Rightarrow \langle N|N+1\rangle = \sqrt{N+1}\,a^*. \tag{21.16}$$

For the time-reversed process, it is convenient to alter the last expression by changing N to $N-1$ in the initial condition. This will bring symmetry to the initial conditions for both processes (absorption and emission will now start from the same number N):

$$\langle N-1|N\rangle = \sqrt{N}\,a^*. \tag{21.17}$$

The probabilities for a photon to be emitted to join a state $|N\rangle$ and for a photon to quit this state due to absorption are, respectively,

$$\mathcal{P}_e = (N+1)|a|^2. \tag{21.18a}$$

$$\mathcal{P}_a = N|a|^2. \tag{21.18b}$$

Equations 21.15 and 21.16 show that the rules governing behavior of bosons are symmetric with respect to time reversal. When applied to photon exchange between atoms and the radiation field, this means that emission and absorption probabilities are proportional ($\sim N+1$ for emission and $\sim N$ for absorption) to the number of photons already existing in this state.

Multiplying (21.18) by $\hbar\omega$, we recover Einstein's coefficients. Equation 21.18a clearly shows that the emission coefficient consists of two parts: the probability of stimulated emission $\mathcal{P}_i(N) = N|a|^2$ and the probability of spontaneous emission $\mathcal{P}_s = |a|^2$. The latter can be considered as a special case of (21.18a) when $N=0$. In this case, stimulated emission dies out, but spontaneous emission survives since its probability does not depend on N. An atom in an excited state can emit a photon even when there are no photons around.

We mentioned at the end of Section 2.2 that atom's instability with respect to photon emission even in the absence of any external radiation can be explained in the language of quantum field theory as a result of stimulation from the "sea" of virtual photons. According to this explanation, there is only stimulated emission and no spontaneous emission. Thus, stimulated emission comes in two varieties: stimulation by really existing photons, whose rate is naturally proportional to their actual amount N, and stimulation by virtual photons, whose rate $|a|^2$ is determined only by the properties of the corresponding atomic eigenstates and the intensity of what we call vacuum fluctuations. Hence, the net proportionality factor for emission is $N+1$.

In the language of QM, the same thing is explained in a simpler way without any reference to virtual photons. The "extra" term 1 in the factor $N+1$ reflects the QM rule that the mere ability of the empty space to accommodate a free photon provokes the excited atom to eventually emit it. Whatever the interpretation, it gives a natural explanation of the Einstein coefficients A_n^m, B_n^m, and B_m^n introduced in Section 2.2. And if we consider probability per photon, then the coefficients of absorption, stimulated emission, and spontaneous emission are all equal to each other.

21.3
BBR Again

It is instructive to rederive the BBR spectrum using Equations 21.15–21.17. Consider a box with N photons in a certain state $\omega, \mathbf{k}$ with some definite polarization. Equilibrium requires that there must be a certain amount of atoms in the box with at least two energy states such that $E_2 - E_1 = \Delta E = \hbar\omega$. Denoting the average number of atoms in the respective states as $\bar{n}_1$ and $\bar{n}_2$, we have

$$\frac{\bar{n}_2}{\bar{n}_1} = e^{-(\hbar\omega/kT)}. \tag{21.19}$$

If $\bar{n}_2$ atoms in state 2 are emitting photons with probability $\mathcal{P}_e = (N+1)|a|^2$ and $\bar{n}_1$ atoms in state 1 are absorbing these photons with probability $\mathcal{P}_a = N|a|^2$, equilibrium will ensue when the rates of both processes become equal:

$$\bar{n}_1 N = \bar{n}_2 (N+1) \tag{21.20}$$

There follows

$$\bar{N} = \frac{\bar{n}_2}{\bar{n}_1 - \bar{n}_2} = \frac{1}{e^{\hbar\omega/kT} - 1}. \tag{21.21}$$

This gives the average number of photons in the optical mode $(\omega, \mathbf{k}, \mathbf{e}) = (\omega, (\omega/c)\hat{\mathbf{n}}, \mathbf{e})$, where $\hat{\mathbf{n}}$ is the unit vector in the direction of propagation, and $\mathbf{e}$ specifies polarization state. We know that this number is determined by ω and T. Multiplying it by $\hbar\omega$ gives the average radiation energy per unit volume stored in one mode:

$$\bar{E}(\omega, \hat{\mathbf{n}}) = \frac{\hbar\omega}{e^{\hbar\omega/kT} - 1}. \tag{21.22}$$

The number of EM modes within a narrow frequency range $\omega, \omega + \Delta\omega$ and with all possible directions of $\hat{\mathbf{n}}$ is given by $\omega^2/2\pi^2 c^3$. In order to include both polarization states, we multiply that number by 2 and, combining it with the last expression, recover Plank's formula for radiation energy density per unit frequency range:

$$\rho(\omega, T) = \frac{\hbar}{\pi^2 c^3} \frac{\omega^3}{e^{\hbar\omega/kT} - 1}. \tag{21.23}$$

We see now that the above result rests on the fact that photons are bosons obeying the rules of quantum statistics.

21.4
Lasers and Masers

Consider Equation 21.19 again. It describes the ratio of occupancy numbers of two energy states in thermal equilibrium. The higher the temperature, the closer this ratio to 1. At $T \to \infty$, we have $\bar{n}_2 = \bar{n}_1$ for any pair of levels: At infinite temperature, all energy levels are occupied uniformly. This is the most one can get under the equilibrium conditions.

But thermal equilibrium is only a special case. Nonequilibrium states are routinely observed in natural phenomena. In particular, the functioning of lasers is based on light interaction with atoms whose energy distribution is different from (21.19). Namely, at least for two distinct energy levels, the population of the higher level E_2 is *greater* than that of the lower level E_1. A population satisfying $\bar{n}_2 > \bar{n}_1$ is usually called *inverse population*.

This seems impossible as it would make the ratio (21.19) greater than 1 and, according to the formula, 1 is the maximum that we can get. But think outside the box! Go beyond $T = \infty$ and you will get $T < 0$!

To see what it means "to go beyond $T = \infty$" and how such a trespassing leads to negative T, think of T formally as a specific function of some parameter x, for instance, $T \sim 1/x$. In the discussed situation, $x = n_1 - n_2$ and the relation between T and x is more complicated, but $T = 1/x$ is sufficient to illustrate the main point. If you graph the function $T(x)$ thus defined, you will see that it is positive for all positive x and it goes to $+\infty$ as x approaches zero while remaining positive (Figure 21.2). But as x goes beyond zero (gets negative), T goes "beyond infinity," that is, makes a jump from $+\infty$ to $-\infty$, and remains negative while decreasing in magnitude as x recedes from $x = 0$ in the negative direction.

So we can obtain $\bar{n}_2 > \bar{n}_1$ from (21.19) by formally ascribing negative sign to temperature. Reversing the argument, we can introduce a physically meaningful notion of negative temperature that can be interpreted as associated with the inverse population. And in this case, it may have some significance as describing "abnormal" distributions over energy levels, when at least one higher level is more populated than the lower level.

How will the inverse population affect the frequency distribution of photons? In equilibrium, the distribution $\rho(\omega, T)$ does not change with time or changes very slowly. An interesting feature of the nonequilibrium states is that even when we manage to maintain the inverse *atomic* population constant, the *photon* population generally does not stay constant. Since the number of light-emitting atoms exceeds that of absorbing ones, the rate of photon emission exceeds that of absorption. The result is permanent growth of the photon population N in a given and maintained nonequilibrium state. In other words, we expect $dN/dt > 0$.

How can we find $N(t)$? Equations 21.19 and 21.21 with positive T do not hold at $\bar{n}_2 > \bar{n}_1$. In this case, we need to derive a new equation that would describe nonequilibrium states with inverse population. In order to do that, we have to drop the requirement that the rates of absorption and emission be equal. Instead, we

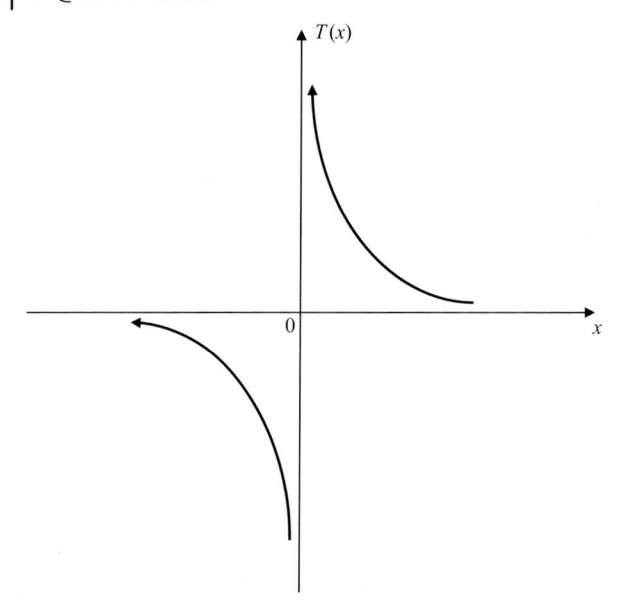

Figure 21.2 Illustration of the concept of negative temperatures. The temperature in equilibrium is related to population distribution over the energy levels. For two selected levels 1 and 2, it may be considered as a function of $x = n_1 - n_2$. In equilibrium, x is positive and T increases unboundedly as $x \to 0$. As x passes the zero point (the system goes beyond equilibrium and the inverse population is formed), T goes beyond positive infinity – it jumps to negative infinity and then approaches zero remaining negative as x is shifted in negative direction.

write

$$\frac{d\bar{N}_a}{dt} = \bar{n}_1 \bar{N} \omega_\gamma, \tag{21.24}$$

$$\frac{d\bar{N}_e}{dt} = \bar{n}_2 (\bar{N} + 1) \omega_\gamma. \tag{21.25}$$

Suppose there are no other processes that could change the net average photon number $\bar{N}$, so this number depends only on the rates (21.24) and (21.25). Then,

$$\frac{d\bar{N}}{dt} = \frac{d\bar{N}_e}{dt} - \frac{d\bar{N}_a}{dt} = \bar{n}_2 (\bar{N} + 1) \omega_\gamma - \bar{n}_1 \bar{N} \omega_\gamma. \tag{21.26}$$

Rearranging the expression on the right-hand side as $(\bar{n}_2 - \bar{n}_1) \bar{N} \omega_\gamma + \bar{n}_2 \omega_\gamma$, we get

$$\frac{d\bar{N}}{dt} - \omega_\gamma (\bar{n}_2 - \bar{n}_1) \bar{N} = \omega_\gamma \bar{n}_2. \tag{21.27}$$

Assuming constant atomic population, we have a nonhomogeneous differential equation with constant coefficients. Its general solution is

$$\bar{N}(t) = \tilde{N} e^{\omega_\gamma (\bar{n}_2 - \bar{n}_1) t} - \frac{\bar{n}_2}{\bar{n}_2 - \bar{n}_1}. \tag{21.28}$$

Here, $\tilde{N}$ is an adjustable parameter whose value is determined by the initial conditions. Suppose that initially we have a known number N_0 of photons. Then,

$$\bar{N}(0) = N_0 = \tilde{N} - \frac{\bar{n}_2}{\bar{n}_2 - \bar{n}_1} \quad \text{and} \quad \tilde{N} = N_0 + \frac{\bar{n}_2}{\bar{n}_2 - \bar{n}_1}. \quad (21.29)$$

Putting this back into the general solution, (21.28) gives

$$\bar{N}(t) = \left(N_0 + \frac{\bar{n}_2}{\bar{n}_2 - \bar{n}_1}\right) e^{\omega_\gamma (\bar{n}_2 - \bar{n}_1)t} - \frac{\bar{n}_2}{\bar{n}_2 - \bar{n}_1}. \quad (21.30)$$

According to (21.30), even if there are initially no photons ($N_0 = 0$) in the system, the system will generate photons at an increasing rate starting from the initial moment. This process will inevitably be triggered by the spontaneous emission that has a finite rate ω_γ and then, since at least one photon is already around, stimulated emission will kick in and progress at an increasing rate at the cost of the energy taken from the higher energy level (whose population must be maintained constant by some external source!). The result will be generation of predominantly coherent light. The coherence is ensured by the condition that most of emitted photons are identical to the very first one that caused the stimulated emission. The word "predominantly" reflects the irreducible spontaneous emission throughout the whole process.

If $N_0 \neq 0$ (e.g., a certain number of identical photons entered the system from the outside before the onset of generation), this initial number will be exponentially amplified after the initial moment. This process is called light amplification. Thus, the laser can work in the *generation* regime (photon production starts of its own accord when $N_0 = 0$) or in the *amplification* regime (enhancing the already existing radiation when $N_0 \neq 0$). The majority of emitted photons will be exact clones of the one that induced the emission. But overall the cloning is never perfect (its fidelity is less than 1) due to spontaneous emission.

In both cases, the output will be *predominantly* a pure ensemble, at least after sufficiently long t. The difference is that in the generation regime some characteristics (e.g., polarization) even in the pure part of the ensemble are generally unpredictable – they will replicate those of the spontaneously emitted photon(s). In the amplification regime, we can predict that such characteristics will replicate those in the input.

Problems

21.1 Suppose we have an initial state $|100, 0\rangle$ of 100 identical R-polarized photons in a box. An H-polarized photon and then all 101 photons are passed through a circular polarizer. Find the probabilities of outcomes $|101, 0\rangle$ and $|100, 1\rangle$.

21.2 Consider two identical fermions in a state $|q_1, q_2\rangle_{\downarrow\downarrow}$, where q is a collective notation for a set of compatible observables other than spin.

a) What can you say about such state when $q_1 = q_2$? Prove your answer.
b) What would you say about such state in case of two identical bosons?

21.3 Consider an excited state of a He atom in the zero approximation:

$$\Psi(\mathbf{r}_1, \mathbf{r}_2) = \frac{1}{\sqrt{2}} \left(\psi_{1,0,0}(\mathbf{r}_1)\psi_{2,0,0}(\mathbf{r}_2) - \psi_{1,0,0}(\mathbf{r}_2)\psi_{2,0,0}(\mathbf{r}_1) \right)$$

(the spin dimension is dropped for the sake of simplicity).

a) Prove that $\Psi(\mathbf{r}_1, \mathbf{r}_2)$ is normalized.
b) Find the corresponding zero-approximation energy of this state.

References

1 Streater, R.F. and Wightman, A.S. (1964) *PCT, Spin and Statistics, and All That*, W. A. Benjamin, Inc., New York.

2 Nielsen, M.A. and Chuang, I.L. (2000) *Quantum Computation and Quantum Information*, Cambridge University Press.

22
Second Quantization

We start this chapter by revisiting the quantum oscillator. We will find in its treatment the embryo of a broader view and new description of the multiparticle systems. Instead of traditional treatment of a wave function as a solution of the Schrödinger equation in the position representation or representation of some dynamical observable, we will now find a totally different basis formed by *number states* or *Fock states*. This basis turns out to be a convenient and powerful tool to treat the systems of identical particles.

22.1
Quantum Oscillator Revisited

A quantum oscillator in one dimension is described by the Hamiltonian (9.70). Its energy is quantized and represented by a set of eigenvalues $E_n = (n + 1/2)\hbar\omega$ with $\omega \equiv \sqrt{k/\mu}$. We obtained this result in Section 9.6 by solving the stationary Schrödinger equation with the Hamiltonian (9.70) in the *x*-representation. Here, we will use another approach to this problem [1–3], which opens a new venue in describing systems of identical particles. Let us try to represent the sum of the squares in (9.70) as a product of two factors. This seems possible in view of the algebraic identity:

$$a_1^2 + a_2^2 = (a_1 + ia_2)(a_1 - ia_2) \equiv aa^* \tag{22.1}$$

(in which a_1 and a_2 are real numbers). So we introduce two new operators:

$$a = \frac{1}{\sqrt{2\hbar\omega}}\left(\frac{\hat{p}}{\sqrt{\mu}} - i\sqrt{k}x\right) \quad \text{and} \quad a^\dagger = \frac{1}{\sqrt{2\hbar\omega}}\left(\frac{\hat{p}}{\sqrt{\mu}} + i\sqrt{k}x\right), \tag{22.2}$$

and express $\hat{p}$ and x in terms of a and $a^\dagger$:

$$\hat{p} = \sqrt{\frac{\hbar\mu\omega}{2}}(a + a^\dagger), \quad x = i\sqrt{\frac{\hbar\omega}{2k}}(a - a^\dagger). \tag{22.3}$$

Note that unlike (22.1), $a^\dagger$ in (22.2) is not complex conjugate to a, since $\hat{p}$ in it remains unchanged.

Quantum Mechanics and Quantum Information: A Guide through the Quantum World,
First Edition. Moses Fayngold and Vadim Fayngold.
© 2013 Wiley-VCH Verlag GmbH & Co. KGaA. Published 2013 by Wiley-VCH Verlag GmbH & Co. KGaA.

The operators a and $a^\dagger$ do not commute. It is easy to show that they satisfy the simplest possible commutation relation for noncommutable operators:

$$[a, a^\dagger] = 1. \tag{22.4}$$

The Hamiltonian expressed in terms of $a, a^\dagger$ has the form

$$\hat{H} = \frac{1}{2}\hbar\omega(aa^\dagger + a^\dagger a) \equiv \frac{1}{2}\hbar\omega\{a, a^\dagger\}. \tag{22.5}$$

So it is not just a product of a and $a^\dagger$ as we wanted, but the anticommutator of $a^\dagger$ and a. Still it is a little simpler than (9.70).

Now recall the comment to Equation 9.92. It was suggested there that the quantum number n for an oscillator model can be interpreted as the number of identical particles all in one state that we can denote as $|n\rangle$. Applying this to the current case, we see that $|n\rangle$ is an eigenstate of the Hamiltonian (22.5):

$$\frac{1}{2}\hbar\omega(aa^\dagger + a^\dagger a)|n\rangle = \left(n + \frac{1}{2}\right)\hbar\omega|n\rangle \quad \text{or} \quad (aa^\dagger + a^\dagger a)|n\rangle = (2n+1)|n\rangle. \tag{22.6}$$

In other words, the anticommutator $\{a, a^\dagger\}$ of operators $a, a^\dagger$ is itself an operator with the eigenvalues $2n + 1$ and eigenstates $|n\rangle$. Actually, *it is the Hamiltonian in the $|n\rangle$-basis*. On the other hand, applying the commutator operator (22.4) to the same state, we have

$$[aa^\dagger - a^\dagger a]|n\rangle = |n\rangle. \tag{22.7}$$

Combining Equations 22.6 and 22.7 gives

$$aa^\dagger|n\rangle = (n+1)|n\rangle \quad \text{and} \quad a^\dagger a|n\rangle = n|n\rangle. \tag{22.8}$$

So $|n\rangle$ is also an eigenfunction of the product operators $aa^\dagger$ and $a^\dagger a$. Since applying either of these two products to $|n\rangle$ returns the same state multiplied by the corresponding occupation number, these products are also called the *number operators*. This brings us closer to the main question: What is the physical meaning and the properties of a and $a^\dagger$ themselves?

This question will be addressed in Section 22.2.

22.2
Creation and Annihilation Operators: Bosons

Here we outline the description of a specific mathematical formalism – second quantization, which turns out to be much more effective than the position representation of multiparticle systems considered in Chapter 21. This formalism starts with a very simple observation: According to the results of Section 22.1, the *product* of operators a and $a^\dagger$ acting on a state $|n\rangle$ *returns the same state* unchanged and only multiplied by an integer number associated with the number n of particles in the given state. A product of two operators leaves a state unchanged if the action of one operator cancels that of the other. It is natural therefore to assume that a *single* *operator a or $a^\dagger$* changes a state by decreasing or increasing its number of particles by 1. In order for the product of operators to take the state back to itself (up to a numerical factor), one operator must increase the number of particles by 1 and the other one must decrease this number by 1. We make this assumption specific and consistent with results (22.8) if we *define* the actions of operators a and $a^\dagger$ as

$$a^\dagger|n\rangle = \sqrt{n+1}|n+1\rangle, \quad a|n\rangle = \sqrt{n}|n-1\rangle. \tag{22.9}$$

Indeed, acting on $|n\rangle$ with $aa^\dagger$ and applying (22.9), we have

$$aa^\dagger|n\rangle = a\sqrt{n+1}|n+1\rangle = \sqrt{n+1}a|n+1\rangle = (n+1)|n\rangle,$$

which recovers the first of Equation 22.8. Acting on it with $a^\dagger a$ gives

$$a^\dagger a|n\rangle = a^\dagger\sqrt{n}|n-1\rangle = \sqrt{n}a^\dagger|n-1\rangle = n|n\rangle,$$

which recovers the second of Equation 22.8. Definition (22.9) is also consistent with results of Section 9.6. We had already obtained there (ahead of schedule!) the expression (9.79) for the operator "raising" a state $\psi_n \to \psi_{n+1}$ by one step up the energy scale. It is different from operator $a^\dagger$ introduced here only by the factor i. This just reflects the difference by the corresponding phase factor between $|n\rangle$ and eigenstates obtained in (9.74), that is, $|n\rangle \sim e^{-in(\pi/2)}|\psi_n\rangle$. We can also say that Equations 22.9 and 9.79 are identical, but one of them is written in the Heisenberg representation, while the other in the Schrödinger representation.

Thus, (22.9) gives the basic algorithm for actions of operators a and $a^\dagger$. According to it, each operator transforms the state $|n\rangle$ into a neighboring state. Operator $a^\dagger$ increasing the number of particles n by 1 is called the *creation operator* or, sometimes, the *raising operator*. Operator a decreasing n by 1 is accordingly called the *annihilation* or *lowering operator*. Both of them are also called the *ladder operators*, because by applying each one consecutively to an initial state, we can either increase or decrease the initial number of particles, raising or lowering the state as if taking it up or down the vertical ladder of states.

We can also look at it from another perspective if we note that all states $|n\rangle$ with different n, being the eigenstates of the Hamiltonian, are mutually orthogonal, $\langle n|n'\rangle = \delta_{nn'}$. Therefore, any transformation of states of the type $|n\rangle \to |n\pm 1\rangle$ is a rotation in $\mathcal{H}$ similar to rotation $\mathbf{x} \to \mathbf{y} \to \mathbf{z}$ of Cartesian axes in V. In other words, operator $a^\dagger$ applied to a state $|n\rangle$ rotates it trough 90° to state $|n+1\rangle$. Operator a rotates the same state through the same angle, but in the different plane to state $|n-1\rangle$. Thus, the ladder operators, raising or lowering states in the energy space, rotate these states in the $\mathcal{H}$-space.

Using definition (22.9), we can represent a state $|n\rangle$ as a result of successive raising steps starting from the vacuum state $|0\rangle$. Acting on $|0\rangle$ by the raising operator $a^\dagger$ produces a 1-particle state (mathematically, rotates eigenvector $|0\rangle$ to vector $|1\rangle$): $|1\rangle = a^\dagger|0\rangle$. The second step gives $a^\dagger|1\rangle = (a^\dagger)^2|0\rangle$ or, in view of (22.9), $|2\rangle = \frac{1}{\sqrt{2}}(a^\dagger)^2|0\rangle$. Reiterating the process, we obtain $|n\rangle$ expressed in terms of the vacuum state:

$$|n\rangle = \frac{1}{\sqrt{n!}}(a^\dagger)^n|0\rangle, \tag{22.10}$$

which recovers (9.80). Thus, all set of the number states can be "pulled out of vacuum" by successive action of the creation operator. This is just the image of the same process in position representation in Section 9.6, where we used recurrent relations to obtain all eigenstates from the ground state. Equation 22.10 is just (9.80) written in the basis of the number states.

Now, what happens if, instead of (22.10), we apply the *annihilation operator* to the vacuum state? The answer is obvious from simple consideration: There is nothing to annihilate there, so the result must be zero. Formally, this result follows automatically from (22.9): The lowering operator must turn $|0\rangle$ to $|-1\rangle$, but there are no states with negative number of particles. Accordingly, the rule (22.9) wisely gives the eigenvalue 0 to such state, so we have

$$a|0\rangle = 0. \tag{22.11}$$

This brings in a certain asymmetry to actions of $a^\dagger$ and a: The former, no matter how high up the ladder of states, can produce higher and higher states with no restrictions; the latter, no matter how high it may start, will be stopped at the vacuum state.

Let us see how these rules work when we need to find some observable properties of the system. Suppose we want to find the average $\bar{x}$ in the ground state of the oscillator. The ground state is just the vacuum state in the $|n\rangle$ basis, so we need to find $\bar{x} = \langle 0|x|0\rangle$. Using (22.3) gives

$$\bar{x} = i\sqrt{\frac{\hbar}{2\mu\omega}}\langle 0|(a - a^\dagger)|0\rangle = i\sqrt{\frac{\hbar}{2\mu\omega}}[\langle 0|a|0\rangle - \langle 0|a^\dagger|0\rangle] = 0 \tag{22.12}$$

(the first term in the square bracket is zero by virtue of (22.11), and the second term is zero by virtue of (22.10) at $n = 1$ and orthogonality of eigenstates).

Consider now the average $\overline{x^2}$. In view of (22.12), it is the same as indeterminacy $(\Delta x)^2$. The same algorithm gives

$$\overline{x^2} = -\frac{\hbar}{2\mu\omega}\langle 0|a^2 - aa^\dagger - a^\dagger a + (a^\dagger)^2|0\rangle = \frac{\hbar}{2\mu\omega}. \tag{22.13}$$

Here, only the second term gives a nonzero contribution. In this way, one can calculate the average of an arbitrary power of x or p or of any analytic function $f(x), \phi(p)$.

The considered case of a single oscillator can be generalized onto a system of N coupled oscillators. We know that vibrations in such system can be described as a set of N independent (uncoupled) normal oscillations or normal modes (recall an example at the end of Section 11.11). As an illustration, consider the case of equidistant identical coupled oscillators forming a linear chain of length L.[1] Then each normal mode can be described as a (generally, standing) wave with propagation number k:

$$k \to k_j = \frac{2\pi}{L}j, \quad j = 1, 2, \ldots, N, \tag{22.14}$$

---

1) Linear periodic chain of coupled oscillators in CM.

and the corresponding normal frequency

$$\omega_k \equiv \omega(k_j) = 2\omega_0 \sin \frac{1}{2} D k_j. \qquad (22.15)$$

Here, $D = L/N$ is the distance between two neighboring oscillators and ω_0 is the characteristic vibration frequency for an isolated oscillator. As seen from (22.14), the discrete variable k_j and accordingly $\omega(k_j)$ takes on N possible values. In the limit $L \to \infty$, variables k, ω become continuous.

In QM, the energy of each such oscillation is quantized according to

$$E(n_k) = \hbar \omega_k \left(n_k + \frac{1}{2} \right), \quad n_k = 0, 1, 2, \ldots. \qquad (22.16)$$

We must be careful not to confuse notations j in (22.14) and n_k in (22.16). The former is number of normal modes, whereas the latter is the number of excitations, that is, the number of (quasi)particles in the given mode. Since normal oscillations are independent, all previous results of this section apply to each one of them separately. In the same way as before, we can describe each oscillator in the basis of its own number states, with two distinctions: first, now we may have up to N different types (or families) of quanta with respective energies $\hbar \omega_k$ and variable number n_k of identical quanta in each family: while all the quanta (considered as particles) of the same family are identical, the different families are characterized by different ω_k. Accordingly, we will have N pairs of ladder operators $\left(a_k, a_k^\dagger \right)$ – one for each family. Second, the displacement operator that was a real number x in case of one oscillator can now be a complex number X_k because it is a component in Fourier expansion of regular displacement x_l [2,4]. Accordingly, the complex conjugates of operators of normal modes satisfy the condition:

$$X_k^* = X_{-k}; \quad \hat{P}_k^* = \hat{P}_{-k}. \qquad (22.17)$$

Otherwise, the mathematical structure of the theory remains the same. For each type of vibrations, we introduce the respective couple of its ladder operators:

$$a_k = \frac{1}{\sqrt{2\mu\omega_k}} \left(\hat{P}_k - i\mu\omega_k X_k^* \right), \quad a_k^\dagger = \frac{1}{\sqrt{2\mu\omega_k}} \left(\hat{P}_k^* + i\mu\omega_k X_k \right). \qquad (22.18)$$

Here, we have complex conjugate of $\hat{P}_k$ in the second equation (22.17) in order to compensate for the complex conjugate of X_k in the first equation. Conversely, using the rules (22.17), we find

$$\hat{P}_k = \sqrt{\frac{\mu \hbar \omega_k}{2}} (a_k^* + a_{-k}), \quad X_k = i \sqrt{\frac{\hbar}{2\mu\omega_k}} (a_k^* - a_{-k}). \qquad (22.19)$$

The commutation relations for the ladder operators now depend on whether they belong to the same family of normal modes ($k = k'$) or not ($k \neq k'$):

$$\left[a_k, a_{k'}^\dagger \right] = \delta_{kk'}; \quad [a_k, a_{k'}] = \left[a_k^\dagger, a_{k'}^\dagger \right] = 0. \qquad (22.20)$$

Using (22.19) and (22.20), we can express the Hamiltonian in terms of $a_k, a_{k'}$ in the same way as in (22.5):

$$\hat{H} = \frac{1}{2}\sum_k \hbar\omega_k \left(a_k^\dagger a_k + a_k a_k^\dagger\right). \qquad (22.21)$$

In all these relations, we must now specify the ladder operators with respect to the corresponding state k. In the considered situation, they create or annihilate quanta (the *phonons* in the language of solid state physics) with the wave number $k = k_j$. The number states of the Hamiltonian will now include N variables:

$$|n\rangle \rightarrow |n_1, n_2, \ldots, n_N\rangle. \qquad (22.22)$$

Here, n_1 is the number of phonons with $k = k_1$, n_2 is the number of phonons with $k = k_2$, and so on, up to $k = k_N$. Each of these numbers can be an arbitrary integer, $n_j = 0, 1, 2, \ldots$, with $j = 1, 2, \ldots, N$. Then a ladder operator acts on number state (22.22) according to

$$a_k^\dagger |n_1, n_2, \ldots, n_k, \ldots, n_N\rangle = \sqrt{n_k + 1}|n_1, n_2, \ldots n_k + 1, \ldots, n_N\rangle,$$
$$a_k |n_1, n_2, \ldots, n_k, \ldots, n_N\rangle = \sqrt{n_k}|n_1, n_2, \ldots, n_k - 1, \ldots, n_N\rangle. \qquad (22.23)$$

22.3
Creation and Annihilation Operators: Fermions

The theoretical scheme described in Section 22.2 works for bosons. For fermions, it must be modified in accordance with the exclusion principle, which reduces the number of members in each normal family down to zero or one. That is, in contrast to (22.23), each n_k is allowed to have only one of the two values, $n_j = 0, 1$.

Denote the raising and lowering operators for fermions as $b_k^\dagger$ and b_k, respectively. Their action on a number state is determined similar to (22.23), but with population numbers only 0 or 1. Therefore, for the raising operators, we must have

$$b_k^\dagger |n_1, n_2, \ldots, n_{k-1}, 0, n_{k+1}, \ldots, n_N\rangle = |n_1, n_2, \ldots, n_{k-1}, 1, n_{k+1}, \ldots, n_N\rangle, \qquad (22.24a)$$

$$\left(b_k^\dagger\right)^2 |n_1, n_2, \ldots, n_{k-1}, 0, n_{k+1}, \ldots, n_N\rangle = 0. \qquad (22.24b)$$

Equation 22.24b is imposed by the exclusion principle: For bosons, the repeated application of the raising operator would create two identical particles in the same state $|k\rangle$, but this is forbidden for fermions. Equation 22.24b is mathematical expression of this ban. It just means that for fermions

$$b_k^\dagger |1\rangle_k = 0. \qquad (22.25)$$

In this respect, (22.24b) or (22.25) is a mirror image of (22.11). The fermion state $n_k = 1$ for the raising operator is analogous to the state $n_k = 0$ for the lowering operator.

Similar to (22.25), for the lowering operators b_k, we have

$$b_k|n_1, n_2, \ldots, n_{k-1}, 1, n_{k+1}, \ldots, n_N\rangle = |n_1, n_2, \ldots, n_{k-1}, 0, n_{k+1}, \ldots, n_N\rangle, \tag{22.26a}$$

$$(b_k)^2|n_1, n_2, \ldots, n_{k-1}, 1, n_{k+1}, \ldots, n_N\rangle = 0. \tag{22.26b}$$

Relations equivalent to (22.24a, 22.24b, 22.26a, and 22.26b) can be obtained formally from (22.9) if we just change there the sign $n \to -n$ in the numerical factor for the raising operator:

$$b_k^\dagger|n_k\rangle = \sqrt{1 - n_k}|n_k + 1\rangle, \quad b_k|n_k\rangle = \sqrt{n_k}|n_k - 1\rangle. \tag{22.27}$$

These results are obvious: The first application of lowering operator to a state with $n_k = 1$ converts it to the vacuum state, so the second application produces nothing. Actually, this is just a more fancy way to write already familiar requirement (22.11), which is universal for all kinds of particles. We can summarize this as a simple property of both operators for any k:

$$\left(b_k^\dagger\right)^2 = (b_k)^2 = 0. \tag{22.28}$$

In this respect, the properties of fermions display a certain symmetry with respect to the ladder of states and the corresponding operators: The exclusion principle forbids motion up the ladder beyond the state with $n_k = 1$, so that motions in both directions are equally restricted and can occur only between the states $|0\rangle$ and $|1\rangle$. Is it not because of this symmetry that only the fermions build up the known matter, and bosons are only carriers of interaction forces between the fermions?

Using (22.27), we can reconstruct the properties of the number operators $b_k^\dagger b_k$ and $b_k b_k^\dagger$:

$$b_k^\dagger b_k|n_k\rangle = \sqrt{n_k(2 - n_k)}|n_k\rangle = \begin{cases} 0|n_k\rangle, & n_k = 0, \\ 1|n_k\rangle, & n_k = 1. \end{cases} \tag{22.29}$$

In the set of allowed values $n_k = 0, 1$, this is equivalent to $b_k^\dagger b_k|n_k\rangle = n_k|n_k\rangle$. In other words, $b_k^\dagger b_k$ is the number operator with the eigenvalues equal to population n_k.

Similarly,

$$b_k b_k^\dagger|n_k\rangle = \sqrt{1 - n_k^2}|n_k\rangle = \begin{cases} 1|n_k\rangle, & n_k = 0, \\ 0|n_k\rangle, & n_k = 1. \end{cases} \tag{22.30}$$

This means that $b_k b_k^\dagger$ is the number operator with the eigenvalues equal to "complementary" population number $1 - n_k$.

It is also instructive to consider the products of the ladder operators with different k.

Let us start with the vacuum state and simplify (22.24a) and (22.24b) to

$$b_k^\dagger|0\rangle = |1\rangle_k, \quad b_{k'}^\dagger|0\rangle = |1\rangle_{k'}. \tag{22.31}$$

Here, we applied two different raising operators to the vacuum state: one "pulling up" from vacuum the number state k and the other "pulling up" the number state k' (operators producing different eigenstates of an observable act independently).

Now we came to the point when the suppressed representation may be important for completeness of the argument. Let it be representation of some observable q compatible with k. Then we can denote the two produced number states in (22.31) more completely as $|1\rangle_k \to |k,q\rangle$ and $|1\rangle_{k'} \to |k',q'\rangle$, and (22.31) will take the form

$$b_k^\dagger |0\rangle = |k,q\rangle, \quad b_{k'}^\dagger |0\rangle = |k',q'\rangle. \tag{22.32}$$

This immediately suggests the way to produce two fermions at once by applying two raising operators in succession (here the coexistence of two fermions will not contradict anything since their states are different):

$$b_{k'}^\dagger b_k^\dagger |0\rangle = |k,q\rangle |k',q'\rangle. \tag{22.33}$$

But this would be only half-way toward the final result. The total state must be antisymmetric! Each raising operator in (22.33) takes care only of its respective k-eigenstate and knows nothing about q. Acting in the same succession, they could also produce the outcome with q and q' swapped:

$$b_{k'}^\dagger b_k^\dagger |0\rangle = |k,q'\rangle |k',q\rangle. \tag{22.34}$$

The total outcome must be their weighted superposition:

$$b_{k'}^\dagger b_k^\dagger |0\rangle = \frac{1}{\sqrt{2}} \left(|k,q\rangle |k',q'\rangle - |k,q'\rangle |k',q\rangle \right). \tag{22.35}$$

If the two operators act in the opposite order, the result will be

$$b_k^\dagger b_{k'}^\dagger |0\rangle = \frac{1}{\sqrt{2}} \left(|k',q\rangle |k,q'\rangle - |k',q'\rangle |k,q\rangle \right) = -b_{k'}^\dagger b_k^\dagger |0\rangle. \tag{22.36}$$

We see that the order matters, and specifically, $\left(b_{k'}^\dagger b_k^\dagger + b_k^\dagger b_{k'}^\dagger \right) |0\rangle = 0$. Note that the expression in parentheses is an anticommutator of $b_{k'}^\dagger$ and $b_k^\dagger$. In view of (22.25), it will produce the same result when acting on the state $|1\rangle$. Thus, it produces zero by acting on any of the two existing number states, which means that it is zero itself:

$$\{ b_{k'}^\dagger, b_k^\dagger \} \equiv b_{k'}^\dagger b_k^\dagger + b_k^\dagger b_{k'}^\dagger = 0. \tag{22.37}$$

The property (22.28) for $b_k^\dagger$ can be considered as a special case of (22.37) at $k = k'$. The same argument leads to a similar property for the lowering operator:

$$\{ b_{k'}, b_k \} \equiv b_{k'} b_k + b_k b_{k'} = 0. \tag{22.38}$$

We see the beautiful symmetry between the commutation relations (22.20) for bosons and anticommutation relations (22.37 and 22.38) for fermions. This reflects the symmetry–antisymmetry properties of the respective wave functions of the two kinds of particles.

In a way, we have made a step toward a powerful procedure in QM, named the *second quantization*. The first quantization acts on observables. The second quantization involves the wave function, because the discrete number of phonons in vibrating lattice or photons in EM field is quantization of the corresponding field, and accordingly the field itself can be represented by the corresponding operator [2–4].

Problems

22.1 Derive the commutation relation for creation and annihilation operators $a^\dagger$ and a

22.2 Express the Hamiltonian for harmonic oscillator in terms of $a^\dagger$ and a.

References

1 Ziman, J.M. (1969) *Elements of Advanced Quantum Theory*, Cambridge University Press.

2 Scadron, M.D. (2007) *Advanced Quantum Theory*, 3rd edn, World Scientific Publishing Co. Pte. Ltd.

3 Sakurai, J.J. (1984) *Advanced Quantum Mechanics*, Benjamin/Cummings Publishing Co.

4 Lipkin, G. (1973) *Quantum Mechanics, New Approaches to Selected Topics*, North-Holland Publishing Co.

23
Quantum Mechanics and Measurements

One of the paradoxes inherent in the quantum theory is what Roger Penrose referred to as the X-mystery – a mystery concerning the measurement problem [1]. In this chapter, we further explore some of the aspects of measurements in QM.

23.1
Collapse or Explosion?

What happens in the $\mathcal{H}$-space when we perform a measurement? We know that the state vector evolves deterministically as a time-continuous solution of the wave equation between measurements. Its norm remains equal to 1, but its tip traces out a continuous curve in the $\mathcal{H}$-space. Its components $c(q_n, t)$ in the chosen basis $|\psi_n\rangle$ are generally functions of time:

$$|\Psi(t)\rangle = \sum_n c(q_n, t)|\psi_n\rangle. \tag{23.1}$$

When the observable q is measured, all these components suddenly vanish except one, which at the same moment suddenly becomes unity. The set of nonzero amplitudes representing the state vector reduces to a single amplitude: $c(q_n) \to \delta_{mn}$ or $c(q) \to \delta(q - q')$, depending on whether the spectrum is discrete or continuous. Since these components defined the state vector at each moment of its continuous evolution, they stored the same information about the system as the vector itself. The discontinuous jump to an unpredictable m or q' means the *loss of all that information*. The system begins a totally new life.

Let us illustrate this with a quasi-free particle from Section 9.1. The most general state for a particle in a box (in the E-basis) is

$$|\Psi\rangle = \sum_n c_n e^{i(E_n/\hbar)t}|\psi_n\rangle, \tag{23.2}$$

with eigenfunctions $\psi_n(x)$ and eigenvalues E_n determined by (9.10) and (9.9), respectively. The spatial probability distribution in (9.11) is a function of time, and the state is not characterized by a definite energy – the energy spectrum is determined by the probability function (not to be confused with density matrix!)

Quantum Mechanics and Quantum Information: A Guide through the Quantum World,
First Edition. Moses Fayngold and Vadim Fayngold.
© 2013 Wiley-VCH Verlag GmbH & Co. KGaA. Published 2013 by Wiley-VCH Verlag GmbH & Co. KGaA.

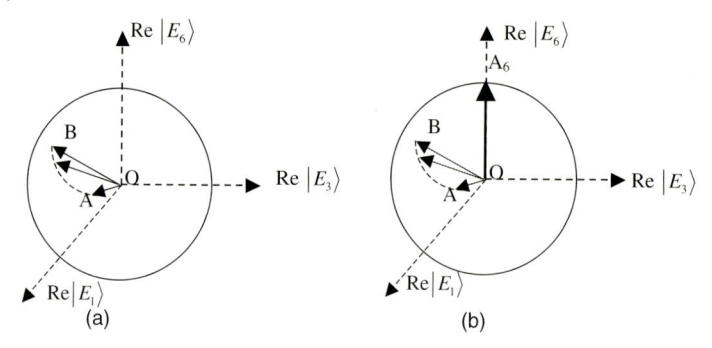

Figure 23.1 The energy measurement in the Hilbert space: (a) right before the measurement, (b) right after the measurement.

$p(E_n) \sim |c(E_n)|^2$ with amplitudes $c_n = c(E_n)$. Suppose that we want to *measure* the energy in state (23.2). If the measurement is performed without affecting the box, it will return one of the energy eigenstates, say, $|\psi_m\rangle$, and the corresponding eigenvalue $E = E_m$, with respective probability $|c_m|^2$. The superposition of eigenstates collapses to a single eigenstate with $n = m$. In the energy space, the set of amplitudes c_n collapses to a single nonzero amplitude δ_{mn}. In the position representation, this will be written as $\Psi(x) \to \psi_m(x)$.

The same process can also be considered in the $\mathcal{H}$-space. Stationary states in (23.2) form the basis for $\mathcal{H}$, so the dimensionality of $\mathcal{H}$ is infinite. In addition, its basis vectors are complex: $|E_n\rangle = \text{Re}\,|E_n\rangle + i\,\text{Im}\,|E_n\rangle$. We cannot represent all this graphically, so let us restrict to three dimensions, by choosing any three arbitrary eigenstates, in a subspace formed by real parts of these eigenvectors. Even this severely truncated picture can illustrate the basic features involved.

Prior to the measurement, the state (23.2) evolves continuously as described by the Schrödinger equation. To watch this evolution in the $\mathcal{H}$-space, one only has to look at the continuous curve traced out by the tip of the state vector $|\Psi(t)\rangle$ on the surface of the unit sphere (Figure 23.1 a). The curve AB in this figure is a historical record of the state evolution. We can call this evolution deterministic because it is determined by an analytical function of time, $\Psi(x, t)$, which is in turn determined by the Schrödinger equation. But even in this case, closest to our classical intuition, what is evolving in a deterministic manner is the *probability amplitude*, which is not a measurable characteristic of a system. It is only a tool for calculating the *expectation value* of any such characteristic. We still cannot say anything definite about the *specific value* of such characteristic (in our case, energy) until we perform a measurement.

Suppose an energy measurement is performed and the system is found in eigenstate No. 6. This outcome is illustrated in Figure 23.1b. The state vector "jumps" from a position it had right before the measurement (point B) to a totally disconnected new position (the tip A_6 of vector $|E_6\rangle$). In the figure, we still preserve the information about the prehistory AB by duplicating the curve AB from Figure 23.1a. But in the *actual measurement*, this information would be irretrievably lost. This type of measurement is irreversible.

It is essential to remember that the displacement from B to A_6 is a *discontinuous* process. The tip of the state vector does not proceed from B to A_6 by following one of the innumerous curves connecting these two points on the sphere. It just disappears as vector OB and immediately reemerges as vector OA_6. Hence, the name of the process is quantum jump or collapse of the wave function. This process *is not described by the Schrödinger equation or any other equation known to us*. It is one of those points where QM most radically departs from classical physics.

Suppose now that instead of energy, we measure x and find the particle at x'. Then the initial state would be converted into a totally different state that is a δ-function:

$$\Psi(x) \to \tilde{\Psi}(x,0) = \delta(x-x'). \tag{23.3}$$

(We assume that the measurement is performed at $t = 0$.) If the measurement (violent as it may be on the particle) does not damage the box, then $0 < x' < a$ and the new state (different as it is from $\Psi(x)$) is still a superposition of the same eigenstates $\Psi_n(x)$ of the box. But now it is a totally different superposition, with the amplitudes *other* than the amplitudes c_n in (23.2). According to Equation B.16, at the initial moment it can be represented as

$$\tilde{\Psi}(x,0) = \delta(x-x') = \sum_n \psi_n(x')\psi_n(x), \tag{23.4}$$

so the new expansion amplitudes are now determined as $\tilde{c}_n = \psi_n(x')$. After that the newly formed state evolves according to

$$\tilde{\Psi}(x,t) = \sum_n \tilde{c}_n \psi_n(x) e^{i(E_n/\hbar)t} = \sum_n \psi_n(x')\psi_n(x) e^{i(E_n/\hbar)t}. \tag{23.5}$$

But recall from Section 11.7 that for such an outrageously discontinuous state as the δ-function, we must use the *relativistic expressions* for energies E_n if we want to get an accurate description of the new evolution! Also note the irony: Whereas $\Psi(x)$ before the position measurement might have been a superposition of only a few (in our example, three) energy eigenstates, the new superposition (23.5) employs a larger (generally, infinite) set of eigenstates. Even if we start with just one eigenstate (e.g., $\Psi(x) = \psi_n(x)$), the x-measurement will yield the infinite set (23.5)! In this sense, the result of the x-measurement is an "anticollapse" or "explosion." As you can see, it is really a matter of which observable's eigenstates we are looking at! An actual measurement is a *combination of collapse and anticollapse* – the state function collapses in the $\mathcal{H}$-space of one observable and "explodes" in the $\mathcal{H}$-space of the other observable.[1] Thus, the measurement of x increases (up to infinity) the number of energy eigenstates but decreases (from infinity to one) the number of position eigenstates. As you think more of it, it should come as no surprise. The two observables in question are incompatible and the feature we stumbled upon is

---

[1] One obvious exception is the measurement of an observable in its own eigenstate, which, if carried out accurately, does not change the state.

yet another illustration of the trade-off between their indeterminacies. (Recall also that, as we saw in Section 13.2, by losing some information about particle's momentum in higher energy eigenstates, one can gain extra information about its position due to the more detailed account of the fine structure in its probability distribution within the box.)

Consider now a position measurement on a free particle. Let the free state at $t = 0$ be described by a "decent" wave packet, for example, a Gaussian function, $\Psi(x, 0) = \Psi_0 e^{-\alpha x^2}$, as in Section 11.3. Its Fourier transform is $\Phi(k) = \Phi_0 e^{-\beta k^2}$, where $\beta \equiv 1/4\alpha$. If position measurement at this moment finds the particle at $x = x'$, then the state $\Psi_0 e^{-\alpha x^2}$ collapses to the δ-function in the configuration space, much like the situation in Figure 23.2 (x-representation). At the same time there will be an "explosion" in the momentum space (p-representation) as the (initially narrow) packet $\Phi_0 e^{-\beta k^2}$ "blows up" into an infinitely wide and uniform packet (recall one of the definitions of δ-function in Appendix B). Both effects are two faces

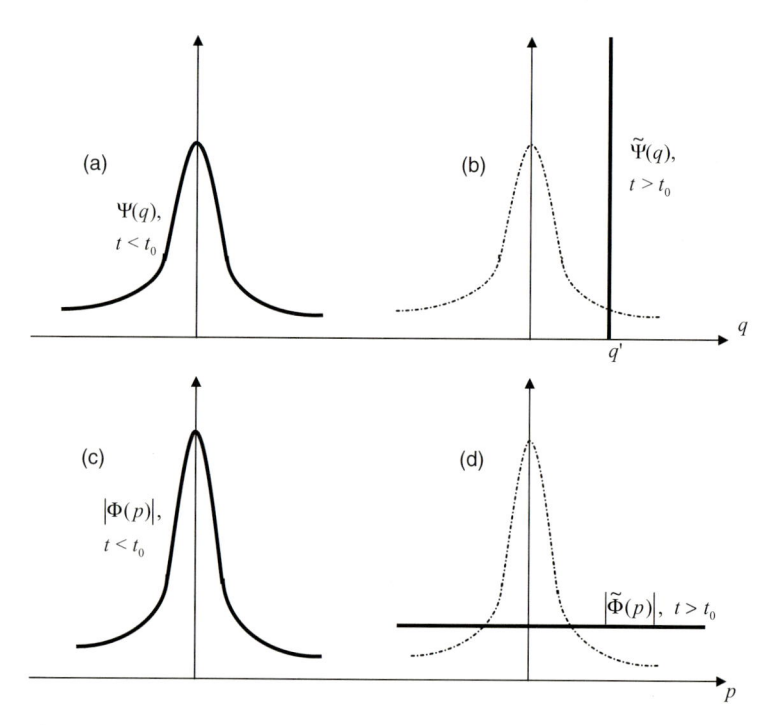

Figure 23.2 Double-faced nature of the state change in a measurement of an observable q. (a and b) The state change in q-space. (a) The observable q is indeterminate before the moment of measurement at $t = t_0$. (b) Right after this moment, we have $q = q'$. The initial state $\Psi(q)$ with finite Δq collapses to $\tilde{\Psi}(q) = \delta(q - q')$ with $\Delta q = 0$. (c and d) The same process now shown in p-space. (c) Right before the measurement. (d) Right after the measurement. The initial state $\Phi(p)$ with finite Δp "explodes" to the state $\tilde{\Phi}(p) = $ const with $\Delta p = \infty$. Both cases (a and b) and (c and d) are the graphical representations of *one* process in *the same* Hilbert space, but with respect to different bases. This result is immediately obvious from the indeterminacy relationship for incompatible observables.

of the same process. Analytically,

$$\Psi(x) = \Psi_0 e^{-\alpha x^2} \to \delta(x - x') \quad \text{(x-representation)}, \tag{23.6}$$

$$\Phi(k) = \Phi_0 e^{-\beta k^2} \to 2\pi \quad \text{(k-representation)}. \tag{23.7}$$

All this is illustrated graphically in Figure 23.2, which uses notations (p, q) instead of (k, x) to show that this effect takes place for any two incompatible variables. One can clearly see how "squeezing" the particle in the configuration space (collapse of the wave function) (Figure 23.2a and b) is accompanied by an explosion in the energy–momentum space (Figure 23.2c and d).

Now suppose that instead of position, we measured *momentum* for the state $\Psi_0 e^{-\alpha x^2}$ (or, which is the same, $\Phi_0 e^{-\beta k^2}$) and found it to be $k = k'$. In this case, the state collapses to a point $k = k'$ in the momentum space and "explodes" to an unbounded monochromatic wave in the configuration space. Analytically,

$$\Phi(k) = \Phi_0 e^{-\beta k^2} \to \delta(k - k') \quad \text{(k-representation)}, \tag{23.8}$$

$$\Psi(x) = \Psi_0 e^{-\alpha x^2} \to \Psi_0 e^{i(kx - \omega t)} \quad \text{(x-representation)}. \tag{23.9}$$

This also follows graphically from the same Figure 23.2, with q, p, $\Psi(q)$, $\Phi(p)$ swapped.

Two points are worth mentioning. First, if at least one of the observables p, q is continuous and measured with infinite accuracy, the measurement also changes the normalization condition. In Figure 23.2, both observables are continuous, and such a change is observed in both spaces. Before the measurement, representations $\Psi(q)$, $\Phi(p)$ are normalized according to (3.3); after the measurement, they are normalized according to (3.7).

Second, when considering the same process in both representations, we have to recall (Section 6.10) that the basis states of noncommuting operators reside in *the same* Hilbert space. Because the state $|x'\rangle$ that emerges from an x-measurement has a *totally indeterminate* momentum, the chances to find each possible value of p in a future p-measurement are equally distributed over the whole range of p. Each basis vector $|x\rangle$ is an equally weighted superposition of all basis vectors $|k\rangle$. Geometrically, it has equal projections onto all of them and thereby has equal angular distance from all of them. So to describe the process (23.7) – an x-measurement in the p-representation – you can just point out that the state vector suddenly jumps from its initial position (which may, generally speaking, be tilted toward – though not merging with – one of the vectors $|k_j\rangle$) to a position *equidistant* from all these vectors. This is one specific example of the general correlation between two sets of basis vectors: The orthonormal sets $|x_n\rangle$ and $|k_j\rangle$ are rotated with respect to one another in such a way that every $|x_n\rangle$ has the same angular distance from each $|k_j\rangle$ and vice versa.[2] As a result, when $|\Psi\rangle$ jumps to an eigenstate $|x_n\rangle$, its initial distribution of projections onto $|k_j\rangle$ (e.g., $\Phi(k_j)$ from (23.8)) jumps to a uniform distribution. The

---

[2] Note that we can restrict ourselves to discrete subsets of x and k without loss of generality.

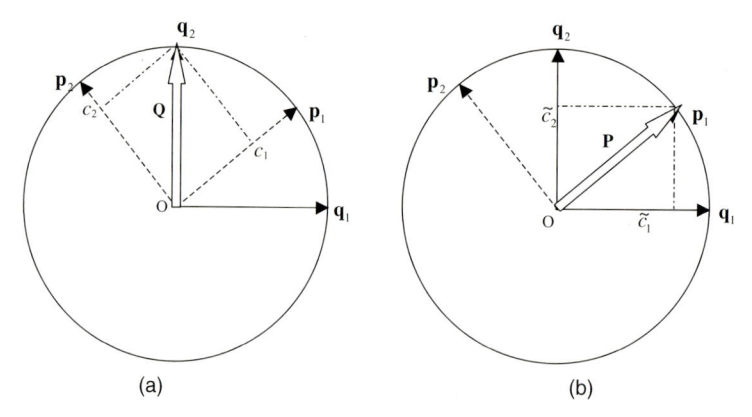

Figure 23.3 Measurement of observable p in a two-state system and its representation in the corresponding Hilbert space. (a) The state before measurement. The system is in an eigenstate $|q_2\rangle$ of observable q. The state vector is coincident with the basis vector $\mathbf{q}_2$. It has only one nonzero projection ($\tilde{c}_2 = 1$) in the q-basis, and two nonzero projections (c_1, c_2) onto the basis vectors ($\mathbf{p}_1, \mathbf{p}_2$), respectively, of the p-basis. (b) The state right after the p-measurement. The state vector is now coincident with the basis vector $\mathbf{p}_1$. It has only one nonzero projection ($c_1 = 1$) in the p-basis (the system "collapsed" to eigenstate $|p_1\rangle$ of observable p). On the other hand, it has now two nonzero projections ($\tilde{c}_1, \tilde{c}_2$) in the q-basis (the system anticollapsed from the single eigenstate $|q_2\rangle$ to a weighted superposition of $|q_1\rangle$ and $|q_2\rangle$). Both the collapse in p-space and the anticollapse in q-space are two faces of the same measurement. In the Hilbert space accommodating both bases, the state vector suddenly disappears as $\mathbf{Q}$ and immediately appears as $\mathbf{P}$.

corresponding value $\Phi = $ const is determined from the normalization condition and the fact that all $|k_j\rangle$ have equal weights[3] in the state $|x_n\rangle$. This gives $1/\sqrt{2}$ for a 2D space, $1/\sqrt{3}$ for a 3D space, and $1/\sqrt{N}$ for N-dimensions (see Figure 23.3). For a large N, the resulting state vector $|x_n\rangle$ will have vanishingly small projections onto any $|k_j\rangle$.

Before finishing this section, let us stop for a moment and have some fun. Imagine a scene from a science fiction movie. A certain agency keeps all secret information engraved on a 10 km straight steel rod so that it is impossible to steal it unnoticeably. An alien comes in, packs the rod into a miniature cubic box with the side of 1 cm, and carries it away unnoticeably for the guard. How is that possible? For the reader returning from a journey to the $\mathcal{H}$-space, the answer must be clear. The alien's box was a multidimensional cube. If such a cube has side a, its main diagonal is $D = (Na^2)^{1/2} = \sqrt{N}a$. This diagonal can be made arbitrarily large for a sufficiently large N (in our example, in order to accommodate 10 km rod, N must be greater than 10^{12}). So the rod was just rotated from 3D subspace to coincide with the diagonal of a 1 cm cube in ND space with $N > 10^{12}$.

---

3) Such uniformity is *not* a universal property of all pairs of incompatible variables. For instance, in a pair of variables (E, x) for a particle in a box, $|E\rangle$ is not an equally weighted superposition of $|x\rangle$, and vice versa.

To summarize, a measurement result is represented differently depending on what representation we use. In the p-representation of p-measurement, the state vector jumps (collapses) to one definite $|k_j\rangle$. In the x-representation of *the same* measurement, the state vector becomes equidistant from all $|x_n\rangle$ (the state "explodes" in the x-space to include all possible $|x_n\rangle$). For a pair of variables (x, p_x) with discrete spectrum (which can be realized in case of dispersed indeterminacy (Section 13.7)), this would mean that all components even out to one common value $1/\sqrt{N}$. At large N, all components become vanishingly small, and yet the norm of the state vector remains equal to 1.[4] Geometrically, the vector as seen by a 3D observer shrinks in its magnitude to a hardly, if at all, visible size. The result we have obtained for the x-measurement in p-representation is obtained now for p-measurement in x-representation. This reflects the equivalence of both representations.

23.2
"Schrödinger's Cat" and Classical Limits of QM

> The observer appears as an integral part of what is observed.
> *Stephen Rosen*

There is ancient problem in philosophy about the point at which accumulated incremental quantitative changes translate into a radical qualitative change. In this sense, QM is a treasure trove for philosophers because the transition between QM and CM is so fuzzy and tricky that one can't help wondering where the one ends and where the other begins. To accentuate the absence of a sharply defined boundary between micro- and macroscale, Schrödinger suggested a lurid *gedanken experiment* involving a macroscopic object – a cat (C) in a dark room. In one version of this experiment, a photon (P) hits a beam splitter (BS) where it can get reflected (state $|R\rangle_P$) or transmitted (state $|T\rangle_P$) with respective probability amplitudes a and b. If reflected, it exits the apparatus with no consequences and the observer who looks inside the room sees a living cat (state $|L\rangle_C$). If transmitted, it hits a detector (D), which then triggers the gun (G), and the observer finds the cat dead (state $|D\rangle_C$). The ultimate fate of the cat depends on the photon's state ($|R\rangle_P$ versus $|T\rangle_P$) after its interaction with the BS. Until the experimenter opens the door and checks what happened to the cat, the composite system must be in a weird entangled state described as

$$|\Psi\rangle = a|R\rangle_P \otimes |L\rangle_C + b|T\rangle_P \otimes |D\rangle_C. \qquad (23.10)$$

The first paradox is that such state of the cat is seemingly inconceivable for a macroscopic object: We are familiar only with live cats and dead cats, but no one can describe what it feels like to be in a superposition of both, let alone the entangled

---
4) In a continuous spectrum, all components acquire one common value $1/\sqrt{2\pi}$. This reflects normalization (3.7).

superposition like (23.10). Second, upon further reflection, we realize that we must introduce other objects into the last equation. In addition to being entangled with the photon, the cat's state cannot be separated from the state of the detector that remains idle in case $|R\rangle_P$ or fires in case $|T\rangle_P$; denote these two D-states as $|I\rangle_D$ and $|F\rangle_D$, respectively. This necessarily involves the states of the gun G that also remains idle in case $|R\rangle_P$ or fires, in synch with D, in case $|T\rangle_P$. Denote these two G-states as $|I\rangle_G$ and $|F\rangle_G$, respectively. We may add more and more objects (e.g., automatic video camera to record the cat's state), but this will only extend the chain of entangled states until the room is opened and the conscious observer peeps in. But the situation is even worse than that! Imagine that the observer himself is locked in the room. Since the observer is a physical object obeying the laws of Nature, he must be also included in the equation: He will either be in state $|L\rangle_O$ of seeing the live cat in case $|R\rangle_P$ or in state $|D\rangle_O$ of seeing the dead cat in case $|T\rangle_P$. Then the equation will tell us that the state of the observer is intimately correlated with all other relevant objects of the experiment, so we cannot in this case ascribe to the observer a separate wave function. Instead, we will have the observer himself in an entangled superposition [1–3]. Using the introduced notations for the corresponding states, we can write

$$|\Psi\rangle = a|R\rangle_P \otimes |I\rangle_D \otimes |I\rangle_G \cdots \otimes |L\rangle_C \otimes |L\rangle_O + b|T\rangle_P \otimes |F\rangle_D \otimes |F\rangle_G \cdots \otimes |D\rangle_C \otimes |D\rangle_O.$$
(23.11)

This chain of objects can never end because as soon as one attempts to identify the last link, one realizes that the state of that link is also in doubt! To untangle it and reduce (23.11) to a single product state, one must employ an apparatus to measure its state, but such an apparatus would immediately add a new link to the chain.

Von Neumann was well aware of this difficulty and his solution was to postulate the existence of an *ad hoc* process (he called it "process 1") whereby some object in the sequence (23.11) known as *von Neumann's chain*, would spontaneously jump into a definite state. Presumably (but not necessarily!) that would be an object far enough down the chain to be regarded as classical. In von Neumann's theory, process 1 is acausal: For example, if the object was the cat and if process 1 took the cat into a $|D\rangle_C$-state, one cannot explain how or why the choice was made. One can only predict the outcome probabilistically. This makes process 1 fundamentally different from "process 2" – deterministic evolution of $|\Psi\rangle$ under the Schrödinger equation. From the standpoint of a single object, process 1 is synonymous with the collapse of $|\Psi\rangle$.

As mentioned, QM does not explain how process 1 works to make the system choose a final state. And there is no real need to explain that. For one thing, the final result would be unaffected if a different object was chosen. If $|\Psi\rangle$ collapses to the second product state on the right, it is immaterial whether it was because the cat jumped into state $|D\rangle_C$ or because the gun jumped into state $|F\rangle_G$ or because the photon from start chose to jump into state $|T\rangle_P$. Either event would make the first product state impossible and the second product state certain. In this way, *any* link at which von Neumann's chain was cut would instantly collapse the entire chain. And, since such collapse happens in no time, it is meaningless to try identifying the cause and the effect here. Von Neumann's model implied that one can make a subjective decision as to where process 1 will "break the chain" or, using another common term, "choose a cut into the system."

Usually the object chosen for the cut is assumed to be "classical," that is, large enough for a human observer to interact with it without perturbing it. So the choice of the cut amounted to deciding which size makes the object classical. Such an object would effectively function as a measuring device (called simply the "apparatus") because as a result of "process 1," the quantum particle (e.g., a photon) starting the chain would jump into a definite state (e.g., $|R\rangle_P$ or $|T\rangle_P$) that can be identified from the readings of the apparatus (e.g., $|D\rangle_C$ versus $|L\rangle_C$). The apparatus is then said to be in a *pointer state*, on the theory that a set of its classically distinguishable states can be represented by the set of marks on the dial of a classical voltmeter. The set of pointer states forms a *pointer basis*, and the apparatus property whose values correspond uniquely to the pointer states is called the *pointer observable*. From this viewpoint, the cat is just another classical apparatus with two pointer states to measure the two quantum states of the photon.

Thus, according to von Neumann, a measurement amounts to an interaction between two systems – quantum and classical – that brings the classical system – the apparatus into one of its pointer states.[5] Classicality of the apparatus means that when *isolated*, it will not jump spontaneously from one pointer state to another. In particular, if a *spontaneous* transition between two distinct states of the apparatus is thermodynamically impossible (i.e., it would involve a large decrease of its entropy in a random fluctuation), they can be thought of as pointer states. Then, any observed transition between such states must be attributed to the interaction.

Von Neumann's model of the measurement was modified by the modern decoherence theory [4–9]. According to it, Nature itself chooses the cut for us. To show how that happens, we turn to another animal analogy. Imagine that the Schrödinger cat C is interacting simultaneously with a large number of mice (labeled M1, M2, ...), whom we, stretching things a little bit, will model by a set of mutually independent objects. Suppose that when a mouse sees a live cat, it runs away (let's call it state $|1\rangle$), and when it sees a dead cat, it rummages through the room looking for food (call it state $|2\rangle$). The states $|1\rangle$ and $|2\rangle$ are not necessarily orthogonal (the cautious mouse keeps its distance from the dead cat!). All we claim is that these states are not identical, so $|\langle 1|2\rangle| < 1$. This is a very reasonable requirement, as having the inner product equal to unity would mean an environment (the mice) totally indifferent to the state of the system (the cat). That's almost an oxymoron, since the state of a system is by definition described by the way it

---

5) For a continuous observable such as position, we can only consider a "practical" measurement, which should allow a small (e.g., ~1 μm) indeterminacy in the pointer's position. Otherwise, the measurement would give the system an infinite energy, which would have a more devastating effect than a bullet on the cat! However, as long as the mass involved is large enough, it would take an incredibly unlikely coincidence to produce a noticeable spontaneous change in our admittedly imperfect devices. (Think of the thermodynamic analogy: Random fluctuations are not likely to have a macroscopic effect that would spontaneously convert a live cat into a dead cat (or vice versa!).)

interacts with the environment. The entangled state of the cat and the mice can be described by

$$|\Psi\rangle = a|1\rangle_{M1} \otimes |1\rangle_{M2} \cdots \otimes |L\rangle_C + b|2\rangle_{M1} \otimes |2\rangle_{M2} \cdots \otimes |D\rangle_C. \quad (23.12)$$

Repeating the procedure outlined in Chapter 12, we will now write the density matrix of the composite system and trace out the mice. In particular, for a single mouse M, we have

$$\rho^{CM} = |\Psi\rangle\langle\Psi| = |a|^2 |L\rangle_C \langle L|_C \otimes |1\rangle_M \langle 1|_M + |b|^2 |D\rangle_C \langle D|_C \otimes |2\rangle_M \langle 2|_M$$
$$+ ab^* |L\rangle_C \langle D|_C \otimes |1\rangle_M \langle 2|_M + ba^* |D\rangle_C \langle L|_C \otimes |2\rangle_M \langle 1|_M,$$

$$\rho^C = \mathrm{Tr}_M \rho^{CM} = |a|^2 |L\rangle_C \langle L|_C \langle 1|1\rangle_M + |b|^2 |D\rangle_C \langle D|_C \langle 2|2\rangle_M$$
$$+ ab^* |L\rangle_C \langle D|_C \langle 2|1\rangle_M + ba^* |D\rangle_C \langle L|_C \langle 1|2\rangle_M,$$

and after a trivial generalization for the multiple mice,

$$\rho^C = \begin{pmatrix} |a|^2 & ab^* \langle 2|1\rangle_{M1} \langle 2|1\rangle_{M2} \cdots \\ ba^* \langle 1|2\rangle_{M1} \langle 1|2\rangle_{M2} \cdots & |b|^2 \end{pmatrix}. \quad (23.13)$$

The diagonal elements are just the probabilities of orthogonal states $|L\rangle_C$ and $|D\rangle_C$. The off-diagonal elements contain multiple inner products whose magnitude is less than 1. As time goes on, the cat will have interacted with more and more mice, and new inner products will appear in (23.13). Hence, the off-diagonal elements will decrease exponentially with time until the cat's state becomes indistinguishable from the mixture:

$$\rho^C = |a|^2 |L\rangle_C \langle L|_C + |b|^2 |D\rangle_C \langle D|_C. \quad (23.14)$$

Though, formally speaking, the system (23.13) remains entangled at all moments, the eventual outcome is classical for all practical purposes: The observer finds either a living cat or a dead cat, and in just the right percentage of trials. In other words, diagonal form of the density matrix is *classically interpretable*. The interaction of a classical system (the cat) with the multiple degrees of freedom (modeled by mice in the above example) of its environment that drives the reduced density matrix to a classically interpretable form is called environment-induced *decoherence*.

We define the *time of decoherence* t_d as the characteristic time needed for the off-diagonal elements to get reduced by a factor of e:

$$\langle 2|1\rangle_{M1} \langle 2|1\rangle_{M2} \cdots \sim e^{-t/t_d}. \quad (23.15)$$

The stronger the system interacts with its environment and the more the environmental degrees of freedom get involved, the faster the loss of coherence between the terms in the superposition (23.12). That's why we should expect t_d to be short for macroscopic objects, which interact with their environment in so many ways (e.g., by intense photon exchange) that the number of degrees of freedom in their environment is practically incalculable.

Models have been developed to estimate t_d for various scenarios. One such model proposed by Caldeira and Leggett in 1981 describes the decohering object as an open system coupled to a heat bath (i.e., a large thermal reservoir that represents the "environment"). Without going into details, we will only mention one important upshot of the model. Let's say, we want to know the position of an object to the accuracy Δx and initially different position states were correlated. For example, think of a system prepared in a superposition $|\psi\rangle = (1/\sqrt{2})|x_0\rangle + (1/\sqrt{2})|x_0 + \Delta x\rangle$ and placed in a parabolic potential so that in the macroscopic limit it will behave as a classical pendulum. Under some realistic assumptions (high temperatures and low potentials involved), the system's coupling with the environment can be modeled as a classical damping force $F_{\text{damping}} = -\eta v$. As the reader remembers, in CM, a damped harmonic oscillator with mass μ, spring constant k, and proper frequency $\omega_0 = \sqrt{k/\mu}$ is described by the equation $\mu(d^2x/dt^2) + \eta(dx/dt) + kx = 0$ with the solution $x = x(0)\cos(\omega t - \phi)e^{-\gamma t}$, where $\gamma = \eta/2\mu$ is the damping factor and $\omega = \sqrt{\omega_0^2 - \gamma^2}$ is the frequency of damped oscillations. The time $\tau = 1/\gamma$ is the characteristic time of damping. Detailed quantum-mechanical analysis of this system in the Caldeira–Leggett model yields the following result for decoherence time:

$$t_d = \tau \frac{\hbar^2}{4\mu k_B T (\Delta x)^2}. \tag{23.16}$$

To put this in perspective, notice that a 10 g pendulum with $\tau = 60$ s and $\Delta x = 1$ μm will decohere in $t_d \simeq 10^{-26}$ s at room temperature. Similar calculations show t_d to be on the order of nanoseconds for many *mesoscopic*[6] objects. Given how tiny t_d really is,[7] the "mystery" of instantaneous collapse of a particle's wave function has a ready explanation: It looks instantaneous to us because the macroscopic measuring device through which we interact with the particle becomes fully classical faster than we can notice.[8] So we can't notice superpositions of dead and alive cats for much the same reason we don't notice relativistic effects at low speeds. Such superpositions are immediately destroyed by a slightest interaction with the environment.

6) Mesoscopic means on the intermediate length scale between "macro" and "micro" – this includes objects that are less than a μm in size, but larger than an atom.
7) There are some notable exceptions: superfluids, superconductors, and laser pulses – all of these are many-particle and thereby macroscopic systems that yet maintain their coherence for a surprisingly long time.
8) There is an important clause to the above argument. In the final analysis, a phrase such as "faster than we can notice" makes an implicit reference to the observer's consciousness. If consciousness is contained within a physical object (e.g., brain neuron), it is reasonable to ask what happens during the short time when this object is entangled with the entire von Neumann chain that includes the measuring device. This is a legitimate question because for all we know, consciousness may well be a quantum phenomenon. Alternatively, if the brain is closer to an ordinary computer, storing and processing knowledge in classical bits, then states of the mind must be classical and t_d must be the time interval during which the mind remains inoperative. Even so, this does not constitute an exact answer as to why we cannot map the intermediate entangled quantum state of the neuron to a final classical state of the mind; in other words, why our brain thinks classically and not quantum-mechanically. It seems plausible, however, that such "quantum thinking" would offer little or no evolutionary advantage to classical life forms.

It is also truly remarkable that the environment helps address causation concerns. An observer who opens the door and sees a dead cat should not blame himself as the odds are overwhelming that the feline's fate was decided on a much lower (e.g., molecular) scale within the first microsecond. The very first sufficiently macroscopic object[9] is the chain (e.g., a large molecule) that decoheres before others affect a cut into the system.

The correspondence principle in the Schrödinger cat experiment is evident in that we only have two product terms in (23.13), one for the alive cat and the other for the dead cat, ensuring that after decoherence, a classical observer will see either the one or the other. Indeed, if we are consistent in applying QM, we have in the described case a superposition of two distinctly different *chains of events*, rather than superposition of just two distinct states of *only one object*.

Box 23.1 Does a system become "classical" after decoherence?

There is a common misconception that t_d is the time for a system to become "classical." The truth is that decoherence is basis dependent. You must make your choice of basis before writing the density matrix. Then decoherence will cause the off-diagonal elements to vanish, but only in that basis!

For example, consider a qubit interacting with its environment via the interaction Hamiltonian $H_{int} = -(1/2)\Lambda\sigma_x$ that is proportional to the Pauli spin operator (here magnetic field has been absorbed into the coupling constant Λ). In this simple case, there is no need to solve the eigenvalue problem: The eigenstates of H_{int} are those of σ_x, and the eigenvalues are the eigenvalues of σ_x, multiplied by $(-\Lambda/2)$. In other words, spin-right and spin-left states form the energy eigenbasis, $\{|R\rangle, |L\rangle\}$, and the respective eigenvalues are $\mp\Lambda/2$. Let us start with a superposition $|\psi\rangle = a|R\rangle + b|L\rangle$ where the amplitudes a and b can be time dependent. After a short while ($t_d \approx 3 - 20$ ns is a typical value for the parameters encountered in current chips), we end up with a mixture $\rho = \mathcal{P}_R|R\rangle\langle R| + \mathcal{P}_L|L\rangle\langle L|$ (where $\mathcal{P}_R$, $\mathcal{P}_L$ are the respective probabilities), which behaves classically with respect to the energy eigenstates. However, just because the off-diagonal elements are now zero in the energy basis does not mean they will be zero, for example, in the $\{|\uparrow\rangle, |\downarrow\rangle\}$ basis. In fact, we will have $\rho = \mathcal{P}_R \frac{|\uparrow\rangle+|\downarrow\rangle}{\sqrt{2}} \frac{\langle\uparrow|+\langle\downarrow|}{\sqrt{2}} + \mathcal{P}_L \frac{|\uparrow\rangle-|\downarrow\rangle}{\sqrt{2}} \frac{\langle\uparrow|-\langle\downarrow|}{\sqrt{2}} = \frac{1}{2}[|\uparrow\rangle\langle\uparrow| + |\downarrow\rangle\langle\downarrow|] + \frac{\mathcal{P}_R - \mathcal{P}_L}{2}[|\uparrow\rangle\langle\downarrow| + |\downarrow\rangle\langle\uparrow|]$ and for the state to be classical in the new basis as well, we need an equally weighted mixture $\mathcal{P}_R = \mathcal{P}_L$.

---

9) Note that we did not invoke the standard quantum-mechanical argument (large mass means short de Broglie wavelength, which means the system is weakly affected by measurements, implying the commutator of incompatible observables is very small and the object behaves classically). We only require that $t_d \to 0$.

Generally speaking, we cannot count on that! We can even make a stronger statement: Even if the superposition was *initially* equally weighted, $|\psi\rangle = (1/\sqrt{2})|R\rangle + (1/\sqrt{2})|L\rangle$, the resulting mixture will not be equally weighted if the amplitudes change with time. The reason is that the role of environment in this problem is twofold: In addition to decoherence, it also induces thermal transitions between energy levels until the qubit attains thermal equilibrium with its environment. The latter process is called relaxation, and is described by its own characteristic time t_r, which is generally different from t_d. The probabilities for a relaxed system are given by the Boltzmann distribution: $\mathcal{P}_R = 1/(1 + e^{-\Lambda/kT})$ and $\mathcal{P}_L = 1/(1 + e^{\Lambda/kT})$.

Suppose now that the qubit has fully relaxed and decohered and the thermal equilibrium state ρ has been attained. This system (we should now call that, instead of "qubit") is classical in the energy basis. No energy measurement will show any correlation between $|R\rangle$ and $|L\rangle$. However, the superposition still exists in the computational basis $\{|\uparrow\rangle, |\downarrow\rangle\}$, as shown by the nonzero off-diagonal elements, so the resulting equilibrium state can be highly quantum-mechanical! For example, an experimenter who measures σ_z may detect an interference pattern. The system may now remain in state ρ forever. Formally, we can explain this outcome by saying that the system has attained a state that is resilient to further decoherence. As far as σ_z-eigenstates are concerned, the system is still a qubit that can easily be used in a quantum computer. Furthermore, this system can become highly entangled by establishing correlations between spin-up and spin-down states and some other environmental observable, and all the while it will maintain its classicality in the energy basis! Upon a second thought, there does seem to be some hope for Schrödinger's cat to remain nonclassical for longer than one might expect, so when one makes a blanket statement about its supposedly "instant" decoherence, some caution is warranted!

As a special case of (23.11), consider an equal ($a = b$) superposition of product states:

$$|\Psi\rangle = \frac{1}{\sqrt{2}}(|0\rangle \otimes |0\rangle \otimes \cdots \otimes |0\rangle) + \frac{1}{\sqrt{2}}(|1\rangle \otimes |1\rangle \otimes \cdots \otimes |1\rangle), \qquad (23.17)$$

where the composite system $|\Psi\rangle$ is made up of many identical *qubits* and $|0\rangle, |1\rangle$ are the qubit eigenstates. After a measurement, all qubits will be found in the same state – either $|0\rangle$ or $|1\rangle$. By analogy with the Schrödinger's cat, states such as (23.16) are called *cat states*.

Recent experiments report successful creation of the Schrödinger's cat in a laboratory setup. Cat states involving up to six atoms have been produced in an experiment by Leibfried et al. [10]. De Martini et al. [11] have demonstrated a quantum superposition for a macroscopic "cat" – a pulse consisting of $N \simeq 10^5$

photons entangled with a faraway single-photon state. In principle, one could also envision a two-cat variant of Schrödinger's thought experiment where the reflected photon sets in motion a similar chain of events in a second chamber, killing a second cat. If one cat is found alive, the other will be found dead and vice versa. Such an experiment would entangle two macroscopic objects, as opposed to the micro–macro entanglement arising in the original configuration. The possibility of producing of such states in practice is discussed in several recent articles [12,13].

Decoherence is theoretically reversible (resulting in *recoherence*). In practice, however, any information stored in the off-diagonal terms dissipates into the environment and then it is hard or impossible to retrieve it – for much the same reason as air is never seen to separate spontaneously into oxygen and nitrogen.

Box 23.2 Decoherence as einselection

In Box 23.1, we saw an example of a system for which superpositions of E-eigenstates are extremely fragile, while superpositions of spin-up and spin-down states are surprisingly robust. There is another important lesson we can draw from this. Note that we did not give a strict proof that decoherence *must* occur in the $\{|R\rangle, |L\rangle\}$-basis. We just assumed it as a working hypothesis and studied the consequences. But these consequences are experimentally testable! Let an assembly of qubits interact with the environment and then do the measurements. If experiments indicate classical behavior with respect to $|R\rangle, |L\rangle$-states and quantum behavior with respect to $|\uparrow\rangle, |\downarrow\rangle$-states, then our assumption was correct; otherwise, the density matrix must be diagonal in some other basis, and we can devise a new series of tests to find out which one. In other words, there is always a *preferred basis* in decoherence process, such that environment will diagonalize the system only in that basis. This selection by the environment of one preferred basis (out of the infinitely many of them spanning the $\mathcal{H}$-space) is known as *einselection* [7–9]. If we call preferred states "most fit" in terms of their ability to remain stable notwithstanding environmental influences, we can conceptualize einselection as Darwinian survival of the fittest.

While Box 23.1 dealt with a single qubit, its real goal was to give a simplified model of the decoherence of a macroscopic system such as a cat or, more generally (and in line with the idea of breaking von Neumann's chain at some link), a measuring apparatus. In this context, the preferred basis is regarded as the *pointer basis* of the apparatus. The corresponding observable (e.g., x-component of spin in the above example) will then be the *pointer observable*. Thus, we have the following model of a classical apparatus: It is a macroscopic system that, when brought into a quantum superposition, will quickly decohere into a diagonal matrix and will be perceived by a conscious observer as existing in one of the pointer states that are "singled out by environmental influence from the larger quantum menu."

There arises a natural question: How does nature perform einselection? The answer is given by the dynamics of the decoherence process, which varies from problem to problem. In Box 23.1, the choice of the pointer basis was dictated by the fact that H_{int} was proportional to the Pauli spin operator. A realistic situation involving a macroscopic system will of course be much more complex; in particular, one should take into account the system's self-Hamiltonian H in addition to any interaction terms H_{int}. Also, the "environment" itself is not always described by a simple model; it may consist of several subenvironments and it need not even be isotropic (case in point: Casimir effect with its reliance on the geometric configuration of the vacuum). The emergence of a preferred basis in complex systems is a topic of ongoing research. In most cases, one would attempt to describe it by solving the *Caldeira–Leggett* master equation. For several relatively simple systems whose self-Hamiltonian is either ignored or is codiagonal (i.e., associated with the same energy eigenbasis) with H_{int}, this process is well understood. It turns out that in many cases environment tends to favor the energy basis as it did in Box 23.1. In particular, it was shown that einselection chooses the energy basis for the free particles and for harmonic oscillators.

On a final note, even though von Neumann's "process 1" is now understood better in the context of einselection, this does not remove its central mystery: God is still playing dice when the time comes to choose the outcome. The most we can get from the theory is the set of probabilities on the main diagonal!

23.3
Von Neumann's Measurement Scheme

> It is impossible to construct any logically consistent mechanical theory for a system made only of quantum-mechanical objects.
>
> L. Landau and E. Lifshitz

As we mentioned above, in von Neumann's measurement scheme, the apparatus is the object that "breaks the chain." It follows that there must exist at least one classical object in the Universe if we are to make any quantum measurements; otherwise, no object's state can be known for sure. Somewhat paradoxically, QM requires the presence of at least one classical object, while classical laws are themselves deduced as a limiting case of QM [14].

The dual role of the apparatus, which interacts with the quantum system and then with the classical observer, gave rise to the concept of a *quasiclassical* object [2,3,14,15]. That is, the apparatus is treated as an ordinary quantum object for the purposes of writing an entangled superposition such as (23.12). At the same time, it is deemed large enough that decoherence occurs almost instantly, and the superposition quickly collapses to a single product state.

Consider a quantum system (e.g., electron) $|\Psi\rangle$ in the proximity of the apparatus $|\Phi\rangle$. Let the electron have eigenstates $|\Psi_n\rangle$ with respective probabilities c_n, and let the apparatus have eigenstates $|\Phi_n\rangle$. Here, eigenstates $|\Psi_n\rangle$ are those of some observable F that we want to measure. Eigenstates $|\Phi_n\rangle$ correspond to some observable G related to the apparatus. For example, G could be the position of a devise pointer (this analogy prompts us to call $|\Phi_n\rangle$ "pointer states"). Before any interaction takes place, we should in principle have

$$|\Psi\rangle = \sum_n c_n |\Psi_n\rangle; \quad |\Phi\rangle = \sum_n d_n |\Phi_n\rangle. \tag{23.18}$$

However, the apparatus is quasiclassical, meaning that it doesn't like being in a superposition state for too long. Before the measurement even starts, the apparatus will have made up its mind, choosing one of its pointer states:[10]

$$|\Psi\rangle = \sum_n c_n |\Psi_n\rangle; \quad |\Phi\rangle = |\Phi_r\rangle. \tag{23.19}$$

When we consider the electron and apparatus as a composite system, our analysis should start with the product state:

$$|\Psi\rangle \otimes |\Phi\rangle = \left(\sum_n c_n |\Psi_n\rangle\right) \otimes |\Phi_r\rangle. \tag{23.20}$$

Next, the electron interacts with the apparatus. This interaction is considered as a quantum-mechanical process that is governed by a certain time-evolution operator $U(t)$. As the interaction Hamiltonian could be extremely complex, we shall not attempt to determine the form of such an operator. Suffice it to say that $U(t)$ is unitary and that it transforms the input state into some entangled superposition of the electron and the apparatus. This unitary evolution is sometimes called *premeasurement*. The apparatus is declared suitable for measuring the observable A if the evolution operator $U(t)$ and the eigenstate $|\Phi_r\rangle$ are such that

$$U(t)(|\Psi\rangle \otimes |\Phi\rangle) = \sum_n c_n U(t)(|\Psi_n\rangle \otimes |\Phi_r\rangle) = \sum_n c_n |\Psi_n\rangle \otimes |\Phi_n\rangle \tag{23.21}$$

or

$$U(t)(|\Psi\rangle \otimes |\Phi\rangle) = \sum_n c_n U(t)(|\Psi_n\rangle \otimes |\Phi_r\rangle) = \sum_n c_n |\Upsilon_n\rangle \otimes |\Phi_n\rangle. \tag{23.22}$$

States $|\Upsilon_n\rangle$ in Equation 23.22 are generally different from $|\Psi_n\rangle$. They are *not* eigenstates of F; they are normalized, but they need not be orthogonal. The exact form of these states depends on the details of the interaction and is inconsequential for our purposes.

After the premeasurement stage, the classical nature of the apparatus again comes into focus. Superposition (23.21) or (23.22) quickly decoheres and only

---

10) Of course, this is an idealization of a real system. Superpositions of eigenstates of the apparatus are excluded because they are statistically unlikely, and not because they contradict QM. In principle, we might need a separate measurement to prepare the apparatus in eigenstate $|\Phi_r\rangle$.

one term survives:

$$\sum_n c_n |\Psi_n\rangle \otimes |\Phi_n\rangle \Rightarrow |\Psi_m\rangle \otimes |\Phi_m\rangle \quad \text{or} \quad \sum_n c_n |\Upsilon_n\rangle \otimes |\Phi_n\rangle \Rightarrow |\Upsilon_m\rangle \otimes |\Phi_m\rangle. \tag{23.23}$$

The apparatus in the eigenstate $|\Phi_m\rangle$ is now independent of electron. The probability of the outcome (23.23) is $|c_m|^2$. (Note that the final state is again normalized!) This is the same process 1, or *wave function collapse*, that we mentioned earlier.[11]

In the final stage, the observer looks at the state $|\Phi_m\rangle$ of the apparatus. This is a classical measurement that leaves $|\Phi_m\rangle$ unchanged. The observer now has the knowledge of the electron state $|\Psi_m\rangle$ or $|\Upsilon_m\rangle$.[12]

Equation 23.21 describes a *von Neumann measurement of the first kind*, which collapses the electron into one of its eigenstates. These measurements are useful as idealized models, but seldom carried out in practice. The more general case (23.22) corresponds to a *von Neumann measurement of the second kind*. These measurements return outcomes $|\Upsilon_m\rangle$ with the same probabilities $|c_m|^2$ that we would expect from measurements of the first kind, but they don't leave the electron in one of the eigenstates of F. One extreme case takes place when we measure a photon's frequency using resonance absorption: We let the photon pass sequentially through a line of detectors, each detector tuned to a certain characteristic frequency. A successful measurement yields the frequency value but at the cost of the photon, which gets absorbed. In this case $|\Upsilon_m\rangle$ is a vacuum state $|N = 0\rangle$ of the EM field. Often, second-kind measurements are less dramatic and destruction of the object is avoided. At the same time, the majority of supposedly first-kind measurements will turn out to be of the second kind when you start reviewing approximations involved in the idealized model.

Conceptually, (23.21) describes a measurement that is "minimally invasive." Let's assume for simplicity that the eigenvalue spectrum of F is discrete. Then we can

---

11) Another name for this stage is *von Neumann projection*. We speak of von Neumann projections in the case (23.21) when $|\Psi\rangle$ gets "projected" onto an eigenstate $|\Psi_m\rangle$. In this context, projection means the destruction of entanglement in the composite state (23.21) or, to use formal language, transition from the pure state $U(t)(|\Psi\rangle \otimes |\Phi\rangle)$ to a mixture $\rho^E = \text{Tr}_A \rho^{EA} = \sum_n |c_n|^2 |\Psi_n\rangle\langle\Psi_n|$ where E stands for the electron and A for the apparatus (recall partial traces from Chapter 12). The reader is encouraged to generalize (23.21) for the case of degeneracy. In this case the term *Luders projection* is sometimes substituted for *von Neumann projection*.

12) The real situation can be more subtle than the one described by (23.21) and (23.22) because for a quasi-classical apparatus, different eigenvalues of the apparatus may be so densely spaced that they would form a practically continuous spectrum. Then it is not feasible to prepare the apparatus in one single state $|\Phi_r\rangle$ and the measurement will begin with the apparatus in some superposition of (infinitely many) states whose eigenvalues lie in a narrow interval $[G_r - \Delta G, G_r + \Delta G]$ around G_r. Such an apparatus is considered "sharply defined" for all practical purposes provided that $\Delta G \ll \langle G \rangle$. One can then use the linearity of $U(t)$ to apply (23.21) and (23.22) separately to each term in such a superposition and after the averaging, still get a result substantially identical to the right-hand side of (23.21) and (23.22).

make three equivalent statements about the measurement (23.21): (a) It is of the first kind; (b) when carried out on an eigenstate of F multiple times, it will always keep returning the same eigenvalue; and (c) it results in a mixture that will not change upon repetition of the measurement, and that is true for any arbitrary input state. The last two statements express *value reproducibility* and *repeatability* of first-kind measurements on discrete observables [16]. In case of continuous observables, the three statements are no longer equivalent. It can be shown formally that one can achieve first-kindness and value reproducibility,[13] but not repeatability [17]. Finally, measurements of the second kind are invasive measurements that are not value reproducible or repeatable. Even an eigenstate at the input will generally be disturbed beyond recognition by such a measurement.[14]

Box 23.3 Special case: nonoptimal measurement

In the above treatment, each time decoherence of the apparatus produces one of the pointer states $|\Phi_m\rangle$, the eigenstate of the system is determined precisely. This is the hallmark of an *optimal* measurement, that is, one designed to measure system's eigenstates, whether in a first- or second-kind measurement. If one and the same pointer state could have resulted from several eigenstates of the system, the measurement is *nonoptimal*. Such a measurement does not allow us to reach a definite conclusion about which eigenstate the system is in.

First we give an example of a measurement that is, technically speaking, optimal although quite useless to the experimenter. Suppose that we tried an electron spin measurement, and the electron got entangled with the apparatus, forming the analogue of a spin singlet state. Let the apparatus be initially in some state $|\Phi\rangle \equiv \sum_{S_z} c_{S_z}|S_z\rangle$ (here S_z is any possible eigenvalue of z-component of spin for the apparatus; the real apparatus is generally in a superposition of such eigenstates, with some average $\bar{S}_z$). After the singlet is formed, we have, dropping the normalizing factors and assuming maximally strong entanglement so that

---

13) In the continuous case, any first-kind measurement is value reproducible, but the converse statement is not true.
14) When reading the literature, it is important to keep in mind the kind of measurement being implied because this has some potential for confusion. For example, compare these two quotations, the former of which implicitly assumes von Neumann measurement of the first kind, and the latter measurement of the second kind:
"It is always possible to choose a measuring device so that the studied properties (of quantum system) will not change in the process of measurement . . . In principle, it is always possible to design an apparatus which could measure a given observable without changing it in the process of measurement." [18]; and ". . . the results of measurements in quantum mechanics cannot be reproduced. If the electron was in a state $\Psi_n(q)$ [which is an eigenstate of an operator f] then a measurement of the quantity f carried out on it, leads with certainty to the value f_n. After the measurement, however, the electron is in a state $\varphi_n(q)$ different from the initial one, and in this state the quantity f does not in general have any definite value. Hence, on carrying out a second measurement on the electron immediately after the first, we should obtain for f a value which does not agree with that obtained from the first measurement" [14].

both summands will be equally weighted, $U(t)(|\Psi\rangle \otimes |\Phi\rangle) = |\uparrow\rangle \otimes |\Phi_+\rangle - |\downarrow\rangle \otimes |\Phi_-\rangle$ where we denoted $|\Phi_\pm\rangle = \sum_{S_z} c_{S_z} |S_z \pm 1/2\rangle$. The correlation characterizing entanglement in this case must be observed in *every* trial, so we must increase/decrease spin by 1/2 in every term of superpositions $|\Phi_\pm\rangle$. Suppose now that, for instance, $|S_z| \gg 1$ and the states $|\Phi_+\rangle$ and $|\Phi_-\rangle$, while being different orthogonal vectors in $\mathcal{H}$, look the same to the classical observer: $|\Phi_+\rangle \approx |\Phi_-\rangle \approx |\Phi\rangle$. The *premeasurement* state is $U(t)(|\Psi\rangle \otimes |\Phi\rangle) \approx |\Phi_\pm\rangle \otimes (|\uparrow\rangle - |\downarrow\rangle)$ and is for all practical purposes a simple superposition. The *measurement* yields a definite result (spin-up or spin-down) after the apparatus decoheres, but the experimenter remains in the dark. This measurement is optimal though poorly designed.

Now suppose the entanglement was not maximal. We introduce, in the spirit of Section 15.3, parameters a, b, c, and d that might be positive or negative, and are not equal: $U(t)(|\Psi\rangle \otimes |\Phi\rangle) = a|\uparrow\rangle \otimes |S_{z,+}\rangle + b|\uparrow\rangle \otimes |S_{z,-}\rangle + c|\downarrow\rangle \otimes |S_{z,+}\rangle + d|\downarrow\rangle \otimes |S_{z,-}\rangle$. Then decoherence of the apparatus leaves the electron in a superposition and the measurement remains *inconclusive*. One also calls such measurements *unsharp* or *inexact*. Such measurements are of great theoretical importance in the derivation of the theory of generalized measurement. We will also encounter in Section 25.6 one situation where they are of greater utility than sharp measurements.

23.3.1
Discussion

Here, we mention a few examples of the first- and second-kind measurements.

a) *A polarizing filter (polarizer)*, say, with a horizontal transmission axis. It will absorb a $|\updownarrow\rangle$ photon, which will ideally be accompanied by scintillation or any other such recordable event. The act of absorption will then give us the observable value, but at the cost of the photon itself. A $|\leftrightarrow\rangle$ photon will pass, leaving the filter idle. The absence of absorption (assuming we knew that the photon *was* emitted) is the signature of the $|\leftrightarrow\rangle$ state. In this experiment, the input eigenstate is preserved half of the time, and half of the time it is destroyed with the photon. Since it is not preserved all of the time, the measurement is of the second kind.

b) *A polarizing beam splitter (PBS)*. This is a more "gentle" version of the previous experiment. Depending on the PBS orientation, one polarization state (say, $|\updownarrow\rangle$) will be transmitted undisturbed, whereas the other (respectively $|\leftrightarrow\rangle$) will be reflected.[15] Generally, the input photon may come in some arbitrary superposi-

---

15) Be sure to distinguish between *polarization direction* (i.e., where the electric field is pointing to) and *polarization state* (relation between the field vector and the direction of propagation). The former is rigidly bound to the photon's momentum and will change whenever the momentum changes. The latter is generally unchanged upon reflection.

tion of linearly polarized states. For example, it could have diagonal or circular polarization according to

$$|\updownarrow\rangle = \frac{1}{\sqrt{2}}(|\updownarrow\rangle + |\leftrightarrow\rangle), \quad |\otimes\rangle = \frac{1}{\sqrt{2}}(|\updownarrow\rangle + i|\leftrightarrow\rangle), \tag{23.24}$$

$$|\updownarrow\rangle = \frac{1}{\sqrt{2}}(|\updownarrow\rangle - |\leftrightarrow\rangle), \quad |\odot\rangle = \frac{1}{\sqrt{2}}(|\updownarrow\rangle - i|\leftrightarrow\rangle). \tag{23.25}$$

Here, $|\otimes\rangle$, $|\odot\rangle$ denote the right and left circular polarization state, respectively. A linear PBS will collapse the photon into a linear state $|\updownarrow\rangle$ or $|\leftrightarrow\rangle$. (This measurement is another example of wave function collapse – this time from the initial state (23.19) or (23.20) with indefinite linear polarization to one of the eigenstates $|\updownarrow\rangle$ or $|\leftrightarrow\rangle$.)

To tell transmission from reflection, we will want the photon to leave its signature by interacting (or not interacting!) with something behind the PBS. For example, we can put an electron in the *reflected* path so that an incident $|\updownarrow\rangle$ photon will leave the electron "idle," while a $|\leftrightarrow\rangle$ photon will make the electron recoil (the Compton effect!). This time the photon survives in either event and we have a measurement of the first kind with respect to polarization.[16] With respect to the momentum observable, this is a measurement of the second kind because momentum state changes in half[17] of the trials – the photon changes direction upon reflection and then again during the scattering.

c) *A modified PBS experiment.* We could in principle eliminate the electron entirely and have the PBS itself signal the measurement outcome to the observer. For instance, transmitted photon leaves the PBS state intact, while reflected photon changes its momentum by some amount $\Delta\mathbf{p}$, causing the PBS to recoil. If we could measure that recoil (a very difficult engineering challenge!), we would know for sure the path chosen by the electron. This time we have a "clean" polarization measurement of the first kind (no scattering to perturb polarization state!), but the momentum measurement is, technically speaking, again of the second kind. (Note that this remains true even though momentum and polarization commute!) But it is a measurement of the second kind where the original eigenstate can be reconstructed with certainty. This feature is due to the one-to-one correspondence between $|\Psi_m\rangle$ and $|\Upsilon_m\rangle$ if one thinks in terms of Equation 23.23. As a result, it is possible to restore the old momentum state simply by adding a mirror in the reflected path. This example shows that one can in principle make a measurement of the first kind on the entire complete set of commuting observables.

d) *Measurement of a spatial coordinate.* This is a notable special case. Such a measurement is always value reproducible if performed accurately *and* right after

---

[16] That is, provided that we neglect the change of photon polarization in Compton scattering. Note that the scatterer (electron) is part and parcel of the measurement procedure, which is not complete until *after* the scattering.

[17] Assuming an equal mixture of $|\updownarrow\rangle$ and $|\leftrightarrow\rangle$ photons; also note that in an N-dimensional $\mathcal{H}$-space, leaving the system completely undisturbed would be possible in a smaller (no more than $1/N$) fraction of trials.

the previous one. This requirement is imposed by special relativity. Two *different* outcomes for coordinate of a particle measured at two infinitesimally close moments of time would mean superluminal velocity of the wave packet between measurements, which is forbidden by SR.[18] On the other hand, we have seen in Section 11.8 that precisely in the case of idealized position measurement on a particle, QM equations predict just that! So where did we go wrong? The answer is that equations in this book describe *nonrelativistic* QM. When relativity is properly taken into account, coordinate measurements will remain reproducible and superluminal velocities in Section 11.8 will be capped by the light barrier (recall the "spikes" in Sections 9.3–9.5).

e) *Measurement of the photon number.* Here it is possible to have a first-kind measurement, which leaves the EM field in a definite number state. Suppose we have two field modes $|N_1\rangle$ and $|N_2\rangle$, where N_2 is known to us and N_1 is the observable we want to measure. We let the modes overlap in the physical space, which causes interaction between them. It can be described by the operator $U = e^{i\chi(N_1 \otimes N_2)}$, where χ is a coupling constant. This operator causes a phase shift in the second mode, which is proportional to the number N_1.

Box 23.4 Erasure and creation of information in a measurement

In von Neumann's model, the role of measurement is to transfer the preexisting information from the quantum level to the classical level with or without distorting the eigenstate. Von Neumann measurements ((23.21–23.23) are often called *selective* because when the input state is a superposition, the measurement tells us exactly which of the allowed possibilities was realized. In this sense, it creates new information. A variant of the selective measurement is measurement with *postselection*. It means that the experimenter will take an ensemble (pure or mixed), run a large number of tests, and out of all possible outcomes, keep only those that produced a desired result, thus obtaining a new pure ensemble. Postselection is usually implied when one speaks of "preparing a system" in a certain state. One can also have a *nonselective* measurement where the outcome is recorded, but we don't force the system to choose a specific path toward that outcome. This is the case in the double-slit experiment when we record the landing position but not the slit through which the particle passed, thereby keeping both amplitudes.

Sometimes the statement is made that such a measurement also completely obliterates the *old* information, destroying our knowledge about the previous state. This is not quite true. Even though after getting outcome $|\updownarrow\rangle$ we cannot tell the actual amplitudes c_1 and c_2 in the previous state $|\Psi\rangle = c_1|\updownarrow\rangle + c_2|\leftrightarrow\rangle$, we do know for sure that c_1 was nonzero. Moreover, if there was no initial reason for us to favor some pair

---

[18] We do not consider here the complicating issues like vacuum fluctuations that are essential in the quantum field theory.

of values (c_1, c_2) over the others, then after obtaining $|\uparrow\rangle$ we can claim that $|c_1|^2$ was probably closer to 1 than to 0. That's some retrospective knowledge about the preexisting state!

Let Alice prepare a quantum system in some (not necessarily eigen-) state $|\Psi_i\rangle$ and let Bob make a measurement (possibly of the second kind) on that system, obtaining an outcome $|\Upsilon_j\rangle$. By a prior arrangement made by both players, each possible preparation $|\Psi_i\rangle$ may be chosen by Alice with probability $P(|\Psi_i\rangle)$. Then the role of measurement is to change Bob's initial knowledge of the system, which amounts to the set of probabilities $\{\mathcal{P}(|\Psi_i\rangle)\}$, to the new set $\{\mathcal{P}'(|\Psi_i\rangle, |\Upsilon_j\rangle)\}$ that accounts for the outcome $|\Upsilon_j\rangle$ obtained. Bob can now make a better guess about the state initially prepared by Alice. Using measurement outcomes for retrospective analysis of the premeasurement state is the central idea of the important branch of QM known as *quantum-state tomography*.

As you see, saying that old information gets destroyed is not the best way to describe what happens in an irreproducible measurement. In case you wonder what is the right way, what you should really say is that quantum measurements are *Markovian* or *memoryless* processes. That is to say, they are the processes in which next stage of the process depends only on the current stage and not on the stages preceding it. In deciding upon its next step, the system only looks at where it is now. How it got there and whatever may have happened in the past are irrelevant to its decision. Some variants of the past may be more likely than others, but even if we could reconstruct the past precisely, that wouldn't matter when it comes to probabilities of future events. The Markovian property is formally expressed in the Schrödinger equation where the time evolution is determined by the current wave function.

23.4
Quantum Information and Measurements

> A measurement always causes the system to jump into an eigenstate of the dynamical variable that is being measured.
>
> P. Dirac

When apparatus A performs a measurement on some quantum system S, we describe A and S by respective sets of variables q_A and q_S. In the absence of interaction, A and S will have independent self-Hamiltonians $\hat{H}(q_A)$ and $\hat{H}(q_S)$. At this first stage, we often face the question: Where should we draw the line between systems A and S? As we saw in Section 23.2, this question is not so trivial.

To illustrate what's involved in this decision, we take a look at one simple example. We have already encountered the Stern–Gerlach (SG) experiment [18–22] in Chapter 7. As we mentioned then, the Ar atom has spin that is for all practical purposes equal to the spin of its outer shell electron. This spin (call it **S**) creates a magnetic moment $\mathcal{M} = -(ge/2\mu_e)\mathbf{S}$ (here $g = 2$ stands for the electron spin g-factor) and this magnetic moment interacts with the field **B**. The classical potential energy of a magnet placed in a magnetic field is $U = -\mathcal{M} \cdot \mathbf{B}$. Thus, interaction between the atom and the field gives rise to the corresponding term in the Hamiltonian: $H_{int} = (ge/2\mu_e)\mathbf{S} \cdot \mathbf{B}$.

This completes the description from the energy viewpoint. What happens in terms of forces? Obviously, field **B** exerts a force $\mathbf{F} = -\vec{\nabla} U = \vec{\nabla}(\mathcal{M} \cdot \mathbf{B}) = -(ge/2\mu_e)\vec{\nabla}(\mathbf{S} \cdot \mathbf{B})$ on the magnetic moment associated with spin [23]. If we align the field along z-axis, the force becomes $\mathbf{F} = \vec{\nabla}(\mathcal{M}_z B_z)$. If in addition B_z is uniform in the x- and y-directions, then $\mathbf{F} = \mathcal{M}_z(\partial B_z/dz)\mathbf{e}_z = -(ge/2\mu_e)(S_z(\partial B_z/dz))\mathbf{e}_z$.

So a force will be exerted on the magnetic moment associated with spin [23] as long as the field is inhomogeneous. The z-component of spin is $S_z = \pm(\hbar/2)$, so the force on the atom may point up or down depending on the sign of the gradient (which we can control) and on whether the electron is in the $|\downarrow\rangle$ or $|\uparrow\rangle$ state (which we cannot predict except probabilistically).

Our goal is to determine the spin state by looking at the deflection of the atomic trajectory. If the atom leaves a mark on the upper half of the screen, we can conclude that the spin was down and vice versa.[19] In this way, information about the microscopic quantum characteristic (atomic spin) is extracted indirectly by observing the macroscopic result (visible mark on the screen). Now, where is the borderline between "quantum" and "classical?" Our first impulse is to identify the atom with the quantum system S, and the rest (macroscopic surroundings) with the classical system A. This, however, would be incorrect [15]. The atom in this experiment leaves a classical mark on the screen, that is, its landing position is a pointer variable. To be sure, spreading of the atomic wave packet introduces some fuzziness in the mark's position, but this fuzziness is usually negligible compared to the apparatus dimensions and the whole atomic trajectory can be described classically to a good approximation. In view of this, the quantum domain reduces to the atomic electron only. The atom sheltering and carrying this electron must be considered as part of the apparatus A, together with the magnet, magnetic field, and the screen. Our first conclusion is that the apparatus need not be macroscopic. Under certain conditions, its role may be taken by a microscopic object [14]. This classical, even though microscopic, part of A serves as a "conveyer belt" for information transfer from the electron to the screen.

Now let us describe the process quantitatively. Before the measurement, the apparatus A and the studied quantum system S do not interact with each other.[20] The total Hamiltonian is the sum

$$\hat{H} = \hat{H}_S(S_z) + \hat{H}_A(z), \tag{23.26}$$

where the first and second terms are contributions from systems S (electron) and A (apparatus), respectively, S_z is the measured spin variable, and z is the vertical deflection of the atom. When the atom enters the field region, there appears an

---

19) That is, assuming $(\partial B_z/\partial z) > 0$. Otherwise, a mark in upper half of the screen will indicate the spin-up state.

20) If, as we said, we consider the nucleus a part of the apparatus A, then they interact by the Coulomb force independent of the atom's position. But this interaction does not count because it is not generating any new information about the spin state or uncovering any old information.

interaction term $H_{int} = (ge/2\mu_e)S_z B_z$ that depends on both variables, S_z and z (dependence on z is baked into the magnetic field: $B_z = B_z(z)$):

$$\hat{H}' = \hat{H}_S(s) + \hat{H}_A(z) + \hat{H}_{Int}(S_z, z). \tag{23.27}$$

The interaction lasts during the time $t_0 = l_0/v$, where l_0 is the length of the magnet and v is the initial speed of the atom. We want the interaction to be strong enough to produce a noticeable deflection of the atom's trajectory. If the interaction term is sufficiently large (or lasts a sufficiently long time), we can even neglect the part $\hat{H}$ in the total Hamiltonian during the interaction. In this case, we can with sufficient accuracy describe the evolution of the total wave function Ψ of the whole system (A+S) by the Schrödinger equation in the form

$$i\hbar \frac{\partial \Psi}{\partial t} \cong \begin{cases} \hat{H}\Psi, & t < 0 \quad (a), \\ \hat{H}_{Int}\Psi, & 0 < t < t_0 \quad (b), \\ \hat{H}\Psi, & t > t_0 \quad (c). \end{cases} \tag{23.28}$$

Here a question pops up: If $\hat{H}_{Int}$ is large, how do we know that we are really measuring the spin observable and not the change to that observable caused by $\hat{H}_{Int}$ itself? In order to have a "clean" experiment, we must ensure that $\hat{H}_{Int}$ is strong enough to change the "classical" coordinate z appreciably, yet not strong enough to damage the extremely fragile QM ("small" part of the experiment). That is to say, it should not cause any atomic transitions between $|\uparrow\rangle$ and $|\downarrow\rangle$ states (such transitions would amount to flipping the spin observable S_z, effectively destroying the original state *before* the information is extracted from it). The electron should be gently urged to choose one of the two spin directions, but not roughly shaken to the extent that the decision is really made by the field. (For example, if the original state involved equal probability amplitudes and in the experiment the majority of atoms drift upward, we will know that something was wrong in our setup.) How is that possible?

The atom's position responds (i.e., atom accelerates) to even a small force. Because of the strong Coulomb attraction between the electron and the nucleus (imagine a very tight spring), magnetic force $\mathbf{F} = -(ge/2\mu_e)(S_z(\partial B_z/dz))\mathbf{e}_z$ on the electron is transferred onto the nucleus without significant distortion of the atomic "shape."[21] The atom responds to this force almost immediately, while the spin state of the electron itself may remain substantially unaffected. So the atom's trajectory is sensitive to the *field gradient*, and not to the field itself.

What about spin? We can picture the electron as a magnet $\mathcal{M}$ placed inside the field $\mathbf{B}$. Classically, it would be subject to the torque $\boldsymbol{\tau} = \mathcal{M} \times \mathbf{B}$, which tends to align $\mathcal{M}$ with $\mathbf{B}$ (this minimizes classical potential energy). Conversely, a state in which $\mathcal{M}$ and $\mathbf{B}$ are antiparallel has the maximum potential energy and is therefore unstable.

---

[21] Of course, this also means the Lorentz forces should be much weaker than the Coulomb force. Otherwise, the atomic shape will deform, it will become difficult to treat the electron cloud as spherically symmetric, and we will have to account for the nonzero orbital angular momentum, introducing all kinds of complications such as spin-orbit coupling. Fortunately, in a practical setup, both v and B_z are small enough to remove this concern.

In terms of QM, this may cause a spin-flip, thus totally changing the spin state we wanted to measure. On the other hand, CM tells us that if the field is not excessively strong, the torque will only cause the spin to process[22] about the direction of **B** (in our case, this means the z-axis), without changing spin component S_z. Thus, spin state is sensitive to the field, but not to the field gradient. The SG experiment should succeed if we provide a large gradient of a weak field![23]

Let us now choose the optimal function $B_z(z)$. The most obvious solution is $B_z(z) = bz$ where b is a constant. To justify this choice, just notice that in the first approximation (first-order term of the Taylor series), *any* function can be written $B_z(z) \cong B_0 + bz$ with $b \equiv (\partial B_z/\partial z)|_{z=0}$; then, we choose $B_0 = 0$ to minimize the torque. However, keep in mind that in the real world, fields such as $\mathbf{B} = (bz)\mathbf{e}_z$ are expressly prohibited by the Maxwell equation $\nabla \cdot \mathbf{B} = 0$. Therefore, the treatment that follows should not be taken at its face value! It is only a convenient approximation of the conditions in a real experiment. With this reservation, let us now write the interaction Hamiltonian responsible for the ultimate vertical deflection:

$$H_{\text{int}} = (geb/2\mu_e) S_z z. \tag{23.29}$$

Before the measurement ($t < 0$, Equation 23.28a), when $B_z = 0$, atomic motion and electron spin (observables z and S_z) are uncorrelated [18]. Let $|\Phi_0\rangle$ be the initial spatial state ($z = 0$) and $|\Phi_+(t)\rangle$, $|\Phi_-(t)\rangle$ denote the two possible spatial states after the measurement ($z > 0$ and $z < 0$, respectively). These states evolve with time as the wave packet is traveling inside the apparatus. As usual, $|\uparrow\rangle$, $|\downarrow\rangle$ will denote spin-up and spin-down states. The composite system is initially in some superposition:

$$|\Psi_0\rangle = |\Phi_0\rangle(c_+|\uparrow\rangle + c_-|\downarrow\rangle), \tag{23.30}$$

where c_+, c_- are the initial amplitudes of spin states, which may or may not be known. The SG apparatus splits $|\Phi_0\rangle$ into spatial states $|\Phi_+(t)\rangle$, $|\Phi_-(t)\rangle$, entangling the classical and quantum parts:

$$|\Psi(t)\rangle = \frac{1}{\sqrt{2}} |\Phi_+(t)\rangle|\downarrow\rangle + \frac{1}{\sqrt{2}} |\Phi_-(t)\rangle|\uparrow\rangle. \tag{23.31}$$

This is a typical cat state and the atom here acts as a miniature version of the Schrödinger cat. In principle, if we could get rid of decoherence, we might consider a highly idealized SG thought experiment with a soccer ball instead of the atom, and with two screens placed in different galaxies, realizing the Schrödinger cat on an astronomic scale. Of course, under realistic conditions, collapse of the cat state will happen very quickly, most likely while the atom is still inside the apparatus.

Formally, one can think of (23.31) as an expansion of the composite system's wave function over the eigenfunctions of a quantum subsystem. Naturally, the expansion

---

22) This will be true when we have a single atom in the field (otherwise, interactions between different atoms complicate the picture). This means our beam of silver atoms should not be overly intense, so the interatomic distances could be kept sufficiently long on the average.

23) One can argue that gradient need not be too strong because we can amplify its effect by giving it a sufficiently long time $t_0 = l_0/v$ to work. Keep in mind, however, that Ag atoms produced by the oven will be moving fast and we cannot extend the length of the apparatus indefinitely.

coefficients $(1/\sqrt{2})|\Phi_+(t)\rangle$ and $(1/\sqrt{2})|\Phi_-(t)\rangle$ depend on the classical variable z, and also on time t that describes their evolution. But we can also think of (23.31) as an expansion over the eigenfunctions of the classical system, with $(1/\sqrt{2})|\downarrow\rangle$, $(1/\sqrt{2})|\uparrow\rangle$ serving as expansion coefficients.[24]

The Schrödinger equation $i\hbar(\partial/\partial t)|\Psi(t)\rangle = H_{\text{int}}|\Psi(t)\rangle$ splits into two separate equations for the orthogonal states $|\uparrow\rangle$ and $|\downarrow\rangle$ and a general solution of the resulting system is found as follows:

$$i\hbar\frac{\partial|\Phi_+(t)\rangle}{\partial t}|\downarrow\rangle + i\hbar\frac{\partial|\Phi_-(t)\rangle}{\partial t}|\uparrow\rangle = \frac{gebz}{2\mu_e}\left(-|\Phi_+(t)\rangle\frac{\hbar}{2}|\downarrow\rangle + |\Phi_-(t)\rangle\frac{\hbar}{2}|\uparrow\rangle\right),$$
(23.32)

$$\begin{cases} i\hbar\dfrac{\partial|\Phi_+(t)\rangle}{\partial t} = -\dfrac{\hbar}{2}\dfrac{gebz}{2\mu_e}|\Phi_+(t)\rangle \\ i\hbar\dfrac{\partial|\Phi_-(t)\rangle}{\partial t} = \dfrac{\hbar}{2}\dfrac{gebz}{2\mu_e}|\Phi_-(t)\rangle \end{cases} \Rightarrow \begin{cases} |\Phi_+(t)\rangle = |\Phi_+(0)\rangle e^{(i/2)(gebz/2\mu_e)t} \\ |\Phi_-(t)\rangle = |\Phi_-(0)\rangle e^{-((i/2)(gebz/2\mu_e)t)} \end{cases}.$$
(23.33)

Stitching together Equations 23.30 and 23.31 at $t = 0$, we see that the initial conditions must be

$$\frac{1}{\sqrt{2}}|\Phi_+(0)\rangle = c_-|\Phi_0\rangle, \quad \frac{1}{\sqrt{2}}|\Phi_-(0)\rangle = c_+|\Phi_0\rangle.$$
(23.34)

Plugging that back into (23.33), we get the total wave function $|\Psi(t)\rangle$:

$$\begin{cases} |\Phi_+\rangle = \sqrt{2}c_-|\Phi_0\rangle e^{(i/2)(gebz/2\mu_e)t} \\ |\Phi_-\rangle = \sqrt{2}c_+|\Phi_0\rangle e^{-((i/2)(gebz/2\mu_e)t)} \end{cases} \Rightarrow |\Psi(t)\rangle = c_-|\Phi_0\rangle e^{(i/2)(gebz/2\mu_e)t}|\downarrow\rangle + c_+|\Phi_0\rangle e^{-((i/2)(gebz/2\mu_e)t)}|\uparrow\rangle.$$
(23.35)

Solution (23.35) describes the system evolving within the time interval $0 \leq t \leq \tau$, where τ is equal to the time of decoherence t_d for the atom or to the time t_0 the atom spends in the field region, whichever is shorter. Strictly speaking, time–energy indeterminacy $\Delta t \geq \hbar/\Delta E$ prevents us from knowing t_0 exactly. However, if l_0 is much larger than the wave packet width, we can treat the atomic longitudinal motion classically, setting $t_0 = l_0/v$:

$$\tau = \min(t_d, t_0) = \min\left(t_d, \frac{l_0}{v}\right).$$
(23.36)

---

24) In presentation [14] of the same problem, the latter expansion is performed. However, expansion over the *quantum-mechanical* eigenstates seems to be more appropriate: First, it is the observable S_z, and not z, that is the ultimate goal of the measurement, and second, the corresponding quantum eigenstates are "primary" or "fundamental" standard functions, while the "classical" eigenstates of the apparatus emerge as a response to the interaction between the quantum and classical parts and depend on the specific details of the experiment. For instance, the range of possible values of z on the screen depends on magnetic field **B**, kinetic energy of the atomic beam, mass of the atom, the length of the magnet, and the distance between the magnet and the screen.

If the decoherence time is long, the atom exits the apparatus as an entangled superposition:

$$|\Psi(\tau=t_0)\rangle = |\Phi_0\rangle\left(c_- e^{ik_z z}|\downarrow\rangle + c_+ e^{-ik_z z}|\uparrow\rangle\right), \quad k_z = \frac{gebl_0}{4\mu_e v}, \qquad (23.37)$$

where k_z characterizes the z-component of the atom's motion (the ultimate transverse component of its propagation vector **k**). This equation tells us that the original wave packet $\Phi_0(z)$ has split into two packets moving away from one another along the z-axis even as they both keep progressing in the x-direction. For an arbitrary moment t within the interval $0 < t < t_0$, when the atom has traveled the distance $l(t) = vt$ inside the apparatus, the wave function (23.35) depends explicitly on time through $k_z(t) = (geb/4\mu_e)t$. Physically, this means that the two split parts of the atom's wave packet depart from each other with acceleration. Accordingly, the split parts trace out parabolic trajectories as if they were projectiles in a gravitational filed, with masses of opposite signs. In actuality, however, it's one and the same atom and it is moving simultaneously along both paths. But we should not confuse this state with an ordinary superposition! Here each spatial part is entangled with a corresponding "splinter" of the original spin state of the electron. If two independent experimenters Alice and Bob measure the atom's position (by using the SG scheme) and the electron's spin (by using some other procedure), their results will always correlate. For instance, when Alice finds the atom deflected upward, Bob finds that the spin is pointing down and vice versa.[25]

When the state (23.37) finally decoheres at time $t_d > t_0$ or when the atom hits the screen (whichever happens earlier), one of the terms in (23.37) actualizes and the other disappears. Accordingly, one of the coefficients c_+, c_- jumps to 1 and the other to zero. The final state will be given by *only one* of the two expressions:

$$|\Psi(t)\rangle = \begin{bmatrix} |\Phi_0\rangle e^{ik_z z}|\downarrow\rangle \\ |\Phi_0\rangle e^{-ik_z z}|\uparrow\rangle \end{bmatrix}. \qquad (23.38)$$

On the other hand, if the decoherence time is short, the atom will "choose" one of the two spatial terms long before it exits the apparatus. This case will again be described by (23.38), with wave number k_z defined by the decoherence time: $k_z(t) = (geb/4\mu_e)t_d$, and with the vertical coordinate also determined at the moment

---

25) In practice, it may be quite difficult to arrange an experiment where separate measurements on the atom and on the electron would be truly independent. The EPR pair of entangled photons would be much easier to handle because they are spatially separated and don't interact at all. Here, we would essentially need to find some way to break the atom, but so "gently" that the electron doesn't notice it. But, at least in theory, our example with Alice and Bob is legitimate because z and S_z are compatible observables.

t_d, that is, with $z = z(t_d) \equiv z_d$:

$$|\Psi(\tau = t_d)\rangle = \begin{bmatrix} |\Phi_0\rangle e^{i(geb/4\mu_e)t_d z_d}|\downarrow\rangle \\ |\Phi_0\rangle e^{-i(geb/4\mu_e)t_d z_d}|\uparrow\rangle \end{bmatrix}. \tag{23.39}$$

During the remaining time $t_d \leq t \leq t_0$, the resulting product state will continue to evolve under the influence of the interaction Hamiltonian:

$$|\Psi(t_d \leq t \leq t_0)\rangle = \begin{bmatrix} |\Phi_0\rangle e^{i(geb/4\mu_e)zt}|\downarrow\rangle = |\Phi_0\rangle e^{ik_z(t)z}|\downarrow\rangle \\ |\Phi_0\rangle e^{-(i(geb/4\mu_e)zt)}|\uparrow\rangle = |\Phi_0\rangle e^{-ik_z(t)z}|\uparrow\rangle \end{bmatrix}, \quad k_z(t) = \frac{geb}{4\mu_e}t. \tag{23.40}$$

After $t = t_0$, the interaction stops, and k_z maintains its final value $k_z(t \geq t_0) = gebt_0/4\mu_e$ from now on. The atom keeps going at its constant final vertical velocity without acceleration. The end result is identical to (23.38), so it will not matter to the observer whether the system left the apparatus and then decohered or it decohered first and then left the apparatus.

If the system is prepared in one of the spin eigenstates, the wave function always remains "packed" in a single packet, which only changes its direction, deflecting toward the top or the bottom of the screen. Unfortunately, the original SG experiment does not preserve the spin eigenstate because the atom collides with the screen and so we have a von Neumann measurement of the second kind. To have a measurement of the first kind with respect to spin, we must improve the SG setup by replacing the screen with a large number of scatterers that absorb only a small fraction of the atom's kinetic energy, leaving the spin state more or less intact. If we then pass the atom through another SG apparatus, we will again get the same value of S_z, meaning that our improved measurement scheme is repeatable. This scheme can even leave the momentum state undisturbed as long as we can neglect the small deflection of the atom (change of **k**), which is acceptable in the approximation $k_z \ll k$. A purist can eliminate this small deflection by employing another SG apparatus with the opposite direction of magnetic field gradient [24]. This will return the atom, without disturbing its internal state, to its initial horizontal path.

In the general case c_+ and c_- are *both* nonzero. The outcome probabilities are determined by $|c_+|^2$ or $|c_-|^2$, and their sum is, of course, equal to 1. The *individual outcome* in this case is principally unpredictable. We can still have a von Neumann measurement of the first kind. However, this measurement now prepares a new state, so in this type of experiment one cannot perform consecutive trials on one and the same sample, because the sample will not be the same after the first trial.

When $(t_d - t_0) \to 0$, we can think of the whole measurement as an instantaneous process. It is sometimes convenient to skip the "intermediate" stage, focusing instead on the system's state before and after the measurement. But we should keep in mind that the entangled state is what connects the initial state (observable S_z indeterminate) and the final state ($S_z = \pm \hbar/2$). The entangled state does not form if only S_z was already defined prior to the measurement. What is really astounding is that QM can so accurately predict probabilities of the possible measurement outcomes *without* any need to watch the evolution of the system between "premeasured" and "postmeasured" states. It is this feature that makes us think that Nature is intrinsically probabilistic. Even if quantum behavior admits any explanation at all, it is in vain that one would look for such an explanation in the specifics of the measurement procedure.

Two important conclusions follow from the above discussion. First, it is not always true that the "quantum-mechanical" should be small and the "classical" big. Second, the quantum system is not always spatially separated from a "classical" apparatus. The quantum and classical parts can be both packed *within* the same object (e.g., an electron within an atom) or two different observables describing that object (e.g., SG experiment for a muon would show that the muon's trajectory is roughly classical, while its spin is perfectly quantum; see also Problem 23.7). One often hears that observation of wave-like and particle-like aspects of an object requires mutually exclusive experimental situations; however, here we observe both aspects in the same experiment!

The only reason this is possible is because we never tried to pinpoint the electron *exactly*. The thickness of the mark left by the electron on the screen and the thickness of a cloudy track in a Wilson chamber are macroscopic objects much bigger than the atomic size. When the path is determined with such low accuracy, the electron is an entirely classical object. At the same time, the electron is *always* quantum-mechanical with respect to its spin [14]. This is to be expected because spin does not lend itself to any classical interpretation.

In terms of the information transfer, the flow of information in such experiments occurs not in the real space, but rather in the $\mathcal{H}$-space of the particle – from the quantum attribute of this particle to the classical one. Sometimes an intimate connection between these two attributes produces a quantum entanglement between them, but the final measurement result is always a pointer state of a classical observable. And it is precisely this peculiarly correlated symbiosis that enables us to infer the quantum face of reality by studying its classical face.

Exercise 23.1

Evaluate the torque on an atom in the SG experiment, assuming the nonzero magnetic field.
Solution:

a) If the electron is in a spin eigenstate, the magnetic moment has a *definite z-component* collinear with **B**. Classically, we represent it by an arrow pointing either strictly up or strictly down, which implies that the angle between the spin and the z-axis is either 0 or 180°. According to (23.30), this would mean zero torque and accordingly no danger of spin-flip during the measurement (ignoring the relatively small instability in the state with $\theta = 180°$). Quantum-mechanically, however, the angle θ between **S** and **B** is given by $\cos\theta = m/\sqrt{s(s+1)}$, where m and s are corresponding quantum numbers of a given spin state. For the electron spin, we have $s = 1/2$ and

$m = \pm 1/2$, which gives us $\sin\theta = \sqrt{2/3}$, and the torque is $\tau = \sqrt{2/3}\mu B$ regardless of the sign of m.

b) If S_z does not initially have a definite value, we can only talk about the average torque for a given initial state of the electron. If the spin-up and spin-down amplitudes have equal amplitudes ($c_+ = c_- = 1/\sqrt{2}$), then the average over the state will be the same as in case (a) (show this).

23.5
Interaction-Free Measurements: Quantum Seeing in the Dark

So far we have seen examples where a measurement perturbs the system being measured, sometimes even changing it beyond recognition as in photon absorption experiments. In this section, we discuss the opposite example demonstrating the possibility of interaction-free measurements [25].

We start with the thought experiment first proposed by Elitzur and Vaidman [26]. The experiment uses a Michelson or Mach–Zehnder interferometer [27]. Light entering the device is divided by the BS into two beams following the two different paths shown in Figure 23.4. They meet at the second BS. After the second splitting, the final transmitted part of the lower beam and the final reflected part of the upper beam interfere destructively, so nothing emerges in this direction; the detector placed in the way of this would-be beam remains idle. Using notations in [25], we denote this detector as D_{dark}. The reflected part of the lower beam and the transmitted part of the upper beam (both directed to go horizontally) interfere constructively, thus restoring the initial beam. If the incident light is dimmed so that we have only one photon at a time, the detector in the way of the horizontal output beam

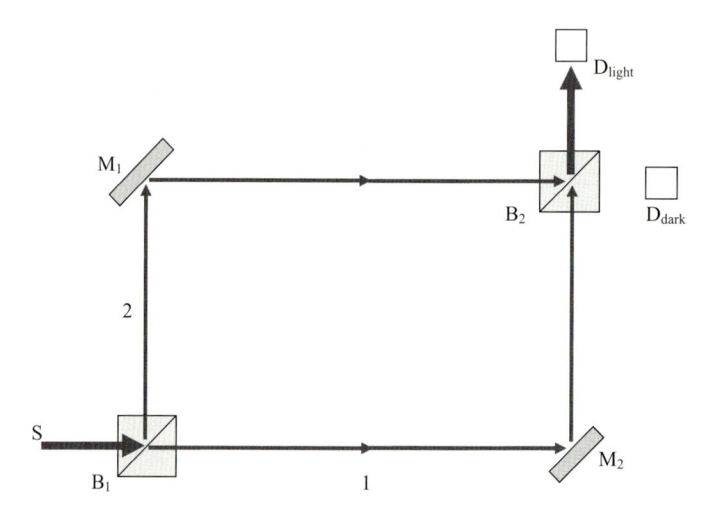

Figure 23.4 Schematic for the interaction-free measurement using the Mach–Zender interferometer. "Quantum seeing" of a macroscopic (classical) object in the dark; S – the source of light; B_1, B_2 – BSs; M_1, M_2 – mirrors; D_{dark}, D_{light} – detectors.

(denoted as D_{light}) will always click recording the arrival of each photon. In a way, the scheme is analogous to the double-slit experiment, with the distinction that here we have only two possible landing positions assigned to each photon. The first position (D_{dark}) is analogous to the center of the dark fringe of the interference pattern; the second position (D_{light}) is analogous to the center of the bright fringe. In order for the scheme to be informative, we must be able to record every photon entry into the device, and this can always be done [25].

Suppose somebody blocks the upper path of the interferometer with an opaque object (Figure 23.5). This is equivalent to an attempt (a very crude one!) to peep in to see the "actual state" of the photon when between the two BSs. We know that in the double-slit case, this destroys the interference, thus enabling the photon to land at a previously forbidden location (at the center of the dark fringe). By analogy, we can now predict the possibility of D_{dark} firing, and will estimate the corresponding probability a little later.

Following [25,26,28], we can imagine a much more dramatic scenario, when the inserted object is so sensitive to light that it explodes even if a single photon hits it. In this case, it becomes vitally important to discover the presence of the object without exposing it even to a single photon. It is equivalent to seeing in total darkness. On the face of it, this seems to be impossible by definition: We can only see an object by recording at least one photon emitted or scattered by it. Without a single photon exchange, there would be no interaction with the object and the object cannot be detected without interaction. And yet, this conclusion turned out to be wrong! In the described procedure, the object can be detected, on the average, at least in one quarter of all trials. All such cases of detection are equivalent to an interaction-free measurement. This can be done by extending functionality of the experiment: Apart from detecting the original photon, we use the photon itself for detecting the object.

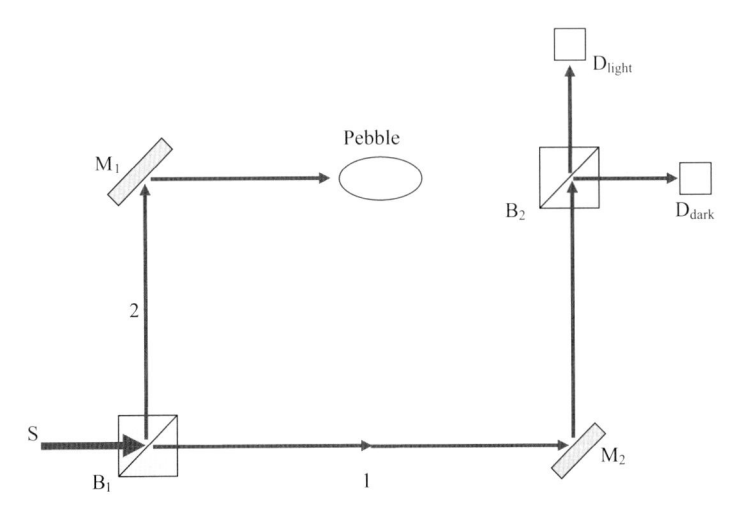

Figure 23.5 The same as Figure 23.4 with a pebble in path 2. Now there is a 50/50 chance for a photon in path 1 to activate D_{dark}. Together with 50/50 chance of choosing this path, this makes D_{dark} activated 25% of the time.

Our ability to do that without the photon getting in direct contact with the object is another manifestation of quantum nonlocality. Let us see how it works.

Assume that our BSs are symmetric – each one divides the incoming beam into two equal parts. If the object is placed in one of the paths, there is 50% chance that the photon will hit the object, ending the game right then and there. Both detectors will remain idle. Even without the explosion, we'll know what happened from the disappearance of the photon (neither detector fires). As there was direct contact between the object and the photon, this is not an interaction-free measurement. We get information, but in a mundane way. So these 50% count as failure.

The remaining possibility (another 50%) is for the photon to choose the unobstructed path. After reaching the second BS, the photon can be either reflected to D_{light} and recorded there or transmitted. The former case (another 25% of trials) will give us no information since this outcome would be observed in the absence of the object as well. Therefore, it cannot even be counted as a measurement. In the second case (the photon is transmitted through the second BS, 25% of all trials), the way is open for it with no interference! The photon is recorded in D_{dar} and this immediately tells us that there is an external object in one of the paths, because D_{dark} could not fire in the absence of the object! And since the photon arrived safely at D_{dark}, it must have traveled along the free path, and the object is in the other path. The scheme works only 0.25 of the time, but when it works, it does so completely.

Now we describe the process quantitatively.

I. Consider first the case when both paths are free.

The choice of which path to take is made by the photon at the BS B1. Now we will consider the general case, without assuming that the BSs are symmetric. Let the probability of passing through B1 (and thereby choosing path 1) be $\mathcal{P}_T = p$. Then the probability of reflection from B1 (and thereby choosing path 2) will be $\mathcal{P}_R = 1 - p$. The corresponding probability amplitudes with their respective phase factors can be written as

$$a_T = p^{1/2} e^{i\delta_T} \quad \text{and} \quad a_R = (1-p)^{1/2} e^{i\delta_R}. \tag{23.41}$$

In either path, the photon is reflected by the corresponding mirror M_1 or M_2 toward the second BS. Upon reaching it, it makes the same choice as before of whether to be transmitted or reflected, and winds up in one of the detectors.

Let us find the corresponding probabilities, assuming both BSs to be identical. The amplitude of arriving at D_{dark} by way 1 is

$$A_1(D_{dark}) = a_T a_R e^{ikl_1} = \sqrt{p(1-p)} e^{i\delta} e^{ikl_1}, \quad \delta = \delta_T + \delta_R. \tag{23.42}$$

The amplitude of doing this by way 2 is

$$A_2(D_{dark}) = a_R a_T e^{ikl_2} = \sqrt{p(1-p)} e^{i\delta} e^{ikl_2}. \tag{23.43}$$

Here, l_1 and l_2 are the lengths of the first and second paths, respectively, and k is the wave vector of the photon. If there is no way to determine which path is chosen, the output is determined by the interference of both possibilities:

$$A(D_{dark}) = A_1(D_{dark}) + A_2(D_{dark}) = \sqrt{p(1-p)} e^{i\delta} \left(e^{ikl_1} + e^{ikl_2}\right). \tag{23.44}$$

23.5 Interaction-Free Measurements: Quantum Seeing in the Dark

Similarly, the amplitudes of being detected in D_{light} after having traveled by way 1 or 2 are, respectively,

$$A_1(D_{light}) = a_T a_T e^{ikl_1} = p e^{2i\delta_T} e^{ikl_1}, \tag{23.45}$$

$$A_2(D_{light}) = a_R a_R e^{ikl_2} = (1-p) e^{2i\delta_R} e^{ikl_2}. \tag{23.46}$$

The total amplitude of being detected in D_{light} is

$$A(D_{light}) = A_1(D_{light}) + A_2(D_{light}) = p e^{2i\delta_T} e^{ikl_1} + (1-p) e^{2i\delta_R} e^{ikl_2}. \tag{23.47}$$

The corresponding probabilities are

$$\mathcal{P}(D_{dark}) = |A(D_{dark})|^2 = 4p(1-p) \cos^2 \frac{k\Delta l}{2}, \quad \Delta l = l_1 - l_2, \tag{23.48}$$

$$\mathcal{P}(D_{light}) = |A(D_{light})|^2 = 1 - 4p(1-p) \sin^2 \left(\Delta\delta + \frac{k\Delta l}{2} \right), \quad \Delta\delta = \delta_T - \delta_R. \tag{23.49}$$

In order for the total probability to be normalized, there must be $\Delta\delta = \pi/2$, so that the last two equations can be written as

$$\begin{aligned} \mathcal{P}(D_{dark}) &= |A(D_{dark})|^2 = 4p(1-p) \cos^2 \frac{k\Delta l}{2}, \\ \mathcal{P}(D_{light}) &= |A(D_{light})|^2 = 1 - 4p(1-p) \cos^2 \frac{k\Delta l}{2}. \end{aligned} \tag{23.50}$$

In order for D_{dark} detector to justify its name, there must be $\mathcal{P}(D_{dark}) = 0$ regardless of what fraction of the incident beam is transmitted, that is, for any p. This can be easily done by making $k\Delta l = \pi$. Alternatively, we can introduce an appropriate phase shifter in one of the paths. For instance, with both path lengths equal ($l_1 = l_2 = l$), the phase shifter must change the phase of the photon passing through it by $\Delta\phi = \pi/2$. Then, including this additional phase into (23.50), we will have

$$\mathcal{P}(D_{dark}) = 0 \quad \text{and} \quad \mathcal{P}(D_{light}) = 1 \tag{23.51}$$

for any p. Thus, in the given setup, the photon will arrive at D_{light} with certainty and cannot be observed in D_{dark}.

II. There is a macroscopic object (a "pebble") in one of the paths, say, path 2. The probability for the photon to hit the object is equal to probability of getting reflected and thereby choosing path 2, that is $\mathcal{P}_R = 1 - p$. The probability for the photon to show up in D_{light} is in this case p^2 (passing through both BSs) and the probability to be detected at D_{dark} is

$$\mathcal{P}_D = p(1-p). \tag{23.52}$$

Comparing with (23.51) shows how the presence of an object in path 2 changes the odds.

The graph of function $\mathcal{P}_D(p)$ is shown in Figure 23.6. It reaches its maximum $\mathcal{P}_D^{max} = 1/4$ at $p = 1/2$ (symmetric BS). The probability to hit the object is in this case equal to $\mathcal{P}_R(1/2) = 1/2$. Thus, even at its maximum, the probability of

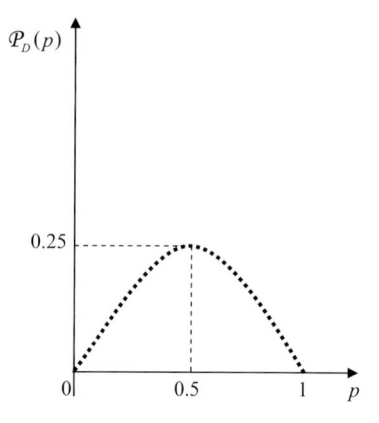

Figure 23.6 Plot of $\mathcal{P}_D(p)$ versus p in the presence of an object in one of the paths.

interaction-free detection is in this case only half of the probability of direct-contact detection. But this ratio can be increased by reducing the reflectivity of the BS (i.e., increasing p). Indeed, taking the ratio $\mathcal{P}_D/\mathcal{P}_R = p$, we see that it approaches 1 at $p \to 1$. It is true that the value of $\mathcal{P}_D$ itself goes to zero in this case, but so does $\mathcal{P}_R$, and the fractional rate of D_{dark} firings increases (note that, as mentioned before, all D_{light} firings are discarded as noninformative, so the ratio $\mathcal{P}_D/\mathcal{P}_R$ rather than just $\mathcal{P}_D$ becomes really meaningful characteristic).

One can show (Problem 23.9) that similar results can be obtained if the object is inserted in path 1, if we *increase* reflectivity of the BS.

The first actual experiment confirming the effect had been performed with a different version of interferometer (the Michelson interferometer)(Figure 23.7). The

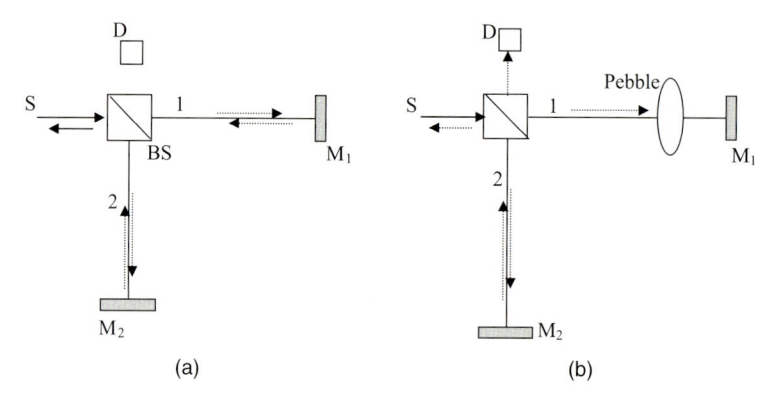

Figure 23.7 Schematic for interaction-free measurement using the Michelson interferometer. (a) The empty Michelson interferometer. The lengths of paths 1, 2 are adjusted so that all the entering photons exit back toward the source, and none of them gets into D. (b) "Quantum seeing" of a macroscopic (classical) object in path 1. Now some photons get into D. This indicates the presence of the object in one of the paths.

source of the initial photons was *parametric down-conversion* effect [29–32] in a nonlinear crystal. When irradiated with an intense laser beam of the UV light, the crystal converted some of the incident UV photons into the pair of photons of lower energy, which flew off in different directions. Detecting one of them indicates the existence of its sister that was directed into the interferometer adjusted so that each entering photon left it by the same way it came in. This is analogous to going to D_{light} in the initial thought experiment or to the center of a bright fringe in the double-slit experiment. The D_{dark} detector is practically always idle due to destructive interference of the two recombined beams at the BS. This detector is analogous to the center of a dark fringe in the double-slit experiment.

The observed results confirmed all basic features of the interaction-free measurement envisioned in Ref. [26] and described above. But this was not the end of the story. Later the researchers found a more sophisticated procedure in which all direct contact cases have been eliminated, and nonlocal detection was observed all the time.

The new procedure employed the evolution of the polarization state in gyrating media, which was discussed in Section 11.13. But in the described experiments, a single gyrator was used, and the photon was passed repeatedly through it, instead of traveling through a set of gyrators. In essence, the interaction-free measurements in such experiments realize a variant of the quantum Zeno effect in which the detected object plays the role of the agent (e.g., a set of polarizers in Section 11.13) stopping the evolution. The experimental setup is shown in Figure 23.8. The interferometer with two paths remains its essential part, but the PBS at its entrance is totally different from the one described earlier. It discriminates between the two

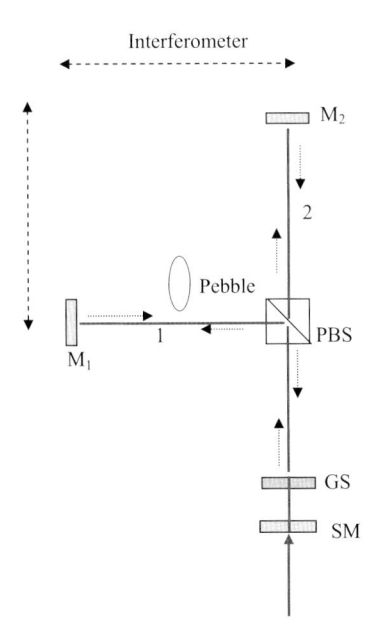

Figure 23.8 Schematic for interaction-free measurement using the Michelson interferometer with polarizing BS; SM – switchable mirror; GS – gyrating slab (polarization rotator).

polarization states of the incident photon, passing the x-polarized photon (in the plane of the figure) and sending it along the (vertical) path 1, but reflecting the y-polarized photon and sending it along the (horizontal) path 2. If polarization is tilted, forming angle θ with the x-direction, the outcome is determined by the analogue of Malus's law: Transmission and reflection have respective probabilities $\cos^2 \theta$ and $\sin^2 \theta$. And there are two new important elements: a gyrating plate and a switchable mirror. The latter is an interference device that can switch very fast between the reflecting and transmitting modes of operation.

Here is how the system works: An x-polarized photon enters through the mirror, which at this moment is transparent. The photon then passes through the gyrator that turns its polarization by a small angle $\Delta\theta$, so that the photon exits it in the state

$$\cos\Delta\theta |x\rangle + \sin\Delta\theta |y\rangle. \tag{23.53}$$

The polarizing BS converts this state into

$$\cos\Delta\theta |x\rangle|1\rangle + \sin\Delta\theta |y\rangle|2\rangle. \tag{23.54}$$

This is a superposition of being x-polarized in path 1 and being y-polarized in path 2. The mirrors at the end of each path reverse the motion back to the splitter. Both paths are made equal in length, so no phase difference has been accumulated between the two amplitudes in (23.54) when back at the splitter, and the photon reemerges in exactly the same polarization state (23.53), except for its direction, now heading toward the entrance mirror. By this time the entrance mirror is made nearly 100% reflecting, so the photon bounces back, and the process reiterates. After each cycle, the amount $\Delta\theta$ is added to the previous polarization angle. After N cycles with N such that $N\Delta\theta = \pi/2$, the photon winds up vertically polarized.[26] By this moment the entrance mirror is made transparent again and the photon is free to exit the system. Its vertical polarization can be checked by placing the polarizing filter (not shown in the figure) with transmission axis along x behind the mirror. The photon will be absorbed, which indicates that its polarization had indeed been evolved from $|x\rangle$ to $|y\rangle$ state.

Now suppose an opaque object (or even a superbomb) is placed, say, in path 2. There will appear a small chance of photon absorption in each cycle. According to (23.53) and (23.54), the amplitude of this process is equal to $\sin\Delta\theta \approx \Delta\theta$. If this small chance is not realized, the photon survives, but its superposition (23.54) collapses to only one state $|x\rangle|1\rangle$, and the probability of this favorable outcome is $\cos^2 \Delta\theta \approx 1 - (1/2)\Delta\theta^2$, which is close to 1 at small $\Delta\theta$.

At the end of cycle, the photon reemerges from the BS x-polarized. The gyrator rotates its polarization direction through $\Delta\theta$ and the next cycle starts from scratches. The process reiterates. According to the general rule, the amplitude $A(N)$ of passing

---

[26] In this respect, the described cycle is similar to a cyclotron in high-energy physics, which replaces a linear set of accelerating units (linear accelerator) by a circular track in the magnetic field (see Ref. [33] and references therein). The systems and physics involved are totally different, but the function (using one unit instead of a long linear set) is the same.

through the system after N cycles is the product of individual passing amplitudes for each cycle. Using the same argument as in Section 11.13, we obtain

$$A(N) = \prod_{j=1}^{N} \cos \Delta\theta_j. \tag{23.55}$$

In our case all cycles are identical, so all $\Delta\theta_j$ are equal to each other, $\Delta\theta_j = \Delta\theta$, and (23.55) reduces to

$$A(N) = (\cos \Delta\theta)^N \approx \left(1 - \frac{(\Delta\theta)^2}{2}\right)^N. \tag{23.56}$$

At $\Delta\theta = \pi/2N$, $N \gg 1$, this gives

$$A(N) \approx \left(1 - \frac{(\pi/2)^2}{2N^2}\right)^N \approx 1 - \frac{(\pi/2)^2}{2N}. \tag{23.57}$$

The corresponding probability is

$$P(N) \approx \left(1 - \frac{(\pi/2)^2}{2N^2}\right)^{2N} \approx 1 - \frac{(\pi/2)^2}{N}. \tag{23.58}$$

At $N \to \infty$, the probability of passing approaches 1 and the probability of collapsing to state $|y\rangle|2\rangle$ and getting absorbed (or worse, exploding the bomb) approaches 0. But the beautiful magic of this is that even with the latter possibility totally negligible, its mere presence dramatically changes the outcome! The polarization state of the photon in the end, instead of being converted to vertical, remains horizontal as if there were no gyrators in the way! The polarization is being frozen because the "boiling pot" had been watched by the object! And the object had been watched by the photon – without both getting into direct contact with one another! The fact that the photon emerges intact and x-polarized indicates the presence of the object, and ideally, this interaction-free detection can happen with certainty all the time!

23.6
QM and the Time Arrow

We know that CM and EM are time-symmetric theories. Any solution of their equations with t changed to $-t$ describes a physically possible situation. In other words, their equations are invariant under the time reversal $t \Rightarrow -t$. Even in the processes involving entropy and heat exchange, association of "natural" flow of time with increase of entropy is only statistical. If we choose a moment when the entropy drops because of some statistical fluctuation and reverse time at that very moment, both directions of time will be associated with entropy increase. In particular, when we see a sufficiently small fluctuation, our odds of seeing the entropy increase or decrease at the next moment are about the same.[27] Generally, it is only less probable but not impossible, for the entropy to be decreasing at a given moment [34].

As to time symmetry in QM, it only holds between measurements, when the system's behavior is described by the time evolution operator. An interrupting measurement that collapses a superposition of eigenstates to a single eigenstate is innately time asymmetric. Time reversal of this process would constitute a reversed measurement in the same basis. But this reversed process would also restore the initial superposition from a single eigenstate (so an eigenstate would no longer return the corresponding eigenvalue!), which is not how measurements work in direct-time QM. Thus, our world and time-reversed world would be described by different QM!

As an example, consider a photon linearly polarized at an angle θ to the vertical. It has the probability $\mathcal{P}(\theta) = \cos^2 \theta$ of collapsing to vertically polarized state after passing through a polarizing filter with a vertical transmission axis. Suppose this potentiality has been actualized. In a time reversal of this process, the incident photon with polarization parallel to the transmission axis would reemerge from the filter polarized at an angle θ to that axis. For $\theta \neq 0$, this would be a total nonsense in direct-time QM. But suppose we accept this weirdness as one of the rules in time-reversed QM. Then there follows another one. Namely, the reversed measurement on two *identical* pure ensembles in the same eigenstate (say, both are vertically polarized) would produce two *different* postensembles. For example, if you run the above experiment for $\theta = 45°$ and $\theta = 30°$ and then reverse the time, you will get two different postensembles with $\theta = 45°$ and $\theta = 30°$ from the same initial ensemble $\theta = 0°$ of vertically polarized photons, and there is nothing in the initial conditions to warrant this divergence of experimental outcomes. In other words, we would observe a totally acausal effect – also unheard – of thing in direct-time QM!

Sometimes an argument is put forth that the change of any state induced by the measurement is always time reversible because the corresponding wave function never ceases to be unitary. This argument is false because unitarity alone is not sufficient for time reversibility. It is true that the system's state function before and after the collapse remains unitary (the tip of its state vector remains on the same sphere of unit radius in the corresponding Hilbert space). However, if we try to reverse it in time *and* require obedience to the regular (direct time) QM, that will be an attempt to combine two mutually exclusive requirements. In direct-time QM, the system will generally not return to the previous state and hence time reversal will not be achieved. (Case in point: The filter changes $\theta = 45°$ input ensemble to a $\theta = 0°$ output ensemble, but if you swap input and output, $\theta = 0°$ ensemble will again emerge as $\theta = 0°$).[28] Conversely, if we still insist that the time reverse must be

---

27) For a large fluctuation, chances are that we are going to see a reversion to the mean, as continuation of the "countertrend" is rather unlikely.

28) We want to emphasize that the QM statement (vertically polarized photon will *always* pass unchanged through an ideal filter with vertical transmission axis) is not probabilistic, but 100% deterministic.

possible, then the corresponding scenario would be the subject of a different QM.[29]

The ultimate verdict is that (at least for some types of measurements) there exists a time arrow in QM. Such measurements are time irreversible and information about the premeasurement state is irretrievably lost. The only way to construct a fully reversible quantum theory would be to find some missing elements in the description of particle interaction with the classical measuring device. This would mean finding the answer to the hotly debated question as to what actually happens during the measurement, which would accordingly change today's formulation of QM. This has not happened so far, and according to all evidence, is not going to happen.

Problems

23.1 Suppose you measure the position of a particle in a box of width a. Assuming that before the measurement the particle was in the sixth eigenstate of the box, find the positions in which it *cannot* be found in a position measurement.

23.2 Suppose that the particle in the previous problem was found at $x' = a/12$. Find the terms that will be missing in expansion (23.5) describing the new state.

23.3 Consider the system discussed in Box 23.1. Identify two regimes of operation for which quantum superposition in the computational basis can be removed by decoherence. Explain.

23.4 Consider the system from the previous problem and try to apply the argument in Box 23.1 in reverse. Assume that complete decoherence produces a diagonal matrix in the $\{|\uparrow\rangle, |\downarrow\rangle\}$ basis, and then make the transition to a nondiagonal density matrix in $\{|R\rangle, |L\rangle\}$ basis.

23.5 Show that the previous result holds for any possible probability distribution between "up" and "down" states.

23.6 Describe qualitatively the possible outcomes of a SG measurement if the screen is replaced with a second magnet to measure the y-component of spin.

---

[29] The subtle point here is that the equivalent argument would not work if the original photon had collapsed from diagonal ($\theta = 45°$) to horizontal ($\theta = 90°$) polarization. In this case, the photon would be absorbed by the filter and the filter would accordingly warm up. (For the sake of the argument, we assume the filter had zero initial temperature.) The entropy of the system would increase. The system would now have just enough thermal energy to emit one photon of the same frequency. In this case, the time reversal of the process would be possible in principle. We would have the warm filter cool down and emit back the diagonally polarized photon. This would mean a spontaneous decrease of entropy – forbidden by pure (nonstatistical) thermodynamics, but allowed by statistical mechanics. In the framework of the latter, the discussed event would be just a relatively rare but totally legitimate fluctuation. The initial information would be restored, and the whole process would be time reversible. All of this does not undermine our main argument. Even one instance of a diagonal photon collapsing to *vertical* ($\theta = 0°$) polarization would suffice to prove that quantum measurements are generally irreversible, and we know that such instances occur 50% of the time.

23.7 Suppose you want to measure S_z of a *free electron* using the SG procedure.

a) What changes in the experimental setup and mathematical description of such measurement would you make?

b) Where would you draw the boundary between the "quantum-mechanical" and "classical" part of the system in this case?

23.8 What requirement should be imposed on the magnetic field gradient in the SG experiment for a free electron? Express the minimal required gradient in terms of the initial width of the electron beam and the time t_0.

23.9 Suppose in the thought experiment [26] described in Section 23.6, an object is inserted into path 1 of the interferometer shown in Figure 23.4, instead of path 2. Find the fractional number $\mathcal{P}_D/\mathcal{P}_R$ of the informative detections of the object as a function of p in this case.

References

1 Penrose, R., Shimony, A., Cartwright, N., and Hawking, S. (1999) *The Large, the Small, and the Human Mind*, Cambridge University Press.

2 Wheeler, J.A. and Zurek, W. (1983) *Quantum Theory and Measurements*, Princeton University Press, p. 157.

3 Davies, P. and Brown, J.R. (1986) *The Ghost in the Atom*, Cambridge University Press.

4 Greenstein, G. and Zajonc, A.G. (1997) *The Quantum Challenge*, Jones and Bartlett Publishers, pp. 157–166.

5 Schlosshauer, M. (2005) Decoherence, the measurement problem, and interpretations of quantum mechanics. *Rev. Mod. Phys.*, **76** (4), 1267–1305.

6 Bacon, D. (2001) Decoherence, control, and symmetry in quantum computers, arXiv:quant-ph/0305025.

7 Zurek, W.H. (2003) Decoherence, einselection, and the quantum origins of the classical. *Rev. Mod. Phys.*, **75**, 715.

8 Zurek, W.H. (1991) Decoherence and the transition from quantum to classical. *Phys. Today*, **44**, 36–44.

9 Zurek, W.H. (2003) Decoherence and the transition from quantum to classical – revisited, arXiv:quant-ph/0306072.

10 Leibfried, D., Knill, E., Seidelin, S., Britton, J., Blakestad, R.B., Chiaverini, J., Hume, D., Itano, W.M., Jost, J.D., Langer, C., Ozeri, R., Reichle, R., and Wineland., D.J. (2005) Creation of a six atom "Schrödinger cat" state. *Nature*, **438**, 639–642.

11 De Martini, F., Sciarrino, F., and Vitelli, C. (2008) Entanglement and Non-locality in a micro-macroscopic system, arXiv:0804.0341.

12 De Martini, F. (2009) Entanglement and quantum superposition of a macroscopic–macroscopic system, arXiv:0903.1992v2.

13 De Martini, F. and Sciarrino, F. (2012) Investigating macroscopic quantum superpositions and the quantum-to-classical transition by optical parametric amplification, arXiv:1202.5518v1.

14 Landau, L. and Lifshits, E. (1965) *Quantum Mechanics: Non-Relativistic Theory*, 2nd edn, Pergamon Press.

15 von Neumann, J. (1955) *Mathematical Foundations of Quantum Mechanics*, Princeton University Press.

16 Busch, P., Grabowski, M., and Lahti, P.J. (1995) *Operational Quantum Physics*, Springer, Berlin.

17 Busch, P., Lahti, P.J., and Mittelstaedt, P. (1996) *The Quantum Theory of Measurement*, vol. 2, Springer, p. 60.

18 Bohm, D. (1989) *Quantum Theory*, Dover Publications, New York.

19 Herbert, Nick (1987) *Quantum Reality: Beyond New Physics*, Anchor.

20 Reinisch, G. (1999) Stern–Gerlach experiment as the pioneer – and probably

the simplest – quantum entanglement test? *Phys. Lett. A*, **259** (6), 427–430.
21 Venugopalan, A. (1997) Decoherence and Schrödinger-cat states in a Stern–Gerlach-type experiment. *Phys. Rev. A*, **56** (5), 4307–4310.
22 Bernstein, Jeremy. (2010) The Stern–Gerlach experiment, arXiv:1007.2435v1 [physics.hist-ph].
23 Griffiths, D. (1999) *Introduction to Electrodynamics*, 3rd edn, Addison-Wesley.
24 Feynman, R., Leighton, R., and Sands, M. (1963) *The Feynman Lectures on Physics*, vol. 1–3, Addison-Wesley, Reading, PA, p. 64.
25 Kwiat, P., Weinfurter, H., and Zeilinger, A. (1996) Quantum seeing in the dark. *Sci. Am.*, **275** (5), 72–78.
26 Elitzur, A.C. and Vaidman, L. (1993) Quantum-mechanical interaction-free measurements. *Found. Phys.*, **23** (7), 987–997.
27 Hecht, E. (2002) *Optics*, 4th edn, Addison-Wesley, p. 385.
28 Kwiat, P.G., Weinfurter, H., Herzog, T., Zeilinger, A., and Kasevich, M.A. (1995) *Phys. Rev. Lett.*, **74** (24), 4763–4766.
29 Gilder, L. (2008) *The Age of Entanglement*, Vintage Books, New York, p. 299.
30 Kwiat, P. et al. (1995) New high-intensity source of polarization-entangled photon pairs. *Phys. Rev. Lett.*, **75** (24), 4337–4341.
31 Zeilinger, A. (2010) *Dance of the Photons*, Farrar, Straus and Giroux, New York, p. 205.
32 Burnham, D.C. and Weinberg, D.L. (1970) Observation of simultaneity in parametric production of optical photon pairs. *Phys. Rev. Lett.*, **25**, 84.
33 Fayngold, M. (2008) *Special Relativity and How It Works*, Wiley-VCH Verlag GmbH, pp. 485–487.
34 Landau, L.D. and Lifshits, E.M. (1976) *Statistical Physics, Part I* (Russian edition), Nauka, Moscow.

24
Quantum Nonlocality

24.1
Entangled Superpositions I

The actual phenomenon of entanglement is much richer than a case study involving 2 qubits. The particles can have more than two states, there can be many particles in the system, and finally, the entangled particles can be nonidentical and accordingly can have different sets of quantum states.

The first possibility is exemplified by a pair of (distinguishable!) photons of frequency ω with the zero net momentum. (In cases when the set of eigenvalues is continuous, we will select only a discrete subset.) If we choose any one axis x and subdivide it by equidistant points separated by δp (an appropriate unit of momentum will be $\delta p = \hbar\omega/Mc$ with integer M), then result $p_x = 3\delta p$ for the first photon will mean $p_x = -3\delta p$ for the second photon. The photons are entangled with respect to p_x (Figure 24.1).

The net momentum is represented by point 0 in the figure; the individual p_x components may range within $-M\delta p \leq p_x \leq M\delta p$. The relevant variables describing our pair are two numbers (j, k), where $j, k = 0, \pm 1, \pm 2, \ldots, \pm M$ identifying two points in the periodic array along p_x. Let $|j\rangle$ be a photon state corresponding to $p_x = j\,\delta p$. Then a state with one photon in $|j\rangle$ and the other in $|k\rangle$ will be $|j, k\rangle_{1,2} = |j\rangle_1 \otimes |k\rangle_2$. The most general state is a superposition of all possible states with different j, k:

$$|\tilde{\Psi}\rangle_{1,2} = \sum_{j,k} c_{jk} |j, k\rangle_{1,2} = \sum_{j,k} c_{jk} |j\rangle_1 \otimes |k\rangle_2, \quad \sum_{j,k} |c_{jk}|^2 = 1. \quad (24.1)$$

This expression would describe two independent particles if it were separable. In our case, this cannot be done because the set c_{jk} is predetermined by the initial condition of the zero net momentum. This correlates j and k by $k = -j$, so that if one photon is in a state $|j\rangle$, the other must be in state $|-j\rangle$, and accordingly, $c_{jk} \to c_{j,-j} \equiv c_j$:

$$|\tilde{\Psi}\rangle_{1,2} \Rightarrow \sum_{j} c_j |j\rangle_1 \otimes |-j\rangle_2. \quad (24.2)$$

Now we have only one summation, and this does not admit factorizing (24.2) into the product of two sums. Equation 24.2 describes an entangled superposition of two-particle states. Each photon knows that if it is in a state $|j\rangle$, then its twin sister is in state $|-j\rangle$, but

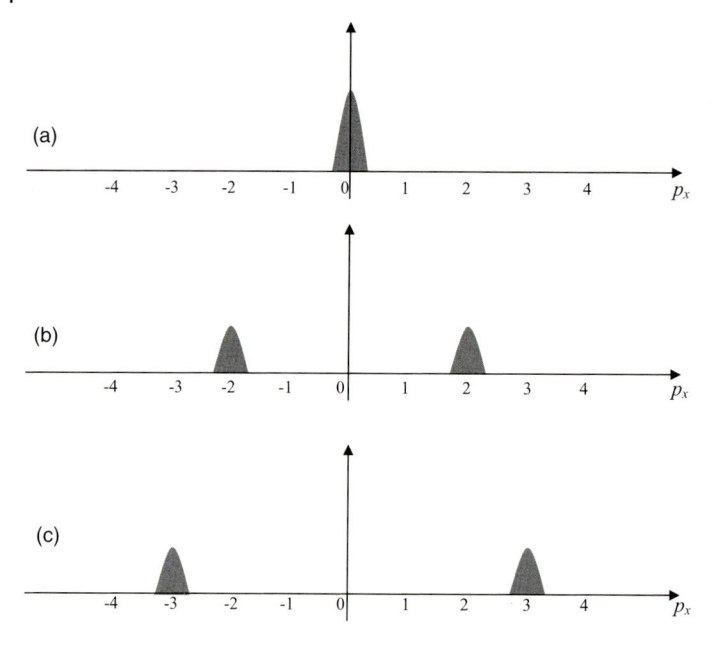

Figure 24.1 Possible states of two identical particles with the zero net momentum. The states can be described by the pairs of quantum numbers (j, k) labeling momentum of the corresponding particle. Shown here is a small subset of possible outcomes of p_x-momentum measurement. (a) $(j, k) = (0, 0)$. (b) $(j, k) = (-2, 2)$. (c) $(j, k) = (-3, 3)$.

neither photon has its own definite state. Only momentum measurement can resolve this "ambiguous" coexistence. Such a measurement collapses state (24.2) to a state with some definite j, for example, $|\tilde{\Psi}\rangle_{1,2} \Rightarrow c_3'|3\rangle_1 \otimes |-3\rangle_2$. Mathematically, the set of coefficients c_j (the entangled state $|\Psi\rangle_{1,2}$ in the $|j\rangle$-basis) collapses to one coefficient $c_3' = e^{i\alpha}$. Since only one term is left, the value of α is immaterial, so we set $\alpha = 0$:

$$|\tilde{\Psi}\rangle_{1,2} = \sum_j c_j |j\rangle_1 \otimes |-j\rangle_2 \Rightarrow |-3\rangle_1 \otimes |3\rangle_2 \quad \text{or} \quad \begin{pmatrix} c_M \\ \vdots \\ c_0 \\ \vdots \\ c_{-M} \end{pmatrix} \Rightarrow \delta_{j,3}, |j| \leq M.$$

(24.3)

The system is now disentangled and accordingly is described by the product of the two individual states with definite p_x, one with $p_x = 3\delta p$ and the other with $p_x = -3\delta p$.

If we have more than two photons with zero net momentum, the constraints on their individual characteristics become more complicated. In case of three photons, we already have different conditions depending on whether their frequencies are all equal or not. If they are equal, the three momenta cannot be collinear. But in either case, they must be coplanar. The latter restriction drops for $N > 3$. And in all cases, there is an additional constraint on frequencies. Altogether, conservation of energy and momentum imposes conditions

$$\hbar \sum_{i=1}^{N} \omega_i = E_0 \quad \text{and} \quad \sum_{i=1}^{N} \mathbf{p}_i = 0, \tag{24.4}$$

where E_0 is the energy of the whole system.

Now we turn to a system of N *arbitrary* particles with the zero net angular momentum $\mathbf{J}$. Then, instead of (24.4), we will have only one condition $J_z = 0$, where z is the chosen basis axis. But instead of two sets of numbers j, k, we have to consider N such sets $m_1, m_2, \ldots, m_N$, where m_n is the net magnetic quantum number of nth particle. Each m_n takes on the values within its own range $|m_n| = 0, 1/2, 1, 3/2, \ldots, M_n$. One particle can be in state $|m_1\rangle$, the other in state $|m_2\rangle$, and so on. The composite state can be written as

$$|m_1, m_2, \ldots, m_n, \ldots, m_N\rangle_{1,2,\ldots,N} = |m_1\rangle_1 \otimes |m_2\rangle_2 \otimes \cdots |m_n\rangle_n \cdots \otimes |m_N\rangle_N. \tag{24.5}$$

The most general state will be a superposition of products (24.5):

$$|\tilde{\Psi}_{1,2,\ldots,N}\rangle = \sum_{m_1, m_2, \ldots, m_N} c_{m_1, m_2, \ldots, m_N} \prod_{n=1}^{N} |m_n\rangle. \tag{24.6}$$

The state is entangled or not, depending on whether (24.6) can be factorized. The condition $\mathbf{J} = 0$ means that $\sum_{n=1}^{N} m_n = 0$. This severely restricts the possible sets of coefficients in (24.6). Then the sum cannot be written as a product state of N particles. The corresponding state becomes entangled. In addition, if all $c_{m_1, m_2, \ldots, m_N}$ have the same modulus, the entanglement is maximally strong. Some cases of such system with $N = 3$ are shown in Figure 24.2.

24.2
Entangled Superpositions II

It is instructive to look at entangled superpositions in the context of conservation laws. For instance, spin measurement collapses an entangled *singlet* state (3.60) to

$$|\tilde{\Psi}_{1,2}\rangle = \alpha|\uparrow\rangle_1 \otimes |\downarrow\rangle_2 + \beta|\downarrow\rangle_1 \otimes |\uparrow\rangle_2 \Rightarrow |\uparrow\rangle_1 \otimes |\downarrow\rangle_2 \quad \text{or} \quad |\downarrow\rangle_1 \otimes |\uparrow\rangle_2. \tag{24.7}$$

Here, the net value of $S_z \equiv (\hbar/2)\sigma_z$ for *the whole system* is always zero on both sides of Equation 24.7, so S_z conserves. In contrast, *one particle* in a simple superposition (3.59) does not show this feature. The spin measurement collapses the state according to

$$|\psi_1\rangle = \alpha_1|\uparrow\rangle_1 + \beta_1|\downarrow\rangle_1 \Rightarrow |\uparrow\rangle_1 \quad \text{or} \quad |\downarrow\rangle_1. \tag{24.8}$$

Here, spin conserves only statistically. A single measurement yields $\sigma_z = +1$ or -1 on the right-hand side, while we have $\langle \sigma_z \rangle = |\alpha_1|^2 - |\beta_1|^2$ on the left-hand side. In

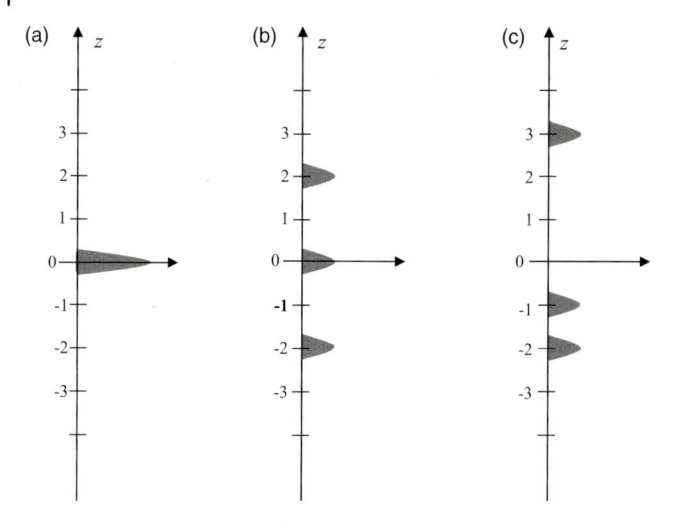

Figure 24.2 Quantum entanglement of three identical particles The states allowed by the initial condition (the net angular momentum **L** = 0) are described by the triplet of quantum numbers (m_1, m_2, m_3) labeling the individual angular momentum of the corresponding particle. Here is shown a small subset of possible outcomes of the angular momentum measurement. (a) $(m_1, m_2, m_3) = (0, 0, 0)$. (b) $(m_1, m_2, m_3) = (2, 0, -2)$. (c) $(m_1, m_2, m_3) = (3, -1, -2)$.

this respect, the measurement does not preserve the variable that it measures. Actually, such nonpreservation is implied from the start in the very definition of a superposition: An observable in question (denote it q) does not have a definite value in a state before the measurement, but acquires such a value in the process of measurement. This, in turn, implies that the reasonable characteristic of the premeasured state is the average $\langle q \rangle$. This means that in a single measurement characterized by conversion of the type (24.8), we *generally* can speak about nonpreservation of the corresponding observable. If the observable is a conserving quantity like spin, then its nonpreservation would mean violation of the corresponding conservation law unless we explicitly include the measuring device as a part of the whole system. This brings in the interaction between the measured particle and measuring device. Then we no longer can consider the particle as an isolated system and its initial state as a simple superposition. It turns out that a simple superposition may actually be only a visible top of an iceberg. And if we take a closer look at this iceberg, we will find an entangled superposition.

In view of this, the apparatus in the Stern–Gerlach experiment should be treated as described earlier in Box 23.3. The electron gets spin-entangled with the apparatus, an effect that we can safely ignore when describing electron behavior, but we must keep it in mind when it comes to conservation laws. Then we realize that our electron is in fact an *open system*, and not not an *isolated system*, for which conservation would hold.

Box 24.1 Measurement and conservation laws

It is instructive to see how conservation laws work in a measurement. To take a familiar example, let the incident photon in the BS experiment (Figure 3.2a) have initial momentum $|\mathbf{k}_1\rangle$ and emerge from the BS in a superposition $(1/\sqrt{2})(|\mathbf{k}_1\rangle + |\mathbf{k}_2\rangle)$, where $|\mathbf{k}_1| = |\mathbf{k}_2| = \omega/c$. Also, let the BS have initial momentum $|\mathbf{K}\rangle_{BS}$. The process of premeasurement entangles the photon with the BS:

$$|\mathbf{k}_1\rangle \otimes |\mathbf{K}\rangle_{BS} \Rightarrow \frac{1}{\sqrt{2}}\left(|\mathbf{K}\rangle_{BS} \otimes |\mathbf{k}_1\rangle + |\mathbf{K} + \mathbf{k}_1 - \mathbf{k}_2\rangle_{BS} \otimes |\mathbf{k}_2\rangle\right). \qquad (24.9)$$

The net momentum of the system (photon + BS) is conserved in each separate trial. On the other hand, if we focus on the photon alone, momentum is conserved only half of the time – when the photon is found in transmitted state $|\mathbf{k}_1\rangle$ (then $|\mathbf{K}\rangle_{BS}$ is unchanged). In the outcomes with $|\mathbf{k}_2\rangle$, the photon's momentum is not conserved. This nonconservation appears to persist even when we take the average of all trials:

$$\langle \mathbf{k} \rangle = (1/2)(\mathbf{k}_1 + \mathbf{k}_2) \neq \omega/c.$$

But it is merely an artifact arising from our neglecting the BS's momentum. Nature keeps track of it, and if we follow this track, we will find no violations.

In a perfect analogy with the situation in Box 23.3, $|\mathbf{K}\rangle_{BS}$ is generally not a momentum eigenstate, but rather a shorthand for the superposition of eigenstates $|\mathbf{k}\rangle_{BS}$: $|\mathbf{K}\rangle_{BS} \equiv \int C(\mathbf{k})|\mathbf{k}\rangle_{BS} d\mathbf{k}$. Even if initially $\langle \mathbf{K} \rangle = 0$ (the natural situation for a laboratory BS!), there is a nonzero indeterminacy $\Delta \mathbf{K} \neq 0$ in state $|\mathbf{K}\rangle_{BS}$. Similarly, $|\mathbf{K} + \mathbf{k}_1 - \mathbf{k}_2\rangle_{BS}$ is the shorthand for $|\mathbf{K} + \mathbf{k}_1 - \mathbf{k}_2\rangle_{BS} \equiv \int C(\mathbf{k})|\mathbf{k} + \mathbf{k}_1 - \mathbf{k}_2\rangle_{BS} d\mathbf{k}$ with the "shifted" eigenstates $|\mathbf{k}\rangle_{BS} \to |\mathbf{k} + \mathbf{k}_1 - \mathbf{k}_2\rangle_{BS}$ after the interaction with the photon.

If the photon is reflected, it changes its momentum, $|\mathbf{k}_1\rangle \to |\mathbf{k}_2\rangle$. This sets the BS into translational motion and it will now have, on the average, $\langle \mathbf{K} \rangle = \mathbf{k}_1 - \mathbf{k}_2$ instead of the initial $\langle \mathbf{K} \rangle = 0$. But in view of its huge mass $M_{BS} \gg \hbar\omega$, the velocity is so small that the classical observer does not notice the difference. Using the same reasoning as in Box 23.3, we can write $|\mathbf{K} + \mathbf{k}_1 - \mathbf{k}_2\rangle_{BS} \approx |\mathbf{K}\rangle_{BS}$, so that

$$|\mathbf{K}\rangle_{BS} \otimes |\mathbf{k}_1\rangle + |\mathbf{K} + \mathbf{k}_1 - \mathbf{k}_2\rangle_{BS} \otimes |\mathbf{k}_2\rangle \underset{M_{BS} \gg \hbar\omega}{\approx} |\mathbf{K}\rangle_{BS} \otimes (|\mathbf{k}_1\rangle + |\mathbf{k}_2\rangle). \qquad (24.10)$$

Even if $\langle \mathbf{K} \rangle = 0$ (which is usually the case), the *indeterminacy* in the BS's momentum can be much greater than $|\mathbf{k}_2|$, that is, $|\mathbf{k}_2| = |\mathbf{k}_1| \ll |\Delta \mathbf{K}|$. This happens if indeterminacy Δx in the BS's position is $\Delta x \ll \lambda \equiv c/2\pi\omega$. In the optical region, $\Delta x \cong 10^{-3}\lambda$. In all such cases, the difference between $|\mathbf{K}\rangle_{BS}$ and $|\mathbf{K} - \mathbf{k}_2\rangle_{BS}$ is essential on the photon scale but minuscule on the BS scale. Inasmuch as these states refer to the BS only, the difference between them can be neglected and we can consider Equation 24.10 as exact. This justifies the above reduction of entangled superposition to the simple superposition of the photon states alone, but for a fully accurate description of the process, we must use Equation 24.9.

24.2.1
Discussion

The skeptical reader may ask: Does not it look like a double standard? When we discuss the BS's state, we say that the BS acquires additional momentum when reflecting the photon, but as it comes to the photon, we do not even mention that it may acquire the additional momentum due to indeterminacy $\Delta \mathbf{K}$ or just due to thermal "jittering" of BS.

The answer to this is that there is no double standard. The presence of a certain monochromatic component in the BS momentum spectrum does not mandate its automatic transfer to the photon [just as possible absence of $\mathbf{k}_2$ and/or $\mathbf{k}_1$ in this spectrum does not preclude the photon from acquiring $\mathbf{k}_2$ (and losing $\mathbf{k}_1$!) in the process of reflection]. The change $\mathbf{k}_1 \rightarrow \mathbf{k}_2$ in momentum of reflected photon has nothing to do with momentum indeterminacy $\Delta \mathbf{K}$ of BS. Even if we assume $\mathbf{K} = \Delta \mathbf{K} = 0$ (the BS in eigenstate $|\mathbf{K} = 0\rangle$) before collision, we will get the states $(|\mathbf{k}_2\rangle, |\mathbf{k}_1 - \mathbf{k}_2\rangle_{BS})$ after collision. This is due to specific geometry and physical structure of the BS, not due to its Fourier spectrum. Actually, it is another example of mutual deflection of the two colliding objects – recall Chapters 4 and 19. To be sure, we can expect the photon to receive some additional kick from BS due to nonzero $\Delta \mathbf{K}$. But the magnitude of this kick is determined by *velocity* of the BS's motion, rather than by its momentum. The additional kick on the photon is manifest to us as the Doppler shift resulting from the photon's collision with a *moving* target. But the target may be moving fast while having small momentum or slow while having a large momentum, depending on its mass M. In our case M is so huge that the target may be considered as sitting still, even with nonzero momentum spectrum. Thus, there is practically no effect of BS's *momentum spectrum* on *momentum transfer* to the photon. If it were, the BS spectrum would be manifest also in change of the photon frequency, which is not observed in experiments with a stationary BS. The momentum transfer in the given setup is manifest only in rotation of the photon momentum $\mathbf{k}$ through a certain angle without changing its norm (elastic collision!).

The reader can evaluate the contribution $|\Delta \mathbf{k}| \sim |\Delta \omega|/c$ to the photon momentum from the momentum spectrum of the target by assuming $\Delta \mathbf{k} = \Delta \mathbf{K}$, using the Doppler shift formula $\Delta \omega / \omega = \Delta v / c$ and expressing Δv as $\Delta P / M$. We can also consider the contribution due to the thermal jittering of the target. In this case, one can show that $|\Delta \mathbf{k}|/|\mathbf{k}| \leq 10^{-6}$ for any practical situations (Problems 24.1 and 24.2).

24.3
Quantum Teleportation

We look here at a pair of separated but still entangled particles. The correlations between measurement results performed on the particles are *nonlocal*, which seems to demonstrate the existence of faster than light (FTL) communication between distant objects. The whole phenomenon of quantum nonlocality is so impressive and difficult to accept that it appears to be on the verge of mystical. It has been widely discussed from the early days of QM up to this day. Its most characteristic features have been confirmed in experiments by Aspect *et al.* [1].

None of the discussed superluminal motions can be used for FTL communication [2]. Here, we want to show that this also applies to the nonlocal effects. Consider the already familiar entangled electron pair (3.60) in the spin singlet state and with no orbital motion. If the preparation left the spin direction of either particle undefined, all we can say is that (a) in the z-basis the measured electron spin is either up or down and (b) the spin of either electron is always opposite to that of its partner.

Suppose we measure the spin of the first electron, with the result $|\uparrow\rangle$. Then it is immediately known that the second electron is in state $|\downarrow\rangle$. By measuring spin of one electron, we have measured both spins at once. If one performs a measurement on the second electron (which is no longer necessary!), it will only confirm this fact. It seems as if some agent has instantaneously transferred information to the second electron.

In 1935, Einstein, Podolsky, and Rosen (EPR) published their analysis of these kind of phenomena [3]. Their intention was, *first*, to show that a particle's characteristic can, in principle, be measured without performing an actual measurement of it, and, *second*, to point out that such measurement demonstrates an instantaneous physical action (*spooky action at a distance*, as Einstein has called it), which would be incompatible with relativistic causality. The authors interpreted these results as an indication that theoretical framework of QM was incomplete (the *EPR paradox*).

The first of these statements (about possibility of "interaction-free" measurements) is true. The second one (regarding incompleteness of QM) is false. Before explaining why it is false, let us first analyze a possible objection to the interpretation of this thought experiment. Suppose that Alice's friend Tom working with her on Earth has two bottles of wine. One bottle is Burgundy and another one is Chardonnay. Tom prepared one bottle for himself and the other for Bob on Armad. It does not matter who gets which wine, so Tom just puts one bottle into his fridge and the other into the cargo spaceship heading for Armad – without looking – and then goes to hibernation for a few million years. When the spaceship finally arrives at its destination, Tom wakes up completely unaware which wine is on which planet. He opens the fridge and immediately knows that the wine on Armad is Chardonnay.

How could he in an instant get the information about an object in another Galaxy?

The answer is simple. Tom did not receive any signal from Armad. The only signal was the one from his fridge. It has changed his knowledge about the fridge's contents from total uncertainty to 100% certainty that it is Burgundy. Together with the preliminary information about the two bottles that he had, this enables him to conclude with the same dead certainty that the wine on Armad is Chardonnay. The act of observation in this case did not (and could not) physically change the type of wine in either package. The Burgundy in the fridge had been Burgundy before the observation. The Chardonnay on Armad had been Chardonnay long before it arrived there. We can distinguish here between the state of the observed object and the state of the observer. Only the latter has changed in the act of observation. There was no distant communication in this case.

The situation with our pair of particles is fundamentally different. When Alice on Earth performs the experiment and finds her particle's spin pointing up, this does *not* mean that the particle was in a spin-up state before the experiment. This state was *created* in the process of measurement. It was the measurement that converted the initial state (3.60) of the pair into a disentangled state $|\uparrow\rangle_1 \otimes |\downarrow\rangle_2$. Thus, the measurement made on Earth instantaneously changes the situation not only on Earth but also on Armad (and vice versa). The entangled system instantly collapses into one of the two possible independent states.

It appears that some physical agent does indeed carry information about the change on Earth, and this communication occurs instantaneously, changing immediately the situation on Armad. The above scheme could be realized in a more sophisticated version. After the EPR pair was shared, Tom brings to Alice's lab a third electron in an *unknown* state $|\chi\rangle$ and asks if she could send this state to Bob. Alice agrees to try. She acts on *her* two electrons with a specially chosen operator that entangles the free electron with the EPR pair, then applies an appropriate unitary transformation to that electron, and finally, measures both electron spins. The last step leaves Bob's electron untangled, giving it a state of its own. In one quarter of trials, this state will be identical to $|\chi\rangle$. It looks as if $|\chi\rangle$ was packed into a parcel and shipped to Bob with infinite speed. This is an example of *quantum teleportation*. One could think we found a way to move things and signals faster than light!

And yet this conclusion would be wrong. Causality is not violated, because the discussed changes of states are inherently statistical. By definition, causation between events N and M means that a change in M must change some observable property at N from one uniquely defined value to another uniquely defined value or, in the most general case, must produce some noticeable change in the density matrix at N. For example, if the influence from M changes the spin at N from $|\uparrow\rangle_N$ to $|\downarrow\rangle_N$, then a signal has been transmitted. In our case, however, the original spin direction of the electron on Armad was not sharply defined. Suppose that $|a|^2 = |b|^2$. Then from the very beginning there was a 50% chance to find its spin pointing up and a 50% chance to find it pointing down. Therefore, in any individual measurement, when Bob finds his electron's spin pointing down, he can always say: so what? Indeed, if Bob had known with certainty that the spin before the measurement was up, then the actual outcome would strongly suggest that it was due to the effect of some outside agent. But when there had already been a 50% chance to find it in a spin-down state, why should we attribute it to some external influence? We would rather say that this is just the actualization of a preexisting potentiality.

Thus, no individual outcome in the described type of measurement can be the evidence of superluminal or any other telecommunication. If the measurement results are statistical by nature, the only way we can ensure that FTL signaling is taking place is by finding a difference in *statistical distributions* of measurements at N with and without corresponding measurements at M.

Following this program, we must perform the *set* of individual measurements under two different conditions and compare their results.

Condition 1. Prepare a big pure ensemble of electron pairs in state (3.60), assuming, for simplicity, $|a|^2 = |b|^2 = 1/2$. For each pair, send one electron to Earth and the other to Armad. Measure only the electron spin on Armad without disturbing its counterpart on Earth and record the results.

Condition 2. Prepare again the same ensemble. But now, for each pair, measure first the electron spin on Earth and immediately thereafter the electron spin on Armad and record the results.

Next compare the records obtained for the two conditions.

1) *The results for condition 1.* The distribution of results is 50 : 50. Within the margin of statistical fluctuations, half the electrons collapse to state $|\uparrow\rangle$ and the rest to state $|\downarrow\rangle$.
2) *The results for condition 2.* The measurement outcomes are 50 : 50. Half the electrons collapse to state $|\uparrow\rangle$ and the rest to state $|\downarrow\rangle$.

The net results are identical. The measurements on Earth *do not cause* any changes on the collective experimental data on Armad. The whole experiment does not reveal any evidence of any communications or signaling between the distant objects.

This result constitutes the contents of the Eberhard theorem [4], according to which the expectation values for characteristics of two objects separated by a space-like interval are totally independent of one another.

With all that, the persisting fact that for each pair in case 2 the individual outcomes always come out opposite gives an irrefutable evidence of the long-range (nonlocal) correlations. One electron appears to know instantly what happened to its distant counterpart and act accordingly. These increasingly improbable coincidences reveal a new physical phenomenon – nonlocal quantum correlations (or quantum nonlocality). They have no classical analogue and show that a quantum system can keep certain correlations between its parts even when they are separated by voids of space.

Of course, the described situation is only a thought experiment. The real experiments are much more difficult, because it is extremely difficult to maintain the "purity" of the system – to protect it from decoherence that destroys the quantum correlations. The farther apart the constituents of the system, the faster they decohere. And yet such experiments [1] have been successfully carried out. They were performed on a much smaller scale than the thought experiment considered here, but their idea and results are essentially the same. They did show the coordinated behavior of the parts of extended entangled systems. But they did not show any evidence of FTL signaling.

24.4
The "No-Cloning" Theorem

The attempts to uncover FTL signaling between the entangled particles failed. For Bob, finding his electron in a certain spin state does not tell anything about the result of Alice's measurement because of the principal unpredictability of any single quantum event. When Bob measures his electron and finds its spin pointing up, Bob cannot know whether Alice has indeed performed the measurement in the first place. Even without her doing anything, Bob's electron might emerge with spin up from the measurement. So the observed value of its spin can hardly be considered a message. And there is another aspect to this. When Alice sends a message using a certain carrier, she encodes the contents of her message by "imprinting" it into the carrier in a certain way. If the carrier is an EM wave, the "imprints" are the corresponding modulations of the wave's frequency or amplitude. But when she wants to use an entangled system, there is no way for her to encode her message into a spin state of the particle she measures, since the measurement outcome is totally unpredictable. It is not under her or anybody else's control!

Thus, we have trouble at both ends of the line. Bob cannot consider the output of his apparatus as a message, and Alice cannot convert her intended message into a controllable input. In the language of information theory, the two distinct spin states (up or down) cannot, under the described conditions, be used by Alice as 1 bit (0 or 1) of information sent and by Bob as 1 bit of information received.

In 1982, N. Herbert in his paper [5] described an important variation of the above-discussed experiments.[1] Suppose that instead of using the two distinct *spin states in one basis* (the z-axis) as components of 1 bit, Alice chooses *two distinct bases* (x-axis or z-axis) as such components. In other words, Alice and Bob make an arrangement that when Alice measures her electron's spin along the z-direction, it makes the "zero" of the bit; and when she measures spin along the x-direction, this is the "one" of the bit. In contrast to individual spin-up or spin-down states, the use of the corresponding x- or z-setup is totally under Alice's control. Bob, on the other hand, keeps on measuring his electron's spin only along the z-direction (of course, the z-directions of both observers must coincide). In order to receive information from Alice, Bob must have the means to determine from his measurements which option Alice has chosen in each case. Now Bob's concern is to distinguish between Alice's two choices, rather than between the electron's choices. So the question now is as follows: If Alice switches from the z-measurement to x-measurement, will it affect the distribution pattern of Bob's measurement results? If yes, then FTL communication should be possible.

So, suppose Alice does switch to spin measurements along x, obtaining "spin right" ($|\rightarrow\rangle$) or "spin left" ($|\leftarrow\rangle$) state:

$$|\rightarrow\rangle = \frac{1}{\sqrt{2}}(|\uparrow\rangle + |\downarrow\rangle), \tag{24.11a}$$

$$|\leftarrow\rangle = \frac{1}{\sqrt{2}}(|\uparrow\rangle - |\downarrow\rangle). \tag{24.11b}$$

Suppose that in one of such measurements, Alice obtains "spin right." Immediately, Bob's electron state collapses to "spin left," and is now described by Equation (24.11b). But Bob measures spin only along the z-direction. Equations 24.11a and 24.11b predict that in such measurement, there is a 50 : 50 chance for his electron to collapse into either up or down state. The probability distribution for one electron is

---

1) In Herbert's original work, he considered cloning of the photon polarization state in a laser, taking into account the spontaneous emission. In our description here, we use instead the example of cloning the spin state, which can illustrate the same idea.

24.4 The "No-Cloning" Theorem

exactly the same as that in case when Alice performs the z-measurement, although for a different reason. In case of Alice's x-measurement, Bob's electron is in a *superposition* of up and down states; in case of her z-measurement, his electron is either in definite up state or in definite down state, but Bob does not know which one until he performs the measurement. After a succession of such measurements, he will obtain 50:50 outcome distribution because such was the distribution of Alice's results. Thus, Bob's measurement from an x-state *creates a new state*, while measurement from a z-state only confirms a preexisting state. But the experimental pattern is the same in both cases regardless of the difference in the physical mechanism. Therefore, Bob cannot distinguish between the choices made by Alice within this experimental scheme.

Herbert's idea was to modify the experiment in such a way that one could utilize the described difference. Imagine that in each single trial, Bob, before testing his electron, makes many exact copies of it (meaning copies of its spin state, *not the complete set* of its observables, which is impossible for fermions). Note that he copies his electron's spin state (makes its clones) without himself knowing this state. Call it "cloning of an unknown state."

Suppose Alice has chosen the z-measurement, with the result $|\uparrow\rangle$. Immediately, Bob's electron collapses to $|\downarrow\rangle$. *Before measuring this state*, Bob makes many "xerox copies" of it. Then he starts the measurements, passing one electron at a time through his apparatus. All of his cloned electrons will be found in the same state $|\downarrow\rangle$ – too unlikely to happen by accident! Bob concludes that the electron was in state $|\downarrow\rangle$ *before* his measurement, therefore Alice prepared *her* electron in state $|\uparrow\rangle$, in a z-measurement. Bob receives 1 bit of reliable information.

Now let Alice make the x-measurement. She reorients her spin detector from z- to x-direction and finds her electron in state $|\rightarrow\rangle$. Immediately, the entangled electron at Bob's end collapses to state $|\leftarrow\rangle$. Bob again starts with cloning, and gets a bunch of electrons all in state $|\leftarrow\rangle$. Bob passes all these electrons through his apparatus, one at a time, and finds in the end a 50:50 distribution between the two possible z-states. This is quite different from the previous result. Bob immediately concludes that this time Alice must have performed the x-measurement. Again, Bob receives 1 bit of reliable information. And since information about Alice's choice is transmitted instantaneously between two entangled electrons no matter how far apart, we must now have superluminal communication.

The question is whether such a scheme will work. Since its crucial part is cloning, this reduces to the question whether it is possible to clone an unknown state. The answer to this question was given in the same year 1982 [6,7]. It says that exact cloning of an unknown state is impossible. It is forbidden by the requirement that QM operators must be linear. In our case, the operator in question describes an algorithm or a process possessing the following property. Suppose we have a particle in an unknown state $|\Psi\rangle$ and an object in a state $|\Omega\rangle$. Considered as one combined system S, they can be described by a state function $|\Psi\rangle|\Omega\rangle$. But this state may be nonstationary, especially if the parts of the system interact with one another. As a result, within a time interval t, the system S may evolve from the initial state $|S(0)\rangle = |\Psi\rangle|\Omega\rangle$ into a certain final state $S(t)$. The corresponding outcome can be considered as a result of the action of a time evolution operator $\hat{U}(t)$, so that $|S(t)\rangle = \hat{U}(t)|S(0)\rangle$. It depends on the Hamiltonian of the system, but here we will only use the fact that it is linear.

Now imagine a situation when the final state is $S(t) = |\Psi\rangle|\Psi\rangle$. We say that in this case, operator $\hat{U}(t)$ has copied $|\Psi\rangle$ onto system $|\Omega\rangle$. This would constitute the cloning of the initial state – precisely what we need for superluminal signaling. Actually, we need "a little" more – the possibility to clone an *arbitrary* state, since it was essential that Bob could clone both – the state $|\leftarrow\rangle$ of his electron after Alice had performed the x-measurement and state $|\downarrow\rangle$ after she performed the z-measurement.

Let us see if this is possible. Represent $|\Psi\rangle$ as a superposition of some eigenstates $|\psi_j\rangle$ with unknown amplitudes a_j:

$$|\Psi\rangle = \sum_j a_j |\psi_j\rangle. \tag{24.12}$$

Copying this state means that

$$\hat{U}(t)|\Psi\rangle|\Omega\rangle = |\Psi\rangle|\Psi\rangle = \left(\sum_j a_j |\psi_j\rangle\right)^2. \tag{24.13}$$

But if $\hat{U}(t)$ is to be a universal copier able to clone *any* state, then it must be able to clone any single eigenstate in (24.12), that is,

$$\hat{U}(t)|\psi_j\rangle|\Omega\rangle = |\psi_j\rangle|\psi_j\rangle. \tag{24.14}$$

Let us apply again $\hat{U}(t)$ to $|\Psi\rangle|\Omega\rangle$ as we did in (24.13), but now using the property (24.14) and linearity of $\hat{U}(t)$. We obtain

$$\hat{U}(t)|\Psi\rangle|\Omega\rangle = \hat{U}(t)\left(\sum_j a_j |\psi_j\rangle\right)|\Omega\rangle = \sum_j a_j \hat{U}(t)|\psi_j\rangle|\Omega\rangle = \sum_j a_j |\psi_j\rangle^2. \tag{24.15}$$

The right-hand side of this equation is not the same as that in (24.13). Mathematically, the difference is obvious and does not require any comments. Physically, it has a simple interpretation. The postcloning system (24.13) is, as expected, a system of two particles in the same state $|\Psi\rangle$. The system is factorized, that is, each particle is independent of the other. In contrast, the postcloning system (24.15) is an *entangled superposition* of the two particles over a set of states $|\psi_j \psi_j\rangle$. This is immediately seen in the familiar (and the most simple) case of only two eigenstates ($j = 1, 2$), when (24.15) reduces to

$$\hat{U}(t)|\Psi\rangle|\Omega\rangle = \hat{U}(t)(a_1|\psi_2\rangle + a_2|\psi_2\rangle)|\Omega\rangle \Rightarrow (a_1|\psi_1\rangle|\psi_1\rangle + a_2|\psi_2\rangle|\psi_2\rangle). \tag{24.16}$$

This is an entangled superposition of the original and the cloned particle in which there is the amplitude a_1 to find the pair in state $|\psi_1\rangle|\psi_1\rangle$ and the amplitude a_2 to find it in state $|\psi_2\rangle|\psi_2\rangle$ (the subscripts 1, 2 here denote the states, not the particles!). The state (24.16) of the pair is totally different from the initial state (24.12) (with $j = 1, 2$) of the original particle. In the attempt to clone that state, we have destroyed it together with the corresponding information contained in it and obtained instead an entirely different state.

We see that requirements (24.13) and (24.14) are incompatible and therefore our assumption about universal applicability of the evolution operator for cloning was wrong. If it can clone each of the eigenstates in (24.12) separately, it cannot clone their superposition and vice versa. Recall that an electron with its spin, say, to the right is a superposition of the spin-up and spin-down states, and you will immediately see that the cloning procedure as described in the Alice–Bob experiment is not executable.

A similar example is known in optics: We can make a laser with a filter adjusted to amplify, say, vertically polarized light (i.e., to clone $|\updownarrow\rangle$ photon states) and to reduce the contribution of spontaneously emitted photons, but the corresponding orientation of the filter makes it impossible to amplify the $|\leftrightarrow\rangle$ states. On the other hand, removing the filter may enable the laser to amplify any light regardless of its polarization, but it will also uncontrollably amplify the spontaneously emitted light. The cloning will be much farther from the optimal.

We see that the no-cloning theorem prevents not only the perfect cloning of an unknown state but also one important type of imperfect cloning, where we agree to

---

Box 24.2 How rigorous is the no-cloning theorem?

The last example with a laser also illustrates a very important mathematical aspect of the "no-cloning" theorem. As any mathematical theorem, it must be exact: either 100% true or 100% false. If it does have exceptions, their domains must be unambiguously defined. Once this is done, the theorem must work within the remaining domain of applicability *all* of the time. A well-known example is the indeterminacy principle for the angular momentum **L**. It has an obvious exception when **L** = 0, in which case all three components and their indeterminacies are zeros. Similarly, we must be very cautious in formulating the "no-cloning" theorem.

The theorem was interpreted by many as a decisive argument against the possibility of superluminal signaling. However, the actual situation turned out to be more subtle. For instance, van Enk [8] has pointed out that such signaling would not necessarily violate the no-cloning theorem. On the other hand, there are loopholes in the theorem itself, which may disable the ban imposed by it (i.e., remove the word "impossible"), but just do not allow a 100% success rate!

Indeed, the proof (24.13) and (24.15) applies only to the cases when there is no preexisting information about the amplitudes a_j. Now consider the case when we have some (albeit incomplete) information, for example, we know that all $a_j = 0$, except for one, that is, $a_j = \delta_{jk}$ (but we do not know k).

Then Equations 24.13 and 24.15 become equivalent, and the no-cloning theorem actually proves that copying such a system is mathematically allowed, even though the state of the system (the number k) may be unknown. In other words, there must be at least one exception to the no-cloning theorem – namely, when a system is known to be in one of mutually orthogonal eigenstates of a known operator. Thus, $a_j = \delta_{jk}$ constitutes the simple analytical condition for the breach in the no-cloning theorem.

The subtle point here is that even though the ban seems to be lifted in this case (the cloning is mathematically possible), it is still not ascertained physically! This is evident from the above example with the laser amplification: Suppose we know that the photon to be cloned in our laser has a distinct polarization along one of the axes – x or y, but we do not know which one specifically. There is only a 50% chance that we choose the correct orientation of the polarizing filter out of the two possible ones. Generally, if the particle to be cloned is in one out of N distinct eigenstates in the known basis, there is a $1/N$ probability for it to be cloned successfully. In other words, even in the case when we have a partial knowledge of the initial state, the no-cloning theorem allows only a certain chance of success, which stands in proportion to the degree of this knowledge. The case $N = 2$ gives the maximal possible chance. One might say that this chance (50%) is enormously better than the absolute ban. But even this 50% chance of the right guess applies just for 1 bit. For a real message containing $J \gg 1$ bits, the probability to retrieve the information decreases as $\mathcal{P}(J) = (2)^{-J}$. For a relatively short message with $J = 100$ (just about four five-lettered words), this is roughly 10^{-33}. (We show later that even this tiny chance is in fact nonexistent, when discussing imperfect cloning.)

Of course, in practical applications, perfect accuracy is never attainable and usually is not required. A message remains readable even if one or two letters are misspelled. Moreover, by resending the same message several times, we can always distill the information contained in the message, making the chance of an error arbitrarily small as long as each bit can be read with an accuracy better than 50%! It follows that in most practical situations, we should be perfectly satisfied with imperfect cloning. Thus, lowering the claim from "exact" to "imperfect" and from "cloning" to "copying" may bring in a new dimension to our discussion. In this case, a laser with no filters at all *can* amplify a photon even in a *totally unknown* polarization state. This is only an imperfect copying, which is not forbidden.

When applied to fermions, the theorem may at first seem like a tautology, since the fermions, by the exclusion principle, cannot be in one state, so the "no-cloning" is automatically satisfied for them. However, the exclusion principle applies to a state determined by a *complete set* of all relevant observables. But the fermions can still have a common eigenstate of *one* observable provided the values of other observables are different. We can have an intense beam of electrons all in one spin state, but with different positions or linear momenta. Such states do not fall under the jurisdiction of the exclusion principle. Thus, considering what we said earlier about the $\mathcal{H}$-subspaces, the "no cloning" is as relevant to fermions as it is to bosons. We cannot obtain an intense beam of fermions in the same spin state by *exact cloning* of the *spin* state of one fermion if this state is initially unknown.

forgo the cloning of all relevant observables, aiming to clone only the observables that we use to encode bits of information. Indeed, Equations 24.13–24.15 did not specify whether our basis was a *complete* basis, spanning the entire $\mathcal{H}$ space, or if it only spanned a subspace related to some but not all of the observables. For instance, in the communication example considered above, we only encode messages in spin states and we are mainly interested in the possibility of replicating the spin state only (as mentioned above, it is impossible to get even two fermions with the same *complete set* of observables). The no-cloning theorem imposes *additional restrictions*, forbidding ideal copying in a *subspace* of $\mathcal{H}$, for example, by preventing us from cloning an arbitrary spin state.

24.5
Hidden Variables and Bell's Theorem

Here we analyze the hidden variables' theory, which can be considered as an extended EPR argument. According to EPR [3], each particle of the entangled pair must have its own definite state. Since QM cannot describe such states and restricts only to description of the whole system, QM must be incomplete. (Einstein's assertion was that QM is incomplete even when an individual state function is known because its predictions are still probabilistic.) But if QM is in fact incomplete, what is it that we are missing?

If you believe that Einstein was right ("God doesn't play dice!"), you must assume the existence of *hidden variables* – a set of parameters that are, for some yet unknown reasons, inaccessible to us and that determine all future experimental outcomes for a quantum object. As an illustration, consider linearly polarized light in a state

$$|\psi\rangle = \cos\theta|\updownarrow\rangle + e^{i\alpha}\sin\theta|\leftrightarrow\rangle \qquad (24.17)$$

(Figure 24.3), where $|\updownarrow\rangle$ and $|\leftrightarrow\rangle$ denote vertical and horizontal polarization states, respectively. The light passes through a polarizing filter with vertical transmission axis. According to classical theory, light is an EM wave (let its intensity be I_0 and let its polarization vector make angle θ with the vertical). Then, by Malus's law [9,10], the fraction $I = I_0 \cos^2\theta$ of incident light will be transmitted and the rest will be absorbed. As we know, classical theory fails at low I_0, where it must be replaced by quantum description. In QM, the same light forms a pure ensemble of photons and the components E_x, E_y of the electric field vector will now become the (generally complex) projections of state $|\psi\rangle$ onto the eigenvectors $|\updownarrow\rangle, |\leftrightarrow\rangle$. These projections are the amplitudes for the photon to pass through or to get absorbed. From the classical viewpoint, there is an outrageous inconsistency here. On the one hand, we have an ensemble with all photons being in *the same* state. On the other hand, one photon passes through and the other does not. If the states of both incident photons are totally identical, what causes the different results in their interaction with the same filter?

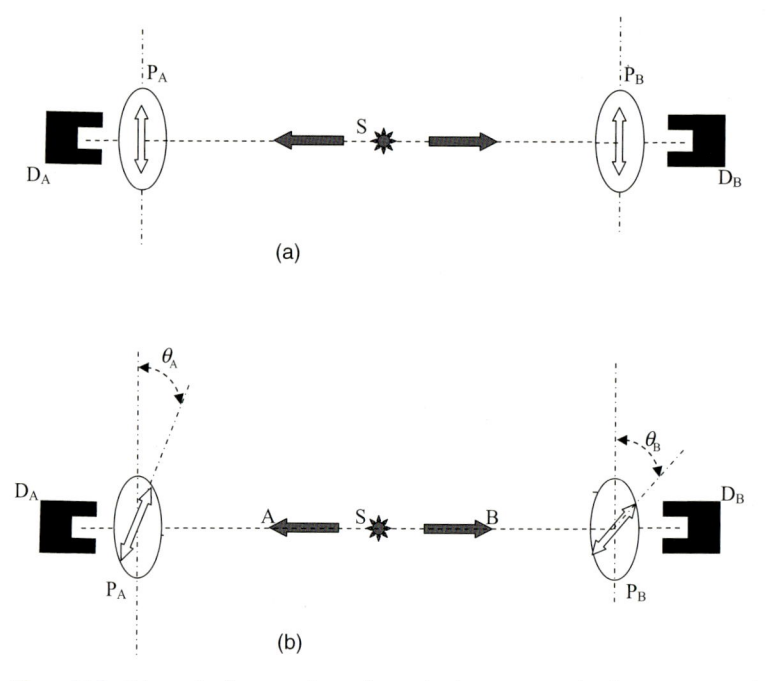

Figure 24.3 Schematic of an experiment for study of quantum nonlocality. S – source of pairs of entangled photons A and B; P_A and P_B – polarizing filters; D_A and D_B – detectors. (a) Polarizers with parallel transmission axes. (b) Polarizers with misaligned transmission axes.

The assertion of the hidden variable theory is that those states were not truly identical. There must be some additional variables that decide the fate of a photon, but they are hidden from QM as we know it today. Obviously, these variables must have different values for the photons that face different outcomes. Then, each photon behaves totally deterministically – the choice is predetermined by the value of the corresponding hidden variable. If such a variable were accessible to observation and its values could be measured, we would give it the status of an observable. Then, knowing θ and having measured this new observable for one specific photon, we could predict exactly whether this photon is going to pass through the filter or be absorbed in it. So the world is not probabilistic and no dice are being tossed. It is only our lack of information about the hidden variables that makes QM a probabilistic theory.

This line of reasoning has a certain appeal, especially to a classical mind. If hidden variables exist, then all chance is eliminated from the world and the clockwork Universe of the earlier centuries is restored. But as often happens when one considers a beautiful theory, one has to wonder if this theory is beautiful enough to be correct. And if Bohr and the majority of contemporary physicists remained skeptical about the idea, their intuition was well grounded in history. As we know from the example of wave and corpuscular theories, when a discarded theory makes a comeback, it tends to come back laden with new concepts and the world never remains as simple as it was under the original theory. But the hidden variable hypothesis would simply restore nineteenth century determinism with little, if any, modification. Even this success would come at a price of introducing a quantum analogue of ether – an "observable" that somehow resists our best effort to detect it, and Einstein should have known better than anyone else that such models have a poor track record. Yet it was not until 1964 that John Bell gave a definitive answer to the question put forth in the EPR paper.

Bell has shown that the hidden variable theory and QM make different predictions that can be tested experimentally. Consider the following experimental setup (Figure 24.3b): A source S produces *entangled* pairs of (distinguishable!) photons. They fly off in the opposite directions, each photon being in indefinite polarization state. The only definite feature is that their states are identical, that is, if we choose the rectilinear basis $\{|\updownarrow\rangle, |\leftrightarrow\rangle\}$, then both photons will be found either in state $|\updownarrow\rangle$ or in state $|\leftrightarrow\rangle$. (The same principle will hold if one chooses any other basis; for example, in case of $\{|R\rangle, |L\rangle\}$ basis, both photons will be right polarized or both will be left polarized.) In the language of QM, they are in a maximally entangled superposition. If A and B denote the photons of the pair, this superposition can be written as

$$|\psi_{AB}\rangle = \frac{1}{\sqrt{2}}\left(|\updownarrow\rangle_A|\updownarrow\rangle_B + e^{i\delta}|\leftrightarrow\rangle_A|\leftrightarrow\rangle_B\right). \tag{24.18}$$

When far apart, each photon is passed through "vertical" polarizer. The respective outcomes are recorded by detectors D_A and D_B. According to QM, there are only two possible outcomes in this experiment: either both photons jump into a definite state $|\updownarrow\rangle$, in which case they will pass through and be detected, or they both jump into a definite state $|\leftrightarrow\rangle$ and are absorbed. In other words, (24.18) collapses into one of the two states:

$$|\psi\rangle \Rightarrow |\updownarrow\rangle_A|\updownarrow\rangle_B \quad \text{or} \quad |\leftrightarrow\rangle_A|\leftrightarrow\rangle_B. \tag{24.19}$$

In the first case, we will have both detectors clicking in tandem; in the second case, both will stay idle. Assuming ideal detectors and polarizers, there will be no other outcomes.

On the average, half of the photons arriving at A will go through, activating D_A, and every such event will be mirrored at B. The second half of the photons will be absorbed thereby leaving D_A idle, and again, every such event will be mirrored at B. The events are rigidly correlated, and we will have the same $(1/2)$ probability for either of the two outcomes (24.19). The correlated response of the detectors would be observed even if the two photons were not maximally entangled, that is, if instead of (24.18) we had

$$|\psi_{AB}\rangle = c_1|\updownarrow\rangle_A|\updownarrow\rangle_B + c_2|\leftrightarrow\rangle_A|\leftrightarrow\rangle_B, \tag{24.20}$$

with the normalized amplitudes c_1, c_2 having different moduli. The only difference would be in the ratio of the number of coincident "vertical" and "horizontal" outcomes $N_{V,V}/N_{H,H}$ – this ratio is a function of c_1 or c_2 (Problem 24.3). Generally, there will be no cases of uncorrelated action of detectors in the described situation. It looks as if the rigid correlation between the two photons (a purely QM phenomenon) is transferred through amplification onto macroscopic systems – the two detectors that can be separated by voids of space and yet react in concord. Of course, the detectors remain independent objects, but the entangled photons make them respond in a maximally coordinated way.

In all hidden variable theories, both photons in the pair are independent objects and each photon has its own polarization state. It is true that the states of both are identical, but this is merely an initial condition imposed by their common origin. Denote this common polarization state as $|\theta\rangle$, which indicates the polarization angle θ with the vertical axis. In other words, each photon is in the state (24.17). This is a definite (not entangled!) state albeit possibly yet unknown to us, so there is nothing to collapse into for the *whole pair*. Once they reach their respective polarizers, each photon separately collapses from superposition (24.17) into either $|\updownarrow\rangle$ or $|\leftrightarrow\rangle$, and since there are no instantaneous correlations between the distant objects, each photon acts *totally independent* of its partner. This brings in the lack of concurrence in the detectors' responses – one may click when the other remains idle or vice versa. The result will be a decrease of the total number of coincident responses (combined clicks or combined nonclicks) – they will no longer be 100% of all cases. The specific fraction of coincident responses will be a function of θ (Problem 24.4). Assuming that the above-mentioned initial condition correlates only *observable variables* – in our case polarizations – and does not affect the hidden variables, we can find the result if we know the distribution $F(\theta)$ of the initial values of θ. For the uniform distribution $F(\theta) = \text{const}$, that is, for the source firing the *pairs* of identically polarized photons with equal probability for all values of θ for each pair, the average fractional number of coincident responses drops from 1 to 3/4 (Problem 24.5). This is different from the QM prediction, and it can be confirmed or refuted by the experiment.

But the experimental setup is not necessarily restricted to the case of the parallel transmission axes of two polarizers, and it is interesting to see the predictions of the two theories in the more general case. So, let the two filters have nonparallel transmission axes, making the angles θ_A and θ_B, respectively, with the vertical (Figure 24.3b).

According to the hidden variable theory, each photon in the pair has a definite state described by (24.17). Applying (24.17) to photons A and B, respectively, and taking their product, we get the state of the pair (with the phase factor dropped) right before they arrive at their respective destinations:

$$|\psi_{AB}\rangle = |\psi_A\rangle|\psi_B\rangle = \left(\cos\theta|\updownarrow\rangle_A + \sin\theta|\leftrightarrow\rangle_A\right)\left(\cos\theta|\updownarrow\rangle_B + \sin\theta|\leftrightarrow\rangle_B\right)$$
$$= \cos^2\theta|\updownarrow\rangle_A|\updownarrow\rangle_B + \sin\theta\cos\theta\left(|\updownarrow\rangle_A|\leftrightarrow\rangle_B + |\leftrightarrow\rangle_A|\updownarrow\rangle_B\right) + \sin^2\theta|\leftrightarrow\rangle_A|\leftrightarrow\rangle_B.$$
(24.21)

24.5 Hidden Variables and Bell's Theorem

The first term on the right-hand side is the amplitude to find both photons in the $|\updownarrow\rangle$-state; the last term is the amplitude to find both in the $|\leftrightarrow\rangle$-state; and the middle term is the amplitude to find one of them in the $|\updownarrow\rangle$-state and the other in the $|\leftrightarrow\rangle$-state. If both filters are oriented along the vertical axis, as in the above-considered case, we would have the probability of both detectors clicking $\mathcal{P}_{AB}^+ = \cos^4\theta$; the probability of both remaining idle $\mathcal{P}_{AB}^- = \sin^4\theta$; and the probability of only one of them clicking $\mathcal{P}_{AB}^\pm = 2\sin^2\theta\cos^2\theta$. The sum of all three is equal to 1 just as it should. But in this sum, the fraction of correlated responses (coincident clicks or coincident nonclicks) is reduced to

$$\mathcal{P}_c(\theta) = 1 - 2\sin^2\theta\cos^2\theta. \tag{24.22}$$

As one would expect, it reaches its maximum 1 (only coincident responses) at $\theta = 0$ or $\theta = \pi/2$; the minimal value is $1/2$ and is reached at $\theta = \pi/4$. If we take the average of (24.22) over all θ, we will get $\langle\mathcal{P}_c(\theta)\rangle = 3/4$.

This is easy to generalize for the case of misaligned filters making angles θ_A and θ_B, respectively, with the vertical axis. The respective probabilities of A and B clicking in this case are

$$\mathcal{P}_A^+(\theta) = \cos^2(\theta - \theta_A), \tag{24.23a}$$

$$\mathcal{P}_B^+(\theta) = \cos^2(\theta - \theta_B). \tag{24.23b}$$

The probability of both clicking in tandem for a fixed polarization $|\theta\rangle$ is

$$\mathcal{P}_{AB}^+(\theta) = \mathcal{P}_A^+(\theta)\mathcal{P}_B^+(\theta) = \cos^2(\theta - \theta_A)\cos^2(\theta - \theta_B). \tag{24.24}$$

For a uniform initial distribution over θ in all trials, the net probability of both detectors firing will be

$$\mathcal{P}_{AB}^+ = \frac{1}{\pi}\int_0^\pi \mathcal{P}_{AB}^+(\theta)d\theta = \frac{1}{8}\left[1 + 2\cos^2(\theta_A - \theta_B)\right]. \tag{24.25}$$

Now we also need to consider the idle responses, since they are as informative as the clicks. A similar procedure (Problem 26.5.4) gives the same result:

$$\mathcal{P}_{AB}^- = \frac{1}{\pi}\int_0^\pi \mathcal{P}_{AB}^-(\theta)d\theta = \mathcal{P}_{AB}^+. \tag{24.26}$$

The total probability of coincident responses is

$$\mathcal{P}_c = \mathcal{P}_{AB}^+ + \mathcal{P}_{AB}^- = (1/4)\left[1 + 2\cos^2(\theta_A - \theta_B)\right]. \tag{24.27}$$

It varies within the range

$$1/4 \leq \mathcal{P}_C \leq 3/4. \tag{24.28}$$

In case of aligned filters, $\theta_A = \theta_B$, (24.28) reaches its maximum value obtained above. The expression (24.28) is a version of *Bell's inequality* [11,12]. It must be satisfied if the object's behavior at each location is governed by the hidden variables of this object alone and is independent of other objects at this instant. These variables, if known to us, would make the outcome of any measurement exactly predictable. The value of this result by Bell is that it is a prediction expressed in terms of measurable characteristics (the probability $\mathcal{P}_C$ is an observable and measurable quantity). There are a few versions of this inequality corresponding to different experimental situations [13–16], and (24.28) is one of them. It remains to compare it with QM predictions for the same arrangement and see which one of them is confirmed by the experiment.

So let us now turn to QM. According to QM, neither of the two photons has a state of its own; the whole pair is in the entangled superposition until both reach their final destinations. In the rectilinear basis, this superposition is described by (24.18), where we can set $\delta = 0$.

This situation is fundamentally different from the scenario with hidden variables. Now, if both detectors fire, it may occur from the state $|\updownarrow\rangle_A|\updownarrow\rangle_B$ or from the state $|\leftrightarrow\rangle_A|\leftrightarrow\rangle_B$. In the former case, the amplitude of the two coincident clicks is $\cos\theta_A \cos\theta_B$; in the latter case, it is $\sin\theta_A \sin\theta_B$. And there is no way to distinguish between the two possibilities. Therefore, they interfere:

$$|\psi_{AB}^+\rangle = \frac{1}{\sqrt{2}}(\cos\theta_A \cos\theta_B + \sin\theta_A \sin\theta_B) = \frac{1}{\sqrt{2}}\cos(\theta_A - \theta_B). \tag{24.29}$$

Similarly, the total amplitude of nonclicks is

$$|\psi_{AB}^-\rangle = \frac{1}{\sqrt{2}}(\sin\theta_A \sin\theta_B + \cos\theta_A \cos\theta_B) = |\psi_{AB}^+\rangle. \tag{24.30}$$

In this respect, the situation is similar to the double-slit experiment, in which the amplitude of arrival at a certain location on the screen is determined by interference of the two virtual paths threading each through the respective slit. In our case, instead of having two paths in the configuration space, we have two paths in the $\mathcal{H}$-space of photon polarizations. Accordingly, instead of an infinite number of possible distinct landing points in the double-slit experiment, here we have only four possible outcomes: coincident firings, coincident nonfirings, firing only at A, and firing only at B. The amplitudes of the latter two possibilities are

$$|\psi_{AB}^{\mp}\rangle = \frac{1}{\sqrt{2}}(\sin\theta_A \cos\theta_B - \cos\theta_A \sin\theta_B) = |\psi_{AB}^{\pm}\rangle. \tag{24.31}$$

It follows from (24.30) that the probabilities of coincident firings and coincident nonfirings are the same:

$$\mathcal{P}^+_{AB} = \mathcal{P}^-_{AB} = (1/2)\cos^2(\theta_A - \theta_B) \Rightarrow \mathcal{P}_c = \mathcal{P}^+_{AB} + \mathcal{P}^-_{AB} = \cos^2(\theta_A - \theta_B). \tag{24.32}$$

$\mathcal{P}_c$ here is the net probability of coincidences (firings or nonfirings). Similarly, the probability of a noncoincident response is the same for ($\pm$) and for ($\mp$) combination:

$$\mathcal{P}^\pm_{AB} = \mathcal{P}^\mp_{AB} = (1/2)\sin^2(\theta_A - \theta_B) \Rightarrow \mathcal{P}^\pm_{AB} = \sin^2(\theta_A - \theta_B), \tag{24.33}$$

where $\mathcal{P}^\pm_{AB}$ is the *net* probability of noncoincident responses. The sum $\mathcal{P}_c + \mathcal{P}^\pm_{AB}$ is, as it should, equal to 1. In case of the two filters lined up ($\theta_A = \theta_B$), Equation 24.32 reduces, as it should, to $\mathcal{P}_c = 1$, in an obvious violation of Bell's inequality. Note that in QM treatment of the case, there is no integration over θ (this variable does not even appear in the equations) because the entangled superposition (24.18) already has *all* θ included as *virtual* polarization states.

The comparison of (24.28) with (24.32) is the essence of *Bell's inequality*, according to which each theory gives its own prediction, the predictions are different and it remains to turn to the experiment as the ultimate judge. The first experiments were carried out in the 1970s [15,16] and unambiguously confirmed the QM prediction. Similar experiments were performed later under a wide range of conditions [17–19] and in all of them Bell's inequality was violated. By now, we have an overwhelming experimental evidence against hidden parameters and in favor of QM.

24.6
Bell-State Measurements

Take two particles A and B that have eigenstates $|0\rangle$ and $|1\rangle$. We can prepare entangled states by taking various combinations of $|0\rangle_A$, $|1\rangle_A$, $|0\rangle_B$, $|1\rangle_B$. Entanglement will manifest itself as a correlation between measurement results for A and B.

We remember that Bell's theorem by itself does not provide a numerical measure of the degree of this correlation. Indeed, Bell's 1964 paper [11] has only shown that the perfect correlations between eigenstates of A and B predicted by QM cannot be realized in any local model via some prearranged agreement stored in a hidden variable. Our task now is to derive such a measure and express the degree of entanglement as a number.

We will adopt the approach taken in Ref. [13] by Clauser, Horne, Shimony, and Holt ("the CHSH paper"), which proposed the following experimental test of the entire family of local hidden variable theories (see also Ref. [15]). Suppose two observers, Alice and Bob, try to determine whether a quantum-mechanical particle can know its future state in advance by storing this information in some hidden variable λ. To this end, they use an ensemble of correlated EPR optical photon pairs

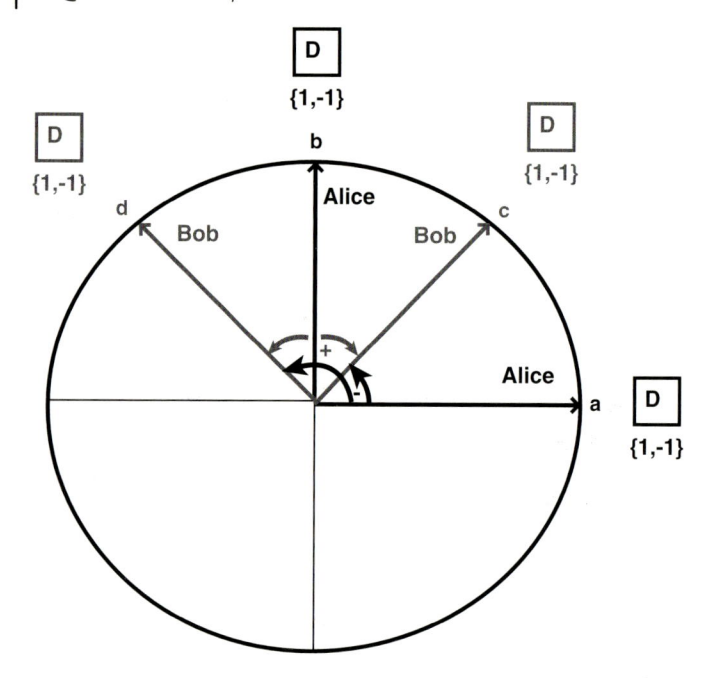

Figure 24.4 CHSH apparatus. Alice and Bob use detectors that either register or fail to register a particle (outcomes 1 and −1, respectively). Test angles a, b, c, d can be selected arbitrarily and quantity Q is composed as the difference of correlation functions for Alice's angle a plus the sum of correlation functions for Alice's angle b. A special case is shown where all four detectors are oriented in the same plane of the Poincaré sphere (see also next diagram). The test angles shown (Bell test angles) are optimal for measuring correlation between Alice's and Bob's particles. Quantum mechanics predicts the correlation between results could be as high as $|Q| = 2\sqrt{2}$; a local hidden variable model would limit that correlation to $|Q| \leq 2$.

and arrange the experiment so that the first photon of a pair enters Alice' apparatus A while the second photon enters Bob's apparatus B. Each apparatus has a polarizing filter followed by a detector. When a photon successfully passes through the polarizer and therefore gets detected, we will represent this outcome as 1; otherwise, we will use −1 to represent the negative outcome.

Furthermore, it is agreed that Alice can orient her polarizer at either angle a or angle b with respect to some chosen reference direction, while Bob will be using polarizer orientations c and d (Figure 24.4). (These angles are unrelated to the coefficients a, b, c, d that we used in Chapter 15 to define concurrence). In this notation,[2] a statement that, say, $A(a) = 1$ (or $A(a) = -1$) will tell us that the photon passed (or failed to pass) Alice's apparatus whose polarizer was oriented at an angle a.

---

2) For the sake of those readers who want to use the original source for reference, let it be mentioned that the CHSH paper used a somewhat less intuitive notation, with angles a and b for apparatus A and angles b and c for apparatus B.

QM predicts a correlation between emergence (or nonemergence) of photons from the apparatus, as discussed earlier in the EPR section.[3] On the other hand, a local hidden variable theory must explain the same correlation in terms of a shared [4] variable λ, which we presumably cannot measure directly, and this variable must provide instructions for the photon such as "if Alice orients her polarizer at a, choose $A(a) = 1$; in the case of orientation b, choose $A(b) = -1$" and so on. Locality of the theory means that when Alice's photon makes its choice, it does not consult Bob's photon, which could be arbitrarily far away since Alice and Bob can control the distance between A and B. Therefore, it can only make its decision based on the orientation of A, because nothing is known whatsoever about the orientation of B. As a result, $A(a)$ must be a function of a and λ only and independent of c and d, so we can write it as $A(a, \lambda)$. The same will of course be true for the other experimental outcomes, which will then be written as $A(b, \lambda)$, $B(c, \lambda)$, and $B(d, \lambda)$. Finally, let λ itself have a normalized probability distribution $\rho(\lambda)$.

Now, when Alice and Bob actually do the experiment, they must operate with the *average* values of $A(a, \lambda)$, $A(b, \lambda)$, $B(c, \lambda)$, and $B(d, \lambda)$. This averaging is required because a hidden variable theory would predict that in the process of measurement of $A(a, \lambda)$ Alice is going to affect (damage) the variable λ, thereby corrupting the possible future result $A(b, \lambda)$ for this particle, so a simultaneous measurement of these two quantities is impossible. Hence, Alice and Bob must use ensemble averages. (This can be quite obvious in the case of photons since when a photon gets absorbed, it cannot be used in any future measurements, but it's not that obvious for particles such as electrons. And upon a second thought, it's not really "obvious" even for photons if one recalls the nondemolition measurements. That's why we had to cite the damage to the hidden variable as the real reason behind the general rule that a particle cannot be reused.)

Let's write the expectation value of the product $A(a)B(c)$:

$$E(a,c) = \int A(a, \lambda) B(c, \lambda) \rho(\lambda) d\lambda. \tag{24.34}$$

(In the CHSH paper, this is called the "emergence correlation function" between apparatus orientations and denoted $P(a, c)$; here, we will stick with the more familiar notation.)

3) In statistics, correlation between random variables X and Y is defined as

$$\rho_{XY} \equiv \text{corr}(X, Y) = \frac{\text{cov}(X, Y)}{\sigma_X \sigma_Y} = \frac{\langle (X - \mu_X)(Y - \mu_Y) \rangle}{\sigma_X \sigma_Y},$$

where μ and σ denote the mean and the variance, respectively. Then, $\rho_{XY} = 1$ is called perfect correlation, $\rho_{XY} = -1$ is called perfect anticorrelation, and $\rho_{XY} = 0$ means X and Y are uncorrelated.where μ and σ denote the mean and the variance, respectively. Then, $\rho_{XY} = 1$ is called perfect correlation, $\rho_{XY} = -1$ is called perfect anticorrelation, and $\rho_{XY} = 0$ means X and Y are uncorrelated.

4) The word *shared* is necessary because here we are trying to explain a perfect correlation where Alice's photon choosing state $|V\rangle$ means that Bob's photon is also in the same state $|V\rangle$, and so on. In a local hidden variable theory, this would only be possible if λ is the same for both photons. So the argument that follows is valid only for the case of perfect correlations as in an EPR photon pair.

If we now do the same experiment, orienting the second apparatus at angle d, then we can similarly write

$$E(a, d) = \int A(a, \lambda) B(d, \lambda) \rho(\lambda) d\lambda \tag{24.35}$$

and then

$$|E(a, c) - E(a, d)| = \left| \int [A(a, \lambda) B(c, \lambda) - A(a, \lambda) B(d, \lambda)] \rho(\lambda) d\lambda \right|. \tag{24.36}$$

However, using the general fact that absolute value of an integral is less than or equal to the integral of the absolute value or

$$\left| \int f(x) dx - \int g(x) dx \right| \leq \int |f(x) - g(x)| dx \tag{24.37}$$

(this is a corollary of the triangle inequality!), we can write

$$|E(a, c) - E(a, d)| \leq \int |A(a, \lambda) B(c, \lambda) - A(a, \lambda) B(d, \lambda)| \rho(\lambda) d\lambda \tag{24.38}$$

or simply

$$|E(a, c) - E(a, d)| \leq \int |A(a, \lambda) B(c, \lambda)[1 - B(c, \lambda) B(d, \lambda)]| \rho(\lambda) d\lambda. \tag{24.39}$$

(Here, we have used the fact that $B(c, \lambda) = \pm 1$ so that $B(c, \lambda)$ is equal to its own reciprocal.)

The product of A and B has absolute value 1, while the term in the brackets is nonnegative, so we can drop the absolute value sign on the right-hand side:

$$|E(a, c) - E(a, d)| \leq \int [1 - B(c, \lambda) B(d, \lambda)] \rho(\lambda) d\lambda. \tag{24.40}$$

Due to the normalization requirement, the first integral will be equal to 1 so that

$$|E(a, c) - E(a, d)| \leq 1 - \int B(c, \lambda) B(d, \lambda) \rho(\lambda) d\lambda. \tag{24.41}$$

We will now prove a short lemma about the second term on the right-hand side. Let us consider the orientation b of the first detector and write the expectation value again:

$$E(b, c) = \int A(b, \lambda) B(c, \lambda) \rho(\lambda) d\lambda. \tag{24.42}$$

Then divide the whole range of λ into two subranges: Γ_+ where both terms in the product have the same sign (i.e., both 1 or both -1) and Γ_- where they have opposite signs. Then, rewrite the last equation as

$$E(b, c) = \int_{\Gamma_+} \rho(\lambda) d\lambda - \int_{\Gamma_-} \rho(\lambda) d\lambda = \int \rho(\lambda) d\lambda - 2 \int_{\Gamma_-} \rho(\lambda) d\lambda = 1 - 2 \int_{\Gamma_-} \rho(\lambda) d\lambda.$$

$$\tag{24.43}$$

Next, replacing $B(c, \lambda)$ with plus or minus $A(b, \lambda)$ depending on the chosen subrange, we can write

$$\int B(c, \lambda) B(d, \lambda) \rho(\lambda) d\lambda = \int_{\Gamma_+} A(b, \lambda) B(d, \lambda) \rho(\lambda) d\lambda - \int_{\Gamma_-} A(b, \lambda) B(d, \lambda) \rho(\lambda) d\lambda$$

$$= E(b, d) - 2 \int_{\Gamma_-} A(b, \lambda) B(d, \lambda) \rho(\lambda) d\lambda$$

(24.44)

and since replacing the integrand with its absolute value would increase the second term, and that absolute value is equal to 1, we have

$$\int B(c, \lambda) B(d, \lambda) \rho(\lambda) d\lambda \geq E(b, d) - 2 \int_{\Gamma_-} \rho(\lambda) d\lambda = E(b, d) + E(b, c) - 1. \quad (24.45)$$

Using this result, we now go back to (24.41) and replace the right-hand side with a larger value:

$$|E(a, c) - E(a, d)| \leq 2 - E(b, d) - E(b, c), \quad (24.46)$$

which also implies (we are now making the theorem less strict!) that

$$E(a, c) - E(a, d) \leq 2 - E(b, d) - E(b, c). \quad (24.47)$$

Let's group all the expectation values on the left-hand side and introduce a new variable Q for the resulting combination:

$$Q \equiv E(a, c) - E(a, d) + E(b, c) + E(b, d) \leq 2. \quad (24.48)$$

This is still only an intermediate result. However, using a similar line of reasoning, one can show that $Q \geq -2$. The combination of these two results gives us the CHSH inequality:

$$|E(a, c) - E(a, d) + E(b, c) + E(b, d)| \leq 2. \quad (24.49)$$

It must hold for any local hidden variable theory, and any violation discovered in the experiment will indicate that Nature does not use hidden variables to prearrange the measured result.[5]

On the other hand, we can use QM to predict Q for any given angles a, b, c, and d. Obviously, we will be most interested in the choice of angles that produces the greatest discrepancy between QM and the CHSH inequality. We shall now consider two physically important cases.

Case 1: Spin-1/2 particles. Let's take an EPR pair of electrons forming a singlet state, that is, $s = 0$, $m = 0$. Since electrons have no polarization, the "polarizer" in this case will be a Stern–Gerlach apparatus. When the first electron of the pair (Alice's electron) encounters the apparatus oriented at angle a, it must decide

---

[5] It is worthwhile to make two remarks. First, even though "CHSH" or "Bell–CHSH inequality" is the preferred name for the last result, you may still sometimes see it mentioned in the literature as "Bell's inequality." Second, because the CHSH authors originally left their inequality in the form (24.48), some derivations you may see in the literature will omit the last step, making you wonder about the origin of the lower bound $Q \geq -2$. Alternatively, there exists a simpler derivation of (24.49) (Problem 24.7).

whether its spin's projection on direction $\vec{a}$ (which now acts as an axis of quantization) will be 1 or −1 (in $\hbar/2$ units). Let us direct our z-axis along **a** and write the singlet state of the EPR pair as

$$\psi = \frac{1}{\sqrt{2}}(\uparrow\downarrow - \downarrow\uparrow). \tag{24.50}$$

Let $\beta = (\mathbf{a}, \mathbf{c})$ be the angle between Alice's and Bob's polarizers. Then, if Alice's electron chooses to align itself with **a**, Bob's electron must be aligned against **a**; with respect to direction **c**, the angle will be $\pi - \beta$. When measured by Bob's apparatus, it will be found in spin-up state with probability $\cos^2((\pi - \beta)/2) = \sin^2(\beta/2)$ and in spin-down state with probability $\sin^2((\pi - \beta)/2) = \cos^2(\beta/2)$. Similarly, if Alice's photon aligns itself against **a**, then $\sin^2(\beta/2)$ and $\cos^2(\beta/2)$ will be the probabilities of getting spin-down and spin-up states for Bob's photon. Therefore, a coincidence where Bob and Alice get the same sign (both spin-up or both spin-down) will be observed in $\sin^2(\beta/2)$ percentage of experiments, and the remaining experiments will produce different signs, so that

$$E(a,c) = \langle A(a)B(c)\rangle = \sin^2\frac{\beta}{2} - \cos^2\frac{\beta}{2} = \frac{1-\cos\beta}{2} - \frac{1+\cos\beta}{2} = -\cos\beta. \tag{24.51}$$

In other words, QM says that in an electron singlet state, we always have

$$E(a,c) = -\mathbf{a}\cdot\mathbf{c} \tag{24.52}$$

for any units vectors **a**, **c**. In particular, when $\beta = 0$, the expectation value will be −1 and Bob's and Alice's results will be perfectly anticorrelated just as one might expect; when the apparatuses are antiparallel ($\beta = \pi$), the results will be perfectly correlated; and when the apparatuses are perpendicular ($\beta = \pi/2$), we get zero correlation, $E(a,c) = 0$. We can now write the quantum-mechanical prediction for Q:

$$|Q| = |E(a,c) - E(a,d) + E(b,c) + E(b,d)| \\ = |-\cos(a,c) + \cos(a,d) - \cos(b,c) - \cos(b,d)|. \tag{24.53}$$

The absolute value is maximized at $a = 0$, $b = 90$, $c = 45$, and $d = 135$, when we get $|Q| = 2\sqrt{2}$.

The angles for which the violation of CHSH inequality is greatest are also called "Bell test angles."

Case 2: Spin-1 particles. Let's consider an EPR pair of photons, whose total spin is again 0. Photons have spin quantum number $s = 1$, which would allow three values of m: −1, 0, 1. However, $m = 0$ is disallowed because of the zero rest mass of a photon. The remaining two eigenvalues, $m = 1$ and $m = -1$, correspond to eigenstates $|R\rangle$ and $|L\rangle$, right and left circularly polarized light. However, we can also create linear combinations of these eigenstates to form a linear basis. For example, $|x\rangle = (1/\sqrt{2})(|R\rangle + |L\rangle)$ and $|y\rangle = (-i/\sqrt{2})(|R\rangle - |L\rangle)$ are suitable choices for linearly polarized photons along x- and y-axes, respectively.

The difference from the electron case is that for a photon, a single polarization state would be defined by indicating an axis without specifying the direction along

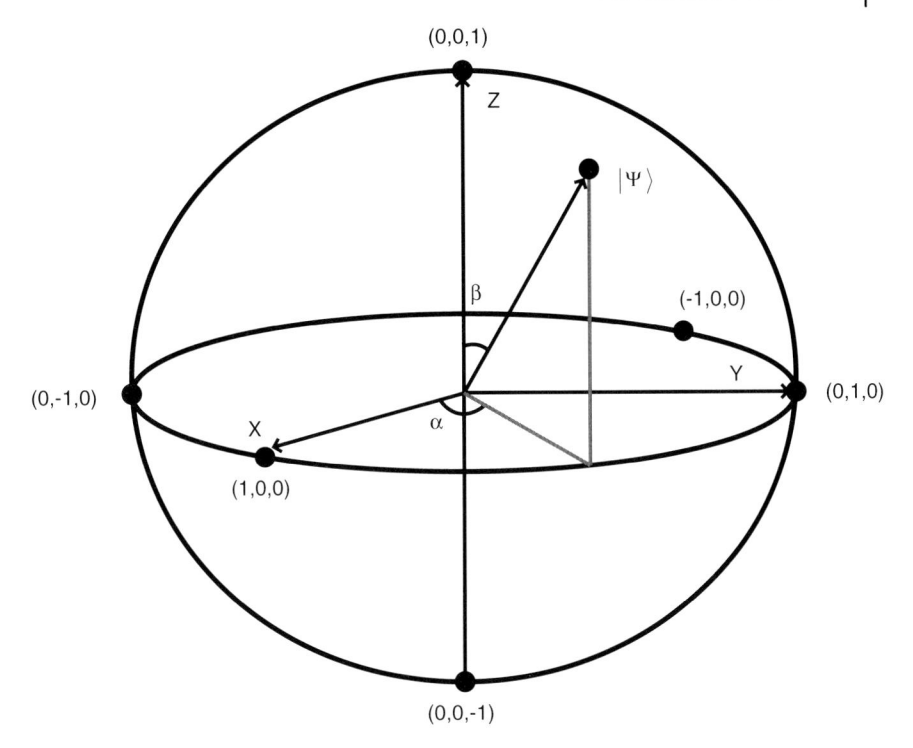

Figure 24.5 Poincaré sphere shows photon polarization states. By convention, the circular component of polarization is represented by the vertical axis. In many situations, it is sufficient to draw a single cross section of this sphere. Thus, when polarization state has no circular component, one can work in the equatorial plane XY.

this axis (for example, by a double-headed arrow). To escape this ambiguity, we can draw the following electron-photon analogy. The analogue of electron's spin-up and spin-down states for a photon is vertical and horizontal polarization states. Alternatively, one could take diagonal-up and diagonal-down states or right and left circularly polarized states, just as for an electron one could direct the quantization axis along x- or y-axis instead of z-axis.

One can easily see that the 180° angle that separated two mutually orthogonal states in the electron case turns into a 90° angle when we switch to photons. To make polarization states behave the same way as electron spin states, it is convenient to represent them by points on the Poincaré sphere shown in Figure 24.5. One can think of it as extension of the Bloch sphere approach to represent photon polarizations.[6] It is defined by the equation

$$X^2 + Y^2 + Z^2 = 1, \tag{24.54}$$

6) Some authors use the terms "Bloch sphere" and "Poincaré sphere" interchangeably. We will reserve the former term for a general two-level system that behaves analogously to a spin-1/2 particle and use the latter term when referring specifically to photon polarizations.

where X, Y, and Z stand for the rectilinear, diagonal, and circular components of polarization. The two opposite points (1, 0, 0) and (−1, 0, 0) will then stand for the horizontal and vertical polarizations, while the point pairs (0, 1, 0), (0, −1, 0) and (0, 0, 1) and (0, 0, −1) will denote the diagonal-up and diagonal-down states and right and left circularly polarized states, respectively.[7]

With this convention, the "real-life" angles are doubled so that the actual 90° angles between orthogonal states are mapped into 180° angles on the Poincaré sphere. To compensate for that, we must also switch from photon formulas, which use the entire angle (as in Malus' law and its QM analogues) to the electron formulas that divide the angle in half. Then, as long as we operate with Poincaré sphere angles, we can use the familiar formulas for electron spins.

It is already clear at this point that the optimal choice of test angles will again be $a = 0$, $b = 90$, $c = 45$, and $d = 135$. When we "translate" these into real-life angles, we will get $a = 0$, $b = 45$, $c = 22.5$, and $d = 67.5$. These are Bell test angles for the EPR photon pair. You may sometimes see a statement that these are the general values of Bell test angles used in *all* practical situations. The truth is that most experiments use entangled pairs of optical photons because they are easier to produce and detect in the laboratory, but other particles have been used as well, so you should be ready to derive the actual Bell test angles for the given experimental setup.[8]

One experiment to test the CHSH inequality was done by Aspect's group in France [17,20]. The measured value was $Q \cong 2.8$, violating the CHSH inequality, but in perfect agreement with the prediction $Q = 2\sqrt{2}$ of QM. This and other experiments strongly suggest that QM *does not* in fact rely on hidden variables to prepare the results of future measurements in advance. Instead, the new state of the photon is created only in the process of measurement and the EPR correlations arising in these experiments must then be a purely quantum-mechanical phenomenon with no underlying deterministic models of the kind envisioned by Einstein. In Aspect's experiment, God certainly plays dice!

The expectation values such as $E(a, c)$ provide us with a numerical measure of the quantum correlation between the two particles in the pair. This correlation is maximum when $E(a, c) = \pm 1$, in other words, when a given measurement outcome for the first particle of the pair rigidly determines the outcome for the second particle. This will be true if the EPR pair is prepared in one of the four states where a measurement of the state of one particle will immediately determine the state of the other:

$$|\Phi^+\rangle = \frac{1}{\sqrt{2}}(|0\rangle_A|0\rangle_B + |1\rangle_A|1\rangle_B), \quad |\Psi^+\rangle = \frac{1}{\sqrt{2}}(|0\rangle_A|1\rangle_B + |1\rangle_A|0\rangle_B),$$

$$|\Phi^-\rangle = \frac{1}{\sqrt{2}}(|0\rangle_A|0\rangle_B - |1\rangle_A|1\rangle_B), \quad |\Psi^-\rangle = \frac{1}{\sqrt{2}}(|0\rangle_A|1\rangle_B - |1\rangle_A|0\rangle_B).$$

(24.55)

7) It should be mentioned that different conventions exist in the literature. Some authors will put the horizontal and vertical polarizations at the north and south poles and diagonal and circular polarizations on the equatorial plane.
8) In addition, we have found the test angles only for the simplest possible case when the pair's total spin is 0, but the EPR pair might as well be prepared, for example, from a triplet state (total spin is 1) or from some superposition of singlet and triplet states, or it might even conceivably involve different particles (e.g., an electron and a deuterium nucleus)

Physically, they correspond to maximum possible entanglement between the particles. These four states are called *Bell states*.

You may also recognize the fourth Bell state as the singlet state of two electrons. So, here is the physical reason that the electron experiment discussed above produced the greatest physically possible violation of CHSH inequality: We were able to squeeze the maximum correlation from this electron pair because it had the maximum entanglement to begin with! The third Bell state should also be familiar to you: It is the triplet state corresponding to $m = 0$. However, the remaining two triplet states for $m = \pm 1$ are *not* Bell states: They are deterministic states where Alice and Bob can anticipate the measurement outcomes in advance, so this experiment cannot tell us anything about the "spooky action at a distance" even though Alice's and Bob's results will show perfect correlation.

It is possible to imagine a QM measurement that *prepares* the two-particle system in one of the four possible Bell states. This is the so-called *Bell-state measurement*. It must be a joint measurement of two particles, and if the particles were in some arbitrary state (i.e., with less than perfect entanglement), the measurement will then act as an entangling operation.

24.7
QM and the Failure of FTL Proposals

We should now say a few words about the relationship between QM and SR. Ever since Einstein came up with SR, physicists, science fiction writers, philosophers, and crackpots have been looking for loopholes that would make FTL signaling possible. In addition to many practical benefits of such signaling, it would be extremely tempting to refute Einstein himself! But assailing relativity was no easy matter: As of this writing, all proposed FTL schemes have failed. It would seem on the face of it that the EPR paradox with its superluminal correlations (confirmed by numerous experiments!) must hold some promise of breaking the light barrier. Of course, it would be preposterous to expect that one could actually move things – physical particles or just energy – across space faster than the speed of light. However, a sophisticated opponent may well point out that changing the physical state of a system does not necessarily involve energy transfer. After all, if we just want to flip a system from one degenerate state to another and we manage to utilize EPR for that purpose, then no change of energy has taken place, but information may have been transferred. Thus, one could design a scheme where Alice, by measuring her particle in a certain basis, causes a change in the state of Bob's particle that Bob will be able to identify – then we would have a FTL exchange of information without violating relativity. It would violate the principle of *causality* since it would be possible to send signals into the past. But difficult as it may be to accept causality violation on the microscale, we cannot reject this possibility outright. Meanwhile, as far as relativity is concerned, Nature always has an alibi: The outcome measured by Bob *could* have been realized in a measurement since it was one of the possibilities already present in the entangled state of the EPR pair. And since preparation of that pair, which involved separating the two particles, occurred at a subluminal speed, there is no outright violation of relativity. Thus, strictly speaking, relatively cannot be used to impose a ban on FTL communication in EPR experiments. Any such ban must be imposed from *outside* of the relativistic framework. It must come from the requirement of causality (if we take causality as a fundamental physical law) or it must follow directly from the structure of the quantum theory. In this connection, one may well ask what exactly QM does to prevent FTL communication. There has been a regrettable tendency among some authors to derive the ban on such communication from the no-cloning theorem as it was originally formulated by Wootters and Zurek in their 1982 paper. To add to the confusion, other authors went one step further and suggested that the no-cloning theorem itself was really the consequence of the ban on superluminal communication – QM's way of complying with the demands of SR. But upon a closer inspection, neither statement can be accepted as accurate. As we said earlier, if there is any such ban, it would come from the causality principle, rather than from SR. Second, the no-cloning theorem, at least in its original formulation, prohibits *perfect* cloning. But even if the cloning apparatus is an imperfect one, producing bad but still readable copies (in other words, the overlap between a state and its "xerox copy" is sufficiently large, but less than 1) or producing some number of perfect copies and a small percentage of noise, Bob can still read the signal. To prevent this possibility, one must require that Bob's success rate in reading each bit be no greater than 50%; otherwise, Alice and Bob can reduce the number of errors exponentially simply by duplicating the same message several times and transmit the text of "Hamlet" to the Andromeda nebula instantly with only a few typos. So it is important to see how QM handles the case of *imperfect* cloning. Our current understanding is that QM prevents FTL signaling by ensuring that while Alice can *send* an instant message to Bob, Bob cannot *read* that message.

24.8
Do Lasers Violate the No-Cloning Theorem?

Since lasers rely on stimulated emission to produce a powerful beam of photons all in the same state, it is easy to think of a laser as a perfect cloning machine. However, their cloning cannot be perfect because of spontaneous emission. Since Einstein's coefficients for stimulated and spontaneous emissions are related by

$$B_{21} = \frac{c^3}{8\pi h \nu^3} A_{21}, \tag{24.56}$$

the corresponding probabilities are proportional, so one cannot just get rid of spontaneous emission without suppressing stimulated emission as well. As a result, any output beam emerging from the laser will be contaminated with spontaneous noise. We may ask then what is the most accurate copy that one can make using a laser?

Let us consider this question first from a classical viewpoint and then from a quantum viewpoint. A real laser is a macroscopic system having a large number of excited atoms. To simplify our analysis, we will take the special case of a "laser" consisting of just one atom that has two energy levels and a transition dipole moment **d**. In this case we should no longer call it "laser", but instead use the term "photon amplifier", because there is considerable difference in the properties of emitted light; also, a "true" laser has more than one excited state. Consider a monochromatic photon passing by the atom positioned at the origin. The photon's electric field can be written as

$$\mathbf{E} = \mathbf{E}_0 e^{i(\vec{k}\cdot\vec{r} - \omega t)}. \tag{24.57}$$

In the vicinity of the atom, the values of $|\mathbf{r}|$ must be on the order of the atomic size a. Then assuming $a \ll \lambda$, the incident field at any given moment of time can be considered as spatially uniform. Mathematically, this is easily seen from the Taylor expansion:

$$\mathbf{E} = \mathbf{E}_0 e^{-i\omega t}\left(1 + \mathbf{k}\cdot\mathbf{r} + \frac{(\mathbf{k}\cdot\mathbf{r})^2}{2} + \cdots\right) \approx \mathbf{E}_0 e^{-i\omega t}, \tag{24.58}$$

where we have dropped the higher order terms because $|\mathbf{k}\cdot\mathbf{r}| \leq 2\pi r/\lambda \ll 1$. Physically, it means that the atom does not "see" individual crests and troughs of the wave, so the atom's interaction with light reduces to the problem of an atomic electric dipole moment placed in a uniform field. The potential energy of a dipole in a uniform field is given by

$$U = -\mathbf{E}\cdot\mathbf{d}. \tag{24.59}$$

The long wavelength assumption, which enables the use of Equation 24.59, is also known as the *dipole approximation*.

Equation 24.59 tells us that the electric field of the photon will interact strongly with the atom when the field is parallel to the dipole moment. We will make an additional assumption that the photon frequency is close to the atomic proper frequency, giving us a chance to produce resonance. We can then expect from classical electrodynamics that the atomic dipole will act as a driven oscillator and emit electromagnetic radiation mostly in its equatorial plane. Accordingly, the polarization of light emitted in this plane will be parallel to **d**. This situation is the classical analogue of stimulated emission.

On the other hand, if $\mathbf{E} \perp \mathbf{d}$, there will be no resonance and the atom will simply not notice this photon. It will continue, however, to emit radiation on its own in a process that is the classical analogue of spontaneous emission. Once again, the polarization of a spontaneous photon will be parallel to the dipole axis, but that makes it orthogonal to the input photon polarization. So, in the case of stimulated emission, we may expect to end up with two identical photons; while in the case of spontaneous emission, we end up with a photon pair whose polarizations are orthogonal.

The quantum-mechanical analysis of photon amplifiers in the context of the no-cloning theorem was first performed by Mandel [21] in 1983. In the *Nature* magazine, Mandel published his conclusions in a short paper titled "Is a photon amplifier always polarization dependent?" His analysis showed that the noise introduced by spontaneous emission imposes a severe limitation on the accuracy of the cloning process.

We can start the quantum description by writing the total Hamiltonian as

$$H = H_{\text{atom}} + H_{\text{light}} + H_{\text{I}}(t). \tag{24.60}$$

The last term is just the energy of interaction of the atom with the field that is given by Equation 24.59; since the classical field equals $\mathbf{E} = \mathbf{E}_0 e^{-i\omega t}$, we can tell at once that this term must depend on time. However, we can no longer use the classical field. Once we decided to think in terms of quanta, we must quantize the field using creation and annihilation operators:

$$\mathbf{E}(t) = \left(\frac{\hbar\omega}{\varepsilon_0 V}\right)^{1/2}(a + a^+)e^{-i\omega t}\boldsymbol{\varepsilon}. \tag{24.61}$$

Here $\boldsymbol{\varepsilon}$ is the (complex!) unit vector that shows the direction of polarization of the photon.[9] We proceed to write the interaction term as

$$H_{\text{I}}(t) = -\mathbf{d}\cdot\mathbf{E} = -\mathbf{d}\cdot\boldsymbol{\varepsilon}\left(\frac{\hbar\omega}{\varepsilon_0 V}\right)^{1/2}(a + a^+)e^{-i\omega t} = d_\varepsilon g(a + a^+), \tag{24.62}$$

where $d_\varepsilon = \mathbf{d}\cdot\boldsymbol{\varepsilon}$ is the dipole moment's projection on the polarization direction, and the constant terms as well as the time dependence[10] have been absorbed into the quantity $g = -(\hbar\omega/\varepsilon_0 V)^{1/2}e^{-i\omega t}$. The second term is also straightforward: It is the free-field Hamiltonian:

$$H_{\text{light}} = \hbar\omega\left(a^+ a + \frac{1}{2}\right). \tag{24.63}$$

The first term describes the energy of the free atom. If $|e\rangle$ and $|g\rangle$ are the excited and ground states, the Hamiltonian can be written in the energy representation as

$$H_{\text{atom}} = E_g|g\rangle\langle g| + E_e|e\rangle\langle e| \tag{24.64}$$

(it is straightforward to show that (24.64) is a diagonal matrix whose matrix elements are the energy eigenvalues).

We now introduce the transition frequency between energy levels:

$$\omega_{eg} = \frac{E_e - E_g}{\hbar} \approx \omega, \tag{24.65}$$

which according to our assumption should be close to the frequency ω of the photon. Furthermore, since the choice of the zero energy level is arbitrary, we can place it

---

9) To see why this vector is generally complex, consider the way we obtain circular polarization. If the polarization vector rotates in a plane, then at any given moment of time it can be written as a complex number. Generally, a circularly polarized EM wave propagating along z-direction will be written as $\mathbf{E} = E_1 e^{i(kz-\omega t)}\mathbf{x} + E_2 e^{i\alpha}e^{i(kz-\omega t)}\mathbf{y}$ and its polarization will be described by the complex unit vector $\boldsymbol{\varepsilon} = (E_1\mathbf{x} + E_2 e^{i\alpha}\mathbf{y})/\sqrt{E_1^2 + E_2^2}$.

10) In the general case, g also includes the spatial dependence that we dropped due to the dipole approximation.

halfway between the ground and excited states to write Equation 24.64 in a more convenient form:

$$H_{\text{atom}} = -\frac{1}{2}\hbar\omega_{eg}|g\rangle\langle g| + \frac{1}{2}\hbar\omega_{eg}|e\rangle\langle e| = \frac{1}{2}\hbar\omega_{eg}\hat{\sigma}_3, \tag{24.66}$$

where we have introduced a new operator:

$$\hat{\sigma}_3 = |e\rangle\langle e| - |g\rangle\langle g|. \tag{24.67}$$

The Hamiltonian now takes the form

$$H = \frac{1}{2}\hbar\omega_{eg}\hat{\sigma}_3 + \hbar\omega\left(a^+ a + \frac{1}{2}\right) + g d_\varepsilon (a + a^+). \tag{24.68}$$

We now have to write the quantum operator $\hat{d}_\varepsilon$ explicitly. To this end, we introduce two auxiliary operators, called *atomic transition operators*:

$$\sigma^+ = |e\rangle\langle g|,$$
$$\sigma^- = |g\rangle\langle e|. \tag{24.69}$$

In the $\{|e\rangle, |g\rangle\}$ basis, these two operators will form matrices:

$$\sigma^+ = \begin{pmatrix} 0 & 1 \\ 0 & 0 \end{pmatrix}, \quad \sigma^- = \begin{pmatrix} 0 & 0 \\ 1 & 0 \end{pmatrix}. \tag{24.70}$$

To elucidate their physical meaning, let us act with these two operators on the ground and excited states, respectively. We see immediately that $\hat{\sigma}^+$ flips the ground state into the excited state, while $\hat{\sigma}^-$ flips the excited state into the ground state. Particularly, in situations where the two energy levels of an atom are determined by the z-component of spin, the above operators become the *spin-flip operators*: $\hat{\sigma}^+$ ("spin-up" operator) flips $|\downarrow\rangle$ into $|\uparrow\rangle$, while $\hat{\sigma}^-$ ("spin-down" operator) flips $|\uparrow\rangle$ into $|\downarrow\rangle$. Note also that the result of $\hat{\sigma}^+$ acting on $|e\rangle$ (respectively $\hat{\sigma}^-$ on $|g\rangle$) is zero.

With this in mind, we now take a fresh look at the dipole moment $\hat{\mathbf{d}}$. In electrostatics, dipole moment is defined as the product of the magnitude of charge and the displacement vector (from the negative to the positive charge):

$$\mathbf{d} = q\mathbf{r}. \tag{24.71}$$

Replacing observables with operators, we can write the expectation value of the dipole moment as follows:

$$\langle\psi|\hat{\mathbf{d}}|\psi\rangle = q\big(c_e^*\langle e| + c_g^*\langle g|\big)\mathbf{r}\big(c_e|e\rangle + c_g|g\rangle\big)$$
$$= q\big\{|c_e|^2\langle e|\mathbf{r}|e\rangle + |c_g|^2\langle g|\mathbf{r}|g\rangle + c_e c_g^*\langle g|\mathbf{r}|e\rangle + c_g c_e^*\langle e|\mathbf{r}|g\rangle\big\}. \tag{24.72}$$

However, the parity-odd operator $\mathbf{r}$ connects states with opposite parity, so the first two terms are equal to zero. Let us denote $\mathbf{M} = q\langle g|\mathbf{r}|e\rangle$ and write the last equation as

$$\langle\psi|\hat{\mathbf{d}}|\psi\rangle = c_e c_g^* \mathbf{M} + c_g c_e^* \mathbf{M}^*. \tag{24.73}$$

Here, **M** and **d** are vectors. Multiplying both sides by the polarization vector ε and denoting $M = \mathbf{M} \cdot \boldsymbol{\varepsilon}$, we can write

$$\langle \psi | \hat{d}_\varepsilon | \psi \rangle = c_e c_g^* M + c_g c_e^* M^*. \tag{24.74}$$

In a broad variety of cases (i.e., when polarization has no circular/elliptical component so that ε is not complex), we can assume M to be a real number, which leads to

$$\langle \psi | \hat{d}_\varepsilon | \psi \rangle = M(c_e c_g^* + c_g c_e^*). \tag{24.75}$$

This is where atomic transition operators come into play. Using (24.69) we can write $c_e c_g^* = \langle \psi | \hat{\sigma}^- | \psi \rangle$ and $c_g c_e^* = \langle \psi | \hat{\sigma}^+ | \psi \rangle$. Then,

$$\langle \psi | \hat{d}_\varepsilon | \psi \rangle = \langle \psi | M(\sigma^+ + \sigma^-) | \psi \rangle \tag{24.76}$$

and the dipole moment operator is

$$\hat{d}_\varepsilon = M(\sigma^+ + \sigma^-). \tag{24.77}$$

Finally, following the suggestion known as the *Jaynes–Cumming model*, we will now drop the $(1/2)\hbar\omega$ term from Equation 24.68. This is justified because this term does not affect the dynamics of the system. This gives us the following Hamiltonian:

$$H = \frac{1}{2}\hbar\omega_{eg}\hat{\sigma}_3 + \hbar\omega a^+ a + gM(\sigma^+ + \sigma^-)(a + a^+). \tag{24.78}$$

But we are not done yet! To complete the analysis, we will now apply the *rotating wave approximation* described below.

First we take note of the fact that the first two terms in the Hamiltonian are time independent, while the interaction term is thought to be relatively weak compared to the first two terms. In such situations it is convenient to work in the *interaction picture*, where the dominant terms in the Hamiltonian are written as operators changing with time as in the Heisenberg picture, while the weak "interaction term" is considered as an operator in the Schrodinger picture, which causes the state to evolve in time. The sum of the first two terms, which we are going to denote H_0, will then obey the Heisenberg equation of motion:

$$\frac{dA^{(H)}(t)}{dt} = \frac{1}{i\hbar}\left[A^{(H)}(t), H_0\right], \tag{24.79}$$

which will be valid for an arbitrary operator $\hat{A}^{(H)}(t)$. In particular, for the annihilation operator,

$$\frac{da(t)}{dt} = \frac{1}{i\hbar}[a, H_0] = \frac{1}{i\hbar}[a, \hbar\omega a^+ a] = \frac{\omega}{i}(aa^+ a - a^+ aa) = -i\omega(aa^+ - a^+ a)a$$
$$= -i\omega a. \tag{24.80}$$

In deriving the last formula, we have used the commutation property $[a, a^+] = 1$. The solution of this differential equation is $a(t) = a(0)e^{-i\omega t}$. In the same manner, writing the Heisenberg equation of motion for a^+, σ^+, and σ^-, we obtain

$$a(t) = a(0)e^{-i\omega t}, \qquad a^+(t) = a^+(0)e^{i\omega t},$$
$$\sigma^+(t) = \sigma^+(0)e^{i\omega_{eg}t}, \qquad \sigma^-(t) = \sigma^-(0)e^{-i\omega_{eg}t}. \tag{24.81}$$

It follows that

$$\sigma^+ a \sim e^{i(\omega_{eg}-\omega)t}, \qquad \sigma^+ a^+ \sim e^{i(\omega_{eg}+\omega)t},$$
$$\sigma^- a \sim e^{-i(\omega_{eg}+\omega)t}, \qquad \sigma^- a^+ \sim e^{-i(\omega_{eg}-\omega)t}. \tag{24.82}$$

The rotation wave approximation (RWA) means dropping the fast oscillating terms $\sigma^+ a^+$ and $\sigma^- a^-$ from the Hamiltonian. This works mathematically, but it is also interesting to see the underlying philosophy: RWA retains only the terms that conserve energy! Indeed, when we act with $\sigma^+ a$ on a system consisting of an atom and a quantum of EM field, the photon gets absorbed due to the annihilation operator and the atomic transition operator flips the atom into the excited state. Conversely, $\sigma^- a^+$ operator causes the atom to fall to the ground state, emitting a photon. Both these processes represent a self-intuitive exchange of energy between the atom and the photon. In contrast, the $\sigma^+ a^+$ and $\sigma^- a$ operators correspond to such "bizarre" situations as emission accompanied by excitation of the atom and absorption accompanied by de-excitation, so these terms do not conserve energy.

We finally write (24.78) in the RWA:

$$H = (1/2)\hbar\omega_{eg}\hat{\sigma}_3 + \hbar\omega a^+ a + gM(\sigma^+ a + \sigma^- a^+). \tag{24.83}$$

In the context of cloning, we are interested in the interaction term of the Hamiltonian. Since M is proportional to $\hat{d}_\varepsilon$, this term is written as

$$H_I = g\mathbf{d} \cdot \boldsymbol{\varepsilon}(\sigma^+ a + \sigma^- a^+). \tag{24.84}$$

In the simple model considered by Mandel, our input photon can be prepared in one of the two polarization states $\boldsymbol{\varepsilon}_1$ and $\boldsymbol{\varepsilon}_2$ and arrives into the amplifier in a given spatial mode a. Then our analysis starting from Equation 24.61 must include separate creation/annihilation operators for each polarization state:[11]

$$H_I = g(\mathbf{d} \cdot \boldsymbol{\varepsilon}_1 \sigma^- a_1^+ + \mathbf{d} \cdot \boldsymbol{\varepsilon}_1 \sigma^+ a_1 + \mathbf{d} \cdot \boldsymbol{\varepsilon}_2 \sigma^- a_2^+ + \mathbf{d} \cdot \boldsymbol{\varepsilon}_2 \sigma^+ a_2). \tag{24.85}$$

Here, a_1^+ is the creation operator for photons in spatial mode a and polarization state $\boldsymbol{\varepsilon}_1$, while a_2^+ is the creation operator for photons in spatial mode a and polarization state $\boldsymbol{\varepsilon}_2$. To simplify matters further, let us take the case when $\boldsymbol{\varepsilon}_1$ and $\boldsymbol{\varepsilon}_2$ represent the vertical and horizontal polarizations that we will now denote as V and H. Let us use notation $|k_V, l_H\rangle$ for a light pulse having k_V vertically and l_H horizontally polarized photons.

Suppose we "feed" one photon, say, vertically polarized photon, into an amplifier, so that the initial state of electromagnetic field is $|1_V, 0_H\rangle$ (we ignore the atomic wave function since we are only interested in the photon output). Its time evolution is

---

[11] Note that the last two terms are Hermitian conjugates of the first two. Therefore, some authors write this equation as $H = g\sum_{s=1}^{2}(\boldsymbol{\mu} \cdot \boldsymbol{\varepsilon}_s \sigma^- a_s^+ + \text{h.c.})$, where h.c. stands for "Hermitian conjugate."

given by $e^{-iHt} \approx 1 - iHt$, where we assumed the exponent to be small enough that we can drop the higher order terms in the Taylor expansion. (This is true for a realistic experimental situation.) Furthermore, we are interested in the case when the atom was initially prepared in the excited state so that it will not absorb photons and the σ^+ operator acting on the atom will not contribute to the output. Therefore, we can ignore the second and fourth terms in (24.85) and write the resulting photon state as follows:

$$e^{-iHt}|1_V, 0_H\rangle \approx |1_V, 0_H\rangle - igt(d_V\sigma^- a_V^+ + d_H\sigma^- a_H^+)|1_V, 0_H\rangle. \tag{24.86}$$

The zeroth order term describes the possibility that there may be no additional photons produced; in other words, our cloning amplifier will sometimes just output the input photon. The first-order term describes a two-photon output beam that may have both photons in the same state or in different states. The user can always discard the trials in which the laser fails to produce the second photon, so we should only be concerned with the two-photon outcomes. We thus get the following very simple equation that describes cloning in the amplifier:[12]

$$|1_V, 0_H\rangle \rightarrow \sqrt{2}d_V|2_V, 0_H\rangle + d_H|1_V, 1_H\rangle. \tag{24.87}$$

We have said earlier that according to the classical picture, stimulated emission occurs when photon polarization is parallel to the dipole moment (i.e., $d_V = |d|$, $d_H = 0$). This situation corresponds to the first term of (24.87). On the other hand, if the dipole moment is perpendicular to polarization ($d_V = 0$, $d_H = |d|$), then (24.87) reduces to the second term. We can thus associate $|2_V, 0_H\rangle$ with stimulated emission and $|1_V, 1_H\rangle$ with spontaneous emission.

Suppose the dipole moment is also vertical. Then, we have from (24.87)

$$|1_V, 0_H\rangle \rightarrow |2_V, 0_H\rangle, \quad |0_V, 1_H\rangle \rightarrow |1_V, 1_H\rangle. \tag{24.88}$$

Equation (24.88) allows perfect cloning of the vertically polarized photon. This should not surprise us: We know that cloning is possible if the initial state is known (but then we don't need a cloning apparatus in the first place.) The whole point of the experiment is that we don't know the photon polarization in advance! And our photon amplifier is asymmetric with respect to initial photon polarization: It succeeds 100% of the time when $\mathbf{d} \cdot \boldsymbol{\varepsilon} = 1$, fails completely when $\mathbf{d} \cdot \boldsymbol{\varepsilon} = 0$, and succeeds some of the time in the intermediate cases. We have to conclude that a simple one-atom amplifier cannot clone an arbitrary photon state.

Is it possible to remove this asymmetry, making a so-called *universal cloning machine* that would produce clones of equal quality for all input states? The answer is yes: One needs to consider only a two-atom amplifier with orthogonal dipole moments $\mathbf{d}_a$ and $\mathbf{d}_b$. Intuitively, it is clear that when the first atom clones the state poorly, the other atom clones it well and vice versa. Formally, we simply augment

---

12) Of course, we have not yet normalized this state, so more accurately it should be written as
$|1_V, 0_H\rangle \rightarrow (\sqrt{2}d_V|2_V, 0_H\rangle + d_H|1_V, 1_H\rangle)/\sqrt{2d_V^2 + d_H^2}$.

the Hamiltonian (24.85) in a self-intuitive way by adding the respective terms for the two atoms:

$$H_1 = g \sum_{s=1}^{2} \left[(\mathbf{d}_a \boldsymbol{\varepsilon}_s + \mathbf{d}_b \boldsymbol{\varepsilon}_s) \sigma^- a_s^+ + \text{h.c.} \right]. \quad (24.89)$$

We again ignore the Hermitian conjugate (h.c.) terms in (24.89) since both atoms are initially excited. Generalizing (24.87) for the two-atom case, we get

$$|1_V, 0_H\rangle \rightarrow \sqrt{2}(d_{a,V} + d_{b,V})|2_V, 0_H\rangle + (d_{a,H} + d_{b,H})|1_V, 1_H\rangle,$$
$$|0_V, 1_H\rangle \rightarrow \sqrt{2}(d_{a,H} + d_{b,H})|0_V, 2_H\rangle + (d_{a,V} + d_{b,V})|1_V, 1_H\rangle, \quad (24.90)$$

and it is easy to see that when $\mathbf{d}_a$ is directed vertically, $\mathbf{d}_b$ is directed horizontally, and $|\mathbf{d}_a| = |\mathbf{d}_b| = d$, Equation 24.88 goes over to the Equation 24.91, which gives us identical cloning behavior for the two possible polarizations:

$$|1_V, 0_H\rangle \rightarrow \sqrt{2}|2_V, 0_H\rangle + |1_V, 1_H\rangle,$$
$$|0_V, 1_H\rangle \rightarrow \sqrt{2}|0_V, 2_H\rangle + |1_V, 1_H\rangle. \quad (24.91)$$

(The reader can show that the same behavior will hold when the input state is a superposition of vertical and horizontal polarizations.) Thus, it is possible to make a polarization-independent amplifier, where stimulated emission is twice more likely than spontaneous emission. The amplitudes $\sqrt{2}$ and 1 give rise (after normalization) to respective probabilities 2/3 and 1/3,[13] and the resulting mixture for a $|1_V, 0_H\rangle$ input will be described by the density operator:

$$\rho = (2/3)|2_V, 0_H\rangle\langle 2_V, 0_H| + (1/3)|1_V, 1_H\rangle\langle 1_V, 1_H|. \quad (24.92)$$

The mixture resulting from the $|0_V, 1_H\rangle$ input state is written similarly.

We can see at once that the no-cloning requirement was not violated. It would only be violated if we get a pure state $\rho = |2_V, 0_H\rangle\langle 2_V, 0_H|$. As it is, our result (24.91) shows that a laser is an *imperfect cloning machine* that does not contradict the no-cloning theorem stated by Wootters and Zurek. Instead, perfect cloning is achieved 2/3 of the time or, to use a more formal language, the signal-to-noise ratio is 2 : 1.

You may wonder if this result could be improved by looking at the *direction* of the output photons. After all, isn't it true that a spontaneous photon should have random direction? If so, then it should be possible to reduce noise by weeding out the photons that deflected from the initial direction. However, our Hamiltonian (24.85) did not take into account *all* spontaneous photons; it only dealt with those emitted in the spatial mode a, the same mode as for the incoming photon. So the space–time wave functions of both output photons are identical; the only possible difference is in their polarization.

---

13) To be more accurate, we should first have the normalized output state that includes the $|1, 0\rangle$ component and *then* write the probabilities; only after that and once single-photon outcomes are discarded, the resulting ensemble will be described by the weights 2/3 and 1/3. Note also that strictly speaking, we should have called (24.92) a *reduced* density matrix, because emerging photons are entangled with the emitting atom(s); by focusing on the photons only, we have implicitly traced out the atomic part of the system.

Knowing the signal-to-noise ratio, one can find the probability to pick the correct input state from the output beam. There is a 2/3 chance that both photons will be in the input state in which case we will certainly pick that state. There is also 1/3 chance of spontaneous emission in which case our odds of picking either a V photon or an H photon are the same. Overall, the probability for a clone to exactly replicate the input is $(2/3) \times 1 + (1/3)(1/2) = 5/6$. This probability is known as the *fidelity* of the cloning process. So to conclude this analysis, at the atomic level, fidelity of universal cloning does not exceed 5/6 because of the possibility of spontaneous emission. If one uses asymmetric cloning, fidelity ranges from 1 in the case of optimal input to 1/2 in the worst possible case. For example, if we make a series of experiments, with half of input photons polarized vertically and half horizontally, then on the average we will get $(1/2) \times 1 + (1/2)(1/2) = 3/4$, doing even worse than before. So a universal quantum cloning machine (UQCM) also turns out to be optimal in terms of fidelity!

24.9
Imperfect Cloning

As we see from the previous discussion, the no-cloning theorem survives the "laser challenge" in a broad variety of experiments involving lasers, one-atom photon amplifiers, and multiatom amplifiers. This conclusion is all the more remarkable that what really precludes perfect cloning is the spontaneous emission – a phenomenon that is not part of standard QM courses and is studied as part of quantum field theory. If it were not for the vacuum fluctuations, Mandel's photon amplifier would be a perfect cloner. But then again we know from Chapter 2 that spontaneous emission is necessary for a self-consistent theory of radiation. This underscores the vital role of quantum field theory as the foundation of QM. Had we been able to construct a perfect cloner, QM as we know it simply would not exist.

But another equally important conclusion is that *imperfect* cloning *can* be realized. The quality of resulting clones is described by fidelity, a measure originally suggested by Schumacher [22]. For the current purposes, we can define it casually as the probability for the output state to match the input state (a more strict definition will be given in Chapter 25). Thus, fidelity can range from 0 to 1. If the output and input states are identical, then fidelity is 1 and the cloning is perfect.

In particular, for the pure states, this definition reduces to the overlap between states, thus acquiring the meaning mentioned above – probability for a clone to exactly replicate the input:

$$F(\psi_{in}, \psi_{out}) = \text{Tr}\left(|\psi_{in}\rangle\langle\psi_{in}|\psi_{out}\rangle\langle\psi_{out}|\right) = \langle\psi_{in}|\psi_{out}\rangle \text{Tr}\left(|\psi_{in}\rangle\langle\psi_{out}|\right)$$
$$= |\langle\psi_{in}|\psi_{out}\rangle|^2.$$

(24.93)

The fact that fidelity can be much greater than 1/2 gives rise to the problem of probabilistic FTL signaling. The question is what kind of information, if any, Bob can extract from Alice's EPR photon after cloning it with a given fidelity F. We will

24.9 Imperfect Cloning

first consider the standard density matrix approach to the problem and then take a more nuanced view, which underscores the importance of quantum compounds.

Let us consider again the universal photon amplifier from Section 24.8. As we saw earlier, its fidelity under ideal conditions can be as high as 5/6. That sounds very promising, but only until we realize that even this ratio is not enough for probabilistic FTL signaling. Indeed, look at the implication of (24.92). First of all, we can surely tell the difference between an input signal consisting of one V photon and a signal consisting of one H photon. In either case we will be picking a perfect clone of the input state 5/6 of the time. But this is nothing new: After all, we know that the no-cloning theorem allows us to clone a photon that we know is going to be prepared in one of the two orthogonal states. We could even have cloned it perfectly as far as the theorem is concerned or, better yet, we could just have measured the incoming photon polarization without involving a cloner. The trouble is that a photon whose state is well defined cannot be sent through space faster than the speed of light. If Alice wants to transmit a V photon to Bob, this photon will travel at the speed of light as a well-behaving classical object. If she wants to harness EPR correlations to send a message faster than the speed of light, then the photon at Bob's side must be a part of an entangled system. As Alice collapses the EPR state, Bob's photon could wind up horizontally or vertically polarized with the same likelihood (as long as we consider an EPR singlet pair). That is to say, if we consider many trials *at once*, then Bob will now be trying to clone a density matrix:

$$\rho_{in} = \frac{1}{2}|1_V, 0_H\rangle\langle 1_V, 0_H| + \frac{1}{2}|0_V, 1_H\rangle\langle 0_V, 1_H|. \tag{24.94}$$

After passing through the amplifier, it will emerge as an equally weighted mixture:

$$\rho_{out} = \frac{1}{2}\left[\frac{2}{3}|2_V, 0_H\rangle\langle 2_V, 0_H| + \frac{1}{3}|1_V, 1_H\rangle\langle 1_V, 1_H|\right] + \frac{1}{2}\left[\frac{2}{3}|0_V, 2_H\rangle\langle 0_V, 2_H| + \frac{1}{3}|1_V, 1_H\rangle\langle 1_V, 1_H|\right]$$

$$= \frac{1}{3}|2_V, 0_H\rangle\langle 2_V, 0_H| + \frac{1}{3}|0_V, 2_H\rangle\langle 0_V, 2_H| + \frac{1}{3}|1_V, 1_H\rangle\langle 1_V, 1_H|, \tag{24.95}$$

where any single photon picked at random will be vertically polarized with probability $(1/3) \times 1 + (1/3)(1/2) = 1/2$ and similarly for horizontal polarization. Suppose now Bob chooses one photon at random from this mixture and measures its polarization in the {V, H} basis. Then this mixture by itself cannot carry any signal. As we discussed earlier, the signal has to be encoded in Alice's choice of the measuring basis. Suppose she will be choosing between the rectilinear basis {V, H} and the circular basis {R, L}. If she chooses the circular basis,

$$\rho_{in} = \frac{1}{2}|1_R, 0_L\rangle\langle 1_R, 0_L| + \frac{1}{2}|0_R, 1_L\rangle\langle 0_R, 1_L|, \tag{24.96}$$

then Bob will end up with the density matrix:

$$\rho_{out} = \frac{1}{3}|2_R, 0_L\rangle\langle 2_R, 0_L| + \frac{1}{3}|0_R, 2_L\rangle\langle 0_R, 2_L| + \frac{1}{3}|1_R, 1_L\rangle\langle 1_R, 1_L|. \tag{24.97}$$

This matrix is indistinguishable from matrix (24.95)! A randomly picked photon will be R-polarized or L-polarized with the same probability 1/2; if Bob uses a {V, H} basis, he will again have the probability 1/2 for either outcome. So Bob cannot derive any information from the resulting density matrix.

The fact that the two input states represented by density matrices (24.94) and (24.96) will produce indistinguishable mixtures at the output is known in the literature as the *no-signaling condition*. It was first put forth by Eberhard [4,23]. Formally, we can state it thus. Equation 24.95 corresponds to the case when Bob feeds into the amplifier an equally weighted mixture of V-polarized and H-polarized photons. As evident from the first line of (24.95), the resulting output matrix can be written as the sum of the corresponding outputs for pure states:

$$\rho_{out} = \rho_{out}(V) + \rho_{out}(H). \tag{24.98}$$

Similarly, Equation 24.97 describes the output matrix:

$$\rho_{out} = \rho_{out}(R) + \rho_{out}(L). \tag{24.99}$$

If these two matrices are equivalent, then, according to Eberhard, message sent by Alice cannot be read. So the no-signaling condition really boils down to the very simple statement that

$$\rho_{out}(V) + \rho_{out}(H) = \rho_{out}(R) + \rho_{out}(L). \tag{24.100}$$

When does Equation 24.100 hold and when does it fail? The answer turns out to be very nontrivial and since it involves some heavy math, we will only give the main conclusions. The interested reader will find the missing details in the references included in the chapter.

Mandel's amplifier was just one example of a UQCM, albeit the first one mentioned in the literature. The general algorithm of such machines was formulated in 1996 by Buzek and Hillery [24,25]. The important fact is that the general Buzek–Hillery machine had fidelity $F = 5/6$ – exactly the same as Mandel's universal photon amplifier! It turned out that this match was not just a coincidence. In fact, it was shown independently by Gisin and Massar [26] and then by Bruss et al. [27] that the Buzek–Hillery UQCM was the optimal one. Thus, QM imposes an upper limit on the quality of clones:

$$F_{max}(\rho_1, \rho_2) = \frac{5}{6}. \tag{24.101}$$

The upper bound (24.101) has been tested experimentally for photons [28] and the experiment showed excellent agreement with the theory. Under optimal conditions, very high-fidelity clones were obtained with $F \approx 0.81$ – just short of the theoretical limit $F_{max} = 5/6 = 0.833$, but never exceeding this limit. Thus, it is clear that the fidelity of Mandel's amplifier cannot be improved any further. The very first UQCM also turned out to be the best one!

We can now ask: Is this upper limit merely a consequence of some features of the cloning process, or does this "magic" number 5/6 reflect something more fundamental? We will argue that there is a deeper reason why Nature decided to restrict the fidelity to $F \leq 5/6$. In what follows, we will give a strict proof that 5/6 is the maximum fidelity of cloning that is compatible with relativistic causality.

24.9 Imperfect Cloning

Let the Bloch vector **m** describe polarization of the input photon, and write the corresponding pure state as $\rho_{\text{in}}(\mathbf{m})$. After the cloning, we have at the output a two-photon system $\rho_{\text{out}}(\mathbf{m})$ that is generally entangled. The entanglement means that outcomes of a *single*-photon measurement on the output system will be described by a partial trace $\text{Tr}_1 \rho_{\text{out}}(\mathbf{m}) = \text{Tr}_2 \rho_{\text{out}}(\mathbf{m})$. It does not matter which photon is traced out because the photons are indistinguishable and we cannot tell whether the photon that arrives to our detector is the original photon or the cloned photon. A pure single-photon state can be written as $|\mathbf{m}\rangle\langle\mathbf{m}| = (I + \mathbf{m} \cdot \boldsymbol{\sigma})/2$, so we'll write $\rho_{\text{in}}(\mathbf{m})$ in that form. Since perfect cloning is impossible, we will assume that polarization state of the clone will coincide with that of the input photon with probability η; in the remaining $1 - \eta$ fraction of experiments, the clone ends up in the orthogonal state. If the cloning succeeds, all output photons will be in the same pure state $(I + \mathbf{m} \cdot \boldsymbol{\sigma})/2$. If it fails, we'll get a mixture where both polarizations are equally likely. Recall that such a mixture is described by the density matrix $I/2$, regardless of the choice of basis. Finally, we have

$$\text{Tr}_1 \rho_{\text{out}}(\mathbf{m}) = \text{Tr}_2 \rho_{\text{out}}(\mathbf{m}) = \eta \frac{I + \mathbf{m} \cdot \boldsymbol{\sigma}}{2} + (1 - \eta)\frac{I}{2} = \frac{I + \eta\mathbf{m} \cdot \boldsymbol{\sigma}}{2}. \quad (24.102)$$

So imperfect cloning is described by the same simple pure state formula, the only change being the multiplication of **m** by a constant multiplier η. This multiplier has a special name: the *shrinking factor* of the Bloch vector. The word "shrinking" refers to the fact that $\eta < 1$.

We now want to write the 4×4 matrix for the actual two-photon output $\rho_{\text{out}}(\mathbf{m})$. In doing so, we avoid any assumptions whatsoever about the direction of the Bloch vector or the quality of cloning. When two particles get entangled, each term in the superposition for the first particle generally gets multiplied by each term in the superposition for the second particle, and all the products (weighed by their respective factors) are added together. Our case is more tricky because the terms I, $m_x \sigma_x$, $m_y \sigma_y$, and $m_z \sigma_z$ in the density matrix are not the basis states for either photon. Nevertheless, the analysis shows that the entangled system can be described by the general formula:

$$\rho_{\text{out}}(\mathbf{m}) = \frac{1}{4}\left(I \otimes I + \eta(\mathbf{m} \cdot \boldsymbol{\sigma} \otimes I + I \otimes \mathbf{m} \cdot \boldsymbol{\sigma}) + \sum_{j,k=x,y,z} t_{jk} \sigma_j \otimes \sigma_k\right). \quad (24.103)$$

The reader can verify that (24.103) indeed gives rise to (24.102) by taking the partial trace of the first three terms and by noticing that contribution from the $t_{jk}\sigma_j \otimes \sigma_k$ terms is zero because Pauli matrices are traceless. This form of density operator is also symmetric, ensuring that $\text{Tr}_1 \rho_{\text{out}}(\mathbf{m}) = \text{Tr}_2 \rho_{\text{out}}(\mathbf{m})$. The coefficients t_{jk} will depend on the details of the cloning process. Note that we cannot predict t_{jk} with the knowledge of **m** alone. For example, if **m** points in the z-direction, the partial trace is $(I + \eta\sigma_z)/2$ for either particle and it is easy to think that only the terms with σ_z should appear in (24.105). But this is not necessarily true! It would only be true if we naively multiplied $[(I + \eta\sigma_z)/2] \otimes [(I + \eta\sigma_z)/2]$. However, by doing so we would be claiming that the photons are mutually independent since what we just wrote is a product state. This is not how entanglement works! We have no right to describe an electron in the entangled pair as an entity that knows its Bloch vector. In fact, it is only *after* the measurement that we can determine the absence of σ_x and σ_y terms for each photon; before the measurement, even Nature itself does not know that. So crossing out these terms from (24.103) is a bad idea! We must at this stage keep all nine terms $t_{jk}\sigma_j \otimes \sigma_k$.

Since the machine is universal, it must act similarly on all input states. The process of cloning a "vertical" photon must be similar in all features to that of cloning a "horizontal" photon. Suppose we rotate the coordinate system through angle $-\varphi$ around some arbitrarily chosen rotation axis **n** (this is equivalent to an active rotation of all vectors through angle φ) and represent this rotation with a transformation matrix $R_\mathbf{n}(\varphi)$. Then the old density matrix $\rho_\text{out}(\mathbf{m})$ will be seen in new coordinates as $\rho_\text{out}(R_\mathbf{n}(\varphi)\mathbf{m})$ because vector **m** is now seen as $R_\mathbf{n}(\varphi)\mathbf{m}$. At the same time, universality of the cloning machine means that rotation should not affect the state $\rho_\text{out}(\mathbf{m})$ in any way *other* than the ordinary (and well-familiar to the reader) similarity transformation. This implies

$$\rho_\text{out}(R_\mathbf{n}(\varphi)\mathbf{m}) = D(R_\mathbf{n}(\varphi))\rho_\text{out}(\mathbf{m})D^+(R_\mathbf{n}(\varphi)). \qquad (24.104)$$

In particular, if **n** coincides with **m**, the Bloch vector remains unchanged and the left side is just equal to $\rho_\text{out}(\mathbf{m})$. Multiplying both sides by $D(R_\mathbf{n}(\varphi))$ from the right, we get

$$\rho_\text{out}(\mathbf{m})D(R_\mathbf{n}(\varphi)) = D(R_\mathbf{n}(\varphi))\rho_\text{out}(\mathbf{m}). \qquad (24.105)$$

Now, since the four-dimensional Hilbert space of the system is a direct product of the two-dimensional Hilbert spaces of the two photons, the rotation operator must also be written as a direct product: $D(R_\mathbf{n}(\varphi)) = e^{-i\mathbf{m}\cdot\sigma(\varphi/2)} \otimes e^{-i\mathbf{m}\cdot\sigma(\varphi/2)}$. After a trivial change of notation $-\varphi/2 = \phi$, we get

$$\left[\rho_\text{out}(\mathbf{m}), e^{i\mathbf{m}\cdot\sigma\phi} \otimes e^{i\mathbf{m}\cdot\sigma\phi}\right] = 0. \qquad (24.106)$$

That's the universality condition and it must hold for any UQCM. To make the math look as easy as possible, let us consider the last equation with **m** pointing in the z-direction. The rotation operator can then be written as a simple 4×4 matrix:

$$e^{i\sigma_z\phi} \otimes e^{i\sigma_z\phi} = \begin{pmatrix} e^{2i\phi} & 0 & 0 & 0 \\ 0 & 1 & 0 & 0 \\ 0 & 0 & 1 & 0 \\ 0 & 0 & 0 & e^{-2i\phi} \end{pmatrix}. \qquad (24.107)$$

As for $\rho_\text{out}(\mathbf{m})$, it is easy to check that the first three terms commute with the above operator. Also, writing the nine matrices for the $t_{jk}\sigma_j \otimes \sigma_k$ terms and adding them, we obtain the following after tedious but straightforward algebra:

$$\sum_{i,j=x,y,z} t_{jk}\sigma_j \otimes \sigma_k$$
$$= \begin{pmatrix} t_{33} & t_{31}-it_{32} & t_{13}-it_{23} & t_{11}-t_{22}-i(t_{12}+t_{21}) \\ t_{31}+it_{32} & -t_{33} & t_{11}+t_{22}+i(t_{12}-t_{21}) & -t_{13}+it_{23} \\ t_{13}+it_{23} & t_{11}+t_{22}+i(t_{21}-t_{12}) & -t_{33} & -t_{31}+it_{32} \\ t_{11}-t_{22}+i(t_{12}+t_{21}) & -t_{13}-it_{23} & -t_{31}-it_{32} & t_{33} \end{pmatrix}.$$

In order for this matrix to commute with (24.107) for all rotation angles, it must hold

$$t_{jk} = \begin{pmatrix} t_{11} & t_{12} & 0 \\ -t_{12} & t_{11} & 0 \\ 0 & 0 & t_{33} \end{pmatrix}. \tag{24.108}$$

The analysis for the case of **m** pointing in the $-z$-direction will be exactly the same, the only difference being the change of sign of ϕ. We will again get the result in the same form:

$$t'_{jk} = \begin{pmatrix} t'_{11} & t'_{12} & 0 \\ -t'_{12} & t'_{11} & 0 \\ 0 & 0 & t'_{33} \end{pmatrix} \tag{24.109}$$

(primes were used to distinguish between the above two cases). However, this matrix algebra does not tell us how t'_{jk} relates to t_{jk}. The relation follows easily once we recall how spin operators change under rotations. If we rotate the coordinates $180°$ about the x-axis (we could take the y- axis as well, with the same conclusions) so that z changes to $-z$, then operators transform according to $\sigma_x \to \sigma_x$, $\sigma_y \to -\sigma_y$, and $\sigma_z \to -\sigma_z$. As a result, to keep the form of the terms $t_{jk}\sigma_j \otimes \sigma_k$ unchanged, we only have to set $t'_{xy} = -t_{xy}$, $t'_{xz} = -t_{xz}$, $t'_{zx} = -t_{zx}$, and $t'_{zx} = -t_{zx}$ and leave the other five primed coefficients equal to the unprimed ones. So the matrix (24.109) is now written as

$$t'_{jk} = \begin{pmatrix} t_{11} & -t_{12} & 0 \\ t_{12} & t_{11} & 0 \\ 0 & 0 & t_{33} \end{pmatrix} \tag{24.110}$$

and the sum of output matrices $\rho_{out}(\uparrow) + \rho_{out}(\downarrow)$ gives us a diagonal matrix for the coefficients:

$$t_{jk} + t'_{jk} = \begin{pmatrix} 2t_{11} & 0 & 0 \\ 0 & 2t_{11} & 0 \\ 0 & 0 & 2t_{33} \end{pmatrix}. \tag{24.111}$$

If we now repeat the same procedure for the directions $+x$ and $-x$, the resulting matrix $\rho_{out}(\rightarrow) + \rho_{out}(\leftarrow)$ will have the same diagonal form, as the reader can verify:

$$\tilde{t}_{jk} + \tilde{t}'_{jk} = \begin{pmatrix} 2\tilde{t}_{11} & 0 & 0 \\ 0 & 2\tilde{t}_{22} & 0 \\ 0 & 0 & 2\tilde{t}_{22} \end{pmatrix}. \tag{24.112}$$

The tilde here was used to distinguish between the input mixtures (24.111) and (24.112). Note that the *first* two diagonal elements are equal in the former matrix and the *last* two in the latter. The no-signaling condition implies that $\rho_{out}(\uparrow) + \rho_{out}(\downarrow) = \rho_{out}(\rightarrow) + \rho_{out}(\leftarrow)$ (see also Equation 24.100). Setting the last two matrices equal to each other, we come to a remarkable conclusion: To prevent FTL signaling, Nature must ensure that $t_{11} = t_{22} = t_{33} = t$. With that requirement met, the array of coefficients (24.108) will be written as

$$t_{jk} = \begin{pmatrix} t & t_{12} & 0 \\ -t_{12} & t & 0 \\ 0 & 0 & t \end{pmatrix}. \tag{24.113}$$

We are now in a position to determine the limitations imposed by Nature on the quality of cloning. First we'll use the obtained values of t_{jk} to write the density operator (24.103) as a matrix:

$$\rho_{\text{out}}(\uparrow) = \frac{1}{4} \left\{ \begin{pmatrix} 1 & 0 & 0 & 0 \\ 0 & 1 & 0 & 0 \\ 0 & 0 & 1 & 0 \\ 0 & 0 & 0 & 1 \end{pmatrix} + \eta \begin{pmatrix} 2 & 0 & 0 & 0 \\ 0 & 0 & 0 & 0 \\ 0 & 0 & 0 & 0 \\ 0 & 0 & 0 & -2 \end{pmatrix} + \begin{pmatrix} t & 0 & 0 & 0 \\ 0 & -t & 2t+2it_{12} & 0 \\ 0 & 2t-2it_{12} & -t & 0 \\ 0 & 0 & 0 & t \end{pmatrix} \right\}.$$

The eigenvalues of this operator are found by solving the eigenvalue problem:

$$\frac{1}{4} \begin{vmatrix} 1+2\eta+t-4\lambda & 0 & 0 & 0 \\ 0 & 1-t-4\lambda & 2t+2it_{12} & 0 \\ 0 & 2t-2it_{12} & 1-t-4\lambda & 0 \\ 0 & 0 & 0 & 1-2\eta+t-4\lambda \end{vmatrix} = 0. \tag{24.114}$$

The solutions are

$$\lambda_{1,2} = (1/4)(1 \pm 2\eta + t), \quad \lambda_{3,4} = (1/4)\left(1 - t \pm 2\sqrt{t^2 + t_{xy}^2}\right). \tag{24.115}$$

But in order for $\rho_{\text{out}}(\uparrow)$ to be a valid QM operator, it must be positive semidefinite, which implies that all its eigenvalues have to be nonnegative. The maximum value of η compatible with this condition is obtained for $t_{xy} = 0$ and $t = 1/3$. Then, the quality of cloning is limited by $\eta_{\max} = 2/3$ and $F_{\max} = (1+\eta_{\max})/2 = 5/6$. In this optimal case, the density operator is

$$\rho_{\text{out}}(\uparrow) = \frac{1}{4 \times 3} \begin{pmatrix} 8 & 0 & 0 & 0 \\ 0 & 2 & 2 & 0 \\ 0 & 2 & 2 & 0 \\ 0 & 0 & 0 & 0 \end{pmatrix}. \tag{24.116}$$

We can now see the true meaning of the upper limit $F_{\max} = 5/6$. This is the maximum fidelity that keeps probabilities nonnegative *and* does not violate the no-signaling condition (24.100). This is as far as we can go in theory, and, as evidenced by (24.101), QM allows us to reach this exact limit but not to go any further! As long as one defines superluminal signaling in terms of obtaining different density matrices at the receiver's end, the verdict is clear: The theory of vacuum fluctuations saves causality by providing for just enough spontaneous emission to bring fidelity of photon clones under the allowed limit. This conclusion, first obtained by Gisin [29], is so important that a special term was invented to describe it – "peaceful coexistence between QM and relativity" [30]. Actually, what was at stake is not relativity *per se*, but only the causality principle. It seems that QM is on the verge of violating causality, but stops just short of doing so.

24.10
The FLASH Proposal and Quantum Compounds

We can now discuss how QM prevents superluminal FLASH signaling. In the original FLASH scheme, Alice would prepare input photons in a rectilinear {V, H} basis, resulting in an equal mixture of V- and H-polarized states, or in a circular {R, L} basis, resulting in an equal mixture of R- and L-polarized states. The FLASH apparatus would then amplify each photon by a *nonselective* (with respect to polarization) laser-gain tube (LGT) and attempt to identify the polarization of the input photon by analyzing the photon bursts that emerge from the LGT. In the absence of noise, the problem would reduce to telling a multiphoton plane unpolarized (PUP) pulse from a similar circularly unpolarized (CUP) pulse. This could be easily accomplished by splitting the output N-photon pulse into two halves and subjecting each half to a measurement in the appropriate basis. In this case, each input state will have a unique signature. For instance, if we have a V-polarized pulse at the input, then the $N/2$ photons measured in the {V, H} basis will all come out V-polarized, while the $N/2$ photons measured in the {R, L} basis will split equally between R- and L-polarized states, so the unique signature of this input pulse will be $(0, N/2, N/4, N/4)$. In the same manner, we can obtain the signatures of the other three input pulses. So, as long as N is sufficiently large, Bob could read Alice's message unmistakably.

At the time of FLASH publication, the no-cloning theorem was still unknown,[14] but after it was stated by Wootters and Zurek a few months later, many commentators rushed to dismiss the whole idea of FLASH for the wrong reasons. Of course, it was only perfect cloning that was shown to be impossible. But a closer look at the FLASH paper shows that Herbert was well aware that his output photon bursts would be masked by the noise. His actual proposal was to look at the *macroscopic* outcome – a powerful beam emerging from the LGT and identifying the dominant component. To quote Herbert, "By looking at many bursts we can average out the noise and extract the appropriate signal signature (either an $|H - V|$ excess in the case of a PUP signal, or an $|R - L|$ excess in the case of a CUP signal)" [5]. In other words, the scheme was intended from the very beginning to use imperfect cloning only.

It was only later, when theoretical research focused on imperfect cloning, that the second type of objections to FLASH emerged, based on the density matrix arguments. As we saw in Section 24.9, the two mixtures Bob gets at this end of the

---

14) In retrospect, it is now clear that several physicists had been tacitly using the no-cloning assumption prior to 1982. As Asher Peres, who was the reviewer of the FLASH paper, put it many years later, "As it often happens in science, these things were well known to those who knew things well." Though Peres felt the FLASH paper was wrong, he recommended it for publication because he thought that "finding the error would lead to significant progress in our understanding of physics" [31].

apparatus are indistinguishable. If Alice's signal is to be encoded in such a mixture, then there can be no FTL signaling.

However, one can make a compelling argument that this kind of proof cannot be considered decisive. To see why it falls short, consider this simple conceptual analogy. Let us say, one is looking at two sequences of digits. The first sequence is 1010101010... and the second one is 1100110011.... The statistical characteristics – mean value of a digit and standard deviation – are the same for both sequences; so, if one applies statistical analysis only, the sequences will be equivalent. But if one pays attention to the repetitive pattern of ones and zeros, one immediately notices the difference.

Equations 24.95–24.100 describe the *statistical average* of many independent trials. But it is not the average assembly resulting from many identically prepared EPR pairs that Bob is trying to clone! As far as Bob is concerned, his experiment consists of a large number of individual trials, and in each trial the input to the UQCM is a pure state $|1_V, 0_H\rangle$ or $|0_V, 1_H\rangle$ (for a rectilinear basis) or $|1_R, 0_L\rangle$ or $|0_R, 1_L\rangle$ (for the circular basis).[15] And it is these individual outcomes rather than a statistical mixture averaged over all trials that FLASH proposal aims to use! Therefore, a no-signaling proof cannot rely on density matrix arguments alone. To strengthen the proof, it is necessary to show that one cannot glimpse any additional information from the pattern formed by $|V\rangle$ and $|H\rangle$ (respectively $|R\rangle$ and $|L\rangle$) input states.

Another way to put it is by saying that Herbert's FLASH proposal involves *quantum compounds* rather than ordinary quantum mixtures. Stated simply, Bob's sequence is not completely random in the sense that Alice, when she measures the state of her photon, knows exactly the outcome that Bob will get if he chooses the same measuring basis. To quote Herbert again, "Certainly if someone can predict the polarization of every single one of your photons, one would hesitate to call such a beam "unpolarized" [5]. This is the way a compound differs from an ordinary mixture even though they are described by the same density matrix.

Fortunately, as it often happens in physics, by glossing over the distinction between individual outcomes and statistical averages, we would still not get a different answer. The current consensus among physicists is that FTL signaling appears to be impossible even if one attempts to employ quantum compounds. However, in order to explain the failure of FLASH correctly, one must look carefully at the dynamics of the actual amplification process and the resulting photon statistics in the output beam. The exact features of the process will depend on the type of amplifier used.[16]

Suppose, for example, we try the one-photon amplifier described by Mandel. Because this amplifier also happens to be the optimal UQCM, it is natural to assume that if a signaling attempt fails with this amplifier, it must fail with all other amplifiers as well. The output of this amplifier is two photons in the same space–time mode – a

---

15) This fact has caused some confusion in the 1980s as some authors mistakenly reduced the argument against FTL signaling to Equations 24.98–24.100.

16) The statistics of photons produced by different types of amplifiers is well studied, but their suitability in the context of FLASH signaling has not received as much theoretical attention as various no-signaling proofs based on the density matrix approach. As of this writing, the amount of research in this area remains limited, and some of the conclusions may be tentative.

situation Howard Yuen calls "biphoton." Let's say Alice prepared the photon in state $|V\rangle$ so that Bob's UQCM produces a biphoton $|2_V, 0_H\rangle$ two times out of three. One time out of three the result will be $|1_V, 1_H\rangle$, but Bob hopes this noise level will not distort his results too much. By the same token, if the input photon came in state $|H\rangle$, the output biphoton will likely have state $|0_V, 2_H\rangle$. So Bob is betting that an output sequence dominated by $|2_V, 0_H\rangle$ and $|0_V, 2_H\rangle$ biphotons will be distinguishable from a similar sequence dominated by $|2_R, 0_L\rangle$ and $|0_R, 2_L\rangle$ biphotons.

The weak point in this scheme is the assumption that photons would behave *independently* of each other. If we recall the discussion of number states in Chapter 21 and the HOM experiment analogies, we can show that input mixtures $(1/2)|V\rangle\langle V| + (1/2)|H\rangle\langle H|$ and $(1/2)|R\rangle\langle R| + (1/2)|L\rangle\langle L|$ will look indistinguishable to Bob on a *per trial* basis.

Let Bob use the circular basis to measure his cloned photons. Suppose Alice chooses the linear basis and measures $|V\rangle$. Then, Bob will be looking at biphoton $|2_V, 0_H\rangle$ with probability 2/3. This biphoton will indeed behave as a coin pair, tossing one "heads" and one "tails" half of the time. Bob also has a 1/3 chance to obtain a different biphoton, $|1_V, 1_H\rangle$. This biphoton will behave as in the HOM experiment, always resulting in coincident detections (either two "heads" or two "tails"). So overall Bob obtains each of the three possible measurement results with the same probability 1/3. The same conclusion holds if Alice chooses the linear basis and measures $|H\rangle$. Obviously, no amount of "patterning" such as sending 10 signals in a row in the same linear basis can help in this case. Now let's say, Alice chooses the circular basis and measures $|R\rangle$. Bob will then obtain two right-polarized photons with probability 2/3 and oppositely polarized photons with probability 1/3. Let's say, Alice chooses the circular basis and measures $|L\rangle$. Now Bob will obtain two left-polarized photons with probability 2/3 and oppositely polarized photons with probability 1/3. But Alice cannot control the state, $|R\rangle$ or $|L\rangle$, she measures! Both of them are equally likely. It is again pointless to arrange signals in a pattern when each separate trial yields the same probability distribution 1/3, 1/3, 1/3 at the receiver end. Thus, the FTL scheme is destroyed in the end by the very same phenomenon of correlation between outcomes that offered the hope of FTL signaling in the first place.

One experiment to test the FLASH proposal has been done recently [32] using a quantum injected-optical parametric amplifier (QI-OPA) as the cloning machine. This amplifier is different from the one-photon amplifier considered by Mandel in that the interaction Hamiltonian responsible for amplification has the form

$$H \sim i(a_H^+ a_V^+) + \text{h.c.}, \tag{24.117}$$

where the two creation operators are associated with the H-polarized mode and V-polarized mode, respectively. Let angle φ describe the input photon polarization in the equatorial plane of the Poincaré sphere so that $\varphi = 0$, $\varphi = \pi$ correspond to S-polarized (respectively D-polarized) photon and $\varphi = -\pi/2$, $\varphi = \pi/2$ correspond to R-polarized (respectively L-polarized) photon. Any optical amplifier is characterized by *gain g*, which is just the amplification factor (ratio of output power and input power). A QI-OPA whose g is given generates a multiphoton output beam whose average photon numbers in the two diagonal polarization modes are given by

$$M_{S,D}(\varphi) = \sinh^2 g + \frac{2\sinh^2 g + 1}{2}(1 \pm \cos\varphi). \tag{24.118}$$

In the experiment, the amplifier gain was $g \approx 4.45$, corresponding to the average photon number on the order of 4000. The researchers tried to realize the FLASH scheme by setting the input photon polarization and then measuring the excess of output intensity in one polarization mode over the other ($|I_S - I_D|$ and $|I_R - I_L|$) in line with the FLASH proposal. To imitate the density matrix of the output state, the measurements were averaged over a large number (2500) of independent trials. Once again, the output photons produced by the QI-OPA are not mutually independent and photon statistics predicts, somewhat counterintuitively, that the result should not depend on Alice's choice of the coding basis. But the photon statistics in this experiment is super-Poissonian, that is, the standard deviation of the photon number is greater than the mean. Physically, it means that photons will tend to bunch together instead of arriving at the detector independent of one another. As a result, in individual trials, we should constantly observe an excess of photons in one polarization mode (S instead of D or vice versa), even when *on the average* there is no reason to prefer one mode over the other.

The experimental data convincingly demonstrated the impossibility of FTL signaling. As one would expect, *without* taking the absolute value, the average of $I_S - I_D$ and $I_R - I_L$ depended on φ. The reason for that is clear. Suppose that Bob uses the $\{S, D\}$ measuring bases in all trials. Then, an S-polarized input photon would, on the average, cause a predominance of S-polarized photons in the output, which means $\langle I_S - I_D \rangle > 0$, and a D-polarized photon would skew the output toward D-photons so that $\langle I_S - I_D \rangle < 0$. Meanwhile, circularly polarized output photons due to an R/L-polarized photon at the input will *on the average* be equally likely to collapse into either of the two diagonal states: $\langle I_S - I_D \rangle = 0$. However, we are interested in the average of the *absolute value,* $\langle |I_S - I_D| \rangle = 0$. This is because the message sent by Alice consists in the choice of the coding basis, so an output containing mostly S-photons and an output containing mostly D-photons should both correspond to the same bit (diagonal basis). The same principle is true if Alice chooses the circular basis. So Bob has to record the absolute value first and only then take the statistical average. This is where the results become really "strange." In the trials where Alice chooses the $\{S, D\}$ basis, we would expect always (or at least most of the time) to see an excess of photons along one or the other diagonal, and this is what actually happens. So far, so good. The surprising thing happens when Alice chooses the $\{R, L\}$ basis. We would naturally think that photons would split about equally between the S- and D-directions in each individual trial. But this is where our common sense proves misleading. As experiment shows, the photons will behave *exactly* the same way as before! In a typical trial, we will again see the same statistics of photons, with the majority of photons choosing one or the other diagonal. The absolute value of photon excess along one diagonal as measured by Bob turns out to be the same in all cases, regardless of the angle φ:

$$\langle |I_S - I_D| \rangle = \text{const.} \tag{24.119}$$

As a result, Bob finds himself in an upsetting situation: The information is in the output beam sent by Alice faster than the speed of light, but he cannot read this information! Or, to put it in a formal way, Bob cannot perform *quantum state estimation* for the resulting output beam with sufficient accuracy that would let him distinguish between two initial preparations {S, D} and {R, L}. And the irony does not stop there. In fact, one can show [33] that the output states produced by *optimal* universal cloning machines are the *worst* ones from the point of view of state estimation on the resulting correlated copies. The moral of the story is that Bob has lost his ability to extract information from the "quantum compound" because of the very same optimality of cloning that enabled Alice to send that information to him! It looks like QM first creates the possibility of FTL signaling due to the entanglement between EPR photons and then removes this possibility because of the very same entanglement between the photons in the output. This irony evidences the crucial role of nonlocal correlations in *preventing* FTL signaling. While the fidelity-bound (24.102) is sufficient to prevent the type of FTL signaling where the parties attempt to exchange a general density matrix, a more subtle attempt to make use of the preexisting pattern in that matrix (quantum compound) is thwarted by the presence of *high-order correlations* among different clones that were introduced in the cloning process. The reader must keep in mind that if it were possible to write the output as a product state, then FLASH would succeed even in the presence of spontaneous noise!

Box 24.3 The HOM effect

Since a detailed quantum optics treatment of high-order correlations between photons that predetermine the failure of FLASH is outside the scope of this text, we can offer a somewhat less rigorous proof, which involves the following modified FTL signaling proposal. Imagine that Alice sends to Bob *two* signals in a row, prepared in one and the same (say, rectilinear) basis. Then, Bob will have at his disposal two independent photons, each photon polarized either vertically or horizontally. Suppose he sends these photons toward a polarization-neutral beam splitter, ensuring they arrive at respective input ports a and b simultaneously. Thus, there are four equally likely possibilities for the beam splitter input: $|V\rangle_a|V\rangle_b, |V\rangle_a|H\rangle_b, |H\rangle_a|V\rangle_b, |H\rangle_a|H\rangle_b$.

Suppose the two photons turn out to have the same polarization. Assuming they also arrive at the same time and have the same frequency, this makes them identical and therefore indistinguishable. This situation is described by the HOM effect (already discussed briefly in Section 4.6), which is observed when two identical photons arrive simultaneously at two *different* input ports a and b of a 50:50 beam splitter (Figure 4.7). In this case quantum optics predicts that both photons will exit from the *same* output port – either c or d – with 50% probability for each port. This phenomenon is due to the way amplitudes of indistinguishable events are added together. In this case we have four possible events: both photons are reflected; both are transmitted; a is reflected and b is transmitted; and a is transmitted and b is reflected. Since the photons are identical, the first two events look exactly the same and we must add their amplitudes. But upon each reflection, the amplitude undergoes a 90° phase shift, so the respective amplitude gets multiplied by i; two reflections result in a factor $i^2 = -1$ appearing before the first amplitude. As a result, the first two amplitudes cancel each other,[17] and only the two remaining possibilities survive.

Thus, the HOM effect makes it possible for Bob to weed out all instances of $|V\rangle_a|V\rangle_b$ and $|H\rangle_a|H\rangle_b$ by discarding all experiments where both photons emerge from the same output port. This will remove 50% of all trials. On the other hand, if the input is $|V\rangle_a|H\rangle_b$ or $|H\rangle_a|V\rangle_b$, then photons will behave "classically," and all four possibilities will be equally likely. Only half of such trials (or 25% of the total) will be discarded then. The remaining 25% are the trials where the photons with orthogonal polarizations emerge from different ports. Measuring these photons in the rectilinear basis will produce zero coincidences (i.e., if one photon is vertical, the other is horizontal and vice versa), and if measurements in a circular basis resulted in coincidences half of the time as one would expect for independent photons, then we would have FTL signaling.

Unfortunately, this modified FTL signaling proposal fails for a similar reason: the entanglement between the photons in the output. In this case, the beam splitter also acts as an entangler and the actual output state is

$$|\Psi\rangle = \left(1/\sqrt{2}\right)\left(|V\rangle_c|H\rangle_d - |H\rangle_c|V\rangle_d\right). \tag{24.120}$$

This state will look exactly the same way in the circular basis, and again there will be zero coincidences.

So 25% of the total that were selected by Bob do not contain any extractable information that could be used for FTL signaling. Obviously, the initial sequence of single photons does not carry such information either. Therefore, the discarded component is also useless for FTL signaling. However, two-thirds of that component is pairs of photons with the same polarization and one-third – pairs with different polarizations, in other words, the discarded component, simulates the output of Mandel's amplifier, which therefore cannot be used for superluminal signaling.

---

17) This assumes the beam splitter is symmetric, that is, treating input beams a and b identically. Another type of splitter is a reflecting surface such that reflection off the upper side does not change the phase, while reflection off the bottom shifts the phase by 180°. The end result is the same, only this time the third and the fourth amplitudes will cancel each other. A more detailed treatment of the problem involves photon algebra (creation and annihilation operators).

Problems

24.1 Find the expected magnitude of the photon's momentum in the state (24.9).

24.2 Consider a photon in the optical range of spectrum, interacting with a BS. Assume that the molecular weight of molecules of BS is about 50. Evaluate the magnitude of possible momentum transfer to the incident photon, which results from thermal jittering of BS at normal temperature.

24.3 Suppose that the two photons described in Section 24.5 are not necessarily maximally entangled, and the state of the pair is described by Equation 24.20. Find the ratio $N_{V,V}/N_{H,H}$ in terms of c_1.

24.4 Consider ensemble of photon pairs emitted by source S toward A and B, respectively (Figure 24.3). In all trials, both photons in a pair are in the same state $|\theta\rangle$ of *definite* linear polarization. The transmission axes of polarizing filters at the ends A and B are parallel to each other. Find (a) the fractional number of coincident clicks of both detectors and (b) the fractional number of coincident nonclicks in both detectors.

24.5 Consider the same situation as in the previous problem, and assume that the probability for a pair to be polarized along a direction θ is the same for all θ. Find the percentage of coincident responses at the ends A and B.

24.6 Derive the expression (24.26) for coincident nonfirings of both detectors if the behavior of quantum systems were governed by hidden parameters

24.7 Show that $A(a)B(c) - A(a)B(d) + A(b)B(c) + A(b)B(d) = \pm 2$. Once you prove it, the CHSH inequality is derived easily. (Hint: you may want to factor out the common terms).

24.8 Show that all 4 Bell states have the maximum possible concurrence.

References

1 Aspect, A., Grangier, P., and Roger, C. (1982) Experimental test of Bell's inequalities using time-varying analyzers. *Phys. Rev. Lett.*, **49**, 91–94.

2 Fayngold, M. (2008) *Special Relativity and How It Works*, Wiley-VCH Verlag GmbH, pp. 485–487.

3 Einstein, A., Podolsky, B., and Rosen, N. (1935) Can quantum-mechanical description of physical reality be considered complete? *Phys. Rev.*, **47**, 777–780.

4 Eberhard, P. (1978) Bell's theorem and the different concepts of locality. *Nuovo Cim.*, **46**, 392.

5 Herbert, N. (1982) FLASH: a superluminal communicator based upon a new kind of quantum measurement. *Found. Phys.*, **12**, 1171.

6 Wootters, W.K. and Zurek, W.H. (1982) A single quantum cannot be cloned. *Nature*, **299**, 802.

7 Dieks, D. (1982) Communication by EPR devices. *Phys. Lett. A*, **92** (6), 271.

8 van Enk, S.J. (1998) No-cloning and superluminal signalling. arXiv: quant-ph/9803030 v.1, 13 Mar.

9 Hecht, E. (2002) *Optics*, 4th edn, Addison-Wesley, p. 385.

10 Born, M. and Wolf, E. (1999) *Principles of Optics: Electromagnetic Theory of Propagation, Interference and Diffraction of Light*, 7th ed., Cambridge University Press, Cambridge, UK.

11 Bell, J.S. (1964) On the Einstein–Podolsky–Rosen paradox. *Physics*, **1**, 195–200.

12 Bell, J.S. (1966) On the problem of hidden variables in QM. *Rev. Mod. Phys.*, **38**, 447.

13 Clauser, J.F., Horne, M.A., Shimony, A., and Holt, R.A. (1969) Proposed experiment to test local hidden-variable theories. *Phys. Rev. Lett.*, **23**, 880–884.

14 Bell, J.S. (1987) *Speakable and Unspeakable in QM*, Cambridge University Press, Cambridge.

15 Clauser, J.F. and Horne, M.A. (1974) Experimental consequences of objective local theories. *Phys. Rev. D*, **10**, 526–535.

16 Clauser, J.F. and Shimony, A. (1978) Bell's theorem: experimental tests and implications. *Rep. Prog. Phys.*, **41**, 1881.

17 Aspect, A., Grangier, P., and Roger, G. (1982) Experimental realization of Einstein–Podolsky–Rosen–Bohm Gedanken experiment: a new violation of Bell's inequalities. *Phys. Rev. Lett.*, **49**, 91–94.

18 Aspect, A., Dalibard, J., and Roger, G. (1982) Experimental test of Bell's inequalities using time-varying analyzers. *Phys. Rev. Lett.*, **49**, 1804.

19 Greenberger, D., Horne, M., and Zeilinger, A. (2005) Bell theorem without inequalities for two particles: I. Efficient detectors. *Phys. Rev. A*, **78**, 022110.

20 Aspect, A. (2007) To be or not to be local. *Nature*, **446**, 866–867.

21 Mandel, L. (1983) Is a photon amplifier always polarization-dependent? *Nature*, **304**, 188.

22 Schumacher, B. (1995) Quantum coding. *Phys. Rev. A*, **51**, 2738.

23 Eberhard, P.H. and Ross, R.R. (1989) Quantum field theory cannot provide faster-than-light communication. *Found. Phys. Lett.*, **2**, 127–149.

24 Buzek, V. and Hillery, M. (1996) Quantum copying: beyond the no-cloning theorem. *Phys. Rev. A*, **54**, 1844.

25 Buzek, V. and Hillery, M. (1996) Universal optimal cloning of arbitrary quantum states: from qubits to quantum registers. *Phys. Rev. A*, **54**, 1844.

26 Gisin, N. and Massar, S. (1997) Optimal quantum cloning machines. *Phys. Rev. Lett.*, **79**, 2153.

27 Bruss, D., DiVincenzo, D.P., Ekert, A., Fuchs, C.A., Macchiavello, C., and Smolin, J.A. (1998) Optimal universal and state-dependent quantum cloning. *Phys. Rev. A*, **57**, 2368.

28 Lamas-Linares, A., Simon, C., Howell, J.C., and Bouwmeester, D. Experimental quantum cloning of single photons. *Science*, **296**, 712.

29 Gisin, N. (1998) Quantum cloning without signaling. *Phys. Lett. A*, **242** (1–2), 1–3.

30 Shimony, A. (1983) *Foundations of Quantum Mechanics in the Light of New Technology* (ed. S. Kamefuchi), Physical Society of Japan, Tokyo, Japan.

31 Peres, A. (2002) How the no-cloning theorem got its name. *Fortschr. Phys.*, **51**, 458–461.

32 DeAngelis, T., De Martini, F., Nagali, E., and Sciarrino, F. (2007) Experimental test of the no-signaling theorem. *Phys. Rev. Lett.*, **99**, 193601.

33 Demkowicz-Dobrzanski, R. (2005) State estimation on correlated copies. *Phys. Rev. A*, **71**, 062321.

25
Quantum Measurements and POVMs

25.1
Projection Operator and Its Properties

We have found earlier that to understand a system's behavior with respect to measurements, it is convenient to work in an orthonormal basis. Suppose observable A of a given system has a set of eigenvalues a_i. Let there be N such eigenvalues and let each eigenvalue a_i define an eigenstate $|a_i\rangle$. We can then compose from N eigenstates $|a_i\rangle$ an orthonormal basis:

$$\{|a_i\rangle\} = \{|1\rangle, |2\rangle, \ldots, |N\rangle\}. \tag{25.1}$$

So the system will "live" in some N-dimensional $\mathcal{H}$-space that has basis (25.1).

Suppose now the system is in some arbitrary state $|\psi\rangle$. It can be written as a superposition of eigenstates:

$$|\psi\rangle = \sum_i |a_i\rangle \langle a_i|\psi\rangle. \tag{25.2}$$

Projection operator $\hat{P}_{a_i}$ is then defined as

$$\hat{P}_{a_i} = |a_i\rangle\langle a_i|, \tag{25.3}$$

so that $|\psi\rangle = \sum_i \hat{P}_{a_i}|\psi\rangle$. The name "projection" arises from a simple geometric analogy: When applied to an arbitrary state $|\psi\rangle$, this operator selects the projection $|\psi_{a_i}\rangle \equiv |a_i\rangle\langle a_i|\psi\rangle$ along the "unit vector" $|a_i\rangle$:

$$\hat{P}_{a_i}|\psi\rangle = |a_i\rangle\langle a_i|\psi\rangle. \tag{25.4}$$

In particular, if the system is already prepared in state $|a_i\rangle$, the inner product is unity and projection operator just leaves the system unchanged:

$$\hat{P}_{a_i}|a_i\rangle = |a_i\rangle. \tag{25.5}$$

So it is clear that operator $\hat{P}_{a_i}$ has the eigenstate $|a_i\rangle$ and the eigenvalue 1.

We must now say a few words about normalization – a topic that often causes confusion. The eigenstates $|a_i\rangle$ in Equations 25.1–25.3 are assumed to be normalized. This cannot be said about the left-hand side of Equation 25.4. Indeed, as is clear

Quantum Mechanics and Quantum Information: A Guide through the Quantum World,
First Edition. Moses Fayngold and Vadim Fayngold.
© 2013 Wiley-VCH Verlag GmbH & Co. KGaA. Published 2013 by Wiley-VCH Verlag GmbH & Co. KGaA.

from the fact that $\langle a_i|\psi\rangle \le 1$, the projections $|\psi_{a_i}\rangle$ are vectors whose norm is less than unity. This is fine when we are writing an equation like (25.4). The current state of the system is the normalized state $|\psi\rangle$ and the length (norm) of vector $|\psi_{a_i}\rangle$ expresses the amplitude of one possible outcome (eigenvalue a_i) of a measurement that we plan to perform in the future. But what happens *after* we performed that measurement and obtained the eigenvalue a_i? The vector $|\psi_{a_i}\rangle$ cannot describe the postmeasurement state because it has not been normalized. Indeed, the system is now in the eigenstate $|a_i\rangle$, and Equation 25.4 tells us that

$$|a_i\rangle = \frac{\hat{P}_{a_i}|\psi\rangle}{\langle a_i|\psi\rangle}. \tag{25.6}$$

Hence, if for some reason we don't like notation $|a_i\rangle$ and instead want to write the final state of the system in terms of the projection operator, we can do so by using the right-hand side of (25.6).

One may ask a legitimate question: What's wrong with the conventional eigenstate language that we would instead choose the longer and more abstract expression? It turns out that discussing the outcomes of measurements in terms of projection operators makes it possible to embrace those situations where the eigenstate language is no longer sufficient. These are the so-called POVMs that we are going to cover in the rest of this chapter. These are the nonoptimal measurements, employing operators that are *not* designed to return an eigenstate with certainty, so describing the outcomes in terms of $|a_i\rangle$ is no longer feasible. At the same time, the operator language works perfectly in such cases. This is why we want to develop this language now, with a view to future applications.

In the measurement theory, we no longer say "eigenvalue a_i" but rather "outcome a_i," mindful of the fact that in general, the outcomes will not necessarily be equivalent to eigenvalues. Let us therefore adopt this term, which for now will of course mean the same thing as eigenvalue, but in the following sections will take on an extended meaning.

When measuring the arbitrary state $|\psi\rangle$, the probability of outcome a_i is equal to

$$p(a_i) = \langle\psi|\hat{P}_{a_i}|\psi\rangle. \tag{25.7}$$

The proof of this statement is elementary: One just needs to write the operator explicitly, and the right-hand side of (25.7) will turn into the square of the probability amplitude. Here we have abandoned the old symbol for probability and adopted lowercase p to stress the fact that "probability" is now that of an *outcome*. Since linear momentum is not part of this chapter, this should not cause any confusion.

Students should learn the last formula by heart. This is one of the most important properties of the projection operator. Today, most articles on measurement theory will just write the expression $\langle\psi|\hat{P}_{a_i}|\psi\rangle$ instead of the traditional notation such as $p(a_i)$ or $\Pr(a_i)$. Yet it means exactly the same thing: probability that a measuring apparatus oriented along direction a_i is going to register an event if the system was prepared in state $|\psi\rangle$. Of course, one can also obtain this probability by squaring the modulus of the corresponding probability amplitude $\langle a_i|\psi\rangle$. Therefore, the following ways of writing Equation 25.6 are equivalent:

$$|a_i\rangle = \frac{\hat{P}_{a_i}|\psi\rangle}{\langle a_i|\psi\rangle} = \frac{\hat{P}_{a_i}|\psi\rangle}{\sqrt{p(a_i)}} = \frac{\hat{P}_{a_i}|\psi\rangle}{\sqrt{\langle\psi|\hat{P}_{a_i}|\psi\rangle}}. \tag{25.8}$$

As a corollary, we can now prove the closure (completeness) relation

$$\sum_i \hat{P}_{a_i} = I \tag{25.9}$$

by requiring that the probabilities of outcomes add to 1. Indeed, from

$$1 = \sum_i p(a_i) = \sum_i \langle\psi|\hat{P}_{a_i}|\psi\rangle = \langle\psi|\sum_i \hat{P}_{a_i}|\psi\rangle = \langle\psi|I|\psi\rangle = 1, \tag{25.10}$$

there follows that the sum of projection operators must be equal to the identity operator I.

Of course, we could also have come to the same conclusion by simply comparing the left- and right-hand sides of (25.2), but the proof we chose underscores the physical justification of (25.9).

With this result, the spectral decomposition theorem for Hermitian operators, mentioned earlier in Chapter 12, can be written in terms of projectors. Then, operator $\hat{A}$ corresponding to the observable A will take the form

$$\hat{A} = \sum_i a_i \hat{P}_{a_i}. \tag{25.11}$$

The fact that the probabilities of outcomes given by (25.7) are all positive or zero tells us that the projection operator is positive semidefinite. We already encountered this term in Chapter 12 in the context of density matrices. This is not a coincidence because $\hat{P}_{a_i}$ is mathematically identical to the density matrix of a pure state $|a_i\rangle$. Alternatively, turning the argument on its head, we might say that probabilities are nonnegative because of the positive semidefiniteness of $\hat{P}_{a_i}$. As was shown then, all eigenvalues of a positive semidefinite Hermitian operator are nonnegative, and conversely, if all eigenvalues are nonnegative, then the Hermitian operator is positive semidefinite (Problem 25.1).

As the density operator for a pure state, $\hat{P}_{a_i}$ naturally possesses all properties of such operators, including Hermiticity and idempotence:

$$\hat{P}_{a_i} = \hat{P}_{a_i}^+, \tag{25.12}$$

$$\hat{P}_{a_i}^2 = \hat{P}_{a_i}. \tag{25.13}$$

Idempotence of $\hat{P}_{a_i}$ simply means that the second (and all subsequent) application of the projection operator does not change the resulting state. In this sense, measurements described by $\hat{P}_{a_i}$ are *reproducible*.

Projection operators are mutually orthogonal:

$$\hat{P}_{a_i}\hat{P}_{a_j} = \delta_{ij}\hat{P}_{a_i}. \tag{25.14}$$

It is also easy to see that $\hat{P}_{a_i}$ has no inverse. The reason is simple: It is always possible to choose at least two different states $|\psi_1\rangle$ and $|\psi_2\rangle$ that will have the same projection on a given unit vector $|a_i\rangle$; in other words, $\hat{P}_{a_i}|\psi_1\rangle = \hat{P}_{a_i}|\psi_2\rangle$. Then the existence of the inverse operator $\hat{P}_{a_i}^{-1}$ would imply $|\psi_1\rangle = |\psi_2\rangle$, contradicting our initial assumption.

The hermiticity and idempotence properties are important to us because we can now rewrite our previous results in a more general form. The closure relation and spectral decomposition theorem can be written as

$$\sum_i \hat{P}_{a_i}^+ \hat{P}_{a_i} = I \tag{25.15}$$

and

$$\hat{A} = \sum_i a_i \hat{P}_{a_i}^+ \hat{P}_{a_i}. \tag{25.16}$$

The probability of an outcome (25.7) and the postmeasurement state (25.8) will now take the following symmetrical form:

$$p(a_i) = \langle \psi | \hat{P}_{a_i}^+ \hat{P}_{a_i} | \psi \rangle \tag{25.17}$$

and

$$|a_i\rangle = \frac{\hat{P}_{a_i}|\psi\rangle}{\langle a_i|\psi\rangle} = \frac{\hat{P}_{a_i}|\psi\rangle}{\sqrt{p(a_i)}} = \frac{\hat{P}_{a_i}|\psi\rangle}{\sqrt{\langle \psi | \hat{P}_{a_i}^+ \hat{P}_{a_i} | \psi \rangle}}. \tag{25.18}$$

The reader should keep in mind that there exists some ambiguity in the literature about how the term "projection operator" (or simply "projector") is used. You may sometimes encounter the phrase "projector onto a subspace." Now, strictly speaking, there *is* a definition for a projector onto a subspace consisting of several states: Such a projector is just a sum of projectors for the individual states:

$$\hat{P}_{\text{subspace}} = \sum_i \hat{P}_{a_i} = \sum_i |a_i\rangle\langle a_i|. \tag{25.19}$$

In particular, the identity operator can be described as a projection operator that projects onto the entire Hilbert space $\mathcal{H}$. In fact, problems that call for this type of "generalized projectors" do exist. However, there is a tendency among physicists to use terms casually, and as a rule of thumb, when someone says "projector onto a subspace," he usually means an ordinary projection operator (25.3). Sometimes you will get the clue from the fact that the "subspace" really consists of a single state (e.g., when one say something like "projector onto a subspace orthogonal to $|a_i\rangle$"). But even when you encounter a phrase like "$\hat{P}_{a_i}$ is the projector onto the eigenspace of A with eigenvalue a_i," you should start with the working assumption that $\hat{P}_{a_i}$ here just means one term $\hat{P}_{a_i} = |a_i\rangle\langle a_i|$ rather than the sum of such terms. It is usually possible to see from the context what type of projectors is implied in each specific situation.

25.2
Projective Measurements

The projection operator formalism describes the most common type of measurement in QM called *projective* or *von Neumann measurements*.[1] These are the measurements where the observable A is measured in an orthonormal basis composed of eigenstates of $\hat{A}$. The measurement projects the system onto one of these eigenstates (with probability given by the square of the length of the projection) and returns the corresponding eigenvalue as shown by Equations 25.4 and 25.11.

Before the measurement is performed, one cannot predict (except probabilistically) which one of the N eigenstates the system will choose to collapse into. So if we are given a single projector $\hat{P}_{a_i}$, we will only be able to tell the probability and the resulting postmeasurement state for only a single outcome a_i. To account for all N possible outcomes, in other words, to describe the nature of the measurement completely, we need to have all N projectors. The complete set of operators $\{\hat{P}_{a_i}\}$ written in either ket–bra or matrix form will tell us all that we need to know about the measurement. For example, we can find the expectation value of A:

$$\langle A \rangle = \sum_i a_i p(a_i) = \sum_i a_i \langle \psi | \hat{P}_{a_i}^+ \hat{P}_{a_i} | \psi \rangle = \langle \psi | \sum_i a_i \hat{P}_{a_i} | \psi \rangle = \langle \psi | \hat{A} | \psi \rangle. \tag{25.20}$$

The last step follows from the spectral decomposition theorem. We have recovered the result $\langle A \rangle = \langle \psi | \hat{A} | \psi \rangle$ that is familiar to every student of QM, using only the projection operator formalism. In particular, the special case of (25.20),

$$\langle \hat{P}_{a_i} \rangle = \langle \psi | \hat{P}_{a_i} | \psi \rangle, \tag{25.21}$$

convinces us that probability of an outcome is identical to the expectation value of the projector:

$$p(a_i) = \langle \hat{P}_{a_i} \rangle. \tag{25.22}$$

We can also apply (25.18) and (25.19) to find the probabilities of outcomes and the final states. In particular, by imagining a large ensemble of systems, we can describe the measurement in the language of density matrices. Since a postmeasurement state $|\psi'\rangle = (\hat{P}_{a_i}|\psi\rangle)/\left(\sqrt{\langle \psi | \hat{P}_{a_i} | \psi \rangle}\right)$ occurs with probability $p(a_i) = \langle \psi | \hat{P}_{a_i}^+ \hat{P}_{a_i} | \psi \rangle$, the projective measurement on a *pure* system will result in the following mixed state:

---

1) The relation between projection operators and projective measurements merits a short remark about the terminology we are using. Even though formula $\hat{P}_{a_i} = |a_i\rangle\langle a_i|$ that defines projection operator has been introduced for the orthonormal basis (25.1), most physicists will still call an expression of the form $|a_i\rangle\langle a_i|$ a "projection operator" onto the direction $|a_i\rangle$ even if the basis $\{|a_i\rangle\}$ is nonorthogonal. It is OK to speak of "projection operators" in such a case as long as one keeps in mind that the measurement in question is not a projective measurement.

$$\xi = \sum_i p(a_i)|\psi'\rangle\langle\psi'| = \sum_i \langle\psi|\hat{P}_{a_i}|\psi\rangle \frac{\hat{P}_{a_i}|\psi\rangle\langle\psi|\hat{P}_{a_i}}{\left(\sqrt{\langle\psi|\hat{P}_{a_i}|\psi\rangle}\right)^2} = \sum_i \hat{P}_{a_i}|\psi\rangle\langle\psi|\hat{P}_{a_i}.$$

(25.23)

This equation should be used in cases when we are not told the outcome of the projective measurement, but we do know that the measurement has taken place.

Suppose now the initial state was itself some mixture ρ,

$$\rho = \sum_j w_j |\psi_j\rangle\langle\psi_j|.$$

(25.24)

Then the net probability of outcome a_i is found by taking the trace in one of the three ways:

$$p(a_i) = \sum_j w_j p_j(a_i) = \sum_j w_j \langle\psi_j|\hat{P}_{a_i}^+\hat{P}_{a_i}|\psi_j\rangle$$

$$= \begin{cases} \mathrm{Tr}\sum_j w_j \hat{P}_{a_i}^+\hat{P}_{a_i}|\psi_j\rangle\langle\psi_j| = \mathrm{Tr}\left(\hat{P}_{a_i}^+\hat{P}_{a_i}\rho\right) \\ \mathrm{Tr}\sum_j w_j \hat{P}_{a_i}|\psi_j\rangle\langle\psi_j|\hat{P}_{a_i}^+ = \mathrm{Tr}\left(\hat{P}_{a_i}\rho\hat{P}_{a_i}^+\right), \\ \mathrm{Tr}\sum_j w_j |\psi_j\rangle\langle\psi_j|\hat{P}_{a_i}^+\hat{P}_{a_i} = \mathrm{Tr}\left(\rho\hat{P}_{a_i}^+\hat{P}_{a_i}\right) \end{cases}$$

(25.25)

where $p_j(a_i)$ is the probability that outcome a_i will occur from the *pure* state $|\psi_j\rangle$.

Let us find the resulting density matrix ρ_i provided that outcome a_i was obtained in a measurement on ρ. Applying (25.23) to each $|\psi_j\rangle$ and summing over j gives

$$\tilde{\rho}'_i = \sum_j p_j(a_i)|\psi'_j\rangle\langle\psi'_j| = \left(\sum_j p_j(a_i) \frac{\hat{P}_{a_i}|\psi_j\rangle\langle\psi_j|\hat{P}_{a_i}^+}{\sqrt{p_j(a_i)}\sqrt{p_j(a_i)}}\right) = \hat{P}_{a_i}\rho\hat{P}_{a_i}^+. \quad (25.26)$$

We put the tilde over the resulting matrix to indicate that it is not normalized. Indeed, its norm is not equal to 1 because

$$\sum_j p_j(a_i) \equiv p(a_i) \leq 1.$$

(25.27)

Thus, to normalize ρ, we divide it by the probability (25.27):

$$\rho'_i = \frac{\hat{P}_{a_i}\rho\hat{P}_{a_i}^+}{p(a_i)}.$$

(25.28)

Equations 25.25–25.28 give the most general rule for the transformation of a system under a projective measurement.

Thus, if our goal is to describe the complete statistics of outcomes, all we need to know in addition to the wave function $|\psi\rangle$ and the observable eigenvalues a_i is the set of projectors $\{\hat{P}_{a_i}\}$. This set is said to *define* the corresponding projective measurement.

Note that once we have this information, we can always switch back to the familiar version of QM and calculate things "the old way." The spectral theorem $\hat{A} = \sum_i a_i \hat{P}_{a_i}$ will give us the corresponding operator $\hat{A}$ and from $|a_i\rangle = (\hat{P}_{a_i}|\psi\rangle)/\left(\sqrt{\langle\psi|\hat{P}_{a_i}|\psi\rangle}\right)$, we can always find the eigenstates. (This is useful in case the projectors were given to us in the matrix form!). The probability amplitudes can then be found directly as $\langle a_i|\psi\rangle$ or as $(\hat{P}_{a_i}|\psi\rangle)/|a_i\rangle$. This procedure will take the problem back to the conventional form.

One important property of projective measurements is that the number of possible outcomes cannot exceed the dimensionality of the Hilbert space. This is clear from the fact that we associate each basis state $|a_i\rangle$ with a separate outcome, so we cannot have more outcomes than the number of basis states.

Box 25.1 Spectral theorem for density operators

The spectral decomposition theorem, $\hat{A} = \sum_i a_i \hat{P}_{a_i}$, applies for *any* positive semidefinite Hermitian operator. Since the density matrix operator is both Hermitian and positive semidefinite, we can always decompose it as

$$\rho = \sum_i v_i \hat{P}_{v_i} = \sum_i v_i |v_i\rangle\langle v_i|, \tag{25.29}$$

where v_i are the eigenvalues of ρ and vectors $\{|v_i\rangle\}$ form an orthonormal basis. One can easily show that ρ is diagonal in that basis, and that eigenvalues v_i are nonnegative.

Therefore, a density matrix can always be diagonalized:

$$\rho = \begin{pmatrix} v_1 & 0 & \cdots & 0 \\ 0 & v_2 & \cdots & 0 \\ 0 & 0 & \cdots & v_n \end{pmatrix} \tag{25.30}$$

and from the trace condition for density matrices, it follows that $\text{Tr}\,\rho = \sum_i v_i = 1$. We want to stress that eigenstates $\{|v_i\rangle\}$ should never be confused with the pure states comprising the density matrix: In terms of pure states, this matrix will be written as $\rho = \sum_j w_j |\psi_j\rangle\langle\psi_j|$ where pure states $\{|\psi_j\rangle\}$ are generally different from the eigenstates $\{|v_i\rangle\}$.

To illustrate the importance of the spectral theorem, let us consider the problem of finding the *matrix exponential* e^ρ, which is defined as the Taylor series, $e^\rho = \sum_{k=0}^\infty \rho^k/k!$. If we chose an inconvenient basis in which ρ is not diagonal, then we must perform an infinite number of matrix operations. However, the spectral theorem guarantees the existence of a diagonal representation (25.30).

Then, working in this new basis, we can immediately see that

$$\rho^k = \begin{pmatrix} v_1^k & 0 & \cdots & 0 \\ 0 & v_2^k & \cdots & 0 \\ 0 & 0 & \cdots & v_n^k \end{pmatrix}$$

and therefore

$$e^\rho = \sum_{k=0}^{\infty} \frac{\rho^k}{k!} = \begin{pmatrix} e^{v_1} & 0 & \cdots & 0 \\ 0 & e^{v_2} & \cdots & 0 \\ 0 & 0 & \cdots & e^{v_n} \end{pmatrix}.$$

Having obtained the matrix exponential in this way, we can then return to the old basis. This result has an important corollary that will be used later in the discussion of quantum entropy. Let us define the *matrix logarithm* as the inverse operation of the exponential; in other words, $\sigma = \ln \rho$, when $e^\sigma = \rho$. Then it is easy to see that

$$\sigma = \begin{pmatrix} \ln v & 0 & \cdots & 0 \\ 0 & \ln v_2 & \cdots & 0 \\ 0 & 0 & \cdots & \ln v_n \end{pmatrix}.$$

Therefore, in order to find the logarithm of a matrix, we first diagonalize that matrix and then take the logarithm of each diagonal element.

25.3
POVMs

Projective measurements require that unit vectors $|a_i\rangle$ along which the outcomes are being measured must form a complete orthogonal set. It turns out that this type of measurements is not always the best choice. There exists a whole range of problems such as pure state discrimination or joint measurement on several qubits where it is more advantageous to use a more general measurement procedure that tries to detect outcomes along the unit vectors $|a_i\rangle$ where the set $\{|a_i\rangle\}$ is *not* orthogonal. These measurements are known by the acronym POVM that stands for positive operator-valued measure. The POVM formalism modifies the concepts from the previous section to adapt them for the nonorthogonal case.

Consider the measurement scheme shown in Figure 25.1. As before, we are interested in the outcome statistics for observable A. Suppose the system was prepared in state $|a_1\rangle$ or $|a_2\rangle$. But this time we intend to perform a measurement that will leave the system in one of the three states $|b_1\rangle, |b_2\rangle, |b_3\rangle$. If these states were orthogonal, we could accomplish this with a projective measurement $\{\hat{P}_{b_i}\} = \{|b_1\rangle\langle b_1|, |b_2\rangle\langle b_2|, |b_3\rangle\langle b_3|\}$. However, in this example, the Hilbert space of the system is only two dimensional, that is, any two orthogonal vectors will form

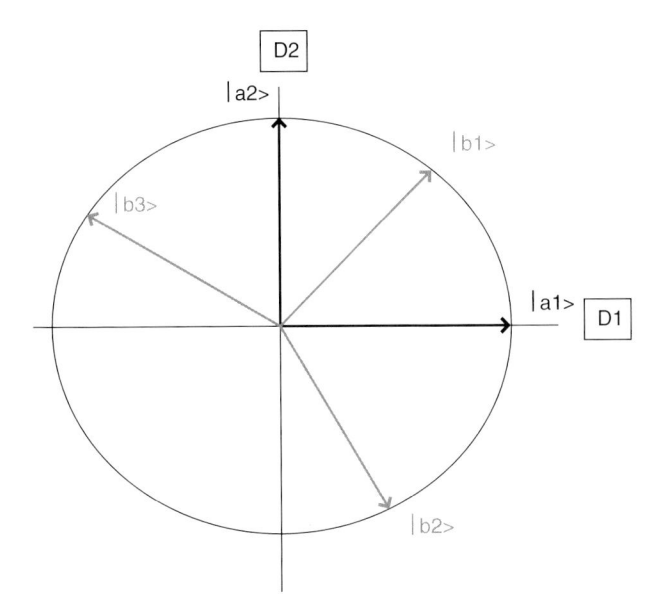

Figure 25.1 Two-dimensional Hilbert space $\{|a_1\rangle, |a_2\rangle\}$ makes it impossible to have three mutually orthogonal states $|b_1\rangle, |b_2\rangle, |b_3\rangle$. One cannot use more than two detectors at the same time.

a complete basis for this space. If we try to squeeze in a third b-vector, the resulting set will necessarily be nonorthogonal. This is generally the kind of situation that motivates the POVM: We want the ability to detect three outcomes, but the system only allows for two outcomes at a time. It must be already clear that the measurement in this case cannot be projective. We will therefore represent this measurement by a set of operators $\{\hat{F}_{b_i}\}$, which, generally speaking, are not identical to the orthogonal projectors $\{\hat{P}_{b_i}\}$. Let us see now which properties of projective measurements will hold in this case and which need to be adapted to the new situation.

First of all, whatever we measure now, generally, is *not* the value of observable A in one of the states $|b_1\rangle, |b_2\rangle, |b_3\rangle$. There are several ways to explain this. One way is to simply say that such values would result from a projective measurement (if such a measurement were possible), while our measurement is of a different type. Alternatively, we can note that measuring these values would mean asking the system a set of incompatible questions. One can ask an electron two simultaneous questions: (i) Is your spin pointing up? (ii) Is your spin pointing down? But one cannot add the third question, "Is your spin pointing to the right?" at the same time. This is impossible because the corresponding operators do not commute. Therefore, it would be a mistake to think of this POVM as a collection of "questions" $\{|b_i\rangle\langle b_i|\}$, that is, as a set of projectors $\{\hat{P}_{b_i}\}$. Finally, to convince yourself that the system cannot take all these questions at once, imagine as an illustration that we try to measure the spin state with a Stern–Gerlach apparatus. Obviously, we can only have one direction of magnetic field at any given point, so the quantization axis as well as the field gradient is well defined. The electron can only choose between deflecting along the gradient or against the gradient, so there are just two possible landing

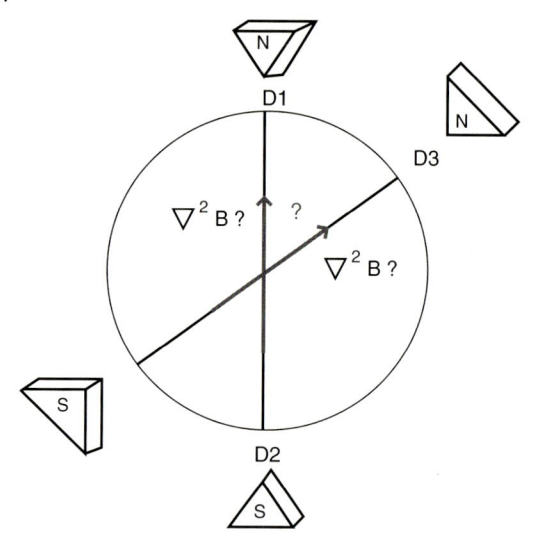

Figure 25.2 Electron spin measurement with a Stern–Gerlach apparatus is a projective measurement in $\mathcal{H}^2$-space. The two possible outcomes (spin-up and spin-down) are registered by detectors D_1 and D_2. The arrow shows magnetic field gradient. Any attempt to insert a third detector D_3 would be equivalent to asking the field to have *two* different gradients at the same point, which clearly cannot happen.

points on the screen and the experiment will only resolve two orthogonal spin states. Then again we can try to ask two questions first and the third one later, but that would be two different measurements that require two different apparatuses. So one should never try to visualize a POVM by imagining a series of detectors as in Figure 25.1 placed along directions $|b_i\rangle$. Figure 25.2 illustrates this conceptual fallacy by showing a failed attempt at spin measurement using three detectors.

Second, one may well ask: "But if we cannot do that, then how does one implement a POVM at all?" The answer is that instead of working with an isolated system, we are going to couple it with another system called the *ancilla* and then let the joint system evolve in time until the states that were initially nonorthogonal become orthogonal in the larger Hilbert space of the joint system. In this way, a POVM reduces to a projective measurement on a different Hilbert space. (However, $\{\hat{F}_{b_i}\}$ still describes the final effect on the given system only, ignoring the ancilla!).[2] But this achievement comes at a price: As shown in Figure 25.3, an operator associated with direction $|b_i\rangle$ is physically implemented by a detector set up along a different direction $|B_i\rangle$. A conceptual "measuring operator" is no longer coincident with the actual apparatus carrying out the measurement!

Third, this feature of POVMs underscores the following basic distinction: Projective measurements are repetitive, but POVMs are not. When a projective

---

2) The effect on the ancilla will depend on the way in which the POVM was implemented. Generally, there are infinitely many possible implementations. Anyway, the final state of the ancilla is interesting to us only as a way to get a glimpse of the system's state.

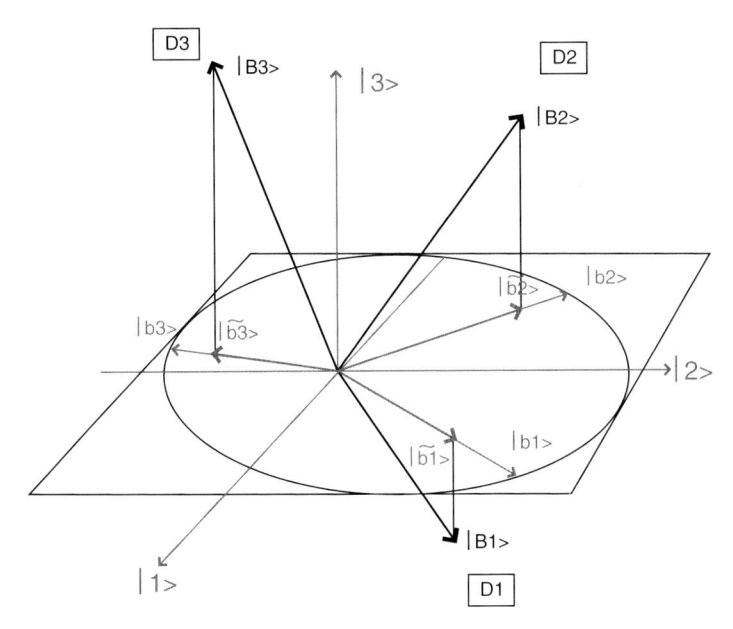

Figure 25.3 Three orthogonal detectors can only be squeezed in by "raising" them above a 2D plane, placing them in a 3D or higher dimensional space. Extension of the system's 2D $\mathcal{H}^2$-space is achieved by combining the system with an ancilla. The resulting space is generally multidimensional; for illustrative purposes, only three dimensions are shown. The initial state of the joint system is in a plane spanned by vectors $\{|1\rangle, |2\rangle\}$ and orthogonal to vector $|3\rangle$. The gray arrows show directions in that plane that define a given POVM. A unitary transformation of the joint system is, from the system's viewpoint, equivalent to a rotation of coordinate axis. If such a rotation will transform the unit vectors according to $\{|1\rangle, |2\rangle, |3\rangle\} \Rightarrow \{|B_1\rangle, |B_2\rangle, |B_3\rangle\}$, then we have the physical implementation of this POVM.

measurement collapses the system into one of its eigenstates, all subsequent applications of that measurement will leave it in the same eigenstate.[3] In contrast, a POVM may correctly indicate the initial state $|a_i\rangle$, but the system itself is already in state $|b_j\rangle$. When the next application of this POVM indicates state $|b_j\rangle$, the system will already be in some new state, and so on. (This by itself would amount to a von Neumann measurement of the 2nd kind, but in addition as we will see later, a POVM will sometimes *fail* to indicate the state, in which case we say the measurement outcome was *inconclusive*.)

Fourth, the reader can now see why physicists use the word "outcome" instead of "eigenvalue" or "eigenstate." The outcome of a POVM is the retrospective identification of the preparation state as $|a_i\rangle$ or, in some cases, the outcome is failure to identify the state. However, $|a_i\rangle$ is generally not an eigenstate of observable A. (Usually states $|a_i\rangle$ are nonorthogonal and therefore they cannot *all* be eigenstates.) If it is not an eigenstate, it leaves the value of the observable in that state

---

3) From the formal viewpoint, this is the consequence of idempotence and mutual orthogonality of projection operators.

indeterminate. We will also know the postmeasurement state $|b_j\rangle$, which in most problems we don't really care about;[4] and in the event it coincides with one of the eigenstates, we will know the postmeasurement value of A, which again is usually of little practical importance.

Fifth, since POVM is actually also a projective measurement, only on a larger system, it is natural that some basic properties of projective measurements must hold. For one, the probabilities of outcomes must still add to 1. This is a natural requirement since each of the "nonorthogonal outcomes" $|b_i\rangle$ corresponds to an outcome of a projective measurement on the entangled bipartite system. Therefore, we will require that the equivalent of the closure relation must hold for POVMs:

$$\sum_i \hat{F}_{b_i} = 1. \tag{25.31}$$

Sixth, the same line of reasoning will also convince us that the fundamental property 25.7 of projective measurements must also be retained:

$$p(b_i) = \langle \psi | \hat{F}_{b_i} | \psi \rangle. \tag{25.32}$$

Seventh, since probabilities are obviously nonnegative, all operators $\hat{F}_{b_i}$ making up a POVM – we call them *POVM elements* – are positive semidefinite.

The last property together with the closure relation is sufficient to *define* a general POVM. Thus, a POVM is defined as a set of positive semidefinite operators $\{\hat{F}_{b_i}\}$ that partition unity. Once we are given such a set, we can determine the corresponding outcomes and their probabilities according to (25.32), and it also possible to show that the physical implementation (i.e., the right ancilla and the suitable transformation of the resulting bipartite system) will always be available.

As we said earlier, as long as our attention is restricted to the system only, POVM elements $\hat{F}_{b_i}$ are not projection operators, so $\hat{F}_{b_i} \neq |b_i\rangle\langle b_i|$. Fortunately, they are closely related and finding them will only require a small adjustment. Let us multiply each "quasiprojector" $|b_i\rangle\langle b_i|$ by a corresponding nonnegative number λ_i. Then by a suitable choice of λ_i, we can compose a POVM:

$$\hat{F}_{b_i} = \lambda_i |b_i\rangle\langle b_i|. \tag{25.33}$$

Equation 25.33 will easily satisfy the POVM definition. The positivity requirement is satisfied automatically as long as we have

$$\langle \psi | \hat{F}_{b_i} | \psi \rangle = \langle \psi | \lambda_i | b_i \rangle \langle b_i | \psi \rangle = \lambda_i |\langle b_i | \psi \rangle|^2 \geq 0. \tag{25.34}$$

The closure relation will be satisfied if we require that

$$\sum_i \hat{F}_{b_i} = \sum_i \lambda_i |b_i\rangle\langle b_i| = I. \tag{25.35}$$

The restrictions on λ_i follow from (25.34). Since $p(b_i) = \langle \psi | \hat{F}_{b_i} | \psi \rangle$ has the meaning of probability, it cannot become greater than 1. The inner product $\langle b_i | \psi \rangle$ can become

---

4) The only practical role of $|b_j\rangle$ is to help identify the initial state $|a_i\rangle$ and thereafter the system is usually discarded.

as large as 1 when $|\psi\rangle$ is collinear with $|b_i\rangle$ (both are assumed to have been normalized). Therefore, 1 is the upper bound for all λ_i:

$$0 \leq \lambda_i \leq 1. \tag{25.36}$$

The number of possible POVMs is equal to the number of sets $\{\lambda_i\}$ satisfying (25.35). The case $\lambda_i = 1$ corresponds to the projective measurement.

Once a POVM has been constructed, we can simplify it by absorbing the constant multiplier into the state function:

$$\left|\tilde{b}_i\right\rangle = \sqrt{\lambda_i}|b_i\rangle. \tag{25.37}$$

Then, the POVM takes a simple form that should be already familiar to the reader from the discussion of unitary freedom for density matrices:

$$\hat{F}_{b_i} = \left|\tilde{b}_i\right\rangle\left\langle\tilde{b}_i\right|. \tag{25.38}$$

This similarity is not just a coincidence. When we act on the pure state $|\psi\rangle$ with the POVM defined by $\{\hat{F}_{b_i}\}$, the result is a mixture formed by pure states $|b_i\rangle$ with respective weights determined by (25.34):

$$w_i = \langle\psi|\hat{F}_{b_i}|\psi\rangle = \lambda_i|\langle b_i|\psi\rangle|^2. \tag{25.39}$$

This results in the following density matrix for a POVM:[5]

$$\xi_{\text{POVM}} = \sum_i w_i|b_i\rangle\langle b_i| = \sum_i \lambda_i|\langle b_i|\psi\rangle|^2|b_i\rangle\langle b_i|. \tag{25.40}$$

Compare this with the density matrix (see Equation 25.23) that we would obtain in the case of a projective measurement:

$$\xi = \sum_i w_i|b_i\rangle\langle b_i| = \sum_i \hat{P}_{b_i}|\psi\rangle\langle\psi|\hat{P}_{b_i}^\dagger = \sum_i |\langle b_i|\psi\rangle|^2|b_i\rangle\langle b_i|. \tag{25.41}$$

Coefficients λ_i acquire a simple physical meaning: They attenuate the relative weights of outcomes – pure states $|b_i\rangle$ – in this mixture in cases when we have more outcomes than that can be accommodated by dimensionality of the Hilbert space. Conceptually, what happens is this: In a projective measurement, wave function $|\psi\rangle$ of the system chooses to collapse into state $|b_i\rangle$ with probability $|\langle b_i|\psi\rangle|^2$. The sum of these probabilities is $\sum_i|\langle b_i|\psi\rangle|^2 = 1$ and states $|b_i\rangle$ are orthogonal and span the Hilbert space. Suppose now that we try to add one more state $|b_j\rangle$, which forms an acute or obtuse angle with some of the states $|b_i\rangle$; in other words, $\langle b_j|b_i\rangle \neq 0$. This state is now going to "grab" some percentage of the outcomes. But since the total sum is limited to 100%, it will have to take some outcomes away from the other states. This means that each state is going to get a

---

[5] To avoid confusion, we use ξ for the density matrix after the measurement. It represents the mixture that forms after *all* POVM elements were given their chance to act on the input state. In contrast, we will be using ρ when discussing the effect of an *individual* POVM element $\hat{F}_{b_i}$ on the input state.

smaller piece of the pie than it seems to be entitled to if you just look at the value of the inner product. The function of multipliers λ_i is to modify each share $|\langle b_i|\psi\rangle|^2$ accordingly to make room for the newcomer. Just like in real life, such redistribution can be achieved in multiple ways. While the simplest way is just to spread the sacrifices equally:

$$\lambda_1 = \lambda_1 = \cdots = \lambda_N = \frac{1}{\sum_i |\langle b_i|\psi\rangle|^2}, \qquad (25.42)$$

they can also be made disproportional, with some losing smaller share than others. The only thing limiting this real-life analogy is that it is not possible for any state $|b_i\rangle$ to *gain* at the expense of the others: λ_i cannot be greater than 1.

25.4
POVM as a Generalized Measurement

As we have seen, a set of nonorthogonal unit vectors $|b_i\rangle$ defines a POVM $\{\hat{F}_{b_i}\}$ with elements $\hat{F}_{b_i} = |\tilde{b}_i\rangle\langle\tilde{b}_i| = \lambda_i|b_i\rangle\langle b_i|$. The fact that there is no orthogonality condition for $\hat{F}_a$ similar to the requirement $\hat{P}_{a_i}\hat{P}_{a_j} = \delta_{i,j}\hat{P}_{a_i}$ for projection operators is the distinguishing characteristic of POVMs. In particular, it explains why POVMs are not repetitive: Even as one measurement collapses the system into some state $|b_i\rangle$, the next measurement can change that – There is a finite probability that the system will respond to one of the elements $\hat{F}_{b_j}, j \neq i$.

The reader remembers that projective operators had the property $\hat{P}_{b_i}\hat{P}_{b_i} = \hat{P}_{b_i}$. The condition $0 < \lambda_i \leq 1$ means that we cannot write a similar identity for POVMs. Instead, we have

$$\hat{F}_{b_i}\hat{F}_{b_i} = \lambda_i^2|b_i\rangle\langle b_i|b_i\rangle\langle b_i| = \lambda_i\hat{F}_{b_i}. \qquad (25.43)$$

However, the hermiticity property still holds for POVM elements because λ_i are real:

$$\hat{F}_{b_i}^+ = \lambda_i^+|b_i\rangle\langle b_i| = \hat{F}_{b_i}. \qquad (25.44)$$

As a consequence of (25.43) and (25.44),

$$\hat{F}_{b_i} = \frac{1}{\lambda_i}\hat{F}_{b_i}^+\hat{F}_{b_i} = \left(\frac{\hat{F}_{b_i}}{\sqrt{\lambda_i}}\right)^+\left(\frac{\hat{F}_{b_i}}{\sqrt{\lambda_i}}\right). \qquad (25.45)$$

In the special case $\lambda_i = 1$, this reduces to the familiar result for projective operators:

$$\hat{P}_{b_i} = \left(\hat{P}_{b_i}\right)^+\left(\hat{P}_{b_i}\right). \qquad (25.46)$$

The probabilities of outcomes can now be written most generally as

$$p(b_i) = \langle\psi|\left(\frac{\hat{F}_{b_i}}{\sqrt{\lambda_i}}\right)^+\left(\frac{\hat{F}_{b_i}}{\sqrt{\lambda_i}}\right)|\psi\rangle. \qquad (25.47)$$

Let us also determine the postmeasurement state of the system. Suppose the system chose to respond to operator $\hat{F}_{b_i}$ so that $\hat{F}_{b_i}|\psi\rangle = \lambda_i|b_i\rangle\langle b_i|\psi\rangle$. Since we must end up with the normalized state $|b_i\rangle$, we therefore have

$$|\psi'\rangle = \frac{\hat{F}_{b_i}|\psi\rangle}{\lambda_i\langle b_i|\psi\rangle}, \tag{25.48}$$

which is the POVM analogue of (25.8). But as noted earlier and expressed by (25.40), the expression $\lambda_i|\langle b_i|\psi\rangle|^2$ corresponds to the probability of outcome $|b_i\rangle$, which is also given by (25.47). Therefore,

$$|\psi'\rangle = \frac{\left(\hat{F}_{b_i}/\sqrt{\lambda_i}\right)|\psi\rangle}{\sqrt{p(b_i)}} = \frac{\left(\hat{F}_{b_i}/\sqrt{\lambda_i}\right)|\psi\rangle}{\sqrt{\langle\psi|\left(\hat{F}_{b_i}/\sqrt{\lambda_i}\right)^+\left(\hat{F}_{b_i}/\sqrt{\lambda_i}\right)|\psi\rangle}}. \tag{25.49}$$

Equations 25.47 and 25.49 generalize Equations 25.17 and 25.18 and reduce to them when $\lambda_i = 1$. It is natural to think of the operators $\left(\hat{F}_{b_i}/\sqrt{\lambda_i}\right)^+$ and $\left(\hat{F}_{b_i}/\sqrt{\lambda_i}\right)$ as the square roots[6] of the operator $\left(\hat{F}_{b_i}/\sqrt{\lambda_i}\right)^+\left(\hat{F}_{b_i}/\sqrt{\lambda_i}\right)$, much in the same way as we think of functions ψ^* and ψ as square roots of the function $|\psi|^2 = \psi^*\psi$. Then, if we denote

$$\hat{M}_i \equiv \left(\frac{\hat{F}_{b_i}}{\sqrt{\lambda_i}}\right) \quad \text{and} \quad \hat{M}_i^+ \equiv \left(\frac{\hat{F}_{b_i}}{\sqrt{\lambda_i}}\right)^+, \tag{25.50}$$

the POVM measurement will take a familiar form:

$$p(b_i) = \langle\psi|\hat{M}_i^+\hat{M}_i|\psi\rangle, \tag{25.51}$$

$$|\psi'\rangle = \frac{\hat{M}_i|\psi\rangle}{\sqrt{p(b_i)}} = \frac{\hat{M}_i|\psi\rangle}{\sqrt{\langle\psi|\hat{M}_i^+\hat{M}_i|\psi\rangle}}, \tag{25.52}$$

$$p(b_i) = \sum_j w_j p_j(b_i) = \text{Tr}\left(\hat{M}_{b_i}^+\hat{M}_{b_i}\rho\right) = \text{Tr}\left(\hat{M}_{b_i}\rho\hat{M}_{b_i}^+\right) = \text{Tr}\left(\rho\hat{M}_{b_i}^+\hat{M}_{b_i}\right). \tag{25.53}$$

---

6) In linear algebra, matrix B is said to be a *square root* of matrix A if the square of B is equal to A, that is, $BB = A$. If the matrices are defined in a Hilbert space, then we must tweak this definition a little by replacing the first matrix in the product with a Hermitian adjoint: $B^+B = A$. We also remember that a matrix is equivalent to an operator. Thus, the operator equation $\hat{B}^+\hat{B} = \hat{A}$ defines the square root of operator $\hat{A}$ as $\sqrt{\hat{A}} = \hat{B}$ or $\sqrt{\hat{A}} = \hat{B}^+$. In general, an operator can have many square roots. This can be seen from the fact that any operator $\hat{B} = \hat{U}\sqrt{\hat{A}}$, where $\hat{U}$ is unitary, will also be the square root because of the defining property of the unitary operator, $\hat{B}^+\hat{B} = \left(\hat{U}^+\hat{U}\right)\hat{A} = \left(\hat{U}^{-1}\hat{U}\right)\hat{A} = \hat{A}$. Therefore, the root is generally known up to an arbitrary "multiplier" $\hat{U}$. However, one can show that a positive-definite operator has only one positive-definite root, which is called *principal square root* (Problem 25.4).

$$\tilde{\rho}'_i = \hat{M}_{b_i} \rho \hat{M}_{b_i}^+, \qquad (25.54)$$

$$\rho'_i = \frac{\tilde{\rho}'_i}{p(b_i)}, \qquad (25.55)$$

with

$$\hat{M}_i \equiv \sqrt{\hat{M}_i^+ \hat{M}_i} = \sqrt{\frac{\hat{F}_{b_i}^+ \hat{F}_{b_i}}{\lambda_i}} = \sqrt{\hat{F}_{b_i}}. \qquad (25.56)$$

The last equation follows immediately from (25.45). Operators $\hat{M}_{b_i}$, $\hat{M}_{b_i}^+$ are known as *Kraus operators, operation elements, decomposition operators*, or sometimes as *noise operators* or *error operators* when the operation in question is due to environmental interactions.

Equations 25.51–25.55 are similar to their projective analogues[7] and are often said to *define* the POVM measurement. In fact, it is possible to construct an axiomatic formulation of QM by taking (25.51–25.53) as a postulate. This top–down approach was taken, for example, in the well-known Ref. [1]. Here we wanted to show how this formalism follows from the "ordinary" rules of QM. Once the student has a solid grasp of the concepts leading to (25.51–25.53), these equations can well be taken as the starting point for further study.

25.5
POVM Examples

Equation 25.35 gives us a general recipe for constructing a POVM out of a given set of N pure states, according to

$$\hat{F}_{b_i} = \lambda_i |b_i\rangle \langle b_i|. \qquad (25.57)$$

However, in each specific situation, making the right choice λ_i constitutes a separate problem. We always have to restrict the range of λ_i to (25.36) and then also ensure that POVM elements partition unity:

$$\sum_{i=1}^{N} \hat{F}_{b_i} = \sum_{i=1}^{N} \lambda_i |b_i\rangle \langle b_i| = I. \qquad (25.58)$$

Even then, there will generally be an infinite number of possible solutions, corresponding to different measurements one can perform on the given system. The optimal POVM will be chosen based on the desired characteristics of such a measurement.

For example, suppose we want to make a POVM measurement of electron spin. Let the N spin states $|b_i\rangle$ appearing in (25.58) be defined by respective unit vectors $\vec{b}_i$. Recalling the discussion leading to (24.102), we can write for such a state

$$\frac{I + \vec{b} \cdot \vec{\sigma}}{2} = |\vec{b}\rangle \langle \vec{b}|, \qquad (25.59)$$

---

[7] Projective measurements are sometimes called projection value measures (PVMs) since they follow from POVMs as a limiting case.

where $\sigma_x, \sigma_y, \sigma_z$ are the Pauli matrices and $\vec{\sigma} = \vec{i}\sigma_x + \vec{j}\sigma_y + \vec{k}\sigma_z$. As a result, the general requirement (25.58) applied to electron spin states can be written as follows:

$$\sum_{i=1}^{N} \lambda_i \frac{I + \vec{b}_i \cdot \vec{\sigma}}{2} = I. \tag{25.60}$$

Or, equivalently,

$$\sum_{i=1}^{N} \left(\frac{\lambda_i}{2}\right) I + \sum_{i=1}^{N} \left(\frac{1}{2}\lambda_i \vec{b}_i\right) \cdot \vec{\sigma} = I. \tag{25.61}$$

This condition will be met automatically if we choose the constants in such a way that

$$\sum_{i=1}^{N} \lambda_i = 2 \quad \text{and} \quad \sum_{i=1}^{N} \lambda_i \vec{b}_i = 0. \tag{25.62}$$

The number of POVMs of type (25.62) we can create by varying the directions of unit vectors and the respective set of λ_i is going to depend on N. Thus, for $N = 2$, Equation 25.62 gives us one single choice $\lambda_1 = \lambda_2 = 1$ and $\vec{b}_2 = -\vec{b}_1$. Note that the antiparallel unit vectors (such as spin-up and spin-down directions) correspond to orthogonal states: $\langle \vec{b}_2 | \vec{b}_1 \rangle = 0$. In this special case, the resulting POVM $\hat{F}_i = |\vec{b}_i\rangle\langle\vec{b}_i|$, $i = 1, 2$ reduces to an ordinary projective spin measurement as it should. But for larger N, we have much greater freedom in choosing the constants.

Of particular interest is the symmetric case where we satisfy the first equation (25.62) by taking $\lambda_1 = \lambda_2 = \cdots = \lambda_N = 2/N$. Then, the POVM becomes

$$\hat{F}_i = \frac{2}{N} |\vec{b}_i\rangle\langle\vec{b}_i|, \tag{25.63}$$

and the second equation (25.62) reduces to the statement that the sum of N unit vectors is zero. (In other words, not every choice of unit vectors allows for this symmetric solution!). For example, if $N = 3$, we have

$$\hat{F}_i = \frac{2}{3} |\vec{b}_i\rangle\langle\vec{b}_i|, \tag{25.64}$$

and the three unit vectors make angles of 120° with one another, for example, $\beta_1 = 0°$, $\beta_2 = 120°$, and $\beta_3 = 240°$ as shown in Figure 25.4a:

$$\vec{b}_1 = (0, 0, 1), \quad \vec{b}_2 = \left(\frac{\sqrt{3}}{2}, 0, -\frac{1}{2}\right), \quad \vec{b}_3 = \left(-\frac{\sqrt{3}}{2}, 0, -\frac{1}{2}\right). \tag{25.65}$$

While $\vec{b}_i$ is a unit vector, $|\vec{b}_i\rangle$ has the physical meaning of a spin state where spin $\vec{S}$ is directed along $\vec{b}_i$; this is the same state that we wrote in earlier sections of the book as $|\vec{S} \cdot \vec{b}_i\rangle$. It can be written as

$$|\vec{b}_i\rangle = \begin{pmatrix} \cos\frac{\beta_i}{2} \\ \sin\frac{\beta_i}{2} e^{i\alpha} \end{pmatrix}, \tag{25.66}$$

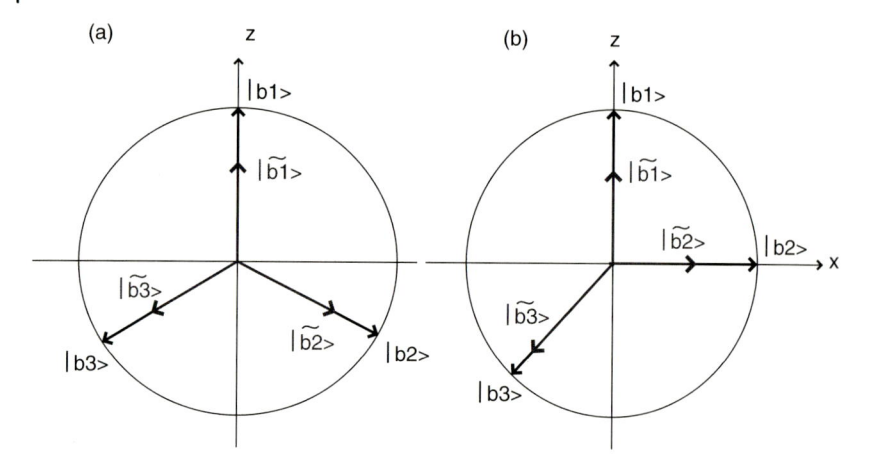

Figure 25.4 (a) $N=3$ symmetric $(\lambda_1 = \lambda_2 = \lambda_3 = 2/3)$ case where the z-axis was chosen to coincide with b_1 and b_1, b_2, b_3 lie in the xz-plane, that is, $\alpha = 0$ and $\beta = 0°, 120°,$ and $240°$. (b) POVM measurement of an electron spin given unit vectors $\vec{b}_1, \vec{b}_2, \vec{b}_3$ where $\vec{b}_1 \perp \vec{b}_2$ and $\lambda_1 = \lambda_2 = \lambda$. Note that the value of λ follows from normalization requirement for $\vec{b}_3$. As a result, in this case $\lambda = \sqrt{2}/1+\sqrt{2} \approx 0.5858$ and $\lambda_1 = \lambda_2 = \lambda,\ \lambda_3 = \sqrt{2}\lambda$.

where α and β are the azimuthal and polar angles, respectively. If we allow our three unit vectors to lie in the xz-plane (Figure 25.4a), making $\alpha = 0$, then

$$|\vec{b}_1\rangle = \begin{pmatrix} 1 \\ 0 \end{pmatrix}, \quad |\vec{b}_2\rangle = \begin{pmatrix} \frac{1}{2} \\ \frac{\sqrt{3}}{2} \end{pmatrix}, \quad |\vec{b}_3\rangle = \begin{pmatrix} -\frac{1}{2} \\ \frac{\sqrt{3}}{2} \end{pmatrix}. \tag{25.67}$$

We then have for POVM elements (25.64)

$$\hat{F}_i = |\tilde{b}_i\rangle\langle\tilde{b}_i| = \frac{2}{3}|b_i\rangle\langle b_i|, \tag{25.68}$$

where

$$|\tilde{b}_1\rangle = \begin{pmatrix} \sqrt{\frac{2}{3}} \\ 0 \end{pmatrix}, \quad |\tilde{b}_2\rangle = \begin{pmatrix} \sqrt{\frac{1}{6}} \\ \sqrt{\frac{1}{2}} \end{pmatrix}, \quad |\tilde{b}_3\rangle = \begin{pmatrix} -\sqrt{\frac{1}{6}} \\ \sqrt{\frac{1}{2}} \end{pmatrix}. \tag{25.69}$$

The resulting set $\{\hat{F}_1, \hat{F}_2, \hat{F}_3\}$ satisfies all mathematical requirements for POVMs. Again, we remind the reader that the physical realization of such a measurement *cannot* be achieved by simply setting three detectors along the directions $\beta_1 = 0°$, $\beta_2 = 120°$, and $\beta_3 = 240°$. The electron must be coupled with an ancilla, and the detectors will be measuring joint states in the enlarged Hilbert space. As a side note, phrases like "setting a detector along $|b_i\rangle$" are quite common in the literature; this is fine as long as one remembers not to take them too literally. All they mean is that the detector in question will realize the corresponding POVM element $\hat{F}_{b_i}$.

Consider another interesting case with $N = 3$ when $\vec{b}_1$ and $\vec{b}_2$ form a 90° angle:

$$\vec{b}_1 = \hat{z} = (0,0,1), \quad \vec{b}_2 = \hat{x} = (1,0,0), \quad \vec{b}_3 = \frac{1}{\lambda_1 + \lambda_2 - 2}\left(\lambda_1 \vec{b}_1 + \lambda_2 \vec{b}_2\right). \quad (25.70)$$

The last equation follows from (25.62). We see that Equation 25.70 gives us an infinite variety of ways to select λ_1 and λ_2, thus defining $\vec{b}_3$. For example, let us choose $\lambda_1 = \lambda_2 = \lambda$. Then, we have $\vec{b}_3 = -(\lambda/(2 - 2\lambda))(\vec{b}_1 + \vec{b}_2)$, and to make its norm equal to 1, we must require that $\lambda/(2 - 2\lambda) = 1/\sqrt{2}$, hence $\lambda = \sqrt{2}/(1 + \sqrt{2})$.

How do we write the corresponding POVM elements? The first two states are straightforward: These are just eigenstates with spin pointing in the z- and x-directions, respectively. The last state is given by the vector $\vec{b}_3 = -(1/\sqrt{2})(\vec{b}_1 + \vec{b}_2)$ – a vector for which $\beta = 225°$. Therefore,

$$\left|\vec{b}_1\right\rangle = |+\rangle, \quad \left|\vec{b}_2\right\rangle = \frac{1}{\sqrt{2}}(|+\rangle + |-\rangle),$$

$$\left|\vec{b}_3\right\rangle = \cos\frac{225°}{2}|+\rangle + \frac{\sin 225°}{2}|-\rangle = -\frac{1}{\sqrt{2}}\left(\sqrt{1 - \frac{1}{\sqrt{2}}}|+\rangle - \sqrt{1 + \frac{1}{\sqrt{2}}}|-\rangle\right). \quad (25.71)$$

(The reader can verify the last equation using the half-angle formulas.) With $\lambda_1 = \lambda_2 = \sqrt{2}/(1 + \sqrt{2})$ and $\lambda_3 = 2 - \lambda_1 - \lambda_2 = 2/(1 + \sqrt{2})$, we get the POVM shown in Figure 25.4b.

$$\hat{F}_1 = \lambda_1 \left|\vec{b}_1\right\rangle\left\langle\vec{b}_1\right| = \frac{\sqrt{2}}{1 + \sqrt{2}}|+\rangle\langle+|, \quad \hat{F}_2 = \lambda_2 \left|\vec{b}_2\right\rangle\left\langle\vec{b}_2\right| = \frac{\sqrt{2}}{1 + \sqrt{2}}\frac{(|+\rangle + |-\rangle)(\langle+| + \langle-|)}{2},$$

$$\hat{F}_3 = \lambda_3 \left|\vec{b}_3\right\rangle\left\langle\vec{b}_3\right| = \frac{2}{1 + \sqrt{2}}\left(\cos\frac{225°}{2}|+\rangle + \sin\frac{225°}{2}|-\rangle\right)\left(\cos\frac{225°}{2}\langle+| + \sin\frac{225°}{2}\langle-|\right). \quad (25.72)$$

Box 25.2 The last POVM element

As we have seen, listing the POVM elements $\hat{F}_{b_i} = \lambda_i |b_i\rangle\langle b_i|$ will tell the reader unambiguously which POVM has been chosen. In practice, however, many authors don't bother to list all N elements. Instead, they will list the first $N - 1$ elements, in view of the fact that the last element will be determined from the closure relation (25.58):

$$\hat{F}_{b_N} = I - \sum_{i=1}^{N-1} \hat{F}_{b_i}. \quad (25.73)$$

We personally feel that it's better to write the last element in the same form as the other elements. As physicists, we always want to be able to visualize that element by learning the corresponding direction $|b_N\rangle$. This direction is given explicitly by the ket–bra form $\hat{F}_{b_N} = \lambda_N |b_N\rangle\langle b_N|$. In contrast, the form (25.73) does not give us that direction, although we could, if necessary, do the matrix subtraction, obtain $\hat{F}_{b_N}$ in the matrix form, and then recover $|b_N\rangle$ (think how one would find the constant λ_N from this matrix!) Besides, Equation 25.73 often makes an unprepared reader wonder, "Is the last element somehow different from the rest since it was not written in the same ket–bra notation as the other elements?" But in fact, this form does not mean that the author intended this element to stand out from the crowd. Essentially, all it tells you is that the author just did not bother to derive $\hat{F}_{b_N} = \lambda_N |b_N\rangle\langle b_N|$ explicitly from (25.73) and leaves it to the reader to finish the job if the reader will find it worthwhile. Mathematically, no element of a POVM is "special." But physically, the last element often has a special meaning in that it corresponds to the "inconclusive" outcome of the measurement. This is the case, for example, in many state discrimination problems where the first $N - 1$ directions corresponding to conclusive outcomes are clear from the problem's condition, and the task is to construct an optimal POVM by making the best possible choice of the last remaining direction (the one for the inconclusive outcome!) and of the coefficients λ_i. In such cases one has no choice but to use (25.73) first while solving the optimization problem, and only then, after the optimal POVM has been found, one can, if one wants to do so, write the resulting direction explicitly.

25.6
Discrimination of Two Pure States

Consider the following problem. Two parties – Alice and Bob – have agreed to measure photon polarization in the default computational basis $\{|\leftrightarrow\rangle, |\updownarrow\rangle\}$. Alice prepares and then sends to Bob a photon in one of the two nonorthogonal polarization states $|a_1\rangle$ and $|a_2\rangle$ with angle θ between them:

$$|\langle a_1 | a_2 \rangle| = \cos\theta, \qquad (25.74)$$

where $\theta \neq \pm(\pi/2)$. The choice between $|a_1\rangle$ and $|a_2\rangle$ is made randomly so that each one occurs with a 50% probability. Bob's task is to determine the photon's state correctly.

It is natural to start the search for the optimal solution with ordinary projective measurements. However, it is clear at once that there is no way one could successfully discriminate these states 100% of the time. Any projective measurement must consist of a set of projectors $\{\hat{P}_{b_i}\}$ onto orthogonal directions $|b_i\rangle$. Namely, for a photon, such a measurement will be described by two projectors $\hat{P}_{b_1} = |b_1\rangle\langle b_1|$ and $\hat{P}_{b_2} = |b_2\rangle\langle b_2|$, where $\langle b_1|b_2\rangle = 0$. Clearly, the orthogonal directions $|b_1\rangle, |b_2\rangle$ cannot be aligned with nonorthogonal directions $|a_1\rangle, |a_2\rangle$. As a consequence, no matter how we arrange the two detectors, at least one of them will have a finite chance to register *either* of the two states so that a click in that detector will mean an inconclusive outcome.

We are typically interested in *unambiguous* discrimination of quantum states, where we may get inconclusive results quite often, but once we jump to a conclusion, this conclusion must be 100% certain.[8] In this case defining success and failure is easy. Let us call conclusive outcome a *success* and denote its probability by P; likewise, an inconclusive outcome will be termed a *failure* and its probability denoted by Q. Obviously, we are going to have $Q = 1 - P$.

The projective algorithm for unambiguous state discrimination works as follows. First, we introduce two (nonorthogonal) directions $|b_1\rangle$, $|b_2\rangle$ defined as follows:

$$|b_1\rangle = |a_2^\perp\rangle,$$
$$|b_2\rangle = |a_1^\perp\rangle, \qquad (25.75)$$

where, intuitively, $|a_1^\perp\rangle, |a_2^\perp\rangle$ denote the states orthogonal to $|a_1\rangle, |a_2\rangle$.

Suppose Bob makes a projective measurement along direction $|b_2\rangle$. This measurement has two possible outcomes, which will be captured by detectors D_1 and D_2, respectively. As shown in Figure 25.5, a click of detector D_1 is conclusive and identifies the photon state as $|a_2\rangle$. On the other hand, a click of detector D_2 does not tell us whether the photon state was $|a_1\rangle$ or $|a_2\rangle$ and is therefore inconclusive. In order for Bob to have a successful measurement, Alice must send the state $|a_2\rangle$ (which happens with probability $1/2$) and this state must collapse into $|b_2\rangle$ (which happens with probability $|\langle b_2|a_2\rangle|^2 = 1 - |\langle b_1|a_2\rangle|^2$). Therefore,

$$P = \frac{1}{2}(1 - |\langle b_1|a_2\rangle|^2) \quad \text{and} \quad Q = \frac{1}{2}(1 + |\langle b_1|a_2\rangle|^2) \qquad (25.76)$$

determine the odds of success in a projective measurement. The same conclusions will hold if Bob chooses direction $|b_1\rangle$; therefore, Equations 25.75 and 25.76 have the general character.

Let us now describe this experiment in terms of projection operators. Success results from operating with $\hat{P}_{b_2}$ on state $|a_2\rangle$:

$$\frac{\hat{P}_{b_2}|a_2\rangle}{\sqrt{\langle a_2|\hat{P}_{b_2}^\dagger \hat{P}_{b_2}|a_2\rangle}} = \frac{|b_2\rangle\langle b_2|a_2\rangle}{\sqrt{|\langle b_2|a_2\rangle|^2}} = |b_2\rangle. \qquad (25.77)$$

What happens when the same projector $\hat{P}_{b_2}$ acts on $|a_1\rangle$? It is easy to answer this question qualitatively: Since this state has zero probability to be registered by D_1, we must have $\langle a_1|\hat{P}_{b_2}|a_1\rangle = 0$, which means

$$\hat{P}_{b_2}|a_1\rangle = 0. \qquad (25.78)$$

---

8) Another facet of the same problem is discriminating the states with *minimum error* when we allow all our measurements to be inconclusive, but require that they give us a pretty good guess about the original state. We will not consider this case here; the interested reader will find a comprehensive analysis of this problem in Ref. [2]. Furthermore, our attention here will be limited to a subclass of the general problem where both states are prepared with equal probabilities and where both states are *pure* (the most general analysis of the problem should replace $|a_1\rangle$ and $|a_2\rangle$ with *mixed* states described by density matrices ρ_1 and ρ_2).

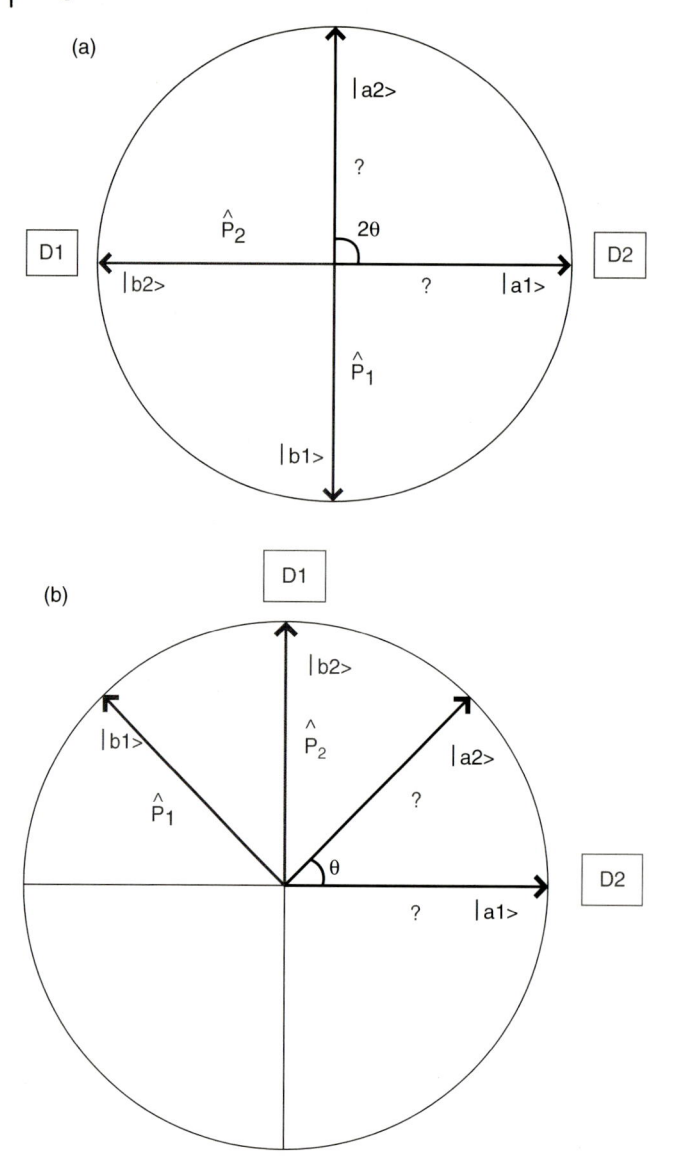

Figure 25.5 Unambiguous discrimination of photon states with a projective measurement. (a) Bloch sphere representation. (b) Physical photon polarizations. Only one pair of detectors (D_1 and D_2) can be used in any single experiment. Measurement shown in figure attempts to identify state $|a_2\rangle$ unambiguously. Angle θ is the physical angle between photon polarizations (case $\theta = \pi/4$ is shown). Question marks indicate inconclusive outcomes.

This result is also apparent if we write the projector explicitly,

$$\hat{P}_{b_2}|a_1\rangle = |a_1^\perp\rangle\langle a_1^\perp|a_1\rangle = 0, \tag{25.79}$$

or if we use the closure relation to write

$$\hat{P}_{b_2}|a_1\rangle = |a_1^+\rangle\langle a_1^+|a_1\rangle = (I - |a_1\rangle\langle a_1|)|a_1\rangle = I|a_1\rangle - |a_1\rangle = 0. \tag{25.80}$$

We thus conclude that projector $\hat{P}_{b_2}$ identifies state $|a_2\rangle$ and *annihilates* state $|a_1\rangle$. To generalize our conclusion, we say that state $|a_i\rangle$ will be annihilated by the projector

$$\hat{P}_{a_i^\perp} = |a_i^\perp\rangle\langle a_i^\perp|. \tag{25.81}$$

Since $|a_1^\perp\rangle$ is not orthogonal to $|a_2^\perp\rangle$, the projective measurement scheme does not allow us to use two corresponding detectors at the same time. So in any given trial, we just use one or the other. Either way, we must hope that Alice sends the "suitable" state that can in principle be identified by the chosen detector. Finally, let us write success and failure probabilities (25.76) in terms of the angle between the two states:

$$P = \frac{1}{2}(1 - \cos^2\theta), \quad Q = \frac{1}{2}(1 + \cos^2\theta). \tag{25.82}$$

For example, if one state is horizontal and the other diagonal ($\theta = 45°$), we get $P = 1/4$ and $Q = 3/4$. This completes the analysis of unambiguous discrimination by projective measurements.

The question arises: Can we use other measurement procedures to discriminate between the states with better chance of success?

Research in this field started with the work of Ivanovic [3] who first suggested the idea that unambiguous discrimination could be achieved if one also allowed for some inconclusive results. In addition to pointing out the possibility of projective measurements (25.81), Ivanovic realized that a sequence of measurements can be more efficient than a single measurement (25.81). The next step was made by Dieks [4] who showed how to replace Ivanovic's sequence with a single POVM measurement. Finally, Peres [5] showed this POVM to be the most efficient type of measurement in terms of minimizing the chance of failure (i.e., an inconclusive result). We are now going to show how one can maximize the success rate with a POVM measurement.

The optimal POVM will consist of three elements: $\{\hat{\Pi}_{b_1}, \hat{\Pi}_{b_2}, \hat{\Pi}_{b_?}\}$. To construct the first two, we will attenuate projection operators $\hat{P}_{b_1} = |b_1\rangle\langle b_1|, \hat{P}_{b_2} = |b_2\rangle\langle b_2|$, with multipliers λ_1 and λ_2 as described earlier. The resulting POVM elements $\hat{\Pi}_{a_1} = \lambda_1 \hat{P}_{b_1}$ and $\hat{\Pi}_{a_2} = \lambda_2 \hat{P}_{b_2}$ (which are no longer projectors!) are sometimes called *quantum detection operators*. These operators will be performing the same function as before – $\hat{\Pi}_{b_1}$ will confirm state $|a_1\rangle$ by annihilating the possibility of $|a_2\rangle$, and $\hat{\Pi}_{b_2}$ will confirm $|a_2\rangle$ by annihilating $|a_1\rangle$. The crucial advantage of the POVM is that both detection operators can be applied simultaneously. This means that instead of committing himself to one single physical detector tailored to a particular state, Bob now has a chance to keep an eye on both possibilities at once. The trade-off is that there will be trials in which neither of these two detectors clicks. Completing the

POVM is the third operator $\hat{\Pi}_{b_?}$ that corresponds to these inconclusive outcomes. The three operators must partition unity:

$$\hat{\Pi}_{b_1} + \hat{\Pi}_{b_2} + \hat{\Pi}_{b_?} = I, \tag{25.83}$$

so our POVM elements are

$$\hat{\Pi}_{b_1} = \lambda_1 |b_1\rangle\langle b_1|, \quad \hat{\Pi}_{b_2} = \lambda_2 |b_2\rangle\langle b_2|, \quad \hat{\Pi}_{b_?} = I - \hat{\Pi}_{b_1} - \hat{\Pi}_{b_2} \tag{25.84}$$

and the task is to find λ_1 and λ_2 that would maximize the success rate.

With a projective measurement, we could only get two possible outcomes: one is failure and the other is unambiguous discrimination of the state (if $\hat{P}_{b_1}$ measurement was used, $|a_1\rangle$, or else state $|a_2\rangle$ for a $\hat{P}_{b_2}$ measurement). In contrast, the POVM measurement (25.83) allows for three possible outcomes: confirmation of $|a_1\rangle$, confirmation of $|a_2\rangle$, and failure. This versatility of POVM (25.83) cannot be matched by a projective measurement whose number of outcomes is limited to two because the Hilbert space of a polarized photon is two-dimensional.

Consider operator $\hat{\Pi}_{b_1}$. If the photon was prepared in state $|a_2\rangle$, the operator will annihilate it, that is,

$$\langle a_2 | \hat{\Pi}_{b_1} | a_2 \rangle = 0. \tag{25.85}$$

But if the photon state was $|a_1\rangle$, the detection operator can either succeed (i.e., detector clicks, identifying the state) or fail (detector fails to click, and we never learn that $|a_1\rangle$ was the correct state; to compensate, the inconclusive operator $\hat{\Pi}_{b_?}$ must then give us $|?\rangle$ outcome). With p_1 being the probability of success, we can write

$$\langle a_1 | \hat{\Pi}_{b_1} | a_1 \rangle = p_1 \tag{25.86}$$

and comparison with (25.84) immediately shows that $\lambda_1 |\langle b_1 | a_1 \rangle|^2 = p_1$. Applying the same reasoning to $\hat{P}_{b_2}$ to obtain $\lambda_2 |\langle b_2 | a_2 \rangle|^2 = p_2$, we then express the unknown multipliers in terms of respective probabilities:

$$\lambda_1 = \frac{p_1}{|\langle b_1 | a_1 \rangle|^2}, \quad \lambda_2 = \frac{p_2}{|\langle b_2 | a_2 \rangle|^2}. \tag{25.87}$$

Also, for operator $\hat{\Pi}_{b_?}$, it is easy to see that

$$\langle a_1 | \hat{\Pi}_{b_?} | a_1 \rangle = q_1, \quad \langle a_2 | \hat{\Pi}_{b_?} | a_2 \rangle = q_2, \tag{25.88}$$

where $q_1 = 1 - p_1$ and $q_2 = 1 - p_2$ are the probabilities of failing to identify the respective states $|a_1\rangle$ and $|a_2\rangle$. As a result, the last multiplier is found from

$$\lambda_? |\langle b_? | a_1 \rangle|^2 = q_1 \quad \text{and} \quad \lambda_? |\langle b_? | a_2 \rangle|^2 = q_2 \tag{25.89}$$

and is equal to

$$\lambda_? = \frac{q_1}{|\langle b_? | a_1 \rangle|^2} = \frac{q_2}{|\langle b_? | a_2 \rangle|^2}. \tag{25.90}$$

25.6 Discrimination of Two Pure States

The POVM elements are

$$\hat{\Pi}_{b_1} = \frac{p_1}{|\langle b_1|a_1\rangle|^2}|b_1\rangle\langle b_1|, \quad \hat{\Pi}_{b_2} = \frac{p_2}{|\langle b_2|a_2\rangle|^2}|b_2\rangle\langle b_2|,$$

$$\hat{\Pi}_{b_?} = \frac{q_1}{|\langle b_?|a_1\rangle|^2}|b_?\rangle\langle b_?| = \frac{q_2}{|\langle b_?|a_2\rangle|^2}|b_?\rangle\langle b_?|,$$

(25.91)

and they obviously fulfill the closure requirement by making the possibilities exhaustive. Indeed, it is easy to see that

$$\langle a_1|\left(\hat{\Pi}_{b_1} + \hat{\Pi}_{b_2} + \hat{\Pi}_{b_?}\right)|a_1\rangle = p_1 + 0 + q_1 = 1,$$

$$\langle a_2|\left(\hat{\Pi}_{b_1} + \hat{\Pi}_{b_2} + \hat{\Pi}_{b_?}\right)|a_2\rangle = 0 + p_2 + q_2 = 1,$$

(25.92)

which immediately imply that our POVM elements add to unity. Also, it is easy to see that (25.85) and the similar condition for the other element express the requirement of unambiguous discrimination (detection operator should not react positively to the wrong state).

The fact that all probabilities in (25.79 and 25.80) are real, nonnegative numbers means that diagonal elements of matrices representing $\hat{\Pi}_{b_1}$, $\hat{\Pi}_{b_2}$, and $\hat{\Pi}_{b_?}$ are real (i.e., operators are Hermitian) and nonnegative (i.e., operators are positive semi-definite). To simplify these formulas, let us recall that $|\langle a_1|a_2\rangle| = \cos\theta$. Now, since $|b_1\rangle$ is orthogonal to $|a_2\rangle$ and $|b_2\rangle$ is orthogonal to $|a_1\rangle$, it is easy to see that

$$|\langle a_1|b_1\rangle| = |\langle b_2|a_2\rangle| = \sin\theta,$$

(25.93)

and therefore

$$\hat{\Pi}_{b_1} = \frac{p_1}{\sin^2\theta}|b_1\rangle\langle b_1|, \quad \hat{\Pi}_{b_2} = \frac{p_2}{\sin^2\theta}|b_2\rangle\langle b_2|,$$

$$\hat{\Pi}_{b_?} = \frac{q_1}{|\langle b_?|a_1\rangle|^2}|b_?\rangle\langle b_?| = \frac{q_2}{|\langle b_?|a_2\rangle|^2}|b_?\rangle\langle b_?| = I - \hat{\Pi}_{b_1} - \hat{\Pi}_{b_2}.$$

(25.94)

The attentive reader will notice that while we have determined $\hat{\Pi}_{b_1}$, $\hat{\Pi}_{b_2}$ uniquely in terms of initial states and success probabilities, $\hat{\Pi}_{b_?}$ is yet to be determined explicitly. But we can use the last equation (25.94) to impose a constraint upon $\hat{\Pi}_{b_?}$ by requiring that it must be positive-semidefinite. This means it must have nonnegative eigenvalues. As you know already, the corresponding eigenvalue problem is solved by writing the operator in the matrix form and setting up the characteristic equation. How does one obtain the matrix form of the detection operators? The "obvious" (and wrong) way is to surround the operators with $\langle a_1|$ on the left and $|a_2\rangle$ on the right. But that would be a serious mistake because these states are not orthogonal! The eigenvalue problem must be solved in an orthogonal basis. Therefore, let us choose the basis $\{|a_1\rangle, |a_1^\perp\rangle\}$. We leave it to the reader to verify the following results:

$$\hat{\Pi}_{b_1} = \frac{p_1}{\sin^2\theta}\begin{pmatrix} \sin^2\theta & \sin\theta\cos\theta \\ \sin\theta\cos\theta & \cos^2\theta \end{pmatrix}, \quad \hat{\Pi}_{b_2} = \frac{p_2}{\sin^2\theta}\begin{pmatrix} 0 & 0 \\ 0 & 1 \end{pmatrix},$$

$$\hat{\Pi}_{b_?} = \begin{pmatrix} 1 & 0 \\ 0 & 1 \end{pmatrix} - \hat{\Pi}_1 - \hat{\Pi}_2 = \begin{pmatrix} 1-p_1 & -p_1\cot\theta \\ -p_1\cot\theta & 1 - p_1\cot^2\theta - p_2\csc^2\theta \end{pmatrix}.$$

(25.95)

From the resulting characteristic equation, we obtain

$$\det \begin{vmatrix} 1 - p_1 - \lambda & -p_1 \cot\theta \\ -p_1 \cot\theta & 1 - p_1 \cot^2\theta - p_2 \csc^2\theta - \lambda \end{vmatrix} = 0 \qquad (25.96)$$

or

$$(q_1 - \lambda)\left[1 - (1 - q_1)\cot^2\theta - (1 - q_2)\csc^2\theta - \lambda\right] = (1 - q_1)\cot^2\theta, \qquad (25.97)$$

where we have switched from probabilities of success p_1 and p_2 to the respective failure probabilities $q_1 = 1 - p_1$ and $q_2 = 1 - p_2$. (Also, don't confuse eigenvalue λ with the POVM element multipliers λ_i).

After some algebra, this gives us a quadratic equation for the eigenvalue λ. To ensure both eigenvalues are nonnegative, it is enough to satisfy this requirement for the smaller eigenvalue. After some simplification, the resulting inequality leads to

$$q_1 q_2 \geq \cos^2\theta = |\langle a_1|a_2\rangle|^2. \qquad (25.98)$$

As a corollary, in the optimal POVM measurement, we have

$$q_2 = \frac{\cos^2\theta}{q_1}. \qquad (25.99)$$

As we see, there is a limit to what a POVM can do for us as the product of error rates cannot drop below the minimum threshold defined by the overlap between the initial states.

The total failure probability is given by

$$Q^{POVM} = \frac{1}{2}q_1 + \frac{1}{2}q_2 \qquad (25.100)$$

and is minimized when (25.98) becomes a strict equality. Then differentiating

$$Q^{POVM} = \frac{1}{2}q_1 + \frac{1}{2}\frac{\cos^2\theta}{q_1} \qquad (25.101)$$

with respect to q_1, we find the minimum at

$$q_1 = q_2 = \cos\theta \qquad (25.102)$$

for which the probability of total failure is

$$Q^{POVM}_{min} = \cos\theta. \qquad (25.103)$$

This is the theoretical minimum for the failure rate that was first established in the works of Ivanovic, Dieks, and Peres and is sometimes called the *IDP limit*. Recalling our definition of $\cos\theta$, we finally write

$$Q_{IDP} = Q^{POVM}_{min} = |\langle a_1|a_2\rangle| \qquad (25.104)$$

and thereby the maximum theoretically possible success rate is

$$P_{IDP} = P^{POVM}_{max} = 1 - |\langle a_1|a_2\rangle|. \qquad (25.105)$$

Compare this result with (25.82). It is easy to see that POVM offers a much lower failure rate, especially for strongly nonorthogonal states. For example, let's look at the special case of $\theta = 45$, which plays important role in quantum cryptography applications such as B92 protocol. For this angle, projective measurement gives

$$Q_{min}^{proj}(\theta = 45) = \frac{1}{2}\left(1 + |\langle a_2|a_1\rangle|^2\right) = 0.75, \qquad (25.106)$$

while the above-described POVM reduces the failure rate to

$$Q_{min}^{POVM}(\theta = 45) = |\langle a_2|a_1\rangle| = \frac{1}{\sqrt{2}} \approx 0.707. \qquad (25.107)$$

The advantage of POVM as compared to a traditional projective measurement is illustrated in Figure 25.6.

But there is an important catch. It is easy to think from the above graph that POVM should be chosen as the optimal measurement regardless of the initial conditions. But this is not quite true. In deriving equations starting from (25.83) and all the way to (25.104 and 25.105), we have tacitly assumed that the right POVM will exist. This assumption needs to be checked. One obvious sign that we are outstepping the legitimate domain of the POVM is when our POVM starts giving us failure rates exceeding 1. Thus, we should always require that

$$q_1 \leq 1, \quad q_2 = \frac{\cos^2\theta}{q_1} \leq 1. \qquad (25.108)$$

In view of (25.102), this requirement is clearly fulfilled in our case. However, consider a more general situation when states $|a_1\rangle$ and $|a_2\rangle$ are chosen with probabilities n_1 and n_2. Then equation (25.100) should be written as

$$Q^{POVM} = n_1 q_1 + n_2 q_2. \qquad (25.109)$$

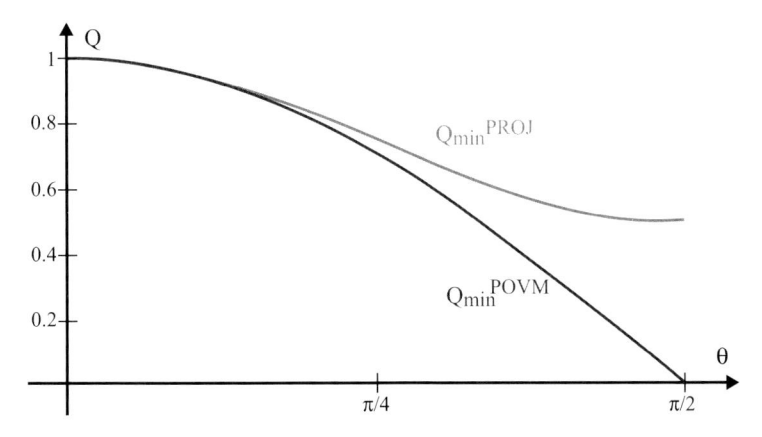

Figure 25.6 Minimum failure rate Q_{min} as a function of θ according to Equations 25.84 and 25.106. The upper graph corresponds to a projective measurement $Q_{min}^{proj} = (1/2)(1 + \cos^2\theta)$, the lower graph – to POVM $Q_{min}^{POVM} = \cos\theta$. States $|a_1\rangle$ and $|a_2\rangle$ are chosen with the same probability $n_1 = n_2 = 1/2$.

Using the optimum condition $q_2 = \cos^2\theta/q_1$ as before and minimizing the resulting Q^{POVM}, we obtain the following general result:

$$q_1 = \sqrt{\frac{n_2}{n_1}}\cos\theta, \quad q_2 = \sqrt{\frac{n_1}{n_2}}\cos\theta, \quad Q^{POVM}_{min} = 2\sqrt{n_1 n_2}\cos\theta. \tag{25.110}$$

Then the first condition (25.108) becomes

$$\cos^2\theta \leq \frac{n_1}{n_2} = \frac{n_1}{1-n_1} \Rightarrow n_1 \geq \frac{\cos^2\theta}{1+\cos^2\theta} \tag{25.111}$$

and the second condition (25.108) becomes

$$\cos^2\theta \leq \frac{n_2}{n_1} = \frac{1-n_1}{n_1} \Rightarrow n_1 \leq \frac{1}{1+\cos^2\theta}. \tag{25.112}$$

Combining them together, we get the following inequality that must be satisfied so that there exist an optimal POVM procedure:

$$\frac{\cos^2\theta}{1+\cos^2\theta} \leq n_1 \leq \frac{1}{1+\cos^2\theta}. \tag{25.113}$$

We show this requirement in Figure 25.7. The reader can see that for any given overlap $|\langle a_1|a_2\rangle| = \cos\theta$ between the nonorthogonal states, the optimal POVM will exist only as long as n_1 and n_2 fit into the allowed range. In fact, the choice $n_1 = n_2 = 1/2$ is the only one for which the POVM exists in all cases. What happens if the lower bound is breached? Since this bound arises from the first condition (25.108), it will mean that q_1 has exceeded 1; in other words, the probability p_1 that POVM, acting through its element $\hat{\Pi}_{b_1}$, will successfully detect $|a_1\rangle$ state has dropped to 0. To put it in different terms, the experiment is no longer capable of discriminating $|a_1\rangle$; the remaining outcomes it can only produce are $|a_2\rangle$ and $|?\rangle$. This corresponds to the projective measurement (25.79) that discriminates state $|a_2\rangle$ unambiguously when successful or, in case of failure, gives an inconclusive answer. This measurement will be described by Equation 25.76 generalized for the case of arbitrary n_1 and n_2:

$$Q(\hat{P}_{b_2}) = n_1 + n_2|\langle a_2|a_1\rangle|^2. \tag{25.114a}$$

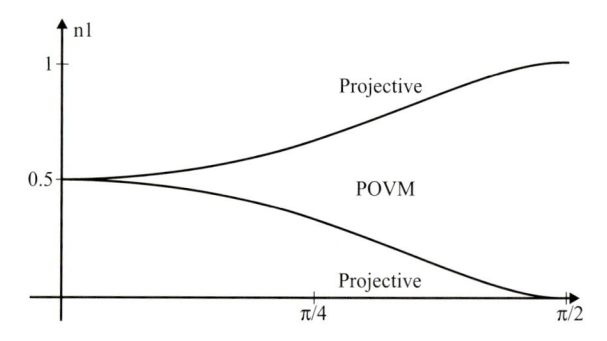

Figure 25.7 The allowed range of n_1 is bounded by the upper and lower curve. Outside this range, POVM goes over to a projective measurement.

The above equation expresses the fact that operator $\hat{P}_{b_2} = I - |a_1\rangle\langle a_1|$ is guaranteed to be inconclusive in n_1 percentage of cases when Alice sends to Bob state $|a_1\rangle$ and succeeds in the remaining n_2 percentage of cases (when Alice sends state $|a_2\rangle$) with probability $\cos^2\theta$.

With the same reasoning, we can show that if the upper bound is breached, the experiment reduces to a projective measurement described by

$$Q(\hat{P}_{b_1}) = n_2 + n_1 |\langle a_2|a_1\rangle|^2. \qquad (25.114b)$$

Combining our results, we finally write

$$Q_{min} = \begin{cases} Q_{min}^{POVM} = 2\sqrt{n_1 n_2}\cos\theta, & \dfrac{\cos^2\theta}{1+\cos^2\theta} \leq n_1 \leq \dfrac{1}{1+\cos^2\theta}, \\[6pt] Q(\hat{P}_{b_2}) = n_1 + n_2\cos^2\theta, & n_1 < \dfrac{\cos^2\theta}{1+\cos^2\theta}, \\[6pt] Q(\hat{P}_{b_1}) = n_2 + n_1\cos^2\theta, & n_1 > \dfrac{1}{1+\cos^2\theta}. \end{cases} \qquad (25.115)$$

This equation has a simple physical meaning. The main trouble with projective measurements $\hat{P}_{b_2}$ and $\hat{P}_{b_1}$ is that they are *guaranteed* to fail when encountering the "wrong" state. This inconvenient feature of projective measurements is manifest in the terms n_1 and n_2 in (25.115). If one of the two states is assigned a predominant probability, then projective measurements offer the optimal detection efficiency because we can always choose the projector designed specifically with the more likely state in mind. Such a measurement will be inconclusive when encountering the other state, but the chance of that is relatively low. But when the two probabilities are more or less equal, projective measurement becomes inefficient because it cannot anticipate the state in advance. In other words, we will too often apply projector $\hat{P}_{b_2}$ when $\hat{P}_{b_1}$ would have offered a much better chance of conclusive result or apply $\hat{P}_{b_1}$ when we should have applied $\hat{P}_{b_2}$. This is when a POVM comes in handy because of its ability to "listen" to all three possible outcomes. By offering the experimenter a nonzero chance of success in *all* trials, a POVM then does a better job than either $\hat{P}_{b_2}$ or $\hat{P}_{b_1}$, which have a nonzero chance of success only in *some* of the trials. The trade-off is that with a POVM, we must accept a greater chance of failure for a given state $|a_i\rangle$ ($i = 1, 2$) than if we used the "right" projection operator:

$$q_i = \sqrt{\dfrac{n_j}{n_i}}\cos\theta \approx \cos\theta \qquad (25.116)$$

for a POVM (where $j \neq i$ and the probabilities are assumed to be about equal), and

$$q_i = \cos^2\theta < \cos\theta \qquad (25.117)$$

for a projective measurement provided that the experimenter correctly guessed the more likely state and tailor-suited his projector for that state. Equation 25.115 establishes the range of optimum efficiency for projective measurements versus POVMs.

With these conclusions, we are now in a position to provide the experimenter with instructions about building the optimal POVM. First of all, let us write our POVM explicitly. We have from (25.94)

$$\hat{\Pi}_{b_1} = \frac{1 - \sqrt{(n_2/n_1)}\cos\theta}{\sin^2\theta}|b_1\rangle\langle b_1|, \quad \hat{\Pi}_{b_2} = \frac{1 - \sqrt{(n_1/n_2)}\cos\theta}{\sin^2\theta}|b_2\rangle\langle b_2|,$$

$$\hat{\Pi}_{b_?} = I - \hat{\Pi}_{b_1} - \hat{\Pi}_{b_2},$$

(25.118)

which defines the POVM uniquely for the given states $|a_1\rangle, |a_2\rangle$ and probabilities n_1, n_2. Figure 25.8 shows this POVM for the case $\theta = \pi/4$. When looking at the diagram, one should keep in mind the difference between the Bloch sphere representation and the actual polarization angles for photons. Thus, the physical angle θ goes over to 2θ on the Bloch sphere. Then the figure becomes identical to Figure 25.4(b) once we set $n_1 = n_2 = 1/2$, which ensures that projectors $|b_1\rangle\langle b_1|, |b_2\rangle\langle b_2|$ get attenuated by the same factor $\lambda_1 = \lambda_2 = \lambda$.

Last but not least, note that since detector $D_?$ is set symmetrically between the two states, there is no reason to attribute an inconclusive outcome to one state instead of

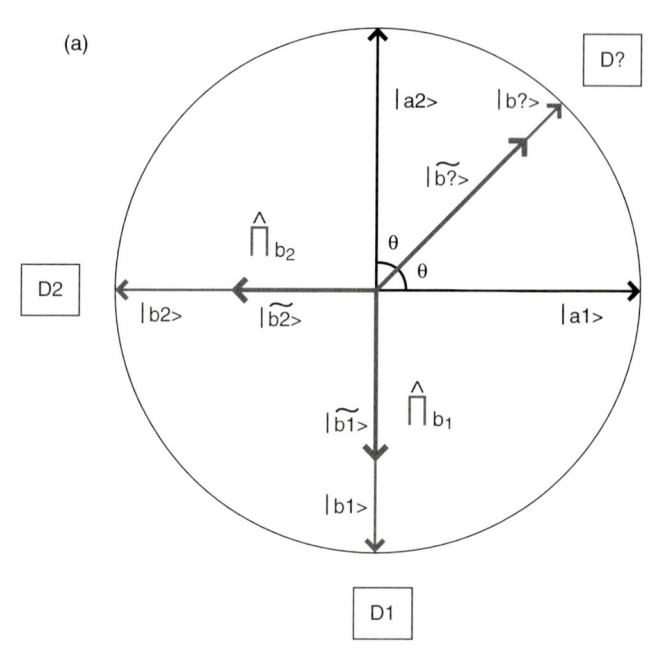

Figure 25.8 POVM realizing the optimal quantum state discrimination experiment. (a) The Bloch sphere representation. (b) Physical angles between photons. When detector D_1 or D_2 clicks, we know for sure that the corresponding state has been prepared. The third detector, $D_?$ is set up symmetrically between $|a_1\rangle$ and $|a_2\rangle$. When it clicks, it corresponds to an inconclusive result. Note that if neither D_1 nor D_2 clicks, then $D_?$ *will* click, so the experiment will *always* produce some answer $(|a_1\rangle, |a_2\rangle, |?\rangle)$. $D_?$ will click with the probability $Q_{min}^{POVM} = 2\sqrt{n_1 n_2}\cos\theta$. The thick arrows give a schematic representation of the three POVM elements.

(b)

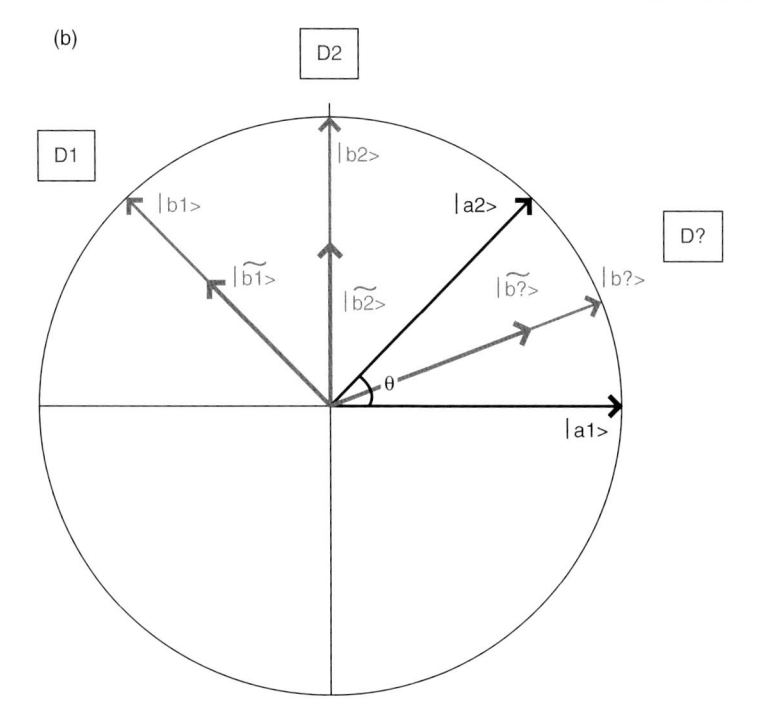

Figure 25.8 (Continued).

the other. In other words, inconclusive outcomes are inconclusive in the full sense of the word: We cannot glimpse any probabilistic information from them, so both states remain equally probable.

25.7
Neumark's Theorem

It is time to say a few words about the physical meaning of POVMs, which up to now has remained hidden in the rather abstract formal definition from Section 25.2. Let an arbitrary system with wave function $|\psi\rangle$ exist in an n-dimensional Hilbert space $\mathcal{H}^n$ and suppose we perform a POVM measurement on that system. Furthermore, let this POVM be described by N elements $\vec{F}_i = |\tilde{b}_i\rangle\langle\tilde{b}_i|$, and suppose that $N > n$. Though these measurements are not projective, one still needs some way to visualize them. To that end, one can imagine a set of N quasiprojective measurements onto the corresponding directions $|\tilde{b}_i\rangle = \sqrt{\lambda_i}|b_i\rangle$, as seen from Equation 25.119:

$$|\psi\rangle = I|\psi\rangle = \left(\sum_{i=1}^{N}\vec{F}_i\right)|\psi\rangle = \sum_{i=1}^{N}\lambda_i|b_i\rangle\langle b_i|\psi\rangle. \qquad (25.119)$$

However, this equation conceals a serious conceptual pitfall for the unsuspecting reader. It may look *as if* we had found a way to arrange N detectors along $|b_1\rangle, \ldots, |b_N\rangle$ in such a way that they would not interfere with one another and carry out N simultaneous projective measurements. Of course, a POVM does not avoid the physical ban on performing more than n projective measurements simultaneously, and directions $|b_i\rangle$ do not tell us the physical placement of detectors, as we have mentioned in Section 25.3. As we said then, we must switch to a three-dimensional Hilbert space prior to making our measurement, so that our three detectors could fit easily along any three orthogonal directions. Let us explore this idea further.

It is quite common in geometry to project a 3D vector onto a 2D plane. This geometric analogy can help us visualize the previous POVM, showing how it can originate from a projective measurement. Consider a 3D Cartesian space R^3 spanned by orthogonal unit vectors $\{\hat{i}, \hat{j}, \hat{k}\}$. Let $\vec{r}$ be an arbitrary vector in this space, defined by its coordinates:

$$\vec{r} = \vec{i}\,(\vec{i}\cdot\vec{r}) + \vec{j}\,(\vec{j}\cdot\vec{r}) + \vec{k}\,(\vec{k}\cdot\vec{r}). \tag{25.120}$$

Equation 25.120 justifies the closure relation in Cartesian space, namely,

$$\vec{i}\,(\vec{i}\cdot) + \vec{j}\,(\vec{j}\cdot) + \vec{k}\,(\vec{k}\cdot) = 1, \tag{25.121}$$

where the dots after the unit vectors indicate these vectors are waiting to form the dot product.

Next, suppose we choose two arbitrary orthogonal unit vectors 1 and 2 starting from the origin and spanning a plane R^2 that is clearly a 2D subspace of R^3. Imagine an observer who "lives" in that plane and can only see the components of vectors parallel to that plane. To this observer, the same vector $\vec{r}$ will appear as

$$\vec{r}^{\,\|} = \vec{i}^{\,\|}(\vec{i}\cdot\vec{r}) + \vec{j}^{\,\|}(\vec{j}\cdot\vec{r}) + \vec{k}^{\,\|}(\vec{k}\cdot\vec{r}) \tag{25.122}$$

and the unit vectors spanning R^3 will form three new directions, $\hat{i}^{\,\|}, \hat{j}^{\,\|}, \hat{k}^{\,\|}$, when projected onto R^2. These new vectors are clearly nonorthogonal since a plane cannot contain more than two mutually orthogonal vectors (with the trivial exception of the third vector having zero length). In addition, the lengths of these vectors are generally less than 1. So the R^2-based observer now sees what appears as a very poor choice of basis for his vector space: three seemingly arbitrary vectors, non-orthogonal and unnormalized. However, these vectors help us model a three-element POVM defined by directions $|b_1\rangle, |b_2\rangle, |b_3\rangle$. We leave it as problem (Problem 25.8) for the reader to show that a closure relation similar to (25.121) also holds for these vectors:

$$\hat{i}^{\,\|}\left(\hat{i}^{\,\|}\cdot\right) + \hat{j}^{\,\|}\left(\hat{j}^{\,\|}\cdot\right) + \hat{k}^{\,\|}\left(\hat{k}^{\,\|}\cdot\right) = 1. \tag{25.123}$$

We have arrived at the equivalent of closure relation (25.120) for the vectors $\vec{r}^{\,\|}$ in the 2D space R^2. Interpreting the three terms on the left-hand side as operators and denoting them by $\hat{F}_1, \hat{F}_2, \hat{F}_3$ and accordingly interpreting 1 on the right-hand side as the identity matrix, we get a sum of operators that add to unity:

$$\hat{F}_1 + \hat{F}_2 + \hat{F}_3 = I. \quad (25.124)$$

It is straightforward to show that these operators are both – positive semidefinite and Hermitian.

The generalization of this result to spaces with an arbitrary number of dimensions is trivial and so when an orthogonal set of unit vectors $\{\hat{e}_i\}$ spanning an N-dimensional Cartesian space R^N projects onto an n-dimensional subspace R^n where $n < N$, it forms a new set of unnormalized, nonorthogonal vectors $\{\hat{e}_i^{\parallel}\}$ in that subspace such that

$$\sum_{i=1}^{N} \hat{F}_i \equiv \sum_{i=1}^{N} \hat{e}_i^{\parallel}\left(\hat{e}_i^{\parallel} \cdot\right) = I, \quad (25.125)$$

where I is the identity operator in R^n and $\hat{F}_i$ are positive semidefinite Hermitian operators. A similar expression can be derived for $\mathcal{H}$-spaces as well. So when an orthonormal basis $\{|B_i\rangle\}$ spanning an N-dimensional space $\mathcal{H}^n$ projects onto an n-dimensional subspace $\mathcal{H}^n$ where $n < N$, it forms a new unnormalized, nonorthogonal basis $\{|\tilde{b}_i\rangle\}$ where $|\tilde{b}_i\rangle$ are components of $|B_i\rangle$ parallel to the subspace:

$$\sum_{i=1}^{N} \hat{F}_i \equiv \sum_{i=1}^{N} |\tilde{b}_i\rangle\langle\tilde{b}_i| = I, \quad (25.126)$$

where I is the identity operator in $\mathcal{H}^n$ and $\hat{F}_i$ are positive semidefinite Hermitian operators. Here, the original orthonormal basis $|B_i\rangle$ defines a corresponding projective measurement on $\mathcal{H}^N$, while Equation 25.126 expresses the defining property of POVMs since operators $\hat{F}_i$ function as nonorthogonal projectors that add to unity. This leads us to express the physical meaning of (25.126) in the following corollary: A projective measurement defined on $\mathcal{H}^N$ by a set of N orthogonal projectors $|B_i\rangle\langle B_i|$ will appear to an observer who lives in subspace $\mathcal{H}^n$ ($n < N$) as an N-element POVM defined by a set of nonorthogonal projectors $\{|\tilde{b}_i\rangle\langle\tilde{b}_i|\}$. We can thus interpret POVM as a usual projective experiment that is viewed from a lower dimensional space (Figure 25.9).

One can also see from the last equation that a click in jth detector will collapse the system into state $|B_j\rangle$. Having N detectors provides us with N possible outcomes $|B_1\rangle, \ldots, |B_N\rangle$ for a POVM measurement. Let a projective measurement on $\mathcal{H}^N$ be associated with the corresponding POVM on $\mathcal{H}^n$. Then, the N detectors carrying out this projective measurement provide the physical implementation of the POVM in question, whereby a click in jth detector corresponds to jth outcome in $\mathcal{H}^n$.

It may seem now, once we associated every $|\tilde{b}_i\rangle$ with a separate POVM outcome, that we have found a sure way of discriminating these states. The reality is more subtle. The correct conclusion is that we have just found a way to fit the extra $N - n$ detectors into the apparatus by "lifting" them above the "overcrowded" $\mathcal{H}^n$ subspace into the other dimensions. But this achievement came at the price of bringing fuzziness into detector readings! Look at an arbitrary initial state $|\tilde{b}_i\rangle$ in $\mathcal{H}^n$.

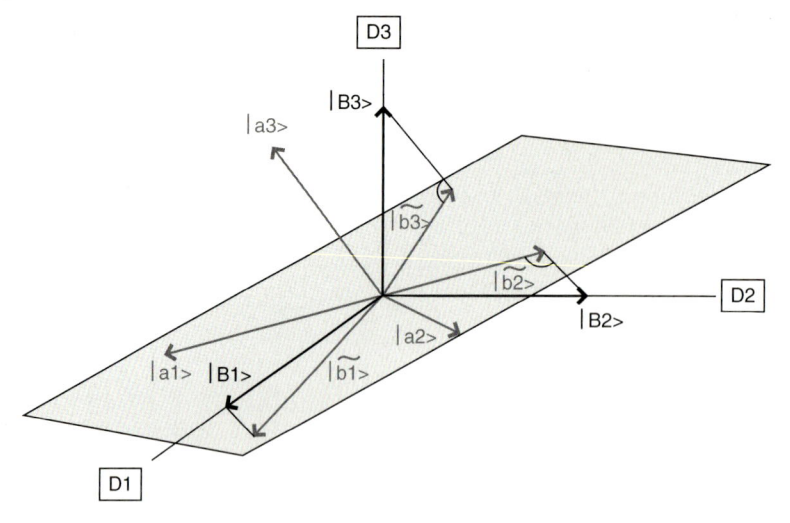

Figure 25.9 Projective measurement defined by $|B_1\rangle, |B_2\rangle, |B_3\rangle$ will look like a POVM defined by $|\tilde{b}_1\rangle, |\tilde{b}_2\rangle, |\tilde{b}_2\rangle$ to an observer in the shaded plane. The plane is spanned by two unit vectors $|a_1\rangle, |a_2\rangle$ (not necessarily orthogonal!). Vector $|a_3\rangle$ is orthogonal to the plane. Switching from $\{|a_1\rangle, |a_2\rangle, |a_3\rangle\}$ basis to $\{|B_1\rangle, |B_2\rangle, |B_3\rangle\}$ basis will implement the POVM.

Physically, it represents the overlap between $|B_i\rangle$ and the corresponding normalized state $|b_i\rangle = (1/\sqrt{\lambda_i})|\tilde{b}_i\rangle$; in other words, it is the probability amplitude for the ith detector to click when a system is in the state $|b_i\rangle$. However, this amplitude is less than 1! Indeed, one can see from the picture that $|b_i\rangle$ generally has finite projections along the other directions as well. The nonzero overlap $\langle \psi_{j\neq i}|b_i\rangle \neq 0$ means that the remaining probability will be shared between the other detectors. In other words, even if the initial state is known, it is impossible to predict with certainty which detector is going to click! That is to say, a POVM measurement is, in a counterintuitive way, intrinsically collective. If N drivers must share N parking spaces, the Dirichlet principle allows each driver to reserve his own space exclusively for himself. This analogy breaks when we want to construct a POVM. We *can* have N detectors listen to N possible outcomes, but we *cannot* assign a specific detector for each outcome. As a consequence, discrimination between nonorthogonal initial states will not be error-free, as we saw in the last section for the case of $N = 3$.[9] So to summarize, POVM realizes the type of measurement where the same initial input can generally result in more than one outcome.

---

9) More specifically, there is always a trade-off between allowed ambiguity and the success rate. As shown in the previous section, it is possible to resolve two states unambiguously (the two detectors are completely "privatized" by the corresponding states) at the cost of lower success rate (the two detectors will sometimes fail, and the clicks that belong to them by right will instead go to the third detector).

Let us now look at the reverse situation: We have a POVM defined on $\mathcal{H}^n$ by a set of N elements $\{\hat{F}_i\}$ where $n < N$ and want to find out if there exists a corresponding H-space $\mathcal{H}^N$ such that to an observer who "lives" in that larger space, this POVM will appear as a projective measurement. The POVM elements $\hat{F}_i = |\tilde{b}_i\rangle\langle\tilde{b}_i|$ will be composed from nonorthogonal states $|\tilde{b}_i\rangle$. Since they are defined in an n-dimensional space, every $|\tilde{b}_i\rangle$ can be written as an $n \times 1$ column matrix.[10] Let us put the resulting N column matrices next to each other, forming an $n \times N$ rectangular matrix:

$$M_{n \times N} = \begin{pmatrix} |\tilde{b}_1\rangle & |\tilde{b}_2\rangle & \cdots & |\tilde{b}_N\rangle \\ \hline \begin{pmatrix} c_1 \\ c_2 \\ \vdots \\ c_n \end{pmatrix}_1 & \begin{pmatrix} c_1 \\ c_2 \\ \vdots \\ c_n \end{pmatrix}_2 & \cdots & \begin{pmatrix} c_1 \\ c_2 \\ \vdots \\ c_n \end{pmatrix}_N \end{pmatrix} \Big| n . \tag{25.127}$$

If we take any state $|\tilde{b}_a\rangle$, the corresponding POVM element $\hat{F}_a = |\tilde{b}_a\rangle\langle\tilde{b}_a|$ will be an $n \times n$ square matrix whose elements will be given by $(F_{ij})_a = (c_i c_j^*)_a$ according to the definition of the ket–bra operator. Due to (25.126), the sum of projectors has to form an identity matrix whose general element is $I_{ij} = \delta_{ij}$. Hence,

$$\sum_\alpha \left(c_i c_j^*\right)_\alpha = \delta_{ij} \tag{25.128}$$

and thus the property of unitary matrices (ith row times complex-conjugated jth row equals Kronecker delta) holds for the *rows* of matrix (25.127). This shows that the rows of (25.127) are mutually orthonormal. Let us now add another $N - n$ rows making them orthonormal (in the sense of Equation 25.128) to each other and to the first n rows. Linear algebra ensures that this can be done (the resulting $N \times N$ matrix defines an N-dimensional Hilbert space $\mathcal{H}^N$ and then a set of n orthogonal states can always be complemented to a complete basis of N orthogonal states). This will give us a unitary $N \times N$ matrix:

$$M_{N \times N} = \begin{pmatrix} |B_1\rangle & |B_2\rangle & \cdots & |B_N\rangle \\ \hline \begin{pmatrix} c_1 \\ \vdots \\ c_n \end{pmatrix}_1 & \begin{pmatrix} c_1 \\ \vdots \\ c_n \end{pmatrix}_2 & \cdots & \begin{pmatrix} c_1 \\ \vdots \\ c_n \end{pmatrix}_N \Big| n \\ \begin{pmatrix} d_1 \\ \vdots \\ d_{N-n} \end{pmatrix}_1 & \begin{pmatrix} d_1 \\ \vdots \\ d_{N-n} \end{pmatrix}_2 & \cdots & \begin{pmatrix} d_1 \\ \vdots \\ d_{N-n} \end{pmatrix}_N \Big| N-n \end{pmatrix} . \tag{25.129}$$

---

10) Of course, by doing that, we have implicitly switched to a new basis for $\mathcal{H}^n$ – an orthonormal basis of n eigenstates, and will now write the matrix in that basis.

By the property of unitary matrices, the columns of (25.129) will also be mutually orthonormal. Thus, the above procedure will augment the initial nonorthonormal states $|\tilde{b}_i\rangle$ in (25.127) to form orthonormal states $|B_i\rangle$ in $\mathcal{H}^N$. We have used uppercase B to indicate that these states belong to the extended $\mathcal{H}$-space, and also removed the tilde because these states are now normalized. This proves that an N-element POVM defined in $\mathcal{H}^n$ where $n < N$ can always be expressed as a projective measurement on $\mathcal{H}^N$, of which space $\mathcal{H}^n$ is a subspace.

We now have an important result that is known as *Neumark's dilation theorem*: One can always express a projective measurement as a POVM on a smaller subspace, and, conversely, one can always express a POVM as a projective measurement on a higher dimensional space.

The geometric meaning of Neumark's theorem becomes clear from looking again at the 3D case at Figure 25.3. Initially, the POVM-generating directions $|\tilde{b}_i\rangle = \sqrt{\lambda_i}|b_i\rangle$ were defined in the $\{|1\rangle, |2\rangle, |3\rangle\}$-basis. Suppose we now rotate the coordinate system to switch to the new basis $\{|B_1\rangle, |B_2\rangle, |B_3\rangle\}$. If the rotation is selected in such a way that $\{|B_1\rangle, |B_2\rangle, |B_3\rangle\}$ will look in the old $\{|1\rangle, |2\rangle, |3\rangle\}$ basis as the sums of "parallel" and "perpendicular" components defined by the c's and d's in (25.129), then detectors D_1, D_2, D_3 set along $\{|B_1\rangle, |B_2\rangle, |B_3\rangle\}$ will implement the POVM. Neumark's theorem guarantees us that we will always be able to find the appropriate rotation.

25.8
How to Implement a Given POVM

> If our qubit is secretly two components of a qutrit, the POVM may be realized as orthogonal measurement of the qutrit. A typical qubit harbors no such secret, though.
>
> J. Preskill

The existence of a process that would realize a given POVM is guaranteed by Neumark's theorem, but the physical implementation of such a process is somewhat nontrivial. We will now discuss the basic steps involved.

a) *Choosing the joint system*. The addition of extra dimensions to the original Hilbert space must be grounded in physical reality. Our apparatus should be able to see possible system states along the third dimension. Sometimes we have situations when our qubit is by default a part of a *qunit* (an N-state system). This will be the case if we have for some reason neglected the extra dimensions available to the qunit and focused on the first two dimensions only, effectively choosing to treat the qunit as a qubit. For example, when we call the photon a two-state system based on its polarization, we ignore the possibility that the photon could be absorbed. If we keep this possibility in mind, then the photon is really a *qutrit* with the basis $\{|\leftrightarrow\rangle, |\updownarrow\rangle, |0\rangle\}$ where $|0\rangle$ is the vacuum state (see Box 25.3).

Having said that, the "secret" third dimension provided by photon absorption within the apparatus is rather a special case that cannot serve as general rule [6]. Generally, we must assume that the qubit will be just that – a two-state system lacking extra dimensions.

We must then make the qubit part of a composite system. Let us therefore combine our qubit Q with another system A called the *ancilla*,[11] and let C be the resulting composite system in the Hilbert product space $\mathcal{H}_C = \mathcal{H}_Q \otimes \mathcal{H}_A$. For example, if both the qubit and the ancilla are (distinguishable!) photons, polarization space of the joint system will be four dimensional, making it possible to implement a POVM with the number of elements $N = 3$ or $N = 4$. The number of POVM elements determines the proper choice of ancilla. For example, a six-element POVM can be realized with a two-state qubit and a three-state ancilla such as an electron and a deuterium nucleus; for an eight-element POVM, one should consider a four-state ancilla, and so on. Here we assume the qubit and ancilla to be distinguishable particles.

Initially, before they had a chance to interact, the qubit and ancilla are just independent parts of the joint system: $|\psi_C\rangle = |\psi_Q\rangle \otimes |\psi_A\rangle$. We shall not go into details as to what physical process could entangle the particles. Suppose such a process has been found.

Box 25.3 Photon as a qutrit

To give you one practical example of this qutrit nature of a photon, we can mention an alternative way to discriminate between two nonorthogonal photon polarizations. There are linear optical devices that can perform polarization-sensitive attenuation of light pulses (Figure 25.10). Due to attenuation, initially nonorthogonal polarization vectors can be orthogonalized and then after leaving the device the state can be determined unambiguously. We would then have a perfect method of discriminating the states, bypassing the complex procedure outlined in Section 25.3. The trouble is that the intensity of a given photon will be reduced in the process. In the language of QM, this means we will have a nonzero probability to discover the photon in a vacuum state. From an experimenter's viewpoint, in $1 - 2\sin^2\theta = \cos 2\theta$ fraction of the trials, neither of the two detectors will click, leading him to conclude that the photon was absorbed and determining its initial state is no longer possible.

---

11) Alternatively, the ancilla could represent another aspect of the *same* system – for example, momentum space of an electron could serve as the ancilla, helping extend the "system" defined by the spin space of the same electron. In the above-mentioned direct sum example, the vacuum state was acting as the ancilla. To put it in most general terms, we say that ancilla is comprised by the extra degrees of freedom that we want to add to a system in question.

Figure 25.11 constructs the Hilbert space of this photon by adding a third basis vector $|0\rangle$ perpendicular to the plane of polarization. Mathematically, this space has the structure of a direct sum $\mathcal{H}^3 = \mathcal{H}^2 \oplus \mathcal{H}^1$ of the constituent subspaces – plane of polarization $\mathcal{H}^2 = \{|\leftrightarrow\rangle, |\updownarrow\rangle\}$ and a one-dimensional subspace with a single vacuum basis state $\mathcal{H}^1 = \{|0\rangle\}$. The 3D basis is comprised from the old basis vectors by filling zeros to the unused entries:

$$|\leftrightarrow\rangle = \begin{pmatrix} 1 \\ 0 \\ 0 \end{pmatrix}, \quad |\updownarrow\rangle = \begin{pmatrix} 0 \\ 1 \\ 0 \end{pmatrix}, \quad |0\rangle = \begin{pmatrix} 0 \\ 0 \\ 1 \end{pmatrix}. \tag{25.130}$$

The two states of the incident photon will be transformed by the apparatus as

$$\hat{U} \begin{pmatrix} \cos\theta \\ \sin\theta \\ 0 \end{pmatrix} = \begin{pmatrix} \sin\theta \\ \sin\theta \\ \sqrt{1 - 2\sin^2\theta} \end{pmatrix}, \quad \hat{U} \begin{pmatrix} \cos\theta \\ -\sin\theta \\ 0 \end{pmatrix} = \begin{pmatrix} \sin\theta \\ -\sin\theta \\ \sqrt{1 - 2\sin^2\theta} \end{pmatrix}, \tag{25.131}$$

where $\hat{U}$ is an operator describing the transformation of a photon state by the apparatus and then a projective measurement can easily be performed upon the output state.

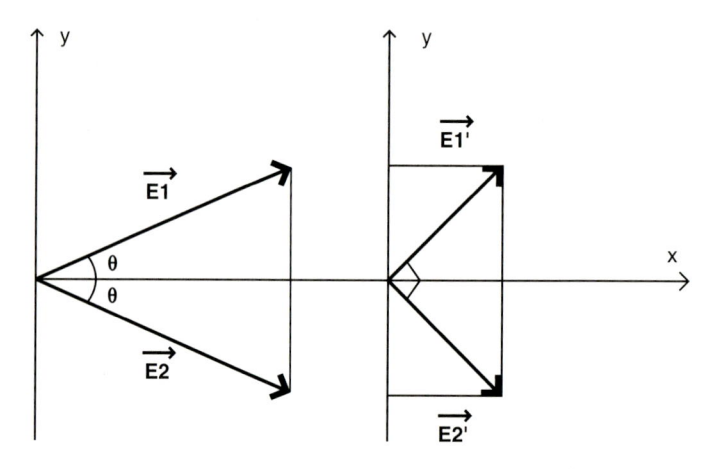

Figure 25.10 Unambiguous photon state discrimination in classical optics. The optical device attenuates the x-component of polarization by $\tan\theta$ without affecting the y-component. The intensity of the resulting pulse is reduced by $2\sin^2\theta$. Two possible polarization states of a photon are given by unit vectors $\vec{E}_1$ and $\vec{E}_2$. The medium attenuates the x-component of the field by a factor of $\tan\theta$, leaving the y-component unchanged. As a result, the two states become orthogonal and can be discriminated by a simple projective measurement. However, intensity changes from $I_{1,2} = |\vec{E}_{1,2}|^2$ to $I'_{1,2} = |\vec{E}'_{1,2}|^2$ so that the intensity ratio is $I'_{1,2}/I_{1,2} = (E_x^2 \tan^2\theta + E_y^2)/(E_x^2 + E_y^2) = (E^2 \sin^2\theta + E^2 \sin^2\theta)/(E^2 \cos^2\theta + E^2 \sin^2\theta) = 2\sin^2\theta$. Then success rate is $p = 2\sin^2\theta = 1 - \cos 2\theta = 1 - |\vec{E}_1 \cdot \vec{E}_2|$ and failure rate is $q = 1 - p = |\vec{E}_1 \cdot \vec{E}_2|$.

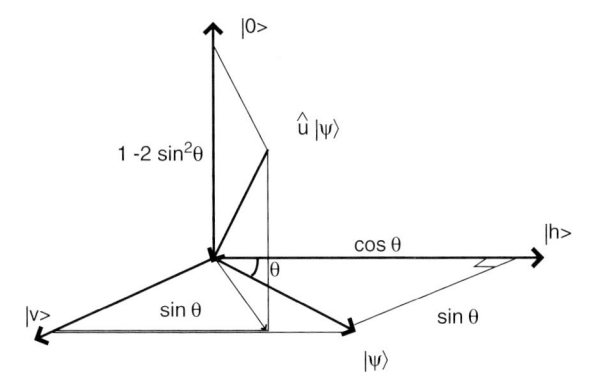

Figure 25.11 Hilbert space of the photon qutrit. The vacuum state is orthogonal to both polarization states. The photon enters the device in a superposition $E\cos\theta|\leftrightarrow\rangle + E\sin\theta|\updownarrow\rangle + 0|0\rangle$ and leaves it in a superposition $E\sin\theta|\leftrightarrow\rangle + E\sin\theta|\updownarrow\rangle + E\sqrt{1-2\sin^2\theta}|0\rangle$.

b) *Initial state preparation.* Initial state of the joint system will be a vector in the new $\mathcal{H}$-space (Figure 25.12). Suppose the qubit and ancilla are photons and we want to implement a three-element POVM The general state of the joint system is

$$|\psi_C\rangle = c_{11}|\updownarrow\rangle_Q \otimes |\updownarrow\rangle_A + c_{12}|\updownarrow\rangle_Q \otimes |\leftrightarrow\rangle_A + c_{21}|\leftrightarrow\rangle_Q \otimes |\updownarrow\rangle_A \qquad (25.132)$$
$$+ c_{22}|\leftrightarrow\rangle_Q \otimes |\leftrightarrow\rangle_A$$

and thus we have a four-dimensional space spanned by the basis (pay attention to the order!):

$$\mathcal{H}_C = \left\{|\updownarrow\rangle_Q|\updownarrow\rangle_A, |\updownarrow\rangle_Q|\leftrightarrow\rangle_A, |\leftrightarrow\rangle_Q|\updownarrow\rangle_A, |\leftrightarrow\rangle_Q|\leftrightarrow\rangle_A\right\}. \qquad (25.133)$$

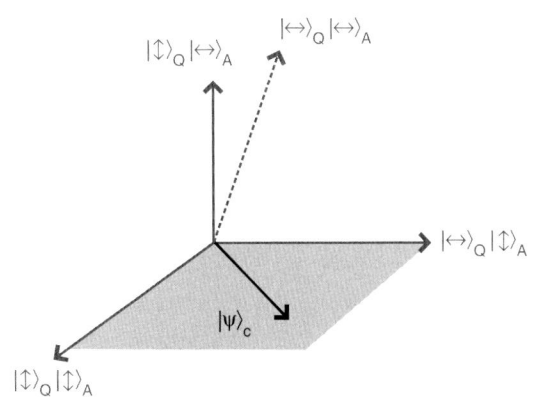

Figure 25.12 Initial state of the joint system. Subscript Q stands for qubit and subscript A stands for ancilla. The qubit is in a superposition $|\psi\rangle_Q = c_1|\updownarrow\rangle_Q + c_2|\leftrightarrow\rangle_Q$ with arbitrary amplitudes. The ancilla is prepared in a vertically polarized state $|\psi\rangle_A = |\updownarrow\rangle_A$. The dashed line represents the unused dimension that will be ignored later.

Let us now form the joint state $|\psi\rangle_C = |\psi\rangle_Q \otimes |\psi\rangle_A$ of the qubit and ancilla prepared in initial states:

$$|\psi\rangle_Q = c_1|\updownarrow\rangle + c_2|\leftrightarrow\rangle, \quad |\psi\rangle_A = d_1|\updownarrow\rangle + d_2|\leftrightarrow\rangle. \tag{25.134}$$

If we choose $d_1 = 1$, $d_2 = 0$, making the ancilla photon V-polarized, the last equation becomes

$$|\psi\rangle_C = |\psi\rangle_Q \otimes |\psi\rangle_A = \begin{pmatrix} c_1 \\ 0 \\ c_2 \\ 0 \end{pmatrix} \tag{25.135}$$

and we can represent the initial state with a vector in the plane formed by the first and third basis vectors (25.133). Here, the ancilla is V-polarized, while the qubit can be in a superposition of polarization states that depends on the initial preparation.

c) *Discarding the extra dimensions.* The new H-space may give us more dimensions than we really need. Thus, in the present example, our extended basis is comprised of four eigenstates, but we only need three dimensions to realize a three-element POVM. Therefore, we can leave the extra dimension "underutilized" by setting the fourth component equal to zero, and let the entire action unfold in the 3D subspace of $\mathcal{H}_C$. Which dimension to discard is up to us. In our case, the initial state already has components along $|\updownarrow\rangle_Q \otimes |\updownarrow\rangle_A$ and $|\leftrightarrow\rangle_Q \otimes |\updownarrow\rangle_A$, so we can discard either of the remaining two dimensions. Suppose we drop the $|\leftrightarrow\rangle_Q \otimes |\leftrightarrow\rangle_A$ dimension and work in the reduced subspace,

$$\mathcal{H}'_C = \{|\updownarrow\rangle_Q \otimes |\updownarrow\rangle_A, |\updownarrow\rangle_Q \otimes |\leftrightarrow\rangle_A, |\leftrightarrow\rangle_Q \otimes |\updownarrow\rangle_A\}. \tag{25.136}$$

The extra detector that could not fit into the two-dimensional space of the qubit has now been assigned a distinct ancilla state with horizontal polarization. It will be convenient then, when generalizing to an N-element POVM on a qubit, to take an ancilla with $N-1$ eigenstates $|0\rangle_A, |1\rangle_A, |2\rangle_A, \ldots, |N-2\rangle_A$ (Figure 25.13). The

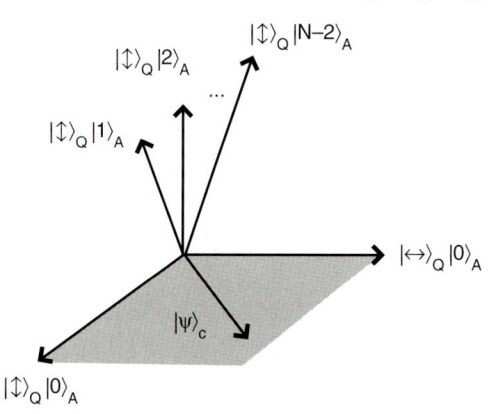

Figure 25.13 N-element POVM on a qubit realized with an ancilla that has $N-1$ eigenstates $|0\rangle_A, |1\rangle_A, |2\rangle_A, \ldots, |N-2\rangle_A$.

joint system's state vector will initially lie in the plane $\{|\updownarrow\rangle_Q \otimes |0\rangle_A, |\leftrightarrow\rangle_Q \otimes |0\rangle_A\}$. The remaining ancilla eigenstates $|1\rangle_A, |2\rangle_A, \ldots, |N-2\rangle_A$ will be assigned to the same qubit state $|\updownarrow\rangle_Q$. The Hilbert space $\mathcal{H}_C = \mathcal{H}_Q \otimes \mathcal{H}_A$ now has $2N-2$ dimensions out of which we will only use N, leaving the other some $N-2$ dimensions in $\mathcal{H}_C$ unutilized. The advantage is that a projective measurement becomes particularly easy. The last $N-2$ detectors will only need to have a polarization analyzer for the ancilla, since the qubit state is guaranteed to be $|\updownarrow\rangle_Q$. The first two detectors must have two polarization analyzers – one for the ancilla and the other for the qubit. The ancilla will be measured first. Outcome $|0\rangle_A$ will collapse the qubit into the horizontal plane, and then a measurement on the qubit will make it choose between $|\updownarrow\rangle_Q$ and $|\leftrightarrow\rangle_Q$. This setup simplifies both the detector design and the structure of the corresponding equations and is in fact taken for granted in most texts.

d) *Constructing the unitary matrix.* Next, we construct matrix (25.129) whose existence is guaranteed by Neumark's theorem. Namely, we compose from vectors $|\tilde{b}_i\rangle$ an $n \times N$ matrix, and then fill the remaining $N - n$ rows with numbers as needed to obtain a unitary $N \times N$ matrix. Components of the "new basis" are now given by the columns of this matrix. The POVM will be realized by a projective measurement in that basis.

As an example, consider the POVM $\hat{F}_i = (2/3)|b_i\rangle\langle b_i|$ that we encountered earlier in Equation 25.64. As we found then, its elements, as defined in the Hilbert space $\mathcal{H}_Q$, are

$$|\tilde{b}_1\rangle = \frac{1}{\sqrt{6}}\begin{pmatrix} 2 \\ 0 \end{pmatrix}, \quad |\tilde{b}_2\rangle = \frac{1}{\sqrt{6}}\begin{pmatrix} 1 \\ \sqrt{3} \end{pmatrix}, \quad |\tilde{b}_3\rangle = \frac{1}{\sqrt{6}}\begin{pmatrix} -1 \\ \sqrt{3} \end{pmatrix}. \qquad (25.137)$$

As we now switch to the joint space $\mathcal{H}_C$, the above vectors will become four-dimensional. We can cross out the last row that corresponds to the (now excluded) $|\leftrightarrow\rangle_Q|\leftrightarrow\rangle_A$ dimension:

$$\{|\tilde{b}_1\rangle \otimes |\psi\rangle_A, |\tilde{b}_2\rangle \otimes |\psi\rangle_A, |\tilde{b}_3\rangle \otimes |\psi\rangle_A\} = \frac{1}{\sqrt{6}}\begin{pmatrix} 2 & 1 & -1 \\ 0 & 0 & 0 \\ 0 & \sqrt{3} & \sqrt{3} \\ \cancel{0} & \cancel{0} & \cancel{0} \end{pmatrix}. \qquad (25.138)$$

Then fill the second row with numbers $\left(\sqrt{1/3}, -\sqrt{1/3}, \sqrt{1/3}\right)$ to make the remaining 3×3 matrix unitary:

$$\{|B_1\rangle, |B_2\rangle, |B_3\rangle\} = \frac{1}{\sqrt{6}}\begin{pmatrix} 2 & 1 & -1 \\ \sqrt{2} & -\sqrt{2} & \sqrt{2} \\ 0 & \sqrt{3} & \sqrt{3} \end{pmatrix}. \qquad (25.139)$$

The second row is now comprised of components $|B_1^\perp\rangle, |B_2^\perp\rangle, |B_3^\perp\rangle$ normal to the plane $\{|\updownarrow\rangle_Q \otimes |\updownarrow\rangle_A, |\leftrightarrow\rangle_Q \otimes |\updownarrow\rangle_A\}$. These components are directed along the axis $|\updownarrow\rangle_Q \otimes |\leftrightarrow\rangle_A$, which corresponds to a horizontally polarized ancilla. We thus have the physical meaning of the third dimension! The POVM $\hat{F}_i = (2/3)|b_i\rangle\langle b_i|$ can now be realized by a projective measurement along the new basis vectors $|B_1\rangle, |B_2\rangle, |B_3\rangle$ given by the columns of this matrix. In a similar manner, we can construct a unitary martix for any other POVM.

e) *Rotating the initial state vector.* By construction, matrix $M_{N \times N}$ consists of the new column vectors $\{|B_1\rangle, |B_2\rangle, |B_3\rangle\}$ written in the *old* basis $\{|1\rangle, |2\rangle, |3\rangle\}$. Hence it is nothing else but the transformation matrix from the new to the old basis, which is described by the similarity transformation studied earlier:

$$\mathbf{r}' = M^+ \mathbf{r}, \quad \hat{A}' = M^+ \hat{A} M \tag{25.140}$$

In this context, it is natural that $M_{N \times N}$ turned out to be unitary, since coordinate system rotations are norm-preserving and hence represented by unitary matrices. Then to see what the initial state looks like in the new basis, we only need to act on it with unitary operator $\hat{U} = M_{N \times N}^+$. It has the mathematical meaning of transformation matrix from the old to the new basis such that

$$\mathbf{r}' = \hat{U}\mathbf{r}, \quad \hat{A}' = \hat{U}\hat{A}\hat{U}^+. \tag{25.141}$$

There are two ways to think of the transformation $\mathbf{r}' = \hat{U}\mathbf{r}$ (Figure 25.14). We could think of it as a *passive* transformation where $\mathbf{r}$ remains unchanged but the axes with detectors affixed to them move to their new positions $\{|B_1\rangle, |B_2\rangle, |B_3\rangle\}$. Or else we could picture an *active* transformation that rotates state vector $\mathbf{r}$ with respect to a fixed basis. Both pictures are mathematically equivalent but radically different in their physical implementation. To "rotate the axes" one would have to build a new measuring device, which, instead of identifying simple states $\{|\updownarrow\rangle_Q \otimes |\updownarrow\rangle_A, |\updownarrow\rangle_Q \otimes |\leftrightarrow\rangle_A, |\leftrightarrow\rangle_Q \otimes |\updownarrow\rangle_A\}$, will be identifying their superpositions (in other words, such a device would have to measure observables B_1, B_2, B_3). To rotate the state vector, one only needs to let the system evolve in time in the "correct" way, $\hat{U} \equiv M_{N \times N}^+ = e^{-i\hat{H}t/\hbar}$; no changes to the lab equipment are necessary then. Any practical POVM implementation will rely on active rotations, and the only challenge to the experimenter will be to produce an external field that will have the required Hamiltonian.

Any n-dimensional vector $\mathbf{r}$ can be augmented to an N-dimensional vector by appending $N - n$ zero components and then be transformed according to (25.141). For example, "old" vector $|\tilde{b}_i\rangle = \sqrt{\lambda_i}|b_i\rangle$ will transform as follows: $|\tilde{b}_i'\rangle = \hat{U}|\tilde{b}_i\rangle$ or

$$\begin{pmatrix} \tilde{b}'_{1i} \\ \vdots \\ \tilde{b}'_{ni} \\ \tilde{b}'_{n+1,i} \\ \vdots \\ \tilde{b}'_{Ni} \end{pmatrix} = \begin{pmatrix} c_{11}^* & \cdots & c_{n1}^* & d_{n+1,1}^* & \cdots & d_{N1}^* \\ \vdots & \vdots & \vdots & \vdots & \vdots & \vdots \\ c_{1n}^* & c_{2n}^* & \cdots & \cdots & \cdots & c_{Nn}^* \\ d_{1,n+1}^* & d_{2,n+1}^* & \cdots & \cdots & \cdots & d_{N,n+1}^* \\ \vdots & \vdots & \vdots & \vdots & \vdots & \vdots \\ d_{1N}^* & d_{2N}^* & \cdots & \cdots & \cdots & d_{NN} \end{pmatrix} \begin{pmatrix} \tilde{b}_{1i} \\ \vdots \\ \tilde{b}_{ni} \\ 0 \\ \vdots \\ 0 \end{pmatrix}$$

$$\tag{25.142}$$

and we will know the new components of this vector. Since a unitary transformation is norm preserving, the length of this vector is still $\sqrt{\lambda_i}$ and the corresponding unit vector is of course $|b_i'\rangle = |\tilde{b}_i'\rangle/\sqrt{\lambda_i}$. By forming dot products between the new vectors $|b_i'\rangle$ and detection directions $|B_i\rangle$, we can find the detection amplitudes for each outcome and hence estimate outcome statistics of the given POVM (Figure 25.14). Thus, the unitary operator for POVM $\hat{F}_i = (2/3)|b_i\rangle\langle b_i|$ will be obtained by taking Hermitian adjoint of matrix (25.139):

$$\hat{U} = \frac{1}{\sqrt{6}} \begin{pmatrix} 2 & \sqrt{2} & 0 \\ 1 & -\sqrt{2} & \sqrt{3} \\ -1 & \sqrt{2} & \sqrt{3} \end{pmatrix}, \qquad (25.143)$$

and then applying it to $|\tilde{b}_1\rangle, |\tilde{b}_2\rangle, |\tilde{b}_3\rangle$ vectors from (25.138) we will obtain

$$\hat{U}|\tilde{b}_1\rangle = \frac{1}{3}\begin{pmatrix} 2 \\ 1 \\ -1 \end{pmatrix}, \quad \hat{U}|\tilde{b}_2\rangle = \frac{1}{3}\begin{pmatrix} 1 \\ 2 \\ 1 \end{pmatrix}, \quad \hat{U}|\tilde{b}_3\rangle = \frac{1}{3}\begin{pmatrix} -1 \\ 1 \\ 2 \end{pmatrix}. \qquad (25.144)$$

These are our new vectors $|\tilde{b}_1'\rangle, |\tilde{b}_2'\rangle, |\tilde{b}_3'\rangle$. As we see, the norm $\sqrt{2/3}$ of $|\tilde{b}_1\rangle, |\tilde{b}_2\rangle, |\tilde{b}_3\rangle$ is preserved by the transformation.

f) Acting with $\hat{U}$ on the initial state. Initially, the three directions $|\tilde{b}_1\rangle, |\tilde{b}_2\rangle, |\tilde{b}_3\rangle$ that generated the POVM were located in one 2D plane. Linearity of QM means that it

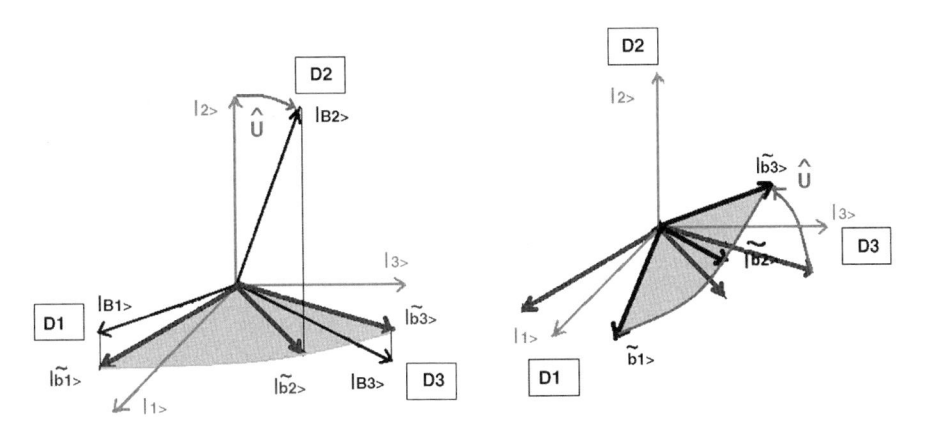

Figure 25.14 Passive vs. active rotations. (a) Passive rotation of coordinate system that would implement a POVM, enabling possible outcomes along $|\tilde{b}_1\rangle, |\tilde{b}_2\rangle, |\tilde{b}_3\rangle$. Unitary transformation $\hat{U} \equiv M_{N\times N}^+$ from the old $\{|1\rangle, |2\rangle, |3\rangle\}$ to the new $\{|B_1\rangle, |B_2\rangle, |B_3\rangle\}$ basis can be deduced. (b) Equivalently, active rotation $\hat{U}$ puts state vectors of the system at the correct angles with respect to the detector directions, while leaving the coordinate system unchanged.

will be fully sufficient for our purposes to describe the effect of $\hat{U}$ on any *two* vectors in that plane. For instance, let us choose two arbitrary normalized initial states $|a_1\rangle$ and $|a_2\rangle$ defined by their coordinates

$$|a_1\rangle = \begin{pmatrix} g_1 \\ 0 \\ g_3 \end{pmatrix}, \quad |a_2\rangle = \begin{pmatrix} h_1 \\ 0 \\ h_3 \end{pmatrix}. \tag{25.145}$$

Acting on these vectors with the unitary operator, we obtain[12]

$$\hat{U}(|a_1\rangle) = \hat{U}\begin{pmatrix} g_1 \\ 0 \\ g_2 \end{pmatrix} = \begin{pmatrix} g'_1 \\ g'_2 \\ g'_3 \end{pmatrix} = g'_1|\updownarrow\rangle_Q \otimes |\updownarrow\rangle_A + g'_2|\updownarrow\rangle_Q \otimes |\leftrightarrow\rangle_A + g'_3|\leftrightarrow\rangle_Q \otimes |\updownarrow\rangle_A,$$

$$\hat{U}(|a_2\rangle) = U\begin{pmatrix} h_1 \\ 0 \\ h_2 \end{pmatrix} = \begin{pmatrix} h'_1 \\ h'_2 \\ h'_3 \end{pmatrix} = h'_1|\updownarrow\rangle_Q \otimes |\updownarrow\rangle_A + h'_2|\updownarrow\rangle_Q \otimes |\leftrightarrow\rangle_A + h'_3|\leftrightarrow\rangle_Q \otimes |\updownarrow\rangle_A.$$

(25.146)

In particular, if the system is initially prepared in either $|a_1\rangle$ or $|a_2\rangle$, then (25.146) gives us the possible final states of this system after the unitary transformation. The effect of $\hat{U}$ on any *other* vector $|\varphi\rangle$ in the plane follows easily from (25.146) since we can always write $|\varphi\rangle$ as a linear combination of $|a_1\rangle$ and $|a_2\rangle$. (In other words, a unitary transformation always maps a plane onto another plane.) This is why description of the POVM provided by (25.146) is complete and in practice, instead of writing the transition matrix explicitly, one usually just says, "the POVM is implemented by a unitary transformation that takes $|a_1\rangle$ into $|a'_1\rangle$ and $|a_2\rangle$ into $|a'_2\rangle$."

g) *Making the projective measurement.* Finally, we make a projective measurement that will collapse the joint system into one of the states $|1\rangle = |\updownarrow\rangle_Q \otimes |\updownarrow\rangle_A$, $|2\rangle = |\updownarrow\rangle_Q \otimes |\leftrightarrow\rangle_A$, $|3\rangle = |\leftrightarrow\rangle_Q \otimes |\updownarrow\rangle_A$. The general measurement (25.144) is fuzzy in the sense that a given detector D_l (where $l = 1, 2, 3$) can react to either state $|a_1\rangle$ or $|a_2\rangle$ with respective probabilities $|g'_l|^2$ and $|h'_l|^2$. If we want to achieve unambiguous state discrimination, we must try to eliminate this fuzziness at least for some detectors.

For example, consider again the state discrimination problem discussed in Section 25.6. We want to realize the POVM 25.118 that will help us discriminate between two distinct input states $|a_1\rangle$ and $|a_2\rangle$. Let us then set $|h'_1|^2 = |g'_3|^2 = 0$ so that D_1 will never react to $|a_2\rangle$ and D_3 will never react to $|a_1\rangle$. Then, (25.146) simplifies to

$$\hat{U}(|a_1\rangle) = g'_1|\updownarrow\rangle_Q \otimes |\updownarrow\rangle_A + g'_2|\updownarrow\rangle_Q \otimes |\leftrightarrow\rangle_A,$$
$$\hat{U}(|a_2\rangle) = h'_2|\updownarrow\rangle_Q \otimes |\leftrightarrow\rangle_A + h'_3|\leftrightarrow\rangle_Q \otimes |\updownarrow\rangle_A. \tag{25.147}$$

---

[12] Of course, if the qubit space H_Q had n dimensions, then (25.146) would be a system of n equations.

Then a click in D_1 identifies the input state unambiguously as $|a_1\rangle$, while a click in D_3 identifies it as $|a_2\rangle$. The respective probabilities $|g_1'|^2$ and $|h_3'|^2$ define success rates p_1 and p_2 for these two input states. On the other hand, a click in D_2 is inconclusive and probabilities $|g_2'|^2$ and $|h_2'|^2$ define the respective failure rates q_1 and q_2. Taking the square root leaves the phases of probability amplitudes undetermined:

$$U(|a_1\rangle) = \sqrt{p_1}e^{i\delta_1}|\updownarrow\rangle_Q \otimes |\updownarrow\rangle_A + \sqrt{q_1}e^{i\delta_2}|\updownarrow\rangle_Q \otimes |\leftrightarrow\rangle_A,$$
$$U(|a_2\rangle) = \sqrt{q_2}e^{i\theta_1}|\updownarrow\rangle_Q \otimes |\leftrightarrow\rangle_A + \sqrt{p_2}e^{i\theta_2}|\leftrightarrow\rangle_Q \otimes |\updownarrow\rangle_A.$$
(25.148)

However, this equation can easily be simplified. Absorbing the phases δ_1 and θ_2 into the initial states, then absorbing the phase difference $\delta_2 - \delta_1$ into the ancilla state $|\updownarrow\rangle_A$, and redenoting the remaining phase, we finally get the following simple algorithm for the unitary transformation:

$$U(|b_1\rangle|\updownarrow\rangle_A) = \sqrt{p_1}|\updownarrow\rangle_Q \otimes |\updownarrow\rangle_A + \sqrt{q_1}|\updownarrow\rangle_Q \otimes |\leftrightarrow\rangle_A,$$
$$U(|b_2\rangle|\updownarrow\rangle_A) = \sqrt{p_2}|\leftrightarrow\rangle_Q \otimes |\updownarrow\rangle_A + \sqrt{q_2}e^{i\theta}|\updownarrow\rangle_Q \otimes |\leftrightarrow\rangle_A.$$
(25.149)

The success and failure rates for an optimal POVM measurement are already known to us from the theoretical analysis in Section 25.6 where we derived them as functions of the overlap $|\langle a_1|a_2\rangle| = \cos\theta$ between the input states. Therefore, (25.149) provides an experimenter with a complete recipe for implementing the optimal state discrimination measurement. Transformation (25.149) is shown schematically in Figure 25.15. Notice that the POVM has now been reduced to an entirely projective measurement on the ancilla exclusively! A click along the $|\leftrightarrow\rangle_A$ direction represents the inconclusive outcome; both input states collapse into the same output and information about the original state is lost. A click along the $|\updownarrow\rangle_A$ direction collapses the qubit state into $|\updownarrow\rangle_Q$ if the input state was $|a_1\rangle$ or into $|\leftrightarrow\rangle_Q$ if the input state was $|a_2\rangle$. From a physicist's viewpoint, $\hat{U}$ will be realized if we can think of some physical process that will couple the system with the ancilla and bring the resulting joint system into a superposition of states given by (25.149). Once this process is completed, an individual measurement on the ancilla will be our best chance to resolve the initial qubit state. Thus, at least in theory, we can always suggest an algorithm for implementing a given POVM. But practical quantum engineering is a different matter. Constructing the appropriate apparatus from standard devices such as linear optical elements is a challenging problem, which requires a lot of creativity from the experimenter.

25.9
Comparison of States and Mixtures

Another problem that makes use of the concept of POVMs is finding how close two given states or mixtures are to each other. We have encountered it in Chapter 24 when discussing fidelity F of the cloning process. Back then, we defined it loosely as

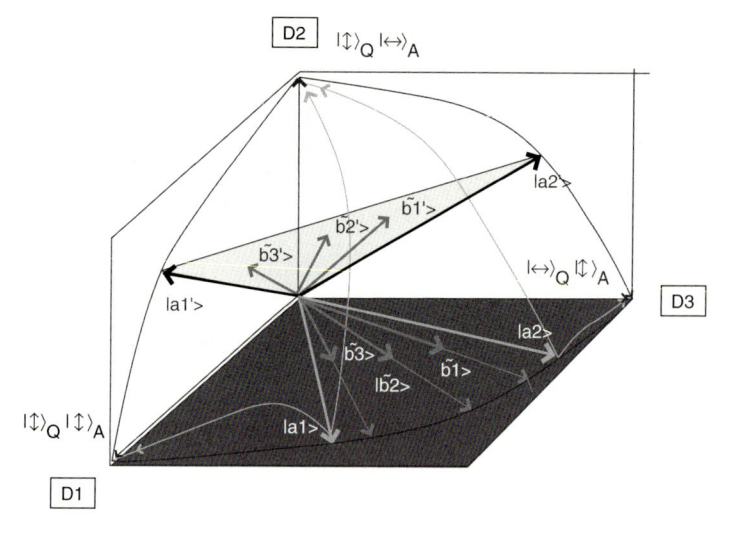

Figure 25.15 State discrimination measurement is physically implemented by the POVM shown below. The system of coordinates corresponds to the initial state of the joint system of the qubit and the ancilla. A unitary transformation takes the horizontal plane (shaded in dark gray) into a tilted plane (shaded in light gray). Outcome D_2 is inconclusive and could indicate either $|a_1\rangle$ or $|a_2\rangle$.

the probability to pick a perfect copy from the mixture produced by the UQCM. It is time to state the rigorous definition.

In classical statistics, one useful measure of difference or, one sometimes says, the *distance* between two random variables is the Bhattacharyya coefficient defined as

$$B = \sum_i \sqrt{P(x_i)Q(x_i)}, \tag{25.150}$$

where P and Q are the respective probabilities for the random variables to assume value x_i and the summation is over all possible x_i. In QM, if we are given normalized pure states $\varphi(x_i), \psi(x_i)$ with x_i being the generalized coordinate, we can also think of them as random variables with $P(x_i) = |\varphi(x_i)|^2$, $Q(x_i) = |\psi(x_i)|^2$. Then the above equation reduces to the absolute value of the inner product:

$$B(\varphi, \psi) = |\langle \varphi | \psi \rangle|. \tag{25.151}$$

However, the Bhattacharyya coefficient thus defined cannot serve as the general measure of distance in QM. Conceptually, the distance between quantum states is as big or small as our ability to distinguish them. The latter is defined by the *optimal* POVM $\{F_i\}$, which returns ith outcome with probability $p_i = \text{Tr}\,[\rho F_i]$ where ρ is a density matrix (see Equation 25.53). This gives us grounds to define the quantum analogue of the Bhattacharyya coefficient, which can handle not only pure states but also mixtures ρ and σ:

$$B(\rho, \sigma) = \sum_i \sqrt{\text{Tr}\,[\rho F_i] \text{Tr}\,[\sigma F_i]}. \tag{25.152}$$

The next step is given without proof. Fuchs and Caves showed definition (25.152) to be equivalent to

$$B(\rho, \sigma) \equiv \text{Tr}\left[\sqrt{\sqrt{\rho}\sigma\sqrt{\rho}}\right] = \text{Tr}\left[\sqrt{\sqrt{\sigma}\rho\sqrt{\sigma}}\right] \equiv B(\sigma, \rho). \tag{25.153}$$

The function is symmetric in its inputs, as one should expect. The root in this formula is the principal square root whose uniqueness is guaranteed by positive semidefiniteness of density matrices (recall Chapter 12). If both input states are pure, this formula goes over to

$$B(\varphi, \psi) = \text{Tr}\left[\sqrt{\sqrt{|\varphi\rangle\langle\varphi|}(|\psi\rangle\langle\psi|)\sqrt{|\varphi\rangle\langle\varphi|}}\right]. \tag{25.154}$$

Following Jozsa [1997], we call the square of the Bhattacharyya coefficient *fidelity*:

$$\mathcal{F}(\rho, \sigma) \equiv (B(\rho, \sigma))^2 = \left\{\text{Tr}\left[\sqrt{\sqrt{\rho}\sigma\sqrt{\rho}}\right]\right\}^2. \tag{25.155}$$

This is the same quantity $\mathcal{F}$ (not to be confused with the POVM) that we encountered earlier in the context of quality of cloning. For two pure states, Equation 25.154 goes over to (25.151) and then $\mathcal{F}$ has the meaning of transition probability between the states.[13]

25.10
Generalized Measurements

In the POVMs we considered so far, there was a one-to-one correspondence between a POVM element $\hat{F}_m$ and the respective outcome m. Such POVMs are governed by the formulas[14]

$$p(m) = \langle\psi|\hat{M}_i^+\hat{M}_i|\psi\rangle, \tag{25.156}$$

$$|\psi'\rangle = \frac{\hat{M}_m|\psi\rangle}{\sqrt{p(m)}} = \frac{\hat{M}_m|\psi\rangle}{\sqrt{\langle\psi|\hat{M}_m^+\hat{M}_m|\psi\rangle}}, \tag{25.157}$$

$$p(m) = \sum_j w_j p_j(m) = \text{Tr}\left(\hat{M}_m^+\hat{M}_m\rho\right) = \text{Tr}\left(\hat{M}_m\rho\hat{M}_m^+\right) = \text{Tr}\left(\rho\hat{M}_m^+\hat{M}_m\right), \tag{25.158}$$

$$\tilde{\rho}_m' = \hat{M}_m\rho\hat{M}_m^+, \tag{25.159}$$

$$\rho_m' = \frac{\tilde{\rho}_m}{p(m)}, \tag{25.160}$$

---

13) Some textbooks refer to the quantum Bhattacharyya coefficient (25.153) as "fidelity." However, definition (25.155) is becoming common in modern publications.

14) To streamline the notations, we will remove the (now redundant) references to the "projection directions" a_i, b_i, $\hat{b}_i$ and also replace i with the more common symbol m.

with

$$\hat{M}_m \equiv \sqrt{\hat{M}_m^+ \hat{M}_m} = \sqrt{\frac{\hat{F}_m^+ \hat{F}_m}{\lambda_m}} = \sqrt{\hat{F}_m}, \qquad (25.161)$$

$$\sum_m \hat{M}_m^+ \hat{M}_m = I. \qquad (25.162)$$

Here, both ρ and $\tilde{\rho}', \rho'$ are density matrices defined in the Hilbert space of the system (e.g., qubit, as in the last section). It is often convenient to introduce a new operator $\hat{\mathcal{E}}_m$ that takes ρ as an input and produces the operator product $\hat{M}_m \rho \hat{M}_m^+$:

$$\hat{\mathcal{E}}_m(\rho) \equiv \hat{M}_m \rho \hat{M}_m^+. \qquad (25.163)$$

This formula, which represents $\hat{\mathcal{E}}_m(\rho)$ in terms of Kraus operators, is known as *operator-sum decomposition*, though in the case (25.163) it obviously has only one summand. Then $\hat{\mathcal{E}}_m(\rho)$ shows how a density matrix is transformed under a POVM, and the probabilities of outcomes are found simply by taking the trace:

$$p(m) = \text{Tr}\left(\hat{\mathcal{E}}_m(\rho)\right), \quad \tilde{\rho}'_m = \hat{\mathcal{E}}_m(\rho), \quad \rho'_m = \frac{\tilde{\rho}'_m}{p(m)} = \frac{\hat{\mathcal{E}}_m(\rho)}{\text{Tr}\left(\hat{\mathcal{E}}_m(\rho)\right)}. \qquad (25.164)$$

Operator $\hat{\mathcal{E}}_m$ is one example of so-called *superoperators*, which transform one operator into another in the same way an ordinary operator transforms functions. It is a linear operator that does not conserve the trace of ρ because $\text{Tr}\hat{\mathcal{E}}_m(\rho) = \text{Tr}\tilde{\rho}' \leq 1$. If we take a set of possible matrices on the (n-dimensional) Hilbert space of the system $\mathcal{H}^n$, then $\hat{\mathcal{E}}_m$ associates with each member ρ of the set a new matrix ρ'. This way it *maps* the set of density matrices onto itself. The mapping is linear, that is, $\hat{\mathcal{E}}_m(\rho_1 + \rho_2) = \hat{\mathcal{E}}_m(\rho_1) + \hat{\mathcal{E}}_m(\rho_2)$ and $\hat{\mathcal{E}}_m(c\rho) = c\hat{\mathcal{E}}_m(\rho)$. A map is also called *positive* if it will transform any positive semidefinite matrix into another positive semidefinite matrix. Since all density matrices are positive semidefinite, $\hat{\mathcal{E}}_m$ is a positive linear map.

Let us now mention another important property of $\hat{\mathcal{E}}_m$. Let ρ be some $n \times n$ density matrix in $\mathcal{H}^n$. Suppose there exists another system (not ancilla!) whose Hilbert space is k-dimensional, and consider some density matrix σ in $\mathcal{H}^k$. We ask: Is there a way to define a POVM that would act on a product state $\sigma \otimes \rho$ that lives in the product space $\mathcal{H}^{kn} = \mathcal{H}^k \otimes \mathcal{H}^n$? This question turns our attention to the map $I \otimes \hat{\mathcal{E}}_m$ defined by analogy with the way we defined the tensor product of operators in Chapter 12. Here, I is a $k \times k$ identity matrix, so that

$$I \otimes \hat{\mathcal{E}}_m \equiv \begin{pmatrix} \hat{\mathcal{E}}_m & 0 & \cdots & 0 \\ 0 & \hat{\mathcal{E}}_m & \cdots & 0 \\ \vdots & \vdots & \vdots & \vdots \\ 0 & 0 & \cdots & \hat{\mathcal{E}}_m \end{pmatrix}. \qquad (25.165)$$

25.10 Generalized Measurements

When applied to $\sigma \otimes \rho$, this map will leave σ unchanged, while ρ is transformed by $\hat{E}_m$. The end result is the tensor product of matrices, which itself is a $kn \times kn$ matrix:

$$(I \otimes \hat{E}_m)(\sigma \otimes \rho) = \sigma \otimes \hat{E}_m(\rho). \tag{25.166}$$

Thus, $I \otimes \hat{E}_m$ defines a new map between density matrices in kn-dimensional Hilbert space $\mathcal{H}^{kn}$. Now, in linear algebra, a positive map on $\mathcal{H}^n$ will not necessarily give rise to a positive map on $\mathcal{H}^{kn}$ (Problem 25.12); when the positivity of the map holds given a $k \times k$ identity matrix I, such map is called k-positive. If positivity of thus derived new maps holds for all k, the map is called *completely positive*. Then since $I \otimes \hat{E}_m$ amounts to a transformation from one density matrix to another, $\hat{E}_m$ is a completely positive map.[15]

The superoperator $\hat{E}_m$ does not give the most general form of a POVM because the latter must include the case when the same outcome m could result from several different POVM elements, namely,

$$\hat{M}_m^+ \hat{M}_m = \sum_i \hat{M}_{m,i}^+ \hat{M}_{m,i}. \tag{25.167}$$

In this case the term "Kraus operators" applies to $\hat{M}_m^+$, $\hat{M}_m$. The above sum is less than unity; we obtain unity only after summation over both indices:

$$\sum_i \hat{M}_{m,i}^+ \hat{M}_{m,i} < I, \quad \sum_{m,i} \hat{M}_{m,i}^+ \hat{M}_{m,i} = I. \tag{25.168}$$

In this case we can introduce another superoperator $\hat{\mathcal{M}}_m$ that generalizes $\hat{E}_m$ and whose operator-sum decomposition has multiple summands:

$$\hat{\mathcal{M}}_m(\rho) = \sum_i \hat{M}_{m,i} \rho \hat{M}_{m,i}^+. \tag{25.169}$$

Then,

$$p(m) = \text{Tr}\left(\hat{\mathcal{M}}_m(\rho)\right), \quad \tilde{\rho}_m' = \hat{\mathcal{M}}_m(\rho), \quad \rho_m' = \frac{\tilde{\rho}_m'}{p(m)} = \frac{\hat{\mathcal{M}}_m(\rho)}{\text{Tr}\left(\hat{\mathcal{M}}_m(\rho)\right)}. \tag{25.170}$$

Once again, $\hat{\mathcal{M}}_m$ is a completely positive map on the Hilbert space of the system.

We can introduce one last level of abstraction. Generalize (25.163) and (25.169) by defining two new superoperators $\hat{E}(\rho) \equiv \sum_m \hat{E}_m \rho \hat{E}_m^+$ and $\hat{\mathcal{M}}(\rho) \equiv \sum_m \hat{\mathcal{M}}_m \rho \hat{\mathcal{M}}_m^+$. These superoperators take a given mixed state as an input, then sandwich it between the Kraus operators, and finally carry out summation. Imagine now that we act with a POVM on a large number of independent samples ρ. Since outcome m occurs with probability $p(m)$ and produces a (normalized) post-measurement density matrix

---

[15] When stated with full mathematical rigor, the definition should consider the general case of maps $\hat{E}_m$ that transform an $n \times n$ matrix ρ into an $l \times l$ matrix ρ'. This should not concern us however, because the Hilbert space of the *system* (ancilla excluded!) retains its dimensionality throughout this chapter, so we are dealing here with the special case $l = n$.

$\tilde{\rho}'_m = \tilde{\rho}'_m/p(m)$, we can average over the outcomes to write the resulting density matrix as follows:

$$\rho' = \sum_m p(m)\rho'_m = \sum_m \tilde{\rho}'_m = \sum_m \hat{E}_m \rho \hat{E}_m^+ \equiv \hat{E}(\rho), \qquad (25.171)$$

$$\rho' = \sum_m p(m)\rho'_m = \sum_m \tilde{\rho}'_m = \sum_m \hat{M}_m \rho \hat{M}_m^+ \equiv \hat{M}(\rho) \qquad (25.172)$$

Equation (25.171) should be sufficient for many applications. If one wishes to use the most general form of Kraus operators, one must write (25.172) instead, and the system state transforms as $\rho \to \rho' = \hat{M}(\rho)$.

The last result is the most general equation one can write for the transformation of a system under a measurement. Moreover, it turns out to be a complete description of the generic *quantum operation*, which means any transformation that takes one density matrix into another (measurement, interaction with environment, unitary evolution, or symmetry transformations).

This property is guaranteed by *Kraus' theorem*, which we shall give without proof. It says that a mapping $\rho \to \rho'$ always describes a certain quantum measurement if the corresponding superoperator that describes this mapping has an operator-sum decomposition (not necessarily unique!) in the form $\sum_i \hat{K}_i \rho \hat{K}_i^+$ where $\hat{K}_i, \hat{K}_i^+$ are arbitrary linear operators such that $\sum_i \hat{K}_i^+ \hat{K}_i \leq 1$. Conversely, if we are given a quantum operation, we can always represent it in this from.

For the derivation of Kraus' theorem, we refer the reader to the literature [7]. What matters to us now is that superoperator $\hat{M}_m(\rho)$ for the POVM satisfies Kraus' theorem. Thus, it can be interpreted as a generic quantum operation. As a result, if we want to describe quantum measurement as broadly as possible, then the POVM formalism provides as general a definition as we can possibly get. Thus, we have the full right to think of POVM as a *generalized quantum measurement*.

Problems

25.1 Show that the necessary and sufficient condition for a Hermitian operator to be positive semidefinite is that all its eigenvalues are nonnegative.

25.2 A projection onto state $|a_i\rangle$ is defined by $\hat{P}_{a_i} = |a_i\rangle\langle a_i|$. Write in terms of eigenkets $|a_i\rangle$ and eigenbras $\langle a_i|$ (a) a projector onto a state orthogonal to $|a_i\rangle$, and (b) a projector onto state $|b\rangle$ forming an angle θ with state $|a_i\rangle$, that is, $|\langle a_i|b\rangle| = \cos\theta$.

25.3 Prove that matrix A is positive semidefinite if and only if there exists a positive semidefinite matrix B such that $BB = A$.

25.4 Show that the principal square root B found in the previous problem is unique.

25.5 Modify POVM (25.71) for the case when $\vec{b}_1 = \hat{z} = (0,0,1)$, $\vec{b}_2 = -\hat{x} = (-1,0,0)$.

25.6 Analyze the case where the states $|a_1\rangle$ and $|a_2\rangle$ are prepared with respective probabilities n_1 and n_2 to derive formula (25.110), $Q^{POVM} = 2\sqrt{n_1 n_2}\cos\theta$.

25.7 Use the closure relation to find the direction $|b_?\rangle$ explicitly.

25.8 Prove (25.123).

25.9 In the last example (Section 25.7), find the angles formed by each of the elements $|\tilde{b}'_1\rangle, |\tilde{b}'_2\rangle, |\tilde{b}'_3\rangle$ with the detector directions $\{|B_1\rangle, |B_2\rangle, |B_3\rangle\}$.

25.10 Show that (25.154) is equivalent to (25.151).

25.11 We have derived (25.143) "the hard way" by ordering the basis vectors according to (25.133). Repeat this derivation, reordering the basis in a way that would swap the 2nd and 3d rows of matrix (25.138). How will that change the Figures 25.12–15?

25.12 Show that the transposition map, which takes the transpose of the input matrix, is not a completely positive map.

References

1 Nielsen, M.A. and Chuang, I.L. (2000) *Quantum Computation and Quantum Information*, Cambridge University Press.
2 Bergou, J.A., Herzog, U., and Hillery, M. (2004) Discrimination of quantum states. *Lect. Notes Phys.*, **649**, 417–465.
3 Ivanovic, I.D. (1987) How to differentiate between non-orthogonal states. *Phys. Lett. A*, **123**, 257.
4 Dieks, D. (1988) Overlap and distinguishability of quantum states. *Phys. Lett. A.*, **126**, 303.
5 Peres, A. (1988) How to differentiate between non-orthogonal states. *Phys. Lett. A*, **128**, 19.
6 Preskill, J. Foundations II: Measurement and evolution (course notes), http://www.theory.caltech.edu/people/preskill/ph229/notes/chap3.pdf.
7 Audretsch, J. (2006) *Entangled Systems*, Wiley-VCH Verlag GmbH, Berlin.

26
Quantum Information

> One feels, for example, that two punched cards should have twice the capacity of one for information storage, and two identical channels twice the capacity of one for transmitting information.
>
> C.E. Shannon, A Mathematical Theory of Communication

In computer science, the basic unit of information is 1 bit. Simply stated, 1 bit is the amount of information we are going to get when we learn the actual state of a system that can be found in one of the two states (state 0 or state 1), where the *a priori* probability of either state is 50%. The physical agent used to store 1 bit of information could be a radio lamp or, in a modern computer, a transistor placed on the surface of a silicon microchip. Modern transistors can be as small as several nanometers; if we try to reduce the scale much further, we are getting into the quantum territory. Ideally, we would like to encode the state of a bit into a system as small as an atom, or, better yet, a single electron (or any other particle). Such a system will possess quantum, rather than classical properties. The unique features of computation on a subatomic scale and practical study of quantum-sized computers is the subject of quantum information science.

Since we are still going to use the binary language of ones and zeros, it is clear that 1 and 0 must now be represented by eigenstates of the chosen quantum system. What we need then is any two-state system such as NH_3 molecule or an electron in a spin-up/spin-down state. Since this quantum system will be used to store 1 bit of information, we will call it a *qubit*. Obviously, qubits will differ from the classical radio lamps or transistors in that the probability amplitudes for the eigenstates will now exhibit quantum behavior.

26.1
Deterministic Information and Shannon Entropy

In classical computer science, information is a very straightforward concept. Using N classical two-state systems (bits), we can create 2^N possible states. For example, to save one letter of the English alphabet in computer memory, we must tell the computer which one of the 26 letters we have chosen,[1] so we need $N = \log_2 26 \approx$ 4.7 bits to store that information (though of course the hardware implementation will require five registers).

But if one thinks about it, the simplicity of this classical concept is only due to the fact that a classical computer is a rigid device that needs to know each bit *exactly*. The computer does not care that the letter typed by an English-speaking user is more likely to be "e" than "r," because in either case it will still use five registers to store the letter, and then later on it will need the same amount of time to retrieve the actual letter from its five registers. A superposition of "e" and "r" is not allowed. As far as the classical computer is concerned, all of the following is true: (a) All letters are equally likely. (b) No guessing is allowed – a letter is either known or unknown. (c) Information is learned in chunks of 4.7 bits. (d) Any 4.7 bit information gain means we have learned one letter with 100% certainty. In such cases, we say that the bits are *deterministic*.

Suppose now that we want to communicate information via a quantum system. Even though the sender can still encode a letter in 5 qubits, it is not certain that the receiver will be using the same measurement basis as the sender. One must therefore prepare for the possibility that the bits sent will be read *probabilistically*, and the actual message will be decoded with some, hopefully small, uncertainty. When the outcomes are probabilistic in nature, any information gain measured in bits will merely imply that we now have greater confidence than before, but not a 100% certainty. We must then introduce *probabilistic bits* that are not to be confused with our familiar deterministic bits. It becomes clear that we now need a more refined definition of information.

Formally, we can say that we have a *random variable W* that can take one of the 26 different values from "a" to "z." Every possible value is an *outcome*. Let us denote outcomes of a random variable with the lowercase symbol and a subindex, so we can have outcomes $\{w_i\}$ where $i = 1, \ldots, 26$. Let's say an outcome w_i can occur with probability $P(w_i)$. Suppose now that we actually look at the text and read some letter w_n. How much knowledge have we actually gained? At first it may be inclined to answer 4.7 bits. But remember, we are now thinking probabilistically! This means we have to consider the actual probabilities $P(w_i)$.

The relative letter frequency in a typical English text is summarized in the following table:

	a	b	c	d	e	f	g	h	i	j	k	l	m
%	8.17	1.5	2.8	4.3	12.7	2.2	2	6.1	7	0.15	0.77	4.03	2.41
	n	o	p	q	r	s	t	u	v	w	x	y	z
%	6.75	7.57	1.93	0.1	5.99	6.33	9.06	2.76	0.98	2.36	0.15	1.97	0.07

It is easy to see that a low-probability outcome must be considered more valuable than a high-probability outcome. Indeed, encountering the letter "e" does not provide us with

---

1) A realistic text will of course contain more than 26 symbols because of the numbers, punctuation marks, and special characters.

much new information: Statistics tells us that this letter is common enough, so if we were smart, we would anticipate this outcome anyway. On the other hand, encountering the letter "q" would be a much greater surprise than encountering "e," thus providing us with information that would be hard for us to anticipate in advance.

We can ascribe to outcome w_n a quantity $I(w_n)$ called *surprisal*, but more often referred to as *self-information* or *information content*. $I(w_n)$ simply tells us how much probabilistic information is learned when this outcome is detected by the observer.

To see what formula can be used to define $I(w_n)$, consider the two limiting cases. As the probability of an event tends to 1, the information content must clearly go to zero (there is no information to be gained from an outcome that was certain to begin with). On the other hand, as the probability goes to zero, the information content is maximized. It will be reasonable to make it infinite (one will be *infinitely* surprised to see a walking dinosaur, and the value of information gained thereby can hardly be overestimated!) Finally, we want $I(w_n)$ to be *additive*, that is, surprisals of two independent events (when the probabilities multiply) should be equal to the sum of surprisals for each event.

The astute reader may already realize that what we need is a logarithmic function. Indeed, a logarithm goes to zero when the argument reaches unity. A zero argument will make a logarithmic function go to *minus* infinity, which is not a problem as long as we are going to take the logarithm with the negative sign. So we hope the next formula will be a high-probability outcome for most readers!

$$I(w_n) = -\log P(w_n) = \log \frac{1}{P(w_n)}. \tag{26.1}$$

Note that self-information is additive due to the property of the logarithm.

There are three flavors of this formula, with the base of the logarithm being 2, 10, and e. Depending on the base chosen, we say that information content is expressed in *bits*, *hartleys*, or *nats*, respectively. Further on, we will be using bits unless specified otherwise.

When outcome w_n happens and we learn the value of W, we thus gain $I(w_n) = -\log_2 P(w_n)$ probabilistic bits of new information. But the probability that we are going to learn that amount of information in a given trial is of course equal to the probability $P(w_n)$ that the outcome in question occurs. So on average we should expect to learn

$$\langle I(w_n) \rangle = -\sum_n P(w_n) \log_2 P(w_n) \tag{26.2}$$

bits of information, where summation is over all possible outcomes.

The expected information content associated with a measurement of W is called Shannon information and denoted H:[2]

$$H(W) = -\sum_n P(w_n) \log_2 P(w_n). \tag{26.3}$$

[2] There exists a certain confusion in terminology. Some people causally say "information content" when really mean Shannon information. We recommend to maintain the distinction between the information content, which characterizes a single outcome, and the Shannon information, which is the expectation value of $I(w_n)$ in a measurement with many possible outcomes w_n. Also, sometimes one would include the subscript s and write H_S to distinguish Shannon information from other information measures. But the notation H (never to be confused with the Hamiltonian!) is by far more common.

Let us see how this definition works in the two limiting cases. First, suppose there are N possible outcomes and the probability distribution is uniform, that is, $P(w_n) = 1/N$. Then, Shannon information becomes equivalent to the ordinary deterministic information:

$$H(W) = -\sum_{n=1}^{N} \frac{1}{N} \log_2 \frac{1}{N} = \log_2 N. \tag{26.4}$$

This is just what we got in the beginning of this section for $N = 26$ letters of the alphabet!

Suppose, on the other hand, that probability distribution is a Kronecker delta, that is, one single outcome has probability 1 and all the rest have zero probability. Since $\lim_{x \to 0} (x \log x) = 0$ (prove it!), Shannon information becomes

$$H(W) = -1 \cdot \log_2 1 = 0 \tag{26.5}$$

In other words, if we were told that all letters in the text were "e," we would not learn anything by reading this text, even though we would still need ceil(4.7) = 5 hardware bits to store an "e" (unless we redesign the hardware to work based on the idea that the alphabet is now reduced to one letter, but that would mean informing the computer that probability distribution is no longer uniform).

We want to emphasize that quantity H defined above is associated with the process of *measurement* in which we obtain more knowledge about the system in question. In other words, H is equal to the difference between our final (and presumably perfect) knowledge that we obtain by learning the outcome and our initial knowledge when the outcome was yet unknown. Suppose that instead of talking in terms of measurements performed on a given state, we now want to define a physical quantity that will be associated with the state itself. Then, we can still use H for that purpose, with the understanding that H will now describe the amount of our *ignorance* about that state. This ignorance is a function of *disorder* or the amount of freedom for the system to choose between the various outcomes. As seen from (26.4), in a completely disorderly system when all outcomes are equiprobable, H attains its maximum value (e.g., $H = 1$ in case of a single bit). Conversely, when there is perfect order in the system, that is, there is no leeway to deviate from a predetermined outcome, then H vanishes according to (26.5). Thus, the more the uncertainty present in the system under study, the higher the H becomes. All this gives us reason to associate H with the classical concept of *entropy*.

Indeed, Equation 26.3 should look familiar to the students who studied statistical mechanics and have seen a very similar definition for thermodynamic entropy (aka Gibbs entropy):

$$S = -k_B \sum_{n} P(w_n) \ln P(w_n). \tag{26.6}$$

The quantity $P(w_n)$ in (26.6) is the probability associated with microstate w_n. This means that apart from the base of the logarithm and the Boltzmann constant k_B in the entropy definition, the only physical difference is that statistical mechanics treats *microstates* as possible outcomes. We remember that entropy is a measure of disorder in a physical system. But disorder tells us the amount of "surprisal" (when there is complete order, the outcome is predictable; the more the chaos, the greater the chance of an outlier event), so entropy has the same physical meaning. For this reason, Shannon information is often referred to as Shannon *entropy*. Like most authors, we also find this term preferable for at least two reasons. First, it avoids the ambiguity that could otherwise confuse an unprepared reader (what "information" are we taking about – the one already present in W or the one we expect to gain in the process of measurement). Second, one could well think of some measurement that fails to eliminate disorder completely. Then we would have to explain to reader that "information" we are taking about refers to the idealized measurement and not to the one that is currently being discussed. "Entropy," on the other hand, has a clear physical meaning, is well familiar to most students, and the thermodynamic analogies translate well to the field of quantum information theory. From now on, we'll be using "Shannon entropy" to describe disorder in a quantum system, and we'll talk of "Shannon information" or "Shannon bits" to describe how much we already know or expect to learn about the system. When we use the word "bit" without a qualifier, it will mean a bit of deterministic information.

Let us now consider several important examples.

- *A single coin toss.* There are $N=2$ equally probable outcomes, so a toss gives us $H(W = \{heads, tails\}) = \log_2 2$ or 1 bit of Shannon information, which also coincides with the deterministic result (one hardware register to store the outcome).
- *Throwing a dice.* Again, $H(W)$ reduces to deterministic information: $H(W = \{1, \ldots, 6\}) = \log_2 6$.
- *Two dice.* Each outcome (say, dice1 = 2 and dice2 = 5) has probability 1/36. With 36 possible outcomes, we get $H(W = \{(1,1), \ldots, (6,6)\}) = \log_2 36$. Note that in this example, the two experiments are independent (one dice does not affect the other). Therefore, we could have defined *two* independent random variables, $W_1 = \{1, \ldots, 6\}$, $W_2 = \{1, \ldots, 6\}$ and written the answer as $H(W_1, W_2) = \log_2 36$. Here, $H(W_1, W_2)$ is the *joint* Shannon information. Note that $H(W_1, W_2) = H(W_1) + H(W_2)$.
- *Binary entropy function.* There are two possible outcomes, for example, heads and tails in a weighted coin toss experiment. If they occur with probabilities p and $1 - p$, we get $H(W) = -p \log_2 p - (1-p) \log_2 (1-p)$.
  Note that in case of $p = 0.5$, the binary entropy function is 1.
- *Spin measurement.* Suppose an electron's spin forms angle β with the z-axis. A measurement of s_z will result in "success" ($|\uparrow\rangle$) with probability $\cos^2 (\beta/2)$ and in failure ($|\downarrow\rangle$) with probability $\sin^2 (\beta/2)$. Then, $H(W = \{|\uparrow\rangle, |\downarrow\rangle\}) = -\cos^2 (\beta/2) \log_2 (\cos^2 (\beta/2)) - \sin^2 (\beta/2) \log_2 (\sin^2 (\beta/2))$. This of course is the same as a weighted coin toss with $p = \cos^2 (\beta/2)$.
- *Collision.* Suppose we define our experiment as consisting of a *pair* of independent measurements where in each measurement we look up a random letter in a text. (Imagine we took two samples of the text from a pure ensemble!) When we get the *same* letter in both measurements, we call such an occurrence a *collision*. We do

not require the occurrence of any *specific* letter; instead, *any* pair of letters, as long as they are the same, will count as a collision. The experiment is deemed a success if it results in a collision, and a failure otherwise. Denoting collision probability by symbol P_c, we can write $P_c = \sum_{n=1}^{26} P_n^2$ and the problem then reduces a binary entropy function experiment.

- *A string of bits.* Suppose we have a string of N bits whose values are completely unknown to us. We further assume that the bit values are independent, so each bit may take values of 0 and 1 regardless of the values of its neighbors. Then, by looking at this bit string, we will obviously acquire N deterministic bits. What about the Shannon information? There are 2^N equiprobable states of this system, so it will be like throwing a dice with that many faces. Thus, we learn $\log_2 2^N = N$ Shannon bits, so in this case both measures of information are equal. Note that our initial information was 0, whether measured in deterministic or Shannon bits, since we started from the state of total ignorance! Since we started from 0 and acquired N bits of new information, we now possess N bits, which is just what our intuition should suggest in this case.

- *A string of bits with prior information.* Suppose now that the value of each bit in the string is known with probability p. When the actual bit values are revealed to us, the information we will learn is $H = -p \log_2 p - (1-p) \log_2 (1-p)$ for each bit. Therefore, the total information gain is $H = -N(p \log_2 p + (1-p) \log_2 (1-p))$ Shannon bits. This will leave us with N Shannon bits of information about the final string, so if someone asks how much we knew initially, the answer is $N - H = N + N(p \log_2 p + (1-p) \log_2 (1-p))$ Shannon bits. As we should expect, the case of $p = 1/2$ reduces to $N - H = 0$, so we have zero initial knowledge about a totally random string of bits.

- *A wave function.* Let a quantum system be prepared in state $|\psi\rangle = c_1 |\varphi_1\rangle + \cdots + c_n |\varphi_n\rangle$. The probabilities of outcomes are given by the squares of amplitudes, so it is easy to write the Shannon entropy function: $H(|\psi\rangle) = -(|c_1|^2 \log_2 |c_1|^2 + \cdots + |c_n|^2 \log_2 |c_n|^2)$. Alternatively, we can write the state in the outer product notation: $|\psi\rangle\langle\psi| = (c_1|\varphi_1\rangle + \cdots + c_n|\varphi_n\rangle)(c_1^*\langle\varphi_1| + \cdots + c_n^*\langle\varphi_n|)$ and have the probabilities lined up along the main diagonal of the resulting density matrix. Then, Shannon entropy becomes $H(\rho) = -(\rho_{11} \log_2 \rho_{11} + \cdots + \rho_{nn} \log_2 \rho_{nn})$.

Just like thermodynamic entropy, Shannon entropy is additive, that is, the entropy of a system is the sum of entropies of its independent parts:

$$H(X, Y) = H(X) + H(Y). \tag{26.7}$$

This additivity holds because products become sums on a logarithmic scale. Or, to borrow a philosophic term, entropies do not behave *holistically* (holism is defined as the phenomenon where a whole is greater than the sum of parts). In fact, one can motivate the definition of information – and thermodynamic entropy! – in the way described above by the desire to avoid holism. This was actually the approach taken by Shannon whose argument in favor of the logarithmic scale we put in the epigraph to this chapter.

Entropy defined by (26.7) is called *joint entropy* of random variables X and Y. One can formally define it in line with (26.3), so that

$$H(X, Y) = -\sum_{m,n} P(x_m, y_n) \log_2 P(x_m, y_n). \tag{26.8}$$

This definition has a quite general character. Here, the symbol $P(x, y)$ stands for the probability that outcomes x and y both occur. In contrast, additivity condition (26.7) is only true as long as X and Y are totally uncorrelated.

26.2
von Neumann Entropy

Consider now a mixed state

$$\rho = \sum_i w_i |\psi_i\rangle\langle\psi_i| \tag{26.9}$$

composed from pure states $|\psi_i\rangle$, which are written in a certain default basis $\{|\varphi_i\rangle\}$. We have seen earlier that by rewriting the density matrix in an appropriate orthonormal basis $\{|v_i\rangle\}$, we can always obtain the diagonal form of the matrix:

$$\rho = \sum_i v_i |v_i\rangle\langle v_i| = \begin{pmatrix} v_1 & 0 & \cdots & 0 \\ 0 & v_2 & \cdots & 0 \\ 0 & 0 & \cdots & v_n \end{pmatrix}. \tag{26.10}$$

Getting Equation 26.10 is straightforward enough: One only needs to find the eigenvalues and have them lined up along the main diagonal. Notice however that we must apply caution if we want to write ρ in terms of pure states in this *new* basis. Since the basis has been changed, pure states will also have to be written in the new form; so, Equation 26.9 will now appear as

$$\rho = \sum_i w_i |\Psi_i\rangle\langle\Psi_i|. \tag{26.11}$$

Note that we have used the uppercase to distinguish (26.11) from (26.9). Density matrices (26.10) and (26.11) are identical, even though $|v_i\rangle$ may, generally speaking, be different from $|\Psi_i\rangle$ as a result of the unitary freedom in the choice of basis.

Let us now go back again to (26.9) and consider the pure states $|\psi_i\rangle$ in their *default* representation. If we now write them in the outer product notation,

$$\rho_i \equiv |\psi_i\rangle\langle\psi_i|, \tag{26.12}$$

these states will have their Shannon entropy given by

$$H(\rho_i) = -\big((\rho_i)_{11} \log_2 (\rho_i)_{11} + \cdots + (\rho_i)_{nn} \log_2 (\rho_i)_{nn}\big), \tag{26.13}$$

as discussed in the previous section. This is easy to see because their diagonal elements have the meaning of probabilities of respective outcomes, so (26.13) is just the definition of Shannon entropy. However, this formula gives us an important hint: When calculating the entropy we can ignore the nondiagonal matrix elements of ρ_i. This gives us the right to replace for the purpose of Shannon entropy calculation all ρ_i with ρ'_i that contain the diagonal elements only:

$$\rho'_i \equiv \begin{pmatrix} \rho_{11} & 0 & \cdots & 0 \\ 0 & \rho_{22} & \cdots & 0 \\ 0 & 0 & \cdots & \rho_{nn} \end{pmatrix}. \tag{26.14}$$

This is *not* a change in basis yet: So far we have just dropped the nondiagonal elements. The reader may remember that for this diagonal case, matrix logarithm takes an especially simple form,[3]

$$\log_2 \rho'_i \equiv \begin{pmatrix} \log_2 \rho_{11} & 0 & \cdots & 0 \\ 0 & \log_2 \rho_{22} & \cdots & 0 \\ 0 & 0 & \cdots & \log_2 \rho_{nn} \end{pmatrix}. \tag{26.15}$$

Combining (26.14) and (26.15), we see that the Shannon entropy of a pure state can be written as

$$H(\rho_i) = -\text{Tr}\left(\rho'_i \log_2 \rho'_i\right). \tag{26.16}$$

Let us now recall that we ought to be working in the *new* basis from the moment we have written (26.10). How will this affect the situation? For the pure states $|\psi_i\rangle$, not much is going to change. They will now be written in the new basis — the one that diagonalizes the density matrix ρ, but it does not necessarily diagonalize the pure states themselves. We can still calculate Shannon entropies of these states by considering the diagonal elements only. Generally speaking, we will then obtain different values since Shannon entropy of a state depends on the choice of basis. But we are now mostly interested in the mixed state ρ that has been diagonalized. Formally, it now looks similar to (26.14). It will be reasonable then to define an entropy measure for this mixture by analogy with (26.16) as

$$S(\rho) = -\text{Tr}\left(\rho' \log_2 \rho'\right). \tag{26.17}$$

The change of notation from H_S to S was needed to make the point that while defined by analogy, this entropy measure is *not* the same as Shannon entropy. Also, the primes have now become redundant because nondiagonal elements of ρ have already been removed in the process of diagonalization. We can finally write the entropy of the mixed state as

$$S(\rho) = -\text{Tr}\left(\rho \log_2 \rho\right). \tag{26.18}$$

The entropy measure defined in this manner is called *von Neumann entropy*.[4] What is its physical meaning? Imagine that state ρ is pure. We are then looking at a pure

---

[3] The generalization of $\ln \rho'_i$ to an arbitrary base of the logarithm is straightforward and left to the reader as an exercise.
[4] Some authors may include the Boltzmann constant in the definition: $S(\rho) = -k_B \text{Tr}\left(\rho \log_2 \rho\right)$, in an effort to draw a parallel with the Gibbs entropy formula.

state whose outer product is diagonal (unlike states ρ_i whose outer products are not necessarily diagonal in $\{|v_i\rangle\}$ basis!). Then ρ must be one of the eigenstates, meaning that only one of the elements $\rho_{11}, \ldots, \rho_{nn}$ is 1 and the rest are zeros. Then the expression

$$S(\rho) = -\left(\rho_{11} \log_2 \rho_{11} + \cdots + \rho_{nn} \log_2 \rho_{nn}\right) \tag{26.19}$$

immediately vanishes and von Neumann entropy of a pure state is zero. Next, let us make ρ a mixture, assigning a nonzero weight to more than one diagonal element $\rho_{11}, \ldots, \rho_{nn}$. Then (26.19) will become a positive number. The farther we depart from a pure state, the greater the value of $S(\rho)$ we will get until we reach a uniform distribution (a completely mixed state) where $\rho_{11} = \cdots = \rho_{nn} = 1/n$ and $S(\rho) = -((1/n) \log_2 (1/n) + \cdots + (1/n) \log_2 (1/n)) = \log_2 n$. Thus, von Neumann entropy represents the departure of the system from a pure state. This is where Shannon entropy analogies become misleading even though $S(\rho)$ was certainly defined "by analogy." As we remember, a pure state will generally have nonzero Shannon entropy, reflecting the uncertainty of outcomes inherent in that state. On the other hand, von Neumann entropy of a pure state is zero, a result that, however, makes perfect sense if we change our paradigm and think of *verification* of that state as the "outcome." We can always run an experiment tailor-suited for this specific state, where a conclusive outcome will mean that unambiguous state identification has occurred. If we know which state has been prepared and also know that it is a pure state, then the result of our experiment is 100% predictable and there are no surprises in store for us. At the other extreme, if the state is perfectly mixed, we cannot predict the outcome of state discrimination experiment at all, and by that measure, the entropy has been maximized.

It is the thermodynamic entropy analogy that seems more appropriate for von Neumann entropy. When the microstate of a system is known, such a system is said to have zero entropy. The entropy quickly increases as we make more microstates available to the system, and reaches its maximum when all microstates become equally probable. In particular, when we mix wine and water, entropy of the system always increases. The analogous property in QM is called *concavity*. What it means is this: If we mix together several (not necessarily pure) states ρ_i to form a new state $\rho = \sum_i P(\rho_i) \rho_i$ where $P(\rho_i)$ is the relative weight of ρ_i in the resulting mixture, then entropy of the mixture cannot be less than the sum of entropies of individual states ρ_i. The inequality

$$S(\rho) \geq \sum_i P(\rho_i) S(\rho_i) \tag{26.20}$$

expresses this concavity property: von Neumann entropy never decreases under a mixing operation!

26.3
Conditional Probability and Bayes's Theorem

The probabilities of outcomes appearing in the above formulas were simply constants defined by the given probability distribution and independent of any other events. There are situations though when one starts a certain probability distribution expressing the current state of knowledge, which then can change if some other fact about the system becomes known.

For example, suppose that in a certain college, 90% of students have taken a Quantum Information course. Unfortunately, the exams are difficult, and only 30% of students who take the course get a passing grade. Then, if Alice is a college student, the probability that she has passed the course is $0.9 \times 0.3 = 0.27$, but if we are told in addition that Alice *did* take the course, the above probability increases to 0.3. The first probability is called *prior* probability of an event (i.e., probability that we assign to an event in the absence of any additional information), and the "updated" one is the *posterior* probability (i.e., expressing our state of knowledge after some additional event has happened).

Let A denote the event of passing the course and B denote the event of taking the course. Then the prior probabilities are $P(A) = 0.27$ and $P(B) = 0.90$. The posterior probability of A expresses the likelihood that A is true *on the condition* that B is true. We call it *conditional probability* and denote as $P(A|B)$. Obviously, in this example, $P(A|B) = 0.30$.

We can also find the *joint probability* $P(A, B)$ of attending and passing the course. Joint probability is defined simply as the probability that both events took place. In this example, it cannot be written simply as $P(A)P(B)$ because the events are not independent (analogy with (26.7) fails in this case!). Instead, the joint probability is obtained from logic. In order to attend and pass the course, Alice has to first get enrolled and then get a passing grade once she is enrolled. This means we must multiply the respective probabilities:

$$P(A, B) = P(B)P(A|B). \tag{26.21}$$

Plugging in the numbers, we readily get $P(A, B) = 0.27$ as before. Notice that the joint probability turned out to be the same as $P(A)$; this is not a big surprise because in the current example, passing the course already assumes enrollment, so to say that A is true is the same as to say that A and B are both true.

We could also run the same logic argument in reverse: In order for A and B to be true, it is necessary that A be true and that B conditioned on A also be true, that is,

$$P(A, B) = P(A)P(B|A). \tag{26.22}$$

Or, in numbers, $0.27 = 0.27 \times P(B|A)$ and the conditional probability of B given A is 1 as it should because all students who passed must have been enrolled in the course.

Obviously, the following two joint probabilities are equal:

$$P(A, B) = P(B, A) \tag{26.23}$$

As a consequence of (26.21) and (26.22), conditional probability can be expressed as

$$P(A|B) = \frac{P(A, B)}{P(B)} \tag{26.24}$$

or

$$P(B|A) = \frac{P(A,B)}{P(A)}, \tag{26.25}$$

and combining these two formulas we get the result known as *Bayes's theorem*:

$$P(A|B) = P(B|A)\frac{P(A)}{P(B)}. \tag{26.26}$$

Let us now describe experimental outcomes in terms of Shannon entropy. To this end, we are going to introduce random variables X and Y with each variable having two possible outcomes: $X = \{\text{passed}, \text{failed}\}$ and $Y = \{\text{enrolled}, \text{declined}\}$. That is, the student has a choice to enroll or decline to enroll, and the student can pass the course or fail for whatever reasons (bad luck in the exam or the student never enrolled in the first place). This allows us to formalize the problem as follows. First, we write the condition of the problem in three lines:

$$\begin{cases} P(Y = \text{enrolled}) = 0.90, \\ P(X = \text{passed}|Y = \text{enrolled}) = 0.30, \\ P(X = \text{passed}|Y = \text{declined}) = 0. \end{cases} \tag{26.27}$$

Then, we can derive all the remaining probabilities. Probability conservation requires that

$$\begin{cases} P(Y = \text{declined}) = 0.10, \\ P(X = \text{failed}|Y = \text{enrolled}) = 0.70, \\ P(X = \text{failed}|Y = \text{declined}) = 1, \end{cases} \tag{26.28}$$

so all conditional probabilities have been determined. Next, we use (26.22) to obtain the joint probabilities:

$$\begin{cases} P(X = \text{passed}, Y = \text{enrolled}) = P(Y = \text{enrolled})P(X = \text{passed}|Y = \text{enrolled}) = 0.27, \\ P(X = \text{passed}, Y = \text{declined}) = P(Y = \text{declined})P(X = \text{passed}|Y = \text{declined}) = 0, \\ P(X = \text{failed}, Y = \text{enrolled}) = P(Y = \text{enrolled})P(X = \text{failed}|Y = \text{enrolled}) = 0.63, \\ P(X = \text{failed}, Y = \text{declined}) = P(Y = \text{declined})P(X = \text{failed}|Y = \text{declined}) = 0.10. \end{cases}$$

Next, let us find the prior probabilities for X by adding the appropriate joint probabilities:

$$\begin{cases} P(X = \text{passed}) = P(X = \text{passed}, Y = \text{enrolled}) + P(X = \text{passed}, Y = \text{declined}) = 0.27, \\ P(X = \text{failed}) = P(X = \text{failed}, Y = \text{enrolled}) + P(X = \text{failed}, Y = \text{declined}) = 0.73. \end{cases}$$

As a final step, we will use Bayes's theorem to list the conditional probabilities with "reversed" order of variables:

$$\begin{cases} P(Y = \text{enrolled}|X = \text{passed}) = P(X = \text{passed}|Y = \text{enrolled}) \dfrac{P(Y = \text{enrolled})}{P(X = \text{passed})} = 1, \\[4pt] P(Y = \text{enrolled}|X = \text{failed}) = P(X = \text{failed}|Y = \text{enrolled}) \dfrac{P(Y = \text{enrolled})}{P(X = \text{failed})} = \dfrac{63}{73}, \\[4pt] P(Y = \text{declined}|X = \text{passed}) = P(X = \text{passed}|Y = \text{declined}) \dfrac{P(Y = \text{declined})}{P(X = \text{passed})} = 0, \\[4pt] P(Y = \text{declined}|X = \text{failed}) = P(X = \text{failed}|Y = \text{declined}) \dfrac{P(Y = \text{declined})}{P(X = \text{failed})} = \dfrac{10}{73}. \end{cases}$$

The following table compiles the joint and conditional probability distributions $P(X, Y) P(X|Y)$, and $P(Y|X)$:

P(X,Y)	Enrolled	Declined	P(X|Y)	Enrolled	Declined	P(Y|X)	Enrolled	Declined
Passed	0.27	0	Passed	0.30	0	Passed	1	0
Failed	0.63	0.10	Failed	0.70	1	Failed	63/73	10/73

The left part of the table illustrates the important general rule:

$$\begin{cases} P(X = x) = \displaystyle\sum_y P(X = x, Y = y), \\[4pt] P(Y = y) = \displaystyle\sum_x P(X = x, Y = y). \end{cases} \qquad (26.29)$$

Notice that we have adapted a common notation where the uppercase letter stands for the random variable and the lowercase letter stands for the possible values of that variable. If the variables are continuous, summation in (26.29) is to be replaced by integration.

The set of probabilities $P(X = x)$ obtained from (26.29) for all possible x forms the probability distribution $P(X)$, which in this context is often called the *marginal distribution* of X. The marginal distribution of Y is defined similarly.

Suppose we now decide to track the progress of a randomly selected student without prior knowledge and see whether the student passes the course. The Shannon entropy of random variables X and Y are equal to

$$\begin{cases} H(X = \{\text{passed, failed}\}) = -0.27 \log_2 0.27 - 0.73 \log_2 0.73 \approx 0.8414, \\ H(X = \{\text{enrolled, declined}\}) = -0.90 \log_2 0.90 - 0.10 \log_2 0.10 \approx 0.469. \end{cases} \qquad (26.30)$$

These results make perfect sense. The odds of passing or failing are more or less comparable, so the experiment is close to a coin toss and the entropy is close to maximum. The enrollment experiment is more ordered, so the entropy is lower, although there is still significant potential for a surprising outcome.

Let us now find the joint and conditional entropies for this experiment. The joint entropy of random variables X and Y is defined as

$$H(X, Y) = -\sum_{x,y} P(x, y) \log_2 P(x, y), \tag{26.31}$$

which naturally extends the definition (26.11) to the case of two variables. Applying this definition, we readily obtain

$$H(X, Y) = -0.27 \log_2 0.27 + 0 \log_2 0 + 0.63 \log_2 0.63 + 0.10 \log_2 0.10 \approx 1.2621. \tag{26.32}$$

What about conditional probabilities? We have previously defined Shannon information as the expected surprisal in an experiment. A conditional experiment for which $H(X|Y)$ is defined should be thought of in the following way. First, we learn the value of Y, with each outcome y having probability $P(Y = y)$. Then, depending on the measured y, we update our probability distribution for X, replacing the prior distribution $P(X = x)$ with the new distribution based on what we now know: $P(X = x|Y = y)$. Then, we actually make a measurement of X, obtain some value x, and record the surprisal $\log_2 P(X = x|Y = y)$. Finally, we take statistical average over the ensemble and define conditional Shannon entropy of X given random variable Y [5] as the expected surprisal:[6]

$$H(X|Y) \equiv -\sum_{x,y} P(Y = y) P(X = x|Y = y) \log_2 P(X = x|Y = y). \tag{26.33}$$

If we now use the shorthand notation to get rid of the uppercase letters and also recall that $P(x|y) = P(x, y)/P(y)$, this gives us two equivalent definitions of conditional entropy:

$$H(X|Y) \equiv -\sum_{x,y} P(y) P(x|y) \log_2 P(x|y) \tag{26.34}$$

and

$$H(X|Y) \equiv -\sum_{x,y} P(x, y) \log_2 \frac{P(x, y)}{P(y)}. \tag{26.35}$$

Yet another definition that also follows from (26.33) expresses $H(X|Y)$ as a weighted sum of $H(X|y)$:

$$H(X|Y) = \sum_{y \in Y} P(y) H(X|y). \tag{26.36}$$

---

5) Yet another common term is *equivocation* of X about Y.
6) The surprisal associated with the result $P(Y = y)$ is not counted for the purpose of defining conditional entropy since this entropy measure was designed to capture the surprisal for X only, with Y being treated as a parameter.

We are now in a position to obtain $H_S(X|Y)$ numerically:

$$H(X|Y) = 0.9H(X|Y = \text{enrolled}) + 0.1H(X|Y = \text{declined})$$
$$= -0.9(0.3\log_2 0.3 + 0.7\log_2 0.7) - 0.1(0\log_2 0 + 1\log_2 1) \approx 0.7932.$$
(26.37)

The reader can see that this number is simply the difference of the two previous results $1.2621 - 0.469$ or, written in symbols,

$$H(X|Y) = H(X, Y) - H(Y).$$
(26.38)

It is easy to see why (26.38) must be true by taking the logarithm of equation $P(x|y) = P(x, y)/P(y)$. Conceptually, when we learn Y, we reduce our total ignorance $H(X, Y)$ about the variable pair by the amount $H(Y)$.

Box 26.1 Conditioning on variable versus conditioning on value

It is very important to keep in mind that the conditional Shannon entropy $H(X|Y)$ is conditioned on a random variable (Y in this case) rather than on that variable having one specific value. Indeed, the summation $H(X|Y) \equiv -\sum_{x,y} P(x, y) \log_2 P(x, y)/P(y)$ is carried out over *all* possible values of Y. The confusion between $H(X|Y)$ and $H(X|y)$ is a common mistake among students. Suppose we tried naively to find $H(X|Y)$ based on the observation that 30% of enrolled students pass the course and 70% fail it: $-0.3\log_2 0.3 - 0.7\log_2 0.7 \approx 0.8813$. However, the number found in this way would be $H(X|Y = \text{enrolled})$, and it is conditioned on random variable Y having one specific value "enrolled." The correct way is to find $H(X|\text{enrolled}) = 0.8813$ and $H(X|\text{declined}) = 0$ and take their sum weighted by the respective probabilities as we did in (26.37): $H(X|Y) = 0.90 \times 0.8813 + 0.10 \times 0 \approx 0.7932$.

Another typical mistake is made when calculating the joint probability. For example, a student may assume that because 27% of students pass and 73% fail, these two probabilities must appear in the joint entropy expression and go on to calculate $H(X, Y) = -0.27\log_2 0.27 - 0.73\log_2 73 \approx 0.8414$. The mistake in this expression is that it artificially combines two outcomes from the table with probabilities 0.63 and 0.10 into a single outcome with probability 0.73. The correct way is to account for all four possible (X, Y) pairs with respective probabilities 0.27, 0.63, 0, and 0.10 as we did in (26.32).

26.4 KL Divergence

Suppose observer Bob knows that a random variable $X = \{x\}$ has a probability distribution $P(x)$. Suppose next that a misinformed observer Eve mistakenly believes the probability distribution to be $Q(x)$. The question is: How much extra information will she have to learn to become as knowledgeable as Bob? To answer this question, we must rethink the meaning of Equations 26.1–26.3.

As we know already, Bob's average surprisal at learning the actual value of X will be $-\sum_{x \in X} P(x) \log_2 P(x)$. On the other hand, while Eve's surprisal at getting a specific outcome x is given by $-\log_2 Q(x)$ just as these formulas suggest, the likelihood of her being surprised in this way is not in fact $Q(x)$ (Eve has been misinformed!), instead it is equal to $P(x)$ because this is the actual probability of this outcome. Thus, Eve's average surprisal is $-\sum_{x \in X} P(x) \log_2 Q(x)$. If Eve's average surprisal turns out to be greater than Bob's, it means she currently has *less* information, and the extra amount that she needs to learn is equal to the difference between Shannon entropies of Eve and Bob:

$$D_{KL}(P||Q) = \left(-\sum_x P(x) \log_2 Q(x)\right) - \left(-\sum_x P(x) \log_2 P(x)\right) = \sum_x P(x) \log_2 \frac{P(x)}{Q(x)}.$$
(26.39)

The symbol $D_{KL}(P \| Q)$ we just introduced is known as the *Kullback–Leibler divergence* or just KL divergence from Q to P. It measures the expected number of Shannon bits required to "upgrade" one's knowledge based on the assumed probability distribution Q to the one based on the correct distribution P. Other names for the KL divergence are *information divergence*, *information gain*, and *relative entropy*.

In particular, when X is known by Bob with certainty, then KL divergence turns into the already familiar information gain by Eve when she learns the actual value of X, with the only nuance that Eve's surprisal must be averaged using the actual, that is, known probabilities.

26.5
Mutual Information

Suppose now that random variable $X = \{x\}$ also depends on another random variable $Y = \{y\}$ so that $P(X)$ is different from the conditional distribution $P(X|Y)$. Suppose further that Eve observes variable Y only. For example, if we again consider our imaginary Quantum Information course, suppose that Eve was allowed to find out whether the student has enrolled in the course. The question is: How much extra information can she learn about X from this observation?

The answer is again given by the KL divergence. Intuitively, it is clear that one who knows the value taken by Y will thus have more information about X than one who has no idea what Y is equal to. Then KL divergence will tell us by how much Eve's information about X was augmented after she learned Y. In other words, the assumed distribution Q of the previous problem now figures as $P(X)$, while the actual distribution P is now $P(X|Y=y)$, where y is the specific outcome (e.g., enrolled or declined) known by Eve. Thus, information obtained by Eve is equal to the following KL divergence:

$$D_{KL}(P(X|y)||P(X)). \tag{26.40}$$

But we want to know how much Eve can learn on the average, so this must be averaged over all possible values of Y:

$$H(X;Y) = \sum_y P(y) D_{KL}(P(X|y)||P(X)). \tag{26.41}$$

The newly introduced symbol $H(X;Y)$ is called *mutual information*[7] and it measures the average amount of knowledge one can obtain about random variable X by observing random variable Y.

This meaning of mutual information immediately implies the following basic property:

$$H(X;Y) = H_S(X) - H_S(X|Y). \tag{26.42}$$

The right-hand side contains the difference between Eve's uncertainly about X before and after the measurement of Y; the resulting change in uncertainty tells us how much she learned, so the last formula should be self-evident. On the other hand, we remember that $H_S(X|Y) = H_S(X,Y) - H_S(Y)$, which gives us the following important corollary:

$$H(X;Y) = H_S(X) + H_S(Y) - H_S(X,Y). \tag{26.43}$$

Hence, mutual information is symmetric:

$$H(X;Y) = H(Y;X) \tag{26.44}$$

and studying Y gives us as much information about X as studying X gives us about Y.

It is easy to see by looking at (26.39) that it is OK to multiply both terms in the KL divergence by a constant. Let this constant be $P(y)$. Then, (26.41) becomes

$$H(X;Y) = \sum_y P(y) D_{KL}(P(y)P(X|y)||P(y)P(X)) = D_{KL}(P(X,Y)||P(X)P(Y)), \tag{26.45}$$

in other words, mutual information is simply the KL divergence from the product of marginal distributions to the joint distribution. And by virtue of (26.39),

$$H(X;Y) = \sum_{x,y} P(x,y) \log_2 \frac{P(x,y)}{P(x)P(y)}. \tag{26.46}$$

The logarithm in (26.46) is sometimes called *pointwise mutual information* of outcomes x and y and written as

$$\text{pmi}(x,y) = \log_2 \frac{P(x,y)}{P(x)P(y)}. \tag{26.47}$$

---

7) Some authors denote mutual information as $I(X;Y)$. Other authors use semicolon, so the reader may also encounter notations such as $H(X:Y)$ and $I(X:Y)$.

The following definitions of pmi are equivalent:

$$\text{pmi}(x, y) = \log_2 \frac{P(x, y)}{P(x)P(y)} = \log_2 \frac{P(x|y)}{P(x)} = \log_2 \frac{P(y|x)}{P(y)}. \tag{26.48}$$

In terms of pmi, mutual information is written

$$H(X; Y) = \sum_{x,y} P(x, y) \, \text{pmi}(x, y). \tag{26.49}$$

We have mentioned above that entropies of independent parts of a system add together to form the total entropy. However, when the parts are mutually dependent, that is, when they are specified by two correlated random variables, the total becomes *less* than the sum of parts. Indeed, as a consequence of (26.43),

$$H(X, Y) = H(X) + H(Y) - H(X; Y), \tag{26.50}$$

and we get the self-intuitive result that the disorder associated with a combined pair of random variables is expressed as the disorder calculated as if these variables were independent minus the term that describes the actual correlation between the variables.

The concept of mutual information is useful in cases when, for example, Bob composes a bit string and Eve (whose apparatus is spying on Bob) makes only an imperfect measurement of Bob's string and can only look at the poor version she thereby obtains. If we treat the entire strings as random variables (with 2^N possible values!), then it is mutual information from Bob's string to Eve's string that tells us how much Eve has learned about Bob's string.

26.6
Rényi Entropy

Until now, we have been using Shannon entropy as a measure of probabilistic information. In 1960, Alfred Rényi defined yet another important measure, which is based on the concept of collision that we mentioned earlier. Rényi proposed to measure the amount of disorder of a random variable in terms of the surprisal of a collision (for this reason, a term *collision entropy* can sometimes be encountered in the literature):

$$R_2(X) = -\log_2 P_c(X) = -\log_2 \sum_x P^2(x). \tag{26.51}$$

The symbol on the left is formally called "Renyi entropy of order 2," for the reason that will be explained later. Here and below, we will just call it "Rényi entropy," implying order 2 by default unless the order is specifically mentioned.

We can find $R_2(X)$ for the few cases considered previously.

- *A single coin toss.* There are $N=2$ equally probable outcomes, so a toss gives us $R_2(X = \{\text{heads}, \text{tails}\}) = -\log_2 P_c = -\log_2((1/4) + (1/4)) = 1$ or 1 bit of Rényi information. In this example, Shannon and Rényi information are the same.

- *Throwing a dice.* Again, Rényi information reduces to deterministic information just as Shannon information does: $R_2(X = \{1, \ldots, 6\}) = -\log_2 \left(\frac{1}{36} + \cdots + \frac{1}{36}\right) = \log_2 6$.
- *Two dice.* Each outcome (say, dice1 = 2 and dice2 = 5) has probability 1/36 and the square of probability 1/1296. With 36 possible outcomes, we get $R_2(X = \{(1, 1), \ldots, (6, 6)\}) = -\log_2 (1/1296 + \cdots + 1/1296) = \log_2 36$. Again, the two entropies agree in this case. Just as before, we can write the answer as the joint Rényi information $R_2(X_1, X_2) = R_2(X_1) + R_2(X_2)$ of two random variables $X_1 = \{1, \ldots, 6\}$, $X_2 = \{1, \ldots, 6\}$. Since the variables are independent, $R_2(X_1, X_2) = R_2(X_1) + R_2(X_2)$ holds in this case, as it does for its Shannon analogue.
- *Binary entropy function.* A coin toss results in heads with probability p: $R_2(X) = -\log_2 (p^2 + (1-p)^2)$.

  This time, Shannon and Rényi entropies differ as long as p is not one-half. As you can see, the equivalence of the two entropies only holds for a uniform probability distribution.
- *Spin measurement.* Again, the experiment is similar to a weighted coin toss with $p = \cos^2 \beta/2$. Rényi entropy is $R_2(X = \{\uparrow, \downarrow\}) = -\log_2 \left(\cos^4 \beta/2 + \sin^4 \beta/2\right)$.
- *A string of bits.* Again, we have a string of N bits whose values are completely unknown to us. With 2^N equiprobable states of this system, probability squares are $1/2^{2N}$ and $R_2(X) = -\log_2 \left((1/2^{2N}) 2^N\right) = N$. Rényi and Shannon entropies coincide once again and all previous conclusions are valid for both entropies.
- *A string of bits with prior information.* When each bit in the string is known with probability p, the Rényi entropy is given by $-\log_2 (p^2 + (1-p)^2)$ for each bit, so the total current uncertainty is N times: $R_2(X) = -N \log_2 (p^2 + (1-p)^2)$. Once again we have $R_2(X) = N$ when $p = 1/2$.

The more general definition for the Rényi entropy of an arbitrary order α is given below:

$$R_\alpha(X) = \frac{1}{1-\alpha} \log_2 \sum_x P_x^\alpha. \tag{26.52}$$

Here, α can be any real number such that $\alpha \geq 0$ and $\alpha \neq 1$ (note that α does not need to be integer!) Just as before, the random variable X is allowed to take values $X = \{x_1, \ldots, x_N\}$, and P_n is the probability to obtain value $X = x_n$. These probabilities are then raised to the power α and added together.

Thus, Rényi entropy of order 2 is really a special case of the more general definition (26.52). However, it is this special case that plays an important role in quantum applications, as we will see later.

One can readily see how Rényi entropy of order 2 is different from Shannon entropy if one rewrites both entropies in terms of expectation values: just compare

$$H(X) = -\langle \log_2 P(x_n) \rangle \tag{26.53}$$

and

$$R_2(X) = -\log_2 \langle P(x_n) \rangle. \tag{26.54}$$

That is, whereas Shannon is averaging the logarithm of probability, Rényi takes logarithm of the average. One can easily see from these equations that both entropies are always positive.

Like Shannon entropy, Rényi entropy is also additive and the equation similar to (26.7) holds for independent variables:

$$R_2(X, Y) = R_2(X) + R_2(Y). \tag{26.55}$$

We will now prove an important theorem about entropy, but first we need to invoke another mathematical result, known as Jensen's inequality. You must be familiar with its shorter version, which says that arithmetic mean is greater than geometric mean or

$$\frac{x_1 + x_2 + \cdots + x_n}{n} \geq \sqrt[n]{x_1 x_2 \ldots x_n}, \tag{26.56}$$

with equality taking place when $x_1 = x_2 \ldots = x_n$. But this is only a special case of the more general inequality that says that if P_i are any numbers that add up to 1 and function f is real, continuous, and concave up, then the weighted function is greater than or equal to the function of the weighted variable:

$$\sum_{i=1}^{n} P_i f(x_i) \geq f\left(\sum_{i=1}^{n} P_i x_i\right). \tag{26.57}$$

You can prove the special case by taking $f(x_i) = -\log x_i$.

Applying Jensen's inequality to entropies, we immediately find that Rényi entropy is upper bounded by the Shannon entropy:

$$H(X) \geq R_2(X) \tag{26.58}$$

and that equality only takes place when the probability distribution $P(x_n)$ is uniform.

26.7
Joint and Conditional Renyi Entropy

When X depends on Y and a specific value $Y = y$ is known, probability of X in (6.1) becomes conditioned on y.

$$R_2(X|y) = -\log_2 P_c(X|y) = -\log_2 \sum_x P^2(x|y). \tag{26.59}$$

We are interested in the entropy averaged over all possible y. This gives us grounds to define conditional Rényi entropy by averaging Equation 26.59:

$$R_2(X|Y) = \sum_y P(y) R_2(X|y), \tag{26.60}$$

which gives us the following formula:

$$R_2(X|Y) = \sum_y P(y)\left(-\log_2 \sum_x P^2(x|y)\right). \tag{26.61}$$

Alternatively, we can express conditional probability in (26.61) in terms of joint probability:

$$R_2(X|Y) = \sum_y P(y)\left(-\log_2 \sum_x \frac{P^2(x,y)}{P^2(y)}\right). \tag{26.62}$$

So far it looks very similar to conditional Shannon entropy defined by ((26.34–26.35). However, there is one crucial difference: This time $P(y)$ appears before the second summation sign and therefore it cannot cancel with $P(y)$ in the denominator. This difference has important implications: Even though $P(x|y) = P(x,y)/P(y)$ is still true, this equation does not give rise to the equivalent of (26.38); in other words,

$$R_2(X|Y) \neq R_2(X, Y) - R_2(Y). \tag{26.63}$$

This leads to a conceptual paradox. It would seem that by learning Y, we should reduce our ignorance $R_2(X, Y)$ by the amount $R_2(Y)$ and then the right-hand side of (26.63) should be equal to the conditional Rényi entropy. In this case our intuition fails us. To see what went wrong with this argument, let us write the joint entropy explicitly and derive the right-hand side:

$$R_2(X, Y) = -\log_2 P_c(X, Y) = -\log_2 \sum_x P^2(x, y), \tag{26.64}$$

$$R_2(X, Y) - R_2(Y) = -\log_2 \sum_x P^2(x, y) + \log_2 \sum_x P^2(y) = -\log_2 \frac{\sum_x P^2(x, y)}{\sum_x P^2(y)}. \tag{26.65}$$

Now compare (26.62) and (26.65). We immediately see that the equality does not hold:

$$R_2(X|Y) = -\sum_y P(y)\left(\log_2 \sum_x \frac{P^2(x,y)}{P^2(y)}\right) \neq -\log_2 \sum_x \frac{P^2(x,y)}{P^2(y)} = R_2(X, Y) - R_2(Y). \tag{26.66}$$

Moreover, an important distinction emerges between conditioning on a variable and conditioning on a specific value. Comparing (26.61) and (26.59), we see at once that the weighted logarithm (26.61) is generally not the same as logarithm (26.59).

Applying Jensen's inequality to (26.61) and (26.59), we see that

$$R_2(X|Y) = \sum_y P(y)\left(-\log_2 \sum_x P^2(x|y)\right) \geq -\log_2 \left(\sum_y P(y) \sum_x P^2(x|y)\right), \tag{26.67}$$

26.7 Joint and Conditional Renyi Entropy

and since the second sum on the right-hand side is equal to $P_c(X|y)$, the last equation in conjunction with (26.58) gives us an important inequality:

$$H(X|Y) \geq R_2(X|Y) \geq -\log_2 \sum_y P(y) P_c(X|y). \tag{26.68}$$

The "paradoxical" behavior of Rényi entropy raises an interesting question: Can one define the KL divergence and the mutual entropy in the same manner as we did for Shannon entropy in the absence of straightforward conceptual analogies? The answer turns out to be negative. For example, one cannot get the divergence by subtracting the entropies of Eve and Bob as we did in Section 26.4. This seems hard to accept, but notice that so far we have failed to give the Rényi entropy any "transparent" physical meaning such as the average surprisal for Shannon entropy. The Rényi entropy was introduced by mathematical formulas (26.51)–(26.52) as a convenient alternative entropy measure, and nothing was said about its physical properties. Then it should come as a no surprise when this quantity exhibits "strange" physical behavior.

The abstract character of Rényi entropy causes much confusion. For example, one may be tempted to "define" D_{KL} using the following argument. Imagine as before that a misinformed Eve believes the probability distribution to be $Q(x)$ when in reality it is $P(x)$. Thus, Eve imagines the collision probability to be $P_c(X) = \sum_x Q^2(x)$, while Bob, who knows better, determines that $P_c(X) = \sum_x P^2(x)$. Subtracting the entropies, get $(-\log_2 \sum_x Q^2(x)) - (-\log_2 \sum_x P^2(x)) = \log_2 \left[(\sum_x P^2(x))/(\sum_x Q^2(x)) \right]$ and this must be sought-for KL divergence. Unfortunately, the simplicity of the argument is deceptive. For example, in Section 26.4, Eve's entropy contained both $Q(x)$ and $P(x)$, the former corresponding to "surprisal" (a subjective characteristic for Eve) and the latter corresponding to the objective picture. Since Shannon entropy had been defined as the average surprisal of Eve, we then had a clear algorithm for constructing D_{KL}. This is certainly not the case for Rényi entropy. What quantity – $Q(x)$ or $P(x)$ or both – should appear in "collision probability" for Eve and why? Besides, we have already seen that addition and subtraction of Rényi entropies of mutually dependent variables is not a straightforward process that one would imagine. One could try to remedy the situation by assuming $P_c(X) = \sum_x P(x) Q(x)$ for Eve, in the spirit of Section 26.4. Alas, this trick does not work any better. The resulting quantity $\log_2 \left[(\sum_x P^2(x))/(\sum_x P(x) Q(x)) \right]$ is a perfectly legitimate mathematical construct, but it is not equal to Bob's information advantage over Eve.

Instead, the KL divergence for Rényi entropy of order α must be defined formally Rényi entropy itself. The general definition is

$$D_\alpha(P||Q) = \frac{1}{\alpha - 1} \log_2 \sum_x P^\alpha(x) Q^{1-\alpha}(x), \tag{26.69}$$

where D_α, the Rényi divergence of order α, generalizes the KL divergence D_{KL}. For $\alpha = 2$, this takes the form

$$D_2(P||Q) = \log_2 \sum_x \frac{P^2(x)}{Q(x)}. \tag{26.70}$$

In case the reader still finds it necessary to justify this definition, let us assume KL divergence formula (26.69) and use it to motivate the entropy definition (26.51). The idea is that Rényi entropy of any arbitrary distribution $P(X)$ should be equal to the KL divergence from the uniform distribution $U(X)$ to $P(X)$, where the uniform distribution means that all x occur with the same probability:

$$U(X) = \left(\frac{1}{N}, \ldots, \frac{1}{N}\right). \tag{26.71}$$

Then, one can write

$$R_\alpha(P) = R_\alpha(U) - D_\alpha(P||U), \tag{26.72}$$

and the reader can verify that this is consistent with Rényi entropy definition:

$$R_\alpha(X) = \frac{1}{1-\alpha} \log_2 \sum_x P^\alpha(x), \tag{26.73}$$

because we then have $R_\alpha(U) = \log_2 N$ and $D_\alpha(P||U) = \log_2 N + (1/\alpha - 1) \log_2 \sum_x P^\alpha(x)$.

There is no intuitive way to define the Rényi equivalent of Shannon mutual information $R_\alpha(X; Y)$. Several different definitions exist, but they are beyond the scope of this text. The interested reader will find details in Refs [1–3]. Let us only mention that once again one must resist the temptation to define mutual Rényi information "by analogy" as $R_2(X; Y) = R_2(X) - R_2(X|Y)$ in the same way as we defined mutual Shannon entropy. This is when the analogy breaks. A "mutual entropy" defined in this way would not be symmetric and, moreover, in some situations it even could turn negative! This would imply that by learning about a correlated random variable Y, we may not only fail to increase our knowledge of X but actually end up with *less* knowledge we had initially! This sounds totally counterintuitive. However, Figure 26.1 shows why it makes perfect sense. Suppose a random variable $X = \{x_1, \ldots, x_N\}$ is correlated with a random variable $U = \{u_1, u_2\}$ and the conditional probability $P(X|U)$ shows a sharp spike when $U = u_1$ but is uniform when $U = u_2$. If Eve gets to know U, she immediately knows the correct distribution. If she is told that $U = u_1$, then it is the first distribution that has low entropy (ideally, zero if the spike tends to a delta function). If she is told that $U = u_2$, then it is the second distribution that has high entropy ($\log_2 N$, to be exact). Then we only need to skew the marginal probability of U toward the outcome $U = u_2$ to ensure that Rényi entropy is close to maximum $R_2(X|U) \equiv \sum_u P(u) R_2(X|u) \approx \log_2 N$. On the other hand, marginal probability of X could well provide for a lower Rényi entropy, $R_2(X) < \log_2 N$. As we can see, both Shannon and Rényi entropies adequately describe the case in Figure 26.1b, identifying it, self-intuitively, as the maximum entropy state. The "paradox" arises because Rényi entropy "understates" the disorder in Figure 26.1a. Of course the "understatement" only happens from the point of view of a physicist looking for direct Shannon analogies. The Rényi theory itself is quite self-consistent and the behavior of $R_2(X|U)$ is simply caused by the way in which Rényi entropy was defined. The mathematical reason for that is the tendency of Rényi entropy to assign less weight to the moderate probability events than Shannon entropy

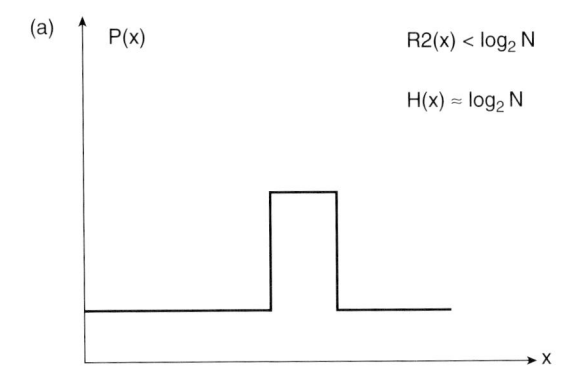

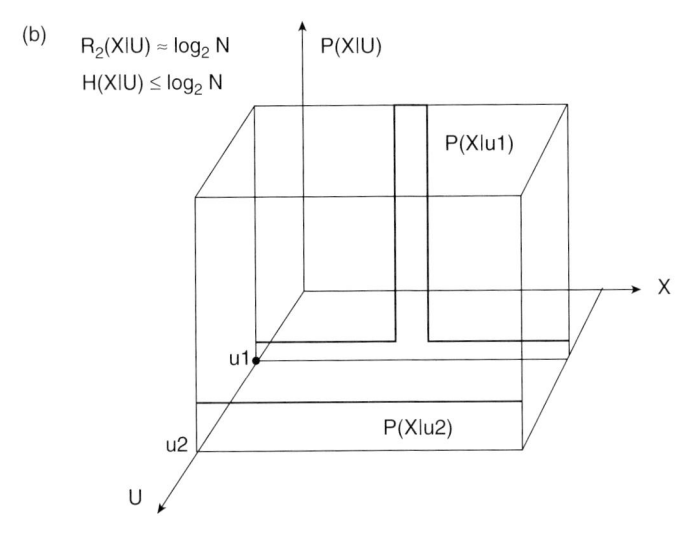

Figure 26.1 Knowledge of U may result in increase of Rényi entropy. (a) Marginal probability distribution P(X). (b) Conditional probability distribution P(X|U). Note that Shannon entropy still behaves "normally," decreasing when conditioned upon U.

does. The distribution in Figure 26.1a belongs precisely to this "moderate probability" category being the intermediate case between the sharp spike and the perfectly flat distribution that one sees in Figure 26.1b. This is where the Rényi entropy "sags," falling below its Shannon counterpart $H(X|U)$. On the other hand, random variable $U = \{u_1, u_2\}$ was specially designed to maximize both $R_2(X|U)$ and $H(X|U)$ in Figure 26.1b. While Shannon entropy definition was robust enough to stand this "stress test" so that $H(X|U)$ is still slightly less than $H(X)$, this is not so for Rényi entropy. To use the term coined by Silvio Micali, we will call information about U *spoiling knowledge* – yes, as far as Rényi entropy is concerned, too much knowledge is sometimes a dangerous thing!

This failure of direct Shannon analogies proves very useful in quantum key distribution applications where we seek to decrease the adversary's information. If the adversary Eve is able to observe a random variable Y that is correlated with X, and hopes to get some hint about X from her observations, she may be disappointed to find out that there is a distinct possibility of

$$R_2(X) - R_2(X|Y) < 0, \tag{26.74}$$

which means that by making observations she has actually *increased* her ignorance about X. To be sure, her spying attempt was not totally useless because she was certainly able to gain some information in Shannon's sense. However, there are situations when Rényi information is more valuable than Shannon's, and then inequality (26.74) suggests an interesting possibility to limit the adversary's eavesdropping powers. So next time when you see a statement in the literature that "Renyi entropy can increase when it is conditioned on a random variable," you will know what property the author is referring to!

26.8
Universal Hashing

A *hash function* is just a fancy term for converting a large set of elements into a smaller set, where every element in the original set will be projected to an element in the smaller set. One example of a hash function is an algorithm that divides the n-bit key x into r (possibly overlapping) shorter substrings and composes the final r-bit key from the substring *parities* (i.e., if the bit values in the substring add to an even/odd number, we say that parity is even/odd and represent this substring by bit 0/1).

Since there is more than one way of subdividing a string into substrings, we in fact have an entire family of hash functions, each of which can accomplish the task. The one that will actually be used can be selected at random. So let us say we have a random variable $G = \{g_1, g_2, g_3, \ldots\}$ corresponding to the choice of some hash function g from this family. Let us also introduce random variable $X = \{x_1, x_2, x_3, \ldots\}$ whose possible values correspond to different strings x that Alice and Bob may have exchanged.

The result of applying G to X will be a shorter (r-bit) string Y, which is also a random variable. In other words,

$$G(X) = Y. \tag{26.75}$$

Since Y consists of r bits, there are 2^r ways to compose this string, and we have many possible values for that random variable:

$$Y = \{y_1, \ldots, y_{2^r}\}. \tag{26.76}$$

We may expect that a well-designed hash function will cause the final string to be chosen from the available options in an unpredictable manner; in other words, to an outside observer, it should appear as if a hash function just threw a dice having 2^r faces to select the specific value of Y. That is to say, the resulting Y will have a uniform probability distribution. As a result, the chance that $g_i(x_i)$ will select any specific value is y_i equal to $1/2^r$. Suppose the choice was made, and then the same hash function was applied to a different string x_j. Clearly, the chance that a collision will happen, that is, the chance to get the same result $y_j = y_i$ is again $1/2^r$. When this is true, we call our family of hash functions *universal*. In particular, the family G of functions that are based on parities of substrings is one example of a universal family of hash functions. Formally, the defining property of a universal family G of hash functions is that for any g in that family, the requirement

$$P(g(x_i) = g(x_j)) \leq \frac{1}{m} \qquad (26.77)$$

holds for any two different $x_i, x_j \in X$, where m is the size of the set of all possible values of Y.

Then applying (26.68) to the conditional probability of Y given G, we can write

$$H(Y|G) \geq R_2(Y|G) \geq -\log_2 \sum_g P(g) P_c(Y|g). \qquad (26.78)$$

The last term on the right-hand side features the total probability of a collision, that is, probability that a given hash function g will produce the same string $G(x_i) = G(x_j)$ in a pair of independent experiments. But this can happen in two ways: Either because the hash function got the *same* inputs, $x_i = x_j$, and consequently generated the same output or because the inputs were different, but the function accidentally produced the same output anyway. So the sum on the right-hand side will now be expressed as

$$\sum_g P(g) P_c(Y|g) = P(x_i = x_j) + P(x_i \neq x_j) P\big(G(x_i) = G(x_j) | x_i \neq x_j\big). \qquad (26.79)$$

But the probability that $x_i = x_j$ is simply $P_c(X)$, while the chance that a universal hash function will output the same result when the inputs were different, is no greater than 2^{-r} as we discussed earlier. Therefore,

$$\sum_g P(g) P_c(Y|g) \leq P_c(X) + (1 - P_c(X)) 2^{-r} \qquad (26.80)$$

and (26.78) now takes the form

$$H(Y|G) \geq R_2(Y|G) \geq -\log_2 \left[P_c(X) + (1 - P_c(X)) 2^{-r} \right]$$
$$= -\log_2 \left[2^{-r} (2^r P_c(X) + 1 - P_c(X)) \right] = r - \log_2 \left[1 + (2^r - 1) P_c(X) \right].$$

Let us replace the right-hand side with a lesser expression, $r - \log_2 (1 + 2^r P_c(X))$, and also use the fact $R_2(X) = -\log_2 P_c(X)$ is equivalent to $P_c(X) = 2^{-R_2(X)}$:

$$H(Y|G) \geq R_2(Y|G) \geq r - \log_2 \left[1 + 2^r 2^{-R_2(X)} \right]. \qquad (26.81)$$

Finally, inequality $\log_2 (1 + a) \leq (a/\ln 2)$, which we leave for the reader to prove, gives us

$$H(Y|G) \geq R_2(Y|G) \geq r - \log_2 \left(1 + 2^{r - R_2(X)} \right) \geq r - \frac{2^{r - R_2(X)}}{\ln 2}. \qquad (26.82)$$

Suppose now that Alice and Bob who hold the same bit string X suspect that an adversary Eve may know some of the bits in that string. To reduce her knowledge, they decided to use some hash function from the family G and reduce their string to Y. The trouble is that Eve is listening to their conversation and therefore knows the hash function. Then, the leftmost term in (26.82) is a measure of Eve's ignorance about Y given the fact that Eve knows G. The right-hand side then establishes the lower bound on that ignorance. The maximum theoretically available information about Y is $H(Y) = r$ bits. Then the mutual information is

$$H(Y;G) = H_S(Y) - H_S(Y|G) \leq \frac{2^{r-R_2(X)}}{\ln 2}, \qquad (26.83)$$

and the knowledge of the hash function gives Eve no more than $2^{r-R_2(X)}/\ln 2$ Shannon bits of information. In particular, if the Rényi entropy of X is much greater than r as is usually the case given that X is a longer string than Y, then Eve's information about the final string gets arbitrarily small.

It should be emphasized that the Shannon entropy on the left-hand side is only the *average* value, where the averaging was carried out over all available hash functions g. In other words, the relative privacy enjoyed by Alice and Bob is only a statistical phenomenon. It is still possible for Eve to learn substantially more information than (26.78) would suggest if she just gets lucky enough so that the hash function g used in one particular experiment reveals more information to her than the average hash function would. Although this privacy risk is mitigated by the low probability of such an outlier event, Alice and Bob may still want to demand a better privacy guarantee for mission-critical applications.

All this may seem straightforward enough. But at this point the attentive reader may ask: "Wait a minute! The left term in (26.82) is conditioned on the hash function only. Shouldn't Eve's information also depend on her actual eavesdropping procedure?" Indeed, we have cut some corners in an effort to get to the conclusion (26.83) quicker. It is time now to add more precision to the proof. In the actual experiment, Eve will know the hash function G and the string Z composed of the bits of X that she either learned or guessed. We can reproduce the argument that led to (26.82), setting Z equal to a specific value z. As a result, we get the following modification of (26.82):

$$H(Y|G,z) \geq R_2(Y|G,z) \geq r - \log_2\left(1 + 2^{r-R_2(X|z)}\right) \geq r - \frac{2^{r-R_2(X|z)}}{\ln 2}. \qquad (26.84)$$

The question arises: Can we average inequality (26.84) over z to switch to entropies conditioned on the random variable Z? Strictly speaking, the answer is negative. The inequality

$$H(Y|G,Z) \geq R_2(Y|G,Z) \geq r - \log_2\left(1 + 2^{r-R_2(X|Z)}\right) \geq r - \frac{2^{r-R_2(X|Z)}}{\ln 2} \qquad (26.85)$$

is generally false. In the following, we derive the correct version of (26.85).

Obviously, Eve would like to minimize $R_2(X|z)$. Let c be the minimum value that $R_2(X|z)$ can take. Then, inequality (26.84) can be rewritten as follows:

$$H(Y|G,z) \geq R_2(Y|G,z) \geq r - \log_2\left(1 + 2^{r-c}\right) \geq r - \frac{2^{r-c}}{\ln 2}. \tag{26.86}$$

Here, c is going to depend on the specific procedure used by Eve to get her string $Z = z$.

To get a concrete estimate of the upper bound of Eve's knowledge of Y for some specific eavesdropping scenario, suppose that Eve does some eavesdropping on the n-bit string X, thereby learning at most t deterministic bits of information, which she then records in a t-bit string Z (note that we treat X and Z as random variables). Here, of course, t must be less than n.[8] Formally, if we introduce an eavesdropping function e defined by

$$e : \{0,1\}^n \to \{0,1\}^t, \tag{26.87}$$

then

$$Z = e(X). \tag{26.88}$$

Note that since Z consists of t bits, there are 2^t possibilities[9] for the specific value $Z = z$ of the random variable Z.

Suppose that a given z can be obtained by eavesdropping in c_z possible ways; in other words, there are c_z different values of x such that $z = e(x)$. Then, if z is known, each legitimate candidate for x must be assigned probability $1/c_z$. Conditional Rényi entropy will be

$$R_2(X|z) = -\log_2 P_c(X|z) = -\log_2 \sum \frac{1}{c_z^2} = -\log_2 \frac{1}{c_z} = \log_2 c_z, \tag{26.89}$$

and then (26.86) becomes

$$H(Y|G,z) \geq R_2(Y|G,z) \geq r - \frac{2^{r-\log_2 c_z}}{\ln 2} = r - \frac{2^r}{c_z \ln 2}. \tag{26.90}$$

Averaging over z yields

$$H(Y|G,Z) \geq \sum_z P(z)\left[r - \frac{2^{r-\log_2 c_z}}{\ln 2}\right] = r - \sum_z P(z)\frac{2^r}{c_z \ln 2}. \tag{26.91}$$

---

[8] In a more typical practical scenario, Eve will compose an n-bit string Z where only t bits will be correct, and the rest will be essentially random guesses. Since random bits don't provide any useful information, Alice and Bob should only be concerned about the t bits that are "meaningful."

[9] Before we move on, we want to emphasize again that t is the number of deterministic bits. This is a very strict requirement: Not only it is necessary that some (integer) number of physical bits have correct values with certainty, but Eve should also know with certainty *which* of her bits are correct. Even if she cannot accurately identify the locations of these correct bits in the string, she will still have only probabilistic information about each specific bit.

But the probability in (26.91) is $P(z) = c_z 2^{-n}$ because string z can be obtained from c_z possible values of x, and each x occurs with probability 2^{-n}. Therefore,

$$H(Y|G,Z) \geq r - \sum_z \frac{2^{r-n}}{\ln 2}, \qquad (26.92)$$

and the mutual Shannon information is

$$H(Y;G,Z) = H(Y) - H(Y|G,Z) \leq \sum_z \frac{2^{r-n}}{\ln 2}. \qquad (26.93)$$

However, there are 2^t different values of z and hence 2^t terms in the sum, so that

$$H(Y;G,Z) = H(Y) - H(Y|G,Z) \leq \frac{2^{t+r-n}}{\ln 2}. \qquad (26.94)$$

Introducing the notation $s \equiv n - r - t$, we can finally write the important result known as the *main privacy amplification theorem*:

$$H(Y;G,Z) \leq \frac{2^{-s}}{\ln 2}. \qquad (26.95)$$

The number s here is the length of the original string X minus the length of the shorted string Y minus the length of the eavesdropped string Z. This number is usually called *safety parameter* because it determines how much information Eve is left with after the privacy amplification process. Alice and Bob can reduce it as close to zero as they like by choosing a sufficiently large safety parameter. This puts only one restriction on the size r of their final string Y:

$$r = n - t - s. \qquad (26.96)$$

That is, they only need to drop from their initial string the number of bits that Eve has managed to eavesdrop, and then some additional s bits where s only depends on the desired level of security.[10]

Up till now we have been assuming that Eve is learning deterministic bits. The question arises: What if instead, she decides to use some other procedure that will give her the bit values probabilistically? Suppose that she managed to obtain t bits of information in the Rényi sense, that is,

$$R_2(X;z) = R_2(X) - R_2(X|z) = t. \qquad (26.97)$$

This immediately implies

$$R_2(X|z) = R_2(X) - t = n - t \qquad (26.98)$$

---

10) A separate important question is whether we can make the security *infinite* by forcing Eve's information to be strictly equal to zero. Inequality (26.95) cannot help us in this case since s cannot exceed $n - t$. However, the problem does have a solution, although it would extract a high cost in terms of the allowed length r of the secret key. For example, to erase completely $t = 2$ deterministic bits of Eve's information, Alice and Bob will have to shorten the string to $r \leq (2/3)n$ as shown in Ref. [4].

and then (26.84) takes the form

$$H(Y|G,z) \geq R_2(Y|G,z) \geq r - \frac{2^{r-(n-t)}}{\ln 2} = r - \frac{2^{-s}}{\ln 2}. \qquad (26.99)$$

Finally,

$$H(Y;G,z) \leq R_2(Y;G,z) \leq \frac{2^{-s}}{\ln 2} \qquad (26.100)$$

and t Rényi bits behave the same way as t deterministic bits with regard to privacy amplification. The result (26.100) is sometimes called the *auxiliary privacy amplification theorem*.

But this still leaves open the question as to what happens if Eve learns t Shannon bits. One might think we should again get a result similar to (26.100). But it turns out not to be the case! In fact, one can demonstrate with concrete examples that if we allow Eve to learn t Shannon bits about X, then, generally speaking, privacy amplification will fail to reduce her information substantially (see Problem 26.1).

26.9
The Holevo Bound

At the end of Chapter 24, we have briefly mentioned the quantum state estimation problem, which arose in the context of no-signaling requirement. In the most general case, Alice selects some (possibly mixed) quantum state ρ out of a given set $\{\rho_1, \rho_2, \ldots, \rho_n\}$, and sends it to Bob. Then, Bob is facing a variation of the state discrimination problem that we encountered in Section 25.6. The difference is that this time, the number of states is greater than 2 and the states themselves are mixtures. We do not discuss here the question as to what POVM will be optimal for the purpose. Instead, we ask: How much information can Bob possibly learn from such a POVM?

Let us describe Alice's choice by a random variable X where $X = \{0, \ldots, n\}$, and denote the state she chooses by the symbol ρ_X. Just as before, we will assign to Alice's choice of each specific ρ_x (note the lowercase subscript!) the probability $P(x)$. Then, we can say that *on the average*, Alice sends to Bob the probability-weighted matrix:

$$\rho = \sum_x P(x)\rho_x. \qquad (26.101)$$

As we remember, in case of density matrices, Shannon entropy is replaced by von Neumann entropy S, so the entropy of the mixture (26.101) will be given by

$$S(\rho) = -\mathrm{Tr}\left(\rho \log_2 \rho\right). \qquad (26.102)$$

Having said that, the mixture (26.101) is still a theoretical construct since Alice never sends to Bob the state ρ in any single trial, but actually sends one of the states ρ_x, which has entropy

$$S(\rho_x) = -\mathrm{Tr}\left(\rho_x \log_2 \rho_x\right). \tag{26.103}$$

We can then find the average entropy:

$$\langle S(\rho_x) \rangle = \sum_x P(x) S(\rho_x) = -\sum_x P(x) \mathrm{Tr}\left(\rho_x \log_2 \rho_x\right). \tag{26.104}$$

During his measurement, Bob can get different outcomes. Let the set of possible outcomes be denoted by a random variable $Y = \{y_1, y_2, \ldots, y_m\}$. (Note that the subscript m here corresponds to the number of POVM elements and is different from n.) Then the question becomes, how much information can Bob learn about X by observing Y? As we remember, the answer is given by the mutual information $H(X;Y)$. But how large can it get? Could it be that Bob obtains perfect knowledge of X and thus performs perfect state estimation?

It turns out that there exists an upper bound on $H(X;Y)$, which is given by the difference of (26.102) and (26.104). This difference is known as *Holevo χ quantity*:

$$\chi = S(\rho) - \langle S(\rho_x) \rangle. \tag{26.105}$$

As a result, information learned by Bob in this experiment obeys the inequality

$$H(X;Y) \leq \chi. \tag{26.106}$$

This inequality, which we state without proof, is called the *Holevo bound*.

As an example, consider the already familiar two-state discrimination measurement where Alice prepares photon states $|u_1\rangle$ and $|u_2\rangle$ with equal probability and $|\langle u_1 | u_2 \rangle| = \cos\theta$. If we choose the basis that that $|u_1\rangle$ coincides with one basis vector, then our two density matrices ρ_1 and ρ_2 are equal to

$$\rho_1 = |u_1\rangle\langle u_1| = \begin{pmatrix} 1 & 0 \\ 0 & 0 \end{pmatrix}, \quad \rho_2 = |u_2\rangle\langle u_2| = \begin{pmatrix} \cos^2\theta & \cos\theta\sin\theta \\ \cos\theta\sin\theta & \sin^2\theta \end{pmatrix}. \tag{26.107}$$

Both these matrices have the same pair of eigenvalues $\lambda_{1,2} = 0, 1$ and therefore their von Neumann entropies are equal to 0. This illustrates the general rule that von Neumann entropy of a pure state always vanishes. (We leave the proof of this statement to the reader as an exercise.) Thus, in this case, χ will be determined by von Neumann entropy (26.102) of the average density matrix:

$$\rho = \frac{1}{2}\begin{pmatrix} 1+\cos^2\theta & \cos\theta\sin\theta \\ \cos\theta\sin\theta & \sin^2\theta \end{pmatrix}. \tag{26.108}$$

Solving for eigenvalues, we get two solutions, $\lambda_{1,2} = (1 \pm \cos\theta)/2$, and then it is easy to write the logarithm and von Neumann entropy:

$$S(\langle\rho\rangle) = -\mathrm{Tr}\begin{pmatrix} (1+\cos\theta)/2 & 0 \\ 0 & (1-\cos\theta)/2 \end{pmatrix}\begin{pmatrix} \log_2(1+\cos\theta)/2 & 0 \\ 0 & \log_2(1-\cos\theta)/2 \end{pmatrix}$$
$$= -(\cos^2\theta/2)\log_2(\cos^2\theta/2) - (\sin^2\theta/2)\log_2(\sin^2\theta/2) \equiv H(\cos^2\theta/2). \tag{26.109}$$

Thus, χ is equal to the right-hand side of (26.109), which is just the binary entropy function, and the Holevo bound is

$$H(X; Y) \leq H\left(\cos^2 \frac{\theta}{2}\right). \tag{26.110}$$

When the states are orthogonal, that is, when $\theta = \pi/2$, this inequality tells us that $H(X; Y) \leq H(1/2) = 1$ and Bob is allowed to learn up to 1 bit of Shannon information about X, which means the state can be determined with certainty. We know that precise state discrimination is in fact possible for orthogonal states and all it takes is a simple projective measurement. In the other limiting case $\theta = 0$, we get $H(X; Y) \leq H(0) = 0$ and the Holevo bound prohibits Bob to know anything about X, which again makes perfect sense since Alice's two choices are now totally indistinguishable. For other values of θ, Bob can learn less than 1 bit of information. Note that it would be a mistake to plug the failure rates Q from Section 25.6 into the binary entropy function to estimate mutual information on the left-hand side of (26.110)! Instead, it must be estimated as described earlier in Section 26.5. To take one example, in the important case $\theta = \pi/4$, the upper bound is $H(X; Y) \leq H(\cos^2(\pi/8)) \approx 0.6$ and the secrecy of Alice's choice remains protected to a reasonable degree no matter what type of POVM Bob tries.

26.10 Entropy of Entanglement

Let ρ^{AB} be the density matrix of a composite system, and $\rho^A = \text{Tr}_B \rho^{AB}$ and $\rho^B = \text{Tr}_A \rho^{AB}$ be the reduced density matrices of its subsystems. Accordingly, we have von Neumann entropies $S(\rho^{AB})$, $S(\rho^A)$, and $S(\rho^B)$. While in the previous sections we were interested in deriving different measures of disorder, our current agenda is to define a new measure of entanglement, alternative to the concurrence. To this end, we are going to consider the simplest case only, when $\rho^A = \rho^B$.

Suppose first the system is in a pure state: $\rho^{AB} \to \psi^{AB}$ and $\rho^A \to \psi^A$, $\rho^B \to \psi^B$. It is clear then from the two limiting cases below that we know as much or as little about its entanglement as we know about its subsystems. When there is no entanglement, the system is in a product state $|\psi^{AB}\rangle = |\psi^A\rangle \otimes |\psi^B\rangle$ so that $S(\psi^A) = S(\psi^B) = 0$. When the system is maximally entangled, it must be in a Bell state where $\rho^A = \rho^B = I/2$. Then, $S(\rho^A) = S(\rho^B) = 1$. We thus have in the intermediate case $0 \leq S(\rho^A) = S(\rho^B) \leq 1$, with both entropies increasing monotonically from 0 to 1 as the subsystems become more entangled. This shows that $S(\rho^A) = S(\rho^B)$ can serve as a measure of entanglement. We define for a pure system

$$\mathcal{E}(\psi^{AB}) \equiv S(\rho^A) = S(\rho^B). \tag{26.111}$$

The quantity $\mathcal{E}$ is called *entropy of entanglement* [5].

We now have two competing measures: entropy of entanglement $\mathcal{E}$ and concurrence C that was defined earlier. A computation that we will not reproduce here shows that we can relate the one with the other as follows:

$$\mathcal{E}(\psi^{AB}) = h\left(\frac{1+\sqrt{1-C^2(\psi^{AB})}}{2}\right), \qquad (26.112)$$

where h is the binary entropy function.

When we turn to the mixed composite state ρ^{AB}, there exists an infinite number of possible pure state decompositions $\rho^{AB} = \sum_i P(i)|\psi_i^{AB}\rangle\langle\psi_i^{AB}|$. Here, each $|\psi_i^{AB}\rangle$ is a pure state that can, however, be entangled and therefore possesses entropy of entanglement (26.112). Then, we can find the *average* entropy of entanglement as a weighted sum of entropies of pure states:

$$\langle\mathcal{E}(\rho^{AB})\rangle = \sum_i P(i)\mathcal{E}(\psi_i^{AB}). \qquad (26.113)$$

However, we note that the average taken over some arbitrary decomposition cannot in principle serve as a meaningful measure! What makes sense is to find the *minimum* average entropy of entanglement if one tries all possible decompositions:

$$\langle\mathcal{E}(\rho^{AB})\rangle_{\min} = \min\left(\sum_i P(i)\mathcal{E}(\psi_i^{AB})\right). \qquad (26.114)$$

Now according to the theorem that we state here without proof, this latter quantity is given by the expression analogous to (26.112) in which we only have to substitute ρ^{AB} for ψ^{AB}. We refer the reader to Ref. [6] for details of the derivation. This gives us grounds to define the entropy of entanglement for a mixture by analogy with (26.112):

$$\mathcal{E}(\rho^{AB}) = \langle\mathcal{E}(\rho^{AB})\rangle_{\min} = \min\left(\sum_i P(i)\mathcal{E}(\psi_i^{AB})\right) = h\left(\frac{1+\sqrt{1-C^2(\rho^{AB})}}{2}\right). \qquad (26.115)$$

The special term for this measure is *entanglement of formation*.

Problems

26.1 Show that for $\alpha = 1$, the Rényi entropy is equivalent to Shannon entropy.

(*Hint*: Use L'Hospital's rule.)

26.2 Derive the special case (26.56) from the general Jensen inequality (26.57).

(*Hint*: Pay special attention to your choice of function f.)

26.3 A random variable X is an n-bit string that is chosen by Bob (with uniform probability distribution) from a given set of such strings. By watching Bob's actions through an imperfect pair of glasses, Eve has recorded a string Z, which does not necessarily have n bits in it and some of the bit value may differ from

Bob's. Eve's conditional probability distribution is defined by

$$P(X|z) = \begin{cases} 2^{-(n/4)}, & x = z, \\ \dfrac{1 - 2^{-(n/4)}}{2^n - 1}, & x \neq z. \end{cases}$$

An oracle gives Eve for free another random variable U, defined by

$$U = \begin{cases} 0, & x = z, \\ 1, & x \neq z. \end{cases}$$

Show that with an overwhelming probability for large n, the conditional Rényi entropy $R_2(X|U, z)$ is about twice as large as the unconditioned entropy $R_2(X|z)$. (If you had solved that problem in 1995, you could have coauthored the article in Ref. [7].)

26.4 Suppose Eve obtains an n-bit string Z that is equal to X with probability t/n (where t is an arbitrary number less than n) and is completely random otherwise, and that Eve does not know which possibility was realized. Then, Alice and Bob proceed to establish a distilled secret key $Y = g(X)$. Prove that $H(X; Z) \leq t$ and $H(Y; Z, g) > (rt/n) - 1$, $\forall g \in G$. The purpose of this problem is to show that Eve's Shannon information can be less than t bits and yet successfully withstand privacy amplification.

References

1 Shayevitz, O. (2010) A note on a characterization of Renyi measures and its relation to composite hypothesis testing, arXiv: 1012.4401v1.

2 Harrimoes, P. (2005) Interpretations of Renyi entropies and divergences, University of Copenhagen. http://arxiv.org/ftp/math-ph/papers/0510/0510002.pdf.

3 Csiszar, I. (1995) Generalized cutoff rates and Renyi's information measures. *IEEE Trans.*, **41**(1), 26–34.

4 Chor, B., Goldreich, O., Histad, J., Friedman, J., Rudich, S., and Smolensky, R. (1985) The bit extraction problem or t-resilient functions. *Proceedings of the 26th IEEE Symposium on Foundations of Computer Science*, pp. 396–407.

5 Audretsch, J. (2006) *Entangled Systems*, Wiley-VCH Verlag GmbH, Berlin.

6 Wootters, W. K. (2001). Entanglement of formation and concurrence. *Quant. Inf. Comput.*, **1**, 27–44.

7 Bennett, C., Brassard, G., Crepeau, C., and Mauer, U.M. (1995) Generalized privacy amplification. *IEEE Trans. Inf. Technol. B*, **41** (6), 1915–1923.

27
Quantum Gates

Logic gates are devices that transform the state of input bits according to some predefined algorithm. The three most common gates used in modern classical computers are NOT, NAND, and NOR. The operation of logic gates is described by Boolean (truth) tables that list all possible input and output states. Implementation of gates can be carried out by electric circuits, transistors, or, as we get to the microscopic level, by individual molecules and atoms. What makes quantum logic gates unique is their ability to accept as input a linear superposition of states in the truth table, and process this input in accordance with linearity of quantum mechanics.

27.1
Truth Tables

Figure 27.1 shows some of the common logic gates used in classical computer science. The operation of these gates should be clear from the following truth tables:

A	B	A AND B	A OR B	A XOR B	A NAND B	A NOR B	A XNOR B
0	0	0	0	0	1	1	1
0	1	0	1	1	1	0	0
1	0	0	1	1	1	0	0
1	1	1	1	0	0	0	1

A	Buffer A	NOT A
0	0	1
1	1	0

A logic gate is called universal if all other gates can be constructed from it. For example, consider the NAND gate. If we fix $A = 1$, the gate will act as a NOT with respect to bit B:

$$\text{NOT}(B) = \text{NAND}(1, B). \tag{27.1}$$

Quantum Mechanics and Quantum Information: A Guide through the Quantum World,
First Edition. Moses Fayngold and Vadim Fayngold.
© 2013 Wiley-VCH Verlag GmbH & Co. KGaA. Published 2013 by Wiley-VCH Verlag GmbH & Co. KGaA.

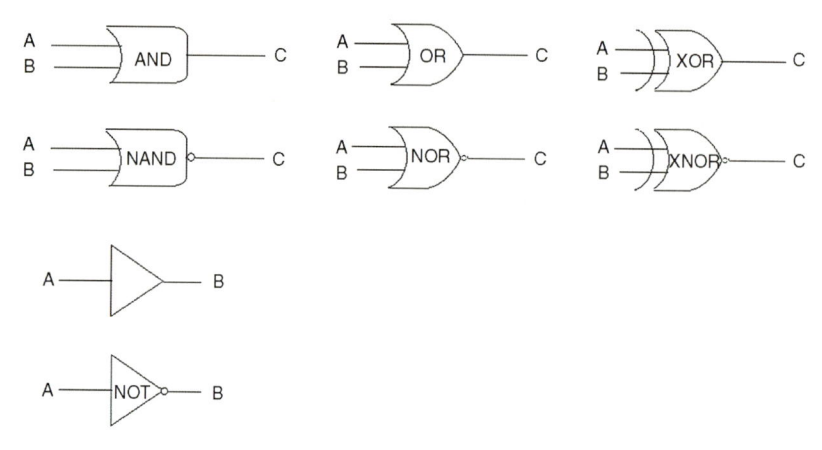

Figure 27.1 Common logic gates shown schematically. The NOT gate reverses the input bit, transforming 0 into 1 and 1 into 0, an operation that is called "taking the complement" in Boolean algebra. Sometimes this gate is called "inverting buffer." Accordingly, if a gate just leaves the bit as it is, we call it simply a buffer. A buffer gate is shown directly above the NOT. It corresponds to the identity operator in QM. The NAND, NOR, XNOR gates are best thought of as AND, OR, XOR gates followed by an inverting buffer. The inversion of output is represented by an empty circle (bubble) in the figure. XOR stands for "exclusive OR," which results in 1 only when the two inputs are different.

To obtain AND, we can use two NAND gates as follows:

$$\text{AND}(A, B) = \text{NAND}(1, \text{NAND}(A, B)). \tag{27.2}$$

The output of the first NAND is used as an input to the second NAND, with the other bit held fixed at 1. This result is easy to verify by using the truth table, or else one can use (27.1) to show that the right-hand side of (27.2) is equal to NOT(NAND(A,B)) and then double inversion results in AND(A,B).

Finally, we can obtain OR using three NAND gates:

$$\text{OR}(A, B) = \text{NAND}(\text{NAND}(1, A), \text{NAND}(1, B)). \tag{27.3}$$

The two NANDs within parenthesis evaluate NOT(A) and NOT(B) and then (27.3) follows from the truth table.

The remaining gates can be constructed similarly (Problem 27.1). One can therefore build a computer using NAND gates only, so this gate is universal. Another universal gate is NOR (Problem 27.3). However, not all gates possess the universality property (Problems 27.4 and 27.5).

The AND and XOR gates play an important role in computer design because they can perform binary addition. In the binary system, the sum of two bits is a binary number that can take values 00, 01, and 10, where the leading bit is the so-called carry bit and the trailing bit is the sum bit. The four possible inputs and the respective outputs are shown in the table below:

A	B	Carry	Sum
0	0	0	0
0	1	0	1
1	0	0	1
1	1	1	0

One can see that the sum bit is equal to A XOR B. Formally, one gets this bit by performing *modulo 2* addition. (Recall that in arithmetic, whenever we have an integer number expression $a = km + b$, we can say that $a = b \bmod m$. For example, $17 = 1 \bmod 8$ and $7 = 1 \bmod 2$.) So the sum bit S can be written $S = (A + B) \bmod 2$, or if one denoted addition modulo 2 by symbol $\oplus$ (not to be confused with direct sum notation for matrices!), $S = A \oplus B$. To get the carry bit, one can use the output of A AND B. Thus, a combination of XOR and AND shown in Figure 27.2 will act as a *half-adder*. The term half-adder refers to the fact that it can add two single-bit numbers, but it cannot be used for addition of two multiple bit numbers. To create a *bona fide* full adder, we must have a way to account for the carry bit created during the previous addition. This is possible if we allow the adder to accept three input bits and add them together. The operation of the full adder is shown in the following table and illustrated in Figure 27.3.

A	B	C_{in}	C_{out}	S
0	0	0	0	0
0	0	1	0	1
0	1	0	0	1
0	1	1	1	0
1	0	0	0	1
1	0	1	1	0
1	1	0	1	0
1	1	1	1	1

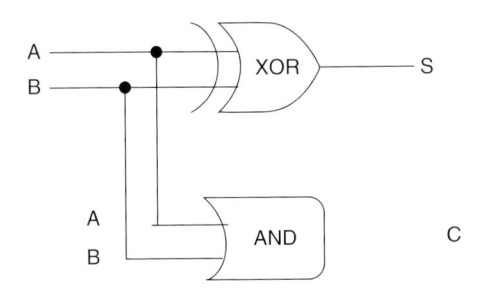

Figure 27.2 A half-adder implemented by a combination of XOR and AND gates. Both gates receive the same inputs A and B.

27 Quantum Gates

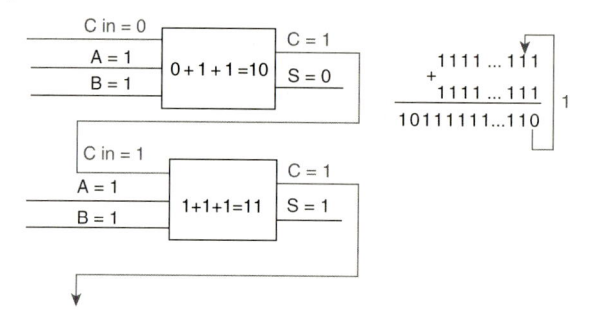

Figure 27.3 Binary addition of two numbers on the right is carried out step by step. The first two steps are shown. The carry bit from the previous step becomes an input for the next step. The final answer is comprised by the bits emerging from the S ports.

Another remarkable property of XOR is that after a small modification, it can act as a CNOT, or controlled NOT gate, if we interpret one input bit, say, A, as the *control bit*, and the other, B, as the *target bit*. When the control bit is set to 0, the gate leaves the target bit unchanged, but when the control is set to 1, the target bit is flipped. One can see from the truth table that this is indeed the behavior of XOR. The only way CNOT differs from XOR is that CNOT has two outputs: the control bit A is output in parallel with the modified target bit $A \oplus B$, as shown in Figure 27.4. The truth table for this gate follows, with subscripts denoting the initial and final (i.e., input and output) states of the two bits:

A_i	B_i	A_f	B_f
0	0	0	0
0	1	0	1
1	0	1	1
1	1	1	0

The next crucial concept related to gates is reversibility. The 2 bit gates AND, OR, NAND, NOR, XOR, and XNOR are all *irreversible*. By this we mean that by looking at the output, we generally cannot reconstruct the input. For example, the input to the AND gate is known with certainty when the output is 1, but is uncertain when the output is 0, so deterministic information about the input states is lost three times out

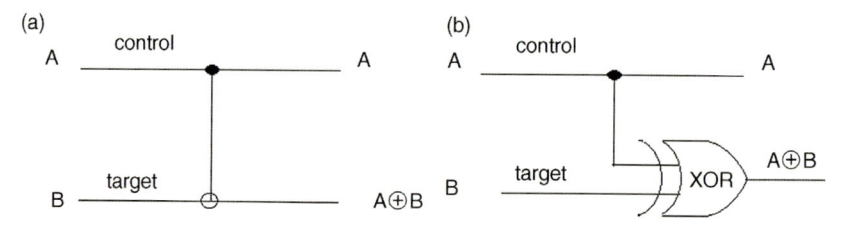

Figure 27.4 CNOT gate flips the target bit if the control bit is set to 1, and leaves it alone otherwise. Schematic (circuit) representation is shown in part (a); implementation with XOR is shown in part (b).

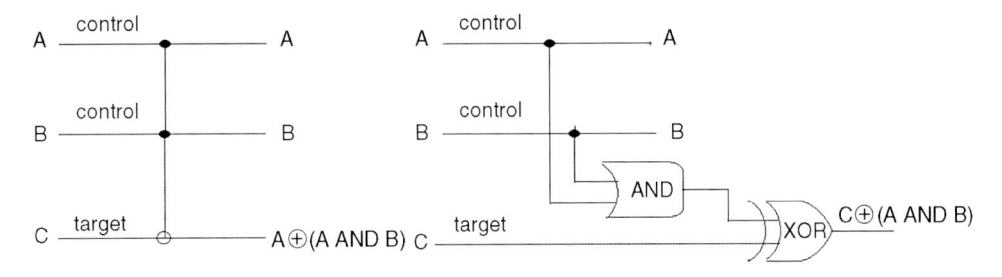

Figure 27.5 Toffoli gate flips the target bit C only if A and B are both 1.

of four. Moreover, *any* gate that inputs more bits than it outputs is irreversible since one cannot assign uniquely one combination of input values to every combination at the output. Reversibility is only possible (although not guaranteed) when the output consists of at least as many bits as the input. For example, the buffer and inverter gates are both reversible. The CNOT gate is also reversible since its truth table provides a one-to-one mapping between the possible inputs and possible outputs. (In fact, reversibility is the main reason we wanted to "upgrade" the XOR gate to the CNOT by adding a second output port.) The physical meaning of reversibility is that no information is lost during the operation of the gate. Thus, as long as we use CNOT gates, the entropy of the system will remain constant.

Though CNOT is not a universal gate, we can achieve universality if we "upgrade" it by adding the third input and the third output. The resulting CCN, or "control–control–NOT" gate was first suggested by Toffoli in 1980. It flips the target bit only when *both* control bits are 1 and leaves it alone otherwise; both control bits remain unchanged. The corresponding equation for the target bit is

$$\text{CCN}(A, B, C) = C \oplus (A \text{ AND } B). \tag{27.4}$$

Constructing the truth table and establishing the reversibility property is left to the reader as an exercise. The reason the Toffoli gate (Figure 27.5) is so important is because it functions as a NAND gate for bits A and B if we set $C = 1$:

$$\text{CCN}(A, B, 1) = \text{NAND}(A, B). \tag{27.5}$$

But the NAND gate is universal and so as long as we keep consistently sending bit 1 to the input port C of every Toffoli gate in the computer, we produce all other gates as needed, and have a fully reversible computer that does not dissipate information.

27.2
Quantum Logic Gates

Consider a two-state system:

$$|\psi\rangle = c_0|0\rangle + c_1|1\rangle, \tag{27.6}$$

with eigenstates $|0\rangle$ and $|1\rangle$ corresponding to energies E_0, E_1. The basis $\{|0\rangle, |1\rangle\}$ formed by these two eigenstates is the most natural basis in which one can describe the system and is usually chosen by default. In the language of quantum information theory, it is called *computational basis*. The two-state system is called a *qubit*.

If we place qubit (27.6) in an external field, it is going to evolve in time; so, the amplitudes c_0 and c_1 should be written as functions of t:

$$|\psi(t)\rangle = \hat{U}|\psi(0)\rangle = c_0 \hat{U}|0\rangle + c_1 \hat{U}|1\rangle = c_0 e^{-i(E_0 t/\hbar)}|0\rangle + c_1 e^{-i(E_1 t/\hbar)}|1\rangle$$
$$= c_0(t)|0\rangle + c_1(t)|1\rangle. \tag{27.7}$$

Here, $\hat{U} = e^{-(i\hat{H}t/\hbar)}$ is the time evolution operator defined by the given Hamiltonian. In the matrix form,

$$\begin{pmatrix} c_0(t) \\ c_1(t) \end{pmatrix} = \begin{pmatrix} u_{11} & u_{12} \\ u_{21} & u_{22} \end{pmatrix} \begin{pmatrix} c_0 \\ c_1 \end{pmatrix}, \tag{27.8}$$

and the only constraint on the operator is the unitarity requirement. Any allowed operator $\hat{U}$ represents a valid quantum gate.

Unlike their classical analogues, quantum gates turn out to have an entirely new property – ability to process a superposition of 1 and 0 bits. That is, while a classical gate could only handle the cases $c_0 = 0$, $c_1 = 1$ and $c_0 = 1$, $c_1 = 0$, a quantum gate will accept any pair of complex amplitudes as long as $|c_0|^2 + |c_1|^2 = 1$.

We now see that one can design a reversible universal quantum gate. Equation 27.7 shows that the operation of time evolution operator $\hat{U} = e^{-(i\hat{H}t/\hbar)}$ on the qubit is reversible because one can reproduce the original state as unitarity guarantees the existence of the inverse operator $\hat{U}^{-1} = \hat{U}^+ = e^{i\hat{H}t/\hbar}$. This tells that at least on the single-qubit level, quantum operations are reversible. Thus, we can make both a buffer by simply applying the identity operator to the input eigenstate and an inverter by applying the time evolution operator that takes $|0\rangle$ into $|1\rangle$ and $|1\rangle$ into $|0\rangle$.

A reversible quantum-mechanical gate can be realized if we are able to construct the Toffoli gate from three qubits A, B, and C, whose joint state will be written as a ket $|ABC\rangle$, where the variables are allowed to take values 0 or 1. In other words, we need a time evolution operator of the form

$$\text{CCN} = |110\rangle\langle 111| + |111\rangle\langle 110| + |101\rangle\langle 101| + |011\rangle\langle 011| + |000\rangle\langle 000|$$
$$+ |001\rangle\langle 001| + |010\rangle\langle 010| + |100\rangle\langle 100|. \tag{27.9}$$

This operator, when applied to $|111\rangle$ state, transforms it into $|110\rangle$, just as expected from the Toffoli gate, and similarly, when applied to $|110\rangle$, it produces state $|111\rangle$. In the remaining six cases, all three bits are left unchanged. Notice that all eight possible input combinations are mutually exclusive and therefore orthogonal. It is straightforward to show that this operator is Hermitian. So at least in theory, one can build a quantum computer with all the classical features; the only thing it takes is to realize the

27.2 Quantum Logic Gates

Hamiltonian $\hat{H}$ satisfying

$$\text{CCN} = e^{-(i\hat{H}t/\hbar)}, \tag{27.10}$$

that is,

$$\hat{H} = \frac{i\hbar}{t}\ln(\text{CCN}), \tag{27.11}$$

where CCN is given by (27.9).

It should be easy now to write the quantum analogues of the common classical gates. The buffer and NOT gates act on a single qubit and are described by 2×2 matrices:

$$\begin{pmatrix} 1 & 0 \\ 0 & 1 \end{pmatrix}\begin{pmatrix} c_1 \\ c_2 \end{pmatrix} = \begin{pmatrix} c_1 \\ c_2 \end{pmatrix}, \quad \begin{pmatrix} 0 & 1 \\ 1 & 0 \end{pmatrix}\begin{pmatrix} c_1 \\ c_2 \end{pmatrix} = \begin{pmatrix} c_2 \\ c_1 \end{pmatrix}. \tag{27.12}$$

The identity matrix leaves the bit unchanged, thus realizing the buffer. The second matrix swaps the amplitudes, which means that input $|0\rangle$ will be flipped to $|1\rangle$ and vice versa, so it realizes the NOT.

Yet another important single-qubit gate is the Hadamard gate [1]:

$$H = \frac{1}{\sqrt{2}}\begin{pmatrix} 1 & 1 \\ 1 & -1 \end{pmatrix}, \tag{27.13}$$

whose operation is described by

$$\frac{1}{\sqrt{2}}\begin{pmatrix} 1 & 1 \\ 1 & -1 \end{pmatrix}\begin{pmatrix} c_1 \\ c_2 \end{pmatrix} = \frac{1}{\sqrt{2}}\begin{pmatrix} c_1 + c_2 \\ c_1 - c_2 \end{pmatrix}. \tag{27.14}$$

In particular, input $|0\rangle$ is transformed by the gate into $(|0\rangle + |1\rangle)/\sqrt{2}$, and input $|1\rangle$ is transformed into $(|0\rangle - |1\rangle)/\sqrt{2}$ (see Figure 27.6). It is easy to see why this gate has no

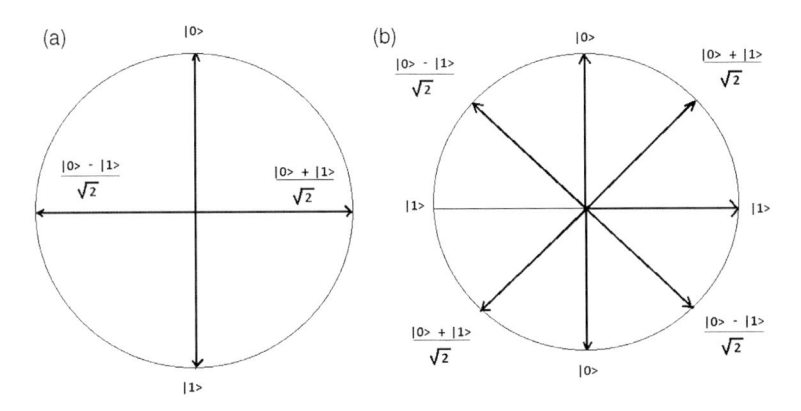

Figure 27.6 Operation of the Hadamard gate (a) on an electron qubit (or a photon qubit in the Poincarè sphere representation) and (b) on a photon qubit with $|0\rangle$ and $|1\rangle$ corresponding to vertical and horizontal polarization, respectively (using double-headed arrows in the diagram on the right to indicate the state; angles are between physical polarization directions). The output photon will have diagonal polarization.

classical analogy. The output qubit is "halfway" between the two eigenstates – a condition that does not exist in Boolean algebra or, for that matter, in classical physics (the best one can do with a classical gate is to prepare a 1 : 1 mixture of zeros and ones). The resulting output states are orthogonal and form the *Hadamard basis* that can be used as an alternative to the standard computational basis.

Pauli spin matrices $\hat{\sigma}_x, \hat{\sigma}_y, \hat{\sigma}_z$ can also be thought of as quantum gates. When used in this context, they are called Pauli-X gate, Pauli-Y gate, and Pauli-Z gate and are written in their usual matrix form:

$$X = \begin{pmatrix} 0 & 1 \\ 1 & 0 \end{pmatrix}, \quad Y = \begin{pmatrix} 0 & -i \\ i & 0 \end{pmatrix}, \quad Z = \begin{pmatrix} 1 & 0 \\ 0 & -1 \end{pmatrix}. \tag{27.15}$$

The reader can see from (27.12) that Pauli-X matrix is equivalent to the quantum NOT gate. Pauli-Y matrix changes the qubit state as follows:

$$\begin{cases} Y|0\rangle = \begin{pmatrix} 0 & -i \\ i & 0 \end{pmatrix}\begin{pmatrix} 1 \\ 0 \end{pmatrix} = \begin{pmatrix} 0 \\ i \end{pmatrix} = i\begin{pmatrix} 0 \\ 1 \end{pmatrix} = i|1\rangle, \\ Y|1\rangle = \begin{pmatrix} 0 & -i \\ i & 0 \end{pmatrix}\begin{pmatrix} 0 \\ 1 \end{pmatrix} = \begin{pmatrix} -i \\ 0 \end{pmatrix} = -i\begin{pmatrix} 1 \\ 0 \end{pmatrix} = -i|0\rangle. \end{cases} \tag{27.16}$$

The resulting mapping of $|0\rangle$ to $i|1\rangle$ and of $|1\rangle$ to $-i|0\rangle$ is equivalent to a rotation by π rad around the Y-axis of the Poincaré sphere. For Pauli-Z matrix, we have

$$\begin{cases} Z|0\rangle = \begin{pmatrix} 1 & 0 \\ 0 & -1 \end{pmatrix}\begin{pmatrix} 1 \\ 0 \end{pmatrix} = \begin{pmatrix} 1 \\ 0 \end{pmatrix} = |0\rangle, \\ Z|1\rangle = \begin{pmatrix} 1 & 0 \\ 0 & -1 \end{pmatrix}\begin{pmatrix} 0 \\ 1 \end{pmatrix} = \begin{pmatrix} 0 \\ -1 \end{pmatrix} = -|1\rangle, \end{cases} \tag{27.17}$$

and the result is equivalent to a phase shift by π. One can generalize the last equation by defining the *phase shift gate*:

$$\hat{R}_\theta = \begin{pmatrix} 1 & 0 \\ 0 & e^{i\theta} \end{pmatrix}, \tag{27.18}$$

which leaves the $|0\rangle$ eigenstate unchanged and changes $|1\rangle$ to $e^{i\theta}|1\rangle$. Or, for an arbitrary input state,

$$\hat{R}_\theta \begin{pmatrix} c_0 \\ c_1 \end{pmatrix} = \begin{pmatrix} c_0 \\ e^{i\theta} c_1 \end{pmatrix}, \tag{27.19}$$

and the gate preserves probabilities while modifying the phase.

Finally, one can have the *rotation gate*:

$$\hat{R}_{\alpha,\beta} = \begin{pmatrix} \cos\beta & -e^{i\alpha}\sin\beta \\ e^{-i\alpha}\sin\beta & \cos\beta \end{pmatrix}, \tag{27.20}$$

which rotates the qubit states by angles α, β. Its action on an arbitrary qubit is described by

$$\begin{pmatrix} \cos\beta & -e^{i\alpha}\sin\beta \\ e^{-i\alpha}\sin\beta & \cos\beta \end{pmatrix} \begin{pmatrix} c_0 \\ c_1 \end{pmatrix} = \begin{pmatrix} c_0\cos\beta - c_1 e^{i\alpha}\sin\beta \\ c_0 e^{-i\alpha}\sin\beta + c_1 \cos\beta \end{pmatrix}. \quad (27.21)$$

For the two-qubit gates that also output a two-qubit state, we must compose 4×4 matrices.[1] Let us work in the basis $\{|00\rangle, |01\rangle, |10\rangle, |11\rangle\}$, with eigenstates following in that order, and derive the matrix for CNOT. The operation of the gate on the four input states is described by a system of four equations:

$$\begin{cases} \hat{U}|00\rangle = |00\rangle, \\ \hat{U}|01\rangle = |01\rangle, \\ \hat{U}|10\rangle = |11\rangle, \\ \hat{U}|11\rangle = |10\rangle. \end{cases} \quad (27.22)$$

Writing the first equation in the matrix form, we get

$$\begin{pmatrix} u_{11} & u_{12} & u_{13} & u_{14} \\ u_{21} & u_{22} & u_{23} & u_{24} \\ u_{31} & u_{32} & u_{33} & u_{34} \\ u_{41} & u_{42} & u_{43} & u_{44} \end{pmatrix} \begin{pmatrix} 1 \\ 0 \\ 0 \\ 0 \end{pmatrix} = \begin{pmatrix} 1 \\ 0 \\ 0 \\ 0 \end{pmatrix}, \quad (27.23)$$

from which $u_{11} = 1$, $u_{21} = u_{31} = u_{41} = 0$. The second equation gives us

$$\begin{pmatrix} 1 & u_{12} & u_{13} & u_{14} \\ 0 & u_{22} & u_{23} & u_{24} \\ 0 & u_{32} & u_{33} & u_{34} \\ 0 & u_{42} & u_{43} & u_{44} \end{pmatrix} \begin{pmatrix} 0 \\ 1 \\ 0 \\ 0 \end{pmatrix} = \begin{pmatrix} 0 \\ 1 \\ 0 \\ 0 \end{pmatrix}, \quad (27.24)$$

from which $u_{12} = 0$, $u_{22} = 1$, $u_{32} = u_{42} = 0$. The next equation,

$$\begin{pmatrix} 1 & 0 & u_{13} & u_{14} \\ 0 & 1 & u_{23} & u_{24} \\ 0 & 0 & u_{33} & u_{34} \\ 0 & 0 & u_{43} & u_{44} \end{pmatrix} \begin{pmatrix} 0 \\ 0 \\ 1 \\ 0 \end{pmatrix} = \begin{pmatrix} 0 \\ 0 \\ 0 \\ 1 \end{pmatrix}, \quad (27.25)$$

---

1) If the gate has 1 bit output, as is the case for AND, OR, NAND, NOR, XOR, and XNOR, we can still write such a matrix with the understanding that information contained in the "extra" bit will be discarded.

determines the third column of the matrix: $u_{13} = u_{23} = u_{33} = 0$, $u_{43} = 1$. Finally,

$$\begin{pmatrix} 1 & 0 & 0 & u_{14} \\ 0 & 1 & 0 & u_{24} \\ 0 & 0 & 0 & u_{34} \\ 0 & 0 & 1 & u_{44} \end{pmatrix} \begin{pmatrix} 0 \\ 0 \\ 0 \\ 1 \end{pmatrix} = \begin{pmatrix} 0 \\ 0 \\ 1 \\ 0 \end{pmatrix} \tag{27.26}$$

gives us the last column: $u_{14} = u_{24} = 0$, $u_{34} = 1$, $u_{44} = 0$. Then,

$$\text{CNOT} = \begin{pmatrix} 1 & 0 & 0 & 0 \\ 0 & 1 & 0 & 0 \\ 0 & 0 & 0 & 1 \\ 0 & 0 & 1 & 0 \end{pmatrix}. \tag{27.27}$$

If the joint system of two qubits was prepared in an arbitrary superposition of its four eigenstates, then CNOT will transform it by swapping amplitudes c_3 and c_4:

$$\begin{pmatrix} 1 & 0 & 0 & 0 \\ 0 & 1 & 0 & 0 \\ 0 & 0 & 0 & 1 \\ 0 & 0 & 1 & 0 \end{pmatrix} \begin{pmatrix} c_1 \\ c_2 \\ c_3 \\ c_4 \end{pmatrix} = \begin{pmatrix} c_1 \\ c_2 \\ c_4 \\ c_3 \end{pmatrix}. \tag{27.28}$$

The matrix for CCN gate can be obtained similarly (Problem 27.6).

27.3
Shor's Algorithm

The ability of quantum gates to work with superpositions of bit values makes it possible to solve certain types of problems that would be intractable for any classical computer. Herein lies the main advantage of quantum computation. Note that it is not the *speed* of number processing that we are after. It is hard to beat a classical computer in the speed race anyway, and the costs of doing so would be prohibitive. Where the quantum computer really shines through is in doing parallel processing of all possible outcomes and eliminating the undesired outcome by destructive interference. Here is one example of a problem that is suitable for a quantum computer.

One of the most popular encryption techniques, the so-called RSA algorithm, is based on the idea that one cannot realistically expect to factorize a large number. To be sure, some large numbers can be factorized easily. For example, if the number is even, the first step is obvious – divide it by 2. In some cases (think of large powers of 2!), a number of consecutive divisions will result in an immediate answer. However, if we think of a large number N that is the product of two primes,

$$N = PQ, \tag{27.29}$$

then all of a sudden factorization of N becomes a formidable problem. In an effort to find P and Q, one must generally try all possible numbers from 1 to $\sqrt{N}$. Sometimes one will find a suitable factor right after the start, and sometimes only at the very end; on the average, it will take about $\sqrt{N}/2$ trials. This problem is intractable, that is, it cannot be done in a realistic time. Case in point: Suppose it takes $L = 200$ bits to write the number N. Unless one is extremely lucky, direct trial-and-error will take about 2^{99} steps and even if a computer spends only 10^{-12} of a second for each step, the average required time of 2^{87} s (longer than the age of the Universe!) should discourage any potential adversary from trying to break the code. The actual time will be somewhat less since one only needs to consider *prime* numbers as potential candidates, but the general conclusion still remains the same: There is no reasonable chance to find P and Q by brute force.

The RSA algorithm makes the number N publicly available, but leaves P and Q hidden. The encryption key e (also publicly available) defines an encryption function $E(M)$ that turns the plain text message M into a ciphertext (cryptogram) C:

$$C = E(M) = M^e \bmod N. \tag{27.30}$$

The decryption key d will define a decryption function $D(C)$ that restores the message from the ciphertext:

$$M = D(C) = C^d \bmod N. \tag{27.31}$$

This decryption key is held privately by the receiving party.

Suppose a sender Alice and a receiver Bob want to exchange a confidential plain text message M. Bob starts the procedure by choosing two large prime numbers P and Q and calculating N. Next, he chooses the decryption key d that he intends to use for decoding Alice's message. This key has to be a large integer $d > \max(P, Q)$ that is also relatively prime[2] to $\phi(N) = (P-1)(Q-1)$.[3] The last condition is needed to ensure the existence of the proper encryption key e for the chosen d and N. Namely, e is the *multiplicative inverse* of d, modulo $\phi(N)$, meaning that multiplication results in 1:

$$e \cdot d = 1 \bmod \phi(N). \tag{27.32}$$

Bob can find this multiplicative inverse by using the well-known classical Euclidean algorithm which we will not discuss here (see, for example, [3]). Thus, he obtains $e = e(d, N)$ and publicly broadcasts e and N to Alice. Upon receiving these two numbers (which can also be overheard by the adversary Eve), Alice represents her message M by an integer number that could range anywhere from 0 to $N-1$ and uses algorithm (27.30) to generate the ciphertext to be sent to Bob. Upon receiving

---

[2] Two numbers are called relatively prime (or coprime) if their greatest common divisor (GCD) is 1.

[3] $\phi(N)$ is called the *totient function* of N. This is the function that tells us how many numbers less than N are relatively prime to N. Since all numbers $1, 2, \ldots, P-1$ are relatively prime to P and all $1, 2, \ldots, Q-1$ to Q, one can see that $\phi(P) = P-1$ and $\phi(Q) = Q-1$. Furthermore, the totient function is *multiplicative*, that is, as long as a and b are relatively prime to each other, $\phi(ab) = \phi(a)\phi(b)$. Hence, $\phi(N) = \phi(PQ) = \phi(P)\phi(Q) = (P-1)(Q-1)$.

the ciphertext, Bob will use (27.31) to decrypt it. Eve too can learn C, but unlike Bob, she does not know d. Knowing e and N, she could find d easily from (27.32) if only she knew $\phi(N)$. But to obtain $\phi(N)$, she must first find P and Q, which we said is an intractable problem for classical computers.

In 1994, Shor [2] suggested a four-step factorization algorithm that runs on a quantum computer and can break the RSA scheme. In the first step, one is to choose a number $a < N$ that is relatively prime to N. This can be done by choosing a at random and checking to see if $GCD(a, N) = 1$. One can show this part to take L trials on average.

In the next step, one tries to find the period T of the function

$$f(x) = a^x \bmod N \tag{27.33}$$

by calculating its value for all x ranging from 0 to N^2. This part involves N^2 operations and is thus intractable for classical computers. Usually, the period will be even; if it is odd, go back to step 1 and find some other a.

The third step is to calculate $z = a^{T/2}$, which is a trivial task. In the final step, one has to find $GCD(Z - 1, N)$ and $GCD(Z + 1, N)$. Both these numbers will be factors of N. Since GCD is easily found even for large numbers, this step does not give trouble to a classical computer. So the task is now to find the period of function (27.33) using quantum parallelism.

To this end, divide the quantum memory into two registers r_1 and r_2, with L bits in each. The first register will store the number x and the second will store the function $f(x)$. Let the corresponding states be denoted by kets $|r_1\rangle$ and $|r_2\rangle$, so the joint state of the two registers will be denoted by $|r_1, r_2\rangle \equiv |r_1\rangle|r_2\rangle$. Initially, both registers will be filled by zeros. Suppose the function (27.33) is realized by an operator $\hat{f}$ that works as follows:

$$\hat{f}|x, 0\rangle = |x, f(x)\rangle. \tag{27.34}$$

Each time we put a certain number x into the first register (in other words, each bit of this register is either 1 or 0), the gate implementing (27.34) will fill the (initially empty) second register with bits of the number $|f(x)\rangle$. However, by making all bits in $|r_1\rangle$ pass independently through a rotation gate, we can prepare each bit as a uniform superposition of 1 and 0:

$$\hat{R}|0\rangle = \frac{1}{\sqrt{2}}(|0\rangle + |1\rangle),$$
$$\vdots \tag{27.35}$$
$$\hat{R}|0\rangle = \frac{1}{\sqrt{2}}(|0\rangle + |1\rangle).$$

Then the joint state of L bits will be a superposition of all 2^L possible values of x:

$$|r1\rangle = \frac{1}{\sqrt{2}}(|0\rangle + |1\rangle) \cdots \frac{1}{\sqrt{2}}(|0\rangle + |1\rangle) = \frac{|00\cdots 0\rangle + |00\cdots 1\rangle + \cdots + |11\cdots 1\rangle}{\sqrt{N}}. \tag{27.36}$$

When gate $\hat{f}$ works on this state, the resulting joint state will be a superposition of all possible pairs of values $|x, f(x)\rangle$:

$$|r_1, r_2\rangle = \hat{f}|r_1, 0\rangle = \hat{f}\sum_x c_x|x, 0\rangle = \sum_{x=0}^{N-1} c_x \hat{f}|x, 0\rangle = \sum_{x=0}^{N-1} c_x|x, f(x)\rangle = \frac{1}{\sqrt{N}}\sum_{x=0}^{N-1}|x\rangle|f(x)\rangle, \quad (27.37)$$

where in the last step we used the fact that all amplitudes are equal. This wave function describes an entangled state of qubits in the $|r_1\rangle$ and $|r_2\rangle$ registers.

Suppose we now decided to measure the state of the second register and found that it contains some specific value of $f(x)$, say, $f(x_i)$. We know that $f(x)$ is periodic with period T, so numbers $x_i, x_i + T, x_i + 2T, \ldots$ are all compatible with this output state. Therefore, such a measurement would instantly collapse the first register into a superposition of states $|x_i\rangle, |x_i + T\rangle, |x_i + 2T\rangle, \ldots, |x_i + m_{\max}T\rangle$, where the sequence contains $m_{\max} + 1$ terms and starts from some integer i. This thought experiment should be enough to convince you that some periodic pattern must be already contained in the entangled state (27.37). Our task is to find this pattern.

In mathematics, periodicities in the input data are often revealed by applying the discrete Fourier transform (DFT). This is the approach we are going to try now. The DFT operation on a multiple qubit state $|x\rangle$ is defined as follows [3]:

$$\text{DFT}|x\rangle = \frac{1}{\sqrt{N}}\sum_{k=0}^{N-1} e^{i(2\pi kx/N)}|k\rangle. \quad (27.38)$$

Here, the numbers x and k are both written in the decimal notation; when put inside a ket, they will denote the corresponding state of the first register. Simply stated, each possible eigenstate $|x\rangle$ will be transformed by the operator into a sum of such eigenstates, each eigenstate taken with its respective amplitude. But we are going to apply DFT (27.38) to the subsystem (first register) of the entangled system (27.37). As long as we don't actually *measure* the state of the subsystem, such an operation will preserve the entanglement. We can now write the result of this operation:

$$\text{DFT}|r_1, r_2\rangle = \text{DFT}\left(\frac{1}{\sqrt{N}}\sum_{x=0}^{N-1}|x\rangle|f(x)\rangle\right) = \frac{1}{N}\sum_{x=0}^{N-1}|f(x)\rangle\sum_{k=0}^{N-1} e^{i(2\pi kx/N)}|k\rangle. \quad (27.39)$$

The amplitudes $(1/N)e^{i(2\pi kx/N)}$ of states $|f(x)\rangle|k\rangle$ can be grouped into the following table (we dropped the $1/N$ factor to reduce clutter):

Amplitude	$|0\rangle$	$|1\rangle$	$\vdots$	$|N-1\rangle$
$|f(0)\rangle$	$e^{i(2\pi 00/N)}$	$e^{i(2\pi 10/N)}$	$\vdots$	$e^{i(2\pi(N-1)0)/N}$
$|f(1)\rangle$	$e^{i(2\pi 01/N)}$	$e^{i(2\pi 11/N)}$	$\vdots$	$e^{i(2\pi(N-1)1)/N}$
$\ldots$			$\vdots$	
$|f(N-1)\rangle$	$e^{i(2\pi 0(N-1))/N}$	$e^{i(2\pi 1(N-1)/N)}$	$\ldots$	$e^{i(2\pi(N-1)(N-0))/N}$

If we now measure the states of both registers, we are going to obtain some result $|x_j\rangle$ in the first register and the second register will be found in some generally, different) state $|f(x_i)\rangle$. The total probability to obtain $|x_j\rangle$ can be found by adding the squares of amplitudes of distinguishable outcomes. But here is where the periodicity of the function comes to the forefront. Since

$$|f(x_i)\rangle = |f(x_i + T)\rangle = \cdots = |f(x_i + m_{max}T)\rangle \tag{27.40}$$

holds for every x_i, the outcomes $|x_j\rangle|f(x_i)\rangle, |x_j\rangle|f(x_i + T)\rangle, \ldots, |x_j\rangle|f(x_i + m_{max}T)\rangle$ will be indistinguishable since they leave combined system in exactly the same final state. Therefore, one must *add* their amplitudes *before* squaring. The resulting probability is

$$\Pr\left(|x_j\rangle|f(x_i)\rangle\right) = \left(\frac{1}{N}\sum_{m=0}^{m_{max}} e^{i[2\pi j(x_i+mT)]/N}\right)^2 \tag{27.41}$$

and the total probability to obtain $|x_j\rangle$ will be the sum of probabilities (27.41) for all $x_i \in [0, T-1]$:

$$\Pr\left(|x_j\rangle\right) = \sum_{x_i=0}^{T-1}\left(\frac{1}{N}\sum_{m=0}^{m_{max}} e^{i[2\pi j(x_i+mT)]/N}\right)^2. \tag{27.42}$$

The phasor approach to determining constructive interference conditions is widely used in undergraduate optics courses. The same method, when applied to (27.42), will tell us which outcomes should have a substantially nonzero probability. When we add the amplitudes within parenthesis, the phase difference between two nearest amplitudes is

$$\Delta\phi = i\frac{2\pi j T}{N}. \tag{27.43}$$

For the amplitudes to interfere constructively, one needs to have

$$\frac{jT}{N} = n, \tag{27.44}$$

where $n = 0, 1, 2, \ldots, T-1$. All other cases can be ignored because their contribution to the final outcome will be negligible. Thus, only numbers j satisfying (27.44) will be selected by this experiment. For example, if $N = 8, T = 2$, only two outcomes will be possible: $j = 0$ and $j = 4$, and the first register will end up with the same probability in state $|000\rangle$ or $|100\rangle$, storing $x = 0$ or $x = 4$, respectively. To conclude the argument, by reading x from the first register, the experimenter will know the value

$$x = \frac{nN}{T}, \quad n = 0, 1, 2, \ldots, T-1. \tag{27.45}$$

In most cases, this result will give the experimenter a sufficiently short list of candidates for T:

$$T = n\frac{N}{x}, \quad n = 0, 1, 2, \ldots, T-1. \tag{27.46}$$

This is especially true if x is small enough so that one gets a limited choice of possibilities that can then easily be checked by a classical computer. If, on the other hand, the outcome is too inconclusive, one only needs to repeat the above procedure until after a (typically very small) number of trials the list of possible values can be narrowed down sufficiently.

Box 27.1 How to realize a DFT?

DFT can be realized by applying several one-qubit and two-qubit operators to the bits in the first register in the following manner. Suppose the register contains number x. The state of the register is

$$|x\rangle = |x_{L-1}\rangle |x_{L-2}\rangle \cdots |x_2\rangle |x_1\rangle |x_0\rangle. \tag{27.47}$$

Act on this state with the following sequence of operators:

$$\underbrace{A_0 B_{0,1} B_{0,2} B_{0,3} \cdots B_{0,L-1}} \ \underbrace{A_1 B_{1,2} B_{1,3} \cdots B_{1,L-1}} \ \underbrace{A_2 B_{2,3} B_{2,4} \cdots B_{2,L-1}} \cdots \underbrace{A_{L-2} B_{L-2,L-1}} \ \underbrace{A_{L-1}}, \tag{27.48}$$

where A_j is an operator acting on qubit $|x_j\rangle$ and $B_{j,k}$ is an operator acting on a pair of qubits $|x_j\rangle$, $|x_k\rangle$. The action of these operators is defined by Equations 27.49 and 27.50:

$$A_j|0_j\rangle = \frac{1}{\sqrt{2}}(|0_j\rangle + |1_j\rangle), \quad A_j|1\rangle = \frac{1}{\sqrt{2}}(|0_j\rangle - |1_j\rangle), \tag{27.49}$$

and

$$B_{j,k}|0_j\rangle|0_k\rangle = |0_j\rangle|0_k\rangle, \quad B_{j,k}|0_j\rangle|1_k\rangle = |0_j\rangle|1_k\rangle, \quad B_{j,k}|1_j\rangle|0_k\rangle$$
$$= |1_j\rangle|0_k\rangle, \quad B_{j,k}|1_j\rangle|1_k\rangle = e^{i\pi/2^{k-j}}|1_j\rangle|1_k\rangle. \tag{27.50}$$

Finally, apply to the resulting state a qubit-reversal operator defined as

$$C_{L-1,L-2,\ldots,0}|x_{L-1}\rangle|x_{L-2}\rangle \cdots |x_0\rangle = |x_0\rangle|x_1\rangle \cdots |x_{L-1}\rangle \tag{27.51}$$

or simply read the qubits forming register state (27.50) in reverse order. One can show [3] that this procedure realizes discrete Fourier transform (27.38).

Problems

27.1 Construct NOR, XOR, XNOR, and buffer gates from NAND only.
27.2 Prove that any binary expression written in terms of AND and OR operations remains unchanged if you invert all bits, replace all ANDs with ORs and all ORs with ANDs, and then invert the entire expression.
27.3 Prove that NOR is a universal gate.

27.4 Show that XOR is not universal.

27.5 Show that no combination of XOR and NOT gates will form a universal gate.

27.6 Derive a 8×8 matrix for the CCN gate defined by (27.9).

27.7 Show for $L = 3$ that the procedure described by Equations 27.48–27.51 amounts to a DFT.

27.8 Show that DFT is an *efficient* algorithm, that is, it can be carried out in a reasonable amount of time (as opposed to computationally infeasible algorithms such as classical factorization of integers).

27.9 Consider a qubit A in arbitrary spin state, and an EPR pair B and C in the first Bell state (24.55). Alice has qubits A and B while Bob has qubit C. Design a scheme to teleport the state of A from Alice to Bob (see Section 24.3). Which gate would you use to entangle A with BC and which gate–for the transformation on A prior to the measurement? Can you modify this scheme to work with the other three Bell states?

References

1. Nielsen, M.A. and Chuang, I.L. (2000) *Quantum Computation and Quantum Information*, Cambridge University Press.
2. Shor, P.W. (1997) Polynomial-time algorithms for prime factorization and discrete logarithms on a quantum computer. *SIAM J. Comput.*, **26**, 5.
3. Berman, G., Doolen, G., Mainieri, R., and Tsifrinovich, V. (1998) *Introduction to Quantum Computers*, World Scientific.

28
Quantum Key Distribution

28.1
Quantum Key Distribution (QKD) with EPR

Consider the following problem. Alice and Bob want to exchange a secret message with a high degree of security. The message is a string of N bits, where N is some relatively small number ($N \approx 1000$). The typical application is to use this secret message as a *key* that will later on allow Alice and Bob to encrypt and decrypt their conversations over an open channel such as a radio broadcast or a (potentially bugged) phone line. As long as the encryption key is known to Alice and Bob only, they can talk to each other safely. The problem is that they did not have a chance to agree on the key in advance. So one of them (say, Alice) now has to generate a key and try to transmit it to Bob. Let's say an eavesdropping party (Eve) wants to learn the bits of the key, and she may be in a position to intercept any information sent between Alice and Bob. The problem is: Can Alice and Bob exchange the key ensuring that Eve does not overhear their conversation?

You may think at first that what Alice and Bob need is *perfect* secrecy of communication, with absolute protection of each and every bit. Fortunately, this requirement can be relaxed somewhat. It is OK if Eve can learn some of the bits in the exchanged key. It is even OK if Eve tampers with the message by changing some of the bit values. The reason we don't need complete perfection is because it is possible for Alice and Bob to perform *key reconciliation*, that is, they can detect the bits that were changed by Eve and drop them from the string. The easiest way to do that is by running a series of *parity checks* on the key. In computer-speak, "parity" simply tells us if the sum of all ones and zeros in a given string is an even or an odd number. Let's say Alice and Bob suspect that some of the bits in the bit string recorded by Bob are the wrong bits. They can then subdivide their respective strings into blocks of bits. The size of the blocks must be small enough that each block is statistically unlikely to contain more than one error. For every such block, Alice and Bob will exchange public messages about the block's parity and retain the block only if the parities match. Of course, Eve could be listening to this public conversation and thus also learn the parity of each block. It is OK because after the conversation, Alice and Bob will drop the last bit of each block whose parity they have announced in public so that parity of the remaining block will become random again. Therefore,

Eve cannot use the knowledge of parity to gain any extra information about the string x. Next, whenever the parities differ, the offending block is subdivided into smaller blocks and the parities are compared again, until the error is localized and eliminated. At some point, when a significant number of consecutive checks performed on ever-larger blocks all result in a parity match, Alice and Eve conclude with a high degree of confidence that all errors have been removed and that they are now looking at the same key.

What about the bits whose values are correct but may be known to Eve? Here, again parity checks come to the rescue. Suppose Eve knows the values of K bits from the reconciled N-bit sequence. Then, Alice and Bob extract $N - K - S$ subblocks from the string (S is an arbitrary number known as safety parameter) without revealing their content, and the union of these subblocks' parity values (1 for odd, 0 for even) will now form a *new* string. Since Eve does not possess complete information about the original string, she cannot guess the output of such a procedure with an acceptable accuracy. This procedure is called *privacy amplification*.

It can be shown formally that key reconciliation followed by privacy amplification can make the reconciled string error-free with an overwhelming probability, while reducing Eve's information essentially to zero provided that Eve has not learned or modified too large a fraction of bits in the original key. The problem is then to determine how much Eve was actually able to learn. Unfortunately, no satisfactory solution exists if you use any classical mode of communication. But with Heisenberg's uncertainty principle, it turns out that Eve cannot learn the key without Alice and Bob noticing it.

The idea is to encode each bit in a quantum state, using the fact that no one can measure the state of a particle without disturbing it. We will be using the same EPR setup that already helped us discuss Bell state measurements in Chapter 24. Let's have a source emit pairs of spin-1/2 particles (e.g., electrons) in a singlet state, and let these electrons fly apart toward Alice and Bob, who then measure the direction of spin in the basis they choose. Each bit sent by Alice will be encoded in her choice of basis (z-axis or x-axis). Bob will also make a random choice of basis, and obtain his result (spin-up or spin-down along the basis chosen). Then, Alice and Bob announce publicly their choice of basis for each measurement, leave only the measurements in which they used the same basis, and discard all the rest. (The bit string that is left after discarding these bits is called the *sifted* key, which is about one half the length of the original key). Then, since the pair was prepared in a singlet state (total spin is zero), whenever Alice finds her electron in a spin-up state, Bob must always find his electron in a spin-down state; in other words, their results must be perfectly *anticorrelated*.

It is easy to see that Eve has no chance to learn the bits while remaining unnoticed. Indeed, if Eve tries to measure the state of an electron (e.g., by performing a Stern–Gerlach measurement), she must disturb the state at least half of the time since she doesn't know in advance which basis – z or x– she should use. The right basis will be announced eventually, but only *after* both electrons have reached their intended recipients. But at the time she makes her measurement, the only way she can guess the basis is by pure luck. Suppose that for a given particle pair, Alice and Bob have both independently chose the z-axis, and Alice's electron was found with $S_z = 1/2$ (in units of $\hbar$). If Eve tries to measure the state of the electron, she must disturb the state at least half of the time since she doesn't know in advance which basis – z or x – she should use. If Eve is lucky enough to choose the z-axis, her measurement leaves the electron anticorrelated with Alice's electron, so Eve's result is $S_z = -(1/2)$ and the electron leaves Eve and reaches Bob in the spin-down state. This is just what Bob would measure if Eve were not present. Suppose however that Eve guessed wrong and chose the x-axis. Now the spin is pointing along x, and when Bob measures its z-projection, he may get either $S_z = 1/2$ or $S_z = -(1/2)$. And since Eve's presence will inevitably introduce errors into the exchanged bit string 25% of the time, it will be easy for Alice and Bob to detect her presence by publicly comparing a part of the bit string and realizing that it contains more errors than one would naturally expect from random noise in the quantum channel. Then, they abort the key and make another attempt.

Consider what is happening here from the formal standpoint. Alice and Bob are performing an EPR experiment with the choice of test angles $a = 0, b = 90, c = 0, d = 90$ with respect to the reference z-axis (in other words, Alice can choose z-axis or x-axis, and so can Bob). Correlation coefficients $E(a,c) = -\vec{a} \cdot \vec{c}$ and $E(a,d) = -\vec{a} \cdot \vec{d}$ are equal to -1 and 0, respectively, meaning perfect anticorrelation of results for the same choice of basis and no correlation when different bases are chosen.

Let us find the quantity Q that we introduced in connection with CHSH inequality:

$$|Q| = |E(a,c) - E(a,d) + E(b,c) + E(b,d)| = |-a \cdot c + a \cdot d - b \cdot c - b \cdot d|$$
$$= |-1 + 0 - 0 - 1| = 2. \tag{28.1}$$

This is obviously not the set of test angles we would choose to test CHSH against predictions of QM! Indeed, if that were our purpose, then we would have made the poorest possible choice of the test angles. However, this choice was attractive because it made the underlying physics look quite simple. But we could also have used a different scheme that would make a better use of CHSH. Let Alice's test angles remain $a = 0$ and $b = 90$ as before, and let Bob orient his polarizer at either angle $c = 45$ or $d = 135$. In other words, we are going to repeat the same experiment, but this time using Bell's test angles. In this case, we will of course get $|Q| = 2\sqrt{2}$ as we should expect. Let us now see what happens when Eve attempts to measure the spin states along directions n_a and n_b for Alice's and Bob's electrons, respectively. If she measures only the spin of Bob's electron, it is the same as choosing $n_a = -n_b$ because once she knows the spin of Bob's electron along n_b, she also knows that Alice's electron has the opposite spin projection on that direction. Then each of the correlation coefficients in the last equation must be modified to account for Eve's intervention:

$$|Q| = \left| -\int [(a \cdot n_a)(c \cdot n_b) - (a \cdot n_a)(d \cdot n_b) + (b \cdot n_a)(c \cdot n_b) + (b \cdot n_a)(d \cdot n_b)]\rho(n_a, n_b) dn_a dn_b \right|, \tag{28.2}$$

where normalized probability measure $\rho(n_a, n_b)$ describes Eve's strategy by giving the probability for Eve to find the spin component along a given direction for a particular measurement. Let α and β be the angles formed by directions n_a and n_b with 0 and 45 directions, respectively (Figure 28.1).

Then, we can rewrite the expression in the square brackets as

$$\cos\alpha\cos\beta - \cos\alpha\sin\beta + \sin\alpha\cos\beta + \sin\alpha\sin\beta$$
$$= \cos(\alpha - \beta) + \sin(\alpha - \beta) = \sqrt{2}\cos(45 - (\alpha - \beta)). \quad (28.3)$$

The difference $45 - \alpha + \beta$ is just the angle $\gamma = n_a \cdot n_b$ between the two directions n_a and n_b. So we can write,

$$|Q| = \sqrt{2}\int (n_a \cdot n_b)\rho(n_a, n_b)dn_a dn_b. \quad (28.4)$$

And since the scalar product is no greater than 1 and $\rho(n_a, n_b)$ was normalized to unity, this implies

$$|Q| \leq \sqrt{2}. \quad (28.5)$$

But this contradicts the QM requirement that $|Q| = 2\sqrt{2}$ for the Bell test angles. Moreover, the obtained result for Q, namely, $|Q| \leq \sqrt{2} < 2$, means the CHSH inequality holds for this case! As we remember, quantum mechanics says that CHSH inequality should not hold for entangled particles. If it does hold, it means the particles were not truly entangled. And thus it is always possible to detect the eavesdropper by measuring Q from a distribution of experimental outcomes.

Physically, by trying to learn the spin states of the electrons, Eve unavoidably destroys the original entangled state of the pair; this is expressed mathematically in

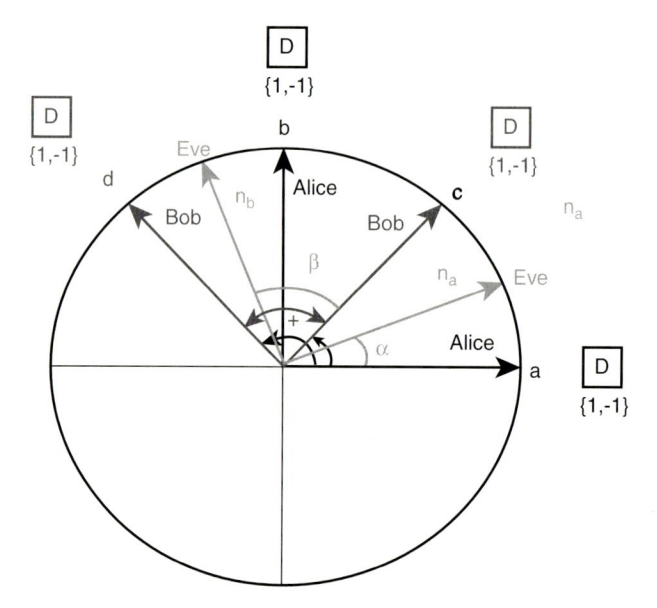

Figure 28.1 Security proof with CHSH. Here, $\gamma = \angle(n_a, n_b) = 45 - \alpha + \beta$. Eve's interference destroys the entanglement as evidenced by $|Q| \leq \sqrt{2} < 2$, and Alice and Bob discover that CHSH holds.

the result $|Q| < 2$ that agrees with the local reality hypothesis. As the author of this scheme, Ekert [1] has put it, "In this case [eavesdropper's] intervention will be equivalent to introducing elements of physical reality to the measurements of the spin components."

Ekert published this result in 1991 with the goal of proving that security of key distribution is ensured by the completeness of quantum mechanics (i.e., absence of hidden variables) that is manifested in the CHSH inequality. But only a few months later, Bennett et al. [2] pointed out that the security proof does not need to invoke CHSH. Indeed, let us now go back to the example we started with, namely, $a = 0, b = 90, c = 0, d = 90$, where both Alice and Bob choose randomly between $0°$ and $90°$, which we take to define the z- and x-axes, respectively. Then Eve's eavesdropping attempt results in Equation 28.6:

$$|Q| = \left| \int [(z \cdot n_a)(z \cdot n_b) - (z \cdot n_a)(x \cdot n_b) + (x \cdot n_a)(z \cdot n_b) + (x \cdot n_a)(x \cdot n_b) \rho(n_a, n_b)] dn_a dn_b \right|.$$

(28.6)

This time α and β will be the angles formed by directions n_a and n_b with the z-axis (Figure 28.2). Then,

$$|Q| = \left| \int [\cos \alpha \cos \beta - \cos \alpha \sin \beta + \sin \alpha \cos \beta + \sin \alpha \sin \beta] \rho(n_a, n_b) dn_a dn_b \right|.$$

(28.7)

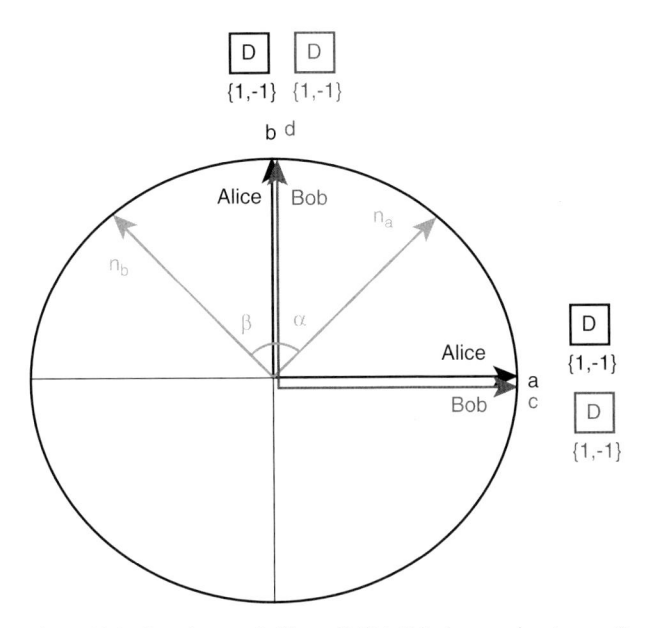

Figure 28.2 Security proof without CHSH. This time we denote $\gamma = \beta - \alpha$. As a result of Eve's meddling, the expected $|Q| = 2$ goes over to $|Q| \leq 1$, which Alice and Bob will easily notice.

You can see that the bracket again reduces to $\sqrt{2}\cos(45-(\alpha-\beta))$. But this time, $\gamma = \beta - \alpha$ (we want to keep the angle positive) so that

$$|Q| = \sqrt{2}\left|\int \cos(45+\gamma)\rho(n_a,n_b)dn_a dn_b\right| = \left|\int(\cos\gamma - \sin\gamma)\rho(n_a,n_b)dn_a dn_b\right| \leq 1. \quad (28.8)$$

(Please note that to keep things simple, we showed n_a and n_b in the first quadrant only, but the proof will work for the general case as well). Unlike in the previous case, this time we have a *difference* of two positive numbers in the parenthesis, that's why we got $\sqrt{2}$ in the previous case, but not now. (The astute reader will point out that we could predict this in advance by dropping the two middle terms from (28.6) because of the property of the scalar product that $(z \cdot n_a)(x \cdot n_b) = (x \cdot n_a)(z \cdot n_b)$. But it always helps to obtain the same answer in two different ways!) As you remember, we should expect $|Q| = 2$ for the *undisturbed* singlet state. So even though we have chosen precisely those test angles for which CHSH is guaranteed to hold in any event, the laws of physics still ensure that Eve cannot remain unnoticed. One only needs to compare the expected versus the actual Q to realize that Eve has disturbed the original singlet state. Ekert's original idea of using Bell's test angles would merely make the effect of her eavesdropping more dramatic.

28.2
BB84 Protocol

In the last section, we discussed the modified Ekert scheme where the two parties choose randomly between z- and x-axes as the basis of measurement. But it could hardly be a practical QKD scheme (just think of the equipment one would need to produce EPR electron pairs and transmit them over large distances, while also preventing decoherence). It is easy to see however that as far as key distribution is concerned, entanglement of electrons is an unnecessary element in the above-described protocol. Imagine that we first place the source of EPR pairs in Alice's laboratory rather than in the middle of the communication line. Obviously, this modification will not change anything whatsoever. Next, imagine that instead of spin measurement on an entangled electron, which is going to produce perfectly random outcomes, Alice just prepares a bunch of unentangled electrons in a random sequence of states ↑, ↓, →,←. Again, as far as Bob is concerned, this will not make any difference in the statistics of measurement outcomes. The result is a totally equivalent scheme, equally secure but without CHSH inequality and without EPR. We have arrived at what is known as the BB84 protocol, after the paper published in 1984 by Bennett et al. [3]. It only remains to overcome the last difficulty, since the original protocol was formulated for photons. As we remember, the photon equivalent of "spin-up" and "spin-down" states are states with horizontal and vertical polarizations, and similarly "spin-left" and 'spin-right" translate into 45° and 135° diagonal polarizations. Therefore, to make the transition to the photon case, we merely have to halve the 0, 180, 90, 270 physical angles (but not the angles on the Poincare sphere!) for the states ↑, ↓, →, ←. (Keep in mind that these four angles are the angles for the particle states, and *not* the two test angles for polarizers that we discussed in connection with Bell-CHSH experiments!)

The BB84 scheme is thus based on four *pairwise nonorthogonal* states of polarization of a single photon (i.e., each rectilinear state is nonorthogonal to both diagonal states) [4]. Alice encodes the 0 bit in the mixture of equally likely nonorthogonal quantum states (0° and 45°) and, similarly, bit 1 is encoded in the mixture of equally likely nonorthogonal states (90° and 135°). The choice of rectilinear versus diagonal basis and the choice of the photon state within the basis are random. Bob also makes a random choice between rectilinear and diagonal basis and measures the photon's polarization in the basis chosen.

Knowing the basis in which the photon was sent and also the result of his measurement, Bob therefore knows the bit originally encoded by Alice in that photon. Just as in Ekert's scheme, Eve would have to disturb the polarization state at least half of the time since she doesn't know in advance which basis – rectilinear of diagonal – she should use, and again, Bob should statistically expect that 1/4 of his bits will be the wrong ones. Communicating with photons is technically more feasible than producing EPR electron pairs, as photons can just be sent by Alice to Bob in the form of coherent laser pulses traveling over a fiber-optic cable.

Formally, BB84 protocol uses photons that exist in a two-dimensional Hilbert space so that the photon state can be described in terms of two unit vectors, $r_1 = (1, 0)$ and $r_2 = (0, 1)$, representing the horizontal and vertical polarizations in the rectilinear basis. The "alternative" diagonal basis is formed by the states $d_1 = (1/\sqrt{2}, 1/\sqrt{2})$, $d_2 = (-1/\sqrt{2}, 1/\sqrt{2})$, corresponding to 45° and 135° photons, respectively. A photon polarized at an arbitrary angle α to the horizontal would be described as $r = (\cos \alpha, \sin \alpha)$. The rectilinear and diagonal bases are said to be "conjugate,' meaning that a photon prepared in one of the rectilinear basis states will behave completely randomly when measured in the diagonal basis and vice versa. Mathematically, we can say that each vector of one basis has equal-length projections onto all vectors of the other basis.

Some versions of the BB84 protocol may use a circular basis instead of the diagonal one, with the basis vectors $c_1 = (1/\sqrt{2}, (1/\sqrt{2})i)$, $c_2 = (1/\sqrt{2}, (-1/\sqrt{2})i)$. This circular basis is conjugate to both the rectilinear and the diagonal bases. This variant of the protocol is for all practical purposes equivalent to the protocol described above.

Since BB84 still remains one of the most popular protocols currently used for QKD, it will be instructive to study it in more detail. For example, what is the best strategy for Eve, and how much information can she learn in the best case? To answer that question, we must first define what exactly constitutes success and failure for Alice, Bob, and Eve.

Start with Eve first. Eve succeeds if she measures (or learns in some other way or just guesses) Alice's photon's state correctly. Otherwise, her eavesdropping attempt is considered a failure. Eve is always successful when she uses the correct basis. She is also successful when she uses the wrong basis but, luckily for her, she still makes the correct guess. Note however that Eve's "success" does not by itself guarantee that the bit she learned will not ultimately be removed from the string by Alice and Bob in the privacy amplification stage. In other words, a "successful" Eve could still inadvertently alert Alice and Bob by corrupting the state of the photon. Of course, Eve has every interest to ensure that Bob receives the photon in the correct state so that he does not grow suspicious, so she will always resend the measured photon to Bob in what she believes to be the correct state originally encoded by Alice.

What about Bob? Bob succeeds if he gets the same state that was sent by Alice. This happens when Eve succeeds in learning the correct state of the photon and resends it in that correct state. It also happens if Eve fails and consequently resends the photon in the wrong state, but fortunately for her, Bob still reads the correct value (of course, in order for that to happen, Eve must resend the photon in a state that is nonorthogonal to the correct state!). Note again that a "successful" Bob is not necessarily aware about eavesdropping that might have taken place.

With these definitions of success and failure, we are now in a position to determine the optimal strategies for Eve and Bob. Before we begin, let us mention that the kind of information Eve receives about Alice's photon depends on the basis she chooses. If Eve uses rectilinear or diagonal (change that to circular if protocol implementation uses circular polarization) basis, then either she chose the basis correctly and will know the exact state of the photon or she used the wrong basis and spoiled the bit so that she knows nothing at all about its original value and can make a random guess with 50% chance of being correct. Since each correct guess counts as success, Eve will have a 75% success rate ($1/2 \times 1 + 1/2 \times 1/2 = 3/4$). But one can show that 75% is not the upper limit because she could also use some *intermediate* basis between the two, gaining some useful information in either case, except of course that this information will be probabilistic.

One can find the maximum possible success rate for Eve in the following way. Consider the Poincaré sphere $X^2 + Y^2 + Z^2 = 1$, where X, Y, and Z are the rectilinear, diagonal, and circular components of an arbitrary polarization state L. As mentioned in the first section, states (1,0,0) and (−1,0,0) corresponding, respectively, to horizontal and vertical polarizations, comprise the rectilinear basis, and similarly, states (0,1,0) and (0, −1,0) for 45° and 135° polarizations, respectively, comprise the diagonal basis, while states (0,0,1) and (0,0, −1) (left- and right-circular polarized light, respectively) comprise the circular basis. These three bases are *mutually conjugate* because the corresponding pairs of points are 90° apart on the Poincaré sphere. Denote one of these bases (say, rectilinear) as $\{-K, K\}$, and the angle between states L and K as viewed from the center of the sphere as α. Just as electron spin states will project onto a given direction with probability amplitude equal to the cosine of the half-angle, the photon polarization states will show a similar behavior. Therefore, a photon collapses into state K with probability $\cos^2(\alpha/2)$ and into $-K$ with probability $\sin^2(\alpha/2)$. As a result, any basis state will behave equiprobably when measured in a conjugate basis because $\cos^2(90/2) = 1/2$.

By definition, in the BB84 protocol, Alice sends a photon from one of the pairs of conjugate bases. Without loss of generality, assume this basis to be rectilinear. Suppose that Eve measures this photon in some arbitrary basis $\{K, -K\}$ where $K = \{X, Y, Z\}$, and this basis forms angle α with Alice's basis (Figure 28.3). The

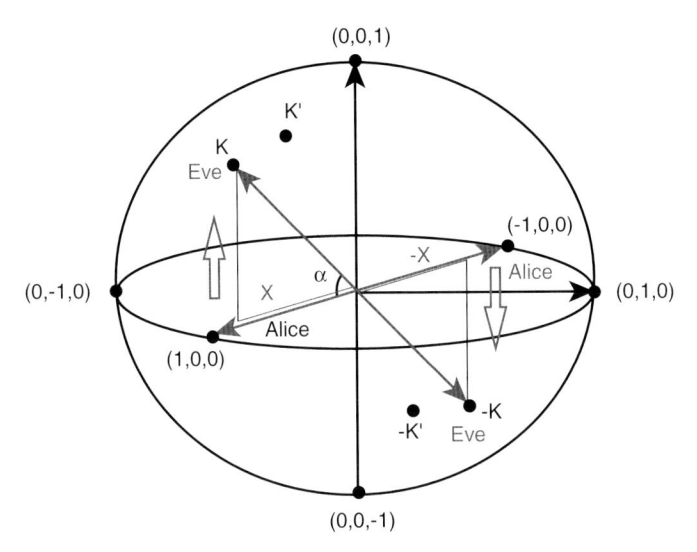

Figure 28.3 Intercept-resend attack by Eve. Poincaré sphere representation of the intercept basis $\{K, -K\}$. Outcome K is interpreted as $(1, 0, 0)$ and $-K$ as $(-1, 0, 0)$. Rectilinear (X) component of polarization is equal to $\cos \alpha$. $\{K', -K'\}$ denotes the basis from which Eve will resend the photons.

rectilinear projection of Eve's basis state K is then given by $\cos \alpha$. We can assume without loss of generality that Eve has chosen K in such a way that α is an *acute* angle. Then, if her reading collapses the photon into state K, she can guess with a better than 50% probability of success that Alice sent the photon in state $(1, 0, 0)$. Likewise, if her result is $-K$, then she will guess that Alice must probably have prepared her photon in state $(-1, 0, 0)$.

The respective probabilities of success and failure for Eve are given by

$$P_E = \cos^2 \frac{\alpha}{2} = \frac{1 + \cos \alpha}{2} = \frac{1 + X}{2}, \quad Q_E = \sin^2 \frac{\alpha}{2} = \frac{1 - \cos \alpha}{2} = \frac{1 - X}{2}. \tag{28.9}$$

Note that if α were an *obtuse* angle, Eve's success rate would be less than 50%. However, since the clever Eve will always use an acute angle to maximize her success rate, we can take $X \geq 0$ without loss of generality. For the photons sent by Alice in a diagonal basis, we can similarly write success and failure probabilities as $(1 + Y)/2$ and $(1 - Y)/2$. Averaging over both bases (half of the photons were sent in the rectilinear basis and half in the diagonal), we then write

$$P_E = \frac{(1 + X) + (1 + Y)}{4}, \quad Q_E = \frac{(1 - X) + (1 - Y)}{4}. \tag{28.10}$$

We see that Eve can minimize the chance of failure by maximizing $X + Y$, given that $X^2 + Y^2 + Z^2 = 1$. This clearly happens when $Z = 0$ and then $X = Y = \sqrt{2}/2$. Then, the minimum error rate for Eve is equal to

$$(Q_E)_{\min} = \frac{2 - \sqrt{2}}{4} \approx 15\%, \tag{28.11}$$

which is much better than the 25% error rate associated with the choice of a conjugate basis. Note however that this error rate describes Eve's best attempt to maximize the knowledge that she expects to gain *after* Alice and Bob have revealed their bases and therefore when resending the photon to Bob, Eve has to operate with lesser knowledge.[1]

Eve resends the photon to Bob using some basis $\{K', -K'\}$, which, generally speaking, can be different from $\{K, -K\}$. The acute angle logic dictates that Eve, who wants to maximize Bob's success rate, will again in each case make the state of the resent photon form an acute angle with what she believes to be the correct polarization. This consideration will determine her choice between states K' and $-K'$ for the resent photon. We further consider only the bits of the sifted key where Alice and Bob use the same basis. With these assumptions, we readily find that Bob succeeds in registering the correct state in two cases: Either Eve makes the correct guess [we give it the chance $(1+X)/2$] and resends the photon at an acute angle to its original polarization, which then lets Bob detect the correct state with a $(1+X')/2$ probability, or Eve is mistaken (that happens in $(1-X)/2$ of trials) and her resent photon is "tailored" for the wrong polarization state, making Bob succeed in $(1-X')/2$ of trials. So Bob's chance of success and his chance of failure respectively are

$$\begin{cases} P_B = \dfrac{1+X}{2}\dfrac{1+X'}{2} + \dfrac{1-X}{2}\dfrac{1-X'}{2} = \dfrac{(1+X)(1+X') + (1-X)(1-X')}{4}, \\ Q_B = \dfrac{1+X}{2}\dfrac{1-X'}{2} + \dfrac{1-X}{2}\dfrac{1+X'}{2} = \dfrac{(1+X)(1-X') + (1-X)(1+X')}{4}. \end{cases}$$

(28.12)

Averaging this result over the two bases, we finally get

$$\begin{cases} P_B = \dfrac{(1+X)(1+X') + (1-X)(1-X') + (1+Y)(1+Y') + (1-Y)(1-Y')}{8}, \\ Q_B = \dfrac{(1+X)(1-X') + (1-X)(1+X') + (1+Y)(1-Y') + (1-Y)(1+Y')}{8}. \end{cases}$$

(28.13)

Bob's error probability is minimized when Eve receives and resends photons from the same basis ($K' = K$). Then, the last equation becomes

$$Q_B = \frac{2\left[(1-X^2) + (1-Y^2)\right]}{8} = \frac{2 - (X^2 + Y^2)}{4} = \frac{1+Z^2}{4}. \tag{28.14}$$

---

1) Reference [4] states that by choosing $X = Y = \sqrt{2}/2$, $Y = 0$, Eve maximizes her *a priori* information, while her *a posteriori* information would be maximized by the choice of a rectilinear or diagonal basis. However, since at the time she receives the photon Eve does not know which basis Alice actually used, her *a priori* information really boils down to this: "If Alice chose the rectilinear basis, *then* I can tell the correct polarization with error probability $(1-X)/2$, and if she chose the diagonal basis, *then* I can tell the correct polarization with error probability $(1-Y)/2$, and only *after* the basis is announced, I will know which one of these is the case." Hence, the error rate expression $[(1-X) + (1-Y)]/4$ does not represent Eve's *a priori* knowledge, but rather represents her best attempt to maximize her *a posteriori* knowledge based on what she knows *a priori*.

To ensure Bob's error rate attains its minimum value of $1/4$, the clever Eve will make the circular component equal to zero; in other words, she can use *any* basis $K' = K$ in the XY-plane. If $K' \neq K$ or if $Z \neq 0$, Bob will get additional errors. Thus, to minimize both Bob's *and* her own error rate, Eve should choose $X = Y = \sqrt{2}/2$, $Z = 0$, which is known as the *Breidbart basis* (Figure 28.4). With this choice of basis, Eve can learn the value of Alice's bits with probability at most $(2 + \sqrt{2})/4 \approx 85\%$, while introducing at least a 25% error probability on Bob's side for each resent photon.

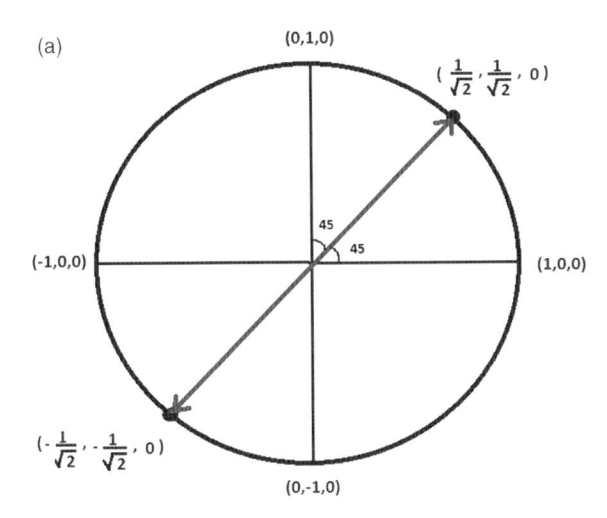

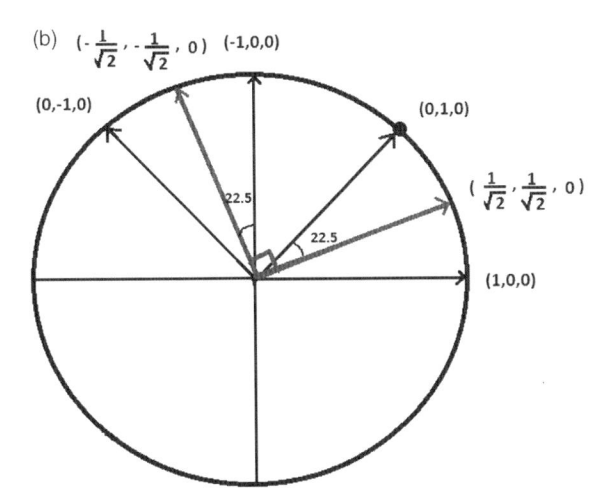

Figure 28.4 Breidbart basis. (a) Equatorial ($Z = 0$) plane of the Poincaré sphere. Breidbart basis $\{((1/\sqrt{2}), (1/\sqrt{2}), 0), ((-1/\sqrt{2}), (-1/\sqrt{2}), 0)\}$ is halfway between rectilinear and diagonal bases. Antipodal points on the x-axis denote horizontal and vertical polarizations. Antipodal points on the y-axis denote two diagonal polarizations. (b) Physical angles between polarization states.

Let us see what kind of information advantage Eve can obtain by choosing the Breidbart basis over a canonical (i.e. rectilinear or diagonal) basis. A measurement in a canonical basis would provide (on the average) 1/2 bit of deterministic information, as Eve would know about half of the bits with certainty while being completely in the dark with regard to the other half. Moreover, Eve knows *which* bits are certain and which ones are random. Suppose the sifted key was prepared based on n photons. Then there are $n/2$ photons whose polarization states are known to Eve with certainty; they contribute zero Shannon and Rényi entropy. For the remaining $n/2$ photons, the uncertainty is perfect, so the Shannon and Rényi entropies are both equal to $n/2$. So by choosing this basis, Eve will have

$$H = R_2 = 0.5n \qquad (28.15)$$

and her knowledge of the key will be equal to $0.5n$ bits. Strictly speaking, we should have written the entropies in (28.15) as conditional entropies, since Eve is trying to guess Alice's string A based on the knowledge of her own string E that she recorded after her eavesdropping attack:

$$H(A|E) = R_2(A|E) = 0.5n. \qquad (28.16)$$

Then Eve's knowledge of A in terms of Shannon bits will be given by mutual information:

$$\begin{cases} H(A;E) = H(A) - H(A|E) = n - H(A|E) = 0.5n \\ R_2(A) - R_2(A|E) = n - R_2(A|E) = 0.5n \end{cases}. \qquad (28.17)$$

Notice that we refrained from calling a similar quantity in the second equation "mutual information" because we have seen that definition of mutual Rényi information is tricky. We can still say, however, that Eve has learned $0.5n$ bits about A in Rényi's sense.

What if Eve chooses the Breidbart basis? Each photon's state will be known with 85% certainty so that

$$\begin{cases} H(A|E) = -n(0.85 \log_2 0.85 + 0.15 \log_2 0.15) \approx 0.601n, \\ R_2(A|E) = -n \log_2 (0.85^2 + 0.15^2) \approx 0.415n, \end{cases} \qquad (28.18)$$

and Eve's actual knowledge of the key is

$$\begin{cases} H(A;E) = H(A) - H(A|E) = 0.399n \\ R_2(A) - R_2(A|E) = 0.585n. \end{cases} \qquad (28.19)$$

Note that our findings (28.18) are in agreement with the result (26.58), since $H \geq R_2$ holds for the canonical basis (as an equality) as well as for the Breidbart basis.

As evidenced by (28.17) and (28.19), as far as Shannon entropy is concerned, the Breidbart basis will actually put Eve at a disadvantage! Knowing all the bits with 85% probability is not quite as informative as knowing half of the bits with certainty, although that dynamics changes if we look at it from the Rényi entropy perspective. So what is the incentive for Eve to choose that basis if her goal is to learn as much information as possible? The answer is that n bits known with 85% probability can resist privacy amplification better than $0.5n$ bits known with certainty [5].[2] Clearly, there is not much advantage for Eve in learning information that is going to be erased by her opponents at a later stage! So it makes a lot of sense for her to sacrifice some of the information for a better chance to withstand privacy amplification.[3]

Finally, what is the amount of Shannon and Rényi knowledge that Bob gets about Alice's key A by looking at his own key B? Assuming that Eve has been actively listening to *every* passing bit, Bob's error rate will be 25% under both scenarios, so we just need to write an equation similar to (28.18):

$$\begin{cases} H(A|B) = -n(0.75\log_2 0.75 + 0.25\log_2 0.25) \approx 0.811n, \\ R_2(A|B) = -n\log_2(0.75^2 + 0.25^2) \approx 0.678n, \end{cases} \quad (28.20)$$

and then

$$\begin{cases} H(A;B) = H(A) - H(A|B) \approx 0.189n \\ \qquad\quad = R_2(A) - R_2(A|B) \approx 0.322n. \end{cases} \quad (28.21)$$

It turns out that Bob has considerably less knowledge than Eve! Then again, it's only natural that an attacker always has an informational advantage over the legitimate user because she can use the same technology and, in addition, she is dealing with the original, uncorrupted version of the message. The point is, however, that she

---

2) There is a limit to their tenacity, however, and it is given by the Rényi entropy. It can be shown that $0.399n$ Shannon bits learned by Eve can resist privacy amplification no better than $0.585n$ deterministic or Rényi bits would.

3) Once we established that the Breidbart basis looks like the optimal choice for Eve, you might think she is always going to choose that basis. But this does not necessarily describe the dynamics of a guessing game in which the opponents are trying to outsmart each other. The reason is that while Eve wants to be efficient, she certainly does not want to be too predictable! After all, if Eve's strategy were known in advance, then Alice could easily send a few decoy photons prepared in the Breidbart basis. This would make Eve's response entirely deterministic as she would always resend the photon to Bob in the same (e.g. vertically polarized) state and thus reveal herself, because without her tampering with the message, Bob would expect to get some dispersion of results. Being clever, Eve must realize that, so there is some incentive for her to choose another basis (possibly somewhere halfway between Breidbart and canonical) even if it's somewhat less than optimal.

> **Box 28.1 Is the Breidbart basis really optimal?**
>
> An attentive reader may ask: What if Eve tried a POVM measurement? Would it then be possible to make the error rate less than 15%?
>
> In general, you should always be asking this question when presented with a new QKD protocol. But for BB84, the answer is no. The reason is that BB84 is a four-state protocol employing symmetric states that are linearly dependent (the last part follows from the fact that Hilbert space of a photon is two-dimensional). We have previously considered the problem of unambiguous discrimination of two states, where the POVM approach does prove optimal. Unambiguous discrimination is out of question in the case of four states. There is a general theorem proved by Chefles [6] that states that are only linearly independent states can be unambiguously discriminated. For photons, this means two states. If the number of states is greater, only ambiguous discrimination is possible. Note that Breidbart basis measurement is also a case of ambiguous discrimination (all results are probabilistic), as is the canonical basis measurement since there is no way to ensure that the bit value inferred by Eve is actually the one that was prepared by Alice.
>
> The general problem of ambiguous discrimination is a complex one, but fortunately, BB84 protocol falls under the special case category. There is a Helstrom formula [7] for two-state ambiguous discrimination, which gives error rate $\Pr_{F,E} = 1/2 \left(1 - \sqrt{1 - 4\eta_1 \eta_2 |\langle \psi_1 | \psi_2 \rangle|^2}\right)$ and when the two states are equiprobable ($\eta_1 = \eta_2 = 0.5$), it results in $Q_E = (1 - \sin\theta)/2$ where $|\langle \psi_1 | \psi_2 \rangle| = \cos\theta$. For $\theta = 45$, we get $Q_E = 15\%$. In general, we should expect discrimination of four states to be harder than discrimination of two states and thus $Q_E \geq 15\%$ should be expected. Breidbart basis measurement with its 15% error rate turns out to be optimal then. The reason four states could be discriminated as easily as two is because they were pairwise orthogonal: $\psi_1 \perp \psi_3$, $\psi_2 \perp \psi_4$. Hence, the optimal POVM in this case reduces to a simple projective Breidbart measurement as far as Eve is concerned. However, we will show later that a POVM can improve efficiency when additional information is provided to Eve.

cannot avoid causing a certain error rate at the receiver end and thus a stealth attack is impossible.

28.3
QKD as Communication Over a Channel

The technical term for the transmission line (a fiber-optic cable or any other physical medium that serves as a conduit for the signal) is *communication channel*. Such a channel can be public (accessible to anyone) or private (it is intended to be shared by the legitimate users only).[4] When a channel transmits just two possible bit values (0 or 1), it is called *binary*. The private channel used in QKD is clearly binary because Alice sends either 1 or 0 into the channel and Bob also retrieves either 1 or 0. The error rate Q in such a channel is the probability that when Alice sends 1, Bob will retrieve 0 or when Alice sends 0, Bob will retrieve 1, in other words, that the bit sent will flip while in transit. In the language of communication theory, this is called *crossover probability* and a channel with $Q > 0$ is called *noisy*.[5] Furthermore, our channel is *symmetric* because crossover probability Q for bit 1 is the same as for bit 0. So the cable carrying polarized photons from Alice to Bob is one example of a noisy binary symmetric channel with crossover probability Q.

We already know from the previous discussion Bob's uncertainty about the value of the bit he receives that is then given by the binary entropy function $h(Q)$. Initially Bob has no prior information about the bit. The Shannon information he learns upon receiving the bit is equal to

$$1 - h(Q) = 1 + [Q \log_2 Q + (1-Q) \log_2 (1-Q)]. \tag{28.22}$$

That is, for every bit sent by Alice, Bob receives only $1 - h(Q)$ bits of information in Shannon's sense. The ratio C between received and sent information is called *capacity* of the channel:

$$C = \frac{\text{(bits received)}}{\text{(bits sent)}} = 1 - h(Q). \tag{28.23}$$

Formula (28.23) establishes the theoretical upper bound on channel capacity. In practice, a channel can be less than ideal (e.g., Bob has an inefficient photon detector, uses a suboptimal algorithm to deduce the bit value from his measurement results, etc.). Then Alice will need to send more information to Bob than is theoretically required. The actual capacity achieved in a real-life experiment is sometimes called R_{comp}:

$$R_{comp} \leq C. \tag{28.24}$$

To summarize, in formal terms, QKD distribution is a special case of communication carried out over two channels, one of which (public) is deemed secure and the other (private) insecure.

---

4) The word public in this context means not only that anybody can overhear the conversation but also that an adversary can listen to, but cannot modify, block, or substitute the signals exchanged over such a channel. In contrast, private channel allows Eve not only to eavesdrop on the signals but also to modify them or even fabricate her own signals and send these bogus signals to legitimate users.

5) It is assumed without loss of generality that $Q < 0.5$ because otherwise Bob can simply flip the retrieved bit value, effectively obtaining a channel with crossover probability $1 - Q$.

Box 28.2 Why QKD?

Perhaps a short historical note is warranted. When Shannon [8] first investigated communication involving two channels in 1948, he envisaged a classical algorithm where Alice "feeds" a message M and a key K to an encryption function E, which produces the encrypted ciphertext C:

$$E(M, K) = C. \tag{28.25}$$

Then any user who knows E, C, and K can decode the message by feeding C and K to a *decryption function* $D \equiv E^{-1}$ that is uniquely determined for a given E:

$$M = D(C, K). \tag{28.26}$$

Shannon simulated the effect of noise by introducing a noise random variable e that is simply a string of bits having the same length as the ciphertext C, as shown in Figure 28.5. Each bit of e has value 1 with probability Q and 0 with probability $1 - Q$. Then, we can model channel noise by adding e to C:

$$C \rightarrow C \oplus e. \tag{28.27}$$

Here, the $\oplus$ sign should be read as modulo 2 addition. As a result, the decoding function now has to take the ciphertext corrupted by noise, and the chance of retrieving the original message M becomes uncertain:

$$M' = D(C \oplus e, K), \tag{28.28}$$

with M' generally different from M. But Shannon was able to prove his *noisy channel coding theorem* that states one can always select a pair of encoding and decoding functions to guarantee that $M' = M$, with chance of error exponentially small in n, where n is the length of string C and δ is a constant:

$$\Pr(M' \neq M) \leq 2^{-\delta n}. \tag{28.29}$$

Moreover, it is OK to choose the encoding function *randomly* from a list of all possible encoding functions for the given message. Only the decoding function

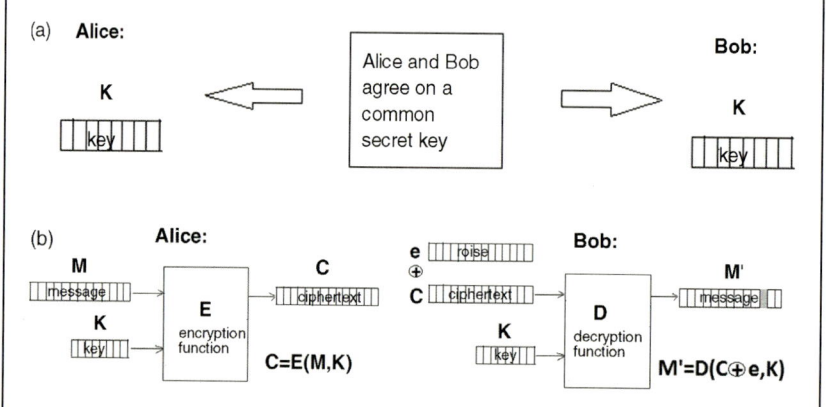

Figure 28.5 (a) Alice and Bob establish a secret key. (b) Alice encrypts her message to Bob, and Bob decrypts it despite the presence of noise in the channel. The decrypted message may contain a small percentage of errors.

must be tailor-suited for the specific choice of M, E, and K. Thus, even classical cryptography allows one in principle to exchange coded messages over a noisy channel. The question arises, however, how to establish the key K common for both parties, and how long K should be.

In Shannon's model, both E and C are communicated publicly (which is also true for modern QKD protocols), but the key K had to be either established by Alice and Bob during a physical meeting or sent secretly with a trusted courier. Thus, Eve is allowed to know everything except K (this is true for QKD as well). Shannon's pessimistic conclusion was that if Alice and Bob wanted perfect secrecy, meaning that Eve's knowledge of E would be totally useless in decoding C, then their secret key K had to be at least as long as M. In view of this, they might as well give the message itself to the courier!

A critic might well point out that the traditional approach to security goes circular: The insecurity of a communication channel is addressed by postulating the existence of another communication channel with perfect security. Clearly, in order to be of any use, K had to be much shorter than M, which meant that communication could not be perfectly secure. But there was still hope that even with a small hint provided by the knowledge of E and C, Eve would find it very difficult to guess this shorter K. And thus the goal of classical cryptography was to make determination of K *computationally infeasible*. This can be achieved, for example, by RSA cryptographic scheme, which is used heavily in today's applications.

Wiesner was the first to realize that quantum mechanics offered a better approach to security than any classical scheme. If computational infeasibility of finding the key could be guaranteed by the laws of nature rather than some mathematical algorithm, then the practical security of communication could not be compromised as long as QM was thought to be correct. Thus, QKD offers the prospect of *unconditional security*, that is, security that can hold against all attacks allowed by QM.

28.4
Postprocessing of the Key

After Alice and Bob have exchanged a sifted key X, they must perform key reconciliation and privacy amplification. Key reconciliation is necessary for three reasons: First, there may be errors introduced by Eve inadvertently when she tried to eavesdrop on X. (We have seen that in BB84 protocol, Eve must spoil no less than 25% of bits she listens to.) Second, random noise in the channel can introduce additional errors. Third, Eve could deliberately tamper with the message, sending some meaningless bits to Bob. (Why do that? One simple answer is because by knowing a higher percentage of bits in X, Eve would be in a better position to resist privacy amplification.) The ratio of the wrong bits in the sifted key to the total length of the key is known as QBER, or quantum bit error rate. Thus, the first task of Alice and Bob is to get rid of the wrong bits and establish beyond reasonable doubt that from this point on, they will be working with the *same* string.

The next step, privacy amplification, is when Alice and Bob will try to remove any information that Eve may have learned by using a publicly announced procedure that will transform their reconciled string X into a different string. Since Eve possesses only partial information about X, she cannot guess the output of such a procedure with an acceptable accuracy. In other words, we want a procedure such that a small uncertainty in the input will cause a large uncertainty in the output. If the procedure is sufficiently well designed, it should reduce Eve's information essentially to zero.

Suppose Alice sent Bob an n-bit string $X = \{0,1\}^n$, and Bob received n-bit string $Y = \{0,1\}^n$, which may differ from X due to inadvertent or malicious tampering by Eve or due to natural random noise. Then their task falls into three parts: (a) Check whether $X \neq Y$, ensuring that the chance of error is arbitrarily small. (b) If the answer is positive, reconcile the strings until $X = Y$ with high probability. (c) Estimate how much information was leaked to Eve via public communication in steps (a) and (b) and then address this problem in the privacy amplification stage.

a) This part is quite simple. Alice chooses at random a hash function $G : \{0,1\}^n \to \{0,1\}^r$, where $r < n$, and sends to Bob the description of the function together with the resulting r-bit string $G(X)$ produced by that function. If Bob finds that $G(X) = G(Y)$, it is strong evidence that X and Y are probably the same. Indeed, there are 2^r possible r-bit strings, so the chance for a hash function to return the same output when $X \neq Y$ is less than 2^{-r}. On the other hand, if $G(X) \neq G(Y)$, Alice and Bob know for sure that $X \neq Y$.

The amount of information leaked to Eve depends on the parameter r but is independent of n. In the most general case, the knowledge of $G(X)$ reveals to Eve up to r Shannon bits of information [9] about X. Its removal would be a formidable problem in the most general case. (We have found earlier that privacy amplification is not very efficient at reducing the adversary's Shannon information.) Fortunately, there exists at least one class of hash functions $G(X)$ such that the conclusions derived previously for r deterministic or r Rényi bits will hold for these functions [10]. It is safe to assume then that r Shannon bits leaked to Eve can be effectively erased later by sacrificing $r + s$ bits where s is the safety parameter.

It is essential for the success of this procedure that the hash function be chosen *after* X has been transmitted to Bob. Otherwise, Eve would know in advance how $G(X)$ works and could tamper just with those bits that will not particularly affect the output of $G(X)$.

b) If the hash function test indicates $X \neq Y$, then key reconciliation is called upon. At this stage, the chance of success depends on how much X differs from Y. It is clear from the limiting case that there is a limit to what a reconciliation procedure can accomplish: If, unbeknownst to Alice and Bob, *all* the bits of X have been tampered with, then after eliminating all errors they will simply end up with an empty string. The number of bits that differ between X and Y is called the *Hamming distance* and is denoted $d(X,Y)$. If this number turns out to be too large, Alice and Bob have no choice but to abandon the effort, restart the protocol from the beginning, and exchange a new key.[6]

If it is believed that $d(X, Y) = 1$, then the simplest approach is to try the hash function method of part (a) for all possible Y that differ from X in 1 bit only. This approach, known as *bit twiddling*, quickly becomes intractable for the larger $d(X, Y)$ and therefore other techniques should be used if Alice and Bob expect more than a few errors. Even so, it is good to keep this option in mind because once the majority of errors have been eliminated by the more powerful methods, bit twiddling could finish the remaining job in a short amount of time.

The most common general technique used when $d(X, Y)$ is large is for Alice to generate (possibly with a hash function) an m-bit *check string* $C(X) = \{0, 1\}^m$, which depends on X, but is shorter than X (i.e., $m < n$), and transmit it to Bob through a public channel so that Eve learns it as well, but cannot tamper with it. This guarantees that, unlike X itself, $C(X)$ will reach Bob unchanged. The property of the check string is that it allows Bob to recover the original X despite the occasional errors introduced by Eve,[7] while revealing no more than m Shannon bits of information about X. Once again, it is shown in error correction code theory that it will be possible to select the check string in a way that will allow later to treat these m bits as if they were deterministic of Rényi bits when it comes to privacy amplification.

c) Alice and Bob now have a reconciled n-bit string X. The final task is to erase Eve's knowledge about X as thoroughly as possible. Overall, $r + m$ Shannon bits (now treated as deterministic or Rényi information) have been leaked to Eve during the reconciliation procedure. In addition, Eve will know t physical or Rényi bits that she learned during the key distribution stage. (If she learned t Shannon bits, privacy amplification theorem cannot be used directly. In this case, we must first find the upper bound on the amount of Rényi information she learned and then use that number instead of t). As a result, Eve now has her own n-bit string $Z = \{0, 1\}^n$, which, generally speaking, is correlated with X.

The main privacy amplification theorem then guarantees that Alice and Bob can distill a shorted k-bit string $K = \{0, 1\}^k$ and make Eve's average expected information about K arbitrarily small by choosing a sufficiently large safety parameter s:

$$H(K; G, Z) \leq \frac{2^{-s}}{\ln 2}. \tag{28.30}$$

---

[6] Of course, Eve always has a choice to block their communication for good by just eavesdropping in an ostentatious manner so that they have to restart from scratch continuously, but the point is that she cannot eavesdrop to the conversation *and* remain undetected. Furthermore, if Alice and Bob could in principle resort to a different communication channel instead of the one that has been compromised (or simply if learning the key is more important for Eve than blocking the conversation), then one can argue that Eve has an incentive not to eavesdrop *too* rudely. Thus, even when faced with a malicious adversary, one should still assume there will be a reasonable chance to reconcile the strings.

[7] Obviously, the feasibility of this procedure depends on how much information Eve has actually learned about X. If her version of X is too close to Bob's version Y, then by publicly broadcasting $C(X)$, we would essentially give her the instructions how to gain the perfect knowledge of X.

The even more strict auxiliary theorem (more strict because its claim is valid for every specific instance z of random variable Z) guarantees the same security in case Eve has learned t Rényi bits:

$$H(K; G, z) \leq R_2(K; G, z) \leq \frac{2^{-s}}{\ln 2}. \tag{28.31}$$

Note that hash functions G appearing in these equations should not be confused with the hash functions used in step (a). The final key length is determined by Equation 28.32:

$$k = n - t - r - m - s, \tag{28.32}$$

and Eve is left with no more than $2^{-s}/\ln 2$ Shannon bits of information. The keys held by Alice and Bob are now identical with error probability 2^{-r}. The value of m will be determined by the initial error rate Q at the receiver end. Postprocessing of the key is successful if k is positive and not too small.

There are several reasons why the above requirement may in fact be overly conservative. Information that was leaked to Eve in the process of eavesdropping, error checking, and error correction will most likely at least partially overlap, making it possible to somewhat relax the restrictions suggested by that expression. Also note that, strictly speaking, we have obtained estimates only for the *lower bound* on distilled key length. While we know that a $k = n - t - r - m - s$ bit key will divulge to Eve *less* than $2^{-s}/\ln 2$ Shannon bits, we don't really know by how much less, so we can still hope that some extra gain in key length could be eked out without exceeding the $2^{-s}/\ln 2$ limit. Yet another, and more serious, reason why the distilled key length could be greater than what is suggested by (28.32) is that Alice and Bob may be willing to put up with a small chance of leaking extra information to Eve in return for a longer key. It can be shown [5] that in exchange for a very low risk (that tends to zero for large n), one can make k as large as $k \approx 0.6n$ in the Breidbart basis experiment. This pleasant surprise is actually a consequence of the ability of Rényi energy to increase when conditioned on an auxiliary random variable. Still, it's better to err on the safe side, and requirement (28.32) can be used as a rough guide.

The above results make it possible for Alice and Bob to decide if they should go ahead with key reconciliation and privacy amplification or abort the attempt and start from scratch. Key distribution is safe as long as the right-hand side of (28.32) is positive. A negative number means that the error rate Q was too high and to eliminate Eve's knowledge it would be necessary to sacrifice the entire length of the key. In particular, for a sufficiently long string $n \to \infty$, the parameters r and s become negligible and a key can be distilled as long as

$$k(Q) = n - t(Q) - m(Q) > 0, \tag{28.33}$$

where the number of bits leaked to Eve in the previous stages is written as a function of Bob's error rate Q.

One can legitimately ask: Where is the assurance that Q will be low enough to make safe communication possible? We have seen that Eve's chance of flipping the bit during the measurement is at least 25% in BB84 protocol. However, a less intrusive Eve may decide to measure only a fraction δ of bits. Of course, this means she would also obtain less information. Eve can also, if she wants to, decrease Q to zero by refraining from eavesdropping completely or increase Q potentially all the way to 50% by inserting random noise into the message. Therefore, Q should be treated as a parameter entirely under Eve's control.

The fact that Eve could set Q high enough to make secure communication impossible does not in any way invalidate the idea of QKD. One can see that by realizing that if that was Eve's intention, she could just intercept all messages and never resend anything, in effect completely blocking the quantum channel. Needless to say, no cryptographic scheme can work when the communication channel is cut off, so the fact that legitimate users cannot exchange the message if the adversary won't let them to should be considered as a given. The point of cryptography is not to get the message across, defying all obstacles, but to make sure that when and if the message is exchanged, it is exchanged confidentially. The final key length expressed as a function of error rate Q thus serves as the formal safety criterion: QKD remains safe as long as $k(Q) > 0$. Turning the argument on its head, it is now Eve's problem to keep eavesdropping below the critical threshold Q_c that will force Alice and Bob to abandon the attempt due to $k(Q_c) = 0$ depriving Eve of her chance to learn anything!

On the other hand, once Q has been chosen by Eve, she definitely wants to adopt the most efficient eavesdropping strategy compatible with that Q. This should be clear from the game theory viewpoint because whatever opinion Alice and Bob may have about her eavesdropping capabilities, she cannot influence their assumptions except by changing Q. An eavesdropping strategy is called *optimal* if for a given Q at the receiving end, the eavesdropper obtains the maximum information allowed by the laws of QM. To fully describe the safety of QKD, one therefore needs to (a) establish which strategy is optimal for each given error rate, and (b) obtain the maximum error rate Q_c at which secure communication is possible assuming Eve chooses the optimal strategy for that error rate.[8]

---

8) We should keep in mind that the "optimal" strategy is optimal in a limited sense only, because we have three players with multiple and conflicting goals. Insomuch as Eve wants Alice and Bob to keep talking, she must reduce her intrusion to keep the error rate under the threshold, but how she should choose the error rate in a real-life situation is really the subject of game theory where the answer will depend on the players' priorities: the importance of transmitting the message versus importance of privacy and the allowed number of attempts for Alice and Bob, and the benefits of knowing the key versus the benefits of frustrating communication between the opponents for Eve. Furthermore, it is not at all clear what kind of information is more valuable to Eve – physical bits, Shannon or Rényi information, or perhaps some other information measure, and it is not even clear how to define the value of information about the transmitted qubits versus information about the final distilled key because Eve does not know precisely how far her opponents are willing to go with privacy amplification. For example, neglecting r and s in the derivation of Q_c works in the idealized case of $n \to \infty$, but for a finite key length n, one can only guess what values of r and s will be considered acceptably safe by Alice and Bob. But once the information measure has been chosen (Shannon information, say) and all the necessary assumptions have been made, the definition becomes straightforward: Eve's attack on the protocol is considered optimal if for a given error rate at the receiving end, she can learn the maximum theoretically possible amount of information about the key.

The most conservative estimate for Q_c is derived as follows. Normally, one would find $t(Q)$ in terms of deterministic or Rényi bits by guessing Eve's measurement strategy and the fraction δ of bits she measures. However, to be completely on the safe side, one may want to apply the strongest measure of uncertainty called *min-entropy* and defined as the surprisal of the most probable outcome:

$$H_\infty(X) = -\log_2 (\max_x P(x)). \tag{28.34}$$

This measure tells us how hard it would be for someone to *guess* the correct value of the random variable X. It is very easy to see why this measure makes a lot of sense in cryptographic applications: When we are uncertain about the best way to proceed, we will usually just choose the strategy where success is deemed most likely. For example, if we are allowed one attempt to guess an unknown letter of the English alphabet, of course we are going to assume that it's an "e" and discard all other possibilities. Shannon or Rényi entropy is not particularly good at describing this simple commonsense strategy since they both assume a player who will give consideration to all possibilities.

Now, Bob's ignorance (in Shannon's sense) about his key is given by the binary entropy function:

$$h(Q) = -Q \log_2 Q - (1 - Q) \log_2 (1 - Q). \tag{28.35}$$

Then it can be shown [11] that in terms of min-entropy, Eve's uncertainty about Alice's key X is asymptotically bounded by

$$H_\infty(X) \geq n - h(Q)n \tag{28.36}$$

and for privacy amplification purposes, this entropy measure can be considered equivalent to knowing $h(Q)n$ deterministic bits. Furthermore, under certain assumptions of the coding theory [12], we can also approximate the check string length as $m(Q) \approx h(Q)n$.[9] Then critical error rate follows from

$$n - 2h(Q_c)n = 0. \tag{28.37}$$

Q_c is found as a root of the transcendental equation:

$$h(Q) = \frac{1}{2}. \tag{28.38}$$

The graph of this function in Figure 28.6 shows that

$$Q_c \approx 0.11 \tag{28.39}$$

and thus BB84 protocol remains secure as long as Bob's QBER is under 11%.

---

9) This assumes Alice and Bob use the Shannon random code for key reconciliation. When other codes are used, more bits of the string will be consumed in the process and the critical error rate will be much lower. For example, codes based on checking parities of substrings (so-called cascade codes) would result in critical error rate $Q_c \approx 8.9\%$.

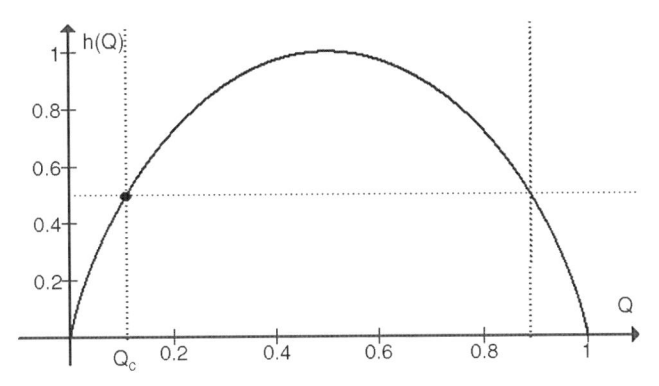

Figure 28.6 Graphical solution of the transcendental equation for critical error rate Q_c.

Alternatively, by noticing that the quantity $n - m(Q) = n - h(Q)n$ is nothing else but the mutual information between Alice's and Bob's strings, while $t(Q) = h(Q)n$ can be thought of as mutual information between Eve's and Alice strings,[10] we can now write (28.37) in a particularly simple form:[11]

$$H(A;B)(Q_c) = H(A;E)(Q_c). \tag{28.40}$$

The communication effort has to be abandoned when Bob infers from his error rate that Eve may know as much about the string as he does.[12] At lower error rates,

$$H(A;B)(Q) > H(A;E)(Q), \tag{28.41}$$

and a secret key can in principle be distilled. Conceptually, the more knowledgeable players will always be able by sacrificing some of their information to erase the knowledge of the less knowledgeable player. What could be easier than that? But keep in mind that derivation of this "simple" condition involves some very heavy math, especially with regard to the left-hand side that makes use of $m(Q) = h(Q)n$. The interested reader may find details of this derivation in Ref. [13]. Practically speaking, however, Eve's knowledge of the key at 11% error rate is likely to be way less than the theoretical maximum.

---

10) What is meant here under A, B, E is of course the *sifted* key since the trials in which Alice and Bob choose different bases will be discarded after a public conversation; Eve will overhear the conversation and drop the corresponding bits from her string as well.

11) Some authors will denote mutual information $H(A;B)(Q)$ as $H_{A;B}(Q)$ or $H_{AB}(Q)$; this notation can improve the appearance of the expression in cases when mutual information is written as a function of another parameter.

12) When we write mutual information $H(A;B)(Q)$, $H(A;E)(Q)$, $H(B;E)(Q)$ as functions of Q, the implied assumption is that joint probability distribution P_{ABE} is known equally well by all three parties. This is, strictly speaking, not necessarily true, since Alice and Bob, as cautious players, ascribe *all* errors to Eve's eavesdropping, while in fact some of them may be caused by ordinary natural noise; that is, Alice and Bob may well believe that Eve knows more than she actually does. Also, due to noise, Eve's success rate will be less than that allowed in principle by the protocol. In this case, P_{ABE} assumed by Alice and Bob may be different from P_{ABE} assumed by Eve.

In practice, it is more convenient to write the critical error condition in a "hybrid" form, with binary entropy on the left:

$$n(1 - h(Q_c)) = H(A; E)(Q_c). \qquad (28.42)$$

When written in this form, the left-hand side is considered to be independent of the eavesdropping strategy. As for the right-hand side, the strongest entropy measure – min-entropy – is in practice often replaced with the familiar entropy measures. For example, consider a canonical basis measurement where Eve learns $t(Q) = (n/2)\delta$ deterministic bits, and knows nothing at all about the other bits. Clearly, in this case, the min-entropy approach gives no additional advantages to Eve and we can then write

$$n - nh(Q_c) = \frac{n}{2}\delta. \qquad (28.43)$$

Bob, of course, will observe an error rate that depends on the number δ chosen by Eve: $Q_c = 0.25\delta$. Equation 28.43 reduces to

$$1 - h(Q_c) = 2Q_c \qquad (28.44)$$

and $Q_c \approx 17\%$.

28.5
B92 Protocol

In 1992, Bennett [14] suggested an alternative to the BB84 protocol. Instead of the four different photon states (0°, 90°, 45°, and 135° polarization angles) utilized in BB84, the new protocol, now known as B92, uses just two states. The only requirement is that these states must be *nonorthogonal*. One simple example is a 0° and 45° photon polarizations used by Alice, encoding, respectively, bits 0 and 1. We shall denote these states $|u_1\rangle$ and $|u_2\rangle$ respectively, with

$$|\langle u_1 | u_2 \rangle| \equiv \cos\theta. \qquad (28.45)$$

(Here, again, θ is the physical angle!) Bob now tries to "read" the state sent by Alice. If the result is conclusive, he records it. The inconclusive results will be discarded by Alice and Bob when they compare their keys. If Eve is actively listening on the line, she will have to flip or otherwise disturb some bits in the process. Note that this time there is no need for Alice and Bob to compare the bases. Instead, Bob only has to announce publicly which of his measurements were inconclusive.

In formal terms, Bob and Eve both face a nonorthogonal state discrimination problem that was discussed in Section 25.6. One difference between them is that Bob is interested in *unambiguous* discrimination so that his string B can exactly match Alice's string A when Eve does not tamper with the message. In this case, he has two attractive strategies at his disposal. One is projective measurement described by

$$Q_B = \frac{1}{2}(1 + \cos^2\theta), \qquad (28.46)$$

which makes 25% of all outcomes conclusive for $\theta = 45°$. The other is the optimal POVM:

$$Q_B^{\text{POVM}}(\theta) = \cos\theta, \tag{28.47}$$

which boosts the rate of conclusive outcomes to 29.3% for $\theta = 45°$. The POVM method can enhance throughput by a small margin, but is extremely difficult technologically and does not offer any significant advantages, so any practical B92 protocol will be implemented by Bob carrying out projective measurement (28.46). This is what we will assume in this section. This also implies that the "inconclusive" outcomes discarded from the sifted string will not in fact be totally inconclusive since Bob could still guess the correct state with 2/3 probability.

Eve, on the other hand, will usually find unambiguous discrimination a poor strategy, and will mostly be interested in the optimal ambiguous discrimination measurement given by the Helstrom formula:

$$P_E = \frac{1 + \sin\theta}{2}, \quad Q_E = \frac{1 - \sin\theta}{2}, \tag{28.48}$$

which for $\theta = 45°$ gives her the same 85% success rate as the Breidbart basis measurement for BB84. The more simple-minded strategy of listening in the rectilinear or diagonal basis would only let her find 25% of the bits exactly and 75% of bits probabilistically with probability 2/3, for a total success rate of 75%.

We can now establish the optimal intercept and resend bases for Eve by the following simple argument. Disturbance introduced by Eve may possibly affect Bob in three ways: first, by changing his percentage of conclusive outcomes; second, by introducing a positive error rate among his conclusive outcomes; and last, by changing the statistics of inconclusive outcomes in cases when Bob uses a measurement procedure that makes these outcomes suggestive, albeit in a probabilistic way. Suppose that after her ambiguous discrimination measurement, Eve resends the photons to Bob according to her best guess. That is, if she believes the photon was prepared in state $|u_1\rangle$, she also resends it in state $|u_1\rangle$. Likewise, a suspected $|u_2\rangle$ photon will be resent as $|u_2\rangle$. Since Eve is right 85% of the time, the distribution of outcomes in 85% of Bob's measurements will be exactly the same as it would have been without Eve, namely, one quarter will be correct and conclusive, and three quarters, though suggestive of the correct state with probability 2/3, will be discarded as inconclusive. However, 15% of the time photons will arrive to Bob's detector in the wrong state and for these photons, the statistics of outcomes will work the same way but with a nuance: One quarter will be "conclusive" but wrong, and three quarters will be discarded, which will not be a big mistake because they would suggest the wrong bit values by assigning them a 2/3 probability. Overall, Bob will get conclusive results in 25% of the trials just as he should, but 15% of his "conclusive" results will be wrong. Hence, Bob's error rate is 15%.

It is clear that Eve cannot reduce it still further by resending the photons in some intermediate basis (otherwise, she could resend them to herself and get her own error rate under the 15% theoretical minimum). Hence, the above strategy is optimal in terms of minimizing Q_B. It is also easy to see that the overall dispersion of outcomes at Bob's side will be exactly the same as it would be without Eve's interference, since deflecting 15% of the photons in the $|u_1\rangle$ beam to $|u_2\rangle$ beam and 15% of photons in the $|u_2\rangle$ beam to $|u_1\rangle$ beam leave the overall statistics of the beam unchanged. Therefore, Bob cannot detect the eavesdropping attack by looking at his percentage of conclusive outcomes or at the statistics of inconclusive outcomes. Here too, Eve's strategy of resending bits in state $|u_1\rangle$ or $|u_2\rangle$ proves to be optimal. Since in all three categories the $\{|u_1\rangle, |u_2\rangle\}$ resend basis does the best theoretically possible job of hiding Eve's attack, it must be the optimal basis. This is in sharp contrast with BB84 protocol where Eve's best bet was to resend photons in the intermediate basis halfway between their original preparation states. This time, it turns out that there is no advantage in that; Eve's best chance is simply to act upon her best knowledge, resending each photon in what she believes to have been its original preparation state.[13] Thus, the only indication that eavesdropping may be taking place is unusually high error rate.

Summarizing, we get $Q_E = Q_B = 15\%$ and Eve causes lower error rate at the receiver end than what we saw in BB84. By the same reasoning, if Eve chooses the canonical (rectilinear of diagonal) basis, the error rates will be $Q_E = Q_B = 25\%$ – in full analogy with the choice of canonical basis in BB84.

As Eve now knows all Alice's bits probabilistically with 85% confidence, her Shannon and Rényi entropies are just the same as in the case of BB84 for Breidbart basis measurements:

$$H(A|E) \approx 0.601n, \quad R_2(A|E) \approx 0.415n. \tag{28.49}$$

As for Bob, his success rate now is also equal to 85%, so

$$H(A|B) \approx 0.601n, \quad R_2(A|B) \approx 0.415n. \tag{28.50}$$

Tables 28.1 and 28.2 summarize the results we got for BB84 and B92 protocols.

Table 28.1 BB84.

	Eve		Bob	
	Canonical	Breidbart	Canonical	Breidbart
P	0.75	$(2+\sqrt{2})/4 \approx 0.85$	0.75	0.75
Q	0.25	$(2-\sqrt{2})/4 \approx 0.15$	0.25	0.25
H	$0.5n$	$0.601n$	$0.811n$	$0.811n$
R_2	$0.5n$	$0.415n$	$0.678n$	$0.678n$

Alice sends to Bob n-bit string A, and Eve obtains n-bit string E by eavesdropping. Success rates for Eve and Bob are defined as the fraction of correct physical bits recorded in the respective strings E and B. Mutual information for Eve and Bob, $H(A; E)$ and $H(A; B)$, can be found from the respective Shannon entropies.

---

13) This has another important corollary. After the experiment, inconclusive measurements will be discarded by Alice and Bob and hence also by Eve. However, with this choice of the resend basis, Bob's measurement has the same 25% chance to be conclusive, whether or not Eve guessed the state correctly. Therefore, by publicly announcing to Alice which of his trials were inconclusive, Bob does not provide any new information to Eve.

Table 28.2 B92.

	Alice	Bob
P_E	0.85	0.85
Q_E	0.15	0.15
H	$0.601n$	$0.601n$
R_2	$0.415n$	$0.415n$

28.6 Experimental Implementation of QKD Schemes

It is now time to take our discussion of QKD from concrete examples to a more abstract level. So far we have considered two ways for Alice and Bob to exchange a secret key – first, by encoding bit values in four polarization states (BB84), and then, by encoding them in two nonorthogonal polarization states (B92). But photon polarization is not the only possible mechanism to implement QKD protocols. For example, one could encode bit values in electron spin states, or use some other observable that is allowed to take only two values. Therefore, to avoid giving the impression that we cannot think beyond one specific example, we often want to describe the protocol in general terms, dropping the implementation details and specifying only the number of states employed by the protocol, the inner products of these states, and the types of measurements performed by all the parties involved. Thus, BB84 protocol is often referred to as a *four-state protocol* in scientific literature, and B92 is simply described as using two nonorthogonal states.

To convince the reader that there are multiple ways to implement one and the same protocol, we are now going to discuss an important example where BB84 and B92 are realized by encoding bit values in photon phase shifts. (To increase the relevance of this example, let us point out that it describes the original implementation of B92 that was proposed by Bennett when he suggested this protocol.)

Consider the setup shown in Figure 28.7. Alice produces a photon ψ that goes toward the beam splitter BSA – a half-reflecting mirror that sends one-half of the

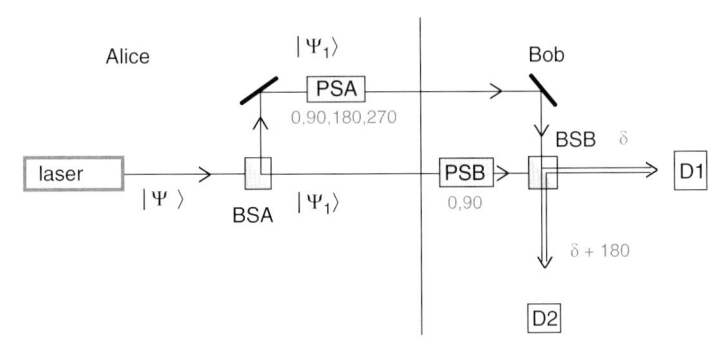

Figure 28.7 BB84 with phase shift.

wave function (ψ_1) into the upper arm of the interferometer and the other half (ψ_2) into the lower arm. After traveling through long fiber-optic cables, both parts of the wave function finally reach Bob's laboratory and recombine in his part of the interferometer. Bob also has a beam splitter BSB – another half-reflecting mirror that splits each half of the wave function into transmitted and reflected parts. Finally, a pair of detectors D_1 and D_2 record the experimental results.

A key part of the design is that when ψ_1 and ψ_2 approach Bob's beam splitter with phase difference δ, the wave parts that reach D_1 have the same phase difference δ while the parts that reach D_2 have phase shift $\delta + 180°$. In other words, when we have constructive interference at D_1, the interference at D_2 is destructive and vice versa. This means that the result of the experiment is going to be *deterministic* when $\delta = 0$ or $\delta = 180°$, that is, one detector always "clicks," while other always stays idle. In order to achieve this effect for the setup in Figure 28.7, we have made the shaded parts of the beam splitters from optically denser glass. (Recall that a photon reflection is accompanied by 180° phase change only when it reflects from a denser medium.)

This setup is known as the Mach–Zender interferometer. The main difference from the Michelson interferometer that you encountered in introductory physics courses is that Mach–Zender has two detectors, which allow us to set the arm lengths so as to make experimental outcomes perfectly deterministic. Then if determinism is violated, we know that there was an additional phase shift in one of the arms, and can estimate that phase shift from the distribution of outcomes.

Imagine now that Alice inserts a phase shifter PSA in her arm of the interferometer and introduces a 0, 90, 180, or 270 shift chosen at random. Likewise, Bob inserts another shifter PSB in the other arm and applies a randomly chosen 0° or 90° shift. If Bob chooses 0° (PSB = 0), the outcomes will behave deterministically when PSA = 0 and 180 and will be distributed randomly for PSA = 90 and 270. If he chooses 90° (PSB = 90), then the outcomes for PSA = 0 and 180 will behave randomly and the outcomes for PSA = 0 and 270 deterministically. Finally, if we do not see deterministic behavior in either detector (e.g., if neither detector clicks), this means Eve is actively listening on the line and resends the photons in some intermediate basis (0 < PSF < 90). We have arrived at one possible implementation of BB84 (Table 28.3).

Table 28.3 Phase shift implementation of BB84.

BB84 basis	Polarization angle (physical)	Polarization angle (on Poincaré sphere)	PSA	PSE	PSB
Rectilinear	0	0	0	0	0
	90	180	180		
Diagonal	45	90	90	90	90
	135	270	270		
Breidbart	22.5	45		45	
	112.5	225			

PSB = 0 simulates the choice of rectilinear basis by Bob. PSB = 90 simulates the choice of the diagonal basis. In order to overhear the conversation, Eve will want to introduce her own phase shift PSE (and then change it to PSE' when re-emitting the signal). A choice of PSE = 45 is equivalent to Eve choosing the Breidbart basis.

Indeed, if we let the four values of PSA play the same role as Alice's four polarization states in BB84, and the two values of PSB play the same role as the two choices of polarization basis by Bob, we will obtain the same statistics of experimental outcomes.

At this point we should say a few words about the source of light used in QKD experiments. Up till now, we have assumed that Alice will always generate single photons in a desired state. In principle, a workable QKD scheme could be designed in this way. But in practice the cost and complexity of such systems has motivated experimenters to adopt a less ideal solution. A laser source is used to generate a bright monochromatic beam, which then gets attenuated (e.g., by a lens made of opaque glass) so that the resulting beam contains less than one photon on the average.

Light produced by a laser consists of photons that are essentially all in the same state except for a small and ideally negligible noise component. Alice can thus easily choose the desired frequency, polarization, and phase, which will apply to all photons in the laser pulse. However, once the phase is known, the exact number of photons in the pulse cannot be arranged in advance. Such a pulse is described by a superposition of photon number states:

$$|\alpha\rangle = C_0|0\rangle + C_1|1\rangle + C_2|2\rangle + \cdots + C_i|i\rangle + \cdots, \tag{28.51}$$

and the probabilities to find 0,1,2, or more photons in that pulse will be equal to

$$|\langle 0|\alpha\rangle|^2, |\langle 1|\alpha\rangle|^2, |\langle 2|\alpha\rangle|^2, \ldots, |\langle i|\alpha\rangle|^2. \tag{28.52}$$

Here, $|i\rangle$ denotes a state with i photons. Note that the *vacuum* state (zero-photon state $|0\rangle$) is also possible. Of course, the *average* number of photons in the pulse is under Alice's control (the averaging is over the ensemble of pulses prepared under identical conditions). But she cannot guarantee that $|\alpha\rangle = |1\rangle$ would hold for any specific pulse.

In order to perform QKD, Alice must attenuate her pulse to the point where the average number of photons μ would be much less than 1. You may ask: What is wrong with using a more powerful beam? The answer is in the photon number statistics describing laser-generated pulses. The probability to find n photons in a laser pulse with the mean photon number μ is given by a Poisson distribution:

$$p(n,\mu) = \frac{\mu^n e^{-\mu}}{n!}. \tag{28.53}$$

Thus, we find 0, 1, 2, or more photons with respective probabilities:

$$p(0,\mu) = e^{-\mu}, \quad p(1,\mu) = \mu e^{-\mu}, \quad p(2,\mu) = \frac{1}{2}\mu^2 e^{-\mu}, \ldots \tag{28.54}$$

Note that while the probability of finding two or more photons falls off rapidly as μ goes to zero, it can be far from negligible for $\mu \approx 1$. But that would create a security breach because Eve could then divert the extra photons, make a measurement on them, and thus learn valuable information while leaving one photon undisturbed.[14] This strategy is known as PNS or photon number splitting attack. This explains why in real-life experiments, laser beam is usually attenuated to the order of $\mu \approx 0.05$.

This addresses the PNS threat, but the side effect is that in the majority of trials, Bob's detectors will not register any photons.[15] This drastically decreases the throughput as most pulses are just wasted. The actual transmission rate (i.e., fraction of pulses that finds its way to the sifted key) is given by

$$t^{(BB84)} = \frac{1}{2}\left(1 - |\langle 0|\alpha\rangle|^2\right) = \frac{1}{2}(1 - e^{-\mu}), \tag{28.55}$$

which also accounts for the fact that half of the photons will be discarded after comparing the bases.

The reader can see from this discussion that practical implementation of BB84 is quite different from an idealized thought experiment. In theory, Alice and Bob could communicate with single photons, ridding themselves of vacuum states for good. In practice, generation of single photons is often too difficult, and the actual experimental prototypes rely on attenuated laser pulses. This creates another subtle difficulty: The "orthogonal" states such as vertically and horizontally polarized light are in fact nonorthogonal because of the vacuum component! Indeed, consider the inner product of vertical and horizontal laser pulse states:

$$|\alpha_\updownarrow\rangle = C_0|0\rangle + C_1|1_\updownarrow\rangle + C_2|2_\updownarrow\rangle + \cdots + C_i|i_\updownarrow\rangle + \cdots,$$
$$|\alpha_\leftrightarrow\rangle = C_0|0\rangle + C_1|1_\leftrightarrow\rangle + C_2|2_\leftrightarrow\rangle + \cdots + C_i|i_\leftrightarrow\rangle + \cdots.$$

Polarization was not indicated for the first term because it corresponds to the absence of a photon. Canceling the zero inner products of mutually orthogonal states, we see that

$$\langle \alpha_\updownarrow | \alpha_\leftrightarrow \rangle = |C_0|^2 \equiv p(0,\mu) = e^{-\mu}. \tag{28.56}$$

In other words, orthogonality can only be preserved for sufficiently bright pulses. Moreover, if the pulses are infinitely faint, they become essentially "parallel!" Clearly,

---

14) The question arises, what should Eve do with these diverted photons. Resending them to Bob alongside the undisturbed photon is not a good idea because Bob will get suspicious if he gets too many mixed states. It is best for Eve to keep the diverted photons to herself without resending them. This will slightly decrease the average light intensity observed by Bob. But Eve could easily compensate Bob for that loss by slightly increasing her μ in those cases when she carries out an ordinary intercept-resend attack on individual photons so that her strategy of siphoning off any available extra information will remain undiscovered.

15) There is a small chance that a detector will click even when no photons have been sent. Such events, known as *dark counts*, are mostly of thermal origin and can be strongly suppressed at low temperatures. Dark count rate is an important characteristic describing noise in the detector.

the problem indicated by (28.56) will also persist in the interferometric implementation considered above:

$$\begin{cases} \langle a|e^{i\pi}a\rangle = [C_0^*\langle 0|+C_1^*\langle 1|+\cdots][C_0|0\rangle+C_1|e^{i\pi}1\rangle+\cdots] = |C_0|^2 = e^{-\mu} \\ \langle e^{i(\pi/2)}a|e^{i(3\pi/2)}a\rangle = [C_0^*\langle 0|+C_1^*\langle e^{i(\pi/2)}1|+\cdots][C_0|0\rangle+C_1|e^{i(3\pi/2)}1\rangle+\cdots] = |C_0|^2 = e^{-\mu} \end{cases}$$
(28.57)

Now the supposedly orthogonal states 0, 180 or 90, 270 have overlap between them that is due to the vacuum state. It follows from (28.57) that a state with phase 0 projects onto a state with phase 180 with probability

$$|\langle a|e^{i\pi}a\rangle|^2 = e^{-2\mu}. \tag{28.58}$$

As a result, these states fail to destruct completely in interferometry experiments. Suppose indeed that state $|a\rangle$ is split in half, then one part acquires a 180° phase shift to become $|e^{i\pi}a\rangle$. We now have two states, $(1/\sqrt{2})|a\rangle$ and $(1/\sqrt{2})|e^{i\pi}a\rangle$. Then, both states are split in half again, becoming $(1/\sqrt{2})|a\rangle$ and $(1/2)|e^{i\pi}a\rangle$ and recombine in the detector. Then we can write,

$$\left|\frac{1}{2}|a\rangle+\frac{1}{2}|e^{i\pi}a\rangle\right|^2 = \frac{1}{4}(C_0^*\langle 0|+C_0^*\langle 0|+C_1^*\langle 1|+C_1^*\langle e^{i\pi}1|+\cdots)(C_0|0\rangle+C_0|0\rangle+C_1|1\rangle+C_1|e^{i\pi}1\rangle+\cdots)$$

$$= \frac{1}{4}(4|C_0|^2\langle 0|0\rangle+|C_1|^2[\langle 1|1\rangle+e^{i\pi}\langle 1|1\rangle+e^{-i\pi}\langle 1|1\rangle+e^{i\pi}e^{-i\pi}\langle 1|1\rangle]+\cdots) = |C_0|^2 = e^{-\mu}.$$
(28.59)

In other words, even as all the other states cancel each other, the vacuum term common to both "orthogonal" states interferes constructively.

One may think this is where the second detector would come in handy. Normally, destructive interference in one detector will show as constructive interference in the other and it seems at first that if both detectors fail to click, we could take it as a signature of the vacuum state. The reality is that there is no way to make *all* measurements conclusive even if one uses two detectors. Surely, if one detector clicks, this means constructive interference took place in that detector, hence interference in the second detector must be destructive. But if neither detector clicks, we do not have any conclusive result because the effect of a vacuum state is indistinguishable from the effect of destructive interference. All we can say is that in one detector, the pulses interfered destructively, and in the other, interference was constructive and we would have observed a click but unfortunately the vacuum state was realized. This does not tell us in which detector the pulses interfered constructively. Moreover, we cannot even determine whether the vacuum state was in fact realized in the detector with destructive interference. So the second detector does not help us in this case. Simply stated, we cannot generally distinguish 0 photons from 1 photon.

Indeed, suppose two pulses arrive in the detector D_1 in phase and interfere constructively. This case is straightforward. We have

$$\left|\frac{1}{2}|a\rangle+\frac{1}{2}|a\rangle\right|^2 = |\langle a|a\rangle|^2 = |C_0|^2+|C_1|^2+|C_2|^2+\cdots = 1 \tag{28.60}$$

and a detector count is registered with probability

$$P_{D_1,\text{count}} = 1 - |C_0|^2 = 1 - e^{-\mu}, \tag{28.61}$$

while nothing is registered in detector D_2:

$$P_{D_2,\text{count}} = 0. \tag{28.62}$$

Therefore, our chance to have a conclusive result is $1 - e^{-\mu}$, and there is always a $e^{-\mu}$ chance that our measurement will be inconclusive.

In the earlier QKD experiments, the restriction (28.56) on orthogonality was considered a weakness. At low pulse intensities, the states will have too much of an overlap; on the other hand, set the intensity too high and the scheme will become vulnerable to the PNS attack. Therefore, if we were 100% realistic in our analysis in Section 28.2, we should have mentioned that the majority of experiments will fail to register a photon and the statistics of outcomes that we discussed applies only to those rare cases when the photon was registered. Thus, a feature of B92 protocol (the discarding of inconclusive results) is present already in real-life BB84 experiments. In this sense, Bennett's B92 proposal was virtue borne out of necessity.[16] If we cannot make the pulses orthogonal enough, then nonorthogonality must be accepted and turned into advantage. This is why, even though B92 *can* in principle work with single photons, it was designed specifically with the intention to harness the nonorthogonality of dim laser pulses.

Let us now see how interferometry can help us implement B92 protocol. The basic scheme is shown in Figure 28.8. This time the interferometer is of traditional variety, with only one detector. Once again Alice starts with a bright beam, but this time she uses an *unsymmetric beam splitter* (UBS) (where the mirror is mostly transparent and only a small part of intensity is reflected) to split a dim *signal pulse* $\mu \ll 1$ from the initial beam. The remaining bright beam is time-delayed by Δt so that the dim pulse goes into the fiber-optic cable first, followed by the bright beam. The bright beam is called *reference pulse*. At his end of the interferometer, Bob also

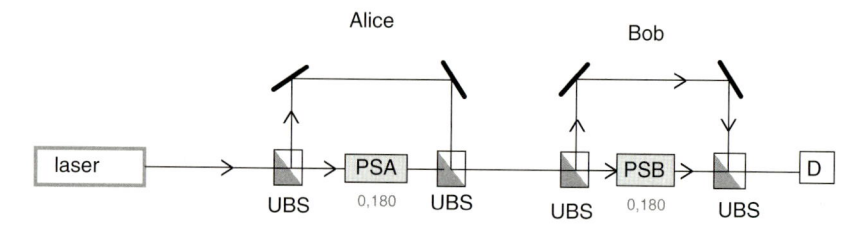

Figure 28.8 B92 with phase shift. The interfering states are nonorthogonal due to the vacuum component. The overlap is given by $|\langle u_1|u_2\rangle|^2 = |C_0|^4 = e^{-2\mu} \approx 1 - 2\mu$, where μ is the average photon number in a pulse.

---

16) In this respect, the Ekert scheme based on EPR photon pairs is more reliable than laser pulse-based BB84. Unfortunately, when it comes to practice, transmission of EPR pairs over large distances turns out to be too great a challenge.

subjects both signals to the same UBS splitting followed by the same time delaying of the brighter part that Alice accomplished at her end of the apparatus. The dim pulse split off from the dim pulse is so weak that it can be safely neglected. The greater part of the dim pulse is time-delayed by Δt, thus losing its initial time advantage compared to the reference pulse. But the reference pulse gets split as well, with the brighter part falling behind in time by another Δt. As a result, Bob now has two simultaneous dim pulses, each traveling in a separate arm of the interferometer. At the last beam splitter, these pulses recombine to produce constructive or destructive interference at the detector (Figure 28.9). The bright pulse arrives in the detector Δt seconds later.

In the original version of the protocol, Alice's two nonorthogonal states $|u_1\rangle$, $|u_2\rangle$ were realized by 0° and 45° polarizations of a single photon. This time they will be realized by phase shifts PSA = 0 and PSA = 180:

$$|u_1\rangle = |\alpha\rangle, \quad |u_2\rangle = |e^{i\pi}\alpha\rangle. \tag{28.63}$$

The seemingly orthogonal pulses in fact have a significant overlap because both of them contain a heavy admixture of the vacuum state.

As described in Section 25.6, Bob's operators $\hat{P}_1$ and $\hat{P}_2$ are described by

$$\hat{P}_1 = I - |u_2\rangle\langle u_2| = I - |e^{i\pi}\alpha\rangle\langle e^{i\pi}\alpha|, \quad \hat{P}_2 = I - |u_1\rangle\langle u_1| = I - |\alpha\rangle\langle\alpha|. \tag{28.64}$$

In the current implementation, they will be realized by the choice of PSB = 0 and PSB = 180 by Bob at his end of the interferometer:

$$\hat{P}_1 \Rightarrow \text{PSB} = 0, \quad \hat{P}_2 \Rightarrow \text{PSB} = 180. \tag{28.65}$$

One can see that the choice of PSB = 0 does in fact act as a projection operator $\hat{P}_1$ onto a subspace orthogonal to $|u_2\rangle$. A positive result (i.e., detector clicks) tells Bob that Alice must have chosen PSA = 0. Hence, state $|u_1\rangle$ is confirmed and state $|u_2\rangle$ is annihilated.

By the same token, the choice of PSB = 180 acts as a projection operator $\hat{P}_2$ onto a subspace orthogonal to $|u_1\rangle$. A detector count tells Bob that PSB = 180, thus confirming state $|u_2\rangle$ and annihilating state $|u_1\rangle$.

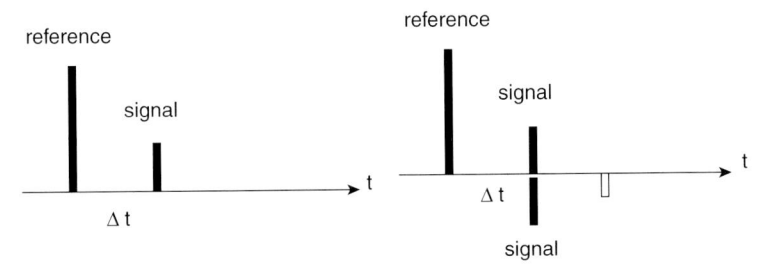

Figure 28.9 Reference and signal pulses. Signal pulses interfere constructively or destructively depending on the phase shift. The empty rectangle marks the very faint pulse that can be neglected.

Once we have written our projection operators, operator $\hat{P}_?$ follows immediately:

$$\hat{P}_? = I - \hat{P}_1 - \hat{P}_2 = |u_1\rangle\langle u_1| + |u_2\rangle\langle u_2| - I = |a\rangle\langle a| + |e^{i\pi}a\rangle\langle e^{i\pi}a| - I. \tag{28.66}$$

If Bob chooses the right basis in which to measure the state prepared by Alice, his chances to get conclusive versus inconclusive result are

$$P = 1 - |\langle u_1|u_2\rangle|^2, \quad Q = |\langle u_1|u_2\rangle|^2. \tag{28.67}$$

If he chooses the wrong basis, these chances become

$$P = 0, \quad Q = 1. \tag{28.68}$$

Overall, averaging over the choices of basis, we can write

$$P = \frac{1}{2}\left(1 - |\langle u_1|u_2\rangle|^2\right), \quad Q = \frac{1}{2}\left(1 + |\langle u_1|u_2\rangle|^2\right) \tag{28.69}$$

and since (28.56) gives us $|\langle u_1|u_2\rangle|^2 = |C_0|^4 = e^{-2\mu}$, the last equation reduces to

$$P = \frac{1}{2}(1 - e^{-2\mu}), \quad Q = \frac{1}{2}(1 + 2^{-2\mu}). \tag{28.70}$$

Since only conclusive measurements contribute to the sifted key, the transmission rate is simply

$$t^{(B92)} = \frac{1}{2}(1 - e^{-2\mu}). \tag{28.71}$$

Comparing it with (28.55), we see that B92 protocol offers a higher transmission rate for a given intensity of light pulses. In particular, for low-intensity pulses,

$$t^{(BB84)} \approx \frac{\mu}{2}, \quad t^{(B92)} \approx \mu. \tag{28.72}$$

Suppose we want to imitate the effect of a 45° angle between photon polarizations that we had in Section 28.5. Then, we must set the inner product equal to

$$\langle u_1|u_2\rangle = \frac{1}{\sqrt{2}}. \tag{28.73}$$

According to (28.57), this requires

$$e^{-\mu} = \frac{1}{\sqrt{2}}. \tag{28.74}$$

With this choice of μ, transmission rate becomes

$$t^{(B92)} = \frac{1}{2}\left(1 - \left(\frac{1}{\sqrt{2}}\right)^2\right) = 0.25 \tag{28.75}$$

and the result from Section 28.5 has been recovered. Thus, interferometric implementation becomes equivalent to the scheme with 0° and 45° photon polarizations at

$$\mu = \frac{\ln 2}{2} \approx 0.347, \tag{28.76}$$

which, however, may be too high to prevent a successful PNS attack (at this intensity, Eve has a 4.26% chance to discover two photons in a given pulse). Reducing the intensity still further has the same effect as decreasing the angle between photon polarizations.

One important feature of the above scheme is that the absence of a photon count in Bob's detector amounts to an inconclusive result. As a corollary, if Eve tries to replace some legitimate signals with vacuum states, Alice and Bob will discard all such instances as inconclusive. If Eve tampers with the conclusive part of the message, she would better know the actual phase states!

Finally, the reader may wonder about the function of the reference beam. Its role is auxiliary, but also important. It is assumed that the moments of time Alice is going to emit her photons have not been established in advance. The arrival of the reference beam tells Bob that the dim signal beam should have arrived Δt seconds ago. If Eve tries to suppress the signal pulse while allowing the reference pulse to arrive, Bob will soon realize that he is getting too many vacuum states. Suppressing the reference signal is dangerous for Eve because Alice and Bob will realize during subsequent public discussion that too many emitted reference signals failed to arrive. So Eve has no choice but to allow the reference signal to go through and then she must also resend the signal pulse to Bob, which creates a serious difficulty for Eve because just like Bob, she cannot hope to learn the correct PSA conclusively from her measurement. On the other hand, if there were no reference pulses, Bob would have no reason to expect any specific signal from Alice to arrive at a specific time. This would allow Eve to subvert B92 in the following simple way: She could perform an unambiguous discrimination measurement of each passing photon, resend the signal to Bob only when she gets a conclusive result, and suppress the rest of the signals without causing suspicion. Of course, she would have to fabricate additional photons to compensate Bob for the loss of intensity caused by signal suppression. This can be done by increasing the intensity of signals that she does resend (but still keeping the intensity well below $\mu \approx 1$). An increased chance to detect the photons whose state Eve has learned will keep the energy flux at the receiver end unchanged and then the zero error rate will convince Alice and Bob that no eavesdropping is taking place.[17] The reference beam prevents this by forcing Eve to resend all photons, including those whose state she failed to learn conclusively.[18]

17) However, she must apply some extra effort to replicate the statistics of the original pulses that were characterized by a Poissonian distribution of lesser mean. Since Eve decided to increase the intensity, a Poissonian curve would now have a different shape, which could set off an alarm when Bob notices that time correlations between detection events are different from what he expected. So in order for this trick to work perfectly, Eve would need to generate pulses with non-Poissonian photon number statistics.

18) In principle, reference pulses will not be necessary if Alice and Bob agree in advance (or establish *post factum* through public discussion) about the time interval between the pulses. To be sure, the exact moment a laser is going to fire is never known with perfect precision, but the approximate time window can well be controlled by the experimenter. These ideas are explored further in time-shift protocols. One might suggest that the reception of a bright pulse is a more reliable mark on the timescale than some prearranged time window. But this argument is dubious: After all, Eve could also resend the reference pulse with some time delay as long as it is not long enough to be counted as *suppression* of the pulse.

28.7
Advanced Eavesdropping Strategies

The simplest strategy Eve can choose is to use projective measurements to analyze one qubit at a time so that the outcome of measurement on one qubit will not affect the outcomes of measurements on other qubits. This lack of correlation between Eve's measurement results gives us the grounds to call such an eavesdropping strategy *incoherent*. Other terms you may see in the literature are *single qubit attack* – it means the same thing, namely, that Eve attacks one qubit at a time – and *intercept-resend attack* – each bit is resent right after the interception. But there are other strategies available to Eve: POVMs on a single qubit, which were described earlier, and, even more generally, collective measurements on multiple qubits. For instance, Eve could keep incoming qubits in a storage room without measuring them until she accumulates enough of them and then performs a measurement of some collective property of the whole group, which would leave the qubits in an entangled state. This strategy is called *coherent eavesdropping* or *collective attack*. This means we have not proved the security of a protocol until we investigate the possibility of two-qubit, three-qubit, or, generally, N-qubit coherent eavesdropping.

Perhaps the most promising strategy for Eve is the following interesting variation on the collective measurement topic [15,16]. Alice's qubit $|\Psi\rangle$ goes past Eve. Eve appends to this qubit an ancilla $|A\rangle$. Eve then performs a unitary[19] transformation U whose purpose is to entangle the two systems. Thereafter $|\Psi\rangle$ is allowed to go toward Bob's detector, while $|A\rangle$ is retained by Eve. When Bob measures the state of $|\Psi\rangle$, the entangled state will be destroyed and $|A\rangle$ will collapse into a definite state. Eve repeats this procedure for each qubit sent by Alice. After Bob finishes his measurements and the bases are announced, Eve looks at her ancillas and figures the outcomes obtained by Bob. The advantage of this scheme is that instead of trying to learn the information on her own, Eve will allow Bob to reveal the qubit state to her.

With this scheme, the error rate (QBER) on Bob's side is completely determined by the entangling operation U. Alice and Bob will know this error rate (henceforth we will denote it by Q) after comparing bit values in a small subset of the string. The question arises: How much information could Eve possibly learn for a given Q if she uses this entanglement strategy?

In what follows, we will impose only two constraints on the scheme: Eve interacts with only one qubit at a time, and for each qubit, she uses only one ancilla. However, we still allow Eve, after the bases are announced, to make a collective measurement on several ancillas at once if she wishes to do so.

Let us denote the horizontal and vertical polarization states by $|h\rangle$ and $|v\rangle$, and the two diagonal polarization states by $|s\rangle$ and $|d\rangle$. By definition of BB84, Alice's qubit was prepared in one of these four states, using either h,v or s,d basis. Suppose, for the sake of example, it was the h,v basis. It seems natural to assume that if Eve carries out an *optimal* eavesdropping attack, Bob will also receive some mixture of the same

---

19) Is possible to show that any nonunitary interaction between the qubit and the ancilla is equivalent to a unitary interaction between the qubit and a higher dimensional ancilla.

basis eigenstates $|h\rangle$ and $|v\rangle$, and similarly, if the s,d basis was chosen by Alice, Bob will receive a mixture of $|s\rangle$ and $|d\rangle$. We will accept this without proof, referring the interested readers to Section III of Ref. [17] where this assumption results in Eve attaining the maximum theoretical bound on information gain. A choice of U that mixes the basis states in this way is called *symmetric* strategy on Eve's part.

Let us say Alice prepares her qubit in state $|h\rangle$. When $|h\rangle$ comes into contact with $|A\rangle$, they are initially independent so that the joint system should be written $|h\rangle \otimes |A\rangle$.

After the symmetric entangling operation U has run its course, a measurement performed in the same h,v basis will find Alice's qubit in state $|h\rangle$ with ancilla in the corresponding state $|\varphi_h\rangle$, or qubit in state $|v\rangle$ with ancilla in the corresponding state $|\theta_h\rangle$. (Subindex h refers to the initial state of the qubit.) In the former case, Bob will record the correct state $|h\rangle$; in the latter case, his result will be wrong because $|v\rangle$ is not the state that Alice prepared originally. The probability with which wrong qubit values occur at Bob's side is just the error rate Q. Then probability of success is $1 - Q$ and the probability amplitudes are given by the square root of the respective probabilities and finally we can write

$$U(|h\rangle \otimes |A\rangle) \equiv |H\rangle = \sqrt{1-Q}|h\rangle \otimes |\varphi_h\rangle + \sqrt{Q}|v\rangle \otimes |\theta_h\rangle, \qquad (28.77)$$

$$U(|v\rangle \otimes |A\rangle) \equiv |V\rangle = \sqrt{1-Q}|v\rangle \otimes |\varphi_v\rangle + \sqrt{Q}|h\rangle \otimes |\theta_v\rangle. \qquad (28.78)$$

(Equation 28.78 is obtained in a similar manner).

These equations look surprisingly similar to the ones we encountered for earlier POVMS. This is not an accident. The above scheme really amounts to a POVM conducted together by Eve and Bob where Eve performs the unitary transformation and then Bob unwittingly launches the measurement part of the POVM. The difference is that the qubit and ancilla switch roles: The POVM is now on the ancilla, while at Bob's end of the apparatus, this is still a projective measurement on the qubit.

One property of a symmetric attack is that eavesdropping-induced error must be the same in h,v and s,d bases. The orthogonality condition

$$\langle H|V\rangle = 0 \qquad (28.79)$$

must hold because a unitary transformation preserves inner products and $|h\rangle \otimes |A\rangle$ was orthogonal to $|v\rangle \otimes |A\rangle$. Also, if the Hilbert space $\mathcal{H}_A$ of the ancilla has dimensionality 4 or higher, we can have four mutually orthogonal ancilla states [16]:

$$|\varphi_{h,v}\rangle \perp |\theta_{h,v}\rangle. \qquad (28.80)$$

In other words, with the appropriate choice of the entangling operator,

$$\langle \varphi_h|\theta_h\rangle = \langle \varphi_h|\theta_v\rangle = \langle \varphi_v|\theta_h\rangle = \langle \varphi_v|\theta_v\rangle = 0. \qquad (28.81)$$

In this case U will be represented by an 8×8 or higher dimensional matrix according to the dimensionality of the vector space $\mathcal{H}_Q \otimes \mathcal{H}_A$. (Note that although we assumed a four-state ancilla to streamline the argument, generally speaking, a four-dimensional ancilla[20] may not be necessary and in fact in the simplest cases, a

---

20) Language purists would sometimes call a two-state system "two dimensional," referring of course to the dimensionality of the system's Hilbert space.

two-state ancilla will be sufficient.)[21] We will state without a proof that at least in the case of *symmetric attack* (the reader will find details in Ref. [17]), the scalar products $\langle \varphi_h | \varphi_v \rangle$ and $\langle \theta_h | \theta_v \rangle$ are real, and that Eve's information is maximized when they are equal:

$$\langle \varphi_h | \varphi_v \rangle = \langle \theta_h | \theta_v \rangle = \cos \alpha, \tag{28.82}$$

where parameter α is defined by the equation

$$1 - Q = \frac{1 + \langle \theta_h | \theta_h \rangle}{2 - \langle \varphi_h | \varphi_v \rangle + \langle \theta_h | \theta_v \rangle} = \frac{1 + \cos \alpha}{2}. \tag{28.83}$$

Note that α has nothing to do with the 45° angle between qubit polarizations in BB84; rather, it gives the overlap between ancilla states and is entirely determined by the desired error rate Q:

$$\cos \alpha = 1 - 2Q. \tag{28.84}$$

What about Eve's error rate then? After the bases have been announced, Eve must decide if she wants to measure her ancilla states individually or to perform collective measurements on them as a group. Suppose she decides to measure them individually. In that case the whole experiment boils down to individual POVM measurements on the ancillas conducted by Eve and Bob in tandem. For example, suppose Bob announces the h,v basis. There are now two possibilities: Either his photon is h-polarized and Eve's ancilla is then in state $|\varphi_h\rangle$ or $|\theta_v\rangle$ (probabilities $1 - Q$ and Q, respectively) or it is v-polarized and the ancilla is in state $|\varphi_v\rangle$ or $|\theta_h\rangle$ (probabilities $1 - Q$ and Q, respectively). Thus, to guess the bit value measured by Bob, Eve has to distinguish between density matrices:

$$\rho_{B,E}^h = (1 - Q)|\varphi_h\rangle\langle\varphi_h| + Q|\theta_v\rangle\langle\theta_v| \tag{28.85}$$

and

$$\rho_{B,E}^v = (1 - Q)|\varphi_v\rangle\langle\varphi_v| + Q|\theta_h\rangle\langle\theta_h|, \tag{28.86}$$

while in order to guess the bit value sent by Alice, she must distinguish between density matrices:

$$\rho_{A,E}^h = (1 - Q)|\varphi_h\rangle\langle\varphi_h| + Q|\theta_h\rangle\langle\theta_h| \tag{28.87}$$

and

$$\rho_{A,E}^v = (1 - Q)|\varphi_v\rangle\langle\varphi_v| + Q|\theta_v\rangle\langle\theta_v|. \tag{28.88}$$

Of course, Eve is more interested in learning Alice's string, so she takes the latter approach. The problem then reduces to that of discriminating between two mixed

---

21) The optimal dimension of the ancilla's Hilbert space is actually an interesting question. It has been known since 1996 that it need not be greater than 4, at least for the case when Alice's signals are two dimensional (i.e., qubits). There is also a more recent proof of existence of an optimal 2D strategy. The interested reader can find additional details in Ref. [18].

states. This problem is described in detail in Ref. [19–22] and we will only use the final result: Eve's minimum error rate is [17]

$$Q_E = \frac{1 - \sin \alpha}{2} = \frac{1 - \sqrt{1 - (1 - 2Q)^2}}{2}. \tag{28.89}$$

How much information has Eve learned about Alice's string? The answer is given by the binary entropy function:

$$H(A; E) = n - nh(Q_E) = n\{1 + Q_E \log_2 Q_E + (1 - Q_E) \log_2 (1 - Q_E)\}$$

$$= n\left\{1 + \frac{1 - \sqrt{1 - (1 - 2Q)^2}}{2} \log_2 \frac{1 - \sqrt{1 - (1 - 2Q)^2}}{2}\right.$$

$$\left. + \frac{1 + \sqrt{1 - (1 - 2Q)^2}}{2} \log_2 \frac{1 + \sqrt{1 - (1 - 2Q)^2}}{2}\right\}$$

$$= n\left\{\frac{1 - \sqrt{1 - (1 - 2Q)^2}}{2} \log_2 \left(1 - \sqrt{1 - (1 - 2Q)^2}\right)\right.$$

$$\left. + \frac{1 + \sqrt{1 - (1 - 2Q)^2}}{2} \log_2 \left(1 + \sqrt{1 - (1 - 2Q)^2}\right)\right\},$$

and then the critical error rate Q_c is found as before, by solving the transcendental equation

$$n(1 - h(Q)) = H(A; E). \tag{28.90}$$

The graphical solution gives us $Q_c \approx 15\%$ (Figure 28.10), and Eve's POVM strategy is thus more efficient than eavesdropping in a canonical or Breidbart basis.[22]

Looking back, we may have anticipated this result because at $Q_E = 1/2(1 - (\sqrt{2}/2)) \approx 15\%$, we have $\alpha = 45$ and $Q = 1/2(1 - \cos \alpha) = 1/2(1 - (\sqrt{2}/2)) \approx 15\% = Q_E$ and Eve and Bob are equally ignorant about Alice's string, which is the critical error rate condition. Or, to put it differently, with this POVM, Eve can achieve the same accuracy as in a Breidbart basis measurement while keeping Bob's error rate at the same 15% level. Alternatively, by keeping Bob's error rate at $Q = 25\%$, we can attain $Q_E = 1/2(1 - \sin \alpha) = 1/2(1 - \sqrt{1 - (1 - 2Q)^2}) = 1/2(1 - (\sqrt{3}/2)) = 6.7\%$. But wait! We said earlier that the Breidbart projective measurement is the best way to discriminate four pairwise orthogonal states of a qubit! How come then Eve is getting only 6.7% of errors instead of 15%, while Bob is

---

22) To be more exact, the answer is $Q_c = 14.6\%$ as the reader can verify either by calculating $Q_c = 1/2(1 - (\sqrt{2}/2))$ or simply by looking attentively at the graphical solution. But remembering the numbers to a decimal point does not add much to one's understanding and the approximation $Q_c = 15\%$ has already become quite common in the literature.

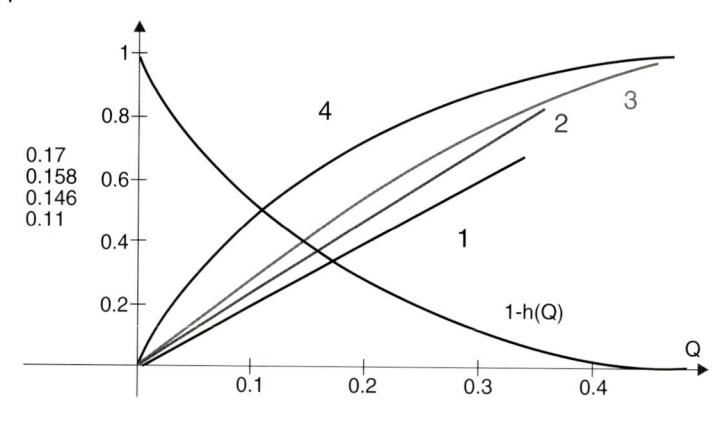

Figure 28.10 Critical error rates in BB84 for different eavesdropping strategies. Critical error rates Q_c are found by soling the corresponding transcendental equation. 1 – Canonical basis measurement discussed at the end of Section 28.5 ($Q_c = 17\%$); 2 – Breidbart basis measurement $Q_c \approx 16\%$; 3 – Eve performs individual measurements on each ancilla, attaining $Q_c \approx 15\%$; 4 – Theoretical limit $Q_c \approx 11\%$ is achieved with a collective attack.

getting the same 25% as in Section 28.2? The factor that is different in the current arrangement is the disclosure of the basis that happens before Eve makes her measurement. So this is not the pure experiment with ambiguous state discrimination envisioned earlier, but rather an experiment in ambiguous state discrimination with additional information. The additional information part comes from the protocol itself. No wonder Eve can optimize performance in that case!

Two limiting cases are of interest. First, when $Q = 0$, Equation 28.89 gives us $Q_E = 1/2$. This happens when Eve leaves Alice's string undisturbed and consequently learns nothing about it. The second case occurs when $Q = 1/2$ and then $Q_E = 0$. This means that Eve simply chose to retain all Alice's bits, sending to Bob some random noise, and then learned all bit values exactly after the bases were announced. Both these cases could obviously occur under the conventional scenario as well and it is in the intermediate error range that the POVM attack shows its full potential.

Eve can learn even more information if she performs collective measurements on her ancillas instead of looking at each ancilla separately [23]. Skipping details of the derivation that can be found in Ref. [15], we only present the final result. Eve's error rate in this case is equal to

$$Q_E = -Q \log_2 Q - (1-Q) \log_2 (1-Q) \equiv h(Q), \qquad (28.91)$$

and this takes us back to the familiar solution $h(Q_c) = 1/2$ or $Q_c \approx 11\%$. Thus, the collective attack not only proves superior to other types of attacks but it also happens to be the optimal strategy for Eve, at least in those cases when Alice prepares her bits independent of one another [24]. Assuming the optimal Shannon random codes are used by Alice and Bob during the reconciliation state, BB84 is theoretically safe as long as $Q_c \leq 11\%$.

Problems

28.1 In the case of deterministic information, Eve knows the values of t bits in X, while the remaining $n - t$ bits remain totally unknown. As a result, her set of candidates for what X might be characterized by a uniform probability distribution $P(x) = \text{const}$. Show that in the case of probabilistic information, her distribution of candidates resists privacy amplification no better than a set of $1/\sum_{x \in X} p^2(x)$ equally weighted candidates.

28.2 Show that unambiguous state discrimination given by $Q_{\min}^{\text{POVM}}(\theta) = \cos \theta$ is a poor eavesdropping strategy.

28.3 Use the same reasoning as in Section 28.2 to establish $\{|u_1\rangle, |u_2\rangle\}$ as the optimal intercept-resend basis.

28.4 Show that when Eve listens in a rectilinear or diagonal basis, B92 protocol gives her less probabilistic information than BB84.

28.5 Derive the critical error rate for B92 assuming Eve performs optimal ambiguous state discrimination measurements.

28.6 Suppose Alice and Bob implementing a phase-shift protocol decided to forgo the interferometry scheme and encode 0 and 1 bits in red and green photons, respectively. Explain why such a scheme would not work.

(*Hint*: A vacuum state of a red pulse can be treated as a superposition of two red photons of opposite phase.)

References

1. Ekert, A. (1991) Quantum cryptography based on Bell's theorem. *Phys. Rev. Lett.*, **67** (6), 661–663.
2. Bennett, C.H., Brassard, G., and Mermin, N.D. (1992) Quantum cryptography without Bell's theorem. *Phys. Rev. Lett.*, **68** (5), 557–559.
3. Bennett, C.H. and Brassard, G. (1984) Quantum cryptography: Public key distribution and coin tossing, Proceedings of IEEE International Conference on Computers, Systems, and Signal Processing, Bangalore, India, p. 175.
4. Bennett, C.H. et al. (1992) Experimental quantum cryptography. *J. Cryptol.*, **5**, 3–28.
5. Bennett, C., Brassard, G., Crepeau, C., and Mauer, U.M. (1995) Generalized privacy amplification. *IEEE Trans. Inf. Technol.*, **41** (6), 1915–1923.
6. Chefles, A. (1998) Unambiguous discrimination between linearly independent quantum states. *Phys. Lett. A*, **239**, 339.
7. Helstrom, C.W. (1976) *Quantum Detection and Estimation Theory*, Academic Press, New York.
8. Shannon, C. (1948) The mathematical theory of communication. *Bell Syst. Tech. J.*, **27**, 379–423, 623–656.
9. Gallager, R.G. (1968) *Information Theory and Reliable Communications*, John Wiley & Sons, Inc., New York.
10. Bennett, C.H., Brassard, G., and Roberts, J.-M. (1988) Privacy amplification by public discussion. *SIAM J. Comp.*, **17** (2), 210–229.
11. Fehr, S. (2010) Quantum cryptography. *Found. Phys.*, **40**, 494–531.
12. Shor, P.W. and Preskill, J. (2000) Simple proof of security of the BB84 quantum key distribution protocol, arXiv:quant-ph/0003004.
13. Csiszar, I. and Korner, J. (1978) Broadcast channels with confidential messages. *IEEE Trans. Inf. Theory*, **24**, 339.
14. Bennett, C.H. (1992) Quantum cryptography using any two nonorthogonal states. *Phys. Rev. Lett.*, **68** (21), 3121–3124.

15 Molotkov, S.N. and Timofeev, A.N. (2007) Explicit attack on the key in quantum cryptography (BB84 protocol) reaching the theoretical error limit $Q_c \approx 11\%$. *JETP Lett.*, **85**, 524.

16 Kronberg, D.N. and Molotkov, S.N. (2010) Quantum scheme for an optimal attack on quantum key distribution protocol BB84. *Bull. Russ. Acad. Sci. Phys.*, **74** (7), 912–918.

17 Fuchs, C.A., Gisin, N., Griffiths, R., Niu, C.-H., and Peres, A. (1997) Optimal eavesdropping in quantum cryptography. I. Information bound and optimal strategy. *Phys. Rev. A*, **56**, 1163.

18 Fuchs, C.A. and Peres, A. (1996) Quantum-state disturbance versus information gain: uncertainty relations for quantum information. *Phys. Rev. A*, **53**, 2038.

19 Kholevo, A.S. (2002) *Introduction to the Quantum Theory of Information* (in Russian), MCCME Publishers, Moscow.

20 Kholevo, A.S. (1998) *Quantum Coding Theorems*, vol. **53** (N6), Russian Math Surveys, pp. 1295–1331.

21 Kholevo, A.S. (1991) Quantum probability and quantum statistics. Probability theory, 8 (in Russian), 5–132, 266–270, 276, Itogi Nauki I Tekhniki, Acad. Nauk SSSR, Vsesoyuz. Inst. Nauchn. I Tekhn. Inform., Moscow.

22 Holevo, A.S. (1979) On capacity of a quantum communications channel, Probl. Peredachi Inf., 3–11.

23 Kholevo, A.S. (1998) Modern mathematical physics: introduction to the quantum theory of information. *Usp. Mat. Nauk*, **53**, 193.

24 Renner, R. (2005) Security of quantum key distribution, PhD dissertation, Swiss Federal Institute of Technology, arXiv: quant-ph/0512258.

Appendix A
Classical Oscillator

Consider equation for a classical harmonic oscillator of mass μ on a spring under external driving force:

$$\left(\mu \frac{d^2}{dt^2} + \beta \frac{d}{dt} + k\right) x(t) = f_0 e^{i\omega t}. \tag{A.1}$$

Here k is the spring constant, β is the dissipative constant responsible for the damping force, and f_0 and ω are the amplitude and frequency, respectively, of the external force (sometimes referred to as the source). It is inhomogeneous linear differential equation of the second order for unknown function $x(t)$. Its *general* solution includes two contributions: the first is a *special* solution of (A.1) and the second one is the general solution of the corresponding *homogeneous* equation, that is, of (A.1) without the source:

$$\left(\mu \frac{d^2}{dt^2} + \beta \frac{d}{dt} + k\right) x(t) = 0. \tag{A.2}$$

Let us start with the special solution of (A.1). It is natural to look for it as a periodic function with the same frequency ω as in the source:

$$x_\omega(t) = x_0 e^{i\omega t}. \tag{A.3}$$

We only need to adjust yet unknown amplitude x_0 so that this expression becomes a solution. Putting it into (A.1) produces the expression on the left:

$$\left(\frac{d^2}{dt^2} + \gamma \frac{d}{dt} + \omega_0^2\right) x_0 e^{i\omega t} = (\omega_0^2 - \omega^2 + i\gamma\omega) x_0 e^{i\omega t}. \tag{A.4}$$

Here

$$\omega_0 \equiv \sqrt{\frac{k}{\mu}} \quad \text{(the proper frequency)} \quad \text{and} \quad \gamma \equiv \frac{\beta}{\mu}. \tag{A.5}$$

Thus, the operation in parentheses

$$\hat{L}(t) \equiv \left(\frac{d^2}{dt^2} + \gamma \frac{d}{dt} + \omega_0^2\right) \tag{A.6}$$

Quantum Mechanics and Quantum Information: A Guide through the Quantum World,
First Edition. Moses Fayngold and Vadim Fayngold.
© 2013 Wiley-VCH Verlag GmbH & Co. KGaA. Published 2013 by Wiley-VCH Verlag GmbH & Co. KGaA.

acting on function (A.3) returns this function multiplied by a numerical (in this case – complex) factor

$$L(\omega) = \omega_0^2 - \omega^2 + i\gamma\omega, \tag{A.7}$$

so that

$$\hat{L}(t)x_\omega(t) = L(\omega)x_\omega(t). \tag{A.8}$$

The capped symbol $\hat{L}$ indicating an operation on a function is called the *operator*. The corresponding function (A.3) and the factor (A.7) are called, respectively, the eigenfunction and the eigenvalue of operator (A.6). The equations of the type (A.8) (but with real eigenvalues) play the fundamental role in the mathematical structure of QM.

Returning to our case and comparing (A.4) with (A.1), we find

$$x_0 = \frac{a_0}{\omega_0^2 - \omega^2 + i\gamma\omega}, \quad a_0 \equiv f_0/\mu. \tag{A.9}$$

Thus,

$$x_\omega(t) = \frac{a_0}{\omega_0^2 - \omega^2 + i\gamma\omega} e^{i\omega t}. \tag{A.10}$$

The imaginary term in the denominator indicates the phase shift (proportional to friction at sufficiently small γ) between the driving force and displacement. At $\omega = \omega_0$ (exact resonance), they are phase shifted by $\pi/2$. In the absence of friction ($\gamma = 0$), the phase shift is π if $\omega > \omega_0$.

If $\gamma = 0$ and $\omega = \omega_0$, Equation A.1 reduces to

$$\left(\frac{d^2}{dt^2} + \omega_0^2\right)x(t) = a_0 e^{i\omega_0 t}. \tag{A.11}$$

The expression (A.10) diverges in this limit, and must be replaced with a non-stationary solution. The existence of the single (second-order) derivative in (A.7) admits, for this special case, such a solution in the form

$$x_s(t) = x_0 t e^{i\omega_0 t}. \tag{A.12}$$

Putting it into (A.7) shows that it satisfies the equation if $x_0 = a_0/2i\omega_0$, so that

$$x_s(t) = \frac{a_0}{2i\omega_0} t e^{i\omega_0 t}. \tag{A.13}$$

The imaginary unit here, again, means the phase shift by $\pi/2$ between the driving force and displacement.

It is interesting to note that there exists also a solution that remains finite and formally stationary even at resonance and in the absence of friction, if the *source* in (A.1) has a complex frequency:

$$\omega \to \tilde{\omega} = \omega + \frac{1}{2}i\gamma, \quad \gamma > 0. \tag{A.14}$$

If we drop the term describing friction in (A.1) and use (A.14) instead, we will come to the equation

$$\left(\frac{d^2}{dt^2} + \omega_0^2\right) x(t) = a_0\, e^{i\tilde{\omega}t}, \tag{A.15}$$

with solution

$$x_{\tilde{\omega}}(t) = \frac{a_0}{\omega_0^2 - \omega^2 - i\gamma\omega + (1/4)\gamma^2} e^{i(\omega+(1/2)i\gamma)t}. \tag{A.16}$$

At $t > 0$, it describes damped oscillations even though the driving force keeps acting, because the force itself is exponentially decreasing with time. In practically important cases, we have $\gamma \ll \omega, \omega_0$, and (A.16) takes the form

$$x_{\tilde{\omega}}(t) = \frac{a_0}{\omega_0^2 - \omega^2 - i\gamma\omega} e^{i(\omega+(1/2)i\gamma)t}. \tag{A.17}$$

This is different from (A.10) only by the sign of imaginary term in the denominator. At exact resonance $\omega = \omega_0$, it reduces to

$$x_s(t) = i\frac{a_0}{\gamma\omega} e^{i(\omega_0+(1/2)i\gamma)t}. \tag{A.18}$$

It remains finite at any finite t.

Now we turn to the solution of *homogeneous* equation (A.2), which describes oscillations without driving force. In a trivial case, there are no oscillations at all. But we are interested in a nontrivial case with oscillations. Physically this happens, for example, after the action of an instantaneous force at a certain moment taken as $t = 0$, after which the oscillator is left alone. So we look for a nonzero solution of (A.2) at $t > 0$. Formally, such solution can be considered as an eigenfunction of (A.6) corresponding to the eigenvalue $L(\omega) = 0$. In this case, ω, instead of being a given frequency of the driving force, becomes the characteristic frequency emerging as the solution of equation

$$L(\omega) = \omega_0^2 - \omega^2 + i\gamma\omega = 0. \tag{A.19}$$

Since (A.19) is a quadratic equation, we have two solutions

$$\omega_1 = \omega_\gamma + i\frac{\gamma}{2} \quad \text{and} \quad \omega_2 = -\omega_\gamma + i\frac{\gamma}{2}, \quad \omega_\gamma \equiv \sqrt{\omega_0^2 - \frac{1}{4}\gamma^2}. \tag{A.20}$$

Accordingly, the general solution of homogeneous equation (A.2) is the sum of the two terms

$$x_{\text{hom}}(t) = x_1 e^{i\omega_1 t} + x_2 e^{i\omega_2 t} = e^{-(1/2)\gamma t}\left(x_1 e^{i\omega_\gamma t} + x_2 e^{-i\omega_\gamma t}\right), \quad t > 0. \tag{A.21}$$

Now we can select (by adjusting the amplitudes x_1 and x_2) a solution satisfying some physical conditions. For instance, we may be interested what happens if an oscillator isolated at $t < 0$ is suddenly hit at $t = 0$ by a huge but a very fleeting force f, so that the impulse $I = \int f\,dt = p$ is finite and accordingly imparts the oscillator with the finite initial speed $v_0 = p/\mu$. Since finite speed cannot produce any finite displacement during the infinitesimally short time interval dt, the initial displacement of the oscillator remains zero. Thus, we have the conditions defining (A.21) at the zero moment $x_{\text{hom}}(0) = 0$ and $\dot{x}_{\text{hom}}(0) = v_0$ (the dot on x means time derivative). Putting here (A.22) gives two simultaneous equations for two amplitudes

$$x_1 + x_2 = 0,$$
$$-\frac{1}{2}\gamma(x_1 + x_2) + i\omega_\gamma(x_1 - x_2) = v_0. \tag{A.22}$$

It follows that $x_2 = -x_1 = -v_0/2i\omega_\gamma$. If $\gamma > 2\omega_0$, then ω_γ determined in (A.20) is imaginary, and (A.21) becomes the sum of two real exponential functions of time. We must retain both because the increasing term is asymptotically counterbalanced by the factor $e^{-(1/2)\gamma t}$, so the resulting displacement will be decreasing with time at large t. The motion is not oscillatory in this case (the medium is too viscous to allow even one oscillation around the center).

Combining all these results, we have the following expression for $x_{\text{hom}}(t)$:

$$x_{\text{hom}}(t) = \begin{cases} 0, & t < 0, \\ \dfrac{v_0}{\omega_\gamma} e^{-(1/2)\gamma t} \sin \omega_\gamma t, & 0 < \gamma < 2\omega_0 \\ -\dfrac{v_0}{|\omega_\gamma|} e^{-(1/2)\gamma t} \sinh(|\omega_\gamma|t), & \gamma > 2\omega_0 \end{cases} \quad t > 0. \tag{A.23}$$

The graph of $x_{\text{hom}}(t)$ is shown in Figure 2.3. We recognize the familiar behavior of a damped oscillator after the action of an instant force. The oscillator is stationary at $t < 0$, and then starts oscillating after being instantly hit at $t = 0$; these oscillations gradually die out due to the damping factor γ. If the medium is so viscous that $\gamma > 2\omega_0$, the oscillator, after being excited by an instant infinite force (as mentioned above, the finite instant force would not produce any momentum), gradually returns to its equilibrium position without vibrations.

A more elegant way to describe the action of an instant force is to use the concept of δ-function: $\delta(t) = 0$ at all times except for $t = 0$, where it is infinite. When representing a force, we can start with a force acting on a system only during a short time Δt around, say, $t = 0$, and imparting to it one unit of momentum, $p = 1$. Let $\Delta t \to 0$, but under condition that the imparted momentum does not change. Then the force f must go to infinity so that $f \Delta t = \int_{-\infty}^{\infty} f(t)dt = 1$, and we can describe it as $f(t) = \delta(t)$. In general case, an instant force producing p units of momentum can be represented as $f(t) = p\delta(t)$ (a detailed description of δ-function is given in Appendix B). In a model of a real physical situation, imparting of a nonzero momentum to an oscillator in the absence of any finite continuous force requires at least one hit, that is, a discontinuous infinite force that is described by δ-function. In this respect, a homogeneous equation (A.2) must be replaced with an equation only "partially homogeneous": there is at least one moment of time, when its source must be nonzero, and even infinite. In other words, (A.23) is, strictly speaking, a special solution to the inhomogeneous equation

$$\left(\mu \frac{d^2}{dt^2} + \beta \frac{d}{dt} + k\right) x(t) = p\delta(t), \tag{A.24}$$

with δ-function as a source. The solution to such equation with normalized source (i.e., with $p = 1$) is called the Green function of the corresponding differential operator. The Green function (A.23) describes natural behavior: the system responds *after* the action of an external agent. The effect follows the cause. Such Green function is called the *retarded Green function*.

Mathematically, one can consider the case when a system responds in advance to action of the cause. The corresponding Green function is called the *advanced Green function* [1].

Reference

1 Fayngold, M. (2008) *Special Relativity and How It Works*, Wiley-VCH Verlag GmbH, Weinheim, pp. 485–487

Appendix B
Delta Function

Dirac's delta function is defined by the following properties:

(a) $\delta(x) = 0, \quad x \neq 0,$

(b) $\int_{-\infty}^{\infty} \delta(x)dx = 1.$ (B.1)

(c) $0 \cdot \delta(0) = 0.$

The shape of this function is obvious from this definition: it is zero everywhere except one point $x = 0$, where it suddenly produces an infinitely sharp spike. It must be infinite at $x = 0$, for otherwise the integral (b) would be zero. However, the infinity at $x = 0$ is a special kind of infinity: it is an infinity that, when multiplied by zero, gives zero.

As a consequence, we can derive some other properties of this function.

(d) $\int_{-\infty}^{\infty} f(x)\delta(x)dx = f(0).$ (B.2)

This property holds for any "well-behaved"[1] function $f(x)$ continuous at $x = 0$ (we require continuity so that some definite value $f(0)$ exists). The proof is not strict but quite intuitive. Only the point $x = 0$ can give any contribution to the integral, because the product is zero everywhere else. At $x = 0$, we have the right to replace $f(x)$ by $f(0)$ because of the continuity of $f(x)$ at this point. Taking $f(0)$ outside the integral and recalling (b), we get our result.[2]

(e) $\delta(-x) = \delta(x).$ (B.3)

---

[1] Well behaved here means finite and continuous.

[2] ome textbooks define the delta function by properties (a) and (B.2). Property (b) then follows as a special case when $f(x) = 1$. This might be a better definition from the mathematical viewpoint. The reader may be dissatisfied with our definition because the proof of (B.2) is not very strict. We chose this definition because it is more illustrative of the shape of the delta function. Also, some textbooks skip property (c) in the definition, but then, as the textbook goes on to prove property (f), one can see that $0 \cdot \delta(0) = 0$ is silently implied.

This is proved by noticing that $\delta(-x)$ satisfies both properties (a) and (b), so it's also a delta function. Property (a) is satisfied as $\delta(-x) = 0$ for all x except zero; property (b) is satisfied because

$$\int_{-\infty}^{\infty} \delta(-x)dx = -\int_{-\infty}^{\infty} \delta(-x)d(-x) = -\int_{\infty}^{-\infty} \delta(u)du = \int_{-\infty}^{\infty} \delta(u)du = 1.$$

(f) $x\delta(x) = 0.$ (B.4)

This is proved by noticing that the product is zero at $x \neq 0$; property (c) finishes the proof.

(g) $\delta(ax) = \dfrac{\delta(x)}{|a|}$ for any real a. (B.5)

We prove it by comparing the integrands on the left and right sides of the following expression:

$$\int_{-\infty}^{\infty} \delta(ax)dx = \frac{1}{a}\int_{-\infty}^{\infty} \delta(ax)d(ax) = \frac{1}{a}\begin{cases} \int_{-\infty}^{\infty} \delta(u)du, & a \geq 0 \\ \int_{\infty}^{-\infty} \delta(u)du, & a < 0 \end{cases}$$

$$= \frac{1}{a}\begin{cases} 1, & a \geq 0 \\ -1, & a < 0 \end{cases} = \frac{1}{|a|} = \frac{1}{|a|} \cdot 1 = \frac{1}{|a|}\int_{-\infty}^{\infty} \delta(x)dx = \int_{-\infty}^{\infty} \frac{1}{|a|}\delta(x)dx.$$

Comparing here the left and right sides yields (B.5).[3]

(h) $\displaystyle\int_{-\infty}^{\infty} f(x')\delta(x' - x)dx' = f(x).$ (B.6)

This is simply a generalization of property (d).

(i) $f(x)\delta(x - x') = f(x')\delta(x - x').$ (B.7)

This is self-obvious; just verify it separately for $x = x'$ and $x \neq x'$.

Dirac's delta function of a vector argument (here we mean a "conventional" Cartesian vector!) is *defined* as the product of delta functions of components:

$$\delta(\vec{a}) = \delta(a_x)\delta(a_y)\delta(a_z). \tag{B.8}$$

Thus, Dirac's delta function of a position vector in rectangular coordinates will be

$$\delta(\vec{r} - \vec{r}') = \delta(x - x')\delta(y - y')\delta(z - z'). \tag{B.8a}$$

If the coordinate system is orthogonal but not Cartesian, an additional multiplier will appear. Thus, in cylindrical coordinates (ρ, ϕ, z) the previous result will appear as

$$\delta(\vec{r} - \vec{r}') = \frac{1}{\rho}\delta(\rho - \rho')\delta(\varphi - \varphi')\delta(z - z'), \tag{B.9}$$

3) $\delta(x - x')$ can be used in this expression just as well due to (e).

and in spherical coordinates (r, θ, ϕ) as

$$\delta(\vec{r} - \vec{r}') = \frac{1}{r^2 \sin \theta} \delta(r - r') \delta(\theta - \theta') \delta(\varphi - \varphi'). \tag{B.10}$$

Kronecker delta can be considered as the analogue of δ-function whose arguments are integer values. It is defined as a function that takes two integer numbers, returning 1 if they are equal, and 0 otherwise. It is denoted by symbol $\delta_{nn'}$, where n and n' are assumed to be any two integers (of course, we can use any letters instead of n and n').

$$\delta_{nn'} = \begin{cases} 1, & n = n', \\ 0, & n \neq n'. \end{cases} \tag{B.11}$$

Now consider some applications. Take the eigenstate expansion of a wave function

$$\Psi(q) = a_1 \psi_1(q) + a_2 \psi_2(q) + \cdots + a_n \psi_n(q) = \sum_{k=1}^{n} a_k \psi_k(q). \tag{B.12}$$

Recall the expression (4.7) for probability amplitudes a_k,

$$a_k = \int_{-\infty}^{\infty} \psi_k^*(q) \Psi(q) dq, \tag{B.13}$$

and substitute (B.12) into (B.13):

$$a_k = \int_{-\infty}^{\infty} \psi_k^*(q) \Psi(q) dq = \int_{-\infty}^{\infty} \psi_k^*(q) \sum_{l=1}^{n} a_l \psi_l(q) dq = \sum_{l=1}^{n} a_l \int_{-\infty}^{\infty} \psi_k^*(q) \psi_l(q) dq. \tag{B.14}$$

The only possible way to make this true for all k is to require that the integral in (B.14) equals 1 for $l = k$ and zero for all other l's. In other words,

$$\int_{-\infty}^{\infty} \psi_k^* \psi_l \, dq = \langle \psi_k | \psi_l \rangle = \delta_{kl}. \tag{B.15}$$

This proves the orthogonality of eigenstates corresponding to *different* eigenvalues. This is precisely the result we predicted in the previous sections purely on philosophical considerations.

Now, put (B.13) into (B.12):

$$\Psi(q) = \sum_{k=1}^{n} \int_{-\infty}^{\infty} (\psi_k^*(q') \Psi(q') dq') \psi_k(q)$$

$$= \int_{-\infty}^{\infty} \sum_{k=1}^{n} \psi_k^*(q') \Psi(q') \psi_k(q) dq' = \int_{-\infty}^{\infty} \Psi(q') \left(\sum_{k=1}^{n} \psi_k^*(q') \psi_k(q) \right) dq'.$$

In view of (B.6), this immediately proves that

$$\sum_{k=1}^{n} \psi_k^*(q') \psi_k(q) = \delta(q - q'). \tag{B.16}$$

We can consider this either as general property of eigenstates or (reading (B.16) from right to left) as a general property of δ-function – it can always be represented as expansion over the set of eigenstates of any differential operator.

Strictly speaking, delta function is not a function, or at least, as physicists say, not well-behaved function[4]; it is really an idealization of function, that is, what we get in the limit. Mathematicians call it a pseudo-function. Sometimes we also call it a *distribution*, which is the mathematical name for an object defined in terms of its integral properties.

Here are some additional useful properties of Dirac's delta function.

(j) $$\int_{-\infty}^{\infty} \delta(a-x)\delta(x-b)dx = \delta(a-b) \quad (a \neq b). \tag{B.17}$$

Just consider one of the delta functions here as $f(x)$. Since $a \neq b$, one function is "well behaved" at the singular point of the other, and therefore we can prove (B.17) by applying (B.6).

(k) $$\delta(x^2 - a^2) = \frac{\delta(x-a) + \delta(x+a)}{|2a|}. \tag{B.18}$$

We just write

$$\delta(x^2 - a^2) = \delta[(x-a)(x+a)] = \begin{cases} \delta[(x-a) \cdot 2a], & x = a, \\ \delta[-2a \cdot (x+a)], & x = -a, \end{cases}$$

and apply property (g).

(l) $$\delta'(x) = -\delta'(-x). \tag{B.19}$$

This is proved by using the formal definition of derivative together with property (e):

$$\delta'(x) = \lim_{\varepsilon \to 0} \frac{\delta(x+\varepsilon) - \delta(x)}{\varepsilon} = \lim_{\varepsilon \to 0} \frac{\delta(-x-\varepsilon) - \delta(-x)}{\varepsilon}$$

$$= -\lim_{\varepsilon \to 0} \frac{\delta(-x) - \delta(-x-\varepsilon)}{\varepsilon} = -\delta'(-x).$$

(Note that at $x \neq 0$ the delta function is uniformly zero, so its derivative is also zero.)

(m) $$x\delta'(x) = -\delta(x). \tag{B.20}$$

The property obviously holds for $x \neq 0$. Now, integrating by parts, we get

$$\int_{-\infty}^{\infty} x\delta'(x)dx = \int_{-\infty}^{\infty} x\,d(\delta(x)) = x\delta(x)\Big|_{-\infty}^{\infty} - \int_{-\infty}^{\infty} \delta(x)dx = 0 - 1 = -1$$

$$= -\int_{-\infty}^{\infty} \delta(x)dx = \int_{-\infty}^{\infty} -\delta(x)dx.$$

---

[4] It violates the classical definition of function that requires that, given (a), the integral (b) *must* be zero. So, as already mentioned, the word "function" is, strictly speaking, a misnomer. We can still use it, but keep in mind that it's only an approximation of a strict definition of a function.

Since the integrals are equal and the integrands are equal everywhere except the point $x=0$, they must be equal at this point as well.

$$\text{(n)} \quad \delta(f(x)) = \sum_i \frac{\delta(x - x_i)}{|f'(x_i)|}, \tag{B.21}$$

where $x = x_i$ are roots of the equation $f(x) = 0$.

For the proof, we recall the identity $dx = df(x)/f'(x)$ and then compare two expressions:

$$\int_{-\infty}^{\infty} \delta(f(x))dx = \int_{x=-\infty}^{x=\infty} \delta(f(x)) \frac{df(x)}{f'(x)} = \int_{f(x)=-\infty}^{f(x)=\infty} \delta(f(x)) \frac{d(f(x))}{|f'(x)|} = \sum_i \frac{1}{|f'(x_i)|}$$

(the modulus appears here because we change from integration variable x to $u \equiv f(x)$; on those intervals where increasing x means increasing u, that is, the derivative is positive, it's fine, while on the intervals where increasing x means decreasing u, the sign changes, and we take the modulus to account for that change). The sum on the right appears because any nonzero result of integration here corresponds to $u \equiv f(x) = 0$, and this equation may have more than one solution. The summation takes care of all of them. In the final step, we write

$$\int_{-\infty}^{\infty} \left(\sum_i \frac{\delta(x-x_i)}{|f(x_i)|} \right) dx = \sum_i \int_{-\infty}^{\infty} \frac{\delta(x-x_i)}{|f'(x_i)|} dx = \sum_i \frac{1}{|f'(x_i)|} \int_{-\infty}^{\infty} \delta(x-x_i) dx$$

$$= \sum_i \frac{1}{|f'(x_i)|}$$

and compare the result with the previous expression.

$$\text{(o)} \quad \int_{-\infty}^{\infty} g(x)\delta(f(x))dx = \sum_i \frac{g(x_i)}{|f'(x_i)|}, \quad \text{where} \quad f(x_i) = 0. \tag{B.22}$$

This follows directly from (n):

$$\int_{-\infty}^{\infty} g(x)\delta(f(x))dx = \int_{-\infty}^{\infty} g(x) \sum_i \frac{\delta(x-x_i)}{|f'(x_i)|} dx = \sum_i \int_{-\infty}^{\infty} g(x) \frac{\delta(x-x_i)}{|f'(x_i)|} dx$$

$$= \sum_i \frac{1}{|f'(x_i)|} g(x_i).$$

$$\text{(p)} \quad \int g(x)\delta[f(x) - a]dx = \frac{g(x_0)}{|f'(x_0)|}, \quad \text{where} \quad f(x_0) = a. \tag{B.23}$$

This is just a special case of (o).

$$\text{(q)} \quad \delta(x) = \frac{1}{2\pi} \int_{-\infty}^{\infty} e^{ikx} dk. \tag{B.24}$$

The proof is not trivial. You can rewrite the *complex form of Fourier's formula* $f(x) = (1/2\pi) \int_{-\infty}^{\infty} \int_{-\infty}^{\infty} f(\xi) e^{i(k\xi-kx)} \, d\xi \, dk$ as $f(x) = \int_{-\infty}^{\infty} ((1/2\pi) \int_{-\infty}^{\infty} e^{i(k\xi-kx)} \, dk) f(\xi) \, d\xi$ and, recalling that $f(x) = \int_{-\infty}^{\infty} f(\xi) \delta(\xi - x) d\xi$, conclude that $\delta(\xi - x) = (1/2\pi) \int_{-\infty}^{\infty} e^{ik(\xi-x)} \, dk$. This is our property (q), after the obvious substitution of variables.

Or, if you don't know the complex form of Fourier's formula, you have to go for a longer derivation, applying the generalized summation [1] and writing the right-hand side as

$$\frac{1}{2\pi} \int_{-\infty}^{\infty} e^{ikx} \, dk = \lim_{\eta \to 0} \frac{1}{2\pi} \int_{-\infty}^{\infty} e^{ikx - \eta k^2} \, dk = \lim_{\eta \to 0} \frac{1}{2\pi} \int_{-\infty}^{\infty} e^{-\eta(k^2 - (ikx/\eta))} \, dk$$

$$= \lim_{\eta \to 0} \frac{1}{2\pi} \int_{-\infty}^{\infty} e^{-\eta[(k-(ix/2\eta))^2 - (ix/2\eta)^2]} \, dk$$

$$= \lim_{\eta \to 0} \frac{1}{2\pi} \int_{-\infty}^{\infty} e^{-\eta(k-(ix/2\eta))^2 + \eta(ix/2\eta)^2} \, dk$$

$$= \frac{1}{2\pi} \lim_{\eta \to 0} \int_{-\infty}^{\infty} e^{-\eta(k-(ix/2\eta))^2} e^{\eta(-x^2/4\eta^2)} \, dk$$

$$= \frac{1}{2\pi} \lim_{\eta \to 0} e^{\eta(-x^2/4\eta^2)} \int_{-\infty}^{\infty} e^{-\eta(k-(ix/2\eta))^2} \, dk$$

$$= \frac{1}{2\pi} \lim_{\eta \to 0} e^{-x^2/4\eta} \int_{-\infty}^{\infty} e^{-\eta(k-(ix/2\eta))^2} \, d\left(k - \frac{ix}{2\eta}\right) = \cdots.$$

Recalling the well-known formula $\int_{-\infty}^{\infty} e^{-u^2} \, du = \sqrt{\pi}$ and its obvious corollary $\int_{-\infty}^{\infty} e^{-au^2} \, du = \sqrt{\pi/a}$, we continue the above expression as

$$\cdots = \frac{1}{2\pi} \lim_{\eta \to 0} e^{-x^2/4\eta} \frac{\sqrt{\pi}}{\sqrt{\eta}} = \lim_{\eta \to 0} \frac{1}{2\sqrt{\pi\eta}} e^{-x^2/4\eta}.$$

We notice that it satisfies the first property of delta function because at $x \neq 0$ the limit goes to zero (exponential function goes to zero faster than $1/\sqrt{\eta}$ goes to infinity!). Then we notice that it also satisfies the second property because

$$\int_{-\infty}^{\infty} \lim_{\eta \to 0} \frac{1}{2\sqrt{\pi\eta}} e^{-x^2/4\eta} \, dx = \lim_{\eta \to 0} \frac{1}{2\sqrt{\pi\eta}} \int_{-\infty}^{\infty} e^{-x^2/4\eta} \, dx = \lim_{\eta \to 0} \frac{1}{2\sqrt{\pi\eta}} \frac{\sqrt{\pi}}{\sqrt{1/4\eta}}$$

$$= \lim_{\eta \to 0} \frac{\sqrt{4\pi\eta}}{2\sqrt{\pi\eta}} = 1$$

and thus conclude it's a delta function.

One of the reasons we included this second, and much longer, proof is for its valuable by-product:

$$(r) \quad \delta(x) = \frac{1}{2\sqrt{\pi}} \lim_{\eta \to 0} \frac{e^{-x^2/4\eta}}{\sqrt{\eta}}, \qquad (B.25)$$

$$(s) \quad \delta(x) = \frac{1}{\pi} \lim_{\eta \to \infty} \frac{\sin \eta x}{x}. \qquad (B.26)$$

We use the result (B.24):

$$\delta(x) = \frac{1}{2\pi} \int_{-\infty}^{\infty} e^{ikx}\, dk = \frac{1}{2\pi ix} e^{ikx}\Big|_{-\infty}^{\infty} = \lim_{\eta \to \infty} \frac{1}{2\pi ix}(e^{i\eta x} - e^{-i\eta x})$$

$$= \lim_{\eta \to \infty} \frac{2i \sin \eta x}{2\pi ix} = \lim_{\eta \to \infty} \frac{\sin \eta x}{\pi x}.$$

Reference

1 Fikhtengol'ts, G.M. (1970) *Course of Differential and Integral Calculus* (in Russian), vol. 2, Nauka, Moscow.

Appendix C
Representation of Observables by Operators

The mean position (an average $\langle \mathbf{r} \rangle$) in a state $\Psi(\mathbf{r})$ is determined by the expression

$$\langle \mathbf{r} \rangle = \int \mathbf{r} \mathcal{P}(\mathbf{r}) d\mathbf{r} = \int \Psi^*(\mathbf{r}) \mathbf{r} \Psi(\mathbf{r}) d\mathbf{r}. \tag{C.1}$$

Consider now the *momentum* measurement in the same state. The state $\Psi(\mathbf{r})$ can be represented as a superposition of monochromatic plane waves (wave packet)

$$\Psi(\mathbf{r}) = \int \Phi(\mathbf{p}) e^{i(\mathbf{p}/\hbar)\mathbf{r}} d\mathbf{p}. \tag{C.2}$$

From the mathematical viewpoint, this is a Fourier expansion of $\Psi(\mathbf{r})$. Physically, the expansion coefficient $\Phi(\mathbf{p})$ is the probability amplitude of finding the particle's *momentum* in the vicinity of $\mathbf{p}$. The set of values $\Phi(\mathbf{p})$ (or just the function $\Phi(\mathbf{p})$) completely specifies the physical state $\Psi(\mathbf{r})$, and vice versa. Once we know $\Psi(\mathbf{r})$, we can uniquely determine $\Phi(\mathbf{p})$:

$$\Phi(\mathbf{p}) = \int \Psi(\mathbf{r}) e^{-i(\mathbf{p}/\hbar)\mathbf{r}} d\mathbf{r} \tag{C.3}$$

(we drop the proportionality coefficient $(2\pi)^{-3}$ since it will not affect the final result). Since $\Phi(\mathbf{p})$ is explicitly expressed in terms of momentum $\mathbf{p}$, it can be called the state function in **p**-representation. It describes *the same* physical state as $\Psi(\mathbf{r})$, but in different representation. In this respect, $\Psi(\mathbf{r})$ and $\Phi(\mathbf{p})$ (when related to one another by (C.2) and (C.3)) are like the two linguistic versions of the *same* story – for example, one in English and one in Russian.

Now, we may be interested in another characteristic of the system – its average momentum $\langle \mathbf{p} \rangle$. Then the **p**-representation of the given state (the state function in the form $\Phi(\mathbf{p})$) provides us with the most natural tool for its calculation. In complete analogy with (C.1), we can write

$$\langle \mathbf{p} \rangle = \int \Phi^*(\mathbf{p}) \mathbf{p} \Phi(\mathbf{p}) d\mathbf{p}. \tag{C.4}$$

But, on the other hand, if both $\Phi(\mathbf{p})$ and $\Psi(\mathbf{r})$ are only different but equivalent representations of the same state, then either of them stores exactly the same (and maximal possible) information about the state. In this case, the rules of QM must have a procedure for calculation of the mean *momentum* from $\Psi(\mathbf{r})$ as well. Such procedure does exist. Indeed, in view of (C.3), the previous equation can be written as

$$\begin{aligned}\langle \mathbf{p} \rangle &= \int \left(\int \Psi^*(\mathbf{r}') e^{i(\mathbf{p}/\hbar)\mathbf{r}'} \, d\mathbf{r}' \, \mathbf{p} \int \Psi(\mathbf{r}) e^{-i(\mathbf{p}/\hbar)\mathbf{r}} \, d\mathbf{r} \right) d\mathbf{p} \\ &= \int \left(\int \Psi^*(\mathbf{r}') e^{i(\mathbf{p}/\hbar)\mathbf{r}'} \, d\mathbf{r}' \int \Psi(\mathbf{r}) \mathbf{p} \, e^{-i(\mathbf{p}/\hbar)\mathbf{r}} \, d\mathbf{r} \right) d\mathbf{p} \\ &= -i\hbar \int \left(\int \Psi^*(\mathbf{r}') e^{i(\mathbf{p}/\hbar)\mathbf{r}'} \, d\mathbf{r}' \int \Psi(\mathbf{r}) \vec{\nabla} \, e^{-i(\mathbf{p}/\hbar)\mathbf{r}} \, d\mathbf{r} \right) d\mathbf{p} \end{aligned} \quad (C.5)$$

and then rearranged as

$$\begin{aligned}\langle \mathbf{p} \rangle &= i\hbar \int d\mathbf{p} \iint d\mathbf{r} \, d\mathbf{r}' \Psi^*(\mathbf{r}') \Psi(\mathbf{r}) \vec{\nabla} \, e^{i\mathbf{k}(\mathbf{r}'-\mathbf{r})} \\ &= i\hbar \iint \Psi^*(\mathbf{r}') \Psi(\mathbf{r}) d\mathbf{r} \, d\mathbf{r}' \, \vec{\nabla} \int e^{i\mathbf{k}(\mathbf{r}'-\mathbf{r})} \, d\mathbf{p}. \end{aligned} \quad (C.6)$$

But, as shown in Appendix B, the last integral in (C.6) is one of the definitions of δ-function. Therefore, integrating by parts, assuming that $\Psi(\mathbf{r})$ and its derivatives approach zero at the boundaries of the region of integration, and using the properties of δ-function, we finally obtain

$$\begin{aligned}\langle \mathbf{p} \rangle &= i\hbar \iint \Psi^*(\mathbf{r}') \Psi(\mathbf{r}) \, d\mathbf{r} \, d\mathbf{r}' \, \vec{\nabla} \delta(\mathbf{r} - \mathbf{r}') \\ &= -i\hbar \iint \Psi^*(\mathbf{r}') \vec{\nabla} \Psi(\mathbf{r}) \delta(\mathbf{r} - \mathbf{r}') d\mathbf{r} \, d\mathbf{r}' = \int \Psi^*(\mathbf{r}) \mathbf{p} \Psi(\mathbf{r}) \, d\mathbf{r}, \end{aligned} \quad (C.7)$$

where $\hat{\mathbf{p}}$ is the operator determined by (3.24).

This approach can be straightforwardly generalized onto situations when we need to find the mean of a function of p^n (with integer n). Thus, in finding $\langle p_x^n \rangle$, the same calculations lead us to expressions analogous to (C.5) with the only distinction that now we have one dimension and p_x^n with arbitrary integer n:

$$\begin{aligned}\langle p_x^n \rangle &= \int \left(\int \Psi^*(x') e^{i(p_x/\hbar)x'} \, dx' \, p_x^n \int \Psi(x) e^{-i(p_x/\hbar)x} \, dx \right) dp_x \\ &= \int \left(\int \Psi^*(x') e^{i(p_x/\hbar)x'} \, dx' \int \Psi(x) p_x^n \, e^{-i(p_x/\hbar)x} \, dx \right) dp_x \\ &= (-i\hbar)^n \int \left(\int \Psi^*(x') e^{i(p_x/\hbar)x'} \, dx' \int \Psi(x) \left(\frac{\partial}{\partial x}\right)^n e^{-i(p_x/\hbar)x} \, dx \right) dp_x. \end{aligned} \quad (C.8)$$

Integrating n times by parts, and using again the properties of δ-function, we obtain

$$\langle p_x^n \rangle = \int \Psi^*(x) \left(-i\hbar \frac{\partial}{\partial x} \right)^n \Psi(x) dx. \quad (C.9)$$

This enables us to find the average $\langle F(p_x) \rangle$, where $\langle F(p_x) \rangle$ is a polynomial:

$$\langle F(p_x) \rangle = \int \Psi^*(x) F\left(-i\hbar \frac{\partial}{\partial x} \right) \Psi(x) dx. \quad (C.10)$$

We can apply this result also to an arbitrary analytic function F that can be expanded into the Taylor series. The generalization onto three dimensions is technically involved, but conceptually trivial – it just requires to increase the number of integrations.

Reversing the argument, we can find the expectation value of arbitrary analytic function of x when the quantum state is given in the momentum representation $\Phi(p_x)$:

$$\langle F(x) \rangle = \int \Phi^*(p_x) \left(i\hbar \frac{\partial}{\partial p_x} \right)^n \Phi(p_x) dp_x. \tag{C.11}$$

This, too, can be easily written in the vector form.

Appendix D
Elements of Matrix Algebra

This appendix helps to elucidate the meaning of Dirac's notation. Suppose we have a set of states such that being found in one state of the set totally excludes the possibility to be found next time in some other state. Such a set forms an orthogonal basis and it only remains to normalize it if we want to follow the convention. Now every possible state can be expressed in terms of the basis states, and we can write the coefficients (i.e., probability amplitudes) as a column matrix. This is, of course, a different representation of the wave function – representation corresponding to the observable whose eigenstates have formed the basis, but we know that it's just another "face" of the same physical entity. We call that entity (i.e., that state) the ket vector. Now we take complex conjugates of these coefficients, write them in a row, and call the new state "the bra vector." What is the physical meaning of this matrix row? If we started with the wave function

$$\psi = \sum_l c_l \psi_l, \tag{D.1}$$

which is our ket, then the corresponding bra is $\psi^* = \sum_l c_l^* \psi_l^*$. If we multiply the row (c_i^*) by the column (c_l), we get a 1×1 matrix, that is, a single number. This number is equal to the sum $\sum |c_l|^2$ of probabilities of each eigenstate, which, of course, is equal to 1. (Note, however, that it was the orthogonality of the basis that allows us to interpret the sum of squares as the total probability!) Of course, we have to maintain the correct order, putting the bra to the left of the ket; otherwise, the result will be $n \times n$ matrix instead of a number.[1] This is why the bra–ket order is required in Dirac's notation.

The dependence of matrix product on the order of multiplied matrices gives rather a figurative explanation why the product $\langle \phi | \varphi \rangle$ (bra–ket) is a number while the product $|\phi\rangle\langle\varphi|$ (ket–bra) is an operator.

The bra vector is more than just the complex conjugate of the ket. The complex conjugation is not enough: the column has to be changed to a row, which is achieved by transposition. One can readily see that the ket vector is changed to a bra and vice

---

[1] We could still obtain the total probability by taking the *spur* (or the *trace*) – the sum of diagonal elements of such a matrix.

Quantum Mechanics and Quantum Information: A Guide through the Quantum World,
First Edition. Moses Fayngold and Vadim Fayngold.
© 2013 Wiley-VCH Verlag GmbH & Co. KGaA. Published 2013 by Wiley-VCH Verlag GmbH & Co. KGaA.

versa by taking the adjoint:

$$\begin{pmatrix} c_1 \\ \vdots \\ c_n \end{pmatrix}^\dagger = \begin{pmatrix} c_1^* & \cdots & c_n^* \end{pmatrix}. \tag{D.2}$$

Thus, we have the following situation. When we study the wave function ψ in analytic form, for instance, when we are looking for the probability of finding the particle somewhere and ψ appears under the integral sign, as in $\int \psi^*(x)\psi(x)dx$, we can take the complex conjugate ψ^* and think of it as the bra vector (although the term itself is reserved for Dirac's notation). But strictly speaking, one should write $|\psi\rangle^\dagger = \langle\psi|$. Whenever you imply that ψ is to be regarded as a matrix, only the latter form is acceptable.

If we have two wave functions, $\Psi = c_1\psi_1 + c_2\psi_2 + \cdots + c_n\psi_n$ and $\Phi = d_1\psi_1 + d_2\psi_2 + \cdots + d_n\psi_n$, then

$$\langle\Psi|\Phi\rangle = \begin{pmatrix} c_1^* & \cdots & c_n^* \end{pmatrix} \begin{pmatrix} d_1 \\ \vdots \\ d_n \end{pmatrix} = c_1^* d_1 + \cdots + c_n^* d_n, \tag{D.3}$$

while

$$\langle\Phi|\Psi\rangle = \begin{pmatrix} d_1^* & \cdots & d_n^* \end{pmatrix} \begin{pmatrix} c_1 \\ \vdots \\ c_n \end{pmatrix} = c_1 d_1^* + \cdots + c_n d_n^* = \langle\Psi|\Phi\rangle^*, \tag{D.4}$$

and you now see why the scalar product in the Hilbert space depends on the order of multipliers the way it does. So, while $\langle\Psi|\Phi\rangle^* = \langle\Phi|\Psi\rangle$ is always correct, keep in mind that conjugation here relates to *whole scalar product*! If you want to apply it separately to each multiplier, retaining their initial order, you will get an expression

$$\langle\Psi|\Phi\rangle^* = \langle\Psi|^*|\Phi\rangle^* = |\Psi\rangle\langle\Phi| = \begin{pmatrix} c_1 \\ \vdots \\ c_n \end{pmatrix} \begin{pmatrix} d_1^* & \cdots & d_n^* \end{pmatrix} = \begin{pmatrix} c_1 d_1^* & \cdots & c_n d_n^* \\ \vdots & \ddots & \vdots \\ c_n d_1^* & \cdots & c_n d_n^* \end{pmatrix} \tag{D.5}$$

with a totally different physical meaning. In contrast to matrix algebra, where complex conjugation does not change the order: $(AB)^* = A^* B^*$, we must change it "by hand" for the inner product (D.4) in the Hilbert space.

This is why we use transposition to get "the real bra vector." This operation has the same effect as reversing the order of multipliers. Indeed, one can show that $(AB)^T = B^T A^T$ and therefore

$$(AB)^\dagger = (B^T A^T)^* = B^\dagger A^\dagger, \tag{D.6}$$

so that

$$\langle\Psi|\Phi\rangle^\dagger = |\Phi\rangle^\dagger \langle\Psi|^\dagger = \langle\Phi||\Psi\rangle = \langle\Phi|\Psi\rangle = \begin{pmatrix} d_1^* & \cdots & d_n^* \end{pmatrix} \begin{pmatrix} c_1 \\ \vdots \\ c_n \end{pmatrix}$$
$$= c_1 d_1^* + \cdots + c_n d_n^*. \tag{D.7}$$

This is just what we should expect, so we have

$$\langle\Phi|\Psi\rangle^\dagger = \langle\Psi|\Phi\rangle, \tag{D.8}$$

where taking the adjoint of the scalar product can now be "decomposed" into smaller operations in full agreement with the rules of matrix algebra.

If there is a constant multiplier, it will not be affected by transposition, but will be affected by complex conjugation, so $|\lambda\Psi\rangle^\dagger = \lambda^*\langle\Psi|$. In our convention, when we see something like $\langle\lambda\Psi|$, we read it as "bra vector corresponding to the ket $|\lambda\Psi\rangle$" and not as "bra corresponding to $|\Psi\rangle$ that was multiplied by λ." Therefore, $\langle\lambda\Psi| = |\lambda\Psi\rangle^\dagger = \lambda^*|\Psi\rangle^\dagger = \lambda^*\langle\Psi|$. Hence,

$$\langle\Phi|\lambda\Psi\rangle^\dagger = \lambda\langle\Phi|\Psi\rangle^\dagger = \lambda\langle\Psi|\Phi\rangle, \qquad \langle\lambda\phi|\varphi\rangle^\dagger = \lambda^*\langle\phi|\varphi\rangle^\dagger = \lambda^*\langle\varphi|\phi\rangle$$

as we pointed out when discussing scalar products in $\mathcal{H}$.

Any Hermitian matrix C can be sandwiched between a column matrix X and an adjoint of X – a row matrix $X^\dagger$:

$$X^\dagger C X = \begin{pmatrix} x_1^* & x_2^* & \cdots & x_N^* \end{pmatrix} \begin{pmatrix} c_{11} & c_{12} & \cdots & c_{1N} \\ c_{21} & c_{22} & \cdots & c_{2N} \\ \cdots & \cdots & \cdots & \cdots \\ c_{N1} & c_{N2} & \cdots & c_{NN} \end{pmatrix} \begin{pmatrix} x_1 \\ x_2 \\ \cdots \\ x_N \end{pmatrix} = \sum_{i,j=1}^N x_i^* c_{ij} x_j \tag{D.9}$$

The sum on the right is called *quadratic form*. In particular, if instead of an arbitrary C we place in the middle a *diagonal* matrix Λ, we get a *diagonal* quadratic form containing the squares only:

$$X^\dagger \Lambda X = \begin{pmatrix} x_1^* & x_2^* & \cdots & x_N^* \end{pmatrix} \begin{pmatrix} \lambda_{11} & 0 & \cdots & 0 \\ 0 & \lambda_{22} & \cdots & 0 \\ \cdots & \cdots & \cdots & \cdots \\ 0 & 0 & \cdots & \lambda_{NN} \end{pmatrix} \begin{pmatrix} x_1 \\ x_2 \\ \cdots \\ x_N \end{pmatrix} = \sum_{i=1}^N \lambda_{ii}|x_i|^2 \tag{D.10}$$

A change from an old to a new basis can be accomplished by a *similarity transformation* that is defined by the transformation (transition) matrix T and transforms state vectors and operator matrices as follows:

$$\begin{cases} X' = T^\dagger X \\ C' = T^\dagger C T \end{cases} \tag{D.11}$$

One can see that this transformation works as intended by comparing state vector $Y = CX$ in the old and new basis: $Y' = C'X' = (T^\dagger CT)(T^\dagger X) = T^\dagger CX = T^\dagger Y = Y'$.

Matrix $T^\dagger$ takes X to X' and is therefore called a transformation (transition) matrix from the *old* to the *new* basis. Then T must be the transition matrix from the *new* to the *old* basis, as seen from the fact that $TX' = T(T^\dagger X) = X$. There exists an easy rule to obtain these transition matrices, namely: A transition matrix from the old to the new basis is constructed by writing the *orts* (basis vectors) of the old basis as column vectors in the new basis and putting the columns next to each other; a transition matrix from the new to the old basis is constructed similarly by writing the new *orts* as column matrices in the old basis. In most problems it is easier to write new orts in the old basis, which then will give matrix T. Then vectors in the new representation will be derived using $T^\dagger$. Note that in the Hilbert space, all transition matrices are unitary.

Some older textbooks use an alternative notation where our T is denoted as $T^\dagger$ and our $T^\dagger$ as T. In that case the similarity transformation will be written as $X' = TX$ and $C' = TCT^\dagger$.

For any given C there always exists a suitable similarity transformation that will *diagonalize* it, transforming it into a diagonal matrix $C' \equiv \Lambda = T^\dagger CT$.

One can then sandwich Λ between T and $T^\dagger$ to write $T\Lambda T^\dagger = TT^\dagger CTT^\dagger = C$. Now let vector X_k have all zero components, except component x_k, which is 1. Then it is easy to see that $\Lambda X_k = \lambda_{kk} X_k$, that is, X_k is an eigenvector and λ_{kk} is its eigenvalue. We arrive at the following important result: a Hermitian matrix C can always be written in an alternative form,

$$C = T\Lambda T^\dagger \tag{D.12}$$

where Λ is a diagonal matrix with eigenvalues along the main diagonal and T is the unitary matrix of a similarity transformation. This result is called the *spectral theorem for matrices*.

Appendix E
Eigenfunctions and Eigenvalues of the Orbital Angular Momentum Operator

Here we have to find the eigenfunctions of the operator $\hat{L}^2$. These eigenfunctions depend on the angles θ and φ. Thus, we need to solve the equation $\hat{L}^2 Y(\theta, \varphi) = L^2 Y(\theta, \varphi)$, or

$$\left[\frac{1}{\sin\theta} \frac{\partial}{\partial\theta}\left(\sin\theta \frac{\partial}{\partial\theta}\right) + \frac{1}{\sin^2\theta} \frac{\partial^2}{\partial\varphi^2} \right] Y(\theta, \varphi) = \Lambda^2 Y(\theta, \varphi), \quad \Lambda \equiv \frac{L}{\hbar}. \tag{E.1}$$

The solution is subject to the usual conditions: it must be finite, continuous, and single-valued within the regions $0 \leq \theta \leq \pi$ and $0 \leq \varphi \leq 2\pi$. This is a well-known problem in mathematical physics. The first step in its solution is separation of variables. Namely, we seek for a solution in the form

$$Y(\theta, \varphi) = \Theta(\theta)\Phi(\varphi). \tag{E.2}$$

Putting this into (E.1) yields

$$\frac{\Phi(\varphi)}{\sin\theta} \frac{\partial}{\partial\theta}\left(\sin\theta \frac{\partial\Theta(\theta)}{\partial\theta}\right) + \frac{\Theta(\theta)}{\sin^2\theta} \frac{\partial^2 \Phi(\varphi)}{\partial\varphi^2} = \Lambda^2 \Theta(\theta)\Phi(\varphi), \tag{E.3}$$

or, after dividing through by $Y(\theta, \varphi)$,

$$\frac{(1/\sin\theta)[\partial(\sin\theta(\partial\Theta(\theta)/\partial\theta))/\partial\theta]}{\Theta(\theta)} + \frac{(1/\sin^2\theta)(\partial^2 \Phi(\varphi)/\partial\varphi^2)}{\Phi(\theta)} = \Lambda^2. \tag{E.4}$$

We notice that if

$$\frac{\partial^2 \Phi}{\partial\varphi^2} = -m^2 \Phi, \tag{E.5}$$

then Equation E.4 reduces to

$$\frac{1}{\sin\theta} \frac{\partial}{\partial\theta}\left(\sin\theta \frac{\partial\Theta}{\partial\theta}\right) - \frac{m^2}{\sin^2\theta}\Theta + \Lambda^2\Theta = 0. \tag{E.6}$$

We obtained two separate differential equations for two functions, each depending on only one variable. Turn first to the simplest one – (E.5). Its obvious solution is

$$\Phi_m(\varphi) = \Phi_0 e^{\pm im\varphi}, \tag{E.7}$$

with some constant Φ_0. The number m here cannot be arbitrary. Function $\Phi_m(\varphi)$ will be single-valued only for integer m:

$$m = 0, \pm 1, \pm 2, \ldots . \tag{E.8}$$

This result has a very straightforward physical interpretation: (E.7), being a periodic function of φ, describes a specific kind of de Broglie wave moving in a circle around the center. Such motion can occur only if the integral number of the wavelengths fit into the circle. This implies quantization of angular momentum, $L_z = m\hbar$ – the result (2.36) in Section 2.4. Here we obtained it rigorously using the formalism of QM.

Now we turn to Equation E.6. As you look at it, it invites for substitution

$$\cos\theta = \xi, \quad -1 \leq \xi \leq 1, \quad d\xi = -\sin\theta \, d\theta, \tag{E.9}$$

after which it takes the form

$$(1-\xi^2)\Theta'' - 2\xi\Theta' + \left(\Lambda - \frac{m^2}{1-\xi^2}\right)\Theta = 0. \tag{E.10}$$

Here we utilize the conventional notation $d\Theta/d\xi \equiv \Theta'$ (since we are left with only one variable here, we switch from partial derivative to the ordinary one).

The values $\xi = \pm 1$ are points of singularity in (E.10). Therefore, we first consider the behavior of the solution in the vicinity of these points. Starting with $\xi = 1$, switch temporarily to a new variable $s = \xi - 1$. Then Equation E.10 becomes

$$\Theta'' + \frac{2s+1}{s\,s+2}\Theta' - \left[\frac{\Lambda^2}{s(s+2)} + \frac{m^2}{s^2(s+2)^2}\right]\Theta = 0. \tag{E.11}$$

All coefficients here are rational functions of s, so it is natural to look for solution as a power series with respect to s:

$$\Theta(s) = s^q Q(s), \quad Q(s) = \sum_{j=0}^{\infty} a_j s^j. \tag{E.12}$$

At $\xi \to 1$, we have $s \to 0$, $Q(s) \to a_0$, and $\Theta(s) \to a_0 s^q$. Here the terms smaller than s^{q-2} can be neglected and Equation E.11 simplifies to

$$\left[q^2 - (m/2)^2\right] a_0 s^{q-2} = 0. \tag{E.13}$$

The equation is satisfied if $q = \pm m/2$. The same result will be obtained by expanding the solution about the point $\xi = -1$. But since we are looking for the finite solutions, we must discard the possible negative powers of small s in (E.12); in other words, we must leave only positive q in $q = \pm m/2$. Therefore, we take $q = |m|/2$, so that $\Theta(s)$ can be written as

$$\Theta(\xi) \approx \begin{cases} (1-\xi)^{|m|/2} Q(s), & \xi \to 1, \\ (1+\xi)^{|m|/2} Q(s), & \xi \to -1. \end{cases} \tag{E.14}$$

This can be written in a single equation embracing both cases: $\Theta(\xi) = (1-\xi^2)^{|m|/2} Q(s)$. Indeed, when $\xi \to 1$, we have $1 - \xi^2 \approx 2(1-\xi)$, and when

$\xi \to -1$, we have $1 - \xi^2 \approx 2(1 + \xi)$, so in either case the result is just multiplied by 2 as compared with (E.14). But the solutions of a homogeneous differential equation are defined up to an arbitrary numerical factor, which in the end will be absorbed in the normalization procedure, anyway. Finally, since $Q(s)$ is a power series, we can as well write it in terms of the original variable ξ – it will be the similar power series, only with different coefficients:

$$Q(s) = \sum_{j=0}^{\infty} a_j s^j \Rightarrow P(\xi) = \sum_{j=0}^{\infty} b_j \xi^j. \tag{E.15}$$

Thus, we can write (E.14) as

$$\Theta(\xi) = (1 - \xi^2)^{|m|/2} P(\xi). \tag{E.16}$$

Putting this into (E.10) gives

$$(1 - \xi^2) P'' - 2(|m| + 1)\xi P' + (\Lambda^2 - |m| - m^2) P = 0. \tag{E.17}$$

Putting here $P(\xi)$ from (E.15) and collecting equal powers of ξ, we obtain the recurrence relation for coefficients b_j:

$$(j+1)(j+2) b_{j+2} = \left[j(j+1) + 2(|m|+1)j - \Lambda^2 + |m| + m^2 \right] b_j. \tag{E.18}$$

This relation is of paramount importance for the final solution. It provides us with the condition necessary to terminate the series at some $j = k$. Such termination is required for the solution to be a finite and continuous function of ξ in the region given in (E.9) and especially at $\xi = \pm 1$. As seen from (E.18), the series can terminate if the expression in the brackets becomes zero for some $j = k$:

$$k(k+1) + 2(|m|+1)k - \Lambda^2 + |m| + m^2 = 0. \tag{E.19}$$

This can happen only if

$$\Lambda^2 = k(k+1) + 2(|m|+1)k + |m| + m^2 = (k + |m|)(k + |m| + 1). \tag{E.20}$$

Now denote

$$k + |m| \equiv l. \tag{E.21}$$

Since k and $|m|$ are both integers, l is another integer, and (E.20) reduces to

$$\Lambda^2 = l(l+1), \quad l = 0, 1, 2, 3, \ldots. \tag{E.22}$$

It is also seen from definition (E.21) that for each l the $|m|$ can change only within the range $[0, l]$, that is,

$$|m| = 0, 1, 2, \ldots, l-1, l. \tag{E.23}$$

This completes the search for the physical solutions of Equation E.1. It turns out that such solutions form the set of eigenfunctions of $\hat{L}^2$ with eigenvalues determined by Equations E.22 and E.23. The function $\Theta(\xi)$ corresponding to the numbers l and m is usually denoted as $P_l^{|m|}(\xi)$, so that the solution is

$$\Theta(\xi) = P_l^{|m|}(\xi), \quad \xi = \cos\theta. \tag{E.24}$$

Differentiating Equation E.21 for $Q(\xi)$ results in $|m|$ being replaced by $|m|+1$. Therefore, denoting the solution for $|m|=0$ as $P_l(\xi)$, we have

$$P_l^{|m|}(\xi) = \left(1 - \xi^2\right)^{|m|/2} \frac{d^{|m|}}{d\xi^{|m|}} P_l(\xi). \tag{E.25}$$

The function $P_l(\xi)$ is known as the Legendre polynomial. It is normalized so that

$$P_l(\xi = 1) = 1. \tag{E.26}$$

This leads to

$$\int_{-1}^{1} P_l(\xi) P_{l'}(\xi) d\xi = \int_{0}^{\pi} P_l(\cos\theta) P_{l'}(\cos\theta) \sin\theta \, d\theta = \frac{2}{2l+1} \delta_{ll'}. \tag{E.27}$$

The function $P_l^{|m|}(\xi)$ is called the associated Legendre polynomial of the mth order.

The coefficients defining $P_l(\xi)$ are obtained from Equation E.24 at $m = 0$, with Λ determined by (E.28):

$$b_{j+2} = \frac{j(j+1) - l(l+1)}{(j+1)(j+2)} b_j. \tag{E.28}$$

When we start with b_0, $b_1 = 0$, the polynomial $P_l(\xi)$ will contain only even powers of ξ. If we set $b_0 = 0$ and $b_1 \neq 0$, it will contain only the odd powers of ξ. Whatever we choose, the condition (E.26) must be satisfied. Once this is done, we can use (E.28) to find all coefficients of $P_l(\xi)$. The resulting polynomial can also be derived from the formula

$$P_l(\xi) = \frac{1}{2^l \cdot l!} \frac{d^l}{d\xi^l} \left(\xi^2 - 1\right)^l. \tag{E.29}$$

The function $G(\xi) \equiv \xi^2 - 1$ is called the *generating function* or just the *generator* of the Legendre polynomials. The name is self-explanatory. Combining (E.29) with (E.25) allows one to obtain the associated polynomials from the generator as well:

$$P_l^m(\xi) = \frac{1}{2^l \cdot l!} \left(1 - \xi^2\right)^{\frac{|m|}{2}} \frac{d^{l+|m|}}{d\xi^{l+|m|}} \left(\xi^2 - 1\right)^l. \tag{E.30}$$

Combining (E.30) with (E.2), we can write the found expression for the *normalized* eigenfunction of $\hat{L}^2$ as

$$Y_{lm}(\theta, \varphi) = \sqrt{\frac{(2l+1)(l-|m|)!}{4\pi(l+|m|)!}} P_l^{|m|}(\cos\theta) e^{im\varphi}. \tag{E.31}$$

Appendix F
Hermite Polynomials

The time-independent Schrödinger equation for a quantum oscillator can be brought to the form

$$\psi''(\xi) = (\eta - \xi^2)\psi(\xi), \tag{F.1}$$

where

$$\xi = \sqrt{\frac{\mu\omega_0}{\hbar}}x, \qquad \eta = 2\frac{E}{\hbar\omega_0}. \tag{F.2}$$

We have to find the solutions of this equation, which asymptotically go to zero at $|\xi| \to \infty$. To this end, start with the trial solution

$$\psi(\xi) = e^{-(1/2)\xi^2} H(\xi). \tag{F.3}$$

Putting this into (F.1) gives the equation for $H(\xi)$:

$$H''(\xi) - 2\xi H'(\xi) + (\eta - 1)H(\xi) = 0. \tag{F.4}$$

Since one of the coefficients here is a linear function of ξ, it is natural to represent $H(\xi)$ as a Taylor expansion

$$H(\xi) = \sum_{j=0}^{\infty} a_j \xi^j. \tag{F.5}$$

Putting this into (F.4) and collecting the terms with equal powers of ξ gives the recurrent equation for coefficients a_j:

$$a_{j+2} = \frac{2j + 1 - \eta}{(j+2)(j+1)} a_j. \tag{F.6}$$

In the simplest case, the coefficients a_j are such that the infinite series (F.5) is truncated down to polynomials of nth power. Often, the simplest possible solution of an equation is precisely the one that describes a studied physical phenomenon. And this is the case in our situation.

Quantum Mechanics and Quantum Information: A Guide through the Quantum World,
First Edition. Moses Fayngold and Vadim Fayngold.
© 2013 Wiley-VCH Verlag GmbH & Co. KGaA. Published 2013 by Wiley-VCH Verlag GmbH & Co. KGaA.

The infinite series (F.5) with coefficients a_j determined by (F.6) gets truncated at some $j = n$ if

$$\eta = 2n+1, \quad n = 0, 1, 2, \ldots. \tag{F.7}$$

This imposes on η the requirement selecting a discrete set of allowed energies. In view of (F.2), this set is

$$E_n = \left(n + \frac{1}{2}\right)\hbar\omega_0. \tag{F.8}$$

Setting in Equation F.4 $\eta = 2n+1$ gives

$$H_n''(\xi) - 2\xi H_n'(\xi) + 2n H_n(\xi) = 0. \tag{F.9}$$

The introduced subscript in $H(\xi) \to H_n(\xi)$ labels each distinct solution of (F.9), corresponding to its respective integer n. These solutions are called the *Hermite polynomials*.

One can check by direct inspection that these polynomials can be obtained as

$$H_n(\xi) = A_n e^{\xi^2} \frac{d^n}{d\xi^n} e^{-\xi^2}, \tag{F.10}$$

where A_n is an arbitrary constant. Its conventional value, used in applied mathematics, is $A_n = (-1)^n$.

In describing a QM oscillator, A_n must be determined by the normalization condition for a state function $\psi(\xi) \to \psi_n(\xi)$ in (F.3). This condition gives $A_n = \sqrt{2^n n! \sqrt{\pi}}$, so that $\psi_n(\xi)$ can be written as

$$\psi_n(\xi) = e^{-(1/2)\xi^2} H_n(\xi), \quad H_n(\xi) = \frac{(-1)^n}{\sqrt{2^n n! \sqrt{\pi}}} e^{\xi^2} \frac{d^n}{d\xi^n} e^{-\xi^2}. \tag{F.11}$$

After we perform all differentiations in the second of Equation F.11, the two exponential factors cancel each other in each term, and we are left with pure polynomial of ξ. But note that, while $H_n(\xi)$ are polynomials, the corresponding state functions $\psi_n(\xi)$ are not.

Being the normalized eigenfunctions of a Hermitian operator, all state functions $\psi_n(\xi)$ form an orthonormal set, $\int \psi_n(\xi)\psi_{n'}(\xi)d\xi = \delta_{nn'}$. An arbitrary function $\Psi(\xi)$ can be represented as a series

$$\Psi(\xi) = \sum_n b_n \psi_n(\xi), \tag{F.12}$$

with the probability amplitudes

$$b_n = \int \Psi(\xi)\psi_n^*(\xi)d\xi \tag{F.13}$$

for finding the corresponding energy eigenvalues E_n in energy measurement from state (F.12).

Let us now consider ξ as a complex variable z. Then we can apply the Cauchy formula to write

$$\frac{d^n}{d\xi^n} e^{-\xi^2} = \frac{n!}{2\pi i} \oint \frac{e^{-z^2}}{(z-\xi)^{n+1}} dz, \qquad (F.14)$$

where the line integral is taken over a closed loop around point ξ. This allows us to represent the (unnormalized) Hermite polynomials as

$$e^{-\xi^2} H_n(\xi) = (-1)^n \frac{n!}{2\pi i} \oint \frac{e^{-z^2}}{(z-\xi)^{n+1}} dz. \qquad (F.15)$$

A new integration variable $\tau = \xi - z$ brings this to the form

$$H_n(\xi) = \frac{n!}{2\pi i} \oint \frac{e^{-\tau^2 + 2\xi\tau}}{\tau^{n+1}} d\tau. \qquad (F.16)$$

This equation implies that

$$e^{-\tau^2 + 2\xi\tau} = \sum_{j=0}^{\infty} \frac{1}{j!} H_j(\xi) \tau^j. \qquad (F.17)$$

Indeed, putting this into (F.16) and using (F.10) with $A_j = (-1)^j$, we see that in the resulting sum only the term with $j = n$ survives and turns the equation into identity. The relation (F.17) shows that the expression

$$g(\tau, \xi) \equiv e^{-\tau^2 + 2\xi\tau} \qquad (F.18)$$

is the generating function for $H_j(\xi)$.

Differentiating (F.17) with respect to τ yields

$$2(\xi - \tau) e^{-\tau^2 + 2\xi\tau} = \sum_{j=0}^{\infty} \frac{1}{(j-1)!} H_j(\xi) \tau^{j-1}. \qquad (F.19)$$

Eliminating $g(\tau, \xi)$ from (F.17) and (F.19) and collecting the terms with equal powers of τ in the resulting equation, we obtain the important recurrent relation between the Hermite polynomials:

$$\xi H_n(\xi) = n H_{n-1}(\xi) + \frac{1}{2} H_{n+1}(\xi). \qquad (F.20)$$

Here we have replaced $j \to n$ for consistency with the initial labeling of H_n starting from (F.9). Now we can convert this into relationship between the corresponding eigenfunctions $\psi_n(\xi)$. Just multiply (F.20) through by $e^{-\xi^2/2}$, and multiply and divide each term by its respective normalizing factor according to (F.11). Then, after dropping the common factor, we obtain

$$\xi \psi_n(\xi) = \sqrt{\frac{n}{2}} \psi_{n-1}(\xi) + \sqrt{\frac{n+1}{2}} \psi_{n+1}(\xi). \qquad (F.21)$$

This expression is of crucial help when we need to calculate the x-matrix in the energy basis. Simply multiply (F.21) by $\psi_m(\xi)$ (no need for complex conjugation here since all $\psi_m(\xi)$ are real) and integrate over ξ. In view of orthonormality, we will get the result without performing actual integration:

$$\xi_{mn} \equiv \int \psi_m(\xi)\xi\psi_n(\xi)d\xi = \sqrt{\frac{n}{2}}\delta_{m,n-1} + \sqrt{\frac{n+1}{2}}\delta_{m,n+1}. \qquad (F.22)$$

In a similar way, we can obtain the recurrent relations for $\xi^l\psi_n(\xi)$ and use them for calculating the matrix elements of ξ^l with arbitrary integer l.

Appendix G
Solutions of the Radial Schrödinger Equation in Free Space

The radial Schrödinger equation (8.39) in a free space for a state with definite quantum number l is

$$G_l'' + \left[k^2 - \frac{l(l+1)}{r^2}\right] G_l = 0. \tag{G.1}$$

Here $G_l(r) = rR_l(r)$, primes stand for derivatives with respect to r, and subscript l shows that $G_l(r)$ is a solution depending on quantum number l.

In the case $l = 0$, we have

$$G_0'' + k^2 G_0 = 0, \tag{G.2}$$

and the solution finite at the origin is

$$G_0(r) = A \sin kr, \qquad R_0(r) = A \frac{\sin kr}{r} \tag{G.3}$$

(the factor $2i$ is absorbed by coefficient A). Note that this is precisely the radial part of result (8.45), which is quite natural in view of the spherical symmetry of the state with $l = 0$. So, in hindsight we know that the spherically symmetric superposition (8.44) of plane waves describes the state with zero orbital angular momentum.

For arbitrary l, we seek the solution in the form

$$G_l(r) = r^{l+1} F_l(r), \tag{G.4}$$

where $F_l(r)$ is a new function, which is known only for $l = 0$:

$$F_0(r) = \frac{G_0(r)}{r} = R_0(r) = A \frac{\sin kr}{r}. \tag{G.5}$$

Putting (G.4) into (G.1), we find that $F_l(r)$ satisfies the equation

$$F_l'' + 2 \frac{l+1}{r} F_l' + k^2 F_l = 0. \tag{G.6}$$

If we change here $l \to l+1$, the result will be

$$F_{l+1}'' + 2 \frac{l+2}{r} F_{l+1}' + k^2 F_{l+1} = 0. \tag{G.7}$$

Quantum Mechanics and Quantum Information: A Guide through the Quantum World,
First Edition. Moses Fayngold and Vadim Fayngold.
© 2013 Wiley-VCH Verlag GmbH & Co. KGaA. Published 2013 by Wiley-VCH Verlag GmbH & Co. KGaA.

Differentiating (G.6) with respect to r gives

$$F_l''' + 2\frac{l+1}{r}F_l'' + \left[k^2 - 2\frac{l+1}{r^2}\right]F_l' = 0. \tag{G.8}$$

Now we notice a very interesting thing: if we set

$$F_l' = rQ_{l+1}, \tag{G.9}$$

where $Q_l(r)$ is another new function, then

$$F_l'' = Q_{l+1} + rQ_{l+1}', \qquad F_{l+1}''' = 2Q_{l+1}' + rQ_{l+1}'', \tag{G.10}$$

and Equation G.8, expressed in terms of Q, takes the form

$$Q_{l+1}'' + 2\frac{l+2}{r}Q_{l+1}' + k^2 Q_{l+1} = 0. \tag{G.11}$$

But this is identical to Equation G.7. Identical equations have identical solutions. Therefore,

$$Q_{l+1} = F_{l+1}, \tag{G.12}$$

or, in view of (G.9),

$$F_{l+1} = \frac{1}{r}\frac{dF_l}{dr}. \tag{G.13}$$

Thus, we have found that F_{l+1} results from action of operator $d/(r\,dr)$ on F_l. This enables us to immediately obtain all higher F_l from F_0:

$$F_l = \left(\frac{1}{r}\frac{d}{dr}\right)^l F_0. \tag{G.14}$$

Putting here known F_0 from (G.5) and then combining with (G.3), (G.5), and (8.38) gives

$$R_l(r) = 2(-1)^l \left(\frac{r}{k}\right)^l \left(\frac{1}{r}\frac{d}{dr}\right)^l \frac{\sin kr}{r}. \tag{G.15}$$

Here the factor $2(k)^{-l}$ replaces A in (G.3) to satisfy the normalization condition, and the phase factor $e^{i\pi l} = (-1)^l$ is introduced for convenience in future applications.

Appendix H
Bound States in the Coulomb Field

Applying the general theory outlined in Section 10.3 to the Coulomb field gives the equation for radial motion:

$$-\frac{\hbar^2}{2\mu}\left(\frac{d^2 G}{dr^2} - \frac{l(l+1)}{r^2}G\right) - \frac{Ze^2}{r}G = EG. \tag{H.1}$$

Here $e^2 \equiv q_e^2/4\pi\varepsilon_0$, q_e is the electron charge, and Z is the number of protons in a given nucleus. We want to find the spectrum and the radial functions $R(r) = rG(r)$ for the bound states. In this problem, it is convenient to introduce the dimensionless variables

$$\zeta = \frac{r}{a}, \quad \mathcal{E} = \frac{E}{\kappa_1}, \quad a \equiv \frac{\hbar^2}{\mu e^2}, \quad \kappa_1 \equiv \frac{e^2}{2a}. \tag{H.2}$$

This will simplify Equation H.1 to

$$\frac{d^2 G}{d\zeta^2} + \left[\mathcal{E} + \frac{2Z}{\zeta} - \frac{l(l+1)}{\zeta^2}\right]G = 0, \tag{H.3}$$

where all atomic parameters are now hidden in the new variables ζ and $\mathcal{E}$. Let us look for the solution in the form

$$G(\zeta) = e^{-\gamma\zeta}f(\zeta), \quad \gamma \equiv \sqrt{-\mathcal{E}}, \tag{H.4}$$

where $f(\zeta)$ is a new unknown function. Putting this expression into (H.3) gives after some simple algebra

$$\frac{d^2 f}{d\zeta^2} - 2\gamma\frac{df}{d\zeta} + \left[\frac{2Z}{\zeta} - \frac{l(l+1)}{\zeta^2}\right]f = 0. \tag{H.5}$$

Let us look for $f(\zeta)$ in the form of power series expansion. As shown in Section 10.3, the lowest power of ζ for each l must be ζ^{l+1} if we want the solution to be finite at the origin. Therefore, we try the expansion in the form

$$f(\zeta) = \zeta^{l+1}\sum_{j=1}^{\infty} b_j \zeta^j. \tag{H.6}$$

The problem reduces to finding the coefficients b_j. To this end, put the expression (H.6) into (H.5) and collect the terms with equal powers of ζ:

$$\sum_j \{2[Z - \gamma(j+l+1)]b_j + [(j+l+1)(j+l+2) - l(l+1)]b_{j+1}\}\zeta^{j+l} = 0. \tag{H.7}$$

In order to satisfy (H.7) for all ζ, the expressions in the curly brackets must be zero:

$$2[Z - \gamma(j+l+1)]b_j + [(j+l+1)(j+l+2) - l(l+1)]b_{j+1} = 0. \tag{H.8}$$

This gives the system of recurrent relations between any two neighboring coefficients b_j and b_{j+1}:

$$b_{j+1} = 2\frac{\gamma(j+l+1) - Z}{[(j+l+1)(j+l+2) - l(l+1)]}b_j, \quad j = 0, 1, 2, \ldots. \tag{H.9}$$

The first coefficient b_0 is not determined by this system. This is natural since the original equation is homogeneous. The value of b_0 can be determined in the end by the normalization condition. As it is, we can just use (H.9) to express b_1 in terms of b_0, then b_2 in terms of b_1, and so on. Thus, relation (H.9) determines all subsequent coefficients through only one coefficient b_0.

As seen from (H.9), the ratio $b_{j+1}\zeta^{j+1}/b_j\zeta^j \to 2\gamma\zeta/j$ at $j \to \infty$. Therefore, the series (H.6) converges at all finite ζ. However, its sum increases rapidly with ζ – so rapidly, in fact, that it overcomes the exponentially decaying factor $\exp(-\gamma\zeta)$ in (H.4). As a result, the function (H.4) increases unboundedly as $\zeta \to \infty$ [1].

But the sought-after solutions describing a physical system must be finite! Are such solutions contained in the expression (H.4) for $G(\zeta)$?

The answer is yes! The series (H.6) has variable coefficients b_j (recall that γ in (H.9) is, according to (H.4), just disguised energy). Therefore, the series can, at certain values of this variable, truncate to a polynomial. The terms of this polynomial are powers of ζ, and any power, no matter how high, increases with ζ slower than exponential function of ζ. Therefore, the sum of a *finite* number of powers cannot counterbalance the rate of decrease of the exponential factor in (H.4). In other words, $G(\zeta)$ in all such cases converts from asymptotically increasing to asymptotically decreasing function of distance.

Our problem thus reduces to finding such values of γ at which the series with coefficients determined by (H.9) reduces to a polynomial. This happens if the numerator in (H.9) becomes zero for a certain number $j = n_r$. Once b_{n_r+1} becomes zero, all subsequent coefficients with numbers greater than $n_r + 1$ automatically become zero as well, so that the series shrinks down to the polynomial of power n_r.

It remains to find the energies at which this happens. Setting the numerator in (H.9) to zero gives the condition

$$\gamma = \frac{Z}{n_r + l + 1}. \tag{H.10}$$

Returning to variable ε from (H.4) and denoting $n_r + l + 1 \equiv n$, we get $\varepsilon = -Z^2/n^2$. Finally, expressing ε in terms of physical observables (H.2), we obtain already familiar formula (2.40) $E_n = -\mu Z^2 e^4/2\hbar^2 n^2$ with $n = 1, 2, 3, \ldots$ for allowed energies in bound states. The number n here is called the *principal quantum number*. The only difference from (2.40) is the additional factor Z for one-electron ions of atoms heavier than a hydrogen atom. Note that the value $n = 0$ was excluded by the *additional postulate* in the Bohr theory to avoid the collapse of the electron cloud onto the nucleus; here the value $n = 0$ is excluded automatically by relation (H.10).

Now we turn to the eigenstates corresponding to the eigenvalues $\gamma = Z/n$. Putting this into (H.9) and simplifying the denominator gives

$$b_{j+1} = \frac{2Z}{n} \frac{j+l+1-n}{(j+1)(j+2l+2)} b_j. \tag{H.11}$$

We use this to determine one coefficient after another and put them into (H.6):

$$f(\zeta) = b_0 \zeta^{l+1} \left\{ 1 - \frac{n-l-1}{1!(2l+2)} u + \frac{(n-l-1)(n-l-2)}{2!(2l+2)(2l+3)} u^2 + \cdots + (-1)^{n_r} \right.$$
$$\left. \times \frac{(n-l-1)(n-l-2) \cdots 1}{n_r!(2l+2)(2l+3) \cdots (2l+n_r)} u^{n_r} \right\}. \tag{H.12}$$

Here we denoted

$$u \equiv \frac{2Z}{n} \zeta = \frac{2Z}{na} r. \tag{H.13}$$

The polynomial in the curly brackets is completely determined by the numbers n and l, and is usually denoted by $L_{n+l}^{2l+1}(u)$. It is related to the *Laguerre polynomials* $L_k(u)$ known in mathematics [1] and defined as

$$L_k(u) = e^u \frac{d^k}{du^k} \left(e^{-u} u^k \right). \tag{H.14}$$

These polynomials give rise to the *associated* Laguerre polynomials

$$L_k^s(u) = \frac{d^s}{du^s} L_k(u). \tag{H.15}$$

If we set here $k = n+l$ and $s = 2l+1$, we obtain our polynomial $L_{n+l}^{2l+1}(u)$. Finally, we combine all this to obtain the radial function $R_{nl}(r)$ for the state with quantum numbers n and l:

$$R_{nl}(r) = N_{nl} e^{-(1/2)u} u^l L_{n+l}^{2l+1}(u), \quad u \equiv \frac{2Z}{na} r. \tag{H.16}$$

Reference

1 Arfken, G.B. and Weber, H.J. (1995) *Mathematical Methods for Physicists*, Academic Press, New York.

Index

a

absolute value   55, 130, 291, 405, 422, 509, 622–624, 646, 696
absorption modeling, blackbody object   494
ad hoc process   564
algebraic identity   547
Alice–Bob experiment   611
Alice's choice   637, 646, 731, 733
Alice's electron   623, 624, 755
Alice's measurement   607
A-momentum   426–428
amplification
– factor   645
– process   644, 730
– regime   545
amplitudes
– actual vibration   230
– bra and ket vectors   71
– complex vector space   69
– composition of   69–73
– Dirac notation   72
– eigenfunctions   73
– eigenvector   70
– Hermitian conjugate   70, 71
– Hilbert space   69, 71
– interpretation   453
– matrix   70
– momentum eigenstate   72
– nonzero matrix element   70, 71
– notations   70
– phase shift   83
– phenomenon degeneracy   73
– probability   230
– superposition   92
– time-dependent   205
– transition   73, 89
– von Neumann projection   70
ancilla   660, 687
angular distribution   497

angular momentum   34, 119, 149, 274, 291, 382–384, 387, 433, 484, 488, 497
– compound system   183–187
– coupled vs. uncoupled representation   187, 188
– definite   460
– orbital   149, 150
– – eigenstates   151–153
– – eigenvalues   151–153
– – operator, and commutation properties   154–164
– – permutations   154, 155
– and rotations   156
– spin   149, 150
– – an intrinsic angular momentum   164, 168–171
– Stern–Gerlach experiment   150
– total angular momentum   149
– two particles, classical net angular momentum   150
– uncertainty relations for components   157, 158
– vector   156, 208, 286, 287
– visualizing components   158, 159
"angular spreading" of probability   286
anisotropy   268
annihilation operators   548, 550, 552, 633
anticommutator   171, 548, 554
antisymmetric function   528, 529, 530
aperture function   378, 392
asymptotic expression   487, 490
asymptotic perturbations   455–457, 456, 462
atomic currents   287–290
atomic number   290
atomic population constant   543
atomic spectra   1, 31
atomic transition frequency   515
atomic transition operators   631, 632
atoms, and spectra   12

Index

autoionization 457
auxiliary privacy amplification theorem 731
average energy 238
Avogadro number 5
axially symmetric scattering 482
  azimuthal angle 73, 78, 82, 132, 153, 173, 175, 382, 383, 477, 481, 517

b

Balmer series 284
BB84 protocols 758, 759, 760, 772, 774, 778
– Alice's four polarization states 781
– critical error rates 792
– phase shift 779, 780
BBR. *See* blackbody radiation (BBR)
beautiful theory 614
Bell-CHSH experiments 759
Bell's inequality 618
Bell-state measurements 619–627
Bell's test angles 626, 755, 756, 758
Bell's theorem 613–619
Bessel spherical functions 275
B-factor 304, 305
Bhattacharyya coefficient 696
– fidelity 697
binary entropy function 707
bit twiddling 771
blackbody radiation (BBR) 1
– Einstein's light quanta and 24–27
– spectrum
– – according to different approaches treating radiation 4
– – at various temperatures 3
blackbody radiation (BBR) spectrum 542
Bloch function 261
Bloch sphere approach 625
Bloch vector 639, 640
Bob's apparatus 620
Bob's detector 787
– signal beam 787
Bob's electron 755
Bob's error 762, 763
– probability 762
– rate 765, 777
Bob's string 719
Bob's success rate 762
Bohr magneton 289
Bohr radius 34, 285, 398, 519
Bohr rules 418
Bohr's complementarity principle 38
Bohr–Sommerfeld quantization 408
– rules 406–409
Bohr theory 285, 829

Boltzmann constant 2, 9, 526, 706
Boltzmann distribution 22, 23, 356, 569
Born approximation 502, 504
Bose particles 528
Boson clusters 539
Bosonic states 81–89, 531, 532, 536, 539
Bosonic system 539
boundary condition 519
bound particle 406
B92 protocols 677, 778, 779
– Bennett's B92 proposal 784
– nonorthogonal polarization states 779
– phase shift 784
– transmission rate 786
Bragg reflection 266, 267
Breidbart basis 763, 766
– measurements 778
Broglie wave 818
BS experiment 576, 603
BS momentum spectrum 604
Buzek–Hillery machine 638

c

Caldeira–Leggett master equation 571
Caldeira–Leggett model 567
Cartesian coordinates 46, 132, 151, 152, 158, 191, 273
Cartesian space 189, 535, 682, 683
Cartesian tensor 189, 190
cat states 569, 581
Cauchy–Schwarz inequality 385
causality principle 628
center of mass 81, 215, 256, 257, 281, 302, 303, 311, 315, 378, 399, 481, 504
– isolated NH3 molecule 417
– stationary 303
centrifugal energy 209, 281
CHSH apparatus 620
CHSH inequality 623, 624, 626, 627, 755
– security proof 756
CHSH paper 619, 621
CHSH security proof 756, 757
circularly unpolarized (CUP) pulse 643
classical determinism concept 64
classical oscillator 795–799
Clebsch–Gordan coefficients 193
CNOT gate 740, 741
coherence 545
coherent eavesdropping 788
cold emission 241–244
collective attack 788, 792
collision entropy 719
collision frequency 322
collisions 1, 8, 12, 88

– elastic   484
– high-energy   169, 202
column matrix   70, 121, 125–127, 129, 174, 176, 181, 360, 364, 369, 491, 685
common classical gates, quantum analogues   743
commutability   96
– defined   101
complex-conjugated coefficients   421
complex energies   492
complex experimental situation   418
complex function   44, 112, 261, 298, 488, 492, 499
complex propagation vector   325
composite systems   368–376, 421, 566, 569, 572, 581
Compton wavelength   206
computational basis   420, 742
concurrence   421
configurations   535
conservation
– law   46, 301, 601–603
– of probability   202
constant coefficients   544
constant phase factor   528
continuity equation   200–202, 325
continuous "energy bands"   34
conveyer belt   579
corpuscular theories   614
Coulomb attraction   580
Coulomb field   448, 455, 456, 457, 475, 827
– bound states   827–829
Coulomb potential   277, 281, 328, 455
– energy   207
Coulomb repulsion   291, 329
Coulomb scattering   82
– of fermions in same spin state   86
– of two α-particles   84
coupled oscillator analogy   414
CP transformation (CPT)   301
creation operator   633
critical error rate
– hybrid form   776
– transcendental equation, graphical solution   775
current moment   323
cyclic perturbations   470
cyclotron frequency   429

d

D-Alambert inhomogeneous equation   503
D'Alembert equation   424
damping force   567
dark fringes   355, 442, 587

d-coefficients   372
de Broglie waves   31–33, 204, 206, 213, 307, 323, 377, 379, 394, 400, 402, 429
– expansion   487
– phase velocity   216
– revised   218–222
decay constant   320, 514
decaying function   404
decay law   327
decay theory   492
decoherence   399, 568, 569, 570, 607
– as einselection   570, 571
– time   567, 583
deformation   510
degeneracy   207, 457–462
degenerate states, superposition   209–211
degrees of freedom   510
delta spike   308
– evolution   311
density matrices   358, 359, 362, 365, 369, 372, 375, 566, 568, 638
– approach   637
– arguments   644
– properties   361
density operator   358–366, 368, 635
– time evolution   366–368
detuning   467
deviation operators   384
DFT. See discrete Fourier transform (DFT)
diagonal basis   458
diagonal elements   73, 75, 124, 126, 128, 170, 190, 297, 359, 360, 365, 458, 463, 566, 641, 658, 709, 710
diagonal matrix   73, 128, 254, 268, 359, 362, 491, 570, 630, 641
– elements   359
diagonal photon   534
diatomic gases   6, 21
diffraction
– electron   40
– of neutrons by a sodium chloride crystal   41
– scattering   40, 494–498
dipole approximation   629, 630
dipole moment   419, 465, 475, 629, 630
– operator   632
dipole radiation   474–477
Dirac comb function   392
Dirac's delta function   801–807, 810
Dirac's notation   813, 814
direct product formula   371
discontinuous process   39, 345, 559
discrete Fourier transform (DFT)   749, 751
dispersed indeterminacy   264, 390–393

Index

dispersion equations 139, 140, 205, 206, 263, 273, 304, 378, 379, 460
– free particle 205–207
– nonrelativistic 316
– relativistic 316
dispersion formula 327, 328
displacement vector 631
disturbance model 621
diverging flux 325
– nonzero 485
diverging wave 14, 210, 277, 377, 378, 484, 485, 490, 491, 495, 496
double-slit experiment 14–19, 74, 388, 391, 437, 439, 618
double slit revised I 74–77
double slit revised II 77, 78
dynamic momentum 426
Dyson series 469

e

Eberhard theorem 607
effective potential 429
effective size, of atom 285, 286
Ehrenfest theorems 300
eigenfunctions 443
– superposition 356
eigenstates 102–104, 106, 107
– degenerate 457
– momentum 72, 73, 110, 119, 120, 139, 140, 192, 273, 356
– orthogonality 107–110, 143, 173, 612
– perturbed 450
– uncoupled 188
– unperturbed 449, 450
eigenvalues 102–104, 106, 107
– energy 224, 249, 282, 318–320, 338, 434, 443, 453, 518, 822
– equation 463
– nonnegative 675
– specific 104
– vector 178
Einstein coefficients 26, 477, 542
Einstein, Podolsky, and Rosen (EPR)
– correlations 637
– phenomenon 605
Einstein's coefficients 541
Einstein's equation for PEE 30
Einstein's light quanta and BBR 24–27
– classical oscillator 27, 28
– damped oscillations 29
– oscillation amplitude 29
– role of dissipative factor 29
– with spontaneous emission 29, 30
Ekert's scheme 758, 759

– symmetric strategy 789
elastic resonant scattering 500
elastic scattering 81, 484, 493, 496, 497
electric potential 440
electromagnetic
– field 633
– interactions 477
– radiation 283
– waves 1, 12
electron's orbital angular momentum 38
electron spin 37
– magnetic moment 443
– measurement 660
emergence correlation function 621
emission–absorption process 517
EM theory 485
energy bands, origin 257–260
– double well 257
– mechanism for energy level splitting 258
– quasimomentum 259
energy correction 449
energy eigenfunction 319
energy eigenstates 418, 531
energy eigenvector 297
energy flux 337
energy level splitting, in a system of wells 258
ensemble average 358
entangled superpositions I 599–601
entangled superpositions II 601–604
entangled systems 376, 608
entanglement phenomenon 599
entropy of entanglement 733
equations of motion 425
equipartition theorem 2
Euclidian algorithm 747
evanescent de Broglie waves 393–394
evanescent waves 380
Eve's average surprisal 717
Eve's error rate 790
Eve's information 754, 770
– symmetric attack 790
– symmetric strategy 789
Eve's intervention 755, 756
Eve's string 719
evolution factor 304, 314
exactly solvable models (ESMs) 508
exchange operator 527, 528
exclusion principle 552, 612
expansion coefficients 326, 488, 512
expectation values 99, 100, 111, 126, 127, 139, 157, 158, 174, 195, 203, 209, 299, 335, 340, 434, 436, 621

– zero 288
exponential function 44, 102, 238, 296, 468, 798, 806, 828

f

Fabry–Pérot interferometer 227
Fabry–Pérot resonator 313
familiar optical theorem 494
faster than light (FTL)
– communication 604, 608, 628
– signaling 627, 645, 647
– – proposal 647, 648
Fermionic states 81–89
fermions 85, 612
Fermi particles 528
fiber-optic cable 766
fidelity 636
field gradient 580
first-order correction 452
– formula 458
first-order perturbation theory 459, 477
FLASH proposal 643–648, 646
Floquet theorem 261
flux density 46, 201, 202, 309, 425, 426, 484, 485, 493
flux intensity 18, 482, 515
flux of particles 4, 481
Fock space 535
Fock states 534, 547
forward-scattered waves 486
Fourier components 316, 475
Fourier expansion 379, 470, 551
Fourier spectrum 317
Fourier transform 378, 392, 469
free particle in 3D 274–276
Fresnel integrals 305, 312
– asymptotic behavior 313
fringe electric field 440

g

Galilean transformations 219
Gamov's decaying states 514
Gamov state 474, 520, 521
Gamov's theory 470
Gamov-type state 519
Gamow states 327, 336
gauge invariance 424
gauge transformations 424, 430, 432, 441, 509
Gaussian distribution 302, 303
Gaussian function 392, 560
Gaussian packet 301, 302, 398
Gauss theorem 325
Gedanken experiment 563

generalized momentum 426, 431
generalized summation 305
generation regime 545
generic density matrix 360
Gibbs ensemble 367
gravitational filed 583
Green function 304
group velocity 212–218, 216, 217, 312
– poorly defined 218
– of wave packet in vacuum 218
gyrating medium 346

h

Hadamard gate 743
Hamilton equations 300, 424
Hamiltonian 295, 300, 304, 318, 321, 427, 429, 432, 436, 447, 508, 512, 584, 631
Hamming distance 770
Hankel functions 277
Hartree products 529, 530, 533, 538
heat capacity 1, 4–9
Heisenberg equation of motion 632
Heisenberg inequality 380–382
Heisenberg picture 298
Heisenberg's inequality 385
Heisenberg's uncertainty principle 754
helium atom 8
– optical transitions 13
helium ion 281
helium lamp 13
Helmholtz equation 206, 273, 274, 503
– for free particle 205–207
Helstrom formula 777
Hermite polynomials 821–824, 822, 823
Hermitian adjoint 692, 693
Hermitian operators 320, 384, 653, 822
– spectral decomposition theorem 653
hidden variables 95, 613–619
– theory 613, 614, 615, 616
high-fidelity clones 638
high-order correlations 647
Hilbert spaces 69, 70, 109, 129–135, 147, 371, 386, 422, 511, 512, 535, 536, 594, 640, 660, 668, 698, 699, 789, 814
– energy measurement in 558
– operations in 135–142
– of photon qutrit 689
– two-dimensional 652
– vector space 298
Hill's equation, solutions 261
Holevo bound 732
HOM dip 538
HOM effect 88, 647, 648
H-space. *See* Hilbert spaces

Huygens construction 496
Huygens principle 14
hydrogen atom 281, 284

i

ideal gas 4
ideal multiple cloning 516
idempotent operator 362
identity operator 527
impact parameter 481
imperfect cloning 628, 636–643
– machine 635
independent variable 112, 132, 220, 324, 721
indeterminacy
– arbitrary potential and 55
– least possible energy in ground state 54
– principle 53, 54, 378, 385
– QM indeterminacy, determine exact size 54
– quantum-mechanical 53
– "radial fall" 53
– radial standing wave 53
– spherical standing wave 53
indeterminacy revisited 377–394
– of angular momentum 382–384
– dispersed indeterminacy 390–394
– Heisenberg inequality 380–382
– N–ϕ indeterminacy 388–390
– Robertson–Schrödinger relation 384–388
– under scrutiny 377–380
inelastic collision 498
integral of motion 300, 301
integral operator 304
interaction-free detection 590, 593
interaction-free measurement 590, 605
– schematic for 591
interaction potential 508
interferometer 590, 591
intrinsic angular momentum 37, 58, 149
– spin as 164, 168–171
inverse population 518, 543, 544
inverse problem 483
ionization energy 34, 283
irreducible spherical tensors (ISTs) 190

j

Jaynes–Cummings model 468
Jensen's inequality 721, 722

k

kinematic momentum 426, 427
kinetic energy 2, 10, 31, 220, 232, 237, 285
Kirchhoff's law 25
Kraus operators 666, 698, 699
Kraus' theorem 700

Kronecker delta 685, 803
Kronig–Penney potential 261
Krylov–Fock theorem 326
Kullback–Leibler divergence 717, 718

l

ladder operators 549, 551, 552
Laguerre polynomials 829
Landau levels 431
laser-gain tube (LGT) 643
Legendre polynomials 489, 820
light amplification 545
light–matter interactions 473–480
– dipole radiation 474–477
– optical transitions 473, 474
– selection rules 477–480
– – oscillator 478–480
linear momentum 47, 57, 177, 208, 274, 460, 531
linear Stark effect 463
Liouville theorem 367
local continuity equation 310
local phase transformation 425
logic gates 737, 738
– quantum 741
Lorentz condition 424
Lorentz force 423, 429, 437
lowering operator 160, 162, 186, 549, 550, 552–554
Lyman series 284, 465

m

Mach–Zehnder interferometer 439, 586, 780
– interaction-free measurement 586
macroscopic measuring device 397, 567
magnetic field 430, 580
– Aharonov–Bohm effect 437–442
– A-momentum origin 427–429
– charged particle in EM field 423–425
– charge in 423–444
– charge in magnetic field 429–432
– continuity equation in EM field 423–425
– flux 441
– spin precession 432–436
– Zeeman effect 442–444
magnetic moment 431, 434, 443, 579
magnetic quantum number 480
magnetic vector 427
– potential 423
Malus's law 592
Mandel's amplifier 648
Markovian/memoryless processes 578
matrices
– adjoint, or conjugate 123

– algebra, elements of 813–816
– complex conjugate matrix 123
– concepts and definitions 123–125
– continuous 124
– elements 359, 453, 454, 458, 463, 479, 480
– orthogonal 124
– self-adjoint/Hermitian matrix 123
– transpose of matrix 123
– unitary 124
maximal signal velocity 316
Maxwell–Boltzmann statistics 525
Maxwell equations 423, 424, 581
measurement laws 603
π-meson–proton scattering 497
Michelson interferometer 590
micro–macro entanglement 570
min-entropy 774
mixed ensemble 356
mixtures 356–358
momentum 213. *See also* angular momentum
– definite 214
– indeterminacy 382
– space 211, 560
– undefined 214
monochromatic EM wave 486
monochromatic laser 516
multiparticle
– states
– – notation for 536, 537
– systems 525, 547, 548
multiphoton plane unpolarized (PUP) signal 643

n

NAND gates 738, 739, 741
natural frequency 337
net magnetic quantum number 601
Neumann projection 573
Neumann's measurements 573, 655
– scheme 571
Neumark's theorem 681–686, 691
neutrino oscillations 336
neutron scattering in crystals 78–80
Newton's second law 300
N–ϕ indeterminacy 388–390
NH_3 molecule 416
– stationary state 417
no-cloning theorem 607–613
noncommutable operators 548, 561
nonhomogeneous differential equation 544
nonoptimal measurement 574, 652
nonorthogonal states 142–147
– normal projections 146
– – contravariant and covariant components 146
– projection operator 144
– "spin-up" and "spin-down" states 143
nonrelativistic approximation 206, 222, 315, 316, 318
nonrelativistic QM 577
nonselective measurement 577
non-square-integrable function 309
nonstationary quantum oscillator 256
nonzero diverging flux 485
normalization
– constant 533
– of determinant 533, 534
normalizing coefficient 532
normal oscillations 550
no-signaling condition 638
NOT gates 743
– reverses 738
nuclear alpha laser 518
nuclear collision 335
nuclear forces 328, 329
number operators 548, 553
number states 534
numerical approximation 398

o

observables 95, 96
– "complete set" 96
– properties 95, 96
– variables 616
occupancy numbers 525, 543
off-diagonal elements 568
one-slit experiment 379
operators
– algebra 100, 101
– atomic transition 631
– Bob's 785
– differential 105
– functions 101, 102
– Hermitian 100, 384, 653
– Kraus 699
– ladder 551, 553
– mathematical symmetry, and differential 105
– in matrix form 125–128
– momentum 48, 104, 188
– operator-sum decomposition 698
– orbital angular momentum operator
– – eigenfunctions/eigenvalues of 817–820
– parity 106
– quantum-mechanical 97–100
– raising 554
– rotation 192, 640

– tensor   189
optical
– mode   534
– phenomenon   521
– spectroscopy   480
– theorem   490
– transitions   473, 474, 478
oscillating
– exponent   504
– frequency   322, 418, 435
– probability   336
oscillator
– energy   461
– in E-representation   254–257
– – average x-coordinate   255
– – c-amplitudes   254
– – matrix elements   255
– – nonstationary quantum oscillator;   256
– – phase of oscillations   255
– – probability density   256
– – superposition of basis states   254
– – superposition of eigenstates   256
– – temporal factors   255
– model   548
overlapping wave packets   526

p

parallel-plate capacitor   438
parametric down-conversion effect   591
parity-odd operator   631
partial cross section   489
partial trace over system   373
partial wave   489
particle–antiparticle interpretation   306
particle–antiparticle oscillations   331
particle-in-a-box analogy   470
particle, in 3D box   273, 274
particle's actual velocity   213, 216, 218, 219
particle's momentum   378
path integrals   89–93
– double-slit experiment   90
– – virtual paths   90–93
– spherical wave diverging   89
– transition amplitude   89
Pauli exclusion principle   290, 531
Pauli matrices   171, 177, 361, 639
Pauli operator   421
Pauli principle   259
Pauli spin matrices   744
PBS experiment   576
periodic motion   397
periodic structures   260–269
– Bloch function   261
– Brillouin zone in k-space   264

– energy spectrum   263
– expansion coefficients   264
– Hill's equation   261
– Kronig–Penney potential   261
– periodic factor   261
– periodic function   261
– quasimomentum   261, 264
– rectangular periodic potential   262
– transcendental equations   263
– zone structure of a particle's energy spectrum   263
periodic table   290–293
periodic variable   387
perturbations   447–470
– asymptotic perturbations   455–457
– element   469
– frequency   473
– function   511
– Hamiltonian   448
– matrix   459
– method   448, 463
– perturbations and degeneracy   457–460
– Stark effect   462–465
– stationary perturbation theory   447–454
– symmetry, degeneracy, and perturbations   460–462
– theory   455, 473, 475, 503
– time-dependent perturbations   465–470
phase diagram. *See* quantum-mechanical, phase space
phase factors   485
phase shifts   28, 85, 246, 302, 439, 487, 490, 491, 577, 780, 785, 795
– gate   744
phase transformation   425
– global   45
– local   45, 425, 426
phase velocity   212–218
– high-frequency limit   316
phenomenological approach   498, 501
photodetectors   16
photoelectric effect (PEE)   9–12
– first law   10
– revisited   30
photoemission effect   243
photon number splitting (PNS) attack   782, 784
photons   687, 688
– detectors   315
– emission   518
– – probability   540
– energy   485
– interaction   486
– momentum   603, 604

– monochromatic   629
– number   476, 577
– polarizations   88, 629, 634, 645, 670, 759, 760, 779, 786
– – instantaneous   344
– – without involving a laser   637
– states, unambiguous discrimination   672
– trajectory   532
– wave function   518
physical system   416
planar resonant scattering   520
Planck's constant   382, 397, 408
Planck's formula   21–24
Poincaré sphere   625, 645, 760
pointer basis   565
pointer observable   565, 570
Poisson quantum bracket   299, 367
polar angle   211
polarization
– angle   592, 680, 776, 780
– direction   478
– independent amplifier   635
– vector   341, 342, 476, 613
– – dot product   475
polarizers   347, 348
polarizing beam splitter (PBS)   575
polarizing filter   340, 341, 575, 594
position vector operator   479
postcloning system   610
potential
– barrier   11, 236, 243, 329, 404, 416, 457, 525
– box expression   227
– energy   2, 44, 200, 209, 232, 236, 237, 242, 264, 299, 462, 475, 579, 580, 629
– function   257, 400, 404, 407, 423, 438, 509
– perturbations   510, 512, 514
– scattering   500
– well   244, 245
potential threshold   232
– amplitude splitting   233
– energy   232
– evanescent wave   236
– heaviside step function   232
– probability amplitudes   233
– probability flux   235
– Stieltjes integral   234
– time-independent Schrödinger equation   233
– transmitted wave   236
– turning point   233
– wavefront splitting   233
POVMs   652, 658–664
– advantage of   677
– constructing, unitary matrix   691
– discarding, extra dimensions   690, 691
– element   669, 670, 686, 690
– generalized measurement   664–666
– generalized quantum measurement   700
– implementation   686, 687
– initial state preparation   689, 690
– joint system, initial state of   689
– minimum failure rate   677
– Neumark's theorem   681–686
– optimal quantum state discrimination experiment   680, 681
– photon as qutrit   687, 688
– projective measurement   678, 679, 684, 694, 695
– rotating, coordinate system   692–694
– state discrimination measurement   695
– success and failure rates   695
– two pure states, discrimination of   670–681
– unambiguous photon state   688
– unitary transformation operator   693
precession frequency   435
p-representation, state function   809–811
principal quantum number   461, 462, 531, 829
probability   313, 348, 412
– amplitudes   298, 558, 580
– density   307, 320, 357
– distribution   308, 328, 380, 389, 417, 608
– flow   319
– flux   309
– – density   426
– function   557
– of reflection   588
pure ensembles   355, 356, 380

q
QKD. *See* quantum key distribution (QKD)
QM. *See* quantum mechanics (QM)
Q-measurement   297
quadratic equation   384, 413
quantization rule   408
quantum bit   420
quantum bit error rate (QBER)   769
quantum communication theory   364
quantum compounds   643–648, 644
quantum design   508
quantum detection operators   673
quantum engineering   507
quantum ensembles   355–376
– composite systems   368–376
– density operator   358–366
– density operator, time evolution   366–368
– direct problem   355
– inverse problem   355

– mixtures   356–358
– pure ensembles   355, 356
quantum entanglement   602
– and nonlocality   58–61
quantum equations   300
quantum evolution   306
quantum field theory   541
quantum gates   737
– DFT   751
– quantum logic gates   741–746
– Shor's algorithm   746–751
– truth tables   737–741
quantum indeterminacy   389
quantum information   703
– Bayes's theorem   711–716
– conditional probability   711–716
– deterministic information   703–709
– entanglement, entropy of   733, 734
– features   420
– Holevo bound   731–733
– KL divergence   716, 717
– mutual information   717–719
– Rényi entropy   719–721
– – joint and conditional   721–726
– Shannon entropy   703–709
– theory   608
– universal hashing   726–731
– von Neumann entropy   709–711
quantum injected-optical parametric amplifier (QI-OPA)   645
quantum jump   559
quantum key distribution (QKD)
– advanced eavesdropping strategies   788–792
– BB84 protocol   758–766
– B92 protocol   776–779
– communication over channel   766, 767
– distribution   767–769
– with EPR   753–758
– experimental implementation of   779–787
– noise effect   768, 769
– postprocessing of   769–776
quantum measurements
– projective measurements   655–657
quantum-mechanical
– analysis   567, 629
– indeterminacy   49–52
– particle   619
– phase space   62, 63
– – Bohr–Sommerfeld quantization condition   62
– – granular phase space   63
– – squeezed states   63
– – trajectories   62

– prediction   624
– superposition of states   39–42
quantum mechanics (QM)   507
– Bhattacharyya coefficient   696
– and classical mechanics   397–410
– – Bohr–Sommerfeld quantization rules   406–409
– – QM and optics   400, 401
– – quasi-classical state function   401–403
– – relationship between   397–399
– – WKB approximation   404–406
– concavity   711
– counterparts   412
– mathematical structure   307, 428
– and measurements   557–595
– – collapse/explosion   557–563
– – interaction-free measurements, quantum seeing in dark   586–593
– – quantum information and measurements   578–586
– – "Schrödinger's Cat" and classical limits   563–571
– – and time arrow   593–595
– – Von Neumann's measurement scheme   571–578
– operators   610, 642
– postulates   382
– principles   331
– theory of quasi-stationary states   327
quantum nonlocality   521, 599–648, 604, 614
– bell-state measurements   619–627
– entangled superpositions I   599–601
– entangled superpositions II   601–604
– FLASH proposal and quantum compounds   643–648
– hidden variables and Bell's theorem   613–619
– imperfect cloning   636–643
– no-cloning theorem   607–613
– QM and FTL proposals failure   627–636
– quantum teleportation   604–607
quantum numbers   290, 291, 292, 314, 391, 448, 450, 460
quantum operation   700
quantum oscillator   249–254, 430, 431, 547
– eigenstates of quantum oscillator   252
– probability distribution for   253
quantum state estimation   647
quantum states evolution   295–350
– arbitrary state, B-factor and evolution   303–306
– 3D barrier and quasi-stationary states   324–327
– Gaussian packet spreading   301–303

– "illegal" spike, fraudulent life 306–311
– Jinnee out of box 311–315
– nonrelativistic approximation inadequacy 315–317
– operators evolution 299–301
– particle–antiparticle oscillations 331–339
– particle decay theory 327–330
– quasi-stationary states 317–324
– time evolution operator 295–299
– watched pot boils faster 344–350
– watched pot never boils (*See* quantum zeno effect)
quantum-state tomography 578
quantum statistics 525–546
– BBR 542
– bosons and fermions, exclusion principle 525–540
– lasers and masers 543–545
– Planck and Einstein 540–542
quantum superposition 569
quantum system 339, 420, 572, 578, 579
quantum teleportation 604–607
quantum theory of signaling 317
quantum Zeno effect 339–343, 591
quasi classical object 571
quasi-classical state function 401–403
quasi-free particle 227–231, 557
– force on walls 231
– stationary states of a particle 229
quasi-stationary states 322, 324, 329, 331
qutrit 687–688

r
Rabi frequency 467
Rabi oscillations 468
radial
– distance 211, 274, 490, 502
– flux 493, 495
– functions 275, 277, 279, 282, 325, 827
– probability distribution 282, 285
– quantum number 285
– Schrödinger equation 207–209
– – in free space 825–826
radiation energy density 542
radiative friction 12
radioactive system 501
Rayleigh–Jeans formula 3
reduced
– density matrices 375
– matrix element 192
reference pulse 784
refractive index 401, 439, 494
relative motion 8, 281
relativistic dispersion equation 316

relativistic expressions 559
Rényi entropy 719–721
– KL divergence 723, 724
Rényi information 771
Rényi sense 730
replicating symbol 392
resonance half-width 500
resonant
– energies 499
– frequency 517
– scattering 457, 499
reversible quantum theory 595
Ritz's combination principle 13
Robertson–Schrödinger relation 110, 111, 383, 384–388
rotating wave approximation 632
rotational invariance 376
rotation gate 744
rotation wave approximation (RWA) 633
RSA algorithm 746, 747
Rutherford experiments 328
Rutherford scattering 82
Rydberg constant 455
Rydberg states 286

s
sampling symbol 392
scattering 481–504
– amplitude 483, 484 (*See also* amplitudes)
– – calculation 502
– angle 80, 81, 82, 482, 483, 502, 504
– Born approximation 501–505
– of classical particle 482
– differential cross section 482
– diffraction scattering 494–498
– elastic 79
– inelastic 79
– isotropic 78
– matrix 491, 492, 495, 498
– operator 491
– α-particle and a C nucleus 81
– QM description 481–487
– resonant scattering 498–501
– scattering matrix and optical theorem 490–494
– stationary scattering 487–490
– theory 456
Schrödinger cat experiment 568
Schrödinger equation 39, 64, 199, 200, 295, 296, 309, 312, 318, 325, 334, 355, 365, 401, 413, 427, 442, 461, 501, 529, 580
– in arbitrary basis 222–224
– temporal 231
second quantization 547–555

842 | Index

- creation and annihilation
-- operators, bosons 548, 552
-- operators, fermions 552–555
- quantum oscillator revisited 547, 548
secular equation 458, 459
selection rules 477–480, 478
- oscillator 478–480
SG apparatus 584, 585
Shah function 392
Shannon analogies 725
Shannon entropy 707, 719, 721, 725, 728, 731, 765, 774
- function 707, 708, 709, 710, 713, 715, 716
Shannon information 419, 707, 730, 767
Shannon random codes 792
Shannon's sense, information 726, 767, 774
short wavelengths, role 382
shrinking factor 639
signal pulses 785
- interfere 785
signal-to-noise ratio 635
single gyrator 591
single matrix equation 467
single-photon measurement 534, 639
Slater determinant 530, 531, 537
Sommerfeld–Brillouin theorem 316
spatial factor 319, 324
spatial probability 417
- distribution 557
spatial variables, separation of 207
spectral decomposition 362
- theorem 363, 653, 654, 655
spectral energy density 2
spectral theorem
- for density operators 657–658
spectrum of eigenvalues 298
spherical coordinates 274–276
spherical harmonics 188–195
- associated Legendre polynomial 188
- Clebsch–Gordan coefficients 193
- commutation rule 193
- equations 188, 189
- exponent 188
- operators 192
- projection theorem 194, 195
- tensors 189–191
- transition probabilities 189
- vector 189–191
- vector components tranformation 189
- Wigner–Eckart theorem 194, 195
spherically symmetric potential 277
spherical potential well 278–281
spherical standing waves 277

spherical vector 191
spherical wave 89
spin component 581
spin-flip operators 631
spin interactions 79
spin matrix 443
spin measurement 601, 707
spin operators 641
spinors 436
spin-statistics theorem 529
spin wave functions 530
spiral scattering 481
spontaneous emission 25, 283
spread of a wave packet 216
spring constant 337
s-states 286, 291
Stark effect 462–465
state functions 200–202
- in different physical situations 119
state space
- for particles 535
stationary environment 203
stationary perturbations
- crude visual illustration 451
- theory 447–454, 457
stationary states 514
statistical distributions 606
Stephan–Boltzmann law 2
Stern–Gerlach experiments 444, 578, 602, 623, 660, 754
stimulated emission 26
strangeness 332
submissive quantum mechanics 507–522
- inverse problem 507, 508
- playing with evolution 514–521
- playing with quantum states 509–514
superconductor 521
superluminal signaling 611
superoperators 698–700
surprisal 705, 715, 717, 723, 774
survival probability 349
symmetry, of system 460–462
symptotic perturbations 509

t

Taylor expansion 221, 266, 267, 295, 634
Taylor series 27, 275, 330, 340
temporal
- extension 327
- factor 229, 393
tensor
- elements 191
- products 368, 535

thermal de Broglie wavelength   526
thermal equilibrium   1, 543
three-dimensional systems   273
time-dependent factor   448
time-dependent magnetic field   466
time-dependent perturbations   465–470
time derivative operator   300
time–energy indeterminacy   320
time evolution operator   296
time-independent coefficients   469
time-independent Schrödinger equation   320
time of decoherence   566
time-reversed transition   541
time-symmetric theories   593
Toffoli gate   741, 742
torque   433
total energy   211
– density   2, 3
transcendental equations   248
transformation
– function   303
– matrix   640
transition
– amplitudes   73, 89, 92, 195, 298, 333, 334, 453, 454, 469, 473, 525, 541
– frequencies   409, 466, 467, 470, 474, 477, 498, 518, 630
– probabilities   189, 298, 467, 474, 475, 476, 477
translational motion   9, 416, 433, 602, 603
transmission coefficient   240, 244, 245, 322
transmission probability   242, 323, 340, 344, 346
trigonometric function   406
tunneling, through a potential barrier   236–241
– energy gaps   266
two-state systems   411–422, 562
– ammonium molecule   415–419
– double potential well   411–415
– qubits introduced   419–422

u

ultrashort laser pulses   307
unbounded monochromatic wave   561
unitary freedom   363
– theorem   365
unitary matrix   364
unitary transformations   121–123, 297
– column matrix   121
– Dirac's notation   121
– inverse transformation   122, 123
– $N \times N$ transformation matrix   122

universal cloning machine   634
universal photon amplifier   637
universal quantum cloning machine (UQCM)   636, 638, 640, 644, 645, 696
unperturbed atom   474
unsymmetric beam splitter (UBS)   784
UQCM. *See* universal quantum cloning machine (UQCM)

v

vacuum fluctuations   642
vacuum state   520
vector function   128, 517
vector matrix element   476
vector spin operator   177
visualizing probability density   202, 203
von Neumann's density operator   359
von Neumann's model   565, 577, 584
V-space   145, 174

w

watched pot rule   339, 340, 344, 520
wavefront-splitting interference   14
wave functions   39, 363, 403, 513, 526, 532, 554, 564
– collapse   39, 573
– de Broglie wave   45
– in E-representation   43
– exponential function   44
– Hamiltonian operator   44
– Kronecker's delta   44
– Laplacian equation   44
– and measurements   112–116
– monochromatic function   43
– $3N$ components   46
– net probability of finding   43
– in nonrelativistic QM   44
– orthonormal basis   44
– particle in a finite region of volume   42
– phase difference measurement   45
– probability amplitude   44, 46
– probability conservation   46
– probability flux density   46
– in r-representation   42
– – square integrable   43
– satisfying continuity equation   45
– Schrödinger equation   44
– state function   47
– transformation   45
– *vs.* representation   125
– waving, defined   42
wave number   228
wave packet   43, 63, 200, 214, 216, 218, 219, 221

wave–particle duality   38
wave theory   14
weak interactions   301
Wentzel-Kramers-Brillouin (WKB)
    approximation   404–406, 408
Wien formula   27
Wigner–Eckart theorem   193, 195
Wilson chamber   585

x

XOR gates   738, 739
X-ray diffraction   266

z

Zeeman effect   444
zero net momentum   599, 600
zero-order approximation   322, 448, 449, 468